Engineering Mechanics
STATICS &
DYNAMICS

FIFTH EDITION

Anthony Bedford • Wallace Fowler
University of Texas at Austin

PEARSON

Prentice
Hall

Upper Saddle River, New Jersey 07458

Library of Congress Cataloging-in-Publication Data
Bedford, A.
  Engineering mechanics: statics and dynamics / Anthony Bedford, Wallace Fowler.—5th ed.
     p. cm.
  Includes index.
  ISBN 978-0-13-614225-6
  1. Mechanics.   I. Fowler, Wallace L.   II. Title.

CIP Data Available

Vice President and Editorial Director, ECS: *Marcia J. Horton*
Acquisitions Editor: *Tacy Quinn*
Associate Editor: *Dee Bernhard*
Managing Editor: *Scott Disanno*
Media Editor: *David Alick*
Marketing Manager: *Tim Galligan*
Production Editor: *Craig Little*
Media Project Manager: *Rich Barnes*
Director of Creative Services: *Paul Belfanti*
Creative Director: *Juan Lopez*
Art Director: *Jonathan Boylan*
Interior Designer: *Kenny Beck*
Cover Designer: *Jonathan Boylan*
Art Editor: *Xiaohong Zhu*
Manufacturing Manager: *Alexis Heydt-Long*
Manufacturing Buyer: *Lisa McDowell*

About the Cover: Anton Balazh/Sutterstock.com

PEARSON
Prentice Hall
© 2008 by Pearson Education, Inc.
Pearson Prentice Hall
Pearson Education, Inc.
Upper Saddle River, NJ 07458

The author and publisher of this book have used their best efforts in preparing this book. These efforts include the development, research, and testing of the theories and programs to determine their effectiveness. The author and publisher make no warranty of any kind, expressed or implied, with regard to these programs or the documentation contained in this book. The author and publisher shall not be liable in any event for incidental or consequential damages in connection with, or arising out of, the furnishing, performance, or use of these programs.

MATLAB is a registered trademark of The MathWorks, Inc., 3 Apple Hill Drive, Natick, MA 01760-2098. Mathcad is a registered trademark of MathSoft Engineering and Education, 101 Main St., Cambridge, MA 02142-1521. All other product names, brand names, and company names are trademarks or registered trademarks of their respective owners.

Printed in the United States of America

3 2020

ISBN 0-13-614225-7
       978-0-13-614225-6

Pearson Education Ltd., *London*
Pearson Education Australia Pty., Limited, *Sydney*
Pearson Education Singapore, Pte. Ltd
Pearson Education North Asia Ltd., *Hong Kong*
Pearson Education Canada, Ltd., *Toronto*
Pearson Educación de Mexico, S.A. de C.V.
Pearson Education—Japan, *Tokyo*
Pearson Education Malaysia, Pte. Ltd.
Pearson Education, *Upper Saddle River, New Jersey*

Contents

STATICS

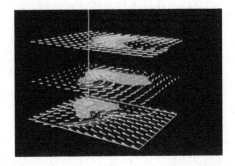

DYNAMICS

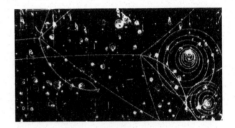

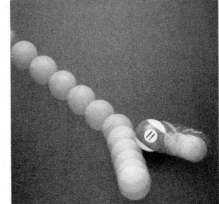

Preface

We began the development of the Fifth Editions of *Engineering Mechanics: Statics* and *Dynamics* by asking ourselves how our textbooks could be restructured to help students learn mechanics more *effectively* and *efficiently*.

From the first editions, our objective has been to present material in a way that emulates the teacher's development of concepts in the classroom and emphasizes visual analysis to enhance student understanding.

Now, based on our classroom experiences and insights provided by colleagues and students over many years, we have designed the fifth editions to conform more closely to the way today's students actually use textbooks in learning mechanics. In developing the new elements described below, we have continued to adhere to our original goals of teaching effective problem-solving procedures and the central importance of free-body diagrams.

New to this Edition

Active Examples

A new example format designed to help students learn concepts and methods and test their understanding. Discussions are visually related to figures and equations in a new *integrated text/art* format for efficient reading. A "Practice Problem" is provided at the end of the Active Example so that students will be motivated to spend more time working with the example and checking whether they understood the material. They can easily assess their understanding by referring to the answer to the Practice Problem that is provided on the page, or by studying the *complete solution* that is presented in an appendix in the same text/art integrated format as the Active Example.

Example-Focused Problems

New homework problems designed to encourage students to study given examples and expand their understanding of concepts. The numbers of these problems are cited at the beginning of each example so that teachers can easily use them to encourage study of selected topics.

Results

Most sections of the text now conclude with a new Results subsection, a self-contained and complete description of the results required to understand the following examples and problems. They are presented in the same integrated text/art format used in the Active Examples for easier comprehension.

Students can efficiently refer to these subsections while studying examples and working problems.

Problem Sets

Thirty percent of the problems are new in the statics text. Problems that are relatively lengthier or more difficult have been marked with an asterisk. Additional problems can be generated using the online homework system with its algorithmic capabilities.

Hallmark Elements of the Text

Examples

In addition to the new Active Examples, we maintain our examples that follow a three-part framework—**Strategy/ Solution/Critical Thinking**—designed to help students develop engineering problem skills. In the Strategy sections, we demonstrate how to plan the solution to a problem. The Solution presents the detailed steps needed to arrive at the required results.

Some of the examples have a focus on design and provide detailed discussions of applications in statics in engineering design.

Computational Mechanics

Some instructors prefer to teach statics without emphasizing the use of the computer. Others use statics as an opportunity to introduce students to the use of computers in engineering, having them either write their own programs in a lower level language of use higher level problem-solving software. Our book is suitable for both approaches. Optional, self-contained Computational Mechanics material is available for this text on the Companion Website, including tutorials using Mathcad and MATLAB. See the supplements section for further information.

Art Program

We recognize the importance of helping students visualize problems in mechanics. Students prefer, and are more motivated by, realistic situations. Our texts include many photographs as well as figures with "photo-realistic" rendering to help students visualize applications and provide a stronger connection to actual engineering practice.

Consistent Use of Color

To help students recognize and interpret elements of figures, we use consistent identifying colors:

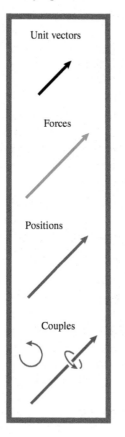

Unit vectors

Forces

Positions

Couples

Triple Accuracy Checking—Commitment to Students and Instructors

It is our commitment to students and instructors to take precautions to ensure the accuracy of the text and its solutions manual to the best of our ability. We use a system of triple accuracy checking in which three parties, in addition to the authors, solve the problems in an effort to be sure that their answers are correct and that they are of an appropriate level of difficulty. Our accuracy team consists of:

- Scott Hendricks of Virginia Polytechnic University
- Karim Nohra of the University of South Florida
- Kurt Norlin of Laurel Technical Services

The parties further verified the text, examples, problems, and solutions manuals to help ensure accuracy. Any errors remain the responsibility of the authors. We welcome communication from students and instructors concerning errors or areas for improvement. Our mailing address is Department of Aerospace Engineering and Engineering Mechanics, University of Texas at Austin, Austin, Texas 78712. Our email address is abedford@mail.utexas.edu

Instructor & Student Resources

Student Resources

The *Statics* and *Dynamics* Study Packs are designed to give students the tools to improve their skills drawing free-body diagrams, and to help them review for tests. Each contains a tutorial on free-body diagrams with fifty practice problems of increasing difficulty with complete solutions. Further strategies and tips help students understand how to use the diagrams in solving the accompanying problems. These supplements and accompanying chapter by chapter review material were prepared by Peter Schiavone of the University of Alberta. An access code for the Companion Website is included inside each Study Pack.

The Study Packs are also available as stand-alone items. Order stand-alone Study Packs with the ISBNs 0-13-614002-5 (Statics) and 0-13-614001-7 (Dynamics).

Web Assessment and Tutorial Resources—Students can access study and review resources such as supplemental practice problems on the Companion Website for this text.

www.prenhall.com/bedford

MATLAB and Mathcad tutorials are available on the Companion Website. Each tutorial discusses a basic mechanics concept, and then shows how to solve a specific problem related to this concept using MATLAB and Mathcad. There are twenty tutorials each for MATLAB and Mathcad. Worksheets were developed by Ronald Larsen and Stephen Hunt of Montana State University–Bozeman.

Additionally, instructors can assign online homework for students using PH GradeAssist. Answers are graded and results are recorded electronically. Online subscription is available or a code can be ordered stand-alone using the ISBN 0-13-601218-3.

Activebook—The Activebook is a subscription-based, full-featured online version of the textbook. Shared comments, note-taking, and text-highlighting make this a powerful tool for study and review. Order access online from www.prenhall.com/bedford or order ISBN 0-13-600348-6 (Statics), 0-13-600349-4 (Dynamics), 0-13-601212-4 (Combined Statics & Dynamics).

Instructor Resources

Instructor's Solutions Manual—This supplement, available to instructors, contains completely worked out solutions. Each solution comes with the problem statement as well as associated artwork. The ISBNs for the printed manuals are 0-13-614003-3 (Statics) and 0-13-614032-7 (Dynamics). Solutions are also available electronically for instructors at www.prenhall.com

Instructor's Resource Center on CD—This CD contains PowerPoint slides and JPEG files of all art from the text. It also contains sets of PowerPoint slides showing each example.

Web Assessment and Tutorial Resources—Through PH (Prentice Hall) GradeAssist, instructors can create online assignments for students using problems from the text. PH GradeAssist also offers every problem in an algorithmic format so that each student can work with slightly different numbers. Students also benefit from an integrated e-book. Answers to problems are recorded in an online gradebook that can be downloaded into Excel. For additional review, study, and practice resources, students should access the Companion Website where they can find supplemental problem sets and information. Contact your Prentice Hall representative for details or a demonstration.

Activebook—Instructors can use the Activebook, a full-featured online version of the textbook, as a powerful tool to annotate and share comments with students. Contact your Prentice Hall representative for details or a demonstration.

Ordering Options

Engineering Mechanics: Statics & Dynamics with Study Packs (ISBN 0-13-241871-1). Engineering Mechanics: Statics & Dynamics with Study Packs and PHGA (ISBN 0-13-235709-7).

Acknowledgments

The following colleagues provided reviews based on their knowledge and teaching experience that greatly helped us in preparing the current and previous editions.

Shaaban Abdallah
University of Cincinnati

Edward E. Adams
Michigan Technological University

George G. Adams
Northeastern University

Raid S. Al-Akkad
University of Dayton

Jerry L. Anderson
Memphis State University

James G. Andrews
University of Iowa

Robert J. Asaro
University of California, San Diego

Leonard B. Baldwin
University of Wyoming

Haim Baruh
Rutgers University

Gautam Batra
University of Nebraska

David M. Bayer
University of North Carolina

Glenn Beltz
University of California–Santa Barbara

Mary Bergs
Marquette University

Don L. Boyer
Arizona State University

Spencer Brinkerhoff
Northern Arizona University

L. M. Brock
University of Kentucky

William (Randy) Burkett
Texas Tech University

Donald Carlson
University of Illinois

Major Robert M. Carpenter
U.S. Military Academy

Douglas Carroll
University of Missouri, Rolla

Paul C. Chan
New Jersey Institute of Technology

Namas Chandra
Florida State University

James Cheney
University of California, Davis

Ravinder Chona
Texas A & M University

Daniel C. Deckler
The University of Akron Wayne College

Anthony DeLuzio
Merrimack College

Mitsunori Denda
Rutgers University

James F. Devine
University of South Florida

Craig Douglas
University of Massachusetts, Lowell

Marijan Dravinski
University of Southern California

S. Olani Durrant
Brigham Young University

Estelle Eke
California State University, Sacramento

Bogdan I. Epureanu
University of Michigan

William Ferrante
University of Rhode Island

Robert W. Fitzgerald
Worcester Polytechnic Institute

George T. Flowers
Auburn University

Mark Frisina
Wentworth Institute

Robert W. Fuessle
Bradley University

Walter Gerstle
University of New Mexico

William Gurley
University of Tennessee, Chattanooga

John Hansberry
University of Massachusetts, Dartmouth

Mark J. Harper
United States Naval Academy

W. C. Hauser
California Polytechnic University, Pomona

Linda Hayes
University of Texas–Austin

R. Craig Henderson
Tennessee Technological University

Paul R. Heyliger
Colorado State University

James Hill
University of Alabama

Robert W. Hinks
Arizona State University

Allen Hoffman
Worcester Polytechnic Institute

Edward E. Hornsey
University of Missouri, Rolla

Robert A. Howland
University of Notre Dame

Joe Ianelli
University of Tennessee, Knoxville

Ali Iranmanesh
Gadsden State Community College

David B. Johnson
Southern Methodist University

E. O. Jones, Jr.
Auburn University

Serope Kalpakjian
Illinois Institute of Technology

Kathleen A. Keil
California Polytechnic University, San Luis Obispo

Yohannes Ketema
University of Minnesota

Seyyed M. H. Khandani
Diablo Valley College

Charles M. Krousgrill
Purdue University

B. Kent Lall
Portland State University

Chad M. Landis
Rice Unversity

Kenneth W. Lau
University of Massachusetts, Lowell

Norman Laws
University of Pittsburgh

William M. Lee
U.S. Naval Academy

Donald G. Lemke
University of Illinois, Chicago

Richard J. Leuba
North Carolina State University

Richard Lewis
Louisiana Technological University

John B. Ligon
Michigan Tech University

Bertram Long
Northeastern University

V. J. Lopardo
U.S. Naval Academy

Frank K. Lu
University of Texas, Arlington

Mark T. Lusk
Colorado School of Mines

K. Madhaven
Christian Brothers College

Nels Madsen
Auburn University

James R. Matthews
University of New Mexico

Gary H. McDonald
University of Tennessee

James McDonald
Texas Technical University

Jim Meagher
California Polytechnic State University, San Luis Obispo

Lee Minardi
Tufts University

Norman Munroe
Florida International University

Shanti Nair
University of Massachusetts, Amherst

Saeed Niku
California Polytechnic State University, San Luis Obispo

Mohammad Noori
North Carolina State University

Harinder Singh Oberoi
Western Washington University

James O'Connor
University of Texas, Austin

Samuel P. Owusu-Ofori
North Carolina A & T State University

Venkata Panchakarla
Florida State University

Assimina A. Pelegri
Rutgers University

Noel C. Perkins
University of Michigan

Corrado Poli
University of Massachusetts–Amherst

David J. Purdy
Rose-Hulman Institute of Technology

Yitshak Ram
Louisiana State University

Colin E. Ratcliffe
U.S. Naval Academy

Daniel Riahi
University of Illinois

Charles Ritz
California Polytechnic State University, Pomona

George Rosborough
University of Colorado, Boulder

Edwin C. Rossow
Northwestern University

Kenneth Sawyers
Lehigh University

Robert Schmidt
University of Detroit

Robert J. Schultz
Oregon State University

Richard A. Scott
University of Michigan

Brian Self
U.S. Air Force Academy

William Semke
University of North Dakota

Patricia M. Shamamy
Lawrence Technological University

Sorin Siegler
Drexel University

Peng Song
Rutgers State University

Candace S. Sulzbach
Colorado School of Mines

L. N. Tao
Illinois Institute of Technology

Craig Thompson
Western Wyoming Community College

John Tomko
Cleveland State University

Kevin Z. Truman
Washington University

John Valasek
Texas A & M University

Christine Valle
Georgia Institute of Technology

Dennis VandenBrink
Western Michigan University

Thomas J. Vasko
University of Hartford

Mark R. Virkler
University of Missouri, Columbia

William H. Walston, Jr.
University of Maryland

Andrew J. Walters
Mississippi University

Reynolds Watkins
Utah State University

Charles White
Northeastern University

Norman Wittels
Worcester Polytechnic Institute

Julius P. Wong
University of Louisville

T. W. Wu
University of Kentucky

Constance Ziemian
Bucknell University

The new elements that make this edition such a departure from our previous ones, particularly the integration of text and art, were developed with the help of students, colleagues and our publisher. Our reviewers of the early samples offered encouragement and suggested helpful refinements. Once the new format had been established, the support we received from Prentice Hall in the development of the books was typically superb. Our editor Tacy Quinn smoothly organized the large team effort that books of this kind require and gave us enthusiastic help and insightful advice. Marcia Horton and Tim Galligan were dedicated to this major revision from our initial conversations about our ideas through the text's publication. Craig Little continued to teach us about the details of book production and was instrumental in keeping the project on schedule. Xiaohong Zhu again provided consummate support on art and photography issues. Dee Bernhard and Mack Patterson managed our communications with reviewers and users of the books. Jennifer Lonschein provided editorial and production support. David Alick, Rich Barnes, Kristin Mayo, Ben Paris, and Kyrce Swenson coordinated the development of the online resources that have become such essential tools for users. Jonathan Boylan designed the covers. We thank Peter Schiavone for developing the study packs that accompany the books, and Stephen Hunt and Ronald Larsen for writing the MATLAB/Mathcad tutorials. Scott Hendricks, Karim Nohra, and Kurt Norlin, valued colleagues from previous campaigns, advised us on style and clarity, corrected many of our errors, and revised the solution manuals. We are responsible for errors that remain. Nancy Bedford gave us editorial advice and help with checking. Many other talented and professional people at Prentice Hall and elsewhere contributed, and we thank them. And once again we thank our families, especially Nancy and Marsha, for their forbearance and understanding of the realities of new editions.

Anthony Bedford and Wallace Fowler
Austin, Texas

About the Authors

Anthony Bedford (*l*) and Wallace T. Fowler

Anthony Bedford is Professor Emeritus of Aerospace Engineering and Engineering Mechanics at the University of Texas at Austin, and has been on the faculty since 1968. A member of the University of Texas Academy of Distinguished Teachers, his main professional activity has been education and research in engineering mechanics. He has written papers on mixture theory, wave propagation, and the mechanics of high velocity impacts and is the author of the books *Hamilton's Principle in Continuum Mechanics, Introduction to Elastic Wave Propagation* (with D. S. Drumheller), and *Mechanics of Materials* (with K. M. Liechti). He has industrial experience at Douglas Aircraft Company, TRW, and Sandia National Laboratories.

Wallace T. Fowler is the Paul D. & Betty Robertson Meek Centennial Professor of Engineering at the University of Texas and is Director of the Texas Space Grant Consortium. He is a fellow of both the American Institute of Aeronautics and Astronautics (AIAA) and the American Society for Engineering Education (ASEE). Dr. Fowler was the recipient of the 1976 General Dynamics Teaching Excellence Award, the 1985 AIAA–ASEE John Leland Atwood Award (for National Outstanding Aerospace Engineering Educator), the 1990–91 University of Texas System Chancellor's Council Teaching Award, and the 1994 ASEE Fred Merryfield Design Teaching Award. In 1997, he was elected to membership in the University of Texas Academy of Distinguished Teachers. Dr. Fowler also served as 2000–2001 national president of the American Society for Engineering Education. Dr. Fowler's research and teaching interests at UT Austin focus on space systems engineering and design.

Photo Credits

Engineering Mechanics
STATICS

CHAPTER
1

Introduction

How do engineers design and construct the devices we use, from simple objects such as chairs and pencil sharpeners to complicated ones such as dams, cars, airplanes, and spacecraft? They must have a deep understanding of the physics underlying the design of such devices and must be able to use mathematical models to predict their behavior. Students of engineering begin to learn how to analyze and predict the behaviors of physical systems by studying mechanics.

◀ Engineers are guided by the principles of statics during each step of the design and assembly of a structure. Statics is one of the sciences underlying the art of structural design.

1.1 Engineering and Mechanics

BACKGROUND

How can engineers design complex systems and predict their characteristics before they are constructed? Engineers have always relied on their knowledge of previous designs, experiments, ingenuity, and creativity to develop new designs. Modern engineers add a powerful technique: They develop mathematical equations based on the physical characteristics of the devices they design. With these mathematical models, engineers predict the behavior of their designs, modify them, and test them prior to their actual construction. Aerospace engineers use mathematical models to predict the paths the space shuttle will follow in flight. Civil engineers use mathematical models to analyze the effects of loads on buildings and foundations.

At its most basic level, mechanics is the study of forces and their effects. Elementary mechanics is divided into *statics*, the study of objects in equilibrium, and *dynamics*, the study of objects in motion. The results obtained in elementary mechanics apply directly to many fields of engineering. Mechanical and civil engineers designing structures use the equilibrium equations derived in statics. Civil engineers analyzing the responses of buildings to earthquakes and aerospace engineers determining the trajectories of satellites use the equations of motion derived in dynamics.

Mechanics was the first analytical science. As a result, fundamental concepts, analytical methods, and analogies from mechanics are found in virtually every field of engineering. Students of chemical and electrical engineering gain a deeper appreciation for basic concepts in their fields, such as equilibrium, energy, and stability, by learning them in their original mechanical contexts. By studying mechanics, they retrace the historical development of these ideas.

Mechanics consists of broad principles that govern the behavior of objects. In this book we describe these principles and provide examples that demonstrate some of their applications. Although it is essential that you practice working problems similar to these examples, and we include many problems of this kind, our objective is to help you understand the principles well enough to apply them to situations that are new to you. Each generation of engineers confronts new problems.

Problem Solving

In the study of mechanics you learn problem-solving procedures that you will use in succeeding courses and throughout your career. Although different types of problems require different approaches, the following steps apply to many of them:

- Identify the information that is given and the information, or answer, you must determine. It's often helpful to restate the problem in your own words. When appropriate, make sure you understand the physical system or model involved.

- Develop a *strategy* for the problem. This means identifying the principles and equations that apply and deciding how you will use them to solve the problem. Whenever possible, draw diagrams to help visualize and solve the problem.

- Whenever you can, try to predict the answer. This will develop your intuition and will often help you recognize an incorrect answer.

- Solve the equations and, whenever possible, interpret your results and compare them with your prediction. This last step is a *reality check*. Is your answer reasonable?

Numbers

Engineering measurements, calculations, and results are expressed in numbers. You need to know how we express numbers in the examples and problems and how to express the results of your own calculations.

Significant Digits   This term refers to the number of meaningful (that is, accurate) digits in a number, counting to the right starting with the first nonzero digit. The two numbers 7.630 and 0.007630 are each stated to four significant digits. If only the first four digits in the number 7,630,000 are known to be accurate, this can be indicated by writing the number in scientific notation as 7.630×10^6.

If a number is the result of a measurement, the significant digits it contains are limited by the accuracy of the measurement. If the result of a measurement is stated to be 2.43, this means that the actual value is believed to be closer to 2.43 than to 2.42 or 2.44.

Numbers may be rounded off to a certain number of significant digits. For example, we can express the value of π to three significant digits, 3.14, or we can express it to six significant digits, 3.14159. When you use a calculator or computer, the number of significant digits is limited by the number of digits the machine is designed to carry.

Use of Numbers in This Book   You should treat numbers given in problems as exact values and not be concerned about how many significant digits they contain. If a problem states that a quantity equals 32.2, you can assume its value is 32.200.... We generally express intermediate results and answers in the examples and the answers to the problems to at least three significant digits. If you use a calculator, your results should be that accurate. Be sure to avoid round-off errors that occur if you round off intermediate results when making a series of calculations. Instead, carry through your calculations with as much accuracy as you can by retaining values in your calculator.

Space and Time

Space simply refers to the three-dimensional universe in which we live. Our daily experiences give us an intuitive notion of space and the locations, or positions, of points in space. The distance between two points in space is the length of the straight line joining them.

Measuring the distance between points in space requires a unit of length. We use both the International System of units, or SI units, and U.S. Customary units. In SI units, the unit of length is the meter (m). In U.S. Customary units, the unit of length is the foot (ft).

Time is, of course, familiar—our lives are measured by it. The daily cycles of light and darkness and the hours, minutes, and seconds measured by our clocks and watches give us an intuitive notion of time. Time is measured by the intervals between repeatable events, such as the swings of a clock pendulum or the vibrations of a quartz crystal in a watch. In both SI units and U.S. Customary units, the unit of time is the second (s). The minute (min), hour (h), and day are also frequently used.

If the position of a point in space relative to some reference point changes with time, the rate of change of its position is called its *velocity*, and the rate of change of its velocity is called its *acceleration*. In SI units, the velocity is expressed in meters per second (m/s) and the acceleration is expressed in meters per second per second, or meters per second squared (m/s^2). In U.S.

Customary units, the velocity is expressed in feet per second (ft/s) and the acceleration is expressed in feet per second squared (ft/s^2).

Newton's Laws

Elementary mechanics was established on a firm basis with the publication in 1687 of *Philosophiae Naturalis Principia Mathematica*, by Isaac Newton. Although highly original, it built on fundamental concepts developed by many others during a long and difficult struggle toward understanding (Fig. 1.1).

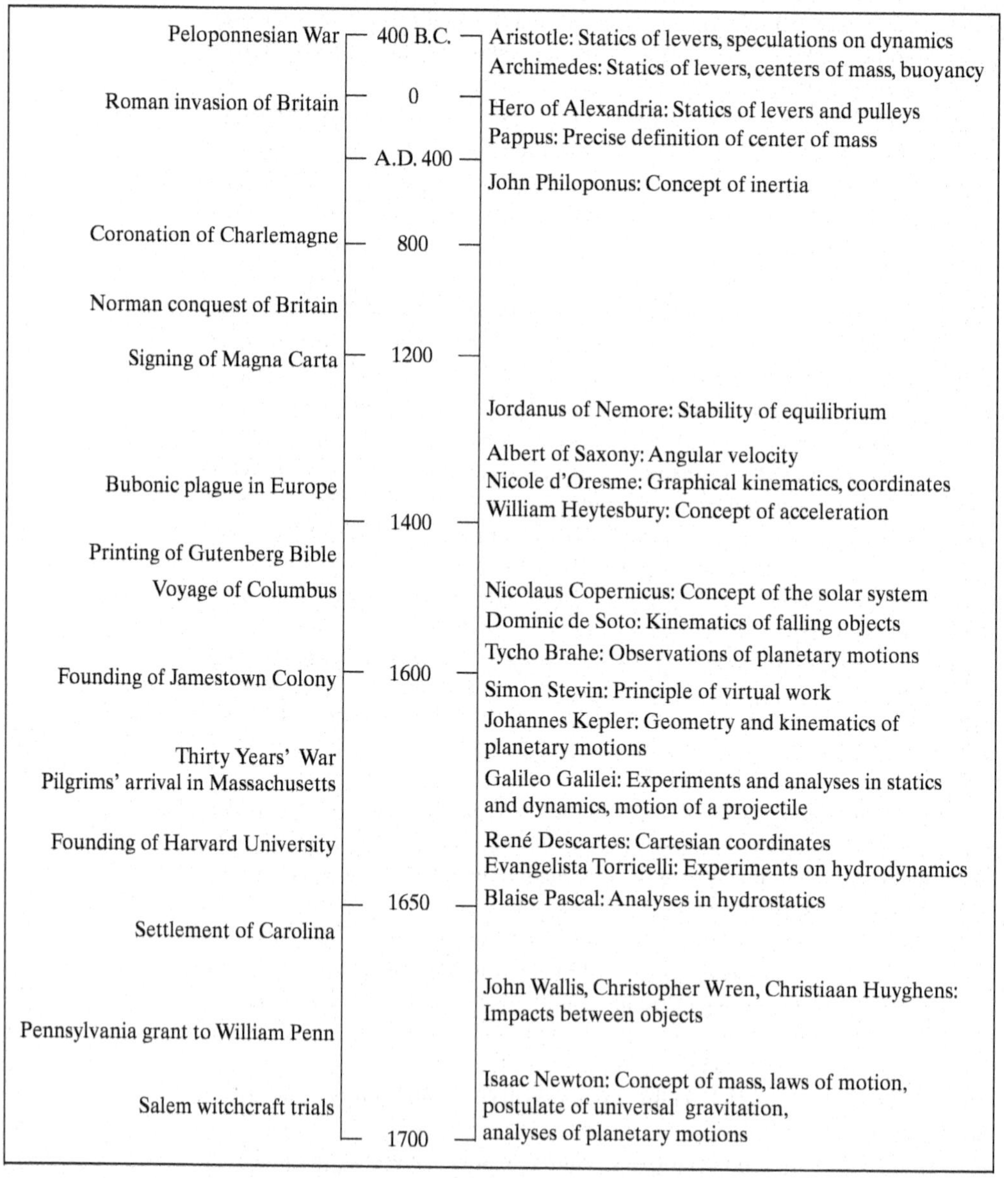

Peloponnesian War	400 B.C.	Aristotle: Statics of levers, speculations on dynamics
		Archimedes: Statics of levers, centers of mass, buoyancy
Roman invasion of Britain	0	Hero of Alexandria: Statics of levers and pulleys
		Pappus: Precise definition of center of mass
	A.D. 400	
		John Philoponus: Concept of inertia
Coronation of Charlemagne	800	
Norman conquest of Britain		
Signing of Magna Carta	1200	
		Jordanus of Nemore: Stability of equilibrium
		Albert of Saxony: Angular velocity
Bubonic plague in Europe		Nicole d'Oresme: Graphical kinematics, coordinates
	1400	William Heytesbury: Concept of acceleration
Printing of Gutenberg Bible		
Voyage of Columbus		Nicolaus Copernicus: Concept of the solar system
		Dominic de Soto: Kinematics of falling objects
		Tycho Brahe: Observations of planetary motions
Founding of Jamestown Colony	1600	Simon Stevin: Principle of virtual work
		Johannes Kepler: Geometry and kinematics of planetary motions
Thirty Years' War		
Pilgrims' arrival in Massachusetts		Galileo Galilei: Experiments and analyses in statics and dynamics, motion of a projectile
Founding of Harvard University		René Descartes: Cartesian coordinates
		Evangelista Torricelli: Experiments on hydrodynamics
	1650	Blaise Pascal: Analyses in hydrostatics
Settlement of Carolina		
		John Wallis, Christopher Wren, Christiaan Huyghens: Impacts between objects
Pennsylvania grant to William Penn		
		Isaac Newton: Concept of mass, laws of motion,
Salem witchcraft trials		postulate of universal gravitation,
	1700	analyses of planetary motions

Figure 1.1

Chronology of developments in mechanics up to the publication of Newton's *Principia* in relation to other events in history.

Newton stated three "laws" of motion, which we express in modern terms:

1. *When the sum of the forces acting on a particle is zero, its velocity is constant. In particular, if the particle is initially stationary, it will remain stationary.*

2. *When the sum of the forces acting on a particle is not zero, the sum of the forces is equal to the rate of change of the linear momentum of the particle. If the mass is constant, the sum of the forces is equal to the product of the mass of the particle and its acceleration.*

3. *The forces exerted by two particles on each other are equal in magnitude and opposite in direction.*

Notice that we did not define force and mass before stating Newton's laws. The modern view is that these terms are defined by the second law. To demonstrate, suppose that we choose an arbitrary object and define it to have unit mass. Then we define a unit of force to be the force that gives our unit mass an acceleration of unit magnitude. In principle, we can then determine the mass of any object: We apply a unit force to it, measure the resulting acceleration, and use the second law to determine the mass. We can also determine the magnitude of any force: We apply it to our unit mass, measure the resulting acceleration, and use the second law to determine the force.

Thus Newton's second law gives precise meanings to the terms *mass* and *force*. In SI units, the unit of mass is the kilogram (kg). The unit of force is the newton (N), which is the force required to give a mass of one kilogram an acceleration of one meter per second squared. In U.S. Customary units, the unit of force is the pound (lb). The unit of mass is the slug, which is the amount of mass accelerated at one foot per second squared by a force of one pound.

Although the results we discuss in this book are applicable to many of the problems met in engineering practice, there are limits to the validity of Newton's laws. For example, they don't give accurate results if a problem involves velocities that are not small compared to the velocity of light (3×10^8 m/s). Einstein's special theory of relativity applies to such problems. Elementary mechanics also fails in problems involving dimensions that are not large compared to atomic dimensions. Quantum mechanics must be used to describe phenomena on the atomic scale.

International System of Units

In SI units, length is measured in meters (m) and mass in kilograms (kg). Time is measured in seconds (s), although other familiar measures such as minutes (min), hours (h), and days are also used when convenient. Meters, kilograms, and seconds are called the *base units* of the SI system. Force is measured in newtons (N). Recall that these units are related by Newton's second law: One newton is the force required to give an object of one kilogram mass an acceleration of one meter per second squared:

$$1 \text{ N} = (1 \text{ kg})(1 \text{ m/s}^2) = 1 \text{ kg-m/s}^2.$$

Because the newton can be expressed in terms of the base units, it is called a *derived unit*.

To express quantities by numbers of convenient size, multiples of units are indicated by prefixes. The most common prefixes, their abbreviations, and the multiples they represent are shown in Table 1.1. For example, 1 km is 1 kilometer, which is 1000 m, and 1 Mg is 1 megagram, which is 10^6 g, or 1000 kg. We frequently use kilonewtons (kN).

Table 1.1   The common prefixes used in SI units and the multiples they represent.

Prefix	Abbreviation	Multiple
nano-	n	10^{-9}
micro-	μ	10^{-6}
milli-	m	10^{-3}
kilo-	k	10^{3}
mega-	M	10^{6}
giga-	G	10^{9}

U.S. Customary Units

In U.S. Customary units, length is measured in feet (ft) and force is measured in pounds (lb). Time is measured in seconds (s). These are the base units of the U.S. Customary system. In this system of units, mass is a derived unit. The unit of mass is the slug, which is the mass of material accelerated at one foot per second squared by a force of one pound. Newton's second law states that

$$1 \text{ lb} = (1 \text{ slug})(1 \text{ ft/s}^2).$$

From this expression we obtain

$$1 \text{ slug} = 1 \text{ lb-s}^2/\text{ft}.$$

We use other U.S. Customary units such as the mile (1 mi = 5280 ft) and the inch (1 ft = 12 in). We also use the kilopound (kip), which is 1000 lb.

Angular Units

In both SI and U.S. Customary units, angles are normally expressed in radians (rad). We show the value of an angle θ in radians in Fig. 1.2. It is defined to be the ratio of the part of the circumference subtended by θ to the radius of the circle. Angles are also expressed in degrees. Since there are 360 degrees (360°) in a complete circle, and the complete circumference of the circle is $2\pi R$, 360° equals 2π rad.

Equations containing angles are nearly always derived under the assumption that angles are expressed in radians. Therefore, when you want to substitute the value of an angle expressed in degrees into an equation, you should first convert it into radians. A notable exception to this rule is that many calculators are designed to accept angles expressed in either degrees or radians when you use them to evaluate functions such as $\sin\theta$.

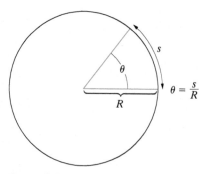

Figure 1.2
Definition of an angle in radians.

Conversion of Units

Many situations arise in engineering practice that require values expressed in one kind of unit to be converted into values in other units. For example, if some of the data to be used in an equation are given in SI units and some are given in U.S. Customary units, they must all be expressed in terms of one system of units before they are substituted into the equation. Converting units is straight-forward, although it must be done with care.

Suppose that we want to express 1 mile per hour (mi/h) in terms of feet per second (ft/s). Because 1 mile equals 5280 feet and 1 hour equals 3600 seconds, we can treat the expressions

$$\left(\frac{5280 \text{ ft}}{1 \text{ mi}}\right) \quad \text{and} \quad \left(\frac{1 \text{ h}}{3600 \text{ s}}\right)$$

as ratios whose values are 1. In this way, we obtain

$$1 \text{ mi/h} = (1 \text{ mi/h})\left(\frac{5280 \text{ ft}}{1 \text{ mi}}\right)\left(\frac{1 \text{ h}}{3600 \text{ s}}\right) = 1.47 \text{ ft/s}.$$

Some useful unit conversions are given in Table 1.2.

Table 1.2 Unit conversions.

Time	1 minute	=	60 seconds
	1 hour	=	60 minutes
	1 day	=	24 hours
Length	1 foot	=	12 inches
	1 mile	=	5280 feet
	1 inch	=	25.4 millimeters
	1 foot	=	0.3048 meters
Angle	2π radians	=	360 degrees
Mass	1 slug	=	14.59 kilograms
Force	1 pound	=	4.448 newtons

RESULTS

- Identify the given information and the answer that must be determined.
- Develop a strategy; identify principles and equations that apply and how they will be used.
- Try to predict the answer whenever possible.
- Obtain the answer and, whenever possible, interpret it and compare it with the prediction.

> Problem Solving: These steps apply to many types of problems.

SI Units—The *base units* are time in seconds (s), length in meters (m), and mass in kilograms (kg). The unit of force is the newton (N), which is the force required to accelerate a mass of one kilogram at one meter per second squared.

> Systems of units.

U.S. Customary Units—The base units are time in seconds (s), length in feet (ft), and force in pounds (lb). The unit of mass is the slug, which is the mass accelerated at one foot per second squared by a force of one pound.

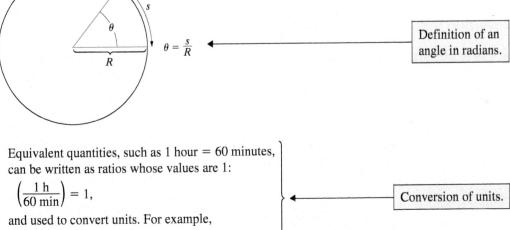

$$\theta = \frac{s}{R}$$

> Definition of an angle in radians.

Equivalent quantities, such as 1 hour = 60 minutes, can be written as ratios whose values are 1:

$$\left(\frac{1\,\text{h}}{60\,\text{min}}\right) = 1,$$

and used to convert units. For example,

$$15\,\text{min} = 15\,\text{min}\left(\frac{1\,\text{h}}{60\,\text{min}}\right) = 0.25\,\text{h}.$$

> Conversion of units.

A comprehensive resource on units has been compiled by Russ Rowlett of the University of North Carolina at Chapel Hill and made available online at www.unc.edu/~rowlett/units.

Active Example 1.1 **Converting Units** (▶ *Related Problem 1.11*)

A man is riding a bicycle at a speed of 6 meters per second (m/s). How fast is he going in kilometers per hour (km/h)?

Strategy
One kilometer is 1000 meters and one hour is 60 minutes × 60 seconds = 3600 seconds. We can use these unit conversions to determine his speed in km/h.

Solution

Convert meters to kilometers.

Convert seconds to hours.

$$6 \text{ m/s} = 6 \text{ m/s} \left(\frac{1 \text{ km}}{1000 \text{ m}} \right) \left(\frac{3600 \text{ s}}{1 \text{ h}} \right)$$

$$= 21.6 \text{ km/h}.$$

Practice Problem A man is riding a bicycle at a speed of 10 feet per second (ft/s). How fast is he going in miles per hour (mi/h)?
Answer: 6.82 mi/h.

Example 1.2 **Converting Units of Pressure** (▶ *Related Problem 1.16*)

Deep Submersible Vehicle.

The pressure exerted at a point of the hull of the deep submersible vehicle is 3.00×10^6 Pa (pascals). A pascal is 1 newton per square meter. Determine the pressure in pounds per square foot.

Strategy
From Table 1.2, 1 pound = 4.448 newtons and 1 foot = 0.3048 meters. With these unit conversions we can calculate the pressure in pounds per square foot.

Solution
The pressure (to three significant digits) is

$$3.00 \times 10^6 \text{ N/m}^2 = (3.00 \times 10^6 \text{ N/m}^2) \left(\frac{1 \text{ lb}}{4.448 \text{ N}} \right) \left(\frac{0.3048 \text{ m}}{1 \text{ ft}} \right)^2$$

$$= 62{,}700 \text{ lb/ft}^2.$$

Critical Thinking
How could we have obtained this result in a more direct way? Notice from the table of unit conversions in the inside front cover that 1 Pa = 0.0209 lb/ft^2. Therefore,

$$3.00 \times 10^6 \text{ N/m}^2 = (3.00 \times 10^6 \text{ N/m}^2) \left(\frac{0.0209 \text{ lb/ft}^2}{1 \text{ N/m}^2} \right)$$

$$= 62{,}700 \text{ lb/ft}^2.$$

Example 1.3	**Determining Units from an Equation** (▶ *Related Problem 1.20*)

Suppose that in Einstein's equation

$$E = mc^2,$$

the mass m is in kilograms and the velocity of light c is in meters per second.

(a) What are the SI units of E?

(b) If the value of E in SI units is 20, what is its value in U.S. Customary base units?

Strategy

(a) Since we know the units of the terms m and c, we can deduce the units of E from the given equation.

(b) We can use the unit conversions for mass and length from Table 1.2 to convert E from SI units to U.S. Customary units.

Solution

(a) From the equation for E,

$$E = (m\text{ kg})(c\text{ m/s})^2,$$

the SI units of E are kg-m^2/s^2.

(b) From Table 1.2, 1 slug = 14.59 kg and 1 ft = 0.3048 m. Therefore,

$$1\text{ kg-m}^2/\text{s}^2 = (1\text{ kg-m}^2/\text{s}^2)\left(\frac{1\text{ slug}}{14.59\text{ kg}}\right)\left(\frac{1\text{ ft}}{0.3048\text{ m}}\right)^2$$

$$= 0.738\text{ slug-ft}^2/\text{s}^2.$$

The value of E in U.S. Customary units is

$$E = (20)(0.738) = 14.8\text{ slug-ft}^2/\text{s}^2.$$

Critical Thinking

In part (a), how did we know that we could determine the units of E by determining the units of mc^2? The dimensions, or units, of each term in an equation must be the same. For example, in the equation $a + b = c$, the dimensions of each of the terms a, b, and c must be the same. The equation is said to be *dimensionally homogeneous*. This requirement is expressed by the colloquial phrase "Don't compare apples and oranges."

Problems

1.1 The value of π is 3.14159265. . . . If C is the circumference of a circle and r is its radius, determine the value of r/C to four significant digits.

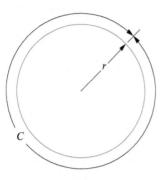

Problem 1.1

1.2 The base of natural logarithms is $e = 2.718281828\ldots$.

(a) Express e to five significant digits.

(b) Determine the value of e^2 to five significant digits.

(c) Use the value of e you obtained in part (a) to determine the value of e^2 to five significant digits.

[Part (c) demonstrates the hazard of using rounded-off values in calculations.]

1.3 A machinist drills a circular hole in a panel with a nominal radius $r = 5$ mm. The actual radius of the hole is in the range $r = 5 \pm 0.01$ mm.

(a) To what number of significant digits can you express the radius?

(b) To what number of significant digits can you express the area of the hole?

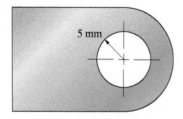

Problem 1.3

1.4 The opening in the soccer goal is 24 ft wide and 8 ft high, so its area is 24 ft $\times$ 8 ft = 192 ft^2. What is its area in m^2 to three significant digits?

Problem 1.4

1.5 The Burj Dubai, scheduled for completion in 2008, will be the world's tallest building with a height of 705 m. The area of its ground footprint will be 8000 m^2. Convert its height and footprint area to U.S. Customary units to three significant digits.

Problem 1.5

1.6 Suppose that you have just purchased a Ferrari F355 coupe and you want to know whether you can use your set of SAE (U.S. Customary unit) wrenches to work on it. You have wrenches with widths w = 1/4 in, 1/2 in, 3/4 in, and 1 in, and the car has nuts with dimensions n = 5 mm, 10 mm, 15 mm, 20 mm, and 25 mm. Defining a wrench to fit if w is no more than 2% larger than n, which of your wrenches can you use?

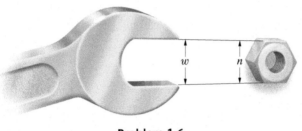

Problem 1.6

1.7 Suppose that the height of Mt. Everest is known to be between 29,032 ft and 29,034 ft. Based on this information, to how many significant digits can you express the height (a) in feet? (b) in meters?

1.8 The maglev (magnetic levitation) train from Shanghai to the airport at Pudong reaches a speed of 430 km/h. Determine its speed (a) in mi/h; (b) in ft/s.

Problem 1.8

1.9 In the 2006 Winter Olympics, the men's 15-km cross-country skiing race was won by Andrus Veerpalu of Estonia in a time of 38 minutes, 1.3 seconds. Determine his average speed (the distance traveled divided by the time required) to three significant digits (a) in km/h; (b) in mi/h.

1.10 The Porsche's engine exerts 229 ft-lb (foot-pounds) of torque at 4600 rpm. Determine the value of the torque in N-m (newton-meters).

Problem 1.10

▶ **1.11** The *kinetic energy* of the man in Active Example 1.1 is defined by $\frac{1}{2}mv^2$, where m is his mass and v is his velocity. The man's mass is 68 kg and he is moving at 6 m/s, so his kinetic energy is $\frac{1}{2}(68\ \text{kg})(6\ \text{m/s})^2 = 1224\ \text{kg-m}^2/\text{s}^2$. What is his kinetic energy in U.S. Customary units?

1.12 The acceleration due to gravity at sea level in SI units is g = 9.81 m/s^2. By converting units, use this value to determine the acceleration due to gravity at sea level in U.S. Customary units.

1.13 A *furlong per fortnight* is a facetious unit of velocity, perhaps made up by a student as a satirical comment on the bewildering variety of units engineers must deal with. A furlong is 660 ft (1/8 mile). A fortnight is 2 weeks (14 nights). If you walk to class at 2 m/s, what is your speed in furlongs per fortnight to three significant digits?

1.14 Determine the cross-sectional area of the beam (a) in m^2; (b) in in^2.

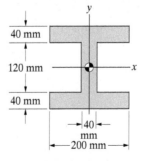

Problem 1.14

1.15 The cross-sectional area of the C12×30 American Standard Channel steel beam is $A = 8.81$ in^2. What is its cross-sectional area in mm^2?

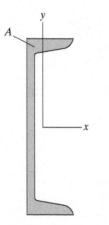

Problem 1.15

▶ **1.16** A pressure transducer measures a value of 300 lb/in^2. Determine the value of the pressure in pascals. A pascal (Pa) is one newton per square meter.

1.17 A horsepower is 550 ft-lb/s. A watt is 1 N-m/s. Determine how many watts are generated by the engines of the passenger jet if they are producing 7000 horsepower.

Problem 1.17

1.18 Chapter 7 discusses distributed loads that are expressed in units of force per unit length. If the value of a distributed load is 400 N/m, what is its value in lb/ft?

1.19 The moment of inertia of the rectangular area about the x axis is given by the equation

$$I = \tfrac{1}{3}bh^3.$$

The dimensions of the area are $b = 200$ mm and $h = 100$ mm. Determine the value of I to four significant digits in terms of (a) mm^4, (b) m^4, and (c) in^4.

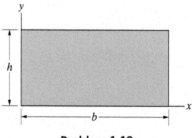

Problem 1.19

▶ **1.20** In Example 1.3, instead of Einstein's equation consider the equation $L = mc$, where the mass m is in kilograms and the velocity of light c is in meters per second. (a) What are the SI units of L? (b) If the value of L in SI units is 12, what is its value in U.S. Customary base units?

1.21 The equation

$$\sigma = \frac{My}{I}$$

is used in the mechanics of materials to determine normal stresses in beams.

(a) When this equation is expressed in terms of SI base units, M is in newton-meters (N-m), y is in meters (m), and I is in meters to the fourth power (m^4). What are the SI units of σ?

(b) If $M = 2000$ N-m, $y = 0.1$ m, and $I = 7 \times 10^{-5}$ m^4, what is the value of σ in U.S. Customary base units?

1.2 Newtonian Gravitation

BACKGROUND

Newton postulated that the gravitational force between two particles of mass m_1 and m_2 that are separated by a distance r (Fig. 1.3) is

$$F = \frac{Gm_1m_2}{r^2},$$ (1.1)

where G is called the universal gravitational constant. The value of G in SI units is 6.67×10^{-11} N-m^2/kg^2. Based on this postulate, he calculated the gravitational force between a particle of mass m_1 and a homogeneous sphere of mass m_2 and found that it is also given by Eq. (1.1), with r denoting the distance from the particle to the center of the sphere. Although the earth is not a homogeneous sphere, we can use this result to approximate the weight of an object of mass m due to the gravitational attraction of the earth. We have

$$W = \frac{Gmm_E}{r^2},$$ (1.2)

where m_E is the mass of the earth and r is the distance from the center of the earth to the object. Notice that the weight of an object depends on its location relative to the center of the earth, whereas the mass of the object is a measure of the amount of matter it contains and doesn't depend on its position.

When an object's weight is the only force acting on it, the resulting acceleration is called the acceleration due to gravity. In this case, Newton's second law states that $W = ma$, and from Eq. (1.2) we see that the acceleration due to gravity is

$$a = \frac{Gm_E}{r^2}.$$ (1.3)

The *acceleration due to gravity at sea level* is denoted by g. Denoting the radius of the earth by R_E, we see from Eq. (1.3) that $Gm_E = gR_E^2$. Substituting this result into Eq. (1.3), we obtain an expression for the acceleration due to gravity at a distance r from the center of the earth in terms of the acceleration due to gravity at sea level:

$$a = g\frac{R_E^2}{r^2}.$$ (1.4)

Since the weight of the object $W = ma$, the weight of an object at a distance r from the center of the earth is

$$W = mg\frac{R_E^2}{r^2}.$$ (1.5)

At sea level $(r = R_E)$, the weight of an object is given in terms of its mass by the simple relation

$$W = mg.$$ (1.6)

The value of g varies from location to location on the surface of the earth. The values we use in examples and problems are $g = 9.81$ m/s^2 in SI units and $g = 32.2$ ft/s^2 in U.S. Customary units.

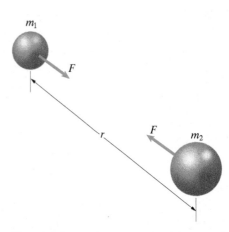

Figure 1.3
The gravitational forces between two particles are equal in magnitude and directed along the line between them.

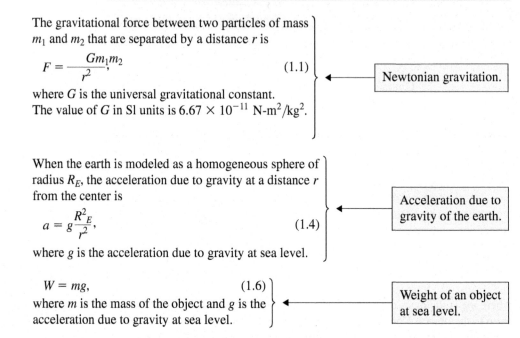

RESULTS

The gravitational force between two particles of mass m_1 and m_2 that are separated by a distance r is

$$F = \frac{Gm_1m_2}{r^2},\qquad(1.1)$$

where G is the universal gravitational constant. The value of G in SI units is 6.67×10^{-11} N-m^2/kg^2.

| Newtonian gravitation. |

When the earth is modeled as a homogeneous sphere of radius R_E, the acceleration due to gravity at a distance r from the center is

$$a = g\frac{R_E^2}{r^2},\qquad(1.4)$$

where g is the acceleration due to gravity at sea level.

| Acceleration due to gravity of the earth. |

$$W = mg,\qquad(1.6)$$

where m is the mass of the object and g is the acceleration due to gravity at sea level.

| Weight of an object at sea level. |

Active Example 1.4   **Weight and Mass** (▶ *Related Problem 1.22*)

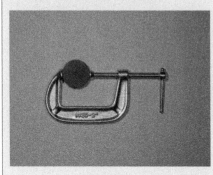

The C-clamp weighs 14 oz at sea level. [16 oz (ounces) = 1 lb.] The acceleration due to gravity at sea level is $g = 32.2$ ft/s^2. What is the mass of the C-clamp in slugs?

Strategy
We must first determine the weight of the C-clamp in pounds. Then we can use Eq. (1.6) to determine the mass in slugs.

Solution

$$14\ oz = 14\ oz\left(\frac{1\ lb}{16\ oz}\right) = 0.875\ lb.$$

| Convert the weight from ounces to pounds. |

$$m = \frac{W}{g} = \frac{0.875\ lb}{32.2\ ft/s^2} = 0.0272\ slug.$$

| Use Eq. (1.6) to calculate the mass in slugs. |

Practice Problem The mass of the C-clamp is 0.397 kg. The acceleration due to gravity at sea level is $g = 9.81$ m/s^2. What is the weight of the C-clamp at sea level in newtons?

Answer: 3.89 N.

Example 1.5	**Determining an Object's Weight** (▶ *Related Problem 1.27*)

When the Mars Exploration Rover was fully assembled, its mass was 180 kg. The acceleration due to gravity at the surface of Mars is 3.68 m/s^2 and the radius of Mars is 3390 km.

(a) What was the rover's weight when it was at sea level on Earth?

(b) What is the rover's weight on the surface of Mars?

(c) The entry phase began when the spacecraft reached the Mars atmospheric entry interface point at 3522 km from the center of Mars. What was the rover's weight at that point?

Mars Exploration Rover being assembled.

Strategy

The rover's weight at sea level on Earth is given by Eq. (1.6) with $g = 9.81$ m/s^2.

We can determine the weight on the surface of Mars by using Eq. (1.6) with the acceleration due to gravity equal to 3.68 m/s^2.

To determine the rover's weight as it began the entry phase, we can write an equation for Mars equivalent to Eq. (1.5).

Solution

(a) The weight at sea level on Earth is

$$W = mg$$
$$= (180 \text{ kg})(9.81 \text{ m/s}^2)$$
$$= 1770 \text{ N } (397 \text{ lb}).$$

(b) Let $g_M = 3.68$ m/s^2 be the acceleration due to gravity at the surface of Mars. Then the weight of the rover on the surface of Mars is

$$W = mg_M$$
$$= (180 \text{ kg})(3.68 \text{ m/s}^2)$$
$$= 662 \text{ N } (149 \text{ lb}).$$

(c) Let $R_M = 3390$ km be the radius of Mars. From Eq. (1.5), the rover's weight when it is 3522 km above the center of Mars is

$$W = mg_M \frac{R_M^2}{r^2}$$
$$= (180 \text{ kg})(3.68 \text{ m/s}^2)\frac{(3{,}390{,}000 \text{ m})^2}{(3{,}522{,}000 \text{ m})^2}$$
$$= 614 \text{ N } (138 \text{ lb}).$$

Critical Thinking

In part (c), how did we know that we could apply Eq. (1.5) to Mars? Equation (1.5) is applied to Earth based on modeling it as a homogeneous sphere. It can be applied to other celestial objects under the same assumption. The accuracy of the results depends on how aspherical and inhomogeneous the object is.

Problems

▶ **1.22** The acceleration due to gravity on the surface of the moon is 1.62 m/s². (a) What would the mass of the C-clamp in Active Example 1.4 be on the surface of the moon? (b) What would the weight of the C-clamp in newtons be on the surface of the moon?

1.23 The 1 ft × 1 ft × 1 ft cube of iron weighs 490 lb at sea level. Determine the weight in newtons of a 1 m × 1 m × 1 m cube of the same material at sea level.

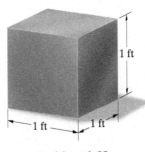

Problem 1.23

1.24 The area of the Pacific Ocean is 64,186,000 square miles and its average depth is 12,925 ft. Assume that the weight per unit volume of ocean water is 64 lb/ft³. Determine the mass of the Pacific Ocean (a) in slugs; (b) in kilograms.

1.25 The acceleration due to gravity at sea level is $g = 9.81$ m/s². The radius of the earth is 6370 km. The universal gravitational constant $G = 6.67 \times 10^{-11}$ N-m²/kg². Use this information to determine the mass of the earth.

1.26 A person weighs 180 lb at sea level. The radius of the earth is 3960 mi. What force is exerted on the person by the gravitational attraction of the earth if he is in a space station in orbit 200 mi above the surface of the earth?

▶ **1.27** The acceleration due to gravity on the surface of the moon is 1.62 m/s². The moon's radius is $R_M = 1738$ km. (See Example 1.5.)
(a) What is the weight in newtons on the surface of the moon of an object that has a mass of 10 kg?
(b) Using the approach described in Example 1.5, determine the force exerted on the object by the gravity of the moon if the object is located 1738 km above the moon's surface.

1.28 If an object is near the surface of the earth, the variation of its weight with distance from the center of the earth can often be neglected. The acceleration due to gravity at sea level is $g = 9.81$ m/s². The radius of the earth is 6370 km. The weight of an object at sea level is mg, where m is its mass. At what height above the surface of the earth does the weight of the object decrease to $0.99mg$?

1.29 The planet Neptune has an equatorial diameter of 49,532 km and its mass is 1.0247×10^{26} kg. If the planet is modeled as a homogeneous sphere, what is the acceleration due to gravity at its surface? (The universal gravitational constant is $G = 6.67 \times 10^{-11}$ N-m²/kg².)

Problem 1.29

1.30 At a point between the earth and the moon, the magnitude of the force exerted on an object by the earth's gravity equals the magnitude of the force exerted on the object by the moon's gravity. What is the distance from the center of the earth to that point to three significant digits? The distance from the center of the earth to the center of the moon is 383,000 km, and the radius of the earth is 6370 km. The radius of the moon is 1738 km, and the acceleration due to gravity at its surface is 1.62 m/s².

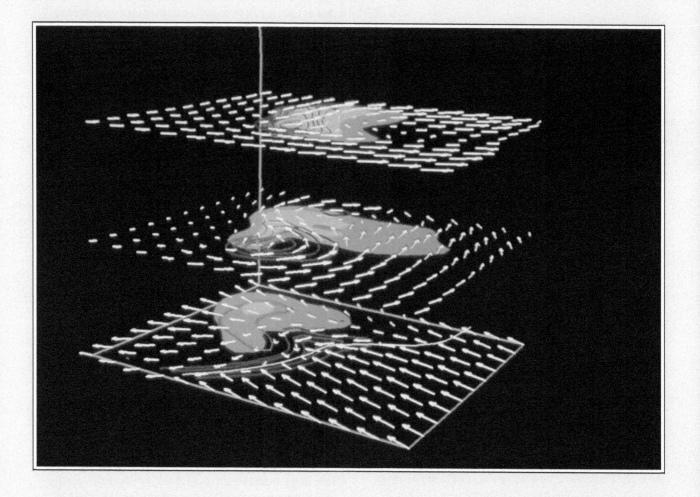

CHAPTER

2

Vectors

If an object is subjected to several forces that have different magnitudes and act in different directions, how can the magnitude and direction of the resulting total force on the object be determined? Forces are vectors and must be added according to the definition of vector addition. In engineering we deal with many quantities that have both magnitude and direction and can be expressed and analyzed as vectors. In this chapter we review vector operations, express vectors in terms of components, and present examples of engineering applications of vectors.

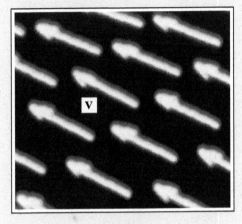

◀ Fields of vectors show the velocities and directions of a gas flow at three vertical positions. Vectors are used to describe and analyze quantities that have magnitude and direction, including positions, forces, moments, velocities, and accelerations.

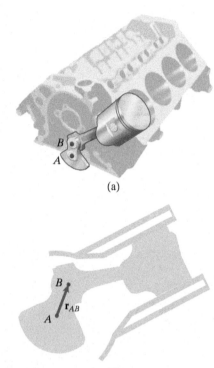

Figure 2.1
(**a**) Two points A and B of a mechanism.
(**b**) The vector $\mathbf{r}_{AB}$ from A to B.

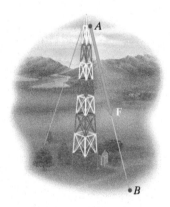

Figure 2.2
Representing the force cable AB exerts on the tower by a vector $\mathbf{F}$.

2.1 Scalars and Vectors

BACKGROUND

A physical quantity that is completely described by a real number is called a *scalar*. Time is a scalar quantity. Mass is also a scalar quantity. For example, you completely describe the mass of a car by saying that its value is 1200 kg.

In contrast, you have to specify both a nonnegative real number, or *magnitude,* and a direction to describe a vector quantity. Two vector quantities are equal only if both their magnitudes and their directions are equal.

The position of a point in space relative to another point is a vector quantity. To describe the location of a city relative to your home, it is not enough to say that it is 100 miles away. You must say that it is 100 miles west of your home. Force is also a vector quantity. When you push a piece of furniture across the floor, you apply a force of magnitude sufficient to move the furniture and you apply it in the direction you want the furniture to move.

We will represent vectors by boldfaced letters, $\mathbf{U}, \mathbf{V}, \mathbf{W}, \ldots$, and will denote the magnitude of a vector $\mathbf{U}$ by $|\mathbf{U}|$. A vector is represented graphically by an arrow. The direction of the arrow indicates the direction of the vector, and the length of the arrow is defined to be proportional to the magnitude. For example, consider the points A and B of the mechanism in Fig. 2.1a. We can specify the position of point B relative to point A by the vector $\mathbf{r}_{AB}$ in Fig. 2.1b. The direction of $\mathbf{r}_{AB}$ indicates the direction from point A to point B. If the distance between the two points is 200 mm, the magnitude $|\mathbf{r}_{AB}| = 200$ mm.

The cable AB in Fig. 2.2 helps support the television transmission tower. We can represent the force the cable exerts on the tower by a vector $\mathbf{F}$ as shown. If the cable exerts an 800-N force on the tower, $|\mathbf{F}| = 800$ N. (A cable suspended in this way will exhibit some sag, or curvature, and the tension will vary along its length. For now, we assume that the curvature in suspended cables and ropes and the variations in their tensions can be neglected. This assumption is approximately valid if the weight of the rope or cable is small in comparison to the tension. We discuss and analyze suspended cables and ropes in more detail in Chapter 10.)

Vectors are a convenient means for representing physical quantities that have magnitude and direction, but that is only the beginning of their usefulness. Just as real numbers are manipulated with the familiar rules for addition, subtraction, multiplication, and so forth, there are rules for manipulating vectors. These rules provide powerful tools for engineering analysis.

Vector Addition

When an object moves from one location in space to another, we say it undergoes a *displacement*. If we move a book (or, speaking more precisely, some point of a book) from one location on a table to another, as shown in Fig. 2.3a, we can represent the displacement by the vector $\mathbf{U}$. The direction of $\mathbf{U}$ indicates the direction of the displacement, and $|\mathbf{U}|$ is the distance the book moves.

Suppose that we give the book a second displacement $\mathbf{V}$, as shown in Fig. 2.3b. The two displacements $\mathbf{U}$ and $\mathbf{V}$ are equivalent to a single displacement of the book from its initial position to its final position, which we represent by the vector $\mathbf{W}$ in Fig. 2.3c. Notice that the final position of the book is the same whether we first give it the displacement $\mathbf{U}$ and then the displacement $\mathbf{V}$ or we first give it the displacement $\mathbf{V}$ and then the displacement $\mathbf{U}$ (Fig. 2.3d). The displacement $\mathbf{W}$ is defined to be the sum of the displacements $\mathbf{U}$ and $\mathbf{V}$:

$$\mathbf{U} + \mathbf{V} = \mathbf{W}.$$

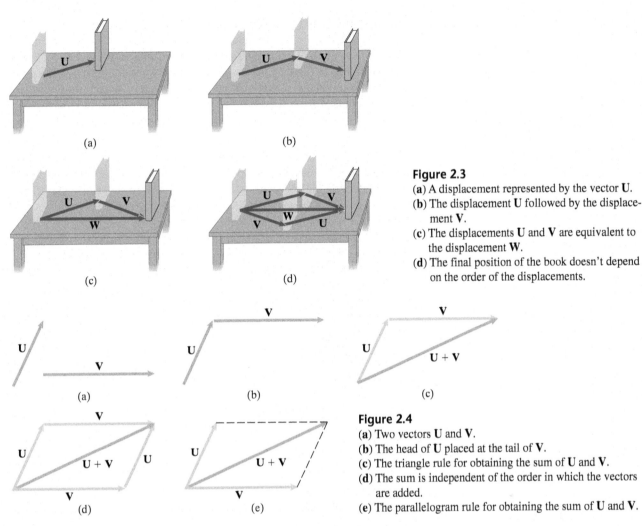

Figure 2.3
(**a**) A displacement represented by the vector **U**.
(**b**) The displacement **U** followed by the displacement **V**.
(**c**) The displacements **U** and **V** are equivalent to the displacement **W**.
(**d**) The final position of the book doesn't depend on the order of the displacements.

Figure 2.4
(**a**) Two vectors **U** and **V**.
(**b**) The head of **U** placed at the tail of **V**.
(**c**) The triangle rule for obtaining the sum of **U** and **V**.
(**d**) The sum is independent of the order in which the vectors are added.
(**e**) The parallelogram rule for obtaining the sum of **U** and **V**.

The definition of vector addition is motivated by the addition of displacements. Consider the two vectors **U** and **V** shown in Fig. 2.4a. If we place them head to tail (Fig. 2.4b), their sum is defined to be the vector from the tail of **U** to the head of **V** (Fig. 2.4c). This is called the *triangle rule* for vector addition. Figure 2.4d demonstrates that the sum is independent of the order in which the vectors are placed head to tail. From this figure we obtain the *parallelogram rule* for vector addition (Fig. 2.4e).

The definition of vector addition implies that

$$\mathbf{U} + \mathbf{V} = \mathbf{V} + \mathbf{U} \quad \text{Vector addition is commutative.} \tag{2.1}$$

and

$$(\mathbf{U} + \mathbf{V}) + \mathbf{W} = \mathbf{U} + (\mathbf{V} + \mathbf{W}) \quad \text{Vector addition is associative.} \tag{2.2}$$

for any vectors **U**, **V**, and **W**. These results mean that when two or more vectors are added, the order in which they are added doesn't matter. The sum can be obtained by placing the vectors head to tail in any order, and the vector from the tail of the first vector to the head of the last one is the sum (Fig. 2.5). If the sum of two or more vectors is zero, they form a closed polygon when they are placed head to tail (Fig. 2.6).

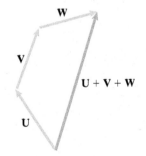

Figure 2.5
Sum of the three vectors **U**, **V**, and **W**.

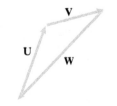

Figure 2.6
Three vectors **U**, **V**, and **W** whose sum is zero.

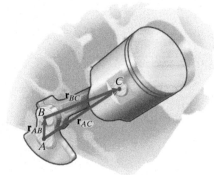

Figure 2.7
Arrows denoting the relative positions of
points are vectors.

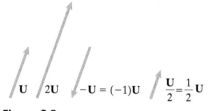

Figure 2.8
A vector **U** and some of its scalar multiples.

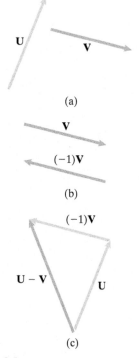

Figure 2.9
(a) Two vectors **U** and **V**.
(b) The vectors **V** and (-1)**V**.
(c) The sum of **U** and (-1)**V** is the vector
   difference **U** − **V**.

A physical quantity is called a vector if it has magnitude and direction and obeys the definition of vector addition. We have seen that displacement is a vector. The position of a point in space relative to another point is also a vector quantity. In Fig. 2.7, the vector $\mathbf{r}_{AC}$ from A to C is the sum of $\mathbf{r}_{AB}$ and $\mathbf{r}_{BC}$. A force has direction and magnitude, but do forces obey the definition of vector addition? For now we will assume that they do. When we discuss dynamics, we will show that Newton's second law implies that force is a vector.

Product of a Scalar and a Vector

The product of a scalar (real number) a and a vector **U** is a vector written as a**U**. Its magnitude is $|a||\mathbf{U}|$, where $|a|$ is the absolute value of the scalar a. The direction of a**U** is the same as the direction of **U** when a is positive and is opposite to the direction of **U** when a is negative.

The product (-1)**U** is written as $-$**U** and is called "the negative of the vector **U**." It has the same magnitude as **U** but the opposite direction. The division of a vector **U** by a scalar a is defined to be the product

$$\frac{\mathbf{U}}{a} = \left(\frac{1}{a}\right)\mathbf{U}.$$

Figure 2.8 shows a vector **U** and the products of **U** with the scalars 2, -1, and 1/2.

The definitions of vector addition and the product of a scalar and a vector imply that

$$a(b\mathbf{U}) = (ab)\mathbf{U}, \quad \text{The product is associative with} \tag{2.3}$$
$$\text{respect to scalar multiplication.}$$

$$(a + b)\mathbf{U} = a\mathbf{U} + b\mathbf{U}, \quad \text{The products are distributive} \tag{2.4}$$
$$\text{with respect to scalar addition.}$$

and

$$a(\mathbf{U} + \mathbf{V}) = a\mathbf{U} + a\mathbf{V}, \quad \text{The products are distributive} \tag{2.5}$$
$$\text{with respect to vector addition.}$$

for any scalars a and b and vectors **U** and **V**. We will need these results when we discuss components of vectors.

Vector Subtraction

The difference of two vectors **U** and **V** is obtained by adding **U** to the vector (-1)**V**:

$$\mathbf{U} - \mathbf{V} = \mathbf{U} + (-1)\mathbf{V}. \tag{2.6}$$

Consider the two vectors **U** and **V** shown in Fig. 2.9a. The vector (-1)**V** has the same magnitude as the vector **V** but is in the opposite direction (Fig. 2.9b). In Fig. 2.9c, we add the vector **U** to the vector (-1)**V** to obtain **U** − **V**.

Unit Vectors

A *unit vector* is simply a vector whose magnitude is 1. A unit vector specifies a direction and also provides a convenient way to express a vector that has a particular direction. If a unit vector **e** and a vector **U** have the same direction, we can write **U** as the product of its magnitude $|\mathbf{U}|$ and the unit vector **e** (Fig. 2.10),

$$\mathbf{U} = |\mathbf{U}|\mathbf{e}.$$

Any vector **U** *can be regarded as the product of its magnitude and a unit vector that has the same direction as* **U**. Dividing both sides of this equation by $|\mathbf{U}|$ yields

$$\frac{\mathbf{U}}{|\mathbf{U}|} = \mathbf{e},$$

so *dividing any vector by its magnitude yields a unit vector that has the same direction.*

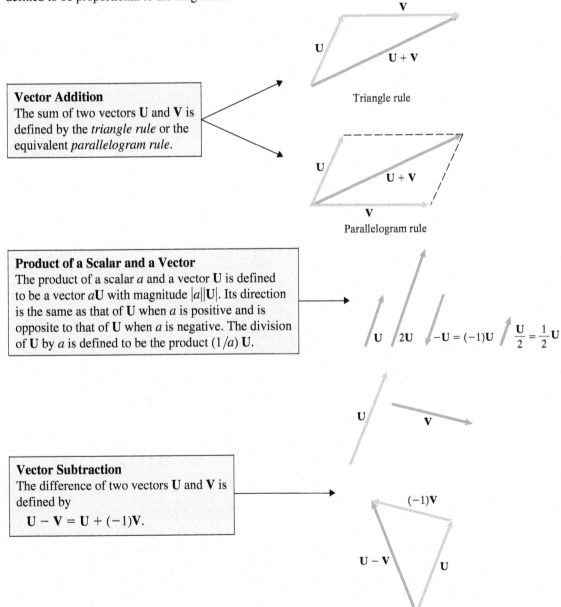

Figure 2.10
Since **U** and **e** have the same direction, the vector **U** equals the product of its magnitude with **e**.

RESULTS

A physical quantity that is completely described by a real number is called a *scalar*. A *vector* has both magnitude and direction and satisfies a defined rule of addition. A vector is represented graphically by an arrow whose length is defined to be proportional to the magnitude.

Vector Addition
The sum of two vectors **U** and **V** is defined by the *triangle rule* or the equivalent *parallelogram rule*.

Triangle rule

Parallelogram rule

Product of a Scalar and a Vector
The product of a scalar *a* and a vector **U** is defined to be a vector *a***U** with magnitude $|a||\mathbf{U}|$. Its direction is the same as that of **U** when *a* is positive and is opposite to that of **U** when *a* is negative. The division of **U** by *a* is defined to be the product $(1/a)\,\mathbf{U}$.

Vector Subtraction
The difference of two vectors **U** and **V** is defined by
$$\mathbf{U} - \mathbf{V} = \mathbf{U} + (-1)\mathbf{V}.$$

Unit Vectors
A unit vector is a vector whose magnitude is 1.
Any vector **U** can be expressed as $|\mathbf{U}|\mathbf{e}$, where
e is a unit vector with the same direction as **U**.
Dividing a vector **U** by its magnitude yields a
unit vector with the same direction as **U**.

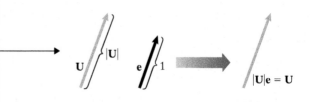

Active Example 2.1 Vector Operations (▶ *Related Problem 2.1*)

The magnitudes of the vectors shown are $|\mathbf{U}| = 8$ and $|\mathbf{V}| = 3$. The vector **V** is vertical. Graphically determine the magnitude of the vector $\mathbf{U} + 2\mathbf{V}$.

Strategy
By drawing the vectors to scale and applying the triangle rule for addition, we can measure the magnitude of the vector $\mathbf{U} + 2\mathbf{V}$.

Solution

Drawing the vectors **U** and **2V** to scale, place them head to tail.

The measured value of $|\mathbf{U} + 2\mathbf{V}|$ is 13.0.

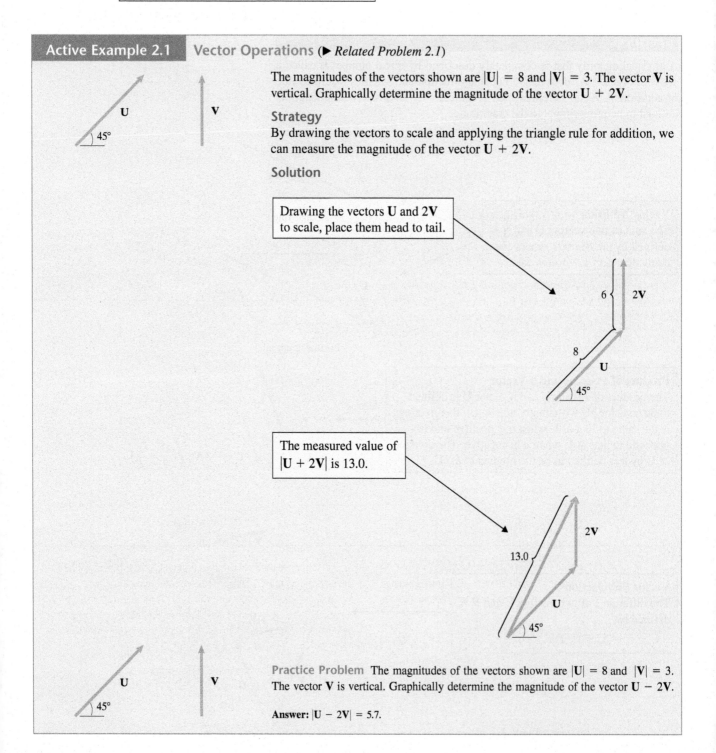

Practice Problem The magnitudes of the vectors shown are $|\mathbf{U}| = 8$ and $|\mathbf{V}| = 3$. The vector **V** is vertical. Graphically determine the magnitude of the vector $\mathbf{U} - 2\mathbf{V}$.

Answer: $|\mathbf{U} - 2\mathbf{V}| = 5.7$.

| Example 2.2 | **Adding Vectors** (▶ *Related Problem 2.2*) |

Part of the roof of a sports stadium is to be supported by the cables AB and AC. The forces the cables exert on the pylon to which they are attached are represented by the vectors $\mathbf{F}_{AB}$ and $\mathbf{F}_{AC}$. The magnitudes of the forces are $|\mathbf{F}_{AB}| = 100$ kN and $|\mathbf{F}_{AC}| = 60$ kN. Determine the magnitude and direction of the sum of the forces exerted on the pylon by the cables.

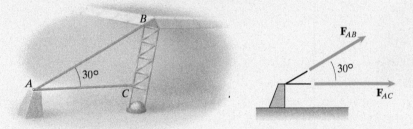

Strategy

By drawing the parallelogram rule for adding the two forces *with the vectors drawn to scale,* we can measure the magnitude and direction of their sum.

Solution

We graphically construct the parallelogram rule for obtaining the sum of the two forces with the lengths of $\mathbf{F}_{AB}$ and $\mathbf{F}_{AC}$ proportional to their magnitudes (Fig. a). By measuring the figure, we estimate the magnitude of the vector $\mathbf{F}_{AB} + \mathbf{F}_{AC}$ to be 155 kN and its direction to be 19° above the horizontal.

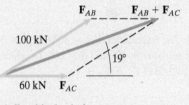

(a) Graphical solution.

Critical Thinking

In engineering applications, vector operations are nearly always done analytically. So why is it worthwhile to gain experience with graphical methods? Doing so enhances your intuition about vectors and helps you understand vector operations. Also, sketching out a graphical solution can often help you formulate an analytical solution.

Problems

▶ **2.1** In Active Example 2.1, suppose that the vectors $\mathbf{U}$ and $\mathbf{V}$ are reoriented as shown. The vector $\mathbf{V}$ is vertical. The magnitudes are $|\mathbf{U}| = 8$ and $|\mathbf{V}| = 3$. Graphically determine the magnitude of the vector $\mathbf{U} + 2\mathbf{V}$.

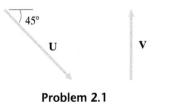

Problem 2.1

▶ **2.2** Suppose that the pylon in Example 2.2 is moved closer to the stadium so that the angle between the forces $\mathbf{F}_{AB}$ and $\mathbf{F}_{AC}$ is 50°. Draw a sketch of the new situation. The magnitudes of the forces are $|\mathbf{F}_{AB}| = 100$ kN and $|\mathbf{F}_{AC}| = 60$ kN. Graphically determine the magnitude and direction of the sum of the forces exerted on the pylon by the cables.

Refer to the following diagram when solving Problems 2.3 through 2.5. The force vectors F_A, F_B, and F_C lie in the same plane.

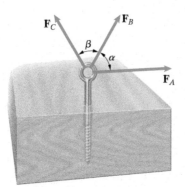

Problems 2.3–2.5

2.3 The magnitude $|\mathbf{F}_A| = 80$ lb and the angle $\alpha = 65°$. The magnitude $|\mathbf{F}_A + \mathbf{F}_B| = 120$ lb. Graphically determine the magnitude of $\mathbf{F}_B$.

2.4 The magnitudes $|\mathbf{F}_A| = 40$ N, $|\mathbf{F}_B| = 50$ N, and $|\mathbf{F}_C| = 40$ N. The angles $\alpha = 50°$ and $\beta = 80°$. Graphically determine the magnitude of $\mathbf{F}_A + \mathbf{F}_B + \mathbf{F}_C$.

2.5 The magnitudes $|\mathbf{F}_A| = |\mathbf{F}_B| = |\mathbf{F}_C| = 100$ lb, and the angle $\alpha = 30°$. Graphically determine the value of the angle β for which the magnitude $|\mathbf{F}_A + \mathbf{F}_B + \mathbf{F}_C|$ is a minimum and the minimum value of $|\mathbf{F}_A + \mathbf{F}_B + \mathbf{F}_C|$.

2.6 The angle $\theta = 50°$. Graphically determine the magnitude of the vector $\mathbf{r}_{AC}$.

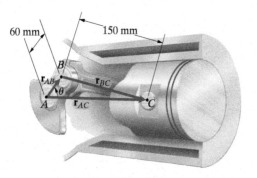

Problem 2.6

2.7 The vectors $\mathbf{F}_A$ and $\mathbf{F}_B$ represent the forces exerted on the pulley by the belt. Their magnitudes are $|\mathbf{F}_A| = 80$ N and $|\mathbf{F}_B| = 60$ N. Graphically determine the magnitude of the total force the belt exerts on the pulley.

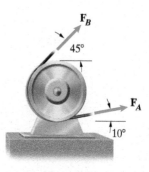

Problem 2.7

2.8 The sum of the forces $\mathbf{F}_A + \mathbf{F}_B + \mathbf{F}_C = \mathbf{0}$. The magnitude $|\mathbf{F}_A| = 100$ N and the angle $\alpha = 60°$. Graphically determine the magnitudes $|\mathbf{F}_B|$ and $|\mathbf{F}_C|$.

2.9 The sum of the forces $\mathbf{F}_A + \mathbf{F}_B + \mathbf{F}_C = \mathbf{0}$. The magnitudes $|\mathbf{F}_A| = 100$ N and $|\mathbf{F}_B| = 80$ N. Graphically determine the magnitude $|\mathbf{F}_C|$ and the angle α.

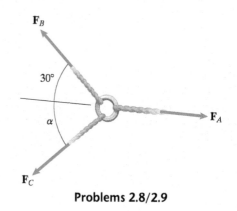

Problems 2.8/2.9

2.10 The forces acting on the sailplane are represented by three vectors. The lift **L** and drag **D** are perpendicular. The magnitude of the weight **W** is 500 lb. The sum of the forces **W** + **L** + **D** = **0**. Graphically determine the magnitudes of the lift and drag.

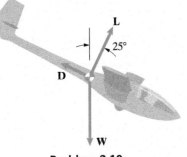

Problem 2.10

2.11 A spherical storage tank is suspended from cables. The tank is subjected to three forces, the forces **F**$_A$ and **F**$_B$ exerted by the cables and its weight **W**. The weight of the tank is $|\mathbf{W}| = 600$ lb. The vector sum of the forces acting on the tank equals zero. Graphically determine the magnitudes of **F**$_A$ and **F**$_B$.

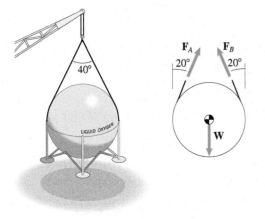

Problem 2.11

2.12 The rope ABC exerts forces $\mathbf{F}_{BA}$ and $\mathbf{F}_{BC}$ of equal magnitude on the block at B. The magnitude of the total force exerted on the block by the two forces is 200 lb. Graphically determine $|\mathbf{F}_{BA}|$.

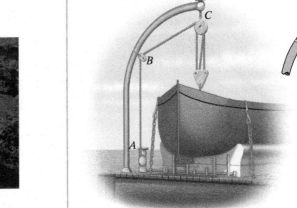

Problem 2.12

2.13 Two snowcats tow an emergency shelter to a new location near McMurdo Station, Antarctica. (The top view is shown. The cables are horizontal.) The total force $\mathbf{F}_A + \mathbf{F}_B$ exerted on the shelter is in the direction parallel to the line L and its magnitude is 400 lb. Graphically determine the magnitudes of $\mathbf{F}_A$ and $\mathbf{F}_B$.

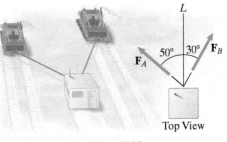

Problem 2.13

2.14 A surveyor determines that the horizontal distance from A to B is 400 m and the horizontal distance from A to C is 600 m. Graphically determine the magnitude of the vector $\mathbf{r}_{BC}$ and the angle α.

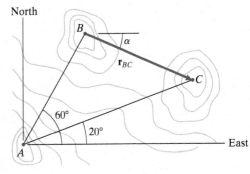

Problem 2.14

2.15 The vector **r** extends from point A to the midpoint between points B and C. Prove that

$$\mathbf{r} = \tfrac{1}{2}(\mathbf{r}_{AB} + \mathbf{r}_{AC}).$$

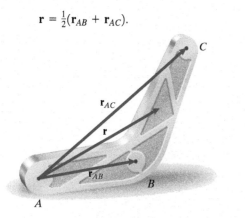

Problem 2.15

2.16 By drawing sketches of the vectors, explain why

$$\mathbf{U} + (\mathbf{V} + \mathbf{W}) = (\mathbf{U} + \mathbf{V}) + \mathbf{W}.$$

2.2 Components in Two Dimensions

BACKGROUND

Vectors are much easier to work with when they are expressed in terms of mutually perpendicular vector components. Here we explain how to express vectors in cartesian components and give examples of vector manipulations using components.

Consider the vector **U** in Fig. 2.11a. By placing a cartesian coordinate system so that the vector **U** is parallel to the x–y plane, we can write it as the sum of perpendicular *vector components* $\mathbf{U}_x$ and $\mathbf{U}_y$ that are parallel to the x and y axes (Fig. 2.11b):

$$\mathbf{U} = \mathbf{U}_x + \mathbf{U}_y.$$

Then by introducing a unit vector **i** defined to point in the direction of the positive x axis and a unit vector **j** defined to point in the direction of the positive y axis (Fig. 2.11c), we can express the vector **U** in the form

$$\mathbf{U} = U_x\mathbf{i} + U_y\mathbf{j}. \tag{2.7}$$

The scalars U_x and U_y are called *scalar components of* **U**. *When we refer simply to the components of a vector, we will mean its scalar components.* We will refer to U_x and U_y as the x and y components of **U**.

The components of a vector specify both its direction relative to the cartesian coordinate system and its magnitude. From the right triangle formed by the vector **U** and its vector components (Fig. 2.11c), we see that the magnitude of **U** is given in terms of its components by the Pythagorean theorem:

$$|\mathbf{U}| = \sqrt{U_x^2 + U_y^2}. \tag{2.8}$$

With this equation the magnitude of a vector can be determined when its components are known.

Manipulating Vectors in Terms of Components

The sum of two vectors **U** and **V** in terms of their components is

$$\begin{aligned}
\mathbf{U} + \mathbf{V} &= (U_x\mathbf{i} + U_y\mathbf{j}) + (V_x\mathbf{i} + V_y\mathbf{j}) \\
&= (U_x + V_x)\mathbf{i} + (U_y + V_y)\mathbf{j}. \tag{2.9}
\end{aligned}$$

(a)

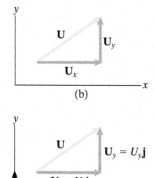

(b)

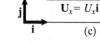

(c)

Figure 2.11
(a) A vector **U**.
(b) The vector components $\mathbf{U}_x$ and $\mathbf{U}_y$.
(c) The vector components can be expressed in terms of **i** and **j**.

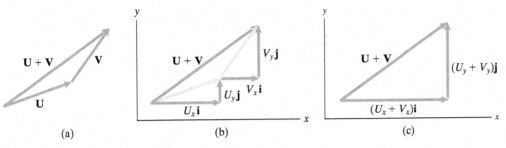

Figure 2.12
(a) The sum of **U** and **V**. (b) The vector components of **U** and **V**. (c) The sum of the components in each coordinate direction equals the component of **U** + **V** in that direction.

The components of **U** + **V** are the sums of the components of the vectors **U** and **V**. Notice that in obtaining this result we used Eqs. (2.2), (2.4), and (2.5).

It is instructive to derive Eq. (2.9) graphically. The summation of **U** and **V** is shown in Fig. 2.12a. In Fig. 2.12b we introduce a coordinate system and show the components **U** and **V**. In Fig. 2.12c we add the x and y components, obtaining Eq. (2.9).

The product of a number a and a vector **U** in terms of the components of **U** is

$$a\mathbf{U} = a(U_x\mathbf{i} + U_y\mathbf{j}) = aU_x\mathbf{i} + aU_y\mathbf{j}.$$

The component of $a\mathbf{U}$ in each coordinate direction equals the product of a and the component of **U** in that direction. We used Eqs. (2.3) and (2.5) to obtain this result.

Position Vectors in Terms of Components

We can express the position vector of a point relative to another point in terms of the cartesian coordinates of the points. Consider point A with coordinates (x_A, y_A) and point B with coordinates (x_B, y_B). Let $\mathbf{r}_{AB}$ be the vector that specifies the position of B relative to A (Fig. 2.13a). That is, we denote the vector *from* a point A *to* a point B by $\mathbf{r}_{AB}$. We see from Fig. 2.13b that $\mathbf{r}_{AB}$ is given in terms of the coordinates of points A and B by

$$\mathbf{r}_{AB} = (x_B - x_A)\mathbf{i} + (y_B - y_A)\mathbf{j}. \tag{2.10}$$

Notice that the x component of the position vector from a point A to a point B is obtained by subtracting the x coordinate of A from the x coordinate of B, and the y component is obtained by subtracting the y coordinate of A from the y coordinate of B.

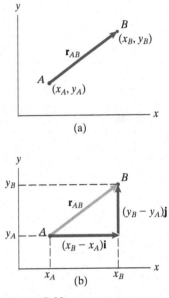

Figure 2.13
(a) Two points A and B and the position vector $\mathbf{r}_{AB}$ from A to B.
(b) The components of $\mathbf{r}_{AB}$ can be determined from the coordinates of points A and B.

RESULTS

A vector **U** that is parallel to the x–y plane can be expressed as

$$\mathbf{U} = U_x\mathbf{i} + U_y\mathbf{j}, \tag{2.7}$$

where **i** is a unit vector that points in the positive x axis direction and **j** is a unit vector that points in the positive y axis direction.

The magnitude of **U** is given by

$$|\mathbf{U}| = \sqrt{U_x^2 + U_y^2}. \tag{2.8}$$

Manipulating Vectors in Terms of Components

Vector addition (or subtraction) and multiplication of a vector by a number can be carried out in terms of components.

$$\begin{cases} \mathbf{U} + \mathbf{V} = (U_x\mathbf{i} + U_y\mathbf{j}) + (V_x\mathbf{i} + V_y\mathbf{j}) \\ \qquad = (U_x + V_x)\mathbf{i} + (U_y + V_y)\mathbf{j}, \quad (2.9) \\ \quad a\mathbf{U} = a(U_x\mathbf{i} + U_y\mathbf{j}) \\ \qquad = aU_x\mathbf{i} + aU_y\mathbf{j}. \end{cases}$$

Position Vectors in Terms of Components

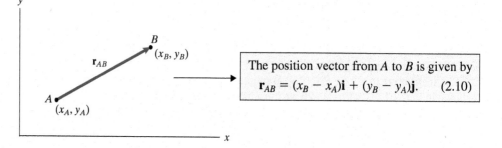

The position vector from A to B is given by
$$\mathbf{r}_{AB} = (x_B - x_A)\mathbf{i} + (y_B - y_A)\mathbf{j}. \quad (2.10)$$

Active Example 2.3   Determining Components (▶ *Related Problem 2.31*)

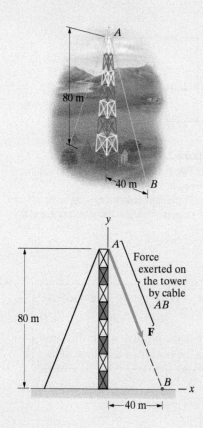

The cable from point A to point B exerts a 900-N force on the top of the television transmission tower that is represented by the vector $\mathbf{F}$. Express $\mathbf{F}$ in terms of components using the coordinate system shown.

Strategy

We will determine the components of the vector $\mathbf{F}$ in two ways. In the first method, we will determine the angle between $\mathbf{F}$ and the y axis and use trigonometry to determine the components. In the second method, we will use the given slope of the cable AB and apply similar triangles to determine the components of $\mathbf{F}$.

Solution

First Method

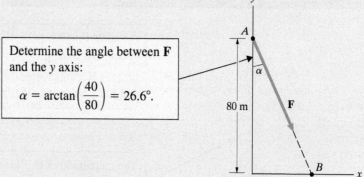

Determine the angle between $\mathbf{F}$ and the y axis:

$$\alpha = \arctan\left(\frac{40}{80}\right) = 26.6°.$$

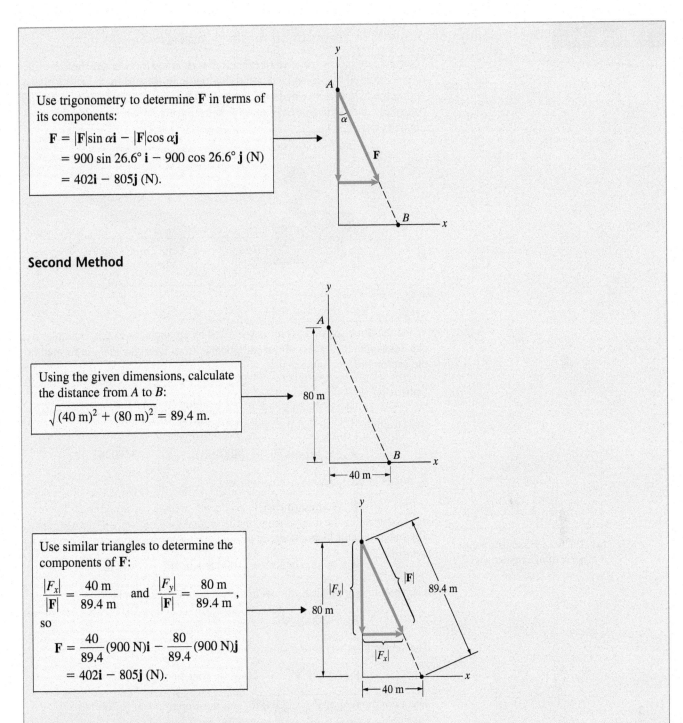

Use trigonometry to determine **F** in terms of its components:

$$\mathbf{F} = |\mathbf{F}|\sin\alpha\mathbf{i} - |\mathbf{F}|\cos\alpha\mathbf{j}$$
$$= 900\sin 26.6°\,\mathbf{i} - 900\cos 26.6°\,\mathbf{j}\ (\mathrm{N})$$
$$= 402\mathbf{i} - 805\mathbf{j}\ (\mathrm{N}).$$

Second Method

Using the given dimensions, calculate the distance from A to B:

$$\sqrt{(40\ \mathrm{m})^2 + (80\ \mathrm{m})^2} = 89.4\ \mathrm{m}.$$

Use similar triangles to determine the components of **F**:

$$\frac{|F_x|}{|\mathbf{F}|} = \frac{40\ \mathrm{m}}{89.4\ \mathrm{m}} \quad\text{and}\quad \frac{|F_y|}{|\mathbf{F}|} = \frac{80\ \mathrm{m}}{89.4\ \mathrm{m}},$$

so

$$\mathbf{F} = \frac{40}{89.4}(900\ \mathrm{N})\mathbf{i} - \frac{80}{89.4}(900\ \mathrm{N})\mathbf{j}$$
$$= 402\mathbf{i} - 805\mathbf{j}\ (\mathrm{N}).$$

Practice Problem The cable from point A to point B exerts a 900-N force on the top of the television transmission tower that is represented by the vector **F**. Suppose that you change the placement of point B so that the magnitude of the y component of **F** is three times the magnitude of the x component of **F**. Express **F** in terms of its components. How far along the x axis from the origin of the coordinate system should B be placed?

Answer: $\mathbf{F} = 285\mathbf{i} - 854\mathbf{j}$ (N). Place point B at 26.7 m from the origin.

Example 2.4 **Determining Components in Terms of an Angle** (▶ *Related Problem 2.33*)

Hydraulic cylinders are used to exert forces in many mechanical devices. The force is exerted by pressurized liquid (hydraulic fluid) pushing against a piston within the cylinder. The hydraulic cylinder *AB* exerts a 4000-lb force **F** on the bed of the dump truck at *B*. Express **F** in terms of components using the coordinate system shown.

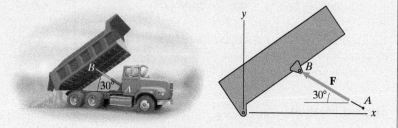

Strategy

When the direction of a vector is specified by an angle, as in this example, we can determine the values of the components from the right triangle formed by the vector and its components.

Solution

We draw the vector **F** and its vector components in Fig. a. From the resulting right triangle, we see that the magnitude of $\mathbf{F}_x$ is

$$|\mathbf{F}_x| = |\mathbf{F}| \cos 30° = (4000\ \text{lb})\cos 30° = 3460\ \text{lb}.$$

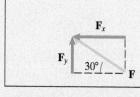

(a) The force **F** and its components form a right triangle.

$\mathbf{F}_x$ points in the negative *x* direction, so

$$\mathbf{F}_x = -3460\mathbf{i}\ (\text{lb}).$$

The magnitude of $\mathbf{F}_y$ is

$$|\mathbf{F}_y| = |\mathbf{F}| \sin 30° = (4000\ \text{lb})\sin 30° = 2000\ \text{lb}.$$

The vector component $\mathbf{F}_y$ points in the positive *y* direction, so

$$\mathbf{F}_y = 2000\mathbf{j}\ (\text{lb}).$$

The vector **F**, in terms of its components, is

$$\mathbf{F} = \mathbf{F}_x + \mathbf{F}_y = -3460\mathbf{i} + 2000\mathbf{j}\ (\text{lb}).$$

The *x* component of **F** is −3460 lb, and the *y* component is 2000 lb.

Critical Thinking

When you have determined the components of a given vector, you should make sure they appear reasonable. In this example you can see from the vector's direction that the *x* component should be negative and the *y* component positive. You can also make sure that the components yield the correct magnitude. In this example,

$$|\mathbf{F}| = \sqrt{(-3460\ \text{lb})^2 + (2000\ \text{lb})^2} = 4000\ \text{lb}.$$

| Example 2.5 | Determining an Unknown Vector Magnitude (▶ *Related Problem 2.47*) |

The cables A and B exert forces $\mathbf{F}_A$ and $\mathbf{F}_B$ on the hook. The magnitude of $\mathbf{F}_A$ is 100 lb. The tension in cable B has been adjusted so that the total force $\mathbf{F}_A + \mathbf{F}_B$ is perpendicular to the wall to which the hook is attached.

(a) What is the magnitude of $\mathbf{F}_B$?

(b) What is the magnitude of the total force exerted on the hook by the two cables?

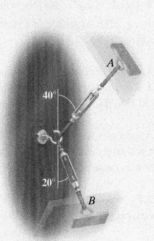

Strategy

The vector sum of the two forces is perpendicular to the wall, so the sum of the components parallel to the wall equals zero. From this condition we can obtain an equation for the magnitude of $\mathbf{F}_B$.

Solution

(a) In terms of the coordinate system shown in Fig. a, the components of $\mathbf{F}_A$ and $\mathbf{F}_B$ are

$$\mathbf{F}_A = |\mathbf{F}_A| \sin 40°\mathbf{i} + |\mathbf{F}_A| \cos 40°\mathbf{j},$$

$$\mathbf{F}_B = |\mathbf{F}_B| \sin 20°\mathbf{i} - |\mathbf{F}_B| \cos 20°\mathbf{j}.$$

The total force is

$$\mathbf{F}_A + \mathbf{F}_B = (|\mathbf{F}_A| \sin 40° + |\mathbf{F}_B| \sin 20°)\mathbf{i}$$
$$+ (|\mathbf{F}_A| \cos 40° - |\mathbf{F}_B| \cos 20°)\mathbf{j}.$$

Now we set the component of the total force parallel to the wall (the y component) equal to zero:

$$|\mathbf{F}_A| \cos 40° - |\mathbf{F}_B| \cos 20° = 0,$$

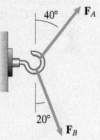

We thus obtain an equation for the magnitude of $\mathbf{F}_B$:

$$|\mathbf{F}_B| = \frac{|\mathbf{F}_A| \cos 40°}{\cos 20°} = \frac{(100 \text{ lb})\cos 40°}{\cos 20°} = 81.5 \text{ lb}.$$

(b) Since we now know the magnitude of $\mathbf{F}_B$, we can determine the total force acting on the hook:

$$\mathbf{F}_A + \mathbf{F}_B = (|\mathbf{F}_A| \sin 40° + |\mathbf{F}_B| \sin 20°)\mathbf{i}$$
$$= [(100 \text{ lb})\sin 40° + (81.5 \text{ lb})\sin 20°]\mathbf{i} = 92.2\mathbf{i} \text{ (lb)}.$$

The magnitude of the total force is 92.2 lb.

Critical Thinking

We can obtain the solution to (a) in a less formal way. If the component of the total force parallel to the wall is zero, we see in Fig. a that the magnitude of the vertical component of $\mathbf{F}_A$ must equal the magnitude of the vertical component of $\mathbf{F}_B$:

$$|\mathbf{F}_A| \cos 40° = |\mathbf{F}_B| \cos 20°.$$

Therefore the magnitude of $\mathbf{F}_B$ is

$$|\mathbf{F}_B| = \frac{|\mathbf{F}_A| \cos 40°}{\cos 20°} = \frac{(100 \text{ lb}) \cos 40°}{\cos 20°} = 81.5 \text{ lb}.$$

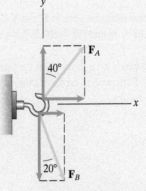

(a) Resolving $\mathbf{F}_A$ and $\mathbf{F}_B$ into components parallel and perpendicular to the wall.

Problems

2.17 A force $\mathbf{F} = 40\mathbf{i} - 20\mathbf{j}$ (N). What is its magnitude $|\mathbf{F}|$?

Strategy: The magnitude of a vector in terms of its components is given by Eq. (2.8).

2.18 An engineer estimating the components of a force $\mathbf{F} = F_x\mathbf{i} + F_y\mathbf{j}$ acting on a bridge abutment has determined that $F_x = 130$ MN, $|\mathbf{F}| = 165$ MN, and F_y is negative. What is F_y?

2.19 A support is subjected to a force $\mathbf{F} = F_x\mathbf{i} + 80\mathbf{j}$ (N). If the support will safely support a force of magnitude 100 N, what is the allowable range of values of the component F_x?

2.20 If $\mathbf{F}_A = 600\mathbf{i} - 800\mathbf{j}$ (kip) and $\mathbf{F}_B = 200\mathbf{i} - 200\mathbf{j}$ (kip), what is the magnitude of the force $\mathbf{F} = \mathbf{F}_A - 2\mathbf{F}_B$?

2.21 The forces acting on the sailplane are its weight $\mathbf{W} = -500\mathbf{j}$ (lb), the drag $\mathbf{D} = -200\mathbf{i} + 100\mathbf{j}$ (lb), and the lift $\mathbf{L}$. The sum of the forces $\mathbf{W} + \mathbf{L} + \mathbf{D} = \mathbf{0}$. Determine the components and the magnitude of $\mathbf{L}$.

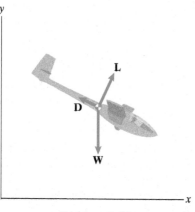

Problem 2.21

2.22 Two perpendicular vectors $\mathbf{U}$ and $\mathbf{V}$ lie in the x–y plane. The vector $\mathbf{U} = 6\mathbf{i} - 8\mathbf{j}$ and $|\mathbf{V}| = 20$. What are the components of $\mathbf{V}$?

2.23 A fish exerts a 10-lb force on the line that is represented by the vector $\mathbf{F}$. Express $\mathbf{F}$ in terms of components using the coordinate system shown.

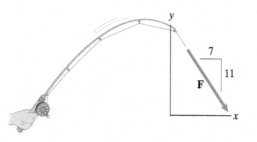

Problem 2.23

2.24 A man exerts a 60-lb force $\mathbf{F}$ to push a crate onto a truck. (a) Express $\mathbf{F}$ in terms of components using the coordinate system shown. (b) The weight of the crate is 100 lb. Determine the magnitude of the sum of the forces exerted by the man and the crate's weight.

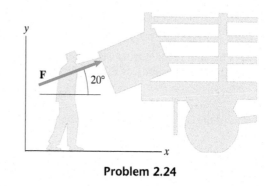

Problem 2.24

2.25 The missile's engine exerts a 260-kN force $\mathbf{F}$. (a) Express $\mathbf{F}$ in terms of components using the coordinate system shown. (b) The mass of the missile is 8800 kg. Determine the magnitude of the sum of the forces exerted by the engine and the missile's weight.

Problem 2.25

2.26 For the truss shown, express the position vector $\mathbf{r}_{AD}$ from point A to point D in terms of components. Use your result to determine the distance from point A to point D.

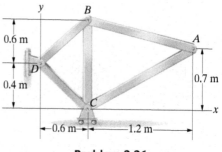

Problem 2.26

2.27 The points $A, B, \ldots$ are the joints of the hexagonal structural element. Let $\mathbf{r}_{AB}$ be the position vector from joint A to joint B, $\mathbf{r}_{AC}$ the position vector from joint A to joint C, and so forth. Determine the components of the vectors $\mathbf{r}_{AC}$ and $\mathbf{r}_{AF}$.

2.28 Determine the components of the vector $\mathbf{r}_{AB} - \mathbf{r}_{BC}$.

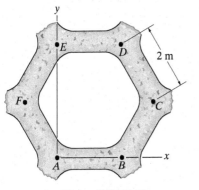

Problems 2.27/2.28

2.29 The coordinates of point A are $(1.8, 3.0)$ ft. The y coordinate of point B is 0.6 ft. The vector $\mathbf{r}_{AB}$ has the same direction as the unit vector $\mathbf{e}_{AB} = 0.616\mathbf{i} - 0.788\mathbf{j}$. What are the components of $\mathbf{r}_{AB}$?

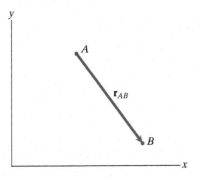

Problem 2.29

2.30 (a) Express the position vector from point A of the front-end loader to point B in terms of components.

(b) Express the position vector from point B to point C in terms of components.

(c) Use the results of (a) and (b) to determine the distance from point A to point C.

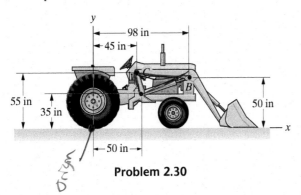

Problem 2.30

▶ **2.31** In Active Example 2.3, the cable AB exerts a 900-N force on the top of the tower. Suppose that the attachment point B is moved in the horizontal direction farther from the tower, and assume that the magnitude of the force $\mathbf{F}$ the cable exerts on the top of the tower is proportional to the length of the cable. (a) What is the distance from the tower to point B if the magnitude of the force is 1000 N? (b) Express the 1000-N force $\mathbf{F}$ in terms of components using the coordinate system shown.

2.32 Determine the position vector $\mathbf{r}_{AB}$ in terms of its components if (a) $\theta = 30°$; (b) $\theta = 225°$.

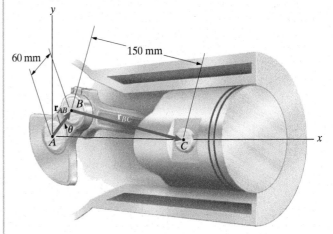

Problem 2.32

▶ **2.33** In Example 2.4, the coordinates of the fixed point A are $(17, 1)$ ft. The driver lowers the bed of the truck into a new position in which the coordinates of point B are $(9, 3)$ ft. The magnitude of the force $\mathbf{F}$ exerted on the bed by the hydraulic cylinder when the bed is in the new position is 4800 lb. Draw a sketch of the new situation. Express $\mathbf{F}$ in terms of components.

2.34 A surveyor measures the location of point A and determines that $\mathbf{r}_{OA} = 400\mathbf{i} + 800\mathbf{j}$ (m). He wants to determine the location of a point B so that $|\mathbf{r}_{AB}| = 400$ m and $|\mathbf{r}_{OA} + \mathbf{r}_{AB}| = 1200$ m. What are the cartesian coordinates of point B?

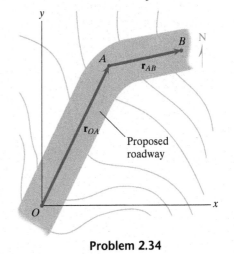

Problem 2.34

2.35 The magnitude of the position vector $\mathbf{r}_{BA}$ from point B to point A is 6 m and the magnitude of the position vector $\mathbf{r}_{CA}$ from point C to point A is 4 m. What are the components of $\mathbf{r}_{BA}$?

2.36 In Problem 2.35, determine the components of a unit vector $\mathbf{e}_{CA}$ that points from point C toward point A.

Strategy: Determine the components of $\mathbf{r}_{CA}$ and then divide the vector $\mathbf{r}_{CA}$ by its magnitude.

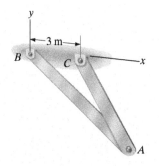

Problems 2.35/2.36

2.37 The x and y coordinates of points A, B, and C of the sailboat are shown.

(a) Determine the components of a unit vector that is parallel to the forestay AB and points from A toward B.

(b) Determine the components of a unit vector that is parallel to the backstay BC and points from C toward B.

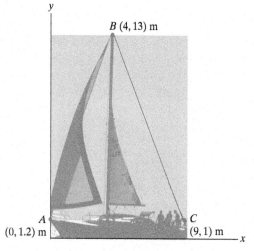

Problem 2.37

2.38 The length of the bar AB is 0.6 m. Determine the components of a unit vector $\mathbf{e}_{AB}$ that points from point A toward point B.

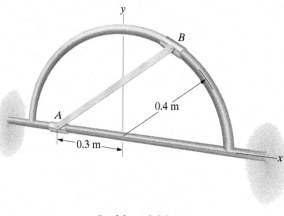

Problem 2.38

2.39 Determine the components of a unit vector that is parallel to the hydraulic actuator BC and points from B toward C.

2.40 The hydraulic actuator BC exerts a 1.2-kN force $\mathbf{F}$ on the joint at C that is parallel to the actuator and points from B toward C. Determine the components of $\mathbf{F}$.

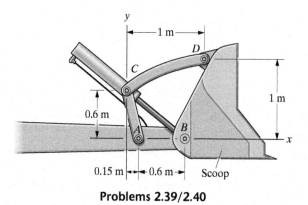

Problems 2.39/2.40

2.41 A surveyor finds that the length of the line *OA* is 1500 m and the length of the line *OB* is 2000 m.

(a) Determine the components of the position vector from point *A* to point *B*.

(b) Determine the components of a unit vector that points from point *A* toward point *B*.

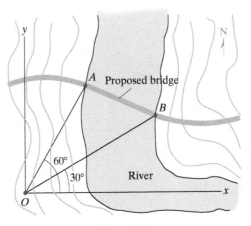

Problem 2.41

2.42 The magnitudes of the forces exerted by the cables are $|T_1| = 2800$ lb, $|T_2| = 3200$ lb, $|T_3| = 4000$ lb, and $|T_4| = 5000$ lb. What is the magnitude of the total force exerted by the four cables?

2.43 The tensions in the four cables are equal: $|T_1| = |T_2| = |T_3| = |T_4| = T$. Determine the value of T so that the four cables exert a total force of 12,500-lb magnitude on the support.

Problems 2.42/2.43

2.44 The rope *ABC* exerts forces $\mathbf{F}_{BA}$ and $\mathbf{F}_{BC}$ on the block at *B*. Their magnitudes are equal: $|\mathbf{F}_{BA}| = |\mathbf{F}_{BC}|$. The magnitude of the total force exerted on the block at *B* by the rope is $|\mathbf{F}_{BA} + \mathbf{F}_{BC}| = 920$ N. Determine $|\mathbf{F}_{BA}|$ by expressing the forces $\mathbf{F}_{BA}$ and $\mathbf{F}_{BC}$ in terms of components.

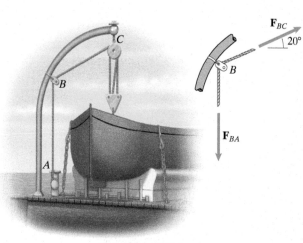

Problem 2.44

2.45 The magnitude of the horizontal force $\mathbf{F}_1$ is 5 kN and $\mathbf{F}_1 + \mathbf{F}_2 + \mathbf{F}_3 = \mathbf{0}$. What are the magnitudes of $\mathbf{F}_2$ and $\mathbf{F}_3$?

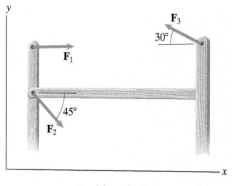

Problem 2.45

2.46 Four groups engage in a tug-of-war. The magnitudes of the forces exerted by groups *B*, *C*, and *D* are $|\mathbf{F}_B| = 800$ lb, $|\mathbf{F}_C| = 1000$ lb, and $|\mathbf{F}_D| = 900$ lb. If the vector sum of the four forces equals zero, what is the magnitude of $\mathbf{F}_A$ and the angle α?

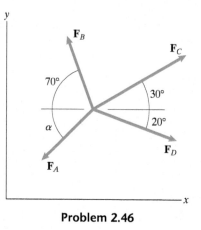

Problem 2.46

▶ **2.47** In Example 2.5, suppose that the attachment point of cable *A* is moved so that the angle between the cable and the wall increases from 40° to 55°. Draw a sketch showing the forces exerted on the hook by the two cables. If you want the total force $\mathbf{F}_A + \mathbf{F}_B$ to have a magnitude of 200 lb and be in the direction perpendicular to the wall, what are the necessary magnitudes of $\mathbf{F}_A$ and $\mathbf{F}_B$?

2.48 The bracket must support the two forces shown, where $|\mathbf{F}_1| = |\mathbf{F}_2| = 2$ kN. An engineer determines that the bracket will safely support a total force of magnitude 3.5 kN in any direction. Assume that $0 \le \alpha \le 90°$. What is the safe range of the angle α?

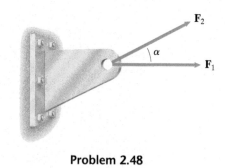

Problem 2.48

2.49 The figure shows three forces acting on a joint of a structure. The magnitude of $\mathbf{F}_C$ is 60 kN, and $\mathbf{F}_A + \mathbf{F}_B + \mathbf{F}_C = \mathbf{0}$. What are the magnitudes of $\mathbf{F}_A$ and $\mathbf{F}_B$?

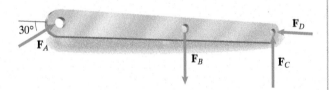

Problem 2.49

2.50 Four coplanar forces act on a beam. The forces $\mathbf{F}_B$ and $\mathbf{F}_C$ are vertical. The vector sum of the forces is zero. The magnitudes $|\mathbf{F}_B| = 10$ kN and $|\mathbf{F}_C| = 5$ kN. Determine the magnitudes of $\mathbf{F}_A$ and $\mathbf{F}_D$.

Problem 2.50

2.51 Six forces act on a beam that forms part of a building's frame. The vector sum of the forces is zero. The magnitudes $|\mathbf{F}_B| = |\mathbf{F}_E| = 20$ kN, $|\mathbf{F}_C| = 16$ kN, and $|\mathbf{F}_D| = 9$ kN. Determine the magnitudes of $\mathbf{F}_A$ and $\mathbf{F}_G$.

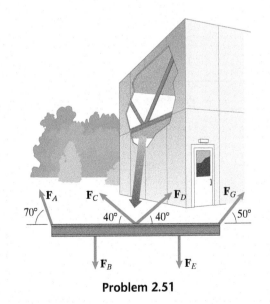

Problem 2.51

2.52 The total weight of the man and parasail is $|\mathbf{W}| = 230$ lb. The drag force $\mathbf{D}$ is perpendicular to the lift force $\mathbf{L}$. If the vector sum of the three forces is zero, what are the magnitudes of $\mathbf{L}$ and $\mathbf{D}$?

Problem 2.52

2.53 The three forces acting on the car are shown. The force $\mathbf{T}$ is parallel to the x axis and the magnitude of the force $\mathbf{W}$ is 14 kN. If $\mathbf{T} + \mathbf{W} + \mathbf{N} = \mathbf{0}$, what are the magnitudes of the forces $\mathbf{T}$ and $\mathbf{N}$?

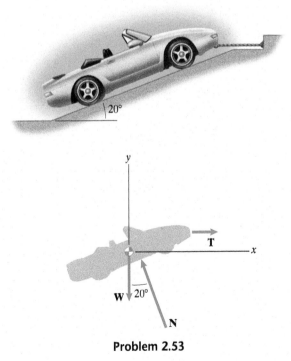

Problem 2.53

2.54 The cables A, B, and C help support a pillar that forms part of the supports of a structure. The magnitudes of the forces exerted by the cables are equal: $|\mathbf{F}_A| = |\mathbf{F}_B| = |\mathbf{F}_C|$. The magnitude of the vector sum of the three forces is 200 kN. What is $|\mathbf{F}_A|$?

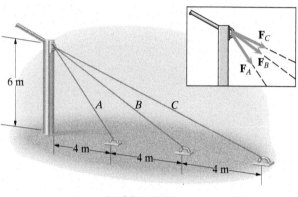

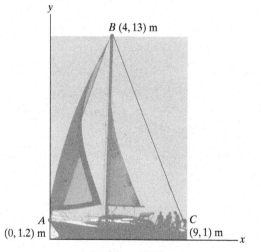

Problem 2.54

2.55 The total force exerted on the top of the mast B by the sailboat's forestay AB and backstay BC is $180\mathbf{i} - 820\mathbf{j}$ (N). What are the magnitudes of the forces exerted at B by the cables AB and BC?

2.56 The structure shown forms part of a truss designed by an architectural engineer to support the roof of an orchestra shell. The members AB, AC, and AD exert forces $\mathbf{F}_{AB}$, $\mathbf{F}_{AC}$, and $\mathbf{F}_{AD}$ on the joint A. The magnitude $|\mathbf{F}_{AB}| = 4$ kN. If the vector sum of the three forces equals zero, what are the magnitudes of $\mathbf{F}_{AC}$ and $\mathbf{F}_{AD}$?

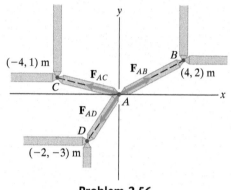

Problem 2.56

2.57 The distance $s = 45$ in.

(a) Determine the unit vector $\mathbf{e}_{BA}$ that points from B toward A.

(b) Use the unit vector you obtained in (a) to determine the coordinates of the collar C.

2.58 Determine the x and y coordinates of the collar C as functions of the distance s.

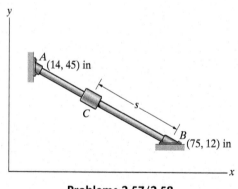

Problems 2.57/2.58

Problem 2.55

2.59 The position vector **r** goes from point A to a point on the straight line between B and C. Its magnitude is $|\mathbf{r}| = 6$ ft. Express **r** in terms of components.

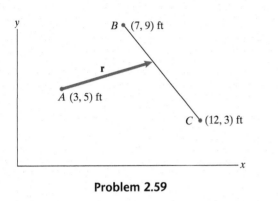

Problem 2.59

2.60 Let **r** be the position vector from point C to the point that is a distance s meters from point A along the straight line between A and B. Express **r** in terms of components. (Your answer will be in terms of s.)

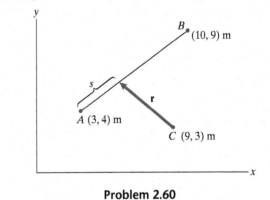

Problem 2.60

2.3 Components in Three Dimensions

BACKGROUND

Many engineering applications require vectors to be expressed in terms of components in a three-dimensional coordinate system. In this section we explain this technique and demonstrate vector operations in three dimensions.

We first review how to draw objects in three dimensions. Consider a three-dimensional object such as a cube. If we draw the cube as it appears when the point of view is perpendicular to one of its faces, we obtain Fig. 2.14a. In this view, the cube appears two dimensional. The dimension perpendicular to the page cannot be seen. To remedy this, we move the point of view upward and to the right, obtaining Fig. 2.14b. In this *oblique* view, the third dimension is visible. The hidden edges of the cube are shown as dashed lines.

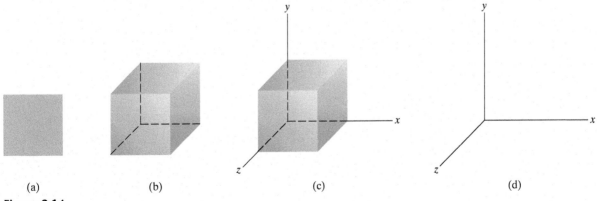

(a)   (b)   (c)   (d)

Figure 2.14
(**a**) A cube viewed with the line of sight perpendicular to a face.
(**b**) An oblique view of the cube.
(**c**) A cartesian coordinate system aligned with the edges of the cube.
(**d**) Three-dimensional representation of the coordinate system.

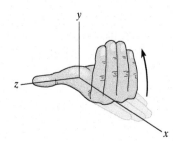

Figure 2.15
Recognizing a right-handed coordinate system.

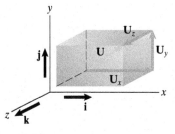

Figure 2.16
A vector **U** and its vector components.

We can use this approach to draw three-dimensional coordinate systems. In Fig. 2.14c we align the x, y, and z axes of a three-dimensional cartesian coordinate system with the edges of the cube. The three-dimensional representation of the coordinate system alone is shown in Fig. 2.14d. The coordinate system shown is said to be *right handed*. If the fingers of the right hand are pointed in the direction of the positive x axis and then bent (as in preparing to make a fist) toward the positive y axis, the thumb points in the direction of the positive z axis (Fig. 2.15). Otherwise, the coordinate system is left handed. Because some equations used in mathematics and engineering do not yield correct results when they are applied using a left-handed coordinate system, we use only right-handed coordinate systems.

We can express a vector **U** in terms of vector components $\mathbf{U}_x$, $\mathbf{U}_y$, and $\mathbf{U}_z$ parallel to the x, y, and z axes, respectively (Fig. 2.16), as

$$\mathbf{U} = \mathbf{U}_x + \mathbf{U}_y + \mathbf{U}_z. \tag{2.11}$$

(We have drawn a box around the vector to help in visualizing the directions of the vector components.) By introducing unit vectors **i**, **j**, and **k** that point in the positive x, y, and z directions, we can express **U** in terms of scalar components as

$$\mathbf{U} = U_x\mathbf{i} + U_y\mathbf{j} + U_z\mathbf{k}. \tag{2.12}$$

We will refer to the scalars U_x, U_y, and U_z as the x, y, and z components of **U**.

Magnitude of a Vector in Terms of Components

Consider a vector **U** and its vector components (Fig. 2.17a). From the right triangle formed by the vectors $\mathbf{U}_y$, $\mathbf{U}_z$, and their sum $\mathbf{U}_y + \mathbf{U}_z$ (Fig. 2.17b), we can see that

$$|\mathbf{U}_y + \mathbf{U}_z|^2 = |\mathbf{U}_y|^2 + |\mathbf{U}_z|^2. \tag{2.13}$$

The vector **U** is the sum of the vectors $\mathbf{U}_x$ and $\mathbf{U}_y + \mathbf{U}_z$. These three vectors form a right triangle (Fig. 2.17c), from which we obtain

$$|\mathbf{U}|^2 = |\mathbf{U}_x|^2 + |\mathbf{U}_y + \mathbf{U}_z|^2.$$

Substituting Eq. (2.13) into this result yields the equation

$$|\mathbf{U}|^2 = |\mathbf{U}_x|^2 + |\mathbf{U}_y|^2 + |\mathbf{U}_z|^2 = U_x^2 + U_y^2 + U_z^2.$$

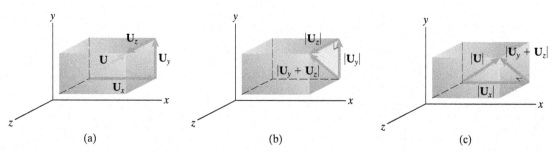

Figure 2.17
(**a**) A vector **U** and its vector components.
(**b**) The right triangle formed by the vectors $\mathbf{U}_y$, $\mathbf{U}_z$, and $\mathbf{U}_y + \mathbf{U}_z$.
(**c**) The right triangle formed by the vectors **U**, $\mathbf{U}_x$, and $\mathbf{U}_y + \mathbf{U}_z$.

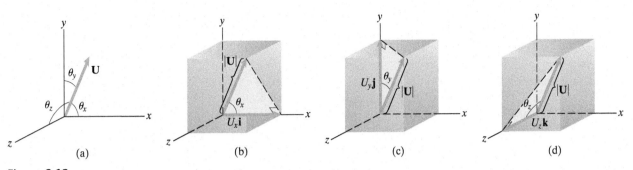

Figure 2.18
(a) A vector $\mathbf{U}$ and the angles θ_x, θ_y, and θ_z.
(b)–(d) The angles θ_x, θ_y, and θ_z and the vector components of $\mathbf{U}$.

Thus, the magnitude of a vector $\mathbf{U}$ is given in terms of its components in three dimensions by

$$|\mathbf{U}| = \sqrt{U_x^2 + U_y^2 + U_z^2}. \tag{2.14}$$

Direction Cosines

We described the direction of a vector relative to a two-dimensional cartesian coordinate system by specifying the angle between the vector and one of the coordinate axes. One of the ways we can describe the direction of a vector in three dimensions is by specifying the angles θ_x, θ_y, and θ_z between the vector and the positive coordinate axes (Fig. 2.18a).

In Figs. 2.18b–d, we demonstrate that the components of the vector $\mathbf{U}$ are respectively given in terms of the angles θ_x, θ_y, and θ_z, by

$$U_x = |\mathbf{U}| \cos \theta_x, \quad U_y = |\mathbf{U}| \cos \theta_y, \quad U_z = |\mathbf{U}| \cos \theta_z. \tag{2.15}$$

The quantities $\cos \theta_x$, $\cos \theta_y$, and $\cos \theta_z$ are called the *direction cosines* of $\mathbf{U}$. The direction cosines of a vector are not independent. If we substitute Eqs. (2.15) into Eq. (2.14), we find that the direction cosines satisfy the relation

$$\cos^2 \theta_x + \cos^2 \theta_y + \cos^2 \theta_z = 1. \tag{2.16}$$

Suppose that $\mathbf{e}$ is a unit vector with the same direction as $\mathbf{U}$, so that

$$\mathbf{U} = |\mathbf{U}|\mathbf{e}.$$

In terms of components, this equation is

$$U_x\mathbf{i} + U_y\mathbf{j} + U_z\mathbf{k} = |\mathbf{U}|(e_x\mathbf{i} + e_y\mathbf{j} + e_z\mathbf{k}).$$

Thus the relations between the components of $\mathbf{U}$ and $\mathbf{e}$ are

$$U_x = |\mathbf{U}|e_x, \quad U_y = |\mathbf{U}|e_y, \quad U_z = |\mathbf{U}|e_z.$$

By comparing these equations to Eqs. (2.15), we see that

$$\cos \theta_x = e_x, \quad \cos \theta_y = e_y, \quad \cos \theta_z = e_z.$$

The direction cosines of a vector $\mathbf{U}$ are the components of a unit vector with the same direction as $\mathbf{U}$.

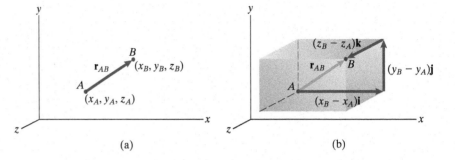

Figure 2.19
(a) The position vector from point A to point B.
(b) The components of $\mathbf{r}_{AB}$ can be determined from the coordinates of points A and B.

Position Vectors in Terms of Components

Generalizing the two-dimensional case, we consider a point A with coordinates (x_A, y_A, z_A) and a point B with coordinates (x_B, y_B, z_B). The position vector $\mathbf{r}_{AB}$ from A to B, shown in Fig. 2.19a, is given in terms of the coordinates of A and B by

$$\mathbf{r}_{AB} = (x_B - x_A)\mathbf{i} + (y_B - y_A)\mathbf{j} + (z_B - z_A)\mathbf{k}. \qquad (2.17)$$

The components are obtained by subtracting the coordinates of point A from the coordinates of point B (Fig. 2.19b).

Components of a Vector Parallel to a Given Line

In three-dimensional applications, the direction of a vector is often defined by specifying the coordinates of two points on a line that is parallel to the vector. This information can be used to determine the components of the vector.

Suppose that we know the coordinates of two points A and B on a line parallel to a vector $\mathbf{U}$ (Fig. 2.20a). We can use Eq. (2.17) to determine the

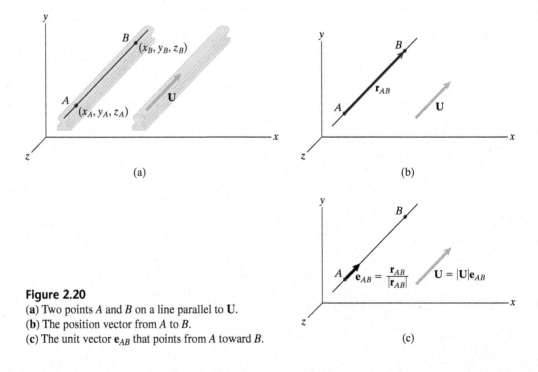

Figure 2.20
(a) Two points A and B on a line parallel to $\mathbf{U}$.
(b) The position vector from A to B.
(c) The unit vector $\mathbf{e}_{AB}$ that points from A toward B.

position vector $\mathbf{r}_{AB}$ from A to B (Fig. 2.20b). We can divide $\mathbf{r}_{AB}$ by its magnitude to obtain a unit vector $\mathbf{e}_{AB}$ that points from A toward B (Fig. 2.20c). Since $\mathbf{e}_{AB}$ has the same direction as $\mathbf{U}$, we can determine $\mathbf{U}$ in terms of its scalar components by expressing it as the product of its magnitude and $\mathbf{e}_{AB}$.

More generally, suppose that we know the magnitude of a vector $\mathbf{U}$ and the components of any vector $\mathbf{V}$ that has the same direction as $\mathbf{U}$. Then $\mathbf{V}/|\mathbf{V}|$ is a unit vector with the same direction as $\mathbf{U}$, and we can determine the components of $\mathbf{U}$ by expressing it as $\mathbf{U} = |\mathbf{U}|(\mathbf{V}/|\mathbf{V}|)$.

RESULTS

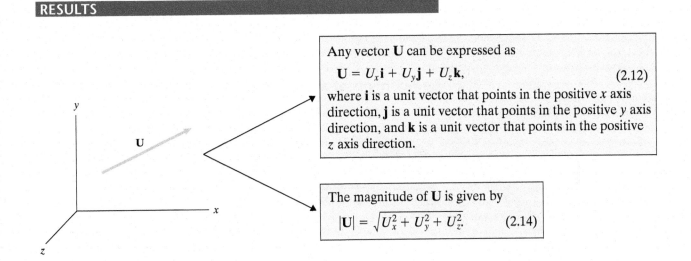

Any vector $\mathbf{U}$ can be expressed as

$$\mathbf{U} = U_x\mathbf{i} + U_y\mathbf{j} + U_z\mathbf{k}, \tag{2.12}$$

where $\mathbf{i}$ is a unit vector that points in the positive x axis direction, $\mathbf{j}$ is a unit vector that points in the positive y axis direction, and $\mathbf{k}$ is a unit vector that points in the positive z axis direction.

The magnitude of $\mathbf{U}$ is given by

$$|\mathbf{U}| = \sqrt{U_x^2 + U_y^2 + U_z^2}. \tag{2.14}$$

Direction Cosines

The direction of a vector $\mathbf{U}$ relative to a given coordinate system can be specified by the angles $\theta_x, \theta_y,$ and θ_z between the vector and the positive coordinate axes.

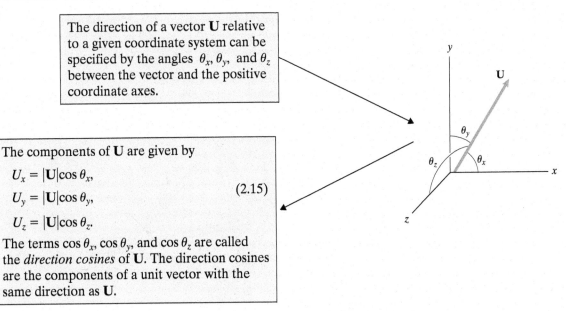

The components of $\mathbf{U}$ are given by

$$U_x = |\mathbf{U}|\cos\theta_x,$$
$$U_y = |\mathbf{U}|\cos\theta_y, \tag{2.15}$$
$$U_z = |\mathbf{U}|\cos\theta_z.$$

The terms $\cos\theta_x$, $\cos\theta_y$, and $\cos\theta_z$ are called the *direction cosines* of $\mathbf{U}$. The direction cosines are the components of a unit vector with the same direction as $\mathbf{U}$.

Position Vectors in Terms of Components

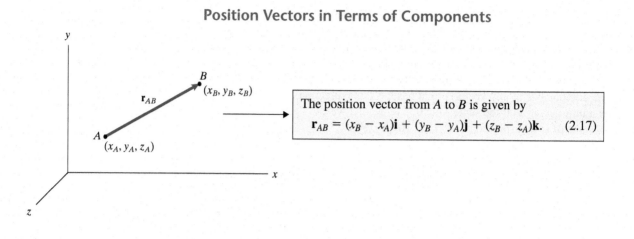

The position vector from A to B is given by

$$\mathbf{r}_{AB} = (x_B - x_A)\mathbf{i} + (y_B - y_A)\mathbf{j} + (z_B - z_A)\mathbf{k}. \qquad (2.17)$$

Components of a Vector Parallel to a Given Line

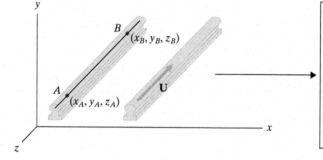

The vector $\mathbf{U}$ is parallel to the line through points A and B. Obtain the position vector $\mathbf{r}_{AB}$ from A to B in terms of its components. Divide $\mathbf{r}_{AB}$ by its magnitude to obtain a unit vector $\mathbf{e}_{AB}$ that is parallel to the line. Then the vector $\mathbf{U}$ in terms of its components is given by

$$\mathbf{U} = |\mathbf{U}|\mathbf{e}_{AB}.$$

Active Example 2.6   **Direction Cosines** (▶ *Related Problem 2.67*)

The coordinates of point C of the truss are $x_C = 4$ m, $y_C = 0$, $z_C = 0$, and the coordinates of point D are $x_D = 2$ m, $y_D = 3$ m, $z_D = 1$ m. What are the direction cosines of the position vector $\mathbf{r}_{CD}$ from point C to point D?

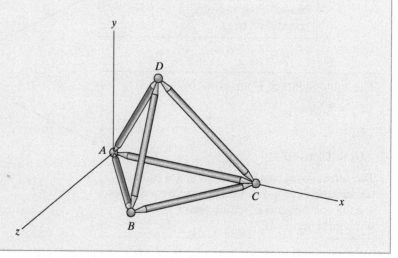

Strategy
Knowing the coordinates of points C and D, we can determine $\mathbf{r}_{CD}$ in terms of its components. Then we can calculate the magnitude of $\mathbf{r}_{CD}$ (the distance from C to D) and use Eqs. (2.15) to obtain the direction cosines.

Solution

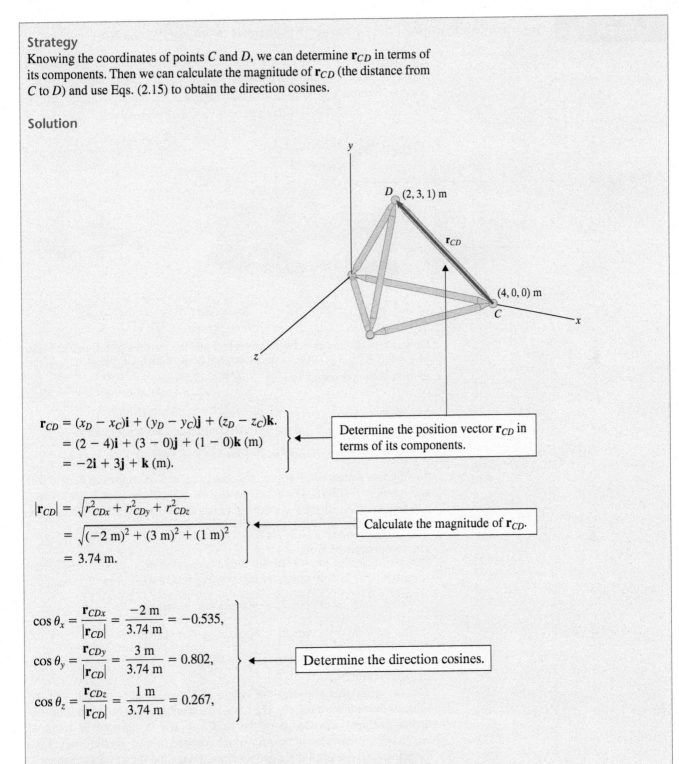

$$\mathbf{r}_{CD} = (x_D - x_C)\mathbf{i} + (y_D - y_C)\mathbf{j} + (z_D - z_C)\mathbf{k}.$$
$$= (2 - 4)\mathbf{i} + (3 - 0)\mathbf{j} + (1 - 0)\mathbf{k} \ (\text{m})$$
$$= -2\mathbf{i} + 3\mathbf{j} + \mathbf{k} \ (\text{m}).$$

Determine the position vector $\mathbf{r}_{CD}$ in terms of its components.

$$|\mathbf{r}_{CD}| = \sqrt{r_{CDx}^2 + r_{CDy}^2 + r_{CDz}^2}$$
$$= \sqrt{(-2 \ \text{m})^2 + (3 \ \text{m})^2 + (1 \ \text{m})^2}$$
$$= 3.74 \ \text{m}.$$

Calculate the magnitude of $\mathbf{r}_{CD}$.

$$\cos \theta_x = \frac{\mathbf{r}_{CDx}}{|\mathbf{r}_{CD}|} = \frac{-2 \ \text{m}}{3.74 \ \text{m}} = -0.535,$$

$$\cos \theta_y = \frac{\mathbf{r}_{CDy}}{|\mathbf{r}_{CD}|} = \frac{3 \ \text{m}}{3.74 \ \text{m}} = 0.802,$$

$$\cos \theta_z = \frac{\mathbf{r}_{CDz}}{|\mathbf{r}_{CD}|} = \frac{1 \ \text{m}}{3.74 \ \text{m}} = 0.267,$$

Determine the direction cosines.

Practice Problem The coordinates of point B of the truss are $x_B = 2.4$ m, $y_B = 0$, $z_B = 3$ m. Determine the components of a unit vector $\mathbf{e}_{BD}$ that points from point B toward point D.

Answer: $\mathbf{e}_{BD} = -0.110\mathbf{i} + 0.827\mathbf{j} - 0.551\mathbf{k}.$

| Example 2.7 | Determining Components in Three Dimensions (▶ *Related Problem 2.76*) |

The crane exerts a 600-lb force **F** on the caisson. The angle between **F** and the *x* axis is 54°, and the angle between **F** and the *y* axis is 40°. The *z* component of **F** is positive. Express **F** in terms of components.

Strategy

Only two of the angles between the vector and the positive coordinate axes are given, but we can use Eq. (2.16) to determine the third angle. Then we can determine the components of **F** by using Eqs. (2.15).

Solution

The angles between **F** and the positive coordinate axes are related by

$$\cos^2 \theta_x + \cos^2 \theta_y + \cos^2 \theta_z = (\cos 54°)^2 + (\cos 40°)^2 + \cos^2 \theta_z = 1.$$

Solving this equation for $\cos \theta_z$, we obtain the two solutions $\cos \theta_z = 0.260$ and $\cos \theta_z = -0.260$, which tells us that $\theta_z = 74.9°$ or $\theta_z = 105.1°$. The *z* component of the vector **F** is positive, so the angle between **F** and the positive *z* axis is less than 90°. Therefore $\theta_z = 74.9°$.

The components of **F** are

$$F_x = |\mathbf{F}| \cos \theta_x = 600 \cos 54° = 353 \text{ lb},$$

$$F_y = |\mathbf{F}| \cos \theta_y = 600 \cos 40° = 460 \text{ lb},$$

$$F_z = |\mathbf{F}| \cos \theta_z = 600 \cos 74.9° = 156 \text{ lb}.$$

Critical Thinking

You are aware that knowing the square of a number does not tell you the value of the number uniquely. If $a^2 = 4$, the number a can be either 2 or −2. In this example, knowledge of the angles θ_x and θ_y allowed us to solve Eq. (2.16) for the value of $\cos^2 \theta_z$, which resulted in two possible values of the angle θ_z. There is a simple geometrical explanation for why this happened. The two angles θ_x and θ_y are sufficient to define a line parallel to the vector **F**, *but not the direction of* **F** *along that line.* The two values of θ_z we obtained correspond to the two possible directions of **F** along the line. Additional information is needed to indicate the direction. In this example, the additional information was supplied by stating that the *z* component of **F** is positive.

| Example 2.8 | Determining Components in Three Dimensions (▶ *Related Problem 2.86*) |

The tether of the balloon exerts an 800-N force **F** on the hook at O. The vertical line AB intersects the x–z plane at point A. The angle between the z axis and the line OA is 60°, and the angle between the line OA and **F** is 45°. Express **F** in terms of components.

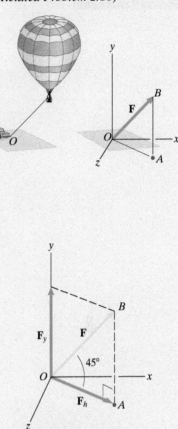

Strategy

We can determine the components of **F** from the given geometric information in two steps. First, we express **F** as the sum of two vector components parallel to the lines OA and AB. The component parallel to AB is the vector component $\mathbf{F}_y$. Then we can use the component parallel to OA to determine the vector components $\mathbf{F}_x$ and $\mathbf{F}_z$.

Solution

In Fig. a, we express **F** as the sum of its y component $\mathbf{F}_y$ and the component $\mathbf{F}_h$ parallel to OA. The magnitude of $\mathbf{F}_y$ is

$$|\mathbf{F}_y| = |\mathbf{F}| \sin 45° = (800 \text{ N}) \sin 45° = 566 \text{ N},$$

and the magnitude of $\mathbf{F}_h$ is

$$|\mathbf{F}_h| = |\mathbf{F}| \cos 45° = (800 \text{ N}) \cos 45° = 566 \text{ N}.$$

In Fig. b, we express $\mathbf{F}_h$ in terms of the vector components $\mathbf{F}_x$ and $\mathbf{F}_z$. The magnitude of $\mathbf{F}_x$ is

$$|\mathbf{F}_x| = |\mathbf{F}_h| \sin 60° = (566 \text{ N}) \sin 60° = 490 \text{ N},$$

and the magnitude of $\mathbf{F}_z$ is

$$|\mathbf{F}_z| = |\mathbf{F}_h| \cos 60° = (566 \text{ N}) \cos 60° = 283 \text{ N}.$$

The vector components $\mathbf{F}_x$, $\mathbf{F}_y$, and $\mathbf{F}_z$ all point in the positive axis directions, so the scalar components of **F** are positive:

$$\mathbf{F} = 490\mathbf{i} + 566\mathbf{j} + 283\mathbf{k} \text{ (N)}.$$

(a) Resolving **F** into vector components parallel to OA and OB.

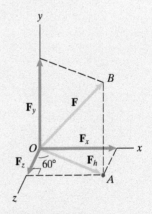

(b) Resolving $\mathbf{F}_h$ into vector components parallel to the x and z axes.

Critical Thinking

As this example demonstrates, two angles are required to specify a vector's direction relative to a three-dimensional coordinate system. The two angles used may not be defined in the same way as in the example, but however they are defined, you can determine the components of the vector in terms of the magnitude and the two specified angles by a procedure similar to the one we used here.

Example 2.9 **Determining Components in Three Dimensions** (▶ *Related Problem 2.90*)

The rope extends from point B through a metal loop attached to the wall at A to point C. The rope exerts forces $\mathbf{F}_{AB}$ and $\mathbf{F}_{AC}$ on the loop at A with magnitudes $|\mathbf{F}_{AB}| = |\mathbf{F}_{AC}| = 200$ lb. What is the magnitude of the total force $\mathbf{F} = \mathbf{F}_{AB} + \mathbf{F}_{AC}$ exerted on the loop by the rope?

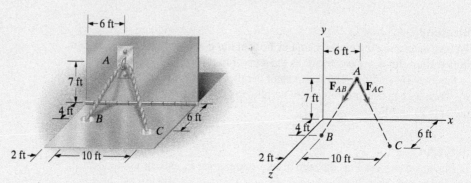

Strategy

The force $\mathbf{F}_{AB}$ is parallel to the line from A to B, and the force $\mathbf{F}_{AC}$ is parallel to the line from A to C. Since we can determine the coordinates of points A, B, and C from the given dimensions, we can determine the components of unit vectors that have the same directions as the two forces and use them to express the forces in terms of scalar components.

Solution

Let $\mathbf{r}_{AB}$ be the position vector from point A to point B and let $\mathbf{r}_{AC}$ be the position vector from point A to point C (Fig. a). From the given dimensions, the coordinates of points A, B, and C are

$$A: (6, 7, 0) \text{ ft}, \qquad B: (2, 0, 4) \text{ ft}, \qquad C: (12, 0, 6) \text{ ft}.$$

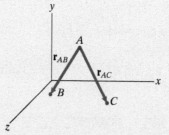

(a) The position vectors $\mathbf{r}_{AB}$ and $\mathbf{r}_{AC}$.

Therefore, the components of $\mathbf{r}_{AB}$ and $\mathbf{r}_{AC}$, with the coordinates in ft, are given by

$$\mathbf{r}_{AB} = (x_B - x_A)\mathbf{i} + (y_B - y_A)\mathbf{j} + (z_B - z_A)\mathbf{k}$$
$$= (2 - 6)\mathbf{i} + (0 - 7)\mathbf{j} + (4 - 0)\mathbf{k}$$
$$= -4\mathbf{i} - 7\mathbf{j} + 4\mathbf{k} \text{ (ft)}$$

and

$$\mathbf{r}_{AC} = (x_C - x_A)\mathbf{i} + (y_C - y_A)\mathbf{j} + (z_C - z_A)\mathbf{k}$$
$$= (12 - 6)\mathbf{i} + (0 - 7)\mathbf{j} + (6 - 0)\mathbf{k}$$
$$= 6\mathbf{i} - 7\mathbf{j} + 6\mathbf{k} \text{ (ft)}.$$

Their magnitudes are $|\mathbf{r}_{AB}| = 9$ ft and $|\mathbf{r}_{AC}| = 11$ ft. By dividing $\mathbf{r}_{AB}$ and $\mathbf{r}_{AC}$ by their magnitudes, we obtain unit vectors $\mathbf{e}_{AB}$ and $\mathbf{e}_{AC}$ that point in the directions of $\mathbf{F}_{AB}$ and $\mathbf{F}_{AC}$ (Fig. b):

$$\mathbf{e}_{AB} = \frac{\mathbf{r}_{AB}}{|\mathbf{r}_{AB}|} = -0.444\mathbf{i} - 0.778\mathbf{j} + 0.444\mathbf{k},$$

$$\mathbf{e}_{AC} = \frac{\mathbf{r}_{AC}}{|\mathbf{r}_{AC}|} = 0.545\mathbf{i} - 0.636\mathbf{j} + 0.545\mathbf{k}.$$

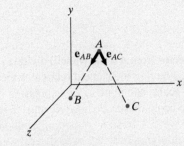

(b) The unit vectors $\mathbf{e}_{AB}$ and $\mathbf{e}_{AC}$.

The forces $\mathbf{F}_{AB}$ and $\mathbf{F}_{AC}$ are

$$\mathbf{F}_{AB} = (200 \text{ lb})\mathbf{e}_{AB} = -88.9\mathbf{i} - 155.6\mathbf{j} + 88.9\mathbf{k} \text{ (lb)},$$
$$\mathbf{F}_{AC} = (200 \text{ lb})\mathbf{e}_{AC} = 109.1\mathbf{i} - 127.3\mathbf{j} + 109.1\mathbf{k} \text{ (lb)}.$$

The total force exerted on the loop by the rope is

$$\mathbf{F} = \mathbf{F}_{AB} + \mathbf{F}_{AC} = 20.2\mathbf{i} - 282.8\mathbf{j} + 198.0\mathbf{k} \text{ (lb)},$$

and its magnitude is

$$|\mathbf{F}| = \sqrt{(20.2)^2 + (-282.8)^2 + (198.0)^2} = 346 \text{ lb}.$$

Critical Thinking

How do you know that the magnitude and direction of the total force exerted on the metal loop at A by the rope is given by the magnitude and direction of the vector $\mathbf{F} = \mathbf{F}_{AB} + \mathbf{F}_{AC}$? At this point in our development of mechanics, we assume that force is a vector, but have provided no proof. In the study of dynamics it is shown that Newton's second law implies that force is a vector.

Example 2.10 | **Determining Components of a Force** (▶ *Related Problem 2.95*)

The cable AB exerts a 50-N force **T** on the collar at A. Express **T** in terms of components.

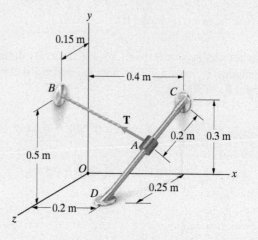

Strategy

Let $\mathbf{r}_{AB}$ be the position vector from A to B. We will divide $\mathbf{r}_{AB}$ by its magnitude to obtain a unit vector $\mathbf{e}_{AB}$ having the same direction as the force **T**. Then we can obtain **T** in terms of scalar components by expressing it as the product of its magnitude and $\mathbf{e}_{AB}$. To begin this procedure, we must first determine the coordinates of the collar A. We will do so by obtaining a unit vector $\mathbf{e}_{CD}$ pointing from C toward D and multiplying it by 0.2 m to determine the position of the collar A relative to C.

Solution

Determining the Coordinates of Point A   The position vector from C to D, with the coordinates in meters, is

$$\mathbf{r}_{CD} = (0.2 - 0.4)\mathbf{i} + (0 - 0.3)\mathbf{j} + (0.25 - 0)\mathbf{k}$$
$$= -0.2\mathbf{i} - 0.3\mathbf{j} + 0.25\mathbf{k} \ (\text{m}).$$

Dividing this vector by its magnitude, we obtain the unit vector $\mathbf{e}_{CD}$ (Fig. a):

$$\mathbf{e}_{CD} = \frac{\mathbf{r}_{CD}}{|\mathbf{r}_{CD}|} = \frac{-0.2\mathbf{i} - 0.3\mathbf{j} + 0.25\mathbf{k}}{\sqrt{(-0.2)^2 + (-0.3)^2 + (0.25)^2}}$$
$$= -0.456\mathbf{i} - 0.684\mathbf{j} + 0.570\mathbf{k}.$$

Using this vector, we obtain the position vector from C to A:

$$\mathbf{r}_{CA} = (0.2 \ \text{m})\mathbf{e}_{CD} = -0.091\mathbf{i} - 0.137\mathbf{j} + 0.114\mathbf{k} \ (\text{m}).$$

The position vector from the origin of the coordinate system to C is $\mathbf{r}_{OC} = 0.4\mathbf{i} + 0.3\mathbf{j}$ (m), so the position vector from the origin to A is

$$\mathbf{r}_{OA} = \mathbf{r}_{OC} + \mathbf{r}_{CA} = (0.4\mathbf{i} + 0.3\mathbf{j}) + (-0.091\mathbf{i} - 0.137\mathbf{j} + 0.114\mathbf{k})$$
$$= 0.309\mathbf{i} + 0.163\mathbf{j} + 0.114\mathbf{k} \ (\text{m}).$$

The coordinates of A are $(0.309, 0.163, 0.114)$ m.

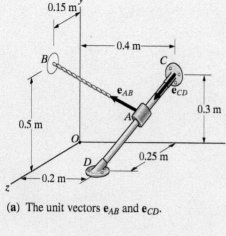

(a) The unit vectors $\mathbf{e}_{AB}$ and $\mathbf{e}_{CD}$.

Determining the Components of T Using the coordinates of point A, we find that the position vector from A to B is

$$\mathbf{r}_{AB} = (0 - 0.309)\mathbf{i} + (0.5 - 0.163)\mathbf{j} + (0.15 - 0.114)\mathbf{k}$$

$$= -0.309\mathbf{i} + 0.337\mathbf{j} + 0.036\mathbf{k} \ (\text{m}).$$

Dividing this vector by its magnitude, we obtain the unit vector $\mathbf{e}_{AB}$ (Fig. a).

$$\mathbf{e}_{AB} = \frac{\mathbf{r}_{AB}}{|\mathbf{r}_{AB}|} = \frac{-0.309\mathbf{i} + 0.337\mathbf{j} + 0.036\mathbf{k} \ (\text{m})}{\sqrt{(-0.309 \ \text{m})^2 + (0.337 \ \text{m})^2 + (0.036 \ \text{m})^2}}$$

$$= -0.674\mathbf{i} + 0.735\mathbf{j} + 0.079\mathbf{k}.$$

The force $\mathbf{T}$ is

$$\mathbf{T} = |\mathbf{T}|\mathbf{e}_{AB} = (50 \ \text{N})(-0.674\mathbf{i} + 0.735\mathbf{j} + 0.079\mathbf{k})$$

$$= -33.7\mathbf{i} + 36.7\mathbf{j} + 3.9\mathbf{k} \ (\text{N}).$$

Critical Thinking

Look at the two ways unit vectors were used in this example. The unit vector $\mathbf{e}_{CD}$ was used to obtain the components of the position vector $\mathbf{r}_{CA}$, which made it possible to determine the coordinates of point A. The coordinates of point A were then used to determine the unit vector $\mathbf{e}_{AB}$, which was used to express the force $\mathbf{T}$ in terms of its components.

Problems

2.61 A vector $\mathbf{U} = 3\mathbf{i} - 4\mathbf{j} - 12\mathbf{k}$. What is its magnitude?

Strategy: The magnitude of a vector is given in terms of its components by Eq. (2.14).

2.62 The vector $\mathbf{e} = \frac{1}{3}\mathbf{i} + \frac{2}{3}\mathbf{j} + e_z\mathbf{k}$ is a unit vector. Determine the component e_z.

2.63 An engineer determines that the attachment point will be subjected to a force $\mathbf{F} = 20\mathbf{i} + F_y\mathbf{j} - 45\mathbf{k}$ (kN). If the attachment point will safely support a force of 80-kN magnitude in any direction, what is the acceptable range of values of F_y?

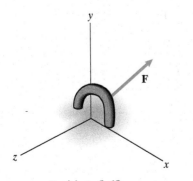

Problem 2.63

2.64 A vector $\mathbf{U} = U_x\mathbf{i} + U_y\mathbf{j} + U_z\mathbf{k}$. Its magnitude $|\mathbf{U}| = 30$. Its components are related by the equations $U_y = -2U_x$ and $U_z = 4U_y$. Determine the components.

2.65 An object is acted upon by two forces $\mathbf{F}_1 = 20\mathbf{i} + 30\mathbf{j} - 24\mathbf{k}$ (kN) and $\mathbf{F}_2 = -60\mathbf{i} + 20\mathbf{j} + 40\mathbf{k}$ (kN). What is the magnitude of the total force acting on the object?

2.66 Two vectors $\mathbf{U} = 3\mathbf{i} - 2\mathbf{j} + 6\mathbf{k}$ and $\mathbf{V} = 4\mathbf{i} + 12\mathbf{j} - 3\mathbf{k}$.

(a) Determine the magnitudes of $\mathbf{U}$ and $\mathbf{V}$.

(b) Determine the magnitude of the vector $3\mathbf{U} + 2\mathbf{V}$.

▶ **2.67** In Active Example 2.6, suppose that you want to redesign the truss, changing the position of point D so that the magnitude of the vector $\mathbf{r}_{CD}$ from point C to point D is 3 m. To accomplish this, let the coordinates of point D be $(2, y_D, 1)$ m, and determine the value of y_D so that $|\mathbf{r}_{CD}| = 3$ m. Draw a sketch of the truss with point D in its new position. What are the new direction cosines of $\mathbf{r}_{CD}$?

2.68 A force vector is given in terms of its components by $\mathbf{F} = 10\mathbf{i} - 20\mathbf{j} - 20\mathbf{k}$ (N).

(a) What are the direction cosines of $\mathbf{F}$?

(b) Determine the components of a unit vector $\mathbf{e}$ that has the same direction as $\mathbf{F}$.

2.69 The cable exerts a force **F** on the hook at O whose magnitude is 200 N. The angle between the vector **F** and the x axis is 40°, and the angle between the vector **F** and the y axis is 70°.

(a) What is the angle between the vector **F** and the z axis?

(b) Express **F** in terms of components.

    Strategy: (a) Because you know the angles between the vector **F** and the x and y axes, you can use Eq. (2.16) to determine the angle between **F** and the z axis. (Observe from the figure that the angle between **F** and the z axis is clearly within the range $0 < \theta_z < 180°$.) (b) The components of **F** can be obtained with Eqs. (2.15).

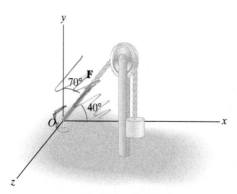

Problem 2.69

2.70 A unit vector has direction cosines $\cos \theta_x = -0.5$ and $\cos \theta_y = 0.2$. Its z component is positive. Express it in terms of components.

2.71 The airplane's engines exert a total thrust force **T** of 200-kN magnitude. The angle between **T** and the x axis is 120°, and the angle between **T** and the y axis is 130°. The z component of **T** is positive.

(a) What is the angle between **T** and the z axis?

(b) Express **T** in terms of components.

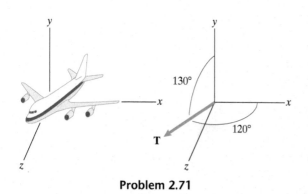

Problem 2.71

Refer to the following diagram when solving Problems 2.72 through 2.75.

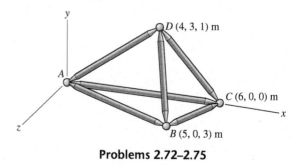

Problems 2.72–2.75

2.72 Determine the components of the position vector $\mathbf{r}_{BD}$ from point B to point D. Use your result to determine the distance from B to D.

2.73 What are the direction cosines of the position vector $\mathbf{r}_{BD}$ from point B to point D?

2.74 Determine the components of the unit vector $\mathbf{e}_{CD}$ that points from point C toward point D.

2.75 What are the direction cosines of the unit vector $\mathbf{e}_{CD}$ that points from point C toward point D?

▶ **2.76** In Example 2.7, suppose that the caisson shifts on the ground to a new position. The magnitude of the force **F** remains 600 lb. In the new position, the angle between the force **F** and the x axis is 60° and the angle between **F** and the z axis is 70°. Express **F** in terms of components.

2.77 Astronauts on the space shuttle use radar to determine the magnitudes and direction cosines of the position vectors of two satellites A and B. The vector $\mathbf{r}_A$ from the shuttle to satellite A has magnitude 2 km and direction cosines $\cos \theta_x = 0.768$, $\cos \theta_y = 0.384$, $\cos \theta_z = 0.512$. The vector $\mathbf{r}_B$ from the shuttle to satellite B has magnitude 4 km and direction cosines $\cos \theta_x = 0.743$, $\cos \theta_y = 0.557$, $\cos \theta_z = -0.371$. What is the distance between the satellites?

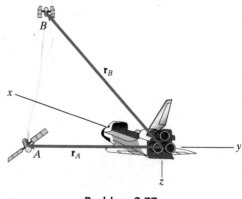

Problem 2.77

2.78 Archaeologists measure a pre-Columbian ceremonial structure and obtain the dimensions shown. Determine (a) the magnitude and (b) the direction cosines of the position vector from point A to point B.

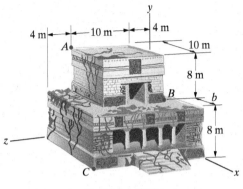

Problem 2.78

2.79 Consider the structure described in Problem 2.78. After returning to the United States, an archaeologist discovers that a graduate student has erased the only data file containing the dimension b. But from recorded GPS data he is able to calculate that the distance from point B to point C is 16.61 m.

(a) What is the distance b?

(b) Determine the direction cosines of the position vector from B to C.

2.80 Observers at A and B use theodolites to measure the direction from their positions to a rocket in flight. If the coordinates of the rocket's position at a given instant are (4, 4, 2) km, determine the direction cosines of the vectors $\mathbf{r}_{AR}$ and $\mathbf{r}_{BR}$ that the observers would measure at that instant.

2.81* Suppose that the coordinates of the rocket's position are unknown. At a given instant, the person at A determines that the direction cosines of $\mathbf{r}_{AR}$ are $\cos\theta_x = 0.535$, $\cos\theta_y = 0.802$, and $\cos\theta_z = 0.267$, and the person at B determines that the direction cosines of $\mathbf{r}_{BR}$ are $\cos\theta_x = -0.576$, $\cos\theta_y = 0.798$, and $\cos\theta_z = -0.177$. What are the coordinates of the rocket's position at that instant?

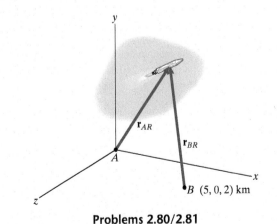

Problems 2.80/2.81

2.82* The height of Mount Everest was originally measured by a surveyor in the following way. He first measured the altitudes of two points and the horizontal distance between them. For example, suppose that the points A and B are 3000 m above sea level and are 10,000 m apart. He then used a theodolite to measure the direction cosines of the vector $\mathbf{r}_{AP}$ from point A to the top of the mountain P and the vector $\mathbf{r}_{BP}$ from point B to P. Suppose that the direction cosines of $\mathbf{r}_{AP}$ are $\cos\theta_x = 0.5179$, $\cos\theta_y = 0.6906$, and $\cos\theta_z = 0.5048$, and the direction cosines of $\mathbf{r}_{BP}$ are $\cos\theta_x = -0.3743$, $\cos\theta_y = 0.7486$, and $\cos\theta_z = 0.5472$. Using this data, determine the height of Mount Everest above sea level.

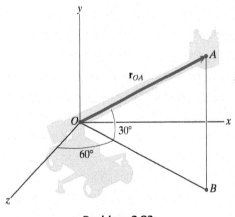

Problem 2.82

2.83 The distance from point O to point A is 20 ft. The straight line AB is parallel to the y axis, and point B is in the x–z plane. Express the vector $\mathbf{r}_{OA}$ in terms of components.

Strategy: You can express $\mathbf{r}_{OA}$ as the sum of a vector from O to B and a vector from B to A. You can then express the vector from O to B as the sum of vector components parallel to the x and z axes. See Example 2.8.

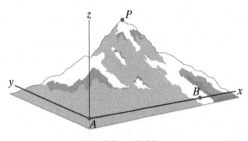

Problem 2.83

2.84 The magnitudes of the two force vectors are $|\mathbf{F}_A| = 140$ lb and $|\mathbf{F}_B| = 100$ lb. Determine the magnitude of the sum of the forces $\mathbf{F}_A + \mathbf{F}_B$.

2.85 Determine the direction cosines of the vectors $\mathbf{F}_A$ and $\mathbf{F}_B$.

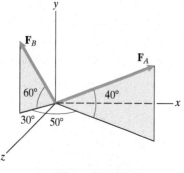

Problems 2.84/2.85

▶**2.86** In Example 2.8, suppose that a change in the wind causes a change in the position of the balloon and increases the magnitude of the force $\mathbf{F}$ exerted on the hook at O to 900 N. In the new position, the angle between the vector component $\mathbf{F}_h$ and $\mathbf{F}$ is 35°, and the angle between the vector components $\mathbf{F}_h$ and $\mathbf{F}_z$ is 40°. Draw a sketch showing the relationship of these angles to the components of $\mathbf{F}$. Express $\mathbf{F}$ in terms of its components.

2.87 An engineer calculates that the magnitude of the axial force in one of the beams of a geodesic dome is $|\mathbf{P}| = 7.65$ kN. The cartesian coordinates of the endpoints A and B of the straight beam are $(-12.4, 22.0, -18.4)$ m and $(-9.2, 24.4, -15.6)$ m, respectively. Express the force $\mathbf{P}$ in terms of components.

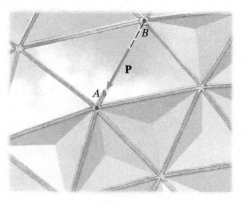

Problem 2.87

2.88 The cable BC exerts an 8-kN force $\mathbf{F}$ on the bar AB at B.
(a) Determine the components of a unit vector that points from point B toward point C.
(b) Express $\mathbf{F}$ in terms of components.

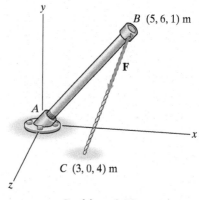

Problem 2.88

2.89 A cable extends from point C to point E. It exerts a 50-lb force $\mathbf{T}$ on the plate at C that is directed along the line from C to E. Express $\mathbf{T}$ in terms of components.

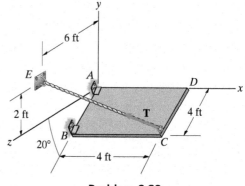

Problem 2.89

▶**2.90** In Example 2.9, suppose that the metal loop at A is moved upward so that the vertical distance to A increases from 7 ft to 8 ft. As a result, the magnitudes of the forces $\mathbf{F}_{AB}$ and $\mathbf{F}_{AC}$ increase to $|\mathbf{F}_{AB}| = |\mathbf{F}_{AC}| = 240$ lb. What is the magnitude of the total force $\mathbf{F} = \mathbf{F}_{AB} + \mathbf{F}_{AC}$ exerted on the loop by the rope?

2.91 The cable *AB* exerts a 200-lb force $\mathbf{F}_{AB}$ at point *A* that is directed along the line from *A* to *B*. Express $\mathbf{F}_{AB}$ in terms of components.

2.92 Cable *AB* exerts a 200-lb force $\mathbf{F}_{AB}$ at point *A* that is directed along the line from *A* to *B*. The cable *AC* exerts a 100-lb force $\mathbf{F}_{AC}$ at point *A* that is directed along the line from *A* to *C*. Determine the magnitude of the total force exerted at point *A* by the two cables.

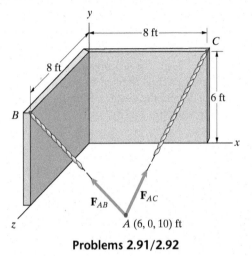

Problems 2.91/2.92

2.93 The 70-m-tall tower is supported by three cables that exert forces $\mathbf{F}_{AB}$, $\mathbf{F}_{AC}$, and $\mathbf{F}_{AD}$ on it. The magnitude of each force is 2 kN. Express the total force exerted on the tower by the three cables in terms of components.

2.94 The magnitude of the force $\mathbf{F}_{AB}$ is 2 kN. The *x* and *z* components of the vector sum of the forces exerted on the tower by the three cables are zero. What are the magnitudes of $\mathbf{F}_{AC}$ and $\mathbf{F}_{AD}$?

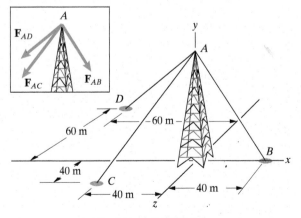

Problems 2.93/2.94

▶ 2.95 In Example 2.10, suppose that the distance from point *C* to the collar *A* is increased from 0.2 m to 0.3 m, and the magnitude of the force **T** increases to 60 N. Express **T** in terms of its components.

2.96 The cable *AB* exerts a 32-lb force **T** on the collar at *A*. Express **T** in terms of components.

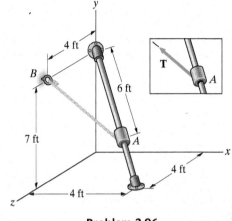

Problem 2.96

2.97 The circular bar has a 4-m radius and lies in the *x–y* plane. Express the position vector from point *B* to the collar at *A* in terms of components.

2.98 The cable *AB* exerts a 60-N force **T** on the collar at *A* that is directed along the line from *A* toward *B*. Express **T** in terms of components.

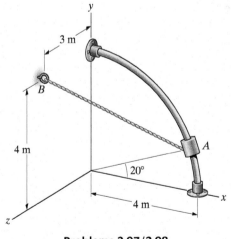

Problems 2.97/2.98

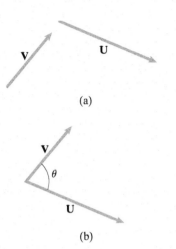

(a)

(b)

Figure 2.21
(a) The vectors **U** and **V**.
(b) The angle θ between **U** and **V** when the two vectors are placed tail to tail.

2.4 Dot Products

Two kinds of products of vectors, the dot and cross products, have been found to have applications in science and engineering, especially in mechanics and electromagnetic field theory. We use both of these products in Chapter 4 to evaluate moments of forces about points and lines.

The dot product of two vectors has many uses, including determining the components of a vector parallel and perpendicular to a given line and determining the angle between two lines in space.

Definition

Consider two vectors **U** and **V** (Fig. 2.21a). The *dot product* of **U** and **V**, denoted by **U** · **V** (hence the name "dot product"), is defined to be the product of the magnitude of **U**, the magnitude of **V**, and the cosine of the angle θ between **U** and **V** when they are placed tail to tail (Fig. 2.21b):

$$\mathbf{U} \cdot \mathbf{V} = |\mathbf{U}||\mathbf{V}| \cos \theta. \tag{2.18}$$

Because the result of the dot product is a scalar, the dot product is sometimes called the scalar product. The units of the dot product are the product of the units of the two vectors. *Notice that the dot product of two nonzero vectors is equal to zero if and only if the vectors are perpendicular.*

The dot product has the properties

$$\mathbf{U} \cdot \mathbf{V} = \mathbf{V} \cdot \mathbf{U}, \quad \text{The dot product is commutative.} \tag{2.19}$$

$$a\,(\mathbf{U} \cdot \mathbf{V}) = (a\mathbf{U}) \cdot \mathbf{V} = \mathbf{U} \cdot (a\mathbf{V}), \quad \text{The dot product is associative with respect to scalar multiplication.} \tag{2.20}$$

and

$$\mathbf{U} \cdot (\mathbf{V} + \mathbf{W}) = \mathbf{U} \cdot \mathbf{V} + \mathbf{U} \cdot \mathbf{W}, \quad \text{The dot product is associative with respect to vector addition.} \tag{2.21}$$

for any scalar a and vectors **U**, **V**, and **W**.

Dot Products in Terms of Components

In this section we derive an equation that allows you to determine the dot product of two vectors if you know their scalar components. The derivation also results in an equation for the angle between the vectors. The first step is to determine the dot products formed from the unit vectors **i**, **j**, and **k**. Let us evaluate the dot product **i** · **i**. The magnitude $|\mathbf{i}| = 1$, and the angle between two identical vectors placed tail to tail is zero, so we obtain

$$\mathbf{i} \cdot \mathbf{i} = |\mathbf{i}||\mathbf{i}| \cos(0) = (1)(1)(1) = 1.$$

The dot product of **i** and **j** is

$$\mathbf{i} \cdot \mathbf{j} = |\mathbf{i}||\mathbf{j}| \cdot \cos(90°) = (1)(1)(0) = 0.$$

Continuing in this way, we obtain

$$
\begin{aligned}
\mathbf{i} \cdot \mathbf{i} = 1, \quad & \mathbf{i} \cdot \mathbf{j} = 0, \quad & \mathbf{i} \cdot \mathbf{k} = 0, \\
\mathbf{j} \cdot \mathbf{i} = 0, \quad & \mathbf{j} \cdot \mathbf{j} = 1, \quad & \mathbf{j} \cdot \mathbf{k} = 0, \\
\mathbf{k} \cdot \mathbf{i} = 0, \quad & \mathbf{k} \cdot \mathbf{j} = 0, \quad & \mathbf{k} \cdot \mathbf{k} = 1.
\end{aligned}
\tag{2.22}
$$

The dot product of two vectors **U** and **V**, expressed in terms of their components, is

$$
\begin{aligned}
\mathbf{U} \cdot \mathbf{V} &= (U_x\mathbf{i} + U_y\mathbf{j} + U_z\mathbf{k}) \cdot (V_x\mathbf{i} + V_y\mathbf{j} + V_z\mathbf{k}) \\
&= U_xV_x(\mathbf{i} \cdot \mathbf{i}) + U_xV_y(\mathbf{i} \cdot \mathbf{j}) + U_xV_z(\mathbf{i} \cdot \mathbf{k}) \\
&\quad + U_yV_x(\mathbf{j} \cdot \mathbf{i}) + U_yV_y(\mathbf{j} \cdot \mathbf{j}) + U_yV_z(\mathbf{j} \cdot \mathbf{k}) \\
&\quad + U_zV_x(\mathbf{k} \cdot \mathbf{i}) + U_zV_y(\mathbf{k} \cdot \mathbf{j}) + U_zV_z(\mathbf{k} \cdot \mathbf{k}).
\end{aligned}
$$

In obtaining this result, we used Eqs. (2.20) and (2.21). Substituting Eqs. (2.22) into this expression, we obtain an equation for the dot product in terms of the scalar components of the two vectors:

$$
\mathbf{U} \cdot \mathbf{V} = U_xV_x + U_yV_y + U_zV_z. \tag{2.23}
$$

To obtain an equation for the angle θ in terms of the components of the vectors, we equate the expression for the dot product given by Eq. (2.23) to the definition of the dot product, Eq. (2.18), and solve for $\cos\theta$:

$$
\cos\theta = \frac{\mathbf{U} \cdot \mathbf{V}}{|\mathbf{U}||\mathbf{V}|} = \frac{U_xV_x + U_yV_y + U_zV_z}{|\mathbf{U}||\mathbf{V}|}. \tag{2.24}
$$

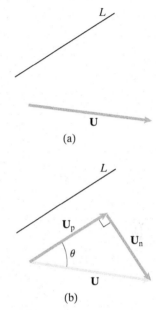

(a)

(b)

Figure 2.22
(a) A vector **U** and line L.
(b) Resolving **U** into components parallel and normal to L.

Vector Components Parallel and Normal to a Line

In some engineering applications a vector must be expressed in terms of vector components that are parallel and normal (perpendicular) to a given line. The component of a vector parallel to a line is called the *projection* of the vector onto the line. For example, when the vector represents a force, the projection of the force onto a line is the component of the force in the direction of the line.

We can determine the components of a vector parallel and normal to a line by using the dot product. Consider a vector **U** and a straight line L (Fig. 2.22a). We can express **U** as the sum of vector components $\mathbf{U}_p$ and $\mathbf{U}_n$ that are parallel and normal to L (Fig. 2.22b).

The Parallel Component   In terms of the angle θ between **U** and the vector component $\mathbf{U}_p$, the magnitude of $\mathbf{U}_p$ is

$$
|\mathbf{U}_p| = |\mathbf{U}| \cos\theta. \tag{2.25}
$$

Let **e** be a unit vector parallel to L (Fig. 2.23). The dot product of **e** and **U** is

$$
\mathbf{e} \cdot \mathbf{U} = |\mathbf{e}||\mathbf{U}| \cos\theta = |\mathbf{U}| \cos\theta.
$$

Comparing this result with Eq. (2.25), we see that the magnitude of $\mathbf{U}_p$ is

$$
|\mathbf{U}_p| = \mathbf{e} \cdot \mathbf{U}.
$$

Therefore the parallel vector component, or projection of **U** onto L, is

$$
\mathbf{U}_p = (\mathbf{e} \cdot \mathbf{U})\mathbf{e}. \tag{2.26}
$$

(This equation holds even if **e** doesn't point in the direction of $\mathbf{U}_p$. In that case, the angle $\theta > 90°$ and $\mathbf{e} \cdot \mathbf{U}$ is negative.) When the components of a vector and the components of a unit vector **e** parallel to a line L are known, we can use Eq. (2.26) to determine the component of the vector parallel to L.

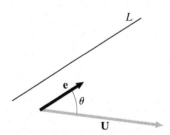

Figure 2.23
The unit vector **e** is parallel to L.

The Normal Component   Once the parallel vector component has been determined, we can obtain the normal vector component from the relation $\mathbf{U} = \mathbf{U}_p + \mathbf{U}_n$:

$$
\mathbf{U}_n = \mathbf{U} - \mathbf{U}_p. \tag{2.27}
$$

RESULTS

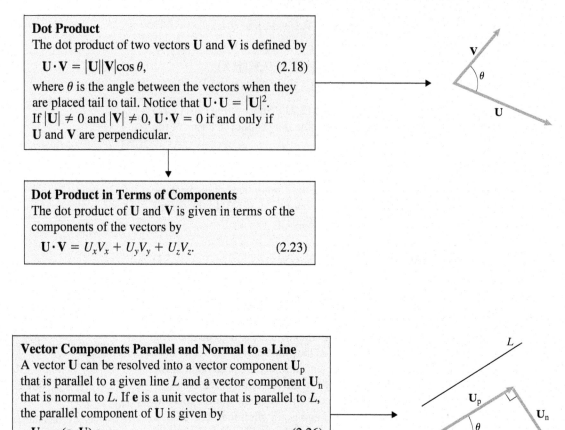

Dot Product
The dot product of two vectors **U** and **V** is defined by

$$\mathbf{U} \cdot \mathbf{V} = |\mathbf{U}||\mathbf{V}|\cos\theta, \qquad (2.18)$$

where θ is the angle between the vectors when they are placed tail to tail. Notice that $\mathbf{U} \cdot \mathbf{U} = |\mathbf{U}|^2$. If $|\mathbf{U}| \neq 0$ and $|\mathbf{V}| \neq 0$, $\mathbf{U} \cdot \mathbf{V} = 0$ if and only if **U** and **V** are perpendicular.

Dot Product in Terms of Components
The dot product of **U** and **V** is given in terms of the components of the vectors by

$$\mathbf{U} \cdot \mathbf{V} = U_x V_x + U_y V_y + U_z V_z. \qquad (2.23)$$

Vector Components Parallel and Normal to a Line
A vector **U** can be resolved into a vector component $\mathbf{U}_p$ that is parallel to a given line L and a vector component $\mathbf{U}_n$ that is normal to L. If **e** is a unit vector that is parallel to L, the parallel component of **U** is given by

$$\mathbf{U}_p = (\mathbf{e} \cdot \mathbf{U})\,\mathbf{e}. \qquad (2.26)$$

The normal component can be obtained from the relation

$$\mathbf{U}_n = \mathbf{U} - \mathbf{U}_p. \qquad (2.27)$$

Active Example 2.11   **Dot Products** (▶ *Related Problem 2.99*)

The components of two vectors **U** and **V** are $\mathbf{U} = 6\mathbf{i} - 5\mathbf{j} - 3\mathbf{k}$ and $\mathbf{V} = 4\mathbf{i} + 2\mathbf{j} + 2\mathbf{k}$. (a) What is the value of $\mathbf{U} \cdot \mathbf{V}$? (b) What is the angle between **U** and **V** when they are placed tail to tail?

Strategy
Knowing the components of **U** and **V**, we can use Eq. (2.23) to determine the value of $\mathbf{U} \cdot \mathbf{V}$. Then we can use the definition of the dot product, Eq. (2.18), to calculate the angle between the vectors.

Solution

$$\mathbf{U} \cdot \mathbf{V} = U_x V_x + U_y V_y + U_z V_z$$
$$= (6)(4) + (-5)(2) + (-3)(2)$$
$$= 8.$$

Use the components of the vectors to determine the value of $\mathbf{U} \cdot \mathbf{V}$.

$$\mathbf{U} \cdot \mathbf{V} = |\mathbf{U}||\mathbf{V}|\cos \theta,$$

so

$$\cos \theta = \frac{\mathbf{U} \cdot \mathbf{V}}{|\mathbf{U}||\mathbf{V}|}$$

$$= \frac{8}{\sqrt{(6)^2 + (-5)^2 + (-3)^2} \ \sqrt{(4)^2 + (2)^2 + (2)^2}}$$

$$= 0.195.$$

Therefore $\theta = 78.7°$.

Use the definition of $\mathbf{U} \cdot \mathbf{V}$ to determine θ.

Practice Problem The components of two vectors $\mathbf{U}$ and $\mathbf{V}$ are $\mathbf{U} = 6\mathbf{i} - 5\mathbf{j} - 3\mathbf{k}$ and $\mathbf{V} = V_x\mathbf{i} + 2\mathbf{j} + 2\mathbf{k}$. Determine the value of the component V_x so that the vectors $\mathbf{U}$ and $\mathbf{V}$ are perpendicular.

Answer: $V_x = 2.67$.

Example 2.12   Using the Dot Product to Determine an Angle (▶ *Related Problem 2.100*)

What is the angle θ between the lines AB and AC?

Strategy
We know the coordinates of the points A, B, and C, so we can determine the components of the vector $\mathbf{r}_{AB}$ from A to B and the vector $\mathbf{r}_{AC}$ from A to C (Fig. a). Then we can use Eq. (2.24) to determine θ.

Solution
The vectors $\mathbf{r}_{AB}$ and $\mathbf{r}_{AC}$, with the coordinates in meters, are

$$\mathbf{r}_{AB} = (6 - 4)\mathbf{i} + (1 - 3)\mathbf{j} + (-2 - 2)\mathbf{k} = 2\mathbf{i} - 2\mathbf{j} - 4\mathbf{k} \ (\text{m}),$$
$$\mathbf{r}_{AC} = (8 - 4)\mathbf{i} + (8 - 3)\mathbf{j} + (4 - 2)\mathbf{k} = 4\mathbf{i} + 5\mathbf{j} + 2\mathbf{k} \ (\text{m}).$$

Their magnitudes are

$$|\mathbf{r}_{AB}| = \sqrt{(2 \text{ m})^2 + (-2 \text{ m})^2 + (-4 \text{ m})^2} = 4.90 \text{ m},$$
$$|\mathbf{r}_{AC}| = \sqrt{(4 \text{ m})^2 + (5 \text{ m})^2 + (2 \text{ m})^2} = 6.71 \text{ m}.$$

The dot product of $\mathbf{r}_{AB}$ and $\mathbf{r}_{AC}$ is

$$\mathbf{r}_{AB} \cdot \mathbf{r}_{AC} = (2 \text{ m})(4 \text{ m}) + (-2 \text{ m})(5 \text{ m}) + (-4 \text{ m})(2 \text{ m}) = -10 \text{ m}^2.$$

Therefore,

$$\cos \theta = \frac{\mathbf{r}_{AB} \cdot \mathbf{r}_{AC}}{|\mathbf{r}_{AB}||\mathbf{r}_{AC}|} = \frac{-10 \text{ m}^2}{(4.90 \text{ m})(6.71 \text{ m})} = -0.304.$$

The angle $\theta = \arccos(-0.304) = 107.7°$.

Critical Thinking
What does it mean if the dot product of two vectors is negative? From Eq. (2.18) and the graph of the cosine (Fig. b), you can see that the dot product is negative, as it is in this example, only if the enclosed angle between the two vectors is greater than $90°$.

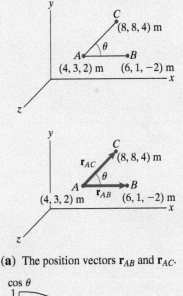

(a) The position vectors $\mathbf{r}_{AB}$ and $\mathbf{r}_{AC}$.

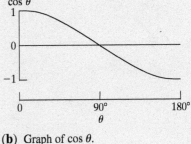

(b) Graph of $\cos \theta$.

| Example 2.13 | Vector Components Parallel and Normal to a Line (▶ *Related Problem 2.111*) |

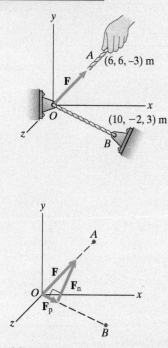

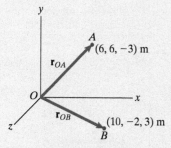

(a) The components of **F** parallel and normal to *OB*.

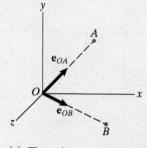

(b) The position vectors **r**$_{OA}$ and **r**$_{OB}$.

Suppose that you pull on the cable *OA*, exerting a 50-N force **F** at *O*. What are the vector components of **F** parallel and normal to the cable *OB*?

Strategy

Expressing **F** as the sum of vector components parallel and normal to *OB* (Fig. a), we can determine the vector components by using Eqs. (2.26) and (2.27). But to apply them, we must first express **F** in terms of scalar components and determine the scalar components of a unit vector parallel to *OB*. We can obtain the components of **F** by determining the components of the unit vector pointing from *O* toward *A* and multiplying them by $|\mathbf{F}|$.

Solution

The position vectors from *O* to *A* and from *O* to *B* are (Fig. b)

$$\mathbf{r}_{OA} = 6\mathbf{i} + 6\mathbf{j} - 3\mathbf{k} \ (\text{m}),$$

$$\mathbf{r}_{OB} = 10\mathbf{i} - 2\mathbf{j} + 3\mathbf{k} \ (\text{m}).$$

Their magnitudes are $|\mathbf{r}_{OA}| = 9$ m and $|\mathbf{r}_{OB}| = 10.6$ m. Dividing these vectors by their magnitudes, we obtain unit vectors that point from the origin toward *A* and *B* (Fig. c):

$$\mathbf{e}_{OA} = \frac{\mathbf{r}_{OA}}{|\mathbf{r}_{OA}|} = \frac{6\mathbf{i} + 6\mathbf{j} - 3\mathbf{k} \ (\text{m})}{9 \ \text{m}} = 0.667\mathbf{i} + 0.667\mathbf{j} - 0.333\mathbf{k},$$

$$\mathbf{e}_{OB} = \frac{\mathbf{r}_{OB}}{|\mathbf{r}_{OB}|} = \frac{10\mathbf{i} - 2\mathbf{j} + 3\mathbf{k} \ (\text{m})}{10.6 \ \text{m}} = 0.941\mathbf{i} - 0.188\mathbf{j} + 0.282\mathbf{k}.$$

The force **F** in terms of scalar components is

$$\mathbf{F} = |\mathbf{F}|\mathbf{e}_{OA} = (50 \ \text{N})(0.667\mathbf{i} + 0.667\mathbf{j} - 0.333\mathbf{k})$$

$$= 33.3\mathbf{i} + 33.3\mathbf{j} - 16.7\mathbf{k} \ (\text{N}).$$

Taking the dot product of $\mathbf{e}_{OB}$ and **F**, we obtain

$$\mathbf{e}_{OB} \cdot \mathbf{F} = (0.941)(33.3 \ \text{N}) + (-0.188)(33.3 \ \text{N}) + (0.282)(-16.7 \ \text{N})$$

$$= 20.4 \ \text{N}.$$

The parallel vector component of **F** is

$$\mathbf{F}_\text{p} = (\mathbf{e}_{OB} \cdot \mathbf{F})\mathbf{e}_{OB} = (20.4 \ \text{N})(0.941\mathbf{i} - 0.188\mathbf{j} + 0.282\mathbf{k})$$

$$= 19.2\mathbf{i} - 3.83\mathbf{j} + 5.75\mathbf{k} \ (\text{N}),$$

and the normal vector component is

$$\mathbf{F}_\text{n} = \mathbf{F} - \mathbf{F}_\text{p} = 14.2\mathbf{i} + 37.2\mathbf{j} - 22.4\mathbf{k} \ (\text{N}).$$

Critical Thinking

How can you confirm that two vectors are perpendicular? It is clear from Eq. (2.18) that the dot product of two nonzero vectors is zero if and only if the enclosed angle between them is 90°. We can use this diagnostic test to confirm that the components of **F** determined in this example are perpendicular. Evaluating the dot product of $\mathbf{F}_\text{p}$ and $\mathbf{F}_\text{n}$ in terms of their components in newtons, we obtain

$$\mathbf{F}_\text{p} \cdot \mathbf{F}_\text{n} = (19.2)(14.2) + (-3.83)(37.2) + (5.75)(-22.4) = 0.$$

(c) The unit vectors $\mathbf{e}_{OA}$ and $\mathbf{e}_{OB}$.

Problems

▶ **2.99** In Active Example 2.11, suppose that the vector **V** is changed to **V** = 4**i** − 6**j** − 10**k**.

(a) What is the value of **U** · **V**?

(b) What is the angle between **U** and **V** when they are placed tail to tail?

▶ **2.100** In Example 2.12, suppose that the coordinates of point *B* are changed to (6, 4, 4) m. What is the angle θ between the lines *AB* and *AC*?

2.101 What is the dot product of the position vector **r** = −10**i** + 25**j** (m) and the force vector **F** = 300**i** + 250**j** + 300**k** (N)?

2.102 Suppose that the dot product of two vectors **U** and **V** is **U** · **V** = 0. If $|\mathbf{U}| \neq 0$, what do you know about the vector **V**?

2.103 Two *perpendicular* vectors are given in terms of their components by **U** = U_x**i** − 4**j** + 6**k** and **V** = 3**i** + 2**j** − 3**k**. Use the dot product to determine the component U_x.

2.104 The three vectors

$$\mathbf{U} = U_x\mathbf{i} + 3\mathbf{j} + 2\mathbf{k},$$

$$\mathbf{V} = -3\mathbf{i} + V_y\mathbf{j} + 3\mathbf{k},$$

$$\mathbf{W} = -2\mathbf{i} + 4\mathbf{j} + W_z\mathbf{k}$$

are mutually perpendicular. Use the dot product to determine the components U_x, V_y, and W_z.

2.105 The magnitudes $|\mathbf{U}| = 10$ and $|\mathbf{V}| = 20$.

(a) Use Eq. (2.18) to determine **U** · **V**.

(b) Use Eq. (2.23) to determine **U** · **V**.

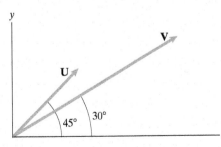

Problem 2.105

2.106 By evaluating the dot product **U** · **V**, prove the identity $\cos(\theta_1 - \theta_2) = \cos\theta_1 \cos\theta_2 + \sin\theta_1 \sin\theta_2$.

  Strategy: Evaluate the dot product both by using Eq. (2.18) and by using Eq. (2.23).

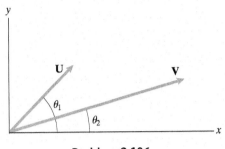

Problem 2.106

2.107 Use the dot product to determine the angle between the forestay (cable *AB*) and the backstay (cable *BC*) of the sailboat.

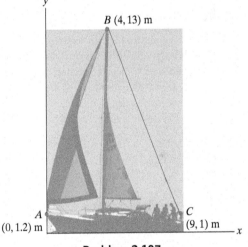

Problem 2.107

2.108 Determine the angle θ between the lines *AB* and *AC*

(a) by using the law of cosines (see Appendix A);

(b) by using the dot product.

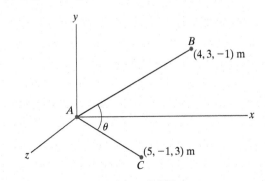

Problem 2.108

2.109 The ship O measures the positions of the ship A and the airplane B and obtains the coordinates shown. What is the angle θ between the lines of sight OA and OB?

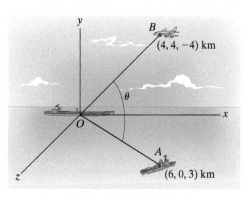

(4, 4, −4) km

θ

(6, 0, 3) km

Problem 2.109

2.110 Astronauts on the space shuttle use radar to determine the magnitudes and direction cosines of the position vectors of two satellites A and B. The vector $\mathbf{r}_A$ from the shuttle to satellite A has magnitude 2 km and direction cosines $\cos \theta_x = 0.768$, $\cos \theta_y = 0.384$, $\cos \theta_z = 0.512$. The vector $\mathbf{r}_B$ from the shuttle to satellite B has magnitude 4 km and direction cosines $\cos \theta_x = 0.743$, $\cos \theta_y = 0.557$, $\cos \theta_z = -0.371$. What is the angle θ between the vectors $\mathbf{r}_A$ and $\mathbf{r}_B$?

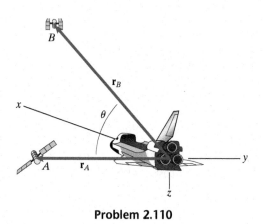

Problem 2.110

▶ **2.111** In Example 2.13, if you shift your position and the coordinates of point A where you apply the 50-N force become $(8, 3, -3)$ m, what is the vector component of $\mathbf{F}$ parallel to the cable OB?

2.112 The person exerts a force $\mathbf{F} = 60\mathbf{i} - 40\mathbf{j}$ (N) on the handle of the exercise machine. Use Eq. (2.26) to determine the vector component of $\mathbf{F}$ that is parallel to the line from the origin O to where the person grips the handle.

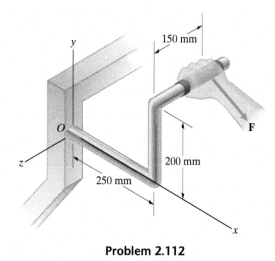

150 mm

200 mm

250 mm

Problem 2.112

2.113 At the instant shown, the Harrier's thrust vector is $\mathbf{T} = 17{,}000\mathbf{i} + 68{,}000\mathbf{j} - 8{,}000\mathbf{k}$ (N) and its velocity vector is $\mathbf{v} = 7.3\mathbf{i} + 1.8\mathbf{j} - 0.6\mathbf{k}$ (m/s). The quantity $P = |\mathbf{T}_p||\mathbf{v}|$, where $\mathbf{T}_p$ is the vector component of $\mathbf{T}$ parallel to $\mathbf{v}$, is the power currently being transferred to the airplane by its engine. Determine the value of P.

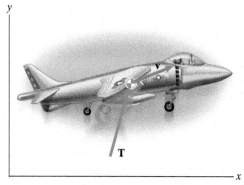

Problem 2.113

2.114 Cables extend from A to B and from A to C. The cable AC exerts a 1000-lb force $\mathbf{F}$ at A.

(a) What is the angle between the cables AB and AC?

(b) Determine the vector component of $\mathbf{F}$ parallel to the cable AB.

2.115 Let $\mathbf{r}_{AB}$ be the position vector from point A to point B. Determine the vector component of $\mathbf{r}_{AB}$ parallel to the cable AC.

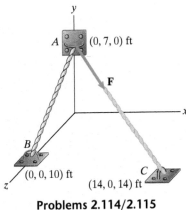

Problems 2.114/2.115

2.116 The force $\mathbf{F} = 10\mathbf{i} + 12\mathbf{j} - 6\mathbf{k}$ (N). Determine the vector components of $\mathbf{F}$ parallel and normal to the line OA.

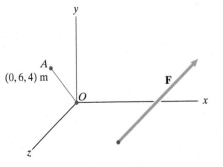

Problem 2.116

2.117 The rope AB exerts a 50-N force $\mathbf{T}$ on collar A. Determine the vector component of $\mathbf{T}$ parallel to the bar CD.

2.118 In Problem 2.117, determine the vector component of $\mathbf{T}$ normal to the bar CD.

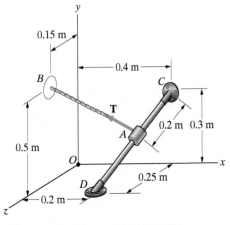

Problems 2.117/2.118

2.119 The disk A is at the midpoint of the sloped surface. The string from A to B exerts a 0.2-lb force $\mathbf{F}$ on the disk. If you express $\mathbf{F}$ in terms of vector components parallel and normal to the sloped surface, what is the component normal to the surface?

2.120 In Problem 2.119, what is the vector component of $\mathbf{F}$ parallel to the surface?

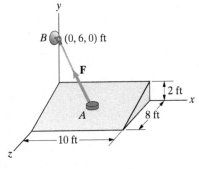

Problems 2.119/2.120

2.121 An astronaut in a maneuvering unit approaches a space station. At the present instant, the station informs him that his position relative to the origin of the station's coordinate system is $\mathbf{r}_G = 50\mathbf{i} + 80\mathbf{j} + 180\mathbf{k}$ (m) and his velocity is $\mathbf{v} = -2.2\mathbf{j} - 3.6\mathbf{k}$ (m/s). The position of an airlock is $\mathbf{r}_A = -12\mathbf{i} + 20\mathbf{k}$ (m). Determine the angle between his velocity vector and the line from his position to the airlock's position.

2.122 In Problem 2.121, determine the vector component of the astronaut's velocity parallel to the line from his position to the airlock's position.

2.123 Point P is at longitude 30°W and latitude 45°N on the Atlantic Ocean between Nova Scotia and France. Point Q is at longitude 60°E and latitude 20°N in the Arabian Sea. Use the dot product to determine the shortest distance along the surface of the earth from P to Q in terms of the radius of the earth R_E.

Strategy: Use the dot product to determine the angle between the lines OP and OQ; then use the definition of an angle in radians to determine the distance along the surface of the earth from P to Q.

Problems 2.121/2.122

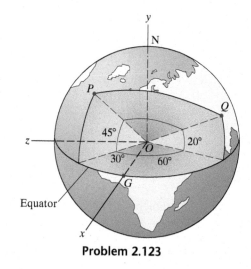

Problem 2.123

2.5 Cross Products

BACKGROUND

Like the dot product, the cross product of two vectors has many applications, including determining the rate of rotation of a fluid particle and calculating the force exerted on a charged particle by a magnetic field. Because of its usefulness for determining moments of forces, the cross product is an indispensable tool in mechanics. In this section we show you how to evaluate cross products and give examples of simple applications.

Definition

Consider two vectors $\mathbf{U}$ and $\mathbf{V}$ (Fig. 2.24a). The *cross product* of $\mathbf{U}$ and $\mathbf{V}$, denoted $\mathbf{U} \times \mathbf{V}$, is defined by

$$\mathbf{U} \times \mathbf{V} = |\mathbf{U}||\mathbf{V}| \sin \theta\, \mathbf{e}. \tag{2.28}$$

The angle θ is the angle between $\mathbf{U}$ and $\mathbf{V}$ when they are placed tail to tail (Fig. 2.24b). The vector $\mathbf{e}$ is a unit vector defined to be perpendicular to both $\mathbf{U}$ and $\mathbf{V}$. Since this leaves two possibilities for the direction of $\mathbf{e}$, the vectors $\mathbf{U}$, $\mathbf{V}$, and $\mathbf{e}$ are defined to be a right-handed system. The *right-hand rule* for determining the direction of $\mathbf{e}$ is shown in Fig. 2.24c. If the fingers of the right hand are pointed in the direction of the vector $\mathbf{U}$ (the first vector in the cross product) and then bent toward the vector $\mathbf{V}$ (the second vector in the cross product), the thumb points in the direction of $\mathbf{e}$.

Because the result of the cross product is a vector, it is sometimes called the vector product. The units of the cross product are the product of the units of the two vectors. Notice that the cross product of two nonzero vectors is equal to zero if and only if the two vectors are parallel.

An interesting property of the cross product is that it is *not* commutative. Eq. (2.28) implies that the magnitude of the vector $\mathbf{U} \times \mathbf{V}$ is equal to the magnitude of the vector $\mathbf{V} \times \mathbf{U}$, but the right-hand rule indicates that they are opposite in direction (Fig. 2.25). That is,

$$\mathbf{U} \times \mathbf{V} = -\mathbf{V} \times \mathbf{U}. \quad \text{The cross product is } not \text{ commutative.} \quad (2.29)$$

The cross product also satisfies the relations

$$a\,(\mathbf{U} \times \mathbf{V}) = (a\mathbf{U}) \times \mathbf{V} = \mathbf{U} \times (a\mathbf{V}) \quad \begin{array}{l}\text{The cross product is}\\\text{associative with}\\\text{respect to scalar}\\\text{multiplication.}\end{array} \quad (2.30)$$

and

$$\mathbf{U} \times (\mathbf{V} + \mathbf{W}) = (\mathbf{U} \times \mathbf{V}) + (\mathbf{U} \times \mathbf{W}) \quad \begin{array}{l}\text{The cross product is}\\\text{distributive with}\\\text{respect to vector}\\\text{addition.}\end{array} \quad (2.31)$$

for any scalar a and vectors $\mathbf{U}$, $\mathbf{V}$, and $\mathbf{W}$.

Cross Products in Terms of Components

To obtain an equation for the cross product of two vectors in terms of their components, we must determine the cross products formed from the unit vectors $\mathbf{i}$, $\mathbf{j}$, and $\mathbf{k}$. Since the angle between two identical vectors placed tail to tail is zero, it follows that

$$\mathbf{i} \times \mathbf{i} = |\mathbf{i}||\mathbf{i}|\sin(0)\mathbf{e} = \mathbf{0}.$$

The cross product $\mathbf{i} \times \mathbf{j}$ is

$$\mathbf{i} \times \mathbf{j} = |\mathbf{i}||\mathbf{j}|\sin 90°\mathbf{e} = \mathbf{e},$$

where $\mathbf{e}$ is a unit vector perpendicular to $\mathbf{i}$ and $\mathbf{j}$. Either $\mathbf{e} = \mathbf{k}$ or $\mathbf{e} = -\mathbf{k}$. Applying the right-hand rule, we find that $\mathbf{e} = \mathbf{k}$ (Fig. 2.26). Therefore,

$$\mathbf{i} \times \mathbf{j} = \mathbf{k}.$$

Continuing in this way, we obtain

$$\begin{array}{lll} \mathbf{i} \times \mathbf{i} = \mathbf{0}, & \mathbf{i} \times \mathbf{j} = \mathbf{k}, & \mathbf{i} \times \mathbf{k} = -\mathbf{j}, \\ \mathbf{j} \times \mathbf{i} = -\mathbf{k}, & \mathbf{j} \times \mathbf{j} = \mathbf{0}, & \mathbf{j} \times \mathbf{k} = \mathbf{i}, \\ \mathbf{k} \times \mathbf{i} = \mathbf{j}, & \mathbf{k} \times \mathbf{j} = -\mathbf{i}, & \mathbf{k} \times \mathbf{k} = \mathbf{0}. \end{array} \quad (2.32)$$

These results can be remembered easily by arranging the unit vectors in a circle, as shown in Fig. 2.27a. The cross product of adjacent vectors is equal to the third vector with a positive sign if the order of the vectors in the cross product is the order indicated by the arrows and a negative sign otherwise. For example, in Fig. 2.27b we see that $\mathbf{i} \times \mathbf{j} = \mathbf{k}$, but $\mathbf{i} \times \mathbf{k} = -\mathbf{j}$.

The cross product of two vectors $\mathbf{U}$ and $\mathbf{V}$, expressed in terms of their components, is

$$\begin{aligned} \mathbf{U} \times \mathbf{V} &= (U_x\mathbf{i} + U_y\mathbf{j} + U_z\mathbf{k}) \times (V_x\mathbf{i} + V_y\mathbf{j} + V_z\mathbf{k}) \\ &= U_xV_x(\mathbf{i} \times \mathbf{i}) + U_xV_y(\mathbf{i} \times \mathbf{j}) + U_xV_z(\mathbf{i} \times \mathbf{k}) \\ &\quad + U_yV_x(\mathbf{j} \times \mathbf{i}) + U_yV_y(\mathbf{j} \times \mathbf{j}) + U_yV_z(\mathbf{j} \times \mathbf{k}) \\ &\quad + U_zV_x(\mathbf{k} \times \mathbf{i}) + U_zV_y(\mathbf{k} \times \mathbf{j}) + U_zV_z(\mathbf{k} \times \mathbf{k}). \end{aligned}$$

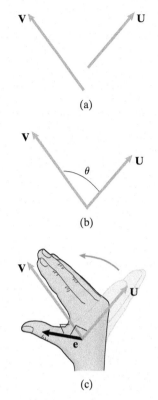

(a)

(b)

(c)

Figure 2.24
(a) The vectors $\mathbf{U}$ and $\mathbf{V}$.
(b) The angle θ between the vectors when they are placed tail to tail.
(c) Determining the direction of $\mathbf{e}$ by the right-hand rule.

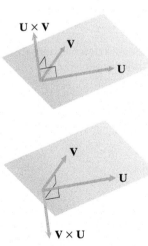

Figure 2.25
Directions of $\mathbf{U} \times \mathbf{V}$ and $\mathbf{V} \times \mathbf{U}$.

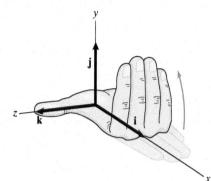

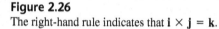

Figure 2.26
The right-hand rule indicates that $\mathbf{i} \times \mathbf{j} = \mathbf{k}$.

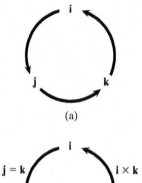

(a)

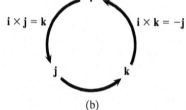

(b)

Figure 2.27
(a) Arrange the unit vectors in a circle with arrows to indicate their order.
(b) You can use the circle to determine their cross products.

By substituting Eqs. (2.32) into this expression, we obtain the equation

$$\mathbf{U} \times \mathbf{V} = (U_y V_z - U_z V_y)\mathbf{i} - (U_x V_z - U_z V_x)\mathbf{j}$$
$$+ (U_x V_y - U_y V_x)\mathbf{k}. \tag{2.33}$$

This result can be compactly written as the determinant

$$\mathbf{U} \times \mathbf{V} = \begin{vmatrix} \mathbf{i} & \mathbf{j} & \mathbf{k} \\ U_x & U_y & U_z \\ V_x & V_y & V_z \end{vmatrix}. \tag{2.34}$$

This equation is based on Eqs. (2.32), which we obtained using a right-handed coordinate system. It gives the correct result for the cross product only if a right-handed coordinate system is used to determine the components of $\mathbf{U}$ and $\mathbf{V}$.

Evaluating a 3 × 3 Determinant

A 3 × 3 determinant can be evaluated by repeating its first two columns and evaluating the products of the terms along the six diagonal lines:

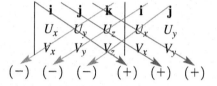

Adding the terms obtained from the diagonals that run downward to the right (blue arrows) and subtracting the terms obtained from the diagonals that run downward to the left (red arrows) gives the value of the determinant:

$$\begin{vmatrix} \mathbf{i} & \mathbf{j} & \mathbf{k} \\ U_x & U_y & U_z \\ V_x & V_y & V_z \end{vmatrix} = \begin{array}{l} U_y V_z \mathbf{i} + U_z V_x \mathbf{j} + U_x V_y \mathbf{k} \\ -U_y V_x \mathbf{k} - U_z V_y \mathbf{i} - U_x V_z \mathbf{j}. \end{array}$$

A 3 × 3 determinant can also be evaluated by expressing it as

$$\begin{vmatrix} \mathbf{i} & \mathbf{j} & \mathbf{k} \\ U_x & U_y & U_z \\ V_x & V_y & V_z \end{vmatrix} = \mathbf{i} \begin{vmatrix} U_y & U_z \\ V_y & V_z \end{vmatrix} - \mathbf{j} \begin{vmatrix} U_x & U_z \\ V_x & V_z \end{vmatrix} + \mathbf{k} \begin{vmatrix} U_x & U_y \\ V_x & V_y \end{vmatrix}.$$

The terms on the right are obtained by multiplying each element of the first row of the 3 × 3 determinant by the 2 × 2 determinant obtained by crossing out that element's row and column. For example, the first element of the first row, $\mathbf{i}$, is multiplied by the 2 × 2 determinant

$$\begin{vmatrix} \mathbf{i} & \mathbf{j} & \mathbf{k} \\ U_x & U_y & U_z \\ V_x & V_y & V_z \end{vmatrix}.$$

Be sure to remember that the second term is subtracted. Expanding the 2 × 2 determinants, we obtain the value of the determinant:

$$\begin{vmatrix} \mathbf{i} & \mathbf{j} & \mathbf{k} \\ U_x & U_y & U_z \\ V_x & V_y & V_z \end{vmatrix} = (U_y V_z - U_z V_y)\mathbf{i} - (U_x V_z - U_z V_x)\mathbf{j} + (U_x V_y - U_y V_x)\mathbf{k}.$$

Mixed Triple Products

In Chapter 4, when we discuss the moment of a force about a line, we will use an operation called the *mixed triple product*, defined by

$$\mathbf{U} \cdot (\mathbf{V} \times \mathbf{W}). \tag{2.35}$$

In terms of the scalar components of the vectors,

$$
\mathbf{U} \cdot (\mathbf{V} \times \mathbf{W}) = (U_x\mathbf{i} + U_y\mathbf{j} + U_z\mathbf{k}) \cdot \begin{vmatrix} \mathbf{i} & \mathbf{j} & \mathbf{k} \\ V_x & V_y & V_z \\ W_x & W_y & W_z \end{vmatrix}
$$

$$
= (U_x\mathbf{i} + U_y\mathbf{j} + U_z\mathbf{k}) \cdot [(V_yW_z - V_zW_y)\mathbf{i}
$$
$$
- (V_xW_z - V_zW_x)\mathbf{j} + (V_xW_y - V_yW_x)\mathbf{k}]
$$

$$
= U_x(V_yW_z - V_zW_y) - U_y(V_xW_z - V_zW_x)
$$
$$
+ U_z(V_xW_y - V_yW_x).
$$

This result can be expressed as the determinant

$$
\mathbf{U} \cdot (\mathbf{V} \times \mathbf{W}) = \begin{vmatrix} U_x & U_y & U_z \\ V_x & V_y & V_z \\ W_x & W_y & W_z \end{vmatrix}. \tag{2.36}
$$

Interchanging any two of the vectors in the mixed triple product changes the sign but not the absolute value of the result. For example,

$$
\mathbf{U} \cdot (\mathbf{V} \times \mathbf{W}) = -\mathbf{W} \cdot (\mathbf{V} \times \mathbf{U}).
$$

If the vectors $\mathbf{U}$, $\mathbf{V}$, and $\mathbf{W}$ in Fig. 2.28 form a right-handed system, it can be shown that the volume of the parallelepiped equals $\mathbf{U} \cdot (\mathbf{V} \times \mathbf{W})$.

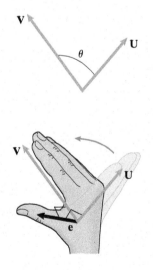

Figure 2.28
Parallelepiped defined by the vectors $\mathbf{U}$, $\mathbf{V}$, and $\mathbf{W}$.

RESULTS

Cross Product
The cross product of two vectors $\mathbf{U}$ and $\mathbf{V}$ is defined by

$$
\mathbf{U} \times \mathbf{V} = |\mathbf{U}||\mathbf{V}|\sin\theta\,\mathbf{e}. \tag{2.28}
$$

As in the dot product, θ is the angle between the vectors when they are placed tail to tail. The unit vector $\mathbf{e}$ is defined to be perpendicular to $\mathbf{U}$, perpendicular to $\mathbf{V}$, and directed so that $\mathbf{U}$, $\mathbf{V}$, $\mathbf{e}$ form a right-handed system. If $|\mathbf{U}| \neq 0$ and $|\mathbf{V}| \neq 0$, $\mathbf{U} \times \mathbf{V} = \mathbf{0}$ if and only if $\mathbf{U}$ and $\mathbf{V}$ are parallel.

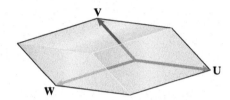

Cross Product in Terms of Components
The cross product of $\mathbf{U}$ and $\mathbf{V}$ is given in terms of the components of the vectors by

$$
\mathbf{U} \times \mathbf{V} = (U_yV_z - U_zV_y)\mathbf{i} - (U_xV_z - U_zV_x)\mathbf{j}
$$
$$
+ (U_xV_y - U_yV_x)\mathbf{k} \tag{2.33}
$$

$$
= \begin{vmatrix} \mathbf{i} & \mathbf{j} & \mathbf{k} \\ U_x & U_y & U_z \\ V_x & V_y & V_z \end{vmatrix} \tag{2.34}
$$

Mixed Triple Product
The operation $\mathbf{U} \cdot (\mathbf{V} \times \mathbf{W})$ is called the mixed triple product of the vectors $\mathbf{U}$, $\mathbf{V}$, and $\mathbf{W}$. It can be expressed in terms of the components of the vectors by the determinant

$$\mathbf{U} \cdot (\mathbf{V} \times \mathbf{W}) = \begin{vmatrix} U_x & U_y & U_z \\ V_x & V_y & V_z \\ W_x & W_y & W_z \end{vmatrix}. \tag{2.36}$$

When $\mathbf{U}$, $\mathbf{V}$, $\mathbf{W}$ form a right-handed system, the volume of the parallelepiped shown equals $\mathbf{U} \cdot (\mathbf{V} \times \mathbf{W})$.

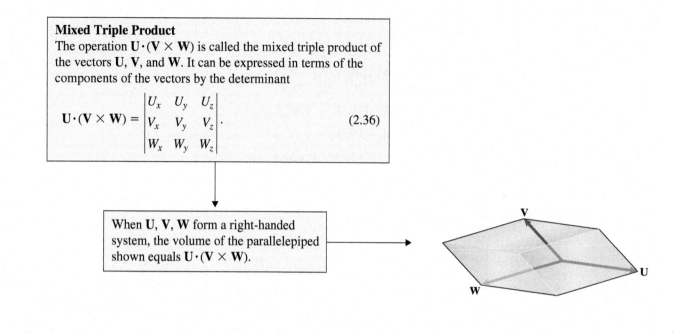

Active Example 2.14   Cross Products (▶ *Related Problem 2.124*)

The components of two vectors $\mathbf{U}$ and $\mathbf{V}$ are $\mathbf{U} = 6\mathbf{i} - 5\mathbf{j} - \mathbf{k}$ and $\mathbf{V} = 4\mathbf{i} + 2\mathbf{j} + 2\mathbf{k}$. (a) Determine the cross product $\mathbf{U} \times \mathbf{V}$. (b) Use the dot product to prove that $\mathbf{U} \times \mathbf{V}$ is perpendicular to $\mathbf{U}$.

Strategy
(a) Knowing the components of $\mathbf{U}$ and $\mathbf{V}$, we can use Eq. (2.33) to determine $\mathbf{U} \times \mathbf{V}$. (b) Once we have determined the components of the vector $\mathbf{U} \times \mathbf{V}$, we can prove that it is perpendicular to $\mathbf{U}$ by showing that $(\mathbf{U} \times \mathbf{V}) \cdot \mathbf{U} = 0$.

Solution

$$
\begin{aligned}
\mathbf{U} \times \mathbf{V} &= (U_y V_z - U_z V_y)\mathbf{i} - (U_x V_z - U_z V_x)\mathbf{j} \\
&\quad + (U_x V_y - U_y V_x)\mathbf{k} \\
&= [(-5)(2) - (-1)(2)]\mathbf{i} - [(6)(2) - (-1)(4)]\mathbf{j} \\
&\quad + [(6)(2) - (-5)(4)]\mathbf{k} \\
&= -8\mathbf{i} - 16\mathbf{j} + 32\mathbf{k}.
\end{aligned}
$$

(a) Use the components of the vectors to determine $\mathbf{U} \times \mathbf{V}$.

$$
\begin{aligned}
(\mathbf{U} \times \mathbf{V}) \cdot \mathbf{U} &= (\mathbf{U} \times \mathbf{V})_x U_x + (\mathbf{U} \times \mathbf{V})_y U_y + (\mathbf{U} \times \mathbf{V})_z U_z \\
&= (-8)(6) + (-16)(-5) + (32)(-1) \\
&= 0.
\end{aligned}
$$

(b) Show that $(\mathbf{U} \times \mathbf{V}) \cdot \mathbf{U} = 0$.

Practice Problem The components of two vectors $\mathbf{U}$ and $\mathbf{V}$ are $\mathbf{U} = 3\mathbf{i} + 2\mathbf{j} - \mathbf{k}$ and $\mathbf{V} = 5\mathbf{i} - 3\mathbf{j} - 4\mathbf{k}$. Determine the components of a unit vector that is perpendicular to $\mathbf{U}$ and perpendicular to $\mathbf{V}$.

Answer: $\mathbf{e} = -0.477\mathbf{i} + 0.304\mathbf{j} - 0.825\mathbf{k}$ or $\mathbf{e} = 0.477\mathbf{i} - 0.304\mathbf{j} + 0.825\mathbf{k}$.

Example 2.15	**Minimum Distance from a Point to a Line** (▶ *Related Problem 2.133*)

Consider the straight lines *OA* and *OB*.
(a) Determine the components of a unit vector that is perpendicular to both *OA* and *OB*.
(b) What is the minimum distance from point *A* to the line *OB*?

Strategy
(a) Let $\mathbf{r}_{OA}$ and $\mathbf{r}_{OB}$ be the position vectors from *O* to *A* and from *O* to *B* (Fig. a). Since the cross product $\mathbf{r}_{OA} \times \mathbf{r}_{OB}$ is perpendicular to $\mathbf{r}_{OA}$ and $\mathbf{r}_{OB}$, we will determine it and divide it by its magnitude to obtain a unit vector perpendicular to the lines *OA* and *OB*.
(b) The minimum distance from *A* to the line *OB* is the length *d* of the straight line from *A* to *OB* that is perpendicular to *OB* (Fig. b). We can see that $d = |\mathbf{r}_{OA}| \sin \theta$, where θ is the angle between $\mathbf{r}_{OA}$ and $\mathbf{r}_{OB}$. From the definition of the cross product, the magnitude of $\mathbf{r}_{OA} \times \mathbf{r}_{OB}$ is $|\mathbf{r}_{OA}||\mathbf{r}_{OB}| \sin \theta$, so we can determine *d* by dividing the magnitude of $\mathbf{r}_{OA} \times \mathbf{r}_{OB}$ by the magnitude of $\mathbf{r}_{OB}$.

Solution
(a) The components of $\mathbf{r}_{OA}$ and $\mathbf{r}_{OB}$ are

$$\mathbf{r}_{OA} = 10\mathbf{i} - 2\mathbf{j} + 3\mathbf{k} \text{ (m)},$$
$$\mathbf{r}_{OB} = 6\mathbf{i} + 6\mathbf{j} - 3\mathbf{k} \text{ (m)}.$$

By using Eq. (2.34), we obtain $\mathbf{r}_{OA} \times \mathbf{r}_{OB}$:

$$\mathbf{r}_{OA} \times \mathbf{r}_{OB} = \begin{vmatrix} \mathbf{i} & \mathbf{j} & \mathbf{k} \\ 10 & -2 & 3 \\ 6 & 6 & -3 \end{vmatrix} = -12\mathbf{i} + 48\mathbf{j} + 72\mathbf{k} \text{ (m}^2\text{)}.$$

This vector is perpendicular to $\mathbf{r}_{OA}$ and $\mathbf{r}_{OB}$. Dividing it by its magnitude, we obtain a unit vector $\mathbf{e}$ that is perpendicular to the lines *OA* and *OB*:

$$\mathbf{e} = \frac{\mathbf{r}_{OA} \times \mathbf{r}_{OB}}{|\mathbf{r}_{OA} \times \mathbf{r}_{OB}|} = \frac{-12\mathbf{i} + 48\mathbf{j} + 72\mathbf{k} \text{ (m}^2\text{)}}{\sqrt{(-12 \text{ m}^2)^2 + (48 \text{ m}^2)^2 + (72 \text{ m}^2)^2}}$$

$$= -0.137\mathbf{i} + 0.549\mathbf{j} + 0.824\mathbf{k}.$$

(b) From Fig. b, the minimum distance *d* is

$$d = |\mathbf{r}_{OA}| \sin \theta.$$

The magnitude of $\mathbf{r}_{OA} \times \mathbf{r}_{OB}$ is

$$|\mathbf{r}_{OA} \times \mathbf{r}_{OB}| = |\mathbf{r}_{OA}||\mathbf{r}_{OB}| \sin \theta.$$

Solving this equation for $\sin \theta$, we find that the distance *d* is

$$d = |\mathbf{r}_{OA}| \left(\frac{|\mathbf{r}_{OA} \times \mathbf{r}_{OB}|}{|\mathbf{r}_{OA}||\mathbf{r}_{OB}|} \right) = \frac{|\mathbf{r}_{OA} \times \mathbf{r}_{OB}|}{|\mathbf{r}_{OB}|}$$

$$= \frac{\sqrt{(-12 \text{ m}^2)^2 + (48 \text{ m}^2)^2 + (72 \text{ m}^2)^2}}{\sqrt{(6 \text{ m})^2 + (6 \text{ m})^2 + (-3 \text{ m})^2}} = 9.71 \text{ m}.$$

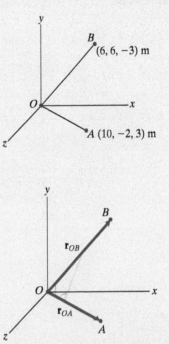

(a) The vectors $\mathbf{r}_{OA}$ and $\mathbf{r}_{OB}$.

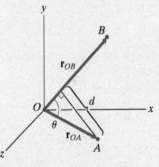

(b) The minimum distance *d* from *A* to the line *OB*.

Critical Thinking
This example is an illustration of the power of vector methods. Determining the minimum distance from point *A* to the line *OB* can be formulated as a minimization problem in differential calculus, but the vector solution we present is far simpler.

| Example 2.16 | Component of a Vector Perpendicular to a Plane (▶ *Related Problem 2.139*) |

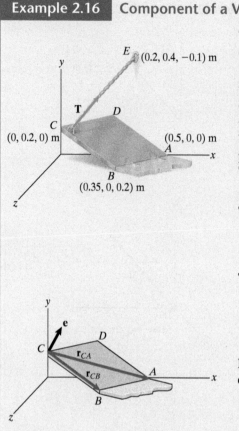

(a) Determining a unit vector perpendicular to the door.

The rope CE exerts a 500-N force $\mathbf{T}$ on the door $ABCD$. What is the magnitude of the component of $\mathbf{T}$ perpendicular to the door?

Strategy

We are given the coordinates of the corners A, B, and C of the door. By taking the cross product of the position vector $\mathbf{r}_{CB}$ from C to B and the position vector $\mathbf{r}_{CA}$ from C to A, we will obtain a vector that is perpendicular to the door. We can divide the resulting vector by its magnitude to obtain a unit vector perpendicular to the door and then apply Eq. (2.26) to determine the component of $\mathbf{T}$ perpendicular to the door.

Solution

The components of $\mathbf{r}_{CB}$ and $\mathbf{r}_{CA}$ are

$$\mathbf{r}_{CB} = 0.35\mathbf{i} - 0.2\mathbf{j} + 0.2\mathbf{k} \text{ (m)},$$
$$\mathbf{r}_{CA} = 0.5\mathbf{i} - 0.2\mathbf{j} \text{ (m)}.$$

Their cross product is

$$\mathbf{r}_{CB} \times \mathbf{r}_{CA} = \begin{vmatrix} \mathbf{i} & \mathbf{j} & \mathbf{k} \\ 0.35 & -0.2 & 0.2 \\ 0.5 & -0.2 & 0 \end{vmatrix} = 0.04\mathbf{i} + 0.1\mathbf{j} + 0.03\mathbf{k} \text{ (m}^2\text{)}.$$

Dividing this vector by its magnitude, we obtain a unit vector $\mathbf{e}$ that is perpendicular to the door (Fig. a):

$$\mathbf{e} = \frac{\mathbf{r}_{CB} \times \mathbf{r}_{CA}}{|\mathbf{r}_{CB} \times \mathbf{r}_{CA}|} = \frac{0.04\mathbf{i} + 0.1\mathbf{j} + 0.03\mathbf{k} \text{ (m}^2\text{)}}{\sqrt{(0.04 \text{ m}^2)^2 + (0.1 \text{ m}^2)^2 + (0.03 \text{ m}^2)^2}}$$
$$= 0.358\mathbf{i} + 0.894\mathbf{j} + 0.268\mathbf{k}.$$

To use Eq. (2.26), we must express $\mathbf{T}$ in terms of its scalar components. The position vector from C to E is

$$\mathbf{r}_{CE} = 0.2\mathbf{i} + 0.2\mathbf{j} - 0.1\mathbf{k} \text{ (m)},$$

so we can express the force $\mathbf{T}$ as

$$\mathbf{T} = |\mathbf{T}|\frac{\mathbf{r}_{CE}}{|\mathbf{r}_{CE}|} = (500 \text{ N})\frac{0.2\mathbf{i} + 0.2\mathbf{j} - 0.1\mathbf{k} \text{ (m)}}{\sqrt{(0.2 \text{ m})^2 + (0.2 \text{ m})^2 + (-0.1 \text{ m})^2}}$$
$$= 333\mathbf{i} + 333\mathbf{j} - 167\mathbf{k} \text{ (N)}.$$

The component of $\mathbf{T}$ parallel to the unit vector $\mathbf{e}$, which is the component of $\mathbf{T}$ perpendicular to the door, is

$$(\mathbf{e} \cdot \mathbf{T})\mathbf{e} = [(0.358)(333 \text{ N}) + (0.894)(333 \text{ N}) + (0.268)(-167 \text{ N})]\mathbf{e}$$
$$= 373\mathbf{e} \text{ (N)}.$$

The magnitude of the component of $\mathbf{T}$ perpendicular to the door is 373 N.

Critical Thinking

Why is it useful to determine the component of the force $\mathbf{T}$ perpendicular to the door? If the y axis is vertical and the rope CE is the only thing preventing the hinged door from falling, you can see intuitively that it is the component of the force perpendicular to the door that holds it in place. We analyze problems of this kind in Chapter 5.

Problems

▶ **2.124** In Active Example 2.14, suppose that the vector **V** is changed to **V** = 4**i** − 6**j** − 10**k**. (a) Determine the cross product **U** × **V**. (b) Use the dot product to prove that **U** × **V** is perpendicular to **V**.

2.125 Two vectors **U** = 3**i** + 2**j** and **V** = 2**i** + 4**j**.
(a) What is the cross product **U** × **V**?
(b) What is the cross product **V** × **U**?

2.126 The two segments of the L-shaped bar are parallel to the x and z axes. The rope AB exerts a force of magnitude $|\mathbf{F}| = 500$ lb on the bar at A. Determine the cross product $\mathbf{r}_{CA} \times \mathbf{F}$, where $\mathbf{r}_{CA}$ is the position vector from point C to point A.

2.127 The two segments of the L-shaped bar are parallel to the x and z axes. The rope AB exerts a force of magnitude $|\mathbf{F}| = 500$ lb on the bar at A. Determine the cross product $\mathbf{r}_{CB} \times \mathbf{F}$, where $\mathbf{r}_{CB}$ is the position vector from point C to point B. Compare your answer to the answer to Problem 2.126.

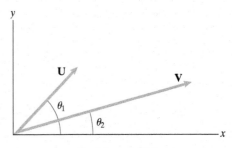

Problems 2.126/2.127

2.128 Suppose that the cross product of two vectors **U** and **V** is **U** × **V** = **0**. If $|\mathbf{U}| \neq 0$, what do you know about the vector **V**?

2.129 The cross product of two vectors **U** and **V** is **U** × **V** = −30**i** + 40**k**. The vector **V** = 4**i** − 2**j** + 3**k**. The vector **U** = 4**i** + U_y**j** + U_z**k**. Determine U_y and U_z.

2.130 The magnitudes $|\mathbf{U}| = 10$ and $|\mathbf{V}| = 20$.
(a) Use the definition of the cross product to determine **U** × **V**.
(b) Use the definition of the cross product to determine **V** × **U**.
(c) Use Eq. (2.34) to determine **U** × **V**.
(d) Use Eq. (2.34) to determine **V** × **U**.

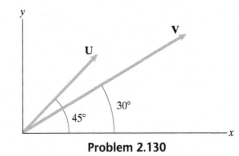

Problem 2.130

2.131 The force **F** = 10**i** − 4**j** (N). Determine the cross product $\mathbf{r}_{AB} \times \mathbf{F}$.

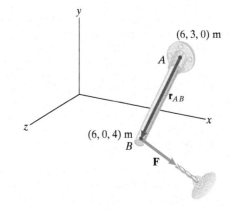

Problem 2.131

2.132 By evaluating the cross product **U** × **V**, prove the identity $\sin(\theta_1 - \theta_2) = \sin\theta_1 \cos\theta_2 - \cos\theta_1 \sin\theta_2$.

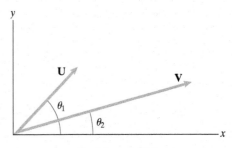

Problem 2.132

▶ **2.133** In Example 2.15, what is the minimum distance from point B to the line OA?

2.134 (a) What is the cross product $\mathbf{r}_{OA} \times \mathbf{r}_{OB}$? (b) Determine a unit vector $\mathbf{e}$ that is perpendicular to $\mathbf{r}_{OA}$ and $\mathbf{r}_{OB}$.

2.135 Use the cross product to determine the length of the shortest straight line from point B to the straight line that passes through points O and A.

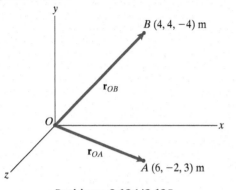

Problems 2.134/2.135

2.136 The cable BC exerts a 1000-lb force $\mathbf{F}$ on the hook at B. Determine $\mathbf{r}_{AB} \times \mathbf{F}$.

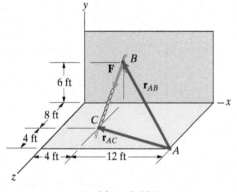

Problem 2.136

2.137 The force vector $\mathbf{F}$ points along the straight line from point A to point B. Its magnitude is $|\mathbf{F}| = 20$ N. The coordinates of points A and B are $x_A = 6$ m, $y_A = 8$ m, $z_A = 4$ m and $x_B = 8$ m, $y_B = 1$ m, $z_B = -2$ m.

(a) Express the vector $\mathbf{F}$ in terms of its components.

(b) Use Eq. (2.34) to determine the cross products $\mathbf{r}_A \times \mathbf{F}$ and $\mathbf{r}_B \times \mathbf{F}$.

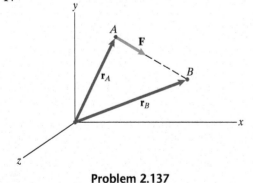

Problem 2.137

2.138 The rope AB exerts a 50-N force $\mathbf{T}$ on the collar at A. Let $\mathbf{r}_{CA}$ be the position vector from point C to point A. Determine the cross product $\mathbf{r}_{CA} \times \mathbf{T}$.

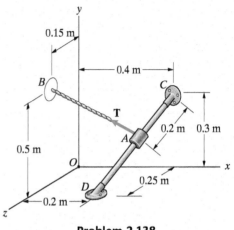

Problem 2.138

▶ **2.139** In Example 2.16, suppose that the attachment point E is moved to the location $(0.3, 0.3, 0)$ m and the magnitude of $\mathbf{T}$ increases to 600 N. What is the magnitude of the component of $\mathbf{T}$ perpendicular to the door?

2.140 The bar AB is 6 m long and is perpendicular to the bars AC and AD. Use the cross product to determine the coordinates x_B, y_B, z_B of point B.

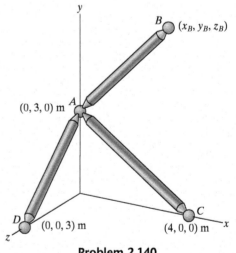

Problem 2.140

2.141* Determine the minimum distance from point P to the plane defined by the three points A, B, and C.

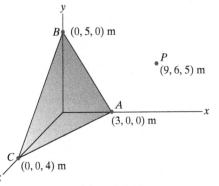

Problem 2.141

2.142* The force vector $\mathbf{F}$ points along the straight line from point A to point B. Use Eqs. (2.28)–(2.31) to prove that

$$\mathbf{r}_B \times \mathbf{F} = \mathbf{r}_A \times \mathbf{F}.$$

Strategy: Let $\mathbf{r}_{AB}$ be the position vector from point A to point B. Express $\mathbf{r}_B$ in terms of $\mathbf{r}_A$ and $\mathbf{r}_{AB}$. Notice that the vectors $\mathbf{r}_{AB}$ and $\mathbf{F}$ are parallel.

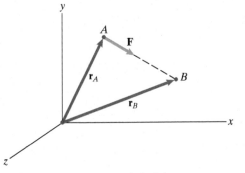

Problem 2.142

2.143 For the vectors $\mathbf{U} = 6\mathbf{i} + 2\mathbf{j} - 4\mathbf{k}$, $\mathbf{V} = 2\mathbf{i} + 7\mathbf{j}$, and $\mathbf{W} = 3\mathbf{i} + 2\mathbf{k}$, evaluate the following mixed triple products:
(a) $\mathbf{U} \cdot (\mathbf{V} \times \mathbf{W})$;
(b) $\mathbf{W} \cdot (\mathbf{V} \times \mathbf{U})$;
(c) $\mathbf{V} \cdot (\mathbf{W} \times \mathbf{U})$.

2.144 Use the mixed triple product to calculate the volume of the parallelepiped.

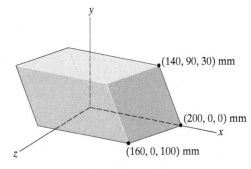

Problem 2.144

2.145 By using Eqs. (2.23) and (2.34), show that

$$\mathbf{U} \cdot (\mathbf{V} \times \mathbf{W}) = \begin{vmatrix} U_x & U_y & U_z \\ V_x & V_y & V_z \\ W_x & W_y & W_z \end{vmatrix}.$$

2.146 The vectors $\mathbf{U} = \mathbf{i} + U_y\mathbf{j} + 4\mathbf{k}$, $\mathbf{V} = 2\mathbf{i} + \mathbf{j} - 2\mathbf{k}$, and $\mathbf{W} = -3\mathbf{i} + \mathbf{j} - 2\mathbf{k}$ are coplanar (they lie in the same plane). What is the component U_y?

Review Problems

2.147 The magnitude of $\mathbf{F}$ is 8 kN. Express $\mathbf{F}$ in terms of scalar components.

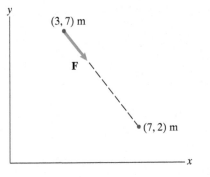

Problem 2.147

2.148 The magnitude of the vertical force $\mathbf{W}$ is 600 lb, and the magnitude of the force $\mathbf{B}$ is 1500 lb. Given that $\mathbf{A} + \mathbf{B} + \mathbf{W} = 0$, determine the magnitude of the force $\mathbf{A}$ and the angle α.

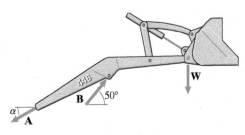

Problem 2.148

2.149 The magnitude of the vertical force vector **A** is 200 lb. If **A** + **B** + **C** = **0**, what are the magnitudes of the force vectors **B** and **C**?

2.150 The magnitude of the horizontal force vector **D** is 280 lb. If **D** + **E** + **F** = **0**, what are the magnitudes of the force vectors **E** and **F**?

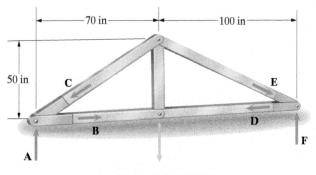
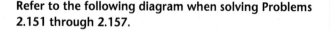
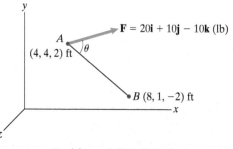

Problems 2.149/2.150

Refer to the following diagram when solving Problems 2.151 through 2.157.

Problems 2.151–2.157

2.151 What are the direction cosines of **F**?

2.152 Determine the components of a unit vector parallel to line AB that points from A toward B.

2.153 What is the angle θ between the line AB and the force **F**?

2.154 Determine the vector component of **F** that is parallel to the line AB.

2.155 Determine the vector component of **F** that is normal to the line AB.

2.156 Determine the vector $\mathbf{r}_{BA} \times \mathbf{F}$, where $\mathbf{r}_{BA}$ is the position vector from B to A.

2.157 (a) Write the position vector $\mathbf{r}_{AB}$ from point A to point B in terms of components.

(b) A vector **R** has magnitude $|\mathbf{R}| = 200$ lb and is parallel to the line from A to B. Write **R** in terms of components.

2.158 The rope exerts a force of magnitude $|\mathbf{F}| = 200$ lb on the top of the pole at B.

(a) Determine the vector $\mathbf{r}_{AB} \times \mathbf{F}$, where $\mathbf{r}_{AB}$ is the position vector from A to B.

(b) Determine the vector $\mathbf{r}_{AC} \times \mathbf{F}$, where $\mathbf{r}_{AC}$ is the position vector from A to C.

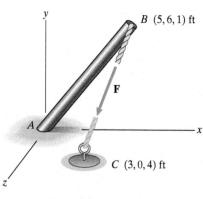

Problem 2.158

2.159 The pole supporting the sign is parallel to the x axis and is 6 ft long. Point A is contained in the y–z plane. (a) Express the vector **r** in terms of components. (b) What are the direction cosines of **r**?

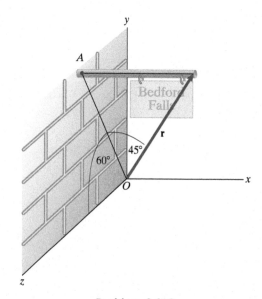

Problem 2.159

2.160 The z component of the force **F** is 80 lb. (a) Express **F** in terms of components. (b) What are the angles θ_x, θ_y, and θ_z between **F** and the positive coordinate axes?

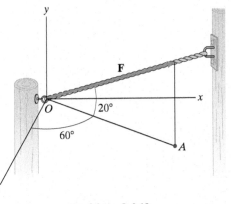

Problem 2.160

2.161 The magnitude of the force vector $\mathbf{F}_B$ is 2 kN. Express it in terms of components.

2.162 The magnitude of the vertical force vector **F** is 6 kN. Determine the vector components of **F** parallel and normal to the line from B to D.

2.163 The magnitude of the vertical force vector **F** is 6 kN. Given that $\mathbf{F} + \mathbf{F}_A + \mathbf{F}_B + \mathbf{F}_C = \mathbf{0}$, what are the magnitudes of $\mathbf{F}_A$, $\mathbf{F}_B$, and $\mathbf{F}_C$?

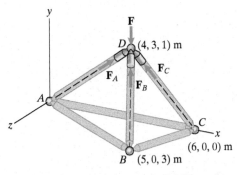

Problems 2.161–2.163

2.164 The magnitude of the vertical force **W** is 160 N. The direction cosines of the position vector from A to B are $\cos \theta_x = 0.500$, $\cos \theta_y = 0.866$, and $\cos \theta_z = 0$, and the direction cosines of the position vector from B to C are $\cos \theta_x = 0.707$, $\cos \theta_y = 0.619$, and $\cos \theta_z = -0.342$. Point G is the midpoint of the line from B to C. Determine the vector $\mathbf{r}_{AG} \times \mathbf{W}$, where $\mathbf{r}_{AG}$ is the position vector from A to G.

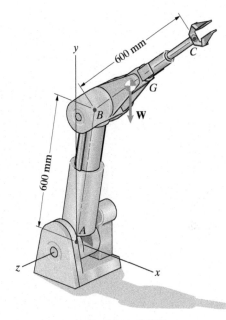

Problem 2.164

2.165 The rope CE exerts a 500-N force **T** on the hinged door.
(a) Express **T** in terms of components.
(b) Determine the vector component of **T** parallel to the line from point A to point B.

2.166 In Problem 2.165, let $\mathbf{r}_{BC}$ be the position vector from point B to point C. Determine the cross product $\mathbf{r}_{BC} \times \mathbf{T}$.

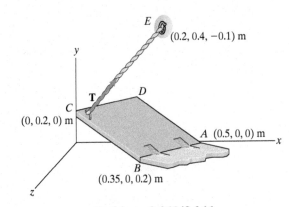

Problems 2.165/2.166

CHAPTER
3

Forces

In Chapter 2 we represented forces by vectors and used vector addition to sum forces. In this chapter we discuss forces in more detail and introduce two of the most important concepts in mechanics, equilibrium and the free-body diagram. We will use free-body diagrams to identify the forces on objects and use equilibrium to determine unknown forces.

◀ The forces due to the weight of the bridge are transferred to the vertical support towers by cables. In this chapter we use free-body diagrams to analyze the forces acting on objects in equilibrium.

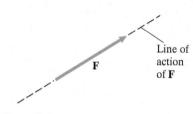

Figure 3.1
A force **F** and its line of action.

3.1 Forces, Equilibrium, and Free-Body Diagrams

BACKGROUND

Force is a familiar concept, as is evident from the words push, pull, and lift used in everyday conversation. In engineering we deal with different types of forces having a large range of magnitudes. In this section we define some terms used to describe forces, discuss particular forces that occur frequently in engineering applications, and introduce the concepts of equilibrium and free-body diagrams.

Terminology

Line of Action   When a force is represented by a vector, the straight line collinear with the vector is called the *line of action* of the force (Fig. 3.1).

Systems of Forces   A *system of forces* is simply a particular set of forces. A system of forces is *coplanar,* or *two dimensional,* if the lines of action of the forces lie in a plane. Otherwise it is *three dimensional.* A system of forces is *concurrent* if the lines of action of the forces intersect at a point (Fig. 3.2a) and *parallel* if the lines of action are parallel (Fig. 3.2b).

External and Internal Forces   We say that a given object is subjected to an *external force* if the force is exerted by a different object. When one part of a given object is subjected to a force by another part of the same object, we say it is subjected to an *internal force*. These definitions require that you clearly define the object you are considering. For example, suppose that you are the object. When you are standing, the floor—a different object—exerts an external force on your feet. If you press your hands together, your left hand exerts an internal force on your right hand. However, if your right hand is the object you are considering, the force exerted by your left hand is an external force.

Body and Surface Forces   A force acting on an object is called a *body force* if it acts on the volume of the object and a *surface force* if it acts on its surface. The gravitational force on an object is a body force. A surface force can be exerted on an object by contact with another object. Both body and surface forces can result from electromagnetic effects.

Gravitational Forces

You are aware of the force exerted on an object by the earth's gravity whenever you pick up something heavy. We can represent the gravitational force, or weight, of an object by a vector (Fig. 3.3).

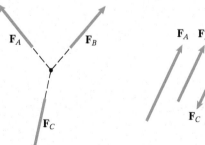

Figure 3.2
(**a**) Concurrent forces.
(**b**) Parallel forces.

(a)

(b)

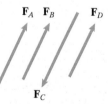
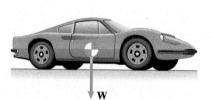

Figure 3.3
Representing an object's weight by a vector.

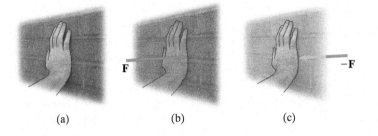

(a)　　　　　　(b)　　　　　　(c)

Figure 3.4
(a) Exerting a contact force on a wall by pushing on it.
(b) The vector **F** represents the force you exert on the wall.
(c) The wall exerts a force −**F** on your hand.

The magnitude of an object's weight is related to its mass m by

$$|\mathbf{W}| = mg, \tag{3.1}$$

where g is the acceleration due to gravity at sea level. We will use the values $g = 9.81$ m/s^2 in SI units and $g = 32.2$ ft/s^2 in U.S. Customary units.

Gravitational forces, and also electromagnetic forces, act at a distance. The objects they act on are not necessarily in contact with the objects exerting the forces. In the next section we discuss forces resulting from contacts between objects.

Contact Forces

Contact forces are the forces that result from contacts between objects. For example, you exert a contact force when you push on a wall (Fig. 3.4a). The surface of your hand exerts a force on the surface of the wall that can be represented by a vector **F** (Fig. 3.4b). The wall exerts an equal and opposite force −**F** on your hand (Fig. 3.4c). (Recall Newton's third law: The forces exerted on each other by any two particles are equal in magnitude and opposite in direction. If you have any doubt that the wall exerts a force on your hand, try pushing on the wall while standing on roller skates.)

We will be concerned with contact forces exerted on objects by contact with the surfaces of other objects and by ropes, cables, and springs.

Surfaces　Consider two plane surfaces in contact (Fig. 3.5a). We represent the force exerted on the right surface by the left surface by the vector **F** in Fig. 3.5b. We can resolve **F** into a component **N** that is normal to the surface and a component **f** that is parallel to the surface (Fig. 3.5c). The component **N** is called the *normal force*, and the component **f** is called the *friction force*. We sometimes assume that the friction force between two surfaces is negligible in comparison to the normal force, a condition we describe by saying that the surfaces are *smooth*. In this case we show only the normal force (Fig. 3.5d). When the friction force cannot be neglected, we say the surfaces are *rough*.

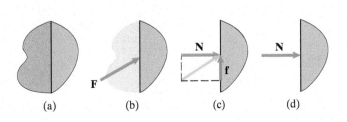

(a)　　　　(b)　　　　(c)　　　　(d)

Figure 3.5
(a) Two plane surfaces in contact.
(b) The force **F** exerted on the right surface.
(c) The force **F** resolved into components normal and parallel to the surface.
(d) Only the normal force is shown when friction is neglected.

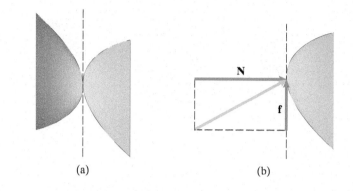

(a)   (b)

Figure 3.6
(a) Curved contacting surfaces. The dashed line indicates the plane tangent to the surfaces at their point of contact.
(b) The normal force and friction force on the right surface.

If the contacting surfaces are curved (Fig. 3.6a), the normal force and the friction force are perpendicular and parallel to the plane tangent to the surfaces at their point of contact (Fig. 3.6b).

Ropes and Cables   A contact force can be exerted on an object by attaching a rope or cable to the object and pulling on it. In Fig. 3.7a, the crane's cable is attached to a container of building materials. We can represent the force the cable exerts on the container by a vector **T** (Fig. 3.7b). The magnitude of **T** is called the *tension* in the cable, and the line of action of **T** is collinear with the cable. The cable exerts an equal and opposite force −**T** on the crane (Fig. 3.7c).

Notice that we have assumed that the cable is straight and that the tension where the cable is connected to the container equals the tension near the crane. This is approximately true if the weight of the cable is small compared to the tension. Otherwise, the cable will sag significantly and the tension will vary along its length. In Chapter 9 we will discuss ropes and cables whose weights are not small in comparison to their tensions. For now, we assume that ropes and cables are straight and that their tensions are constant along their lengths.

A *pulley* is a wheel with a grooved rim that can be used to change the direction of a rope or cable (Fig. 3.8a). For now, we assume that the tension is

(a)

(b)

(c)

Figure 3.7
(a) A crane with its cable attached to a container.
(b) The force **T** exerted on the container by the cable.
(c) The force −**T** exerted on the crane by the cable.

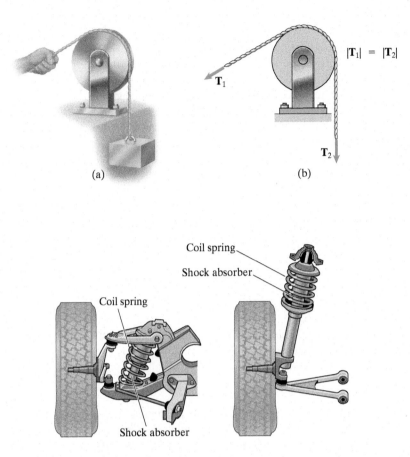

$|\mathbf{T}_1| = |\mathbf{T}_2|$

Figure 3.8
(a) A pulley changes the direction of a rope or cable.
(b) For now, you should assume that the tensions on each side of the pulley are equal.

Figure 3.9
Coil springs in car suspensions. The arrangement on the right is called a MacPherson strut.

the same on both sides of a pulley (Fig. 3.8b). This is true, or at least approximately true, when the pulley can turn freely and the rope or cable either is stationary or turns the pulley at a constant rate.

Springs   Springs are used to exert contact forces in mechanical devices, for example, in the suspensions of cars (Fig. 3.9). Let's consider a coil spring whose unstretched length, the length of the spring when its ends are free, is L_0 (Fig. 3.10a). When the spring is stretched to a length L greater than L_0 (Fig. 3.10b), it pulls on the object to which it is attached with a force $\mathbf{F}$ (Fig. 3.10c). The object exerts an equal and opposite force $-\mathbf{F}$ on the spring (Fig. 3.10d). When the spring is compressed to a length L less than L_0 (Figs. 3.11a, b), the spring pushes on the object with a force $\mathbf{F}$ and the object exerts an equal and opposite force $-\mathbf{F}$ on the spring (Figs. 3.11c, d). If a spring is compressed too much, it may buckle (Fig. 3.11e). A spring designed to exert a force by being compressed is often provided with lateral support to prevent buckling, for example, by enclosing it in a cylindrical sleeve. In the car suspensions shown in Fig. 3.9, the shock absorbers within the coils prevent the springs from buckling.

The magnitude of the force exerted by a spring depends on the material it is made of, its design, and how much it is stretched or compressed relative to its unstretched length. When the change in length is not too large compared to the unstretched length, the coil springs commonly used in mechanical devices exert a force approximately proportional to the change in length:

$$|\mathbf{F}| = k|L - L_0|. \tag{3.2}$$

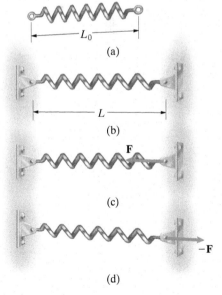

Figure 3.10
(a) A spring of unstretched length L_0.
(b) The spring stretched to a length $L > L_0$.
(c, d) The force $\mathbf{F}$ exerted by the spring and the force $-\mathbf{F}$ on the spring.

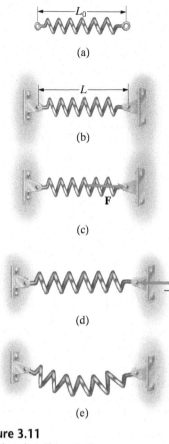

(a)

(b)

(c)

(d)

(e)

Figure 3.11
(a) A spring of length L_0.
(b) The spring compressed to a length
   $L < L_0$.
(c, d) The spring pushes on an object with a
   force **F**, and the object exerts a force
   $-$**F** on the spring.
(e) A coil spring will buckle if it is compressed too much.

Because the force is a linear function of the change in length (Fig. 3.12), a spring that satisfies this relation is called a *linear spring*. The value of the *spring constant k* depends on the material and design of the spring. Its dimensions are (force)/(length). Notice from Eq. (3.2) that k equals the magnitude of the force required to stretch or compress the spring a unit of length.

Suppose that the unstretched length of a spring is $L_0 = 1$ m and $k = 3000$ N/m. If the spring is stretched to a length $L = 1.2$ m, the magnitude of the pull it exerts is

$$k|L - L_0| = 3000(1.2 - 1) = 600 \text{ N}.$$

Although coil springs are commonly used in mechanical devices, we are also interested in them for a different reason. Springs can be used to *model* situations in which forces depend on displacements. For example, the force necessary to bend the steel beam in Fig. 3.13a is a linear function of the displacement δ, or

$$|\mathbf{F}| = k\delta,$$

if δ is not too large. Therefore we can model the force-deflection behavior of the beam with a linear spring (Fig. 3.13b).

Equilibrium

In everyday conversation, equilibrium means an unchanging state—a state of balance. Before we state precisely what this term means in mechanics, let us consider some familiar examples. If you are in a building as you read this, objects you observe around you that are *at rest (stationary) relative to the building*, such as pieces of furniture, are in equilibrium. A person sitting or standing at rest relative to the building is also in equilibrium. If a train travels at constant speed on a straight track, objects within the train that are at rest relative to the train, such as the passenger seats or a passenger standing in the aisle (Fig. 3.14a), are in equilibrium. *The person at rest relative to the building and also the passenger at rest relative to the train are not accelerating.*

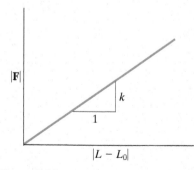

Figure 3.12
The graph of the force exerted by a linear spring as a function of its stretch or compression is a straight line with slope k.

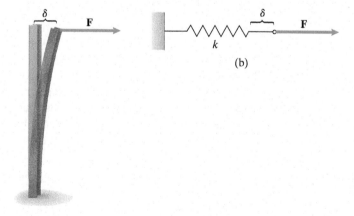

(a)

(b)

Figure 3.13
(a) A steel beam deflected by a force.
(b) Modeling the beam's behavior with a linear spring.

However, if the train should begin increasing or decreasing its speed, the person standing in the aisle of the train would no longer be in equilibrium and might lose his balance (Fig. 3.14b).

We define an object to be in *equilibrium* only if each point of the object has the same constant velocity, which is referred to as *steady translation*. The velocity must be measured relative to a frame of reference in which Newton's laws are valid. Such a frame is called a *Newtonian* or *inertial reference frame*. In many engineering applications, a frame of reference that is fixed with respect to the earth can be regarded as inertial. Therefore, objects in steady translation relative to the earth can be assumed to be in equilibrium. We make this assumption throughout this book. In the examples cited in the previous paragraph, the furniture and person at rest in a building and also the passenger seats and passenger at rest within the train moving at constant speed are in steady translation relative to the earth and so are in equilibrium.

The vector sum of the external forces acting on an object in equilibrium is zero. We will use the symbol $\Sigma\mathbf{F}$ to denote the sum of the external forces. Thus, when an object is in equilibrium,

$$\Sigma\mathbf{F} = \mathbf{0}. \tag{3.3}$$

In some situations we can use this *equilibrium equation* to determine unknown forces acting on an object in equilibrium. The first step will be to draw a *free-body diagram* of the object to identify the external forces acting on it.

Free-Body Diagrams

A free-body diagram serves to focus attention on the object of interest and helps identify the external forces acting on it. Although in statics we are concerned only with objects in equilibrium, free-body diagrams are also used in dynamics to study the motions of objects.

Although it is one of the most important tools in mechanics, a free-body diagram is a simple concept. It is a drawing of an object and the external forces acting on it. Otherwise, nothing other than the object of interest is included. The drawing shows the object *isolated*, or *freed*, from its surroundings.

Drawing a free-body diagram involves three steps:

1. *Identify the object you want to isolate*—As the following examples show, your choice is often dictated by particular forces you want to determine.

2. *Draw a sketch of the object isolated from its surroundings, and show relevant dimensions and angles*—Your drawing should be reasonably accurate, but it can omit irrelevant details.

3. *Draw vectors representing all of the external forces acting on the isolated object, and label them*—Don't forget to include the gravitational force if you are not intentionally neglecting it.

A coordinate system is necessary to express the forces on the isolated object in terms of components. Often it is convenient to choose the coordinate system before drawing the free-body diagram, but in some situations the best choice of a coordinate system will not be apparent until after it has been drawn.

A simple example demonstrates how you can choose free-body diagrams to determine particular forces and also that you must distinguish carefully between external and internal forces. Two stationary blocks of equal weight W are suspended by cables in Fig. 3.15. The system is in equilibrium. Suppose that we want to determine the tensions in the two cables.

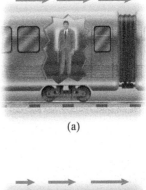

(a)

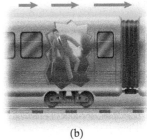

(b)

Figure 3.14
(a) While the train moves at a constant speed, a person standing in the aisle is in equilibrium.
(b) If the train starts to speed up, the person is no longer in equilibrium.

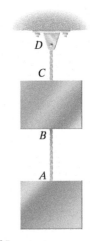

Figure 3.15
Stationary blocks suspended by cables.

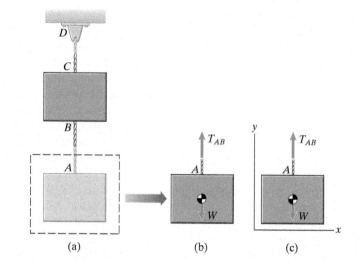

Figure 3.16
(a) Isolating the lower block and part of cable AB.
(b) Indicating the external forces completes the free-body diagram.
(c) Introducing a coordinate system.

To determine the tension in cable AB, we first isolate an "object" consisting of the lower block and part of cable AB (Fig. 3.16a). We then ask ourselves what forces can be exerted on our isolated object by objects not included in the diagram. The earth exerts a gravitational force of magnitude W on the block. Also, where we "cut" cable AB, the cable is subjected to a contact force equal to the tension in the cable (Fig. 3.16b). The arrows in this figure indicate the directions of the forces. The scalar W is the weight of the block and T_{AB} is the tension in cable AB. We assume that the weight of the part of cable AB included in the free-body diagram can be neglected in comparison to the weight of the block.

Since the free-body diagram is in equilibrium, the sum of the external forces equals zero. In terms of a coordinate system with the y axis upward (Fig. 3.16c), we obtain the equilibrium equation

$$\Sigma \mathbf{F} = T_{AB}\mathbf{j} - W\mathbf{j} = (T_{AB} - W)\mathbf{j} = \mathbf{0}.$$

Thus, the tension in cable AB is $T_{AB} = W$.

We can determine the tension in cable CD by isolating the upper block (Fig. 3.17a). The external forces are the weight of the upper block and the tensions in the two cables (Fig. 3.17b). In this case we obtain the equilibrium equation

$$\Sigma \mathbf{F} = T_{CD}\mathbf{j} - T_{AB}\mathbf{j} - W\mathbf{j} = (T_{CD} - T_{AB} - W)\mathbf{j} = \mathbf{0}.$$

Since $T_{AB} = W$, we find that $T_{CD} = 2W$.

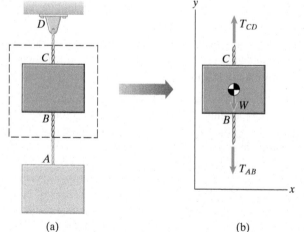

Figure 3.17
(a) Isolating the upper block to determine the tension in cable CD.
(b) Free-body diagram of the upper block.

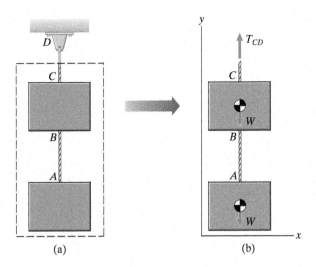

(a)   (b)

Figure 3.18
(a) An alternative choice for determining the tension in cable *CD*.
(b) Free-body diagram including both blocks and cable *AB*.

We could also have determined the tension in cable *CD* by treating the two blocks and the cable *AB* as a single object (Figs. 3.18a, b). The equilibrium equation is

$$\Sigma \mathbf{F} = T_{CD}\mathbf{j} - W\mathbf{j} - W\mathbf{j} = (T_{CD} - 2W)\mathbf{j} = \mathbf{0},$$

and we again obtain $T_{CD} = 2W$.

Why doesn't the tension in cable *AB* appear on the free-body diagram in Fig. 3.18b? Remember that only external forces are shown on free-body diagrams. Since cable *AB* is part of the free-body diagram in this case, the forces it exerts on the upper and lower blocks are internal forces.

RESULTS

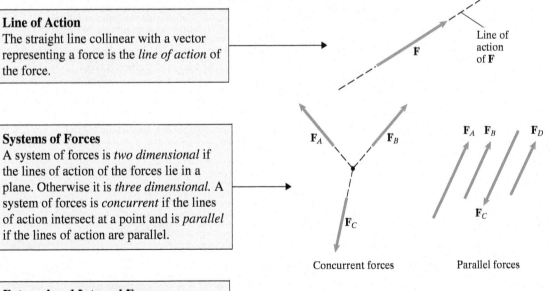

Line of Action
The straight line collinear with a vector representing a force is the *line of action* of the force.

Line of action of **F**

Systems of Forces
A system of forces is *two dimensional* if the lines of action of the forces lie in a plane. Otherwise it is *three dimensional*. A system of forces is *concurrent* if the lines of action intersect at a point and is *parallel* if the lines of action are parallel.

Concurrent forces   Parallel forces

External and Internal Forces
An object is subjected to an *external force* if the force is exerted by a different object. A force exerted on part of an object by a different part of the same object is an *internal force*.

Gravitational Forces

The weight of an object can be represented by a vector. Its magnitude at sea level is related to the mass m of the object by

$$|\mathbf{W}| = mg, \qquad (3.1)$$

where g is the acceleration due to gravity at sea level.

Contact Forces

Contacting objects exert equal and opposite forces on each other.

Ropes and Cables

If the weight of a rope or cable connecting two objects is negligible in comparison to its tension, it exerts equal and opposite forces on the objects that are parallel to the rope or cable.

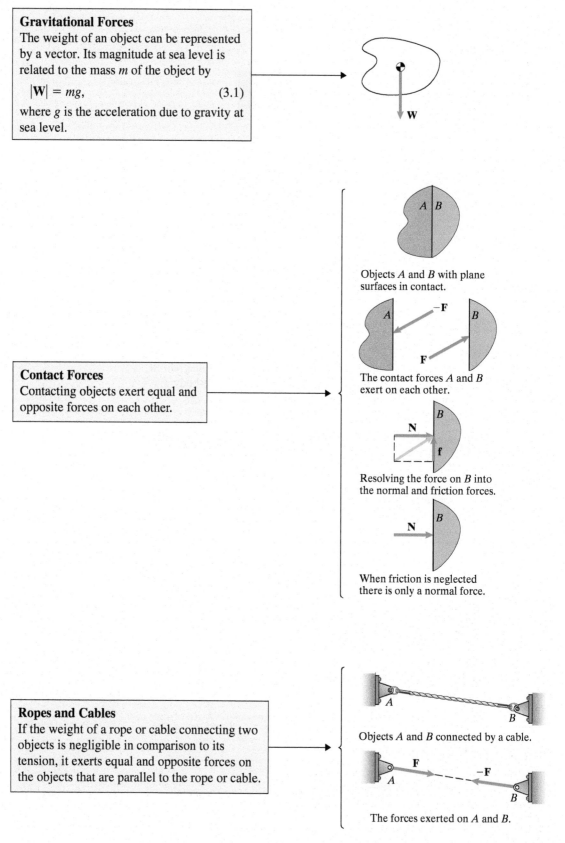

Objects A and B with plane surfaces in contact.

The contact forces A and B exert on each other.

Resolving the force on B into the normal and friction forces.

When friction is neglected there is only a normal force.

Objects A and B connected by a cable.

The forces exerted on A and B.

Linear Springs
The magnitude of the equal and opposite forces exerted on two objects connected by a *linear springs* is

$$|\mathbf{F}| = k|L - L_0|, \qquad (3.2)$$

where k is the *spring constant*, L is the length of the spring, and L_0 is its unstretched length.

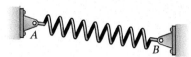

Objects A and B connected by a spring.

The forces exerted on A and B.

Equilibrium
An object is in equilibrium if it is in *steady translation* (each point of the object has the same constant velocity) relative to an inertial reference frame. The sum of the external forces acting on an object in equilibrium is zero:

$$\Sigma\mathbf{F} = \mathbf{0}. \qquad (3.3)$$

Free-Body Diagrams
A *free-body diagram* is a drawing of an object, isolated from its surroundings, that shows the external forces acting on it. Drawing a free-body diagram involves three steps.

1. Identify the object you want to isolate.
2. Draw a sketch of the object isolated from its surroundings.
3. Draw vectors representing the external forces acting on the object.

3.2 Two-Dimensional Force Systems

Suppose that the system of external forces acting on an object in equilibrium is two dimensional (coplanar). By orienting a coordinate system so that the forces lie in the *x–y* plane, we can express the sum of the external forces as

$$\Sigma\mathbf{F} = (\Sigma F_x)\mathbf{i} + (\Sigma F_y)\mathbf{j} = \mathbf{0},$$

where ΣF_x and ΣF_y are the sums of the x and y components of the forces. Since a vector is zero only if each of its components is zero, we obtain two scalar equilibrium equations:

$$\Sigma F_x = 0, \qquad \Sigma F_y = 0. \qquad (3.4)$$

The sums of the x and y components of the external forces acting on an object in equilibrium must each equal zero.

Active Example 3.1 | Using Equilibrium to Determine Forces (▶ *Related Problem 3.1*)

The 1440-kg car is held in place on the inclined ramp by the horizontal cable from *A* to *B*. The car's brakes are not engaged, so the tires exert only normal forces on the ramp. Determine the magnitude of the force exerted on the car by the cable.

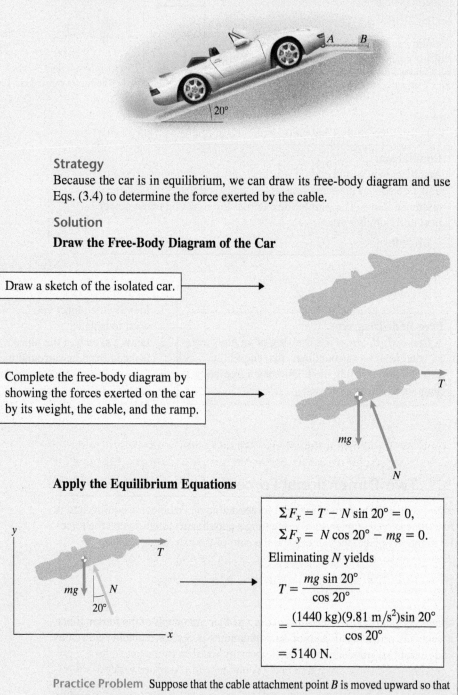

Strategy

Because the car is in equilibrium, we can draw its free-body diagram and use Eqs. (3.4) to determine the force exerted by the cable.

Solution

Draw the Free-Body Diagram of the Car

Draw a sketch of the isolated car. ──────▶

Complete the free-body diagram by showing the forces exerted on the car by its weight, the cable, and the ramp. ──────▶

T

mg

N

Apply the Equilibrium Equations

$$\Sigma F_x = T - N \sin 20° = 0,$$
$$\Sigma F_y = N \cos 20° - mg = 0.$$

Eliminating N yields

$$T = \frac{mg \sin 20°}{\cos 20°}$$

$$= \frac{(1440 \text{ kg})(9.81 \text{ m/s}^2)\sin 20°}{\cos 20°}$$

$$= 5140 \text{ N}.$$

Practice Problem Suppose that the cable attachment point *B* is moved upward so that the cable is parallel to the ramp. Determine the magnitude of the force exerted on the car by the cable.

Answer: 4830 N.

Example 3.2 | **Choosing a Free-Body Diagram** (▶ *Related Problem 3.3*)

The automobile engine block is suspended by a system of cables. The mass of the block is 200 kg. The system is stationary. What are the tensions in cables AB and AC?

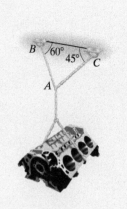

Strategy
We need a free-body diagram that is subjected to the forces we want to determine. By isolating part of the cable system near point A where the cables are joined, we can obtain a free-body diagram that is subjected to the weight of the block and the unknown tensions in cables AB and AC.

Solution
Draw the Free-Body Diagram Isolating part of the cable system near point A (Fig. a), we obtain a free-body diagram subjected to the weight of the block $W = mg = (200 \text{ kg}) (9.81 \text{ m/s}^2) = 1962 \text{ N}$ and the tensions in cables AB and AC (Fig. b).

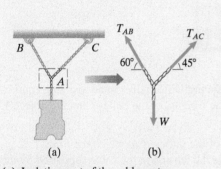

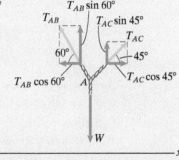

(a) Isolating part of the cable system.
(b) The completed free-body diagram.

(c) Selecting a coordinate system and re-solving the forces into components.

Apply the Equilibrium Equations We select the coordinate system shown in Fig. c and resolve the cable tensions into x and y components. The resulting equilibrium equations are

$$\Sigma F_x = T_{AC} \cos 45° - T_{AB} \cos 60° = 0,$$

$$\Sigma F_y = T_{AC} \sin 45° + T_{AB} \sin 60° - 1962 \text{ N} = 0.$$

Solving these equations, we find that the tensions in the cables are $T_{AB} = 1436 \text{ N}$ and $T_{AC} = 1016 \text{ N}$.

Critical Thinking
How can you choose a free-body diagram that permits you to determine particular unknown forces? There are no definite rules for choosing free-body diagrams. You will learn what to do in many cases from the examples we present, but you will also encounter new situations. It may be necessary to try several free-body diagrams before finding one that provides the information you need. Remember that forces you want to determine should appear as external forces on your free-body diagram, and your objective is to obtain a number of equilibrium equations equal to the number of unknown forces.

| Example 3.3 | Applying Equilibrium to a System of Pulleys (▶ *Related Problem 3.54*) |

The mass of each pulley of the system is m, and the mass of the suspended object A is m_A. Determine the force T necessary for the system to be in equilibrium.

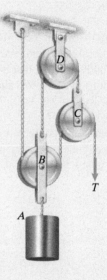

Strategy

By drawing free-body diagrams of the individual pulleys and applying equilibrium, we can relate the force T to the weights of the pulleys and the object A.

Solution

We first draw a free-body diagram of the pulley C to which the force T is applied (Fig. a). Notice that we assume the tension in the cable supported by the pulley to equal T on both sides (see Fig. 3.8). From the equilibrium equation

$$T_D - T - T - mg = 0,$$

we determine that the tension in the cable supported by pulley D is

$$T_D = 2T + mg.$$

We now know the tensions in the cables extending from pulleys C and D to pulley B in terms of T. Drawing the free-body diagram of pulley B (Fig. b), we obtain the equilibrium equation

$$T + T + 2T + mg - mg - m_A g = 0.$$

Solving, we obtain $T = m_A g/4$.

Critical Thinking

Notice that the objects we isolate in Figs. a and b include parts of the cables. The weights of those parts of cable are external forces acting on the free-body diagrams. Why didn't we include them? We tacitly assumed that the weights of those parts of cable could be neglected in comparison to the weights of the pulleys and the suspended object A. You will notice throughout the book that weights of objects are often neglected in analyzing the forces acting on them. This is a valid approximation for a given object if its weight is small compared to the other forces acting on it. But in any real engineering application, this assumption must be carefully evaluated. We discuss the weights of objects in more detail in Chapter 7.

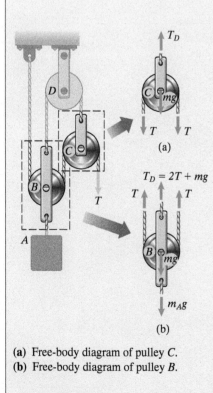

(a) Free-body diagram of pulley C.
(b) Free-body diagram of pulley B.

| Example 3.4 | Forces on an Airplane in Equilibrium (▶ *Related Problems 3.60–3.62*) |

The figure shows an airplane flying in the vertical plane and its free-body diagram. The forces acting on the airplane are its weight W, the thrust T exerted by its engines, and aerodynamic forces resulting from the pressure distribution on the airplane's surface. The dashed line indicates the path along which the airplane is moving. The aerodynamic forces are resolved into a component perpendicular to the path, the lift L, and a component parallel to the path, the drag D. The angle γ between the horizontal and the path is called the flight path angle, and α is the angle of attack. If the airplane remains in equilibrium for an interval of time, it is said to be in steady flight. If $\gamma = 6°$, $D = 125$ kN, $L = 680$ kN, and the mass of the airplane is 72,000 kg, what values of T and α are necessary to maintain steady flight?

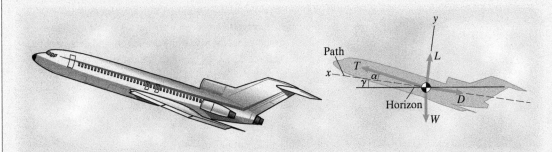

Strategy
The airplane is assumed to be in equilibrium. By applying Eqs. (3.4) to the given free-body diagram, we will obtain two equations with which to determine T and α.

Solution
In terms of the coordinate system in the figure, the equilibrium equations are

$$\Sigma F_x = T \cos \alpha - D - W \sin \gamma = 0, \qquad (1)$$
$$\Sigma F_y = T \sin \alpha + L - W \cos \gamma = 0, \qquad (2)$$

where the airplane's weight is $W = (72{,}000 \text{ kg})(9.81 \text{ m/s}^2) = 706{,}000$ N. We solve Eq. (2) for $\sin \alpha$, solve Eq. (1) for $\cos \alpha$, and divide to obtain an equation for $\tan \alpha$:

$$\tan \alpha = \frac{W \cos \gamma - L}{W \sin \gamma + D}$$

$$= \frac{(706{,}000 \text{ N}) \cos 6° - 680{,}000 \text{ N}}{(706{,}000 \text{ N}) \sin 6° + 125{,}000 \text{ N}} = 0.113.$$

The angle of attack $\alpha = \arctan(0.113) = 6.44°$. Now we use Eq. (1) to determine the thrust:

$$T = \frac{W \sin \gamma + D}{\cos \alpha}$$

$$= \frac{(706{,}000 \text{ N}) \sin 6° + 125{,}000 \text{ N}}{\cos 6.44°} = 200{,}000 \text{ N}.$$

Notice that the thrust necessary for steady flight is 28% of the airplane's weight.

Problems

▶ 3.1 In Active Example 3.1, suppose that the angle between the ramp supporting the car is increased from 20° to 30°. Draw the free-body diagram of the car showing the new geometry. Suppose that the cable from A to B must exert a 1900-lb horizontal force on the car to hold it in place. Determine the car's weight in pounds.

3.2 The ring weighs 5 lb and is in equilibrium. The force $F_1 = 4.5$ lb. Determine the force F_2 and the angle α.

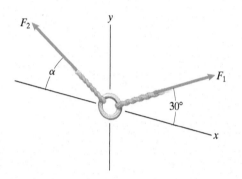

Problem 3.2

▶ 3.3 In Example 3.2, suppose that the attachment point C is moved to the right and cable AC is extended so that the angle between cable AC and the ceiling decreases from 45° to 35°. The angle between cable AB and the ceiling remains 60°. What are the tensions in cables AB and AC?

3.4 The 200-kg engine block is suspended by the cables AB and AC. The angle $\alpha = 40°$. The free-body diagram obtained by isolating the part of the system within the dashed line is shown. Determine the forces T_{AB} and T_{AC}.

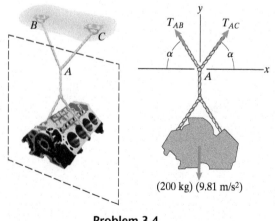

Problem 3.4

3.5 A heavy rope used as a mooring line for a cruise ship sags as shown. If the mass of the rope is 90 kg, what are the tensions in the rope at A and B?

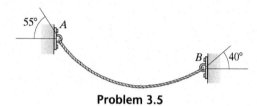

Problem 3.5

3.6 A physiologist estimates that the masseter muscle of a predator, *Martes*, is capable of exerting a force M as large as 900 N. Assume that the jaw is in equilibrium and determine the necessary force T that the temporalis muscle exerts and the force P exerted on the object being bitten.

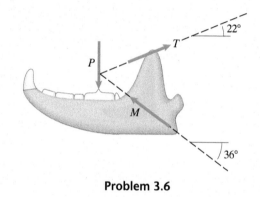

Problem 3.6

3.7 The two springs are identical, with unstretched lengths 250 mm and spring constants $k = 1200$ N/m.

(a) Draw the free-body diagram of block A.

(b) Draw the free-body diagram of block B.

(c) What are the masses of the two blocks?

3.8 The two springs are identical, with unstretched lengths of 250 mm. Suppose that their spring constant k is unknown and the sum of the masses of blocks A and B is 10 kg. Determine the value of k and the masses of the two blocks.

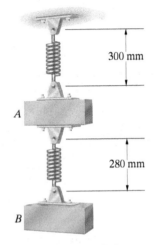

300 mm

A

280 mm

B

Problems 3.7/3.8

3.9 The inclined surface is smooth. (Remember that "smooth" means that friction is negligible.) The two springs are identical, with unstretched lengths of 250 mm and spring constants $k = 1200$ N/m. What are the masses of blocks A and B?

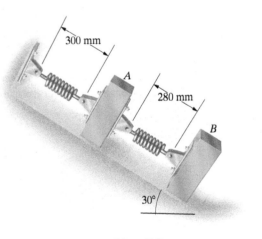

300 mm

A

280 mm

B

30°

Problem 3.9

3.10 The mass of the crane is 20,000 kg. The crane's cable is attached to a caisson whose mass is 400 kg. The tension in the cable is 1 kN.

(a) Determine the magnitudes of the normal and friction forces exerted on the crane by the level ground.

(b) Determine the magnitudes of the normal and friction forces exerted on the caisson by the level ground.

Strategy: To do part (a), draw the free-body diagram of the crane and the part of its cable within the dashed line.

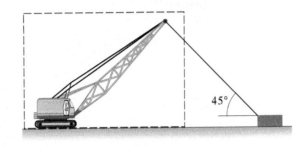

45°

Problem 3.10

3.11 The inclined surface is smooth. The 100-kg crate is held stationary by a force T applied to the cable.

(a) Draw the free-body diagram of the crate.

(b) Determine the force T.

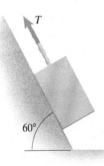

T

60°

Problem 3.11

3.24 The person wants to cause the 200-lb crate to start sliding toward the right. To achieve this, the *horizontal component* of the force exerted on the crate by the rope must equal 0.35 times the normal force exerted on the crate by the floor. In Fig. a, the person pulls on the rope in the direction shown. In Fig. b, the person attaches the rope to a support as shown and pulls upward on the rope. What is the magnitude of the force he must exert on the rope in each case?

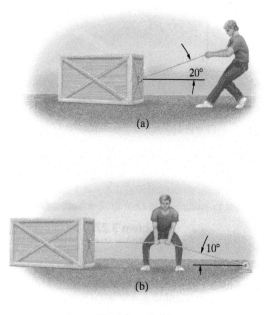

(a)

(b)

Problem 3.24

3.25 A traffic engineer wants to suspend a 200-lb traffic light above the center of the two right lanes of a four-lane thoroughfare as shown. Points *A* and *C* are at the same height. Determine the tensions in the cables *AB* and *BC*.

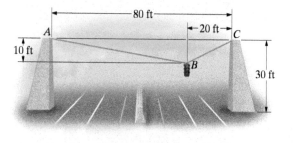

Problem 3.25

3.26 Cable *AB* is 3 m long and cable *BC* is 4 m long. Points *A* and *C* are at the same height. The mass of the suspended object is 350 kg. Determine the tensions in cables *AB* and *BC*.

3.27 The length of cable *AB* is adjustable. Cable *BC* is 4 m long. If you don't want the tension in either cable *AB* or cable *BC* to exceed 3 kN, what is the minimum acceptable length of cable *AB*?

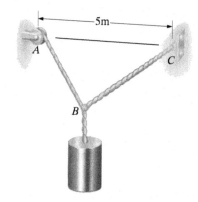

Problems 3.26/3.27

3.28 What are the tensions in the upper and lower cables? (Your answers will be in terms of *W*. Neglect the weight of the pulley.)

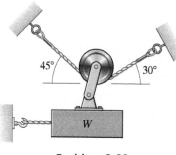

Problem 3.28

3.29 Two tow trucks lift a 660-lb motorcycle out of a ravine following an accident. If the motorcycle is in equilibrium in the position shown, what are the tensions in cables *AB* and *AC*?

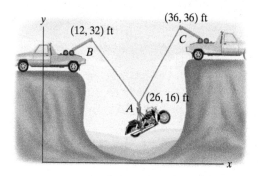

Problem 3.29

3.30 An astronaut candidate conducts experiments on an airbear-ing platform. While she carries out calibrations, the platform is held in place by the horizontal tethers AB, AC, and AD. The forces exerted by the tethers are the only horizontal forces acting on the platform. If the tension in tether AC is 2 N, what are the tensions in the other two tethers?

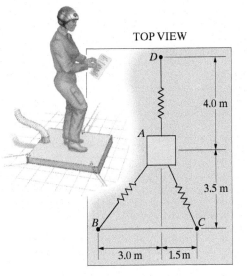

TOP VIEW

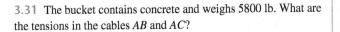

Problem 3.30

3.31 The bucket contains concrete and weighs 5800 lb. What are the tensions in the cables AB and AC?

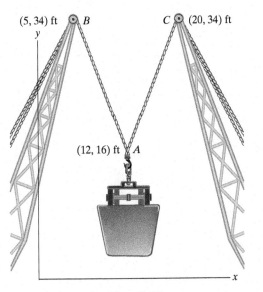

Problem 3.31

3.32 The slider A is in equilibrium and the bar is smooth. What is the mass of the slider?

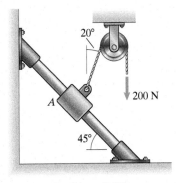

Problem 3.32

3.33 The 20-kg mass is suspended from three cables. Cable AC is equipped with a turnbuckle so that its tension can be adjusted and a strain gauge that allows its tension to be measured. If the tension in cable AC is 40 N, what are the tensions in cables AB and AD?

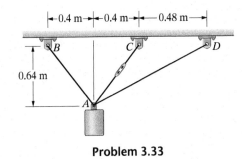

Problem 3.33

3.34 The structural joint is in equilibrium. If $F_A = 1000$ lb and $F_D = 5000$ lb, what are F_B and F_C?

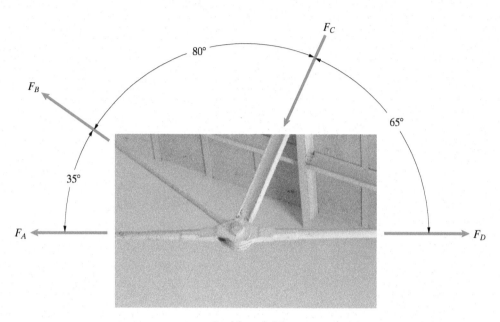

Problem 3.34

3.35 The collar A slides on the smooth vertical bar. The masses $m_A = 20$ kg and $m_B = 10$ kg. When $h = 0.1$ m, the spring is unstretched. When the system is in equilibrium, $h = 0.3$ m. Determine the spring constant k.

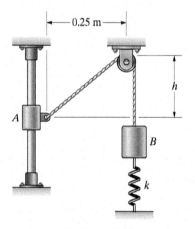

Problem 3.35

3.36* Suppose that you want to design a cable system to suspend an object of weight W from the ceiling. The two wires must be identical, and the dimension b is fixed. The ratio of the tension T in each wire to its cross-sectional area A must equal a specified value $T/A = \sigma$. The "cost" of your design is the total volume of material in the two wires, $V = 2A\sqrt{b^2 + h^2}$. Determine the value of h that minimizes the cost.

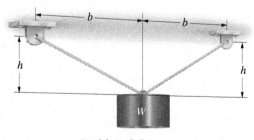

Problem 3.36

3.37 The system of cables suspends a 1000-lb bank of lights above a movie set. Determine the tensions in cables *AB*, *CD*, and *CE*.

3.38 A technician changes the position of the 1000-lb bank of lights by removing the cable *CE*. What is the tension in cable *AB* after the change?

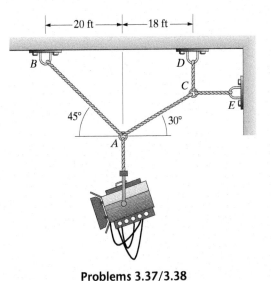

Problems 3.37/3.38

3.39 While working on another exhibit, a curator at the Smithsonian Institution pulls the suspended *Voyager* aircraft to one side by attaching three horizontal cables as shown. The mass of the aircraft is 1250 kg. Determine the tensions in the cable segments *AB*, *BC*, and *CD*.

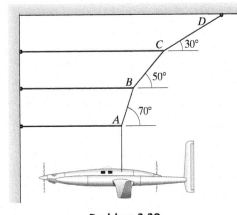

Problem 3.39

3.40 A truck dealer wants to suspend a 4000-kg truck as shown for advertising. The distance $b = 15$ m, and the sum of the lengths of the cables *AB* and *BC* is 42 m. Points *A* and *C* are at the same height. What are the tensions in the cables?

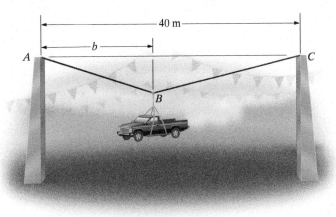

Problem 3.40

3.41 The distance $h = 12$ in, and the tension in cable AD is 200 lb. What are the tensions in cables AB and AC?

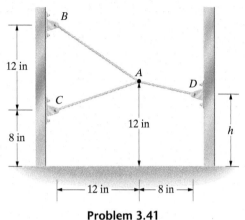

Problem 3.41

3.42 You are designing a cable system to support a suspended object of weight W. Because your design requires points A and B to be placed as shown, you have no control over the angle α, but you can choose the angle β by placing point C wherever you wish. Show that to minimize the tensions in cables AB and BC, you must choose $\beta = \alpha$ if the angle $\alpha \geq 45°$.

Strategy: Draw a diagram of the sum of the forces exerted by the three cables at A.

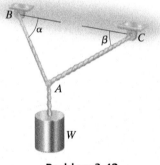

Problem 3.42

3.43* The length of the cable ABC is 1.4 m. The 2-kN force is applied to a small pulley. The system is stationary. What is the tension in the cable?

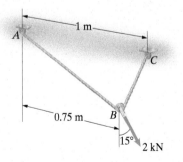

Problem 3.43

3.44 The masses $m_1 = 12$ kg and $m_2 = 6$ kg are suspended by the cable system shown. The cable BC is horizontal. Determine the angle α and the tensions in the cables AB, BC, and CD.

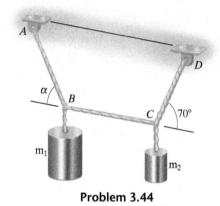

Problem 3.44

3.45 The weights $W_1 = 50$ lb and W_2 are suspended by the cable system shown. Determine the weight W_2 and the tensions in the cables AB, BC, and CD.

3.46 Assume that $W_2 = W_1/2$. If you don't want the tension anywhere in the supporting cable to exceed 200 lb, what is the largest acceptable value of W_1?

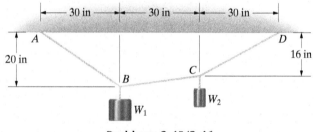

Problems 3.45/3.46

3.47 The hydraulic cylinder is subjected to three forces. An 8-kN force is exerted on the cylinder at B that is parallel to the cylinder and points from B toward C. The link AC exerts a force at C that is parallel to the line from A to C. The link CD exerts a force at C that is parallel to the line from C to D.

(a) Draw the free-body diagram of the cylinder. (The cylinder's weight is negligible.)

(b) Determine the magnitudes of the forces exerted by the links AC and CD.

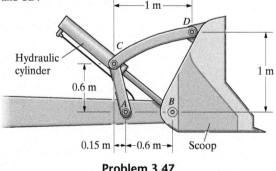

Problem 3.47

3.48 The 50-lb cylinder rests on two smooth surfaces.

(a) Draw the free-body diagram of the cylinder.

(b) If $\alpha = 30°$, what are the magnitudes of the forces exerted on the cylinder by the left and right surfaces?

3.49 Obtain an equation for the force exerted on the 50-lb cylinder by the left surface in terms of the angle α in two ways: (a) using a coordinate system with the y axis vertical, (b) using a coordinate system with the y axis parallel to the right surface.

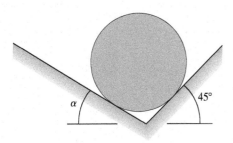

Problems 3.48/3.49

3.50 The two springs are identical, with unstretched length 0.4 m. When the 50-kg mass is suspended at B, the length of each spring increases to 0.6 m. What is the spring constant k?

Problem 3.50

3.51 The cable AB is 0.5 m in length. The *unstretched* length of the spring is 0.4 m. When the 50-kg mass is suspended at B, the length of the spring increases to 0.45 m. What is the spring constant k?

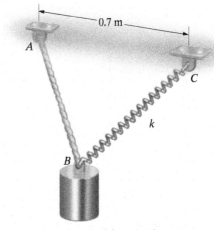

Problem 3.51

3.52* The small sphere of mass m is attached to a string of length L and rests on the smooth surface of a fixed sphere of radius R. The center of the sphere is directly below the point where the string is attached. Obtain an equation for the tension in the string in terms of m, L, h, and R.

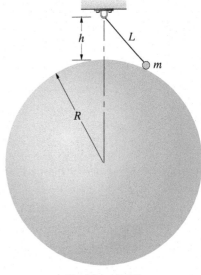

Problem 3.52

3.53 The inclined surface is smooth. Determine the force T that must be exerted on the cable to hold the 100-kg crate in equilibrium and compare your answer to the answer of Problem 3.11.

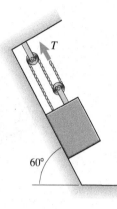

Problem 3.53

▶ **3.54** In Example 3.3, suppose that the mass of the suspended object is m_A and the masses of the pulleys are $m_B = 0.3m_A$, $m_C = 0.2m_A$, and $m_D = 0.2m_A$. Show that the force T necessary for the system to be in equilibrium is $0.275m_A g$.

3.55 The mass of each pulley of the system is m and the mass of the suspended object A is m_A. Determine the force T necessary for the system to be in equilibrium.

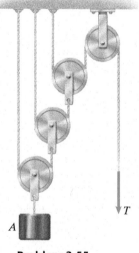

Problem 3.55

3.56 The suspended mass $m_1 = 50$ kg. Neglecting the masses of the pulleys, determine the value of the mass m_2 necessary for the system to be in equilibrium.

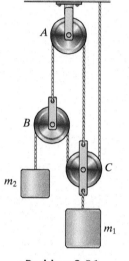

Problem 3.56

3.57 The boy is lifting himself using the block and tackle shown. If the weight of the block and tackle is negligible, and the combined weight of the boy and the beam he is sitting on is 120 lb, what force does he have to exert on the rope to raise himself at a constant rate? (Neglect the deviation of the ropes from the vertical.)

Problem 3.57

3.58 Pulley systems containing one, two, and three pulleys are shown. Neglecting the weights of the pulleys, determine the force T required to support the weight W in each case.

3.59 The number of pulleys in the type of system shown could obviously be extended to an arbitrary number N.

(a) Neglecting the weights of the pulleys, determine the force T required to support the weight W as a function of the number of pulleys N in the system.

(b) Using the result of part (a), determine the force T required to support the weight W for a system with 10 pulleys.

▶ 3.60 A 14,000-kg airplane is in steady flight in the vertical plane. The flight path angle is $\gamma = 10°$, the angle of attack is $\alpha = 4°$, and the thrust force exerted by the engine is $T = 60$ kN. What are the magnitudes of the lift and drag forces acting on the airplane? (See Example 3.4.)

▶ 3.61 An airplane is in steady flight, the angle of attack $\alpha = 0$, the thrust-to-drag ratio $T/D = 2$, and the lift-to-drag ratio $L/D = 4$. What is the flight path angle γ? (See Example 3.4.)

▶ 3.62 An airplane glides in steady flight ($T = 0$), and its lift-to-drag ratio is $L/D = 4$.

(a) What is the flight path angle γ?

(b) If the airplane glides from an altitude of 1000 m to zero altitude, what horizontal distance does it travel? (See Example 3.4.)

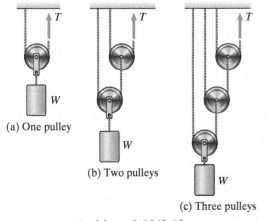

(a) One pulley

(b) Two pulleys

(c) Three pulleys

Problems 3.58/3.59

3.3 Three-Dimensional Force Systems

The equilibrium situations we have considered so far have involved only coplanar forces. When the system of external forces acting on an object in equilibrium is three dimensional, we can express the sum of the external forces as

$$\Sigma \mathbf{F} = (\Sigma F_x)\mathbf{i} + (\Sigma F_y)\mathbf{j} + (\Sigma F_z)\mathbf{k} = \mathbf{0}.$$

Each component of this equation must equal zero, resulting in three scalar equilibrium equations:

$$\Sigma F_x = 0, \quad \Sigma F_y = 0, \quad \Sigma F_z = 0. \tag{3.5}$$

The sums of the x, y, and z components of the external forces acting on an object in equilibrium must each equal zero.

Active Example 3.5   (▶ *Related Problem 3.63*)

The 100-kg cylinder is suspended from the ceiling by cables attached at points B, C, and D. What are the tensions in cables AB, AC, and AD?

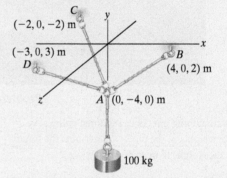

Strategy

By isolating part of the cable system near point A, we will obtain a free-body diagram subjected to forces due to the tensions in the cables. Because the sums of the external forces in the x, y, and z directions must each equal zero, we can obtain three equilibrium equations for the three unknown tensions. To do so, we must express the forces exerted by the tensions in terms of their components.

Solution

Draw the Free-Body Diagram and Apply Equilibrium

Isolate part of the cable system near point A and show the forces exerted due to the tensions in the cables. The sum of the forces must equal zero:

$$\Sigma \mathbf{F} = \mathbf{T}_{AB} + \mathbf{T}_{AC} + \mathbf{T}_{AD} - (981 \text{ N})\mathbf{j} = \mathbf{0}.$$

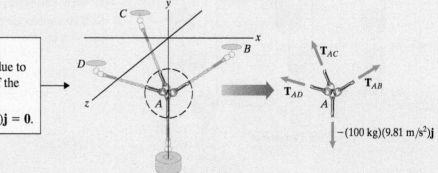

Write the Forces in Terms of Their Components

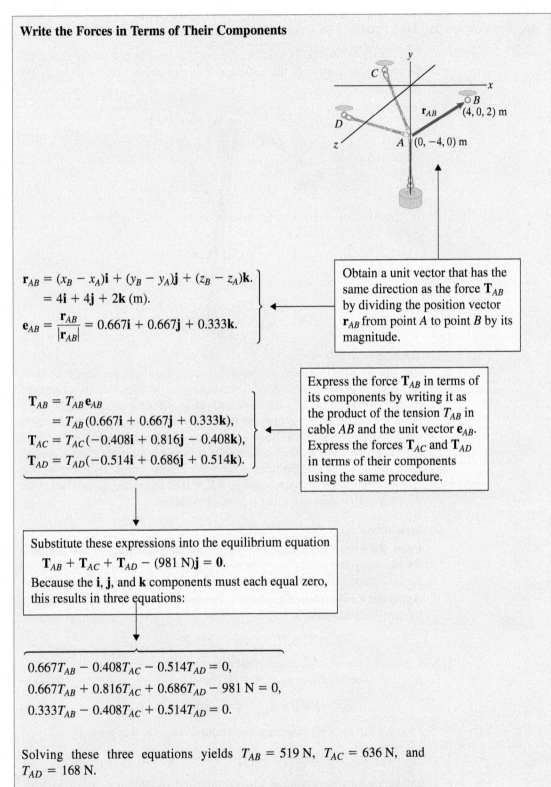

Obtain a unit vector that has the same direction as the force $\mathbf{T}_{AB}$ by dividing the position vector $\mathbf{r}_{AB}$ from point A to point B by its magnitude.

$$\mathbf{r}_{AB} = (x_B - x_A)\mathbf{i} + (y_B - y_A)\mathbf{j} + (z_B - z_A)\mathbf{k}.$$
$$= 4\mathbf{i} + 4\mathbf{j} + 2\mathbf{k} \text{ (m)}.$$
$$\mathbf{e}_{AB} = \frac{\mathbf{r}_{AB}}{|\mathbf{r}_{AB}|} = 0.667\mathbf{i} + 0.667\mathbf{j} + 0.333\mathbf{k}.$$

Express the force $\mathbf{T}_{AB}$ in terms of its components by writing it as the product of the tension T_{AB} in cable AB and the unit vector $\mathbf{e}_{AB}$. Express the forces $\mathbf{T}_{AC}$ and $\mathbf{T}_{AD}$ in terms of their components using the same procedure.

$$\mathbf{T}_{AB} = T_{AB}\mathbf{e}_{AB}$$
$$= T_{AB}(0.667\mathbf{i} + 0.667\mathbf{j} + 0.333\mathbf{k}),$$
$$\mathbf{T}_{AC} = T_{AC}(-0.408\mathbf{i} + 0.816\mathbf{j} - 0.408\mathbf{k}),$$
$$\mathbf{T}_{AD} = T_{AD}(-0.514\mathbf{i} + 0.686\mathbf{j} + 0.514\mathbf{k}).$$

Substitute these expressions into the equilibrium equation
$$\mathbf{T}_{AB} + \mathbf{T}_{AC} + \mathbf{T}_{AD} - (981 \text{ N})\mathbf{j} = \mathbf{0}.$$
Because the $\mathbf{i}$, $\mathbf{j}$, and $\mathbf{k}$ components must each equal zero, this results in three equations:

$$0.667T_{AB} - 0.408T_{AC} - 0.514T_{AD} = 0,$$
$$0.667T_{AB} + 0.816T_{AC} + 0.686T_{AD} - 981 \text{ N} = 0,$$
$$0.333T_{AB} - 0.408T_{AC} + 0.514T_{AD} = 0.$$

Solving these three equations yields $T_{AB} = 519$ N, $T_{AC} = 636$ N, and $T_{AD} = 168$ N.

Practice Problem Suppose that cables AB, AC, and AD are lengthened so that the attachment point A is located at the point $(0, -6, 0)$ m. What are the tensions in the cables?

Answer: $T_{AB} = 432$ N, $T_{AC} = 574$ N, $T_{AD} = 141$ N.

| Example 3.6 | Application of the Dot Product (▶ *Related Problem 3.79*) |

The 100-lb "slider" C is held in place on the smooth bar by the cable AC. Determine the tension in the cable and the force exerted on the slider by the bar.

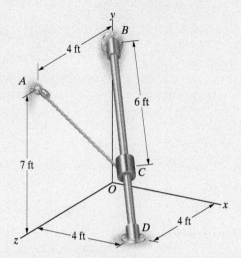

Strategy

Because we want to determine forces that act on the slider, we need to draw its free-body diagram. The external forces acting on the slider are its weight and the forces exerted on it by the cable and the bar. If we approached this example as we did the previous one, our next step would be to express the forces in terms of their components. However, we don't know the direction of the force exerted on the slider by the bar. Since the smooth bar exerts negligible friction force, we do know that the force is normal to the bar's axis. Therefore we can eliminate this force from the equation $\Sigma \mathbf{F} = \mathbf{0}$ by taking the dot product of the equation with a unit vector that is parallel to the bar.

Solution

Draw the Free-Body Diagram We isolate the slider (Fig. a) and complete the free-body diagram by showing the weight of the slider, the force $\mathbf{T}$ exerted by the tension in the cable, and the normal force $\mathbf{N}$ exerted by the bar (Fig. b).

Apply the Equilibrium Equations The sum of the external forces acting on the free-body diagram is

$$\Sigma \mathbf{F} = \mathbf{T} + \mathbf{N} - (100 \text{ lb})\mathbf{j} = \mathbf{0}. \tag{1}$$

Let $\mathbf{e}_{BD}$ be the unit vector pointing from point B toward point D. Since $\mathbf{N}$ is perpendicular to the bar, $\mathbf{e}_{BD} \cdot \mathbf{N} = 0$. Therefore,

$$\mathbf{e}_{BD} \cdot (\Sigma \mathbf{F}) = \mathbf{e}_{BD} \cdot [\mathbf{T} - (100 \text{ lb})\mathbf{j}] = 0. \tag{2}$$

Determining $\mathbf{e}_{BD}$: We determine the vector from point B to point D,

$$\mathbf{r}_{BD} = (4 - 0)\mathbf{i} + (0 - 7)\mathbf{j} + (4 - 0)\mathbf{k} = 4\mathbf{i} - 7\mathbf{j} + 4\mathbf{k}(\text{ft}),$$

and divide it by its magnitude to obtain the unit vector $\mathbf{e}_{BD}$:

$$\mathbf{e}_{BD} = \frac{\mathbf{r}_{BD}}{|\mathbf{r}_{BD}|} = \frac{4}{9}\mathbf{i} - \frac{7}{9}\mathbf{j} + \frac{4}{9}\mathbf{k}.$$

Expressing **T** *in terms of components:* We need to determine the coordinates of the slider C. We can write the vector from B to C in terms of the unit vector $\mathbf{e}_{BD}$,

$$\mathbf{r}_{BC} = 6\mathbf{e}_{BD} = 2.67\mathbf{i} - 4.67\mathbf{j} + 2.67\mathbf{k} \ (\text{ft}),$$

and then add it to the vector from the origin O to B to obtain the vector from O to C:

$$\mathbf{r}_{OC} = \mathbf{r}_{OB} + \mathbf{r}_{BC} = 7\mathbf{j} + (2.67\mathbf{i} - 4.67\mathbf{j} + 2.67\mathbf{k})$$

$$= 2.67\mathbf{i} + 2.33\mathbf{j} + 2.67\mathbf{k} \ (\text{ft}).$$

The components of this vector are the coordinates of point C. Now we can determine a unit vector with the same direction as **T**. The vector from C to A is

$$\mathbf{r}_{CA} = (0 - 2.67)\mathbf{i} + (7 - 2.33)\mathbf{j} + (4 - 2.67)\mathbf{k}$$

$$= -2.67\mathbf{i} + 4.67\mathbf{j} + 1.33\mathbf{k} \ (\text{ft}),$$

and the unit vector that points from point C toward point A is

$$\mathbf{e}_{CA} = \frac{\mathbf{r}_{CA}}{|\mathbf{r}_{CA}|} = -0.482\mathbf{i} + 0.843\mathbf{j} + 0.241\mathbf{k}.$$

Let T be the tension in the cable AC. Then we can write the vector **T** as

$$\mathbf{T} = T\mathbf{e}_{CA} = T(-0.482\mathbf{i} + 0.843\mathbf{j} + 0.241\mathbf{k}).$$

Determining **T** *and* **N**: Substituting our expressions for $\mathbf{e}_{BD}$ and **T** in terms of their components into Eq. (2) yields

$$0 = \mathbf{e}_{BD} \cdot [\mathbf{T} - (100 \text{ lb})\mathbf{j}]$$

$$= \left(\frac{4}{9}\mathbf{i} - \frac{7}{9}\mathbf{j} + \frac{4}{9}\mathbf{k}\right) \cdot [-0.482T\mathbf{i} + (0.843T - 100 \text{ lb})\mathbf{j} + 0.241T\mathbf{k}]$$

$$= -0.762T + 77.8 \text{ lb},$$

and we obtain the tension $T = 102$ lb.

Now we can determine the force exerted on the slider by the bar by using Eq. (1):

$$\mathbf{N} = -\mathbf{T} + (100 \text{ lb})\mathbf{j}$$

$$= -(102 \text{ lb})(-0.482\mathbf{i} + 0.843\mathbf{j} + 0.241\mathbf{k}) + (100 \text{ lb})\mathbf{j}$$

$$= 49.1\mathbf{i} + 14.0\mathbf{j} - 24.6\mathbf{k} \ (\text{lb}).$$

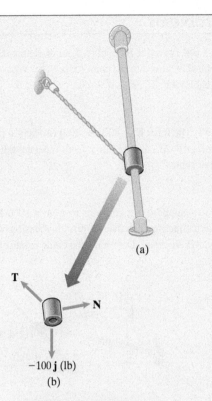

(a)

$-100\,\mathbf{j}$ (lb)

(b)

(a) Isolating the slider.
(b) Free-body diagram of the slider showing the forces exerted by its weight, the cable, and the bar.

Critical Thinking

By taking the dot product of the equilibrium equation for the slider with a unit vector $\mathbf{e}_{BD}$ that is parallel to the smooth bar BD, we obtained Eq. (2), which does not contain the normal force **N**. Why does this happen? The formal answer is that $\mathbf{e}_{BD}$ is perpendicular to **N**, and so $\mathbf{e}_{BD} \cdot \mathbf{N} = 0$. But the physical interpretation of Eq. (2) provides a more compelling explanation: It states that *the component of the slider's weight parallel to the bar is balanced by the component of* **T** *parallel to the bar.* The normal force exerted on the slider by the smooth bar has no component parallel to the bar. We were therefore able to solve for the tension in the cable without knowing the normal force **N**.

Problems

▶ **3.63** In Active Example 3.5, suppose that the attachment point *B* is moved to the point $(5, 0, 0)$ m. What are the tensions in cables *AB*, *AC*, and *AD*?

3.64 The force $\mathbf{F} = 800\mathbf{i} + 200\mathbf{j}$ (lb) acts at point *A* where the cables *AB*, *AC*, and *AD* are joined. What are the tensions in the three cables?

3.65* Suppose that you want to apply a 1000-lb force $\mathbf{F}$ at point *A* in a direction such that the resulting tensions in cables *AB*, *AC*, and *AD* are equal. Determine the components of $\mathbf{F}$.

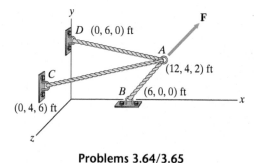

Problems 3.64/3.65

3.66 The 10-lb metal disk *A* is supported by the smooth inclined surface and the strings *AB* and *AC*. The disk is located at coordinates $(5, 1, 4)$ ft. What are the tensions in the strings?

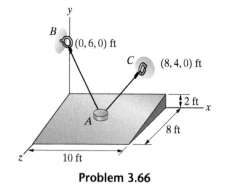

Problem 3.66

3.67 The bulldozer exerts a force $\mathbf{F} = 2\mathbf{i}$ (kip) at *A*. What are the tensions in cables *AB*, *AC*, and *AD*?

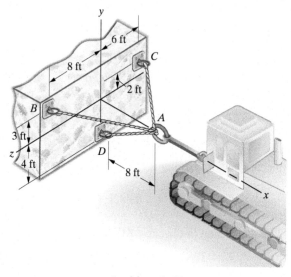

Problem 3.67

3.68 Prior to its launch, a balloon carrying a set of experiments to high altitude is held in place by groups of student volunteers holding the tethers at *B*, *C*, and *D*. The mass of the balloon, experiments package, and the gas it contains is 90 kg, and the buoyancy force on the balloon is 1000 N. The supervising professor conservatively estimates that each student can exert at least a 40-N tension on the tether for the necessary length of time. Based on this estimate, what minimum numbers of students are needed at *B*, *C*, and *D*?

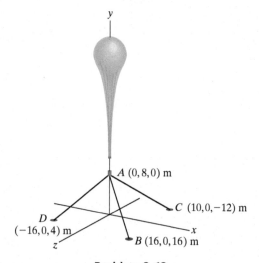

Problem 3.68

3.69 The 20-kg mass is suspended by cables attached to three vertical 2-m posts. Point A is at (0, 1.2, 0) m. Determine the tensions in cables AB, AC, and AD.

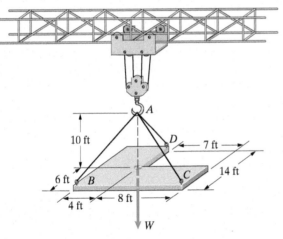

Problem 3.69

3.70 The weight of the horizontal wall section is $W = 20{,}000$ lb. Determine the tensions in the cables AB, AC, and AD.

Problem 3.70

3.71 The car in Fig. a and the pallet supporting it weigh 3000 lb. They are supported by four cables AB, AC, AD, and AE. The locations of the attachment points on the pallet are shown in Fig. b. The tensions in cables AB and AE are equal. Determine the tensions in the cables.

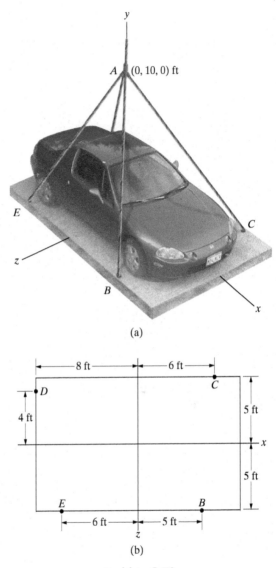

(a)

(b)

Problem 3.71

3.72 The 680-kg load suspended from the helicopter is in equilibrium. The aerodynamic drag force on the load is horizontal. The y axis is vertical, and cable OA lies in the x–y plane. Determine the magnitude of the drag force and the tension in cable OA.

3.73 The coordinates of the three cable attachment points B, C, and D are $(-3.3, -4.5, 0)$ m, $(1.1, -5.3, 1)$ m, and $(1.6, -5.4, -1)$ m, respectively. What are the tensions in cables OB, OC, and OD?

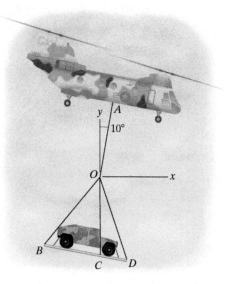

Problems 3.72/3.73

3.74 If the mass of the bar AB is negligible compared to the mass of the suspended object E, the bar exerts a force on the "ball" at B that points from A toward B. The mass of the object E is 200 kg. The y axis points upward. Determine the tensions in the cables BC and BD.

Strategy: Draw a free-body diagram of the ball at B. (The weight of the ball is negligible.)

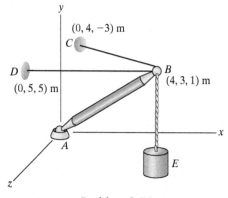

Problem 3.74

3.75* The 3400-lb car is at rest on the plane surface. The unit vector $\mathbf{e}_n = 0.456\mathbf{i} + 0.570\mathbf{j} + 0.684\mathbf{k}$ is perpendicular to the surface. Determine the magnitudes of the total normal force $\mathbf{N}$ and the total friction force $\mathbf{f}$ exerted on the surface by the car's wheels.

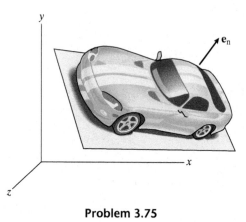

Problem 3.75

3.76 The system shown anchors a stanchion of a cable-suspended roof. If the tension in cable AB is 900 kN, what are the tensions in cables EF and EG?

3.77* The cables of the system will each safely support a tension of 1500 kN. Based on this criterion, what is the largest safe value of the tension in cable AB?

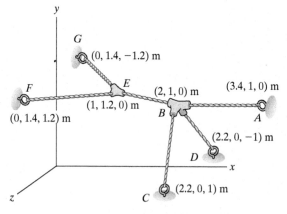

Problems 3.76/3.77

3.78 The 200-kg slider at A is held in place on the smooth vertical bar by the cable AB.

(a) Determine the tension in the cable.

(b) Determine the force exerted on the slider by the bar.

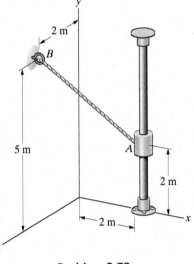

Problem 3.78

▶ **3.79** In Example 3.6, suppose that the cable AC is replaced by a longer one so that the distance from point B to the slider C increases from 6 ft to 8 ft. Determine the tension in the cable.

3.80 The cable AB keeps the 8-kg collar A in place on the smooth bar CD. The y axis points upward. What is the tension in the cable?

3.81* Determine the magnitude of the normal force exerted on the collar A by the smooth bar.

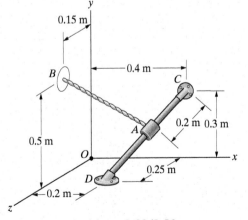

Problems 3.80/3.81

3.82* The 10-kg collar A and 20-kg collar B are held in place on the smooth bars by the 3-m cable from A to B and the force F acting on A. The force F is parallel to the bar. Determine F.

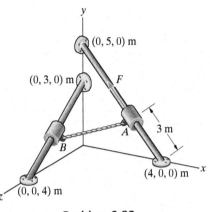

Problem 3.82

Review Problems

3.83 The 100-lb crate is held in place on the smooth surface by the rope AB. Determine the tension in the rope and the magnitude of the normal force exerted on the crate by the surface.

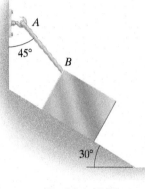

Problem 3.83

3.84 The system shown is called Russell's traction. If the sum of the downward forces exerted at A and B by the patient's leg is 32.2 lb, what is the weight W?

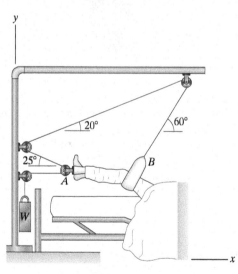

Problem 3.84

3.85 The 400-lb engine block is suspended by the cables AB and AC. If you don't want either T_{AB} or T_{AC} to exceed 400 lb, what is the smallest acceptable value of the angle α?

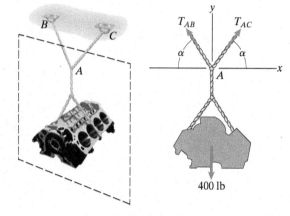

Problem 3.85

3.86 The cable AB is horizontal, and the box on the right weighs 100 lb. The surfaces are smooth.
(a) What is the tension in the cable?
(b) What is the weight of the box on the left?

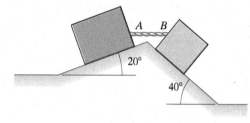

Problem 3.86

3.87 Assume that the forces exerted on the 170-lb climber by the slanted walls of the "chimney" are perpendicular to the walls. If he is in equilibrium and is exerting a 160-lb force on the rope, what are the magnitudes of the forces exerted on him by the left and right walls?

Problem 3.87

3.88 The mass of the suspended object A is m_A and the masses of the pulleys are negligible. Determine the force T necessary for the system to be in equilibrium.

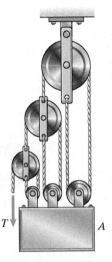

Problem 3.88

3.89 The assembly A, including the pulley, weighs 60 lb. What force F is necessary for the system to be in equilibrium?

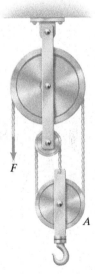

Problem 3.89

3.90 The mass of block A is 42 kg, and the mass of block B is 50 kg. The surfaces are smooth. If the blocks are in equilibrium, what is the force F?

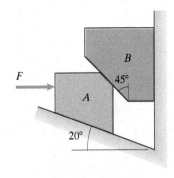

Problem 3.90

3.91 The climber A is being helped up an icy slope by two friends. His mass is 80 kg, and the direction cosines of the force exerted on him by the slope are $\cos \theta_x = -0.286$, $\cos \theta_y = 0.429$, and $\cos \theta_z = 0.857$. The y axis is vertical. If the climber is in equilibrium in the position shown, what are the tensions in the ropes AB and AC and the magnitude of the force exerted on him by the slope?

3.92 Consider the climber A being helped by his friends in Problem 3.91. To try to make the tensions in the ropes more equal, the friend at B moves to the position (4, 2, 0) m. What are the new tensions in the ropes AB and AC and the magnitude of the force exerted on the climber by the slope?

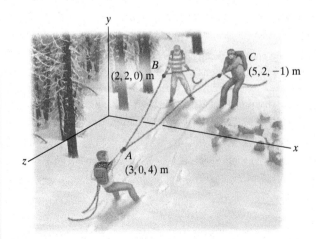

Problems 3.91/3.92

3.93 A climber helps his friend up an icy slope. His friend is hauling a box of supplies. If the mass of the friend is 90 kg and the mass of the supplies is 22 kg, what are the tensions in the ropes *AB* and *CD*? Assume that the slope is smooth. That is, only normal forces are exerted on the man and the box by the slope.

Problem 3.93

3.94 The 2800-lb car is moving at constant speed on a road with the slope shown. The aerodynamics forces on the car are the drag *D* = 270 lb, which is parallel to the road, and the lift *L* = 120 lb, which is perpendicular to the road. Determine the magnitudes of the total normal and friction forces exerted on the car by the road.

Problem 3.94

3.95 An engineer doing preliminary design studies for a new radio telescope envisions a triangular receiving platform suspended by cables from three equally spaced 40-m towers. The receiving platform has a mass of 20 Mg (megagrams) and is 10 m below the tops of the towers. What tension would the cables be subjected to?

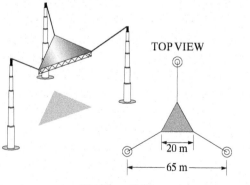

Problem 3.95

3.96 To support the tent, the tension in the rope *AB* must be 35 lb. What are the tensions in the ropes *AC*, *AD*, and *AE*?

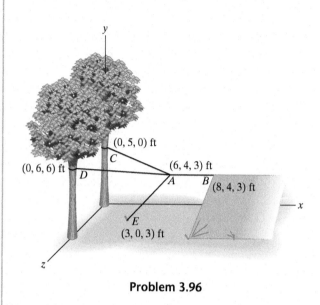

Problem 3.96

3.97 Cable *AB* is attached to the top of the vertical 3-m post, and its tension is 50 kN. What are the tensions in cables *AO*, *AC*, and *AD*?

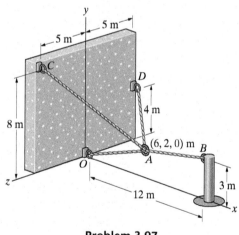

Problem 3.97

3.98* The 1350-kg car is at rest on a plane surface with its brakes locked. The unit vector $\mathbf{e}_n = 0.231\mathbf{i} + 0.923\mathbf{j} + 0.308\mathbf{k}$ is perpendicular to the surface. The y axis points upward. The direction cosines of the cable from A to B are $\cos\theta_x = -0.816$, $\cos\theta_y = 0.408$, $\cos\theta_z = -0.408$, and the tension in the cable is 1.2 kN. Determine the magnitudes of the normal and friction forces the car's wheels exert on the surface.

3.99* The brakes of the car are released, and the car is held in place on the plane surface by the cable AB. The car's front wheels are aligned so that the tires exert no friction forces parallel to the car's longitudinal axis. The unit vector $\mathbf{e}_p = -0.941\mathbf{i} + 0.131\mathbf{j} + 0.314\mathbf{k}$ is parallel to the plane surface and aligned with the car's longitudinal axis. What is the tension in the cable?

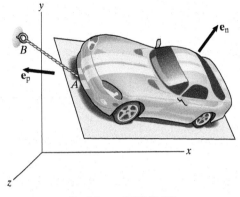

Problems 3.98/3.99

Design Project 1 A possible design for a simple scale to weigh objects is shown. The length of the string AB is 0.5 m. When an object is placed in the pan, the spring stretches and the string AB rotates. The object's weight can be determined by observing the change in the angle α.

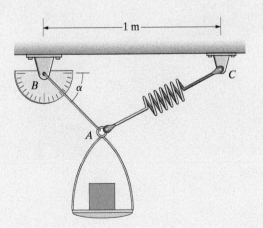

(a) Assume that objects with masses in the range 0.2–2 kg are to be weighed. Choose the unstretched length and spring constant of the spring in order to obtain accurate readings for weights in the desired range. (Neglect the weights of the pan and spring. Notice that a significant change in the angle α is needed to determine the weight accurately.)

(b) Suppose that you can use the same components—the pan, protractor, a spring, string—and also one or more pulleys. Suggest another possible configuration for the scale. Use statics to analyze your proposed configuration and compare its accuracy with that of the configuration shown for objects with masses in the range 0.2–2 kg.

Design Project 2 Suppose that the positions of points A, C, and D of the system of cables suspending the 100-kg mass are fixed, but you are free to choose the x and z coordinates of point B. Investigate the effects of different choices of the location of point B on the tensions in the cables. If the cost of cable AB is proportional to the product of the tension in the cable and its length, investigate the effect of different choices of the location of point B on the cost of the cable. Write a brief report describing the results of your investigations and recommending a location for point B.

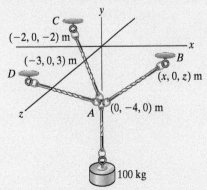

Systems of Forces and Moments

The effects of forces can depend not only on their magnitudes and directions but also on the moments, or torques, they exert. The rotations of objects such as the wheels of a vehicle, the crankshaft of an engine, and the rotor of an electric generator result from the moments of the forces exerted on them. If an object is in equilibrium, the moment about any point due to the forces acting on the object is zero. Before continuing our discussion of free-body diagrams and equilibrium, we must explain how to calculate moments and introduce the concept of equivalent systems of forces and moments.

◀ The counterweight of the building crane exerts a large moment that the crane's structure must support during assembly. In this chapter we calculate moments of forces and analyze systems of forces and moments.

4.1 Two-Dimensional Description of the Moment

BACKGROUND

Consider a force of magnitude F and a point P, and let's view them in the direction perpendicular to the plane containing the force vector and the point (Fig. 4.1a). The *magnitude of the moment* of the force about P is the product DF, where D is the perpendicular distance from P to the line of action of the force (Fig. 4.1b). In this example, the force would tend to cause counterclockwise rotation about point P. That is, if we imagine that the force acts on

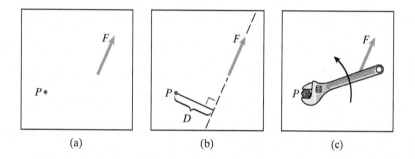

(a)　　　　(b)　　　　(c)

Figure 4.1
(a) The force and point P.
(b) The perpendicular distance D from point P to the line of action of F.
(c) The direction of the moment is counterclockwise.

an object that can rotate about point P, the force would cause counterclockwise rotation (Fig. 4.1c). We say that the *direction of the moment* is counterclockwise. *We define counterclockwise moments to be positive and clockwise moments to be negative.* (This is the usual convention, although we occasionally encounter situations in which it is more convenient to define clockwise moments to be positive.) Thus, the moment of the force about P is

$$M_P = DF. \tag{4.1}$$

Notice that if the line of action of F passes through P, the perpendicular distance $D = 0$ and the moment of F about P is zero.

The dimensions of the moment are (distance) $\times$ (force). For example, moments can be expressed in newton-meters in SI units and in foot-pounds in U.S. Customary units.

Suppose that you want to place a television set on a shelf, and you aren't certain the attachment of the shelf to the wall is strong enough to support it. Intuitively, you place it near the wall (Fig. 4.2a), knowing that the attachment is more likely to fail if you place it away from the wall (Fig. 4.2b). What is the difference in the two cases? The magnitude and direction of the force exerted on the shelf by the weight of the television are the same in each case, but the moments exerted on the attachment are different. The moment exerted about P by its weight when it is near the wall, $M_P = -D_1 W$, is smaller in magnitude than the moment about P when it is placed away from the wall, $M_P = -D_2 W$.

The method we describe in this section can be used to determine the sum of the moments of a system of forces about a point if the forces are two-dimensional (coplanar) and the point lies in the same plane. For example, consider the construction crane shown in Fig. 4.3. The sum of the moments exerted about point P by the load W_1 and the counterweight W_2 is

$$\Sigma M_P = D_1 W_1 - D_2 W_2.$$

This moment tends to cause the top of the vertical tower to rotate and could cause it to collapse. If the distance D_2 is adjusted so that $D_1 W_1 = D_2 W_2$, the moment about point P due to the load and the counterweight is zero.

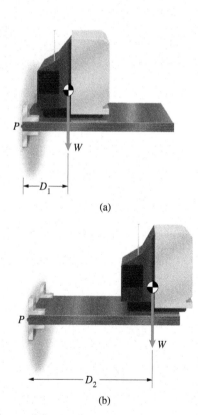

(a)

(b)

Figure 4.2
(a) Placing the television near the wall minimizes the moment exerted on the support of the shelf at P.
(b) Placing the television far from the wall exerts a large moment on the support at P and could cause it to fail.

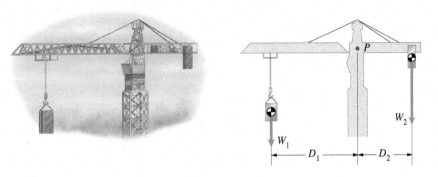

Figure 4.3
A tower crane used in the construction of
high-rise buildings.

*If a force is expressed in terms of components, the moment of the force
about a point P is equal to the sum of the moments of its components about P.*
We prove this very useful result in the next section.

RESULTS

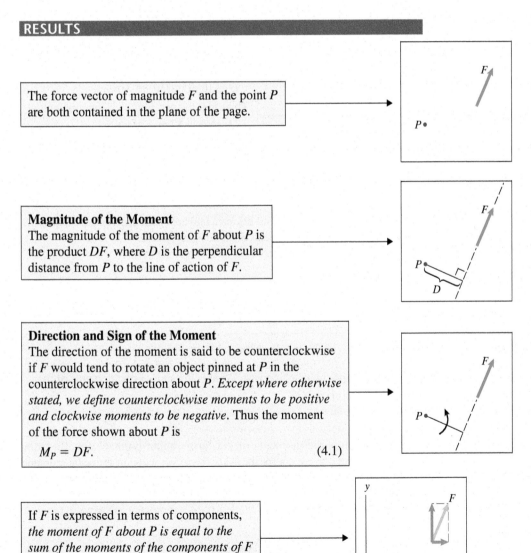

The force vector of magnitude F and the point P
are both contained in the plane of the page.

Magnitude of the Moment
The magnitude of the moment of F about P is
the product DF, where D is the perpendicular
distance from P to the line of action of F.

Direction and Sign of the Moment
The direction of the moment is said to be counterclockwise
if F would tend to rotate an object pinned at P in the
counterclockwise direction about P. *Except where otherwise
stated, we define counterclockwise moments to be positive
and clockwise moments to be negative.* Thus the moment
of the force shown about P is

$$M_P = DF. \tag{4.1}$$

If F is expressed in terms of components,
*the moment of F about P is equal to the
sum of the moments of the components of F
about P.*

Active Example 4.1 **Determining a Moment** (▶ *Related Problem 4.1*)

What is the moment of the 40-kN force about point A?

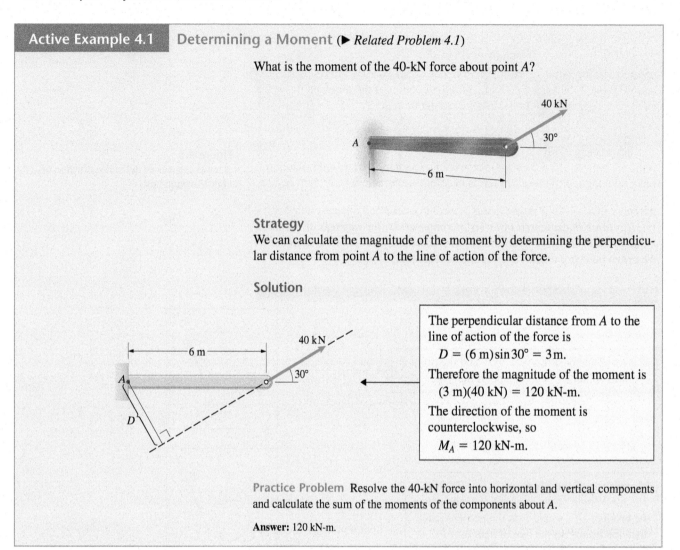

Strategy
We can calculate the magnitude of the moment by determining the perpendicular distance from point A to the line of action of the force.

Solution

> The perpendicular distance from A to the line of action of the force is
> $$D = (6 \text{ m})\sin 30° = 3 \text{ m}.$$
> Therefore the magnitude of the moment is
> $$(3 \text{ m})(40 \text{ kN}) = 120 \text{ kN-m}.$$
> The direction of the moment is counterclockwise, so
> $$M_A = 120 \text{ kN-m}.$$

Practice Problem Resolve the 40-kN force into horizontal and vertical components and calculate the sum of the moments of the components about A.

Answer: 120 kN-m.

Example 4.2 **Moment of a System of Forces** (▶ *Related Problem 4.12*)

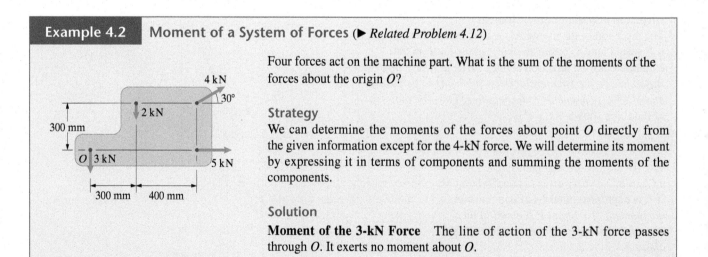

Four forces act on the machine part. What is the sum of the moments of the forces about the origin O?

Strategy
We can determine the moments of the forces about point O directly from the given information except for the 4-kN force. We will determine its moment by expressing it in terms of components and summing the moments of the components.

Solution

Moment of the 3-kN Force The line of action of the 3-kN force passes through O. It exerts no moment about O.

Moment of the 5-kN Force   The line of action of the 5-kN force also passes through O. It too exerts no moment about O.

Moment of the 2-kN Force   The perpendicular distance from O to the line of action of the 2-kN force is 0.3 m, and the direction of the moment about O is clockwise. The moment of the 2-kN force about O is

$$-(0.3 \text{ m})(2 \text{ kN}) = -0.600 \text{ kN-m}.$$

(Notice that we converted the perpendicular distance from millimeters into meters, obtaining the result in terms of kilonewton-meters.)

Moment of the 4-kN Force   In Fig. a, we introduce a coordinate system and express the 4-kN force in terms of x and y components. The perpendicular distance from O to the line of action of the x component is 0.3 m, and the direction of the moment about O is clockwise. The moment of the x component about O is

$$-(0.3 \text{ m})(4 \cos 30° \text{ kN}) = -1.039 \text{ kN-m}.$$

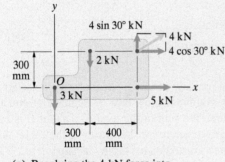

(a) Resolving the 4-kN force into components.

The perpendicular distance from point O to the line of action of the y component is 0.7 m, and the direction of the moment about O is counterclockwise. The moment of the y component about O is

$$(0.7 \text{ m})(4 \sin 30° \text{ kN}) = 1.400 \text{ kN-m}.$$

The sum of the moments of the four forces about point O is

$$\Sigma M_0 = -0.600 - 1.039 + 1.400 = -0.239 \text{ kN-m}.$$

The four forces exert a 0.239 kN-m clockwise moment about point O.

Critical Thinking

If an object is subjected to a system of known forces, why is it useful to determine the sum of the moments of the forces about a given point? As we discuss in Chapter 5, the object is in equilibrium only if the sum of the moments about *any* point is zero, so calculating the sum of the moments provides a test for equilibrium. (Notice that the object in this example is not in equilibrium.) Furthermore, in dynamics the sum of the moments of the forces acting on objects must be determined in order to analyze their angular motions.

| **Example 4.3** | **Summing Moments to Determine an Unknown Force** (▶ *Related Problem 4.23*) |

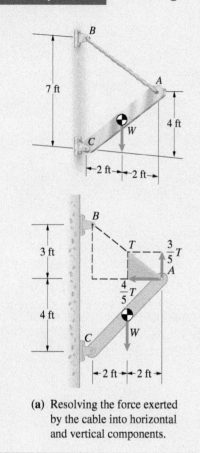

(a) Resolving the force exerted by the cable into horizontal and vertical components.

The weight $W = 300$ lb. The sum of the moments about C due to the weight W and the force exerted on the bar CA by the cable AB is zero. What is the tension in the cable?

Strategy
Let T be the tension in cable AB. Using the given dimensions, we can express the horizontal and vertical components of the force exerted on the bar by the cable in terms of T. Then by setting the sum of the moments about C due to the weight of the bar and the force exerted by the cable equal to zero, we can obtain an equation for T.

Solution
Using similar triangles, we express the force exerted on the bar by the cable in terms of horizontal and vertical components (Fig. a). The sum of the moments about C due to the weight of the bar and the force exerted by the cable AB is

$$\Sigma M_C = 4\left(\frac{4}{5}T\right) + 4\left(\frac{3}{5}T\right) - 2W = 0.$$

Solving for T, we obtain

$$T = 0.357W = 107.1 \text{ lb}.$$

Critical Thinking
This example is a preview of the applications we consider in Chapter 5 and demonstrates why you must know how to calculate moments of forces. If the bar is in equilibrium, the sum of the moments about C is zero. Applying this condition allowed us to determine the tension in the cable. Why didn't we need to consider the force exerted on the bar by its support at C? Because we know that the moment of that force about C is zero.

Problems

▶ **4.1** In Active Example 4.1, the 40-kN force points 30° above the horizontal. Suppose that the force points 30° below the horizontal instead. Draw a sketch of the beam with the new orientation of the force. What is the moment of the force about point A?

4.2 The mass $m_1 = 20$ kg. The magnitude of the total moment about B due to the forces exerted on bar AB by the weights of the two suspended masses is 170 N-m. What is the magnitude of the total moment due to the forces about point A?

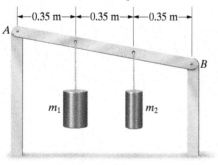

Problem 4.2

4.3 The wheels of the overhead crane exert downward forces on the horizontal I-beam at B and C. If the force at B is 40 kip and the force at C is 44 kip, determine the sum of the moments of the forces on the beam about (a) point A, (b) point D.

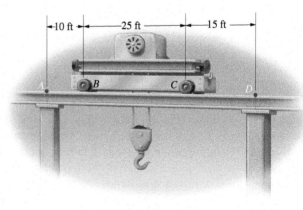

Problem 4.3

4.4 What force F applied to the pliers is required to exert a 4 N-m moment about the center of the bolt at P?

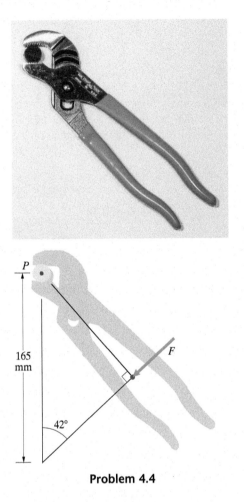

Problem 4.4

4.5 Two forces of equal magnitude F are applied to the wrench as shown. If a 50 N-m moment is required to loosen the nut, what is the necessary value of F?

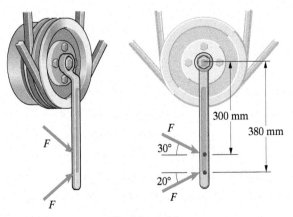

Problem 4.5

4.6 The force $F = 8$ kN. What is the moment of the force about point P?

4.7 If the magnitude of the moment due to the force F about Q is 30 kN-m, what is F?

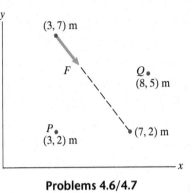

Problems 4.6/4.7

4.8 The support at the left end of the beam will fail if the moment about A of the 15-kN force F exceeds 18 kN-m. Based on this criterion, what is the largest allowable length of the beam?

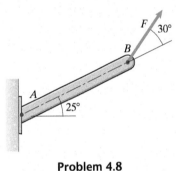

Problem 4.8

4.9 The length of the bar AP is 650 mm. The radius of the pulley is 120 mm. Equal forces $T = 50$ N are applied to the ends of the cable. What is the sum of the moments of the forces (a) about A; (b) about P?

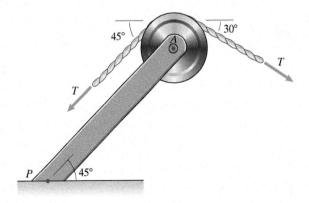

Problem 4.9

4.10 The force $F = 12$ kN. A structural engineer determines that the magnitude of the moment due to F about P should not exceed 5 kN-m. What is the acceptable range of the angle α? Assume that $0 \leq \alpha \leq 90°$.

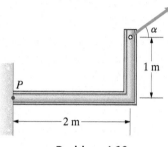

Problem 4.10

4.11 The length of bar AB is 350 mm. The moments exerted about points B and C by the vertical force F are $M_B = -1.75$ kN-m and $M_C = -4.20$ kN-m. Determine the force F and the length of bar AC.

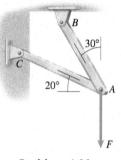

Problem 4.11

▶ **4.12** In Example 4.2, suppose that the 2-kN force points upward instead of downward. Draw a sketch of the machine part showing the orientations of the forces. What is the sum of the moments of the forces about the origin O?

4.13 Two equal and opposite forces act on the beam. Determine the sum of the moments of the two forces (a) about point P; (b) about point Q; (c) about the point with coordinates $x = 7$ m, $y = 5$ m.

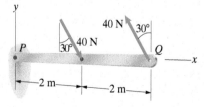

Problem 4.13

4.14 The moment exerted about point E by the weight is 299 in-lb. What moment does the weight exert about point S?

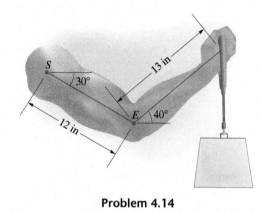

Problem 4.14

4.15 The magnitudes of the forces exerted on the pillar at D by the cables A, B, and C are equal: $F_A = F_B = F_C$. The magnitude of the total moment about E due to the forces exerted by the three cables at D is 1350 kN-m. What is F_A?

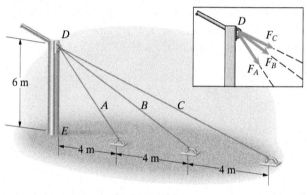

Problem 4.15

4.16 Three forces act on the piping. Determine the sum of the moments of the three forces about point P.

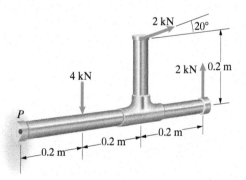

Problem 4.16

4.17 The forces $F_1 = 30$ N, $F_2 = 80$ N, and $F_3 = 40$ N. What is the sum of the moments of the forces about point A?

4.18 The force $F_1 = 30$ N. The vector sum of the three forces is zero. What is the sum of the moments of the forces about point A?

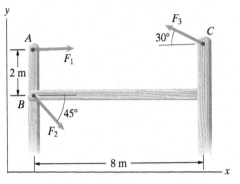

Problems 4.17/4.18

4.19 The forces $F_A = 30$ lb, $F_B = 40$ lb, $F_C = 20$ lb, and $F_D = 30$ lb. What is the sum of the moments of the forces about the origin of the coordinate system?

4.20 The force $F_A = 30$ lb. The vector sum of the forces on the beam is zero, and the sum of the moments of the forces about the origin of the coordinate system is zero. (a) Determine the forces F_B, F_C, and F_D. (b) Determine the sum of the moments of the forces about the right end of the beam.

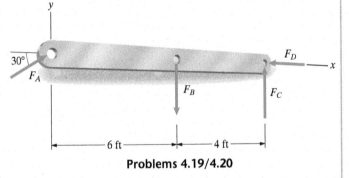

Problems 4.19/4.20

4.21 Three forces act on the car. The sum of the forces is zero and the sum of the moments of the forces about point P is zero.
(a) Determine the forces A and B.
(b) Determine the sum of the moments of the forces about point Q.

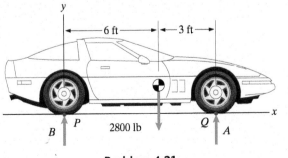

Problem 4.21

4.22 Five forces act on the piping. The vector sum of the forces is zero and the sum of the moments of the forces about point P is zero.
(a) Determine the forces A, B, and C.
(b) Determine the sum of the moments of the forces about point Q.

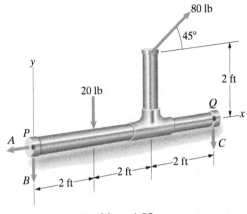

Problem 4.22

▶ **4.23** In Example 4.3, suppose that the attachment point B is moved upward and the cable is lengthened so that the vertical distance from C to B is 9 ft. (The positions of points C and A are unchanged.) Draw a sketch of the system with the cable in its new position. What is the tension in the cable?

4.24 The tension in the cable is the same on both sides of the pulley. The sum of the moments about point A due to the 800-lb force and the forces exerted on the bar by the cable at B and C is zero. What is the tension in the cable?

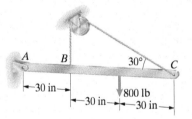

Problem 4.24

4.25 The 160-N weights of the arms AB and BC of the robotic manipulator act at their midpoints. Determine the sum of the moments of the three weights about A.

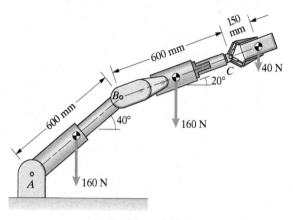

Problem 4.25

4.26 The space shuttle's attitude thrusters exert two forces of magnitude $F = 7.70$ kN. What moment do the thrusters exert about the center of mass G?

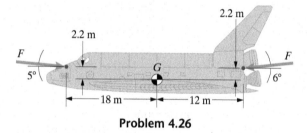

Problem 4.26

4.27 The force F exerts a 200 ft-lb counterclockwise moment about A and a 100 ft-lb clockwise moment about B. What are F and θ?

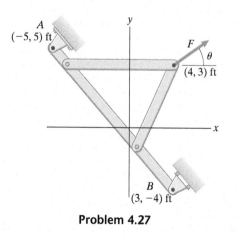

Problem 4.27

4.28 Five forces act on a link in the gear-shifting mechanism of a lawn mower. The vector sum of the five forces on the bar is zero. The sum of their moments about the point where the forces A_x and A_y act is zero.

(a) Determine the forces A_x, A_y, and B.

(b) Determine the sum of the moments of the forces about the point where the force B acts.

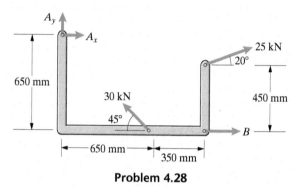

Problem 4.28

4.29 Five forces act on a model truss built by a civil engineering student as part of a design project. The dimensions are $b = 300$ mm and $h = 400$ mm and $F = 100$ N. The sum of the moments of the forces about the point where A_x and A_y act is zero. If the weight of the truss is negligible, what is the force B?

4.30 The dimensions are $b = 3$ ft and $h = 4$ ft and $F = 300$ lb. The vector sum of the forces acting on the truss is zero, and the sum of the moments of the forces about the point where A_x and A_y act is zero.

(a) Determine the forces A_x, A_y, and B.

(b) Determine the sum of the moments of the forces about the point where the force B acts.

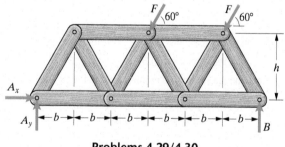

Problems 4.29/4.30

4.31 The mass $m = 70$ kg. What is the moment about A due to the force exerted on the beam at B by the cable?

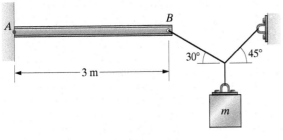

Problem 4.31

4.32 The weights W_1 and W_2 are suspended by the cable system shown. The weight $W_1 = 12$ lb. The cable BC is horizontal. Determine the moment about point P due to the force exerted on the vertical post at D by the cable CD.

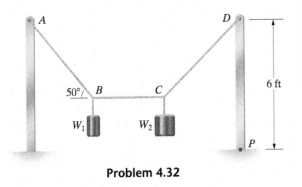

Problem 4.32

4.33 The bar AB exerts a force at B that helps support the vertical retaining wall. The force is parallel to the bar. The civil engineer wants the bar to exert a 38 kN-m moment about O. What is the magnitude of the force the bar must exert?

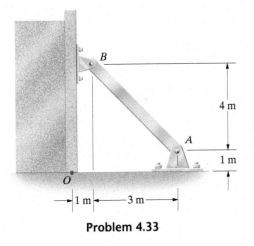

Problem 4.33

4.34 A contestant in a fly-casting contest snags his line in some grass. If the tension in the line is 5 lb, what moment does the force exerted on the rod by the line exert about point H, where he holds the rod?

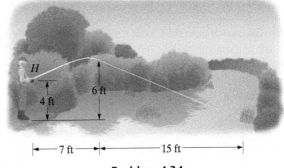

Problem 4.34

4.35 The cables AB and AC help support the tower. The tension in cable AB is 5 kN. The points A, B, C, and O are contained in the same vertical plane. (a) What is the moment about O due to the force exerted on the tower by cable AB? (b) If the sum of the moments about O due to the forces exerted on the tower by the two cables is zero, what is the tension in cable AC?

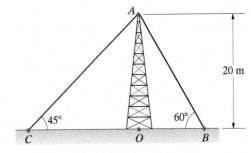

Problem 4.35

4.36 The cable from B to A (the sailboat's forestay) exerts a 230-N force at B. The cable from B to C (the backstay) exerts a 660-N force at B. The bottom of the sailboat's mast is located at $x = 4$ m, $y = 0$. What is the sum of the moments about the bottom of the mast due to the forces exerted at B by the forestay and backstay?

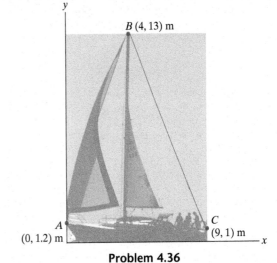

Problem 4.36

4.37 The cable AB exerts a 290-kN force on the crane's boom at B. The cable AC exerts a 148-kN force on the boom at C. Determine the sum of the moments about P due to the forces the cables AB and AC exert on the boom.

4.38 The mass of the crane's boom is 9000 kg. Its weight acts at G. The sum of the moments about P due to the boom's weight, the force exerted at B by the cable AB, and the force exerted at C by the cable AC is zero. Assume that the tensions in cables AB and AC are equal. Determine the tension in the cables.

4.39 The mass of the luggage carrier and the suitcase combined is 12 kg. Their weight acts at A. The sum of the moments about the origin of the coordinate system due to the weight acting at A and the vertical force F applied to the handle of the luggage carrier is zero. Determine the force F (a) if $\alpha = 30°$; (b) if $\alpha = 50°$.

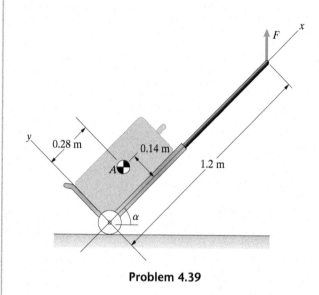

Problem 4.39

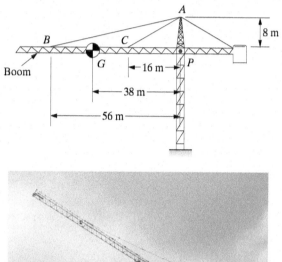

Problems 4.37/4.38

4.40 The hydraulic cylinder BC exerts a 300-kN force on the boom of the crane at C. The force is parallel to the cylinder. What is the moment of the force about A?

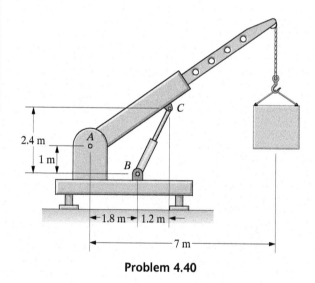

Problem 4.40

4.41 The hydraulic piston *AB* exerts a 400-lb force on the ladder at *B* in the direction parallel to the piston. The sum of the moments about *C* due to the force exerted on the ladder by the piston and the weight *W* of the ladder is zero. What is the weight of the ladder?

4.42 The hydraulic cylinder exerts an 8-kN force at *B* that is parallel to the cylinder and points from *C* toward *B*. Determine the moments of the force about points *A* and *D*.

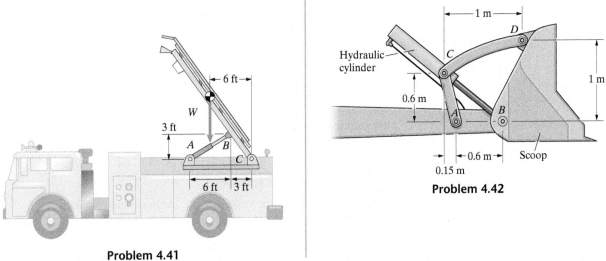

Problem 4.41

Problem 4.42

4.43 The structure shown in the diagram is one of two identical structures that support the scoop of the excavator. The bar *BC* exerts a 700-N force at *C* that points from *C* toward *B*. What is the moment of this force about *K*?

4.44 The bar *BC* exerts a force at *C* that points from *C* toward *B*. The hydraulic cylinder *DH* exerts a 1550-N force at *D* that points from *D* toward *H*. The sum of the moments of these two forces about *K* is zero. What is the magnitude of the force that bar *BC* exerts at *C*?

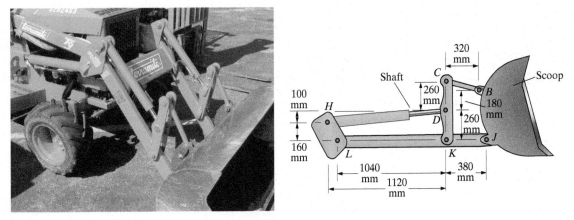

Problems 4.43/4.44

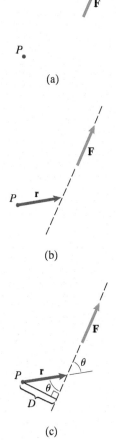

(a)

(b)

(c)

Figure 4.4
(a) The force **F** and point *P*.
(b) A vector **r** from *P* to a point on the line
    of action of **F**.
(c) The angle θ and the perpendicular
    distance *D*.

4.2 The Moment Vector

BACKGROUND

The moment of a force about a point is a vector. In this section we define this vector and explain how it is evaluated. We then show that when we use the two-dimensional description of the moment described in Section 4.1, we are specifying the magnitude and direction of the moment vector.

Consider a force vector **F** and point *P* (Fig. 4.4a). The *moment* of **F** about *P* is the vector

$$\mathbf{M}_P = \mathbf{r} \times \mathbf{F}, \tag{4.2}$$

where **r** is a position vector from *P* to *any* point on the line of action of **F** (Fig. 4.4b).

Magnitude of the Moment

From the definition of the cross product, the magnitude of $\mathbf{M}_P$ is

$$|\mathbf{M}_P| = |\mathbf{r}||\mathbf{F}| \sin \theta,$$

where θ is the angle between the vectors **r** and **F** when they are placed tail to tail. The perpendicular distance from *P* to the line of action of **F** is $D = |\mathbf{r}| \sin \theta$ (Fig. 4.4c). Therefore the magnitude of the moment $\mathbf{M}_P$ equals the product of the perpendicular distance from *P* to the line of action of **F** and the magnitude of **F**:

$$|\mathbf{M}_P| = D|\mathbf{F}|. \tag{4.3}$$

Notice that if we know the vectors $\mathbf{M}_P$ and **F**, this equation can be solved for the perpendicular distance *D*.

Direction of the Moment

We know from the definition of the cross product that $\mathbf{M}_P$ is perpendicular to both **r** and **F**. That means that $\mathbf{M}_P$ is perpendicular to the plane containing *P* and **F** (Fig. 4.5a). Notice in this figure that we denote a moment by a circular arrow around the vector.

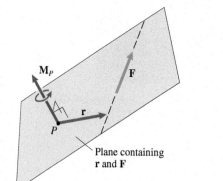

(a)

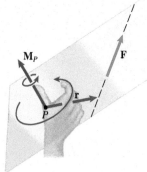

(b)

Figure 4.5
(a) $\mathbf{M}_P$ is perpendicular to the plane
    containing *p* and **F**.
(b) The direction of $\mathbf{M}_P$ indicates the
    direction of the moment.

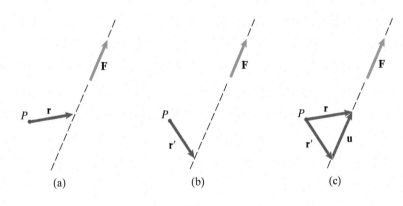

Figure 4.6
(a) A vector **r** from P to the line of action of **F**.
(b) A different vector **r'**.
(c) **r** = **r'** + **u**.

The direction of $\mathbf{M}_P$ also indicates the direction of the moment: Pointing the thumb of the right hand in the direction of $\mathbf{M}_P$, the "arc" of the fingers indicates the direction of the rotation that **F** tends to cause about P (Fig. 4.5b).

The result obtained from Eq. (4.2) doesn't depend on where the vector **r** intersects the line of action of **F**. Instead of using the vector **r** in Fig. 4.6a, we could use the vector **r'** in Fig. 4.6b. The vector **r** = **r'** + **u**, where **u** is parallel to **F** (Fig. 4.6c). Therefore,

$$\mathbf{r} \times \mathbf{F} = (\mathbf{r'} + \mathbf{u}) \times \mathbf{F} = \mathbf{r'} \times \mathbf{F}$$

because the cross product of the parallel vectors **u** and **F** is zero.

In summary, the moment of a force **F** about a point P has three properties:

1. The magnitude of $\mathbf{M}_P$ is equal to the product of the magnitude of **F** and the perpendicular distance from P to the line of action of **F**. If the line of action of **F** passes through P, $\mathbf{M}_P = \mathbf{0}$.
2. $\mathbf{M}_P$ is perpendicular to the plane containing P and **F**.
3. The direction of $\mathbf{M}_P$ indicates the direction of the moment through a right-hand rule (Fig. 4.5b). Since the cross product is not commutative, it is essential to maintain the correct sequence of the vectors in the equation $\mathbf{M}_P = \mathbf{r} \times \mathbf{F}$.

Let us determine the moment of the force **F** in Fig. 4.7a about the point P. Since the vector **r** in Eq. (4.2) can be a position vector to any point on the line of action of **F**, we can use the vector from P to the point of application of **F** (Fig. 4.7b):

$$\mathbf{r} = (12 - 3)\mathbf{i} + (6 - 4)\mathbf{j} + (-5 - 1)\mathbf{k} = 9\mathbf{i} + 2\mathbf{j} - 6\mathbf{k} \text{ (ft)}.$$

The moment is

$$\mathbf{M}_P = \mathbf{r} \times \mathbf{F} = \begin{vmatrix} \mathbf{i} & \mathbf{j} & \mathbf{k} \\ 9 & 2 & -6 \\ 4 & 4 & 7 \end{vmatrix} = 38\mathbf{i} - 87\mathbf{j} + 28\mathbf{k} \text{ (ft-lb)}.$$

The magnitude of $\mathbf{M}_P$,

$$|\mathbf{M}_P| = \sqrt{(38)^2 + (-87)^2 + (28)^2} = 99.0 \text{ ft-lb},$$

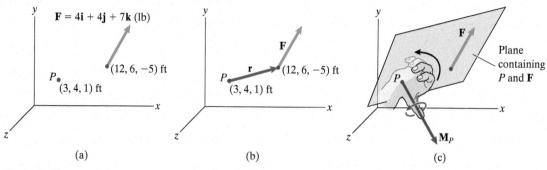

Figure 4.7
(a) A force **F** and point P.
(b) The vector **r** from P to the point of application of **F**.
(c) $\mathbf{M}_P$ is perpendicular to the plane containing P and **F**.
     The right-hand rule indicates the direction of the moment.

equals the product of the magnitude of **F** and the perpendicular distance D from point P to the line of action of **F**. Therefore,

$$D = \frac{|\mathbf{M}_P|}{|\mathbf{F}|} = \frac{99.0 \text{ ft-lb}}{9 \text{ lb}} = 11.0 \text{ ft.}$$

The direction of $\mathbf{M}_P$ tells us both the orientation of the plane containing P and **F** and the direction of the moment (Fig. 4.7c).

Relation to the Two-Dimensional Description

If our view is perpendicular to the plane containing the point P and the force **F**, the two-dimensional description of the moment we used in Section 4.1 specifies both the magnitude and direction of the vector $\mathbf{M}_P$. In this situation, $\mathbf{M}_P$ is perpendicular to the page, and the right-hand rule indicates whether it points out of or into the page.

For example, in Fig. 4.8a, the view is perpendicular to the x–y plane and the 10-N force is contained in the x–y plane. Suppose that we want to determine the

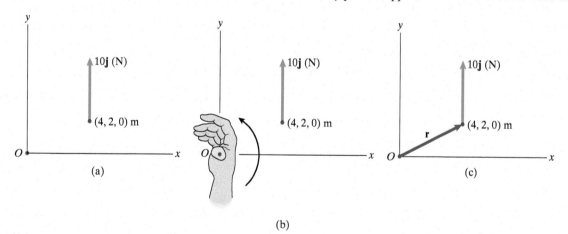

Figure 4.8
(a) The force is contained in the x–y plane.
(b) The counterclockwise direction of the moment indicates that $\mathbf{M}_O$ points out of the page.
(c) The vector **r** from O to the point of application of **F**.

moment of the force about the origin O. The perpendicular distance from O to the line of action of the force is 4 m. The two-dimensional description of the moment of the force about O is that its magnitude is $(4 \text{ m})(10 \text{ N}) = 40$ N-m and its direction is counterclockwise, or

$$M_O = 40 \text{ N-m}.$$

That tells us that the magnitude of the vector $\mathbf{M}_O$ is 40 N-m, and the right-hand rule (Fig. 4.8b) indicates that it points out of the page. Therefore,

$$\mathbf{M}_O = 40\mathbf{k} \text{ (N-m)}.$$

We can confirm this result by using Eq. (4.2). If we let $\mathbf{r}$ be the vector from O to the point of application of the force (Fig. 4.8c),

$$\mathbf{M}_O = \mathbf{r} \times \mathbf{F} = (4\mathbf{i} + 2\mathbf{j}) \times 10\mathbf{j} = 40\mathbf{k} \text{ (N-m)}.$$

As this example illustrates, the two-dimensional description of the moment determines the moment vector. The converse is also true. The magnitude of $\mathbf{M}_O$ equals the product of the magnitude of the force and the perpendicular distance from O to the line of action of the force, 40 N-m, and the direction of the vector $\mathbf{M}_O$ indicates that the moment is counterclockwise (Fig. 4.8b).

Varignon's Theorem

Let $\mathbf{F}_1, \mathbf{F}_2, \ldots, \mathbf{F}_N$ be a concurrent system of forces whose lines of action intersect at a point Q. The moment of the system about a point P is

$$(\mathbf{r}_{PQ} \times \mathbf{F}_1) + (\mathbf{r}_{PQ} \times \mathbf{F}_2) + \cdots + (\mathbf{r}_{PQ} \times \mathbf{F}_N)$$

$$= \mathbf{r}_{PQ} \times (\mathbf{F}_1 + \mathbf{F}_2 + \cdots + \mathbf{F}_N),$$

where $\mathbf{r}_{PQ}$ is the vector from P to Q (Fig. 4.9). This result, known as *Varignon's theorem*, follows from the distributive property of the cross product, Eq. (2.31). It confirms that the moment of a force about a point P is equal to the sum of the moments of its components about P.

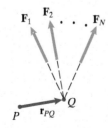

Figure 4.9
A system of concurrent forces and a point P.

RESULTS

Moment

The moment of a force **F** about a point P is defined by

$$\mathbf{M}_P = \mathbf{r} \times \mathbf{F}, \tag{4.2}$$

where **r** is a position vector from P to *any* point on the line of action of **F**.

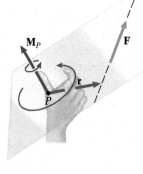

Magnitude of the Moment

The magnitude of the vector $\mathbf{M}_P$ is

$$|\mathbf{M}_P| = D|\mathbf{F}|, \tag{4.3}$$

where D is the perpendicular distance from P to the line of action of **F**.

Direction of the Moment

The vector $\mathbf{M}_P$ is perpendicular to the plane containing the point P and the vector **F**. Pointing the thumb of the right hand in the direction of $\mathbf{M}_P$, the fingers point in the direction of the rotation that **F** tends to cause about P.

Active Example 4.4   **Determining a Moment** (▶ *Related Problem 4.45*)

Determine the moment of the 90-lb force **F** about point A.

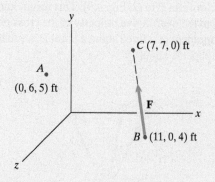

Strategy

To apply Eq. (4.2), we must express the force **F** in terms of its components. The vector **r** is a vector from point A to any point on the line of action of **F**, so we can use the vector from point A to point B.

Solution

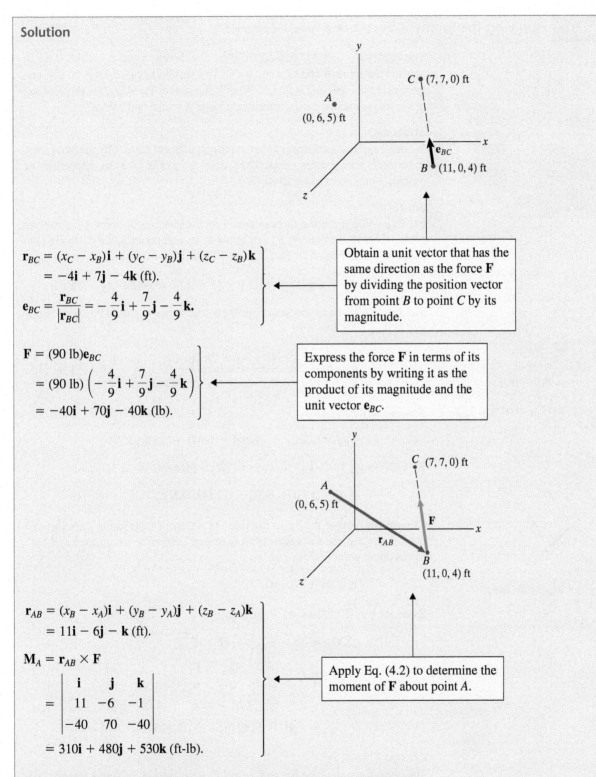

$$\mathbf{r}_{BC} = (x_C - x_B)\mathbf{i} + (y_C - y_B)\mathbf{j} + (z_C - z_B)\mathbf{k}$$
$$= -4\mathbf{i} + 7\mathbf{j} - 4\mathbf{k} \text{ (ft).}$$

$$\mathbf{e}_{BC} = \frac{\mathbf{r}_{BC}}{|\mathbf{r}_{BC}|} = -\frac{4}{9}\mathbf{i} + \frac{7}{9}\mathbf{j} - \frac{4}{9}\mathbf{k}.$$

Obtain a unit vector that has the same direction as the force $\mathbf{F}$ by dividing the position vector from point B to point C by its magnitude.

$$\mathbf{F} = (90 \text{ lb})\mathbf{e}_{BC}$$
$$= (90 \text{ lb})\left(-\frac{4}{9}\mathbf{i} + \frac{7}{9}\mathbf{j} - \frac{4}{9}\mathbf{k}\right)$$
$$= -40\mathbf{i} + 70\mathbf{j} - 40\mathbf{k} \text{ (lb).}$$

Express the force $\mathbf{F}$ in terms of its components by writing it as the product of its magnitude and the unit vector $\mathbf{e}_{BC}$.

$$\mathbf{r}_{AB} = (x_B - x_A)\mathbf{i} + (y_B - y_A)\mathbf{j} + (z_B - z_A)\mathbf{k}$$
$$= 11\mathbf{i} - 6\mathbf{j} - \mathbf{k} \text{ (ft).}$$

$$\mathbf{M}_A = \mathbf{r}_{AB} \times \mathbf{F}$$
$$= \begin{vmatrix} \mathbf{i} & \mathbf{j} & \mathbf{k} \\ 11 & -6 & -1 \\ -40 & 70 & -40 \end{vmatrix}$$
$$= 310\mathbf{i} + 480\mathbf{j} + 530\mathbf{k} \text{ (ft-lb).}$$

Apply Eq. (4.2) to determine the moment of $\mathbf{F}$ about point A.

Practice Problem (a) Use Eq. (4.2) to determine the moment of $\mathbf{F}$ about point A, letting the vector $\mathbf{r}$ be the position vector from point A to point C. (b) Determine the perpendicular distance from point A to the line of action of $\mathbf{F}$.

Answer: (a) $\mathbf{M}_A = 310\mathbf{i} + 480\mathbf{j} + 530\mathbf{k}$ (ft-lb). (b) 8.66 ft.

<table>
<tr><td>**Example 4.5**</td><td>**Applying the Moment Vector** (▶ *Related Problem 4.57*)</td></tr>
</table>

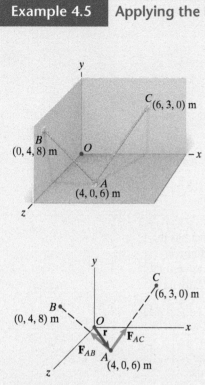

The cables AB and AC extend from an attachment point A on the floor to attachment points B and C in the walls. The tension in cable AB is 10 kN, and the tension in cable AC is 20 kN. What is the sum of the moments about O due to the forces exerted on the attachment point A by the two cables?

Strategy

We must express the forces exerted on the attachment point A by the two cables in terms of their components. Then we can use Eq. (4.2) to determine the moments the forces exert about O.

Solution

Let $\mathbf{F}_{AB}$ and $\mathbf{F}_{AC}$ be the forces exerted on the attachment point A by the two cables (Fig. a). To express $\mathbf{F}_{AB}$ in terms of its components, we determine the position vector from A to B,

$$(0 - 4)\mathbf{i} + (4 - 0)\mathbf{j} + (8 - 6)\mathbf{k} = -4\mathbf{i} + 4\mathbf{j} + 2\mathbf{k} \ (\text{m}),$$

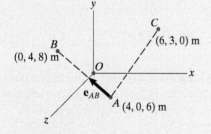

(a) The forces $\mathbf{F}_{AB}$ and $\mathbf{F}_{AC}$ exerted at A by the cables.

and divide it by its magnitude to obtain a unit vector $\mathbf{e}_{AB}$ with the same direction as $\mathbf{F}_{AB}$ (Fig. b):

$$\mathbf{e}_{AB} = \frac{-4\mathbf{i} + 4\mathbf{j} + 2\mathbf{k} \ (\text{m})}{\sqrt{(-4 \ \text{m})^2 + (4 \ \text{m})^2 + (2 \ \text{m})^2}} = -\frac{2}{3}\mathbf{i} + \frac{2}{3}\mathbf{j} + \frac{1}{3}\mathbf{k}.$$

Now we write $\mathbf{F}_{AB}$ as

$$\mathbf{F}_{AB} = 10\mathbf{e}_{AB} = -6.67\mathbf{i} + 6.67\mathbf{j} + 3.33\mathbf{k} \ (\text{kN}).$$

We express the force $\mathbf{F}_{AC}$ in terms of its components in the same way:

$$\mathbf{F}_{AC} = 5.71\mathbf{i} + 8.57\mathbf{j} - 17.14\mathbf{k} \ (\text{kN}).$$

Choose the Vector r   Since the lines of action of both forces pass through point A, we can use the vector from O to A to determine the moments of both forces about point O (Fig. a):

$$\mathbf{r} = 4\mathbf{i} + 6\mathbf{k} \ (\text{m}).$$

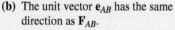

(b) The unit vector $\mathbf{e}_{AB}$ has the same direction as $\mathbf{F}_{AB}$.

Evaluate r × F   The sum of the moments is

$$\Sigma\mathbf{M}_O = (\mathbf{r} \times \mathbf{F}_{AB}) + (\mathbf{r} \times \mathbf{F}_{AC})$$

$$= \begin{vmatrix} \mathbf{i} & \mathbf{j} & \mathbf{k} \\ 4 & 0 & 6 \\ -6.67 & 6.67 & 3.33 \end{vmatrix} + \begin{vmatrix} \mathbf{i} & \mathbf{j} & \mathbf{k} \\ 4 & 0 & 6 \\ 5.71 & 8.57 & -17.14 \end{vmatrix}$$

$$= -91.4\mathbf{i} + 49.5\mathbf{j} + 61.0\mathbf{k} \ (\text{kN-m}).$$

Critical Thinking

The lines of action of the forces $\mathbf{F}_{AB}$ and $\mathbf{F}_{AC}$ intersect at A. Notice that, according to Varignon's theorem, we could have summed the forces first, obtaining

$$\mathbf{F}_{AB} + \mathbf{F}_{AC} = -0.952\mathbf{i} + 15.24\mathbf{j} - 13.81\mathbf{k} \ (\text{kN}),$$

and then determined the sum of the moments of the two forces about O by calculating the moment of the sum of the two forces about O:

$$\Sigma \mathbf{M}_O = \mathbf{r} \times (\mathbf{F}_{AB} + \mathbf{F}_{AC})$$

$$= \begin{vmatrix} \mathbf{i} & \mathbf{j} & \mathbf{k} \\ 4 & 0 & 6 \\ -0.952 & 15.24 & -13.81 \end{vmatrix}$$

$$= -91.4\mathbf{i} + 49.5\mathbf{j} + 61.0\mathbf{k} \ (\text{kN-m}).$$

Problems

▶ **4.45** In Active Example 4.4, what is the moment of $\mathbf{F}$ about the origin of the coordinate system?

4.46 Use Eq. (4.2) to determine the moment of the 80-N force about the origin O letting $\mathbf{r}$ be the vector (a) from O to A; (b) from O to B.

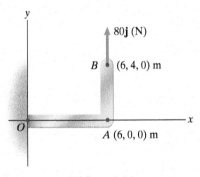

Problem 4.46

4.47 A bioengineer studying an injury sustained in throwing the javelin estimates that the magnitude of the maximum force exerted was $|\mathbf{F}| = 360$ N and the perpendicular distance from O to the line of action of $\mathbf{F}$ was 550 mm. The vector $\mathbf{F}$ and point O are contained in the x–y plane. Express the moment of $\mathbf{F}$ about the shoulder joint at O as a vector.

Problem 4.47

4.48 Use Eq. (4.2) to determine the moment of the 100-kN force (a) about A; (b) about B.

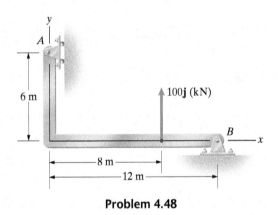

Problem 4.48

4.49 The cable AB exerts a 200-N force on the support at A that points from A toward B. Use Eq. (4.2) to determine the moment of this force about point P, (a) letting $\mathbf{r}$ be the vector from P to A; (b) letting $\mathbf{r}$ be the vector from P to B.

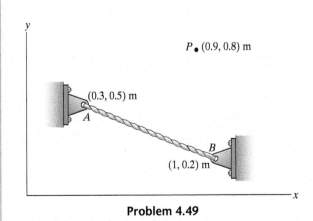

Problem 4.49

4.50 The line of action of **F** is contained in the x–y plane. The moment of **F** about O is 140**k** (N-m), and the moment of **F** about A is 280**k** (N-m). What are the components of **F**?

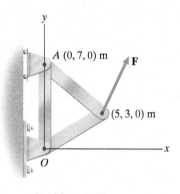

Problem 4.50

4.51 Use Eq. (4.2) to determine the sum of the moments of the three forces (a) about A; (b) about B.

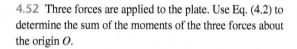

Problem 4.51

4.52 Three forces are applied to the plate. Use Eq. (4.2) to determine the sum of the moments of the three forces about the origin O.

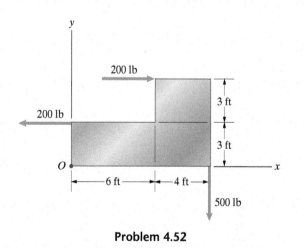

Problem 4.52

4.53 Three forces act on the plate. Use Eq. (4.2) to determine the sum of the moments of the three forces about point P.

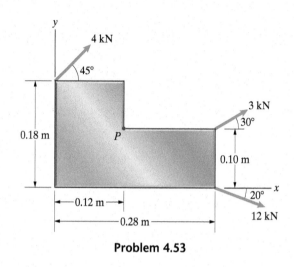

Problem 4.53

4.54 (a) Determine the magnitude of the moment of the 150-N force about A by calculating the perpendicular distance from A to the line of action of the force.

(b) Use Eq. (4.2) to determine the magnitude of the moment of the 150-N force about A.

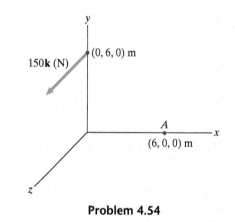

Problem 4.54

4.55 (a) Determine the magnitude of the moment of the 600-N force about A by calculating the perpendicular distance from A to the line of action of the force.

(b) Use Eq. (4.2) to determine the magnitude of the moment of the 600-N force about A.

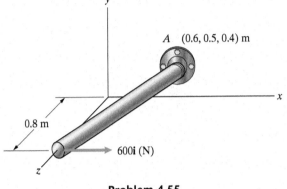

Problem 4.55

4.56 What is the magnitude of the moment of **F** about point B?

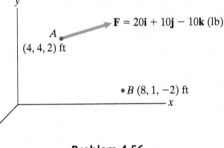

Problem 4.56

▶ **4.57** In Example 4.5, suppose that the attachment point C is moved to the location (8, 2, 0) m and the tension in cable AC changes to 25 kN. What is the sum of the moments about O due to the forces exerted on the attachment point A by the two cables?

4.58 The rope exerts a force of magnitude $|\mathbf{F}| = 200$ lb on the top of the pole at B. Determine the magnitude of the moment of **F** about A.

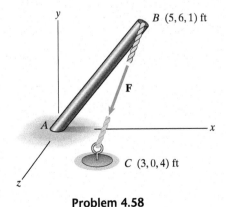

Problem 4.58

4.59 The force $\mathbf{F} = 30\mathbf{i} + 20\mathbf{j} - 10\mathbf{k}$ (N).

(a) Determine the magnitude of the moment of **F** about A.

(b) Suppose that you can change the direction of **F** while keeping its magnitude constant, and you want to choose a direction that maximizes the moment of **F** about A. What is the magnitude of the resulting maximum moment?

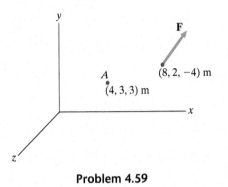

Problem 4.59

4.60 The direction cosines of the force **F** are $\cos \theta_x = 0.818$, $\cos \theta_y = 0.182$, and $\cos \theta_z = -0.545$. The support of the beam at O will fail if the magnitude of the moment of **F** about O exceeds 100 kN-m. Determine the magnitude of the largest force **F** that can safely be applied to the beam.

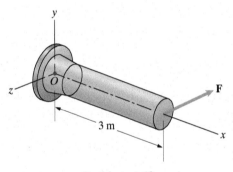

Problem 4.60

4.61 The force **F** exerted on the grip of the exercise machine points in the direction of the unit vector $\mathbf{e} = \frac{2}{3}\mathbf{i} - \frac{2}{3}\mathbf{j} + \frac{1}{3}\mathbf{k}$ and its magnitude is 120 N. Determine the magnitude of the moment of **F** about the origin O.

4.62 The force **F** points in the direction of the unit vector $\mathbf{e} = \frac{2}{3}\mathbf{i} - \frac{2}{3}\mathbf{j} + \frac{1}{3}\mathbf{k}$. The support at O will safely support a moment of 560 N-m magnitude. (a) Based on this criterion, what is the largest safe magnitude of **F**? (b) If the force **F** may be exerted in any direction, what is its largest safe magnitude?

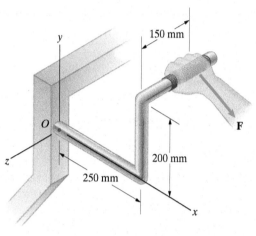

Problems 4.61/4.62

4.63 A civil engineer in Boulder, Colorado, estimates that under the severest expected Chinook winds, the total force on the highway sign will be $\mathbf{F} = 2.8\mathbf{i} - 1.8\mathbf{j}$ (kN). Let $\mathbf{M}_O$ be the moment due to **F** about the base O of the cylindrical column supporting the sign. The y component of $\mathbf{M}_O$ is called the *torsion* exerted on the cylindrical column at the base, and the component of $\mathbf{M}_O$ parallel to the x–z plane is called the *bending moment*. Determine the magnitudes of the torsion and bending moment.

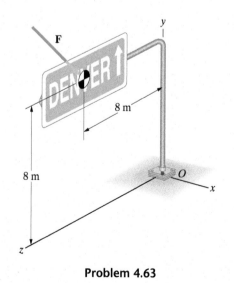

Problem 4.63

4.64 The weights of the arms OA and AB of the robotic manipulator act at their midpoints. The direction cosines of the centerline of arm OA are $\cos \theta_x = 0.500$, $\cos \theta_y = 0.866$, and $\cos \theta_z = 0$, and the direction cosines of the centerline of arm AB are $\cos \theta_x = 0.707$, $\cos \theta_y = 0.619$, and $\cos \theta_z = -0.342$. What is the sum of the moments about O due to the two forces?

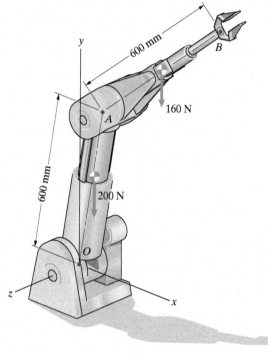

Problem 4.64

4.65 The tension in cable AB is 100 lb. If you want the magnitude of the moment due to the forces exerted on the tree by the two ropes about the base O of the tree to be 1500 ft-lb, what is the necessary tension in rope AC?

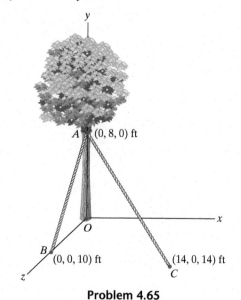

Problem 4.65

4.66* A force **F** acts at the top end A of the pole. Its magnitude is $|\mathbf{F}| = 6$ kN and its x component is $F_x = 4$ kN. The coordinates of point A are shown. Determine the components of **F** so that the magnitude of the moment due to **F** about the base P of the pole is as large as possible.

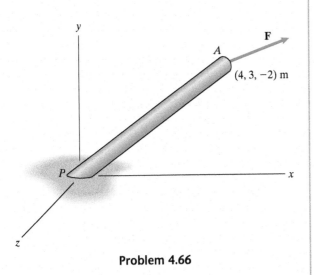

Problem 4.66

4.67 The force $\mathbf{F} = 5\mathbf{i}$ (kN) acts on the ring A where the cables AB, AC, and AD are joined. What is the sum of the moments about point D due to the force **F** and the three forces exerted on the ring by the cables?

Strategy: The ring is in equilibrium. Use what you know about the four forces acting on it.

4.68 In Problem 4.67, determine the moment about point D due to the force exerted on the ring A by the cable AB.

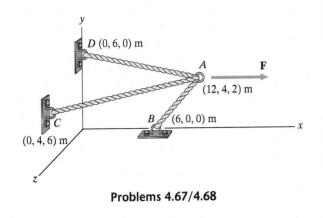

Problems 4.67/4.68

4.69 The tower is 70 m tall. The tensions in cables AB, AC, and AD are 4 kN, 2 kN, and 2 kN, respectively. Determine the sum of the moments about the origin O due to the forces exerted by the cables at point A.

4.70 Suppose that the tension in cable AB is 4 kN, and you want to adjust the tensions in cables AC and AD so that the sum of the moments about the origin O due to the forces exerted by the cables at point A is zero. Determine the tensions.

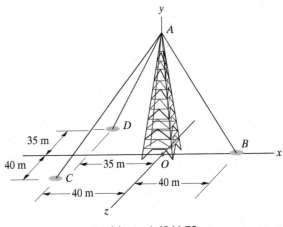

Problems 4.69/4.70

4.71 The tension in cable AB is 150 N. The tension in cable AC is 100 N. Determine the sum of the moments about D due to the forces exerted on the wall by the cables.

4.72 The total force exerted by the two cables in the direction perpendicular to the wall is 2 kN. The magnitude of the sum of the moments about D due to the forces exerted on the wall by the cables is 18 kN-m. What are the tensions in the cables?

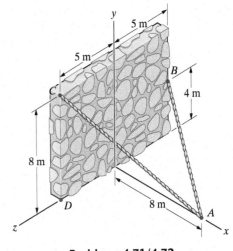

Problems 4.71/4.72

4.73 The tension in the cable *BD* is 1 kN. As a result, cable *BD* exerts a 1-kN force on the "ball" at *B* that points from *B* toward *D*. Determine the moment of this force about point *A*.

4.74* Suppose that the mass of the suspended object *E* is 100 kg and the mass of the bar *AB* is 20 kg. Assume that the weight of the bar acts at its midpoint. If the sum of the moments about point *A* due to the weight of the bar and the forces exerted on the "ball" at *B* by the three cables *BC*, *BD*, and *BE* is zero, determine the tensions in the cables *BC* and *BD*.

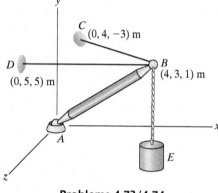

Problems 4.73/4.74

4.75 The 200-kg slider at *A* is held in place on the smooth vertical bar by the cable *AB*. Determine the moment about the bottom of the bar (point *C* with coordinates $x = 2$ m, $y = z = 0$) due to the force exerted on the slider by the cable.

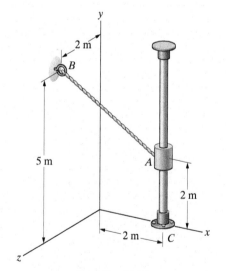

Problem 4.75

4.76 To evaluate the adequacy of the design of the vertical steel post, you must determine the moment about the bottom of the post due to the force exerted on the post at *B* by the cable *AB*. A calibrated strain gauge mounted on cable *AC* indicates that the tension in cable *AC* is 22 kN. What is the moment?

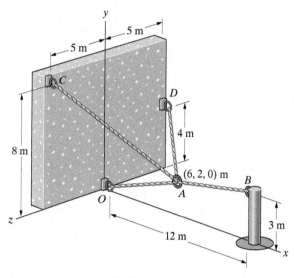

Problem 4.76

4.3 Moment of a Force About a Line

BACKGROUND

The device in Fig. 4.10, called a *capstan*, was used in the days of square-rigged sailing ships. Crewmen turned it by pushing on the handles as shown in Fig. 4.10a, providing power for such tasks as raising anchors and hoisting yards. A vertical force **F** applied to one of the handles as shown in Fig. 4.10b does not cause the capstan to turn, even though the magnitude of the moment about point P is $d|\mathbf{F}|$ in both cases.

The measure of the tendency of a force to cause rotation about a line, or axis, is called the moment of the force about the line. Suppose that a force **F** acts on an object such as a turbine that rotates about an axis L, and we resolve **F** into components in terms of the coordinate system shown in Fig. 4.11. The components F_x and F_z do not tend to rotate the turbine, just as the force parallel to the axis of the capstan did not cause it to turn. It is the component F_y that tends to cause rotation, by exerting a moment of magnitude aF_y about the turbine's axis. In this example we can determine the moment of **F** about L easily because the coordinate system is conveniently placed. We now introduce an expression that determines the moment of a force about any line.

(a) (b)

Figure 4.10
(a) Turning a capstan.
(b) A vertical force does not turn the capstan.

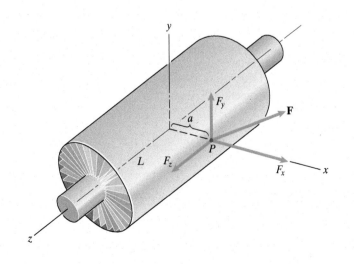

Figure 4.11
Applying a force to a turbine with axis of rotation L.

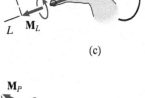

(a)

$\mathbf{M}_P = \mathbf{r} \times \mathbf{F}$

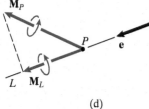

(b)

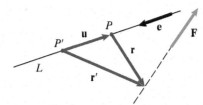

(c)

$\mathbf{M}_P$

(d)

Figure 4.12

(a) The line L and force $\mathbf{F}$.
(b) $\mathbf{M}_P$ is the moment of $\mathbf{F}$ about any point P on L.
(c) The component $\mathbf{M}_L$ is the moment of $\mathbf{F}$ about L.
(d) A unit vector $\mathbf{e}$ along L.

Figure 4.13

Using different points P and P' to determine the moment of $\mathbf{F}$ about L.

Definition

Consider a line L and force $\mathbf{F}$ (Fig. 4.12a). Let $\mathbf{M}_P$ be the moment of $\mathbf{F}$ about an arbitrary point P on L (Fig. 4.12b). The moment of $\mathbf{F}$ about L is the component of $\mathbf{M}_P$ parallel to L, which we denote by $\mathbf{M}_L$ (Fig. 4.12c). The magnitude of the moment of $\mathbf{F}$ about L is $|\mathbf{M}_L|$, and when the thumb of the right hand is pointed in the direction of $\mathbf{M}_L$, the arc of the fingers indicates the direction of the moment about L. In terms of a unit vector $\mathbf{e}$ along L (Fig. 4.12d), $\mathbf{M}_L$ is given by

$$\mathbf{M}_L = (\mathbf{e} \cdot \mathbf{M}_P)\mathbf{e}. \tag{4.4}$$

(The unit vector $\mathbf{e}$ can point in either direction. See our discussion of vector components parallel and normal to a line in Section 2.5.) The moment $\mathbf{M}_P = \mathbf{r} \times \mathbf{F}$, so we can also express $\mathbf{M}_L$ as

$$\mathbf{M}_L = [\mathbf{e} \cdot (\mathbf{r} \times \mathbf{F})]\mathbf{e}. \tag{4.5}$$

The mixed triple product in this expression is given in terms of the components of the three vectors by

$$\mathbf{e} \cdot (\mathbf{r} \times \mathbf{F}) = \begin{vmatrix} e_x & e_y & e_z \\ r_x & r_y & r_z \\ F_x & F_y & F_z \end{vmatrix}. \tag{4.6}$$

Notice that the value of the scalar $\mathbf{e} \cdot \mathbf{M}_P = \mathbf{e} \cdot (\mathbf{r} \times \mathbf{F})$ determines both the magnitude and direction of $\mathbf{M}_L$. The absolute value of $\mathbf{e} \cdot \mathbf{M}_P$ is the magnitude of $\mathbf{M}_L$. If $\mathbf{e} \cdot \mathbf{M}_P$ is positive, $\mathbf{M}_L$ points in the direction of $\mathbf{e}$, and if $\mathbf{e} \cdot \mathbf{M}_P$ is negative, $\mathbf{M}_L$ points in the direction opposite to $\mathbf{e}$.

The result obtained with Eq. (4.4) or (4.5) doesn't depend on which point on L is chosen to determine $\mathbf{M}_P = \mathbf{r} \times \mathbf{F}$. If we use point P in Fig. 4.13 to determine the moment of $\mathbf{F}$ about L, we get the result given by Eq. (4.5). If we use P' instead, we obtain the same result,

$$\begin{aligned}[\mathbf{e} \cdot (\mathbf{r}' \times \mathbf{F})]\mathbf{e} &= \{\mathbf{e} \cdot [(\mathbf{r} + \mathbf{u}) \times \mathbf{F}]\}\mathbf{e} \\ &= [\mathbf{e} \cdot (\mathbf{r} \times \mathbf{F}) + \mathbf{e} \cdot (\mathbf{u} \times \mathbf{F})]\mathbf{e} \\ &= [\mathbf{e} \cdot (\mathbf{r} \times \mathbf{F})]\mathbf{e},\end{aligned}$$

because $\mathbf{u} \times \mathbf{F}$ is perpendicular to $\mathbf{e}$.

Applications

To demonstrate that $\mathbf{M}_L$ is the measure of the tendency of $\mathbf{F}$ to cause rotation about L, we return to the turbine in Fig. 4.11. Let Q be a point on L at an arbitrary distance b from the origin (Fig. 4.14a). The vector $\mathbf{r}$ from Q to P is $\mathbf{r} = a\mathbf{i} - b\mathbf{k}$, so the moment of $\mathbf{F}$ about Q is

$$\mathbf{M}_Q = \mathbf{r} \times \mathbf{F} = \begin{vmatrix} \mathbf{i} & \mathbf{j} & \mathbf{k} \\ a & 0 & -b \\ F_x & F_y & F_z \end{vmatrix} = bF_y\mathbf{i} - (aF_z + bF_x)\mathbf{j} + aF_y\mathbf{k}.$$

Since the z axis is coincident with L, the unit vector $\mathbf{k}$ is along L. Therefore the moment of $\mathbf{F}$ about L is

$$\mathbf{M}_L = (\mathbf{k} \cdot \mathbf{M}_Q)\mathbf{k} = aF_y\mathbf{k}.$$

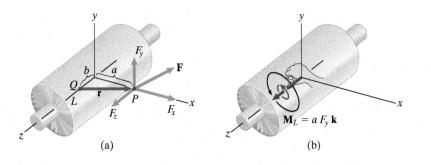

Figure 4.14
(a) An arbitrary point Q on L and the vector $\mathbf{r}$ from Q to P.
(b) $\mathbf{M}_L$ and the direction of the moment about L.

The components F_x and F_z exert no moment about L. If we assume that F_y is positive, it exerts a moment of magnitude aF_y about the turbine's axis in the direction shown in Fig. 4.14b.

Now let us determine the moment of a force about an arbitrary line L (Fig. 4.15a). The first step is to choose a point on the line. If we choose point A (Fig. 4.15b), the vector $\mathbf{r}$ from A to the point of application of $\mathbf{F}$ is

$$\mathbf{r} = (8 - 2)\mathbf{i} + (6 - 0)\mathbf{j} + (4 - 4)\mathbf{k} = 6\mathbf{i} + 6\mathbf{j} \text{ (m)}.$$

The moment of $\mathbf{F}$ about A is

$$\mathbf{M}_A = \mathbf{r} \times \mathbf{F} = \begin{vmatrix} \mathbf{i} & \mathbf{j} & \mathbf{k} \\ 6 & 6 & 0 \\ 10 & 60 & -20 \end{vmatrix}$$

$$= -120\mathbf{i} + 120\mathbf{j} + 300\mathbf{k} \text{ (N-m)}.$$

The next step is to determine a unit vector along L. The vector from A to B is

$$(-7 - 2)\mathbf{i} + (6 - 0)\mathbf{j} + (2 - 4)\mathbf{k} = -9\mathbf{i} + 6\mathbf{j} - 2\mathbf{k} \text{ (m)}.$$

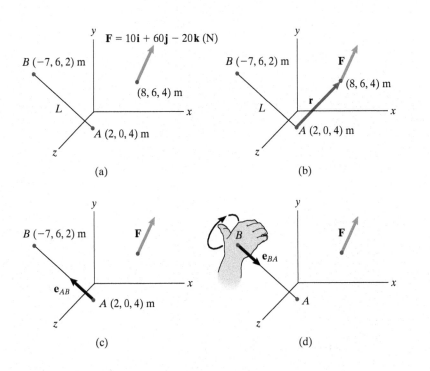

Figure 4.15
(a) A force $\mathbf{F}$ and line L.
(b) The vector $\mathbf{r}$ from A to the point of application of $\mathbf{F}$.
(c) $\mathbf{e}_{AB}$ points from A toward B.
(d) The right-hand rule indicates the direction of the moment.

Dividing this vector by its magnitude, we obtain a unit vector $\mathbf{e}_{AB}$ that points from A toward B (Fig. 4.15c):

$$\mathbf{e}_{AB} = -\frac{9}{11}\mathbf{i} + \frac{6}{11}\mathbf{j} - \frac{2}{11}\mathbf{k}.$$

The moment of $\mathbf{F}$ about L is

$$\mathbf{M}_L = (\mathbf{e}_{AB} \cdot \mathbf{M}_A)\mathbf{e}_{AB}$$

$$= \left[\left(-\frac{9}{11}\right)(-120 \text{ N-m}) + \left(\frac{6}{11}\right)(120 \text{ N-m}) + \left(-\frac{2}{11}\right)(300 \text{ N-m})\right]\mathbf{e}_{AB}$$

$$= 109\mathbf{e}_{AB} \text{ (N-m)}.$$

The magnitude of $\mathbf{M}_L$ is 109 N-m; pointing the thumb of the right hand in the direction of $\mathbf{e}_{AB}$ indicates the direction.

If we calculate $\mathbf{M}_L$ using the unit vector $\mathbf{e}_{BA}$ that points from B toward A instead, we obtain

$$\mathbf{M}_L = -109\mathbf{e}_{BA} \text{ (N-m)}.$$

We obtain the same magnitude, and the minus sign indicates that $\mathbf{M}_L$ points in the direction opposite to $\mathbf{e}_{BA}$, so the direction of $\mathbf{M}_L$ is the same. Therefore the right-hand rule indicates the same direction (Fig. 4.15d).

The preceding examples demonstrate three useful results that we can state in more general terms:

- When the line of action of $\mathbf{F}$ is perpendicular to a plane containing L (Fig. 4.16a), the magnitude of the moment of $\mathbf{F}$ about L is equal to the product of the magnitude of $\mathbf{F}$ and the perpendicular distance D from L to the point where the line of action intersects the plane: $|\mathbf{M}_L| = |\mathbf{F}|D$.

- When the line of action of $\mathbf{F}$ is parallel to L (Fig. 4.16b), the moment of $\mathbf{F}$ about L is zero: $\mathbf{M}_L = 0$. Since $\mathbf{M}_P = \mathbf{r} \times \mathbf{F}$ is perpendicular to $\mathbf{F}$, $\mathbf{M}_P$ is perpendicular to L and the vector component of $\mathbf{M}_P$ parallel to L is zero.

- When the line of action of $\mathbf{F}$ intersects L (Fig. 4.16c), the moment of $\mathbf{F}$ about L is zero. Since we can choose any point on L to evaluate $\mathbf{M}_P$, we can use the point where the line of action of $\mathbf{F}$ intersects L. The moment $\mathbf{M}_P$ about that point is zero, so its vector component parallel to L is zero.

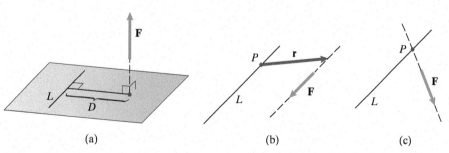

(a)   (b)   (c)

Figure 4.16
(a) $\mathbf{F}$ is perpendicular to a plane containing L.
(b) $\mathbf{F}$ is parallel to L.
(c) The line of action of $\mathbf{F}$ intersects L at P.

Determining the Moment of a Force F About a Line L

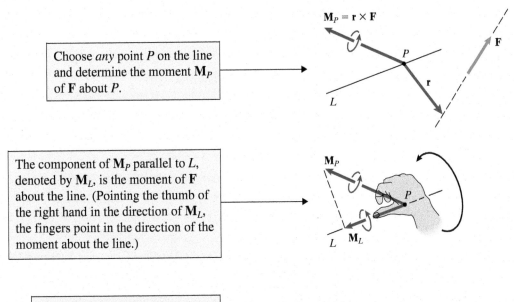

Choose *any* point P on the line and determine the moment $\mathbf{M}_P$ of $\mathbf{F}$ about P.

The component of $\mathbf{M}_P$ parallel to L, denoted by $\mathbf{M}_L$, is the moment of $\mathbf{F}$ about the line. (Pointing the thumb of the right hand in the direction of $\mathbf{M}_L$, the fingers point in the direction of the moment about the line.)

If $\mathbf{e}$ is a unit vector parallel to L,
$$\mathbf{M}_L = (\mathbf{e} \cdot \mathbf{M}_P)\, \mathbf{e}. \qquad (4.4)$$

Special Cases

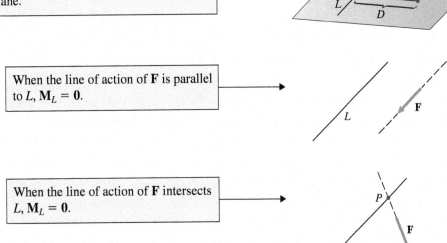

When the line of action of $\mathbf{F}$ is perpendicular to a plane containing L, $|\mathbf{M}_L| = |\mathbf{F}|D$, where D is the perpendicular distance from L to the point where the line of action intersects the plane.

When the line of action of $\mathbf{F}$ is parallel to L, $\mathbf{M}_L = \mathbf{0}$.

When the line of action of $\mathbf{F}$ intersects L, $\mathbf{M}_L = \mathbf{0}$.

Active Example 4.6   Moment of a Force About a Line (▶ *Related Problem 4.87*)

What is the moment of the force **F** about the axis of the bar *BC*?

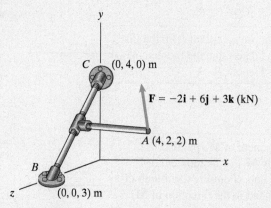

Strategy

Because we know the coordinates of points *A*, *B*, and *C*, we can determine the moment due to **F** about a point on the axis of the bar. We will determine the moment about point *B*. The component of that moment parallel to the axis *BC* is the moment of **F** about the axis. By obtaining a unit vector parallel to the axis, we can use Eq. (4.4) to determine the parallel component.

Solution

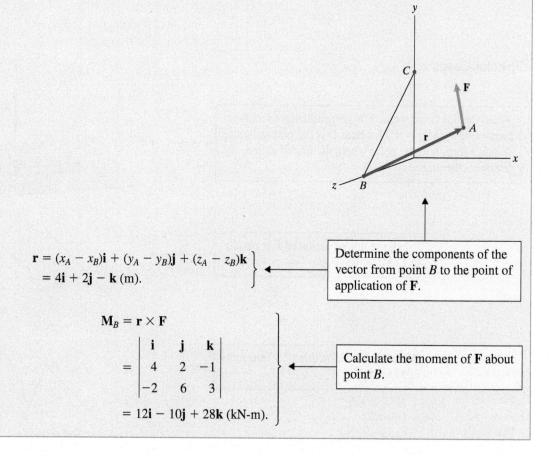

$$\mathbf{r} = (x_A - x_B)\mathbf{i} + (y_A - y_B)\mathbf{j} + (z_A - z_B)\mathbf{k}$$
$$= 4\mathbf{i} + 2\mathbf{j} - \mathbf{k} \text{ (m).}$$

Determine the components of the vector from point *B* to the point of application of **F**.

$$\mathbf{M}_B = \mathbf{r} \times \mathbf{F}$$
$$= \begin{vmatrix} \mathbf{i} & \mathbf{j} & \mathbf{k} \\ 4 & 2 & -1 \\ -2 & 6 & 3 \end{vmatrix}$$
$$= 12\mathbf{i} - 10\mathbf{j} + 28\mathbf{k} \text{ (kN-m).}$$

Calculate the moment of **F** about point *B*.

$$\mathbf{r}_{BC} = (x_C - x_B)\mathbf{i} + (y_C - y_B)\mathbf{j} + (z_C - z_B)\mathbf{k}$$
$$= 4\mathbf{j} - 3\mathbf{k} \text{ (m)}.$$
$$\mathbf{e}_{BC} = \frac{\mathbf{r}_{BC}}{|\mathbf{r}_{BC}|} = 0.8\mathbf{j} - 0.6\mathbf{k}.$$

Obtain a unit vector parallel to the axis BC by dividing the position vector from point B to point C by its magnitude.

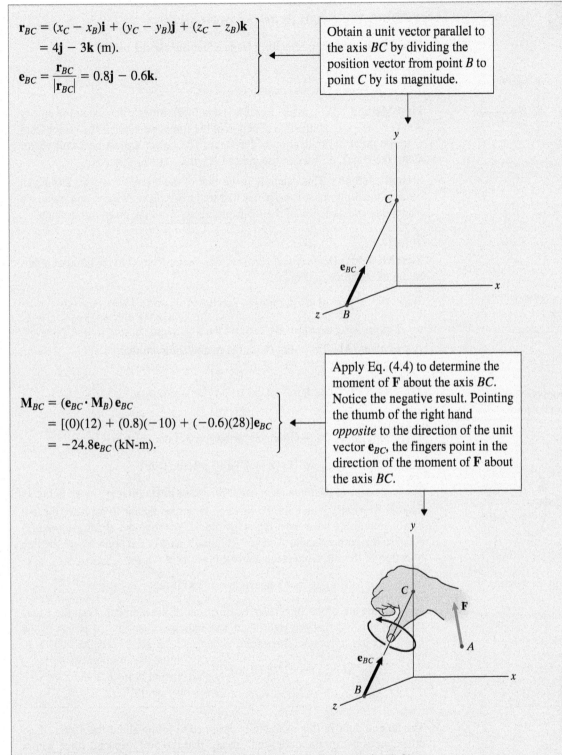

$$\mathbf{M}_{BC} = (\mathbf{e}_{BC} \cdot \mathbf{M}_B)\,\mathbf{e}_{BC}$$
$$= [(0)(12) + (0.8)(-10) + (-0.6)(28)]\mathbf{e}_{BC}$$
$$= -24.8\mathbf{e}_{BC} \text{ (kN-m)}.$$

Apply Eq. (4.4) to determine the moment of $\mathbf{F}$ about the axis BC. Notice the negative result. Pointing the thumb of the right hand *opposite* to the direction of the unit vector $\mathbf{e}_{BC}$, the fingers point in the direction of the moment of $\mathbf{F}$ about the axis BC.

Practice Problem Determine the moment $\mathbf{M}_C$ of the force $\mathbf{F}$ about point C. Use it to calculate the moment of $\mathbf{F}$ about the axis BC by determining the component of $\mathbf{M}_C$ parallel to the axis.

Answer: $\mathbf{M}_{BC} = -24.8\mathbf{e}_{BC}$ (kN-m).

| Example 4.7 | Moment of a Force About the *x* Axis (▶ *Related Problem 4.77*) |

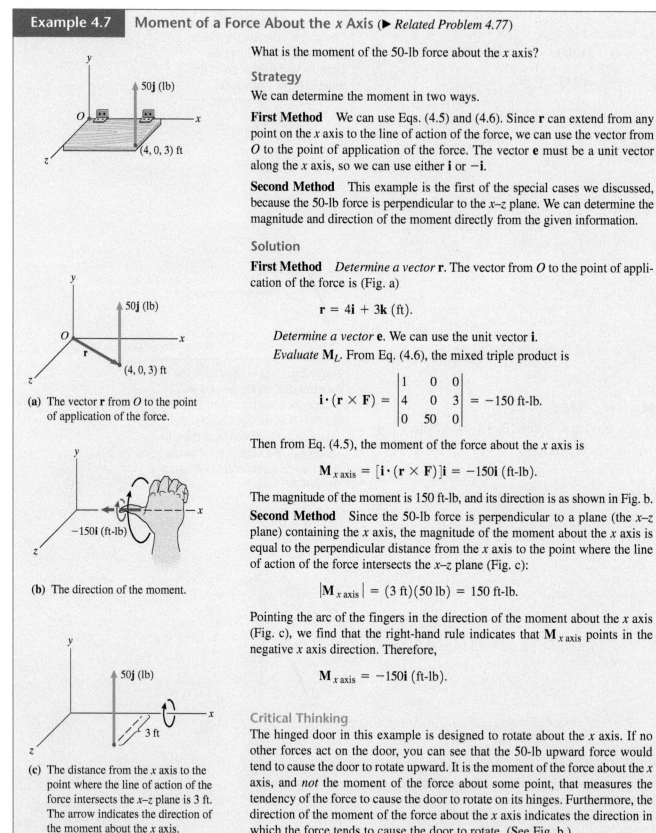

What is the moment of the 50-lb force about the *x* axis?

Strategy

We can determine the moment in two ways.

First Method   We can use Eqs. (4.5) and (4.6). Since **r** can extend from any point on the *x* axis to the line of action of the force, we can use the vector from *O* to the point of application of the force. The vector **e** must be a unit vector along the *x* axis, so we can use either **i** or −**i**.

Second Method   This example is the first of the special cases we discussed, because the 50-lb force is perpendicular to the *x*–*z* plane. We can determine the magnitude and direction of the moment directly from the given information.

Solution

First Method   *Determine a vector* **r**. The vector from *O* to the point of application of the force is (Fig. a)

$$\mathbf{r} = 4\mathbf{i} + 3\mathbf{k} \text{ (ft).}$$

Determine a vector **e**. We can use the unit vector **i**.

Evaluate $\mathbf{M}_L$. From Eq. (4.6), the mixed triple product is

$$\mathbf{i} \cdot (\mathbf{r} \times \mathbf{F}) = \begin{vmatrix} 1 & 0 & 0 \\ 4 & 0 & 3 \\ 0 & 50 & 0 \end{vmatrix} = -150 \text{ ft-lb.}$$

Then from Eq. (4.5), the moment of the force about the *x* axis is

$$\mathbf{M}_{x\text{ axis}} = [\mathbf{i} \cdot (\mathbf{r} \times \mathbf{F})]\mathbf{i} = -150\mathbf{i} \text{ (ft-lb).}$$

The magnitude of the moment is 150 ft-lb, and its direction is as shown in Fig. b.

Second Method   Since the 50-lb force is perpendicular to a plane (the *x*–*z* plane) containing the *x* axis, the magnitude of the moment about the *x* axis is equal to the perpendicular distance from the *x* axis to the point where the line of action of the force intersects the *x*–*z* plane (Fig. c):

$$|\mathbf{M}_{x\text{ axis}}| = (3 \text{ ft})(50 \text{ lb}) = 150 \text{ ft-lb.}$$

Pointing the arc of the fingers in the direction of the moment about the *x* axis (Fig. c), we find that the right-hand rule indicates that $\mathbf{M}_{x\text{ axis}}$ points in the negative *x* axis direction. Therefore,

$$\mathbf{M}_{x\text{ axis}} = -150\mathbf{i} \text{ (ft-lb).}$$

(a) The vector **r** from *O* to the point of application of the force.

(b) The direction of the moment.

(c) The distance from the *x* axis to the point where the line of action of the force intersects the *x*–*z* plane is 3 ft. The arrow indicates the direction of the moment about the *x* axis.

Critical Thinking

The hinged door in this example is designed to rotate about the *x* axis. If no other forces act on the door, you can see that the 50-lb upward force would tend to cause the door to rotate upward. It is the moment of the force about the *x* axis, and *not* the moment of the force about some point, that measures the tendency of the force to cause the door to rotate on its hinges. Furthermore, the direction of the moment of the force about the *x* axis indicates the direction in which the force tends to cause the door to rotate. (See Fig. b.)

Example 4.8   **Rotating Machines** (▶ *Related Problem 4.100*)

The crewman exerts the forces shown on the handles of the coffee grinder winch, where $\mathbf{F} = 4\mathbf{j} + 32\mathbf{k}$ N. Determine the total moment he exerts (a) about point O; (b) about the axis of the winch, which coincides with the x axis.

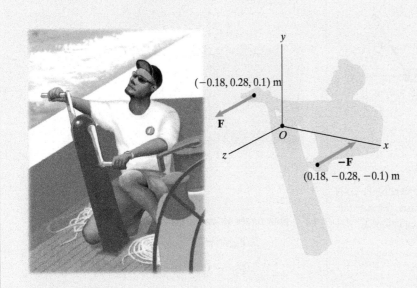

Strategy

(a) To obtain the total moment about point O, we must sum the moments of the two forces about O. Let the sum be denoted by $\Sigma\mathbf{M}_O$.

(b) Because point O is on the x axis, the total moment about the x axis is the component of $\Sigma\mathbf{M}_O$ parallel to the x axis, which is the x component of $\Sigma\mathbf{M}_O$.

Solution

(a) The total moment about point O is

$$\Sigma\mathbf{M}_O = \begin{vmatrix} \mathbf{i} & \mathbf{j} & \mathbf{k} \\ -0.18 & 0.28 & 0.1 \\ 0 & 4 & 32 \end{vmatrix} + \begin{vmatrix} \mathbf{i} & \mathbf{j} & \mathbf{k} \\ 0.18 & -0.28 & -0.1 \\ 0 & -4 & -32 \end{vmatrix}$$

$$= 17.1\mathbf{i} + 11.5\mathbf{j} - 1.4\mathbf{k} \text{ (N-m)}.$$

(b) The total moment about the x axis is the x component of $\Sigma\mathbf{M}_O$ (Fig. a):

$$\Sigma\mathbf{M}_{x\text{ axis}} = 17.1 \text{ (N-m)}.$$

Notice that this is the result given by Eq. (4.4): Since $\mathbf{i}$ is a unit vector parallel to the x axis,

$$\Sigma\mathbf{M}_{x\text{ axis}} = (\mathbf{i} \cdot \Sigma\mathbf{M}_O)\mathbf{i} = 17.1 \text{ (N-m)}.$$

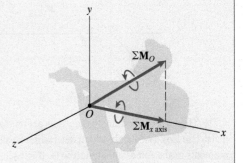

(a) The total moment about the x axis.

Problems

▶ **4.77** The force $\mathbf{F} = 20\mathbf{i} + 40\mathbf{j} - 10\mathbf{k}$ (N). Use both of the procedures described in Example 4.7 to determine the moment due to $\mathbf{F}$ about the z axis.

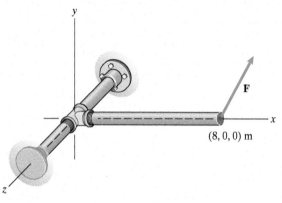

Problem 4.77

4.78 Use Eqs. (4.5) and (4.6) to determine the moment of the 20-N force about (a) the x axis, (b) the y axis, (c) the z axis. (First see if you can write down the results without using the equations.)

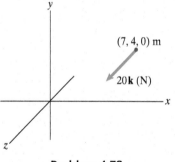

Problem 4.78

4.79 Three forces parallel to the y axis act on the rectangular plate. Use Eqs. (4.5) and (4.6) to determine the sum of the moments of the forces about the x axis. (First see if you can write down the result without using the equations.)

4.80 The three forces are parallel to the y axis. Determine the sum of the moments of the forces (a) about the y axis; (b) about the z axis.

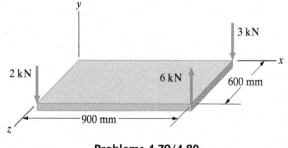

Problems 4.79/4.80

4.81 The person exerts a force $\mathbf{F} = 0.2\mathbf{i} - 0.4\mathbf{j} + 1.2\mathbf{k}$ (lb) on the gate at C. Point C lies in the x–y plane. What moment does the person exert about the gate's hinge axis, which is coincident with the y axis?

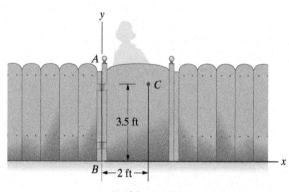

Problem 4.81

4.82 Four forces act on the plate. Their components are

$$\mathbf{F}_A = -2\mathbf{i} + 4\mathbf{j} + 2\mathbf{k} \text{ (kN)},$$

$$\mathbf{F}_B = 3\mathbf{j} - 3\mathbf{k} \text{ (kN)},$$

$$\mathbf{F}_C = 2\mathbf{j} + 3\mathbf{k} \text{ (kN)},$$

$$\mathbf{F}_D = 2\mathbf{i} + 6\mathbf{j} + 4\mathbf{k} \text{ (kN)}.$$

Determine the sum of the moments of the forces (a) about the x axis; (b) about the z axis.

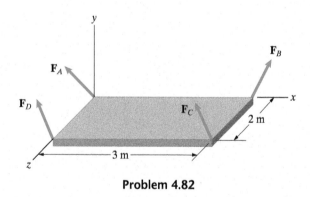

Problem 4.82

4.83 The force $\mathbf{F} = 30\mathbf{i} + 20\mathbf{j} - 10\mathbf{k}$ (lb).

(a) What is the moment of $\mathbf{F}$ about the y axis?

(b) Suppose that you keep the magnitude of $\mathbf{F}$ fixed, but you change its direction so as to make the moment of $\mathbf{F}$ about the y axis as large as possible. What is the magnitude of the resulting moment?

4.84 The moment of the force $\mathbf{F}$ about the x axis is $-80\mathbf{i}$ (ft-lb), the moment about the y axis is zero, and the moment about the z axis is $160\mathbf{k}$ (ft-lb). If $F_y = 80$ lb, what are F_x and F_z?

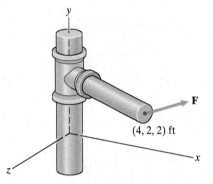

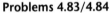

Problems 4.83/4.84

4.85 The robotic manipulator is stationary. The weights of the arms AB and BC act at their midpoints. The direction cosines of the centerline of arm AB are $\cos \theta_x = 0.500$, $\cos \theta_y = 0.866$, $\cos \theta_z = 0$, and the direction cosines of the centerline of arm BC are $\cos \theta_x = 0.707$, $\cos \theta_y = 0.619$, $\cos \theta_z = -0.342$. What total moment is exerted about the z axis by the weights of the arms?

4.86 In Problem 4.85, what total moment is exerted about the x axis by the weights of the arms?

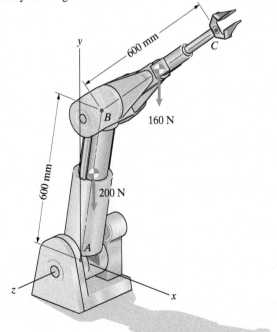

Problems 4.85/4.86

▶ **4.87** In Active Example 4.6, suppose that the force changes to $\mathbf{F} = -2\mathbf{i} + 3\mathbf{j} + 6\mathbf{k}$ (kN). Determine the magnitude of the moment of the force about the axis of the bar BC.

4.88 Determine the moment of the 20-N force about the line AB. Use Eqs. (4.5) and (4.6), letting the unit vector $\mathbf{e}$ point (a) from A toward B; (b) from B toward A.

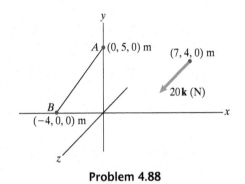

Problem 4.88

4.89 The force $\mathbf{F} = -10\mathbf{i} + 5\mathbf{j} - 5\mathbf{k}$ (kip). Determine the moment of $\mathbf{F}$ about the line AB. Draw a sketch to indicate the direction of the moment.

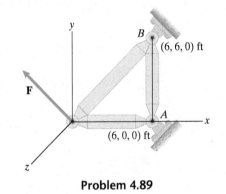

Problem 4.89

4.90 The force $\mathbf{F} = 10\mathbf{i} + 12\mathbf{j} - 6\mathbf{k}$ (N). What is the moment of $\mathbf{F}$ about the line AO? Draw a sketch to indicate the direction of the moment.

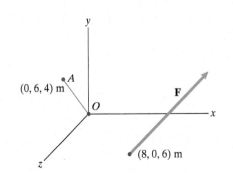

Problem 4.90

4.91 The tension in the cable AB is 1 kN. Determine the moment about the x axis due to the force exerted on the hatch by the cable at point B. Draw a sketch to indicate the direction of the moment.

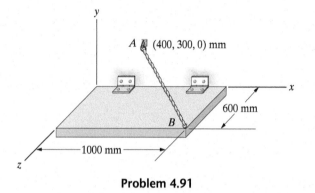

Problem 4.91

4.92 Determine the moment of the force applied at D about the straight line through the hinges A and B. (The line through A and B lies in the y–z plane.)

4.93 The tension in the cable CE is 160 lb. Determine the moment of the force exerted by the cable on the hatch at C about the straight line through the hinges A and B.

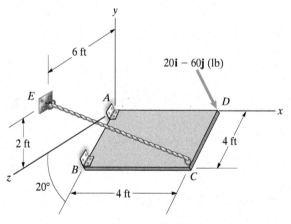

Problems 4.92/4.93

4.94 The coordinates of A are $(-2.4, 0, -0.6)$ m, and the coordinates of B are $(-2.2, 0.7, -1.2)$ m. The force exerted at B by the sailboat's main sheet AB is 130 N. Determine the moment of the force about the centerline of the mast (the y axis). Draw a sketch to indicate the direction of the moment.

Problem 4.94

4.95 The tension in cable AB is 200 lb. Determine the moments about each of the coordinate axes due to the force exerted on point B by the cable. Draw sketches to indicate the direction of the moments.

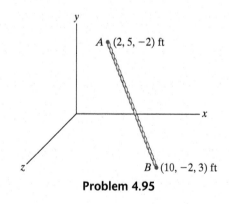

Problem 4.95

4.96 The total force exerted on the blades of the turbine by the steam nozzle is $\mathbf{F} = 20\mathbf{i} - 120\mathbf{j} + 100\mathbf{k}$ (N), and it effectively acts at the point (100, 80, 300) mm. What moment is exerted about the axis of the turbine (the x axis)?

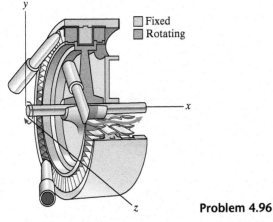

Fixed
Rotating

x

y

z

Problem 4.96

4.97 The pneumatic support AB holds a trunk lid in place. It exerts a 35-N force on the fixture at B that points in the direction from A toward B. Determine the magnitude of the moment of the force about the hinge axis of the lid, which is the z axis.

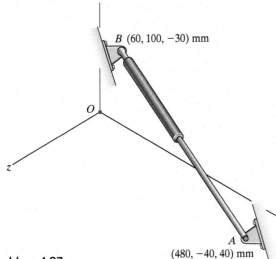

y

B (60, 100, −30) mm

O

z

x

A
(480, −40, 40) mm

Problem 4.97

4.98 The tension in cable AB is 80 lb. What is the moment about the line CD due to the force exerted by the cable on the wall at B?

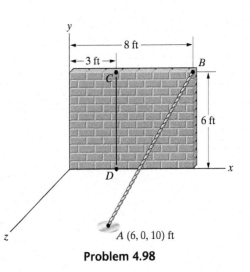

y

8 ft

3 ft

C

B

6 ft

D

x

z

A (6, 0, 10) ft

Problem 4.98

4.99 The magnitude of the force $\mathbf{F}$ is 0.2 N and its direction cosines are $\cos \theta_x = 0.727$, $\cos \theta_y = -0.364$, and $\cos \theta_z = 0.582$. Determine the magnitude of the moment of $\mathbf{F}$ about the axis AB of the spool.

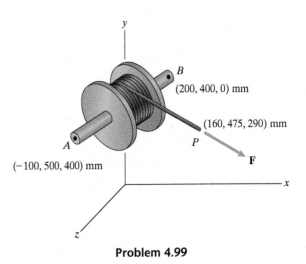

y

B
(200, 400, 0) mm

(160, 475, 290) mm

A

P

F

(−100, 500, 400) mm

x

z

Problem 4.99

▶ 4.100 A motorist applies the two forces shown to loosen a lug nut. The direction cosines of $\mathbf{F}$ are $\cos\theta_x = \frac{4}{13}$, $\cos\theta_y = \frac{12}{13}$, and $\cos\theta_z = \frac{3}{13}$. If the magnitude of the moment about the x axis must be 32 ft-lb to loosen the nut, what is the magnitude of the forces the motorist must apply? (See Example 4.8.)

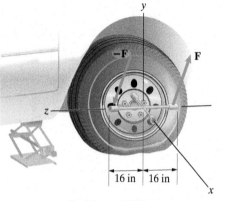

Problem 4.100

4.101 The tension in cable AB is 2 kN. What is the magnitude of the moment about the shaft CD due to the force exerted by the cable at A? Draw a sketch to indicate the direction of the moment about the shaft.

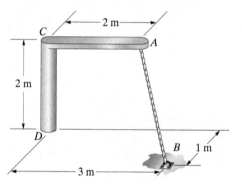

Problem 4.101

4.102 The axis of the car's wheel passes through the origin of the coordinate system and its direction cosines are $\cos\theta_x = 0.940$, $\cos\theta_y = 0$, $\cos\theta_z = 0.342$. The force exerted on the tire by the road effectively acts at the point $x = 0$, $y = -0.36$ m, $z = 0$ and has components $\mathbf{F} = -720\mathbf{i} + 3660\mathbf{j} + 1240\mathbf{k}$ (N). What is the moment of $\mathbf{F}$ about the wheel's axis?

Problem 4.102

4.103 The direction cosines of the centerline OA are $\cos\theta_x = 0.500$, $\cos\theta_y = 0.866$, and $\cos\theta_z = 0$, and the direction cosines of the line AG are $\cos\theta_x = 0.707$, $\cos\theta_y = 0.619$, and $\cos\theta_z = -0.342$. What is the moment about OA due to the 250-N weight? Draw a sketch to indicate the direction of the moment about the shaft.

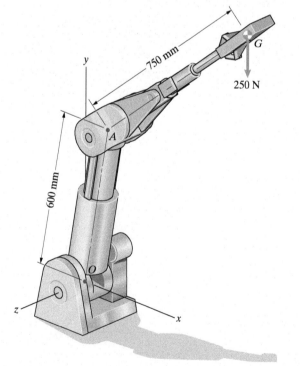

Problem 4.103

4.104 The radius of the steering wheel is 200 mm. The distance from O to C is 1 m. The center C of the steering wheel lies in the x–y plane. The driver exerts a force $\mathbf{F} = 10\mathbf{i} + 10\mathbf{j} - 5\mathbf{k}$ (N) on the wheel at A. If the angle $\alpha = 0$, what is the magnitude of the moment about the shaft OC? Draw a sketch to indicate the direction of the moment about the shaft.

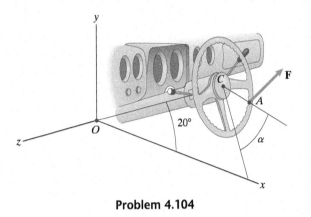

Problem 4.104

4.105* The magnitude of the force $\mathbf{F}$ is 10 N. Suppose that you want to choose the direction of the force $\mathbf{F}$ so that the magnitude of its moment about the line L is a maximum. Determine the components of $\mathbf{F}$ and the magnitude of its moment about L. (There are two solutions for $\mathbf{F}$.)

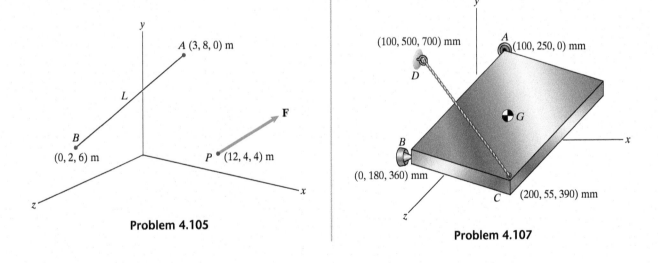

Problem 4.105

4.106 The weight W causes a tension of 100 lb in cable CD. If $d = 2$ ft, what is the moment about the z axis due to the force exerted by the cable CD at point C?

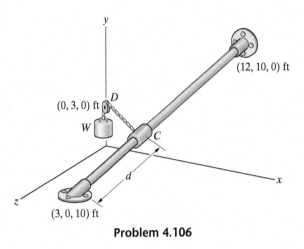

Problem 4.106

4.107* The y axis points upward. The weight of the 4-kg rectangular plate acts at the midpoint G of the plate. The sum of the moments about the straight line through the supports A and B due to the weight of the plate and the force exerted on the plate by the cable CD is zero. What is the tension in the cable?

Problem 4.107

4.4 Couples

Now that we have described how to calculate the moment due to a force, consider this question: Is it possible to exert a moment on an object without subjecting it to a net force? The answer is yes, and it occurs when a compact disk begins rotating or a screw is turned by a screwdriver. Forces are exerted on these objects, but in such a way that the net force is zero while the net moment is not zero.

Two forces that have equal magnitudes, opposite directions, and different lines of action are called a *couple* (Fig. 4.17a). A couple tends to cause rotation of an object even though the vector sum of the forces is zero, and it has the remarkable property that *the moment it exerts is the same about any point.*

The moment of a couple is simply the sum of the moments of the forces about a point P (Fig. 4.17b):

$$\mathbf{M} = [\mathbf{r}_1 \times \mathbf{F}] + [\mathbf{r}_2 \times (-\mathbf{F})] = (\mathbf{r}_1 - \mathbf{r}_2) \times \mathbf{F}.$$

The vector $\mathbf{r}_1 - \mathbf{r}_2$ is equal to the vector $\mathbf{r}$ shown in Fig. 4.17c, so we can express the moment as

$$\mathbf{M} = \mathbf{r} \times \mathbf{F}.$$

Since $\mathbf{r}$ doesn't depend on the position of P, the moment $\mathbf{M}$ is the same for *any* point P.

Because a couple exerts a moment but the sum of the forces is zero, it is often represented in diagrams simply by showing the moment (Fig. 4.17d). Like the Cheshire cat in *Alice's Adventures in Wonderland*, which vanished except for its grin, the forces don't appear; only the moment they exert is visible. But we recognize the origin of the moment by referring to it as a *moment of a couple,* or simply a *couple*.

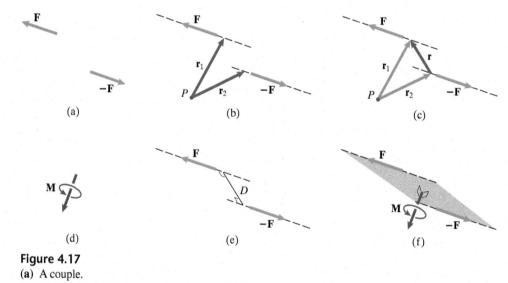

Figure 4.17
(a) A couple.
(b) Determining the moment about P.
(c) The vector $\mathbf{r} = \mathbf{r}_1 - \mathbf{r}_2$.
(d) Representing the moment of the couple.
(e) The distance D between the lines of action.
(f) $\mathbf{M}$ is perpendicular to the plane containing $\mathbf{F}$ and $-\mathbf{F}$.

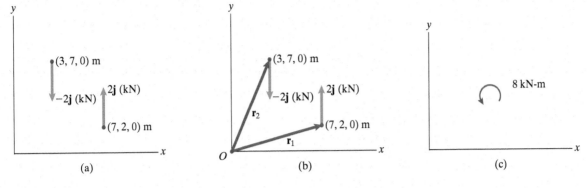

Figure 4.18
(a) A couple consisting of 2-kN forces.
(b) Determining the sum of the moments of the forces about O.
(c) Representing a couple in two dimensions.

Notice in Fig. 4.17c that $\mathbf{M} = \mathbf{r} \times \mathbf{F}$ is the moment of $\mathbf{F}$ about a point on the line of action of the force $-\mathbf{F}$. The magnitude of the moment of a force about a point equals the product of the magnitude of the force and the perpendicular distance from the point to the line of action of the force, so $|\mathbf{M}| = D|\mathbf{F}|$, where D is the perpendicular distance between the lines of action of the two forces (Fig. 4.17e). The cross product $\mathbf{r} \times \mathbf{F}$ is perpendicular to $\mathbf{r}$ and $\mathbf{F}$, which means that $\mathbf{M}$ is perpendicular to the plane containing $\mathbf{F}$ and $-\mathbf{F}$ (Fig. 4.17f). Pointing the thumb of the right hand in the direction of $\mathbf{M}$, the arc of the fingers indicates the direction of the moment.

In Fig. 4.18a, our view is perpendicular to the plane containing the two forces. The distance between the lines of action of the forces is 4 m, so the magnitude of the moment of the couple is $|\mathbf{M}| = (4\text{ m})(2\text{ kN}) = 8$ kN-m. The moment $\mathbf{M}$ is perpendicular to the plane containing the two forces. Pointing the arc of the fingers of the right hand counterclockwise, we find that the right-hand rule indicates that $\mathbf{M}$ points out of the page. Therefore, the moment of the couple is

$$\mathbf{M} = 8\mathbf{k} \text{ (kN-m)}.$$

We can also determine the moment of the couple by calculating the sum of the moments of the two forces about *any* point. The sum of the moments of the forces about the origin O is (Fig. 4.18b)

$$\mathbf{M} = [\mathbf{r}_1 \times (2\mathbf{j})] + [\mathbf{r}_2 \times (-2\mathbf{j})]$$
$$= [(7\mathbf{i} + 2\mathbf{j}) \times (2\mathbf{j})] + [(3\mathbf{i} + 7\mathbf{j}) \times (-2\mathbf{j})]$$
$$= 8\mathbf{k} \text{ (kN-m)}.$$

In a two-dimensional situation like this example, it isn't convenient to represent a couple by showing the moment vector, because the vector is perpendicular to the page. Instead, we represent the couple by showing its magnitude and a circular arrow that indicates its direction (Fig. 4.18c).

By grasping a bar and twisting it (Fig. 4.19a), a moment can be exerted about its axis (Fig. 4.19b). Although the system of forces exerted is distributed over the surface of the bar in a complicated way, the effect is the same as if two equal and opposite forces are exerted (Fig. 4.19c). When we represent a couple as in Fig. 4.19b, or by showing the moment vector $\mathbf{M}$, we imply that some system of forces exerts that moment. The system of forces (such as the forces exerted in twisting the bar, or the forces on the crankshaft that exert a moment on

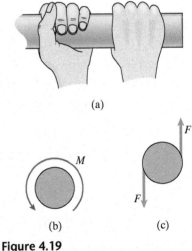

Figure 4.19
(a) Twisting a bar.
(b) The moment about the axis of the bar.
(c) The same effect is obtained by applying two equal and opposite forces.

the drive shaft of a car) is nearly always more complicated than two equal and opposite forces, but the effect is the same. For this reason, we can *model* the actual system as a simple system of two forces.

RESULTS

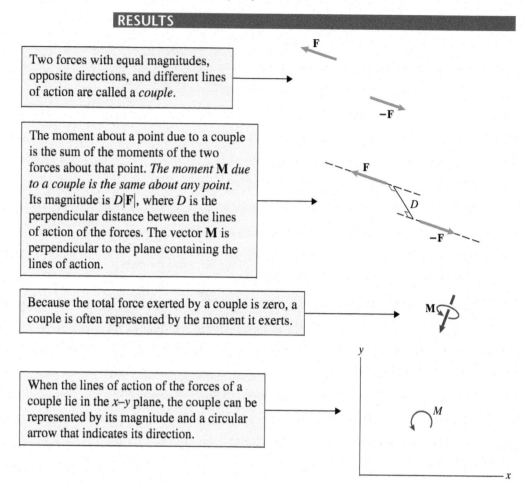

Two forces with equal magnitudes, opposite directions, and different lines of action are called a *couple*.

The moment about a point due to a couple is the sum of the moments of the two forces about that point. *The moment **M** due to a couple is the same about any point.* Its magnitude is $D|\mathbf{F}|$, where D is the perpendicular distance between the lines of action of the forces. The vector **M** is perpendicular to the plane containing the lines of action.

Because the total force exerted by a couple is zero, a couple is often represented by the moment it exerts.

When the lines of action of the forces of a couple lie in the x–y plane, the couple can be represented by its magnitude and a circular arrow that indicates its direction.

Active Example 4.9   Moment of a Couple (▶ *Related Problem 4.108*)

The force $\mathbf{F} = 10\mathbf{i} - 4\mathbf{j}$ (N). Determine the moment due to the couple. Represent the moment by its magnitude and a circular arrow indicating its direction.

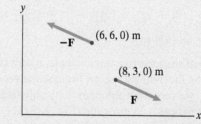

Strategy

We will determine the moment in two ways. In the first method, we will choose a point and calculate the sum of the moments of the two forces about that point. Because the moment due to a couple is the same about any point, we can choose any convenient point. In the second method, we will sum the moments of the two couples formed by the x and y components of the forces.

Solution

First Method

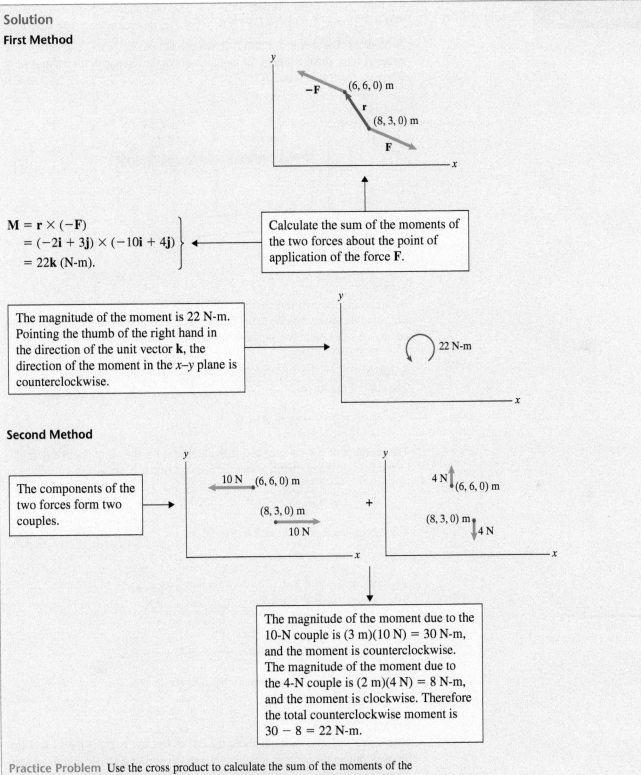

$$\mathbf{M} = \mathbf{r} \times (-\mathbf{F})$$
$$= (-2\mathbf{i} + 3\mathbf{j}) \times (-10\mathbf{i} + 4\mathbf{j})$$
$$= 22\mathbf{k} \text{ (N-m)}.$$

Calculate the sum of the moments of the two forces about the point of application of the force **F**.

The magnitude of the moment is 22 N-m. Pointing the thumb of the right hand in the direction of the unit vector **k**, the direction of the moment in the x–y plane is counterclockwise.

22 N-m

Second Method

The components of the two forces form two couples.

10 N   (6, 6, 0) m

(8, 3, 0) m
10 N

+

4 N (6, 6, 0) m

(8, 3, 0) m
4 N

The magnitude of the moment due to the 10-N couple is (3 m)(10 N) = 30 N-m, and the moment is counterclockwise. The magnitude of the moment due to the 4-N couple is (2 m)(4 N) = 8 N-m, and the moment is clockwise. Therefore the total counterclockwise moment is 30 − 8 = 22 N-m.

Practice Problem Use the cross product to calculate the sum of the moments of the forces **F** and −**F** about the point *P* with coordinates (10, 7, 3) m. Represent the moment by its magnitude and a circular arrow indicating its direction.

Answer: 22**k** (N-m), or 22 N-m counterclockwise.

Example 4.10	Determining Unknown Forces (▶ *Related Problem 4.113*)

Two forces A and B and a 200 ft-lb couple act on the beam. The sum of the forces is zero, and the sum of the moments about the left end of the beam is zero. What are the forces A and B?

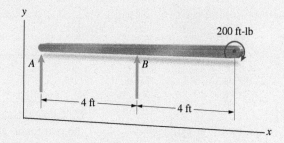

Strategy

By summing the two forces (the couple exerts no net force on the beam) and summing the moments due to the forces and the couple about the left end of the beam, we will obtain two equations in terms of the two unknown forces.

Solution

The sum of the forces is

$$\Sigma F_y = A + B = 0.$$

The moment of the couple (200 ft-lb clockwise) is the same about any point, so the sum of the moments about the left end of the beam is

$$\Sigma M_{\text{left end}} = (4 \text{ ft}) B - 200 \text{ ft-lb} = 0.$$

The forces are $B = 50$ lb and $A = -50$ lb.

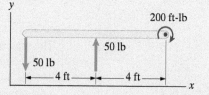

The forces on the beam form a couple.

Critical Thinking

Notice that the total moment about the left end of the beam is the sum of the moment due to the force B and the moment due to the 200 ft-lb couple. As we observe in Chapter 5, if an object subjected to forces and couples is in equilibrium, the sum of the forces is zero and the sum of the moments about any point, *including moments due to couples*, is zero. In this example we needed both these conditions to determine the unknown forces A and B.

Example 4.11   **Sum of the Moments Due to Two Couples** (▶ *Related Problem 4.119*)

Determine the sum of the moments exerted on the pipe by the two couples.

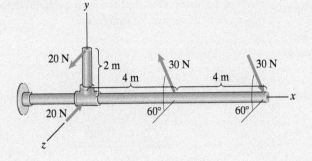

Strategy
We will express the moment exerted by each couple as a vector. To express the 30-N couple in terms of a vector, we will express the forces in terms of their components. We can then sum the moment vectors to determine the sum of the moments exerted by the couples.

Solution
Consider the 20-N couple. The magnitude of the moment of the couple is $(2 \text{ m})(20 \text{ N}) = 40$ N-m. The direction of the moment vector is perpendicular to the y–z plane, and the right-hand rule indicates that it points in the positive x axis direction. The moment of the 20-N couple is $40\mathbf{i}$ (N-m).

By resolving the 30-N forces into y and z components, we obtain the two couples in Fig. a. The moment of the couple formed by the y components is $-(30 \sin 60°)(4)\mathbf{k}$ (N-m), and the moment of the couple formed by the z components is $(30 \cos 60°)(4)\mathbf{j}$ (N-m).

The sum of the moments is therefore

$$\Sigma\mathbf{M} = 40\mathbf{i} + (30 \cos 60°)(4)\mathbf{j} - (30 \sin 60°)(4)\mathbf{k} \text{ (N-m)}$$

$$= 40\mathbf{i} + 60\mathbf{j} - 104\mathbf{k} \text{ (N-m)}.$$

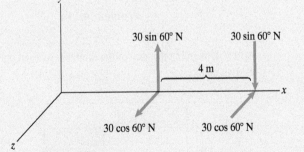

(a) Resolving the 30-N forces into y and z components.

Critical Thinking
Although the method we used in this example helps you recognize the contributions of the individual couples to the sum of the moments, it is convenient only when the orientations of the forces and their points of application relative to the coordinate system are fairly simple. When that is not the case, you can determine the sum of the moments by choosing any point and calculating the sum of the moments of the forces about that point.

Problems

▶ **4.108** In Active Example 4.9, suppose that the point of application of the force **F** is moved from (8, 3, 0) m to (8, 8, 0) m. Draw a sketch showing the new position of the force. From your sketch, will the moment due to the couple be clockwise or counterclockwise? Calculate the moment due to the couple. Represent the moment by its magnitude and a circular arrow indicating its direction.

4.109 The forces are contained in the x–y plane.

(a) Determine the moment of the couple and represent it as shown in Fig. 4.18c.

(b) What is the sum of the moments of the two forces about the point $(10, -40, 20)$ ft?

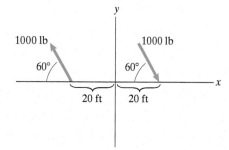

Problem 4.109

4.110 The moment of the couple is 600**k** (N-m). What is the angle α?

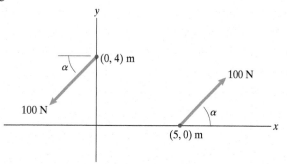

Problem 4.110

4.111 Point P is contained in the x–y plane, $|\mathbf{F}| = 100$ N, and the moment of the couple is $-500\mathbf{k}$ (N-m). What are the coordinates of P?

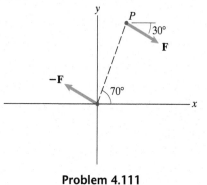

Problem 4.111

4.112 Three forces of equal magnitude are applied parallel to the sides of an equilateral triangle. (a) Show that the sum of the moments of the forces is the same about any point. (b) Determine the magnitude of the sum of the moments.

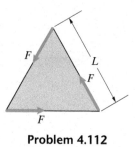

Problem 4.112

▶ **4.113** In Example 4.10, suppose that the 200 ft-lb couple is counterclockwise instead of clockwise. Draw a sketch of the beam showing the forces and couple acting on it. What are the forces A and B?

4.114 The moments of two couples are shown. What is the sum of the moments about point P?

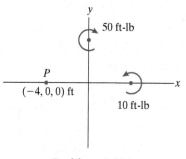

Problem 4.114

4.115 Determine the sum of the moments exerted on the plate by the two couples.

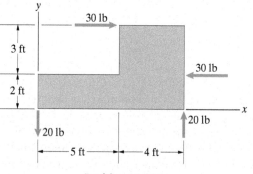

Problem 4.115

4.116 Determine the sum of the moments exerted about A by the couple and the two forces.

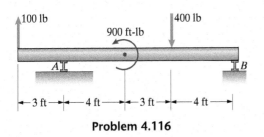

100 lb

400 lb

900 ft-lb

A   B

←3 ft→←4 ft→←3 ft→←4 ft→

Problem 4.116

4.117 Determine the sum of the moments exerted about A by the couple and the two forces.

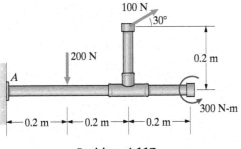

100 N

30°

200 N

0.2 m

A

300 N-m

←0.2 m→←0.2 m→←0.2 m→

Problem 4.117

4.118 The sum of the moments about point A due to the forces and couples acting on the bar is zero.

(a) What is the magnitude of the couple C?

(b) Determine the sum of the moments about point B due to the forces and couples acting on the bar.

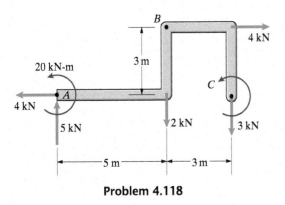

B

4 kN

3 m

20 kN-m

C

A

4 kN

5 kN

2 kN

3 kN

←5 m→←3 m→

Problem 4.118

▶ **4.119** In Example 4.11, suppose that instead of acting in the positive z axis direction, the upper 20-N force acts in the positive x axis direction. Instead of acting in the negative z axis direction, let the lower 20-N force act in the negative x axis direction. Draw a sketch of the pipe showing the forces acting on it. Determine the sum of the moments exerted on the pipe by the two couples.

4.120 (a) What is the moment of the couple?

(b) Determine the perpendicular distance between the lines of action of the two forces.

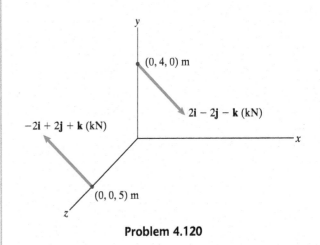

y

(0, 4, 0) m

2i − 2j − k (kN)

−2i + 2j + k (kN)

x

(0, 0, 5) m

z

Problem 4.120

4.121 Determine the sum of the moments exerted on the plate by the three couples. (The 80-lb forces are contained in the x–z plane.)

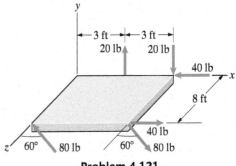

y

←3 ft→←3 ft→

20 lb   20 lb

40 lb

x

8 ft

40 lb

z 60°   80 lb   60°   80 lb

Problem 4.121

4.122 What is the magnitude of the sum of the moments exerted on the T-shaped structure by the two couples?

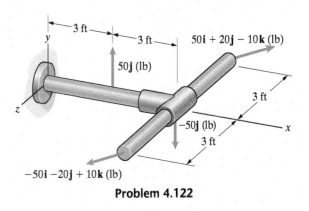

←3 ft→

←3 ft→

50i + 20j − 10k (lb)

50j (lb)

y

3 ft

z

−50j (lb)

x

3 ft

−50i − 20j + 10k (lb)

Problem 4.122

4.123 The tension in cables AB and CD is 500 N.

(a) Show that the two forces exerted by the cables on the rectangular hatch at B and C form a couple.

(b) What is the moment exerted on the plate by the cables?

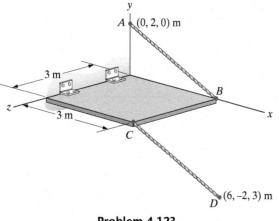

Problem 4.123

4.124 The cables AB and CD exert a couple on the vertical pipe. The tension in each cable is 8 kN. Determine the magnitude of the moment the cables exert on the pipe.

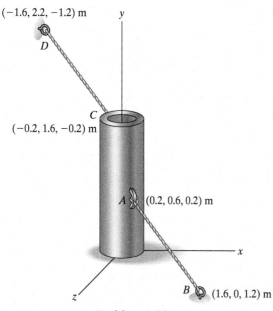

Problem 4.124

4.125 The bar is loaded by the forces
$$\mathbf{F}_B = 2\mathbf{i} + 6\mathbf{j} + 3\mathbf{k} \ (\text{kN}),$$
$$\mathbf{F}_C = \mathbf{i} - 2\mathbf{j} + 2\mathbf{k} \ (\text{kN}),$$
and the couple
$$\mathbf{M}_C = 2\mathbf{i} + \mathbf{j} - 2\mathbf{k} \ (\text{kN-m}).$$
Determine the sum of the moments of the two forces and the couple about A.

4.126 The forces
$$\mathbf{F}_B = 2\mathbf{i} + 6\mathbf{j} + 3\mathbf{k} \ (\text{kN}),$$
$$\mathbf{F}_C = \mathbf{i} - 2\mathbf{j} + 2\mathbf{k} \ (\text{kN}),$$
and the couple
$$\mathbf{M}_C = M_{Cy}\mathbf{j} + M_{Cz}\mathbf{k} \ (\text{kN-m}).$$
Determine the values of M_{Cy} and M_{Cz} so that the sum of the moments of the two forces and the couple about A is zero.

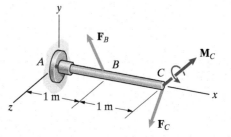

Problems 4.125/4.126

4.127 Two wrenches are used to tighten an elbow fitting. The force $\mathbf{F} = 10\mathbf{k}$ (lb) on the right wrench is applied at $(6, -5, -3)$ in, and the force $-\mathbf{F}$ on the left wrench is applied at $(4, -5, 3)$ in.

(a) Determine the moment about the x axis due to the force exerted on the right wrench.

(b) Determine the moment of the couple formed by the forces exerted on the two wrenches.

(c) Based on the results of (a) and (b), explain why two wrenches are used.

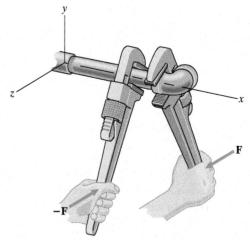

Problem 4.127

4.5 Equivalent Systems

BACKGROUND

A *system of forces and moments* is simply a particular set of forces and moments of couples. The systems of forces and moments dealt with in engineering can be complicated. This is especially true in the case of distributed forces, such as the pressure forces exerted by water on a dam. Fortunately, if we are concerned only with the total force and moment exerted, we can represent complicated systems of forces and moments by much simpler systems.

Conditions for Equivalence

We define two systems of forces and moments, designated as system 1 and system 2, to be *equivalent* if the sums of the forces are equal, or

$$(\Sigma \mathbf{F})_1 = (\Sigma \mathbf{F})_2, \tag{4.7}$$

and the sums of the moments about a point P are equal, or

$$(\Sigma \mathbf{M}_P)_1 = (\Sigma \mathbf{M}_P)_2. \tag{4.8}$$

To see what the conditions for equivalence mean, consider the systems of forces and moments in Fig. 4.20a. In system 1, an object is subjected to two forces $\mathbf{F}_A$ and $\mathbf{F}_B$ and a couple $\mathbf{M}_C$. In system 2, the object is subjected to a force $\mathbf{F}_D$ and two couples $\mathbf{M}_E$ and $\mathbf{M}_F$. The first condition for equivalence is

$$(\Sigma \mathbf{F})_1 = (\Sigma \mathbf{F})_2:$$
$$\mathbf{F}_A + \mathbf{F}_B = \mathbf{F}_D. \tag{4.9}$$

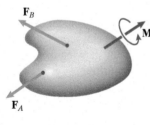
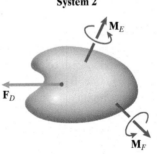

(a)

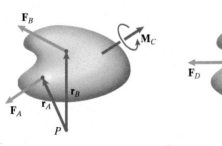

(b)

Figure 4.20
(a) Different systems of forces and moments applied to an object.
(b) Determining the sum of the moments about a point P for each system.

If we determine the sums of the moments about the point P in Fig. 4.20b, the second condition for equivalence is

$$(\Sigma M_P)_1 = (\Sigma M_P)_2:$$

$$(\mathbf{r}_A \times \mathbf{F}_A) + (\mathbf{r}_B \times \mathbf{F}_B) + \mathbf{M}_C = (\mathbf{r}_D \times \mathbf{F}_D) + \mathbf{M}_E + \mathbf{M}_F. \quad (4.10)$$

If these conditions are satisfied, systems 1 and 2 are equivalent.

We will use this example to demonstrate that *if the sums of the forces are equal for two systems of forces and moments and the sums of the moments about one point P are equal, then the sums of the moments about any point are equal.* Suppose that Eq. (4.9) is satisfied, and Eq. (4.10) is satisfied for the point P in Fig. 4.20b. For a different point P' (Fig. 4.21), we will show that

$$(\Sigma M_{P'})_1 = (\Sigma M_{P'})_2:$$

$$(\mathbf{r}'_A \times \mathbf{F}_A) + (\mathbf{r}'_B \times \mathbf{F}_B) + \mathbf{M}_C = (\mathbf{r}'_D \times \mathbf{F}_D) + \mathbf{M}_E + \mathbf{M}_F. \quad (4.11)$$

In terms of the vector $\mathbf{r}$ from P' to P, the relations between the vectors $\mathbf{r}'_A$, $\mathbf{r}'_B$, and $\mathbf{r}'_D$ in Fig. 4.21 and the vectors $\mathbf{r}_A$, $\mathbf{r}_B$, and $\mathbf{r}_D$ in Fig. 4.20b are

$$\mathbf{r}'_A = \mathbf{r} + \mathbf{r}_A, \qquad \mathbf{r}'_B = \mathbf{r} + \mathbf{r}_B, \qquad \mathbf{r}'_D = \mathbf{r} + \mathbf{r}_D.$$

Substituting these expressions into Eq. (4.11), we obtain

$$[(\mathbf{r} + \mathbf{r}_A) \times \mathbf{F}_A] + [(\mathbf{r} + \mathbf{r}_B) \times \mathbf{F}_B] + \mathbf{M}_C$$
$$= [(\mathbf{r} + \mathbf{r}_D) \times \mathbf{F}_D] + \mathbf{M}_E + \mathbf{M}_F.$$

Rearranging terms, we can write this equation as

$$[\mathbf{r} \times (\Sigma \mathbf{F})_1] + (\Sigma \mathbf{M}_P)_1 = [\mathbf{r} \times (\Sigma \mathbf{F})_2] + (\Sigma \mathbf{M}_P)_2,$$

which holds in view of Eqs. (4.9) and (4.10). The sums of the moments of the two systems about any point are equal.

Representing Systems by Equivalent Systems

If we are concerned only with the total force and total moment exerted on an object by a given system of forces and moments, we can *represent* the system by an equivalent one. By this we mean that instead of showing the actual forces and couples acting on an object, we would show a different system that exerts the same total force and moment. In this way, we can replace a given system by a less complicated one to simplify the analysis of the forces and moments acting on an object and to gain a better intuitive understanding of their effects on the object.

Representing a System by a Force and a Couple   Let us consider an arbitrary system of forces and moments and a point P (system 1 in Fig. 4.22). We can represent this system by one consisting of a single force acting at P and a single couple (system 2). The conditions for equivalence are

$$(\Sigma \mathbf{F})_2 = (\Sigma \mathbf{F})_1:$$

$$\mathbf{F} = (\Sigma \mathbf{F})_1$$

and

$$(\Sigma \mathbf{M}_P)_2 = (\Sigma \mathbf{M}_P)_1:$$

$$\mathbf{M} = (\Sigma \mathbf{M}_P)_1.$$

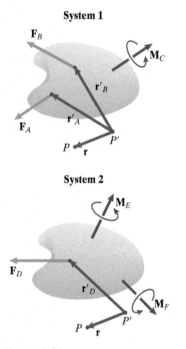

System 1

System 2

Figure 4.21
Determining the sums of the moments about a different point P'.

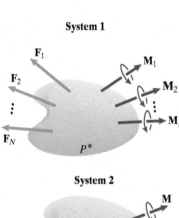

System 1

System 2

Figure 4.22
(a) An arbitrary system of forces and moments.
(b) A force acting at P and a couple.

These conditions are satisfied if **F** equals the sum of the forces in system 1 and **M** equals the sum of the moments about P in system 1.

Thus *no matter how complicated a system of forces and moments may be, it can be represented by a single force acting at a given point and a single couple.*

Representing a Force by a Force and a Couple

A force $\mathbf{F}_P$ acting at a point P (system 1 in Fig. 4.23a) can be represented by a force **F** acting at a different point Q and a couple **M** (system 2). The moment of system 1 about point Q is $\mathbf{r} \times \mathbf{F}_P$, where **r** is the vector from Q to P (Fig. 4.23b). The conditions for equivalence are

$$(\Sigma \mathbf{F})_2 = (\Sigma \mathbf{F})_1:$$
$$\mathbf{F} = \mathbf{F}_P$$

and

$$(\Sigma \mathbf{M}_Q)_2 = (\Sigma \mathbf{M}_Q)_1:$$
$$\mathbf{M} = \mathbf{r} \times \mathbf{F}_P.$$

The systems are equivalent if the force **F** equals the force $\mathbf{F}_P$ and the couple **M** equals the moment of $\mathbf{F}_P$ about Q.

Concurrent Forces Represented by a Force

A system of concurrent forces whose lines of action intersect at a point P (system 1 in Fig. 4.24) can be represented by a single force whose line of action intersects P (system 2). The sums of the forces in the two systems are equal if

$$\mathbf{F} = \mathbf{F}_1 + \mathbf{F}_2 + \cdots + \mathbf{F}_N.$$

The sum of the moments about P equals zero for each system, so the systems are equivalent if the force **F** equals the sum of the forces in system 1.

Parallel Forces Represented by a Force

A system of parallel forces whose sum is not zero can be represented by a single force **F** (Fig. 4.25). We demonstrate this result in Example 4.14.

Representing a System by a Wrench

We have shown that *any* system of forces and moments can be represented by a single force acting at a given point and a single couple. This raises an interesting question: What is the simplest system that can be equivalent to any system of forces and moments?

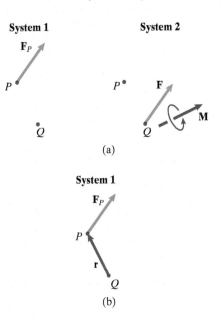

System 1   **System 2**

(a)

System 1

(b)

Figure 4.23
(a) System 1 is a force $\mathbf{F}_P$ acting at point P. System 2 consists of a force **F** acting at point Q and a couple **M**.
(b) Determining the moment of system 1 about point Q.

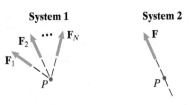

System 1   **System 2**

Figure 4.24
A system of concurrent forces and a system consisting of a single force **F**.

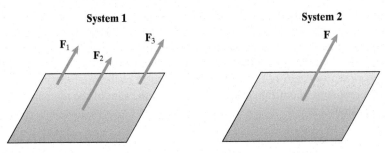

System 1   **System 2**

Figure 4.25
A system of parallel forces and a system consisting of a single force **F**.

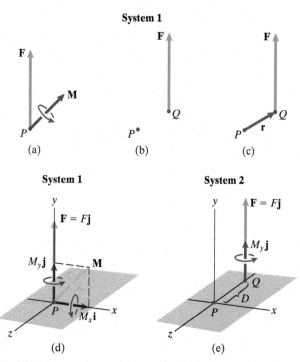

Figure 4.26

(a) System 1 is a single force and a single couple.

(b) Can system 1 be represented by a single force and no couple?

(c) The moment of $\mathbf{F}$ about P is $\mathbf{r} \times \mathbf{F}$.

(d) $\mathbf{F}$ is along the y axis, and $\mathbf{M}$ is contained in the x–y plane.

(e) System 2 is the force $\mathbf{F}$ and the component of $\mathbf{M}$ parallel to $\mathbf{F}$.

To consider this question, let us begin with an arbitrary force $\mathbf{F}$ acting at a point P and an arbitrary couple $\mathbf{M}$ (system 1 in Fig. 4.26a) and see whether we can represent this system by a simpler one. For example, can we represent it by the force $\mathbf{F}$ acting at a different point Q and no couple (Fig. 4.26b)? The sum of the forces is the same as in system 1. If we can choose the point Q so that $\mathbf{r} \times \mathbf{F} = \mathbf{M}$, where $\mathbf{r}$ is the vector from P to Q (Fig. 4.26c), the sum of the moments about P is the same as in system 1 and the systems are equivalent. But the vector $\mathbf{r} \times \mathbf{F}$ is perpendicular to $\mathbf{F}$, so it can equal $\mathbf{M}$ only if $\mathbf{M}$ is perpendicular to $\mathbf{F}$. That means that, in general, we can't represent system 1 by the force $\mathbf{F}$ alone.

However, we can represent system 1 by the force $\mathbf{F}$ acting at a point Q and the component of $\mathbf{M}$ that is parallel to $\mathbf{F}$. Figure 4.26d shows system 1 with a coordinate system placed so that $\mathbf{F}$ is along the y axis and $\mathbf{M}$ is contained in the x–y plane. In terms of this coordinate system, we can express the force and couple as $\mathbf{F} = F\mathbf{j}$ and $\mathbf{M} = M_x\mathbf{i} + M_y\mathbf{j}$. System 2 in Fig. 4.26e consists of the force $\mathbf{F}$ acting at a point on the z axis and the component of $\mathbf{M}$ parallel to $\mathbf{F}$. If we choose the distance D so that $D = M_x/F$, system 2 is equivalent to system 1. The sum of the forces in each system is $\mathbf{F}$. The sum of the moments about P in system 1 is $\mathbf{M}$, and the sum of the moments about P in system 2 is

$$(\Sigma \mathbf{M}_P)_2 = [(-D\mathbf{k}) \times (F\mathbf{j})] + M_y\mathbf{j} = M_x\mathbf{i} + M_y\mathbf{j} = \mathbf{M}.$$

A force $\mathbf{F}$ and a couple $\mathbf{M}_p$ that is parallel to $\mathbf{F}$ is called a *wrench. It is the simplest system that can be equivalent to an arbitrary system of forces and moments.*

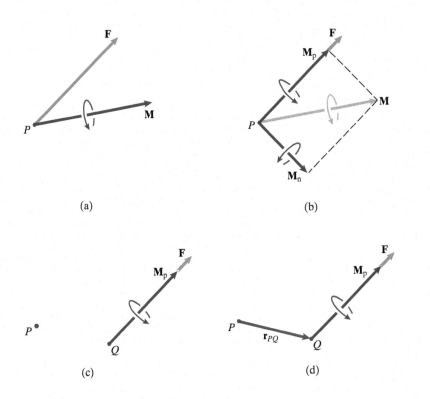

Figure 4.27
(a) If necessary, first represent the system by a single force and a single couple.
(b) The components of $\mathbf{M}$ parallel and normal to $\mathbf{F}$.
(c) The wrench.
(d) Choose Q so that the moment of $\mathbf{F}$ about P equals the normal component of $\mathbf{M}$.

How can we represent a given system of forces and moments by a wrench? If the system is a single force or a single couple or if it consists of a force $\mathbf{F}$ and a couple that is parallel to $\mathbf{F}$, it is a wrench, and we can't simplify it further. If the system is more complicated than a single force and a single couple, we can begin by choosing a convenient point P and representing the system by a force $\mathbf{F}$ acting at P and a couple $\mathbf{M}$ (Fig. 4.27a). Then representing this system by a wrench requires two steps:

1. Determine the components of $\mathbf{M}$ parallel and normal to $\mathbf{F}$ (Fig. 4.27b).
2. The wrench consists of the force $\mathbf{F}$ acting at a point Q and the parallel component $\mathbf{M}_P$ (Fig. 4.27c). To achieve equivalence, the point Q must be chosen so that the moment of $\mathbf{F}$ about P equals the normal component $\mathbf{M}_n$ (Fig. 4.27d)—that is, so that $\mathbf{r}_{PQ} \times \mathbf{F} = \mathbf{M}_n$.

RESULTS

Equivalent Systems of Forces and Moments

A *system of forces and moments* is simply a particular set of forces and moments due to couples. We define two systems of forces and moments, designated as system 1 and system 2, to be *equivalent* if two conditions are satisfied:

1. The sum of the forces in system 1 is equal to the sum of the forces in system 2.
2. The sum of the moments about any point P due to the forces and moments in system 1 is equal to the sum of the moments about *the same point P* due to the forces and moments in system 2.

Representing Systems of Forces and Moments by Equivalent Systems

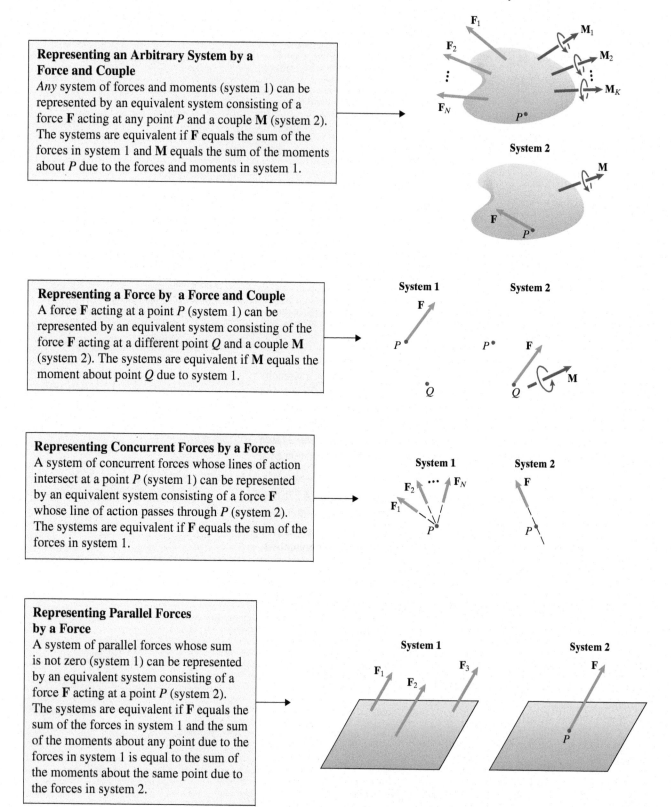

System 1

System 2

Representing an Arbitrary System by a Force and Couple

Any system of forces and moments (system 1) can be represented by an equivalent system consisting of a force **F** acting at any point P and a couple **M** (system 2). The systems are equivalent if **F** equals the sum of the forces in system 1 and **M** equals the sum of the moments about P due to the forces and moments in system 1.

Representing a Force by a Force and Couple

A force **F** acting at a point P (system 1) can be represented by an equivalent system consisting of the force **F** acting at a different point Q and a couple **M** (system 2). The systems are equivalent if **M** equals the moment about point Q due to system 1.

Representing Concurrent Forces by a Force

A system of concurrent forces whose lines of action intersect at a point P (system 1) can be represented by an equivalent system consisting of a force **F** whose line of action passes through P (system 2). The systems are equivalent if **F** equals the sum of the forces in system 1.

Representing Parallel Forces by a Force

A system of parallel forces whose sum is not zero (system 1) can be represented by an equivalent system consisting of a force **F** acting at a point P (system 2). The systems are equivalent if **F** equals the sum of the forces in system 1 and the sum of the moments about any point due to the forces in system 1 is equal to the sum of the moments about the same point due to the forces in system 2.

Active Example 4.12   (▶ *Related Problem 4.151*)

System 1 consists of the following forces and couples:

$$\mathbf{F}_A = -10\mathbf{i} + 10\mathbf{j} - 15\mathbf{k} \text{ (kN)},$$
$$\mathbf{F}_B = 30\mathbf{i} + 5\mathbf{j} + 10\mathbf{k} \text{ (kN)},$$
$$\mathbf{M}_C = -90\mathbf{i} + 150\mathbf{j} + 60\mathbf{k} \text{ (kN-m)}.$$

Suppose that you want to represent system 1 by an equivalent system consisting of a force $\mathbf{F}$ acting at the point P with coordinates $(4, 3, -2)$ m and a couple $\mathbf{M}$ (system 2). Determine $\mathbf{F}$ and $\mathbf{M}$.

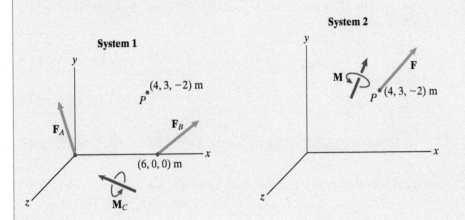

Strategy

The conditions for equivalence are satisfied if $\mathbf{F}$ equals the sum of the forces in system 1 and $\mathbf{M}$ equals the sum of the moments about point P due to the forces and moments in system 1. We can use these conditions to determine $\mathbf{F}$ and $\mathbf{M}$.

Solution

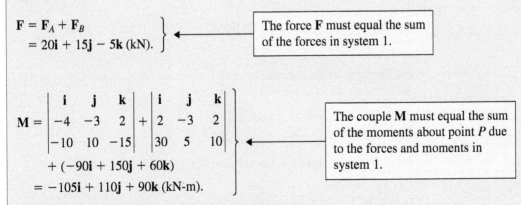

$$\mathbf{F} = \mathbf{F}_A + \mathbf{F}_B$$
$$= 20\mathbf{i} + 15\mathbf{j} - 5\mathbf{k} \text{ (kN).}$$

> The force $\mathbf{F}$ must equal the sum of the forces in system 1.

$$\mathbf{M} = \begin{vmatrix} \mathbf{i} & \mathbf{j} & \mathbf{k} \\ -4 & -3 & 2 \\ -10 & 10 & -15 \end{vmatrix} + \begin{vmatrix} \mathbf{i} & \mathbf{j} & \mathbf{k} \\ 2 & -3 & 2 \\ 30 & 5 & 10 \end{vmatrix}$$
$$+ (-90\mathbf{i} + 150\mathbf{j} + 60\mathbf{k})$$
$$= -105\mathbf{i} + 110\mathbf{j} + 90\mathbf{k} \text{ (kN-m).}$$

> The couple $\mathbf{M}$ must equal the sum of the moments about point P due to the forces and moments in system 1.

Practice Problem Suppose that you want to represent system 2 by an equivalent system consisting of a force $\mathbf{F}'$ acting at the origin of the coordinate system and a couple $\mathbf{M}'$ (system 3). Determine $\mathbf{F}'$ and $\mathbf{M}'$.

Answer: $\mathbf{F}' = 20\mathbf{i} + 15\mathbf{j} - 5\mathbf{k}$ (kN), $\mathbf{M}' = -90\mathbf{i} + 90\mathbf{j} + 90\mathbf{k}$ (kN-m).

Example 4.13 Representing a System by a Simpler Equivalent System (▶ *Related Problem 4.137*)

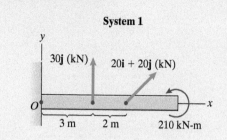

System 1

System 1 consists of two forces and a couple acting on a pipe. Represent system 1 by (a) a single force acting at the origin O of the coordinate system and a single couple and (b) a single force.

Strategy
(a) We can represent system 1 by a force $\mathbf{F}$ acting at the origin and a couple M (system 2 in Fig. a) and use the conditions for equivalence to determine $\mathbf{F}$ and $\mathbf{M}$.
(b) Suppose that we place the force $\mathbf{F}$ with its point of application a distance D along the x axis (system 3 in Fig. b). The sums of the forces in systems 2 and 3 are equal. If we can choose the distance D so that the moment about O in system 3 equals $\mathbf{M}$, system 3 will be equivalent to system 2 and therefore equivalent to system 1.

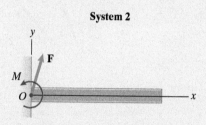

System 2

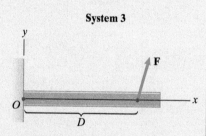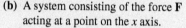

System 3

(a) A force $\mathbf{F}$ acting at O and a couple M.

(b) A system consisting of the force $\mathbf{F}$ acting at a point on the x axis.

Solution
(a) The conditions for equivalence are

$$(\Sigma \mathbf{F})_2 = (\Sigma \mathbf{F})_1:$$
$$\mathbf{F} = 30\mathbf{j} + (20\mathbf{i} + 20\mathbf{j}) \ (\text{kN}) = 20\mathbf{i} + 50\mathbf{j} \ (\text{kN}),$$

and

$$(\Sigma M_O)_2 = (\Sigma M_O)_1:$$
$$M = (30 \ \text{kN})(3 \ \text{m}) + (20 \ \text{kN})(5 \ \text{m}) + 210 \ \text{kN-m}$$
$$= 400 \ \text{kN-m}.$$

(b) The sums of the forces in systems 2 and 3 are equal. Equating the sums of the moments about O yields

$$(\Sigma M_O)_3 = (\Sigma M_O)_2:$$
$$(50 \ \text{kN})D = 400 \ \text{kN-m},$$

and we find that system 3 is equivalent to system 2 if $D = 8$ m.

Critical Thinking
In part (b), why did we assume that the point of application of the force is on the x axis? In order to represent the system in Fig. a by a single force, we needed to place the line of action of the force so that the force would exert a 400 kN-m counterclockwise moment about O. Placing the point of application of the force a distance D along the x axis was simply a convenient way to accomplish that.

Example 4.14   Representing Parallel Forces by a Single Force (▶ *Related Problem 4.154*)

System 1 consists of parallel forces. Suppose you want to represent it by a force
F (system 2). What is **F**, and where does its line of action intersect the x–z plane?

Strategy
We can determine **F** from the condition that the sums of the forces in the two systems must be equal. For the two systems to be equivalent, we must choose the point of application P so that the sums of the moments about a point are equal. This condition will tell us where the line of action intersects the x–z plane.

Solution
The sums of the forces must be equal.

$$(\Sigma \mathbf{F})_2 = (\Sigma \mathbf{F})_1:$$
$$\mathbf{F} = 30\mathbf{j} + 20\mathbf{j} - 10\mathbf{j} \ (\text{lb}) = 40\mathbf{j} \ (\text{lb}).$$

The sums of the moments about an arbitrary point must be equal: Let the coordinates of point P be (x, y, z). The sums of the moments about the origin O must be equal.

$$(\Sigma \mathbf{M}_O)_2 = (\Sigma \mathbf{M}_O)_1:$$

$$\begin{vmatrix} \mathbf{i} & \mathbf{j} & \mathbf{k} \\ x & y & z \\ 0 & 40 & 0 \end{vmatrix} = \begin{vmatrix} \mathbf{i} & \mathbf{j} & \mathbf{k} \\ 6 & 0 & 2 \\ 0 & 30 & 0 \end{vmatrix} + \begin{vmatrix} \mathbf{i} & \mathbf{j} & \mathbf{k} \\ 2 & 0 & 4 \\ 0 & -10 & 0 \end{vmatrix} + \begin{vmatrix} \mathbf{i} & \mathbf{j} & \mathbf{k} \\ -3 & 0 & -2 \\ 0 & 20 & 0 \end{vmatrix}$$

Expanding the determinants, we obtain

$$[20 \text{ ft-lb} + (40 \text{ lb})z]\mathbf{i} + [100 \text{ ft-lb} - (40 \text{ lb})x]\mathbf{k} = \mathbf{0}.$$

The sums of the moments about the origin are equal if

$$x = 2.5 \text{ ft},$$
$$z = -0.5 \text{ ft}.$$

The systems are equivalent if $\mathbf{F} = 40\mathbf{j}$ (lb) and its line of action intersects the x–z plane at $x = 2.5$ ft and $z = -0.5$ ft. Notice that we did not obtain an equation for the y coordinate of P. The systems are equivalent if **F** is applied at any point along the line of action.

Critical Thinking
In this example we could have determined the x and z coordinates of point P in a simpler way. Since the sums of the moments about any point must be equal for the systems to be equivalent, the sums of the moments about any *line* must also be equal. Equating the sums of the moments about the x axis yields

$$(\Sigma M_{x \text{ axis}})_2 = (\Sigma M_{x \text{ axis}})_1:$$
$$-(40 \text{ lb})z = -(30 \text{ lb})(2 \text{ ft}) + (10 \text{ lb})(4 \text{ ft}) + (20 \text{ lb})(2 \text{ ft}),$$

and we obtain $z = -0.5$ ft. Also, equating the sums of the moments about the z axis gives

$$(\Sigma M_{z \text{ axis}})_2 = (\Sigma M_{z \text{ axis}})_1:$$
$$(40 \text{ lb})x = (30 \text{ lb})(6 \text{ ft}) - (10 \text{ lb})(2 \text{ ft}) - (20 \text{ lb})(3 \text{ ft}),$$

and we obtain $x = 2.5$ ft.

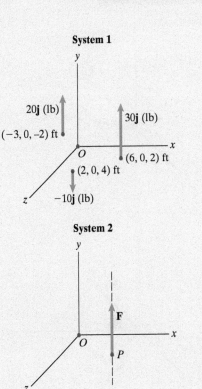

System 1

System 2

| Example 4.15 | Representing a Force and Couple by a Wrench (▶ *Related Problems 4.170, 4.171*) |

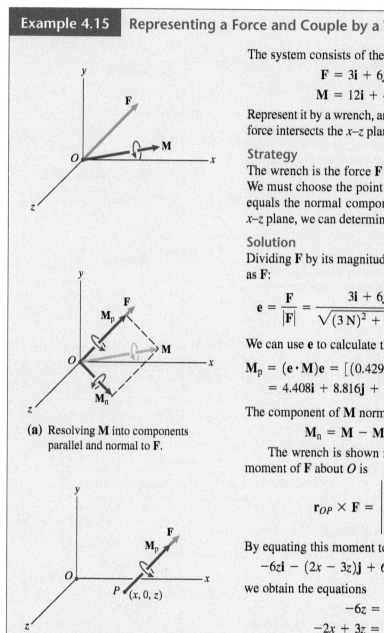

(a) Resolving **M** into components parallel and normal to **F**.

(b) The wrench acting at a point in the x–z plane.

The system consists of the force and couple

$$\mathbf{F} = 3\mathbf{i} + 6\mathbf{j} + 2\mathbf{k} \ (\text{N}),$$

$$\mathbf{M} = 12\mathbf{i} + 4\mathbf{j} + 6\mathbf{k} \ (\text{N-m}).$$

Represent it by a wrench, and determine where the line of action of the wrench's force intersects the x–z plane.

Strategy

The wrench is the force **F** and the component of **M** parallel to **F** (Figs. a, b). We must choose the point of application P so that the moment of **F** about O equals the normal component $\mathbf{M_n}$. By letting P be an arbitrary point of the x–z plane, we can determine where the line of action of **F** intersects that plane.

Solution

Dividing **F** by its magnitude, we obtain a unit vector **e** with the same direction as **F**:

$$\mathbf{e} = \frac{\mathbf{F}}{|\mathbf{F}|} = \frac{3\mathbf{i} + 6\mathbf{j} + 2\mathbf{k} \ (\text{N})}{\sqrt{(3\,\text{N})^2 + (6\,\text{N})^2 + (2\,\text{N})^2}} = 0.429\mathbf{i} + 0.857\mathbf{j} + 0.286\mathbf{k}.$$

We can use **e** to calculate the component of **M** parallel to **F**:

$$\mathbf{M_p} = (\mathbf{e} \cdot \mathbf{M})\mathbf{e} = [(0.429)(12\,\text{N-m}) + (0.857)(4\,\text{N-m}) + (0.286)(6\,\text{N-m})]\mathbf{e}$$

$$= 4.408\mathbf{i} + 8.816\mathbf{j} + 2.939\mathbf{k} \ (\text{N-m}).$$

The component of **M** normal to **F** is

$$\mathbf{M_n} = \mathbf{M} - \mathbf{M_p} = 7.592\mathbf{i} - 4.816\mathbf{j} + 3.061\mathbf{k} \ (\text{N-m}).$$

The wrench is shown in Fig. b. Let the coordinates of P be $(x, 0, z)$. The moment of **F** about O is

$$\mathbf{r}_{OP} \times \mathbf{F} = \begin{vmatrix} \mathbf{i} & \mathbf{j} & \mathbf{k} \\ x & 0 & z \\ 3 & 6 & 2 \end{vmatrix} = -6z\mathbf{i} - (2x - 3z)\mathbf{j} + 6x\mathbf{k} \ (\text{N-m}).$$

By equating this moment to $\mathbf{M_n}$, or

$$-6z\mathbf{i} - (2x - 3z)\mathbf{j} + 6x\mathbf{k} \ (\text{N-m}) = 7.592\mathbf{i} - 4.816\mathbf{j} + 3.061\mathbf{k} \ (\text{N-m}),$$

we obtain the equations

$$-6z = 7.592,$$

$$-2x + 3z = -4.816,$$

$$6x = 3.061.$$

Solving these equations, we find the coordinates of point P are $x = 0.510$ m, $z = -1.265$ m.

Critical Thinking

Why did we place point P at an arbitrary point $(x, 0, z)$ in the x–z plane? Our objective was to place the line of action of the force **F** of the wrench so as to satisfy the condition that the moment of **F** about O would equal $\mathbf{M_n}$. Placing the point of application of **F** at a point $(x, 0, z)$ and then using this condition to determine x and z was a convenient way to determine the necessary location of the line of action. The point $(x, 0, z) = (0.510, 0, -1.265)$ m is the intersection of the line of action with the x–z plane.

Problems

4.128 Two systems of forces act on the beam. Are they equivalent?

Strategy: Check the two conditions for equivalence. The sums of the forces must be equal, and the sums of the moments about an arbitrary point must be equal.

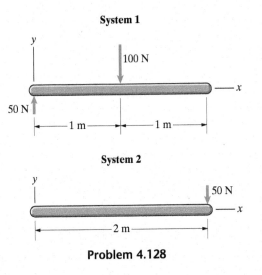

Problem 4.128

4.129 Two systems of forces and moments act on the beam. Are they equivalent?

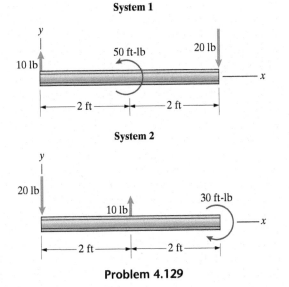

Problem 4.129

4.130 Four systems of forces and moments act on an 8-m beam. Which systems are equivalent?

4.131 The four systems can be made equivalent by adding a couple to one of the systems. Which system is it, and what couple must be added?

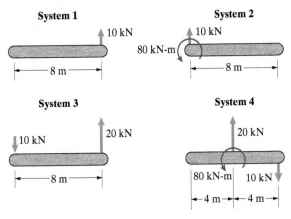

Problems 4.130/4.131

4.132 System 1 is a force **F** acting at a point *O*. System 2 is the force **F** acting at a different point *O'* along the same line of action. Explain why these systems are equivalent. (This simple result is called the *principle of transmissibility*.)

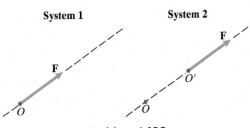

Problem 4.132

4.133 The vector sum of the forces exerted on the log by the cables is the same in the two cases. Show that the systems of forces exerted on the log are equivalent.

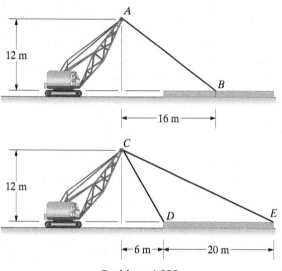

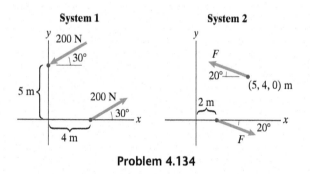

Problem 4.133

4.134 Systems 1 and 2 each consist of a couple. If they are equivalent, what is *F*?

System 1 System 2

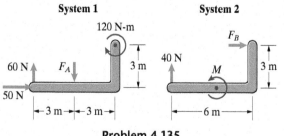

Problem 4.134

4.135 Two equivalent systems of forces and moments act on the L-shaped bar. Determine the forces F_A and F_B and the couple *M*.

System 1 System 2

Problem 4.135

4.136 Two equivalent systems of forces and moments act on the plate. Determine the force *F* and the couple *M*.

System 1 System 2

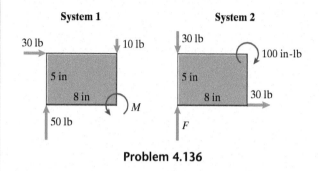

Problem 4.136

▶ **4.137** In Example 4.13, suppose that the 30-kN vertical force in system 1 is replaced by a 230-kN vertical force. Draw a sketch of the new system 1. If you represent system 1 by a single force **F** as in system 3, at what position *D* on the *x* axis must the force be placed?

4.138 Three forces and a couple are applied to a beam (system 1).
(a) If you represent system 1 by a force applied at *A* and a couple (system 2), what are **F** and *M*?
(b) If you represent system 1 by the force **F** (system 3), what is the distance *D*?

System 1

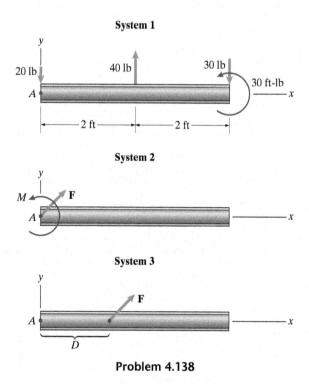

System 2

System 3

Problem 4.138

4.139 Represent the two forces and couple acting on the beam by a force **F**. Determine **F** and determine where its line of action intersects the x axis.

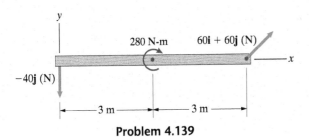

Problem 4.139

4.140 The bracket is subjected to three forces and a couple. If you represent this system by a force **F**, what is **F** and where does its line of action intersect the x axis?

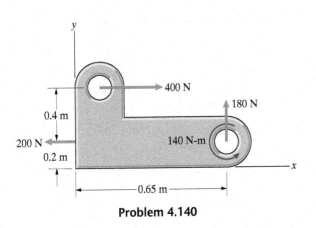

Problem 4.140

4.141 The vector sum of the forces acting on the beam is zero, and the sum of the moments about the left end of the beam is zero.

(a) Determine the forces A_x and A_y, and the couple M_A.

(b) Determine the sum of the moments about the right end of the beam.

(c) If you represent the 600-N force, the 200-N force, and the 30 N-m couple by a force **F** acting at the left end of the beam and a couple M, what are **F** and M?

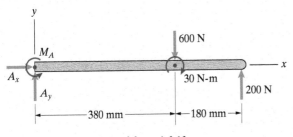

Problem 4.141

4.142 The vector sum of the forces acting on the truss is zero, and the sum of the moments about the origin O is zero.

(a) Determine the forces A_x, A_y, and B.

(b) If you represent the 2-kip, 4-kip, and 6-kip forces by a force **F**, what is **F**, and where does its line of action intersect the y axis?

(c) If you replace the 2-kip, 4-kip, and 6-kip forces by the force you determined in (b), what are the vector sum of the forces acting on the truss and the sum of the moments about O?

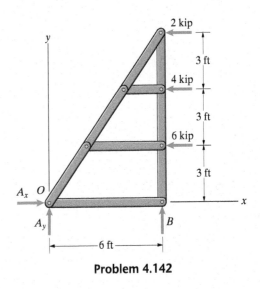

Problem 4.142

4.143 The distributed force exerted on part of a building foundation by the soil is represented by five forces. If you represent them by a force **F**, what is **F**, and where does its line of action intersect the x axis?

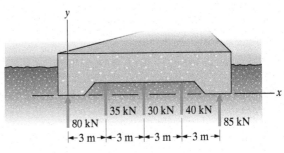

Problem 4.143

4.144 At a particular instant, aerodynamic forces distributed over the airplane's surface exert the 88-kN and 16-kN vertical forces and the 22 kN-m counterclockwise couple shown. If you represent these forces and couple by a system consisting of a force **F** acting at the center of mass G and a couple **M**, what are **F** and **M**?

4.145 If you represent the two forces and couple acting on the airplane by a force **F**, what is **F**, and where does its line of action intersect the x axis?

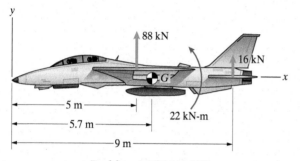

Problems 4.144/4.145

4.146 The system is in equilibrium. If you represent the forces $\mathbf{F}_{AB}$ and $\mathbf{F}_{AC}$ by a force **F** acting at A and a couple **M**, what are **F** and **M**?

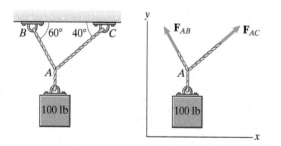

Problem 4.146

4.147 Three forces act on the beam.

(a) Represent the system by a force **F** acting at the origin O and a couple **M**.

(b) Represent the system by a single force. Where does the line of action of the force intersect the x axis?

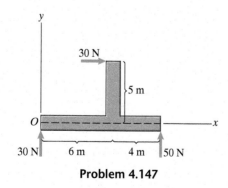

Problem 4.147

4.148 The tension in cable AB is 400 N, and the tension in cable CD is 600 N.

(a) If you represent the forces exerted on the left post by the cables by a force **F** acting at the origin O and a couple **M**, what are **F** and **M**?

(b) If you represent the forces exerted on the left post by the cables by the force **F** alone, where does its line of action intersect the y axis?

4.149 The tension in each of the cables AB and CD is 400 N. If you represent the forces exerted on the right post by the cables by a force **F**, what is **F**, and where does its line of action intersect the y axis?

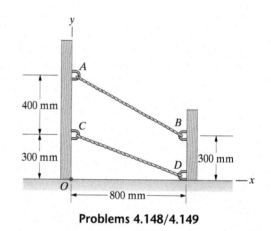

Problems 4.148/4.149

4.150 If you represent the three forces acting on the beam cross section by a force **F**, what is **F**, and where does its line of action intersect the x axis?

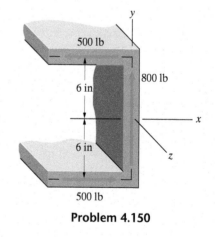

Problem 4.150

▶ **4.151** In Active Example 4.12, suppose that the force $\mathbf{F}_B$ is changed to $\mathbf{F}_B = 20\mathbf{i} - 15\mathbf{j} + 30\mathbf{k}$ (kN), and you want to represent system 1 by an equivalent system consisting of a force **F** acting at the point P with coordinates $(4, 3, -2)$ m and a couple **M** (system 2). Determine **F** and **M**.

4.152 The wall bracket is subjected to the force shown.

(a) Determine the moment exerted by the force about the z axis.

(b) Determine the moment exerted by the force about the y axis.

(c) If you represent the force by a force **F** acting at O and a couple **M**, what are **F** and **M**?

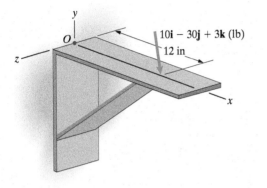

Problem 4.152

4.153 A basketball player executes a "slam dunk" shot, then hangs momentarily on the rim, exerting the two 100-lb forces shown. The dimensions are $h = 14\frac{1}{2}$ in and $r = 9\frac{1}{2}$ in, and the angle $\alpha = 120°$.

(a) If you represent the forces he exerts by a force **F** acting at O and a couple **M**, what are **F** and **M**?

(b) The glass backboard will shatter if $|\mathbf{M}| > 4000$ in-lb. Does it break?

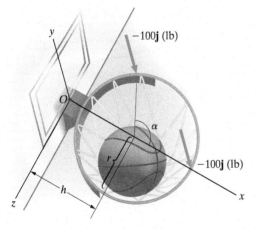

Problem 4.153

▶ **4.154** In Example 4.14, suppose that the 30-lb upward force in system 1 is changed to a 25-lb upward force. If you want to represent system 1 by a single force **F** (system 2), where does the line of action of **F** intersect the x–z plane?

4.155 The normal forces exerted on the car's tires by the road are

$$\mathbf{N}_A = 5104\mathbf{j} \ (\text{N}),$$
$$\mathbf{N}_B = 5027\mathbf{j} \ (\text{N}),$$
$$\mathbf{N}_C = 3613\mathbf{j} \ (\text{N}),$$
$$\mathbf{N}_D = 3559\mathbf{j} \ (\text{N}).$$

If you represent these forces by a single equivalent force **N**, what is **N** and where does its line of action intersect the x–z plane?

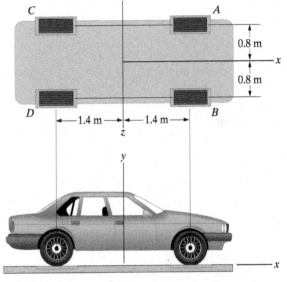

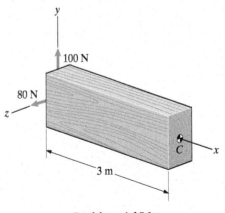

Problem 4.155

4.156 Two forces act on the beam. If you represent them by a force **F** acting at C and a couple **M**, what are **F** and **M**?

Problem 4.156

4.157 An axial force of magnitude P acts on the beam. If you represent it by a force $\mathbf{F}$ acting at the origin O and a couple $\mathbf{M}$, what are $\mathbf{F}$ and $\mathbf{M}$?

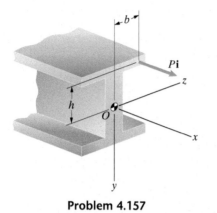

Problem 4.157

4.158 The brace is being used to remove a screw.

(a) If you represent the forces acting on the brace by a force $\mathbf{F}$ acting at the origin O and a couple $\mathbf{M}$, what are $\mathbf{F}$ and $\mathbf{M}$?

(b) If you represent the forces acting on the brace by a force $\mathbf{F}'$ acting at a point P with coordinates (x_P, y_P, z_P) and a couple $\mathbf{M}'$, what are $\mathbf{F}'$ and $\mathbf{M}'$?

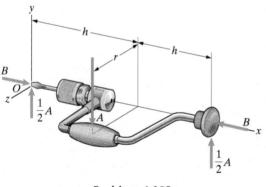

Problem 4.158

4.159 Two forces and a couple act on the cube. If you represent them by a force $\mathbf{F}$ acting at point P and a couple $\mathbf{M}$, what are $\mathbf{F}$ and $\mathbf{M}$?

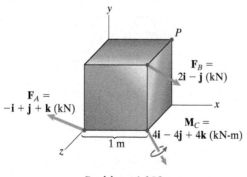

Problem 4.159

4.160 The two shafts are subjected to the torques (couples) shown.

(a) If you represent the two couples by a force $\mathbf{F}$ acting at the origin O and a couple $\mathbf{M}$, what are $\mathbf{F}$ and $\mathbf{M}$?

(b) What is the magnitude of the total moment exerted by the two couples?

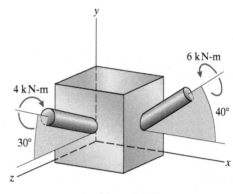

Problem 4.160

4.161 The two systems of forces and moments acting on the bar are equivalent. If

$$\mathbf{F}_A = 30\mathbf{i} + 30\mathbf{j} - 20\mathbf{k} \ (\text{kN}),$$
$$\mathbf{F}_B = 40\mathbf{i} - 20\mathbf{j} + 25\mathbf{k} \ (\text{kN}),$$
$$\mathbf{M}_B = 10\mathbf{i} + 40\mathbf{j} - 10\mathbf{k} \ (\text{kN-m}),$$

what are $\mathbf{F}$ and $\mathbf{M}$?

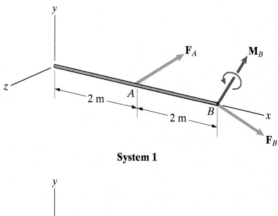

System 1

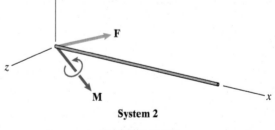

System 2

Problem 4.161

4.162 Point G is at the center of the block. The forces are

$$\mathbf{F}_A = -20\mathbf{i} + 10\mathbf{j} + 20\mathbf{k} \text{ (lb)},$$

$$\mathbf{F}_B = 10\mathbf{j} - 10\mathbf{k} \text{ (lb)}.$$

If you represent the two forces by a force $\mathbf{F}$ acting at G and a couple $\mathbf{M}$, what are $\mathbf{F}$ and $\mathbf{M}$?

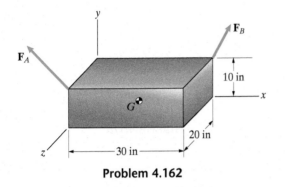

Problem 4.162

4.163 The engine above the airplane's fuselage exerts a thrust $T_0 = 16$ kip, and each of the engines under the wings exerts a thrust $T_U = 12$ kip. The dimensions are $h = 8$ ft, $c = 12$ ft, and $b = 16$ ft. If you represent the three thrust forces by a force $\mathbf{F}$ acting at the origin O and a couple $\mathbf{M}$, what are $\mathbf{F}$ and $\mathbf{M}$?

4.164 Consider the airplane described in Problem 4.163 and suppose that the engine under the wing to the pilot's right loses thrust.

(a) If you represent the two remaining thrust forces by a force $\mathbf{F}$ acting at the origin O and a couple $\mathbf{M}$, what are $\mathbf{F}$ and $\mathbf{M}$?

(b) If you represent the two remaining thrust forces by the force $\mathbf{F}$ alone, where does its line of action intersect the x–y plane?

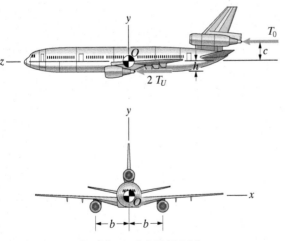

Problems 4.163/4.164

4.165 The tension in cable AB is 100 lb, and the tension in cable CD is 60 lb. Suppose that you want to replace these two cables by a single cable EF so that the force exerted on the wall at E is equivalent to the two forces exerted by cables AB and CD on the walls at A and C. What is the tension in cable EF, and what are the coordinates of points E and F?

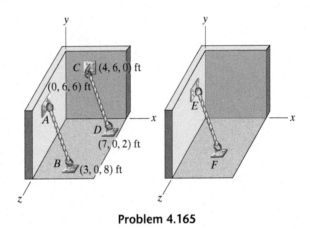

Problem 4.165

4.166 The distance $s = 4$ m. If you represent the force and the 200-N-m couple by a force $\mathbf{F}$ acting at the origin O and a couple $\mathbf{M}$, what are $\mathbf{F}$ and $\mathbf{M}$?

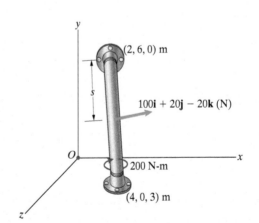

Problem 4.166

4.167 The force **F** and couple **M** in system 1 are

$$F = 12i + 4j - 3k \text{ (lb)},$$

$$M = 4i + 7j + 4k \text{ (ft-lb)}.$$

Suppose you want to represent system 1 by a wrench (system 2). Determine the couple M_p and the coordinates x and z where the line of action of the force intersects the x–z plane.

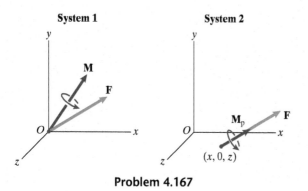

Problem 4.167

4.168 A system consists of a force **F** acting at the origin O and a couple **M**, where

$$F = 10i \text{ (lb)}, \qquad M = 20j \text{ (ft-lb)}.$$

If you represent the system by a wrench consisting of the force **F** and a parallel couple M_p, what is M_p, and where does the line of action of **F** intersect the y–z plane?

4.169 A system consists of a force **F** acting at the origin O and a couple **M**, where

$$F = i + 2j + 5k \text{ (N)}, \qquad M = 10i + 8j - 4k \text{ (N-m)}.$$

If you represent it by a wrench consisting of the force **F** and a parallel couple M_p, (a) determine M_p, and determine where the line of action of **F** intersects (b) the x–z plane, (c) the y–z plane.

▶ **4.170** Consider the force **F** acting at the origin O and the couple **M** given in Example 4.15. If you represent this system by a wrench, where does the line of action of the force intersect the x–y plane?

▶ **4.171** Consider the force **F** acting at the origin O and the couple **M** given in Example 4.15. If you represent this system by a wrench, where does the line of action of the force intersect the plane $y = 3$ m?

4.172 A wrench consists of a force of magnitude 100 N acting at the origin O and a couple of magnitude 60 N-m. The force and couple point in the direction from O to the point $(1, 1, 2)$ m. If you represent the wrench by a force **F** acting at the point $(5, 3, 1)$ m and a couple **M**, what are **F** and **M**?

4.173 System 1 consists of two forces and a couple. Suppose that you want to represent it by a wrench (system 2). Determine the force **F**, the couple M_p, and the coordinates x and z where the line of action of **F** intersects the x–z plane.

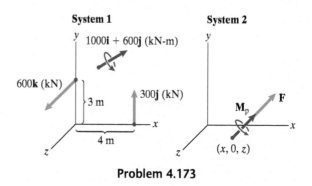

Problem 4.173

4.174 A plumber exerts the two forces shown to loosen a pipe.

(a) What total moment does he exert about the axis of the pipe?

(b) If you represent the two forces by a force **F** acting at O and a couple **M**, what are **F** and **M**?

(c) If you represent the two forces by a wrench consisting of the force **F** and a parallel couple M_p, what is M_p, and where does the line of action of **F** intersect the x–y plane?

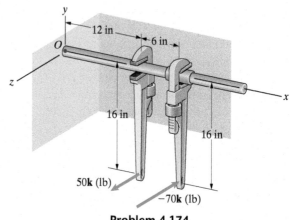

Problem 4.174

Review Problems

4.175 The Leaning Tower of Pisa is approximately 55 m tall and 7 m in diameter. The horizontal displacement of the top of the tower from the vertical is approximately 5 m. Its mass is approximately 3.2×10^6 kg. If you model the tower as a cylinder and assume that its weight acts at the center, what is the magnitude of the moment exerted by the weight about the point at the center of the tower's base?

Problem 4.175

4.176 The cable AB exerts a 300-N force on the support A that points from A toward B. Determine the magnitude of the moment the force exerts about point P.

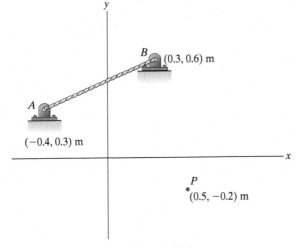

Problem 4.176

4.177 Three forces act on the structure. The sum of the moments due to the forces about A is zero. Determine the magnitude of the force F.

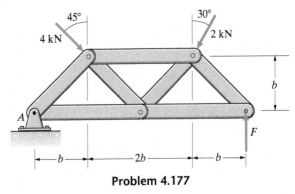

Problem 4.177

4.178 Determine the moment of the 400-N force (a) about A, (b) about B.

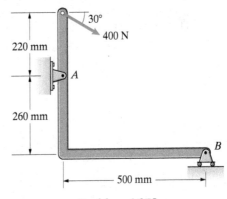

Problem 4.178

4.179 Determine the sum of the moments exerted about A by the three forces and the couple.

4.180 If you represent the three forces and the couple by an equivalent system consisting of a force $\mathbf{F}$ acting at A and a couple $\mathbf{M}$, what are the magnitudes of $\mathbf{F}$ and $\mathbf{M}$?

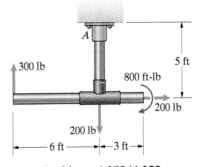

Problems 4.179/4.180

4.181 The vector sum of the forces acting on the beam is zero, and the sum of the moments about A is zero.

(a) What are the forces A_x, A_y, and B?

(b) What is the sum of the moments about B?

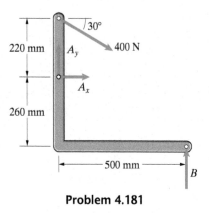

Problem 4.181

4.182 The hydraulic piston BC exerts a 970-lb force on the boom at C in the direction parallel to the piston. The angle $\alpha = 40°$. The sum of the moments about A due to the force exerted on the boom by the piston and the weight of the suspended load is zero. What is the weight of the suspended load?

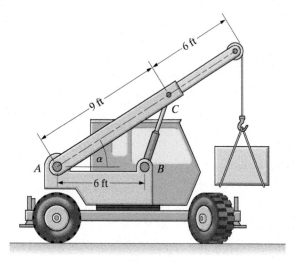

Problem 4.182

4.183 The force $\mathbf{F} = -60\mathbf{i} + 60\mathbf{j}$ (lb).

(a) Determine the moment of **F** about point A.

(b) What is the perpendicular distance from point A to the line of action of **F**?

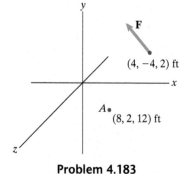

Problem 4.183

4.184 The 20-kg mass is suspended by cables attached to three vertical 2-m posts. Point A is at (0, 1.2, 0) m. Determine the moment about the base E due to the force exerted on the post BE by the cable AB.

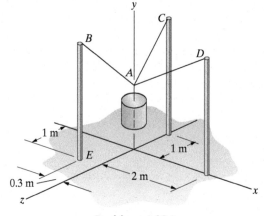

Problem 4.184

4.185 What is the total moment due to the two couples?

(a) Express the answer by giving the magnitude and stating whether the moment is clockwise or counterclockwise.

(b) Express the answer as a vector.

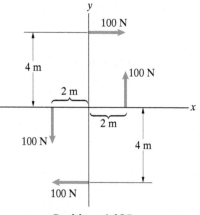

Problem 4.185

4.186 The bar AB supporting the lid of the grand piano exerts a force $\mathbf{F} = -6\mathbf{i} + 35\mathbf{j} - 12\mathbf{k}$ (lb) at B. The coordinates of B are $(3, 4, 3)$ ft. What is the moment of the force about the hinge line of the lid (the x axis)?

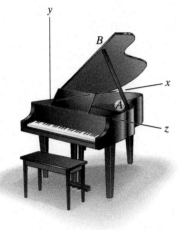

Problem 4.186

4.187 Determine the moment of the vertical 800-lb force about point C.

4.188 Determine the moment of the vertical 800-lb force about the straight line through points C and D.

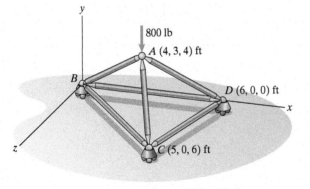

Problems 4.187/4.188

4.189 The system of cables and pulleys supports the 300-lb weight of the work platform. If you represent the upward force exerted at E by cable EF and the upward force exerted at G by cable GH by a single equivalent force $\mathbf{F}$, what is $\mathbf{F}$, and where does its line of action intersect the x axis?

4.190 The system of cables and pulleys supports the 300-lb weight of the work platform.

(a) What are the tensions in cables AB and CD?

(b) If you represent the forces exerted on the work platform by the cables at A and C by a single equivalent force $\mathbf{F}$, what is $\mathbf{F}$ and where does its line of action intersect the x axis?

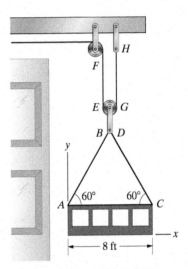

Problems 4.189/4.190

4.191 The two systems are equivalent. Determine the forces A_x and A_y, and the couple M_A.

4.192 If you represent the equivalent systems in Problem 4.191 by a force **F** acting at the origin and a couple M, what are **F** and M?

4.193 If you represent the equivalent systems in Problem 4.191 by a force **F**, what is **F**, and where does its line of action intersect the x axis?

System 1

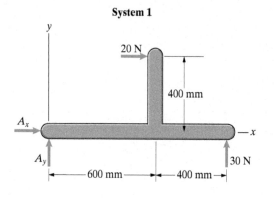

System 2

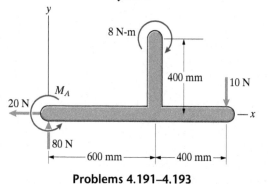

Problems 4.191–4.193

4.194 The two systems are equivalent. If
$$\mathbf{F} = -100\mathbf{i} + 40\mathbf{j} + 30\mathbf{k} \text{ (lb)},$$
$$\mathbf{M}' = -80\mathbf{i} + 120\mathbf{j} + 40\mathbf{k} \text{ (in-lb)},$$
determine **F'** and **M**.

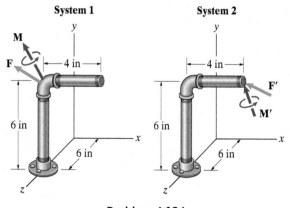

Problem 4.194

4.195 The tugboats A and B exert forces $F_A = 1$ kN and $F_B = 1.2$ kN on the ship. The angle $\theta = 30°$. If you represent the two forces by a force **F** acting at the origin O and a couple M, what are **F** and M?

4.196 The tugboats A and B exert forces $F_A = 600$ N and $F_B = 800$ N on the ship. The angle $\theta = 45°$. If you represent the two forces by a force **F**, what is **F**, and where does its line of action intersect the y axis?

4.197 The tugboats A and B want to exert two forces on the ship that are equivalent to a force **F** acting at the origin O of 2-kN magnitude. If $F_A = 800$ N, determine the necessary values of F_B and θ.

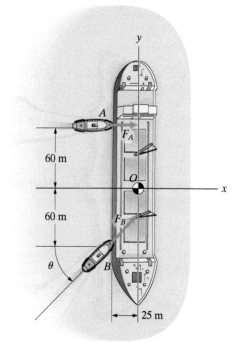

Problems 4.195–4.197

4.198 If you represent the forces exerted by the floor on the table legs by a force **F** acting at the origin *O* and a couple **M**, what are **F** and **M**?

4.199 If you represent the forces exerted by the floor on the table legs by a force **F**, what is **F**, and where does its line of action intersect the *x–z* plane?

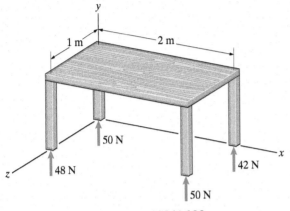

Problems 4.198/4.199

4.200 Two forces are exerted on the crankshaft by the connecting rods. The direction cosines of $\mathbf{F}_A$ are $\cos \theta_x = -0.182$, $\cos \theta_y = 0.818$, and $\cos \theta_z = 0.545$, and its magnitude is 4 kN. The direction cosines of $\mathbf{F}_B$ are $\cos \theta_x = 0.182$, $\cos \theta_y = 0.818$, and $\cos \theta_z = -0.545$, and its magnitude is 2 kN. If you represent the two forces by a force **F** acting at the origin *O* and a couple **M**, what are **F** and **M**?

4.201 If you represent the two forces exerted on the crankshaft in Problem 4.200 by a wrench consisting of a force **F** and a parallel couple $\mathbf{M_p}$, what are **F** and $\mathbf{M_p}$, and where does the line of action of **F** intersect the *x–z* plane?

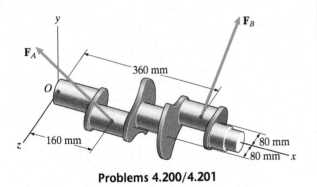

Problems 4.200/4.201

Design Project   A relatively primitive device for exercising the biceps muscle is shown. Suggest an improved configuration for the device. You can use elastic cords (which behave like linear springs), weights, and pulleys. Seek a design such that the variation of the moment about the elbow joint as the device is used is small in comparison to the design shown. Give consideration to the safety of your device, its reliability, and the requirement to accommodate users having a range of dimensions and strengths. Choosing specific dimensions, determine the range of the magnitude of the moment exerted about the elbow joint as your device is used.

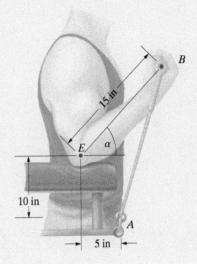

CHAPTER
5

Objects in Equilibrium

Building on concepts developed in Chapters 3 and 4, we first state the general equilibrium equations. We describe various ways that structural members can be supported, or held in place. Using free-body diagrams and equilibrium equations, we then show how to determine unknown forces and couples exerted on structural members by their supports. The principal motivation for this procedure is that it is the initial step in answering an essential question in structural analysis: How do engineers design structural elements so that they will support the loads to which they are subjected?

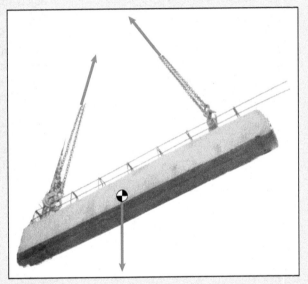

◄ The beam is in equilibrium under the actions of its weight and the forces exerted by the chains. In this chapter we apply the equilibrium equations to determine unknown forces and couples acting on objects.

5.1 Two-Dimensional Applications

BACKGROUND

When an object acted upon by a system of forces and moments is in equilibrium, the following conditions are satisfied:

1. The sum of the forces is zero:

$$\Sigma \mathbf{F} = \mathbf{0}. \tag{5.1}$$

2. The sum of the moments about any point is zero:

$$\Sigma \mathbf{M}_{\text{any point}} = \mathbf{0}. \tag{5.2}$$

From our discussion of equivalent systems of forces and moments in Chapter 4, Eqs. (5.1) and (5.2) imply that the system of forces and moments acting on an object in equilibrium is equivalent to a system consisting of no forces and no couples. This provides insight into the nature of equilibrium. From the standpoint of the total force and total moment exerted on an object in equilibrium, the effects are the same as if no forces or couples acted on the object. This observation also makes it clear that if the sum of the forces on an object is zero and the sum of the moments about one point is zero, then the sum of the moments about every point is zero.

The Scalar Equilibrium Equations

When the loads and reactions on an object in equilibrium form a two-dimensional system of forces and moments, they are related by three scalar equilibrium equations:

$$\Sigma F_x = 0, \tag{5.3}$$

$$\Sigma F_y = 0, \tag{5.4}$$

$$\Sigma M_{\text{any point}} = 0. \tag{5.5}$$

A natural question is whether more than one equation can be obtained from Eq. (5.5) by evaluating the sum of the moments about more than one point. The answer is yes, and in some cases it is convenient to do so. But there is a catch—the additional equations will not be independent of Eqs. (5.3)–(5.5). In other words, *more than three independent equilibrium equations cannot be obtained from a two-dimensional free-body diagram, which means we can solve for at most three unknown forces or couples.* We discuss this point further in Section 5.2.

Supports

When you are standing, the floor supports you. When you sit in a chair with your feet on the floor, the chair and floor support you. In this section we are concerned with the ways objects can be supported, or held in place. Forces and couples exerted on an object by its supports are called *reactions*, expressing the fact that the supports "react" to the other forces and couples, or *loads*, acting on the object. For example, a bridge is held up by the reactions exerted by its supports, and the loads are the forces exerted by the weight of the bridge itself, the traffic crossing it, and the wind.

Some very common kinds of supports are represented by stylized models called *support conventions*. Actual supports often closely resemble the support

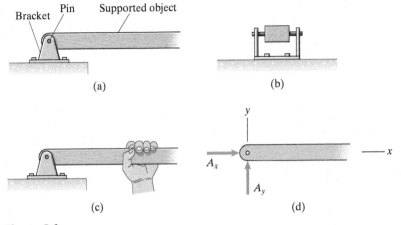

Figure 5.1
(**a**) A pin support.
(**b**) Side view showing the pin passing through the beam.
(**c**) Holding a supported bar.
(**d**) The pin support is capable of exerting two components of force.

conventions, but even when they don't, we represent them by these conventions if the actual supports exert the same (or approximately the same) reactions as the models.

The Pin Support   Figure 5.1a shows a *pin support*. The diagram represents a bracket to which an object (such as a beam) is attached by a smooth pin that passes through the bracket and the object. The side view is shown in Fig. 5.1b.

   To understand the reactions that a pin support can exert, it's helpful to imagine holding a bar attached to a pin support (Fig. 5.1c). If you try to move the bar without rotating it (that is, translate the bar), the support exerts a reactive force that prevents this movement. However, you can rotate the bar about the axis of the pin. The support cannot exert a couple about the pin axis to prevent rotation. Thus a pin support can't exert a couple about the pin axis, but it can exert a force on an object in any direction, which is usually expressed by representing the force in terms of components (Fig. 5.1d). The arrows indicate the directions of the reactions if A_x and A_y are positive. If you determine A_x or A_y to be negative, the reaction is in the direction opposite to that of the arrow.

   The pin support is used to represent any real support capable of exerting a force in any direction but not exerting a couple. Pin supports are used in many common devices, particularly those designed to allow connected parts to rotate relative to each other (Fig. 5.2).

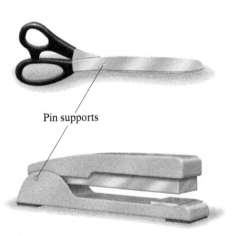

Pin supports

Figure 5.2
Pin supports in a pair of scissors and a stapler.

The Roller Support   The convention called a *roller support* (Fig. 5.3a) represents a pin support mounted on wheels. Like the pin support, it cannot exert a couple about the axis of the pin. Since it can move freely in the direction parallel to the surface on which it rolls, it can't exert a force parallel to the surface but can only exert a force normal (perpendicular) to this surface (Fig. 5.3b). Figures 5.3c–e are other commonly used conventions equivalent to the roller support. The wheels of vehicles and wheels supporting parts of machines are roller supports if the friction forces exerted on them are negligible in comparison to the normal forces. A plane smooth surface can also be modeled by a roller

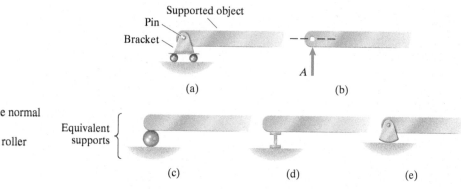

Figure 5.3
(a) A roller support.
(b) The reaction consists of a force normal to the surface.
(c)–(e) Supports equivalent to the roller support.

Figure 5.4
Supporting an object with a plane smooth surface.

support (Fig. 5.4). Beams and bridges are sometimes supported in this way so that they will be free to undergo thermal expansion and contraction.

The supports shown in Fig. 5.5 are similar to the roller support in that they cannot exert a couple and can only exert a force normal to a particular direction. (Friction is neglected.) In these supports, the supported object is attached to a pin or slider that can move freely in one direction but is constrained in the perpendicular direction. Unlike the roller support, these supports can exert a normal force in either direction.

The Fixed Support The *fixed support* shows the supported object literally built into a wall (Fig. 5.6a). This convention is also called a *built-in* support. To understand the reactions, imagine holding a bar attached to a fixed support (Fig. 5.6b). If you try to translate the bar, the support exerts a reactive force that

Figure 5.5
Supports similar to the roller support except that the normal force can be exerted in either direction.

(a) Pin in a slot.   (b) Slider in a slot.   (c) Slider on a shaft.

Figure 5.6
(a) Fixed support.
(b) Holding a supported bar.
(c) The reactions a fixed support is capable of exerting.

prevents translation, and if you try to rotate the bar, the support exerts a reactive couple that prevents rotation. A fixed support can exert two components of force and a couple (Fig. 5.6c). The term M_A is the couple exerted by the support, and the curved arrow indicates its direction. Fence posts and lampposts have fixed supports. The attachments of parts connected so that they cannot move or rotate relative to each other, such as the head of a hammer and its handle, can be modeled as fixed supports.

Table 5.1 summarizes the support conventions commonly used in two-dimensional applications, including those we discussed in Chapter 3. Although

Table 5.1 Supports used in two-dimensional applications.

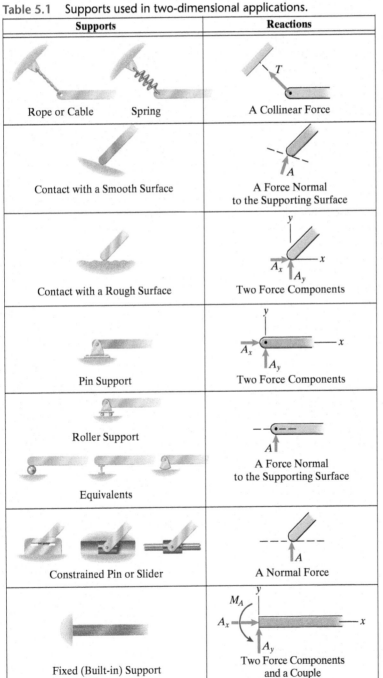

Supports	Reactions
Rope or Cable Spring	A Collinear Force
Contact with a Smooth Surface	A Force Normal to the Supporting Surface
Contact with a Rough Surface	Two Force Components
Pin Support	Two Force Components
Roller Support Equivalents	A Force Normal to the Supporting Surface
Constrained Pin or Slider	A Normal Force
Fixed (Built-in) Support	Two Force Components and a Couple

the number of conventions may appear daunting, the examples and problems will help you become familiar with them. You should also observe how various objects you see in your everyday experience are supported and think about whether each support could be represented by one of the conventions.

Free-Body Diagrams

We introduced free-body diagrams in Chapter 3 and used them to determine forces acting on simple objects in equilibrium. By using the support conventions, we can model more elaborate objects and construct their free-body diagrams in a systematic way.

For example, the beam in Fig. 5.7a has a pin support at the left end and a roller support at the right end and is loaded by a force F. The roller support rests on a surface inclined at 30° to the horizontal. To obtain the free-body diagram of the beam, we first isolate it from its supports (Fig. 5.7b), since the free-body diagram must contain no object other than the beam. We complete the free-body diagram by showing the reactions that may be exerted on the beam by the supports (Fig. 5.7c). Notice that the reaction B exerted by the roller support is normal to the surface on which the support rests.

The object in Fig. 5.8a has a fixed support at the left end. A cable passing over a pulley is attached to the object at two points. We isolate it from its supports (Fig. 5.8b) and complete the free-body diagram by showing the reactions at the fixed support and the forces exerted by the cable (Fig. 5.8c). *Don't forget the couple at a fixed support.* Since we assume the tension in the cable is the same on both sides of the pulley, the two forces exerted by the cable have the same magnitude T.

Once you have obtained the free-body diagram of an object in equilibrium to identify the loads and reactions acting on it, you can apply the equilibrium equations.

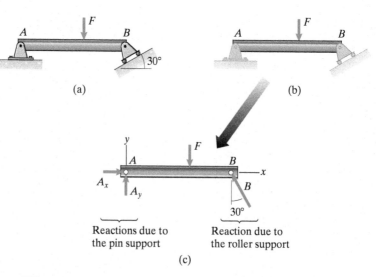

Reactions due to the pin support

Reaction due to the roller support

(c)

Figure 5.7
(**a**) A beam with pin and roller supports.
(**b**) Isolating the beam from its supports.
(**c**) The completed free-body diagram.

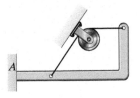

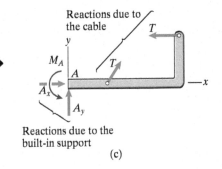

(a) (b) (c)

Figure 5.8
(a) An object with a fixed support.
(b) Isolating the object.
(c) The completed free-body diagram.

Equilibrium Equations

When an object is in equilibrium, the system of forces and moments acting on it satisfies two conditions.

The sum of the forces is zero:

$$\Sigma \mathbf{F} = \mathbf{0}. \qquad (5.1)$$

The sum of the moments about any point is zero:

$$\Sigma \mathbf{M}_{\text{any point}} = \mathbf{0}. \qquad (5.2)$$

When the system of forces and moments acting on an object in equilibrium is two dimensional, it satisfies three scalar equilibrium equations.

$$\Sigma F_x = 0, \qquad (5.3)$$
$$\Sigma F_y = 0, \qquad (5.4)$$
$$\Sigma M_{\text{any point}} = 0. \qquad (5.5)$$

Supports

To draw the free-body diagram of an object, isolate it from its supports and show the *reactions*, the forces and moments that the supports may exert (Table 5.1).

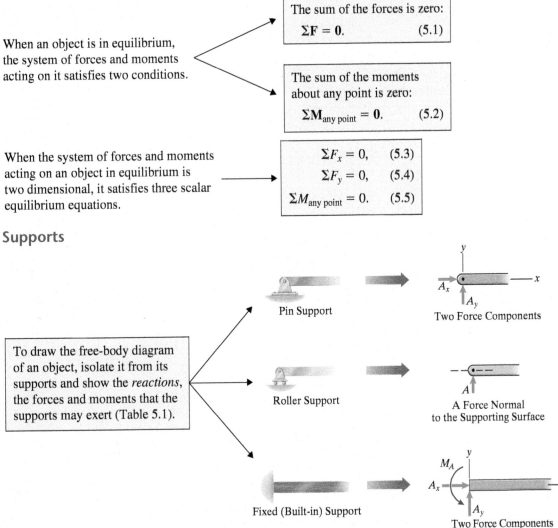

Pin Support

Two Force Components

Roller Support

A Force Normal to the Supporting Surface

Fixed (Built-in) Support

Two Force Components and a Couple

Reactions at a Fixed Support (▶ *Related Problem 5.1*)

The beam has a fixed support at A and is subjected to a 4-kN force. (a) Draw the free-body diagram of the beam. (b) Determine the reactions at the fixed support.

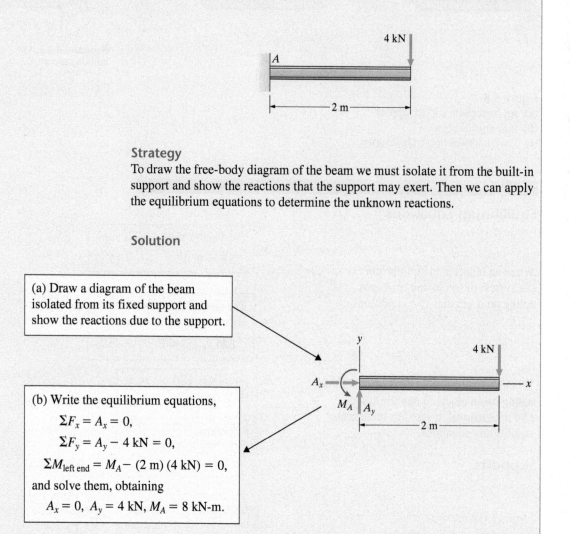

Strategy

To draw the free-body diagram of the beam we must isolate it from the built-in support and show the reactions that the support may exert. Then we can apply the equilibrium equations to determine the unknown reactions.

Solution

(a) Draw a diagram of the beam isolated from its fixed support and show the reactions due to the support.

(b) Write the equilibrium equations,

$$\Sigma F_x = A_x = 0,$$

$$\Sigma F_y = A_y - 4 \text{ kN} = 0,$$

$$\Sigma M_{\text{left end}} = M_A - (2 \text{ m})(4 \text{ kN}) = 0,$$

and solve them, obtaining

$$A_x = 0, \ A_y = 4 \text{ kN}, \ M_A = 8 \text{ kN-m}.$$

Practice Problem The beam has pin and roller supports and is subjected to a 4-kN force. (a) Draw the free-body diagram of the beam. (b) Determine the reactions at the supports.

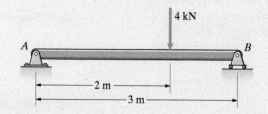

Answer: $A_x = 0, A_y = 1.33$ kN, $B = 2.67$ kN.

| Example 5.2 | Reactions at a Fixed Support (▶ *Related Problem 5.9*) |

The object has a fixed support at *A* and is subjected to two forces and a couple. What are the reactions at the support?

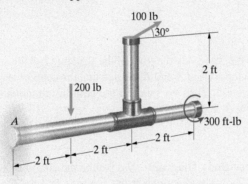

Strategy
We will obtain a free-body diagram by isolating the object from the fixed support at *A* and showing the reactions exerted at *A*, *including the couple that may be exerted by a fixed support*. Then we can determine the unknown reactions by applying the equilibrium equations.

Solution

Draw the Free-Body Diagram   We isolate the object from its support and show the reactions at the fixed support (Fig. a). There are three unknown reactions: two force components A_x and A_y and a couple M_A. (Remember that we can choose the directions of these arrows arbitrarily.) We also resolve the 100-lb force into its components.

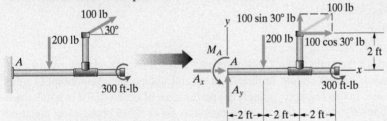

(a) Drawing the free-body diagram.

Apply the Equilibrium Equations   Summing the moments about point *A*, the equilibrium equations are

$$\Sigma F_x = A_x + 100 \cos 30° \text{ lb} = 0,$$
$$\Sigma F_y = A_y - 200 \text{ lb} + 100 \sin 30° \text{ lb} = 0,$$
$$\Sigma M_{\text{point } A} = M_A + 300 \text{ ft-lb} - (2\text{ft})(200 \text{ lb}) - (2 \text{ ft})(100 \cos 30° \text{ lb})$$
$$+ (4 \text{ ft})(100 \sin 30° \text{ lb}) = 0.$$

Solving these equations, we obtain the reactions $A_x = -86.6$ lb, $A_y = 150$ lb, and $M_A = 73.2$ ft-lb.

Critical Thinking
Why don't the 300 ft-lb couple and the couple M_A exerted by the fixed support appear in the first two equilibrium equations? Remember that a couple exerts no net force. Also, because the moment due to a couple is the same about any point, the moment about *A* due to the 300 ft-lb counterclockwise couple is 300 ft-lb counterclockwise.

| Example 5.3 | Choosing the Point About Which to Evaluate Moments (▶ *Related Problem 5.15*) |

The structure AB supports a suspended 2-Mg (megagram) mass. The structure is attached to a slider in a vertical slot at A and has a pin support at B. What are the reactions at A and B?

Strategy
We will draw the free-body diagram of the structure and the suspended mass by removing the supports at A and B. Notice that the support at A can exert only a horizontal reaction. Then we can use the equilibrium equations to determine the reactions at A and B.

Solution
Draw the Free-Body Diagram We isolate the structure and mass from the supports and show the reactions at the supports and the force exerted by the weight of the 2000-kg mass (Fig. a). The slot at A can exert only a horizontal force on the slider.

Apply the Equilibrium Equations Summing moments about point B, we find that the equilibrium equations are

$$\Sigma F_x = A + B_x = 0,$$

$$\Sigma F_y = B_y - (2000)(9.81) \text{ N} = 0,$$

$$\Sigma M_{\text{point } B} = (3 \text{ m})A + (2 \text{ m})[(2000)(9.81) \text{ N}] = 0.$$

The reactions are $A = -13.1$ kN, $B_x = 13.1$ kN, and $B_y = 19.6$ kN.

(a) Drawing the free-body diagram.

Critical Thinking
Although the point about which moments are evaluated in writing equilibrium equations can be chosen arbitrarily, a careful choice can often simplify your solution. In this example, point B lies on the lines of action of the two unknown reactions B_x and B_y. By evaluating moments about B, we obtained an equation containing only one unknown, the reaction at A.

| Example 5.4 | Analysis of a Luggage Carrier (▶ *Related Problems 5.65–5.68*) |

The figure shows an airport luggage carrier and its free-body diagram when it is held in equilibrium in the tilted position. If the luggage carrier supports a weight $W = 50$ lb, the angle $\alpha = 30°$, $a = 8$ in, $b = 16$ in, and $d = 48$ in, what force F must the user exert?

Strategy
The unknown reactions on the free-body diagram are the force F and the normal force N exerted by the floor. If we sum moments about the center of the wheel C, we obtain an equation in which F is the only unknown reaction.

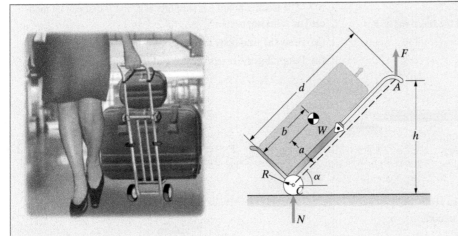

Solution
Summing moments about C,

$$\Sigma M_{(\text{point } C)} = d(F \cos \alpha) + a(W \sin \alpha) - b(W \cos \alpha) = 0,$$

and solving for F, we obtain

$$F = \frac{(b - a \tan \alpha)W}{d}.$$

Substituting the values of W, α, a, b, and d yields the solution $F = 11.9$ lb.

Problems

Assume that objects are in equilibrium. In the statements of the answers, x components are positive to the right and y components are positive upward.

▶ 5.1 In Active Example 5.1, suppose that the beam is subjected to a 6 kN-m counterclockwise couple at the right end in addition to the 4-kN downward force. Draw a sketch of the beam showing its new loading. Draw the free-body diagram of the beam and apply the equilibrium equations to determine the reactions at A.

5.2 The beam has a fixed support at A and is loaded by two forces and a couple. Draw the free-body diagram of the beam and apply equilibrium to determine the reactions at A.

5.3 The beam is subjected to a load $F = 400$ N and is supported by the rope and the smooth surfaces at A and B.

(a) Draw the free-body diagram of the beam.

(b) What are the magnitudes of the reactions at A and B?

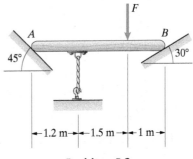

Problem 5.3

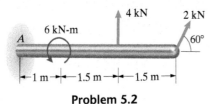

Problem 5.2

5.4 (a) Draw the free-body diagram of the beam.
(b) Determine the tension in the rope and the reactions at B.

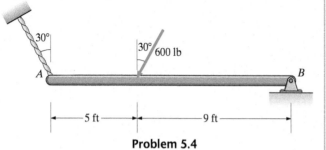

Problem 5.4

5.5 (a) Draw the free-body diagram of the 60-lb drill press, assuming that the surfaces at A and B are smooth.
(b) Determine the reactions at A and B.

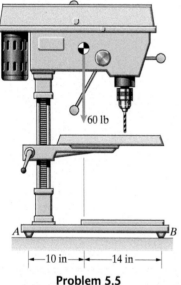

Problem 5.5

5.6 The masses of the person and the diving board are 54 kg and 36 kg, respectively. Assume that they are in equilibrium.
(a) Draw the free-body diagram of the diving board.
(b) Determine the reactions at the supports A and B.

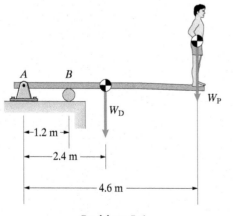

Problem 5.6

5.7 The ironing board has supports at A and B that can be modeled as roller supports.
(a) Draw the free-body diagram of the ironing board.
(b) Determine the reactions at A and B.

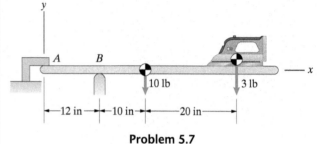

Problem 5.7

5.8 The distance $x = 9$ m.
(a) Draw the free-body diagram of the beam.
(b) Determine the reactions at the supports.

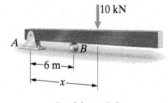

Problem 5.8

▶ **5.9** In Example 5.2, suppose that the 200-lb downward force and the 300 ft-lb counterclockwise couple change places; the 200-lb downward force acts at the right end of the horizontal bar, and the 300 ft-lb counterclockwise couple acts on the horizontal bar 2 ft to the right of the support A. Draw a sketch of the object showing the new loading. Draw the free-body diagram of the object and apply the equilibrium equations to determine the reactions at A.

5.10 (a) Draw the free-body diagram of the beam.
(b) Determine the reactions at the supports.

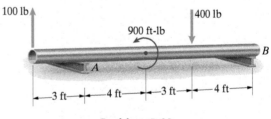

Problem 5.10

5.11 The person exerts 20-N forces on the pliers. The free-body diagram of one part of the pliers is shown. Notice that the pin at *C* connecting the two parts of the pliers behaves like a pin support. Determine the reactions at *C* and the force *B* exerted on the pliers by the bolt.

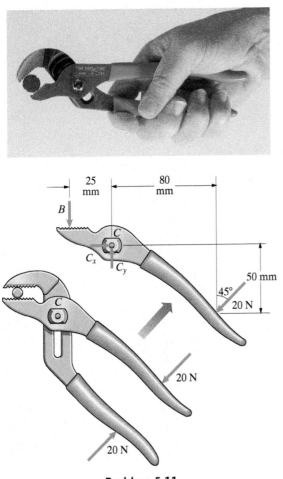

Problem 5.11

5.12 (a) Draw the free-body diagram of the beam.
(b) Determine the reactions at the pin support *A*.

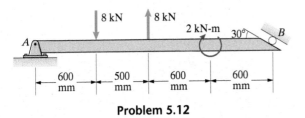

Problem 5.12

5.13 (a) Draw the free-body diagram of the beam.
(b) Determine the reactions at the supports.

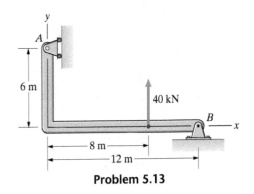

Problem 5.13

5.14 (a) Draw the free-body diagram of the beam.
(b) If $F = 4$ kN, what are the reactions at *A* and *B*?

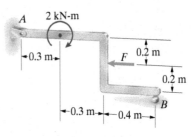

Problem 5.14

▶ **5.15** In Example 5.3, suppose that the attachment point for the suspended mass is moved toward point *B* such that the horizontal distance from *A* to the attachment point increases from 2 m to 3 m. Draw a sketch of the beam *AB* showing the new geometry. Draw the free-body diagram of the beam and apply the equilibrium equations to determine the reactions at *A* and *B*.

5.16 A person doing push-ups pauses in the position shown. His 180-lb weight *W* acts at the point shown. The dimensions $a = 15$ in, $b = 42$ in, and $c = 16$ in. Determine the normal force exerted by the floor on each of his hands and on each of his feet.

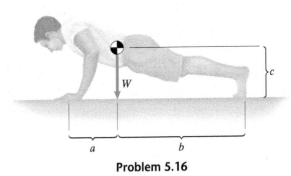

Problem 5.16

5.17 The hydraulic piston *AB* exerts a 400-lb force on the ladder at *B* in the direction parallel to the piston. Determine the weight of the ladder and the reactions at *C*.

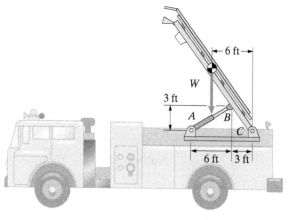

Problem 5.17

5.18 Draw the free-body diagram of the structure by isolating it from its supports at *A* and *E*. Determine the reactions at *A* and *E*.

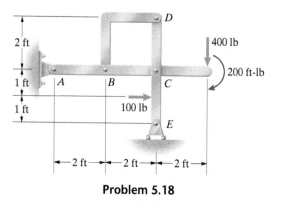

Problem 5.18

5.19 (a) Draw the free-body diagram of the beam.

(b) Determine the tension in the cable and the reactions at *A*.

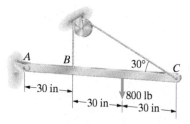

Problem 5.19

5.20 The unstretched length of the spring *CD* is 350 mm. Suppose that you want the lever *ABC* to exert a 120-N normal force on the smooth surface at *A*. Determine the necessary value of the spring constant *k* and the resulting reactions at *B*.

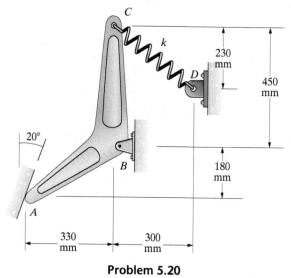

Problem 5.20

5.21 The mobile is in equilibrium. The fish *B* weighs 27 oz. Determine the weights of the fish *A*, *C*, and *D*. (The weights of the crossbars are negligible.)

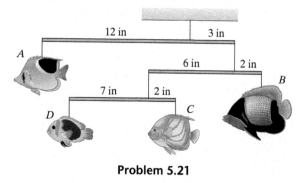

Problem 5.21

5.22 The car's wheelbase (the distance between the wheels) is 2.82 m. The mass of the car is 1760 kg and its weight acts at the point $x = 2.00$ m, $y = 0.68$ m. If the angle $\alpha = 15°$, what is the total normal force exerted on the two rear tires by the sloped ramp?

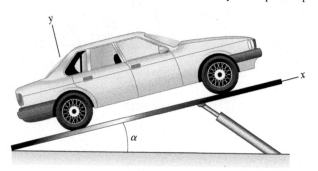

Problem 5.22

5.23 The link AB exerts a force on the bucket of the excavator at A that is parallel to the link. The weight $W = 1500$ lb. Draw the free-body diagram of the bucket and determine the reactions at C. (The connection at C is equivalent to a pin support of the bucket.)

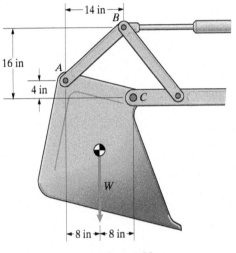

Problem 5.23

5.24 The 14.5-lb chain saw is subjected to the loads at A by the log it cuts. Determine the reactions R, B_x, and B_y that must be applied by the person using the saw to hold it in equilibrium.

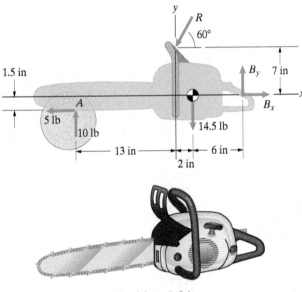

Problem 5.24

5.25 The mass of the trailer is 2.2 Mg (megagrams). The distances $a = 2.5$ m and $b = 5.5$ m. The truck is stationary, and the wheels of the trailer can turn freely, which means the road exerts no horizontal force on them. The hitch at B can be modeled as a pin support.

(a) Draw the free-body diagram of the trailer.

(b) Determine the total normal force exerted on the rear tires at A and the reactions exerted on the trailer at the pin support B.

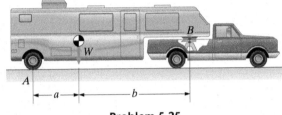

Problem 5.25

5.26 The total weight of the wheelbarrow and its load is $W = 100$ lb. (a) What is the magnitude of the upward force F necessary to lift the support at A off the ground? (b) What is the magnitude of the downward force necessary to raise the wheel off the ground?

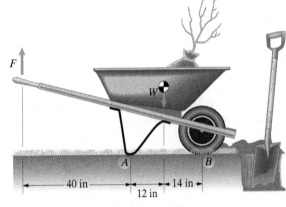

Problem 5.26

5.27 The airplane's weight is $W = 2400$ lb. Its brakes keep the rear wheels locked. The front (nose) wheel can turn freely, and so the ground exerts no horizontal force on it. The force T exerted by the airplane's propeller is horizontal.

(a) Draw the free-body diagram of the airplane. Determine the reaction exerted on the nose wheel and the total normal reaction exerted on the rear wheels

(b) when $T = 0$;

(c) when $T = 250$ lb.

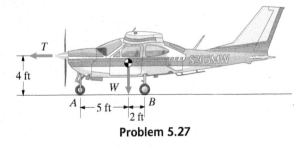

Problem 5.27

5.28 A safety engineer establishing limits on the loads that can be carried by a forklift analyzes the situation shown. The dimensions are $a = 32$ in, $b = 30$ in, and $c = 26$ in. The combined weight of the forklift and operator is $W_F = 1200$ lb. As the weight W_L supported by the forklift increases, the normal force exerted on the floor by the rear wheels at B decreases. The forklift is on the verge of tipping forward when the normal force at B is zero. Determine the value of W_L that will cause this condition.

Problem 5.28

5.29 Paleontologists speculate that the stegosaur could stand on its hind limbs for short periods to feed. Based on the free-body diagram shown and assuming that $m = 2000$ kg, determine the magnitudes of the forces B and C exerted by the ligament–muscle brace and vertebral column, and determine the angle α.

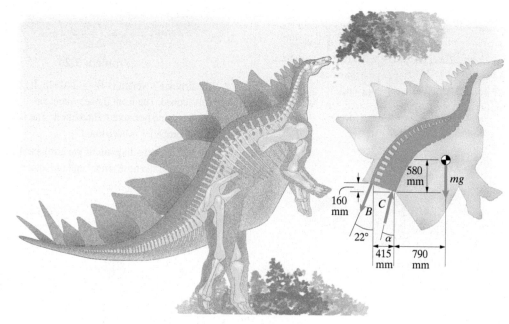

Problem 5.29

5.30 The weight of the fan is $W = 20$ lb. Its base has four equally spaced legs of length $b = 12$ in. Each leg has a pad near the end that contacts the floor and supports the fan. The height $h = 32$ in. If the fan's blade exerts a thrust $T = 2$ lb, what total normal force is exerted on the two legs at A?

5.31 The weight of the fan is $W = 20$ lb. Its base has four equally spaced legs of length $b = 12$ in. Each leg has a pad near the end that contacts the floor and supports the fan. The height $h = 32$ in. As the thrust T of the fan increases, the normal force supported by the two legs at A decreases. When the normal force at A is zero, the fan is on the verge of tipping over. Determine the value of T that will cause this condition.

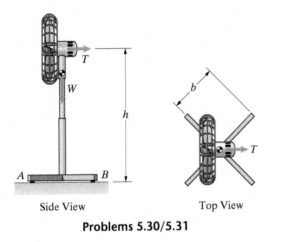

Side View   Top View

Problems 5.30/5.31

5.32 In a measure to decrease costs, the manufacturer of the fan described in Problem 5.31 proposes to support the fan with three equally spaced legs instead of four. An engineer is assigned to analyze the safety implications of the change. The weight of the fan decreases to $W = 19.6$ lb. The dimensions b and h are unchanged. What thrust T will cause the fan to be on the verge of tipping over in this case? Compare your answer to the answer to Problem 5.31.

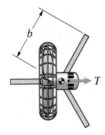

Problem 5.32

5.33 A force $F = 400$ N acts on the bracket. What are the reactions at A and B?

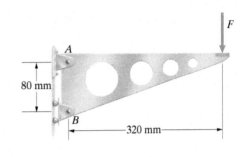

Problem 5.33

5.34 The sign's weight $W_s = 32$ lb acts at the point shown. The 10-lb weight of bar AD acts at the midpoint of the bar. Determine the tension in the cable AE and the reactions at D.

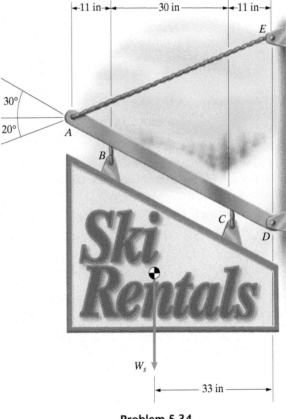

Problem 5.34

5.35 The device shown, called a *swape* or *shadoof*, helps a person lift a heavy load. (Devices of this kind were used in Egypt at least as early as 1550 B.C. and are still in use in various parts of the world.) The dimensions $a = 3.6$ m and $b = 1.2$ m. The mass of the bar and counterweight is 90 kg, and their weight W acts at the point shown. The mass of the load being lifted is 45 kg. Determine the vertical force the person must exert to support the stationary load (a) when the load is just above the ground (the position shown); (b) when the load is 1 m above the ground. Assume that the rope remains vertical.

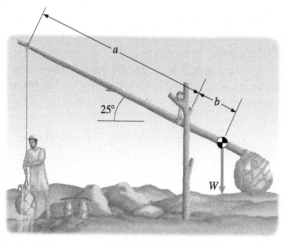

Problem 5.35

5.36 This structure, called a *truss*, has a pin support at A and a roller support at B and is loaded by two forces. Determine the reactions at the supports.

Strategy: Draw a free-body diagram treating the entire truss as a single object.

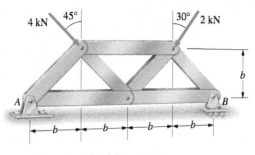

Problem 5.36

5.37 An Olympic gymnast is stationary in the "iron cross" position. The weight of his left arm and the weight of his body *not including his arms* are shown. The distances are $a = b = 9$ in and $c = 13$ in. Treat his shoulder S as a fixed support, and determine the magnitudes of the reactions at his shoulder. That is, determine the force and couple his shoulder must support.

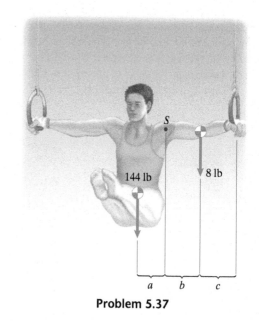

Problem 5.37

5.38 Determine the reactions at A.

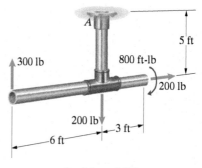

Problem 5.38

5.39 The car's brakes keep the rear wheels locked, and the front wheels are free to turn. Determine the forces exerted on the front and rear wheels by the road when the car is parked (a) on an upslope with $\alpha = 15°$; (b) on a downslope with $\alpha = -15°$.

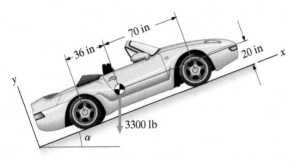

Problem 5.39

5.40 The length of the bar is $L = 4$ ft. Its weight $W = 6$ lb acts at the midpoint of the bar. The floor and wall are smooth. The spring is unstretched when the angle $\alpha = 0$. If the bar is in equilibrium when $\alpha = 40°$, what is the spring constant k?

5.41 The weight W of the bar acts at its midpoint. The floor and wall are smooth. The spring is unstretched when the angle $\alpha = 0$. Determine the angle α at which the bar is in equilibrium in terms of W, k, and L.

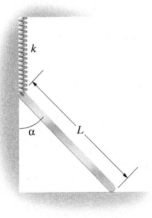

Problems 5.40/5.41

5.42 The plate is supported by a pin in a smooth slot at B. What are the reactions at the supports?

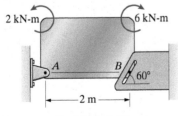

Problem 5.42

5.43 Determine the reactions at the fixed support A.

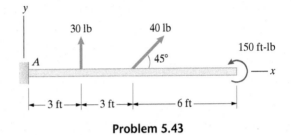

Problem 5.43

5.44 Suppose that you want to represent the two forces and couple acting on the beam in Problem 5.43 by an equivalent force $\mathbf{F}$ as shown. (a) Determine $\mathbf{F}$ and the distance D at which its line of action crosses the x axis. (b) Assume that $\mathbf{F}$ is the only load acting on the beam and determine the reactions at the fixed support A. Compare your answers to the answers to Problem 5.43.

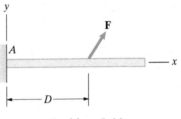

Problem 5.44

5.45 The bicycle brake on the right is pinned to the bicycle's frame at A. Determine the force exerted by the brake pad on the wheel rim at B in terms of the cable tension T.

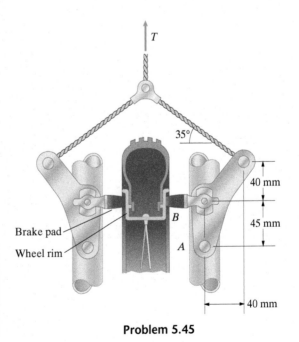

Problem 5.45

5.46 The mass of each of the suspended weights is 80 kg. Determine the reactions at the supports at A and E.

5.47 The suspended weights are each of mass m. The supports at A and E will each safely support a force of 6 kN magnitude. Based on this criterion, what is the largest safe value of m?

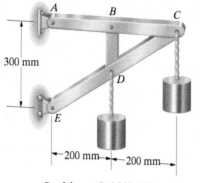

300 mm

200 mm — 200 mm

Problems 5.46/5.47

5.48 The tension in cable BC is 100 lb. Determine the reactions at the fixed support.

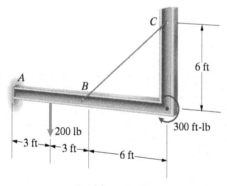

6 ft

200 lb

300 ft-lb

3 ft — 3 ft — 6 ft

Problem 5.48

5.49 The tension in cable AB is 2 kN. What are the reactions at C in the two cases?

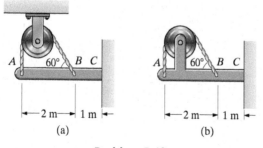

60°   B C

2 m — 1 m

(a)

60°   B C

2 m — 1 m

(b)

Problem 5.49

5.50 Determine the reactions at the supports.

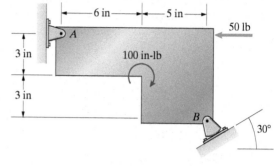

6 in — 5 in

50 lb

3 in

100 in-lb

3 in

B

30°

Problem 5.50

5.51 The weight $W = 2$ kN. Determine the tension in the cable and the reactions at A.

5.52 The cable will safely support a tension of 6 kN. Based on this criterion, what is the largest safe value of the weight W?

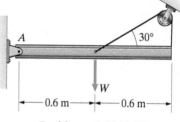

A

30°

W

0.6 m — 0.6 m

Problems 5.51/5.52

5.53 The blocks being compressed by the clamp exert a 200-N force on the pin at D that points from A toward D. The threaded shaft BE exerts a force on the pin at E that points from B toward E.

(a) Draw a free-body diagram of the arm DCE of the clamp, assuming that the pin at C behaves like a pin support.

(b) Determine the reactions at C.

5.54 The blocks being compressed by the clamp exert a 200-N force on the pin at A that points from D toward A. The threaded shaft BE exerts a force on the pin at B that points from E toward B.

(a) Draw a free-body diagram of the arm ABC of the clamp, assuming that the pin at C behaves like a pin support.

(b) Determine the reactions at C.

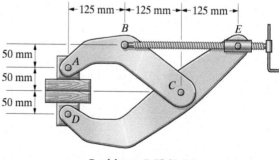

125 mm — 125 mm — 125 mm

B

E

50 mm

A

50 mm

C

50 mm

D

Problems 5.53/5.54

5.55 Suppose that you want to design the safety valve to open when the difference between the pressure p in the circular pipe (diameter = 150 mm) and atmospheric pressure is 10 MPa (megapascals; a pascal is 1 N/m^2). The spring is compressed 20 mm when the valve is closed. What should the value of the spring constant be?

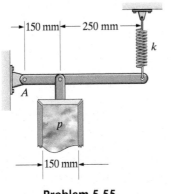

Problem 5.55

5.56 The 10-lb weight of the bar AB acts at the midpoint of the bar. The length of the bar is 3 ft. Determine the tension in the string BC and the reactions at A.

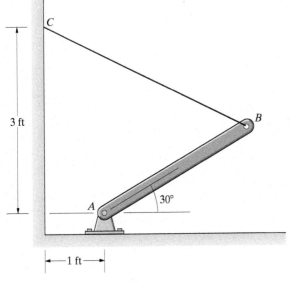

Problem 5.56

5.57 The crane's arm has a pin support at A. The hydraulic cylinder BC exerts a force on the arm at C in the direction parallel to BC. The crane's arm has a mass of 200 kg, and its weight can be assumed to act at a point 2 m to the right of A. If the mass of the suspended box is 800 kg and the system is in equilibrium, what is the magnitude of the force exerted by the hydraulic cylinder?

5.58 In Problem 5.57, what is the magnitude of the force exerted on the crane's arm by the pin support at A?

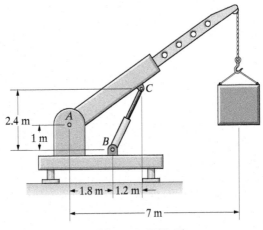

Problems 5.57/5.58

5.59 A speaker system is suspended by the cables attached at D and E. The mass of the speaker system is 130 kg, and its weight acts at G. Determine the tensions in the cables and the reactions at A and C.

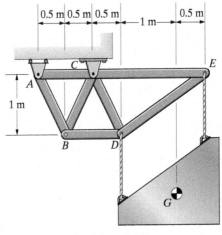

Problem 5.59

5.60 The weight $W_1 = 1000$ lb. Neglect the weight of the bar AB. The cable goes over a pulley at C. Determine the weight W_2 and the reactions at the pin support A.

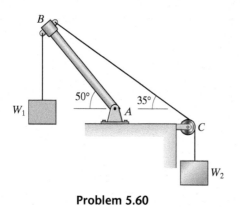

Problem 5.60

5.61 The dimensions $a = 2$ m and $b = 1$ m. The couple $M = 2400$ N-m. The spring constant is $k = 6000$ N/m, and the spring would be unstretched if $h = 0$. The system is in equilibrium when $h = 2$ m and the beam is horizontal. Determine the force F and the reactions at A.

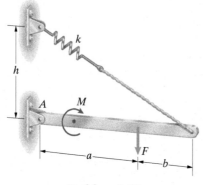

Problem 5.61

5.62 The bar is 1 m long, and its weight W acts at its midpoint. The distance $b = 0.75$ m, and the angle $\alpha = 30°$. The spring constant is $k = 100$ N/m, and the spring is unstretched when the bar is vertical. Determine W and the reactions at A.

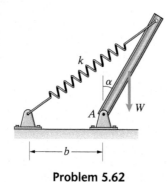

Problem 5.62

5.63 The boom derrick supports a suspended 15-kip load. The booms BC and DE are each 20 ft long. The distances are $a = 15$ ft and $b = 2$ ft, and the angle $\theta = 30°$. Determine the tension in cable AB and the reactions at the pin supports C and D.

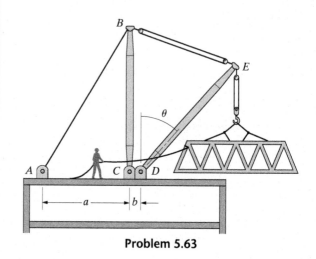

Problem 5.63

5.64 The arrangement shown controls the elevators of an airplane. (The elevators are the horizontal control surfaces in the airplane's tail.) The elevators are attached to member EDG. Aerodynamic pressures on the elevators exert a clockwise couple of 120 in-lb. Cable BG is slack, and its tension can be neglected. Determine the force F and the reactions at the pin support A.

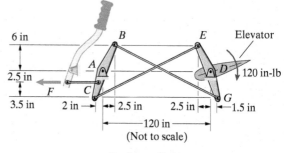

Problem 5.64

▶ **5.65** In Example 5.4, suppose that $\alpha = 40°$, $d = 1$ m, $a = 200$ mm, $b = 500$ mm, $R = 75$ mm, and the mass of the luggage is 40 kg. Determine F and N.

▶ **5.66** In Example 5.4, suppose that $\alpha = 35°$, $d = 46$ in, $a = 10$ in, $b = 14$ in, $R = 3$ in, and you don't want the user to have to exert a force F larger than 20 lb. What is the largest luggage weight that can be placed on the carrier?

▶ 5.67 One of the difficulties in making design decisions is that you don't know how the user will place the luggage on the carrier in Example 5.4. Suppose you assume that the point where the weight acts may be anywhere within the "envelope" $R \leq a \leq 0.75c$ and $0 \leq b \leq 0.75d$. If $\alpha = 30°$, $c = 14$ in, $d = 48$ in, $R = 3$ in, and $W = 80$ lb, what is the largest force F the user will have to exert for any luggage placement?

▶ 5.68 In Example 5.4, assume a user that would hold the carrier's handle at $h = 36$ in above the floor. Assume that $R = 3$ in, $a = 6$ in, $b = 12$ in, and $d = 4$ ft. The resulting ratio of the force the user must exert to the weight of the luggage is $F/W = 0.132$. Suppose that people with a range of heights use this carrier. Obtain a graph of F/W as a function of h for $24 \leq h \leq 36$ in.

5.2 Statically Indeterminate Objects

BACKGROUND

In Section 5.1 we discussed examples in which we were able to use the equilibrium equations to determine unknown forces and couples acting on objects in equilibrium. It is important to be aware of two common situations in which this procedure doesn't lead to a solution. First, the free-body diagram of an object can have more unknown forces or couples than the number of independent equilibrium equations that can be obtained. For example, because no more than three independent equilibrium equations can be obtained from a given free-body diagram in a two-dimensional problem, if there are more than three unknowns they can't all be determined from the equilibrium equations alone. This occurs, for example, when an object has more supports than the minimum number necessary to maintain it in equilibrium. Such an object is said to have *redundant supports*. The second situation is when the supports of an object are improperly designed such that they cannot maintain equilibrium under the loads acting on it. The object is said to have *improper supports*. In either situation, the object is said to be *statically indeterminate*.

Engineers use redundant supports whenever possible for strength and safety. Some designs, however, require that the object be incompletely supported so that it is free to undergo certain motions. These two situations—more supports than necessary for equilibrium or not enough—are so common that we consider them in detail.

Redundant Supports

Consider a beam with a fixed support (Fig. 5.9a). From its free-body diagram (Fig. 5.9b), we obtain the equilibrium equations

$$\Sigma F_x = A_x = 0,$$
$$\Sigma F_y = A_y - F = 0,$$
$$\Sigma M_{\text{point } A} = M_A - \left(\frac{L}{2}\right)F = 0.$$

Assuming we know the load F, we have three equations and three unknown reactions, for which we obtain the solutions $A_x = 0$, $A_y = F$, and $M_A = FL/2$.

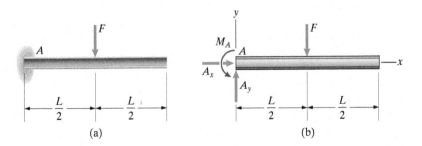

(a)   (b)

Figure 5.9
(a) A beam with a fixed support.
(b) The free-body diagram has three unknown reactions.

Figure 5.10
(a) A beam with fixed and roller supports.
(b) The free-body diagram has four unknown reactions.

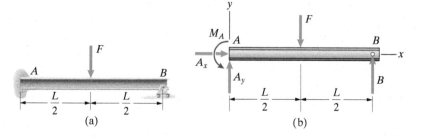

(a)

(b)

Now suppose we add a roller support at the right end of the beam (Fig. 5.10a). From the new free-body diagram (Fig. 5.10b), we obtain the equilibrium equations

$$\Sigma F_x = A_x = 0, \tag{5.6}$$

$$\Sigma F_y = A_y - F + B = 0, \tag{5.7}$$

$$\Sigma M_{\text{point } A} = M_A - \left(\frac{L}{2}\right)F + LB = 0. \tag{5.8}$$

Now we have three equations and four unknown reactions. Although the first equation tells us that $A_x = 0$, we can't solve the two equations (5.7) and (5.8) for the three reactions A_y, B, and M_A.

When faced with this situation, students often attempt to sum the moments about another point, such as point B, to obtain an additional equation:

$$\Sigma M_{\text{point } B} = M_A + \left(\frac{L}{2}\right)F - LA_y = 0.$$

Unfortunately, this doesn't help. This is not an independent equation, but is a linear combination of Eqs. (5.7) and (5.8):

$$\Sigma M_{\text{point } B} = M_A + \left(\frac{L}{2}\right)F - LA_y$$

$$= \underbrace{M_A - \left(\frac{L}{2}\right)F + LB}_{\text{Eq. (5.8)}} - \underbrace{L(A_y - F + B)}_{\text{Eq. (5.7)}}.$$

As this example demonstrates, each support added to an object results in additional reactions. The difference between the number of reactions and the number of independent equilibrium equations is called the *degree of redundancy*. Even if an object is statically indeterminate due to redundant supports, it may be possible to determine some of the reactions from the equilibrium equations. Notice that in our previous example we were able to determine the reaction A_x even though we could not determine the other reactions.

Since redundant supports are so ubiquitous, you may wonder why we devote so much effort to teaching you how to analyze objects whose reactions can be determined with the equilibrium equations. We want to develop your understanding of equilibrium and give you practice writing equilibrium equations. The reactions on an object with redundant supports *can* be determined by supplementing the equilibrium equations with additional equations that relate the forces and couples acting on the object to its deformation, or change in shape. Thus obtaining the equilibrium equations is the first step of the solution.

Improper Supports

We say that an object has improper supports if it will not remain in equilibrium under the action of the loads exerted on it. Thus an object with improper supports will move when the loads are applied. In two-dimensional problems, this can occur in two ways:

1. *The supports can exert only parallel forces.* This leaves the object free to move in the direction perpendicular to the support forces. If the loads exert a component of force in that direction, the object is not in equilibrium. Figure 5.11a shows an example of this situation. The two roller supports can exert only vertical forces, while the force F has a horizontal component. The beam will move horizontally when F is applied. This is particularly apparent from the free-body diagram (Fig. 5.11b). The sum of the forces in the horizontal direction cannot be zero because the roller supports can exert only vertical reactions.

2. *The supports can exert only concurrent forces.* If the loads exert a moment about the point where the lines of action of the support forces intersect, the object is not in equilibrium. For example, consider the beam in Fig. 5.12a. From its free-body diagram (Fig. 5.12b) we see that the reactions A and B exert no moment about the point P, where their lines of action intersect, but the load F does. The sum of the moments about point P is not zero, and the beam will rotate when the load is applied.

Except for problems that deal explicitly with improper supports, objects in our examples and problems have proper supports. You should develop the habit of examining objects in equilibrium and thinking about why they are properly supported for the loads acting on them.

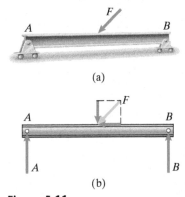

Figure 5.11
(a) A beam with two roller supports is not in equilibrium when subjected to the load shown.
(b) The sum of the forces in the horizontal direction is not zero.

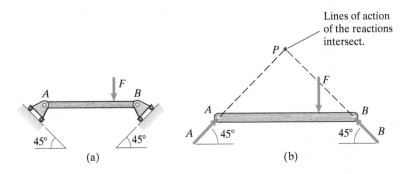

Figure 5.12
(a) A beam with roller supports on sloped surfaces.
(b) The sum of the moments about point P is not zero.

RESULTS

A supported object is said to be *statically indeterminate* in two circumstances:

> **Redundant Supports**
> The object has more supports than the minimum number necessary to maintain equilibrium. The difference between the number of reactions due to the supports and the number of independent equilibrium equations is called the *degree of redundancy*.

> **Improper Supports**
> The supports cannot maintain the object in equilibrium under the loads acting on it.

Active Example 5.5   Recognizing a Statically Indeterminate Object (▶ *Related Problem 5.69*)

The beam has two pin supports and is loaded by a 2-kN force.

(a) Show that the beam is statically indeterminate and determine the degree of redundancy.
(b) Determine as many reactions as possible.

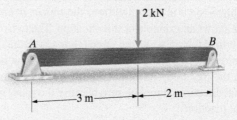

Strategy

The beam is statically indeterminate if its free-body diagram has more unknown reactions than the number of independent equilibrium equations we can obtain. The difference between the number of reactions and the number of equilibrium equations is the degree of redundancy. Even if the beam is statically indeterminate, it may be possible to solve the equilibrium equation for some of the reactions.

Solution

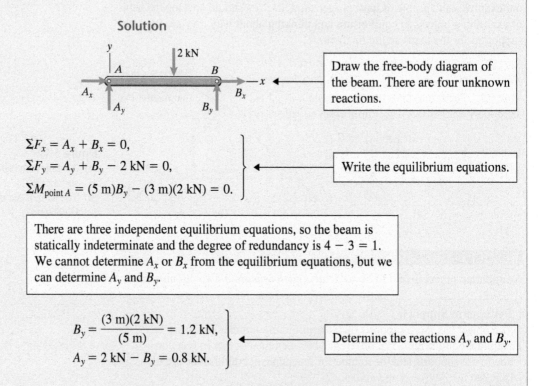

Draw the free-body diagram of the beam. There are four unknown reactions.

$$\Sigma F_x = A_x + B_x = 0,$$
$$\Sigma F_y = A_y + B_y - 2 \text{ kN} = 0,$$
$$\Sigma M_{\text{point } A} = (5 \text{ m})B_y - (3 \text{ m})(2 \text{ kN}) = 0.$$

Write the equilibrium equations.

There are three independent equilibrium equations, so the beam is statically indeterminate and the degree of redundancy is $4 - 3 = 1$. We cannot determine A_x or B_x from the equilibrium equations, but we can determine A_y and B_y.

$$B_y = \frac{(3 \text{ m})(2 \text{ kN})}{(5 \text{ m})} = 1.2 \text{ kN},$$
$$A_y = 2 \text{ kN} - B_y = 0.8 \text{ kN}.$$

Determine the reactions A_y and B_y.

Practice Problem Suppose that the pin support at point A of the beam is replaced by a fixed support. (a) Show that the beam is statically indeterminate and determine the degree of redundancy. (b) Determine as many reactions as possible.

Answer: (a) Degree of redundancy is 2. (b) No reactions can be determined.

Example 5.6 | Proper and Improper Supports (▶ *Related Problems 5.75, 5.76*)

State whether each L-shaped bar is properly or improperly supported. If a bar is properly supported, determine the reactions at its supports.

Strategy

By drawing the free-body diagram of each bar, we can determine whether the reactions of the supports can exert only parallel or concurrent forces on it. If so, we can then recognize whether the applied load results in the bar not being in equilibrium.

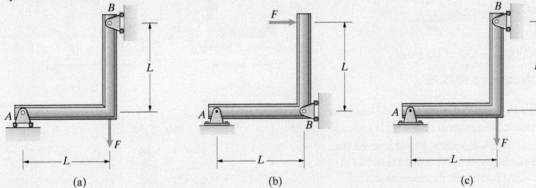

(a)   (b)   (c)

Solution

Consider the free-body diagrams of the bars (shown below):

Bar (a) The lines of action of the reactions due to the two roller supports intersect at P, and the load F exerts a moment about P. This bar is improperly supported.

Bar (b) The lines of action of the reactions intersect at A, and the load F exerts a moment about A. This bar is also improperly supported.

Bar (c) The three support forces are neither parallel nor concurrent. This bar is properly supported. The equilibrium equations are

$$\Sigma F_x = A_x - B = 0,$$
$$\Sigma F_y = A_y - F = 0,$$
$$\Sigma M_{\text{point } A} = BL - FL = 0.$$

Solving these equations, the reactions are $A_x = F$, $A_y = F$, and $B = F$.

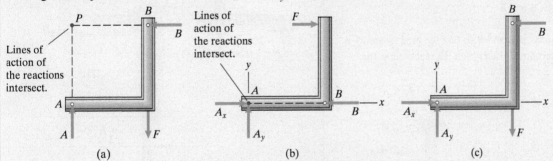

(a)   (b)   (c)

Critical Thinking

An essential part of learning mechanics is developing your intuition about the behaviors of the physical systems we study. In this example, think about the effects of the loads on the three systems, and see if you can predict whether they are properly supported. Will the loads cause the bars to move or not? Then see if your judgment is confirmed by the analyses given in the example.

Problems

▶ 5.69 (a) Draw the free-body diagram of the beam and show that it is statically indeterminate. (See Active Example 5.5.)

(b) Determine as many of the reactions as possible.

5.70 Choose supports at A and B so that the beam is not statically indeterminate. Determine the reactions at the supports.

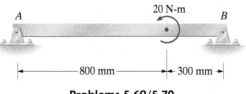

Problems 5.69/5.70

5.71 (a) Draw the free-body diagram of the beam and show that it is statically indeterminate. (The external couple M_0 is known.)

(b) By an analysis of the beam's deflection, it is determined that the vertical reaction B exerted by the roller support is related to the couple M_0 by $B = 2M_0/L$. What are the reactions at A?

5.72 Choose supports at A and B so that the beam is not statically indeterminate. Determine the reactions at the supports.

Problems 5.71/5.72

5.73 Draw the free-body diagram of the L-shaped pipe assembly and show that it is statically indeterminate. Determine as many of the reactions as possible.

Strategy: Place the coordinate system so that the x axis passes through points A and B.

5.74 Choose supports at A and B so that the pipe assembly is not statically indeterminate. Determine the reactions at the supports.

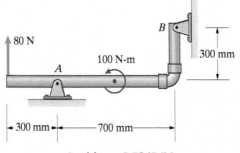

Problems 5.73/5.74

▶ 5.75 State whether each of the L-shaped bars shown is properly or improperly supported. If a bar is properly supported, determine the reactions at its supports. (See Active Example 5.6.)

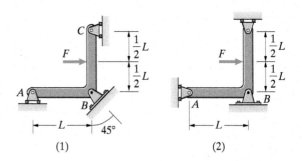

(1)   (2)

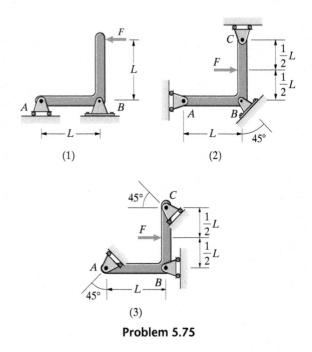

(3)

Problem 5.75

▶ 5.76 State whether each of the L-shaped bars shown is properly or improperly supported. If a bar is properly supported, determine the reactions at its supports. (See Active Example 5.6.)

(1)   (2)

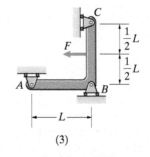

(3)

Problem 5.76

5.3 Three-Dimensional Applications

We have seen that when an object in equilibrium is subjected to a two-dimensional system of forces and moments, no more than three independent equilibrium equations can be obtained. In the case of a three-dimensional system of forces and moments, up to six independent equilibrium equations can be obtained. The three components of the sum of the forces must equal zero and the three components of the sum of the moments about a point must equal zero. The procedure for determining the reactions on an object subjected to a three-dimensional system of forces and moments—drawing a free-body diagram and applying the equilibrium equations—is the same as in two dimensions.

The Scalar Equilibrium Equations

When an object is in equilibrium, the system of forces and couples acting on it satisfy Eqs. (5.1) and (5.2). The sum of the forces is zero and the sum of the moments about any point is zero. Expressing these equations in terms of cartesian components in three dimensions yields the six scalar equilibrium equations.

$$\Sigma F_x = 0, \tag{5.9}$$

$$\Sigma F_y = 0, \tag{5.10}$$

$$\Sigma F_z = 0, \tag{5.11}$$

$$\Sigma M_x = 0, \tag{5.12}$$

$$\Sigma M_y = 0, \tag{5.13}$$

$$\Sigma M_z = 0. \tag{5.14}$$

The sums of the moments can be evaluated about any point. Although more equations can be obtained by summing moments about other points, they will not be independent of these equations. *More than six independent equilibrium equations cannot be obtained from a given free-body diagram, so at most six unknown forces or couples can be determined.*

   The steps required to determine reactions in three dimensions are familiar from the two-dimensional applications we have discussed. First obtain a free-body diagram by isolating an object and showing the loads and reactions acting on it, then use Eqs. (5.9)–(5.14) to determine the reactions.

Supports

We present five conventions frequently used in three-dimensional problems. Even when actual supports do not physically resemble these models, we represent them by the models if they exert the same (or approximately the same) reactions.

The Ball and Socket Support   In the *ball and socket support*, the supported object is attached to a ball enclosed within a spherical socket (Fig. 5.13a). The socket permits the ball to rotate freely (friction is neglected) but prevents it from translating in any direction.

   Imagine holding a bar attached to a ball and socket support (Fig. 5.13b). If you try to translate the bar (move it without rotating it) in any direction, the support exerts a reactive force to prevent the motion. However, you can rotate

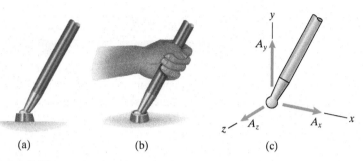

Figure 5.13
(**a**) A ball and socket support.
(**b**) Holding a supported bar.
(**c**) The ball and socket support can exert three components of force.

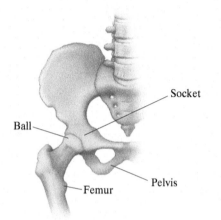

Figure 5.14
The human femur is attached to the pelvis by a ball and socket support.

the bar about the support. The support cannot exert a couple to prevent rotation. Thus a ball and socket support can't exert a couple but can exert three components of force (Fig. 5.13c). It is the three-dimensional analog of the two-dimensional pin support.

The human hip joint is an example of a ball and socket support (Fig. 5.14). The support of the gear shift lever of a car can be modeled as a ball and socket support within the lever's range of motion.

The Roller Support   The *roller support* (Fig. 5.15a) is a ball and socket support that can roll freely on a supporting surface. A roller support can exert only a force normal to the supporting surface (Fig. 5.15b). The rolling "casters" sometimes used to support furniture legs are supports of this type.

The Hinge   The hinge support is the familiar device used to support doors. It permits the supported object to rotate freely about a line, the *hinge axis*. An object is attached to a hinge in Fig. 5.16a. The z axis of the coordinate system is aligned with the hinge axis.

If you imagine holding a bar attached to a hinge (Fig. 5.16b), notice that you can rotate the bar about the hinge axis. The hinge cannot exert a couple about the hinge axis (the z axis) to prevent rotation. However, you can't rotate the bar about the x or y axis because the hinge can exert couples about those

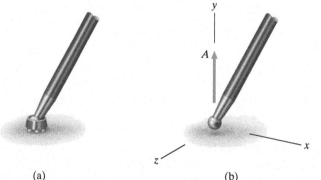

Figure 5.15
(**a**) A roller support.
(**b**) The reaction is normal to the supporting surface.

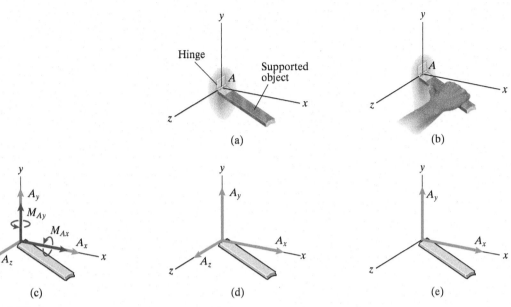

Figure 5.16
(a) A hinge. The z axis is aligned with the hinge axis.
(b) Holding a supported bar.
(c) In general, a hinge can exert five reactions: three force components and two couple components.
(d) The reactions when the hinge exerts no couples.
(e) The reactions when the hinge exerts neither couples nor a force parallel to the hinge axis.

axes to resist the motion. In addition, you can't translate the bar in any direction. The reactions a hinge can exert on an object are shown in Fig. 5.16c. There are three components of force, A_x, A_y, and A_z, and couples about the x and y axes, M_{Ax} and M_{Ay}.

In some situations, either a hinge exerts no couples on the object it supports, or they are sufficiently small to neglect. An example of the latter case is when the axes of the hinges supporting a door are properly aligned (the axes of the individual hinges coincide). In these situations the hinge exerts only forces on an object (Fig. 5.16d). Situations also arise in which a hinge exerts no couples on an object and exerts no force in the direction of the hinge axis. (The hinge may actually be designed so that it cannot support a force parallel to the hinge axis.) Then the hinge exerts forces only in the directions perpendicular to the hinge axis (Fig. 5.16e). In examples and problems, we indicate when a hinge does not exert all five of the reactions in Fig. 5.16c.

The Bearing   The type of bearing shown in Fig. 5.17a supports a circular shaft while permitting it to rotate about its axis. The reactions are identical to those exerted by a hinge. In the most general case (Fig. 5.17b), the bearing can exert a force on the supported shaft in each coordinate direction and can exert couples about axes perpendicular to the shaft but cannot exert a couple about the axis of the shaft.

As in the case of the hinge, situations can occur in which the bearing exerts no couples (Fig. 5.17c) or exerts no couples and no force parallel to the shaft axis (Fig. 5.17d). Some bearings are designed in this way for specific

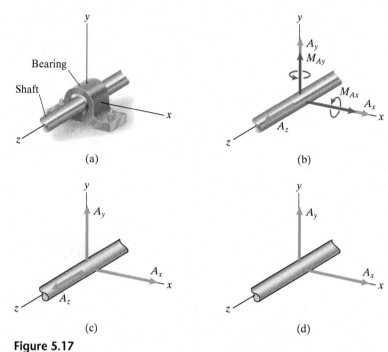

Figure 5.17
(a) A bearing. The z axis is aligned with the axis of the shaft.
(b) In general, a bearing can exert five reactions: three force components and two couple components.
(c) The reactions when the bearing exerts no couples.
(d) The reactions when the bearing exerts neither couples nor a force parallel to the axis of the shaft.

applications. In examples and problems, we indicate when a bearing does not exert all of the reactions in Fig. 5.17b.

The Fixed Support   You are already familiar with the fixed, or built-in, support (Fig. 5.18a). Imagine holding a bar with a fixed support (Fig. 5.18b). You cannot translate it in any direction, and you cannot rotate it about any axis.

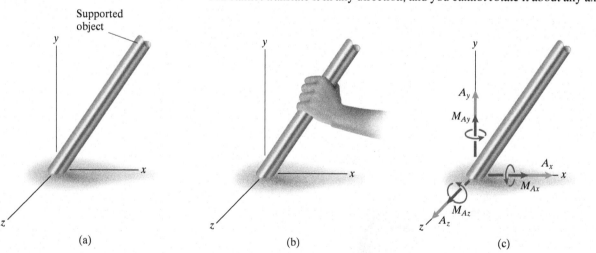

Figure 5.18
(a) A fixed support.
(b) Holding a supported bar.
(c) A fixed support can exert six reactions: three force components and three couple components.

The support is capable of exerting forces A_x, A_y, and A_z in each coordinate direction and couples M_{Ax}, M_{Ay}, and M_{Az} about each coordinate axis (Fig. 5.18c).

Table 5.2 summarizes the support conventions commonly used in three-dimensional applications.

Table 5.2   Supports used in three-dimensional applications.

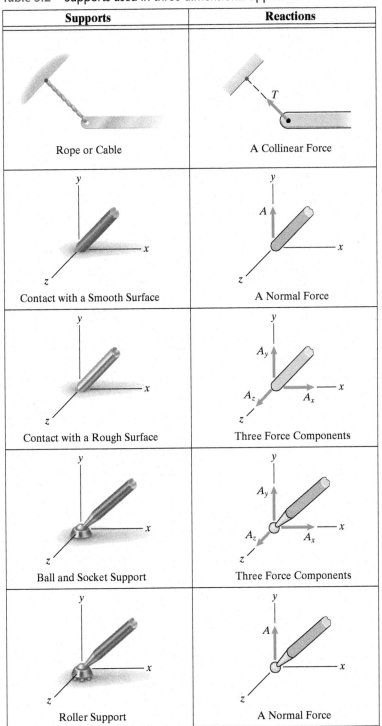

Supports	Reactions
Rope or Cable	A Collinear Force
Contact with a Smooth Surface	A Normal Force
Contact with a Rough Surface	Three Force Components
Ball and Socket Support	Three Force Components
Roller Support	A Normal Force

Table 5.2 *continued*

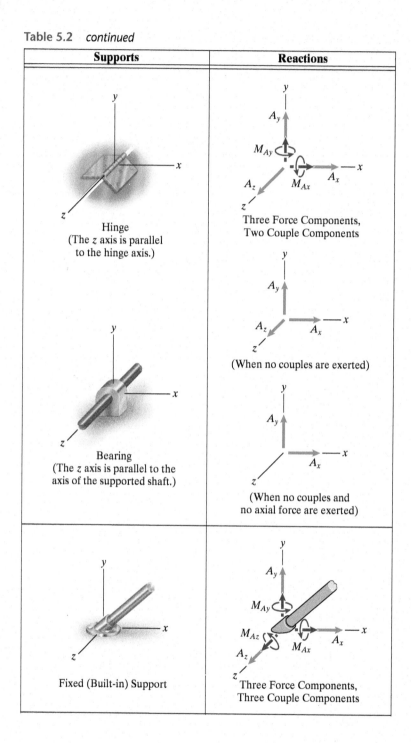

Supports	Reactions
Hinge (The z axis is parallel to the hinge axis.)	Three Force Components, Two Couple Components
Bearing (The z axis is parallel to the axis of the supported shaft.)	(When no couples are exerted) (When no couples and no axial force are exerted)
Fixed (Built-in) Support	Three Force Components, Three Couple Components

RESULTS

Equilibrium Equations

If an object is in equilibrium, the sum of the external forces acting on it equals zero,

$$\Sigma \mathbf{F} = \mathbf{0}$$

$$\Sigma F_x = 0, \qquad (5.9)$$
$$\Sigma F_y = 0, \qquad (5.10)$$
$$\Sigma F_z = 0, \qquad (5.11)$$

and the sum of the moments about any point due to the forces and couples acting on it is zero,

$$\Sigma \mathbf{M}_{\text{any point}} = \mathbf{0}$$

$$\Sigma M_x = 0, \qquad (5.12)$$
$$\Sigma M_y = 0, \qquad (5.13)$$
$$\Sigma M_z = 0. \qquad (5.14)$$

Supports

Examples of supports used in three-dimensional applications. (See Table 5.2.)

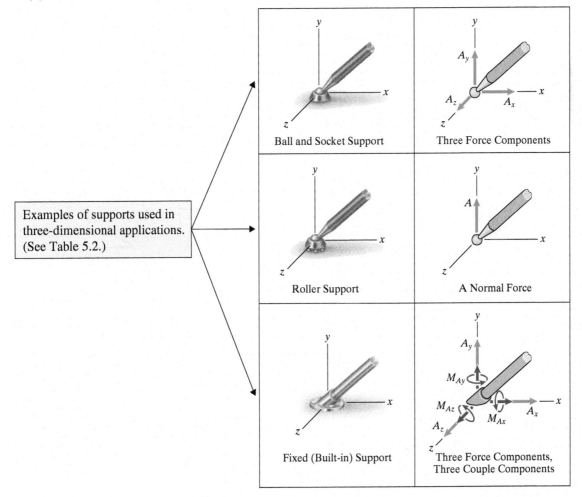

Ball and Socket Support

Three Force Components

Roller Support

A Normal Force

Fixed (Built-in) Support

Three Force Components, Three Couple Components

Active Example 5.7 | Determining Reactions in Three Dimensions (▶ *Related Problem 5.86*)

The bar *AB* is supported by the cables *BC* and *BD* and a ball and socket support at *A*. Cable *BC* is parallel to the *z* axis and cable *BD* is parallel to the *x* axis. The 200-N force acts at the midpoint of the bar. Determine the tensions in the cables and the reactions at *A*.

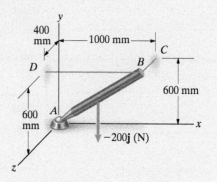

Strategy

We must obtain the free-body diagram of the bar by isolating it and showing the reactions exerted by the cables and the ball and socket support. Then we can apply the equilibrium equations to determine the reactions.

Solution

Draw the Free-Body Diagram of the Bar

Isolate the bar and show the reactions exerted by the cables and the ball and socket support.

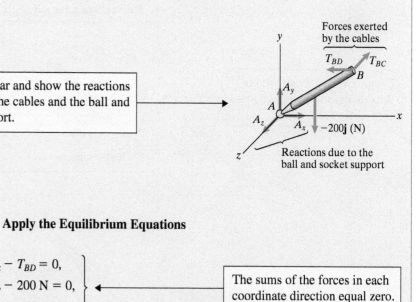

Forces exerted by the cables

Reactions due to the ball and socket support

Apply the Equilibrium Equations

$$\Sigma F_x = A_x - T_{BD} = 0,$$
$$\Sigma F_y = A_y - 200 \text{ N} = 0,$$
$$\Sigma F_z = A_z - T_{BC} = 0.$$

The sums of the forces in each coordinate direction equal zero.

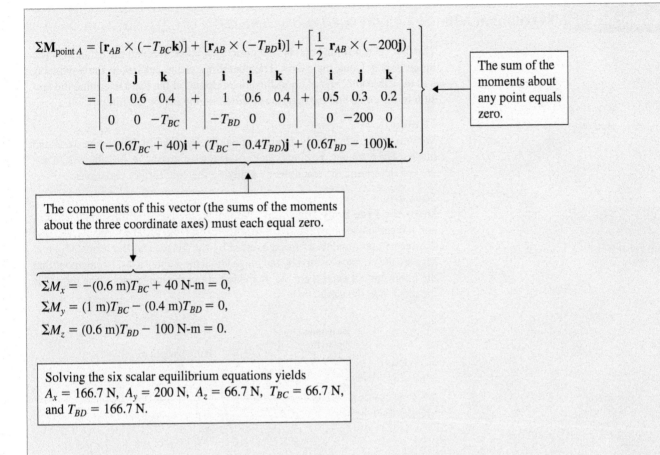

$$\Sigma\mathbf{M}_{\text{point } A} = [\mathbf{r}_{AB} \times (-T_{BC}\mathbf{k})] + [\mathbf{r}_{AB} \times (-T_{BD}\mathbf{i})] + \left[\frac{1}{2}\mathbf{r}_{AB} \times (-200\mathbf{j})\right]$$

$$= \begin{vmatrix} \mathbf{i} & \mathbf{j} & \mathbf{k} \\ 1 & 0.6 & 0.4 \\ 0 & 0 & -T_{BC} \end{vmatrix} + \begin{vmatrix} \mathbf{i} & \mathbf{j} & \mathbf{k} \\ 1 & 0.6 & 0.4 \\ -T_{BD} & 0 & 0 \end{vmatrix} + \begin{vmatrix} \mathbf{i} & \mathbf{j} & \mathbf{k} \\ 0.5 & 0.3 & 0.2 \\ 0 & -200 & 0 \end{vmatrix}$$

$$= (-0.6T_{BC} + 40)\mathbf{i} + (T_{BC} - 0.4T_{BD})\mathbf{j} + (0.6T_{BD} - 100)\mathbf{k}.$$

> The sum of the moments about any point equals zero.

> The components of this vector (the sums of the moments about the three coordinate axes) must each equal zero.

$$\Sigma M_x = -(0.6 \text{ m})T_{BC} + 40 \text{ N-m} = 0,$$
$$\Sigma M_y = (1 \text{ m})T_{BC} - (0.4 \text{ m})T_{BD} = 0,$$
$$\Sigma M_z = (0.6 \text{ m})T_{BD} - 100 \text{ N-m} = 0.$$

> Solving the six scalar equilibrium equations yields
> $A_x = 166.7$ N, $A_y = 200$ N, $A_z = 66.7$ N, $T_{BC} = 66.7$ N, and $T_{BD} = 166.7$ N.

Practice Problem Suppose that the cables BC and BD are removed and the ball and socket joint at A is replaced by a fixed support. Determine the reactions at A.

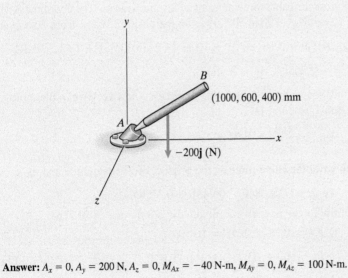

Answer: $A_x = 0, A_y = 200$ N, $A_z = 0, M_{Ax} = -40$ N-m, $M_{Ay} = 0, M_{Az} = 100$ N-m.

Example 5.8 | **Reactions at a Hinge Support** (▶ *Related Problem 5.104*)

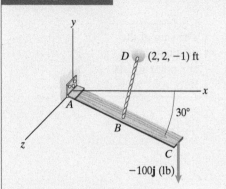

The bar AC is 4 ft long and is supported by a hinge at A and the cable BD. The hinge axis is along the z axis. The centerline of the bar lies in the x–y plane, and the cable attachment point B is the midpoint of the bar. Determine the tension in the cable and the reactions exerted on the bar by the hinge.

Strategy

We will obtain a free-body diagram of bar AC by isolating it from the cable and hinge. (The reactions the hinge can exert on the bar are shown in Table 5.2.) Then we can determine the reactions by applying the equilibrium equations.

Solution

Draw the Free-Body Diagram   We isolate the bar from the hinge support and the cable and show the reactions they exert (Fig. a). The terms A_x, A_y, and A_z are the components of force exerted by the hinge, and the terms M_{Ax} and M_{Ay} are the couples exerted by the hinge about the x and y axes. (Remember that the hinge cannot exert a couple on the bar about the hinge axis.) The term T is the tension in the cable.

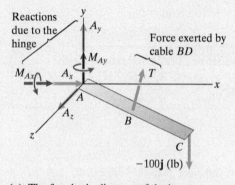

(a) The free-body diagram of the bar.

Apply the Equilibrium Equations   To write the equilibrium equations, we must first express the cable force in terms of its components. The coordinates of point B are $(2 \cos 30°, -2 \sin 30°, 0)$ ft, so the position vector from B to D is

$$\mathbf{r}_{BD} = (2 - 2 \cos 30°)\mathbf{i} + [2 - (-2 \sin 30°)]\mathbf{j} + (-1 - 0)\mathbf{k}$$

$$= 0.268\mathbf{i} + 3\mathbf{j} - \mathbf{k} \text{ (ft)}.$$

We divide this vector by its magnitude to obtain a unit vector $\mathbf{e}_{BD}$ that points from point B toward point D:

$$\mathbf{e}_{BD} = \frac{\mathbf{r}_{BD}}{|\mathbf{r}_{BD}|} = 0.084\mathbf{i} + 0.945\mathbf{j} - 0.315\mathbf{k}.$$

Now we can write the cable force as the product of its magnitude and $\mathbf{e}_{BD}$:

$$T\mathbf{e}_{BD} = T(0.084\mathbf{i} + 0.945\mathbf{j} - 0.315\mathbf{k}).$$

The sums of the forces in each coordinate direction must equal zero:

$$\Sigma F_x = A_x + 0.084T = 0,$$

$$\Sigma F_y = A_y + 0.945T - 100 \text{ lb} = 0, \tag{1}$$

$$\Sigma F_z = A_z - 0.315T = 0.$$

If we sum moments about A, the resulting equations do not contain the unknown reactions A_x, A_y, and A_z. The position vectors from A to B and from A to C are

$$\mathbf{r}_{AB} = 2 \cos 30°\mathbf{i} - 2 \sin 30°\mathbf{j} \text{ (ft)},$$

$$\mathbf{r}_{AC} = 4 \cos 30°\mathbf{i} - 4 \sin 30°\mathbf{j} \text{ (ft)}.$$

The sum of the moments about A, with forces in lb and distances in ft, is

$$\Sigma\mathbf{M}_{\text{point }A} = M_{Ax}\mathbf{i} + M_{Ay}\mathbf{j} + [\mathbf{r}_{AB} \times (T\mathbf{e}_{BD})] + [\mathbf{r}_{AC} \times (-100\mathbf{j})]$$

$$= M_{Ax}\mathbf{i} + M_{Ay}\mathbf{j} + \begin{vmatrix} \mathbf{i} & \mathbf{j} & \mathbf{k} \\ 1.732 & -1 & 0 \\ 0.084T & 0.945T & -0.315T \end{vmatrix}$$

$$+ \begin{vmatrix} \mathbf{i} & \mathbf{j} & \mathbf{k} \\ 3.464 & -2 & 0 \\ 0 & -100 & 0 \end{vmatrix}$$

$$= (M_{Ax} + 0.315T)\mathbf{i} + (M_{Ay} + 0.546T)\mathbf{j}$$

$$+ (1.72T - 346)\mathbf{k} = 0.$$

From this vector equation, we obtain the scalar equations

$$\Sigma M_x = M_{Ax} + (0.315 \text{ ft})T = 0,$$

$$\Sigma M_y = M_{Ay} + (0.546 \text{ ft})T = 0,$$

$$\Sigma M_z = (1.72 \text{ ft})T_{BD} - 346 \text{ ft-lb} = 0.$$

Solving these equations yields the reactions

$$T = 201 \text{ lb}, \quad M_{Ax} = -63.4 \text{ ft-lb}, \quad M_{Ay} = -109.8 \text{ ft-lb}.$$

Then from Eqs. (1) we obtain the forces exerted on the bar by the hinge:

$$A_x = -17.0 \text{ lb}, \quad A_y = -90.2 \text{ lb}, \quad A_z = 63.4 \text{ lb}.$$

Critical Thinking

Notice in Table 5.2 that there are three possibilities for the reactions exerted by a hinge or bearing. How do you know which one to choose? Under certain circumstances, a hinge may not exert significant couples on the object to which it is connected, and it also may not exert a significant force in the direction of the hinge axis. For example, when an object has two hinge supports and their axes are aligned (see Example 5.9), you can often assume that each individual hinge does not exert couples on the object. But in general, it requires experience to make such judgments. In upcoming examples and problems, we will indicate the reactions that you can assume are exerted by a hinge. Whenever you are in doubt, you should assume that a hinge may exert the most general set of reactions shown in Table 5.2 (three force components and two couple components).

Example 5.9 | **Reactions at Properly Aligned Hinges** (▶ *Related Problem 5.112*)

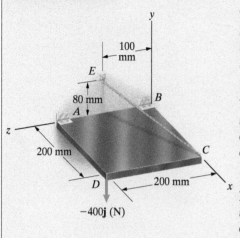

−400**j** (N)

The plate is supported by hinges at A and B and the cable CE. The properly aligned hinges do not exert couples on the plate, and the hinge at A does not exert a force on the plate in the direction of the hinge axis. Determine the reactions at the hinges and the tension in the cable.

Strategy

We will draw the free-body diagram of the plate, using the given information about the reactions exerted by the hinges at A and B. Before the equilibrium equations can be applied, we must express the force exerted on the plate by the cable in terms of its components.

Solution

Draw the Free-Body Diagram   We isolate the plate and show the reactions at the hinges and the force exerted by the cable (Fig. a). The term T is the force exerted on the plate by cable CE.

Apply the Equilibrium Equations   Since we know the coordinates of points C and E, we can express the cable force as the product of its magnitude T and a unit vector directed from C toward E. The result is

$$T(-0.842\mathbf{i} + 0.337\mathbf{j} + 0.421\mathbf{k}).$$

The sums of the forces in each coordinate direction equal zero:

$$\Sigma F_x = A_x + B_x - 0.842T = 0,$$
$$\Sigma F_y = A_y + B_y + 0.337T - 400 = 0, \tag{1}$$
$$\Sigma F_z = B_z + 0.421T = 0.$$

If we sum the moments about B, the resulting equations will not contain the three unknown reactions at B. The sum of the moments about B, with forces in N and distances in m, is

$$\Sigma\mathbf{M}_{\text{point }B} = \begin{vmatrix} \mathbf{i} & \mathbf{j} & \mathbf{k} \\ 0.2 & 0 & 0 \\ -0.842T & 0.337T & 0.421T \end{vmatrix} + \begin{vmatrix} \mathbf{i} & \mathbf{j} & \mathbf{k} \\ 0 & 0 & 0.2 \\ A_x & A_y & 0 \end{vmatrix}$$

$$+ \begin{vmatrix} \mathbf{i} & \mathbf{j} & \mathbf{k} \\ 0.2 & 0 & 0.2 \\ 0 & -400 & 0 \end{vmatrix}$$

$$= (-0.2A_y + 80)\mathbf{i} + (-0.0842T + 0.2A_x)\mathbf{j}$$
$$+ (0.0674T - 80)\mathbf{k} = 0.$$

The scalar equations are

$$\Sigma M_x = -(0.2\text{ m})A_y + 80\text{ N-m} = 0,$$
$$\Sigma M_y = -(0.0842\text{ m})T + (0.2\text{ m})A_x = 0,$$
$$\Sigma M_z = (0.0674\text{ m})T - 80\text{ N-m} = 0.$$

Solving these equations, we obtain the reactions

$$T = 1187\text{ N}, \qquad A_x = 500\text{ N}, \qquad A_y = 400\text{ N}.$$

Then from Eqs. (1), the reactions at B are

$$B_x = 500\text{ N}, \qquad B_y = -400\text{ N}, \qquad B_z = -500\text{ N}.$$

Reactions due to hinge A. It exerts no axial force.

Reactions due to hinge B

Force exerted by cable CE

−400**j** (N)

(a) The free-body diagram of the plate.

"Properly aligned hinges" means hinges that are mounted on an object so that their axes are aligned. When this is the case, as in this example, it can usually be assumed that each individual hinge does not exert couples on the object. Notice that it is also assumed in this example that the hinge at A exerts no reaction parallel to the hinge axis but the hinge at B does. The hinges can be intentionally designed so that this is the case, or it can result from the way they are installed.

If our only objective in this example had been to determine the tension T, we could have done so easily by evaluating the sum of the moments about the line AB (the z axis). Because the reactions at the hinges exert no moment about the z axis, we obtain the equation

$$(0.2 \text{ m})(0.337T) - (0.2 \text{ m})(400 \text{ N}) = 0,$$

which yields $T = 1187$ N.

Problems

5.77 The bar AB has a fixed support at A and is loaded by the forces

$$\mathbf{F}_B = 2\mathbf{i} + 6\mathbf{j} + 3\mathbf{k} \text{ (kN)},$$
$$\mathbf{F}_C = \mathbf{i} - 2\mathbf{j} + 2\mathbf{k} \text{ (kN)}.$$

(a) Draw the free-body diagram of the bar.

(b) Determine the reactions at A.

Strategy: (a) Draw a diagram of the bar isolated from its supports. Complete the free-body diagram of the bar by adding the two external forces and the reactions due to the fixed support (see Table 5.2). (b) Use the scalar equilibrium equations (5.9)–(5.14) to determine the reactions.

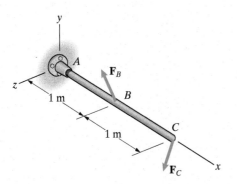

Problem 5.77

5.78 The bar AB has a fixed support at A. The tension in cable BC is 8 kN. Determine the reactions at A.

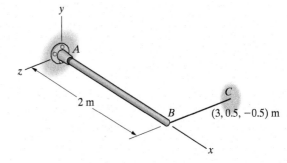

Problem 5.78

5.79 The bar AB has a fixed support at A. The collar at B is fixed to the bar. The tension in the rope BC is 300 lb. (a) Draw the free-body diagram of the bar. (b) Determine the reactions at A.

5.80 The bar AB has a fixed support at A. The collar at B is fixed to the bar. Suppose that you don't want the support at A to be subjected to a couple of magnitude greater than 3000 ft-lb. What is the largest allowable tension in the rope BC?

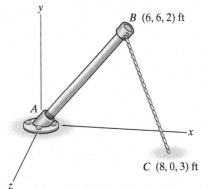

Problems 5.79/5.80

5.81 The total force exerted on the highway sign by its weight and the most severe anticipated winds is $\mathbf{F} = 2.8\mathbf{i} - 1.8\mathbf{j}$ (kN). Determine the reactions at the fixed support.

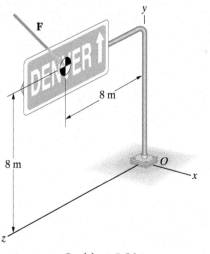

Problem 5.81

5.82 The tension in cable AB is 800 lb. Determine the reactions at the fixed support C.

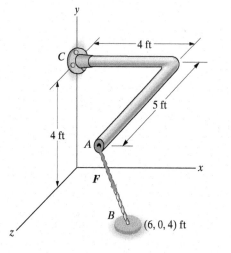

Problem 5.82

5.83 The tension in cable AB is 24 kN. Determine the reactions at the fixed support D.

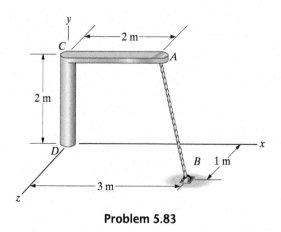

Problem 5.83

5.84 The robotic manipulator is stationary and the y axis is vertical. The weights of the arms AB and BC act at their midpoints. The direction cosines of the centerline of arm AB are $\cos \theta_x = 0.174$, $\cos \theta_y = 0.985$, $\cos \theta_z = 0$, and the direction cosines of the centerline of arm BC are $\cos \theta_x = 0.743$, $\cos \theta_y = 0.557$, $\cos \theta_z = -0.371$. The support at A behaves like a fixed support.

(a) What is the sum of the moments about A due to the weights of the two arms?

(b) What are the reactions at A?

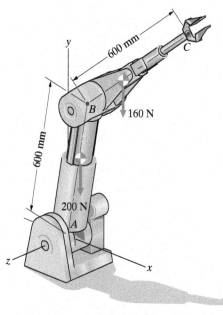

Problem 5.84

5.85 The force exerted on the grip of the exercise machine is $\mathbf{F} = 260\mathbf{i} - 130\mathbf{j}$ (N). What are the reactions at the fixed support at O?

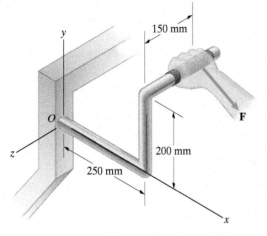

Problem 5.85

▶ **5.86** In Active Example 5.7, suppose that cable BD is lengthened and the attachment point D moved from $(0, 600, 400)$ mm to $(0, 600, 600)$ mm. (The end B of bar AB remains where it is.) Draw a sketch of the bar and its supports showing cable BD in its new position. Draw the free-body diagram of the bar and apply equilibrium to determine the tensions in the cables and the reactions at A.

5.87 The force $\mathbf{F}$ acting on the boom ABC at C points in the direction of the unit vector $0.512\mathbf{i} - 0.384\mathbf{j} + 0.768\mathbf{k}$ and its magnitude is 8 kN. The boom is supported by a ball and socket at A and the cables BD and BE. The collar at B is fixed to the boom.

(a) Draw the free-body diagram of the boom.

(b) Determine the tensions in the cables and the reactions at A.

5.88 The cables BD and BE in Problem 5.87 will each safely support a tension of 25 kN. Based on this criterion, what is the largest acceptable magnitude of the force $\mathbf{F}$?

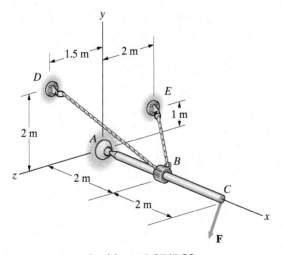

Problems 5.87/5.88

5.89 The suspended load exerts a force $F = 600$ lb at A, and the weight of the bar OA is negligible. Determine the tensions in the cables and the reactions at the ball and socket support O.

5.90 The suspended load exerts a force $F = 600$ lb at A and bar OA weighs 200 lb. Assume that the bar's weight acts at its midpoint. Determine the tensions in the cables and the reactions at the ball and socket support O.

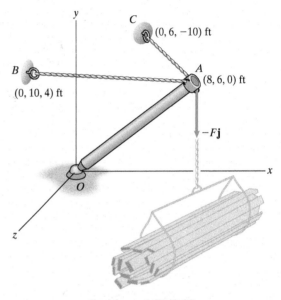

Problems 5.89/5.90

5.91 The 158,000-kg airplane is at rest on the ground ($z = 0$ is ground level). The landing gear carriages are at A, B, and C. The coordinates of the point G at which the weight of the plane acts are $(3, 0.5, 5)$ m. What are the magnitudes of the normal reactions exerted on the landing gear by the ground?

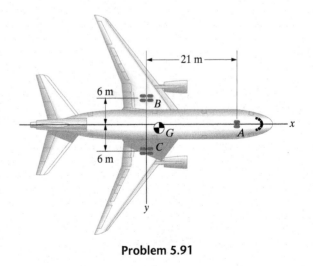

Problem 5.91

5.92 The horizontal triangular plate is suspended by the three vertical cables A, B, and C. The tension in each cable is 80 N. Determine the x and z coordinates of the point where the plate's weight effectively acts.

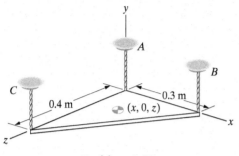

Problem 5.92

5.93 The 800-kg horizontal wall section is supported by the three vertical cables A, B, and C. What are the tensions in the cables?

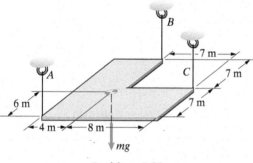

Problem 5.93

5.94 The bar AC is supported by the cable BD and a bearing at A that can rotate about the z axis. The person exerts a force $\mathbf{F} = 10\mathbf{j}$ (lb) at C. Determine the tension in the cable and the reactions at A.

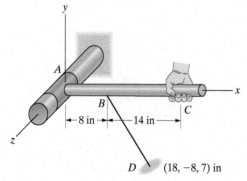

Problem 5.94

5.95 The L-shaped bar is supported by a bearing at A and rests on a smooth horizontal surface at B. The vertical force $F = 4$ kN and the distance $b = 0.15$ m. Determine the reactions at A and B.

5.96 The vertical force $F = 4$ kN and the distance $b = 0.15$ m. If you represent the reactions at A and B by an equivalent system consisting of a single force, what is the force and where does its line of action intersect the x–z plane?

5.97 The vertical force $F = 4$ kN. The bearing at A will safely support a force of 2.5-kN magnitude and a couple of 0.5 kN-m magnitude. Based on these criteria, what is the allowable range of the distance b?

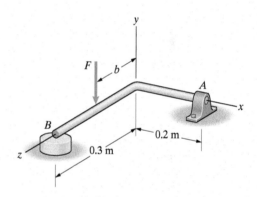

Problems 5.95–5.97

5.98 The 1.1-m bar is supported by a ball and socket support at A and the two smooth walls. The tension in the vertical cable CD is 1 kN.

(a) Draw the free-body diagram of the bar.

(b) Determine the reactions at A and B.

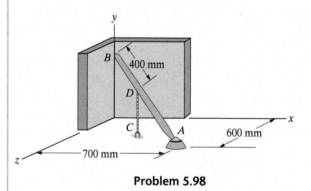

Problem 5.98

5.99 The 8-ft bar is supported by a ball and socket support at A, the cable BD, and a roller support at C. The collar at B is fixed to the bar at its midpoint. The force $\mathbf{F} = -50\mathbf{k}$ (lb). Determine the tension in cable BD and the reactions at A and C.

5.100 The bar is 8 ft in length. The force $\mathbf{F} = F_y\mathbf{j} - 50\mathbf{k}$ (lb). What is the largest value of F_y for which the roller support at C will remain on the floor?

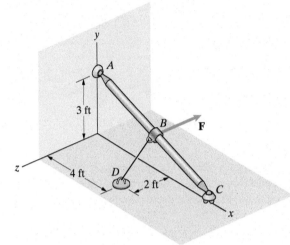

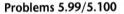

Problems 5.99/5.100

5.101 The tower is 70 m tall. The tension in each cable is 2 kN. Treat the base of the tower A as a fixed support. What are the reactions at A?

5.102 The tower is 70 m tall. If the tension in cable BC is 2 kN, what must the tensions in cables BD and BE be if you want the couple exerted on the tower by the fixed support at A to be zero? What are the resulting reactions at A?

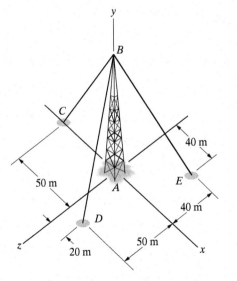

Problems 5.101/5.102

5.103 The space truss has roller supports at B, C, and D and is subjected to a vertical force $F = 20$ kN at A. What are the reactions at the roller supports?

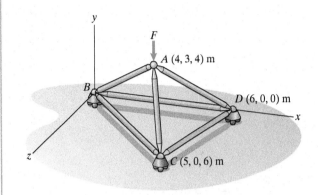

Problem 5.103

▶ **5.104** In Example 5.8, suppose that the cable BD is lengthened and the attachment point B is moved to the end of the bar at C. The positions of the attachment point D and the bar are unchanged. Draw a sketch of the bar showing cable BD in its new position. Draw the free-body diagram of the bar and apply equilibrium to determine the tension in the cable and the reactions at A.

5.105 The 40-lb door is supported by hinges at A and B. The y axis is vertical. The hinges do not exert couples on the door, and the hinge at B does not exert a force parallel to the hinge axis. The weight of the door acts at its midpoint. What are the reactions at A and B?

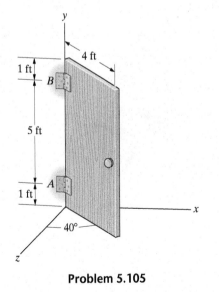

Problem 5.105

5.106 The vertical cable is attached at A. Determine the tension in the cable and the reactions at the bearing B due to the force $\mathbf{F} = 10\mathbf{i} - 30\mathbf{j} - 10\mathbf{k}$ (N).

5.107 Suppose that the z component of the force $\mathbf{F}$ is zero, but otherwise $\mathbf{F}$ is unknown. If the couple exerted on the shaft by the bearing at B is $\mathbf{M}_B = 6\mathbf{j} - 6\mathbf{k}$ N-m, what are the force $\mathbf{F}$ and the tension in the cable?

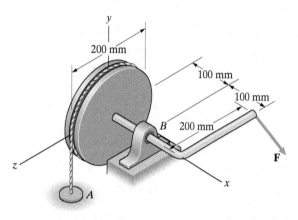

Problems 5.106/5.107

5.108 The device in Problem 5.106 is badly designed because of the couples that must be supported by the bearing at B, which would cause the bearing to "bind." (Imagine trying to open a door supported by only one hinge.) In this improved design, the bearings at B and C support no couples, and the bearing at C does not exert a force in the x direction. If the force $\mathbf{F} = 10\mathbf{i} - 30\mathbf{j} - 10\mathbf{k}$ (N), what are the tension in the vertical cable and the reactions at the bearings B and C?

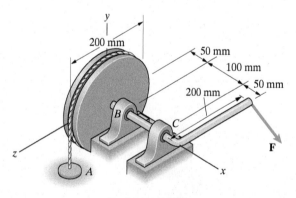

Problem 5.108

5.109 The rocket launcher is supported by the hydraulic jack DE and the bearings A and B. The bearings lie on the x axis and support shafts parallel to the x axis. The hydraulic cylinder DE exerts a force on the launcher that points along the line from D to E. The coordinates of D are $(7, 0, 7)$ ft, and the coordinates of E are $(9, 6, 4)$ ft. The weight $W = 30$ kip acts at $(4.5, 5, 2)$ ft. What is the magnitude of the reaction on the launcher at E?

5.110 Consider the rocket launcher described in Problem 5.109. The bearings at A and B do not exert couples, and the bearing B does not exert a force in the x direction. Determine the reactions at A and B.

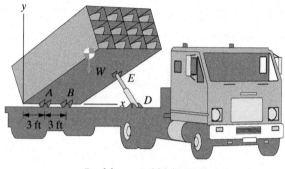

Problems 5.109/5.110

5.111 The crane's cable CD is attached to a stationary object at D. The crane is supported by the bearings E and F and the horizontal cable AB. The tension in cable AB is 8 kN. Determine the tension in the cable CD.

Strategy: Since the reactions exerted on the crane by the bearings do not exert moments about the z axis, the sum of the moments about the z axis due to the forces exerted on the crane by the cables AB and CD equals zero.

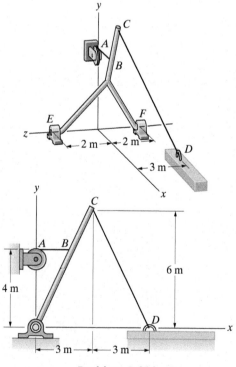

Problem 5.111

▶ **5.112** In Example 5.9, suppose that the cable *CE* is shortened and its attachment point *E* is moved to the point (0, 80, 0) mm. The plate remains in the same position. Draw a sketch of the plate and its supports showing the new position of cable *CE*. Draw the free-body diagram of the plate and apply equilibrium to determine the reactions at the hinges and the tension in the cable.

5.113 The plate is supported by hinges at *A* and *B* and the cable *CE*, and it is loaded by the force at *D*. The edge of the plate to which the hinges are attached lies in the *y–z* plane, and the axes of the hinges are parallel to the line through points *A* and *B*. The hinges do not exert couples on the plate. What is the tension in cable *CE*?

5.114 In Problem 5.113, the hinge at *B* does not exert a force on the plate in the direction of the hinge axis. What are the magnitudes of the forces exerted on the plate by the hinges at *A* and *B*?

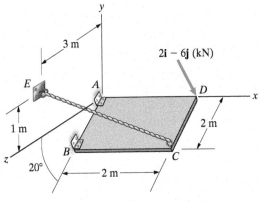

Problems 5.113/5.114

5.115 The bar *ABC* is supported by ball and socket supports at *A* and *C* and the cable *BD*. The suspended mass is 1800 kg. Determine the tension in the cable.

5.116* In Problem 5.115, assume that the ball and socket support at *A* is designed so that it exerts no force parallel to the straight line from *A* to *C*. Determine the reactions at *A* and *C*.

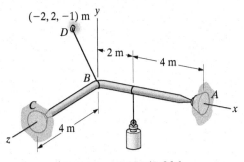

Problems 5.115/5.116

5.117 The bearings at *A*, *B*, and *C* do not exert couples on the bar and do not exert forces in the direction of the axis of the bar. Determine the reactions at the bearings due to the two forces on the bar.

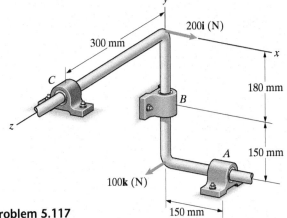

Problem 5.117

5.118 The support that attaches the sailboat's mast to the deck behaves like a ball and socket support. The line that attaches the spinnaker (the sail) to the top of the mast exerts a 200-lb force on the mast. The force is in the horizontal plane at 15° from the centerline of the boat. (See the top view.) The spinnaker pole exerts a 50-lb force on the mast at *P*. The force is in the horizontal plane at 45° from the centerline. (See the top view.) The mast is supported by two cables, the backstay *AB* and the port shroud *ACD*. (The forestay *AE* and the starboard shroud *AFG* are slack, and their tensions can be neglected.) Determine the tensions in the cables *AB* and *CD* and the reactions at the bottom of the mast.

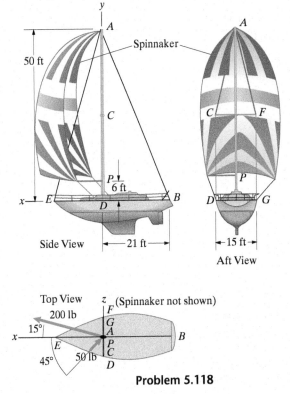

Problem 5.118

5.119* The bar AC is supported by the cable BD and a bearing at A that can rotate about the axis AE. The person exerts a force $\mathbf{F} = 50\mathbf{j}$ (N) at C. Determine the tension in the cable.

   Strategy: Use the fact that the sum of the moments about the axis AE due to the forces acting on the free-body diagram of the bar must equal zero.

5.120* In Problem 5.119, determine the reactions at the bearing A.

   Strategy: Write the couple exerted on the free-body diagram of the bar by the bearing as $\mathbf{M}_A = M_{Ax}\mathbf{i} + M_{Ay}\mathbf{j} + M_{Az}\mathbf{k}$. Then, in addition to the equilibrium equations, obtain an equation by requiring the component of $\mathbf{M}_A$ parallel to the axis AE to equal zero.

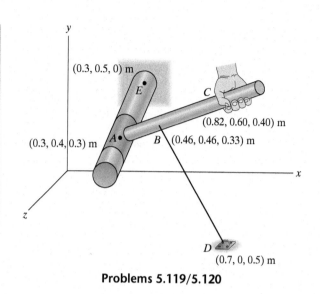

Problems 5.119/5.120

5.4 Two-Force and Three-Force Members

BACKGROUND

We have shown how the equilibrium equations are used to analyze objects that are supported and loaded in different ways. Here we discuss two particular types of loading that occur so frequently they deserve particular attention. The first type, the two-force member, is especially important and plays an important role in our analysis of structures in Chapter 6.

Two-Force Members

If the system of forces and moments acting on an object is equivalent to two forces acting at different points, we refer to the object as a *two-force member*. For example, the object in Fig. 5.19a is subjected to two sets of concurrent forces whose lines of action intersect at A and B. Since we can represent them by single forces acting at A and B (Fig. 5.19b), where $\mathbf{F} = \mathbf{F}_1 + \mathbf{F}_2 + \cdots + \mathbf{F}_N$ and $\mathbf{F}' = \mathbf{F}'_1 + \mathbf{F}'_2 + \cdots + \mathbf{F}'_M$, this object is a two-force member.

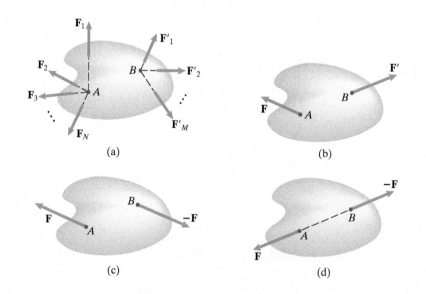

Figure 5.19
(a) An object subjected to two sets of concurrent forces.
(b) Representing the concurrent forces by two forces $\mathbf{F}$ and $\mathbf{F}'$.
(c) If the object is in equilibrium, the forces must be equal and opposite.
(d) The forces form a couple unless they have the same line of action.

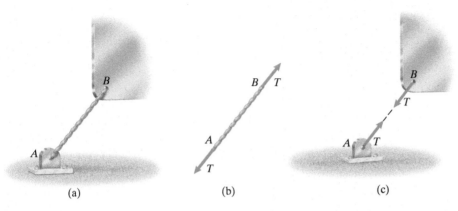

Figure 5.20
(a) A cable attached at A and B.
(b) The cable is a two-force member.
(c) The forces exerted by the cable.

If the object is in equilibrium, what can we infer about the forces $\mathbf{F}$ and $\mathbf{F}'$? The sum of the forces equals zero only if $\mathbf{F}' = -\mathbf{F}$ (Fig. 5.19c). Furthermore, the forces $\mathbf{F}$ and $-\mathbf{F}$ form a couple, so the sum of the moments is not zero unless the lines of action of the forces lie along the line through the points A and B (Fig. 5.19d). Thus equilibrium tells us that *the two forces are equal in magnitude, are opposite in direction, and have the same line of action*. However, without additional information, we cannot determine their magnitude.

A cable attached at two points (Fig. 5.20a) is a familiar example of a two-force member (Fig. 5.20b). The cable exerts forces on the attachment points that are directed along the line between them (Fig. 5.20c).

A bar that has two supports that exert only forces on it (no couples) and is not subjected to any loads is a two-force member (Fig. 5.21a). Such bars are often used as supports for other objects. Because the bar is a two-force member, the lines of action of the forces exerted on the bar must lie along the line between the supports (Fig. 5.21b). Notice that, unlike the cable, the bar can exert forces at A and B either in the directions shown in Fig. 5.21c or in the opposite directions. (In other words, the cable can only pull on its supports, while the bar can either pull or push.)

In these examples we assumed that the weights of the cable and the bar could be neglected in comparison with the forces exerted on them by their supports. When that is not the case, they are clearly not two-force members.

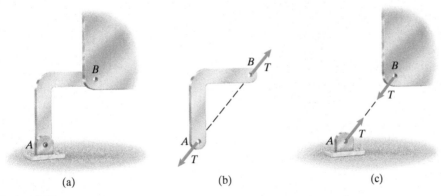

Figure 5.21
(a) The bar AB attaches the object to the pin support.
(b) The bar AB is a two-force member.
(c) The force exerted on the supported object by the bar AB.

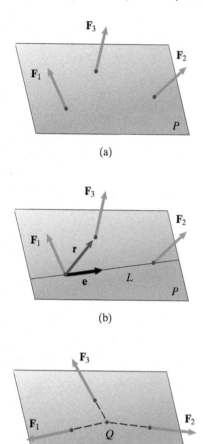

(a)

(b)

(c)

Figure 5.22
(a) The three forces and the plane P.
(b) Determining the moment due to force $\mathbf{F}_3$ about L.
(c) If the forces are not parallel, they must be concurrent.

Three-Force Members

If the system of forces and moments acting on an object is equivalent to three forces acting at different points, we call it a *three-force member*. We can show that if a three-force member is in equilibrium, the three forces are coplanar and are either parallel or concurrent.

We first prove that the forces are coplanar. Let them be called $\mathbf{F}_1$, $\mathbf{F}_2$, and $\mathbf{F}_3$, and let P be the plane containing the three points of application (Fig. 5.22a). Let L be the line through the points of application of $\mathbf{F}_1$ and $\mathbf{F}_2$. Since the moments due to $\mathbf{F}_1$ and $\mathbf{F}_2$ about L are zero, the moment due to $\mathbf{F}_3$ about L must equal zero (Fig. 5.22b):

$$[\mathbf{e} \cdot (\mathbf{r} \times \mathbf{F}_3)]\mathbf{e} = [\mathbf{F}_3 \cdot (\mathbf{e} \times \mathbf{r})]\mathbf{e} = \mathbf{0}.$$

This equation requires that $\mathbf{F}_3$ be perpendicular to $\mathbf{e} \times \mathbf{r}$, which means that $\mathbf{F}_3$ is contained in P. The same procedure can be used to show that $\mathbf{F}_1$ and $\mathbf{F}_2$ are contained in P, so the forces are coplanar. (A different proof is required if the points of application lie on a straight line, but the result is the same.)

If the three coplanar forces are not parallel, there will be points where their lines of action intersect. Suppose that the lines of action of two of the forces intersect at a point Q. Then the moments of those two forces about Q are zero, and the sum of the moments about Q is zero only if the line of action of the third force also passes through Q. Therefore, either the forces are parallel or they are concurrent (Fig. 5.22c).

The analysis of an object in equilibrium can often be simplified by recognizing that it is a two-force or three-force member. However, in doing so we are not getting something for nothing. Once the free-body diagram of a two-force member is drawn, as shown in Figs. 5.20b and 5.21b, no further information can be obtained from the equilibrium equations. And when we require that the lines of action of nonparallel forces acting on a three-force member be coincident, we have used the fact that the sum of the moments about a point must be zero and cannot obtain further information from that condition.

RESULTS

Two-Force Member
If an object in equilibrium is subjected to two forces acting at different points and *no other forces or couples*, it is called a two-force member. Equilibrium requires that the two forces be equal and opposite and parallel to the line between the two points.

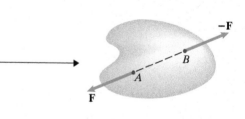

Three-Force Member
If an object in equilibrium is subjected to three forces acting at different points and *no other forces or couples*, it is called a three-force member. Equilibrium requires that the three forces be coplanar and either parallel or concurrent.

Active Example 5.10   Two- and Three-Force Members (▶ *Related Problem 5.121*)

The 100-lb weight of the rectangular plate acts at its midpoint. Neglect the weight of the link *AB*. Determine the reactions exerted on the plate at *B* and *C*.

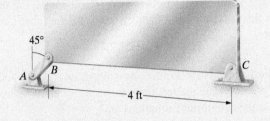

Strategy

The plate is subjected to its weight and the reactions exerted by the pin supports at *B* and *C*, so it is a three-force member. The link *BC* is a two-force member, so the line of action of the reaction it exerts on the plate at *B* must be directed along the line between *A* and *B*. We can use this information to simplify the free-body diagram of the plate.

The reaction exerted on the plate by the two-force member *AB* must be directed along the line between *A* and *B*.

Solution

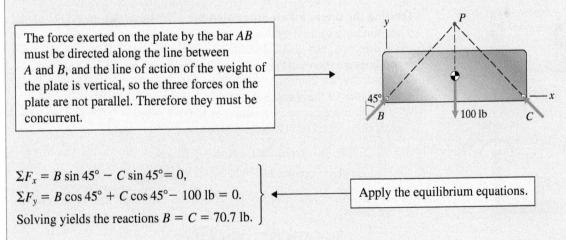

The force exerted on the plate by the bar *AB* must be directed along the line between *A* and *B*, and the line of action of the weight of the plate is vertical, so the three forces on the plate are not parallel. Therefore they must be concurrent.

$\Sigma F_x = B \sin 45° - C \sin 45° = 0,$

$\Sigma F_y = B \cos 45° + C \cos 45° - 100 \text{ lb} = 0.$

Solving yields the reactions $B = C = 70.7$ lb.

Apply the equilibrium equations.

Practice Problem Suppose that the plate is replaced with a 100-lb plate whose thickness (the dimension perpendicular to the page) is not uniform. The line of action of the weight of the nonuniform plate is 3 ft to the right of point *B*. Determine the reactions exerted on the plate at *B* and *C*.

Answer: $B = 35.4$ lb, $C = 79.1$ lb.

Example 5.11   A Two-Force Member (▶ *Related Problem 5.122*)

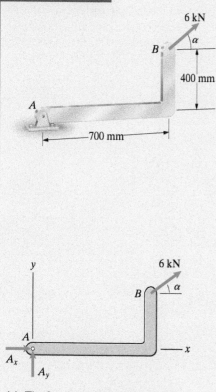

(a) The free-body diagram of the bar.

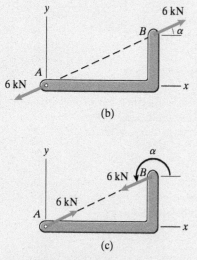

(b)

(c)

(b), (c) The possible directions of the forces.

The L-shaped bar has a pin support at A and is loaded by a 6-kN force at B. Neglect the weight of the bar. Determine the angle α and the reactions at A.

Strategy

The bar is a two-force member because it is subjected only to the 6-kN force at B and the force exerted by the pin support. (If we could not neglect the weight of the bar, it would not be a two-force member.) We will determine the angle α and the reactions at A in two ways, first by applying the equilibrium equations in the usual way and then by using the fact that the bar is a two-force member.

Solution

Applying the Equilibrium Equations   We draw the free-body diagram of the bar in Fig. a, showing the reactions at the pin support. Summing moments about point A, the equilibrium equations are

$$\Sigma F_x = A_x + 6 \cos \alpha \text{ kN} = 0,$$
$$\Sigma F_y = A_y + 6 \sin \alpha \text{ kN} = 0,$$
$$\Sigma M_{\text{point } A} = (0.7 \text{ m})(6 \sin \alpha \text{ kN}) - (0.4 \text{ m})(6 \cos \alpha \text{ kN}) = 0.$$

From the third equation we see that $\alpha = \arctan(0.4/0.7)$. In the range $0 \le \alpha \le 360°$, this equation has the two solutions $\alpha = 29.7°$ and $\alpha = 209.7°$. Knowing α, we can determine A_x and A_y from the first two equilibrium equations. The solutions for the two values of α are

$$\alpha = 29.7°, \quad A_x = -5.21 \text{ kN}, \quad A_y = -2.98 \text{ kN},$$

and

$$\alpha = 209.7°, \quad A_x = 5.21 \text{ kN}, \quad A_y = 2.98 \text{ kN}.$$

Treating the Bar as a Two-Force Member   We know that the 6-kN force at B and the force exerted by the pin support must be equal in magnitude, opposite in direction, and directed along the line between points A and B. The two possibilities are shown in Figs. b and c. Thus by recognizing that the bar is a two-force member, we immediately know the possible directions of the forces and the magnitude of the reaction at A.

In Fig. b we can see that $\tan \alpha = 0.4/0.7$, so $\alpha = 29.7°$ and the components of the reaction at A are

$$A_x = -6 \cos 29.7° \text{ kN} = -5.21 \text{ kN},$$
$$A_y = -6 \sin 29.7° \text{ kN} = -2.98 \text{ kN}.$$

In Fig. c, $\alpha = 180° + 29.7° = 209.7°$, and the components of the reaction at A are

$$A_x = 6 \cos 29.7° \text{ kN} = 5.21 \text{ kN}$$
$$A_y = 6 \sin 29.7° \text{ kN} = 2.98 \text{ kN}.$$

Critical Thinking

Why is it worthwhile to recognize that an object is a two-force member? Doing so tells you the directions of the forces acting on the object and also that the forces are equal and opposite. As this example demonstrates, such information frequently simplifies the solution of a problem.

Problems

▶ **5.121** In Active Example 5.10, suppose that the support at A is moved so that the angle between the bar AB and the vertical decreases from 45° to 30°. The position of the rectangular plate does not change. Draw the free-body diagram of the plate showing the point P where the lines of action of the three forces acting on the plate intersect. Determine the magnitudes of the reactions on the plate at B and C.

▶ **5.122** The magnitude of the reaction exerted on the L-shaped bar at B is 60 lb. (See Example 5.11.)

(a) What is the magnitude of the reaction exerted on the bar by the support at A?

(b) What are the x and y components of the reaction exerted on the bar by the support at A?

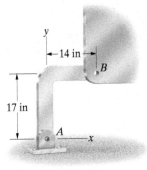

Problem 5.122

5.123 The suspended load weighs 1000 lb. The structure is a three-force member if its weight is neglected. Use this fact to determine the magnitudes of the reactions at A and B.

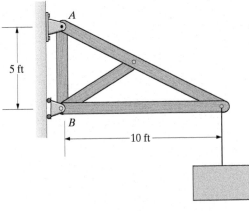

Problem 5.123

5.124 The weight $W = 50$ lb acts at the center of the disk. Use the fact that the disk is a three-force member to determine the tension in the cable and the magnitude of the reaction at the pin support.

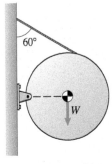

Problem 5.124

5.125 The weight $W = 40$ N acts at the center of the disk. The surfaces are rough. What force F is necessary to lift the disk off the floor?

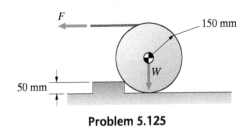

Problem 5.125

5.126 Use the fact that the horizontal bar is a three-force member to determine the angle α and the magnitudes of the reactions at A and B. Assume that $0 \leq \alpha \leq 90°$.

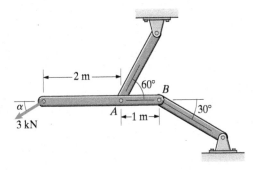

Problem 5.126

5.127 The suspended load weighs 600 lb. Use the fact that *ABC* is a three-force member to determine the magnitudes of the reactions at *A* and *B*.

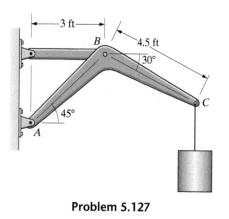

Problem 5.127

5.128 (a) Is the L-shaped bar a three-force member?

(b) Determine the magnitudes of the reactions at *A* and *B*.

(c) Are the three forces acting on the L-shaped bar concurrent?

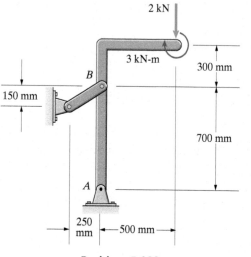

Problem 5.128

5.129 The hydraulic piston exerts a horizontal force at *B* to support the weight *W* = 1500 lb of the bucket of the excavator. Determine the magnitude of the force the hydraulic piston must exert. (The vector sum of the forces exerted at *B* by the hydraulic piston, the two-force member *AB*, and the two-force member *BD* must equal zero.)

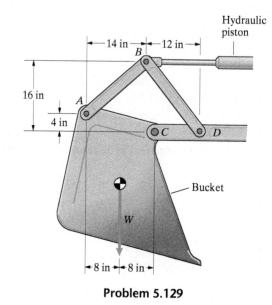

Problem 5.129

5.130 The member *ACG* of the front-end loader is subjected to a load *W* = 2 kN and is supported by a pin support at *A* and the hydraulic cylinder *BC*. Treat the hydraulic cylinder as a two-force member.

(a) Draw the free-body diagrams of the hydraulic cylinder and the member *ACG*.

(b) Determine the reactions on the member *ACG*.

5.131 In Problem 5.130, determine the reactions on the member *ACG* by using the fact that it is a three-force member.

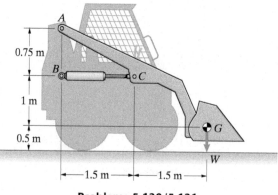

Problems 5.130/5.131

5.132 A rectangular plate is subjected to two forces A and B (Fig. a). In Fig. b, the two forces are resolved into components. By writing equilibrium equations in terms of the components $A_x, A_y, B_x,$ and B_y, show that the two forces A and B are equal in magnitude, opposite in direction, and directed along the line between their points of application.

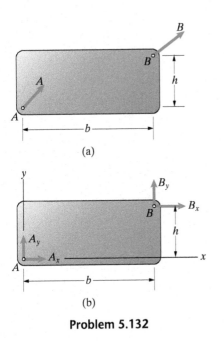

(a)

(b)

Problem 5.132

5.133 An object in equilibrium is subjected to three forces whose points of application lie on a straight line. Prove that the forces are coplanar.

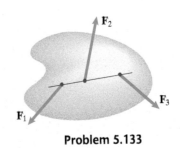

Problem 5.133

Review Problems

5.134 The suspended cable weighs 12 lb.

(a) Draw the free-body diagram of the cable. (The tensions in the cable at A and B are *not* equal.)

(b) Determine the tensions in the cable at A and B.

(c) What is the tension in the cable at its lowest point?

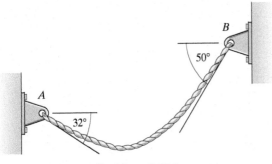

Problem 5.134

5.135 Determine the reactions at the fixed support.

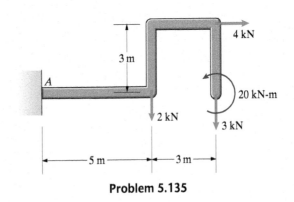

Problem 5.135

5.136 (a) Draw the free-body diagram of the 50-lb plate, and explain why it is statically indeterminate.

(b) Determine as many of the reactions at A and B as possible.

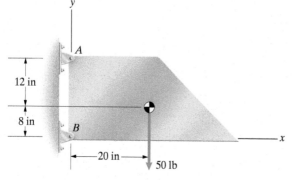

Problem 5.136

5.137 The mass of the truck is 4000 kg. Its wheels are locked, and the tension in its cable is $T = 10$ kN.

(a) Draw the free-body diagram of the truck.

(b) Determine the normal forces exerted on the truck's wheels at A and B by the road.

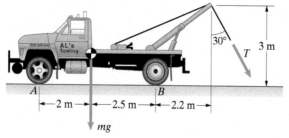

Problem 5.137

5.138 Assume that the force exerted on the head of the nail by the hammer is vertical, and neglect the hammer's weight.

(a) Draw the free-body diagram of the hammer.

(b) If $F = 10$ lb, what are the magnitudes of the force exerted on the nail by the hammer and the normal and friction forces exerted on the floor by the hammer?

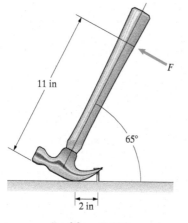

Problem 5.138

5.139 The spring constant is $k = 9600$ N/m and the unstretched length of the spring is 30 mm. Treat the bolt at A as a pin support and assume that the surface at C is smooth. Determine the reactions at A and the normal force at C.

5.140 The engineer designing the release mechanism wants the normal force exerted at C to be 120 N. If the unstretched length of the spring is 30 mm, what is the necessary value of the spring constant k?

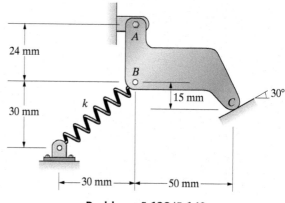

Problems 5.139/5.140

5.141 The truss supports a 90-kg suspended object. What are the reactions at the supports A and B?

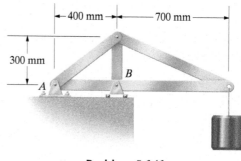

Problem 5.141

5.142 The trailer is parked on a 15° slope. Its wheels are free to turn. The hitch H behaves like a pin support. Determine the reactions at A and H.

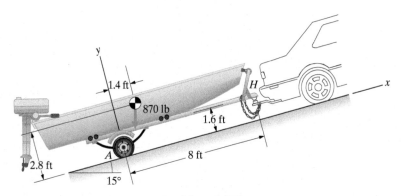

Problem 5.142

5.143 To determine the location of the point where the weight of a car acts (the *center of mass*), an engineer places the car on scales and measures the normal reactions at the wheels for two values of α, obtaining the following results.

α	A_y (kN)	B (kN)
10°	10.134	4.357
20°	10.150	3.677

What are the distances b and h?

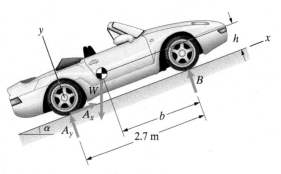

Problem 5.143

5.144 The bar is attached by pin supports to collars that slide on the two fixed bars. Its mass is 10 kg, it is 1 m in length, and its weight acts at its midpoint. Neglect friction and the masses of the collars. The spring is unstretched when the bar is vertical ($\alpha = 0$), and the spring constant is $k = 100$ N/m. Determine the values of α in the range $0 \leq \alpha \leq 60°$ at which the bar is in equilibrium.

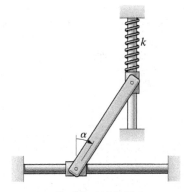

Problem 5.144

5.145 With each of the devices shown you can support a load R by applying a force F. They are called levers of the first, second, and third class.

(a) The ratio R/F is called the *mechanical advantage*. Determine the mechanical advantage of each lever.

(b) Determine the magnitude of the reaction at A for each lever. (Express your answers in terms of F.)

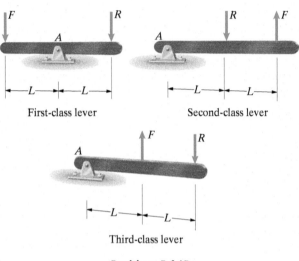

First-class lever   Second-class lever

Third-class lever

Problem 5.145

5.146 The force exerted by the weight of the horizontal rectangular plate is 800 N. The weight of the rectangular plate acts at its midpoint. If you represent the reactions exerted on the plate by the three cables by a single equivalent force, what is the force, and where does its line of action intersect the plate?

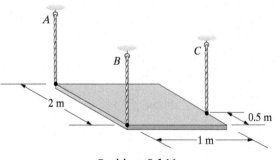

Problem 5.146

5.147 The 20-kg mass is suspended by cables attached to three vertical 2-m posts. Point A is at $(0, 1.2, 0)$ m. Determine the reactions at the fixed support at E.

5.148 In Problem 5.147, the fixed support of each vertical post will safely support a couple of 800 N-m magnitude. Based on this criterion, what is the maximum safe value of the suspended mass?

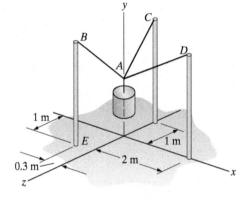

Problems 5.147/148

5.149 The 80-lb bar is supported by a ball and socket support at A, the smooth wall it leans against, and the cable BC. The weight of the bar acts at its midpoint.

(a) Draw the free-body diagram of the bar.

(b) Determine the tension in cable BC and the reactions at A.

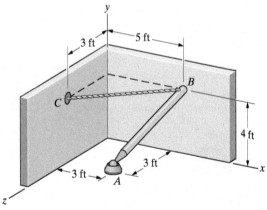

Problem 5.149

5.150 The horizontal bar of weight W is supported by a roller support at A and the cable BC. Use the fact that the bar is a three-force member to determine the angle α, the tension in the cable, and the magnitude of the reaction at A.

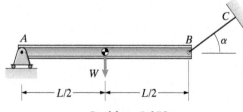

Problem 5.150

Design Project 1 The traditional wheelbarrow shown is designed to transport a load W while being supported by an upward force F applied to the handles by the user. (a) Use statics to analyze the effects of a range of choices of the dimensions a and b on the size of load that could be carried. Also consider the implications of these dimensions on the wheelbarrow's ease and practicality of use. (b) Suggest a different design for this classic device that achieves the same function. Use statics to compare your design to the wheelbarrow with respect to load-carrying ability and ease of use.

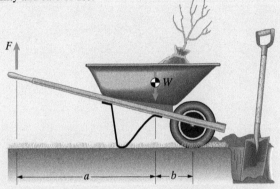

Design Project 2 The figure shows an example of the popular devices called "mobiles," which were introduced as an art form by American artist Alexander Calder (1898–1976). Suppose that you want to design a mobile representing the solar system, and have chosen colored spheres to represent the planets. The masses of the spheres that represent Mercury, Venus, Earth, Mars, Jupiter, Saturn, Uranus, Neptune, and Pluto are 10 g, 25 g, 25 g, 10 g, 50 g, 40 g, 40 g, 40 g, and 10 g. Assume that the cross bars and string you use are of negligible mass. Design your mobile so that the planets are in their correct order relative to the sun. Write a brief report including a drawing of your design and the analysis proving that your mobile is balanced.

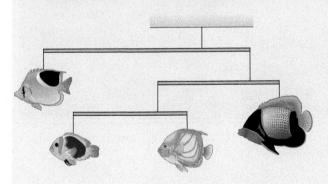

Design Project 3 The bed of the dump truck (Fig. a) is raised by two tandem hydraulic cylinders AB (Fig. b). The mass of the truck's bed and load is 16,000 kg and its weight acts at point G. (Assume that the position of point G *relative to the bed* does not change when the bed is raised.)

(a) Draw a graph of the magnitude of the total force the hydraulic cylinders must exert to support the stationary bed for values of the angle α from zero to 30°.

(b) Consider other choices for the locations of the attachment points A and B that appear to be feasible and investigate how your choices affect the magnitude of the total force the hydraulic cylinders must exert as α varies from zero to 30°. Also compare the costs of your choices of the attachment points to the choices shown in Fig. a, assuming that the cost of the hydraulic cylinders is proportional to the product of the maximum force they must exert as α varies from zero to 30° and their length when $\alpha = 30°$.

(c) Write a brief report presenting your investigations and making a recommendation for the locations of points A and B.

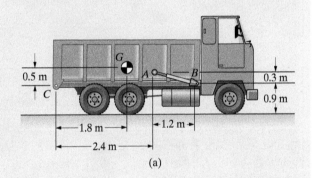

(a)

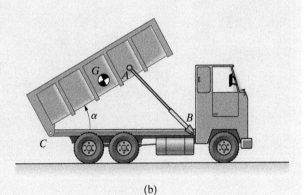

(b)

Structures in Equilibrium

In engineering, the term *structure* can refer to any object that has the capacity to support and exert loads. In this chapter we consider structures composed of interconnected parts, or *members*. To design such a structure, or to determine whether an existing one is adequate, it is necessary to determine the forces and couples acting on the structure as a whole as well as on its individual members. We first demonstrate how this is done for the structures called trusses, which are composed entirely of two-force members. The familiar frameworks of steel members that support some highway bridges are trusses. We then consider other structures, called *frames* if they are designed to remain stationary and support loads and *machines* if they are designed to move and exert loads.

◀ The Neolithic engineers who built Stonehenge set an example for the design of enduring structures. In this chapter we describe techniques for determining the forces and couples acting on individual members of structures.

6.1 Trusses

BACKGROUND

We can explain the nature of truss structures such as the beams supporting a house (Fig. 6.1) by starting with very simple examples. Suppose we pin three bars together at their ends to form a triangle. If we add supports as shown in Fig. 6.2a, we obtain a structure that will support a load F. We can construct more elaborate structures by adding more triangles (Figs. 6.2b and c). The bars are the members of these structures, and the places where the bars are pinned together are called the *joints*. Even though these examples are quite simple, you can see that Fig. 6.2c, which is called a Warren truss, begins to resemble the structures used to support bridges and the roofs of houses (Fig. 6.3). If these structures are supported and loaded at their joints and we neglect the weights of the bars, each bar is a two-force member. We call such a structure a *truss*.

We draw the free-body diagram of a member of a truss in Fig. 6.4a. Because it is a two-force member, the forces at the ends, which are the sums of the forces exerted on the member at its joints, must be equal in magnitude, opposite in direction, and directed along the line between the joints. We call the force T the *axial force* in the member. When T is positive in the direction shown (that is, when the forces are directed away from each other), the member is in *tension*. When the forces are directed toward each other, the member is in *compression*.

In Fig. 6.4b, we "cut" the member by a plane and draw the free-body diagram of the part of the member on one side of the plane. We represent the system of internal forces and moments exerted by the part not included in the free-body diagram by a force $\mathbf{F}$ acting at the point P where the plane intersects

Figure 6.1
A typical house is supported by trusses made of wood beams.

(a)

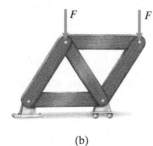

(b)

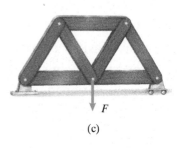

(c)

Figure 6.2
Making structures by pinning bars together to form triangles.

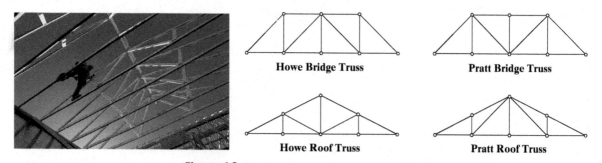

Howe Bridge Truss

Pratt Bridge Truss

Howe Roof Truss

Pratt Roof Truss

Figure 6.3
Simple examples of bridge and roof structures. (The lines represent members, the circles represent joints.)

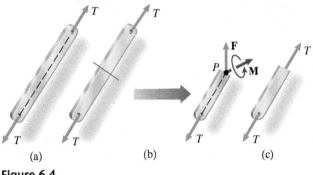

Figure 6.4
(a) Each member of a truss is a two-force member.
(b) Obtaining the free-body diagram of part of the member.
(c) The internal force is equal and opposite to the force acting at the joint, and the internal couple is zero.

the axis of the member and a couple $\mathbf{M}$. The sum of the moments about P must equal zero, so $\mathbf{M} = \mathbf{0}$. Therefore we have a two-force member, which means that $\mathbf{F}$ must be equal in magnitude and opposite in direction to the force T acting at the joint (Fig. 6.4c). The internal force is a tension or compression equal to the tension or compression exerted at the joint. Notice the similarity to a rope or cable, in which the internal force is a tension equal to the tension applied at the ends.

Although many actual structures, including "roof trusses" and "bridge trusses," consist of bars connected at the ends, very few have pinned joints. For example, a joint of a bridge truss is shown in Fig. 6.5. The ends of the members are welded at the joint and are not free to rotate. It is obvious that such a joint can exert couples on the members. Why are these structures called trusses?

The reason is that they are designed to function as trusses, meaning that they support loads primarily by subjecting their members to axial forces. They can usually be *modeled* as trusses, treating the joints as pinned connections under the assumption that couples they exert on the members are small in comparison to axial forces. When we refer to structures with riveted joints as trusses in problems, we mean that you can model them as trusses.

Figure 6.5
A joint of a bridge truss.

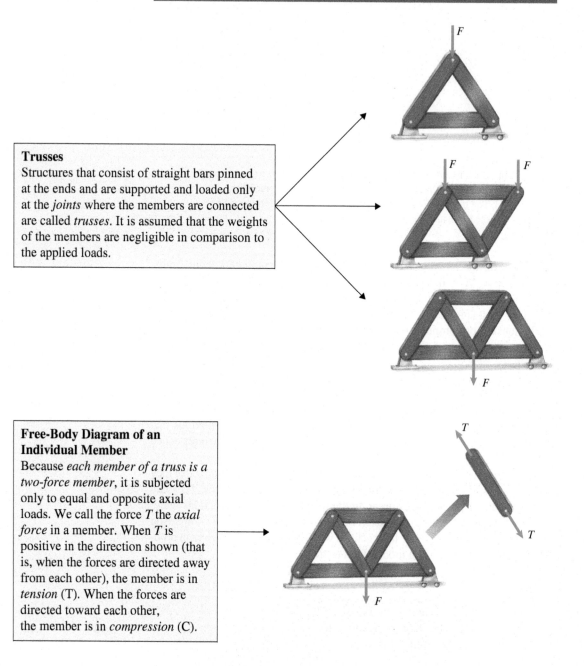

Trusses

Structures that consist of straight bars pinned at the ends and are supported and loaded only at the *joints* where the members are connected are called *trusses*. It is assumed that the weights of the members are negligible in comparison to the applied loads.

Free-Body Diagram of an Individual Member

Because *each member of a truss is a two-force member*, it is subjected only to equal and opposite axial loads. We call the force T the *axial force* in a member. When T is positive in the direction shown (that is, when the forces are directed away from each other), the member is in *tension* (T). When the forces are directed toward each other, the member is in *compression* (C).

6.2 The Method of Joints

BACKGROUND

The method of joints involves drawing free-body diagrams of the joints of a truss one by one and using the equilibrium equations to determine the axial forces in the members. Before beginning, it is usually necessary to draw a free-body diagram of the entire truss (that is, treat the truss as a single object) and determine the reactions at its supports. For example, let's consider the Warren truss in Fig. 6.6a, which has members 2 m in length and supports

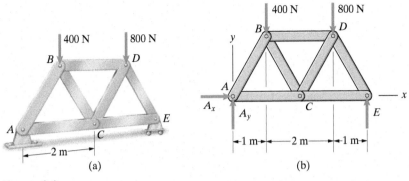

Figure 6.6
(a) A Warren truss supporting two loads.
(b) Free-body diagram of the truss.

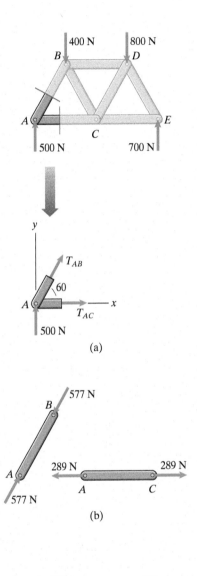

loads at B and D. We draw its free-body diagram in Fig. 6.6b. From the equilibrium equations,

$$\Sigma F_x = A_x = 0,$$
$$\Sigma F_y = A_y + E - 400\text{ N} - 800\text{ N} = 0,$$
$$\Sigma M_{\text{point }A} = -(1\text{ m})(400\text{ N}) - (3\text{ m})(800\text{ N}) + (4\text{ m})E = 0,$$

we obtain the reactions $A_x = 0$, $A_y = 500$ N, and $E = 700$ N.

Our next step is to choose a joint and draw its free-body diagram. In Fig. 6.7a, we isolate joint A by cutting members AB and AC. The terms T_{AB} and T_{AC} are the axial forces in members AB and AC, respectively. Although the directions of the arrows representing the unknown axial forces can be chosen arbitrarily, notice that we have chosen them so that a member is in tension if we obtain a positive value for the axial force. Consistently choosing the directions in this way helps avoid errors.

The equilibrium equations for joint A are

$$\Sigma F_x = T_{AC} + T_{AB}\cos 60° = 0,$$
$$\Sigma F_y = T_{AB}\sin 60° + 500\text{ N} = 0.$$

Solving these equations, we obtain the axial forces $T_{AB} = -577$ N and $T_{AC} = 289$ N. Member AB is in compression, and member AC is in tension (Fig. 6.7b).

Although we use a realistic figure for the joint in Fig. 6.7a to help you understand the free-body diagram, in your own work you can use a simple figure showing only the forces acting on the joint (Fig. 6.7c).

We next obtain a free-body diagram of joint B by cutting members AB, BC, and BD (Fig. 6.8a). From the equilibrium equations for joint B,

$$\Sigma F_x = T_{BD} + T_{BC}\cos 60° + 577\cos 60°\text{ N} = 0,$$
$$\Sigma F_y = -400\text{ N} + 577\sin 60°\text{ N} - T_{BC}\sin 60° = 0,$$

we obtain $T_{BC} = 115$ N and $T_{BD} = -346$ N. Member BC is in tension, and member BD is in compression (Fig. 6.8b). By continuing to draw free-body diagrams of the joints, we can determine the axial forces in all of the members.

In two dimensions, you can obtain only two independent equilibrium equations from the free-body diagram of a joint. Summing the moments about a point does not result in an additional independent equation because the forces are concurrent. Therefore when applying the method of joints, you should choose joints to analyze that are subjected to no more than two unknown forces. In our

Figure 6.7
(a) Obtaining the free-body diagram of joint A.
(b) The axial forces on members AB and AC.
(c) Realistic and simple free-body diagrams of joint A.

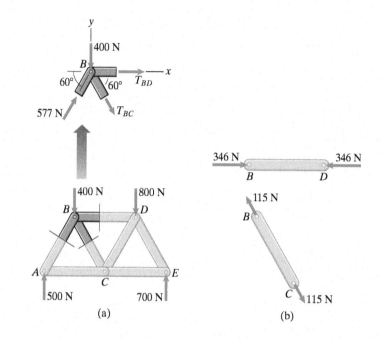

Figure 6.8
(a) Obtaining the free-body diagram of joint B.
(b) Axial forces in members BD and BC.

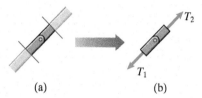

Figure 6.9
(a) A joint with two collinear members and no load.
(b) Free-body diagram of the joint.

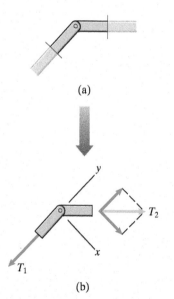

example, we analyzed joint A first because it was subjected to the known reaction exerted by the pin support and two unknown forces, the axial forces T_{AB} and T_{AC} (Fig. 6.7a). We could then analyze joint B because it was subjected to two known forces and two unknown forces, T_{BC} and T_{BD} (Fig. 6.8a). If we had attempted to analyze joint B first, there would have been three unknown forces.

When you determine the axial forces in the members of a truss, your task will often be simpler if you are familiar with three particular types of joints.

- **Truss joints with two collinear members and no load** (Fig. 6.9). The sum of the forces must equal zero, $T_1 = T_2$. The axial forces are equal.
- **Truss joints with two noncollinear members and no load** (Fig. 6.10). Because the sum of the forces in the x direction must equal zero, $T_2 = 0$. Therefore T_1 must also equal zero. The axial forces are zero.
- **Truss joints with three members, two of which are collinear, and no load** (Fig. 6.11). Because the sum of the forces in the x direction must equal zero, $T_3 = 0$. The sum of the forces in the y direction must equal zero, so $T_1 = T_2$. The axial forces in the collinear members are equal, and the axial force in the third member is zero.

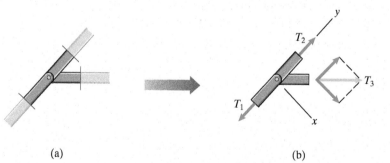

Figure 6.10
(a) A joint with two noncollinear members and no load.
(b) Free-body diagram of the joint.

Figure 6.11
(a) A joint with three members, two of which are collinear, and no load.
(b) Free-body diagram of the joint.

RESULTS

Method of Joints

Before beginning, it is usually necessary to draw the free-body diagram of the entire truss considered as a single object and apply the equilibrium equations to determine the reactions at the supports.

Isolate an individual joint by passing planes through the connected members. Complete the free-body diagram by showing the axial forces in the members. Apply the equilibrium equations $\Sigma F_x = 0$ and $\Sigma F_y = 0$ to the free-body diagram of the joint. Repeat this process for other joints until the desired axial loads have been determined.

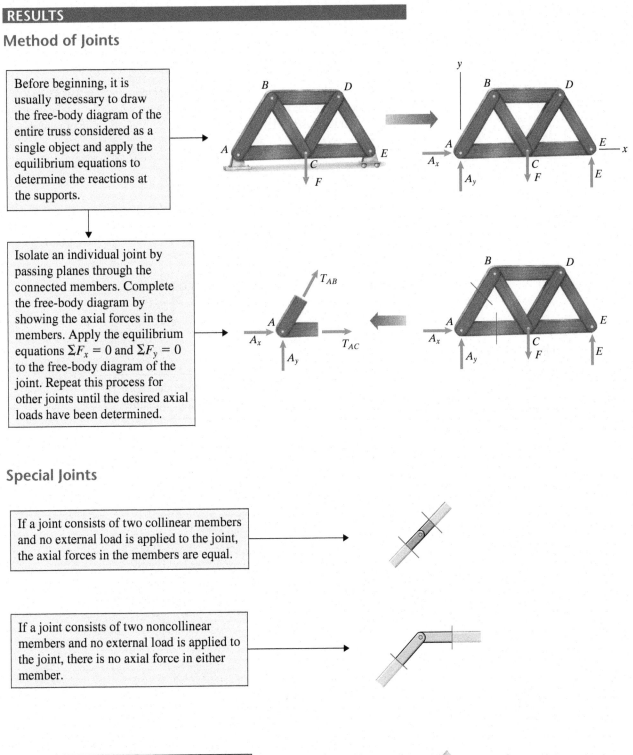

Special Joints

If a joint consists of two collinear members and no external load is applied to the joint, the axial forces in the members are equal.

If a joint consists of two noncollinear members and no external load is applied to the joint, there is no axial force in either member.

If a joint consists of three members, two of which are collinear, and no external load is applied to the joint, the axial forces in the collinear members are equal and the axial force in the third member is zero.

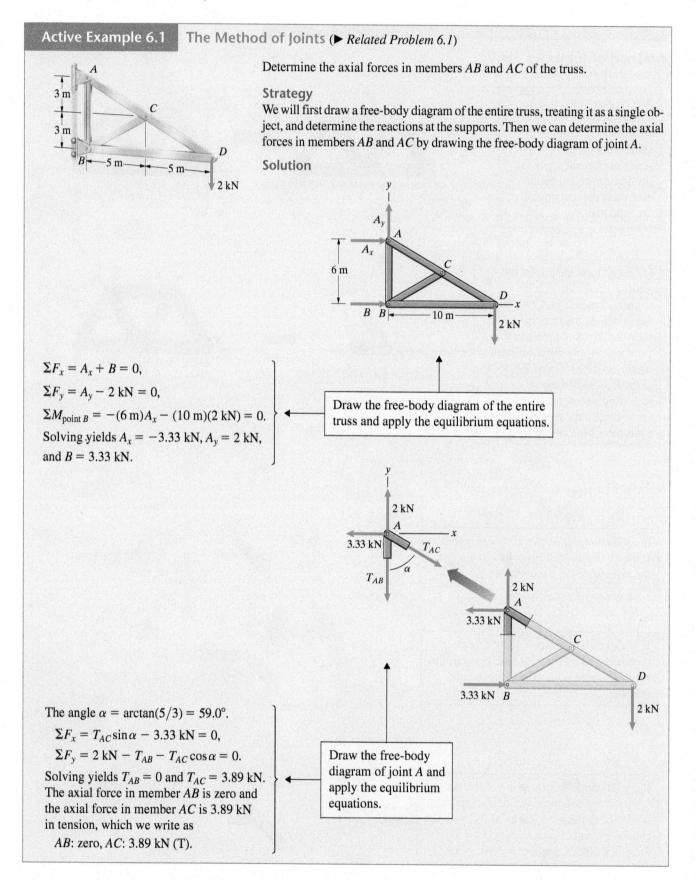

Active Example 6.1   The Method of Joints (▶ *Related Problem 6.1*)

Determine the axial forces in members *AB* and *AC* of the truss.

Strategy

We will first draw a free-body diagram of the entire truss, treating it as a single object, and determine the reactions at the supports. Then we can determine the axial forces in members *AB* and *AC* by drawing the free-body diagram of joint *A*.

Solution

$\Sigma F_x = A_x + B = 0,$

$\Sigma F_y = A_y - 2 \text{ kN} = 0,$

$\Sigma M_{\text{point } B} = -(6 \text{ m})A_x - (10 \text{ m})(2 \text{ kN}) = 0.$

Solving yields $A_x = -3.33$ kN, $A_y = 2$ kN, and $B = 3.33$ kN.

> Draw the free-body diagram of the entire truss and apply the equilibrium equations.

The angle $\alpha = \arctan(5/3) = 59.0°$.

$\Sigma F_x = T_{AC} \sin\alpha - 3.33 \text{ kN} = 0,$

$\Sigma F_y = 2 \text{ kN} - T_{AB} - T_{AC}\cos\alpha = 0.$

Solving yields $T_{AB} = 0$ and $T_{AC} = 3.89$ kN. The axial force in member *AB* is zero and the axial force in member *AC* is 3.89 kN in tension, which we write as

AB: zero, AC: 3.89 kN (T).

> Draw the free-body diagram of joint *A* and apply the equilibrium equations.

Example 6.2   A Bridge Truss (▶ *Related Problem 6.31*)

The loads a bridge structure must support and pin supports where the structure is to be attached are shown in Fig. 1. Assigned to design the structure, a civil engineering student proposes the structure shown in Fig. 2. What are the axial forces in the members?

Strategy
The vertical members AG, BH, CI, DJ, and EK are subjected to compressive forces of magnitude F. Because of the symmetry of the structure, we can determine the axial loads in the remaining members by analyzing joints C and B.

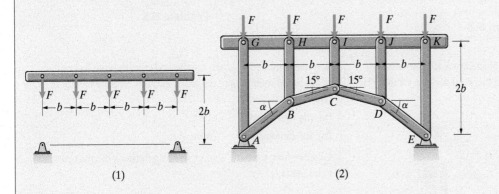

(1)   (2)

Solution
We will leave it as an exercise to show by drawing the free-body diagram of joint C that members BC and CD are subjected to equal compressive loads of magnitude $1.93F$. We draw the free-body diagram of joint B in Fig. a, where $T_{BC} = -1.93F$.

From the equilibrium equations

$$\Sigma F_x = -T_{AB} \cos \alpha + T_{BC} \cos 15° = 0,$$

$$\Sigma F_y = -T_{AB} \sin \alpha + T_{BC} \sin 15° - F = 0,$$

we obtain $T_{AB} = -2.39F$ and $\alpha = 38.8°$. By symmetry, $T_{DE} = T_{AB}$. The axial forces in the members are shown in the table.

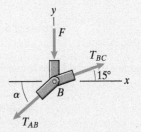

(a) Free-body diagram of joint B.

Axial forces in the members of the bridge structure

Members	Axial Force
AG, BH, CI, DJ, EK	F (C)
AB, DE	$2.39F$ (C)
BC, CD	$1.93F$ (C)

Problems

▶ **6.1** In Active Example 6.1, suppose that in addition to the 2-kN downward force acting at point D, a 2-kN downward force acts at point C. Draw a sketch of the truss showing the new loading. Determine the axial forces in members AB and AC of the truss.

6.2 Determine the axial forces in the members of the truss and indicate whether they are in tension (T) or compression (C).

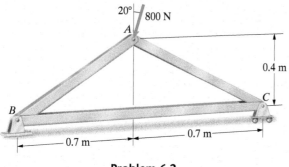

Problem 6.2

6.3 Member AB of the truss is subjected to a 1000-lb tensile force. Determine the weight W and the axial force in member AC.

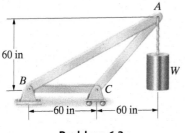

Problem 6.3

6.4 Determine the axial forces in members BC and CD of the truss.

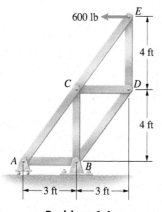

Problem 6.4

6.5 Each suspended weight has mass $m = 20$ kg. Determine the axial forces in the members of the truss and indicate whether they are in tension (T) or compression (C).

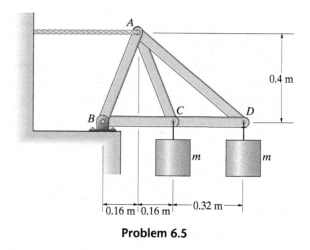

Problem 6.5

6.6 Determine the largest tensile and compressive forces that occur in the members of the truss, and indicate the members in which they occur if

(a) the dimension $h = 0.1$ m;

(b) the dimension $h = 0.5$ m.

Observe how a simple change in design affects the maximum axial loads.

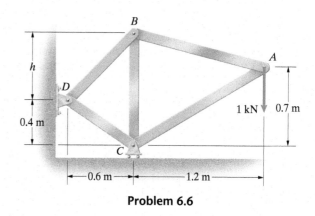

Problem 6.6

6.7 This steel truss bridge is in the Gallatin National Forest south of Bozeman, Montana. Suppose that one of the tandem trusses supporting the bridge is loaded as shown. Determine the axial forces in members *AB*, *BC*, *BD*, and *BE*.

6.8 Determine the largest tensile and compressive forces that occur in the members of the bridge truss, and indicate the members in which they occur.

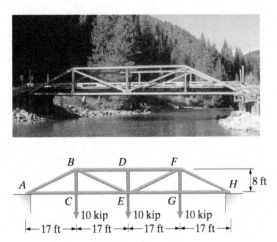

Problems 6.7/6.8

6.9 The trusses supporting the bridge in Problems 6.7 and 6.8 are called Pratt trusses. Suppose that the bridge designers had decided to use the truss shown instead, which is called a Howe truss. Determine the largest tensile and compressive forces that occur in the members, and indicate the members in which they occur. Compare your answers to the answers to Problem 6.8.

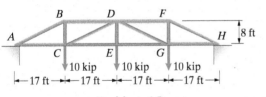

Problem 6.9

6.10 Determine the axial forces in members *BD*, *CD*, and *CE* of the truss.

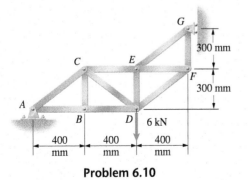

Problem 6.10

6.11 The loads $F_1 = F_2 = 8$ kN. Determine the axial forces in members *BD*, *BE*, and *BG*.

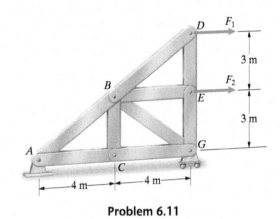

Problem 6.11

6.12 Determine the largest tensile and compressive forces that occur in the members of the truss, and indicate the members in which they occur if

(a) the dimension $h = 5$ in;

(b) the dimension $h = 10$ in.

Observe how a simple change in design affects the maximum axial loads.

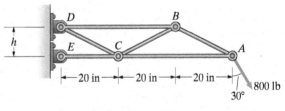

Problem 6.12

6.13 The truss supports loads at *C* and *E*. If $F = 3$ kN, what are the axial forces in members *BC* and *BE*?

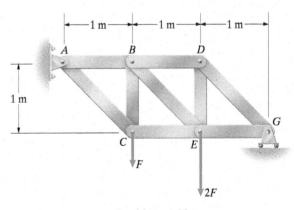

Problem 6.13

6.14 If you don't want the members of the truss to be subjected to an axial load (tension or compression) greater than 20 kN, what is the largest acceptable magnitude of the downward force *F*?

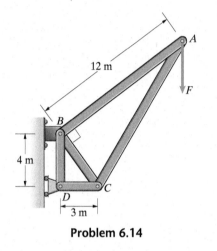

Problem 6.14

6.15 The truss is a preliminary design for a structure to attach one end of a stretcher to a rescue helicopter. Based on dynamic simulations, the design engineer estimates that the downward forces the stretcher will exert will be no greater than 1.6 kN at *A* and at *B*. What are the resulting axial forces in members *CF*, *DF*, and *FG*?

6.16 Upon learning of an upgrade in the helicopter's engine, the engineer designing the truss does new simulations and concludes that the downward forces the stretcher will exert at *A* and at *B* may be as large as 1.8 kN. What are the resulting axial forces in members *DE*, *DF*, and *DG*?

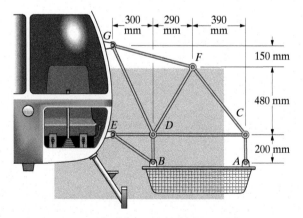

Problems 6.15/6.16

6.17 Determine the axial forces in the members in terms of the weight *W*.

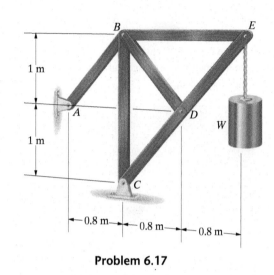

Problem 6.17

6.18 The lengths of the members of the truss are shown. The mass of the suspended crate is 900 kg. Determine the axial forces in the members.

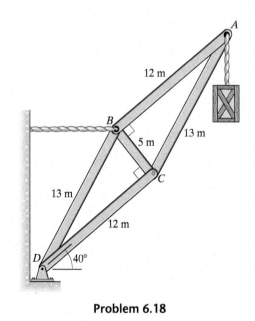

Problem 6.18

6.19 The loads $F_1 = 600$ lb and $F_2 = 300$ lb. Determine the axial forces in members AE, BD, and CD.

6.20 The loads $F_1 = 450$ lb and $F_2 = 150$ lb. Determine the axial forces in members AB, AC, and BC.

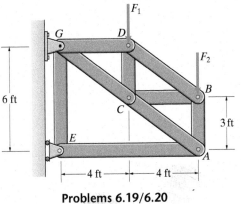

Problems 6.19/6.20

6.21 Determine the axial forces in members BC, CD, and CE of the truss.

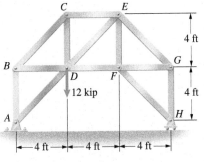

Problem 6.21

6.22 The Warren truss supporting the walkway is designed to support vertical 50-kN loads at B, D, F, and H. If the truss is subjected to these loads, what are the resulting axial forces in members BC, CD, and CE?

6.23 For the Warren truss in Problem 6.22, determine the axial forces in members DF, EF, and FG.

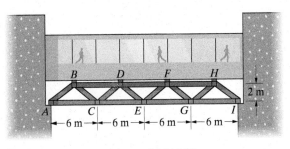

Problems 6.22/6.23

6.24 The Pratt bridge truss supports five forces ($F = 300$ kN). The dimension $L = 8$ m. Determine the axial forces in members BC, BI, and BJ.

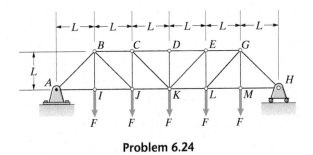

Problem 6.24

6.25 For the roof truss shown, determine the axial forces in members AD, BD, DE, and DG. Model the supports at A and I as roller supports.

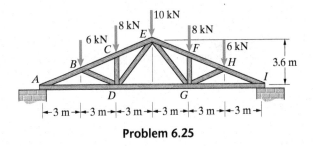

Problem 6.25

6.26 The Howe truss helps support a roof. Model the supports at A and G as roller supports. Determine the axial forces in members AB, BC, and CD.

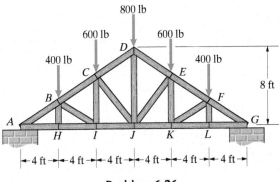

Problem 6.26

6.27   The plane truss forms part of the supports of a crane on an offshore oil platform. The crane exerts vertical 75-kN forces on the truss at B, C, and D. You can model the support at A as a pin support and model the support at E as a roller support that can exert a force normal to the dashed line but cannot exert a force parallel to it. The angle $\alpha = 45°$. Determine the axial forces in the members of the truss.

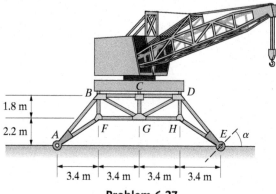

Problem 6.27

6.28   (a) Design a truss attached to the supports A and B that supports the loads applied at points C and D. (b) Determine the axial forces in the members of the truss you designed in (a).

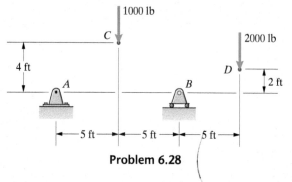

Problem 6.28

6.29   (a) Design a truss attached to the supports A and B that goes over the obstacle and supports the load applied at C.

(b) Determine the axial forces in the members of the truss you designed in (a).

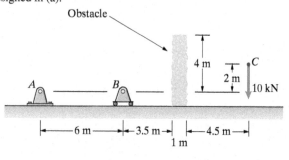

Problem 6.29

6.30   Suppose that you want to design a truss supported at A and B (Fig. a) to support a 3-kN downward load at C. The simplest design (Fig. b) subjects member AC to a 5-kN tensile force. Redesign the truss so that the largest tensile force is less than 3 kN.

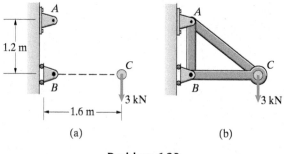

Problem 6.30

▶ **6.31**   The bridge structure shown in Example 6.2 can be given a higher arch by increasing the 15° angles to 20°. If this is done, what are the axial forces in members AB, BC, CD, and DE?

6.3   The Method of Sections

BACKGROUND

When we need to know the axial forces only in certain members of a truss, we often can determine them more quickly using the method of sections than using the method of joints. For example, let's reconsider the Warren truss we used to introduce the method of joints (Fig. 6.12a). It supports loads at B and D, and each member is 2 m in length. Suppose that we need to determine only the axial force in member BC.

Figure 6.12
(a) A Warren truss supporting two loads.
(b) Free-body diagram of the truss, showing the reactions at the supports.

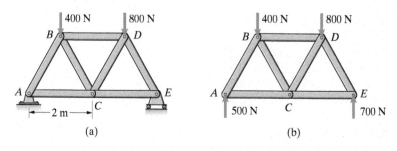

Just as in the method of joints, we begin by drawing a free-body diagram of the entire truss and determining the reactions at the supports. The results of this step are shown in Fig. 6.12b. Our next step is to cut the members AC, BC, and BD to obtain a free-body diagram of a part, or *section*, of the truss (Fig. 6.13). Summing moments about point B, the equilibrium equations for the section are

$$\Sigma F_x = T_{AC} + T_{BD} + T_{BC} \cos 60° = 0,$$
$$\Sigma F_y = 500 \text{ N} - 400 \text{ N} - T_{BC} \sin 60° = 0,$$
$$\Sigma M_{\text{point } B} = (2 \sin 60° \text{ m})T_{AC} - (2 \cos 60° \text{ m})(500 \text{ N}) = 0.$$

Solving them, we obtain $T_{AC} = 289$ N, $T_{BC} = 115$ N, and $T_{BD} = -346$ N.

Notice how similar this method is to the method of joints. Both methods involve cutting members to obtain free-body diagrams of parts of a truss. In the method of joints, we move from joint to joint, drawing free-body diagrams of the joints and determining the axial forces in the members as we go. In the method of sections, we try to obtain a single free-body diagram that allows us to determine the axial forces in specific members. In our example, we obtained a free-body diagram by cutting three members, including the one (member BC) whose axial force we wanted to determine.

In contrast to the free-body diagrams of joints, the forces on the free-body diagrams used in the method of sections are not usually concurrent, and as in our example, we can obtain three independent equilibrium equations. Although there are exceptions, it is usually necessary to choose a section that requires cutting no more than three members, or there will be more unknown axial forces than equilibrium equations.

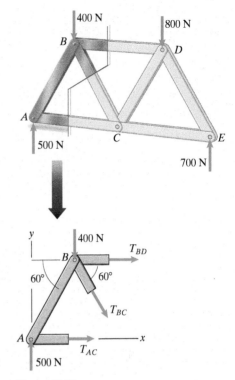

Figure 6.13
Obtaining a free-body diagram of a section of the truss.

RESULTS

The Method of Sections

When the axial forces in particular members of a truss must be determined, the method of sections can often provide the needed results more efficiently than the method of joints.

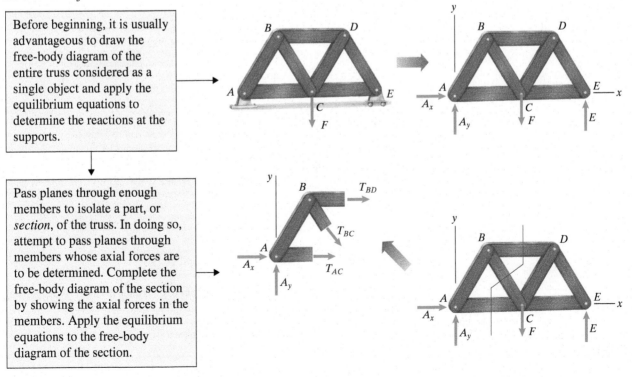

Before beginning, it is usually advantageous to draw the free-body diagram of the entire truss considered as a single object and apply the equilibrium equations to determine the reactions at the supports.

Pass planes through enough members to isolate a part, or *section*, of the truss. In doing so, attempt to pass planes through members whose axial forces are to be determined. Complete the free-body diagram of the section by showing the axial forces in the members. Apply the equilibrium equations to the free-body diagram of the section.

Active Example 6.3   The Method of Sections (▶ *Related Problem 6.32*)

The horizontal members of the truss are each 1 m in length. Determine the axial forces in members *CD*, *CJ*, and *IJ*.

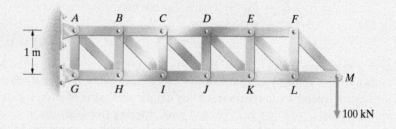

Strategy

By passing planes through members *CD*, *CJ*, and *IJ*, we will obtain a section from which we can obtain the desired axial forces.

Solution

Pass planes through members *CD*, *CJ*, and *IJ* and draw the free-body diagram of the section.

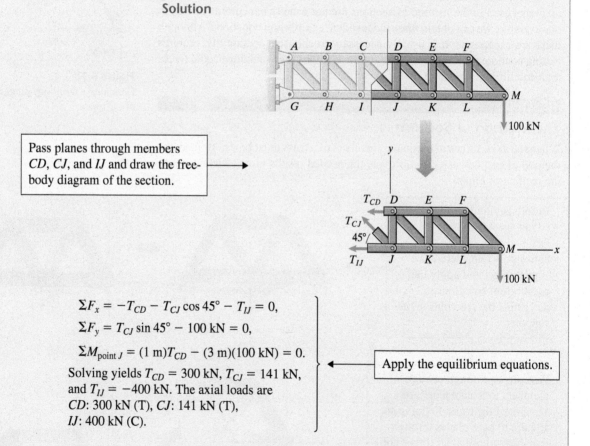

$$\Sigma F_x = -T_{CD} - T_{CJ} \cos 45° - T_{IJ} = 0,$$

$$\Sigma F_y = T_{CJ} \sin 45° - 100 \text{ kN} = 0,$$

$$\Sigma M_{\text{point } J} = (1 \text{ m})T_{CD} - (3 \text{ m})(100 \text{ kN}) = 0.$$

Solving yields $T_{CD} = 300$ kN, $T_{CJ} = 141$ kN, and $T_{IJ} = -400$ kN. The axial loads are
CD: 300 kN (T), *CJ*: 141 kN (T),
IJ: 400 kN (C).

Apply the equilibrium equations.

Practice Problem Use the method of sections to determine the axial forces in members *DE*, *DK*, and *JK* of the truss.

Answer: *DE*: 200 kN (T), *DK*: 141 kN (T), *JK*: 300 kN (C).

| Example 6.4 | **Choosing an Appropriate Section** (▶ *Related Problem 6.33*) |

Determine the axial forces in members *DG* and *BE* of the truss.

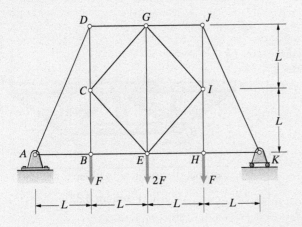

Strategy

We can't obtain a section that involves cutting members *DG* and *BE* without cutting more than three members. However, cutting members *DG*, *BE*, *CD*, and *BC* results in a section with which we can determine the axial forces in members *DG* and *BE*.

Solution

Determine the Reactions at the Supports We draw the free-body diagram of the entire truss in Fig. a. From the equilibrium equations,

$$\Sigma F_x = A_x = 0,$$

$$\Sigma F_y = A_y + K - F - 2F - F = 0,$$

$$\Sigma M_{\text{point } A} = -LF - (2L)(2F) - (3L)F + (4L)K = 0,$$

we obtain the reactions $A_x = 0$, $A_y = 2F$, and $K = 2F$.

Choose a Section

In Fig. b, we obtain a section by cutting members *DG*, *CD*, *BC*, and *BE*. Because the lines of action of T_{BE}, T_{BC}, and T_{CD} pass through point *B*, we can determine T_{DG} by summing moments about *B*:

$$\Sigma M_{\text{point } B} = -L(2F) - (2L)T_{DG} = 0.$$

The axial force $T_{DG} = -F$. Then, from the equilibrium equation

$$\Sigma F_x = T_{DG} + T_{BE} = 0,$$

we see that $T_{BE} = -T_{DG} = F$. Member *DG* is in compression, and member *BE* is in tension.

Critical Thinking

This is a clever example, but not one that is typical of problems faced in practice. The section used to solve it might not be obvious even to a person with experience analyzing structures. Notice that the free-body diagram in Fig. b is statically indeterminate, although it can be used to determine the axial forces in members *DG* and *BE*.

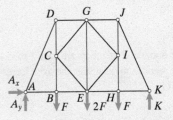

(a) Free-body diagram of the entire truss.

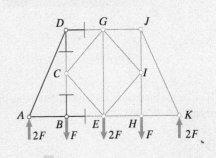

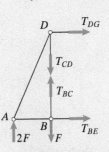

(b) A section of the truss obtained by passing planes through members *DG*, *CD*, *BC*, and *BE*.

Problems

▶ 6.32 In Active Example 6.3, use the method of sections to determine the axial forces in members *BC*, *BI*, and *HI*.

▶ 6.33 In Example 6.4, obtain a section of the truss by passing planes through members *BE*, *CE*, *CG*, and *DG*. Using the fact that the axial forces in members *DG* and *BE* have already been determined, use your section to determine the axial forces in members *CE* and *CG*.

6.34 The truss supports a 100-kN load at *J*. The horizontal members are each 1 m in length.

(a) Use the method of joints to determine the axial force in member *DG*.

(b) Use the method of sections to determine the axial force in member *DG*.

6.35 The horizontal members are each 1 m in length. Use the method of sections to determine the axial forces in members *BC*, *CF*, and *FG*.

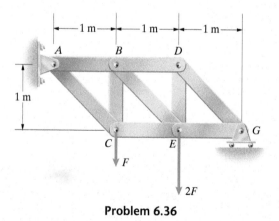

Problems 6.34/6.35

6.36 Use the method of sections to determine the axial forces in members *AB*, *BC*, and *CE*.

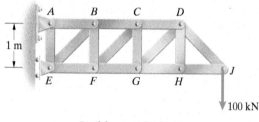

Problem 6.36

6.37 Use the method of sections to determine the axial forces in members *DF*, *EF*, and *EG*.

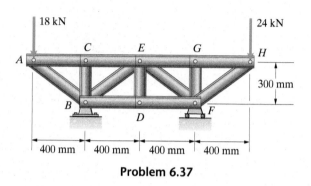

Problem 6.37

6.38 The Pratt bridge truss is loaded as shown. Use the method of sections to determine the axial forces in members *BD*, *BE*, and *CE*.

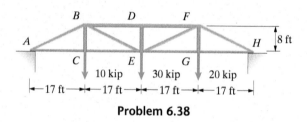

Problem 6.38

6.39 The Howe bridge truss is loaded as shown. Use the method of sections to determine the axial forces in members *BD*, *CD*, and *CE*.

6.40 For the Howe bridge truss, use the method of sections to determine the axial forces in members *DF*, *DG*, and *EG*.

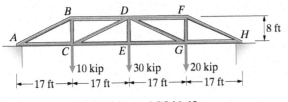
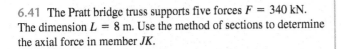
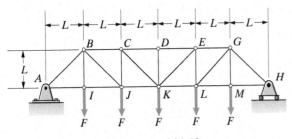

Problems 6.39/6.40

6.41 The Pratt bridge truss supports five forces $F = 340$ kN. The dimension $L = 8$ m. Use the method of sections to determine the axial force in member *JK*.

6.42 For the Pratt bridge truss in Problem 6.41, use the method of sections to determine the axial force in member *EK*.

Problems 6.41/6.42

6.43 The walkway exerts vertical 50-kN loads on the Warren truss at *B*, *D*, *F*, and *H*. Use the method of sections to determine the axial force in member *CE*.

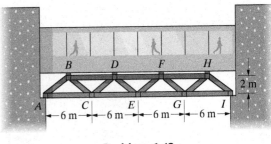

Problem 6.43

6.44 Use the method of sections to determine the axial forces in members *AC*, *BC*, and *BD*.

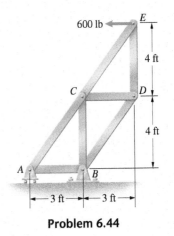

Problem 6.44

6.45 Use the method of sections to determine the axial forces in members *FH*, *GH*, and *GI*.

6.46 Use the method of sections to determine the axial forces in members *DF*, *DG*, and *EG*.

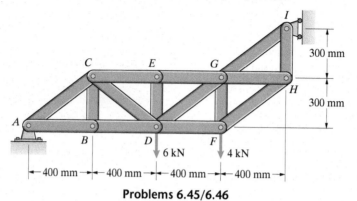

Problems 6.45/6.46

6.47 The Howe truss helps support a roof. Model the supports at *A* and *G* as roller supports.

(a) Use the method of joints to determine the axial force in member *BI*.

(b) Use the method of sections to determine the axial force in member *BI*.

6.48 Use the method of sections to determine the axial force in member *EJ*.

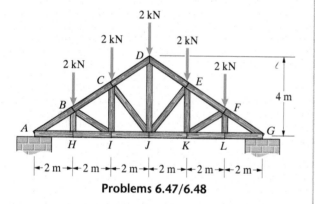

Problems 6.47/6.48

6.49 Use the method of sections to determine the axial forces in members *CE*, *DE*, and *DF*.

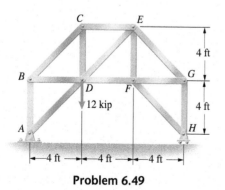

Problem 6.49

6.50 For the bridge truss shown, use the method of sections to determine the axial forces in members *CE*, *CF*, and *DF*.

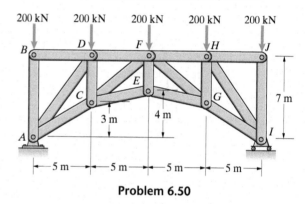

Problem 6.50

6.51 The load $F = 20$ kN and the dimension $L = 2$ m. Use the method of sections to determine the axial force in member *HK*.

Strategy: Obtain a section by cutting members *HK*, *HI*, *IJ*, and *JM*. You can determine the axial forces in members *HK* and *JM* even though the resulting free-body diagram is statically indeterminate.

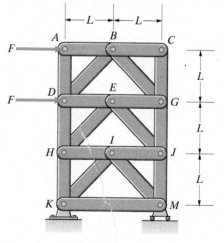

Problem 6.51

6.52 The weight of the bucket is $W = 1000$ lb. The cable passes over pulleys at A and D.

(a) Determine the axial forces in members FG and HI.

(b) By drawing free-body diagrams of sections, explain why the axial forces in members FG and HI are equal.

6.53 The weight of the bucket is $W = 1000$ lb. The cable passes over pulleys at A and D. Determine the axial forces in members IK and JL.

6.54 The truss supports loads at N, P, and R. Determine the axial forces in members IL and KM.

6.55 Determine the axial forces in members HJ and GI.

6.56 By drawing free-body diagrams of sections, explain why the axial forces in members DE, FG, and HI are zero.

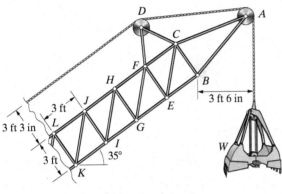

Problems 6.52/6.53

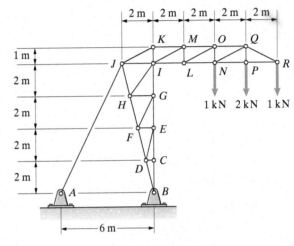

Problems 6.54–6.56

6.4 Space Trusses

BACKGROUND

We can form a simple three-dimensional structure by connecting six bars at their ends to obtain a tetrahedron, as shown in Fig. 6.14a. By adding members, we can obtain more elaborate structures (Figs. 6.14b and c). Three-dimensional structures such as these are called *space trusses* if they have joints that do not exert couples on the members (that is, the joints behave like ball and socket supports) and they are loaded and supported at their joints. Space trusses are analyzed by the same methods we described for two-dimensional trusses. The only difference is the need to cope with the more complicated geometry.

Consider the space truss in Fig. 6.15a. Suppose that the load $\mathbf{F} = -2\mathbf{i} - 6\mathbf{j} - \mathbf{k}$ (kN). The joints A, B, and C rest on the smooth floor. Joint

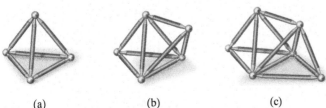

(a)　　　　　(b)　　　　　(c)

Figure 6.14

Space trusses with 6, 9, and 12 members.

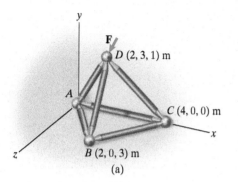

(a)

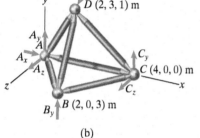

(b)

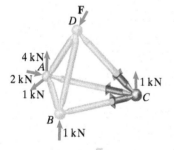

(c)

Figure 6.15
(a) A space truss supporting a load **F**.
(b) Free-body diagram of the entire truss.
(c) Obtaining the free-body diagram of joint C.

A is supported by the corner where the smooth walls meet, and joint C rests against the back wall. We can apply the method of joints to this truss.

First we must determine the reactions exerted by the supports (the floor and walls). We draw the free-body diagram of the entire truss in Fig. 6.15b. The corner can exert three components of force at A, the floor and wall can exert two components of force at C, and the floor can exert a normal force at B. Summing moments about A, we find that the equilibrium equations, with forces in kN and distances in m, are

$$\Sigma F_x = A_x - 2 = 0,$$

$$\Sigma F_y = A_y + B_y + C_y - 6 = 0,$$

$$\Sigma F_z = A_z + C_z - 1 = 0,$$

$$\Sigma M_{\text{point } A} = (\mathbf{r}_{AB} \times B_y \mathbf{j}) + [\mathbf{r}_{AC} \times (C_y \mathbf{j} + C_z \mathbf{k})] + (\mathbf{r}_{AD} \times \mathbf{F})$$

$$= \begin{vmatrix} \mathbf{i} & \mathbf{j} & \mathbf{k} \\ 2 & 0 & 3 \\ 0 & B_y & 0 \end{vmatrix} + \begin{vmatrix} \mathbf{i} & \mathbf{j} & \mathbf{k} \\ 4 & 0 & 0 \\ 0 & C_y & C_z \end{vmatrix} + \begin{vmatrix} \mathbf{i} & \mathbf{j} & \mathbf{k} \\ 2 & 3 & 1 \\ -2 & -6 & -1 \end{vmatrix}$$

$$= (-3B_y + 3)\mathbf{i} + (-4C_z)\mathbf{j}$$

$$+ (2B_y + 4C_y - 6)\mathbf{k} = 0.$$

Solving these equations, we obtain the reactions $A_x = 2$ kN, $A_y = 4$ kN, $A_z = 1$ kN, $B_y = 1$ kN, $C_y = 1$ kN, and $C_z = 0$.

In this example, we can determine the axial forces in members AC, BC, and CD from the free-body diagram of joint C (Fig. 6.15c). To write the equilibrium equations for the joint, we must express the three axial forces in terms of their components. Because member AC lies along the x axis, we express the force exerted on joint C by the axial force T_{AC} as the vector $-T_{AC}\mathbf{i}$. Let $\mathbf{r}_{CB}$ be the position vector from C to B:

$$\mathbf{r}_{CB} = (2 - 4)\mathbf{i} + (0 - 0)\mathbf{j} + (3 - 0)\mathbf{k} = -2\mathbf{i} + 3\mathbf{k} \text{ (m)}.$$

Dividing this vector by its magnitude to obtain a unit vector that points from C toward B yields

$$\mathbf{e}_{CB} = \frac{\mathbf{r}_{CB}}{|\mathbf{r}_{CB}|} = -0.555\mathbf{i} + 0.832\mathbf{k},$$

and we express the force exerted on joint C by the axial force T_{BC} as the vector

$$T_{BC}\mathbf{e}_{CB} = T_{BC}(-0.555\mathbf{i} + 0.832\mathbf{k}).$$

In the same way, we express the force exerted on joint C by the axial force T_{CD} as the vector

$$T_{CD}(-0.535\mathbf{i} + 0.802\mathbf{j} + 0.267\mathbf{k}).$$

Setting the sum of the forces on the joint equal to zero, we obtain

$$-T_{AC}\mathbf{i} + T_{BC}(-0.555\mathbf{i} + 0.832\mathbf{k})$$
$$+T_{CD}(-0.535\mathbf{i} + 0.802\mathbf{j} + 0.267\mathbf{k}) + (1 \text{ kN})\mathbf{j} = 0,$$

and then get the three equilibrium equations

$$\Sigma F_x = -T_{AC} - 0.555T_{BC} - 0.535T_{CD} = 0,$$
$$\Sigma F_y = 0.802T_{CD} + 1 \text{ kN} = 0,$$
$$\Sigma F_z = 0.832T_{BC} + 0.267T_{CD} = 0.$$

Solving these equations, the axial forces are $T_{AC} = 0.444$ kN, $T_{BC} = 0.401$ kN, and $T_{CD} = -1.247$ kN. Members AC and BC are in tension, and member CD is in compression. By continuing to draw free-body diagrams of the joints, we can determine the axial forces in all the members.

As our example demonstrates, three equilibrium equations can be obtained from the free-body diagram of a joint in three dimensions, so it is usually necessary to choose joints to analyze that are subjected to known forces and no more than three unknown forces.

RESULTS

A *space truss* is a truss whose members are not coplanar. Axial forces in the members of a statically determinate space truss can be determined by applying the method of joints.

Before beginning, it is usually necessary to draw the free-body diagram of the entire truss considered as a single object and apply the equilibrium equations to determine the reactions at the supports.

Isolate an individual joint by passing planes through the connected members. Complete the free-body diagram by showing the axial forces in the members. Apply the equilibrium equation $\Sigma\mathbf{F} = \mathbf{0}$ to the free-body diagram of the joint. Repeat this process for other joints until the desired axial loads have been determined.

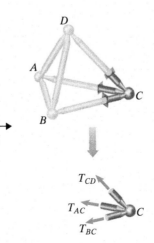

Active Example 6.5   Space Truss (▶ *Related Problem 6.57*)

The space truss has roller supports at *B*, *C*, and *D* and supports a vertical 1200-lb load at *A*. Determine the axial forces in members *AD*, *BD*, and *CD*.

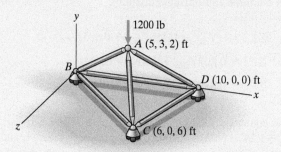

Strategy

We will first draw a free-body diagram of the entire truss, treating it as a single object, and determine the reactions at the supports. Then we can determine the axial forces in members *AD*, *BD*, and *CD* by drawing the free-body diagram of joint *D*.

Solution

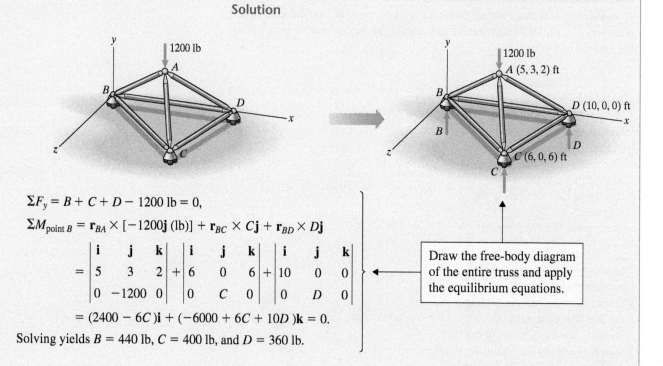

$$\Sigma F_y = B + C + D - 1200 \text{ lb} = 0,$$

$$\Sigma M_{\text{point } B} = \mathbf{r}_{BA} \times [-1200\mathbf{j} \text{ (lb)}] + \mathbf{r}_{BC} \times C\mathbf{j} + \mathbf{r}_{BD} \times D\mathbf{j}$$

$$= \begin{vmatrix} \mathbf{i} & \mathbf{j} & \mathbf{k} \\ 5 & 3 & 2 \\ 0 & -1200 & 0 \end{vmatrix} + \begin{vmatrix} \mathbf{i} & \mathbf{j} & \mathbf{k} \\ 6 & 0 & 6 \\ 0 & C & 0 \end{vmatrix} + \begin{vmatrix} \mathbf{i} & \mathbf{j} & \mathbf{k} \\ 10 & 0 & 0 \\ 0 & D & 0 \end{vmatrix}$$

> Draw the free-body diagram of the entire truss and apply the equilibrium equations.

$$= (2400 - 6C)\mathbf{i} + (-6000 + 6C + 10D)\mathbf{k} = 0.$$

Solving yields *B* = 440 lb, *C* = 400 lb, and *D* = 360 lb.

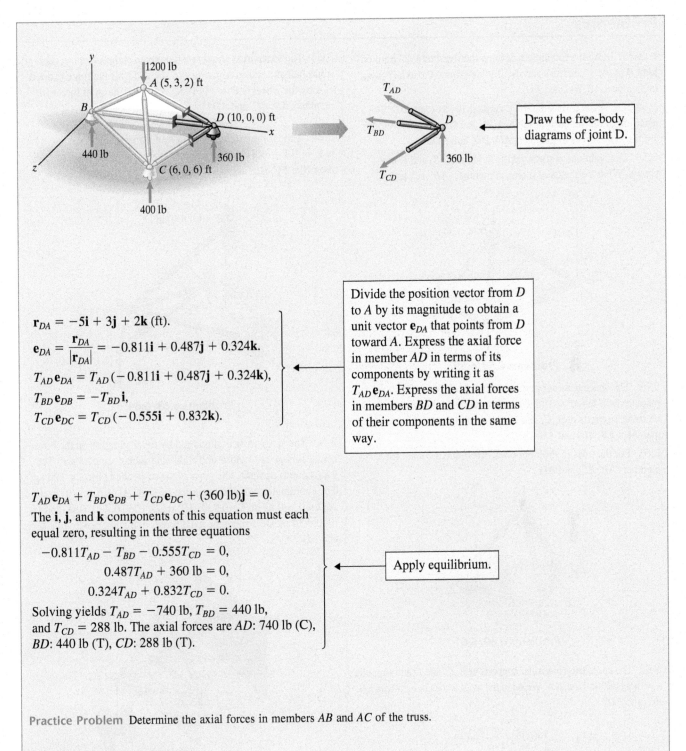

$\mathbf{r}_{DA} = -5\mathbf{i} + 3\mathbf{j} + 2\mathbf{k}$ (ft).

$\mathbf{e}_{DA} = \dfrac{\mathbf{r}_{DA}}{|\mathbf{r}_{DA}|} = -0.811\mathbf{i} + 0.487\mathbf{j} + 0.324\mathbf{k}$.

$T_{AD}\mathbf{e}_{DA} = T_{AD}(-0.811\mathbf{i} + 0.487\mathbf{j} + 0.324\mathbf{k})$,

$T_{BD}\mathbf{e}_{DB} = -T_{BD}\mathbf{i}$,

$T_{CD}\mathbf{e}_{DC} = T_{CD}(-0.555\mathbf{i} + 0.832\mathbf{k})$.

Divide the position vector from D to A by its magnitude to obtain a unit vector $\mathbf{e}_{DA}$ that points from D toward A. Express the axial force in member AD in terms of its components by writing it as $T_{AD}\mathbf{e}_{DA}$. Express the axial forces in members BD and CD in terms of their components in the same way.

$T_{AD}\mathbf{e}_{DA} + T_{BD}\mathbf{e}_{DB} + T_{CD}\mathbf{e}_{DC} + (360 \text{ lb})\mathbf{j} = 0$.

The $\mathbf{i}$, $\mathbf{j}$, and $\mathbf{k}$ components of this equation must each equal zero, resulting in the three equations

$-0.811T_{AD} - T_{BD} - 0.555T_{CD} = 0$,

$\qquad 0.487T_{AD} + 360 \text{ lb} = 0$,

$\qquad 0.324T_{AD} + 0.832T_{CD} = 0$.

Solving yields $T_{AD} = -740$ lb, $T_{BD} = 440$ lb, and $T_{CD} = 288$ lb. The axial forces are AD: 740 lb (C), BD: 440 lb (T), CD: 288 lb (T).

Apply equilibrium.

Practice Problem Determine the axial forces in members AB and AC of the truss.

Answer: AB: 904 lb (C), AC: 680 lb (C).

Problems

▶ **6.57** In Active Example 6.5, draw the free-body diagram of joint B of the space truss and use it to determine the axial forces in members AB, BC, and BD.

6.58 The space truss supports a vertical 10-kN load at D. The reactions at the supports at joints A, B, and C are shown. What are the axial forces in members AD, BD, and CD?

6.59 The reactions at the supports at joints A, B, and C are shown. What are the axial forces in members AB, AC, and AD?

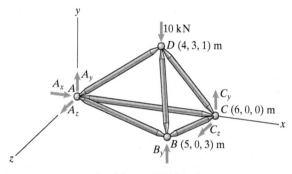
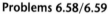

Problems 6.58/6.59

6.60 The space truss supports a vertical load F at A. Each member is of length L, and the truss rests on the horizontal surface on roller supports at B, C, and D. Determine the axial forces in members AB, AC, and AD.

6.61 For the truss in Problem 6.60, determine the axial forces in members AB, BC, and BD.

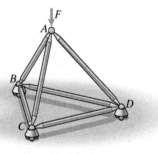

Problems 6.60/6.61

6.62 The space truss has roller supports at B, C, and D and supports a vertical 800-lb load at A. What are the axial forces in members AB, AC, and AD?

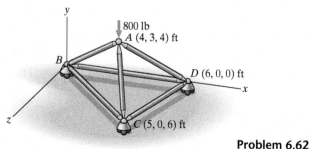

Problem 6.62

6.63 The space truss shown models an airplane's landing gear. It has ball and socket supports at C, D, and E. If the force exerted at A by the wheel is $\mathbf{F} = 40\mathbf{j}$ (kN), what are the axial forces in members AB, AC, and AD?

6.64 If the force exerted at point A of the truss in Problem 6.63 is $\mathbf{F} = 10\mathbf{i} + 60\mathbf{j} + 20\mathbf{k}$ (kN), what are the axial forces in members BC, BD, and BE?

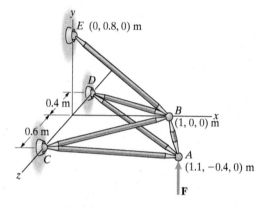

Problems 6.63/6.64

6.65 The space truss is supported by roller supports on the horizontal surface at C and D and a ball and socket support at E. The y axis points upward. The mass of the suspended object is 120 kg. The coordinates of the joints of the truss are A: $(1.6, 0.4, 0)$ m, B: $(1.0, 1.0, -0.2)$ m, C: $(0.9, 0, 0.9)$ m, D: $(0.9, 0, -0.6)$ m, and E: $(0, 0.8, 0)$ m. Determine the axial forces in members AB, AC, and AD.

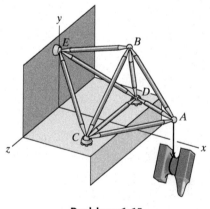

Problem 6.65

6.66 The free-body diagram of the part of the construction crane to the left of the plane is shown. The coordinates (in meters) of the joints A, B, and C are $(1.5, 1.5, 0)$, $(0, 0, 1)$, and $(0, 0, -1)$, respectively. The axial forces P_1, P_2, and P_3 are parallel to the x axis. The axial forces P_4, P_5, and P_6 point in the directions of the unit vectors

$$\mathbf{e}_4 = 0.640\mathbf{i} - 0.640\mathbf{j} - 0.426\mathbf{k},$$

$$\mathbf{e}_5 = 0.640\mathbf{i} - 0.640\mathbf{j} + 0.426\mathbf{k},$$

$$\mathbf{e}_6 = 0.832\mathbf{i} - 0.555\mathbf{k}.$$

The total force exerted on the free-body diagram by the weight of the crane and the load it supports is $-F\mathbf{j} = -44\mathbf{j}$ (kN) acting at the point $(-20, 0, 0)$ m. What is the axial force P_3?

Strategy: Use the fact that the moment about the line that passes through joints A and B equals zero.

6.67 In Problem 6.66, what are the axial forces P_1, P_4, and P_5?

Strategy: Write the equilibrium equations for the entire free-body diagram.

6.68 The mirror housing of the telescope is supported by a 6-bar space truss. The mass of the housing is 3 Mg (megagrams), and its weight acts at G. The distance from the axis of the telescope to points A, B, and C is 1 m, and the distance from the axis to points D, E, and F is 2.5 m. If the telescope axis is vertical ($\alpha = 90°$), what are the axial forces in the members of the truss?

6.69 Consider the telescope described in Problem 6.68. Determine the axial forces in the members of the truss if the angle α between the horizontal and the telescope axis is 20°.

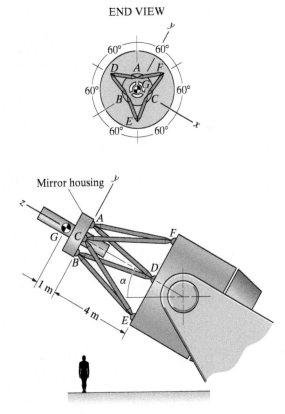

END VIEW

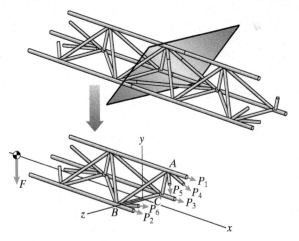

Problems 6.68/6.69

Problems 6.66/6.67

6.5 Frames and Machines

BACKGROUND

Many structures, such as the frame of a car and the human structure of bones, tendons, and muscles (Fig. 6.16), are not composed entirely of two-force members and thus cannot be modeled as trusses. In this section we consider structures of interconnected members that do not satisfy the definition of a truss. Such structures are called *frames* if they are designed to remain stationary and support loads and *machines* if they are designed to move and apply loads.

When trusses are analyzed by cutting members to obtain free-body diagrams of joints or sections, the internal forces acting at the "cuts" are simple axial forces (see Fig. 6.4). This is not generally true for frames or machines, and a different method of analysis is necessary. Instead of cutting members, we isolate entire members, or in some cases combinations of members, from the structure.

To begin analyzing a frame or machine, we draw a free-body diagram of the entire structure (that is, treat the structure as a single object) and determine the reactions at its supports. In some cases the entire structure will be statically indeterminate, but it is helpful to determine as many of the reactions as possible. We then draw free-body diagrams of individual members, or selected combinations of members, and apply the equilibrium equations to determine the forces and couples acting on them. For example, consider the stationary structure in

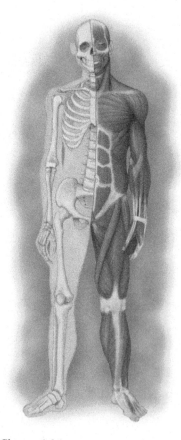

Figure 6.16
The internal structure of a person and a car's frame are not trusses.

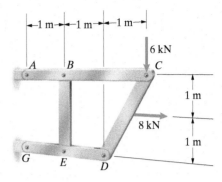

Fig. 6.17. Member *BE* is a two-force member, but the other three members—*ABC*, *CD*, and *DEG*—are not. This structure is a frame. Our objective is to determine the forces on its members.

Analyzing the Entire Structure

We draw the free-body diagram of the entire frame in Fig. 6.18. It is statically indeterminate: There are four unknown reactions, A_x, A_y, G_x, and G_y, whereas we can write only three independent equilibrium equations. However, notice that the lines of action of three of the unknown reactions intersect at A. Summing moments about A yields

$$\Sigma M_{\text{point } A} = (2 \text{ m})G_x + (1 \text{ m})(8 \text{ kN}) - (3 \text{ m})(6 \text{ kN}) = 0,$$

Figure 6.17
A frame supporting two loads.

and we obtain the reaction $G_x = 5$ kN. Then, from the equilibrium equation

$$\Sigma F_x = A_x + G_x + 8 \text{ kN} = 0,$$

we obtain the reaction $A_x = -13$ kN. Although we cannot determine A_y or G_y from the free-body diagram of the entire structure, we can do so by analyzing the individual members.

Analyzing the Members

Our next step is to draw free-body diagrams of the members. To do so, we treat the attachment of a member to another member just as if it were a support. Looked at in this way, we can think of each member as a supported object of the kind analyzed in Chapter 5. Furthermore, the forces and couples the members exert on one another are *equal in magnitude and opposite in direction*. A simple

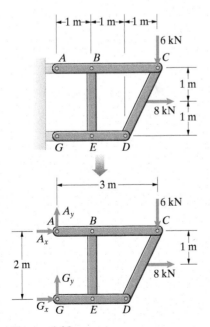

Figure 6.18
Obtaining the free-body diagram of the entire frame.

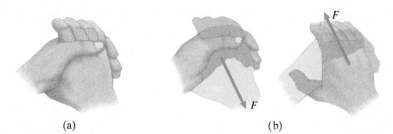

(a)   (b)

Figure 6.19
Demonstrating Newton's third law:
(a) Clasp your hands and pull on your left hand.
(b) Your hands exert equal and opposite forces.

demonstration is instructive. If you clasp your hands as shown in Fig. 6.19a and exert a force on your left hand with your right hand, your left hand exerts an equal and opposite force on your right hand (Fig. 6.19b). Similarly, if you exert a couple on your left hand, your left hand exerts an equal and opposite couple on your right hand.

In Fig. 6.20 we "disassemble" the frame and draw free-body diagrams of its members. Observe that the forces exerted on one another by the members are equal and opposite. For example, at point C on the free-body diagram of member ABC, the force exerted by member CD is denoted by the components

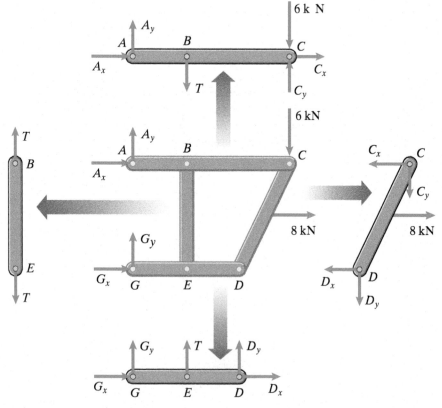

Figure 6.20
Obtaining the free-body diagrams of the members.

C_x and C_y. The forces exerted by member ABC on member CD at point C must be equal and opposite, as shown.

We need to discuss two important aspects of these free-body diagrams before completing the analysis.

Two-Force Members   Member BE is a two-force member, and we have taken this into account in drawing its free-body diagram in Fig. 6.20. The force T is the axial force in member BE, and an equal and opposite force is subjected on member ABC at B and on member GED at E.

Recognizing two-force members in frames and machines and drawing their free-body diagrams as we have done will reduce the number of unknowns and will greatly simplify the analysis. In our example, if we did not treat member BE as a two-force member, its free-body diagram would have four unknown forces (Fig. 6.21a). By treating it as a two-force member (Fig. 6.21b), we reduce the number of unknown forces by three.

Loads Applied at Joints   A question arises when a load is applied at a joint: Where does the load appear on the free-body diagrams of the individual members? The answer is that you can place the load on *any one* of the members attached at the joint. For example, in Fig. 6.17, the 6-kN load acts at the joint where members ABC and CD are connected. In drawing the free-body diagrams of the individual members (Fig. 6.20), we assumed that the 6-kN load acted on member ABC. The force components C_x and C_y on the free-body diagram of member ABC are the forces exerted by the member CD.

To explain why we can draw the free-body diagrams in this way, let us assume that the 6-kN force acts on the pin connecting members ABC and CD, and draw separate free-body diagrams of the pin and the two members (Fig. 6.22a). The force components C'_x and C'_y are the forces exerted by the pin on member ABC, and C_x and C_y are the forces exerted by the pin on member CD. If we superimpose the free-body diagrams of the pin and member ABC, we obtain the two free-body diagrams in Fig. 6.22b, which is the way we drew them in Fig. 6.20. Alternatively, by superimposing the free-body diagrams of the pin and member CD, we obtain the two free-body diagrams in Fig. 6.22c.

Thus if a load acts at a joint, it can be placed on any one of the members attached at the joint when drawing the free-body diagrams of the individual members. Just make sure not to place it on more than one member.

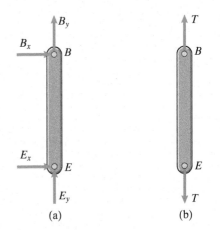

Figure 6.21
Free-body diagram of member BE:
(a) Not treating it as a two-force member.
(b) Treating it as a two-force member.

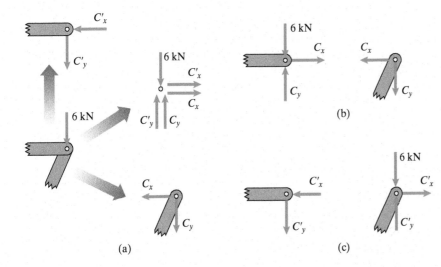

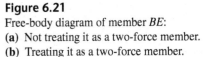

Figure 6.22
(a) Drawing free-body diagrams of the pin and the two members.
(b) Superimposing the pin on member ABC.
(c) Superimposing the pin on member CD.

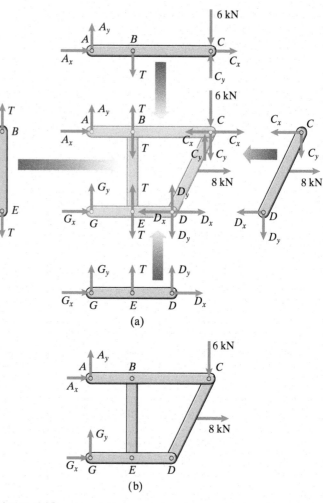

Figure 6.23
(a) "Reassembling" the free-body diagrams of the individual members.
(b) The free-body diagram of the entire frame is recovered.

To detect errors in the free-body diagrams of the members, it is helpful to "reassemble" them (Fig. 6.23a). The forces at the connections between the members cancel (they are internal forces once the members are reassembled), and the free-body diagram of the entire structure is recovered (Fig. 6.23b).

Our final step is to apply the equilibrium equations to the free-body diagrams of the members (Fig. 6.24). In two dimensions, we can obtain three independent equilibrium equations from the free-body diagram of each member of a structure that we do not treat as a two-force member. (By assuming that the forces on a two-force member are equal and opposite axial forces, we have already used the three equilibrium equations for that member.) In this example, there are three members in addition to the two-force member, so we can write $3 \times 3 = 9$ independent equilibrium equations, and there are nine unknown forces: A_x, A_y, C_x, C_y, D_x, D_y, G_x, G_y, and T.

Recall that we determined that $A_x = -13$ kN and $G_x = 5$ kN from our analysis of the entire structure. The equilibrium equations we obtained from the free-body diagram of the entire structure are not independent of the equilibrium

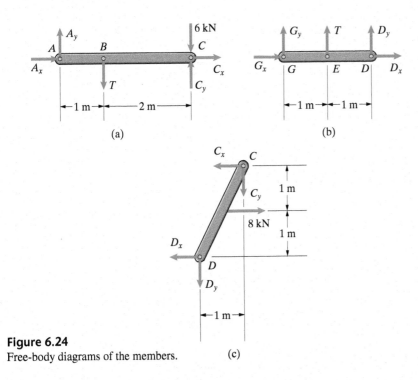

Figure 6.24
Free-body diagrams of the members.

(a)   (b)   (c)

equations obtained from the free-body diagrams of the members, but by using
them to determine A_x and G_x, we get a head start on solving the equations for
the members. Consider the free-body diagram of member ABC (Fig. 6.24a).
Because we know A_x, we can determine C_x from the equation

$$\Sigma F_x = A_x + C_x = 0,$$

obtaining $C_x = -A_x = 13$ kN. Now consider the free-body diagram of GED
(Fig. 6.24b). We can determine D_x from the equation

$$\Sigma F_x = G_x + D_x = 0,$$

obtaining $D_x = -G_x = -5$ kN. Now consider the free-body diagram of member CD (Fig. 6.24c). Because we know C_x, we can determine C_y by summing
moments about D:

$$\Sigma M_{\text{point } D} = (2 \text{ m})C_x - (1 \text{ m})C_y - (1 \text{ m})(8 \text{ kN}) = 0.$$

We obtain $C_y = 18$ kN. Then, from the equation

$$\Sigma F_y = -C_y - D_y = 0,$$

we find that $D_y = -C_y = -18$ kN. Now we can return to the free-body diagrams of members ABC and GED to determine A_y and G_y. Summing moments
about point B of member ABC yields

$$\Sigma M_{\text{point } B} = -(1 \text{ m})A_y + (2 \text{ m})C_y - (2 \text{ m})(6 \text{ kN}) = 0,$$

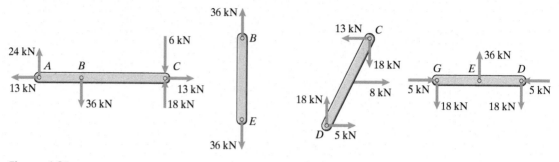

Figure 6.25
Forces on the members of the frame.

and we obtain $A_y = 2C_y - 12$ kN = 24 kN. Then, summing moments about point E of member GED, we have

$$\Sigma M_{\text{point } E} = (1 \text{ m})D_y - (1 \text{ m})G_y = 0,$$

from which we obtain $G_y = D_y = -18$ kN. Finally, from the free-body diagram of member GED, we use the equilibrium equation

$$\Sigma F_y = D_y + G_y + T = 0,$$

which gives us the result $T = -D_y - G_y = 36$ kN. The forces on the members are shown in Fig. 6.25. As this example demonstrates, determination of the forces on the members can often be simplified by carefully choosing the order in which the equations are solved.

We see that determining the forces and couples on the members of frames and machines involves two steps:

1. Determine the reactions at the supports—Draw the free-body diagram of the entire structure, and determine the reactions at its supports. Although this step is not essential, it can greatly simplify your analysis of the members. If the free-body diagram is statically indeterminant, determine as many of the reactions as possible.

2. Analyze the members—Draw free-body diagrams of the members, and apply the equilibrium equations to determine the forces acting on them. You can simplify this step by identifying two-force members. If a load acts at a joint of the structure, you can place the load on the free-body diagram of any one of the members attached at that joint.

RESULTS

A structure of interconnected members that cannot be modeled as a truss is called a *frame* if it is designed to remain stationary and support loads and a *machine* if it is designed to move and apply loads. The forces and couples acting on the individual members of a frame or machine in equilibrium can often be determined by applying the equilibrium equations to the individual members.

It is often advantageous to begin by drawing the free-body diagram of
the entire structure considered as a single object and applying the
equilibrium equations. *Even if the free-body diagram of the entire
structure is statically indeterminate, it may be possible to determine the
reactions from the subsequent analysis of the individual members.*

Draw the free-body diagrams of the individual members and apply the equilibrium
equations to them. Notice that where two members are connected, *the reactions they
exert on each other are equal and opposite*. Notice that member *BD* is a two-force
member. Recognizing two-force members will simplify the analysis of a structure.

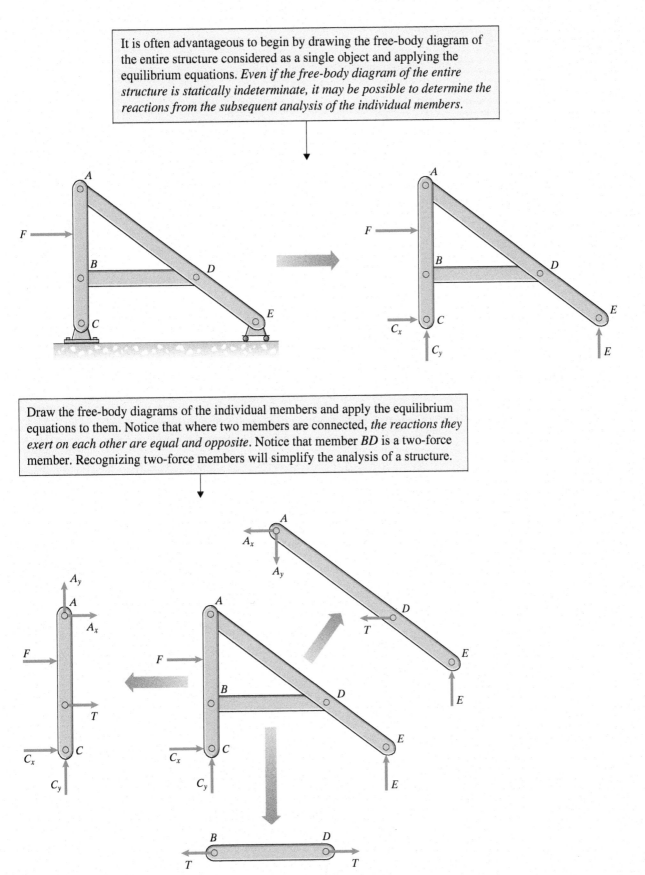

Active Example 6.6 | **Analyzing a Frame** (▶ *Related Problem 6.70*)

Determine the forces and couples acting on the members of the frame.

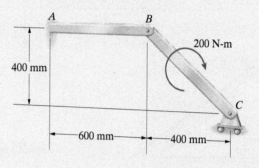

Strategy

We will first draw a free-body diagram of the entire frame, treating it as a single object, and attempt to determine the reactions at the supports. We will then draw free-body diagrams of the individual members and apply the equilibrium equations to determine the forces and couples acting on them.

Solution

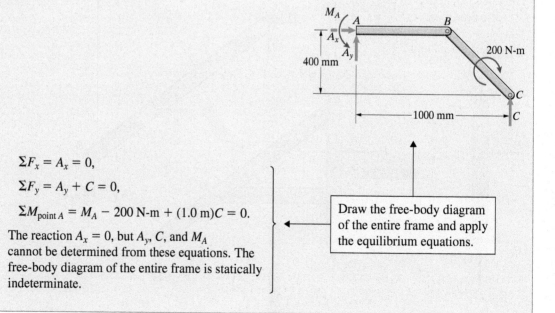

$$\Sigma F_x = A_x = 0,$$

$$\Sigma F_y = A_y + C = 0,$$

$$\Sigma M_{\text{point } A} = M_A - 200 \text{ N-m} + (1.0 \text{ m})C = 0.$$

The reaction $A_x = 0$, but A_y, C, and M_A cannot be determined from these equations. The free-body diagram of the entire frame is statically indeterminate.

Draw the free-body diagram of the entire frame and apply the equilibrium equations.

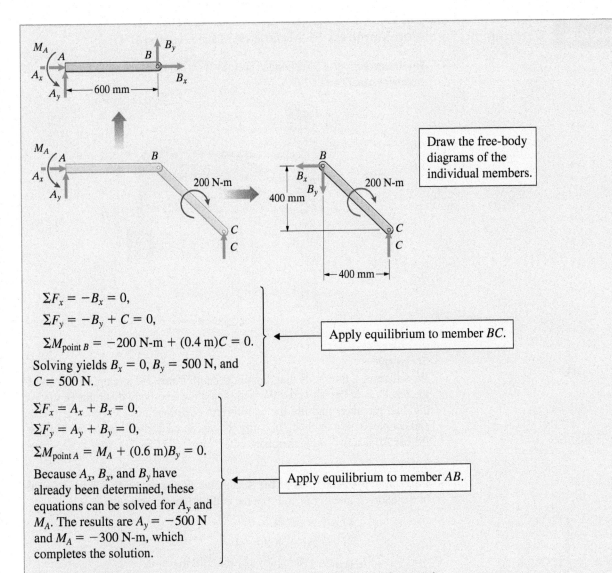

Draw the free-body diagrams of the individual members.

$$\Sigma F_x = -B_x = 0,$$
$$\Sigma F_y = -B_y + C = 0,$$
$$\Sigma M_{\text{point } B} = -200 \text{ N-m} + (0.4 \text{ m})C = 0.$$

Solving yields $B_x = 0$, $B_y = 500$ N, and $C = 500$ N.

Apply equilibrium to member BC.

$$\Sigma F_x = A_x + B_x = 0,$$
$$\Sigma F_y = A_y + B_y = 0,$$
$$\Sigma M_{\text{point } A} = M_A + (0.6 \text{ m})B_y = 0.$$

Because A_x, B_x, and B_y have already been determined, these equations can be solved for A_y and M_A. The results are $A_y = -500$ N and $M_A = -300$ N-m, which completes the solution.

Apply equilibrium to member AB.

Practice Problem The frame has pin supports at A and C. Determine the forces and couples acting on member BC at B and C.

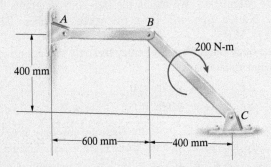

Answer: $B_x = -500$ N, $B_y = 0$, $C_x = 500$ N, $C_y = 0$. (In the statements of the answers, x components are positive to the right and y components are positive upward.)

| Example 6.7 | Determining Forces on Members of a Frame (▶ *Related Problem 6.74*) |

The frame supports a suspended weight $W = 40$ lb. Determine the forces on members *ABCD* and *CEG*.

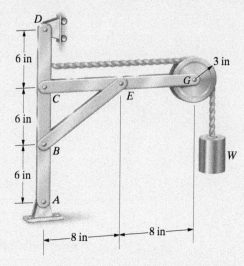

Strategy

We will draw a free-body diagram of the entire frame and attempt to determine the reactions at the supports. We will then draw free-body diagrams of the individual members and use the equilibrium equations to determine the forces and couples acting on them. In doing so, we can take advantage of the fact that the bar *BE* is a two-force member.

Solution

Determine the Reactions at the Supports We draw the free-body diagram of the entire frame in Fig. a. From the equilibrium equations

$$\Sigma F_x = A_x - D = 0,$$
$$\Sigma F_y = A_y - 40 \text{ lb} = 0,$$
$$\Sigma M_{\text{point } A} = (18 \text{ in})D - (19 \text{ in})(40 \text{ lb}) = 0,$$

we obtain the reactions $A_x = 42.2$ lb, $A_y = 40$ lb, and $D = 42.2$ lb.

Analyze the Members We obtain the free-body diagrams of the members in Fig. b. Notice that *BE* is a two-force member. The angle $\alpha = \arctan(6/8) = 36.9°$.

The free-body diagram of the pulley has only two unknown forces. From the equilibrium equations

$$\Sigma F_x = G_x - 40 \text{ lb} = 0,$$
$$\Sigma F_y = G_y - 40 \text{ lb} = 0,$$

we obtain $G_x = 40$ lb and $G_y = 40$ lb. There are now only three unknown forces on the free-body diagram of member *CEG*. From the equilibrium equations

$$\Sigma F_x = -C_x - R \cos \alpha - 40 \text{ lb} = 0,$$
$$\Sigma F_y = -C_y - R \sin \alpha - 40 \text{ lb} = 0,$$
$$\Sigma M_{\text{point } C} = -(8 \text{ in})R \sin \alpha - (16 \text{ in})(40 \text{ lb}) = 0,$$

we obtain $C_x = 66.7$ lb, $C_y = 40$ lb, and $R = -133.3$ lb, completing the solution (Fig. c).

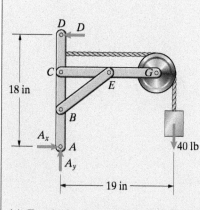

(a) Free-body diagram of the entire frame.

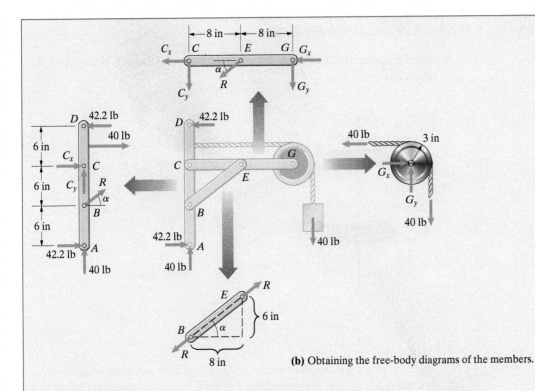

(b) Obtaining the free-body diagrams of the members.

Critical Thinking

In problems of this kind, the reactions on the individual members of the frame can be determined from the free-body diagrams of the members. Why did we draw the free-body diagram of the entire frame and solve the associated equilibrium equations? The reason is that it gave us a head start on solving the equilibrium equations for the members. In this example, when we drew the free-body diagrams of the members we already knew the reactions at A and D, which simplified the remaining analysis. Analyzing the entire frame can also provide a check on your work. Notice that we did not use the equilibrium equations for member $ABCD$. We can check our analysis by confirming that this member is in equilibrium (Fig. c):

$$\Sigma F_x = 42.2 \text{ lb} - 133.3 \cos 36.9° \text{ lb} + 66.7 \text{ lb} + 40 \text{ lb} - 42.2 \text{ lb} = 0,$$

$$\Sigma F_y = 40 \text{ lb} - 133.3 \sin 36.9° \text{ lb} + 40 \text{ lb} = 0,$$

$$\Sigma M_{\text{point } A} = (6 \text{ in})(133.3 \cos 36.9° \text{ lb}) - (12 \text{ in})(66.7 \text{ lb})$$
$$- (15 \text{ in})(40 \text{ lb}) + (18 \text{ in})(42.2 \text{ lb}) = 0.$$

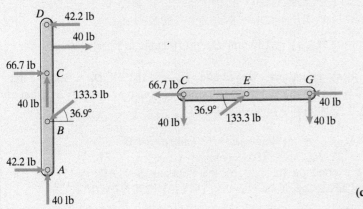

(c) Forces on members $ABCD$ and CEG.

Example 6.8 Analyzing a Machine (▶ *Related Problem 6.103*)

What forces are exerted on the ball at E as a result of the 150-N forces on the pliers?

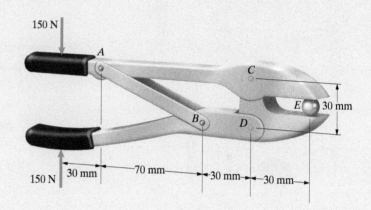

Strategy

A pair of pliers is a simple example of a machine, a structure designed to move and exert forces. The interconnections of the members are designed to create a mechanical advantage, subjecting an object to forces greater than the forces exerted by the user.

In this case there is no information to be gained from the free-body diagram of the entire structure. We must determine the forces exerted on the ball by drawing free-body diagrams of the members.

Solution

We "disassemble" the pliers in Fig. a to obtain the free-body diagrams of the members, labeled (1), (2), and (3). The force R on free-body diagrams (1) and (3) is exerted by the two-force member AB. The angle $\alpha = \arctan(30/70) = 23.2°$. Our objective is to determine the force E exerted by the ball.

The free-body diagram of member (3) has only three unknown forces and the 150-N load, so we can determine R, D_x, and D_y from this free-body diagram alone. The equilibrium equations are

$$\Sigma F_x = D_x + R \cos \alpha = 0,$$

$$\Sigma F_y = D_y - R \sin \alpha + 150 \text{ N} = 0,$$

$$\Sigma M_{\text{point } B} = (30 \text{ mm})D_y - (100 \text{ mm})(150 \text{ N}) = 0.$$

Solving these equations, we obtain $D_x = -1517 \text{ N}$, $D_y = 500 \text{ N}$, and $R = 1650 \text{ N}$. Knowing D_x, we can determine E from the free-body diagram of member (2) by summing moments about C:

$$\Sigma M_{\text{point } C} = -(30 \text{ mm})E - (30 \text{ mm})D_x = 0.$$

The force exerted on the ball by the pliers is $E = -D_x = 1517 \text{ N}$. The mechanical advantage of the pliers is $(1517 \text{ N})/(150 \text{ N}) = 10.1$.

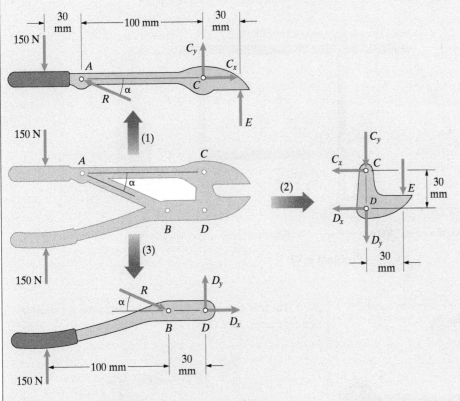

(a) Obtaining the free-body diagrams of the members.

Critical Thinking

What is the motivation for determining the reactions on the members of the pliers? This process is essential for machine and tool design. To design the configuration of the pliers and choose the materials and dimensions of its members, it is necessary to determine all the forces acting on the members, as we have done in this example. Once the forces are known, the methods of mechanics of materials can be used to assess the adequacy of the members to support them.

Problems

Assume that objects are in equilibrium. In the statements of the answers, x components are positive to the right and y components are positive upward.

▶ **6.70** In Active Example 6.6, suppose that in addition to being loaded by the 200 N-m couple, the frame is subjected to a 400-N force at C that is horizontal and points toward the left. Draw a sketch of the frame showing the new loading. Determine the forces and couples acting on member AB of the frame.

6.71 The object suspended at E weighs 200 lb. Determine the reactions on member ACD at A and C.

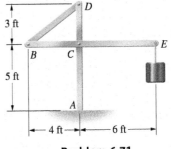

Problem 6.71

6.72 The mass of the object suspended at G is 100 kg. Determine the reactions on member CDE at C and E.

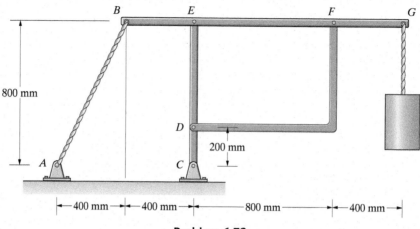

Problem 6.72

6.73 The force $F = 10$ kN. Determine the forces on member ABC, presenting your answers as shown in Fig. 6.25.

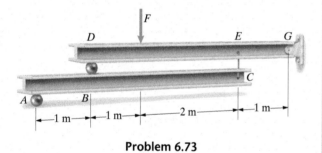

Problem 6.73

▶ **6.74** In Example 6.7, suppose that the frame is redesigned so that the distance from point C to the attachment point E of the two-force member BE is increased from 8 in to 10 in. Determine the forces acting at C on member $ABCD$.

6.75 The tension in cable BD is 500 lb. Determine the reactions at A for cases (1) and (2).

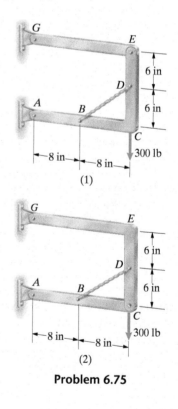

Problem 6.75

6.76 Determine the reactions on member *ABCD* at *A*, *C*, and *D*.

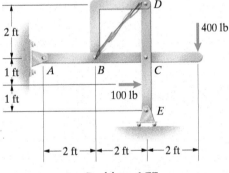

Problem 6.76

6.77 Determine the forces exerted on member *ABC* at *A* and *C*.

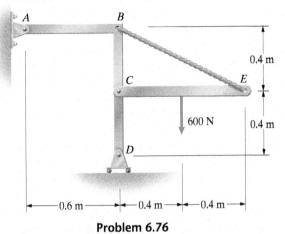

Problem 6.77

6.78 An athlete works out with a squat thrust machine. To rotate the bar *ABD*, she must exert a vertical force at *A* that causes the magnitude of the axial force in the two-force member *BC* to be 1800 N. When the bar *ABD* is on the verge of rotating, what are the reactions on the vertical bar *CDE* at *D* and *E*?

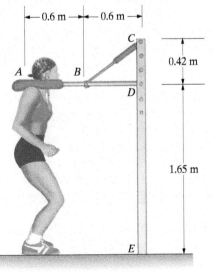

Problem 6.78

6.79 The frame supports a 6-kN vertical load at *C*. The bars *ABC* and *DEF* are horizontal. Determine the reactions on the frame at *A* and *D*.

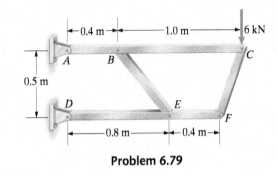

Problem 6.79

6.80 The mass *m* = 120 kg. Determine the forces on member *ABC*, presenting your answers as shown in Fig. 6.25.

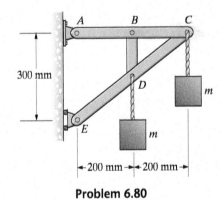

Problem 6.80

6.81 Determine the reactions on member *BCD*.

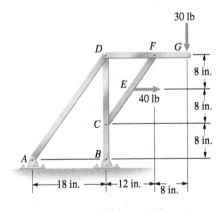

Problem 6.81

6.82 The weight of the suspended object is $W = 50$ lb. Determine the tension in the spring and the reactions at F. (The slotted member DE is vertical.)

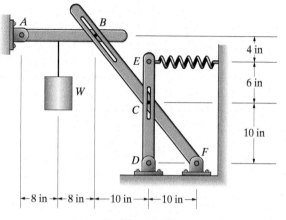

Problem 6.82

6.83 The mass $m = 50$ kg. Bar DE is horizontal. Determine the forces on member $ABCD$, presenting your answers as shown in Fig. 6.25.

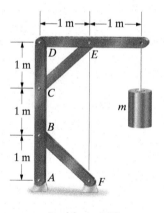

Problem 6.83

6.84 Determine the forces on member BCD.

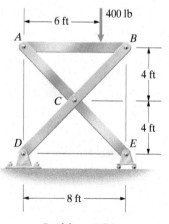

Problem 6.84

6.85 Determine the forces on member ABC.

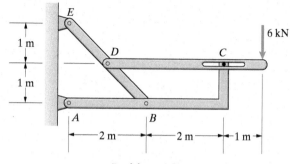

Problem 6.85

6.86 Determine the forces on member ABD.

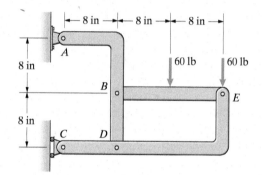

Problem 6.86

6.87 The mass $m = 12$ kg. Determine the forces on member CDE.

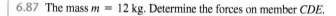
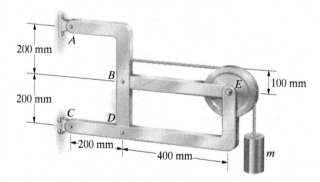

Problem 6.87

6.88 The weight $W = 80$ lb. Determine the forces on member $ABCD$.

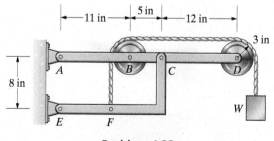

Problem 6.88

6.89 The woman using the exercise machine is holding the 80-lb weight stationary in the position shown. What are the reactions at the fixed support E and the pin support F? (A and C are pinned connections.)

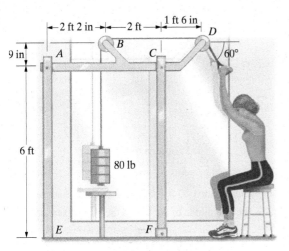

Problem 6.89

6.90 Determine the reactions on member ABC at A and B.

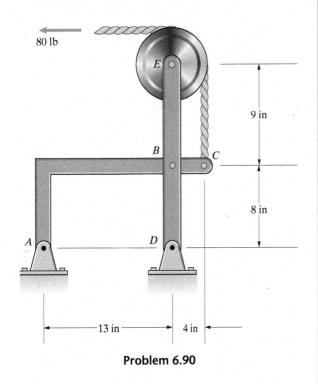

Problem 6.90

6.91 The mass of the suspended object is $m = 50$ kg. Determine the reactions on member ABC.

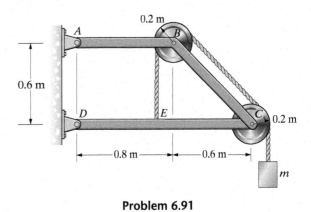

Problem 6.91

6.92 The unstretched length of the spring is L_0. Show that when the system is in equilibrium the angle α satisfies the relation $\sin \alpha = 2(L_0 - 2F/k)/L$.

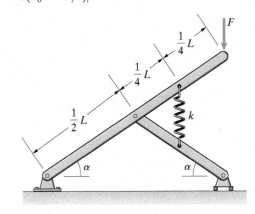

Problem 6.92

6.93 The pin support B will safely support a force of 24-kN magnitude. Based on this criterion, what is the largest mass m that the frame will safely support?

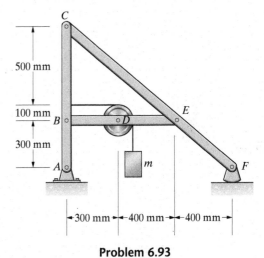

Problem 6.93

6.94 Determine the reactions at *A* and *C*.

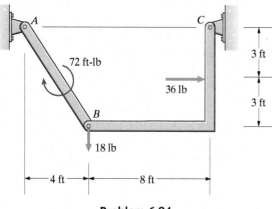

Problem 6.94

6.95 Determine the forces on member *AD*.

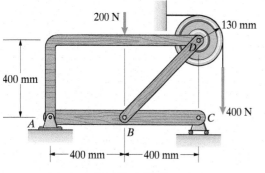

Problem 6.95

6.96 The frame shown is used to support high-tension wires. If $b = 3$ ft, $\alpha = 30°$, and $W = 200$ lb, what is the axial force in member *HJ*?

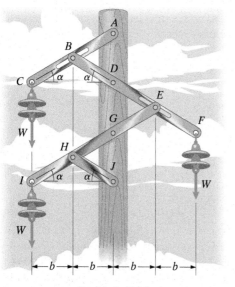

Problem 6.96

6.97 Determine the force exerted on the ball by the bolt cutters and the magnitude of the axial force in the two-force member *AB*.

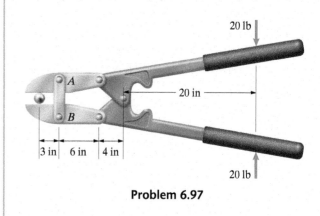

Problem 6.97

6.98 The woman exerts 20-N forces to the pliers as shown.

(a) What is the magnitude of the forces the pliers exert on the bolt at *B*?

(b) Determine the magnitude of the force the members of the pliers exert on each other at the pinned connection *C*.

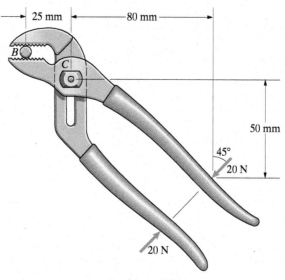

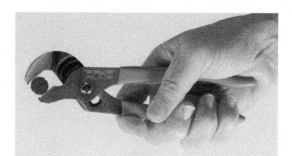

Problem 6.98

6.99 Figure a is a diagram of the bones and biceps muscle of a person's arm supporting a mass. Tension in the biceps muscle holds the forearm in the horizontal position, as illustrated in the simple mechanical model in Fig. b. The weight of the forearm is 9 N, and the mass $m = 2$ kg.

(a) Determine the tension in the biceps muscle AB.

(b) Determine the magnitude of the force exerted on the upper arm by the forearm at the elbow joint C.

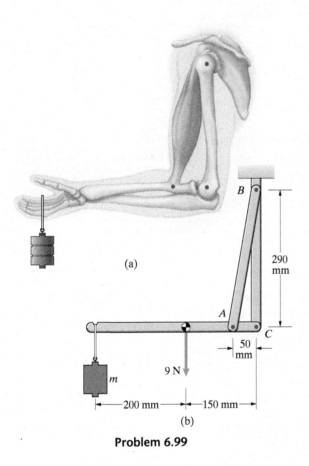

(a)

(b)

Problem 6.99

6.100 The bones and tendons in a horse's rear leg are shown in Fig. a. A biomechanical model of the leg is shown in Fig. b. If the horse is stationary and the normal force exerted on its leg by the ground is $N = 1200$ N, determine the tensions in the superficial digital flexor BC and the patellar ligament DF.

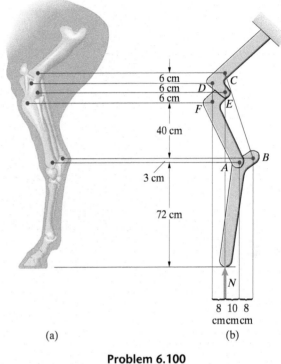

(a)

(b)

Problem 6.100

6.101 The pressure force exerted on the piston is 2 kN toward the left. Determine the couple M necessary to keep the system in equilibrium.

6.102 In Problem 6.101, determine the forces on member AB at A and B.

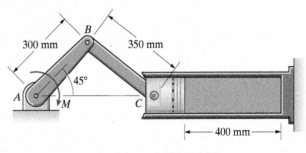

Problems 6.101/6.102

▶ 6.103 In Example 6.8, suppose that the object being held by the pliers is moved to the left so that the horizontal distance from D to the object at E decreases from 30 mm to 20 mm. Draw a sketch of the pliers showing the new position of the object. What forces are exerted on the object at E as a result of the 150-N forces on the pliers?

6.104 The shovel of the excavator is supported by a pin support at E and the two-force member BC. The 300-lb weight W of the shovel acts at the point shown. Determine the reactions on the shovel at E and the magnitude of the axial force in the two-force member BC.

6.105 The shovel of the excavator has a pin support at E. The position of the shovel is controlled by the horizontal hydraulic piston AB, which is attached to the shovel through a linkage of the two-force members BC and BD. The 300-lb weight W of the shovel acts at the point shown. What is the magnitude of the force the hydraulic piston must exert to hold the shovel in equilibrium?

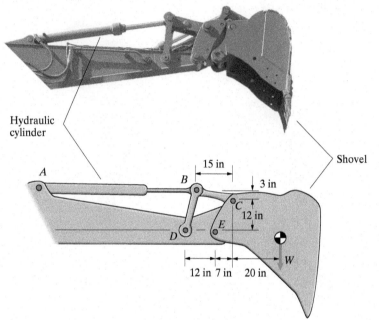

Problems 6.104/6.105

6.106 The woman exerts 20-N forces on the handles of the shears. Determine the magnitude of the forces exerted on the branch at A.

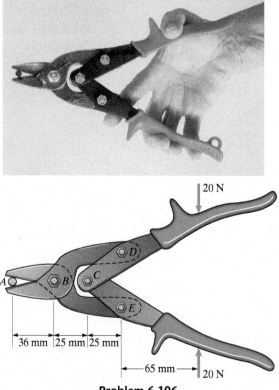

Problem 6.106

6.107 The person exerts 40-N forces on the handles of the locking wrench. Determine the magnitude of the forces the wrench exerts on the bolt at *A*.

6.108 Determine the magnitude of the force the members of the wrench exert on each other at *B* and the axial force in the two-force member *DE*.

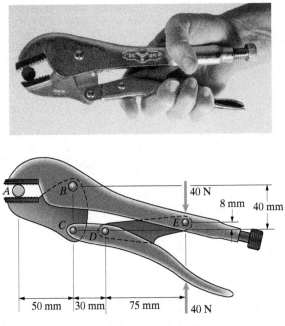

Problems 6.107/6.108

6.109 This device is designed to exert a large force on the horizontal bar at *A* for a stamping operation. If the hydraulic cylinder *DE* exerts an axial force of 800 N and $\alpha = 80°$, what horizontal force is exerted on the horizontal bar at *A*?

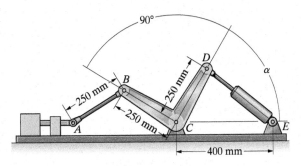

Problem 6.109

6.110 This device raises a load *W* by extending the hydraulic actuator *DE*. The bars *AD* and *BC* are 4 ft long, and the distances $b = 2.5$ ft and $h = 1.5$ ft. If $W = 300$ lb, what force must the actuator exert to hold the load in equilibrium?

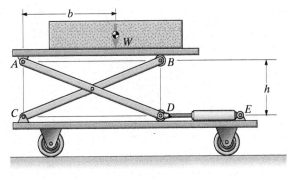

Problem 6.110

6.111 The four-bar linkage operates the forks of a fork lift truck. The force supported by the forks is $W = 8$ kN. Determine the reactions on member *CDE*.

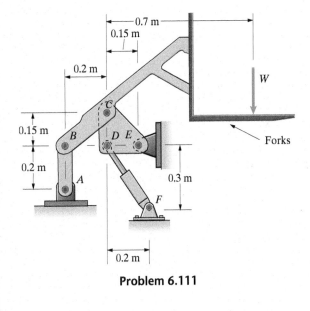

Problem 6.111

6.112 If the horizontal force on the scoop is $F = 2000$ lb, what is the magnitude of the axial force in the hydraulic actuator AC?

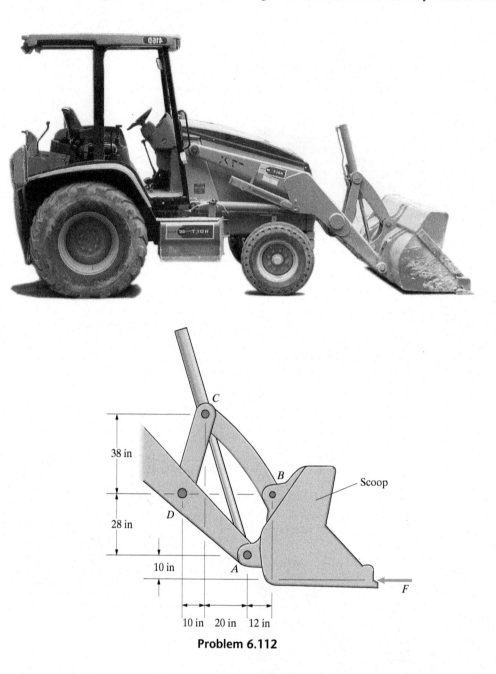

Problem 6.112

6.113 A 10-kip horizontal force acts on the bucket of the excavator. Determine the reactions on member *ACF* at *A* and *F*.

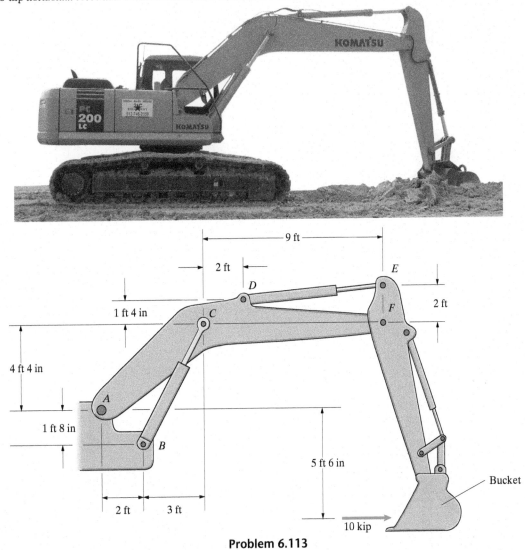

Problem 6.113

6.114 The structure shown in the diagram (one of the two identical structures that support the scoop of the excavator) supports a downward force *F* = 1800 N at *G*. Members *BC* and *DH* can be treated as two-force members. Determine the reactions on member *CDK* at *K*.

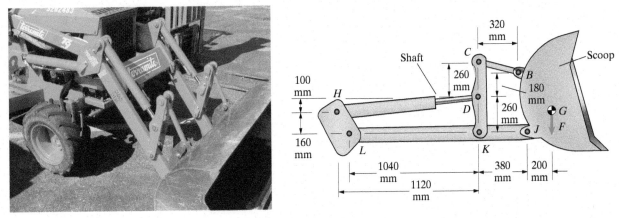

Problem 6.114

Review Problems

6.115 The loads $F_1 = 440$ N and $F_2 = 160$ N. Determine the axial forces in the members. Indicate whether they are in tension (T) or compression (C).

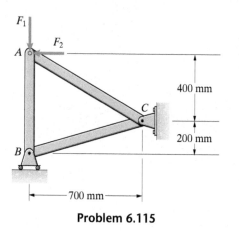

Problem 6.115

6.116 The truss supports a load $F = 10$ kN. Determine the axial forces in members AB, AC, and BC.

6.117 Each member of the truss will safely support a tensile force of 40 kN and a compressive force of 32 kN. Based on this criterion, what is the largest downward load F that can safely be applied at C?

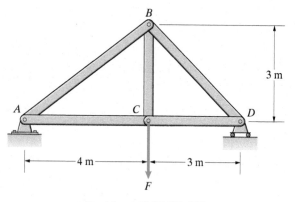

Problems 6.116/6.117

6.118 The Pratt bridge truss supports loads at F, G, and H. Determine the axial forces in members BC, BG, and FG.

6.119 Determine the axial forces in members CD, GD, and GH.

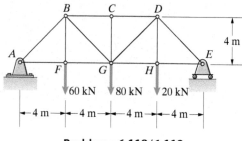

Problems 6.118/6.119

6.120 The truss supports loads at F and H. Determine the axial forces in members AB, AC, BC, BD, CD, and CE.

6.121 Determine the axial forces in members EH and FH.

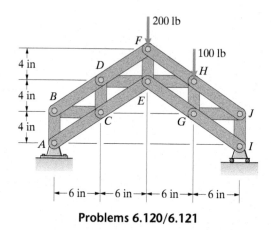

Problems 6.120/6.121

6.122 Determine the axial forces in members *BD*, *CD*, and *CE*.

6.123 Determine the axial forces in members *DF*, *EF*, and *EG*.

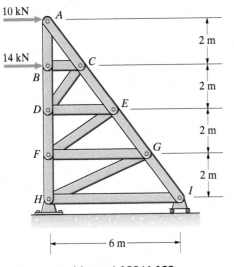

Problems **6.122/6.123**

6.124 The truss supports a 400-N load at *G*. Determine the axial forces in members *AC*, *CD*, and *CF*.

6.125 Determine the axial forces in members *CE*, *EF*, and *EH*.

6.126 Which members have the largest tensile and compressive forces, and what are their values?

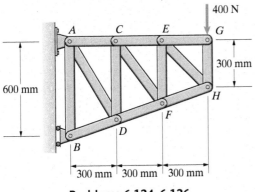

Problems **6.124–6.126**

6.127 The Howe truss helps support a roof. Model the supports at *A* and *G* as roller supports. Use the method of joints to determine the axial forces in members *BC*, *CD*, *CI*, and *CJ*.

6.128 Use the method of sections to determine the axial forces in members *CD*, *CJ*, and *IJ*.

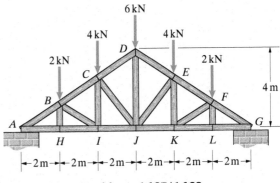

Problems **6.127/6.128**

6.129 A speaker system is suspended from the truss by cables attached at *D* and *E*. The mass of the speaker system is 130 kg, and its weight acts at *G*. Determine the axial forces in members *BC* and *CD*.

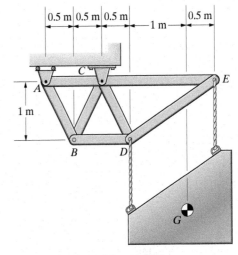

Problem **6.129**

6.130 The mass of the suspended object is 900 kg. Determine the axial forces in the bars AB and AC.

Strategy: Draw the free-body diagram of joint A.

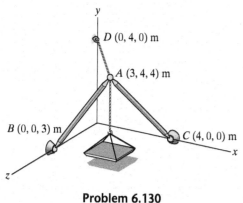

Problem 6.130

6.131 Determine the forces on member ABC, presenting your answers as shown in Fig. 6.25. Obtain the answers in two ways:

(a) When you draw the free-body diagrams of the individual members, place the 400-lb load on the free-body diagram of member ABC.

(b) When you draw the free-body diagrams of the individual members, place the 400-lb load on the free-body diagram of member CD.

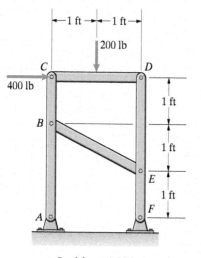

Problem 6.131

6.132 The mass $m = 120$ kg. Determine the forces on member ABC.

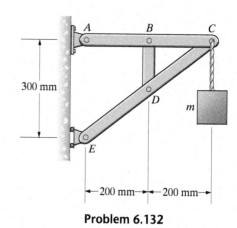

Problem 6.132

6.133 Determine the reactions on member ABC at B and C.

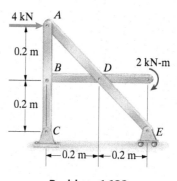

Problem 6.133

6.134 The truck and trailer are parked on a 10° slope. The 14,000-lb weight of the truck and the 8000-lb weight of the trailer act at the points shown. The truck's brakes prevent its rear wheels at B from turning. The truck's front wheels at C and the trailer's wheels at A can turn freely, which means they do not exert friction forces on the road. The trailer hitch at D behaves like a pin support. Determine the forces exerted on the truck at B, C, and D.

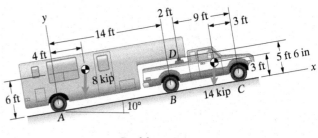

Problem 6.134

6.135 The 600-lb weight of the scoop acts at a point 1 ft 6 in to the right of the vertical line *CE*. The line *ADE* is horizontal. The hydraulic actuator *AB* can be treated as a two-force member. Determine the axial force in the hydraulic actuator *AB* and the forces exerted on the scoop at *C* and *E*.

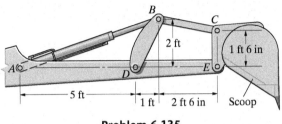

Problem 6.135

6.136 Determine the force exerted on the bolt by the bolt cutters.

6.137 Determine the magnitude of the force the members of the bolt cutters exert on each other at the pin connection *B* and the axial force in the two-force member *CD*.

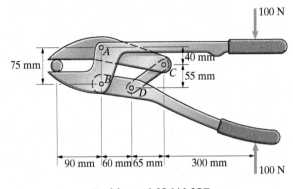

Problems 6.136/6.137

Design Project 1 Design a truss structure to support a foot bridge with an unsupported span (width) of 8 m. Make conservative estimates of the loads the structure will need to support if the pathway supported by the truss is made of wood. Consider two options: (1) Your client wants the bridge to be supported by a truss below the bridge so that the upper surface will be unencumbered by structure. (2) The client wants the truss to be above the bridge and designed so that it can serve as handrails. For each option, use statics to estimate the maximum axial forces to which the members of the structure will be subjected. Investigate alternative designs and compare the resulting axial loads.

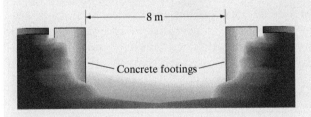

Design Project 2 The truss shown connects one end of a stretcher to a rescue helicopter. Consider alternative truss designs that support the stretcher at *A* and *B* and are supported at *E* and *G*. Compare the maximum tensile and compressive loads in the members of your designs to those in the truss shown. Assuming that the cost of a truss is proportional to the sum of the lengths of its members, compare the costs of your designs to that of the truss shown. Write a brief report describing your analysis and recommending the design you would choose.

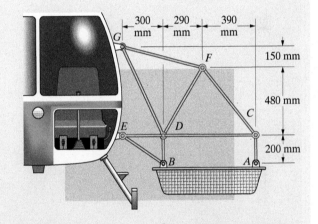

Design Project 3 Go to a fitness center and choose an exercise device that seems mechanically interesting. (For example, it may employ weights, pulleys, and levers.) By measuring dimensions (while the device is not in use), drawing sketches, and perhaps taking photographs, gather the information necessary to analyze the device. Use statics to determine the range of forces a person must exert in using the device.

Suggest changes to the design of the device (other than simply increasing weights) that will increase the maximum force the user must exert.

Prepare a brief report that (1) describes the original device; (2) presents your model and analysis of the device; (3) describes your proposed changes and any analyses supporting them; and (4) recommends the design change you would choose to increase the maximum force the user must employ.

Centroids and Centers of Mass

An object's weight does not act at a single point—it is distributed over the entire volume of the object. But the weight can be represented by a single equivalent force acting at a point called the center of mass. When the equilibrium equations are used to determine the reactions exerted on an object by its supports, the location of the center of mass must be known if the weight of the object is to be included in the analysis. The dynamic behaviors of objects also depend on the locations of their centers of mass. In this chapter we define the center of mass and show how it is determined for various kinds of objects. We also introduce definitions that can be interpreted as the average positions of areas, volumes, and lines. These average positions are called centroids. Centroids coincide with the centers of mass of particular classes of objects, and they also arise in many other engineering applications.

◄ To be balanced, the woman's center of mass—the point at which her weight effectively acts—must be directly above her hands. In this chapter we introduce the concept of an average position, or centroid, and show how to locate the centers of mass of objects.

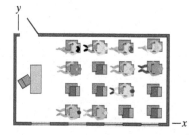

(a)

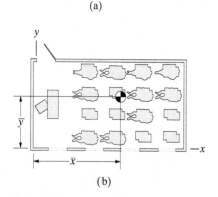

(b)

Figure 7.1
(a) A group of students in a classroom.
(b) Their average position.

7.1 Centroids of Areas

BACKGROUND

Suppose that we want to determine the average position of a group of students sitting in a room. First, we introduce a coordinate system so that we can specify the position of each student. For example, we can align the axes with the walls of the room (Fig. 7.1a). We number the students from 1 to N and denote the position of student 1 by (x_1, y_1), the position of student 2 by (x_2, y_2), and so on. The average x coordinate, which we denote by $\bar{x}$, is the sum of their x coordinates divided by N; that is,

$$\bar{x} = \frac{x_1 + x_2 + \cdots + x_N}{N} = \frac{\sum_i x_i}{N},$$ (7.1)

where the symbol $\sum_i$ means "sum over the range of i." The average y coordinate is

$$\bar{y} = \frac{\sum_i y_i}{N}.$$ (7.2)

We indicate the average position by the symbol shown in Fig. 7.1b.

Now suppose that we pass out some pennies to the students. Let the number of coins given to student 1 be c_1, the number given to student 2 be c_2, and so on. What is the average position of the coins in the room? Clearly, the average position of the coins may not be the same as the average position of the students. For example, if the students in the front of the room have more coins, the average position of the coins will be closer to the front of the room than the average position of the students.

To determine the x coordinate of the average position of the coins, we need to sum the x coordinates of the coins and divide by the number of coins. We can obtain the sum of the x coordinates of the coins by multiplying the number of coins each student has by his or her x coordinate and summing. We can obtain the number of coins by summing the numbers $c_1, c_2, \ldots$. Thus, the average x coordinate of the coins is

$$\bar{x} = \frac{\sum_i x_i c_i}{\sum_i c_i}.$$ (7.3)

We can determine the average y coordinate of the coins in the same way:

$$\bar{y} = \frac{\sum_i y_i c_i}{\sum_i c_i}.$$ (7.4)

By assigning other meanings to $c_1, c_2, \ldots$, we can determine the average positions of other measures associated with the students. For example, we could determine the average position of their age or the average position of their height.

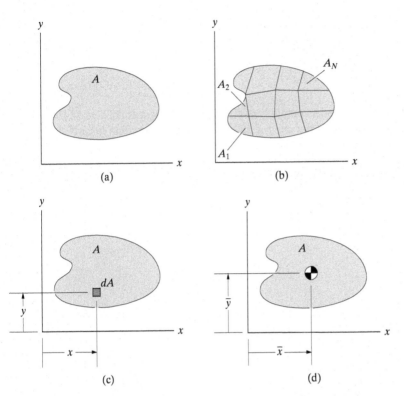

Figure 7.2
(a) The area A.
(b) Dividing A into N parts.
(c) A differential element of area dA with coordinates (x, y).
(d) The centroid of the area.

More generally, we can use Eqs. (7.3) and (7.4) to determine the average position of any set of quantities with which we can associate positions. An average position obtained from these equations is called a *weighted average position*, or *centroid*. The "weight" associated with position (x_1, y_1) is c_1, the weight associated with position (x_2, y_2) is c_2, and so on. In Eqs. (7.1) and (7.2), the weight associated with the position of each student is 1. When the census is taken, the centroid of the population of the United States—the average position of the population—is determined in this way.

Let us consider an arbitrary area A in the x–y plane (Fig. 7.2a). Divide the area into parts $A_1, A_2, \ldots, A_N$ (Fig. 7.2b) and denote the positions of the parts by $(x_1, y_1), (x_2, y_2), \ldots, (x_N, y_N)$. We can obtain the centroid, or average position of the area, by using Eqs. (7.3) and (7.4) with the areas of the parts as the weights:

$$\bar{x} = \frac{\sum_i x_i A_i}{\sum_i A_i}, \qquad \bar{y} = \frac{\sum_i y_i A_i}{\sum_i A_i}. \tag{7.5}$$

A question arises if we try to carry out this procedure: What are the exact positions of the areas $A_1, A_2, \ldots, A_N$? We could reduce the uncertainty in their positions by dividing A into smaller parts, but we would still obtain only approximate values for $\bar{x}$ and $\bar{y}$. To determine the exact location of the centroid, we must take the limit as the sizes of the parts approach zero. We obtain this limit by replacing Eqs. (7.5) by the integrals

$$\bar{x} = \frac{\displaystyle\int_A x \, dA}{\displaystyle\int_A dA}, \tag{7.6}$$

$$\bar{y} = \frac{\displaystyle\int_A y \, dA}{\displaystyle\int_A dA}, \qquad (7.7)$$

where x and y are the coordinates of the differential element of area dA (Fig. 7.2c). The subscript A on the integral signs means the integration is carried out over the entire area. The centroid of the area is shown in Fig. 7.2d.

RESULTS

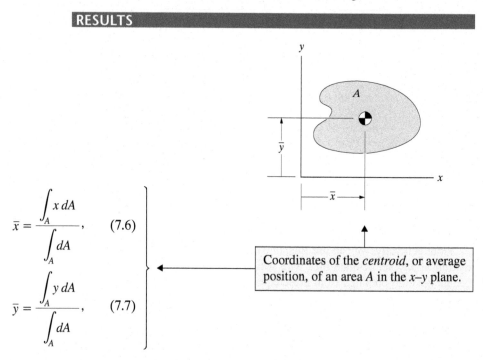

$$\bar{x} = \frac{\displaystyle\int_A x \, dA}{\displaystyle\int_A dA}, \qquad (7.6)$$

$$\bar{y} = \frac{\displaystyle\int_A y \, dA}{\displaystyle\int_A dA}, \qquad (7.7)$$

Coordinates of the *centroid*, or average position, of an area A in the x–y plane.

Keeping in mind that the centroid of an area is its average position will often help in locating it. If an area has "mirror image" symmetry about an axis, its centroid lies on the axis. If an area is symmetric about two axes, the centroid lies at the intersection of the axes.

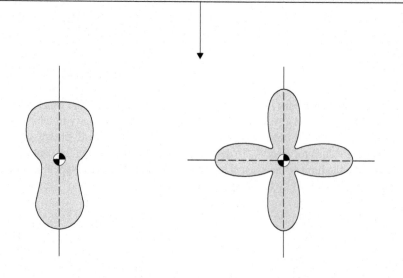

Active Example 7.1   Centroid of an Area by Integration (▶ *Related Problem 7.1*)

Determine the x coordinate of the centroid of the triangular area.

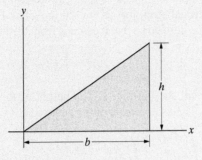

Strategy
We will evaluate Eq. (7.6) using an element of area dA in the form of a vertical "strip" of width dx.

Solution

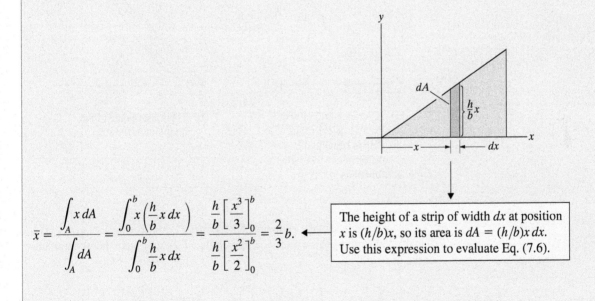

$$\bar{x} = \frac{\displaystyle\int_A x \, dA}{\displaystyle\int_A dA} = \frac{\displaystyle\int_0^b x\left(\frac{h}{b}x\,dx\right)}{\displaystyle\int_0^b \frac{h}{b}x\,dx} = \frac{\dfrac{h}{b}\left[\dfrac{x^3}{3}\right]_0^b}{\dfrac{h}{b}\left[\dfrac{x^2}{2}\right]_0^b} = \frac{2}{3}b.$$

◄ The height of a strip of width dx at position x is $(h/b)x$, so its area is $dA = (h/b)x\,dx$. Use this expression to evaluate Eq. (7.6).

Practice Problem Determine the y coordinate of the centroid of the triangular area. Evaluate Eq. (7.7) by using an element of area dA in the form of a vertical "strip" of width dx, and let y be the height of the midpoint of the strip.

Answer: $\bar{y} = \dfrac{1}{3}h$.

Example 7.2 | **Area Defined by Two Equations** (▶ *Related Problems 7.2, 7.3*)

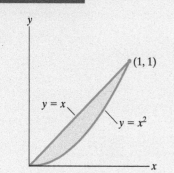

Determine the centroid of the area.

Strategy

We can determine the coordinates of the centroid using an element of area in the form of a vertical strip, just as we did in Active Example 7.1. In this case the strip must be defined so that it extends from the lower curve ($y = x^2$) to the upper curve ($y = x$).

Solution

Let dA be the vertical strip in Fig. a. The height of the strip is $x - x^2$, so $dA = (x - x^2)\,dx$. The x coordinate of the centroid is

$$\bar{x} = \frac{\displaystyle\int_A x\,dA}{\displaystyle\int_A dA} = \frac{\displaystyle\int_0^1 x(x - x^2)\,dx}{\displaystyle\int_0^1 (x - x^2)\,dx} = \frac{\left[\dfrac{x^3}{3} - \dfrac{x^4}{4}\right]_0^1}{\left[\dfrac{x^2}{2} - \dfrac{x^3}{3}\right]_0^1} = \frac{1}{2}.$$

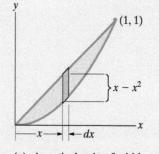

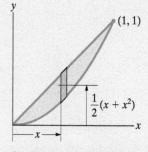

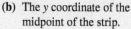

(a) A vertical strip of width dx. The height of the strip is equal to the difference in the two functions.

(b) The y coordinate of the midpoint of the strip.

The y coordinate of the midpoint of the strip is $x^2 + \frac{1}{2}(x - x^2) = \frac{1}{2}(x + x^2)$ (Fig. b). Substituting this expression for y in Eq. (7.7), we obtain the y coordinate of the centroid:

$$\bar{y} = \frac{\displaystyle\int_A y\,dA}{\displaystyle\int_A dA} = \frac{\displaystyle\int_0^1 \left[\frac{1}{2}(x + x^2)\right](x - x^2)\,dx}{\displaystyle\int_0^1 (x - x^2)\,dx} = \frac{\dfrac{1}{2}\left[\dfrac{x^3}{3} - \dfrac{x^5}{5}\right]_0^1}{\left[\dfrac{x^2}{2} - \dfrac{x^3}{3}\right]_0^1} = \frac{2}{5}.$$

Critical Thinking

Notice the generality of the approach we use in this example. It can be used to determine the x and y coordinates of the centroid of any area whose upper and lower boundaries are defined by two functions.

Problems

▶ **7.1** In Active Example 7.1, suppose that the triangular area is oriented as shown. Use integration to determine the x and y coordinates of its centroid. (Notice that you already know the answers based on the results of Active Example 7.1.)

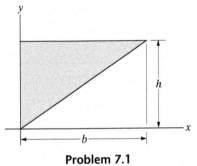
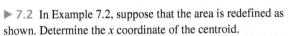

Problem 7.1

▶ **7.2** In Example 7.2, suppose that the area is redefined as shown. Determine the x coordinate of the centroid.

▶ **7.3** In Example 7.2, suppose that the area is redefined as shown. Determine the y coordinate of the centroid.

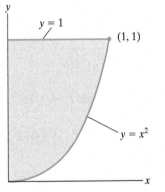

Problems 7.2/7.3

7.4 Determine the centroid of the area.

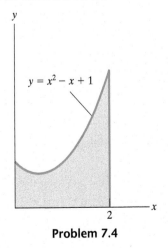

Problem 7.4

7.5 Determine the coordinates of the centroid of the area.

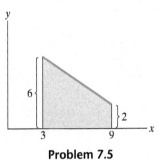

Problem 7.5

7.6 Determine x coordinate of the centroid of the area and compare your answer to the value given in Appendix B.

7.7 Determine the y coordinate of the centroid of the area and compare your answer to the value given in Appendix B.

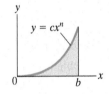

Problems 7.6/7.7

7.8 Suppose that an art student wants to paint a panel of wood as shown, with the horizontal and vertical lines passing through the centroid of the painted area, and asks you to determine the coordinates of the centroid. What are they?

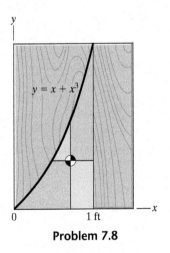

Problem 7.8

7.9 Determine the value of the constant c so that the y coordinate of the centroid of the area is $\bar{y} = 2$. What is the x coordinate of the centroid?

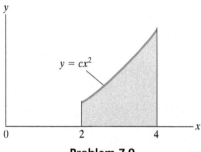

Problem 7.9

7.10 Determine the coordinates of the centroid of the metal plate's cross-sectional area.

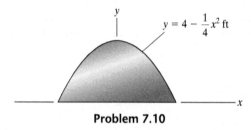

Problem 7.10

7.11 An architect wants to build a wall with the profile shown. To estimate the effects of wind loads, he must determine the wall's area and the coordinates of its centroid. What are they?

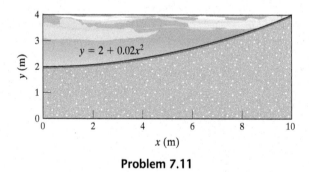

Problem 7.11

7.12 Determine the coordinates of the centroid of the area.

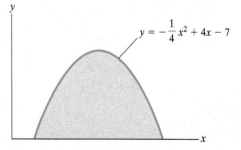

Problem 7.12

7.13 Determine the coordinates of the centroid of the area.

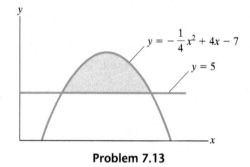

Problem 7.13

7.14 Determine the x coordinate of the centroid of the area.

7.15 Determine the y coordinate of the centroid of the area.

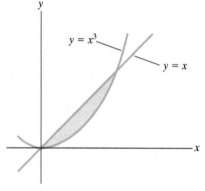

Problems 7.14/7.15

7.16 Determine the x component of the centroid of the area.

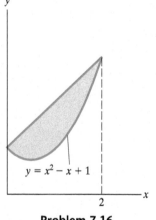

Problem 7.16

7.17 Determine the x coordinate of the centroid of the area.

7.18 Determine the y coordinate of the centroid of the area.

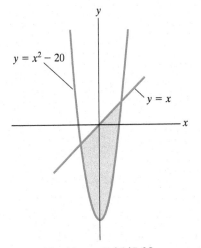

Problems 7.17/7.18

7.19 What is the x coordinate of the centroid of the area?

7.20 What is the y coordinate of the centroid of the area?

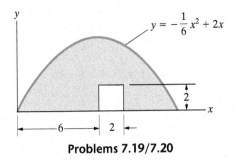

Problems 7.19/7.20

7.21 An agronomist wants to measure the rainfall at the centroid of a plowed field between two roads. What are the coordinates of the point where the rain gauge should be placed?

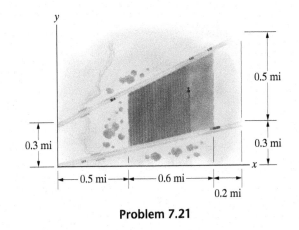

Problem 7.21

7.22 The cross section of an earth-fill dam is shown. Determine the coefficients a and b so that the y coordinate of the centroid of the cross section is 10 m.

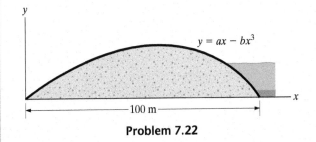

Problem 7.22

7.23 The Supermarine Spitfire used by Great Britain in World War II had a wing with an elliptical profile. Determine the coordinates of its centroid.

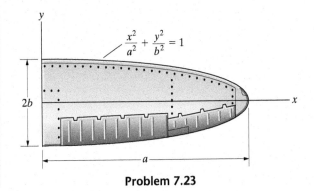

Problem 7.23

7.24 Determine the coordinates of the centroid of the area.

Strategy: Write the equation for the circular boundary in the form $y = (R^2 - x^2)^{1/2}$ and use a vertical "strip" of width dx as the element of area dA.

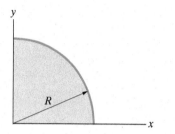

Problem 7.24

7.25* If $R = 6$ and $b = 3$, what is the y coordinate of the centroid of the area?

7.26* What is the x coordinate of the centroid of the area in Problem 7.25?

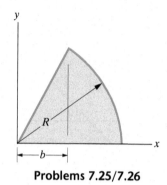

Problems 7.25/7.26

7.2 Composite Areas

BACKGROUND

Although centroids of areas can be determined by integration, the process becomes difficult and tedious for complicated areas. In this section we describe a much easier approach that can be used if an area consists of a combination of simple areas, which we call a *composite area*. We can determine the centroid of a composite area without integration if the centroids of its parts are known.

The area in Fig. 7.3a consists of a triangle, a rectangle, and a semicircle, which we call parts 1, 2, and 3. The x coordinate of the centroid of the composite area is

$$\bar{x} = \frac{\displaystyle\int_A x\, dA}{\displaystyle\int_A dA} = \frac{\displaystyle\int_{A_1} x\, dA + \int_{A_2} x\, dA + \int_{A_3} x\, dA}{\displaystyle\int_{A_1} dA + \int_{A_2} dA + \int_{A_3} dA}. \tag{7.8}$$

The x coordinates of the centroids of the parts are shown in Fig. 7.3b. From the equation for the x coordinate of the centroid of part 1,

$$\bar{x}_1 = \frac{\displaystyle\int_{A_1} x\, dA}{\displaystyle\int_{A_1} dA},$$

we obtain

$$\int_{A_1} x\, dA = \bar{x}_1 A_1.$$

Using this equation and equivalent equations for parts 2 and 3, we can write Eq. (7.8) as

$$\bar{x} = \frac{\bar{x}_1 A_1 + \bar{x}_2 A_2 + \bar{x}_3 A_3}{A_1 + A_2 + A_3}.$$

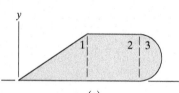

(a)

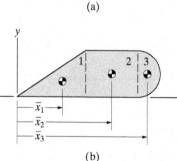

(b)

Figure 7.3
(a) A composite area composed of three simple areas.
(b) The centroids of the parts.

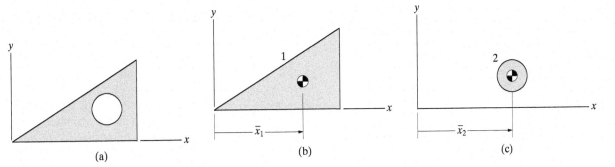

Figure 7.4
(a) An area with a cutout.
(b) The triangular area.
(c) The area of the cutout.

We have obtained an equation for the x coordinate of the composite area in terms of those of its parts. The coordinates of the centroid of a composite area with an arbitrary number of parts are

$$\bar{x} = \frac{\sum_i \bar{x}_i A_i}{\sum_i A_i}, \qquad \bar{y} = \frac{\sum_i \bar{y}_i A_i}{\sum_i A_i}. \tag{7.9}$$

When we can divide an area into parts whose centroids are known, we can use these expressions to determine its centroid. The centroids of some simple areas are tabulated in Appendix B.

We began our discussion of the centroid of an area by dividing an area into finite parts and writing equations for its weighted average position. The results, Eqs. (7.5), are approximate because of the uncertainty in the positions of the parts of the area. The exact Eqs. (7.9) are identical except that the positions of the parts are their centroids.

The area in Fig. 7.4a consists of a triangular area with a circular hole, or cutout. Designating the triangular area (without the cutout) as part 1 of the composite area (Fig. 7.4b) and the area of the cutout as part 2 (Fig. 7.4c), we obtain the x coordinate of the centroid of the composite area:

$$\bar{x} = \frac{\displaystyle\int_{A_1} x\, dA - \int_{A_2} x\, dA}{\displaystyle\int_{A_1} dA - \int_{A_2} dA} = \frac{\bar{x}_1 A_1 - \bar{x}_2 A_2}{A_1 - A_2}.$$

This equation is identical in form to the first of Eqs. (7.9) except that the terms corresponding to the cutout are negative. As this example demonstrates, we can use Eqs. (7.9) to determine the centroids of composite areas containing cutouts by treating the cutouts as negative areas.

We see that determining the centroid of a composite area requires three steps:

1. Choose the parts—Try to divide the composite area into parts whose centroids you know or can easily determine.

2. Determine the values for the parts—Determine the centroid and the area of each part. Watch for instances of symmetry that can simplify your task.

3. Calculate the centroid—Use Eqs. (7.9) to determine the centroid of the composite area.

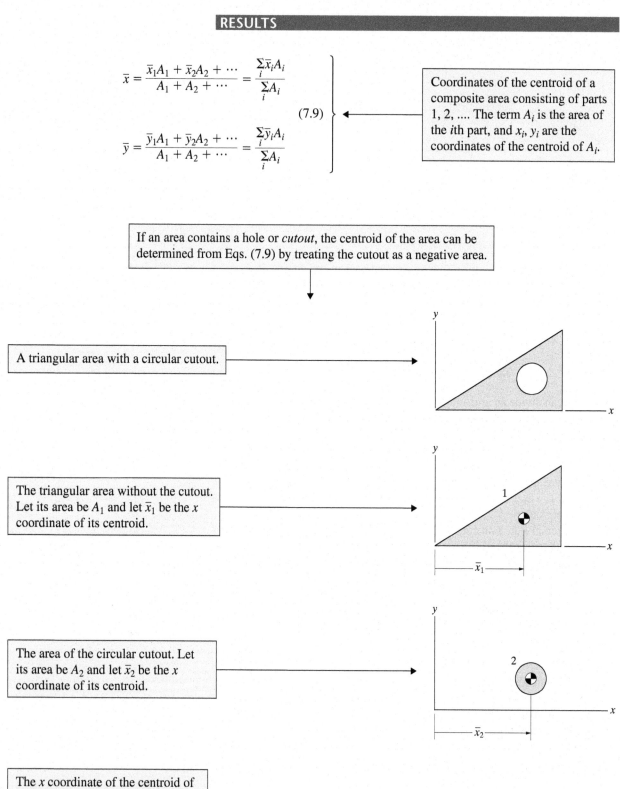

$$\bar{x} = \frac{\bar{x}_1 A_1 + \bar{x}_2 A_2 + \cdots}{A_1 + A_2 + \cdots} = \frac{\sum_i \bar{x}_i A_i}{\sum_i A_i}$$

$$\bar{y} = \frac{\bar{y}_1 A_1 + \bar{y}_2 A_2 + \cdots}{A_1 + A_2 + \cdots} = \frac{\sum_i \bar{y}_i A_i}{\sum_i A_i}$$

(7.9)

Coordinates of the centroid of a composite area consisting of parts 1, 2, ... The term A_i is the area of the ith part, and x_i, y_i are the coordinates of the centroid of A_i.

If an area contains a hole or *cutout*, the centroid of the area can be determined from Eqs. (7.9) by treating the cutout as a negative area.

A triangular area with a circular cutout.

The triangular area without the cutout. Let its area be A_1 and let $\bar{x}_1$ be the x coordinate of its centroid.

The area of the circular cutout. Let its area be A_2 and let $\bar{x}_2$ be the x coordinate of its centroid.

The x coordinate of the centroid of the triangular area with the cutout is

$$\bar{x} = \frac{\bar{x}_1 A_1 - \bar{x}_2 A_2}{A_1 - A_2}.$$

| Active Example 7.3 | Centroid of a Composite Area (▶ *Related Problem 7.27*) |

Determine the x coordinate of the centroid of the composite area.

Strategy
We must divide the area into simple parts (in this example the parts are obvious), determine the areas and centroid locations for the parts, and apply Eq. $(7.9)_1$.

Solution

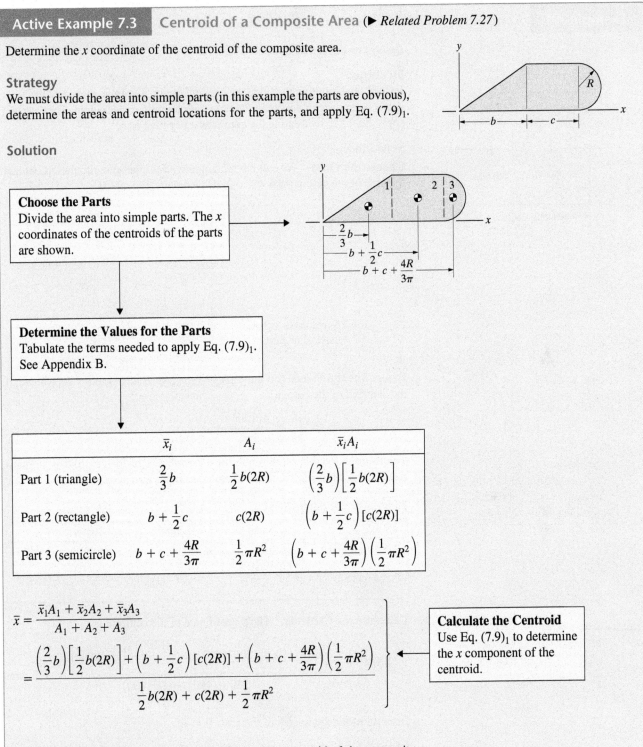

Choose the Parts
Divide the area into simple parts. The x coordinates of the centroids of the parts are shown.

Determine the Values for the Parts
Tabulate the terms needed to apply Eq. $(7.9)_1$.
See Appendix B.

	$\bar{x}_i$	A_i	$\bar{x}_i A_i$
Part 1 (triangle)	$\dfrac{2}{3}b$	$\dfrac{1}{2}b(2R)$	$\left(\dfrac{2}{3}b\right)\left[\dfrac{1}{2}b(2R)\right]$
Part 2 (rectangle)	$b + \dfrac{1}{2}c$	$c(2R)$	$\left(b + \dfrac{1}{2}c\right)[c(2R)]$
Part 3 (semicircle)	$b + c + \dfrac{4R}{3\pi}$	$\dfrac{1}{2}\pi R^2$	$\left(b + c + \dfrac{4R}{3\pi}\right)\left(\dfrac{1}{2}\pi R^2\right)$

$$\bar{x} = \frac{\bar{x}_1 A_1 + \bar{x}_2 A_2 + \bar{x}_3 A_3}{A_1 + A_2 + A_3}$$

$$= \frac{\left(\dfrac{2}{3}b\right)\left[\dfrac{1}{2}b(2R)\right] + \left(b + \dfrac{1}{2}c\right)[c(2R)] + \left(b + c + \dfrac{4R}{3\pi}\right)\left(\dfrac{1}{2}\pi R^2\right)}{\dfrac{1}{2}b(2R) + c(2R) + \dfrac{1}{2}\pi R^2}$$

Calculate the Centroid
Use Eq. $(7.9)_1$ to determine the x component of the centroid.

Practice Problem Determine the y coordinate of the centroid of the composite area.

Answer: $\bar{y} = \dfrac{\left[\dfrac{1}{3}(2R)\right]\left[\dfrac{1}{2}b(2R)\right] + R\left[c(2R)\right] + R\left(\dfrac{1}{2}\pi R^2\right)}{\dfrac{1}{2}b(2R) + c(2R) + \dfrac{1}{2}\pi R^2}$.

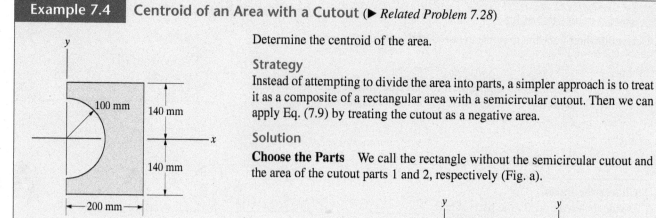

| Example 7.4 | Centroid of an Area with a Cutout (▶ *Related Problem 7.28*) |

Determine the centroid of the area.

Strategy

Instead of attempting to divide the area into parts, a simpler approach is to treat it as a composite of a rectangular area with a semicircular cutout. Then we can apply Eq. (7.9) by treating the cutout as a negative area.

Solution

Choose the Parts   We call the rectangle without the semicircular cutout and the area of the cutout parts 1 and 2, respectively (Fig. a).

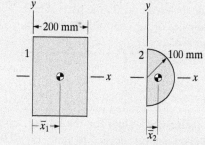

(a) The rectangle and the semicircular cutout.

Determine the Values for the Parts   From Appendix B, the x coordinate of the centroid of the cutout is

$$\bar{x}_2 = \frac{4R}{3\pi} = \frac{4(100)}{3\pi} \text{ mm}.$$

The information for determining the x coordinate of the centroid is summarized in the table. Notice that we treat the cutout as a negative area.

Information for determining $\bar{x}$

	$\bar{x}_i$ (mm)	A_i (mm²)	$\bar{x}_i A_i$ (mm³)
Part 1 (rectangle)	100	(200)(280)	(100)[(200)(280)]
Part 2 (cutout)	$\dfrac{4(100)}{3\pi}$	$-\dfrac{1}{2}\pi(100)^2$	$-\dfrac{4(100)}{3\pi}\left[\dfrac{1}{2}\pi(100)^2\right]$

Calculate the Centroid   The x coordinate of the centroid is

$$\bar{x} = \frac{\bar{x}_1 A_1 + \bar{x}_2 A_2}{A_1 + A_2} = \frac{(100)[(200)(280)] - \dfrac{4(100)}{3\pi}\left[\dfrac{1}{2}\pi(100)^2\right]}{(200)(280) - \dfrac{1}{2}\pi(100)^2} = 122 \text{ mm}$$

Because of the symmetry of the area, $\bar{y} = 0$.

Critical Thinking

If you try to divide the area into simple parts, you will gain appreciation for the approach we used. We were able to determine the centroid by dealing with two simple areas, the rectangular area without the cutout and the semicircular cutout. Determining centroids of areas can often be simplified in this way.

Problems

▶ **7.27** In Active Example 7.3, suppose that the area is placed as shown. Let the dimensions $R = 6$ in, $c = 14$ in, and $b = 18$ in. Use Eq. (7.9)$_1$ to determine the x coordinate of the centroid.

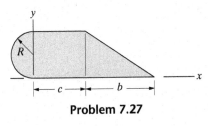

Problem 7.27

▶ **7.28** In Example 7.4, suppose that the area is given a second semicircular cutout as shown. Determine the x coordinate of the centroid.

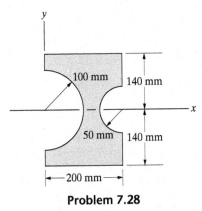

Problem 7.28

For Problems 7.29–7.36, determine the coordinates of the centroids.

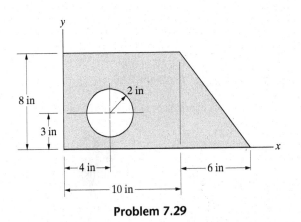

Problem 7.29

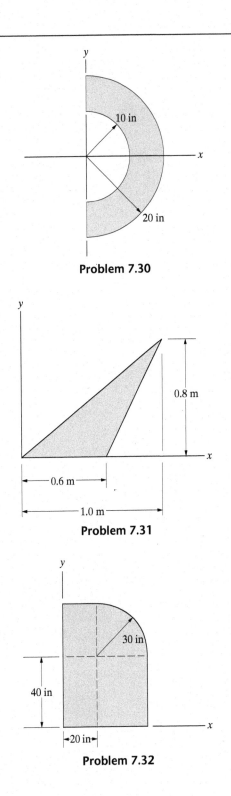

Problem 7.30

Problem 7.31

Problem 7.32

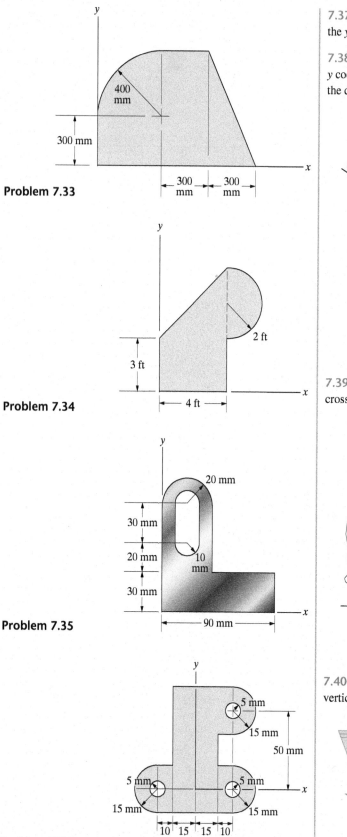

Problem 7.33

Problem 7.34

Problem 7.35

Problem 7.36

7.37 The dimensions $b = 42$ mm and $h = 22$ mm. Determine the y coordinate of the centroid of the beam's cross section.

7.38 If the cross-sectional area of the beam is 8400 mm² and the y coordinate of the centroid of the area is $\bar{y} = 90$ mm, what are the dimensions b and h?

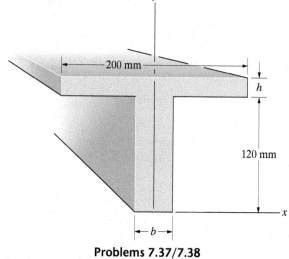

Problems 7.37/7.38

7.39 Determine the y coordinate of the centroid of the beam's cross section.

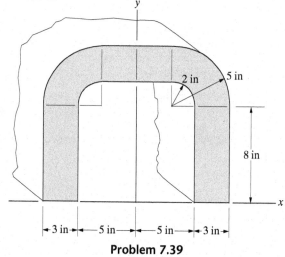

Problem 7.39

7.40 Determine the coordinates of the centroid of the airplane's vertical stabilizer.

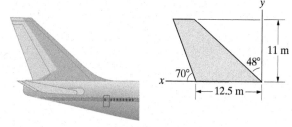

Problem 7.40

7.41 The area has elliptical boundaries. If $a = 30$ mm, $b = 15$ mm, and $\varepsilon = 6$ mm, what is the x coordinate of the centroid of the area?

7.42 By determining the x coordinate of the centroid of the area shown in Problem 7.41 in terms of a, b, and ε, and evaluating its limit as $\varepsilon \to 0$, show that the x coordinate of the centroid of a quarter-elliptical line is

$$\bar{x} = \frac{4a(a + 2b)}{3\pi(a + b)}.$$

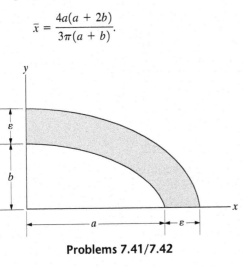

Problems 7.41/7.42

7.43 Three sails of a New York pilot schooner are shown. The coordinates of the points are in feet. Determine the centroid of sail 1.

7.44 Determine the centroid of sail 2.

7.45 Determine the centroid of sail 3.

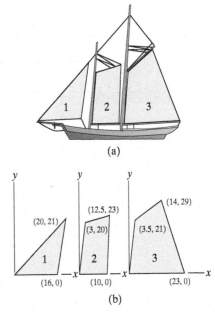

(a)

(b)

Problems 7.43–7.45

7.3 Distributed Loads

BACKGROUND

The load exerted on a beam (stringer) supporting a floor of a building is distributed over the beam's length (Fig. 7.5a). The load exerted by wind on a television transmission tower is distributed along the tower's height (Fig. 7.5b). In many engineering applications, loads are continuously distributed along lines. We will show that the concept of the centroid of an area can be useful in the analysis of objects subjected to such loads.

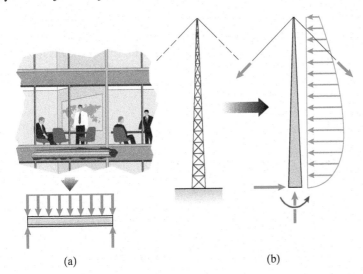

(a)   (b)

Figure 7.5
Examples of distributed forces:
(a) Uniformly distributed load exerted on a beam of a building's frame by the floor.
(b) Wind load distributed along the height of a tower.

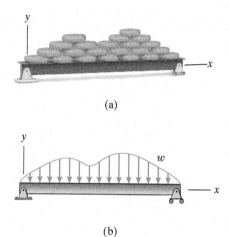

(a)

(b)

Figure 7.6
(a) Loading a beam with bags of sand.
(b) The distributed load w models the load exerted by the bags.

Describing a Distributed Load

We can use a simple example to demonstrate how such loads are expressed analytically. Suppose that we pile bags of sand on a beam, as shown in Fig. 7.6a. It is clear that the load exerted by the bags is distributed over the length of the beam and that its magnitude at a given position x depends on how high the bags are piled at that position. To describe the load, we define a function w such that the *downward* force exerted on an infinitesimal element dx of the beam is $w\,dx$. With this function we can model the varying magnitude of the load exerted by the sand bags (Fig. 7.6b). The arrows in the figure indicate that the load acts in the downward direction. Loads distributed along lines, from simple examples such as a beam's own weight to complicated ones such as the lift distributed along the length of an airplane's wing, are modeled by the function w. Since the product of w and dx is a force, the dimensions of w are (force)/(length). For example, w can be expressed in newtons per meter in SI units or in pounds per foot in U.S. Customary units.

Determining Force and Moment

Let's assume that the function w describing a particular distributed load is known (Fig. 7.7a). The graph of w is called the *loading curve*. Since the force acting on an element dx of the line is $w\,dx$, we can determine the total force F exerted by the distributed load by integrating the loading curve with respect to x:

$$F = \int_L w\,dx. \qquad (7.10)$$

We can also integrate to determine the moment about a point exerted by the distributed load. For example, the moment about the origin due to the force exerted on the element dx is $xw\,dx$, so the total moment about the origin due to the distributed load is

$$M = \int_L xw\,dx. \qquad (7.11)$$

When you are concerned only with the total force and moment exerted by a distributed load, you can represent it by a single equivalent force F (Fig. 7.7b). For equivalence, the force must act at a position $\bar{x}$ on the x axis such that the moment of F about the origin is equal to the moment of the distributed load about the origin:

$$\bar{x}F = \int_L xw\,dx.$$

Therefore the force F is equivalent to the distributed load if we place it at the position

$$\bar{x} = \frac{\int_L xw\,dx}{\int_L w\,dx}. \qquad (7.12)$$

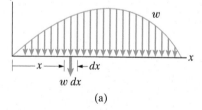

(a)

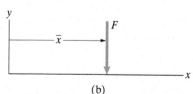

(b)

Figure 7.7
(a) A distributed load and the force exerted on a differential element dx.
(b) The equivalent force.

The Area Analogy

Notice that the term $w\,dx$ is equal to an element of "area" dA between the loading curve and the x axis (Fig. 7.8a). (We use quotation marks because $w\,dx$ is actually a force and not an area.) Interpreted in this way, Eq. (7.10) states that the total force exerted by the distributed load is equal to the "area" A between the loading curve and the x axis:

$$F = \int_L w\,dx = \int_A dA = A. \qquad (7.13)$$

Substituting $w\,dx = dA$ into Eq. (7.12), we obtain

$$\bar{x} = \frac{\int_L xw\,dx}{\int_L w\,dx} = \frac{\int_A x\,dA}{\int_A dA}. \qquad (7.14)$$

The force F is equivalent to the distributed load if it acts at the centroid of the "area" between the loading curve and the x axis (Fig. 7.8b). Using this analogy to represent a distributed load by an equivalent force can be very useful when the loading curve is relatively simple.

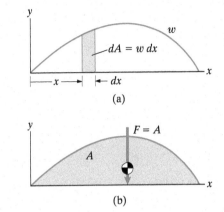

Figure 7.8
(a) Determining the "area" between the function w and the x axis.
(b) The equivalent force is equal to the "area," and the line of action passes through its centroid.

RESULTS

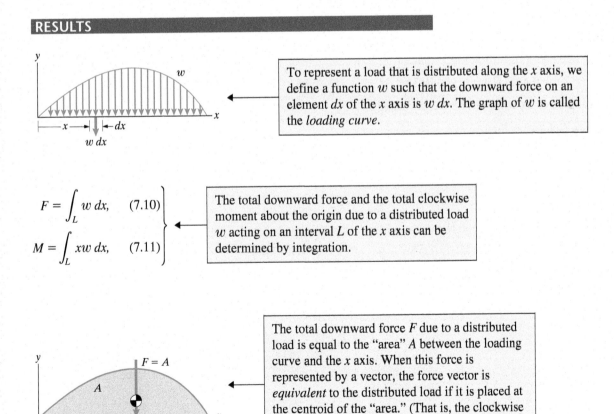

To represent a load that is distributed along the x axis, we define a function w such that the downward force on an element dx of the x axis is $w\,dx$. The graph of w is called the *loading curve*.

$$F = \int_L w\,dx, \qquad (7.10)$$

$$M = \int_L xw\,dx, \qquad (7.11)$$

The total downward force and the total clockwise moment about the origin due to a distributed load w acting on an interval L of the x axis can be determined by integration.

The total downward force F due to a distributed load is equal to the "area" A between the loading curve and the x axis. When this force is represented by a vector, the force vector is *equivalent* to the distributed load if it is placed at the centroid of the "area." (That is, the clockwise moment about the origin due to the force vector is equal to M.) This is called the *area analogy*.

Active Example 7.5 Beam with a Distributed Load (▶ *Related Problem 7.46*)

The beam is subjected to a "triangular" distributed load whose value at B is 100 N/m. (That is, the function w increases linearly from $w = 0$ at A to $w = 100$ N/m at B.) Determine the reactions on the beam at A and B.

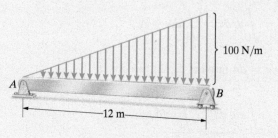

Strategy

We can use the area analogy to represent the distributed load by an equivalent force. Then we can apply the equilibrium equations to determine the reactions at A and B.

Solution

The "area" of the triangular distributed load is one-half its base times its height, or $\frac{1}{2}(12 \text{ m}) \times (100 \text{ N/m}) = 600$ N. The centroid of the triangular "area" is located at $\bar{x} = \frac{2}{3}(12 \text{ m}) = 8$ m.

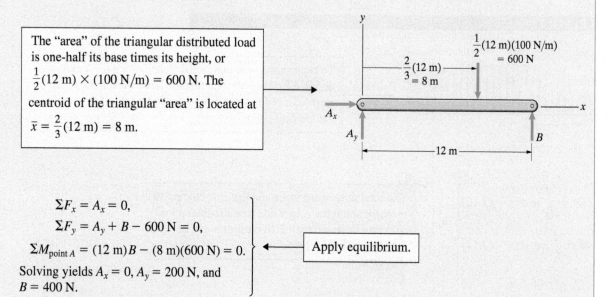

$$\Sigma F_x = A_x = 0,$$
$$\Sigma F_y = A_y + B - 600 \text{ N} = 0,$$
$$\Sigma M_{\text{point } A} = (12 \text{ m})B - (8 \text{ m})(600 \text{ N}) = 0.$$

Apply equilibrium.

Solving yields $A_x = 0$, $A_y = 200$ N, and $B = 400$ N.

Practice Problem (a) Determine w as a function of x for the triangular distributed load in this example. (b) Use Eqs. (7.10) and (7.11) to determine the total downward force and the total clockwise moment about the left end of the beam due to the triangular distributed load.

Answer: (a) $w = \dfrac{100}{12}x$ N/m. (b) $F = 600$ N, $M = 4800$ N-m.

Example 7.6 **Beam Subjected to Distributed Loads** (▶ *Related Problem 7.48*)

The beam is subjected to two distributed loads. Determine the reactions at A and B.

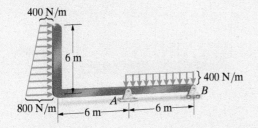

Strategy
We can easily apply the area analogy to the uniformly distributed load between A and B. We will treat the distributed load on the vertical section of the beam as the sum of uniform and triangular distributed loads and use the area analogy to represent each distributed load by an equivalent force.

Solution
We draw the free-body diagram of the beam in Fig. a, expressing the left distributed load as the sum of uniform and triangular loads. In Fig. b, we represent the three distributed loads by equivalent forces. The "area" of the uniform distributed load on the right is $(6\text{ m}) \times (400\text{ N/m}) = 2400\text{ N}$, and its centroid is 3 m from B. The area of the uniform distributed load on the vertical part of the beam is $(6\text{ m}) \times (400\text{ N/m}) = 2400\text{ N}$, and its centroid is located at $y = 3$ m. The area of the triangular distributed load is $\frac{1}{2}(6\text{ m}) \times (400\text{ N/m}) = 1200\text{ N}$, and its centroid is located at $y = \frac{1}{3}(6\text{ m}) = 2$ m.

From the equilibrium equations

$$\Sigma F_x = A_x + 1200\text{ N} + 2400\text{ N} = 0,$$

$$\Sigma F_y = A_y + B - 2400\text{ N} = 0,$$

$$\Sigma M_{\text{point } A} = (6\text{ m})B - (3\text{ m})(2400\text{ N}) - (2\text{ m})(1200\text{ N}) - (3\text{ m})(2400\text{ N}) = 0,$$

we obtain $A_x = -3600$ N, $A_y = -400$ N, and $B = 2800$ N.

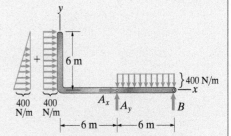

(a) Free-body diagram of the beam.

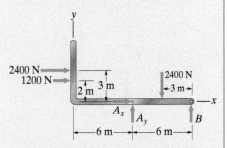

(b) Representing the distributed loads by equivalent forces.

Critical Thinking
When you analyze a problem involving distributed loads, should you always use the area analogy to represent them as we did in this example? The area analogy is useful when a loading curve is sufficiently simple that its area and the location of its centroid are easy to determine. When that is not the case, you can use Eqs. (7.10) and (7.11) to determine the force and moment exerted by a distributed load. We illustrate this approach in Example 7.7.

| Example 7.7 | Beam with a Distributed Load (▶ *Related Problem 7.49*) |

The beam is subjected to a distributed load, a force, and a couple. The distributed load is $w = 300x - 50x^2 + 0.3x^4$ lb/ft. Determine the reactions at the fixed support A.

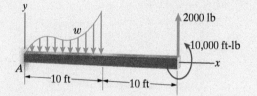

Strategy

Since we know the function w, we can use Eqs. (7.10) and (7.11) to determine the force and moment exerted on the beam by the distributed load. We can then use the equilibrium equations to determine the reactions at A.

Solution

We isolate the beam and show the reactions at the fixed support in Fig. a. The *downward* force exerted by the distributed load is

$$\int_L w\,dx = \int_0^{10} (300x - 50x^2 + 0.3x^4)\,dx = 4330 \text{ lb.}$$

The *clockwise* moment about A exerted by the distributed load is

$$\int_L xw\,dx = \int_0^{10} x(300x - 50x^2 + 0.3x^4)\,dx = 25{,}000 \text{ ft-lb.}$$

From the equilibrium equations

$$\Sigma F_x = A_x = 0,$$

$$\Sigma F_y = A_y - 4330 \text{ lb} + 2000 \text{ lb} = 0,$$

$$\Sigma M_{\text{point } A} = M_A - 25{,}000 \text{ ft-lb} + (20 \text{ ft})(2000 \text{ lb}) + 10{,}000 \text{ ft-lb} = 0,$$

we obtain $A_x = 0$, $A_y = 2330$ lb, and $M_A = -25{,}000$ ft-lb.

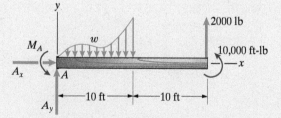

(a) Free-body diagram of the beam.

Critical Thinking

When you use Eq. (7.11), it is important to be aware that you are calculating the *clockwise* moment due to the distributed load w *about the origin* $x = 0$.

Problems

▶ 7.46 In Active Example 7.5, suppose that the distributed load is modified as shown. Determine the reactions on the beam at A and B.

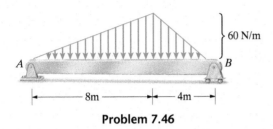

Problem 7.46

7.47 Determine the reactions at A and B.

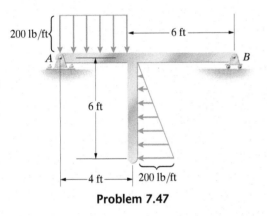

Problem 7.47

▶ 7.48 In Example 7.6, suppose that the distributed loads are modified as shown. Determine the reactions on the beam at A and B.

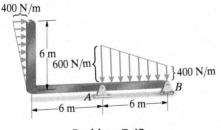

Problem 7.48

▶ 7.49 In Example 7.7, suppose that the distributed load acting on the beam from $x = 0$ to $x = 10$ ft is given by $w = 350 + 0.3x^3$ lb/ft. (a) Determine the downward force and the clockwise moment about A exerted by the distributed load. (b) Determine the reactions at the fixed support.

7.50 Determine the reactions at the fixed support A.

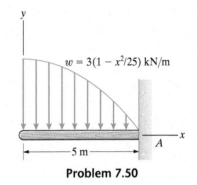

Problem 7.50

7.51 An engineer measures the forces exerted by the soil on a 10-m section of a building foundation and finds that they are described by the distributed load $w = -10x - x^2 + 0.2x^3$ kN/m.

(a) Determine the magnitude of the total force exerted on the foundation by the distributed load.

(b) Determine the magnitude of the moment about A due to the distributed load.

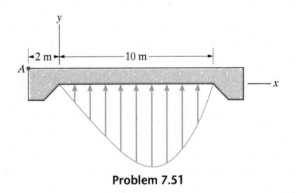

Problem 7.51

7.52 Determine the reactions on the beam at A and B.

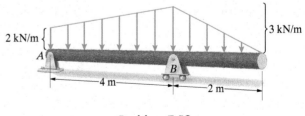

Problem 7.52

7.53 The aerodynamic lift of the wing is described by the distributed load $w = -300\sqrt{1 - 0.04x^2}$ N/m. The mass of the wing is 27 kg, and its center of mass is located 2 m from the wing root R.

(a) Determine the magnitudes of the force and the moment about R exerted by the lift of the wing.

(b) Determine the reactions on the wing at R.

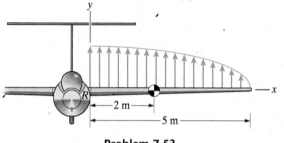

Problem 7.53

7.54 Determine the reactions on the bar at A and B.

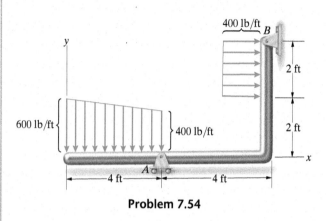

Problem 7.54

7.55 Determine the reactions on member AB at A and B.

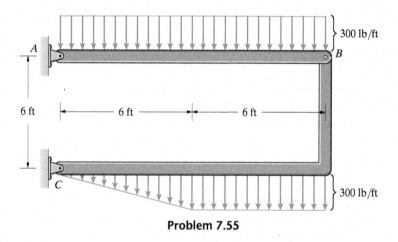

Problem 7.55

7.56 Determine the axial forces in members BD, CD, and CE of the truss and indicate whether they are in tension (T) or compression (C).

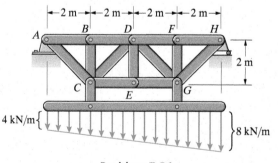

Problem 7.56

7.57 Determine the reactions on member *ABC* at *A* and *B*.

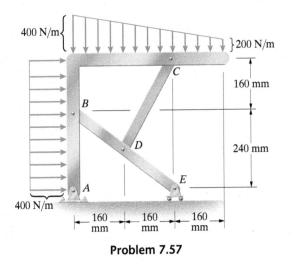

Problem 7.57

7.58 Determine the forces on member *ABC* of the frame.

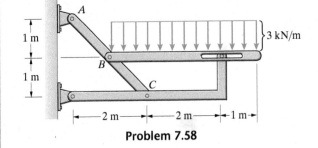

Problem 7.58

7.4 Centroids of Volumes and Lines

BACKGROUND

Here we define the centroids, or average positions, of volumes and lines, and show how to determine the centroids of composite volumes and lines. We will show in Section 7.7 that knowing the centroids of volumes and lines allows you to determine the centers of mass of certain types of objects, which tells you where their weights effectively act.

Volumes   Consider a volume *V*, and let *dV* be a differential element of *V* with coordinates *x*, *y*, and *z* (Fig. 7.9). By analogy with Eqs. (7.6) and (7.7), the coordinates of the centroid of *V* are

$$\bar{x} = \frac{\int_V x \, dV}{\int_V dV}, \qquad \bar{y} = \frac{\int_V y \, dV}{\int_V dV}, \qquad \bar{z} = \frac{\int_V z \, dV}{\int_V dV}. \qquad (7.15)$$

The subscript *V* on the integral signs means that the integration is carried out over the entire volume.

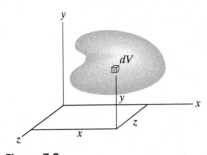

Figure 7.9
A volume *V* and differential element *dV*.

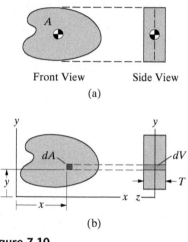

Front View   Side View

(a)

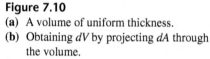

(b)

Figure 7.10
(a) A volume of uniform thickness.
(b) Obtaining dV by projecting dA through the volume.

If a volume has the form of a plate with uniform thickness and cross-sectional area A (Fig. 7.10a), its centroid coincides with the centroid of A and lies at the midpoint between the two faces. To show that this is true, we obtain a volume element dV by projecting an element dA of the cross-sectional area through the thickness T of the volume, so that $dV = T\,dA$ (Fig. 7.10b). Then the x and y coordinates of the centroid of the volume are

$$\bar{x} = \frac{\displaystyle\int_V x\,dV}{\displaystyle\int_V dV} = \frac{\displaystyle\int_A xT\,dA}{\displaystyle\int_A T\,dA} = \frac{\displaystyle\int_A x\,dA}{\displaystyle\int_A dA},$$

$$\bar{y} = \frac{\displaystyle\int_V y\,dV}{\displaystyle\int_V dV} = \frac{\displaystyle\int_A yT\,dA}{\displaystyle\int_A T\,dA} = \frac{\displaystyle\int_A y\,dA}{\displaystyle\int_A dA}.$$

The coordinate $\bar{z} = 0$ by symmetry. Thus you know the centroid of this type of volume if you know (or can determine) the centroid of its cross-sectional area.

Lines   The coordinates of the centroid of a line L are

$$\bar{x} = \frac{\displaystyle\int_L x\,dL}{\displaystyle\int_L dL}, \qquad \bar{y} = \frac{\displaystyle\int_L y\,dL}{\displaystyle\int_L dL}, \qquad \bar{z} = \frac{\displaystyle\int_L z\,dL}{\displaystyle\int_L dL}, \qquad (7.16)$$

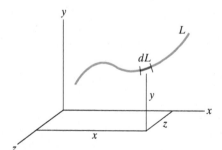

Figure 7.11
A line L and differential element dL.

where dL is a differential length of the line with coordinates x, y, and z. (Fig. 7.11).

RESULTS

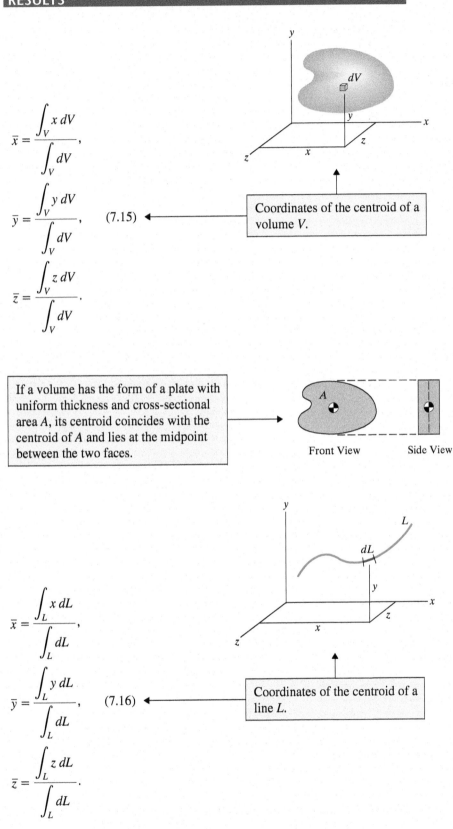

$$\bar{x} = \frac{\int_V x \, dV}{\int_V dV},$$

$$\bar{y} = \frac{\int_V y \, dV}{\int_V dV}, \qquad (7.15)$$

$$\bar{z} = \frac{\int_V z \, dV}{\int_V dV}.$$

Coordinates of the centroid of a volume V.

If a volume has the form of a plate with uniform thickness and cross-sectional area A, its centroid coincides with the centroid of A and lies at the midpoint between the two faces.

Front View   Side View

$$\bar{x} = \frac{\int_L x \, dL}{\int_L dL},$$

$$\bar{y} = \frac{\int_L y \, dL}{\int_L dL}, \qquad (7.16)$$

$$\bar{z} = \frac{\int_L z \, dL}{\int_L dL}.$$

Coordinates of the centroid of a line L.

Active Example 7.8 | Centroid of a Cone by Integration (▶ *Related Problem 7.59*)

Determine the centroid of the cone.

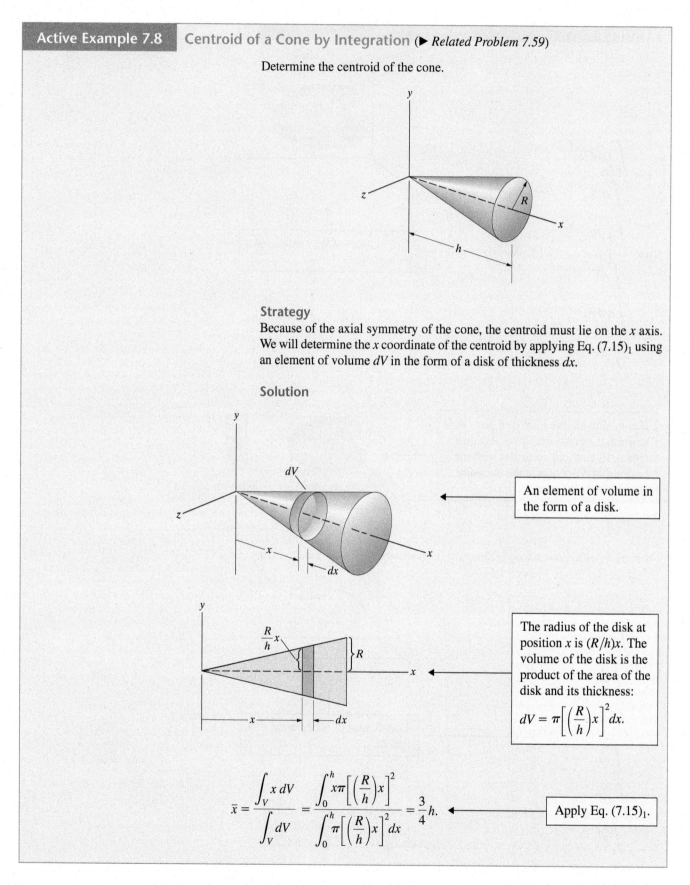

Strategy

Because of the axial symmetry of the cone, the centroid must lie on the x axis. We will determine the x coordinate of the centroid by applying Eq. $(7.15)_1$ using an element of volume dV in the form of a disk of thickness dx.

Solution

An element of volume in the form of a disk.

The radius of the disk at position x is $(R/h)x$. The volume of the disk is the product of the area of the disk and its thickness:

$$dV = \pi \left[\left(\frac{R}{h} \right) x \right]^2 dx.$$

$$\bar{x} = \frac{\int_V x \, dV}{\int_V dV} = \frac{\int_0^h x\pi \left[\left(\frac{R}{h} \right) x \right]^2}{\int_0^h \pi \left[\left(\frac{R}{h} \right) x \right]^2 dx} = \frac{3}{4} h.$$

Apply Eq. $(7.15)_1$.

Practice Problem The radius in feet of the circular cross section of the truncated cone is given as a function of x by $r = 1 + \frac{1}{4}x$. Determine the x coordinate of its centroid.

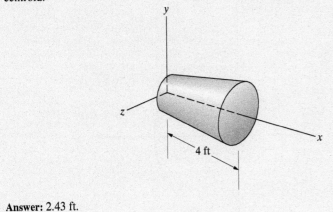

Answer: 2.43 ft.

Example 7.9	Centroid of a Line by Integration (▶ *Related Problem 7.66*)

The line L is defined by the function $y = x^2$. Determine the x coordinate of its centroid.

Strategy
We can express a differential element dL of a line (Fig. a) in terms of dx and dy:

$$dL = \sqrt{dx^2 + dy^2} = \sqrt{1 + \left(\frac{dy}{dx}\right)^2}\, dx.$$

From the equation describing the line, the derivative $dy/dx = 2x$, so we obtain an expression for dL in terms of x:

$$dL = \sqrt{1 + 4x^2}\, dx.$$

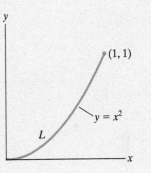

Solution
To integrate over the entire line, we must integrate from $x = 0$ to $x = 1$. The x coordinate of the centroid is

$$\bar{x} = \frac{\displaystyle\int_L x\, dL}{\displaystyle\int_L dL} = \frac{\displaystyle\int_0^1 x\sqrt{1 + 4x^2}\, dx}{\displaystyle\int_0^1 \sqrt{1 + 4x^2}\, dx} = 0.574.$$

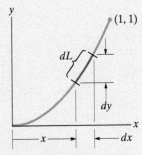

(a) A differential line element dL.

Critical Thinking
Our approach in this example is appropriate to determine the centroid of a line that is described by a function of the form $y = f(x)$. In Example 7.10 we show how to determine the centroid of a line that is described in terms of polar coordinates.

| Example 7.10 | Centroid of a Semicircular Line by Integration (▶ *Related Problem 7.70*) |

Determine the centroid of the semicircular line.

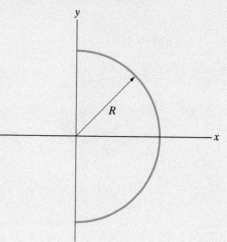

Strategy

Because of the symmetry of the line, the centroid lies on the x axis. To determine $\bar{x}$, we will integrate in terms of polar coordinates. By letting θ change by an amount $d\theta$, we obtain a differential line element of length $dL = R\,d\theta$ (Fig. a). The x coordinate of dL is $x = R\cos\theta$.

Solution

To integrate over the entire line, we must integrate with respect to θ from $\theta = -\pi/2$ to $\theta = +\pi/2$:

$$\bar{x} = \frac{\int_L x\,dL}{\int_L dL} = \frac{\int_{-\pi/2}^{\pi/2}(R\cos\theta)R\,d\theta}{\int_{-\pi/2}^{\pi/2}R\,d\theta} = \frac{R^2[\sin\theta]_{-\pi/2}^{\pi/2}}{R[\theta]_{-\pi/2}^{\pi/2}} = \frac{2R}{\pi}.$$

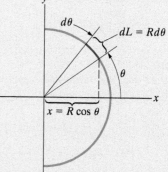

(a) A differential line element $dL = R\,d\theta$.

Critical Thinking

Notice that our integration procedure gives the correct length of the line:

$$\int_L dL = \int_{-\pi/2}^{\pi/2}R\,d\theta = R[\theta]_{-\pi/2}^{\pi/2} = \pi R.$$

Problems

▶ **7.59** Use the method described in Active Example 7.8 to determine the centroid of the truncated cone.

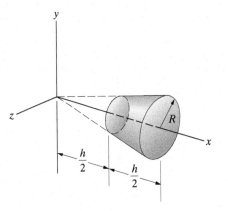

Problem 7.59

7.60 A grain storage tank has the form of a surface of revolution with the profile shown. The height of the tank is 7 m and its diameter at ground level is 10 m. Determine the volume of the tank and the height *above ground level* of the centroid of its volume.

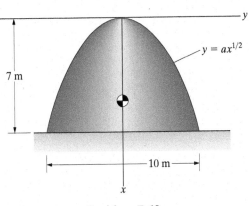

Problem 7.60

7.61 The object shown, designed to serve as a pedestal for a speaker, has a profile obtained by revolving the curve $y = 0.167x^2$ about the x axis. What is the x coordinate of the centroid of the object?

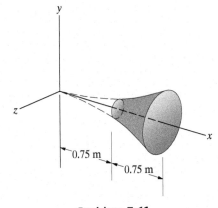

Problem 7.61

7.62 The volume of a nose cone is generated by rotating the function $y = x - 0.2x^2$ about the x axis.

(a) What is the volume of the nose cone?

(b) What is the x coordinate of the centroid of the volume?

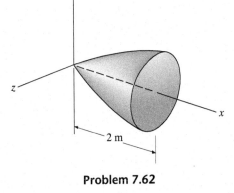

Problem 7.62

7.63 Determine the centroid of the hemispherical volume.

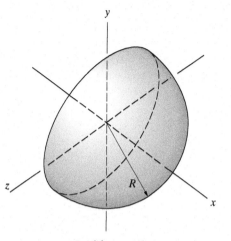

Problem 7.63

7.64 The volume consists of a segment of a sphere of radius R. Determine its centroid.

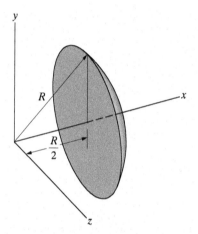

Problem 7.64

7.65 A volume of revolution is obtained by revolving the curve $x^2/a^2 + y^2/b^2 = 1$ about the x axis. Determine its centroid.

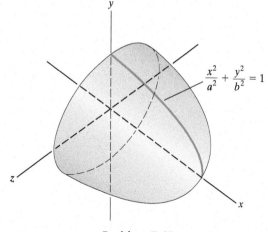

Problem 7.65

▶ **7.66** In Example 7.9, determine the y coordinate of the centroid of the line.

7.67 Determine the coordinates of the centroid of the line.

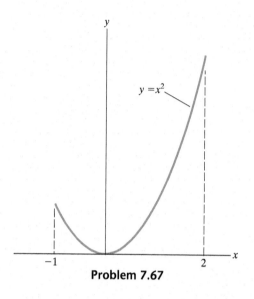

$y = x^2$

Problem 7.67

7.68 Determine the x coordinate of the centroid of the line.

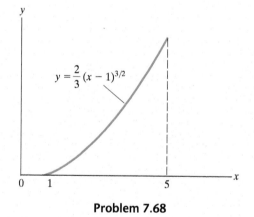

$y = \dfrac{2}{3}(x - 1)^{3/2}$

Problem 7.68

7.69 Determine the x coordinate of the centroid of the line.

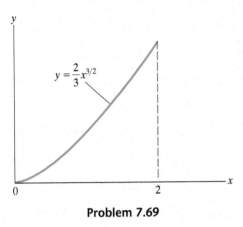

$$y = \frac{2}{3}x^{3/2}$$

Problem 7.69

▶ **7.70** Use the method described in Example 7.10 to determine the centroid of the circular arc.

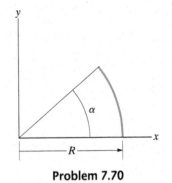

Problem 7.70

7.5 Composite Volumes and Lines

BACKGROUND

The centroids of composite volumes and lines can be derived using the same approach we applied to areas. The coordinates of the centroid of a composite volume are

$$\bar{x} = \frac{\sum_i \bar{x}_i V_i}{\sum_i V_i}, \qquad \bar{y} = \frac{\sum_i \bar{y}_i V_i}{\sum_i V_i}, \qquad \bar{z} = \frac{\sum_i \bar{z}_i V_i}{\sum_i V_i}, \qquad (7.17)$$

and the coordinates of the centroid of a composite line are

$$\bar{x} = \frac{\sum_i \bar{x}_i L_i}{\sum_i L_i}, \qquad \bar{y} = \frac{\sum_i \bar{y}_i L_i}{\sum_i L_i}, \qquad \bar{z} = \frac{\sum_i \bar{z}_i L_i}{\sum_i L_i}. \qquad (7.18)$$

The centroids of some simple volumes and lines are tabulated in Appendices B and C.

Determining the centroid of a composite volume or line requires three steps:

1. **Choose the parts**—Try to divide the composite into parts whose centroids you know or can easily determine.
2. **Determine the values for the parts**—Determine the centroid and the volume or length of each part. Watch for instances of symmetry that can simplify your task.
3. **Calculate the centroid**—Use Eqs. (7.17) or (7.18) to determine the centroid of the composite volume or line.

Active Example 7.11 **Centroid of a Composite Volume** (▶ *Related Problem 7.71*)

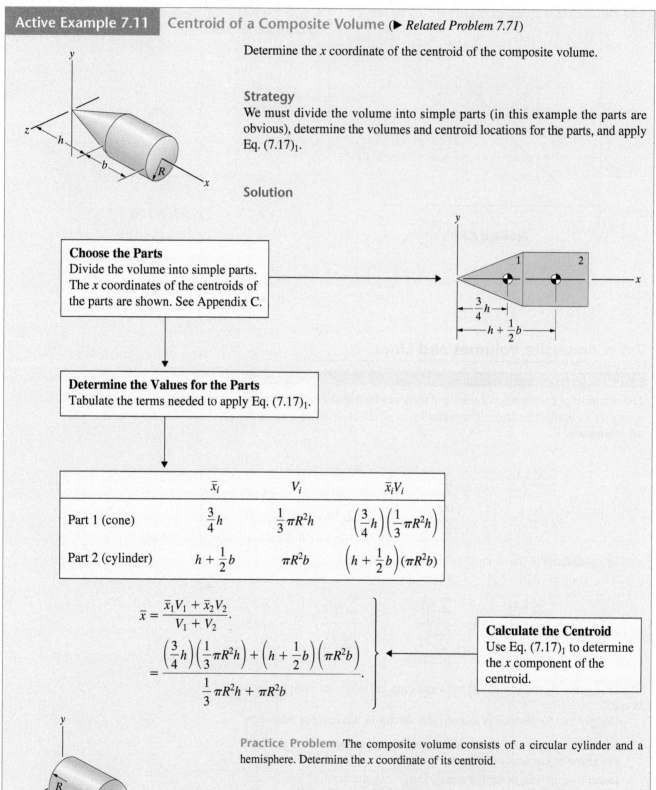

Determine the x coordinate of the centroid of the composite volume.

Strategy
We must divide the volume into simple parts (in this example the parts are obvious), determine the volumes and centroid locations for the parts, and apply Eq. $(7.17)_1$.

Solution

Choose the Parts
Divide the volume into simple parts. The x coordinates of the centroids of the parts are shown. See Appendix C.

Determine the Values for the Parts
Tabulate the terms needed to apply Eq. $(7.17)_1$.

	$\bar{x}_i$	V_i	$\bar{x}_i V_i$
Part 1 (cone)	$\dfrac{3}{4}h$	$\dfrac{1}{3}\pi R^2 h$	$\left(\dfrac{3}{4}h\right)\left(\dfrac{1}{3}\pi R^2 h\right)$
Part 2 (cylinder)	$h + \dfrac{1}{2}b$	$\pi R^2 b$	$\left(h + \dfrac{1}{2}b\right)(\pi R^2 b)$

$$\bar{x} = \frac{\bar{x}_1 V_1 + \bar{x}_2 V_2}{V_1 + V_2}.$$

$$= \frac{\left(\dfrac{3}{4}h\right)\left(\dfrac{1}{3}\pi R^2 h\right) + \left(h + \dfrac{1}{2}b\right)\left(\pi R^2 b\right)}{\dfrac{1}{3}\pi R^2 h + \pi R^2 b}.$$

Calculate the Centroid
Use Eq. $(7.17)_1$ to determine the x component of the centroid.

Practice Problem The composite volume consists of a circular cylinder and a hemisphere. Determine the x coordinate of its centroid.

Answer: $\bar{x} = \dfrac{\left(\frac{1}{2}b\right)\left(\pi R^2 b\right) + \left(b + \frac{3}{8}R\right)\left(\frac{2}{3}\pi R^3\right)}{\pi R^2 b + \frac{2}{3}\pi R^3}.$

| Example 7.12 | Centroid of a Volume Containing a Cutout (▶ *Related Problem 7.72*) |

Determine the centroid of the volume.

Strategy
We can divide this volume into the five simple parts shown in Fig. a. Notice that parts 2 and 3 *do not* have the cutout. It is assumed to be "filled in," which simplifies the geometries of those parts. Part 5, which is the volume of the 20-mm-diameter hole, will be treated as a negative volume in Eqs. (7.17).

Solution

Choose the Parts We can divide the volume into the five simple parts shown in Fig. a. Part 5 is the volume of the 20-mm-diameter hole.

Determine the Values for the Parts The centroids of parts 1 and 3 are located at the centroids of their semicircular cross sections (Fig. b). The information for determining the x-coordinate of the centroid is summarized in the table. Part 5 is a negative volume.

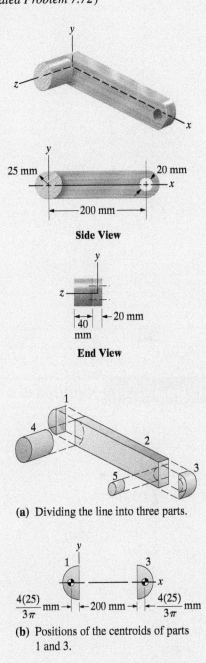

Side View

End View

(a) Dividing the line into three parts.

(b) Positions of the centroids of parts 1 and 3.

Information for determining $\bar{x}$

	$\bar{x}_i$ (mm)	V_i (mm^3)	$\bar{x}_i V_i$ (mm^4)
Part 1	$-\dfrac{4(25)}{3\pi}$	$\dfrac{\pi(25)^2}{2}(20)$	$\left[-\dfrac{4(25)}{3\pi}\right]\left[\dfrac{\pi(25)^2}{2}(20)\right]$
Part 2	100	$(200)(50)(20)$	$(100)[(200)(50)(20)]$
Part 3	$200 + \dfrac{4(25)}{3\pi}$	$\dfrac{\pi(25)^2}{2}(20)$	$\left[200 + \dfrac{4(25)}{3\pi}\right]\left[\dfrac{\pi(25)^2}{2}(20)\right]$
Part 4	0	$\pi(25)^2(40)$	0
Part 5	200	$-\pi(10)^2(20)$	$-(200[\pi(10)^2(20)]$

Calculate the Centroid The x coordinate of the centroid of the composite volume is

$$\bar{x} = \frac{\bar{x}_1 V_1 + \bar{x}_2 V_2 + \bar{x}_3 V_3 + \bar{x}_4 V_4 + \bar{x}_5 V_5}{V_1 + V_2 + V_3 + V_4 + V_5}$$

$$= \frac{\left[-\dfrac{4(25)}{3\pi}\right]\left[\dfrac{\pi(25)^2}{2}(20)\right] + (100)[(200)(50)(20)] + \left[200 + \dfrac{4(25)}{3\pi}\right]\left[\dfrac{\pi(25)^2}{2}(20)\right] + 0 - (200)[\pi(10)^2(20)]}{\dfrac{\pi(25)^2}{2}(20) + (200)(50)(20) + \dfrac{\pi(25)^2}{2}(20) + \pi(25)^2(40) - \pi(10)^2(20)}$$

$$= 72.77 \text{ mm}.$$

The z coordinates of the centroids of the parts are zero except $\bar{z}_4 = 30$ mm. Therefore the z coordinate of the centroid of the composite volume is

$$\bar{z} = \frac{\bar{z}_4 V_4}{V_1 + V_2 + V_3 + V_4 + V_5}$$

$$= \frac{30[\pi(25)^2(40)]}{\dfrac{\pi(25)^2}{2}(20) + (200)(50)(20) + \dfrac{\pi(25)^2}{2}(20) + \pi(25)^2(40) - \pi(10)^2(20)}$$

$$= 7.56 \text{ mm}.$$

Because of symmetry, $\bar{y} = 0$.

Critical Thinking

You can recognize that the volume in this example could be part of a mechanical device. Many manufactured parts have volumes that are composites of simple volumes, and the method used in this example can be used to determine their centroids and, if they are homogeneous, their centers of mass.

Example 7.13   Centroid of a Composite Line (▶ *Related Problem 7.81*)

Determine the centroid of the line. The quarter-circular arc lies in the y–z plane.

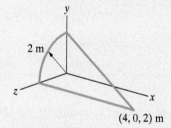

2 m

$(4, 0, 2)$ m

Strategy

We must divide the line into parts (in this case the quarter-circular arc and the two straight segments), determine the centroids of the parts, and apply Eqs. (7.18).

Solution

Choose the Parts   The line consists of a quarter-circular arc and two straight segments, which we call parts 1, 2, and 3 (Fig. a).

Determine the Values for the Parts   From Appendix B, the coordinates of the centroid of the quarter-circular arc are $\bar{x}_1 = 0, \bar{y}_1 = \bar{z}_1 = 2(2)/\pi$ m. The centroids of the straight segments lie at their midpoints. For segment 2, $\bar{x}_2 = 2$ m, $\bar{y}_2 = 0$, and $\bar{z}_2 = 2$ m, and for segment 3, $\bar{x}_3 = 2$ m, $\bar{y}_3 = 1$ m, and $\bar{z}_3 = 1$ m. The length of segment 3 is $L_3 = \sqrt{(4)^2 + (2)^2 + (2)^2} = 4.90$ m. This information is summarized in the table.

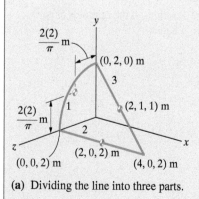

$\dfrac{2(2)}{\pi}$ m

$(0, 2, 0)$ m

3

$\dfrac{2(2)}{\pi}$ m

$(2, 1, 1)$ m

2

$(2, 0, 2)$ m

$(0, 0, 2)$ m $(4, 0, 2)$ m

(a) Dividing the line into three parts.

Information for determining the centroid.

	$\bar{x}_i$ (m)	$\bar{y}_i$ (m)	$\bar{z}_i$ (m)	L_i (m)
Part 1	0	$2(2)/\pi$	$2(2)/\pi$	$\pi(2)/2$
Part 2	2	0	2	4
Part 3	2	1	1	4.90

Calculate the Centroid   The coordinates of the centroid of the composite line are

$$\bar{x} = \frac{\bar{x}_1 L_1 + \bar{x}_2 L_2 + \bar{x}_3 L_3}{L_1 + L_2 + L_3} = \frac{0 + (2)(4) + (2)(4.90)}{\pi + 4 + 4.90} = 1.478 \text{ m},$$

$$\bar{y} = \frac{\bar{y}_1 L_1 + \bar{y}_2 L_2 + \bar{y}_3 L_3}{L_1 + L_2 + L_3} = \frac{[2(2)/\pi][\pi(2)/2] + 0 + (1)(4.90)}{\pi + 4 + 4.90} = 0.739 \text{ m},$$

$$\bar{z} = \frac{\bar{z}_1 L_1 + \bar{z}_2 L_2 + \bar{z}_3 L_3}{L_1 + L_2 + L_3} = \frac{[2(2)/\pi][\pi(2)/2] + (2)(4) + (1)(4.90)}{\pi + 4 + 4.90} = 1.404 \text{ m}.$$

Critical Thinking

What possible reason could you have for wanting to know the centroid (average position) of a line? In Section 7.7 we show that the center of mass of a slender homogeneous bar, which is the point at which the weight of the bar can be represented by an equivalent force, lies approximately at the centroid of the bar's axis.

Problems

▶ 7.71 In Active Example 7.11, suppose that the cylinder is hollow with inner radius $R/2$ as shown. If the dimensions $R = 6$ in, $h = 12$ in, and $b = 10$ in, what is the x coordinate of the centroid of the volume?

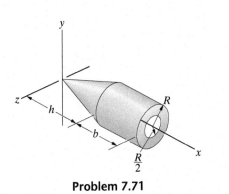

Problem 7.71

▶ 7.72 Use the procedure described in Example 7.12 to determine the x component of the centroid of the volume.

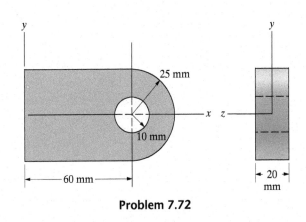

Problem 7.72

For Problems 7.73–7.78, determine the centroids of the volumes.

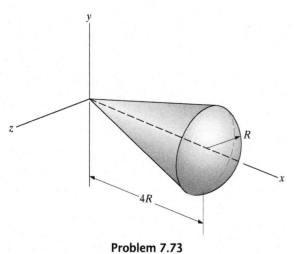

Problem 7.73

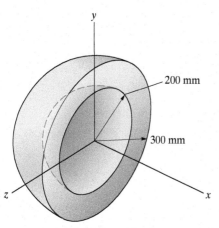

Problem 7.74

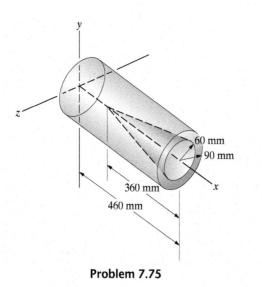

Problem 7.75

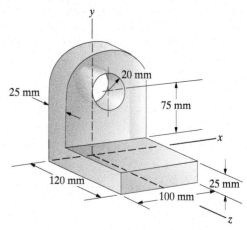

Problem 7.76

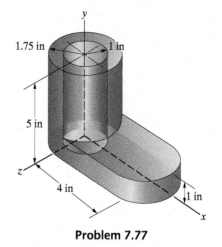

Problem 7.77

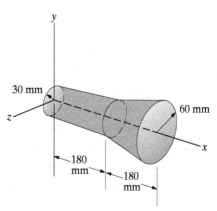

Problem 7.78

7.79 The dimensions of the *Gemini* spacecraft (in meters) were $a = 0.70$, $b = 0.88$, $c = 0.74$, $d = 0.98$, $e = 1.82$, $f = 2.20$, $g = 2.24$, and $h = 2.98$. Determine the centroid of its volume.

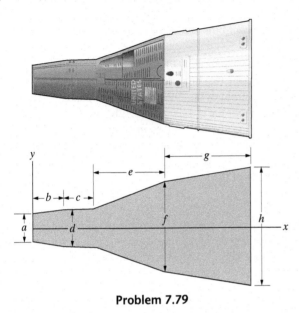

Problem 7.79

7.80 Two views of a machine element are shown. Determine the centroid of its volume.

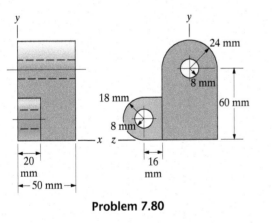

Problem 7.80

▶ **7.81** In Example 7.13, suppose that the circular arc is replaced by a straight line as shown. Determine the centroid of the three-segment line.

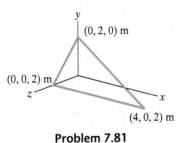

Problem 7.81

For Problems 7.82 and 7.83, determine the centroids of the lines.

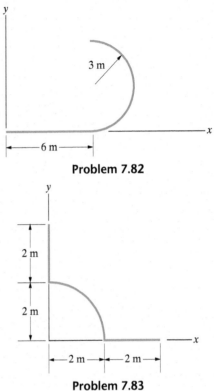

Problem 7.82

Problem 7.83

7.84 The semicircular part of the line lies in the x–z plane. Determine the centroid of the line.

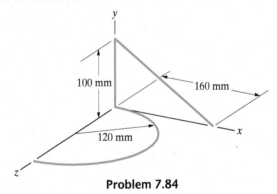

Problem 7.84

7.85 Determine the centroid of the line.

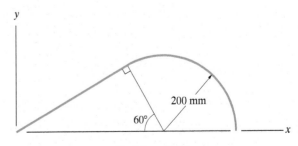

Problem 7.85

7.6 The Pappus–Guldinus Theorems

In this section we discuss two simple and useful theorems relating surfaces and volumes of revolution to the centroids of the lines and areas that generate them.

First Theorem

Consider a line L in the x–y plane that does not intersect the x axis (Fig. 7.12a). Let the coordinates of the centroid of the line be $(\bar{x}, \bar{y})$. We can generate a surface by revolving the line about the x axis (Fig. 7.12b). As the line revolves about the x axis, the centroid of the line moves in a circular path of radius $\bar{y}$.

The first Pappus–Guldinus theorem states that the area of the surface of revolution is equal to the product of the distance through which the centroid of the line moves and the length of the line:

$$A = 2\pi\bar{y}\,L. \qquad (7.19)$$

To prove this result, we observe that as the line revolves about the x axis, the area dA generated by an element dL of the line is $dA = 2\pi y\,dL$, where y is the y coordinate of the element dL (Fig. 7.12c). Therefore, the total area of the surface of revolution is

$$A = 2\pi \int_L y\,dL. \qquad (7.20)$$

From the definition of the y coordinate of the centroid of the line,

$$\bar{y} = \frac{\displaystyle\int_L y\,dL}{\displaystyle\int_L dL},$$

we obtain

$$\int_L y\,dL = \bar{y}L.$$

Substituting this result into Eq. (7.20), we obtain Eq. (7.19).

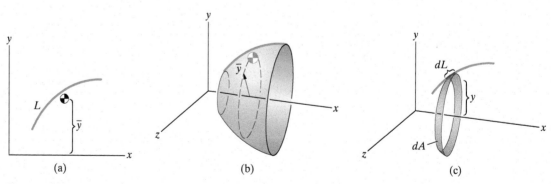

Figure 7.12
(a) A line L and the y coordinate of its centroid.
(b) The surface generated by revolving the line L about the x axis and the path followed by the centroid of the line.
(c) An element dL of the line and the element of area dA it generates.

Second Theorem

Consider an area A in the x–y plane that does not intersect the x axis (Fig. 7.13a). Let the coordinates of the centroid of the area be $(\bar{x}, \bar{y})$. We can generate a volume by revolving the area about the x axis (Fig. 7.13b). As the area revolves about the x axis, the centroid of the area moves in a circular path of length $2\pi\bar{y}$.

The second Pappus–Guldinus theorem states that the volume V of the volume of revolution is equal to the product of the distance through which the centroid of the area moves and the area:

$$V = 2\pi\bar{y}A. \tag{7.21}$$

As the area revolves about the x axis, the volume dV generated by an element dA of the area is $dV = 2\pi y \, dA$, where y is the y coordinate of the element dA (Fig. 7.13c). Therefore, the total volume is

$$V = 2\pi \int_A y \, dA. \tag{7.22}$$

From the definition of the y coordinate of the centroid of the area,

$$\bar{y} = \frac{\displaystyle\int_A y \, dA}{\displaystyle\int_A dA},$$

we obtain

$$\int_A y \, dA = \bar{y}A.$$

Substituting this result into Eq. (7.22), we obtain Eq. (7.21).

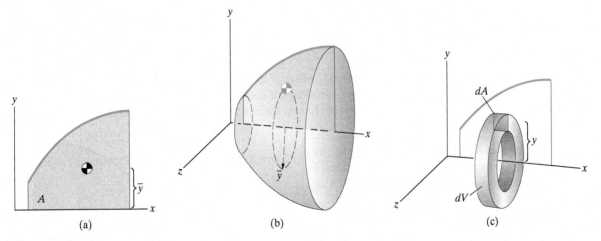

(a)   (b)   (c)

Figure 7.13
(a) An area A and the y coordinate of its centroid.
(b) The volume generated by revolving the area A about the x axis and the path followed by the centroid of the area.
(c) An element dA of the area and the element of volume dV it generates.

Problems

▶ **7.86** Use the method described in Active Example 7.14 to determine the area of the curved part of the surface of the truncated cone.

7.87 Use the second Pappus–Guldinus theorem to determine the volume of the truncated cone.

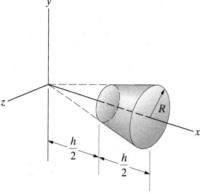

Problems 7.86/7.87

▶ **7.88** The area of the shaded semicircle is $\frac{1}{2}\pi R^2$. The volume of a sphere is $\frac{4}{3}\pi R^3$. Extend the approach described in Example 7.15 to the second Pappus–Guldinus theorem and determine the centroid $\bar{y}_S$ of the semicircular area.

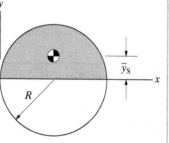

Problem 7.88

7.89 Use the second Pappus–Guldinus theorem to determine the volume generated by revolving the curve about the y axis.

7.90 The length of the curve is $L = 1.479$, and the area generated by rotating it about the x axis is $A = 3.810$. Use the first Pappus–Guldinus theorem to determine the y coordinate of the centroid of the curve.

7.91 Use the first Pappus–Guldinus theorem to determine the area of the surface generated by revolving the curve about the y axis.

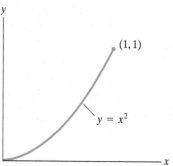

Problems 7.89–7.91

7.92 A nozzle for a large rocket engine is designed by revolving the function $y = \frac{2}{3}(x - 1)^{3/2}$ about the y axis. Use the first Pappus–Guldinus theorem to determine the surface area of the nozzle.

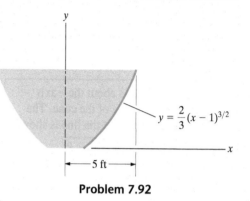

Problem 7.92

7.93 The coordinates of the centroid of the line are $\bar{x} = 332$ mm and $\bar{y} = 118$ mm. Use the first Pappus–Guldinus theorem to determine the area of the surface of revolution obtained by revolving the line about the x axis.

7.94 The coordinates of the centroid of the area between the x axis and the line are $\bar{x} = 355$ mm and $\bar{y} = 78.4$ mm. Use the second Pappus–Guldinus theorem to determine the volume obtained by revolving the area about the x axis.

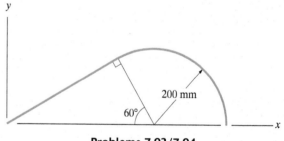

Problems 7.93/7.94

7.95 The volume of revolution contains a hole of radius R.

(a) Use integration to determine its volume.

(b) Use the second Pappus–Guldinus theorem to determine its volume.

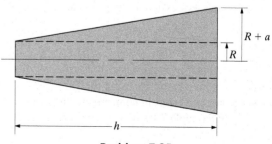

Problem 7.95

7.96 Determine the volume of the volume of revolution.

7.97 Determine the surface area of the volume of revolution.

140 mm

80 mm

Problems 7.96/7.97

7.98 The volume of revolution has an elliptical cross section. Determine its volume.

230 mm

130 mm

180 mm

Problem 7.98

7.7 Centers of Mass of Objects

BACKGROUND

The *center of mass* of an object is the centroid, or average position, of its mass. Here we give the analytical definition of the center of mass and demonstrate one of its most important properties: *An object's weight can be represented by a single equivalent force acting at its center of mass.* We then discuss how to locate centers of mass and show that for particular classes of objects, the center of mass coincides with the centroid of a volume, area, or line.

The center of mass of an object is defined by

$$\bar{x} = \frac{\int_m x \, dm}{\int_m dm}, \qquad \bar{y} = \frac{\int_m y \, dm}{\int_m dm}, \qquad \bar{z} = \frac{\int_m z \, dm}{\int_m dm}, \qquad (7.23)$$

where x, y, and z are the coordinates of the differential element of mass dm (Fig. 7.14). The subscripts m indicate that the integration must be carried out over the entire mass of the object.

Before considering how to determine the center of mass of an object, we will demonstrate that the weight of an object can be represented by a single equivalent force acting at its center of mass. Consider an element of mass dm of an object (Fig. 7.15a). If the y axis of the coordinate system points upward, the weight of dm is $-dm g \, \mathbf{j}$. Integrating this expression over the mass m, we obtain the total weight of the object,

$$\int_m - g\mathbf{j} \, dm = -mg\mathbf{j} = -W\mathbf{j}.$$

The moment of the weight of the element dm about the origin is

$$(x\mathbf{i} + y\mathbf{j} + z\mathbf{k}) \times (-dm g \, \mathbf{j}) = gz\mathbf{i} \, dm - gx\mathbf{k} \, dm.$$

Integrating this expression over m, we obtain the total moment about the origin due to the weight of the object:

$$\int_m (gz\mathbf{i} \, dm - gx\mathbf{k} \, dm) = mg\bar{z}\mathbf{i} - mg\bar{x}\mathbf{k} = W\bar{z}\mathbf{i} - W\bar{x}\mathbf{k}.$$

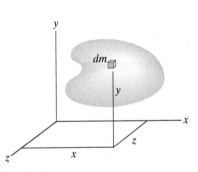

Figure 7.14
An object and differential element of mass dm.

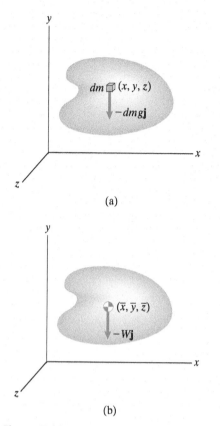

(a)

(b)

Figure 7.15
(a) Weight of the element dm.
(b) Representing the weight by a single force at the center of mass.

If we represent the weight of the object by the force $-W\mathbf{j}$ acting at the center of mass (Fig. 7.15b), the moment of this force about the origin is equal to the total moment due to the weight:

$$(\bar{x}\mathbf{i} + \bar{y}\mathbf{j} + \bar{z}\mathbf{k}) \times (-W\mathbf{j}) = W\bar{z}\mathbf{i} - W\bar{x}\mathbf{k}.$$

This result shows that when we are concerned only with the total force and total moment exerted by the weight of an object, we can assume that its weight acts at the center of mass.

To apply Eqs. (7.23) to specific objects, we will change the variable of integration from mass to volume by introducing the *density*. The density ρ of an object is defined such that the mass of a differential element dV of the volume of the object is $dm = \rho \, dV$. The dimensions of ρ are therefore (mass/volume). For example, it can be expressed in kg/m^3 in SI units or in slug/ft^3 in U.S. Customary units. The total mass of an object is

$$m = \int_m dm = \int_V \rho \, dV. \tag{7.24}$$

An object whose density is uniform throughout its volume is said to be *homogeneous*. In this case, the total mass equals the product of the density and the volume:

$$m = \rho \int_V dV = \rho V. \quad \text{Homogeneous object} \tag{7.25}$$

The *weight density* is defined by $\gamma = g\rho$. It can be expressed in N/m^3 in SI units or in lb/ft^3 in U.S. Customary units. The weight of an element of volume dV of an object is $dW = \gamma \, dV$, and the total weight of a homogeneous object equals γV.

By substituting $dm = \rho \, dV$ into Eq. (7.23), we can express the coordinates of the center of mass in terms of volume integrals:

$$\bar{x} = \frac{\int_V \rho x \, dV}{\int_V \rho \, dV}, \qquad \bar{y} = \frac{\int_V \rho y \, dV}{\int_V \rho \, dV}, \qquad \bar{z} = \frac{\int_V \rho z \, dV}{\int_V \rho \, dV}. \tag{7.26}$$

If ρ is known as a function of position in an object, these expressions determine its center of mass. Furthermore, we can use these expressions to show that the centers of mass of particular classes of objects coincide with centroids of volumes, areas, and lines:

- **The center of mass of a homogeneous object coincides with the centroid of its volume.** If an object is homogeneous, $\rho = $ constant and Eqs. (7.26) become the equations for the centroid of the volume,

$$\bar{x} = \frac{\int_V x \, dV}{\int_V dV}, \qquad \bar{y} = \frac{\int_V y \, dV}{\int_V dV}, \qquad \bar{z} = \frac{\int_V z \, dV}{\int_V dV}.$$

- **The center of mass of a homogeneous plate of uniform thickness coincides with the centroid of its cross-sectional area** (Fig. 7.16). The center of mass of the plate coincides with the centroid of its volume, and we showed in Section 7.4 that the centroid of the volume of a plate of uniform thickness coincides with the centroid of its cross-sectional area.

- **The center of mass of a homogeneous slender bar of uniform cross-sectional area coincides approximately with the centroid of the axis of**

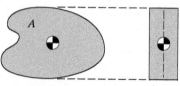

Front View   Side View

Figure 7.16
A plate of uniform thickness.

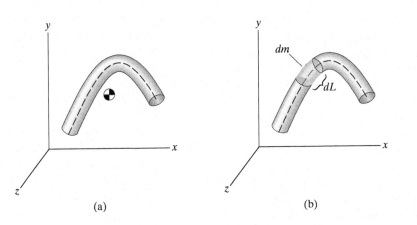

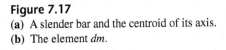

Figure 7.17
(a) A slender bar and the centroid of its axis.
(b) The element dm.

the bar (Fig. 7.17a). The axis of the bar is defined to be the line through the centroid of its cross section. Let $dm = \rho A\, dL$, where A is the cross-sectional area of the bar and dL is a differential element of length of its axis (Fig. 7.17b). If we substitute this expression into Eqs. (7.26), they become the equations for the centroid of the axis:

$$\bar{x} = \frac{\displaystyle\int_L x\, dL}{\displaystyle\int_L dL}, \qquad \bar{y} = \frac{\displaystyle\int_L y\, dL}{\displaystyle\int_L dL}, \qquad \bar{z} = \frac{\displaystyle\int_L z\, dL}{\displaystyle\int_L dL}.$$

This result is approximate because the center of mass of the element dm does not coincide with the centroid of the cross section in regions where the bar is curved.

RESULTS

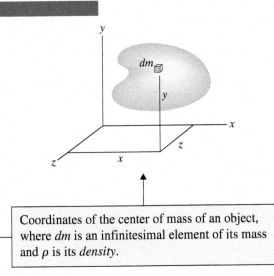

$$\bar{x} = \frac{\displaystyle\int_m x\, dm}{\displaystyle\int_m dm} = \frac{\displaystyle\int_V \rho x\, dV}{\displaystyle\int_i \rho\, dV},$$

$$\bar{y} = \frac{\displaystyle\int_m y\, dm}{\displaystyle\int_m dm} = \frac{\displaystyle\int_V \rho y\, dV}{\displaystyle\int_V \rho\, dV}, \qquad (7.23),\ (7.26).$$

$$\bar{z} = \frac{\displaystyle\int_m z\, dm}{\displaystyle\int_m dm} = \frac{\displaystyle\int_V \rho z\, dV}{\displaystyle\int_V \rho\, dV}.$$

Coordinates of the center of mass of an object, where dm is an infinitesimal element of its mass and ρ is its *density*.

An object is *homogeneous* if its density ρ is constant, or uniform. *The center of mass of a homogeneous object coincides with the centroid of its volume.*

The center of mass of a homogeneous plate of uniform thickness coincides with the centroid of its cross-sectional area.

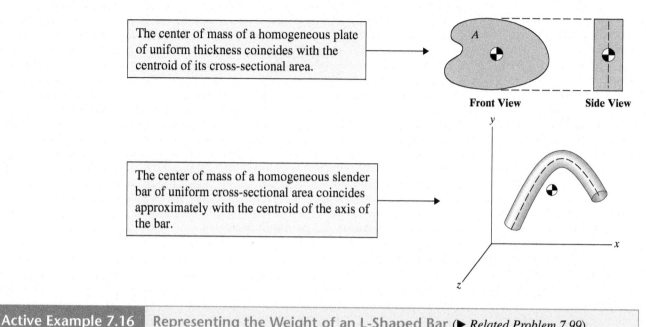

Front View   **Side View**

The center of mass of a homogeneous slender bar of uniform cross-sectional area coincides approximately with the centroid of the axis of the bar.

Active Example 7.16   **Representing the Weight of an L-Shaped Bar** (▶ *Related Problem 7.99*)

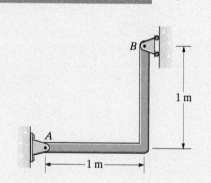

The mass of the homogeneous slender bar is 80 kg. What are the reactions at A and B?

Strategy

We can determine the reactions in two ways.

First Method   By representing the weight of each straight segment of the bar by a force acting at the center of mass of the segment.

Second Method   By determining the center of mass of the entire bar, which is located at the centroid of its axis, and representing the weight of the entire bar by a force acting at its center of mass.

Solution

First Method

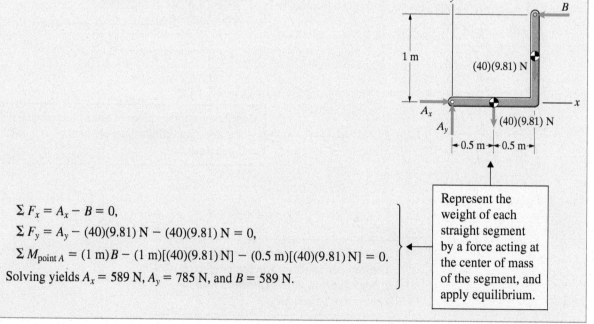

Represent the weight of each straight segment by a force acting at the center of mass of the segment, and apply equilibrium.

$$\Sigma F_x = A_x - B = 0,$$

$$\Sigma F_y = A_y - (40)(9.81) \text{ N} - (40)(9.81) \text{ N} = 0,$$

$$\Sigma M_{\text{point } A} = (1 \text{ m})B - (1 \text{ m})[(40)(9.81) \text{ N}] - (0.5 \text{ m})[(40)(9.81) \text{ N}] = 0.$$

Solving yields $A_x = 589$ N, $A_y = 785$ N, and $B = 589$ N.

Second Method

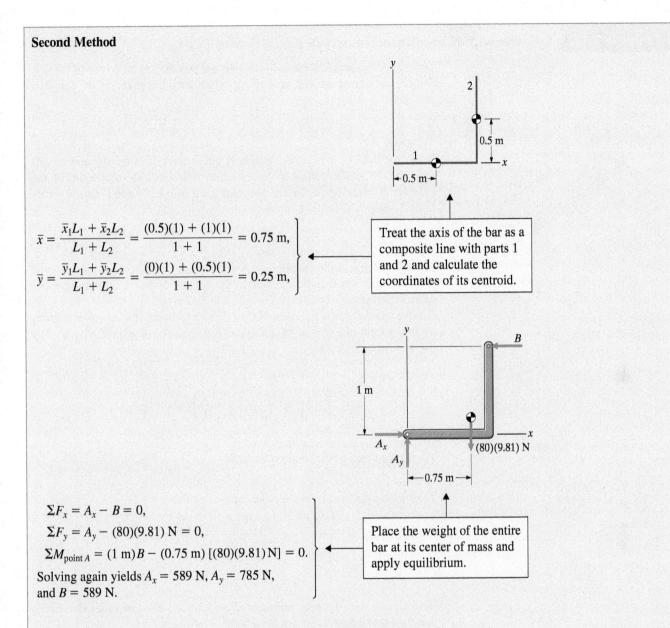

$$\bar{x} = \frac{\bar{x}_1 L_1 + \bar{x}_2 L_2}{L_1 + L_2} = \frac{(0.5)(1) + (1)(1)}{1 + 1} = 0.75 \text{ m,}$$

$$\bar{y} = \frac{\bar{y}_1 L_1 + \bar{y}_2 L_2}{L_1 + L_2} = \frac{(0)(1) + (0.5)(1)}{1 + 1} = 0.25 \text{ m,}$$

Treat the axis of the bar as a composite line with parts 1 and 2 and calculate the coordinates of its centroid.

$$\Sigma F_x = A_x - B = 0,$$

$$\Sigma F_y = A_y - (80)(9.81) \text{ N} = 0,$$

$$\Sigma M_{\text{point } A} = (1 \text{ m})B - (0.75 \text{ m}) [(80)(9.81) \text{ N}] = 0.$$

Solving again yields $A_x = 589$ N, $A_y = 785$ N, and $B = 589$ N.

Place the weight of the entire bar at its center of mass and apply equilibrium.

Practice Problem The mass of the homogeneous circular bar is 80 kg. What are the reactions at A and B?

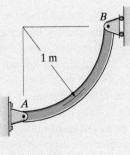

Answer: $A_x = 500$ N, $A_y = 785$ N, $B = 500$ N.

Example 7.17   Cylinder with Nonuniform Density (▶ *Related Problem 7.105*)

Determine the mass of the cylinder and the position of its center of mass if (a) it is homogeneous with density ρ_0; (b) its density is given by the equation $\rho = \rho_0(1 + x/L)$.

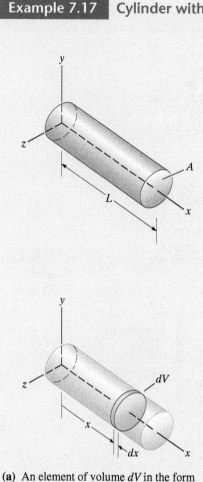

(a) An element of volume dV in the form of a disk.

Strategy

In (a), the mass of the cylinder is simply the product of its density and its volume and the center of mass is located at the centroid of its volume. In (b), the cylinder is inhomogeneous and we must use Eqs. (7.24) and (7.26) to determine its mass and center of mass.

Solution

(a) The volume of the cylinder is LA, so its mass is $\rho_0 LA$. Since the center of mass is coincident with the centroid of the volume of the cylinder, the coordinates of the center of mass are $\bar{x} = \frac{1}{2}L, \bar{y} = 0, \bar{z} = 0$.

(b) We can determine the mass of the cylinder by using an element of volume dV in the form of a disk of thickness dx (Fig. a). The volume $dV = A\,dx$. The mass of the cylinder is

$$m = \int_v \rho\,dV = \int_0^L \rho_0\left(1 + \frac{x}{L}\right)A\,dx = \frac{3}{2}\rho_0 AL.$$

The x coordinate of the center of mass is

$$\bar{x} = \frac{\displaystyle\int_v x\rho\,dV}{\displaystyle\int_v \rho\,dV} = \frac{\displaystyle\int_0^L \rho_0\left(x + \frac{x^2}{L}\right)A\,dx}{\dfrac{3}{2}\rho_0 AL} = \frac{5}{9}L.$$

Because the density does not depend on y or z, we know from symmetry that $\bar{y} = 0$ and $\bar{z} = 0$.

Critical Thinking

Notice that the center of mass of the inhomogeneous cylinder is *not* located at the centroid of its volume. Its density increases from left to right, so the center of mass is located to the right of the midpoint of the cylinder. Many of the objects we deal with in engineering are not homogeneous, but it is not common for an object's density to vary continuously through its volume as in this example. More often, objects consist of assemblies of parts (composites) that have different densities because they consist of different materials. Frequently the individual parts are approximately homogeneous. We discuss the determination of the centers of mass of such composite objects in the next section.

Problems

▶ **7.99** Suppose that the bar in Active Example 7.16 is replaced with this 100-kg homogeneous bar. (a) What is the x coordinate of the bar's center of mass? (b) Determine the reactions at A and B.

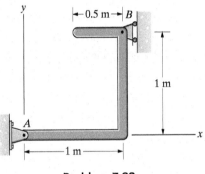

Problem 7.99

7.100 The mass of the homogeneous flat plate is 50 kg. Determine the reactions at the supports A and B.

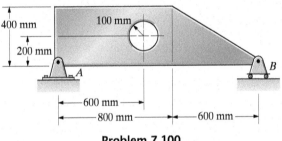

Problem 7.100

7.101 The suspended sign is a homogeneous flat plate that has a mass of 130 kg. Determine the axial forces in members AD and CE. (Notice that the y axis is positive downward.)

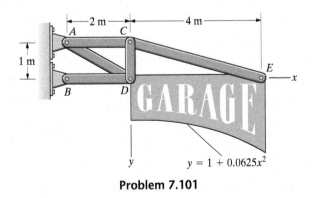

$$y = 1 + 0.0625x^2$$

Problem 7.101

7.102 The bar has a mass of 80 kg. What are the reactions at A and B?

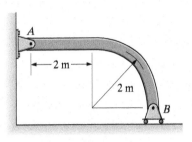

Problem 7.102

7.103 The mass of the bar per unit length is 2 kg/m. Choose the dimension b so that part BC of the suspended bar is horizontal. What is the dimension b, and what are the resulting reactions on the bar at A?

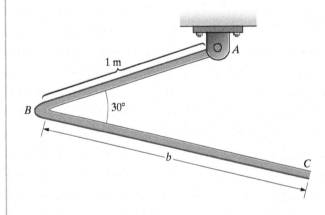

Problem 7.103

7.104 The semicircular part of the homogeneous slender bar lies in the x–z plane. Determine the center of mass of the bar.

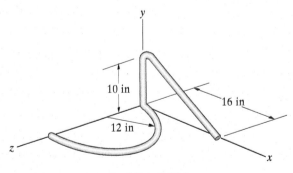

Problem 7.104

▶ 7.105 The density of the cone is given by the equation $\rho = \rho_0(1 + x/h)$, where ρ_0 is a constant. Use the procedure described in Example 7.17 to show that the mass of the cone is given by $m = (7/4)\rho_0 V$, where V is the volume of the cone, and that the x coordinate of the center of mass of the cone is $\bar{x} = (27/35)h$.

7.106 A horizontal cone with 800-mm length and 200-mm radius has a fixed support at A. Its density is $\rho = 6000(1 + 0.4x^2)$ kg/m³, where x is in meters. What are the reactions at A?

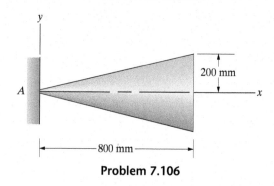

Problem 7.106

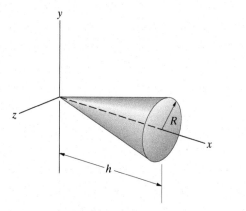

Problem 7.105

7.8 Centers of Mass of Composite Objects

BACKGROUND

The center of mass of an object consisting of a combination of parts can be determined if the centers of mass of its parts are known. The coordinates of the center of mass of a composite object composed of parts with masses $m_1, m_2, \ldots$, are

$$\bar{x} = \frac{\sum\limits_i \bar{x}_i m_i}{\sum\limits_i m_i}, \qquad \bar{y} = \frac{\sum\limits_i \bar{y}_i m_i}{\sum\limits_i m_i}, \qquad \bar{z} = \frac{\sum\limits_i \bar{z}_i m_i}{\sum\limits_i m_i}, \qquad (7.27)$$

where $\bar{x}_i, \bar{y}_i, \bar{z}_i$ are the coordinates of the centers of mass of the parts. Because the weights of the parts are related to their masses by $W_i = gm_i$, Eqs. (7.27) can also be expressed as

$$\bar{x} = \frac{\sum\limits_i \bar{x}_i W_i}{\sum\limits_i W_i}, \qquad \bar{y} = \frac{\sum\limits_i \bar{y}_i W_i}{\sum\limits_i W_i}, \qquad \bar{z} = \frac{\sum\limits_i \bar{z}_i W_i}{\sum\limits_i W_i}. \qquad (7.28)$$

When the masses or weights and the centers of mass of the parts of a composite object are known, these equations determine its center of mass.

Determining the center of mass of a composite object requires three steps:

1. Choose the parts—Try to divide the object into parts whose centers of mass you know or can easily determine.

2. Determine the values for the parts—Determine the center of mass and the mass or weight of each part. Watch for instances of symmetry that can simplify your task.

3. Calculate the center of mass—Use Eqs. (7.27) or (7.28) to determine the center of mass of the composite object.

Active Example 7.18	Center of Mass of a Composite Object (▶ *Related Problem 7.107*)

The L-shaped machine part is composed to two homogeneous bars. Bar 1 is tungsten alloy with a density of 14,000 kg/m³. Bar 2 is steel with a density of 7800 kg/m³. Determine the x coordinate of the center of mass of the machine part.

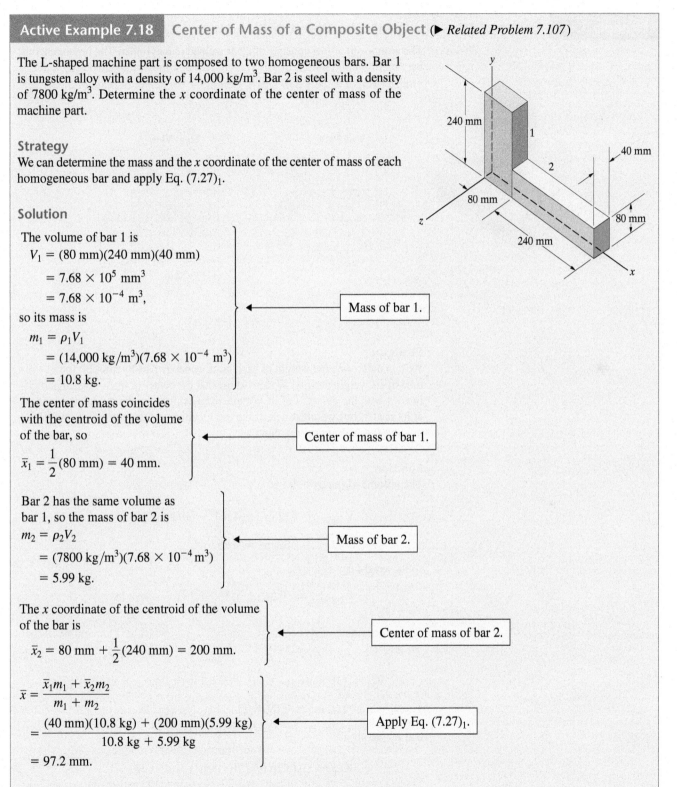

Strategy

We can determine the mass and the x coordinate of the center of mass of each homogeneous bar and apply Eq. $(7.27)_1$.

Solution

The volume of bar 1 is

$V_1 = (80 \text{ mm})(240 \text{ mm})(40 \text{ mm})$

$= 7.68 \times 10^5 \text{ mm}^3$

$= 7.68 \times 10^{-4} \text{ m}^3$,

so its mass is

$m_1 = \rho_1 V_1$

$= (14{,}000 \text{ kg/m}^3)(7.68 \times 10^{-4} \text{ m}^3)$

$= 10.8 \text{ kg}.$

⟵ Mass of bar 1.

The center of mass coincides with the centroid of the volume of the bar, so

$\bar{x}_1 = \dfrac{1}{2}(80 \text{ mm}) = 40 \text{ mm}.$

⟵ Center of mass of bar 1.

Bar 2 has the same volume as bar 1, so the mass of bar 2 is

$m_2 = \rho_2 V_2$

$= (7800 \text{ kg/m}^3)(7.68 \times 10^{-4} \text{ m}^3)$

$= 5.99 \text{ kg}.$

⟵ Mass of bar 2.

The x coordinate of the centroid of the volume of the bar is

$\bar{x}_2 = 80 \text{ mm} + \dfrac{1}{2}(240 \text{ mm}) = 200 \text{ mm}.$

⟵ Center of mass of bar 2.

$\bar{x} = \dfrac{\bar{x}_1 m_1 + \bar{x}_2 m_2}{m_1 + m_2}$

$= \dfrac{(40 \text{ mm})(10.8 \text{ kg}) + (200 \text{ mm})(5.99 \text{ kg})}{10.8 \text{ kg} + 5.99 \text{ kg}}$

⟵ Apply Eq. $(7.27)_1$.

$= 97.2 \text{ mm}.$

Practice Problem Determine the y coordinate of the center of mass of the L-shaped machine part.

Answer: $\bar{y} = 91.4 \text{ mm}.$

| Example 7.19 | Center of Mass of a Composite Object (▶ *Related Problem 7.109*) |

The composite object consists of a bar welded to a cylinder. The homogeneous bar is aluminum (weight density 168 lb/ft³), and the homogeneous cylinder is bronze (weight density 530 lb/ft³). Determine the center of mass of the object.

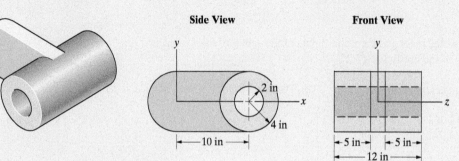

Strategy

We can determine the weight of each homogeneous part by multiplying its volume by its weight density. We also know that the center of mass of each part coincides with the centroid of its volume. The centroid of the cylinder is located at its center, but we must determine the location of the centroid of the bar by treating it as a composite volume.

Solution

The volume of the cylinder is

$$V_{\text{cylinder}} = (12 \text{ in})[\pi(4 \text{ in})^2 - \pi(2 \text{ in})^2]$$

$$= 452 \text{ in}^3 = 0.262 \text{ ft}^3,$$

so its weight is

$$W_{\text{cylinder}} = (0.262 \text{ ft}^3)(530 \text{ lb/ft}^3) = 138.8 \text{ lb}.$$

The x coordinate of its center of mass is $\bar{x}_{\text{cylinder}} = 10$ in. The volume of the bar is

$$V_{\text{bar}} = (10 \text{ in})(8 \text{ in})(2 \text{ in}) + \tfrac{1}{2}\pi(4 \text{ in})^2(2 \text{ in}) - \tfrac{1}{2}\pi(4 \text{ in})^2(2 \text{ in})$$

$$= 160 \text{ in}^3 = 0.0926 \text{ ft}^3,$$

and its weight is

$$W_{\text{bar}} = (0.0926 \text{ ft}^3)(168 \text{ lb/ft}^3) = 15.6 \text{ lb}.$$

We can determine the centroid of the volume of the bar by treating it as a composite volume consisting of three parts (Fig. a). Part 3 is a semicircular "cutout." The centroids of part 1 and the semicircular cutout 3 are located at the

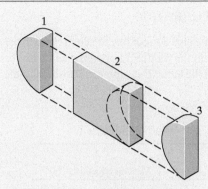

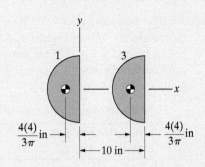

(a) Dividing the bar into three parts.   **(b)** The centroids of the two semicircular parts.

centroids of their semicircular cross sections (Fig b). Using the information summarized in the table, we have

$$\bar{x}_{bar} = \frac{\bar{x}_1 V_1 + \bar{x}_2 V_2 + \bar{x}_3 V_3}{V_1 + V_2 + V_3}$$

$$= \frac{-\dfrac{4(4)}{3\pi}\left[\frac{1}{2}\pi(4)^2(2)\right] + 5\left[(10)(8)(2)\right] - \left[10 - \dfrac{4(4)}{3\pi}\right]\left[\frac{1}{2}\pi(4)^2(2)\right]}{\frac{1}{2}\pi(4)^2(2) + (10)(8)(2) - \frac{1}{2}\pi(4)^2(2)}$$

$$= 1.86 \text{ in.}$$

Information for determining the x coordinate of the centroid of the bar

	$\bar{x}_i$ (in)	V_i (in^3)	$\bar{x}_i V_i$ (in^4)
Part 1	$-\dfrac{4(4)}{3\pi}$	$\frac{1}{2}\pi(4)^2(2)$	$-\dfrac{4(4)}{3\pi}\left[\frac{1}{2}\pi(4)^2(2)\right]$
Part 2	5	$(10)(8)(2)$	$5[(10)(8)(2)]$
Part 3	$10 - \dfrac{4(4)}{3\pi}$	$-\frac{1}{2}\pi(4)^2(2)$	$-\left[10 - \dfrac{4(4)}{3\pi}\right]\left[\frac{1}{2}\pi(4)^2(2)\right]$

Therefore, the x coordinate of the center of mass of the composite object is

$$\bar{x} = \frac{\bar{x}_{bar} W_{bar} + \bar{x}_{cylinder} W_{cylinder}}{W_{bar} + W_{cylinder}}$$

$$= \frac{(1.86 \text{ in})(15.6 \text{ lb}) + (10 \text{ in})(138.8 \text{ lb})}{15.6 \text{ lb} + 138.8 \text{ lb}}$$

$$= 9.18 \text{ in.}$$

Because of the symmetry of the bar, the y and z coordinates of its center of mass are $\bar{y} = 0$ and $\bar{z} = 0$.

Critical Thinking

The composite object in this example is not homogeneous, which means we could not assume that its center of mass coincides with the centroid of its volume. But the bar and the cylinder are each homogeneous, so we *could* determine their individual centers of mass by finding the centroids of their volumes. The primary challenge in this example was determining the centroid of the volume of the bar with its semicircular end and semicircular cutout.

Example 7.20	Centers of Mass of Vehicles (▶ *Related Problems 7.115, 7.116*)

A car is placed on a platform that measures the normal force exerted by each tire independently. Measurements made with the platform horizontal and with the platform tilted at $\alpha = 15°$ are shown in the table. Determine the position of the car's center of mass.

Measurements of the normal forces exerted by the tires

Wheelbase = 2.82 m			
Track = 1.55 m		**Measured Loads (N)**	
		$\alpha = 0$	$\alpha = 15°$
Left front wheel, N_{LF}		5104	4463
Right front wheel, N_{RF}		5027	4396
Left rear wheel, N_{LR}		3613	3956
Right rear wheel, N_{RR}		3559	3898

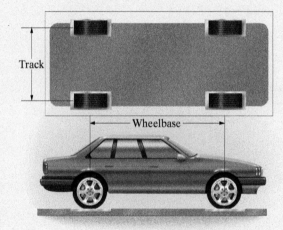

Strategy

The given measurements tell us the normal reactions exerted on the car's tires by the platform. By drawing free-body diagrams of the car in the two positions and applying equilibrium equations, we will obtain equations that can be solved for the unknown coordinates of the car's center of mass.

Solution

We draw the free-body diagram of the car when the platform is in the horizontal position in Figs. a and b. The car's weight is

$$W = N_{LF} + N_{RF} + N_{LR} + N_{RR}$$

$$= 5104 + 5027 + 3613 + 3559$$

$$= 17{,}303 \text{ N.}$$

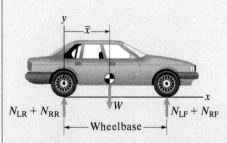

(a) Side view of the free-body diagram with the platform horizontal.

From Fig. a, we obtain the equilibrium equation

$$\Sigma M_{z\text{ axis}} = (\text{wheelbase})(N_{LF} + N_{RF}) - \bar{x}W = 0,$$

which we can solve for $\bar{x}$:

$$\bar{x} = \frac{(\text{wheelbase})(N_{\text{LF}} + N_{\text{RF}})}{W}$$

$$= \frac{(2.82 \text{ m})(5104 \text{ N} + 5027 \text{ N})}{17{,}303 \text{ N}}$$

$$= 1.651 \text{ m}.$$

From Fig. b,

$$\Sigma M_{x \text{ axis}} = \bar{z}W - (\text{track})(N_{\text{RF}} + N_{\text{RR}}) = 0,$$

which we can solve for $\bar{z}$:

$$\bar{z} = \frac{(\text{track})(N_{\text{RF}} + N_{\text{RR}})}{W}$$

$$= \frac{(1.55 \text{ m})(5027 \text{ N} + 3559 \text{ N})}{17{,}303 \text{ N}}$$

$$= 0.769 \text{ m}.$$

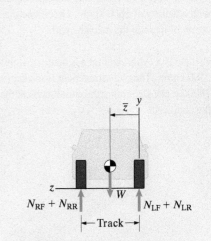

(b) Front view of the free-body diagram with the platform horizontal.

Now that we know $\bar{x}$, we can determine $\bar{y}$ from the free-body diagram of the car when the platform is in the tilted position (Fig. c). From the equilibrium equation

$$\Sigma M_{z \text{ axis}} = (\text{wheelbase})(N_{\text{LF}} + N_{\text{RF}}) + \bar{y}W \sin 15^\circ - \bar{x}W \cos 15^\circ$$

$$= 0,$$

we obtain

$$\bar{y} = \frac{\bar{x}W \cos 15^\circ - (\text{wheelbase})(N_{\text{LF}} + N_{\text{RF}})}{W \sin 15^\circ}$$

$$= \frac{(1.651 \text{ m})(17{,}303 \text{ N}) \cos 15^\circ - (2.82 \text{ m})(4463 \text{ N} + 4396 \text{ N})}{(17{,}303 \text{ N}) \sin 15^\circ}$$

$$= 0.584 \text{ m}.$$

Notice that we could not have determined $\bar{y}$ without the measurements made with the car in the tilted position.

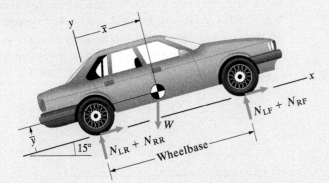

(c) Side view of the free-body diagram with the platform tilted.

Problems

▶ **7.107** In Active Example 7.18, suppose that bar 1 is replaced by a bar with the same dimensions that consists of aluminum alloy with a density of 2600 kg/m³. Determine the x coordinate of the center of mass of the machine part.

7.108 The cylindrical tube is made of aluminum with density 2700 kg/m³. The cylindrical plug is made of steel with density 7800 kg/m³. Determine the coordinates of the center of mass of the composite object.

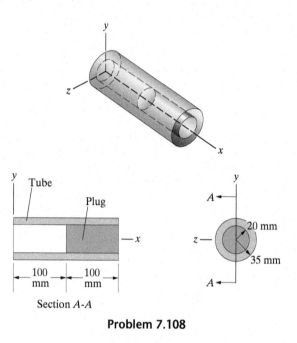

Section A-A

Problem 7.108

▶ **7.109** In Example 7.19, suppose that the object is redesigned so that the radius of the hole in the hollow cylinder is increased from 2 in to 3 in. What is the x coordinate of the center of mass of the object?

7.110 A machine consists of three parts. The masses and the locations of the centers of mass of two of the parts are

Part	Mass (kg)	$\bar{x}$ (mm)	$\bar{y}$ (mm)	$\bar{z}$ (mm)
1	2.0	100	50	−20
2	4.5	150	70	0

The mass of part 3 is 2.5 kg. The design engineer wants to position part 3 so that the center of mass location of the machine is $\bar{x} = 120$ mm, $\bar{y} = 80$ mm, $\bar{z} = 0$. Determine the necessary position of the center of mass of part 3.

7.111 Two views of a machine element are shown. Part 1 is aluminum alloy with density 2800 kg/m³, and part 2 is steel with density 7800 kg/m³. Determine the coordinates of its center of mass.

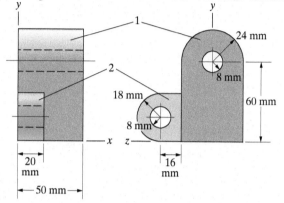

Problem 7.111

7.112 The loads $F_1 = F_2 = 25$ kN. The mass of the truss is 900 kg. The members of the truss are homogeneous bars with the same uniform cross section. (a) What is the x coordinate of the center of mass of the truss? (b) Determine the reactions at A and G.

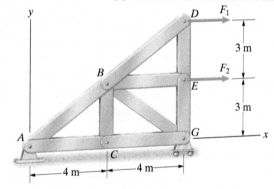

Problem 7.112

7.113 With its engine removed, the mass of the car is 1100 kg and its center of mass is at C. The mass of the engine is 220 kg.

(a) Suppose that you want to place the center of mass E of the engine so that the center of mass of the car is midway between the front wheels A and the rear wheels B. What is the distance b?

(b) If the car is parked on a 15° slope facing up the slope, what total normal force is exerted by the road on the rear wheels B?

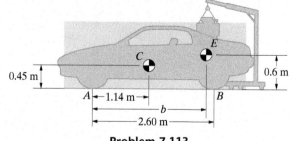

Problem 7.113

7.114 The airplane is parked with its landing gear resting on scales. The weights measured at A, B, and C are 30 kN, 140 kN, and 146 kN, respectively. After a crate is loaded onto the plane, the weights measured at A, B, and C are 31 kN, 142 kN, and 147 kN, respectively. Determine the mass and the x and y coordinates of the center of mass of the crate.

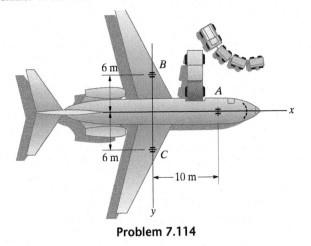
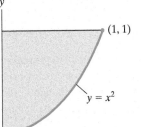

Problem 7.114

▶ **7.115** A suitcase with a mass of 90 kg is placed in the trunk of the car described in Example 7.20. The position of the center of mass of the suitcase is $\bar{x}_s = -0.533$ m, $\bar{y}_s = 0.762$ m, $\bar{z}_s = -0.305$ m. If the suitcase is regarded as part of the car, what is the new position of the car's center of mass?

▶ **7.116** A group of engineering students constructs a miniature device of the kind described in Example 7.20 and uses it to determine the center of mass of a miniature vehicle. The data they obtain are shown in the following table:

Wheelbase = 36 in		
Track = 30 in	Measured Loads (lb)	
	$\alpha = 0$	$\alpha = 10°$
Left front wheel, N_{LF}	35	32
Right front wheel, N_{RF}	36	33
Left rear wheel, N_{LR}	27	34
Right rear wheel, N_{RR}	29	30

Determine the center of mass of the vehicle. Use the same coordinate system as in Example 7.20.

Review Problems

7.117 Determine the centroid of the area by letting dA be a vertical strip of width dx.

7.118 Determine the centroid of the area by letting dA be a horizontal strip of height dy.

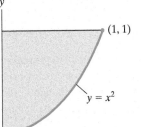

Problems 7.117/7.118

7.119 Determine the centroid of the area.

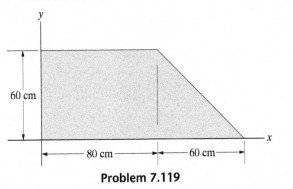

Problem 7.119

7.120 Determine the centroid of the area.

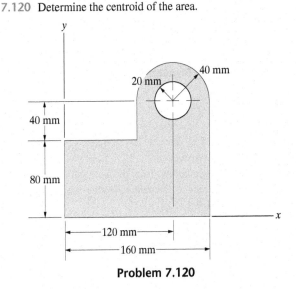

Problem 7.120

7.121 The cantilever beam is subjected to a triangular distributed load. What are the reactions at A?

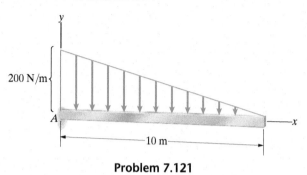

Problem 7.121

7.122 What is the axial load in member BD of the frame?

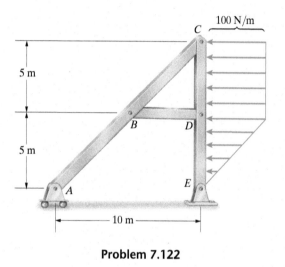

Problem 7.122

7.123 An engineer estimates that the maximum wind load on the 40-m tower in Fig. a is described by the distributed load in Fig. b. The tower is supported by three cables, A, B, and C, from the top of the tower to equally spaced points 15 m from the bottom of the tower (Fig. c). If the wind blows from the west and cables B and C are slack, what is the tension in cable A? (Model the base of the tower as a ball and socket support.)

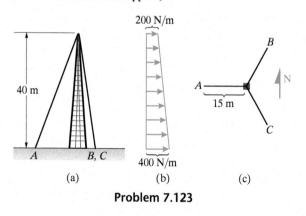

Problem 7.123

7.124 Determine the reactions on member $ABCD$ at A and D.

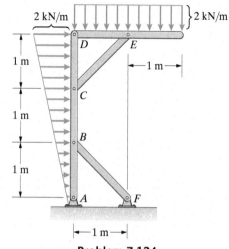

Problem 7.124

7.125 Estimate the centroid of the volume of the *Apollo* lunar return configuration (not including its rocket nozzle) by treating it as a cone and a cylinder.

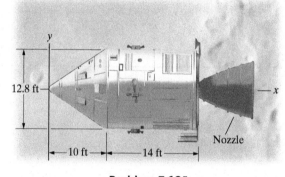

Problem 7.125

7.126 The shape of the rocket nozzle of the *Apollo* lunar return configuration is approximated by revolving the curve shown around the x axis. In terms of the coordinate system shown, determine the centroid of the volume of the nozzle.

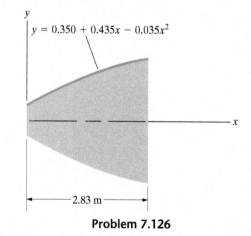

Problem 7.126

7.127 Determine the coordinates of the centroid of the volume.

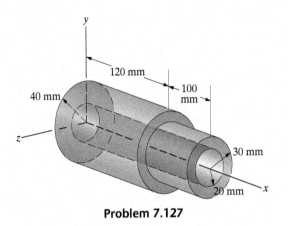

Problem 7.127

7.128 Determine the surface area of the volume of revolution.

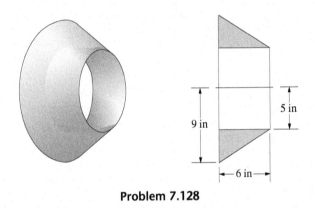

Problem 7.128

7.129 Determine the y coordinate of the center of mass of the homogeneous steel plate.

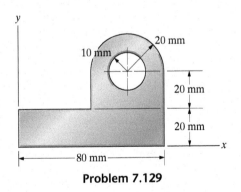

Problem 7.129

7.130 Determine the x coordinate of the center of mass of the homogeneous steel plate.

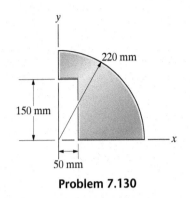

Problem 7.130

7.131 The area of the homogeneous plate is 10 ft². The vertical reactions on the plate at A and B are 80 lb and 84 lb, respectively. Suppose that you want to equalize the reactions at A and B by drilling a 1-ft-diameter hole in the plate. What horizontal distance from A should the center of the hole be? What are the resulting reactions at A and B?

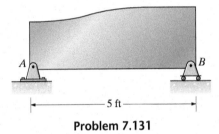

Problem 7.131

7.132 The plate is of uniform thickness and is made of homogeneous material whose mass per unit area of the plate is 2 kg/m². The vertical reactions at A and B are 6 N and 10 N, respectively. What is the x coordinate of the centroid of the hole?

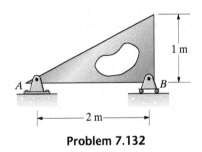

Problem 7.132

7.133 Determine the center of mass of the homogeneous sheet of metal.

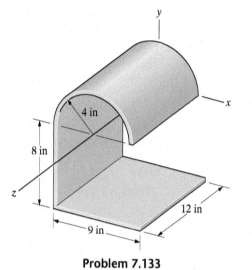

Problem 7.133

7.134 Determine the center of mass of the homogeneous object.

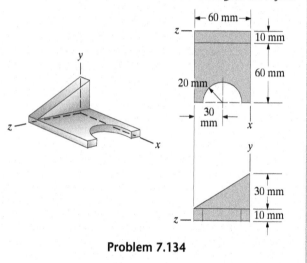

Problem 7.134

7.135 Determine the center of mass of the homogeneous object.

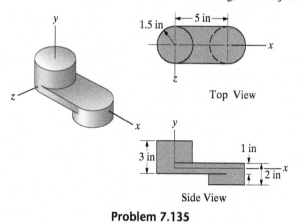

Top View

Side View

Problem 7.135

7.136 The arrangement shown can be used to determine the location of the center of mass of a person. A horizontal board has a pin support at A and rests on a scale that measures weight at B. The distance from A to B is 2.3 m. When the person is not on the board, the scale at B measures 90 N.

(a) When a 63-kg person is in position (1), the scale at B measures 496 N. What is the x coordinate of the person's center of mass?

(b) When the same person is in position (2), the scale measures 523 N. What is the x coordinate of his center of mass?

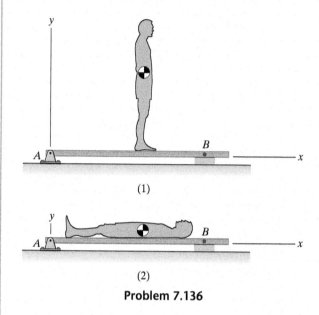

(1)

(2)

Problem 7.136

7.137 If a string is tied to the slender bar at A and the bar is allowed to hang freely, what will be the angle between AB and the vertical?

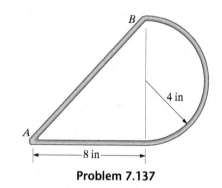

Problem 7.137

7.138 When the truck is unloaded, the total reactions at the front and rear wheels are $A = 54$ kN and $B = 36$ kN. The density of the load of gravel is $\rho = 1600$ kg/m^3. The dimension of the load in the z direction is 3 m, and its surface profile, given by the function shown, does not depend on z. What are the total reactions at the front and rear wheels of the loaded truck?

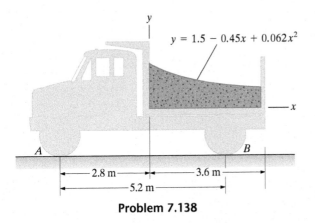

$$y = 1.5 - 0.45x + 0.062x^2$$

2.8 m   3.6 m

5.2 m

Problem 7.138

7.139 The mass of the moon is 0.0123 times the mass of the earth. If the moon's center of mass is 383,000 km from the center of mass of the earth, what is the distance from the center of mass of the earth to the center of mass of the earth–moon system?

Design Project

7.140 Construct a homogeneous thin flat plate with the shape shown in Fig. a. (Use the cardboard back of a pad of paper to construct the plate. Choose your dimensions so that the plate is as large as possible.) Calculate the location of the center of mass of the plate. Measuring as carefully as possible, mark the center of mass clearly on both sides of the plate. Then carry out the following experiments.

(a) Balance the plate on your finger (Fig. b) and observe that it balances at its center of mass. Explain the result of this experiment by drawing a free-body diagram of the plate.

(b) This experiment requires a needle or slender nail, a length of string, and a small weight. Tie the weight to one end of the string and make a small loop at the other end. Stick the needle through the plate at any point other than its center of mass. Hold the needle horizontal so that the plate hangs freely from it (Fig. c). Use the loop to hang the weight from the needle, and let the weight hang freely so that the string lies along the face of the plate. Observe that the string passes through the center of mass of the plate. Repeat this experiment several times, sticking the needle through various points on the plate. Explain the results of this experiment by drawing a free-body diagram of the plate.

(c) Hold the plate so that the plane of the plate is vertical, and throw the plate upward, spinning it like a Frisbee. Observe that the plate spins about its center of mass.

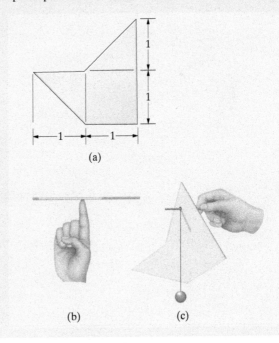

(a)

(b)   (c)

Moments of Inertia

Quantities called moments of inertia arise repeatedly in analyses of engineering problems. Moments of inertia of areas are used in the study of distributed forces and in calculating deflections of beams. The moment exerted by the pressure on a submerged flat plate can be expressed in terms of the moment of inertia of the plate's area. In dynamics, mass moments of inertia are used in calculating the rotational motions of objects. We show how to calculate the moments of inertia of simple areas and objects and then use results called parallel-axis theorems to calculate moments of inertia of more complex areas and objects.

◀ A beam's resistance to bending and ability to support loads depend on a property of its cross section called the moment of inertia. In this chapter we define and show how to calculate moments of inertia of areas.

AREAS

8.1 Definitions

Consider an area A in the x–y plane (Fig. 8.1a). Four moments of inertia of A are defined:

1. Moment of inertia about the x axis:

$$I_x = \int_A y^2 \, dA, \tag{8.1}$$

where y is the y coordinate of the differential element of area dA (Fig. 8.1b). This moment of inertia is sometimes expressed in terms of the *radius of gyration* about the x axis, k_x, which is defined by

$$I_x = k_x^2 A. \tag{8.2}$$

2. Moment of inertia about the y axis:

$$I_y = \int_A x^2 \, dA, \tag{8.3}$$

where x is the x coordinate of the element dA (Fig. 8.1b). The radius of gyration about the y axis, k_y, is defined by

$$I_y = k_y^2 A. \tag{8.4}$$

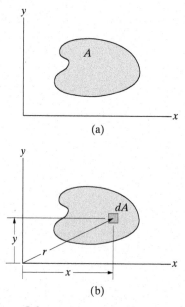

(a)

(b)

Figure 8.1
(a) An area A in the x–y plane.
(b) A differential element of A.

3. Product of inertia:

$$I_{xy} = \int_A xy \, dA. \tag{8.5}$$

4. Polar moment of inertia:

$$J_O = \int_A r^2 \, dA, \tag{8.6}$$

where r is the radial distance from the origin of the coordinate system to dA (Fig. 8.1b). The radius of gyration about the origin, k_O, is defined by

$$J_O = k_O^2 A. \tag{8.7}$$

The polar moment of inertia is equal to the sum of the moments of inertia about the x and y axes:

$$J_O = \int_A r^2 \, dA = \int_A (y^2 + x^2) \, dA = I_x + I_y.$$

Substituting the expressions for the moments of inertia in terms of the radii of gyration into this equation, we obtain

$$k_O^2 = k_x^2 + k_y^2.$$

The dimensions of the moments of inertia of an area are $(\text{length})^4$, and the radii of gyration have dimensions of length. Notice that the definitions of the moments of inertia I_x, I_y, and J_O and the radii of gyration imply that they have positive values for any area. They cannot be negative or zero.

If an area A is symmetric about the x axis, for each element dA with coordinates (x, y), there is a corresponding element dA with coordinates $(x, -y)$, as shown in Fig. 8.2. The contributions of these two elements to the product of inertia I_{xy} of the area cancel: $xy \, dA + (-xy) \, dA = 0$. This means that the product of inertia of the area is zero. The same kind of argument can be used for an area that is symmetric about the y axis. *If an area is symmetric about either the x axis or the y axis, its product of inertia is zero.*

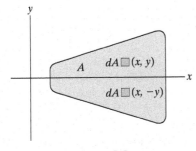

Figure 8.2

Active Example 8.1 Moments of Inertia of a Triangular Area (▶ *Related Problems 8.1–8.3*)

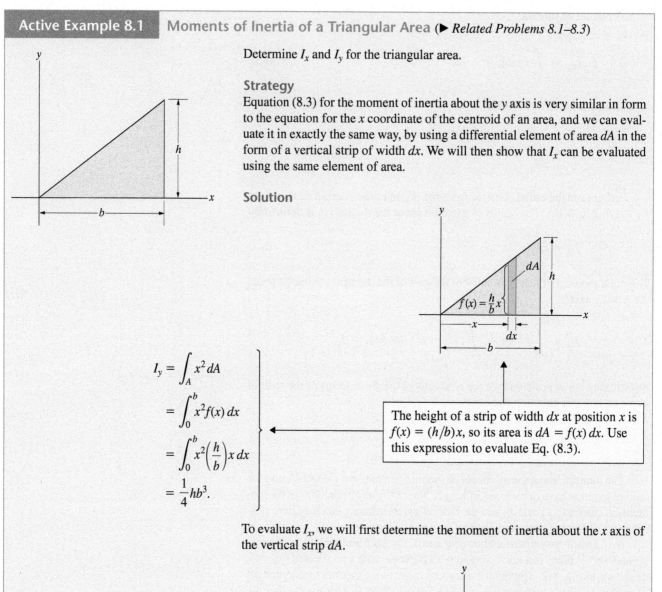

Determine I_x and I_y for the triangular area.

Strategy
Equation (8.3) for the moment of inertia about the y axis is very similar in form to the equation for the x coordinate of the centroid of an area, and we can evaluate it in exactly the same way, by using a differential element of area dA in the form of a vertical strip of width dx. We will then show that I_x can be evaluated using the same element of area.

Solution

$$I_y = \int_A x^2 \, dA$$

$$= \int_0^b x^2 f(x) \, dx$$

$$= \int_0^b x^2 \left(\frac{h}{b}\right) x \, dx$$

$$= \frac{1}{4} h b^3.$$

The height of a strip of width dx at position x is $f(x) = (h/b)x$, so its area is $dA = f(x) \, dx$. Use this expression to evaluate Eq. (8.3).

To evaluate I_x, we will first determine the moment of inertia about the x axis of the vertical strip dA.

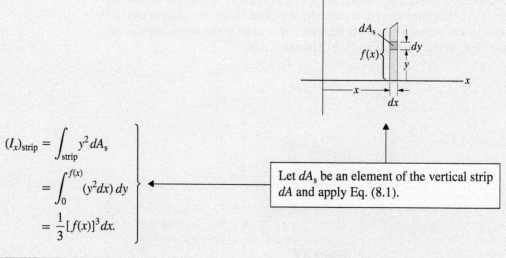

$$(I_x)_{\text{strip}} = \int_{\text{strip}} y^2 \, dA_s$$

$$= \int_0^{f(x)} (y^2 dx) \, dy$$

$$= \frac{1}{3} [f(x)]^3 \, dx.$$

Let dA_s be an element of the vertical strip dA and apply Eq. (8.1).

$$I_x = \int_0^b \frac{1}{3}[f(x)]^3 \, dx$$

$$= \int_0^b \frac{1}{3}\left(\frac{h}{b}x\right)^3 dx$$

$$= \frac{1}{12}bh^3.$$

> Integrate the expression for $(I_x)_{\text{strip}}$ with respect to x from $x = 0$ to $x = b$ to determine I_x for the triangle.

Practice Problem Determine I_{xy} for the triangular area. Do so by determining the product of inertia for the vertical strip dA and then integrating the resulting expression with respect to x from $x = 0$ to $x = b$.

Answer: $I_{xy} = \dfrac{1}{8}b^2h^2$.

Example 8.2	**Moments of Inertia of a Circular Area** (▶ *Related Problem 8.21*)

Determine the moments of inertia and radii of gyration of the circular area.

Strategy

We will first determine the polar moment of inertia J_O by integrating in terms of polar coordinates. We know from the symmetry of the area that $I_x = I_y$, and since $I_x + I_y = J_O$, the moments of inertia I_x and I_y are each equal to $\frac{1}{2}J_O$. We also know from the symmetry of the area that $I_{xy} = 0$.

Solution

By letting r change by an amount dr, we obtain an annular element of area $dA = 2\pi r \, dr$ (Fig. a). The polar moment of inertia is

$$J_O = \int_A r^2 \, dA = \int_0^R 2\pi r^3 \, dr = 2\pi\left[\frac{r^4}{4}\right]_0^R = \frac{1}{2}\pi R^4,$$

and the radius of gyration about O is

$$k_O = \sqrt{\frac{J_O}{A}} = \sqrt{\frac{(1/2)\pi R^4}{\pi R^2}} = \frac{1}{\sqrt{2}}R.$$

The moments of inertia about the x and y axes are

$$I_x = I_y = \frac{1}{2}J_O = \frac{1}{4}\pi R^4,$$

and the radii of gyration about the x and y axes are

$$k_x = k_y = \sqrt{\frac{I_x}{A}} = \sqrt{\frac{(1/4)\pi R^4}{\pi R^2}} = \frac{1}{2}R.$$

The product of inertia is zero:

$$I_{xy} = 0.$$

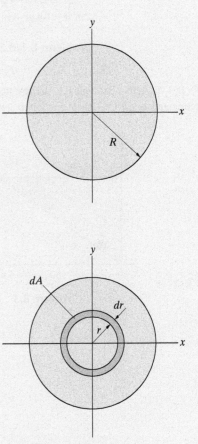

(a) An annular element dA.

Critical Thinking

The symmetry of this example saved us from having to integrate to determine I_x, I_y, and I_{xy}. Be alert for symmetry that can shorten your work. In particular, remember that $I_{xy} = 0$ if the area is symmetric about either the x or the y axis.

Problems

▶ **8.1** Use the method described in Active Example 8.1 to determine I_y and k_y for the rectangular area.

▶ **8.2** Use the method described in Active Example 8.1 to determine I_x and k_x for the rectangular area.

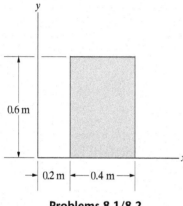

Problems 8.1/8.2

▶ **8.3** In Active Example 8.1, suppose that the triangular area is reoriented as shown. Use integration to determine I_y and k_y.

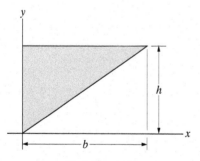

Problem 8.3

8.4 (a) Determine the moment of inertia I_y of the beam's rectangular cross section about the y axis.

(b) Determine the moment of inertia $I_{y'}$ of the beam's cross section about the y' axis. Using your numerical values, show that $I_y = I_{y'} + d_x^2 A$, where A is the area of the cross section.

8.5 (a) Determine the polar moment of inertia J_O of the beam's rectangular cross section about the origin O.

(b) Determine the polar moment of inertia $J_{O'}$ of the beam's cross section about the origin O'. Using your numerical values, show that $J_O = J_{O'} + (d_x^2 + d_y^2)A$, where A is the area of the cross section.

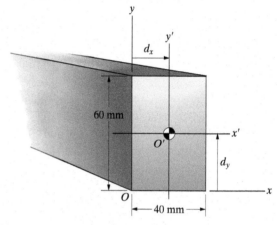

Problems 8.4/8.5

8.6 Determine I_y and k_y.

8.7 Determine J_O and k_O.

8.8 Determine I_{xy}.

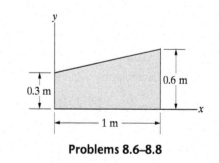

Problems 8.6–8.8

8.9 Determine I_y.

8.10 Determine I_x.

8.11 Determine J_O.

8.12 Determine I_{xy}.

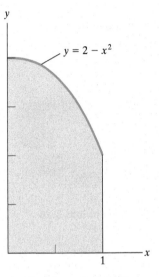

Problems 8.9–8.12

8.13 Determine I_y and k_y.

8.14 Determine I_x and k_x.

8.15 Determine J_O and k_O.

8.16 Determine I_{xy}.

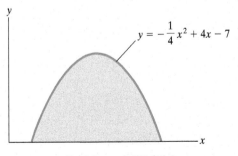

Problems 8.13–8.16

8.17 Determine I_y and k_y.

8.18 Determine I_x and k_x.

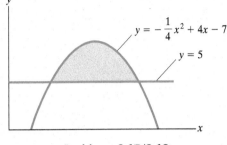

Problems 8.17/8.18

8.19 (a) Determine I_y and k_y by letting dA be a vertical strip of width dx.

(b) The polar moment of inertia of a circular area with its center at the origin is $J_O = \frac{1}{2}\pi R^4$. Explain how you can use this information to confirm your answer to (a).

8.20 (a) Determine I_x and k_x by letting dA be a horizontal strip of height dy.

(b) The polar moment of inertia of a circular area with its center at the origin is $J_O = \frac{1}{2}\pi R^4$. Explain how you can use this information to confirm your answer to (a).

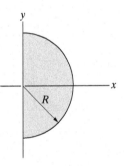

Problems 8.19/8.20

▶ **8.21** Use the procedure described in Example 8.2 to determine the moments of inertia I_x and I_y for the annular ring.

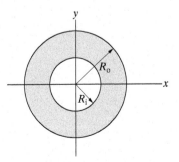

Problem 8.21

8.22 What are the values of I_y and k_y for the elliptical area of the airplane's wing?

8.23 What are the values of I_x and k_x for the elliptical area of the airplane's wing?

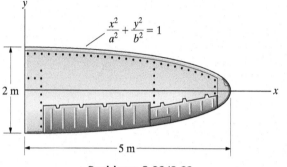

Problems 8.22/8.23

8.24 Determine I_y and k_y.

8.25 Determine I_x and k_x.

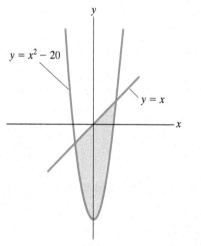

Problems 8.24/8.25

8.26 A vertical plate of area A is beneath the surface of a stationary body of water. The pressure of the water subjects each element dA of the surface of the plate to a force $(p_O + \gamma y)\, dA$, where p_O is the pressure at the surface of the water and γ is the weight density of the water. Show that the magnitude of the moment about the x axis due to the pressure on the front face of the plate is

$$M_{x\,\text{axis}} = p_O \bar{y} A + \gamma I_x,$$

where $\bar{y}$ is the y coordinate of the centroid of A and I_x is the moment of inertia of A about the x axis.

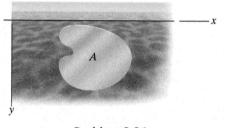

Problem 8.26

8.2 Parallel-Axis Theorems

BACKGROUND

The values of the moments of inertia of an area depend on the position of the coordinate system relative to the area.

In some situations the moments of inertia of an area are known in terms of a particular coordinate system but we need their values in terms of a different coordinate system. When the coordinate systems are parallel, the desired moments of inertia can be obtained by using the theorems we describe in this section. Furthermore, these theorems make it possible for us to determine the moments of inertia of a composite area when the moments of inertia of its parts are known.

Suppose that we know the moments of inertia of an area A in terms of a coordinate system $x'y'$ with its origin at the centroid of the area, and we wish to determine the moments of inertia in terms of a parallel coordinate system xy (Fig. 8.3a). We denote the coordinates of the centroid of A in the xy coordinate system by (d_x, d_y), and $d = \sqrt{d_x^2 + d_y^2}$ is the distance from the origin of the xy coordinate system to the centroid (Fig. 8.3b).

We need two preliminary results before deriving the parallel-axis theorems. In terms of the $x'y'$ coordinate system, the coordinates of the centroid of A are

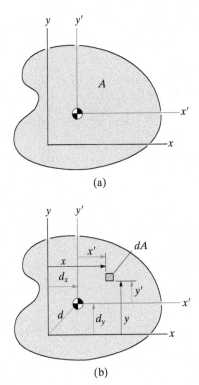

(a)

(b)

Figure 8.3
(a) The area A and the coordinate systems $x'y'$ and xy.
(b) The differential element dA.

$$\bar{x}' = \frac{\int_A x'\,dA}{\int_A dA}, \qquad \bar{y}' = \frac{\int_A y'\,dA}{\int_A dA}.$$

But the origin of the $x'y'$ coordinate system is located at the centroid of A, so $\bar{x}' = 0$ and $\bar{y}' = 0$. Therefore,

$$\int_A x'\,dA = 0, \qquad \int_A y'\,dA = 0. \tag{8.8}$$

Moment of Inertia About the x Axis   In terms of the xy coordinate system, the moment of inertia of A about the x axis is

$$I_x = \int_A y^2\,dA, \tag{8.9}$$

where y is the coordinate of the element of area dA relative to the xy coordinate system. From Fig. 8.3b, we see that $y = y' + d_y$, where y' is the coordinate of dA relative to the $x'y'$ coordinate system. Substituting this expression into Eq. (8.9), we obtain

$$I_x = \int_A (y' + d_y)^2\,dA = \int_A (y')^2\,dA + 2d_y\int_A y'\,dA + d_y^2\int_A dA.$$

The first integral on the right is the moment of inertia of A about the x' axis. From Eq. (8.8), the second integral on the right equals zero. Therefore, we obtain

$$I_x = I_{x'} + d_y^2 A. \tag{8.10}$$

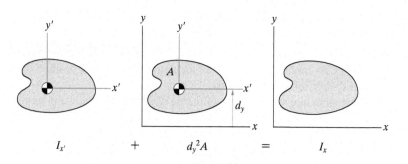

Figure 8.4
The parallel-axis theorem for the moment of inertia about the x axis.

This is a *parallel-axis theorem*. It relates the moment of inertia of A about the x' axis through the centroid to the moment of inertia about the parallel axis x (Fig. 8.4).

Moment of Inertia About the y Axis   In terms of the xy coordinate system, the moment of inertia of A about the y axis is

$$I_y = \int_A x^2 \, dA = \int_A (x' + d_x)^2 \, dA$$

$$= \int_A (x')^2 \, dA + 2d_x \int_A x' dA + d_x^2 \int_A dA.$$

From Eq. (8.8), the second integral on the right equals zero. Therefore, the parallel-axis theorem that relates the moment of inertia of A about the y' axis through the centroid to the moment of inertia about the parallel axis y is

$$I_y = I_{y'} + d_x^2 A. \tag{8.11}$$

Product of Inertia   In terms of the xy coordinate system, the product of inertia is

$$I_{xy} = \int_A xy \, dA = \int_A (x' + d_x)(y' + d_y) \, dA$$

$$= \int_A x'y' dA + d_y \int_A x' dA + d_x \int_A y' dA + d_x d_y \int_A dA.$$

The second and third integrals equal zero from Eq. (8.8). We see that the parallel-axis theorem for the product of inertia is

$$I_{xy} = I_{x'y'} + d_x d_y A. \tag{8.12}$$

Polar Moment of Inertia   The polar moment of inertia $J_O = I_x + I_y$. Summing Eqs. (8.10) and (8.11), the parallel axis theorem for the polar moment of inertia is

$$J_O = J'_O + (d_x^2 + d_y^2)A = J'_O + d^2 A, \tag{8.13}$$

where d is the distance from the origin of the $x'y'$ coordinate system to the origin of the xy coordinate system.

How can the parallel-axis theorems be used to determine the moments of inertia of a composite area? Suppose that we want to determine the moment of inertia about the y axis of the area in Fig. 8.5a. We can divide it into a triangle, a semicircle, and a circular cutout, denoted as parts 1, 2, and 3 (Fig. 8.5b). By using the parallel-axis theorem for I_y, we can determine the moment of inertia of each part about the y axis. For example, the moment of inertia of part 2 (the semicircle) about the y axis is (Fig. 8.5c)

$$(I_y)_2 = (I_{y'})_2 + (d_x)_2^2 A_2.$$

We must determine the values of $(I_{y'})_2$ and $(d_x)_2$. Moments of inertia and centroid locations for some simple areas are tabulated in Appendix B. Once this procedure is carried out for each part, the moment of inertia of the composite area is

$$I_y = (I_y)_1 + (I_y)_2 - (I_y)_3.$$

Notice that the moment of inertia of the circular cutout is subtracted.

We see that determining a moment of inertia of a composite area in terms of a given coordinate system involves three steps:

1. **Choose the parts**—Try to divide the composite area into parts whose moments of inertia you know or can easily determine.

2. **Determine the moments of inertia of the parts**—Determine the moment of inertia of each part in terms of a parallel coordinate system with its origin at the centroid of the part, and then use the parallel-axis theorem to determine the moment of inertia in terms of the given coordinate system.

3. **Sum the results**—Sum the moments of inertia of the parts (or subtract in the case of a cutout) to obtain the moment of inertia of the composite area.

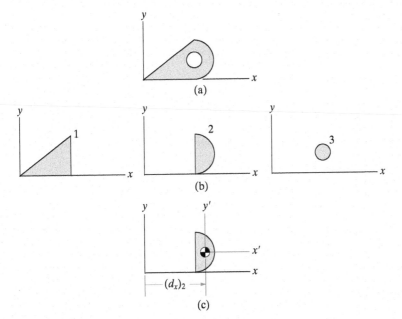

Figure 8.5
(a) A composite area.
(b) The three parts of the area.
(c) Determining $(I_y)_2$.

RESULTS

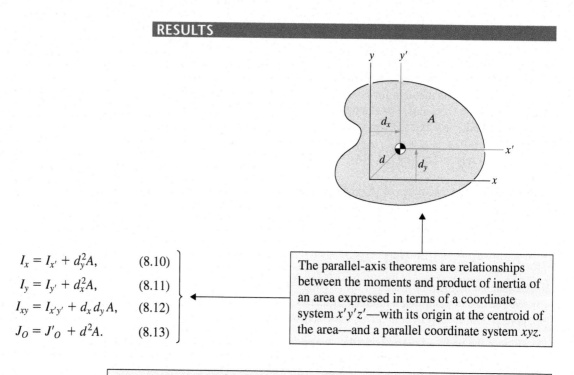

$$I_x = I_{x'} + d_y^2 A, \qquad (8.10)$$

$$I_y = I_{y'} + d_x^2 A, \qquad (8.11)$$

$$I_{xy} = I_{x'y'} + d_x d_y A, \qquad (8.12)$$

$$J_O = J'_O + d^2 A. \qquad (8.13)$$

The parallel-axis theorems are relationships between the moments and product of inertia of an area expressed in terms of a coordinate system $x'y'z'$—with its origin at the centroid of the area—and a parallel coordinate system xyz.

The parallel-axis theorems make it possible to determine the moments and product of inertia of a composite area in terms of a given coordinate system xyz when the moments and products of inertia of each part of the composite area are known in terms of a parallel coordinate system with its origin at the centroid of the part. The values of the moments and product of inertia of the parts in terms of the xyz coordinate system can be summed (or subtracted in the case of a cutout) to obtain the values for the composite area.

Active Example 8.3 Moments of Inertia of a Composite Area (▶ *Related Problem 8.27*)

Determine I_x for the composite area.

Strategy
We can divide this area into two rectangles. We must use the parallel-axis theorems to determine I_x for each rectangle in terms of the xy coordinate system. The values can be summed to determine I_x for the composite area.

Solution

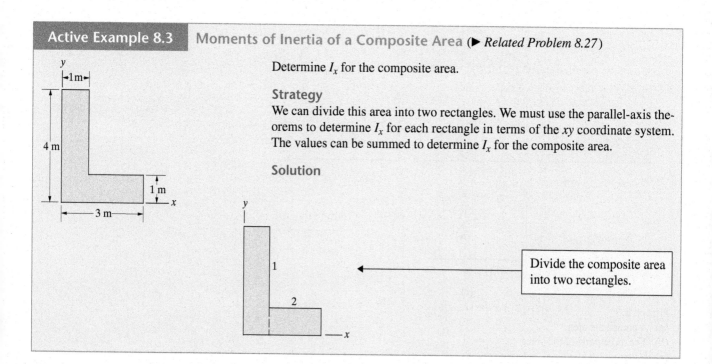

Divide the composite area into two rectangles.

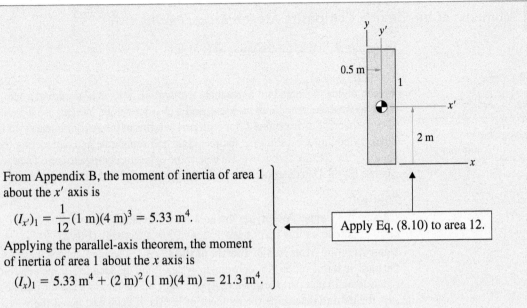

From Appendix B, the moment of inertia of area 1 about the x' axis is

$$(I_{x'})_1 = \frac{1}{12}(1 \text{ m})(4 \text{ m})^3 = 5.33 \text{ m}^4.$$

Applying the parallel-axis theorem, the moment of inertia of area 1 about the x axis is

$$(I_x)_1 = 5.33 \text{ m}^4 + (2 \text{ m})^2 (1 \text{ m})(4 \text{ m}) = 21.3 \text{ m}^4.$$

Apply Eq. (8.10) to area 12.

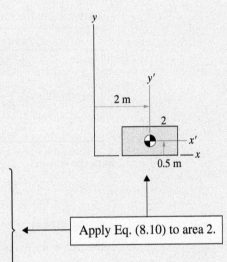

The moment of inertia of area 2 about the x' axis is

$$(I_{x'})_2 = \frac{1}{12}(2 \text{ m})(1 \text{ m})^3 = 0.167 \text{ m}^4.$$

Applying the parallel-axis theorem, the moment of inertia of area 2 about the x axis is

$$(I_x)_2 = 0.167 \text{ m}^4 + (0.5 \text{ m})^2 (2 \text{ m})(1 \text{ m}) = 0.667 \text{ m}^4.$$

Apply Eq. (8.10) to area 2.

The moment of inertia of the composite area about the x axis is

$$I_x = (I_x)_1 + (I_x)_2$$
$$= 21.3 \text{ m}^4 + 0.667 \text{ m}^4$$
$$= 22.0 \text{ m}^4.$$

Sum the values for the parts.

Practice Problem Determine I_{xy} for the composite area.

Answer: $I_{xy} = 6 \text{ m}^4$.

Example 8.4   Moments of Inertia of a Composite Area (▶ *Related Problem 8.30*)

Determine I_y and k_y for the composite area.

Strategy

We can divide this area into a rectangle *without the semicircular cutout*, a semicircle *without the semicircular cutout*, and a circular cutout. We can use a parallel-axis theorem to determine I_y for each part in terms of the xy coordinate system. Then, by adding the values for the rectangle and semicircle and subtracting the value for the circular cutout, we can determine I_y for the composite area. Then we can use Eq. (8.4) to determine the radius of gyration k_y for the composite area.

Solution

Choose the Parts   We divide the area into a rectangle, a semicircle, and the circular cutout, calling them parts 1, 2, and 3, respectively (Fig. a).

Determine the Moments of Inertia of the Parts   The moments of inertia of the parts in terms of the $x'y'$ coordinate systems and the location of the centroid of the semicircular part are given in Appendix B. In the table we use the parallel-axis theorem to determine the moment of inertia of each part about the y axis.

Determining the moments of inertia of the parts

	d_x (mm)	A (mm^2)	$I_{y'}$ (mm^4)	$I_y = I_{y'} + d_x^2 A$ (mm^4)
Part 1	60	$(120)(80)$	$\frac{1}{12}(80)(120)^3$	4.608×10^7
Part 2	$120 + \dfrac{4(40)}{3\pi}$	$\frac{1}{2}\pi(40)^2$	$\left(\dfrac{\pi}{8} - \dfrac{8}{9\pi}\right)(40)^4$	4.744×10^7
Part 3	120	$\pi(20)^2$	$\frac{1}{4}\pi(20)^4$	1.822×10^7

Sum the Results   The moment of inertia of the composite area about the y axis is

$$I_y = (I_y)_1 + (I_y)_2 - (I_y)_3 = (4.608 + 4.744 - 1.822) \times 10^7 \text{ mm}^4$$
$$= 7.530 \times 10^7 \text{ mm}^4.$$

The total area is

$$A = A_1 + A_2 - A_3 = (120 \text{ mm})(80 \text{ mm}) + \frac{1}{2}\pi(40 \text{ mm})^2 - \pi(20 \text{ mm})^2$$
$$= 1.086 \times 10^4 \text{ mm}^2,$$

so the radius of gyration about the y axis is

$$k_y = \sqrt{\frac{I_y}{A}} = \sqrt{\frac{7.530 \times 10^7 \text{ mm}^4}{1.086 \times 10^4 \text{ mm}^2}} = 83.3 \text{ mm}.$$

Critical Thinking

Integration is an additive process, which is why the moments of inertia of composite areas can be determined by adding (or, in the case of a cutout, subtracting) the moments of inertia of the parts. But you can't determine the radii of gyration of composite areas by adding or subtracting the radii of gyration of the parts. This can be seen from the equations relating the moments of inertia, radii of gyration, and area. For this example, we can demonstrate it numerically. The operation

$$(k_y)_1 + (k_y)_2 - (k_y)_3 = \sqrt{\frac{(I_y)_1}{A_1}} + \sqrt{\frac{(I_y)_2}{A_2}} - \sqrt{\frac{(I_y)_3}{A_3}} = 86.3 \text{ mm}$$

does not yield the correct radius of gyration of the composite area.

(a) Parts 1, 2, and 3.

Example 8.5 Beam Cross Sections (▶ *Related Problems 8.81–8.84*)

The equal areas are candidates for the cross section of a beam. (A beam with the second cross section shown is called an I-beam.) Compare their moments of inertia about the *x* axis.

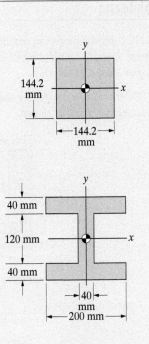

Strategy
We can obtain the moment of inertia of the square cross section from Appendix B. We will divide the I-beam cross section into three rectangles and use the parallel-axis theorem to determine its moment of inertia.

Solution
Square Cross Section   From Appendix B, the moment of inertia of the square cross section about the *x* axis is

$$I_x = \frac{1}{12}(144.2 \text{ mm})(144.2 \text{ mm})^3 = 3.60 \times 10^7 \text{ mm}^4.$$

I-Beam Cross Section   We can divide the area into the rectangular parts shown in Fig. a. Introducing coordinate systems $x'y'$ with their origins at the centroids of the parts (Fig. b), we use the parallel-axis theorem to determine the moments of inertia about the *x* axis (see table). Their sum is

$$I_x = (I_x)_1 + (I_x)_2 + (I_x)_3 = (5.23 + 0.58 + 5.23) \times 10^7 \text{ mm}^4$$

$$= 11.03 \times 10^7 \text{ mm}^4.$$

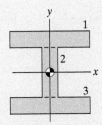

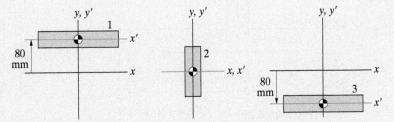

(a) Dividing the I-beam cross section into parts.

(b) Parallel coordinate systems $x'y'$ with origins at the centroids of the parts.

Determining the moments of inertia of the parts about the *x* axis.

	d_y (mm)	A (mm^2)	$I_{x'}$ (mm^4)	$I_x = I_{x'} + d_y^2 A$ (mm^4)
Part 1	80	(200)(40)	$\frac{1}{12}(200)(40)^3$	5.23×10^7
Part 2	0	(40)(120)	$\frac{1}{12}(40)(120)^3$	0.58×10^7
Part 3	−80	(200)(40)	$\frac{1}{12}(200)(40)^3$	5.23×10^7

Critical Thinking
The moment of inertia of the I-beam is 3.06 times that of the square cross section of equal area. Generally a beam with a larger moment of inertia has greater resistance to bending and greater ability to support lateral loads. The cross sections of I-beams are designed to obtain large moments of inertia.

Problems

▶ **8.27** Using the procedure described in Active Example 8.3, determine I_x and k_x for the composite area by dividing it into rectangles 1 and 2 as shown.

8.28 Determine I_y and k_y for the composite area by dividing it into rectangles 1 and 2 as shown.

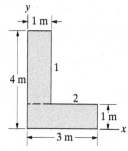

Problems 8.27/8.28

8.29 Determine I_x and k_x.

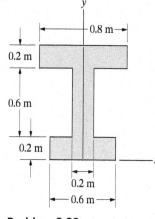

Problem 8.29

▶ **8.30** In Example 8.4, determine I_x and k_x for the composite area.

8.31 Determine I_x and k_x.

8.32 Determine I_y and k_y.

8.33 Determine J_O and k_O.

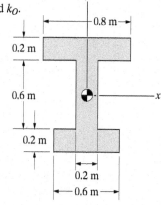

Problems 8.31–8.33

8.34 If you design the beam cross section so that $I_x = 6.4 \times 10^5$ mm⁴, what are the resulting values of I_y and J_O?

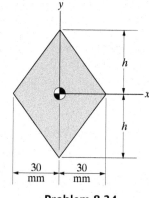

Problem 8.34

8.35 Determine I_y and k_y.

8.36 Determine I_x and k_x.

8.37 Determine I_{xy}.

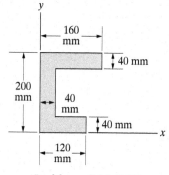

Problems 8.35–8.37

8.38 Determine I_x and k_x.

8.39 Determine I_y and k_y.

8.40 Determine I_{xy}.

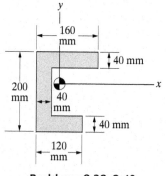

Problems 8.38–8.40

8.41 Determine I_x and k_x.

8.42 Determine J_O and k_O.

8.43 Determine I_{xy}.

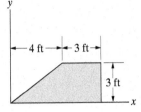

Problems 8.41–8.43

8.44 Determine I_x and k_x.

8.45 Determine J_O and k_O.

8.46 Determine I_{xy}.

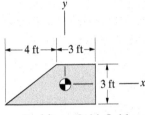

Problems 8.44–8.46

8.47 Determine I_x and k_x.

8.48 Determine J_O and k_O.

8.49 Determine I_{xy}.

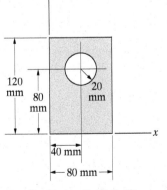

Problems 8.47–8.49

8.50 Determine I_x and k_x.

8.51 Determine I_y and k_y.

8.52 Determine J_O and k_O.

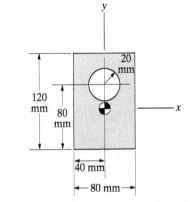

Problems 8.50–8.52

8.53 Determine I_y and k_y.

8.54 Determine J_O and k_O.

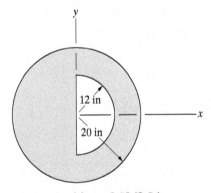

Problems 8.53/8.54

8.55 Determine I_y and k_y if $h = 3$ m.

8.56 Determine I_x and k_x if $h = 3$ m.

8.57 If $I_y = 5$ m^4, what is the dimension h?

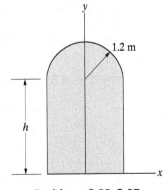

Problems 8.55–8.57

8.58 Determine I_y and k_y.

8.59 Determine I_x and k_x.

8.60 Determine I_{xy}.

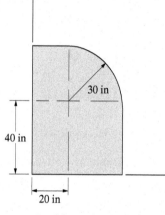

Problems 8.58–8.60

8.61 Determine I_y and k_y.

8.62 Determine I_x and k_x.

8.63 Determine I_{xy}.

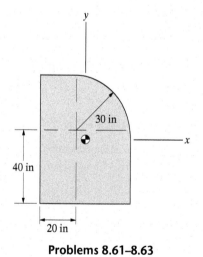

Problems 8.61–8.63

8.64 Determine I_y and k_y.

8.65 Determine I_x and k_x.

8.66 Determine I_{xy}.

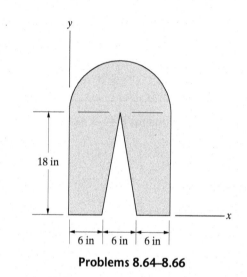

Problems 8.64–8.66

8.67 Determine I_y and k_y.

8.68 Determine J_O and k_O.

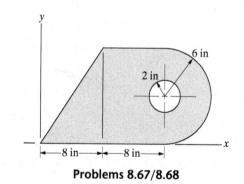

Problems 8.67/8.68

8.69 Determine I_y and k_y.

8.70 Determine I_x and k_x.

8.71 Determine I_{xy}.

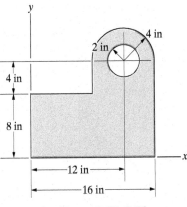

Problems 8.69–8.71

8.72 Determine I_y and k_y.

8.73 Determine I_x and k_x.

8.74 Determine I_{xy}.

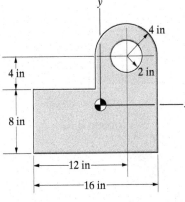

Problems 8.72–8.74

8.75 Determine I_y and k_y.

8.76 Determine J_O and k_O.

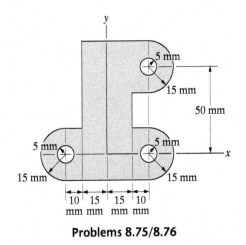

Problems 8.75/8.76

8.77 Determine I_x and I_y for the beam's cross section.

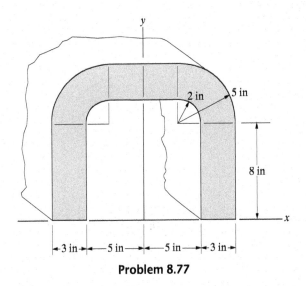

Problem 8.77

8.78 Determine I_x and I_y for the beam's cross section.

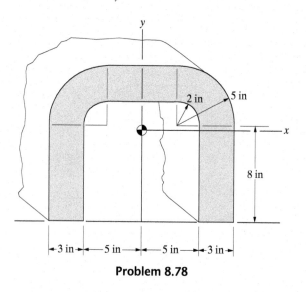

Problem 8.78

8.79 The area $A = 2 \times 10^4$ mm^2. Its moment of inertia about the y axis is $I_y = 3.2 \times 10^8$ mm^4. Determine its moment of inertia about the $\hat{y}$ axis.

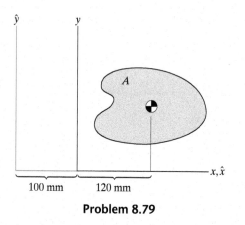

Problem 8.79

8.80 The area $A = 100$ in^2 and it is *symmetric* about the x' axis. The moments of inertia $I_{x'} = 420$ in^4, $I_{y'} = 580$ in^4, $J_O = 11,000$ in^4, and $I_{xy} = 4800$ in^4. What are I_x and I_y?

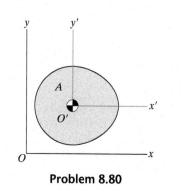

Problem 8.80

▶ **8.81** Determine the moment of inertia of the beam cross section about the x axis. Compare your result with the moment of inertia of a solid square cross section of equal area. (See Example 8.5.)

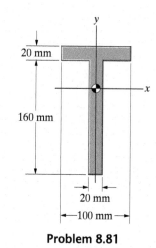

Problem 8.81

▶ **8.82** The area of the beam cross section is 5200 mm^2. Determine the moment of inertia of the beam cross section about the x axis. Compare your result with the moment of inertia of a solid square cross section of equal area. (See Example 8.5.)

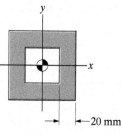

Problem 8.82

▶ 8.83 If the beam in Fig. a is subjected to couples of magnitude M about the x axis (Fig. b), the beam's longitudinal axis bends into a circular arc whose radius R is given by

$$R = \frac{EI_x}{M},$$

where I_x is the moment of inertia of the beam's cross section about the x axis. The value of the term E, which is called the *modulus of elasticity*, depends on the material of which the beam is constructed. Suppose that a beam with the cross section shown in Fig. c is subjected to couples of magnitude $M = 180$ N-m. As a result, the beam's axis bends into a circular arc with radius $R = 3$ m. What is the modulus of elasticity of the beam's material? (See Example 8.5.)

(a) Unloaded.

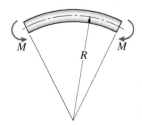

(b) Subjected to couples at the ends.

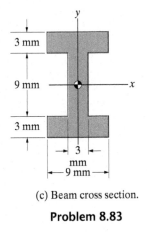

(c) Beam cross section.

Problem 8.83

▶ 8.84 Suppose that you want to design a beam made of material whose density is 8000 kg/m³. The beam is to be 4 m in length and have a mass of 320 kg. Design a cross section for the beam so that $I_x = 3 \times 10^{-5}$ m⁴. (See Example 8.5.)

8.85 The area in Fig. a is a **C230×30** American Standard Channel beam cross section. Its cross-sectional area is $A = 3790$ mm² and its moments of inertia about the x and y axes are $I_x = 25.3 \times 10^6$ mm⁴ and $I_y = 1 \times 10^6$ mm⁴. Suppose that two beams with **C230×30** cross sections are riveted together to obtain a composite beam with the cross section shown in Fig. b. What are the moments of inertia about the x and y axes of the composite beam?

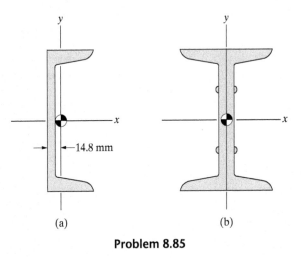

(a)   (b)

Problem 8.85

8.86 The area in Fig. a is an **L152×102×12.7** Angle beam cross section. Its cross-sectional area is $A = 3060$ mm² and its moments of inertia about the x and y axes are $I_x = 7.24 \times 10^6$ mm⁴ and $I_y = 2.61 \times 10^6$ mm⁴. Suppose that four beams with **L152×102×12.7** cross sections are riveted together to obtain a composite beam with the cross section shown in Fig. b. What are the moments of inertia about the x and y axes of the composite beam?

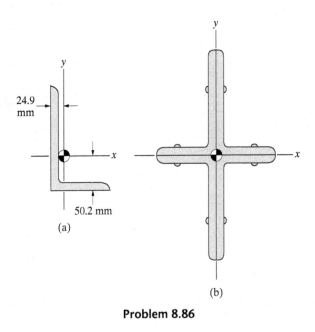

(a)

(b)

Problem 8.86

8.3 Rotated and Principal Axes

BACKGROUND

Suppose that Fig. 8.6a is the cross section of a cantilever beam. If a vertical force is applied to the end of the beam, a larger vertical deflection results if the cross section is oriented as shown in Fig. 8.6b than if it is oriented as shown in Fig. 8.6c. The minimum vertical deflection results when the beam's cross section is oriented so that the moment of inertia I_x is a maximum (Fig. 8.6d).

In many engineering applications we must determine moments of inertia of areas with various angular orientations relative to a coordinate system and also determine the orientation for which the value of a moment of inertia is a maximum or minimum. We discuss these procedures in this section.

Rotated Axes

Consider an area A, a coordinate system xy, and a second coordinate system $x'y'$ that is rotated through an angle θ relative to the xy coordinate system (Fig. 8.7a). Suppose that we know the moments of inertia of A in terms of the xy coordinate system. Our objective is to determine the moments of inertia in terms of the $x'y'$ coordinate system.

In terms of the radial distance r to a differential element of area dA and the angle α in Fig. 8.7b, the coordinates of dA in the xy coordinate system are

$$x = r \cos \alpha, \tag{8.14}$$

$$y = r \sin \alpha. \tag{8.15}$$

Figure 8.6
(a) A beam cross section.
(b)–(d) Applying a lateral load with different orientations of the cross section.

Figure 8.7
(a) The $x'y'$ coordinate system is rotated through an angle θ relative to the xy coordinate system.
(b) A differential element of area dA.

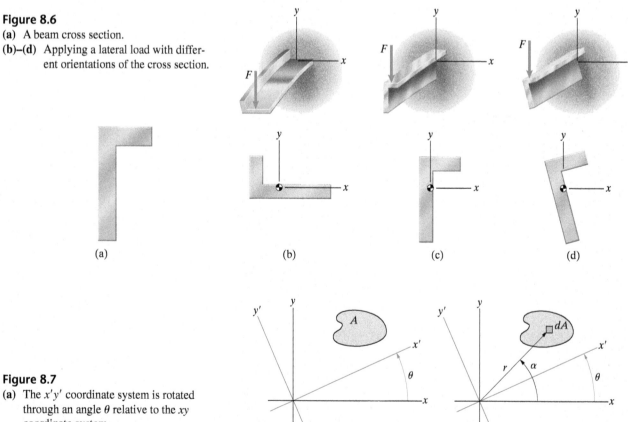

(a) (b) (c) (d)

(a) (b)

The coordinates of dA in the $x'y'$ coordinate system are

$$x' = r\cos(\alpha - \theta) = r(\cos\alpha\cos\theta + \sin\alpha\sin\theta), \qquad (8.16)$$

$$y' = r\sin(\alpha - \theta) = r(\sin\alpha\cos\theta - \cos\alpha\sin\theta). \qquad (8.17)$$

In Eqs. (8.16) and (8.17), we use identities for the cosine and sine of the difference of two angles (Appendix A). By substituting Eqs. (8.14) and (8.15) into Eqs. (8.16) and (8.17), we obtain equations relating the coordinates of dA in the two coordinate systems:

$$x' = x\cos\theta + y\sin\theta, \qquad (8.18)$$

$$y' = -x\sin\theta + y\cos\theta. \qquad (8.19)$$

We can use these expressions to derive relations between the moments of inertia of A in terms of the xy and $x'y'$ coordinate systems.

Moment of Inertia About the x' Axis

$$I_{x'} = \int_A (y')^2\,dA = \int_A (-x\sin\theta + y\cos\theta)^2\,dA$$

$$= \cos^2\theta\int_A y^2\,dA - 2\sin\theta\cos\theta\int_A xy\,dA + \sin^2\theta\int_A x^2\,dA.$$

From this equation we obtain

$$I_{x'} = I_x\cos^2\theta - 2I_{xy}\sin\theta\cos\theta + I_y\sin^2\theta. \qquad (8.20)$$

Moment of Inertia About the y' Axis

$$I_{y'} = \int_A (x')^2\,dA = \int_A (x\cos\theta + y\sin\theta)^2\,dA$$

$$= \sin^2\theta\int_A y^2\,dA + 2\sin\theta\cos\theta\int_A xy\,dA + \cos^2\theta\int_A x^2\,dA.$$

This equation gives us the result

$$I_{y'} = I_x\sin^2\theta + 2I_{xy}\sin\theta\cos\theta + I_y\cos^2\theta. \qquad (8.21)$$

Product of Inertia   In terms of the $x'y'$ coordinate system, the product of inertia of A is

$$I_{x'y'} = (I_x - I_y)\sin\theta\cos\theta + (\cos^2\theta - \sin^2\theta)I_{xy}. \qquad (8.22)$$

Polar Moment of Inertia   From Eqs. (8.20) and (8.21), the polar moment of inertia in terms of the $x'y'$ coordinate system is

$$J'_O = I_{x'} + I_{y'} = I_x + I_y = J_O.$$

Thus the value of the polar moment of inertia is unchanged by a rotation of the coordinate system.

Principal Axes

We have seen that the moments of inertia of A in terms of the $x'y'$ coordinate system depend on the angle θ in Fig. 8.7a. Consider the following question: For what values of θ is the moment of inertia $I_{x'}$ a maximum or minimum?

To answer this question, it is convenient to use the identities

$$\sin 2\theta = 2 \sin \theta \cos \theta,$$

$$\cos 2\theta = \cos^2 \theta - \sin^2 \theta = 1 - 2\sin^2 \theta = 2\cos^2 \theta - 1.$$

With these expressions, we can write Eqs. (8.20)–(8.22) in the forms

$$I_{x'} = \frac{I_x + I_y}{2} + \frac{I_x - I_y}{2} \cos 2\theta - I_{xy} \sin 2\theta, \tag{8.23}$$

$$I_{y'} = \frac{I_x + I_y}{2} - \frac{I_x - I_y}{2} \cos 2\theta + I_{xy} \sin 2\theta, \tag{8.24}$$

$$I_{x'y'} = \frac{I_x - I_y}{2} \sin 2\theta + I_{xy} \cos 2\theta. \tag{8.25}$$

We will denote a value of θ at which $I_{x'}$ is a maximum or minimum by θ_p. To determine θ_p, we evaluate the derivative of Eq. (8.23) with respect to 2θ and equate it to zero, obtaining

$$\tan 2\theta_p = \frac{2I_{xy}}{I_y - I_x}. \tag{8.26}$$

If we set the derivative of Eq. (8.24) with respect to 2θ equal to zero to determine a value of θ for which $I_{y'}$ is a maximum or minimum, we again obtain Eq. (8.26). The second derivatives of $I_{x'}$ and $I_{y'}$ with respect to 2θ are opposite in sign; that is,

$$\frac{d^2 I_{x'}}{d(2\theta)^2} = -\frac{d^2 I_{y'}}{d(2\theta)^2},$$

which means that at an angle θ_p for which $I_{x'}$ is a maximum, $I_{y'}$ is a minimum; and at an angle θ_p for which $I_{x'}$ is a minimum, $I_{y'}$ is a maximum.

A rotated coordinate system $x'y'$ that is oriented so that $I_{x'}$ and $I_{y'}$ have maximum or minimum values is called a set of *principal axes* of the area A. The corresponding moments of inertia $I_{x'}$ and $I_{y'}$ are called the *principal moments of inertia*. In the next section we show that the product of inertia $I_{x'y'}$ corresponding to a set of principal axes equals zero.

Because the tangent is a periodic function, Eq. (8.26) does not yield a unique solution for the angle θ_p. We show, however, that it does determine the orientation of the principal axes within an arbitrary multiple of 90°. Observe in Fig. 8.8 that if $2\theta_0$ is a solution of Eq. (8.26), then $2\theta_0 + n(180°)$ is also a solution for any integer n. The resulting orientations of the $x'y'$ coordinate system are shown in Fig. 8.9.

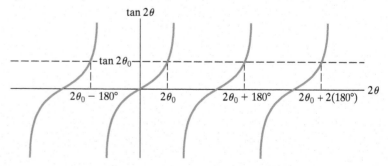

Figure 8.8
For a given value of $\tan 2\theta_0$, there are multiple roots $2\theta_0 + n(180°)$.

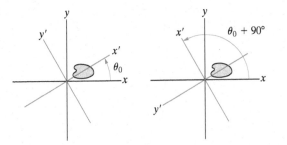
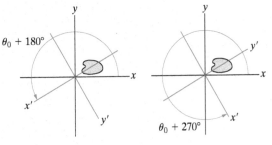

Figure 8.9
The orientation of the $x'y'$ coordinate system
is determined within a multiple of $90°$.

RESULTS

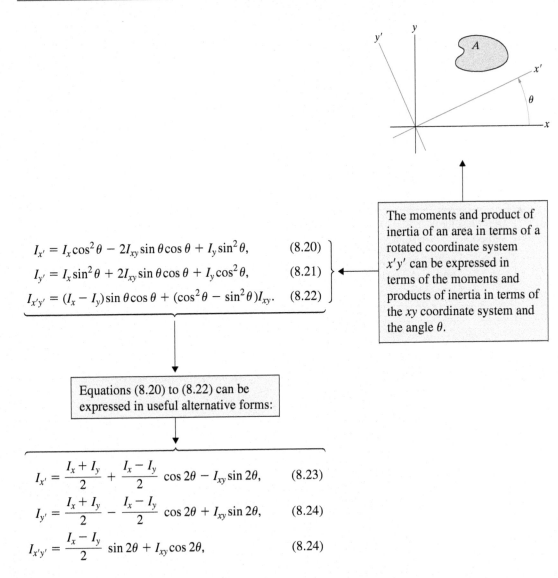

The moments and product of inertia of an area in terms of a rotated coordinate system $x'y'$ can be expressed in terms of the moments and products of inertia in terms of the xy coordinate system and the angle θ.

$$I_{x'} = I_x \cos^2\theta - 2I_{xy}\sin\theta\cos\theta + I_y\sin^2\theta, \quad (8.20)$$

$$I_{y'} = I_x \sin^2\theta + 2I_{xy}\sin\theta\cos\theta + I_y\cos^2\theta, \quad (8.21)$$

$$I_{x'y'} = (I_x - I_y)\sin\theta\cos\theta + (\cos^2\theta - \sin^2\theta)I_{xy}. \quad (8.22)$$

Equations (8.20) to (8.22) can be expressed in useful alternative forms:

$$I_{x'} = \frac{I_x + I_y}{2} + \frac{I_x - I_y}{2}\cos 2\theta - I_{xy}\sin 2\theta, \quad (8.23)$$

$$I_{y'} = \frac{I_x + I_y}{2} - \frac{I_x - I_y}{2}\cos 2\theta + I_{xy}\sin 2\theta, \quad (8.24)$$

$$I_{x'y'} = \frac{I_x - I_y}{2}\sin 2\theta + I_{xy}\cos 2\theta, \quad (8.24)$$

A value of θ for which the moment of inertia $I_{x'}$ obtained from Eq. (8.23) is a maximum or minimum is denoted by θ_{p}. If $I_{x'}$ is a maximum at $\theta = \theta_{\mathrm{p}}$, $I_{y'}$ is a minimum at $\theta = \theta_{\mathrm{p}}$, and if $I_{x'}$ is a minimum, $I_{y'}$ is a maximum. The rotated coordinate system $x'y'$ corresponding to $\theta = \theta_{\mathrm{p}}$ defines the *principal axes* of the area A, and the moments of inertia about the principal axes are the *principal moments of inertia*. The product of inertia $I_{x'y'}$ corresponding to $\theta = \theta_{\mathrm{p}}$ equals zero.

For given values of I_x, I_y, and I_{xy}, the angle θ_{p} can be determined from the equation

$$\tan 2\theta_{\mathrm{p}} = \frac{2I_{xy}}{I_y - I_x}. \tag{8.26}$$

This equation uniquely defines the principal axes, but defines the orientation of the $x'y'$ coordinate system only within a multiple of $90°$. For example, if θ_0 is a solution of Eq. (8.26), then $\theta_0 + 90°$, $\theta_0 + 180°$, and $\theta_0 + 270°$ are also solutions, resulting in four valid orientations of the $x'y'$ coordinate system.

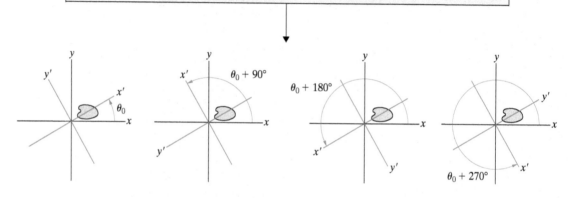

Determining principal axes and principal moments of inertia for a given area A and coordinate system xy involves three steps:

1. Determine I_x, I_y, and I_{xy}.
2. Use Eq. (8.26) to determine θ_{p} to within a multiple of $90°$.
3. Choose the orientation of the $x'y'$ coordinate system and use Eqs.(8.23) and (8.24) to determine the principal moments of inertia.

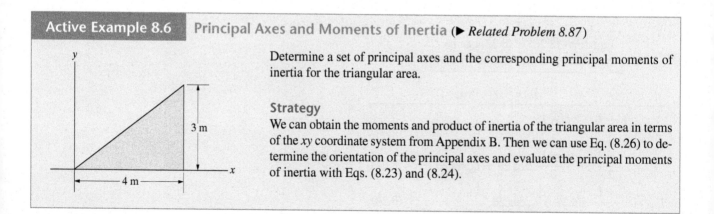

Active Example 8.6 Principal Axes and Moments of Inertia (▶ *Related Problem 8.87*)

Determine a set of principal axes and the corresponding principal moments of inertia for the triangular area.

Strategy

We can obtain the moments and product of inertia of the triangular area in terms of the xy coordinate system from Appendix B. Then we can use Eq. (8.26) to determine the orientation of the principal axes and evaluate the principal moments of inertia with Eqs. (8.23) and (8.24).

Solution

$$I_x = \frac{1}{12}(4\ m)(3\ m)^3 = 9\ m^4,$$

$$I_y = \frac{1}{4}(4\ m)^3(3\ m) = 48\ m^4,$$

$$I_{xy} = \frac{1}{8}(4\ m)^2(3\ m)^2 = 18\ m^4.$$

Determine the moments and products of inertia from Appendix B.

$$\tan 2\theta_p = \frac{2I_{xy}}{I_y - I_x} = \frac{2(18)}{48 - 9} = 0.923.$$

This yields $\theta_p = 21.4°$.

Determine θ_p from Eq. (8.26).

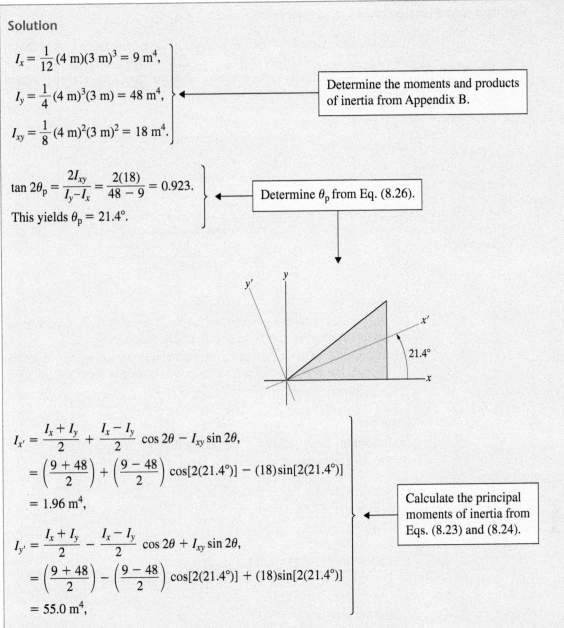

$$I_{x'} = \frac{I_x + I_y}{2} + \frac{I_x - I_y}{2}\cos 2\theta - I_{xy}\sin 2\theta,$$

$$= \left(\frac{9 + 48}{2}\right) + \left(\frac{9 - 48}{2}\right)\cos[2(21.4°)] - (18)\sin[2(21.4°)]$$

$$= 1.96\ m^4,$$

$$I_{y'} = \frac{I_x + I_y}{2} - \frac{I_x - I_y}{2}\cos 2\theta + I_{xy}\sin 2\theta,$$

$$= \left(\frac{9 + 48}{2}\right) - \left(\frac{9 - 48}{2}\right)\cos[2(21.4°)] + (18)\sin[2(21.4°)]$$

$$= 55.0\ m^4,$$

Calculate the principal moments of inertia from Eqs. (8.23) and (8.24).

Practice Problem The moments and product of inertia of the triangular area are $I_x = 9\ m^4$, $I_y = 16\ m^4$, and $I_{xy} = 6\ m^4$. Determine a set of principal axes and the corresponding principal moments of inertia.

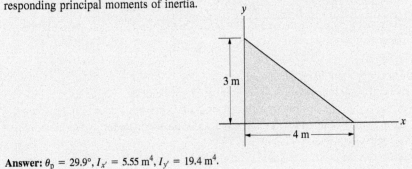

Answer: $\theta_p = 29.9°$, $I_{x'} = 5.55\ m^4$, $I_{y'} = 19.4\ m^4$.

Example 8.7 **Rotated and Principal Axes** (▶ *Related Problems 8.88, 8.89*)

The moments of inertia of the area in terms of the xy coordinate system shown are $I_x = 22$ ft^4, $I_y = 10$ ft^4, and $I_{xy} = 6$ ft^4. (a) Determine $I_{x'}$, $I_{y'}$, and $I_{x'y'}$ for $\theta = 30°$. (b) Determine a set of principal axes and the corresponding principal moments of inertia.

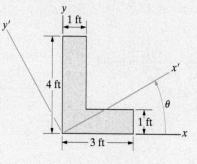

Strategy

(a) We can determine the moments of inertia in terms of the $x'y'$ coordinate system by substituting $\theta = 30°$ into Eqs. (8.23)–(8.25).

(b) The orientation of the principal axes is determined by solving Eq. (8.26) for θ_p. Once θ_p has been determined, the moments of inertia about the principal axes can be determined from Eqs. (8.23) and (8.24).

Solution

(a) Determine $I_{x'}$, $I_{y'}$, and $I_{x'y'}$ By setting $\theta = 30°$ in Eqs. (8.23)–(8.25), we obtain (with moments of inertia in ft^4)

$$I_{x'} = \frac{I_x + I_y}{2} + \frac{I_x - I_y}{2} \cos 2\theta - I_{xy} \sin 2\theta$$

$$= \left(\frac{22 + 10}{2}\right) + \left(\frac{22 - 10}{2}\right) \cos[2(30°)] - (6) \sin[2(30°)] = 13.8 \text{ ft}^4,$$

$$I_{y'} = \frac{I_x + I_y}{2} - \frac{I_x - I_y}{2} \cos 2\theta + I_{xy} \sin 2\theta$$

$$= \left(\frac{22 + 10}{2}\right) - \left(\frac{22 - 10}{2}\right) \cos[2(30°)] + (6) \sin[2(30°)] = 18.2 \text{ ft}^4,$$

$$I_{x'y'} = \frac{I_x - I_y}{2} \sin 2\theta + I_{xy} \cos 2\theta$$

$$= \left(\frac{22 - 10}{2}\right) \sin[2(30°)] + (6) \cos[2(30°)] = 8.2 \text{ ft}^4.$$

(b) Determine θ_p   We substitute the moments of inertia in terms of the xy coordinate system into Eq. (8.26), yielding

$$\tan 2\theta_p = \frac{2I_{xy}}{I_y - I_x} = \frac{2(6)}{10 - 22} = -1.$$

Thus, $\theta_p = -22.5°$. The principal axes corresponding to this value of θ_p are shown in Fig. a.

Calculate $I_{x'}$ and $I_{y'}$ We substitute $\theta_p = -22.5°$ into Eqs. (8.23) and (8.24), obtaining the principal moments of inertia:

$$I_{x'} = 24.5 \text{ ft}^4, \quad I_{y'} = 7.5 \text{ ft}^4.$$

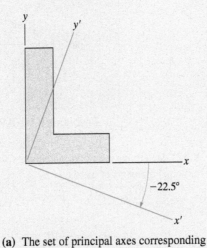

(a) The set of principal axes corresponding to $\theta_p = -22.5°$.

Critical Thinking

Remember that the orientation of the principal axes is only determined within an arbitrary multiple of 90°. In this example we chose to designate the axes in Fig. a as the positive x' and y' axes, but any of these four choices is equally valid.

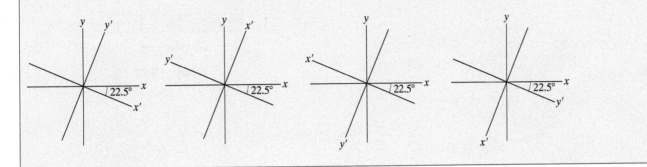

Problems

▶ 8.87 In Active Example 8.6, suppose that the vertical 3-m dimension of the triangular area is increased to 4 m. Determine a set of principal axes and the corresponding principal moments of inertia.

▶ 8.88 In Example 8.7, suppose that the area is reoriented as shown. Determine the moments of inertia $I_{x'}$, $I_{y'}$, and $I_{x'y'}$ if $\theta = 30°$.

▶ 8.89 In Example 8.7, suppose that the area is reoriented as shown. Determine a set of principal axes and the corresponding principal moments of inertia. Based on the results of Example 8.7, can you predict a value of θ_p without using Eq. (8.26)?

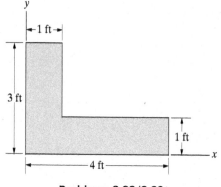

Problems 8.88/8.89

8.90 The moments of inertia of the area are $I_x = 1.26 \times 10^6$ in^4, $I_y = 6.55 \times 10^5$ in^4, and $I_{xy} = -1.02 \times 10^5$ in^4. Determine the moments of inertia of the area $I_{x'}$, $I_{y'}$, and $I_{x'y'}$ if $\theta = 30°$.

8.91 The moments of inertia of the area are $I_x = 1.26 \times 10^6$ in^4, $I_y = 6.55 \times 10^5$ in^4, and $I_{xy} = -1.02 \times 10^5$ in^4. Determine a set of principal axes and the corresponding principal moments of inertia.

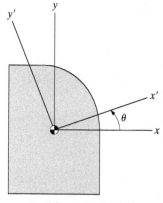

Problems 8.90/8.91

8.92* Determine a set of principal axes and the corresponding principal moments of inertia.

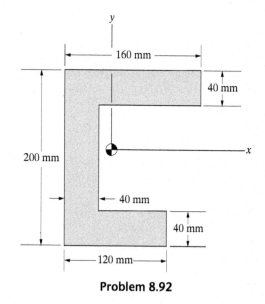

Problem 8.92

8.4 Mohr's Circle

BACKGROUND

Given the moments of inertia of an area in terms of a particular coordinate system, we have presented equations that determine the moments of inertia in terms of a rotated coordinate system, the orientation of the principal axes, and the principal moments of inertia. We can also obtain this information by using a graphical method called *Mohr's circle*, which is very useful for visualizing the solutions of Eqs. (8.23)–(8.25).

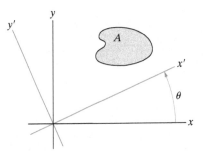

Figure 8.10
The xy coordinate system and the rotated $x'y'$ coordinate system.

Determining $I_{x'}$, $I_{y'}$, and $I_{x'y'}$

We first describe how to construct Mohr's circle and then explain why it works. Suppose we know the moments of inertia I_x, I_y, and I_{xy} of an area in terms of a coordinate system xy and we want to determine the moments of inertia for a rotated coordinate system $x'y'$ (Fig. 8.10). Constructing Mohr's circle involves three steps:

1. Establish a set of horizontal and vertical axes and plot two points: point 1 with coordinates (I_x, I_{xy}) and point 2 with coordinates $(I_y, -I_{xy})$ as shown in Fig. 8.11a.
2. Draw a straight line connecting points 1 and 2. Using the intersection of the straight line with the horizontal axis as the center, draw a circle that passes through the two points (Fig. 8.11b).
3. Draw a straight line through the center of the circle at an angle 2θ measured counterclockwise from point 1. This line intersects the circle at point $1'$ with coordinates $(I_{x'}, I_{x'y'})$ and point $2'$ with coordinates $(I_{y'}, -I_{x'y'})$, as shown in Fig. 8.11c.

Thus, for a given angle θ, the coordinates of points $1'$ and $2'$ determine the moments of inertia in terms of the rotated coordinate system. Why does this graphical construction work? In Fig. 8.12, we show the points 1 and 2 and Mohr's circle. Notice that the horizontal coordinate of the center of the circle is $(I_x + I_y)/2$. The sine and cosine of the angle β are

$$\sin \beta = \frac{I_{xy}}{R}, \qquad \cos \beta = \frac{I_x - I_y}{2R},$$

where R, the radius of the circle, is given by

$$R = \sqrt{\left(\frac{I_x - I_y}{2}\right)^2 + (I_{xy})^2}.$$

Figure 8.13 shows the construction of the points $1'$ and $2'$. The horizontal coordinate of point $1'$ is

$$\frac{I_x + I_y}{2} + R\cos(\beta + 2\theta)$$

$$= \frac{I_x + I_y}{2} + R(\cos \beta \cos 2\theta - \sin \beta \sin 2\theta)$$

$$= \frac{I_x + I_y}{2} + \frac{I_x - I_y}{2}\cos 2\theta - I_{xy}\sin 2\theta = I_{x'},$$

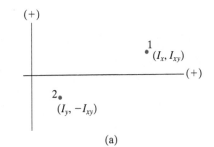

(a)

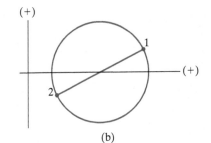

(b)

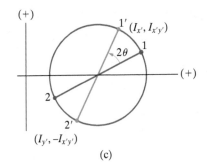

(c)

Figure 8.11
(a) Plotting the points 1 and 2.
(b) Drawing Mohr's circle. The center of the circle is the intersection of the line from 1 to 2 with the horizontal axis.
(c) Finding the points $1'$ and $2'$.

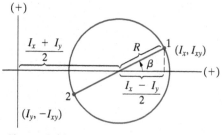

Figure 8.12
The points 1 and 2 and Mohr's circle.

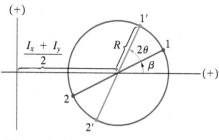

Figure 8.13
The points 1′ and 2′.

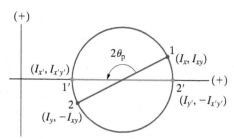

Figure 8.14
To determine the orientation of a set of principal axes, let points 1′ and 2′ be the points where the circle intersects the horizontal axis.

and the horizontal coordinate of point 2′ is

$$\frac{I_x + I_y}{2} - R\cos(\beta + 2\theta)$$

$$= \frac{I_x + I_y}{2} - R(\cos\beta\cos 2\theta - \sin\beta\sin 2\theta)$$

$$= \frac{I_x + I_y}{2} - \frac{I_x - I_y}{2}\cos 2\theta + I_{xy}\sin 2\theta = I_{y'}.$$

The vertical coordinate of point 1′ is

$$R\sin(\beta + 2\theta) = R(\sin\beta\cos 2\theta + \cos\beta\sin 2\theta)$$

$$= I_{xy}\cos 2\theta + \frac{I_x - I_y}{2}\sin 2\theta = I_{x'y'},$$

and the vertical coordinate of point 2′ is

$$-R\sin(\beta + 2\theta) = -I_{x'y'}.$$

We have shown that the coordinates of point 1′ are $(I_{x'}, I_{x'y'})$ and the coordinates of point 2′ are $(I_{y'}, -I_{x'y'})$.

Determining Principal Axes and Principal Moments of Inertia

Because the moments of inertia $I_{x'}$ and $I_{y'}$ are the horizontal coordinates of points 1′ and 2′ of Mohr's circle, their maximum and minimum values occur when points 1′ and 2′ coincide with the intersections of the circle with the horizontal axis (Fig. 8.14). (Which intersection you designate as 1′ is arbitrary. In Fig. 8.14, we have designated the minimum moment of inertia as point 1′.) You can determine the orientation of the principal axes by measuring the angle $2\theta_p$ from point 1 to point 1′, and the coordinates of points 1′ and 2′ are the principal moments of inertia.

Notice that Mohr's circle demonstrates that the product of inertia $I_{x'y'}$ corresponding to a set of principal axes (the vertical coordinate of point 1′ in Fig. 8.14) is always zero. Furthermore, we can use Fig. 8.12 to obtain an analytical expression for the horizontal coordinates of the points where the circle intersects the horizontal axis, which are the principal moments of inertia:

$$\text{Principal moments of inertia} = \frac{I_x + I_y}{2} \pm R$$

$$= \frac{I_x + I_y}{2} \pm \sqrt{\left(\frac{I_x - I_y}{2}\right)^2 + (I_{xy})^2}.$$

RESULTS

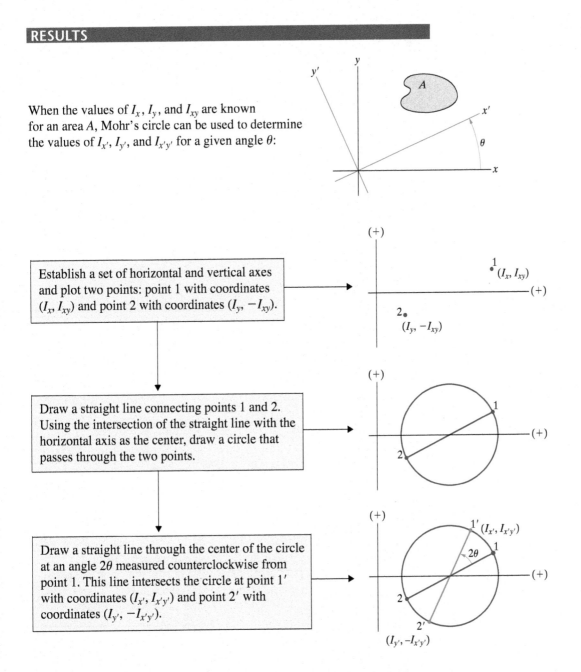

When the values of I_x, I_y, and I_{xy} are known
for an area A, Mohr's circle can be used to determine
the values of $I_{x'}$, $I_{y'}$, and $I_{x'y'}$ for a given angle θ:

> Establish a set of horizontal and vertical axes
> and plot two points: point 1 with coordinates
> (I_x, I_{xy}) and point 2 with coordinates $(I_y, -I_{xy})$.

> Draw a straight line connecting points 1 and 2.
> Using the intersection of the straight line with the
> horizontal axis as the center, draw a circle that
> passes through the two points.

> Draw a straight line through the center of the circle
> at an angle 2θ measured counterclockwise from
> point 1. This line intersects the circle at point 1'
> with coordinates $(I_{x'}, I_{x'y'})$ and point 2' with
> coordinates $(I_{y'}, -I_{x'y'})$.

Mohr's circle can also be used to determine the orientation of the principal axes
and the principal moments of interia:

> Place the point 1' at one of the points where Mohr's
> circle intersects the horizontal axis. Then the values
> of $I_{x'}$ and $I_{y'}$ obtained from points 1' and 2' are
> the principal moments of inertia. The angle
> measured counterclockwise from point 1 to point 1'
> is $2\theta_p$, so the orientation of the principal axes
> can be determined.

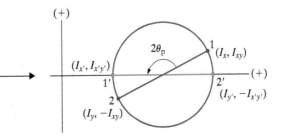

Active Example 8.8 | *Mohr's Circle* (▶ *Related Problems 8.94, 8.95*)

The moments and product of inertia of the area in terms of the xy coordinate system are $I_x = 22 \text{ ft}^4$, $I_y = 10 \text{ ft}^4$, and $I_{xy} = 6 \text{ ft}^4$. Use Mohr's circle to determine the moments of inertia $I_{x'}$, $I_{y'}$, and $I_{x'y'}$ for $\theta = 30°$.

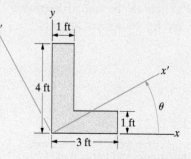

Strategy

By using the given values of I_x, I_y, and I_{xy} to construct Mohr's circle, we can determine $I_{x'}$, $I_{y'}$, and $I_{x'y'}$ for $\theta = 30°$.

Solution

Plot point 1 with coordinates $(I_x, I_{xy}) = (22, 6) \text{ ft}^4$ and point 2 with coordinates $(I_y, -I_{xy}) = (10, -6) \text{ ft}^4$.

Draw a straight line connecting points 1 and 2. Using the intersection of the straight line with the horizontal axis as the center, draw a circle that passes through the two points.

Draw a straight line through the center of the circle at an angle $2\theta = 60°$ measured counterclockwise from point 1. From the coordinates of points 1' and 2', $I_{x'} = 14 \text{ ft}^4$, $I_{y'} = 18 \text{ ft}^4$, and $I_{x'y'} = 8 \text{ ft}^4$.

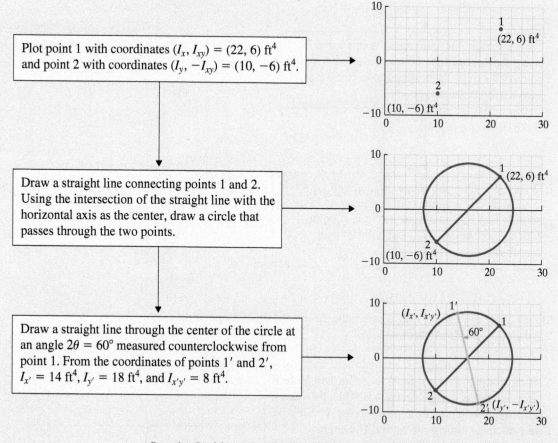

Practice Problem Use Mohr's circle to determine the orientation of the principal axes of the area and the corresponding principal moments of inertia.

Answer: $\theta_p = 67.5°$, $I_{x'} = 7.5 \text{ ft}^4$, $I_{y'} = 24.5 \text{ ft}^4$.

Problems

MASSES

8.5 Simple Objects

BACKGROUND

The acceleration of an object that results from the forces acting on it depends on its mass. The angular acceleration, or rotational acceleration, that results from the forces and couples acting on an object depends on quantities called the mass moments of inertia of the object. In this section we discuss methods for determining mass moments of inertia of particular objects. We show that for special classes of objects, their mass moments of inertia can be expressed in terms of moments of inertia of areas, which explains how the names of those area integrals originated.

An object and a line or "axis" L_O are shown in Fig. 8.15a. The *moment of inertia* of the object about the axis L_O is defined by

$$I_O = \int_m r^2 \, dm, \tag{8.27}$$

where r is the perpendicular distance from the axis to the differential element of mass dm (Fig. 8.15b). Often L_O is an axis about which the object rotates, and the value of I_O is required to determine the angular acceleration, or the rate of change of the rate of rotation, caused by a given couple about L_O. The dimensions of the moment of inertia of an object are $(\text{mass}) \times (\text{length})^2$. Notice that the definition implies that its value must be positive.

Slender Bars

Let us determine the moment of inertia of a straight, slender bar about a perpendicular axis L through the center of mass of the bar (Fig. 8.16a). "Slender" means that we assume that the bar's length is much greater than its width. Let the bar have length l, cross-sectional area A, and mass m. We assume that A is uniform along the length of the bar and that the material is homogeneous.

Consider a differential element of the bar of length dr at a distance r from the center of mass (Fig. 8.16b). The element's mass is equal to the product of its volume and the density: $dm = \rho A \, dr$. Substituting this expression into Eq. (8.27), we obtain the moment of inertia of the bar about a perpendicular axis through its center of mass:

$$I = \int_m r^2 \, dm = \int_{-l/2}^{l/2} \rho A r^2 \, dr = \frac{1}{12}\rho A l^3.$$

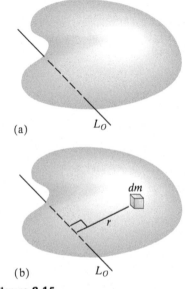
(a)

(b)

Figure 8.15
(a) An object and axis L_O.
(b) A differential element of mass dm.

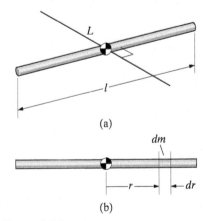
(a)

(b)

Figure 8.16
(a) A slender bar.
(b) A differential element of length dr.

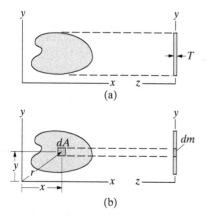

Figure 8.17

(a) A plate of arbitrary shape and uniform thickness T.

(b) An element of volume obtained by projecting an element of area dA through the plate.

The mass of the bar equals the product of the mass density and the volume of the bar, $m = \rho A l$, so we can express the moment of inertia as

$$I = \frac{1}{12}ml^2. \tag{8.28}$$

We have neglected the lateral dimensions of the bar in obtaining this result. That is, we treated the differential element of mass dm as if it were concentrated on the axis of the bar. As a consequence, Eq. (8.28) is an approximation for the moment of inertia of a bar. In the next section we determine the moments of inertia for a bar of finite lateral dimension and show that Eq. (8.28) is a good approximation when the width of the bar is small in comparison to its length.

Thin Plates

Consider a homogeneous flat plate that has mass m and uniform thickness T. We will leave the shape of the cross-sectional area of the plate unspecified. Let a cartesian coordinate system be oriented so that the plate lies in the x–y plane (Fig. 8.17a). Our objective is to determine the moments of inertia of the plate about the x, y, and z axes.

We can obtain a differential element of volume of the plate by projecting an element of area dA through the thickness T of the plate (Fig. 8.17b). The resulting volume is $T\,dA$. The mass of this element of volume is equal to the product of the density and the volume: $dm = \rho T\,dA$. Substituting this expression into Eq. (8.27), we obtain the moment of inertia of the plate about the z axis in the form

$$I_{z\,\text{axis}} = \int_m r^2\,dm = \rho T \int_A r^2\,dA,$$

where r is the distance from the z axis to dA. Since the mass of the plate is $m = \rho T A$, where A is the cross-sectional area of the plate, $\rho T = m/A$. The integral on the right is the polar moment of inertia J_O of the cross-sectional area of the plate. We can therefore write the moment of inertia of the plate about the z axis as

$$I_{z\,\text{axis}} = \frac{m}{A}J_O. \tag{8.29}$$

From Fig. 8.17b, we see that the perpendicular distance from the x axis to the element of area dA is the y coordinate of dA. Therefore, the moment of inertia of the plate about the x axis is

$$I_{x\,\text{axis}} = \int_m y^2\,dm = \rho T \int_A y^2\,dA = \frac{m}{A}I_x, \tag{8.30}$$

where I_x is the moment of inertia of the cross-sectional area of the plate about the x axis. The moment of inertia of the plate about the y axis is

$$I_{y\,\text{axis}} = \int_m x^2\,dm = \rho T \int_A x^2\,dA = \frac{m}{A}I_y, \tag{8.31}$$

where I_y is the moment of inertia of the cross-sectional area of the plate about the y axis.

Because the sum of the area moments of inertia I_x and I_y is equal to the polar moment of inertia J_O, the mass moment of inertia of the thin plate about the z axis is equal to the sum of its moments of inertia about the x and y axes:

$$I_{z\,\text{axis}} = I_{x\,\text{axis}} + I_{y\,\text{axis}}. \qquad \text{Thin plate} \tag{8.32}$$

We have expressed the moments of inertia of a thin homogeneous plate of uniform thickness in terms of the moments of inertia of the cross-sectional area of the plate. In fact, these results explain why the area integrals I_x, I_y, and J_O are called moments of inertia. The use of the same terminology and similar symbols for moments of inertia of areas and moments of inertia of objects can be confusing, but is entrenched in engineering practice. The type of moment of inertia being referred to can be determined either from the context or from the units: $(\text{length})^4$ for moments of inertia of areas and $(\text{mass}) \times (\text{length})^2$ for moments of inertia of masses.

RESULTS

Moment of Inertia of an Object
The moment of inertia of an object about an axis L_O is defined by

$$I_O = \int_m r^2 \, dm, \qquad (8.27)$$

where r is the perpendicular distance from L_O to the differential element of mass dm.

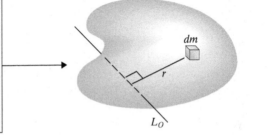

Slender Bars
The differential element of mass $dm = \rho A \, dr$, where ρ is the density of the homogeneous bar and A is its uniform cross-sectional area. The moment of inertia of the slender bar of length l about the perpendicular axis L through its center of mass is

$$I = \int_m r^2 \, dm = \int_{-l/2}^{l/2} \rho A r^2 \, dr = \frac{1}{12} \rho A l^3.$$

In terms of the mass of the bar $m = \rho A l$,

$$I = \frac{1}{12} m l^2. \qquad (8.28)$$

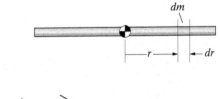
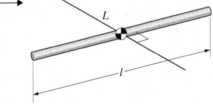

Thin Plates
The moments of inertia of a thin homogeneous plate of uniform thickness and mass m that lies in the x–y plane can be expressed in terms of the moments of inertia of the cross-sectional area A of the plate:

$$I_{x\,\text{axis}} = \frac{m}{A} I_x, \qquad (8.30)$$

$$I_{y\,\text{axis}} = \frac{m}{A} I_y, \qquad (8.31)$$

$$I_{z\,\text{axis}} = \frac{m}{A} J_O = I_{x\,\text{axis}} + I_{y\,\text{axis}}. \qquad (8.29)$$

Here I_x is the moment of inertia of A about the x axis, I_y is the moment of inertia of A about the y axis, and J_O is the polar moment of inertia of A about the origin.

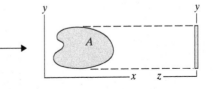

Active Example 8.9   Moment of Inertia of a Triangular Plate (▶ *Related Problem 8.104*)

The thin homogeneous plate is of uniform thickness and mass m. Determine its moment of inertia about the x axis.

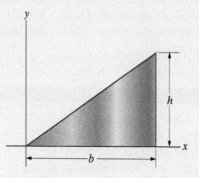

Strategy

The moment of inertia of the plate about the x axis is given by Eq. (8.30) in terms of the moment of inertia of the area of the plate about the x axis. We can obtain the moment of inertia of the area from Appendix B.

Solution

From Appendix B,
$$I_x = \frac{1}{12} bh^3 .$$

> Determine the moment of inertia of the area of the plate about the x axis.

The moment of inertia of the plate about the x axis is

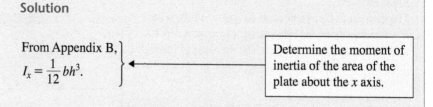

$$I_{x\,\text{axis}} = \frac{m}{A} I_x$$
$$= \frac{m}{\frac{1}{2} bh}\left(\frac{1}{12} bh^3\right)$$
$$= \frac{1}{6} mh^2 .$$

> Apply Eq. (8.30).

Practice Problem Determine the moment of inertia of the plate about the y axis.

Answer: $I_{y\,\text{axis}} = \frac{1}{2} mb^2 .$

Example 8.10   **Moments of Inertia of a Slender Bar** (▶ *Related Problem 8.100*)

Two homogeneous slender bars, each of length l, mass m, and cross-sectional area A, are welded together to form the L-shaped object. Determine the moment of inertia of the object about the axis L_O through point O. (The axis L_O is perpendicular to the two bars.)

Strategy
Using the same integration procedure we used for a single bar, we will determine the moment of inertia of each bar about L_O and sum the results.

Solution
Our first step is to introduce a coordinate system with the z axis along L_O and the x axis collinear with bar 1 (Fig. a). The mass of the differential element of bar 1 of length dx is $dm = \rho A\, dx$. The moment of inertia of bar 1 about L_O is

$$(I_O)_1 = \int_m r^2\, dm = \int_0^l \rho A x^2\, dx = \frac{1}{3}\rho A l^3.$$

In terms of the mass of the bar, $m = \rho A l$, we can write this result as

$$(I_O)_1 = \frac{1}{3}ml^2.$$

The mass of an element of bar 2 of length dy, shown in Fig. b, is $dm = \rho A\, dy$. From the figure we see that the perpendicular distance from L_O to the element is $r = \sqrt{l^2 + y^2}$. Therefore, the moment of inertia of bar 2 about L_O is

$$(I_O)_2 = \int_m r^2\, dm = \int_0^l \rho A(l^2 + y^2)\, dy = \frac{4}{3}\rho A l^3.$$

In terms of the mass of the bar, we obtain

$$(I_O)_2 = \frac{4}{3}ml^2.$$

The moment of inertia of the L-shaped object about L_O is

$$I_O = (I_O)_1 + (I_O)_2 = \frac{1}{3}ml^2 + \frac{4}{3}ml^2 = \frac{5}{3}ml^2.$$

Critical Thinking
In this example we used integration to determine a moment of inertia of an object consisting of two straight bars. The same procedure could be applied to more complicated objects made of such bars, but it would obviously be cumbersome. Once we have used integration to determine a moment of inertia of a single bar, such as Eq. (8.28), it would be very convenient to use that result to determine moments of inertia of composite objects made of bars without having to resort to integration. We show how this can be done in the next section.

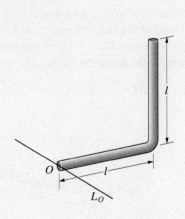

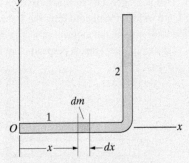

(a) Differential element of bar 1.

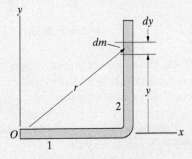

(b) Differential element of bar 2.

Problems

▶ 8.100 The axis L_O is perpendicular to both segments of the L-shaped slender bar. The mass of the bar is 6 kg and the material is homogeneous. Use the method described in Example 8.10 to determine the moment of inertia of the bar about L_O.

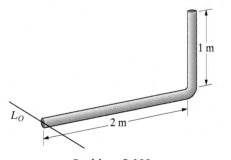

Problem 8.100

8.101 Two homogeneous slender bars, each of mass m and length l, are welded together to form the T-shaped object. Use integration to determine the moment of inertia of the object about the axis through point O that is perpendicular to the bars.

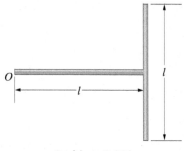

Problem 8.101

8.102 The slender bar lies in the x–y plane. Its mass is 6 kg and the material is homogeneous. Use integration to determine its moment of inertia about the z axis.

8.103 Use integration to determine the moment of inertia of the slender 6-kg bar about the y axis.

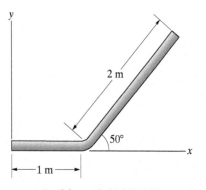

Problems 8.102/8.103

▶ 8.104 The homogeneous thin plate has mass $m = 12$ kg and dimensions $b = 2$ m and $h = 1$ m. Use the procedure described in Active Example 8.9 to determine the moments of inertia of the plate about the x and y axes.

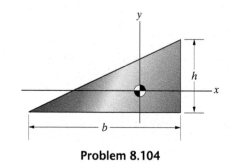

Problem 8.104

8.105 The homogeneous thin plate is of uniform thickness and mass m.

(a) Determine its moments of inertia about the x and z axes.

(b) Let $R_i = 0$ and compare your results with the values given in Appendix C for a thin circular plate.

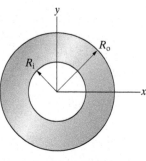

Problem 8.105

8.106 The homogeneous thin plate is of uniform thickness and weighs 20 lb. Determine its moment of inertia about the y axis.

8.107 Determine the moment of inertia of the plate about the x axis.

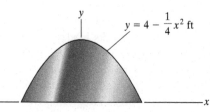

Problems 8.106/8.107

8.6 Parallel-Axis Theorem

BACKGROUND

The parallel-axis theorem allows us to determine the moment of inertia of an object about any axis when the moment of inertia about a parallel axis through the center of mass is known. This theorem can be used to calculate the moment of inertia of a composite object about an axis given the moments of inertia of each of its parts about parallel axes.

Suppose that we know the moment of inertia I about an axis L through the center of mass of an object, and we wish to determine its moment of inertia I_O about a parallel axis L_O (Fig. 8.18a). To determine I_O, we introduce parallel coordinate systems xyz and $x'y'z'$ with the z axis along L_O and the z' axis along L, as shown in Fig. 8.18b. (In this figure the axes L_O and L are perpendicular to the page.) The origin O of the xyz coordinate system is contained in the $x'-y'$ plane. The terms d_x and d_y are the coordinates of the center of mass relative to the xyz coordinate system.

The moment of inertia of the object about L_O is

$$I_O = \int_m r^2 \, dm = \int_m (x^2 + y^2) \, dm, \tag{8.33}$$

where r is the perpendicular distance from L_O to the differential element of mass dm, and x, y are the coordinates of dm in the $x-y$ plane. The coordinates of dm in the two coordinate systems are related by

$$x = x' + d_x, \qquad y = y' + d_y.$$

By substituting these expressions into Eq. (8.33), we can write it as

$$I_O = \int_m [(x')^2 + (y')^2] \, dm + 2d_x \int_m x' \, dm + 2d_y \int_m y' \, dm$$
$$+ \int_m (d_x^2 + d_y^2) \, dm. \tag{8.34}$$

Since $(x')^2 + (y')^2 = (r')^2$, where r' is the perpendicular distance from L to dm, the first integral on the right side of this equation is the moment of inertia I of the object about L. Recall that the x' and y' coordinates of the center of mass of the object relative to the $x'y'z'$ coordinate system are defined by

$$\bar{x}' = \frac{\int_m x' \, dm}{\int_m dm}, \qquad \bar{y}' = \frac{\int_m y' \, dm}{\int_m dm}.$$

Because the center of mass of the object is at the origin of the $x'y'z'$ system, $\bar{x}' = 0$ and $\bar{y}' = 0$. Therefore the integrals in the second and third terms on the right side of Eq. (8.34) are equal to zero. From Fig. 8.18b, we see that $d_x^2 + d_y^2 = d^2$, where d is the perpendicular distance between the axes L and L_O. Therefore, we obtain

$$I_O = I + d^2 m. \tag{8.35}$$

This is the *parallel-axis theorem* for moments of inertia of objects. Equation (8.35) relates the moment of inertia I of an object about an axis *through the center of mass* to its moment of inertia I_O about any parallel axis, where d is the perpendicular distance between the two axes and m is the mass of the object.

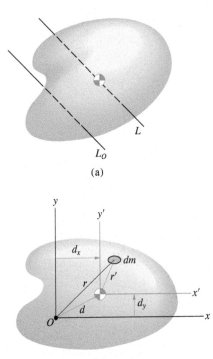

Figure 8.18
(a) An axis L through the center of mass of an object and a parallel axis L_O.
(b) The xyz and $x'y'z'$ coordinate systems.

Determining the moment of inertia of an object about a given axis L_O typically requires three steps:

1. **Choose the parts**—Try to divide the object into parts whose mass moments of inertia you know or can easily determine.

2. **Determine the moments of inertia of the parts**—You must first determine the moment of inertia of each part about the axis through its center of mass parallel to L_O. Then you can use the parallel-axis theorem to determine its moment of inertia about L_O.

3. **Sum the results**—Sum the moments of inertia of the parts (or subtract in the case of a hole or cutout) to obtain the moment of inertia of the composite object.

RESULTS

Parallel-Axis Theorem
The moment of inertia of an object of mass m about an axis L_O is given by

$$I_O = I + d^2 m, \tag{8.35}$$

where I is the moment of inertia of the object about a parallel axis through the center of mass and d is the perpendicular distance between the two axes.

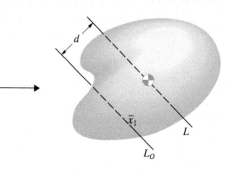

Composite Objects
The parallel-axis theorem makes it possible to determine the moment of inertia of a composite object about a given axis L_O. The moment of inertia of each part must be determined about an axis through the center of mass of the part that is parallel to L_O. Then the parallel-axis theorem can be applied to each part to determine its moment of inertia about L_O. Summing the results gives the moment of inertia of the composite object about L_O.

Active Example 8.11 Parallel-Axis Theorem (▶ *Related Problem 8.111*)

The homogeneous slender bar has mass m and length l. The axis L_O is perpendicular to the bar.
(a) Use integration to determine the moment of inertia of the bar about L_O.
(b) The moment of inertia of the bar about an axis through the center of mass of the bar that is perpendicular to the bar is $I = (1/12)ml^2$. Use this result and the parallel-axis theorem to determine the moment of inertia of the bar about L_O.

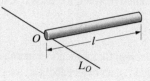

Solution

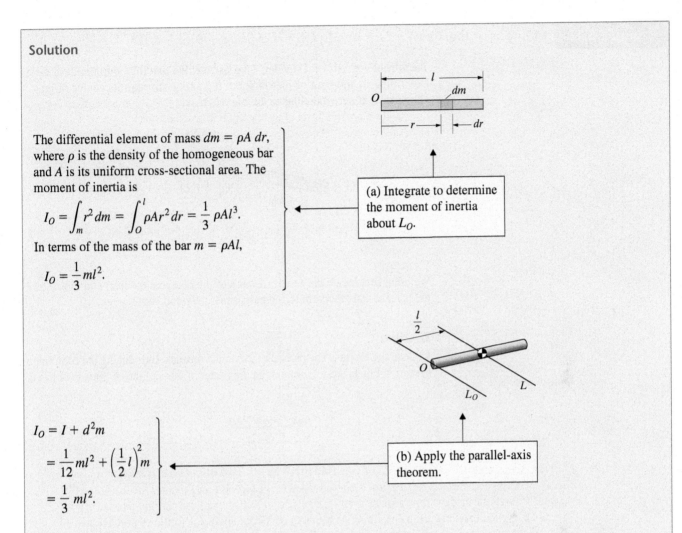

The differential element of mass $dm = \rho A \, dr$, where ρ is the density of the homogeneous bar and A is its uniform cross-sectional area. The moment of inertia is

$$I_O = \int_m r^2 \, dm = \int_0^l \rho A r^2 \, dr = \frac{1}{3} \rho A l^3.$$

In terms of the mass of the bar $m = \rho A l$,

$$I_O = \frac{1}{3} m l^2.$$

(a) Integrate to determine the moment of inertia about L_O.

$$I_O = I + d^2 m$$
$$= \frac{1}{12} m l^2 + \left(\frac{1}{2} l\right)^2 m$$
$$= \frac{1}{3} m l^2.$$

(b) Apply the parallel-axis theorem.

Practice Problem Two homogeneous slender bars, each of length l and mass m, are welded together to form an L-shaped object. Use the parallel-axis theorem to determine the moment of inertia of the object about the axis L_O. (The axis L_O is perpendicular to both bars.)

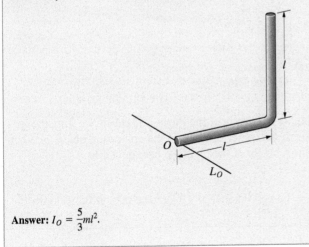

Answer: $I_O = \frac{5}{3} m l^2.$

Example 8.12   Moment of Inertia of a Composite Object (▶ *Related Problem 8.127*)

The object consists of a slender, 3-kg bar welded to a thin, circular 2-kg disk. Determine its moment of inertia about the axis L through its center of mass. (The axis L is perpendicular to the bar and disk.)

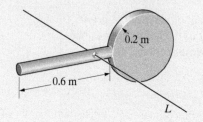

Strategy

We must first locate the center of mass of the composite object and then apply the parallel-axis theorem to the parts separately and sum the results.

Solution

Choose the Parts   The parts are the bar and the disk. Introducing the coordinate system in Fig. a, the x coordinate of the center of mass of the composite object is

$$\bar{x} = \frac{\bar{x}_{bar}m_{bar} + \bar{x}_{disk}m_{disk}}{m_{bar} + m_{disk}}$$

$$= \frac{(0.3 \text{ m})(3 \text{ kg}) + (0.6 \text{ m} + 0.2 \text{ m})(2 \text{ kg})}{(3 \text{ kg}) + (2 \text{ kg})} = 0.5 \text{ m}.$$

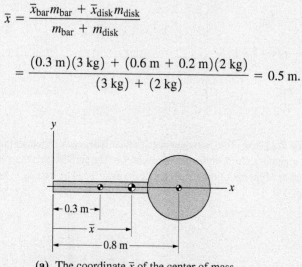

(a) The coordinate $\bar{x}$ of the center of mass of the object.

Determine the Moments of Inertia of the Parts   The distance from the center of mass of the bar to the center of mass of the composite object is 0.2 m (Fig. b). Therefore, the moment of inertia of the bar about L is

$$I_{bar} = \frac{1}{12}(3 \text{ kg})(0.6 \text{ m})^2 + (3 \text{ kg})(0.2 \text{ m})^2 = 0.210 \text{ kg-m}^2.$$

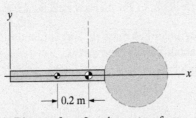

(b) Distance from L to the center of mass of
the bar.

The distance from the center of mass of the disk to the center of mass of the composite object is 0.3 m (Fig. c). The moment of inertia of the disk about L is

$$I_{disk} = \frac{1}{2}(2\ kg)(0.2\ m)^2 + (2\ kg)(0.3\ m)^2 = 0.220\ kg\text{-}m^2.$$

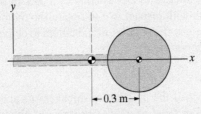

(c) Distance from L to the center of mass of
the disk.

Sum the Results   The moment of inertia of the composite object about L is

$$I = I_{bar} + I_{disk} = 0.430\ kg\text{-}m^2.$$

Critical Thinking

This example demonstrates the most common procedure for determining
moments of inertia of objects in engineering applications. Objects usually
consist of assemblies of parts. The center of mass of each part and its
moment of inertia about the axis through its center of mass must be determined. (It may be necessary to determine this information experimentally,
or it is sometimes supplied by manufacturers of subassemblies.) Then the
center of mass of the composite object is determined and the parallel-axis
theorem is used to determine the moment of inertia of each part about the
axis through the center of mass of the composite object. Finally, the individual moments of inertia are summed to obtain the moment of inertia of the
composite object.

Example 8.13   **Moments of Inertia of a Cylinder** (▶ *Related Problems 8.122, 8.123, 8.125, 8.126*)

The homogeneous cylinder has mass m, length l, and radius R. Determine its moments of inertia about the x, y, and z axes.

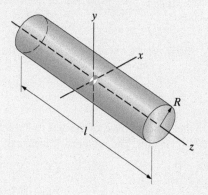

Strategy

We first determine the moments of inertia about the x, y, and z axes of an infinitesimal element of the cylinder consisting of a disk of thickness dz. We then integrate the results with respect to z to obtain the moments of inertia of the cylinder. We must apply the parallel-axis theorem to determine the moments of inertia of the disk about the x and y axes.

Solution

Consider an element of the cylinder of thickness dz at a distance z from the center of the cylinder (Fig. a). (You can imagine obtaining this element by "slicing" the cylinder perpendicular to its axis.) The mass of the element is equal to the product of the mass density and the volume of the element, $dm = \rho(\pi R^2\, dz)$. We obtain the moments of inertia of the element by using the values for a thin circular plate given in Appendix C. The moment of inertia about the z axis is

$$dI_{z\text{ axis}} = \frac{1}{2}dmR^2 = \frac{1}{2}(\rho\pi R^2\, dz)R^2.$$

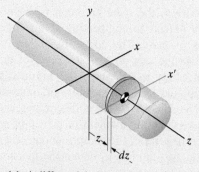

(a) A differential element of the cylinder in the form of a disk.

By integrating this result with respect to z from $-l/2$ to $l/2$, we sum the mass moments of inertia of the infinitesimal disk elements that make up the cylinder. The result is the moment of inertia of the cylinder about the z axis:

$$I_{z\ \text{axis}} = \int_{-l/2}^{l/2} \frac{1}{2}\rho\pi R^4\, dz = \frac{1}{2}\rho\pi R^4 l.$$

We can write this result in terms of the mass of the cylinder, $m = \rho(\pi R^2 l)$, as

$$I_{z\ \text{axis}} = \frac{1}{2}mR^2.$$

The moment of inertia of the disk element about the x' axis is

$$dI_{x'\ \text{axis}} = \frac{1}{4}dm\, R^2 = \frac{1}{4}(\rho\pi R^2\, dz)R^2.$$

We can use this result and the parallel-axis theorem to determine the moment of inertia of the element about the x axis:

$$dI_{x\ \text{axis}} = dI_{x'\ \text{axis}} + z^2\, dm = \frac{1}{4}(\rho\pi R^2\, dz)R^2 + z^2(\rho\pi R^2\, dz).$$

Integrating this expression with respect to z from $-l/2$ to $l/2$, we obtain the moment of inertia of the cylinder about the x axis:

$$I_{x\ \text{axis}} = \int_{-l/2}^{l/2}\left(\frac{1}{4}\rho\pi R^4 + \rho\pi R^2 z^2\right) dz = \frac{1}{4}\rho\pi R^4 l + \frac{1}{12}\rho\pi R^2 l^3.$$

In terms of the mass of the cylinder,

$$I_{x\ \text{axis}} = \frac{1}{4}mR^2 + \frac{1}{12}ml^2.$$

Due to the symmetry of the cylinder,

$$I_{y\ \text{axis}} = I_{x\ \text{axis}}.$$

Critical Thinking

When the cylinder is very long in comparison to its width, $l \gg R$, the first term in the equation for $I_{x\ \text{axis}}$ can be neglected, and we obtain the moment of inertia of a slender bar about a perpendicular axis, Eq. (8.28). Conversely, when the radius of the cylinder is much greater than its length, $R \gg l$, the second term in the equation for $I_{x\ \text{axis}}$ can be neglected, and we obtain the moment of inertia for a thin circular disk about an axis parallel to the disk. This indicates the sizes of the terms you neglect when you use the approximate expressions for the moments of inertia of a "slender" bar and a "thin" disk.

Problems

8.108 The mass of the object is 10 kg. Its moment of inertia about L_1 is 10 kg-m^2. What is its moment of inertia about L_2? (The three axes lie in the same plane.)

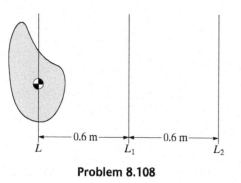

Problem 8.108

8.109 An engineer gathering data for the design of a maneuvering unit determines that the astronaut's center of mass is at $x = 1.01$ m, $y = 0.16$ m and that her moment of inertia about the z axis is 105.6 kg-m^2. Her mass is 81.6 kg. What is her moment of inertia about the z' axis through her center of mass?

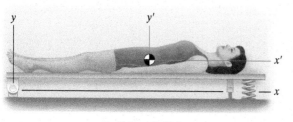

Problem 8.109

8.110 Two homogeneous slender bars, each of mass m and length l, are welded together to form the T-shaped object. Use the parallel-axis theorem to determine the moment of inertia of the object about the axis through point O that is perpendicular to the bars.

▶ **8.111** Use the parallel-axis theorem to determine the moment of inertia of the T-shaped object about the axis through the center of mass of the object that is perpendicular to the two bars. (See Active Example 8.11.)

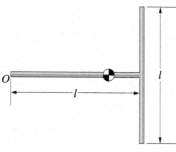

Problems 8.110/8.111

8.112 The mass of the homogeneous slender bar is 20 kg. Determine its moment of inertia about the z axis.

8.113 Determine the moment of inertia of the 20-kg bar about the z' axis through its center of mass.

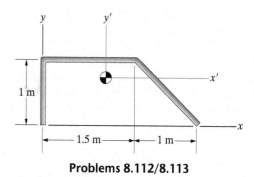

Problems 8.112/8.113

8.114 The homogeneous slender bar weighs 5 lb. Determine its moment of inertia about the z axis.

8.115 Determine the moment of inertia of the 5-lb bar about the z' axis through its center of mass.

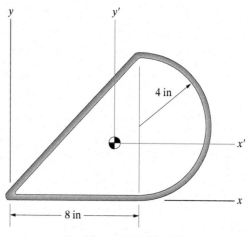

Problems 8.114/8.115

8.116 The rocket is used for atmospheric research. Its weight and its moment of inertia about the z axis through its center of mass (including its fuel) are 10 kip and 10,200 slug-ft^2, respectively. The rocket's fuel weighs 6000 lb, its center of mass is located at $x = -3$ ft, $y = 0$, $z = 0$, and the moment of inertia of the fuel about the axis through the fuel's center of mass parallel to z is 2200 slug-ft^2. When the fuel is exhausted, what is the rocket's moment of inertia about the axis through its new center of mass parallel to z?

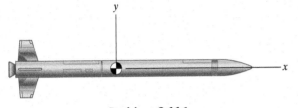

Problem 8.116

8.117 The mass of the homogeneous thin plate is 36 kg. Determine its moment of inertia about the x axis.

8.118 Determine the moment of inertia of the 36-kg plate about the z axis.

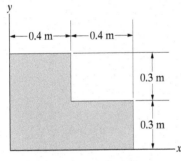

Problems 8.117/8.118

8.119 The homogeneous thin plate weighs 10 lb. Determine its moment of inertia about the x axis.

8.120 Determine the moment of inertia of the 10-lb plate about the y axis.

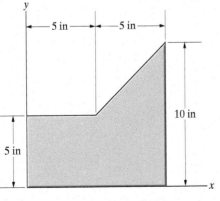

Problems 8.119/8.120

8.121 The thermal radiator (used to eliminate excess heat from a satellite) can be modeled as a homogeneous thin rectangular plate. Its mass is 5 slugs. Determine its moments of inertia about the x, y, and z axes.

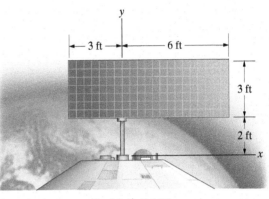

Problem 8.121

▶ **8.122** The homogeneous cylinder has mass m, length l, and radius R. Use integration as described in Example 8.13 to determine its moment of inertia about the x axis.

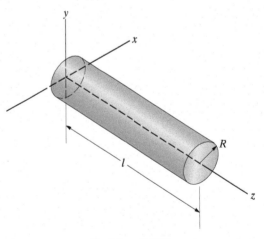

Problem 8.122

▶ 8.123 The homogeneous cone is of mass m. Determine its moment of inertia about the z axis, and compare your result with the value given in Appendix C. (See Example 8.13.)

8.124 Determine the moments of inertia of the homogeneous cone of mass m about the x and y axes, and compare your results with the values given in Appendix C.

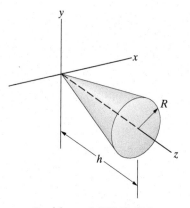

Problems 8.123/8.124

▶ 8.125 The mass of the homogeneous wedge is m. Use integration as described in Example 8.13 to determine its moment of inertia about the z axis. (Your answer should be in terms of m, a, b, and h.)

▶ 8.126 The mass of the homogeneous wedge is m. Use integration as described in Example 8.13 to determine its moment of inertia about the x axis. (Your answer should be in terms of m, a, b, and h.)

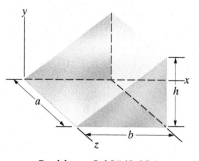

Problems 8.125/8.126

▶ 8.127 In Example 8.12, suppose that part of the 3-kg bar is sawed off so that the bar is 0.4 m long and its mass is 2 kg. Determine the moment of inertia of the composite object about the perpendicular axis L through the center of mass of the modified object.

8.128 The L-shaped machine part is composed of two homogeneous bars. Bar 1 is tungsten alloy with density 14,000 kg/m³, and bar 2 is steel with density 7800 kg/m³. Determine its moment of inertia about the x axis.

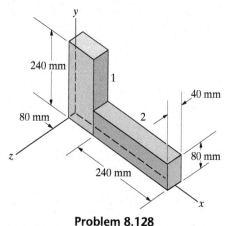

Problem 8.128

8.129 The homogeneous object is a cone with a conical hole. The dimensions $R_1 = 2$ in, $R_2 = 1$ in, $h_1 = 6$ in, and $h_2 = 3$ in. It consists of aluminum alloy with a density of 5 slug/ft³. Determine its moment of inertia about the x axis.

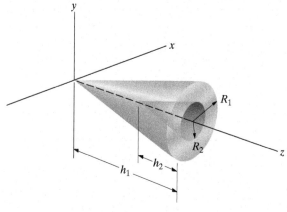

Problem 8.129

8.130 The circular cylinder is made of aluminum (Al) with density 2700 kg/m³ and iron (Fe) with density 7860 kg/m³. Determine its moments of inertia about the x' and y' axes.

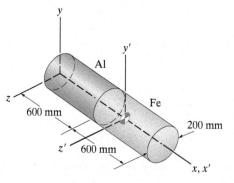

Problem 8.130

8.131 The homogeneous half-cylinder is of mass m. Determine its moment of inertia about the axis L through its center of mass.

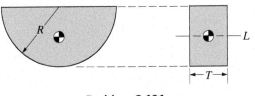

Problem 8.131

8.132 The homogeneous machine part is made of aluminum alloy with density $\rho = 2800$ kg/m³. Determine its moment of inertia about the z axis.

8.133 Determine the moment of inertia of the machine part described in Problem 8.132 about the x axis.

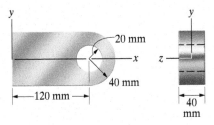

Problems 8.132/8.133

8.134 The object consists of steel of density $\rho = 7800$ kg/m³. Determine its moment of inertia about the axis L_O.

8.135 Determine the moment of inertia of the object in Problem 8.134 about the axis through the center of mass of the object parallel to L_O.

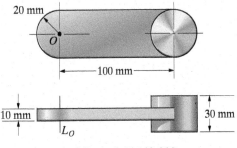

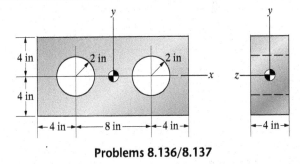

Problems 8.134/8.135

8.136 The thick plate consists of steel of density $\rho = 15$ slug/ft³. Determine its moment of inertia about the z axis.

8.137 Determine the moment of inertia of the plate in Problem 8.136 about the x axis.

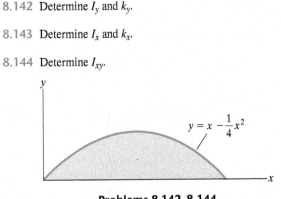

Problems 8.136/8.137

Review Problems

8.138 Determine I_y and k_y.

8.139 Determine I_x and k_x.

8.140 Determine J_O and k_O.

8.141 Determine I_{xy}.

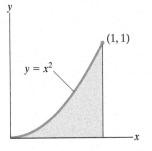

Problems 8.138–8.141

8.142 Determine I_y and k_y.

8.143 Determine I_x and k_x.

8.144 Determine I_{xy}.

$$y = x - \frac{1}{4}x^2$$

Problems 8.142–8.144

8.145 Determine $I_{y'}$ and $k_{y'}$.

8.146 Determine $I_{x'}$ and $k_{x'}$.

8.147 Determine $I_{x'y'}$.

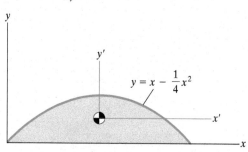

Problems 8.145–8.147

8.148 Determine I_y and k_y.

8.149 Determine I_x and k_x.

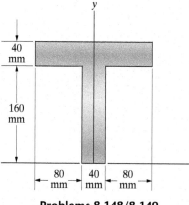

Problems 8.148/8.149

8.150 Determine I_x and k_x.

8.151 Determine J_O and k_O.

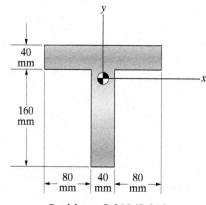

Problems 8.150/8.151

8.152 Determine I_y and k_y.

8.153 Determine J_O and k_O.

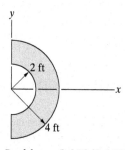

Problems 8.152/8.153

8.154 Determine I_x and k_x.

8.155 Determine I_y and k_y.

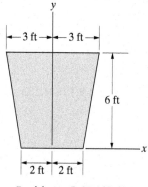

Problems 8.154/8.155

8.156 The moments of inertia of the area are $I_x = 36\ m^4$, $I_y = 145\ m^4$, and $I_{xy} = 44.25\ m^4$. Determine a set of principal axes and the principal moments of inertia.

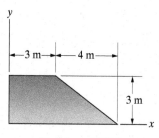

Problem 8.156

8.157 The moment of inertia of the 31-oz bat about a perpendicular axis through point B is 0.093 slug-ft^2. What is the bat's moment of inertia about a perpendicular axis through point A? (Point A is the bat's "instantaneous center," or center of rotation, at the instant shown.)

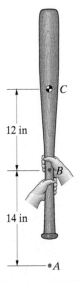

12 in

14 in

C

B

A

Problem 8.157

8.158 The mass of the thin homogeneous plate is 4 kg. Determine its moment of inertia about the y axis.

8.159 Determine the moment of inertia of the 4-kg plate about the z axis.

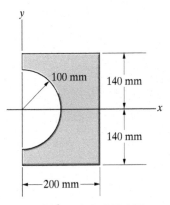

100 mm

140 mm

140 mm

200 mm

Problems 8.158/8.159

8.160 The homogeneous pyramid is of mass m. Determine its moment of inertia about the z axis.

8.161 Determine the moments of inertia of the homogeneous pyramid of mass m about the x and y axes.

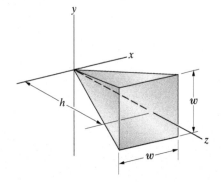

y

x

h

w

z

w

Problems 8.160/8.161

8.162 The homogeneous object weighs 400 lb. Determine its moment of inertia about the x axis.

8.163 Determine the moments of inertia of the 400-lb object about the y and z axes.

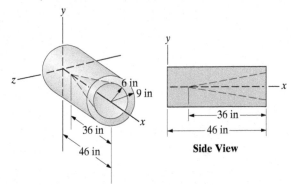

y

z

6 in

9 in

36 in

x

46 in

y

x

36 in

46 in

Side View

Problems 8.162/8.163

8.164 Determine the moment of inertia of the 14-kg flywheel about the axis L.

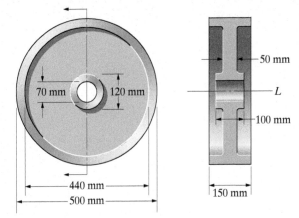

70 mm

120 mm

440 mm

500 mm

50 mm

L

100 mm

150 mm

Problem 8.164

Friction

Friction forces have many important effects, both desirable and undesirable, in engineering applications. The Coulomb theory of friction allows us to estimate the maximum friction forces that can be exerted by contacting surfaces and the friction forces exerted by sliding surfaces. This opens the path to the analysis of important new classes of supports and machines, including wedges (shims), threaded connections, bearings, and belts.

◄ The workpiece exerts normal and friction forces on the grinding wheel. In this chapter we analyze friction forces between contacting surfaces.

9.1 Theory of Dry Friction

BACKGROUND

Suppose that a person climbs a ladder that leans against a smooth wall. Figure 9.1a shows the free-body diagram of the person and ladder. If the person is stationary on the ladder, we can use the equilibrium equations to determine the friction force. But there is another question that we cannot answer using the equilibrium equations alone: Will the ladder remain in place, or will it slip on the floor? If a truck is parked on an incline, the total friction force exerted on its tires by the road prevents it from sliding down the incline (Fig. 9.1b). We can use the equilibrium equations to determine the total friction force. But here too there is another question that we cannot answer: What is the steepest incline on which the truck could be parked without slipping?

To answer these questions, we must examine the nature of friction forces in more detail. Place a book on a table and push it with a small horizontal force, as shown in Fig. 9.2a. If the force you exert is sufficiently small, the book does not move. The free-body diagram of the book is shown in Fig. 9.2b. The force W is the book's weight, and N is the total normal force exerted by the table on the surface of the book that is in contact with the table. The force F is the horizontal force you apply, and f is the total friction force exerted by the table. Because the book is in equilibrium, $f = F$.

Now slowly increase the force you apply to the book. As long as the book remains in equilibrium, the friction force must increase correspondingly, since it equals the force you apply. When the force you apply becomes too large, the book moves. It slips on the table. After reaching some maximum value, the friction force can no longer maintain the book in equilibrium. Also, notice that the force you must apply to keep the book moving on the table is smaller than the force required to cause it to slip. (You are familiar with this phenomenon if you've ever pushed a piece of furniture across a floor.)

Friction force

(a)

Friction force

(b)

Figure 9.1
Objects supported by friction forces.

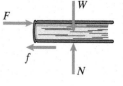

Figure 9.2
(a) Exerting a horizontal force on a book.
(b) The free-body diagram of the book.

(a)

(b)

How does the table exert a friction force on the book? Why does the book slip? Why is less force required to slide the book across the table than is required to start it moving? If the surfaces of the table and the book are magnified sufficiently, they will appear rough (Fig. 9.3). Friction forces arise in part from the interactions of the roughnesses, or *asperities,* of the contacting surfaces. We can gain insight into this mechanism of friction by considering a simple two-dimensional model of the rough surfaces of the book and table.

Suppose that we idealize the asperities of the book and table as the mating two-dimensional "saw-tooth" profiles in Fig. 9.4a. As the horizontal force F increases, the book will remain stationary until the force is sufficiently large to cause the book to slide upward as shown in Fig. 9.4b. What horizontal force is necessary for this to occur? To find out, we must determine the value of F necessary for the book to be in equilibrium in the "slipped" position in Fig. 9.4b. The normal force C_i exerted on the ith saw-tooth asperity of the book is shown in Fig. 9.4c. (Notice that in this simple model we assume the contacting surfaces of the asperities to be smooth.) Denoting the sum of the normal forces exerted on the asperities of the book by the table by $C = \sum_i C_i$, we obtain the equilibrium equations

$$\Sigma F_x = F - C \sin \alpha = 0,$$
$$\Sigma F_y = C \cos \alpha - W = 0.$$

Eliminating C from these equations, we obtain the force necessary to cause the book to slip on the table:

$$F = (\tan \alpha)W.$$

We see that *the force necessary to cause the book to slip is proportional to the force pressing the saw-tooth surfaces together* (the book's weight). Think about stacking increasing numbers of books and applying a horizontal force to them. A progressively larger force is required to cause them to slip as the number of books increases. Also, in our two-dimensional thought experiment, *the angle α is a measure of the roughness of the saw-tooth surfaces.* As $\alpha \to 0$, the surfaces become smooth and the force necessary to cause the book to slip approaches zero. As α increases, the roughness increases and the force necessary to cause the book to slip increases.

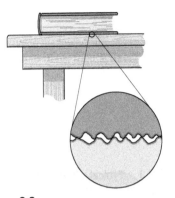

Figure 9.3
The roughnesses of the surfaces can be seen in a magnified view.

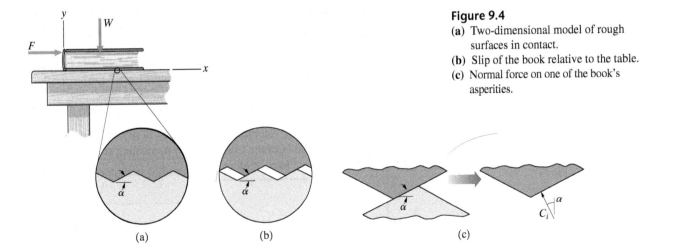

(a)   (b)   (c)

Figure 9.4
(a) Two-dimensional model of rough surfaces in contact.
(b) Slip of the book relative to the table.
(c) Normal force on one of the book's asperities.

| Example 9.2 | Analyzing a Friction Brake (▶ *Related Problem 9.22*) |

The motion of the disk is controlled by the friction force exerted at C by the brake ABC. The hydraulic actuator BE exerts a horizontal force of magnitude F on the brake at B. The coefficients of friction between the disk and the brake are μ_s and μ_k. What couple M is necessary to rotate the disk at a constant rate in the counterclockwise direction?

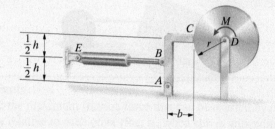

Strategy

We can use the free-body diagram of the disk to obtain a relation between M and the reaction exerted on the disk by the brake, then use the free-body diagram of the brake to determine the reaction in terms of F.

Solution

We draw the free-body diagram of the disk in Fig. a, representing the force exerted by the brake by a single force R. The force R opposes the counterclockwise rotation of the disk, and the friction angle is the angle of kinetic friction $\theta_k = \arctan \mu_k$. Summing moments about D, we obtain

$$\Sigma M_{\text{point } D} = M - (R \sin \theta_k)r = 0.$$

(a) The free-body diagram of the disk.

Then, from the free-body diagram of the brake (Fig. b), we get

$$\Sigma M_{\text{point } A} = -F\left(\frac{1}{2}h\right) + (R \cos \theta_k)h - (R \sin \theta_k)b = 0.$$

We can solve these two equations for M and R. The solution for the couple M is

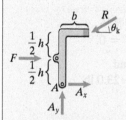

(b) The free-body diagram of the brake.

$$M = \frac{(1/2)hr\, F \sin \theta_k}{h \cos \theta_k - b \sin \theta_k} = \frac{(1/2)hr\, F\mu_k}{h - b\mu_k}.$$

Critical Thinking

If the friction coefficient μ_k is sufficiently small, the denominator in our solution for the couple M, the term $h \cos \theta_k - b \sin \theta_k$, is positive. As μ_k increases, the denominator becomes smaller, because $\cos \theta_k$ decreases and $\sin \theta_k$ increases. As the denominator approaches zero, the couple required to rotate the disk approaches infinity. To understand this result, notice that the denominator equals zero when $\tan \theta_k = h/b$, which means that the line of action of the force R passes through point A (Fig. c). As μ_k increases and the line of action of R approaches point A, the magnitude of R necessary to balance the moment due to F about A approaches infinity. As a result, the *analytical prediction* for M approaches infinity. Of course, at some value of M, the forces F and R would exceed the values the brake could support.

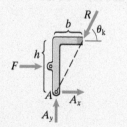

(c) The line of action of R passing through point A.

Example 9.3   Determining Whether an Object Will Tip Over (▶ *Related Problem 9.45*)

Suppose that we want to push the tool chest across the floor by applying the horizontal force F. If we apply the force at too great a height h, the chest will tip over before it slips. If the coefficient of static friction between the floor and the chest is μ_s, what is the largest value of h for which the chest will slip before it tips over?

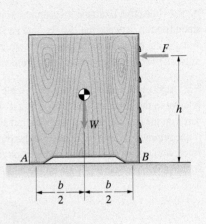

Strategy
When the chest is on the verge of tipping over, it is in equilibrium with no reaction at B. We can use this condition to determine F in terms of h. Then, by determining the value of F that will cause the chest to slip, we will obtain the value of h that causes the chest to be on the verge of tipping over *and* on the verge of slipping.

Solution
We draw the free-body diagram of the chest when it is on the verge of tipping over in Fig. a. Summing moments about A, we obtain

$$\Sigma M_{\text{point } A} = Fh - W\left(\frac{1}{2}b\right) = 0.$$

Equilibrium also requires that $f = F$ and $N = W$.
   When the chest is on the verge of slipping,

$$f = \mu_s N,$$

so

$$F = f = \mu_s N = \mu_s W.$$

Substituting this expression into the moment equation, we obtain

$$\mu_s Wh - W\left(\frac{1}{2}b\right) = 0.$$

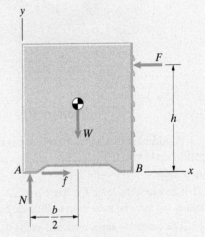

(a) The free-body diagram when the chest is on the verge of tipping over.

Solving this equation for h, we find that when the chest is on the verge of slipping, it is also on the verge of tipping over if it is pushed at the height

$$h = \frac{b}{2\mu_s}.$$

If h is smaller than this value, the chest will begin sliding before it tips over.

Critical Thinking
Notice that the largest value of h for which the chest will slip before it tips over is independent of F. Whether the chest will tip over depends only on where the force is applied, not how large it is. What is the motivation for the solution in this example? The possibility of heavy objects falling over is an obvious safety hazard, and analyses of this kind can influence their design. Once they are in use, safety engineers can establish guidelines (for example, by marking a horizontal line on a vertical cabinet or machine above which it should not be pushed) to prevent tipping.

Problems

▶ **9.1** In Active Example 9.1, suppose that the coefficient of static friction between the 180-lb crate and the ramp is $\mu_s = 0.3$. What is the magnitude of the smallest horizontal force the rope must exert on the crate to prevent it from sliding down the ramp?

9.2 A person places a 2-lb book on a table that is tilted at $15°$ relative to the horizontal. She finds that if she exerts a very small force on the book as shown, the book remains in equilibrium, but if she removes the force, the book slides down the table. What force would she need to exert on the book (in the direction parallel to the table) to cause it to slide up the table?

Problem 9.2

9.3 A student pushes a 200-lb box of books across the floor. The coefficient of kinetic friction between the carpet and the box is $\mu_k = 0.15$.

(a) If he exerts the force F at angle $\alpha = 25°$, what is the magnitude of the force he must exert to slide the box across the floor?

(b) If he bends his knees more and exerts the force F at angle $\alpha = 10°$, what is the magnitude of the force he must exert to slide the box?

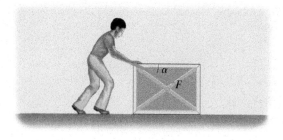

Problem 9.3

9.4 The 2975-lb car is parked on a sloped street. The brakes are applied to both its front and rear wheels.

(a) If the coefficient of static friction between the car's tires and the road is $\mu_s = 0.8$, what is the steepest slope (in degrees relative to the horizontal) on which the car could remain in equilibrium?

(b) If the street were icy and the coefficient of static friction between the car's tires and the road was $\mu_s = 0.2$, what is the steepest slope on which the car could remain in equilibrium?

Problem 9.4

9.5 The truck's winch exerts a horizontal force on the 200-kg crate in an effort to pull it down the ramp. The coefficient of static friction between the crate and the ramp is $\mu_s = 0.6$.

(a) If the winch exerts a 200-N horizontal force on the crate, what is the magnitude of the friction force exerted on the crate by the ramp?

(b) What is the magnitude of the horizontal force the winch must exert on the crate to cause it to start moving down the ramp?

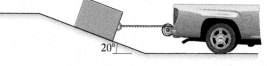

Problem 9.5

9.6 The device shown is designed to position pieces of luggage on a ramp. It exerts a force parallel to the ramp. The suitcase weighs 40 lb. The coefficients of friction between the suitcase and the ramp are $\mu_s = 0.20$ and $\mu_k = 0.18$.

(a) Will the suitcase remain stationary on the ramp when the device exerts no force on it?

(b) What force must the device exert to push the suitcase up the ramp at a constant speed?

Problem 9.6

9.7 The coefficient of static friction between the 50-kg crate and the ramp is $\mu_s = 0.35$. The unstretched length of the spring is 800 mm, and the spring constant is $k = 660$ N/m. What is the minimum value of x at which the crate can remain stationary on the ramp?

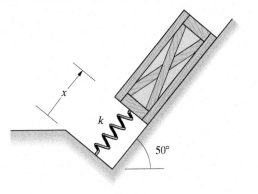

Problem 9.7

9.8 The coefficient of kinetic friction between the 40-kg crate and the slanting floor is $\mu_k = 0.3$. If the angle $\alpha = 20°$, what tension must the person exert on the rope to move the crate at constant speed?

9.9 In Problem 9.8, for what angle α is the tension necessary to move the crate at constant speed a minimum? What is the necessary tension?

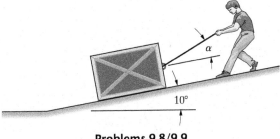

Problems 9.8/9.9

9.10 Box A weighs 100 lb and box B weighs 30 lb. The coefficients of friction between box A and the ramp are $\mu_s = 0.30$ and $\mu_k = 0.28$. What is the magnitude of the friction force exerted on box A by the ramp?

9.11 Box A weighs 100 lb, and the coefficients of friction between box A and the ramp are $\mu_s = 0.30$ and $\mu_k = 0.28$. For what range of weights of the box B will the system remain stationary?

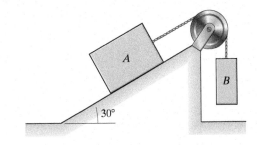

Problems 9.10/9.11

9.12 The mass of the box on the left is 30 kg, and the mass of the box on the right is 40 kg. The coefficient of static friction between each box and the inclined surface is $\mu_s = 0.2$. Determine the minimum angle α for which the boxes will remain stationary.

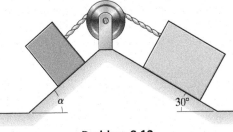

Problem 9.12

9.13 The coefficient of kinetic friction between the 100-kg box and the inclined surface is 0.35. Determine the tension T necessary to pull the box up the surface at a constant rate.

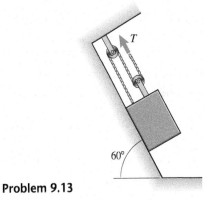

Problem 9.13

9.14 The box is stationary on the inclined surface. The coefficient of static friction between the box and the surface is μ_s.

(a) If the mass of the box is 10 kg, $\alpha = 20°$, $\beta = 30°$, and $\mu_s = 0.24$, what force T is necessary to start the box sliding up the surface?

(b) Show that the force T necessary to start the box sliding up the surface is a minimum when $\tan \beta = \mu_s$.

9.15 In explaining observations of ship launchings at the port of Rochefort in 1779, Coulomb analyzed the system shown to determine the minimum force T necessary to hold the box stationary on the inclined surface. Show that the result is

$$T = \frac{(\sin \alpha - \mu_s \cos \alpha)mg}{\cos \beta - \mu_s \sin \beta}.$$

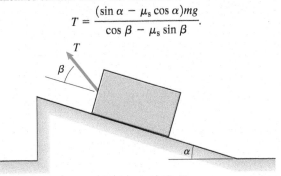

Problems 9.14/9.15

9.16 Two sheets of plywood A and B lie on the bed of the truck. They have the same weight W, and the coefficient of static friction between the two sheets of wood and between sheet B and the truck bed is μ_s.

(a) If you apply a horizontal force to sheet A and apply no force to sheet B, can you slide sheet A off the truck without causing sheet B to move? What force is necessary to cause sheet A to start moving?

(b) If you prevent sheet A from moving by exerting a horizontal force on it, what horizontal force on sheet B is necessary to start it moving?

Problem 9.16

9.17 The weights of the two boxes are $W_1 = 100$ lb and $W_2 = 50$ lb. The coefficients of friction between the left box and the inclined surface are $\mu_s = 0.12$ and $\mu_k = 0.10$. Determine the tension the man must exert on the rope to pull the boxes upward at a constant rate.

9.18 In Problem 9.17, for what range of tensions exerted on the rope by the man will the boxes remain stationary?

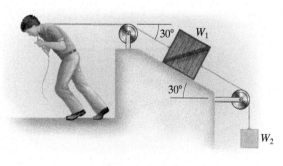

Problems 9.17/9.18

9.19 Each box weighs 10 lb. The coefficient of static friction between box A and box B is 0.24, and the coefficient of static friction between box B and the inclined surface is 0.3. What is the largest angle α for which box B will not slip?

Strategy: Draw individual free-body diagrams of the two boxes and write their equilibrium equations assuming that slip of box B is impending.

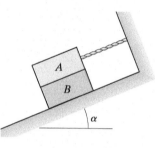

Problem 9.19

9.20 The masses of the boxes are $m_A = 15$ kg and $m_B = 60$ kg. The coefficient of static friction between boxes A and B and between box B and the inclined surface is 0.12. What is the largest force F for which the boxes will not slip?

9.21 In Problem 9.20, what is the smallest force F for which the boxes will not slip?

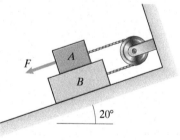

Problems 9.20/9.21

▶ **9.22** In Example 9.2, what clockwise couple M would need to be applied to the disk to cause it to rotate at a constant rate in the clockwise direction?

9.23 The homogeneous horizontal bar AB weighs 20 lb. The homogeneous disk weighs 30 lb. The coefficient of kinetic friction between the disk and the sloping surface is $\mu_k = 0.24$. What is the magnitude of the couple that would need to be applied to the disk to cause it to rotate at a constant rate in the clockwise direction?

9.24 The homogeneous horizontal bar AB weighs 20 lb. The homogeneous disk weighs 30 lb. The coefficient of kinetic friction between the disk and the sloping surface is $\mu_k = 0.24$. What is the magnitude of the couple that would need to be applied to the disk to cause it to rotate at a constant rate in the counterclockwise direction?

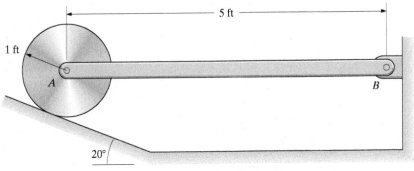

Problems 9.23/9.24

9.25 The mass of the bar is 4 kg. The coefficient of static friction between the bar and the floor is 0.3. Neglect friction between the bar and the wall.

(a) If $\alpha = 20°$, what is the magnitude of the friction force exerted on the bar by the floor?

(b) What is the maximum angle α for which the bar will not slip?

9.26 The coefficient of static friction between the bar and the floor and between the 4-kg bar and the wall is 0.3. What is the maximum angle α for which the bar will not slip?

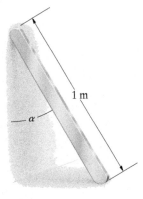

Problems 9.25/9.26

9.27 The ladder and the person weigh 30 lb and 180 lb, respectively. The center of mass of the 12-ft ladder is at its midpoint. The angle $\alpha = 30°$. Assume that the wall exerts a negligible friction force on the ladder.

(a) If $x = 4$ ft, what is the magnitude of the friction force exerted on the ladder by the floor?

(b) What minimum coefficient of static friction between the ladder and the floor is necessary for the person to be able to climb to the top of the ladder without slipping?

9.28 The ladder and the person weigh 30 lb and 180 lb, respectively. The center of mass of the 12-ft ladder is at its midpoint. The coefficient of static friction between the ladder and the floor is $\mu_s = 0.5$. What is the largest value of the angle α for which the person could climb to the top of the ladder without it slipping?

9.29 The ladder and the person weigh 30 lb and 180 lb, respectively. The center of mass of the 12-ft ladder is at its midpoint. The coefficient of static friction between the ladder and the floor is 0.5 and the coefficient of friction between the ladder and the wall is 0.3. What is the largest value of the angle α for which the person could climb to the top of the ladder without it slipping? Compare your answer to the answer to Problem 9.28.

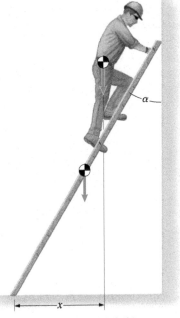

Problems 9.27–9.29

9.30 The disk weighs 50 lb and the bar weighs 25 lb. The coefficients of friction between the disk and the inclined surface are $\mu_s = 0.6$ and $\mu_k = 0.5$.

(a) What is the largest couple M that can be applied to the stationary disk without causing it to start rotating?

(b) What couple M is necessary to rotate the disk at a constant rate?

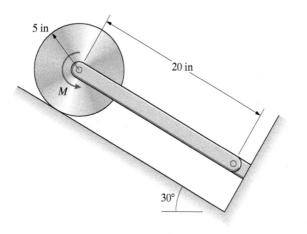

Problem 9.30

9.31 The radius of the 40-kg homogeneous cylinder is $R = 0.15$ m. The slanted wall is smooth and the angle $\alpha = 30°$. The coefficient of static friction between the cylinder and the floor is $\mu_s = 0.2$. What is the largest couple M that can be applied to the cylinder without causing it to slip?

9.32 The homogeneous cylinder has weight W. The coefficient of static friction between the cylinder and both surfaces is μ_s. What is the largest couple M that can be applied to the cylinder without causing it to slip? (Assume that the cylinder slips before rolling up the inclined surface.)

9.33 The homogeneous cylinder has weight W. The coefficient of static friction between the cylinder and both surfaces is μ_s. What is the minimum value of μ_s for which the couple M will cause the cylinder to roll up the inclined surface without slipping?

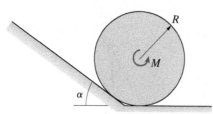

Problems 9.31–9.33

9.34 The coefficient of static friction between the blades of the shears and the object they are gripping is 0.36. What is the largest value of the angle α for which the object will not slip out? Neglect the object's weight.

Strategy: Draw the free-body diagram of the object and assume that slip is impending.

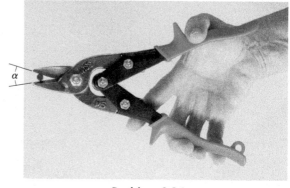

Problem 9.34

9.35 A stationary disk of 300-mm radius is attached to a pin support at D. The disk is held in place by the brake ABC in contact with the disk at C. The hydraulic actuator BE exerts a horizontal 400-N force on the brake at B. The coefficients of friction between the disk and the brake are $\mu_s = 0.6$ and $\mu_k = 0.5$. What couple must be applied to the stationary disk to cause it to slip in the counterclockwise direction?

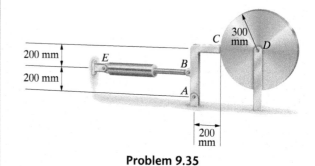

Problem 9.35

9.36 The figure shows a preliminary conceptual idea for a device to exert a braking force on a rope when the rope is pulled downward by the force T. The coefficient of kinetic friction between the rope and the two bars is $\mu_k = 0.28$. Determine the force T necessary to pull the rope downward at a constant rate if $F = 10$ lb and (a) $\alpha = 30°$; (b) $\alpha = 20°$.

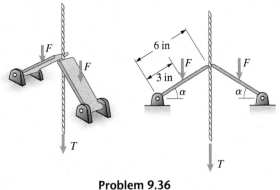

Problem 9.36

9.37 The mass of block B is 8 kg. The coefficient of static friction between the surfaces of the clamp and the block is $\mu_s = 0.2$. When the clamp is aligned as shown, what minimum force must the spring exert to prevent the block from slipping out?

9.38 By altering its dimensions, redesign the clamp in Problem 9.37 so that the minimum force the spring must exert to prevent the block from slipping out is 180 N. Draw a sketch of your new design.

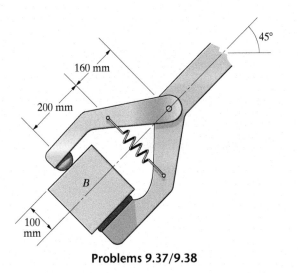

Problems 9.37/9.38

9.39 The horizontal bar is attached to a collar that slides on the smooth vertical bar. The collar at P slides on the smooth horizontal bar. The total mass of the horizontal bar and the two collars is 12 kg. The system is held in place by the pin in the circular slot. The pin contacts only the lower surface of the slot, and the coefficient of static friction between the pin and the slot is 0.8. If the system is in equilibrium and $y = 260$ mm, what is the magnitude of the friction force exerted on the pin by the slot?

9.40 In Problem 9.39, what is the minimum height y at which the system can be in equilibrium?

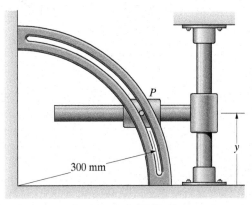

Problems 9.39/9.40

9.41 The rectangular 100-lb plate is supported by the pins A and B. If friction can be neglected at A and the coefficient of static friction between the pin at B and the slot is $\mu_s = 0.4$, what is the largest angle α for which the plate will not slip?

9.42 If you can neglect friction at B and the coefficient of static friction between the pin at A and the slot is $\mu_s = 0.4$, what is the largest angle α for which the 100-lb plate will not slip?

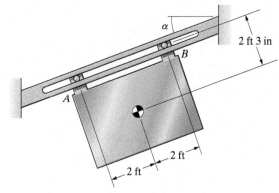

Problems 9.41/9.42

9.43 The airplane's weight is $W = 2400$ lb. Its brakes keep the rear wheels locked, and the coefficient of static friction between the wheels and the runway is $\mu_s = 0.6$. The front (nose) wheel can turn freely and so exerts a negligible friction force on the runway. Determine the largest horizontal thrust force T the plane's propeller can generate without causing the rear wheels to slip.

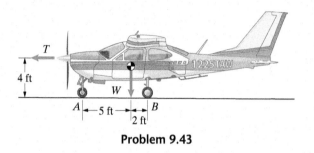

Problem 9.43

9.44 The refrigerator weighs 220 lb. It is supported at A and B. The coefficient of static friction between the supports and the floor is $\mu_s = 0.2$. If you assume that the refrigerator does not tip over before it slips, what force F is necessary for impending slip?

▶ **9.45** The refrigerator weighs 220 lb. It is supported at A and B. The coefficient of static friction between the supports and the floor is $\mu_s = 0.2$. The distance $h = 60$ in and the dimension $b = 30$ in. When the force F is applied to push the refrigerator across the floor, will it tip over before it slips? (See Example 9.3.)

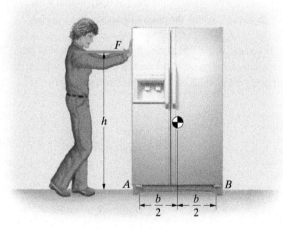

Problems 9.44/9.45

9.46 To obtain a preliminary evaluation of the stability of a turning car, imagine subjecting the stationary car to an increasing lateral force F at the height of its center of mass, and determine whether the car will slip (skid) laterally before it tips over. Show that this will be the case if $b/h > 2\mu_s$. (Notice the importance of the height of the center of mass relative to the width of the car. This reflects on recent discussions of the stability of sport utility vehicles and vans that have relatively high centers of mass.)

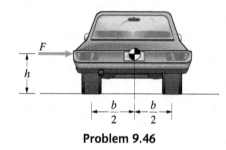

Problem 9.46

9.47 The man exerts a force P on the car at an angle $\alpha = 20°$. The 1760-kg car has front wheel drive. The driver spins the front wheels, and the coefficient of kinetic friction is $\mu_k = 0.02$. Snow behind the rear tires exerts a horizontal resisting force S. Getting the car to move requires overcoming a resisting force $S = 420$ N. What force P must the man exert?

9.48 In Problem 9.47, what value of the angle α minimizes the magnitude of the force P the man must exert to overcome the resisting force $S = 420$ N exerted on the rear tires by the snow? What force must he exert?

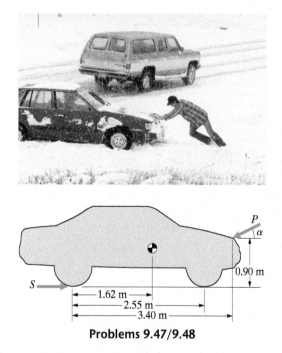

Problems 9.47/9.48

9.49 The coefficient of static friction between the 3000-lb car's tires and the road is $\mu_s = 0.5$. Determine the steepest grade (the largest value of the angle α) the car can drive up at constant speed if the car has (a) rear-wheel drive; (b) front-wheel drive; (c) four-wheel drive.

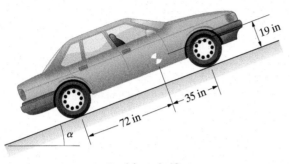

Problem 9.49

9.50 The stationary cabinet has weight W. Determine the force F that must be exerted to cause it to move if (a) the coefficient of static friction at A and at B is μ_s; (b) the coefficient of static friction at A is μ_{sA} and the coefficient of static friction at B is μ_{sB}.

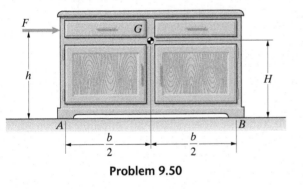

Problem 9.50

9.51 The table weighs 50 lb and the coefficient of static friction between its legs and the inclined surface is 0.7.

(a) If you apply a force at A parallel to the inclined surface to push the table up the inclined surface, will the table tip over before it slips? If not, what force is required to start the table moving up the surface?

(b) If you apply a force at B parallel to the inclined surface to push the table down the inclined surface, will the table tip over before it slips? If not, what force is required to start the table moving down the surface?

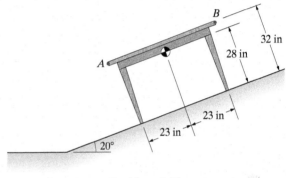

Problem 9.51

9.52 The coefficient of static friction between the right bar and the surface at A is $\mu_s = 0.6$. Neglect the weights of the bars. If $\alpha = 20°$, what is the magnitude of the friction force exerted at A?

9.53 The coefficient of static friction between the right bar and the surface at A is $\mu_s = 0.6$. Neglect the weights of the bars. What is the largest angle α at which the truss will remain stationary without slipping?

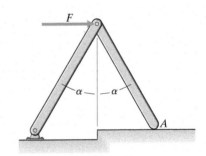

Problems 9.52/9.53

9.54 The bar BC is supported by a rough floor at C. If $F = 2$ kN and bar BC does not slip at C, what is the magnitude of the friction force exerted on the bar at C?

9.55 The bar BC is supported by a rough floor at C. If $F = 2$ kN, what is the minimum coefficient of static friction for which bar BC will not slip at C?

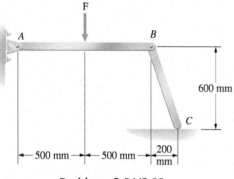

Problems 9.54/9.55

9.56 The weight of the box is 20 lb and the coefficient of static friction between the box and the floor is $\mu_s = 0.65$. Neglect the weights of the bars. What is the largest value of the force F that will not cause the box to slip?

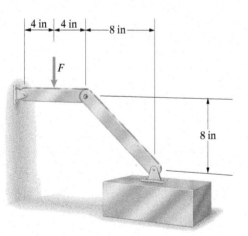

Problem 9.56

9.57 The mass of the suspended object is 6 kg. The structure is supported at B by the normal and friction forces exerted on the plate by the wall. Neglect the weights of the bars.

(a) What is the magnitude of the friction force exerted on the plate at B?

(b) What is the minimum coefficient of static friction at B necessary for the structure to remain in equilibrium?

9.58 Suppose that the lengths of the bars in Problem 9.57 are $L_{AB} = 1.2$ m and $L_{AC} = 1.0$ m and their masses are $m_{AB} = 3.6$ kg and $m_{AC} = 3.0$ kg.

(a) What is the magnitude of the friction force exerted on the plate at B?

(b) What is the minimum coefficient of static friction at B necessary for the structure to remain in equilibrium?

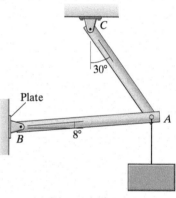

Problems 9.57/9.58

9.59 The frame is supported by the normal and friction forces exerted on the plates at A and G by the fixed surfaces. The coefficient of static friction at A is $\mu_s = 0.6$. Will the frame slip at A when it is subjected to the loads shown?

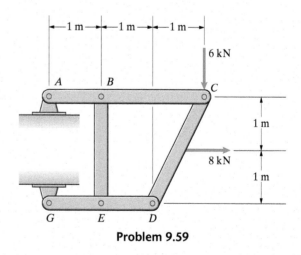

Problem 9.59

9.60 The frame is supported by the normal and friction forces exerted on the plate at A by the wall.

(a) What is the magnitude of the friction force exerted on the plate at A?

(b) What is the minimum coefficient of static friction at A necessary for the structure to remain in equilibrium?

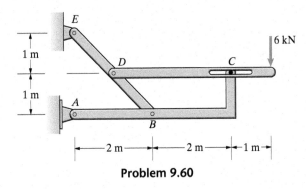

Problem 9.60

9.61 The direction cosines of the crane's cable are $\cos \theta_x = 0.588$, $\cos \theta_y = 0.766$, $\cos \theta_z = 0.260$. The y axis is vertical. The stationary caisson to which the cable is attached weighs 2000 lb and rests on horizontal ground. If the coefficient of static friction between the caisson and the ground is $\mu_s = 0.4$, what tension in the cable is necessary to cause the caisson to slip?

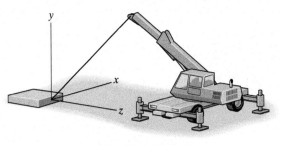

Problem 9.61

9.62* The 10-lb metal disk A is at the center of the inclined surface. The tension in the string AB is 5 lb. What minimum coefficient of static friction between the disk and the surface is necessary to keep the disk from slipping?

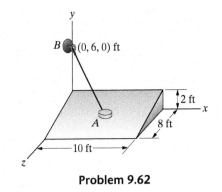

Problem 9.62

9.63* The 5-kg box is at rest on the sloping surface. The y axis points upward. The unit vector $0.557\mathbf{i} + 0.743\mathbf{j} + 0.371\mathbf{k}$ is perpendicular to the sloping surface. What is the magnitude of the friction force exerted on the box by the surface?

9.64* In Problem 9.63, what is the minimum coefficient of static friction necessary for the box to remain at rest on the sloping surface?

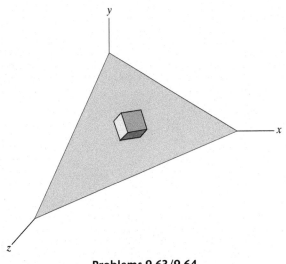

Problems 9.63/9.64

9.2 Wedges

A *wedge* is a bifacial tool with the faces set at a small acute angle (Figs. 9.8a and b). When a wedge is pushed forward, the faces exert large normal forces as a result of the small angle between them (Fig. 9.8c). In various forms, wedges are used in many engineering applications.

The large lateral force generated by a wedge can be used to lift a load (Fig. 9.9a). Let W_L be the weight of the load and W_W the weight of the wedge. To determine the force F necessary to start raising the load, we assume that slip of the load and wedge are impending (Fig. 9.9b). From the free-body diagram of the load, we obtain the equilibrium equations

$$\Sigma F_x = Q - N \sin \alpha - \mu_s N \cos \alpha = 0$$

and

$$\Sigma F_y = N \cos \alpha - \mu_s N \sin \alpha - \mu_s Q - W_L = 0.$$

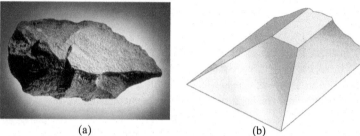

(a)   (b)

Figure 9.8
(a) An early wedge tool—a bifacial "hand axe" from Olduvai Gorge, East Africa.
(b) A modern chisel blade.
(c) The faces of a wedge can exert large lateral forces.

(c)

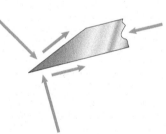

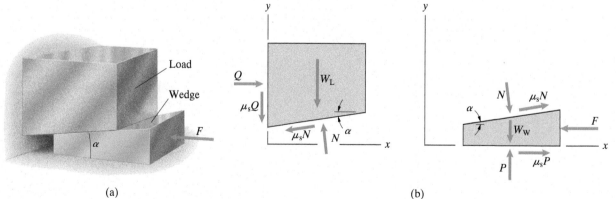

(a)   (b)

Figure 9.9
(a) Raising a load with a wedge.
(b) Free-body diagrams of the load and the wedge when slip is impending.

From the free-body diagram of the wedge, we obtain the equations

$$\Sigma F_x = N \sin \alpha + \mu_s N \cos \alpha + \mu_s P - F = 0$$

and

$$\Sigma F_y = P - N \cos \alpha + \mu_s N \sin \alpha - W_W = 0.$$

These four equations determine the three normal forces Q, N, and P, and the force F. The solution for F is

$$F = \mu_s W_W + \left[\frac{(1 - \mu_s^2) \tan \alpha + 2\mu_s}{(1 - \mu_s^2) - 2\mu_s \tan \alpha} \right] W_L.$$

Suppose that $W_W = 0.2W_L$ and $\alpha = 10°$. If $\mu_s = 0$, the force necessary to lift the load is only $0.176W_L$. But if $\mu_s = 0.2$, the force becomes $0.680W_L$, and if $\mu_s = 0.4$, it becomes $1.44W_L$. From this standpoint, friction is undesirable. But if there were no friction, the wedge would not remain in place when the force F is removed.

Active Example 9.4   Forces on a Wedge (▶ *Related Problems 9.65, 9.66, 9.67*)

A wedge is used to split a log. The angle $\alpha = 10°$. The coefficients of friction between the wedge and the log are $\mu_s = 0.22$ and $\mu_k = 0.20$. If the wedge is driven into the log at a constant speed by a vertical force F, what are the magnitudes of the normal forces exerted on the log by the wedge (that is, what are the magnitudes of the forces causing the log to split)?

Strategy
The friction forces exerted on the wedge by the log resist the motion of the wedge into the log and are of magnitude $\mu_k N$. We can apply equilibrium to the wedge to determine N in terms of F.

Solution

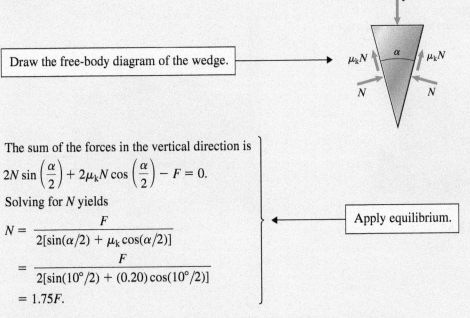

Draw the free-body diagram of the wedge.

The sum of the forces in the vertical direction is

$$2N \sin \left(\frac{\alpha}{2} \right) + 2\mu_k N \cos \left(\frac{\alpha}{2} \right) - F = 0.$$

Solving for N yields

$$N = \frac{F}{2[\sin(\alpha/2) + \mu_k \cos(\alpha/2)]}$$

$$= \frac{F}{2[\sin(10°/2) + (0.20) \cos(10°/2)]}$$

$$= 1.75F.$$

Apply equilibrium.

Practice Problem If the force F is removed, will the wedge remain in place in the log?

Answer: Yes.

Problems

▶ **9.65** In Active Example 9.4, the coefficients of friction between the wedge and the log are $\mu_s = 0.22$ and $\mu_k = 0.20$. What is the largest value of the wedge angle α for which the wedge would remain in place in the log when the force F is removed?

▶ **9.66** The wedge shown is being used to split the log. The wedge weighs 20 lb and the angle α equals 30°. The coefficient of kinetic friction between the faces of the wedge and the log is 0.28. If the normal force exerted by each face of the wedge must equal 150 lb to split the log, what vertical force F is necessary to drive the wedge into the log at a constant rate? (See Active Example 9.4.)

▶ **9.67** The coefficient of static friction between the faces of the wedge and the log in Problem 9.66 is 0.30. Will the wedge remain in place in the log when the vertical force F is removed? (See Active Example 9.4.)

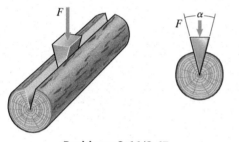

Problems 9.66/9.67

9.68 The weights of the blocks are $W_A = 100$ lb and $W_B = 25$ lb. Between all of the contacting surfaces, $\mu_s = 0.32$ and $\mu_k = 0.30$. What force F is necessary to move B to the left at a constant rate?

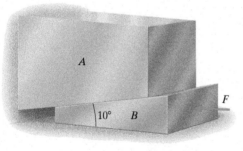

Problem 9.68

9.69 The masses of the blocks are $m_A = 30$ kg and $m_B = 70$ kg. Between all of the contacting surfaces, $\mu_s = 0.1$. What is the largest force F that can be applied without causing the blocks to slip?

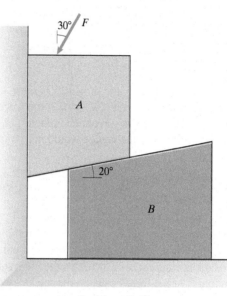

Problem 9.69

9.70 Each block weighs 200 lb. Between all of the contacting surfaces, $\mu_s = 0.1$. What is the largest force F that can be applied without causing block B to slip upward?

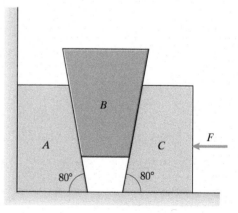

Problem 9.70

9.71 Small wedges called *shims* can be used to hold an object in place. The coefficient of kinetic friction between the contacting surfaces is 0.4. What force F is needed to push the shim downward until the horizontal force exerted on the object A is 200 N?

9.72 The coefficient of static friction between the contacting surfaces is 0.44. If the shims are in place and exert a 200-N horizontal force on the object A, what upward force must be exerted on the left shim to loosen it?

9.75 The box A has a mass of 80 kg, and the wedge B has a mass of 40 kg. Between all contacting surfaces, $\mu_s = 0.15$ and $\mu_k = 0.12$. What force F is required to raise A at a constant rate?

9.76 Suppose that A weighs 800 lb and B weighs 400 lb. The coefficients of friction between all of the contacting surfaces are $\mu_s = 0.15$ and $\mu_k = 0.12$. Will B remain in place if the force F is removed?

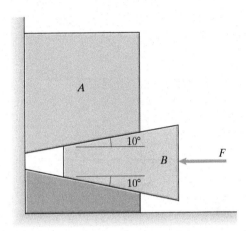

Problems **9.75/9.76**

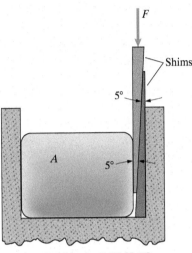

Problems **9.71/9.72**

9.73 The crate A weighs 600 lb. Between all contacting surfaces, $\mu_s = 0.32$ and $\mu_k = 0.30$. Neglect the weights of the wedges. What force F is required to move A to the right at a constant rate?

9.74 Suppose that between all contacting surfaces, $\mu_s = 0.32$ and $\mu_k = 0.30$. Neglect the weights of the 5° wedges. If a force $F = 800$ N is required to move A to the right at a constant rate, what is the mass of A?

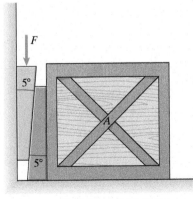

Problems **9.73/9.74**

9.77 Between A and B, $\mu_s = 0.20$, and between B and C, $\mu_s = 0.18$. Between C and the wall, $\mu_s = 0.30$. The weights $W_B = 20$ lb and $W_C = 80$ lb. What force F is required to start C moving upward?

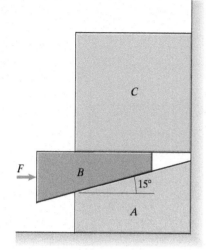

Problem 9.77

9.78 The masses of A, B, and C are 8 kg, 12 kg, and 80 kg, respectively. Between all contacting surfaces, $\mu_s = 0.4$. What force F is required to start C moving upward?

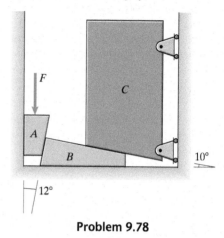

Problem 9.78

9.3 Threads

BACKGROUND

Threads are familiar from their use on wood screws, machine screws, and other machine elements. We show a shaft with square threads in Fig. 9.10a. The axial distance p from one thread to the next is called the *pitch* of the thread, and the angle α is its *slope*. We will consider only the case in which the shaft has a single continuous thread, so the relation between the pitch and slope is

$$\tan \alpha = \frac{p}{2\pi r}, \tag{9.7}$$

where r is the mean radius of the thread.

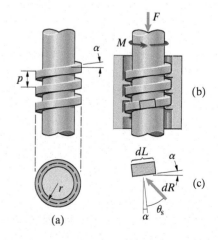

Figure 9.10
(a) A shaft with a square thread.
(b) The shaft within a sleeve with a mating groove and the direction of M that can cause the shaft to start moving in the axial direction opposite to F.
(c) A differential element of the thread when slip is impending.

Suppose that the threaded shaft is enclosed in a fixed sleeve with a mating groove and is subjected to an axial load F (Fig. 9.10b). Applying a couple M in the direction shown will tend to cause the shaft to start rotating and moving in the axial direction opposite to F. Our objective is to determine the couple M necessary to cause the shaft to start rotating.

We draw the free-body diagram of a differential element of the thread of length dL in Fig. 9.10c, representing the reaction exerted by the mating groove by the force dR. If the shaft is on the verge of rotating, dR resists the impending motion and the friction angle is the angle of static friction θ_s. The vertical component of the reaction on the element is $dR\cos(\theta_s + \alpha)$. To determine the total vertical force on the thread, we must integrate this expression over the length L of the thread. For equilibrium, the result must equal the axial force F acting on the shaft:

$$\cos(\theta_s + \alpha) \int_L dR = F. \tag{9.8}$$

The moment about the center of the shaft due to the reaction on the element is $r\,dR\sin(\theta_s + \alpha)$. The total moment must equal the couple M exerted on the shaft:

$$r\sin(\theta_s + \alpha) \int_L dR = M.$$

Dividing this equation by Eq. (9.8), we obtain the couple M necessary for the shaft to be on the verge of rotating and moving in the axial direction opposite to F:

$$M = rF\tan(\theta_s + \alpha). \tag{9.9}$$

Replacing the angle of static friction θ_s in this expression with the angle of kinetic friction θ_k gives the couple required to cause the shaft to rotate at a constant rate.

If the couple M is applied to the shaft in the opposite direction (Fig. 9.11a), the shaft tends to start rotating and moving in the axial direction of the load F. Figure 9.11b shows the reaction on a differential element of the thread of length dL when slip is impending. The direction of the reaction opposes the rotation of the shaft. In this case, the vertical component of the reaction on the element is $dR\cos(\theta_s - \alpha)$. Equilibrium requires that

$$\cos(\theta_s - \alpha) \int_L dR = F. \tag{9.10}$$

The moment about the center of the shaft due to the reaction is $r\,dR\sin(\theta_s - \alpha)$, so

$$r\sin(\theta_s - \alpha) \int_L dR = M.$$

Dividing this equation by Eq. (9.10), we obtain the couple M necessary for the shaft to be on the verge of rotating and moving in the direction of the force F:

$$M = rF\tan(\theta_s - \alpha). \tag{9.11}$$

Replacing θ_s with θ_k in this expression gives the couple necessary to rotate the shaft at a constant rate.

Notice in Eq. (9.11) that the couple required for impending motion is zero when $\theta_s = \alpha$. When the angle of static friction is less than this value, the shaft will rotate and move in the direction of the force F with no couple applied.

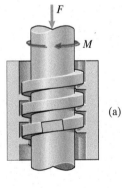

(a)

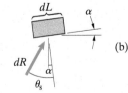

(b)

Figure 9.11
(a) The direction of M that can cause the shaft to move in the axial direction of F.
(b) A differential element of the thread when slip is impending.

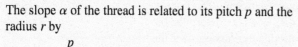

The slope α of the thread is related to its pitch p and the radius r by

$$\tan \alpha = \frac{p}{2\pi r}. \qquad (9.7)$$

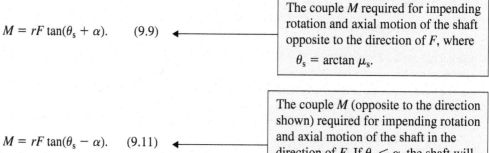

$$M = rF \tan(\theta_s + \alpha). \qquad (9.9)$$

The couple M required for impending rotation and axial motion of the shaft opposite to the direction of F, where

$\theta_s = \arctan \mu_s.$

$$M = rF \tan(\theta_s - \alpha). \qquad (9.11)$$

The couple M (opposite to the direction shown) required for impending rotation and axial motion of the shaft in the direction of F. If $\theta_s < \alpha$, the shaft will rotate and move in the direction of F with no couple applied.

Active Example 9.5   Rotating a Threaded Collar (▶ *Related Problem 9.79*)

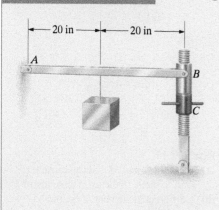

The right end of bar AB is pinned to an unthreaded collar B that rests on a threaded collar C. The mean radius of the threaded vertical shaft is $r = 1.6$ in and its pitch is $p = 0.2$ in. The coefficients of friction between the threads of the collar C and the vertical shaft are $\mu_s = 0.25$ and $\mu_k = 0.22$. The 400-lb suspended object can be raised or lowered by rotating the collar C. When the system is in the position shown, with bar AB horizontal, what is the magnitude of the couple that must be applied to the collar C to cause it to turn at a constant rate and move the suspended object upward?

Strategy

By drawing the free-body diagram of bar AB and the collar B, we can determine the axial force exerted on the collar C. Then we can use Eq. (9.9), with θ_s replaced by θ_k, to determine the required couple.

Solution

Draw the free-body diagram of the bar and collar B.

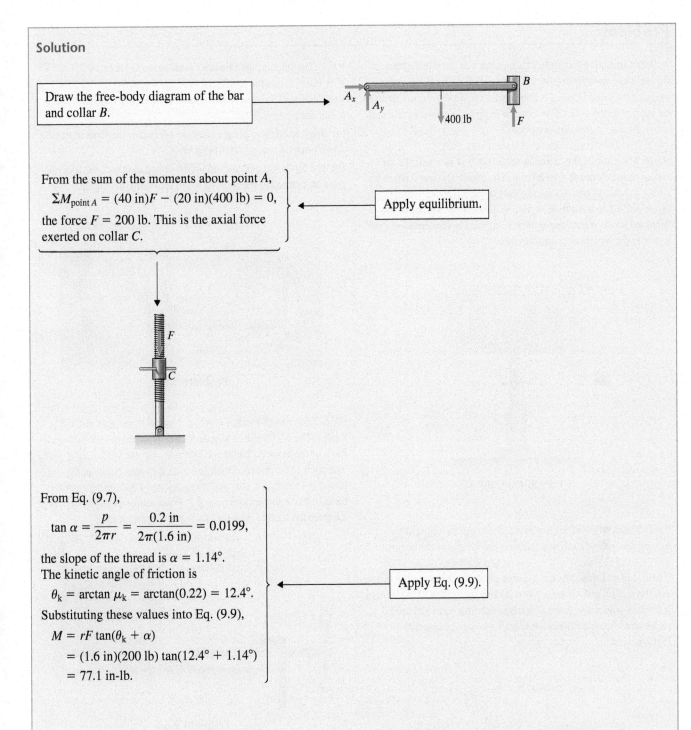

From the sum of the moments about point A,

$$\Sigma M_{\text{point } A} = (40 \text{ in})F - (20 \text{ in})(400 \text{ lb}) = 0,$$

the force $F = 200$ lb. This is the axial force exerted on collar C.

Apply equilibrium.

From Eq. (9.7),

$$\tan \alpha = \frac{p}{2\pi r} = \frac{0.2 \text{ in}}{2\pi(1.6 \text{ in})} = 0.0199,$$

the slope of the thread is $\alpha = 1.14°$.
The kinetic angle of friction is

$$\theta_k = \arctan \mu_k = \arctan(0.22) = 12.4°.$$

Substituting these values into Eq. (9.9),

$$M = rF \tan(\theta_k + \alpha)$$

$$= (1.6 \text{ in})(200 \text{ lb}) \tan(12.4° + 1.14°)$$

$$= 77.1 \text{ in-lb}.$$

Apply Eq. (9.9).

Practice Problem When the system is in the position shown, with bar AB horizontal, what is the magnitude of the couple that must be applied to the collar C to cause it to turn at a constant rate and move the suspended object downward?

Answer: 63.8 in-lb.

Problems

▶ **9.79** In Active Example 9.5, suppose that the pitch of the thread is changed from $p = 0.2$ in to $p = 0.24$ in. What is the slope of the thread? What is the magnitude of the couple that must be applied to the collar C to cause it to turn at a constant rate and move the suspended object upward?

9.80 The pitch of the threaded shaft is $p = 2$ mm and the mean radius of the thread is $r = 20$ mm. The coefficients of friction between the thread and the mating groove are $\mu_s = 0.22$ and $\mu_k = 0.20$. The weight $W = 500$ N. Neglect the weight of the threaded shaft. What couple must be applied to the threaded shaft to lower the weight at a constant rate?

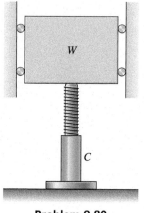

Problem 9.80

9.81 The position of the horizontal beam can be adjusted by turning the machine screw A. Neglect the weight of the beam. The pitch of the screw is $p = 1$ mm, and the mean radius of the thread is $r = 4$ mm. The coefficients of friction between the thread and the mating groove are $\mu_s = 0.20$ and $\mu_k = 0.18$. If the system is initially stationary, determine the couple that must be applied to the screw to cause the beam to start moving (a) upward; (b) downward.

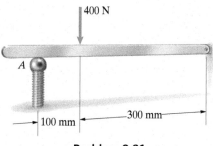

Problem 9.81

9.82 The pitch of the threaded shaft of the C clamp is $p = 0.05$ in, and the mean radius of the thread is $r = 0.15$ in. The coefficients of friction between the threaded shaft and the mating collar are $\mu_s = 0.18$ and $\mu_k = 0.16$.

(a) What maximum couple must be applied to the shaft to exert a 30-lb force on the clamped object?

(b) If a 30-lb force is exerted on the clamped object, what couple must be applied to the shaft to begin loosening the clamp?

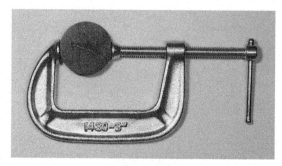

Problem 9.82

9.83 The mass of block A is 60 kg. Neglect the weight of the 5° wedge. The coefficient of kinetic friction between the contacting surfaces of the block A, the wedge, the table, and the wall is $\mu_k = 0.4$. The pitch of the threaded shaft is 5 mm, the mean radius of the thread is 15 mm, and the coefficient of kinetic friction between the thread and the mating groove is 0.2. What couple must be exerted on the threaded shaft to raise the block A at a constant rate?

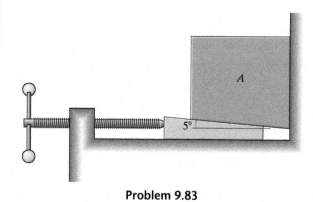

Problem 9.83

9.84 The vise exerts 80-lb forces on A. The threaded shafts are subjected only to axial loads by the jaws of the vise. The pitch of their threads is $p = 1/8$ in, the mean radius of the threads is $r = 1$ in, and the coefficient of static friction between the threads and the mating grooves is 0.2. Suppose that you want to loosen the vise by turning one of the shafts. Determine the couple you must apply (a) to shaft B; (b) to shaft C.

9.85 Suppose that you want to tighten the vise in Problem 9.84 by turning one of the shafts. Determine the couple you must apply

(a) to shaft B;

(b) to shaft C.

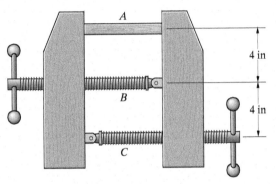

Problems 9.84/9.85

9.86 The threaded shaft has a ball and socket support at B. The 400-lb load A can be raised or lowered by rotating the threaded shaft, causing the threaded collar at C to move relative to the shaft. Neglect the weights of the members. The pitch of the shaft is $p = \frac{1}{4}$ in, the mean radius of the thread is $r = 1$ in, and the coefficient of static friction between the thread and the mating groove is 0.24. If the system is stationary in the position shown, what couple is necessary to start the shaft rotating to raise the load?

9.87 In Problem 9.86, if the system is stationary in the position shown, what couple is necessary to start the shaft rotating to lower the load?

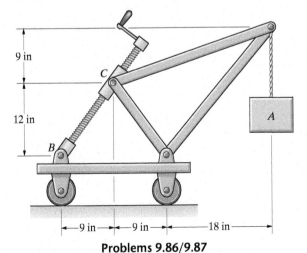

Problems 9.86/9.87

9.88 The car jack is operated by turning the horizontal threaded shaft at A. The threaded shaft fits into a mating threading collar at B. As the shaft turns, points A and B move closer together or farther apart, thereby raising or lowering the jack. The pitch of the threaded shaft is $p = 0.1$ in, the mean radius of the thread is $r = 0.2$ in, and the coefficient of kinetic friction between the threaded shaft and the mating collar at B is 0.15. What couple must be applied at A to rotate the shaft at a constant rate and raise the jack when it is in the position shown if the load $L = 1400$ lb?

9.89 The car jack is operated by turning the horizontal threaded shaft at A. The threaded shaft fits into a mating threading collar at B. As the shaft turns, points A and B move closer together or farther apart, thereby raising or lowering the jack. The pitch of the threaded shaft is $p = 0.1$ in, the mean radius of the thread is $r = 0.2$ in, and the coefficient of kinetic friction between the threaded shaft and the mating collar at B is 0.15. What couple must be applied at A to rotate the shaft at a constant rate and lower the jack when it is in the position shown if the load $L = 1400$ lb?

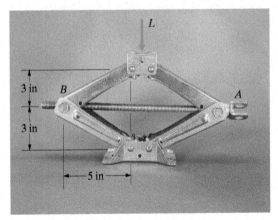

Problems 9.88/9.89

9.90 A *turnbuckle,* used to adjust the length or tension of a bar or cable, is threaded at both ends. Rotating it draws threaded ends of the bar or cable together or moves them apart. Suppose that the pitch of the threads is $p = 0.05$ in, their mean radius is $r = 0.25$ in, and the coefficient of static friction between the threads and the mating grooves is 0.24. If $T = 200$ lb, what couple must be exerted on the turnbuckle to start tightening it?

9.91 Suppose that the pitch of the threads of the turnbuckle is $p = 0.05$ in, their mean radius is $r = 0.25$ in, and the coefficient of static friction between the threads and the mating grooves is 0.24. If $T = 200$ lb, what couple must be exerted on the turnbuckle to start loosening it?

Problems 9.90/9.91

9.92 Member *BE* of the frame has a turnbuckle. (See Problem 9.90.) The threads have pitch $p = 1$ mm, their mean radius is $r = 6$ mm, and the coefficient of static friction between the threads and the mating grooves is 0.2. What couple must be exerted on the turnbuckle to start loosening it?

9.93 In Problem 9.92, what couple must be exerted on the turnbuckle to start tightening it?

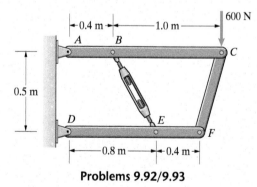

Problems 9.92/9.93

9.94 Members *CD* and *DG* of the truss have turnbuckles. (See Problem 9.90.) The pitch of the threads is $p = 4$ mm, their mean radius is $r = 10$ mm, and the coefficient of static friction between the threads and the mating grooves is 0.18. What couple must be exerted on the turnbuckle of member *CD* to start loosening it?

9.95 In Problem 9.94, what couple must be exerted on the turnbuckle of member *DG* to start loosening it?

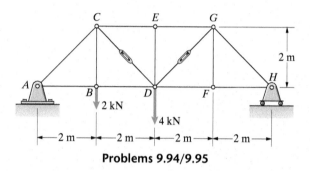

Problems 9.94/9.95

9.96* The load $W = 800$ N can be raised or lowered by rotating the threaded shaft. The distances are $b = 75$ mm and $h = 200$ mm. The pinned bars are each 300 mm in length. The pitch of the threaded shaft is $p = 5$ mm, the mean radius of the thread is $r = 15$ mm, and the coefficient of kinetic friction between the thread and the mating groove is 0.2. When the system is in the position shown, what couple must be exerted to turn the threaded shaft at a constant rate, raising the load?

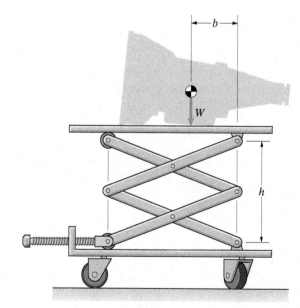

Problem 9.96

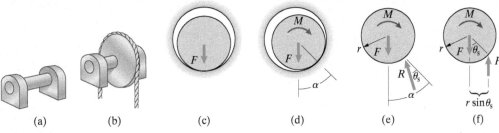

(a) (b) (c) (d) (e) (f)

Figure 9.12
(a) A shaft supported by journal bearings.
(b) A pulley supported by the shaft.
(c) The shaft and bearing when no couple is applied to the shaft.
(d) A couple causes the shaft to roll within the bearing.
(e) Free-body diagram of the shaft.
(f) The two forces on the shaft must be equal and opposite.

9.4 Journal Bearings

BACKGROUND

A *bearing* is a support. This term usually refers to supports designed to allow the supported object to move. For example, in Fig. 9.12a, a horizontal shaft is supported by two *journal bearings*, which allow the shaft to rotate. The shaft can then be used to support a load perpendicular to its axis, such as that subjected by a pulley (Fig. 9.12b).

Here we analyze journal bearings consisting of brackets with holes through which the shaft passes. The radius of the shaft is slightly smaller than the radius of the holes in the bearings. Our objective is to determine the couple that must be applied to the shaft to cause it to rotate in the bearings. Let F be the total load supported by the shaft including the weight of the shaft itself. When no couple is exerted on the shaft, the force F presses it against the bearings as shown in Fig. 9.12c. When a couple M is exerted on the shaft, it rolls up the surfaces of the bearings (Fig. 9.12d). The term α is the angle from the original point of contact of the shaft to its point of contact when M is applied.

In Fig. 9.12e, we draw the free-body diagram of the shaft when M is sufficiently large that slip is impending. The force R is the total reaction exerted on the shaft by the two bearings. Since R and F are the only forces acting on the shaft, equilibrium requires that $\alpha = \theta_s$ and $R = F$ (Fig. 9.12f). The reaction exerted on the shaft by the bearings is displaced a distance $r \sin \theta_s$ from the vertical line through the center of the shaft. By summing moments about the center of the shaft, we obtain the couple M that causes the shaft to be on the verge of slipping:

$$M = rF \sin \theta_s. \qquad (9.12)$$

This is the largest couple that can be exerted on the shaft without causing it to start rotating. Replacing θ_s in this expression by the angle of kinetic friction θ_k gives the couple necessary to rotate the shaft at a constant rate.

The simple type of journal bearing we have described is too primitive for most applications. The surfaces where the shaft and bearing are in contact would quickly become worn. Designers usually incorporate "ball" or "roller" bearings in journal bearings to minimize friction (Fig. 9.13).

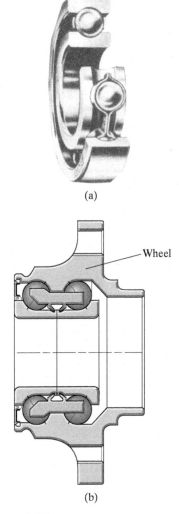

(a)

Wheel

(b)

Figure 9.13
(a) A journal bearing with one row of balls.
(b) Journal bearing assembly of the wheel of a car. There are two rows of balls between the rotating wheel and the fixed inner cylinder.

RESULTS

A journal bearing has a circular hole slightly larger than the circular shaft it supports.

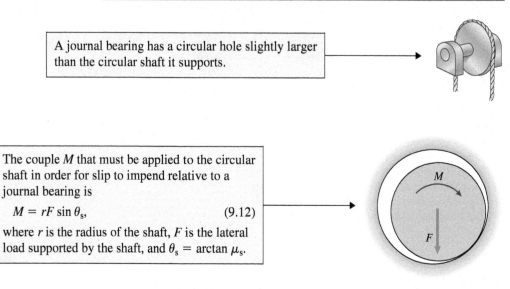

The couple M that must be applied to the circular shaft in order for slip to impend relative to a journal bearing is

$$M = rF \sin \theta_s, \tag{9.12}$$

where r is the radius of the shaft, F is the lateral load supported by the shaft, and $\theta_s = \arctan \mu_s$.

Active Example 9.6   **Pulley Supported by Journal Bearings** (▶ *Related Problem 9.97*)

The weight of the suspended load is $W = 1000$ lb. The pulley P has a 6-in radius and is rigidly attached to a horizontal circular shaft that is supported by journal bearings. The radius of the shaft is 0.5 in, and the coefficient of kinetic friction between the shaft and the bearings is $\mu_k = 0.2$. The weights of the pulley and shaft are negligible. What tension must the winch A exert on the rope to raise the load at a constant rate?

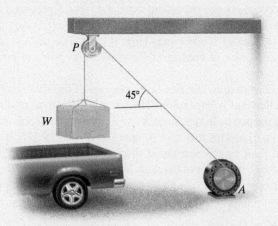

Strategy

Eq. (9.12) with θ_s replaced by θ_k relates the couple M required to turn the pulley at a constant rate to the lateral force F supported by the shaft. By expressing M and F in terms of the forces exerted on the pulley by the rope and applying Eq. (9.12), we can obtain an equation for the tension the winch must exert.

Solution

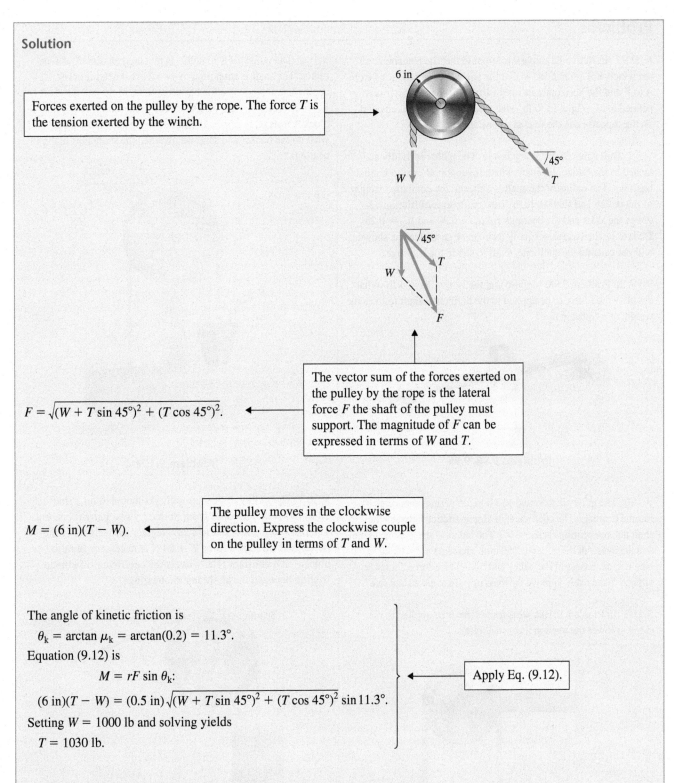

Forces exerted on the pulley by the rope. The force T is the tension exerted by the winch.

The vector sum of the forces exerted on the pulley by the rope is the lateral force F the shaft of the pulley must support. The magnitude of F can be expressed in terms of W and T.

$$F = \sqrt{(W + T \sin 45°)^2 + (T \cos 45°)^2}.$$

The pulley moves in the clockwise direction. Express the clockwise couple on the pulley in terms of T and W.

$$M = (6 \text{ in})(T - W).$$

The angle of kinetic friction is

$$\theta_k = \arctan \mu_k = \arctan(0.2) = 11.3°.$$

Equation (9.12) is

$$M = rF \sin \theta_k:$$

$$(6 \text{ in})(T - W) = (0.5 \text{ in}) \sqrt{(W + T \sin 45°)^2 + (T \cos 45°)^2} \sin 11.3°.$$

Setting $W = 1000$ lb and solving yields

$$T = 1030 \text{ lb}.$$

Apply Eq. (9.12).

Practice Problem What tension must the winch A exert on the rope to lower the load at a constant rate?

Answer: $T = 970$ lb.

Problems

▶ **9.97** In Active Example 9.6, suppose that the placement of the winch at A is changed so that the angle between the rope from A to P and the horizontal increases from 45° to 60°. If the suspended load weighs 1500 lb, what tension must the winch exert on the rope to raise the load at a constant rate?

9.98 The radius of the pulley is 4 in. The pulley is rigidly attached to the horizontal shaft, which is supported by two journal bearings. The radius of the shaft is 1 in, and the combined weight of the pulley and shaft is 20 lb. The coefficients of friction between the shaft and the bearings are $\mu_s = 0.30$ and $\mu_k = 0.28$. Determine the largest weight W that can be suspended as shown without causing the stationary shaft to slip in the bearings.

9.99 In Problem 9.98, suppose that the weight $W = 4$ lb. What couple would have to be applied to the horizontal shaft to raise the weight at a constant rate?

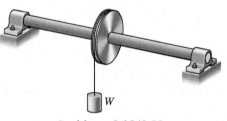

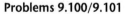
W

Problems 9.98/9.99

9.100 The pulley is mounted on a horizontal shaft supported by journal bearings. The coefficient of kinetic friction between the shaft and the bearings is $\mu_k = 0.3$. The radius of the shaft is 20 mm, and the radius of the pulley is 150 mm. The mass $m = 10$ kg. Neglect the masses of the pulley and shaft. What force T must be applied to the cable to move the mass upward at a constant rate?

9.101 In Problem 9.100, what force T must be applied to the cable to lower the mass at a constant rate?

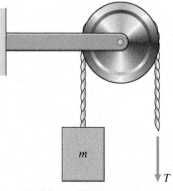

Problems 9.100/9.101

9.102 The pulley of 8-in radius is mounted on a shaft of 1-in radius. The shaft is supported by two journal bearings. The coefficient of static friction between the bearings and the shaft is $\mu_s = 0.15$. Neglect the weights of the pulley and shaft. The 50-lb block A rests on the floor. If sand is slowly added to the bucket B, what do the bucket and sand weigh when the shaft slips in the bearings?

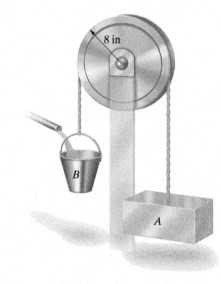

Problem 9.102

9.103 The pulley of 50-mm radius is mounted on a shaft of 10-mm radius. The shaft is supported by two journal bearings. The mass of the block A is 8 kg. Neglect the weights of the pulley and shaft. If a force $T = 84$ N is necessary to raise block A at a constant rate, what is the coefficient of kinetic friction between the shaft and the bearings?

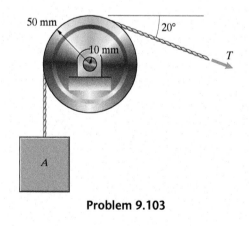

Problem 9.103

9.104 The mass of the suspended object is 4 kg. The pulley has a 100-mm radius and is rigidly attached to a horizontal shaft supported by journal bearings. The radius of the horizontal shaft is 10 mm and the coefficient of kinetic friction between the shaft and the bearings is 0.26. What tension must the person exert on the rope to raise the load at a constant rate?

9.105 In Problem 9.104, what tension must the person exert to lower the load at a constant rate?

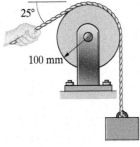

Problems 9.104/9.105

9.106 The radius of the pulley is 200 mm, and it is mounted on a shaft of 20-mm radius. The coefficient of static friction between the pulley and shaft is $\mu_s = 0.18$. If $F_A = 200$ N, what is the largest force F_B that can be applied without causing the pulley to turn? Neglect the weight of the pulley.

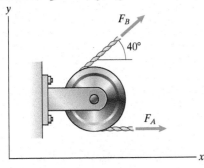

Problem 9.106

9.107 The masses of the boxes are $m_A = 15$ kg and $m_B = 60$ kg. The coefficient of static friction between boxes A and B and between box B and the inclined surface is 0.12. The pulley has a radius of 60 mm and is mounted on a shaft of 10-mm radius. The coefficient of static friction between the pulley and shaft is 0.16. What is the largest force F for which the boxes will not slip?

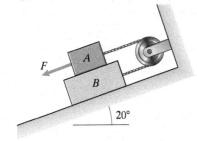

Problem 9.107

9.108 The two pulleys have a radius of 4 in and are mounted on shafts of 1-in radius supported by journal bearings. Neglect the weights of the pulleys and shafts. The tension in the spring is 40 lb. The coefficient of kinetic friction between the shafts and the bearings is $\mu_k = 0.3$. What couple M is required to turn the left pulley at a constant rate?

Problem 9.108

9.109 The weights of the boxes are $W_A = 65$ lb and $W_B = 130$ lb. The coefficient of static friction between boxes A and B and between box B and the floor is 0.12. The pulley has a radius of 4 in and is mounted on a shaft of 0.8-in radius. The coefficient of static friction between the pulley and shaft is 0.16. What is the largest force F for which the boxes will not slip?

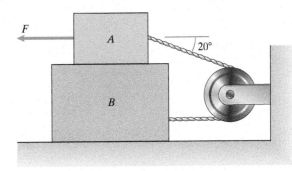

Problem 9.109

9.110 The coefficient of kinetic friction between the 100-kg box and the inclined surface is 0.35. Each pulley has a radius of 100 mm and is mounted on a shaft of 5-mm radius supported by journal bearings. The coefficient of kinetic friction between the shafts and the journal bearings is 0.18. Determine the tension T necessary to pull the box up the surface at a constant rate.

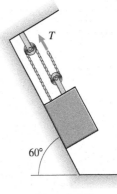

Problem 9.110

9.5 Thrust Bearings and Clutches

BACKGROUND

A *thrust bearing* supports a rotating shaft that is subjected to an axial load. In the type shown in Figs. 9.14a and b, the conical end of the shaft is pressed against the mating conical cavity by an axial load F. Let us determine the couple M necessary to rotate the shaft.

The differential element of area dA in Fig. 9.14c is

$$dA = 2\pi r\, ds = 2\pi r\left(\frac{dr}{\cos\alpha}\right).$$

Integrating this expression from $r = r_i$ to $r = r_o$, we obtain the area of contact:

$$A = \frac{\pi(r_o^2 - r_i^2)}{\cos\alpha}.$$

If we assume that the mating surface exerts a uniform pressure p, the axial component of the total force due to p must equal F: $pA\cos\alpha = F$. Therefore, the pressure is

$$p = \frac{F}{A\cos\alpha} = \frac{F}{\pi(r_o^2 - r_i^2)}.$$

As the shaft rotates about its axis, the moment about the axis due to the friction force on the element dA is $r\mu_k(p\, dA)$. The total moment is

$$M = \int_A \mu_k rp\, dA = \int_{r_i}^{r_o} \mu_k r\left[\frac{F}{\pi(r_o^2 - r_i^2)}\right]\left(\frac{2\pi r\, dr}{\cos\alpha}\right).$$

Integrating, we obtain the couple M necessary to rotate the shaft at a constant rate:

$$M = \frac{2\mu_k F}{3\cos\alpha}\left(\frac{r_o^3 - r_i^3}{r_o^2 - r_i^2}\right). \tag{9.13}$$

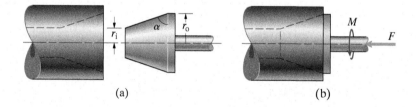

(a)   (b)

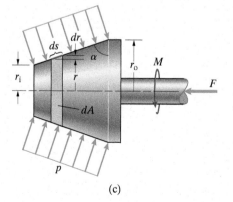

(c)

Figure 9.14

(a), (b) A thrust bearing supports a shaft subjected to an axial load.

(c) The differential element dA and the uniform pressure p exerted by the cavity.

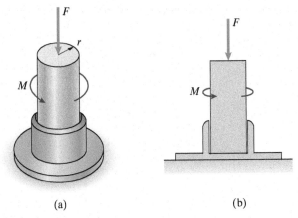

(a)   (b)

Figure 9.15
A thrust bearing that supports a flat-ended shaft.

A simpler thrust bearing is shown in Figs. 9.15a and b. The bracket supports the flat end of a shaft of radius r that is subjected to an axial load F. We can obtain the couple necessary to rotate the shaft at a constant rate from Eq. (9.13) by setting $\alpha = 0$, $r_i = 0$, and $r_o = r$:

$$M = \frac{2}{3}\mu_k Fr. \qquad (9.14)$$

Although they are good examples of the analysis of friction forces, the thrust bearings we have described would become worn too quickly to be used in most applications. The designer of the thrust bearing in Fig. 9.16 minimizes friction by incorporating "roller" bearings.

A *clutch* is a device used to connect and disconnect two coaxial rotating shafts. The type shown in Figs. 9.17a and b consists of disks of radius r attached to the ends of the shafts. When the disks are separated (Fig. 9.17a), the clutch is *disengaged*, and the shafts can rotate freely relative to each other. When the clutch is engaged by pressing the disks together with axial forces F (Fig. 9.17b), the shafts can support a couple M due to the friction forces between the disks. If the couple M becomes too large, the clutch slips.

The friction forces exerted on one face of the clutch by the other face are identical to the friction forces exerted on the flat-ended shaft by the bracket in Fig. 9.15. We can therefore determine the largest couple the clutch can support without slipping by replacing μ_k with μ_s in Eq. (9.14):

$$M = \frac{2}{3}\mu_s Fr. \qquad (9.15)$$

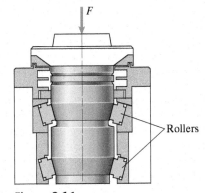

Figure 9.16
A thrust bearing with two rows of cylindrical rollers between the shaft and the fixed support.

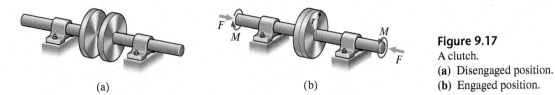

(a)   (b)

Figure 9.17
A clutch.
(a) Disengaged position.
(b) Engaged position.

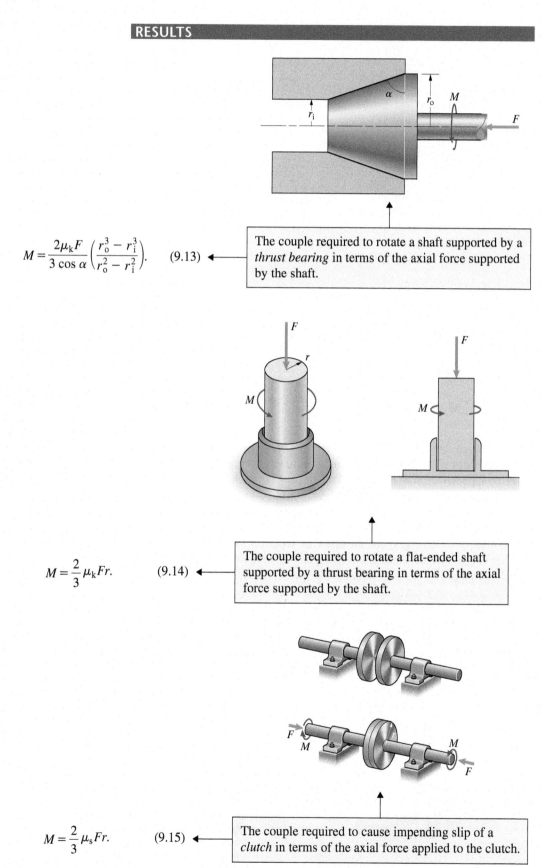

$$M = \frac{2\mu_k F}{3 \cos \alpha} \left(\frac{r_o^3 - r_i^3}{r_o^2 - r_i^2} \right).$$ (9.13)

The couple required to rotate a shaft supported by a *thrust bearing* in terms of the axial force supported by the shaft.

$$M = \frac{2}{3} \mu_k Fr.$$ (9.14)

The couple required to rotate a flat-ended shaft supported by a thrust bearing in terms of the axial force supported by the shaft.

$$M = \frac{2}{3} \mu_s Fr.$$ (9.15)

The couple required to cause impending slip of a *clutch* in terms of the axial force applied to the clutch.

Active Example 9.7 **Thrust Bearing** (▶ *Related Problem 9.111*)

The axial force on the thrust bearing is $F = 200$ lb. The diameters $D_o = 3\frac{1}{2}$ in and $D_i = 1$ in, and the angle $\alpha = 72°$. The coefficient of kinetic friction is $\mu_k = 0.18$. What couple is required to turn the shaft at a constant rate?

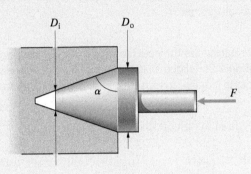

Strategy
The couple is given by Eq. (9.13).

Solution

The radii $r_o = 1.75$ in and $r_i = 0.5$ in.

$$M = \frac{2\mu_k F}{3 \cos \alpha} \left(\frac{r_o^3 - r_i^3}{r_o^2 - r_i^2} \right)$$

$$= \frac{2(0.18)(200 \text{ lb})}{3 \cos 72°} \left[\frac{(1.75 \text{ in})^3 - (0.5 \text{ in})^3}{(1.75 \text{ in})^2 - (0.5 \text{ in})^2} \right]$$

$$= 145 \text{ in-lb}.$$

Apply Eq. (9.13).

Practice Problem The axial force on the thrust bearing is $F = 200$ lb. The diameters $D_o = 3\frac{1}{2}$ in and $D_i = 1$ in, and the dimension $b = 5$ in. The coefficient of kinetic friction is $\mu_k = 0.18$. What couple is required to turn the shaft at a constant rate?

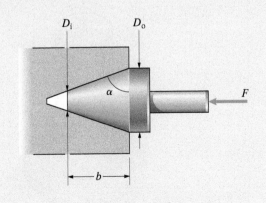

Answer: $M = 184$ in-lb.

Example 9.8 Friction on a Disk Sander (▶ Related Problem 9.118)

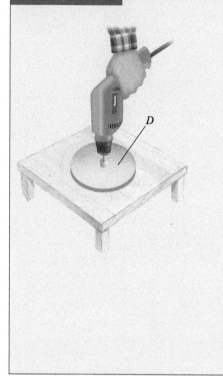

D

The handheld sander has a rotating disk D of 4-in radius with sandpaper bonded to it. The total downward force exerted by the operator and the weight of the sander is 15 lb. The coefficient of kinetic friction between the sandpaper and the surface is $\mu_k = 0.6$. What couple (torque) M must the motor exert to turn the sander at a constant rate?

Strategy
As the disk D rotates, it is subjected to friction forces analogous to the friction forces exerted on a flat-ended shaft supported by a thrust bearing. We can determine the couple required to turn the disk D at a constant rate from Eq. (9.14).

Solution
The couple required to turn the disk at a constant rate is

$$M = \frac{2}{3}\mu_k rF = \frac{2}{3}(0.6)(4 \text{ in})(15 \text{ lb}) = 24 \text{ in-lb}.$$

Critical Thinking
Equations (9.13)–(9.15) were derived under the assumption that the normal force (and consequently the friction force) is uniformly distributed over the contacting surfaces. Evaluating and improving upon this assumption would require analysis of the deformations of the contacting surfaces in specific applications such as the disk sander in this example.

Problems

▶ **9.111** In Active Example 9.7, suppose that the diameters $D_o = 3\frac{1}{2}$ in and $D_i = 1\frac{1}{2}$ in and the angle $\alpha = 72°$. What couple is required to turn the shaft at a constant rate?

9.112 The circular flat-ended shaft is pressed into the thrust bearing by an axial load of 600 lb. The weight of the shaft is negligible. The coefficients of friction between the end of the shaft and the bearing are $\mu_s = 0.20$ and $\mu_k = 0.15$. What is the largest couple M that can be applied to the stationary shaft without causing it to rotate in the bearing?

9.113 The circular flat-ended shaft is pressed into the thrust bearing by an axial load of 600 lb. The weight of the shaft is negligible. The coefficients of friction between the end of the shaft and the bearing are $\mu_s = 0.20$ and $\mu_k = 0.15$. What couple M is required to rotate the shaft at a constant rate?

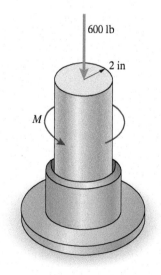

600 lb

2 in

M

Problems 9.112/9.113

9.114 The disk D is rigidly attached to the vertical shaft. The shaft has flat ends supported by thrust bearings. The disk and the shaft together have a mass of 220 kg and the diameter of the shaft is 50 mm. The vertical force exerted on the end of the shaft by the upper thrust bearing is 440 N. The coefficient of kinetic friction between the ends of the shaft and the bearings is 0.25. What couple M is required to rotate the shaft at a constant rate?

9.115 Suppose that the ends of the shaft in Problem 9.114 are supported by thrust bearings of the type shown in Fig. 9.14, where $r_o = 25$ mm, $r_i = 6$ mm, $\alpha = 45°$, and $\mu_k = 0.25$. What couple M is required to rotate the shaft at a constant rate?

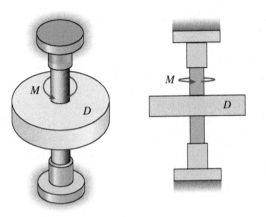

Problems 9.114/9.115

9.116 The shaft is supported by thrust bearings that subject it to an axial load of 800 N. The coefficients of kinetic friction between the shaft and the left and right bearings are 0.20 and 0.26, respectively. What couple is required to rotate the shaft at a constant rate?

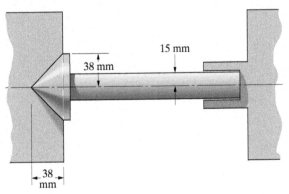

15 mm

38 mm

38 mm

Problem 9.116

9.117 A motor is used to rotate a paddle for mixing chemicals. The shaft of the motor is coupled to the paddle using a friction clutch of the type shown in Fig. 9.17. The radius of the disks of the clutch is 120 mm, and the coefficient of static friction between the disks is 0.6. If the motor transmits a maximum torque of 15 N-m to the paddle, what minimum normal force between the plates of the clutch is necessary to prevent slipping?

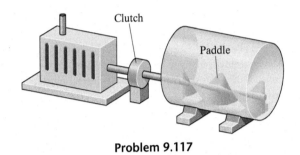

Clutch

Paddle

Problem 9.117

▶ **9.118** The thrust bearing is supported by contact of the collar C with a fixed plate. The area of contact is an annulus with an inside diameter $D_1 = 40$ mm and an outside diameter $D_2 = 120$ mm. The coefficient of kinetic friction between the collar and the plate is $\mu_k = 0.3$. The force $F = 400$ N. What couple M is required to rotate the shaft at a constant rate? (See Example 9.8.)

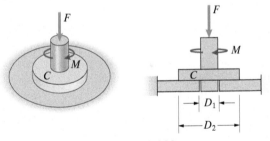

F

M

C

F

M

C

D_1

D_2

Problem 9.118

9.119 An experimental automobile brake design works by pressing the fixed red annular plate against the rotating wheel. If $\mu_k = 0.6$, what force F pressing the plate against the wheel is necessary to exert a couple of 200 N-m on the wheel?

9.120 An experimental automobile brake design works by pressing the fixed red annular plate against the rotating wheel. Suppose that $\mu_k = 0.65$ and the force pressing the plate against the wheel is $F = 2$ kN.

(a) What couple is exerted on the wheel?

(b) What percentage increase in the couple exerted on the wheel is obtained if the outer radius of the brake is increased from 90 mm to 100 mm?

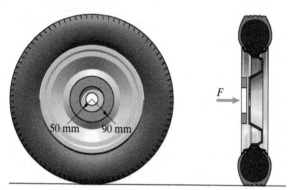

Problems 9.119/9.120

9.121 The coefficient of static friction between the plates of the car's clutch is 0.8. If the plates are pressed together with a force $F = 2.60$ kN, what is the maximum torque the clutch will support without slipping?

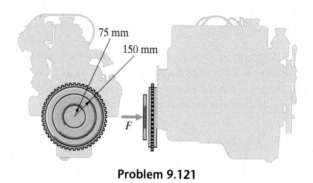

Problem 9.121

9.122* The "Morse taper" is used to support the workpiece on a machinist's lathe. The taper is driven into the spindle and is held in place by friction. If the spindle exerts a uniform pressure $p = 15$ psi on the taper and $\mu_s = 0.2$, what couple must be exerted about the axis of the taper to loosen it?

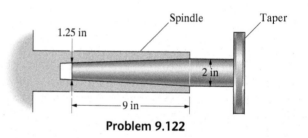

Problem 9.122

9.6 Belt Friction

BACKGROUND

If a rope is wrapped around a fixed post as shown in Fig. 9.18, a large force T_2 exerted on one end can be supported by a relatively small force T_1 applied to the other end. In this section we analyze this familiar phenomenon. It is referred to as *belt friction* because a similar approach can be used to analyze belts used in machines, such as the belts that drive alternators and other devices in a car.

Let us consider a rope wrapped through an angle β around a fixed cylinder (Fig. 9.19a). We will assume that the tension T_1 is known. Our objective is to determine the largest force T_2 that can be applied to the other end of the rope without causing the rope to slip.

We begin by drawing the free-body diagram of an element of the rope whose boundaries are at angles α and $\alpha + \Delta\alpha$ from the point where the rope comes into contact with the cylinder (Figs. 9.19b and c). The force T is the tension in the rope at the position defined by the angle α. We know that the tension in the rope varies with position, because it increases from T_1 at $\alpha = 0$ to T_2 at $\alpha = \beta$. We therefore write the tension in the rope at the position $\alpha + \Delta\alpha$ as $T + \Delta T$. The force ΔN is the normal force exerted on the element by the cylinder. Because we want to determine the largest value of T_2 that will not cause the rope to slip, we assume that the friction force is equal to its maximum possible value $\mu_s \Delta N$, where μ_s is the coefficient of static friction between the rope and the cylinder.

The equilibrium equations in the directions tangential to and normal to the centerline of the rope are

$$\Sigma F_{\text{tangential}} = \mu_s \Delta N + T \cos\left(\frac{\Delta\alpha}{2}\right) - (T + \Delta T) \cos\left(\frac{\Delta\alpha}{2}\right) = 0,$$

$$(9.16)$$

$$\Sigma F_{\text{normal}} = \Delta N - (T + \Delta T) \sin\left(\frac{\Delta\alpha}{2}\right) - T \sin\left(\frac{\Delta\alpha}{2}\right) = 0.$$

Eliminating ΔN, we can write the resulting equation as

$$\left[\cos\left(\frac{\Delta\alpha}{2}\right) - \mu_s \sin\left(\frac{\Delta\alpha}{2}\right)\right]\frac{\Delta T}{\Delta\alpha} - \mu_s T \frac{\sin(\Delta\alpha/2)}{\Delta\alpha/2} = 0.$$

Figure 9.18
A rope wrapped around a post.

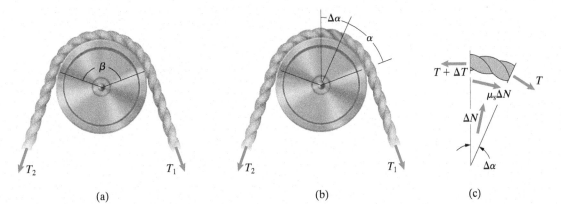

Figure 9.19
(a) A rope wrapped around a fixed cylinder.
(b) A differential element with boundaries at angles α and $\alpha + \Delta\alpha$.
(c) Free-body diagram of the element.

Evaluating the limit of this equation as $\Delta\alpha \to 0$ and observing that

$$\frac{\sin(\Delta\alpha/2)}{\Delta\alpha/2} \to 1,$$

we obtain

$$\frac{dT}{d\alpha} - \mu_s T = 0.$$

This differential equation governs the variation of the tension in the rope. Separating variables yields

$$\frac{dT}{T} = \mu_s \, d\alpha.$$

We can now integrate to determine the tension T_2 in terms of the tension T_1 and the angle β:

$$\int_{T_1}^{T_2} \frac{dT}{T} = \int_0^\beta \mu_s \, d\alpha.$$

Thus, we obtain the largest force T_2 that can be applied without causing the rope to slip when the force on the other end is T_1:

$$T_2 = T_1 e^{\mu_s \beta}. \tag{9.17}$$

The angle β in this equation must be expressed in radians. Replacing μ_s by the coefficient of kinetic friction μ_k gives the force T_2 required to cause the rope to slide at a constant rate.

Equation (9.17) explains why a large force can be supported by a relatively small force when a rope is wrapped around a fixed support. The force required to cause the rope to slip increases exponentially as a function of the angle through which the rope is wrapped. Suppose that $\mu_s = 0.3$. When the rope is wrapped one complete turn around the post $(\beta = 2\pi)$, the ratio $T_2/T_1 = 6.59$. When the rope is wrapped four complete turns around the post $(\beta = 8\pi)$, the ratio $T_2/T_1 = 1880$.

RESULTS

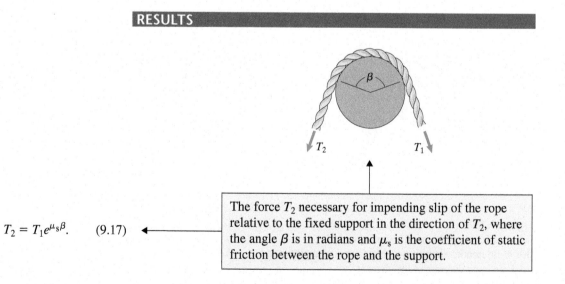

$$T_2 = T_1 e^{\mu_s \beta}. \qquad (9.17)$$

The force T_2 necessary for impending slip of the rope relative to the fixed support in the direction of T_2, where the angle β is in radians and μ_s is the coefficient of static friction between the rope and the support.

Active Example 9.9 Rope Wrapped Around Fixed Cylinders (▶ *Related Problem 9.123*)

The 100-lb box is suspended from a rope that passes over two *fixed* cylinders. The coefficient of static friction is 0.2 between the rope and the left cylinder and 0.4 between the rope and the right cylinder. What is the smallest force the woman needs to exert on the rope to support the stationary box?

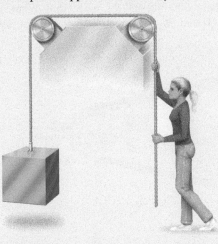

Strategy

She exerts the smallest necessary force when slip of the rope is impending on both cylinders. If we assume that slip is impending and apply Eq. (9.17) to each cylinder, we can determine the force she must apply.

Solution

Let T be the tension in the rope between the two cylinders. The weight $W = 100$ lb and F is the force the woman exerts. The rope is wrapped around each cylinder through an angle (in radians) $\beta = \pi/2$.

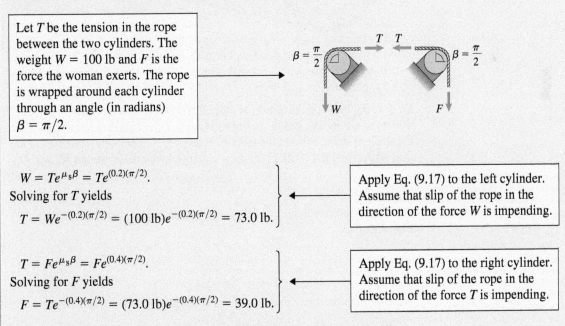

$$W = Te^{\mu_s\beta} = Te^{(0.2)(\pi/2)}.$$

Solving for T yields

$$T = We^{-(0.2)(\pi/2)} = (100 \text{ lb})e^{-(0.2)(\pi/2)} = 73.0 \text{ lb}.$$

Apply Eq. (9.17) to the left cylinder. Assume that slip of the rope in the direction of the force W is impending.

$$T = Fe^{\mu_s\beta} = Fe^{(0.4)(\pi/2)}.$$

Solving for F yields

$$F = Te^{-(0.4)(\pi/2)} = (73.0 \text{ lb})e^{-(0.4)(\pi/2)} = 39.0 \text{ lb}.$$

Apply Eq. (9.17) to the right cylinder. Assume that slip of the rope in the direction of the force T is impending.

Practice Problem What force would the woman need to exert on the rope for slip to be impending in the direction she is pulling? That is, how hard would she have to pull for the box to be on the verge of moving upward? Would she need help?

Answer: 257 lb. Yes.

Example 9.10 **Belts and Pulleys** (▶ *Related Problem 9.134*)

The pulleys turn at a constant rate. The large pulley is attached to a fixed support. The small pulley is supported by a smooth horizontal slot and is pulled to the right by the force $F = 200$ N. The coefficient of static friction between the pulleys and the belt is $\mu_s = 0.8$, the dimension $b = 500$ mm, and the radii of the pulleys are $R_A = 200$ mm and $R_B = 100$ mm. What are the largest values of the couples M_A and M_B for which the belt will not slip?

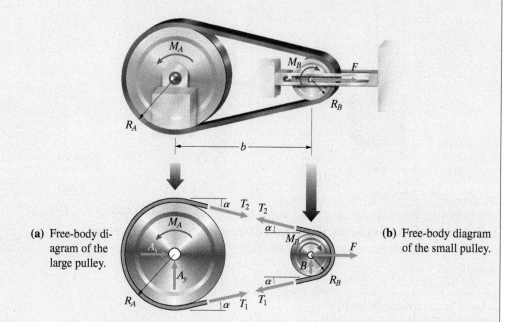

(a) Free-body diagram of the large pulley.

(b) Free-body diagram of the small pulley.

Strategy

By drawing free-body diagrams of the pulleys, we can use the equilibrium equations to relate the tensions in the belt to M_A and M_B and obtain a relation between the tensions in the belt and the force F. When slip is impending, the tensions are also related by Eq. (9.17). From these equations we can determine M_A and M_B.

Solution

From the free-body diagram of the large pulley (Fig. a), we obtain the equilibrium equation

$$M_A = R_A(T_2 - T_1), \tag{1}$$

and from the free-body diagram of the small pulley (Fig. b), we obtain

$$F = (T_1 + T_2) \cos \alpha, \tag{2}$$

$$M_B = R_B(T_2 - T_1). \tag{3}$$

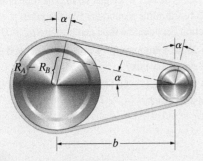

(c) Determining the angle α.

The belt is in contact with the small pulley through the angle $\pi - 2\alpha$ (Fig. c). From the dashed line parallel to the belt, we see that the angle α satisfies the relation

$$\sin \alpha = \frac{R_A - R_B}{b} = \frac{200 \text{ mm} - 100 \text{ mm}}{500 \text{ mm}} = 0.2.$$

Therefore, $\alpha = 11.5° = 0.201$ rad. If we assume that slip impends between the small pulley and the belt, Eq. (9.17) states that

$$T_2 = T_1 e^{\mu_s \beta} = T_1 e^{0.8(\pi - 2\alpha)} = 8.95 T_1.$$

We solve this equation together with Eq. (2) for the two tensions, obtaining $T_1 = 20.5$ N and $T_2 = 183.6$ N. Then from Eqs. (1) and (3), the couples are $M_A = 32.6$ N-m and $M_B = 16.3$ N-m.

 If we assume that slip impends between the large pulley and the belt, we obtain $M_A = 36.3$ N-m and $M_B = 18.1$ N-m, so the belt slips on the small pulley at smaller values of the couples.

Problems

▶ **9.123** In Active Example 9.9, suppose that the left fixed cylinder is replaced by a pulley. Assume that the tensions in the rope on each side of the pulley are approximately equal. What is the smallest force the woman needs to exert on the rope to support the stationary box?

9.124 Suppose that you want to lift a 50-lb crate off the ground by using a rope looped over a tree limb as shown. The coefficient of static friction between the rope and the limb is 0.2, and the rope is wound 135° around the limb. What force must you exert to begin lifting the crate?

Problem 9.124

9.125 *Winches* are used on sailboats to help support the forces exerted by the sails on the ropes (*sheets*) holding them in position. The winch shown is a post that will rotate in the clockwise direction (seen from above), but will not rotate in the counterclockwise direction. The sail exerts a tension $T_S = 800$ N on the sheet, which is wrapped two complete turns around the winch. The coefficient of static friction between the sheet and the winch is $\mu_s = 0.2$. What tension T_C must the crew member exert on the sheet to prevent it from slipping on the winch?

9.126 The coefficient of kinetic friction between the sheet and the winch in Problem 9.125 is $\mu_k = 0.16$. If the crew member wants to let the sheet slip at a constant rate, releasing the sail, what initial tension T_C must he exert on the sheet as it begins slipping?

Problems 9.125/9.126

9.127 The box *A* weighs 20 lb. The rope is wrapped one and one-fourth turns around the fixed wooden post. The coefficients of friction between the rope and post are $\mu_s = 0.15$ and $\mu_k = 0.12$.

(a) What minimum force does the man need to exert to support the stationary box?

(b) What force would the man have to exert to raise the box at a constant rate?

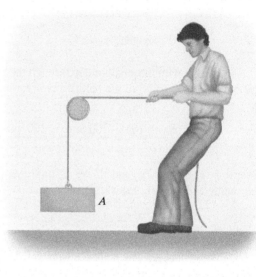

Problem 9.127

9.128 The weight of block *A* is *W*. The disk is supported by a smooth bearing. The coefficient of kinetic friction between the disk and the belt is μ_k. What couple *M* is necessary to turn the disk at a constant rate?

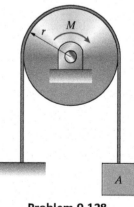

Problem 9.128

9.129 The couple required to turn the wheel of the exercise bicycle is adjusted by changing the weight W. The coefficient of kinetic friction between the wheel and the belt is μ_k. Assume the wheel turns clockwise.

(a) Show that the couple M required to turn the wheel is
$M = WR\,(1 - e^{-3.4\mu_k})$.

(b) If $W = 40$ lb and $\mu_k = 0.2$, what force will the scale S indicate when the bicycle is in use?

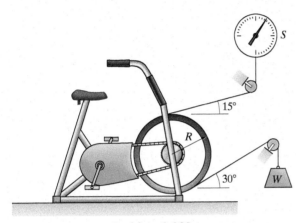

Problem 9.129

9.130 The box B weighs 50 lb. The coefficients of friction between the cable and the fixed round supports are $\mu_s = 0.4$ and $\mu_k = 0.3$.

(a) What is the minimum force F required to support the box?

(b) What force F is required to move the box upward at a constant rate?

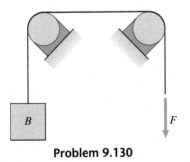

Problem 9.130

9.131 The coefficient of static friction between the 50-lb box and the inclined surface is 0.10. The coefficient of static friction between the rope and the fixed cylinder is 0.05. Determine the force the woman must exert on the rope to cause the box to start moving up the inclined surface.

9.132 In Problem 9.131, what is the minimum force the woman must exert on the rope to hold the box in equilibrium on the inclined surface?

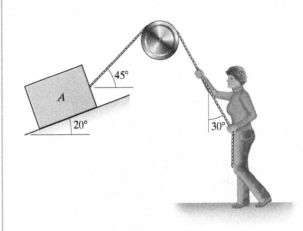

Problems 9.131/9.132

9.133 Blocks B and C each have a mass of 20 kg. The coefficient of static friction at the contacting surfaces is 0.2. Block A is suspended by a rope that passes over a fixed cylinder and is attached to block B. The coefficient of static friction between the rope and the cylinder is 0.3. What is the largest mass block A can have without causing block B to slip to the left?

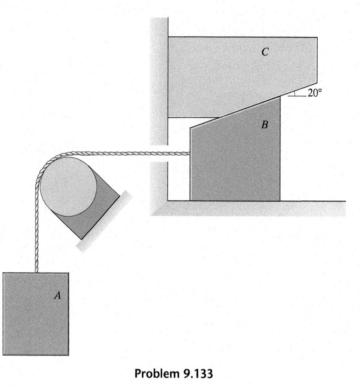

Problem 9.133

▶ **9.134** If the force F in Example 9.10 is increased to 400 N, what are the largest values of the couples M_A and M_B for which the belt will not slip?

9.135 The spring exerts a 320-N force on the left pulley. The coefficient of static friction between the flat belt and the pulleys is $\mu_s = 0.5$. The right pulley cannot rotate. What is the largest couple M that can be exerted on the left pulley without causing the belt to slip?

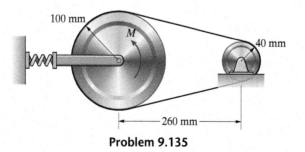

Problem 9.135

Review Problems

9.136 The weight of the box is $W = 30$ lb, and the force F is perpendicular to the inclined surface. The coefficient of static friction between the box and the inclined surface is $\mu_s = 0.2$.

(a) If $F = 30$ lb, what is the magnitude of the friction force exerted on the stationary box?

(b) If $F = 10$ lb, show that the box cannot remain at rest on the inclined surface.

9.137 In Problem 9.136, what is the smallest force F necessary to hold the box stationary on the inclined surface?

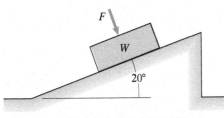

Problems 9.136/9.137

9.138 Blocks A and B are connected by a horizontal bar. The coefficient of static friction between the inclined surface and the 400-lb block A is 0.3. The coefficient of static friction between the surface and the 300-lb block B is 0.5. What is the smallest force F that will prevent the blocks from slipping down the surface?

9.139 What force F is necessary to cause the blocks in Problem 9.138 to start sliding up the plane?

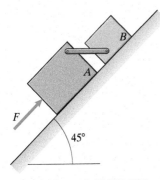

Problems 9.138/9.139

9.140 The masses of crates A and B are 25 kg and 30 kg, respectively. The coefficient of static friction between the contacting surfaces is $\mu_s = 0.34$. What is the largest value of α for which the crates will remain in equilibrium?

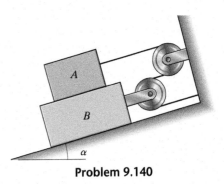

Problem 9.140

9.141 The side of a soil embankment has a 45° slope (Fig. a). If the coefficient of static friction of soil on soil is $\mu_s = 0.6$, will the embankment be stable or will it collapse? If it will collapse, what is the smallest slope that can be stable?

Strategy: Draw a free-body diagram by isolating part of the embankment as shown in Fig. b.

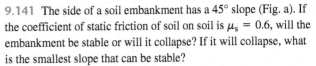

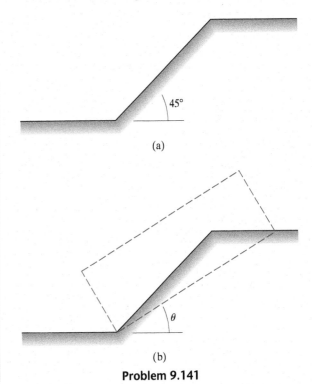

(a)

(b)

Problem 9.141

9.142 The mass of the van is 2250 kg, and the coefficient of static friction between its tires and the road is 0.6. If its front wheels are locked and its rear wheels can turn freely, what is the largest value of α for which it can remain in equilibrium?

9.143 In Problem 9.142, what is the largest value of α for which the van can remain in equilibrium if it points up the slope?

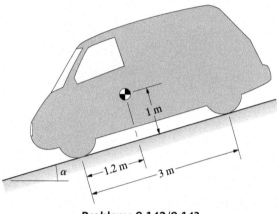

Problems 9.142/9.143

9.144 The shelf is designed so that it can be placed at any height on the vertical beam. The shelf is supported by friction between the two horizontal cylinders and the vertical beam. The combined weight of the shelf and camera is W. If the coefficient of static friction between the vertical beam and the horizontal cylinders is μ_s, what is the minimum distance b necessary for the shelf to stay in place?

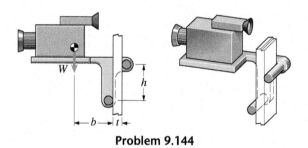

Problem 9.144

9.145 The 20-lb homogeneous object is supported at A and B. The distance $h = 4$ in, friction can be neglected at B, and the coefficient of static friction at A is 0.4. Determine the largest force F that can be exerted without causing the object to slip.

9.146 In Problem 9.145, suppose that the coefficient of static friction at B is 0.36. What is the largest value of h for which the object will slip before it tips over?

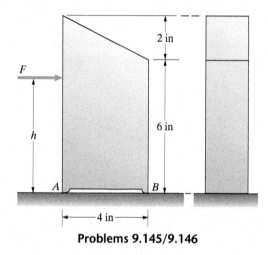

Problems 9.145/9.146

9.147 The 180-lb climber is supported in the "chimney" by the normal and friction forces exerted on his shoes and back. The static coefficients of friction between his shoes and the wall and between his back and the wall are 0.8 and 0.6, respectively. What is the minimum normal force his shoes must exert?

Problem 9.147

9.148 The sides of the 200-lb door fit loosely into grooves in the walls. Cables at A and B raise the door at a constant rate. The coefficient of kinetic friction between the door and the grooves is $\mu_k = 0.3$. What force must the cable at A exert to continue raising the door at a constant rate if the cable at B breaks?

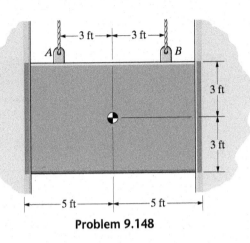

Problem 9.148

9.149 The coefficients of static friction between the tires of the 1000-kg tractor and the ground and between the 450-kg crate and the ground are 0.8 and 0.3, respectively. Starting from rest, what torque must the tractor's engine exert on the rear wheels to cause the crate to move? (The front wheels can turn freely.)

9.150 In Problem 9.149, what is the most massive crate the tractor can cause to move from rest if its engine can exert sufficient torque? What torque is necessary?

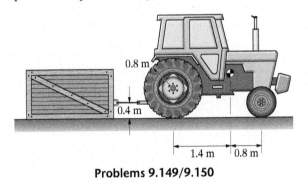

Problems 9.149/9.150

9.151 The mass of the vehicle is 900 kg, it has rear-wheel drive, and the coefficient of static friction between its tires and the surface is 0.65. The coefficient of static friction between the crate and the surface is 0.4. If the vehicle attempts to pull the crate up the incline, what is the largest value of the mass of the crate for which it will slip up the incline before the vehicle's tires slip?

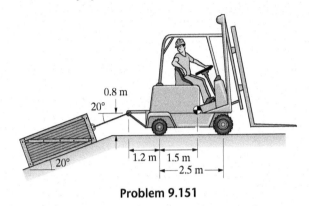

Problem 9.151

9.152 Each 1-m bar has a mass of 4 kg. The coefficient of static friction between the bar and the surface at B is 0.2. If the system is in equilibrium, what is the magnitude of the friction force exerted on the bar at B?

9.153 Each 1-m bar has a mass of 4 kg. What is the minimum coefficient of static friction between the bar and the surface at B necessary for the system to be in equilibrium?

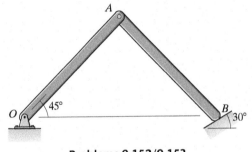

Problems 9.152/9.153

9.154 The collars A and B each have a mass of 2 kg. If friction between collar B and the bar can be neglected, what minimum coefficient of static friction between collar A and the bar is necessary for the collars to remain in equilibrium in the position shown?

9.155 If the coefficient of static friction has the same value μ_s between the 2-kg collars A and B and the bars, what minimum value of μ_s is necessary for the collars to remain in equilibrium in the position shown? (Assume that slip impends at A and B.)

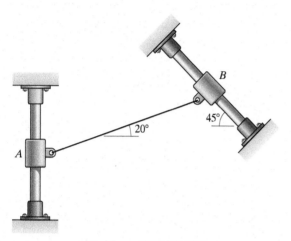

Problems 9.154/9.155

9.156 The clamp presses two pieces of wood together. The pitch of the threads is $p = 2$ mm, the mean radius of the thread is $r = 8$ mm, and the coefficient of kinetic friction between the thread and the mating groove is 0.24. What couple must be exerted on the threaded shaft to press the pieces of wood together with a force of 200 N?

9.157 In Problem 9.156, the coefficient of static friction between the thread and the mating groove is 0.28. After the threaded shaft is rotated sufficiently to press the pieces of wood together with a force of 200 N, what couple must be exerted on the shaft to loosen it?

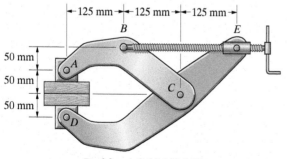

Problems 9.156/9.157

9.158 The axles of the tram are supported by journal bearings. The radius of the wheels is 75 mm, the radius of the axles is 15 mm, and the coefficient of kinetic friction between the axles and the bearings is $\mu_k = 0.14$. The mass of the tram and its load is 160 kg. If the weight of the tram and its load is evenly divided between the axles, what force P is necessary to push the tram at a constant speed?

Problem 9.158

9.159 The two pulleys have a radius of 6 in and are mounted on shafts of 1-in radius supported by journal bearings. Neglect the weights of the pulleys and shafts. The coefficient of kinetic friction between the shafts and the bearings is $\mu_k = 0.2$. If a force $T = 200$ lb is required to raise the man at a constant rate, what is his weight?

9.160 If the man in Problem 9.159 weighs 160 lb, what force T is necessary to lower him at a constant rate?

Problems 9.159/9.160

9.161 If the two cylinders are held fixed, what is the range of W for which the two weights will remain stationary?

9.162 If the system is initially stationary and the left cylinder is slowly rotated, determine the largest weight W that can be

(a) raised;

(b) lowered.

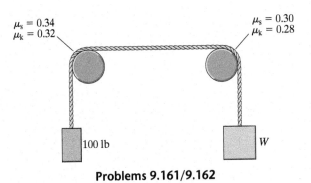

$\mu_s = 0.34$
$\mu_k = 0.32$

$\mu_s = 0.30$
$\mu_k = 0.28$

100 lb

W

Problems 9.161/9.162

Design Project 1

The wedge is used to split firewood by hammering it into a log as shown (see Active Example 9.4). Suppose that you want to design such a wedge to be marketed at hardware stores. Experiments indicate that the coefficient of static friction between the steel wedge and various types of wood varies from 0.2 to 0.4.

(a) Based on the given range of static friction coefficients, determine the maximum wedge angle α for which the wedge would remain in place in a log with no external force acting on it.

(b) Using the wedge angle determined in part (a), and assuming that the coefficient of kinetic friction is 0.9 times the coefficient of static friction, determine the range of vertical forces necessary to drive the wedge into a log at a constant rate.

(c) Write a brief report describing your analysis and recommending a wedge angle for the manufactured product. Consider whether a margin of safety in the chosen wedge angle might be appropriate.

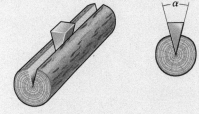

Design Project 2

Design and build a device to measure the coefficient of static friction μ_s between two materials. Use it to measure μ_s for several of the materials listed in Table 9.1 and compare your results with the values in the table. Discuss possible sources of error in your device and determine how closely your values agree when you perform repeated experiments with the same two materials.

CHAPTER 10

Internal Forces and Moments

We began our study of equilibrium by drawing free-body diagrams of individual objects to determine unknown forces and moments acting on them. In this chapter we carry this process one step further and draw free-body diagrams of parts of individual objects to determine internal forces and moments. In doing so, we arrive at the central concern of the structural design engineer: It is the forces within an object that determine whether it will support the external loads to which it is subjected.

◀ The force exerted by the water on the glass window is distributed over the area of the window. In this chapter we analyze distributed forces in beams, in suspended cables, and in stationary liquids.

BEAMS

10.1 Axial Force, Shear Force, and Bending Moment

BACKGROUND

To ensure that a structural member will not fail (break or collapse) due to the forces and moments acting on it, the design engineer must know not only the external loads and reactions acting on the member, but also the forces and moments acting *within* it.

Consider a beam subjected to an external load and reactions (Fig. 10.1a). How can we determine the forces and moments within the beam? In Fig. 10.1b, we "cut" the beam by a plane at an arbitrary cross section and isolate the part of the beam to the left of the plane. It is clear that the isolated part cannot be in equilibrium unless it is subjected to some system of forces and moments at the plane where it joins the other part of the beam. These are the internal forces and moments we seek.

In Chapter 4 we demonstrated that *any* system of forces and moments can be represented by an equivalent system consisting of a force and a couple. Since the system of external loads and reactions on the beam is two-dimensional, we can represent the internal forces and moments by an equivalent system consisting of two components of force and a couple (Fig. 10.1c). The component P parallel to the beam's axis is called the *axial force*. The component V normal to the beam's axis is called the *shear force*, and the couple M is called the

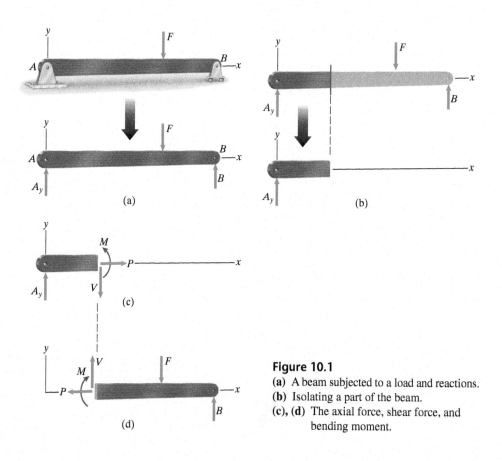

(a)

(b)

(c)

(d)

Figure 10.1
(a) A beam subjected to a load and reactions.
(b) Isolating a part of the beam.
(c), (d) The axial force, shear force, and bending moment.

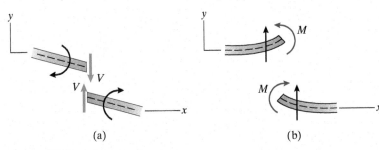

Figure 10.2
(a) Positive shear forces tend to rotate the axis of the beam clockwise.
(b) Positive bending moments tend to bend the axis of the beam upward.

bending moment. The axial force, shear force, and bending moment on the part of the beam to the right of the cutting plane are shown in Fig. 10.1d. Notice that they are equal in magnitude but opposite in direction to the internal forces and moment on the free-body diagram in Fig. 10.1c.

The directions of the axial force, shear force, and bending moment in Figs. 10.1c and 10.1d are the established definitions of the positive directions of these quantities. A positive axial force P subjects the beam to tension. A positive shear force V tends to rotate the axis of the beam clockwise (Fig. 10.2a). A positive bending moment M tends to cause upward curvature of the beam's axis (Fig. 10.2b). Notice that a positive bending moment subjects the upper part of the beam to compression, shortening the beam in the direction parallel to its axis, and subjects the lower part of the beam to tension, lengthening the beam in the direction parallel to its axis.

 RESULTS

The axial force P, shear force V, and bending moment M are an equivalent system representing the internal forces and moment at a cross section of a beam. *These are their defined positive directions.*

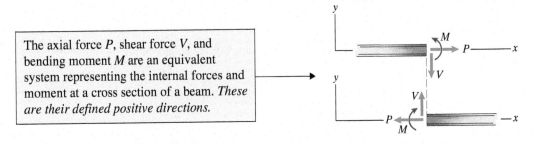

Determining the values of P, V, and M at a particular cross section of a beam typically involves three steps:

1. Draw the free-body diagram of the beam and determine the reactions at its supports.
2. Pass a plane through the beam at the cross section where the internal forces and moment are to be determined. Draw the free-body diagram of one of the resulting parts of the beam, showing P, V, and M in their defined positive directions.
3. Apply the equilibrium equations to determine P, V, and M.

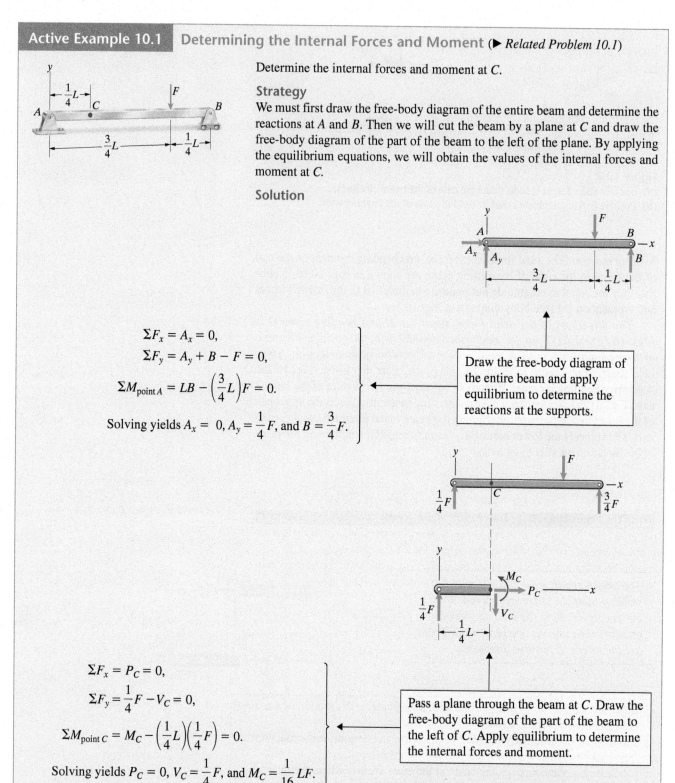

Active Example 10.1 Determining the Internal Forces and Moment (▶ *Related Problem 10.1*)

Determine the internal forces and moment at C.

Strategy

We must first draw the free-body diagram of the entire beam and determine the reactions at A and B. Then we will cut the beam by a plane at C and draw the free-body diagram of the part of the beam to the left of the plane. By applying the equilibrium equations, we will obtain the values of the internal forces and moment at C.

Solution

$$\Sigma F_x = A_x = 0,$$
$$\Sigma F_y = A_y + B - F = 0,$$
$$\Sigma M_{\text{point } A} = LB - \left(\frac{3}{4}L\right)F = 0.$$

Solving yields $A_x = 0$, $A_y = \frac{1}{4}F$, and $B = \frac{3}{4}F$.

> Draw the free-body diagram of the entire beam and apply equilibrium to determine the reactions at the supports.

$$\Sigma F_x = P_C = 0,$$
$$\Sigma F_y = \frac{1}{4}F - V_C = 0,$$
$$\Sigma M_{\text{point } C} = M_C - \left(\frac{1}{4}L\right)\left(\frac{1}{4}F\right) = 0.$$

Solving yields $P_C = 0$, $V_C = \frac{1}{4}F$, and $M_C = \frac{1}{16}LF$.

> Pass a plane through the beam at C. Draw the free-body diagram of the part of the beam to the left of C. Apply equilibrium to determine the internal forces and moment.

Practice Problem Determine the internal forces and moment at C by passing a plane through the beam at C and drawing the free-body diagram of the part of the beam to the right of C.

Answer: $P_C = 0$, $V_C = \frac{1}{4}F$, $M_C = \frac{1}{16}LF$.

| Example 10.2 | Determining the Internal Forces and Moment (▶ *Related Problem 10.8*) |

Determine the internal forces and moment at *B*.

Strategy

To determine the reactions at the supports, we will represent the triangular distributed load by an equivalent force. Then we will determine the internal forces and moment at *B* by cutting the beam by a plane at *B* and drawing the free-body diagram of the part of the beam to the left of the plane, *including the part of the distributed load to the left of the plane.*

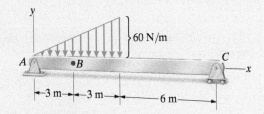

Solution

Determine the External Forces and Moments   We draw the free-body diagram of the beam and represent the distributed load by an equivalent force in Fig. a. The equilibrium equations are

$$\Sigma F_x = A_x = 0,$$

$$\Sigma F_y = A_y + C - 180 \text{ N} = 0$$

$$\Sigma M_{\text{point } A} = (12 \text{ m})C - (4 \text{ m})(180 \text{ N}) = 0$$

Solving them, we obtain $A_x = 0$, $A_y = 120$ N, and $C = 60$ N.

Draw the Free-Body Diagram of Part of the Beam   We cut the beam at *B*, obtaining the free-body diagram in Fig. b. Because point *B* is at the midpoint of the triangular distributed load, the value of the distributed load at *B* is 30 N/m. By representing the distributed load in Fig. b by an equivalent force, we obtain the free-body diagram in Fig. c. From the equilibrium equations

$$\Sigma F_x = P_B = 0,$$

$$\Sigma F_y = 120 \text{ N} - 45 \text{ N} - V_B = 0,$$

$$\Sigma M_{\text{point } B} = M_B + (1 \text{ m})(45 \text{ N}) - (3 \text{ m})(120 \text{ N}) = 0,$$

we obtain $P_B = 0$, $V_B = 75$ N, and $M_B = 315$ N-m.

Critical Thinking

If you attempt to determine the internal forces and moment at *B* by cutting the free-body diagram in Fig. a at *B*, you do *not* obtain correct results. (You can confirm that the resulting free-body diagram of the part of the beam to the left of *B* gives $P_B = 0$, $V_B = 120$ N, and $M_B = 360$ N-m.) The reason is that you do not properly account for the effect of the distributed load on your free-body diagram. You must wait until *after* you have isolated part of the beam before representing distributed loads acting on that part by equivalent forces.

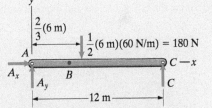

(a) Free-body diagram of the entire beam with the distributed load represented by an equivalent force.

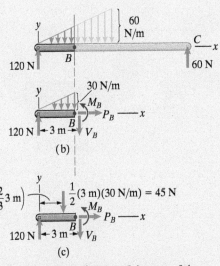

(b), (c) Free-body diagram of the part of the beam to the left of *B*.

Problems

▶ **10.1** In Active Example 10.1, suppose that the distance from point A to point C is increased from $\frac{1}{4}L$ to $\frac{1}{2}L$. Draw a sketch of the beam with C in its new position. Determine the internal forces and moment at C.

10.2 The magnitude of the triangular distributed load is $w_0 = 2$ kN/m. Determine the internal forces and moment at A.

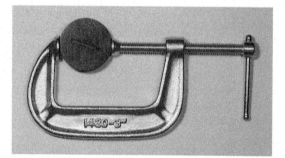

Problem 10.2

10.3 The C clamp exerts 30-lb forces on the clamped object. Determine the internal forces and moment in the clamp at A.

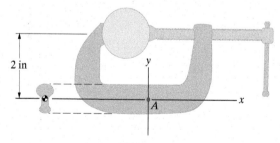

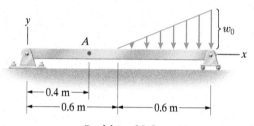

Problem 10.3

10.4 Determine the internal forces and moment at A.

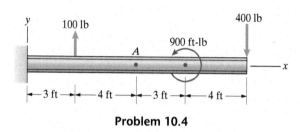

Problem 10.4

10.5 The pipe has a fixed support at the left end. Determine the internal forces and moment at A.

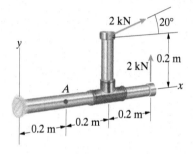

Problem 10.5

10.6 Determine the internal forces and moment at A for each loading.

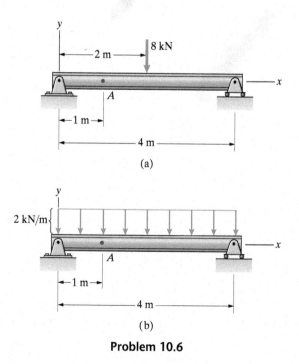

(a)

(b)

Problem 10.6

10.7 Model the ladder rung as a simply supported (pin supported) beam and assume that the 750-N load exerted by the person's shoe is uniformly distributed. Determine the internal forces and moment at A.

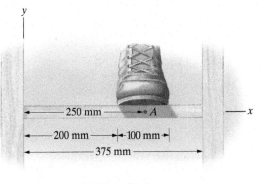

Problem 10.7

▶ **10.8** In Example 10.2, suppose that the distance from point A to point B is increased from 3 m to 4 m. Draw a sketch of the beam with B in its new position. Determine the internal forces and moment at B.

10.9 If x = 3 ft, what are the internal forces and moment at A?

10.10 If x = 4 ft, what are the internal forces and moment at A?

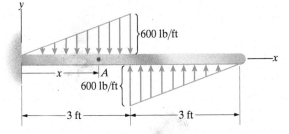

Problems 10.9/10.10

10.11 Determine the internal forces and moment at A for the loadings (a) and (b).

10.12 Determine the internal forces and moment at B for the loadings (a) and (b).

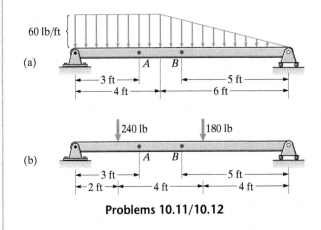
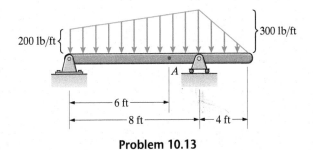

Problems 10.11/10.12

10.13 Determine the internal forces and moment at A.

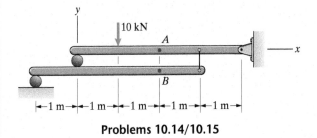

Problem 10.13

10.14 Determine the internal forces and moment at A.

10.15 Determine the internal forces and moment at B.

Problems 10.14/10.15

10.16 Determine the internal forces and moment at A.

10.17 Determine the internal forces and moment at B.

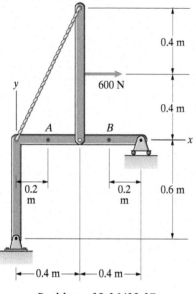

Problems 10.16/10.17

10.18 The tension in the rope is 10 kN. Determine the internal forces and moment at point A.

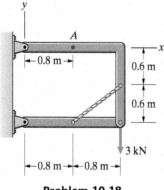

Problem 10.18

10.19 Determine the internal forces and moment at point A of the frame.

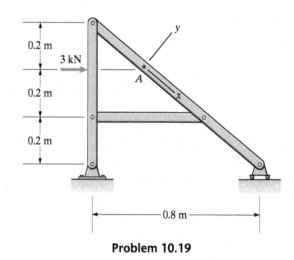

Problem 10.19

10.20 Determine the internal forces and moment at A.

10.21 Determine the internal forces and moment at B.

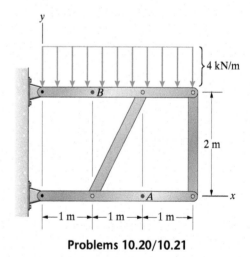

Problems 10.20/10.21

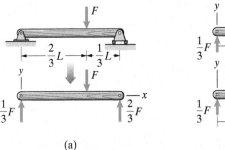

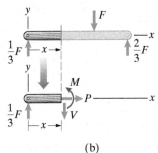

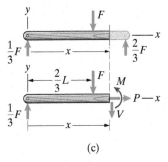

(a)   (b)   (c)

Figure 10.3
(a) A beam loaded by a force F and its free-body diagram.
(b) Cutting the beam at an arbitrary position x to the left of F.
(c) Cutting the beam at an arbitrary position x to the right of F.

10.2 Shear Force and Bending Moment Diagrams

BACKGROUND

To design a beam, an engineer must know the internal forces and moments throughout its length. Of special concern are the maximum and minimum values of the shear force and bending moment and where they occur. In this section we show how the values of P, V, and M can be determined as functions of x and introduce shear force and bending moment diagrams.

Consider a simply supported beam loaded by a force (Fig. 10.3a). Instead of cutting the beam at a specific cross section to determine the internal forces and moment, we cut it at an arbitrary position x between the left end of the beam and the load F (Fig. 10.3b). Applying the equilibrium equations to this free-body diagram, we obtain

$$\left. \begin{array}{l} P = 0 \\[4pt] V = \dfrac{1}{3}F \\[6pt] M = \dfrac{1}{3}Fx \end{array} \right\} \quad 0 < x < \dfrac{2}{3}L.$$

To determine the internal forces and moment for values of x greater than $\frac{2}{3}L$, we obtain a free-body diagram by cutting the beam at an arbitrary position x between the load F and the right end of the beam (Fig. 10.3c). The results are

$$\left. \begin{array}{l} P = 0 \\[4pt] V = -\dfrac{2}{3}F \\[6pt] M = \dfrac{2}{3}F(L - x) \end{array} \right\} \quad \dfrac{2}{3}L < x < L.$$

The *shear force and bending moment diagrams* are simply the graphs of V and M, respectively, as functions of x (Fig. 10.4). They permit you to see the changes in the shear force and bending moment that occur along the beam's length as well as their maximum and minimum values. (By *maximum* we mean the least upper bound of the shear force or bending moment, and by *minimum* we mean the greatest lower bound.)

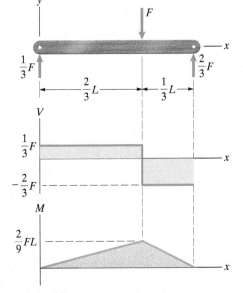

Figure 10.4
The shear force and bending moment diagrams indicating the maximum and minimum values of V and M.

Thus we can determine the distributions of the internal forces and moment in a beam by considering a plane at an arbitrary distance x from the end of the beam and solving for P, V, and M as functions of x. Depending on the complexity of the loading, it may be necessary to draw several free-body diagrams to determine the distributions over the entire length of the beam. The resulting equations for V and M allow us to draw the shear force and bending moment diagrams.

RESULTS

By passing a plane through a beam at an arbitrary position x, the values of P, V, and M can be determined as functions of x. Depending on the loading and supports, it may be necessary to draw several free-body diagrams to determine the distributions for the entire beam.

Shear force and bending moment diagrams for a beam are simply the graphs of V and M as functions of x.

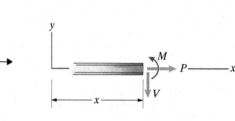

Active Example 10.3   Shear Force and Bending Moment Diagrams (▶ *Related Problem 10.27*)

Determine the shear force V and bending moment M for the beam as functions of x for $0 < x < 2$ m.

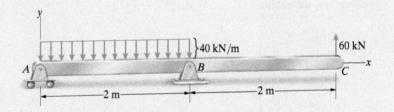

Strategy
We must first draw the free-body diagram of the entire beam and determine the reactions at A and B. Then we will cut the beam by a plane at an arbitrary position x between A and B to obtain functions for V and M that are valid in the range $0 < x < 2$ m.

Solution

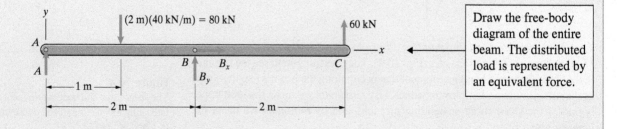

Draw the free-body diagram of the entire beam. The distributed load is represented by an equivalent force.

$$\Sigma F_x = B_x = 0,$$
$$\Sigma F_y = A + B_y - 80 \text{ kN} + 60 \text{ kN} = 0,$$
$$\Sigma M_{\text{point } A} = (2 \text{ m})B_y - (1 \text{ m})(80 \text{ kN}) + (4 \text{ m})(60 \text{ kN}) = 0.$$
Solving yields $A = 100$ kN, $B_x = 0$, and $B_y = -80$ kN.

Apply equilibrium to determine the reactions at A and B.

Pass a plane through the beam at an arbitrary position x between A and B. The distributed load *must not* be represented by an equivalent force before isolating part of the beam.

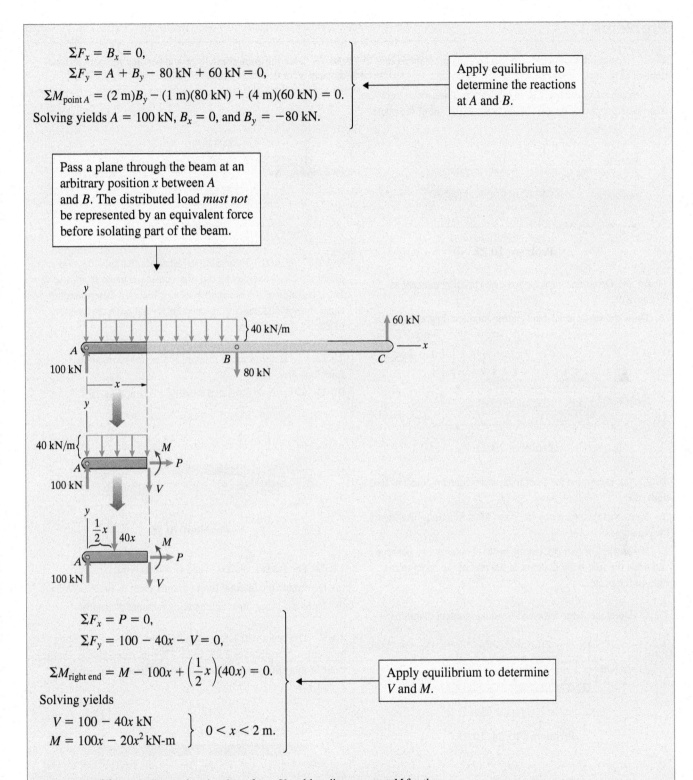

$$\Sigma F_x = P = 0,$$
$$\Sigma F_y = 100 - 40x - V = 0,$$
$$\Sigma M_{\text{right end}} = M - 100x + \left(\frac{1}{2}x\right)(40x) = 0.$$
Solving yields
$$\left.\begin{array}{l} V = 100 - 40x \text{ kN} \\ M = 100x - 20x^2 \text{ kN-m} \end{array}\right\} \quad 0 < x < 2 \text{ m}.$$

Apply equilibrium to determine V and M.

Practice Problem (a) Determine the shear force V and bending moment M for the beam as functions of x for $2 < x < 4$ m. (b) Draw the shear force and bending moment diagrams for the entire beam.

Answer: $V = -60$ kN, $M = 60(4 - x)$ kN-m.

Problems

10.22 Determine the shear force and bending moment as functions of x.

Strategy: Cut the beam at an arbitrary position x and draw the free-body diagram of the part of the beam to the left of the plane.

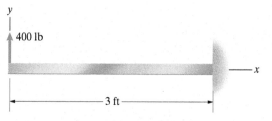

Problem 10.22

10.23 (a) Determine the shear force and bending moment as functions of x.

(b) Draw the shear force and bending moment diagrams.

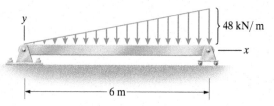

Problem 10.23

10.24 (a) Determine the shear force and bending moment as functions of x.

(b) Show that the equations for V and M as functions of x satisfy the equation $V = dM/dx$.

Strategy: For part (a), cut the beam at an arbitrary position x and draw the free-body diagram of the part of the beam to the right of the plane.

10.25 Draw the shear force and bending moment diagrams.

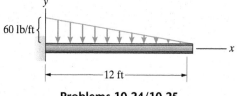

Problems 10.24/10.25

10.26 Determine the shear force and bending moment as functions of x for $0 < x < 2$ m.

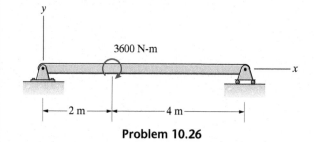

Problem 10.26

▶ **10.27** In Active Example 10.3, suppose that the 40 kN/m distributed load extends all the way across the beam from A to C. Draw a sketch of the beam with its new loading. Determine the shear force V and bending moment M for the beam as functions of x for $2 < x < 4$ m.

10.28 (a) Determine the internal forces and moment as functions of x.

(b) Draw the shear force and bending moment diagrams.

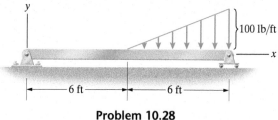

Problem 10.28

10.29 The loads $F = 200$ N and $C = 800$ N-m.

(a) Determine the internal forces and moment as functions of x.

(b) Draw the shear force and bending moment diagrams.

10.30 The beam will safely support shear forces and bending moments of magnitudes 2 kN and 6.5 kN-m, respectively. On the basis of this criterion, can it safely be subjected to the loads $F = 1$ kN, $C = 1.6$ kN-m?

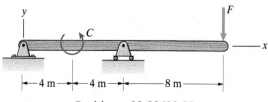

Problems 10.29/10.30

10.31 Model the ladder rung as a simply supported (pin-supported) beam and assume that the 750-N load exerted by the person's shoe is uniformly distributed. Draw the shear force and bending moment diagrams.

10.32 What is the maximum bending moment in the ladder rung in Problem 10.31 and where does it occur?

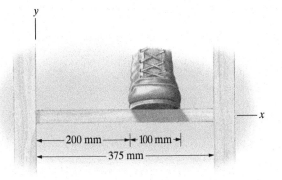

Problems 10.31/10.32

10.33 Assume that the surface the beam rests on exerts a uniformly distributed load. Draw the shear force and bending moment diagrams.

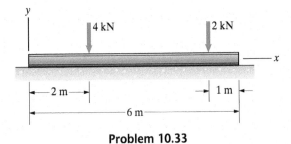

Problem 10.33

10.34 The homogeneous beams AB and CD weigh 600 lb and 500 lb, respectively. Draw the shear force and bending moment diagrams for beam AB.

10.35 The homogeneous beams AB and CD weigh 600 lb and 500 lb, respectively. Draw the shear force and bending moment diagrams for beam CD.

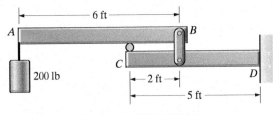

Problems 10.34/10.35

10.36 Determine the shear force V and bending moment M for the beam as functions of x for $0 < x < 3$ ft.

10.37 Draw the shear force and bending moment diagrams for the beam.

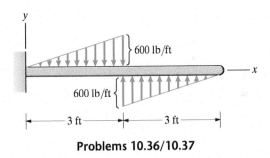

Problems 10.36/10.37

10.38 In preliminary design studies, the vertical forces on an airplane's wing are modeled as shown. The distributed load models aerodynamic forces and the force exerted by the wing's weight. The 80-kN force at $x = 4.4$ m models the force exerted by the weight of the engine. Draw the shear force and bending moment diagrams for the wing for $0 < x < 4.4$ m.

10.39 Draw the shear force and bending moment diagrams for the entire wing.

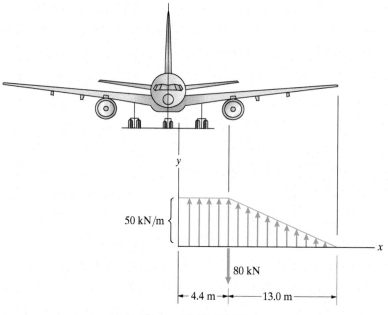

50 kN/m

80 kN

4.4 m — 13.0 m

Problems 10.38/10.39

10.40* Draw the shear force and bending moment diagrams.

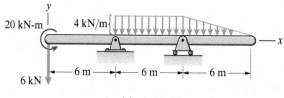

20 kN-m

4 kN/m

6 kN

6 m — 6 m — 6 m

Problem 10.40

10.3 Relations Between Distributed Load, Shear Force, and Bending Moment

BACKGROUND

The shear force and bending moment in a beam subjected to a distributed load are governed by simple differential equations. In this section we derive these equations and show that they provide an interesting and enlightening way to obtain shear force and bending moment diagrams. These equations are also useful for determining deformations of beams.

Suppose that a portion of a beam is subjected to a distributed load w (Fig. 10.5a). In Fig. 10.5b we obtain a free-body diagram by cutting the beam at positions x and $x + \Delta x$. The terms ΔP, ΔV, and ΔM are the changes in the

axial force, shear force, and bending moment, respectively, from x to $x + \Delta x$.
The sum of the forces in the x direction is

$$\Sigma F_x = P + \Delta P - P = 0.$$

Dividing this equation by Δx and taking the limit as $\Delta x \to 0$, we obtain

$$\frac{dP}{dx} = 0,$$

which simply states that the axial force does not depend on x in a portion of a
beam subjected only to a lateral distributed load. To sum the forces on the free-
body diagram in the y direction, we must determine the force exerted by the distributed load. In Fig. 10.5b we introduce a coordinate $\hat{x}$ that measures distance
from the left edge of the free-body diagram. In terms of this coordinate, the
downward force exerted on the free-body diagram by the distributed load is

$$\int_0^{\Delta x} w(x + \hat{x}) \, d\hat{x},$$

where $w(x + \hat{x})$ denotes the value of w at $x + \hat{x}$. To evaluate this integral, we
express $w(x + \hat{x})$ as a Taylor series in terms of $\hat{x}$:

$$w(x + \hat{x}) = w(x) + \frac{dw(x)}{dx}\hat{x} + \frac{1}{2}\frac{d^2w(x)}{dx^2}\hat{x}^2 + \cdots. \quad (10.1)$$

Substituting this equation into the integral expression for the downward force
and integrating term by term, we obtain

$$\int_0^{\Delta x} w(x + \hat{x}) \, d\hat{x} = w(x)\Delta x + \frac{1}{2}\frac{dw(x)}{dx}(\Delta x)^2 + \cdots.$$

The sum of the forces on the free-body diagram in the y direction is therefore

$$\Sigma F_y = V - V - \Delta V - w(x)\Delta x - \frac{1}{2}\frac{dw(x)}{dx}(\Delta x)^2 + \cdots = 0.$$

Dividing by Δx and taking the limit as $\Delta x \to 0$, we obtain

$$\frac{dV}{dx} = -w, \quad (10.2)$$

where $w = w(x)$.

We now want to sum the moments about point Q on the free-body diagram
in Fig. 10.5b. The clockwise moment about Q due to the distributed load is

$$\int_0^{\Delta x} \hat{x} w(x + \hat{x}) \, d\hat{x}.$$

Substituting Eq. (10.1) and integrating term by term, the clockwise moment
about Q is

$$\int_0^{\Delta x} \hat{x} w(x + \hat{x}) \, d\hat{x} = \frac{1}{2}w(x)(\Delta x)^2 + \frac{1}{3}\frac{dw(x)}{dx}(\Delta x)^3 + \cdots.$$

The sum of the moments about Q is therefore

$$\Sigma M_{\text{point } Q} = M + \Delta M - M - (V + \Delta V)\,\Delta x$$

$$-\frac{1}{2}w(x)(\Delta x)^2 - \frac{1}{3}\frac{dw(x)}{dx}(\Delta x)^3 + \cdots = 0.$$

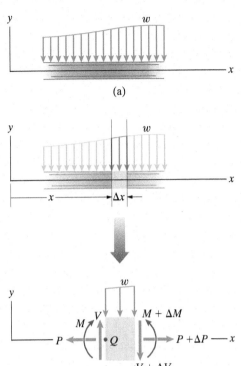

(a)

(b)

Figure 10.5
(a) A portion of a beam subjected to a distributed force w.
(b) Obtaining the free-body diagram of an
element of the beam.

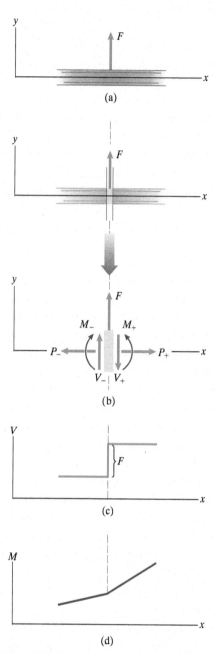

Figure 10.6
(a) A portion of a beam subjected to a distributed force F in the positive y direction.
(b) Obtaining a free-body diagram by cutting the beam to the left and right of F.
(c) The shear force diagram undergoes a positive jump of magnitude F.
(d) The bending moment diagram is continuous.

Dividing by Δx and taking the limit as $\Delta x \rightarrow 0$ gives

$$\frac{dM}{dx} = V. \tag{10.3}$$

In principle, we can use Eqs. (10.2) and (10.3) to determine the shear force and bending moment diagrams for a beam. Equation (10.2) can be integrated to determine V as a function of x, then Eq. (10.3) can be integrated to determine M as a function of x. However, we derived these equations for a segment of beam subjected only to a distributed load. To apply them for a more general loading, we must account for the effects of any point forces and couples acting on the beam.

Let us determine what happens to the shear force and bending moment diagrams where a beam is subjected to a force F in the positive y direction (Fig. 10.6a). By cutting the beam just to the left and just to the right of the force, we obtain the free-body diagram in Fig. 10.6b, where the subscripts $-$ and $+$ denote values to the left and right of the force, respectively. Equilibrium requires that

$$V_+ - V_- = F,$$
$$M_+ - M_- = 0.$$

The shear force diagram undergoes a jump discontinuity of magnitude F (Fig. 10.6c), but the bending moment diagram is continuous (Fig. 10.6d). The jump in the shear force is positive if the force is in the positive y direction.

Now we consider what happens to the shear force and bending moment diagrams when a beam is subjected to a counterclockwise couple C (Fig. 10.7a). Cutting the beam just to the left and just to the right of the couple (Fig. 10.7b), we determine that

$$V_+ - V_- = 0,$$
$$M_+ - M_- = -C.$$

The shear force diagram is continuous (Fig. 10.7c), but the bending moment diagram undergoes a jump discontinuity of magnitude C (Fig. 10.7d), where a beam is subjected to a couple. The jump in the bending moment is *negative* if the couple is in the counterclockwise direction.

We now have the results needed to construct shear force and bending moment diagrams.

Construction of the Shear Force Diagram

In a segment of a beam that is subjected only to a distributed load, we have shown that the shear force is related to the distributed load by

$$\frac{dV}{dx} = -w. \tag{10.4}$$

This equation states that the derivative, or slope, of the shear force with respect to x is equal to the negative of the distributed load. Notice that if there is no distributed load $(w = 0)$ throughout the segment, the slope is zero and the shear force is constant. If w is a constant throughout the segment, the slope of the shear force is constant, which means that the shear force diagram for the segment is a straight line. Integrating Eq. (10.4) with respect to x from a position x_A to a position x_B,

$$\int_{x_A}^{x_B} \frac{dV}{dx}\, dx = -\int_{x_A}^{x_B} w\, dx,$$

yields

$$V_B - V_A = -\int_{x_A}^{x_B} w \, dx.$$

The change in the shear force between two positions is equal to the negative of the area defined by the loading curve between those positions (Fig. 10.8):

$$V_B - V_A = -(\text{area defined by the distributed load from } x_A \text{ to } x_B). \quad (10.5)$$

Where a beam is subjected to a point force of magnitude F in the positive y direction, we have shown that the shear force diagram undergoes an increase of magnitude F. Where a beam is subjected to a couple, the shear force diagram is unchanged (continuous).

Let us demonstrate these results by determining the shear force diagram for the beam in Fig. 10.9. The beam is subjected to a downward force F that results in upward reactions at A and C. Notice that there is no distributed load. Our procedure is to begin at the left end of the beam and construct the diagram from left to right. Figure 10.10a shows the increase in the value of V due to the upward reaction at A. Because there is no distributed load, the value of V remains constant between A and B (Fig. 10.10b). At B, the value of V decreases due to the downward force (Fig. 10.10c). The value of V remains constant between B and C, which completes the shear force diagram (Fig. 10.10d). Compare Fig. 10.10d with the shear force diagram we obtained in Fig. 10.4 by drawing free-body diagrams and applying the equilibrium equations.

Construction of the Bending Moment Diagram

In a segment of a beam subjected only to a distributed load, the bending moment is related to the shear force by

$$\frac{dM}{dx} = V, \quad (10.6)$$

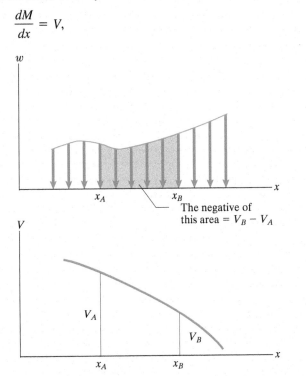

Figure 10.8
The change in the shear force is equal to the negative of the area defined by the loading curve.

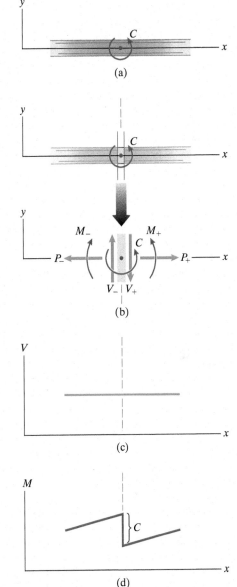

Figure 10.7
(a) A portion of a beam subjected to a counterclockwise couple C.
(b) Obtaining a free-body diagram by cutting the beam to the left and right of C.
(c) The shear force diagram is continuous.
(d) The bending moment diagram undergoes a *negative* jump of magnitude C.

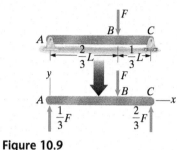

Figure 10.9
Beam loaded by a force F and its free-body diagram.

which states that the slope of the bending moment with respect to x is equal to the shear force. If V is constant throughout the segment, the bending moment diagram for the segment is a straight line. Integrating Eq. (10.6) with respect to x from a position x_A to a position x_B yields

$$M_B - M_A = \int_{x_A}^{x_B} V \, dx.$$

The change in the bending moment between two positions is equal to the area defined by the shear force diagram between those positions (Fig. 10.11):

$$M_B - M_A = \text{area defined by the shear force from } x_A \text{ to } x_B. \tag{10.7}$$

Where a beam is subjected to a counterclockwise couple of magnitude C, the bending moment diagram undergoes a decrease of magnitude C. Where a beam is subjected to a point force, the bending moment diagram is unchanged.

As an example, we will determine the bending moment diagram for the beam in Fig. 10.9. We begin with the shear force diagram we have already determined (Fig. 10.12a) and proceed to construct the bending moment diagram from left to right. The beam is not subjected to a couple at A, so $M_A = 0$. Between A and B, the slope of the bending moment is constant

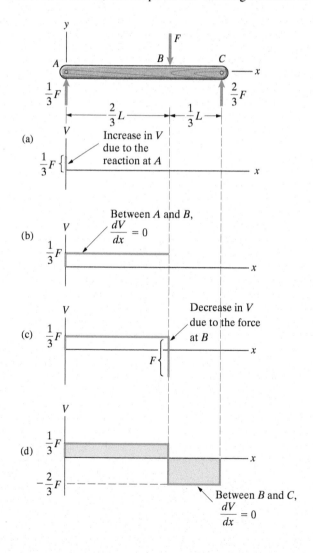

Figure 10.10
Constructing the shear force diagram for the beam in Fig. 10.9.

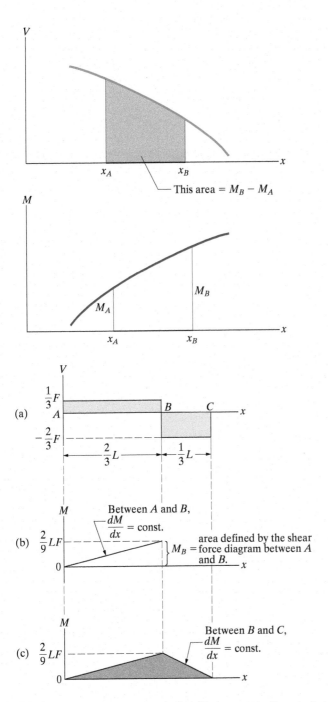

Figure 10.11
The change in the bending moment is equal to the area defined by the shear force diagram.

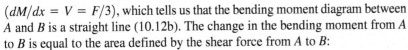

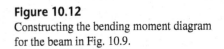

Figure 10.12
Constructing the bending moment diagram for the beam in Fig. 10.9.

$(dM/dx = V = F/3)$, which tells us that the bending moment diagram between A and B is a straight line (10.12b). The change in the bending moment from A to B is equal to the area defined by the shear force from A to B:

$$M_B - M_A = (\tfrac{2}{3}L)(\tfrac{1}{3}F) = \tfrac{2}{9}LF.$$

Therefore, $M_B = 2LF/9$. The slope of the bending moment is also constant between B and C $(dM/dx = V = -2F/3)$, so the bending moment diagram between B and C is a straight line. The change in the bending moment from B to C is equal to the area defined by the shear force from B to C, or

$$M_C - M_B = (\tfrac{1}{3}L)(-\tfrac{2}{3}F) = -\tfrac{2}{9}LF,$$

from which we obtain $M_C = M_B - 2LF/9 = 0$. (Notice that we did not actually need this calculation to conclude that $M_C = 0$, because the beam is not subjected to a couple at C.) The completed bending moment diagram is shown in Fig. 10.12c. Compare it with the bending moment diagram we obtained in Fig. 10.4 by drawing free-body diagrams and applying the equilibrium equations.

RESULTS

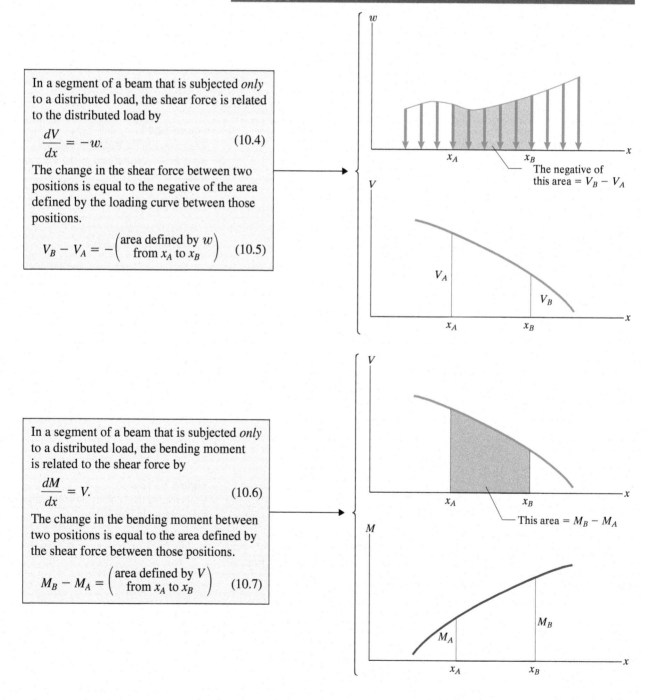

In a segment of a beam that is subjected *only* to a distributed load, the shear force is related to the distributed load by

$$\frac{dV}{dx} = -w. \qquad (10.4)$$

The change in the shear force between two positions is equal to the negative of the area defined by the loading curve between those positions.

$$V_B - V_A = -\left(\begin{array}{c}\text{area defined by } w \\ \text{from } x_A \text{ to } x_B\end{array}\right) \qquad (10.5)$$

In a segment of a beam that is subjected *only* to a distributed load, the bending moment is related to the shear force by

$$\frac{dM}{dx} = V. \qquad (10.6)$$

The change in the bending moment between two positions is equal to the area defined by the shear force between those positions.

$$M_B - M_A = \left(\begin{array}{c}\text{area defined by } V \\ \text{from } x_A \text{ to } x_B\end{array}\right) \qquad (10.7)$$

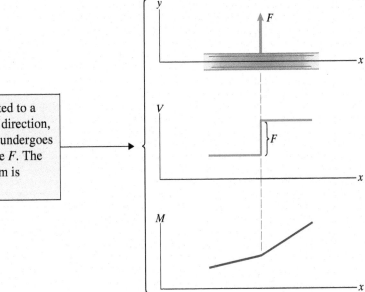

Where a beam is subjected to a force F in the positive y direction, the shear force diagram undergoes an increase of magnitude F. The bending moment diagram is continuous.

Where a beam is subjected to a counterclockwise couple C, the bending moment diagram undergoes a *decrease* of magnitude C. The shear force diagram is continuous.

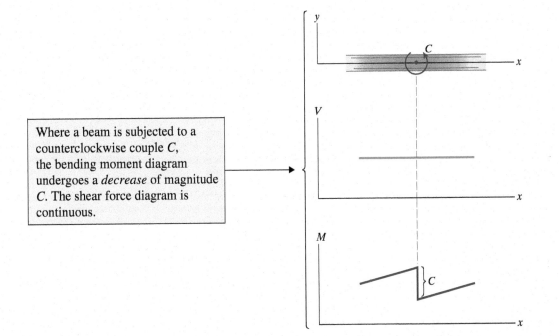

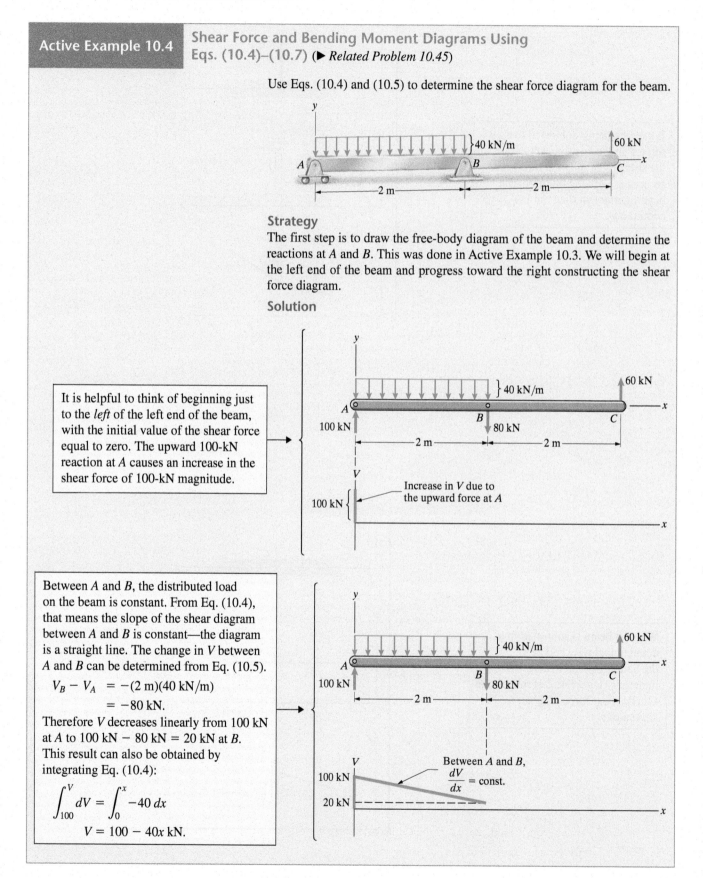

| Active Example 10.4 | Shear Force and Bending Moment Diagrams Using Eqs. (10.4)–(10.7) (▶ *Related Problem 10.45*) |

Use Eqs. (10.4) and (10.5) to determine the shear force diagram for the beam.

Strategy

The first step is to draw the free-body diagram of the beam and determine the reactions at *A* and *B*. This was done in Active Example 10.3. We will begin at the left end of the beam and progress toward the right constructing the shear force diagram.

Solution

It is helpful to think of beginning just to the *left* of the left end of the beam, with the initial value of the shear force equal to zero. The upward 100-kN reaction at *A* causes an increase in the shear force of 100-kN magnitude.

Increase in *V* due to the upward force at *A*

Between *A* and *B*, the distributed load on the beam is constant. From Eq. (10.4), that means the slope of the shear diagram between *A* and *B* is constant—the diagram is a straight line. The change in *V* between *A* and *B* can be determined from Eq. (10.5).

$$V_B - V_A = -(2 \text{ m})(40 \text{ kN/m})$$
$$= -80 \text{ kN.}$$

Therefore *V* decreases linearly from 100 kN at *A* to 100 kN − 80 kN = 20 kN at *B*. This result can also be obtained by integrating Eq. (10.4):

$$\int_{100}^{V} dV = \int_{0}^{x} -40 \, dx$$
$$V = 100 - 40x \text{ kN.}$$

Between *A* and *B*, $\dfrac{dV}{dx}$ = const.

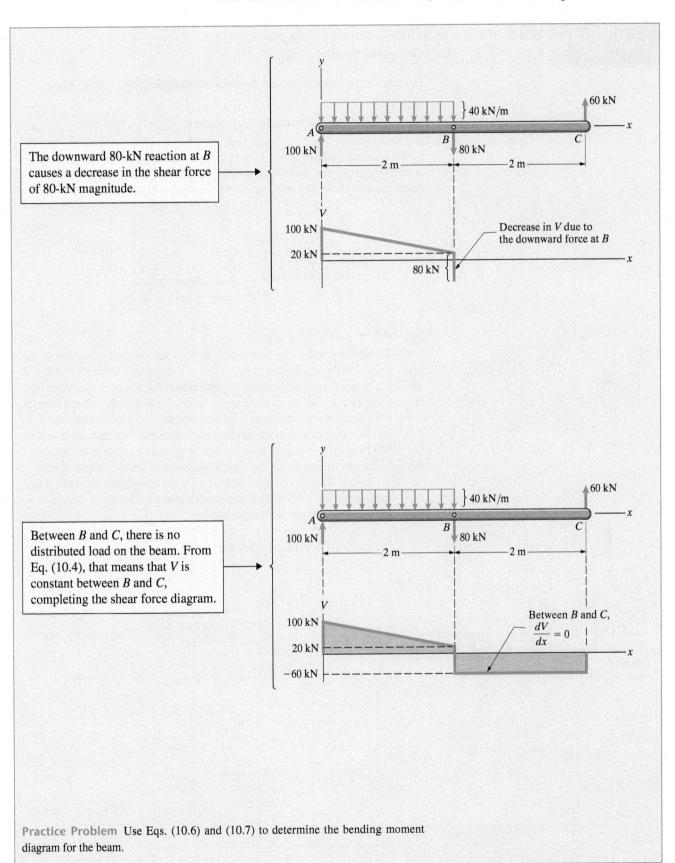

The downward 80-kN reaction at B causes a decrease in the shear force of 80-kN magnitude.

Between B and C, there is no distributed load on the beam. From Eq. (10.4), that means that V is constant between B and C, completing the shear force diagram.

Practice Problem Use Eqs. (10.6) and (10.7) to determine the bending moment diagram for the beam.

Example 10.5	**Shear Force and Bending Moment Diagrams Using Eqs. (10.4)–(10.7)** (▶ *Related Problem 10.44*)

Determine the shear force and bending moment diagrams for the beam.

Strategy

We can begin with the free-body diagram of the beam and use Eqs. (10.4) and (10.5) to construct the shear force diagram. Then we can use the shear force diagram and Eqs. (10.6) and (10.7) to construct the bending moment diagram. In determining both the shear force and bending moment diagrams, we must account for the effects of point forces and couples acting on the beam.

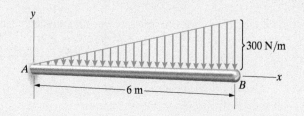

Solution

Shear Force Diagram The first step is to draw the free-body diagram of the beam and determine the reactions at the built-in support A. Using the results of this step, shown in Fig. a, we proceed to construct the shear force diagram from left to right. Figure b shows the increase in the value of V due to the upward force at A. Between A and B, the distributed load on the beam increases linearly from 0 to 300 N/m. Therefore, the slope of the shear force diagram decreases linearly from 0 to −300 N/m. At B, the shear force must be 0, because no force acts there. With this information, we can sketch the shear force diagram qualitatively (Fig. c).

We can also obtain an explicit equation for the shear force between A and B by integrating Eq. (10.4). The distributed load as a function of x is $w = (x/6)300 = 50x$ N/m. We write Eq. (10.4) as

$$dV = -w\,dx = -50x\,dx$$

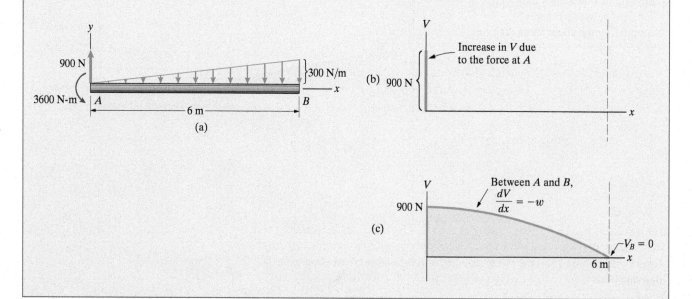

and integrate to determine V at an arbitrary position x:

$$\int_{V_A}^{V} dV = \int_{0}^{x} - 50x \, dx$$

$$V - V_A = -25x^2.$$

Due to the 900-N upward reaction at A, $V_A = 900$ N, so we obtain

$$V = 900 - 25x^2 \text{ N.} \tag{1}$$

Bending Moment Diagram   We construct the bending moment diagram from left to right. Figure d shows the initial decrease in the value of M due to the counterclockwise couple at A. Between A and B, the slope of the bending moment diagram is equal to the shear force V. We see from the shear force diagram (Fig. c) that at A, the slope of the bending moment diagram has a positive value (900 N). As x increases, the slope begins to decrease, and its rate of decrease grows until the value of the slope reaches zero at B. At B, we know that the value of the bending moment is zero, because no couple acts on the beam at B. Using this information, we can sketch the bending moment diagram qualitatively (Fig. e). Notice that its slope decreases from a positive value at A to zero at B, and the rate at which it decreases grows as x increases.

We can obtain an equation for the bending moment between A and B by integrating Eq. (10.6). The shear force as a function of x is given by Eq. (1). We write Eq. (10.6) as

$$dM = V \, dx = (900 - 25x^2) \, dx$$

and integrate:

$$\int_{M_A}^{M} dM = \int_{0}^{x} (900 - 25x^2) \, dx$$

$$M - M_A = 900x - \frac{25}{3}x^3.$$

As a result of the 3600 N-m counterclockwise couple at A, $M_A = -3600$ N-m, yielding the bending moment distribution

$$M = -3600 + 900x - \frac{25}{3}x^3 \text{ N-m.}$$

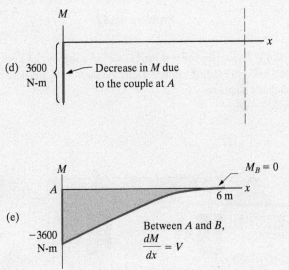

(d)   3600 N-m   ← Decrease in M due to the couple at A

(e) $M_B = 0$ 6 m -3600 N-m   Between A and B, $\dfrac{dM}{dx} = V$

Critical Thinking
As demonstrated in this example, Eqs. (10.4)–(10.7) can be applied in two ways. They provide a basis for rapidly obtaining qualitative sketches of shear force and bending moment diagrams. In addition, explicit equations for the diagrams can be obtained by integrating Eqs. (10.4) and (10.6).

Problems

The following problems are to be solved using Eqs. (10.4)–(10.7).

10.41 Draw the shear force and bending moment diagrams.

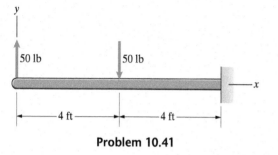

Problem 10.41

10.42 Draw the shear force and bending moment diagrams.

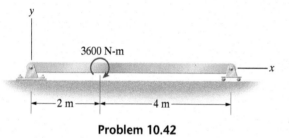

Problem 10.42

10.43 This arrangement is used to subject a segment of a beam to a uniform bending moment. Draw the shear force and bending moment diagrams.

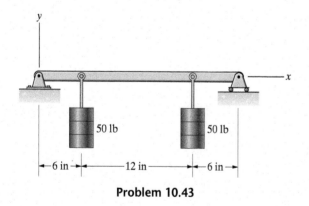

Problem 10.43

▶ **10.44** Use the procedure described in Example 10.5 to draw the shear force and bending moment diagrams for the beam.

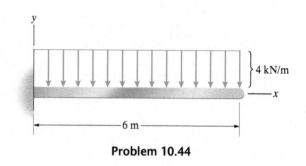

Problem 10.44

▶ **10.45** In Active Example 10.4, suppose that the 40 kN/m distributed load extends all the way across the beam from A to C. Draw a sketch of the beam with its new loading. Draw the shear force diagram for the beam.

10.46 Draw the shear force and bending moment diagrams.

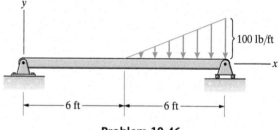

Problem 10.46

10.47 Determine the shear force V and bending moment M for the beam as functions of x.

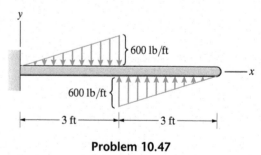

Problem 10.47

10.48* Draw the shear force and bending moment diagrams.

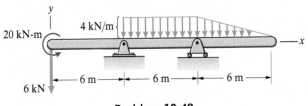

Problem 10.48

10.49 Draw the shear force and bending moment diagrams for the beam *AB*.

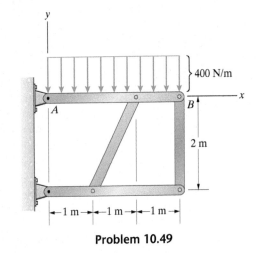

Problem 10.49

CABLES

Because of their unique combination of strength, lightness, and flexibility, ropes and cables are often used to support loads and transmit forces in structures, machines, and vehicles. The great suspension bridges are supported by enormous steel cables. Architectural engineers use cables to create aesthetic structures with open interior spaces (Fig. 10.13). In the following sections we determine the tensions in ropes and cables subjected to distributed and discrete loads.

Figure 10.13
The use of cables to suspend the roof of this sports stadium provides spectators with a view unencumbered by supporting columns.

10.4 Loads Distributed Uniformly Along Straight Lines

BACKGROUND

The main cable of a suspension bridge is the classic example of a cable subjected to a load uniformly distributed along a straight line (Fig. 10.14). The weight of the bridge is (approximately) uniformly distributed horizontally. The load, transmitted to the main cable by the large number of vertical cables, can be modeled as a distributed load. In this section we determine the shape and the variation in the tension of a cable loaded in this way.

Consider a suspended cable subjected to a load distributed uniformly along a horizontal line (Fig. 10.15a). We neglect the weight of the cable. The origin of the coordinate system is located at the cable's lowest point. Let the function $y(x)$ be the curve described by the cable in the x–y plane. Our objective is to determine the curve $y(x)$ and the tension in the cable.

Shape of the Cable

We obtain a free-body diagram by cutting the cable at its lowest point and at an arbitrary position x (Fig. 10.15b). The term T_0 is the tension in the cable at its lowest point, and T is the tension at x. The downward force exerted by the distributed load is wx. From this free-body diagram, we obtain the equilibrium equations

$$T \cos \theta = T_0,$$

$$T \sin \theta = wx. \tag{10.8}$$

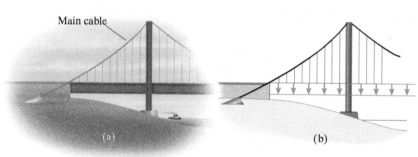

Figure 10.14
(a) Main cable of a suspension bridge.
(b) The load is distributed horizontally.

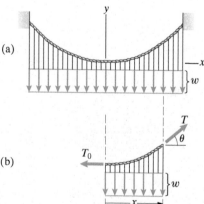

Figure 10.15
(a) A cable subjected to a load uniformly distributed along a horizontal line.
(b) Free-body diagram of the cable between $x = 0$ and an arbitrary position x.

We eliminate the tension T by dividing the second equation by the first one, obtaining

$$\tan \theta = \frac{w}{T_0}x = ax,$$

where

$$a = \frac{w}{T_0}.$$

The slope of the cable at x is $dy/dx = \tan \theta$, so we obtain a differential equation governing the curve described by the cable:

$$\frac{dy}{dx} = ax. \tag{10.9}$$

We have chosen the coordinate system so that $y = 0$ at $x = 0$. Integrating Eq. (10.9),

$$\int_0^y dy = \int_0^x ax \, dx,$$

we find that the curve described by the cable is the parabola

$$y = \frac{1}{2}ax^2. \tag{10.10}$$

Tension of the Cable

To determine the distribution of the tension in the cable, we square both sides of Eqs. (10.8) and then sum them, obtaining

$$T = T_0\sqrt{1 + a^2x^2}. \tag{10.11}$$

The tension is a minimum at the lowest point of the cable and increases monotonically with distance from the lowest point.

Length of the Cable

In some applications it is useful to have an expression for the length of the cable in terms of x. We can write the relation $ds^2 = dx^2 + dy^2$, where ds is an element of length of the cable (Fig. 10.16), in the form

$$ds = \sqrt{1 + \left(\frac{dy}{dx}\right)^2} \, dx.$$

Substituting Eq. (10.9) into this expression and integrating, we obtain an equation for the length s of the cable in the horizontal interval from 0 to x:

$$s = \frac{1}{2}\left\{x\sqrt{1 + a^2x^2} + \frac{1}{a}\ln\left[ax + \sqrt{1 + a^2x^2}\right]\right\}. \tag{10.12}$$

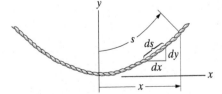

Figure 10.16
The length s of the cable in the horizontal interval from 0 to x.

A suspended cable is subjected to a vertical load uniformly distributed *along a horizontal line*. The origin of the coordinate system is at the cable's lowest point. The curve described by the cable is the parabola

$$y = \frac{1}{2}ax^2. \qquad (10.10)$$

The parameter $a = w/T_0$, where w is the magnitude of the distributed load and T_0 is the tension in the cable at its lowest point.

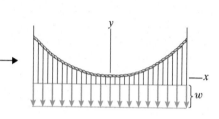

Tension
The tension in the cable in terms of the tension at the lowest point and the horizontal coordinate x relative to the cable's lowest point.

$$T = T_0\sqrt{1 + a^2x^2}. \qquad (10.11)$$

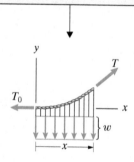

Length
The length of the cable measured from the lowest point $x = 0$ to the point with horizontal coordinate x.

$$s = \frac{1}{2}\left\{x\sqrt{1 + a^2x^2} + \frac{1}{a}\ln\left[ax + \sqrt{1 + a^2x^2}\right]\right\}. \qquad (10.12)$$

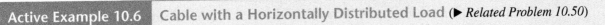

Active Example 10.6 Cable with a Horizontally Distributed Load (▶ *Related Problem 10.50*)

The cable supports a distributed load of 100 lb/ft. What is the tension at its lowest point?

Strategy

The horizontal position of the cable's lowest point is not given. However, the coordinates of each attachment point relative to a coordinate system with its origin at the lowest point must satisfy Eq. (10.10). With those conditions we can determine the horizontal coordinates of the attachment points. Equation (10.10) can be used to determine $a = w/T_0$, which tells us the tension at the lowest point.

Solution

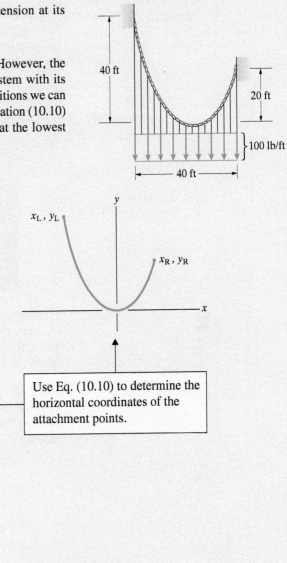

Equation (10.10) must be satisfied for both attachment points:

$$y_L = 40 \text{ ft} = \frac{1}{2}ax_L^2, \quad y_R = 20 \text{ ft} = \frac{1}{2}ax_R^2.$$

Dividing the first equation by the second yields

$$\frac{x_L^2}{x_R^2} = 2.$$

The horizontal span of the cable is

$$x_R - x_L = 40 \text{ ft}.$$

Solving these two equations yields $x_L = -23.4$ ft and $x_R = 16.6$ ft.

> Use Eq. (10.10) to determine the horizontal coordinates of the attachment points.

Substituting the coordinates of the right attachment point into Eq. (10.10),

$$y_R = \frac{1}{2}ax_R^2:$$

$$20 \text{ ft} = \frac{1}{2}a(16.6 \text{ ft})^2,$$

and solving yields $a = 0.146 \text{ ft}^{-1}$. Therefore the tension at the lowest point is

$$T_0 = \frac{w}{a}$$

$$= \frac{100 \text{ lb/ft}}{0.146 \text{ ft}^{-1}}$$

$$= 686 \text{ lb}.$$

> Use Eq. (10.10) to determine the tension at the lowest point.

Practice Problem Determine the maximum tension in the cable.

Answer: 2440 lb.

Example 10.7	Cable with a Horizontally Distributed Load (▶ *Related Problem 10.51*)

The horizontal distance between the supporting towers of the Manhattan Bridge in New York is 1470 ft. The tops of the towers are 145 ft above the lowest point of the main supporting cables. Obtain the equation for the curve described by the cables.

Strategy

We know the coordinates of the cable's attachment points relative to their lowest points. By substituting the coordinates into Eq. (10.10), we can determine the parameter a. Once a is known, Eq. (10.10) describes the shape of the cables.

Solution

The coordinates of the top of the right supporting tower relative to the lowest point of the support cables are $x_R = 735$ ft, $y_R = 145$ ft (Fig. a). By substituting these values into Eq. (10.10),

$$y = \frac{1}{2}ax^2:$$

$$145 \text{ ft} = \frac{1}{2}a(735 \text{ ft})^2,$$

(a) The theoretical curve superimposed on a photograph of the supporting cable.

we obtain $a = 5.37 \times 10^{-4}$ ft^{-1}. The curve described by the supporting cables is

$$y = \frac{1}{2}ax^2 = (2.68 \times 10^{-4})x^2.$$

Figure a compares this parabola with a photograph of the supporting cables.

Critical Thinking

Knowing the relative locations of the cable's highest and lowest points allowed us to determine the value of a. This parameter not only determines the equation describing the cable's shape, as we demonstrated in this example, but is also the ratio of the distributed load w acting on the cable to the tension in the cable at its lowest point. If the value of w was also known, the tension throughout the cable would be determined by Eq. (10.11).

Problems

▶ **10.50** The cable supports a distributed load $w = 12,000$ lb/ft. Using the approach described in Active Example 10.6, determine the maximum tension in the cable.

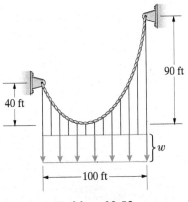

Problem 10.50

▶ **10.51** In Example 10.7, suppose that the tension at the lowest point of one of the main supporting cables of the bridge is two million pounds. What is the maximum tension in the cable?

10.52 A cable is used to suspend a pipeline above a river. The towers supporting the cable are 36 m apart. The lowest point of the cable is 1.4 m below the tops of the towers. The mass of the suspended pipe is 2700 kg.

(a) What is the maximum tension in the cable?

(b) What is the suspending cable's length?

10.53 In Problem 10.52, let the lowest point of the cable be a distance h below the tops of the towers supporting the cable.

(a) If the cable will safely support a tension of 70 kN, what is the minimum safe value of h?

(b) If h has the value determined in part (a), what is the suspending cable's length?

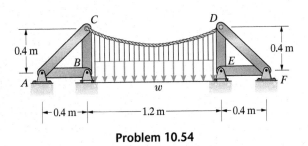

Problems 10.52/10.53

10.54 The cable supports a uniformly distributed load $w = 750$ N/m. The lowest point of the cable is 0.18 m below the attachment points C and D. Determine the axial loads in the truss members AC and BC.

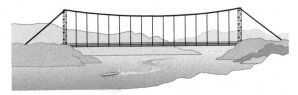

Problem 10.54

10.55 The cable supports a railway bridge between two tunnels. The distributed load is $w = 1$ MN/m, and $h = 40$ m.

(a) What is the maximum tension in the cable?

(b) What is the length of the cable?

10.56 The cable in Problem 10.55 will safely support a tension of 40 MN. What is the shortest cable that can be used, and what is the corresponding value of h?

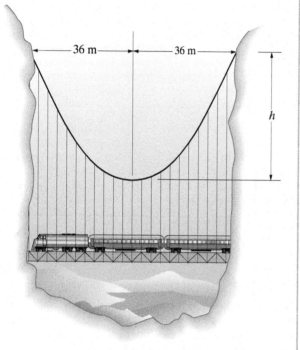

Problems 10.55/10.56

10.57 An oceanographic research ship tows an instrument package from a cable. Hydrodynamic drag subjects the cable to a uniformly distributed force $w = 2$ lb/ft. The tensions in the cable at 1 and 2 are 800 lb and 1300 lb, respectively. Determine the distance h.

10.58 Draw a graph of the shape of the cable in Problem 10.57.

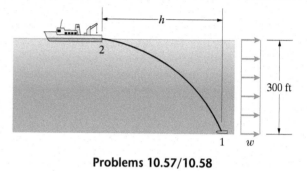

Problems 10.57/10.58

10.5 Loads Distributed Uniformly Along Cables

BACKGROUND

A cable's own weight subjects it to a load that is distributed uniformly along its length. If a cable is subjected to equal, parallel forces spaced uniformly along its length, the load on the cable can often be modeled as a load distributed uniformly along its length. In this section we show how to determine both the cable's resulting shape and the variation in its tension.

Suppose that a cable is acted on by a distributed load that subjects each element ds of its length to a force $w\,ds$, where w is constant. In Fig. 10.17 we show the free-body diagram obtained by cutting the cable at its lowest point and at a point a distance s along its length. The terms T_0 and T are the tensions at the lowest point and at s, respectively. The distributed load exerts a downward force ws. The origin of the coordinate system is located at the lowest point of the cable. Let the function $y(x)$ be the curve described by the cable in the x–y plane. Our objective is to determine $y(x)$ and the tension T.

Shape of the Cable

From the free-body diagram in Fig. 10.17, we obtain the equilibrium equations

$$T \sin \theta = ws, \tag{10.13}$$

$$T \cos \theta = T_0. \tag{10.14}$$

Dividing Eq. (10.13) by Eq. (10.14), we obtain

$$\tan \theta = \frac{w}{T_0} s = as, \tag{10.15}$$

where

$$a = \frac{w}{T_0}. \tag{10.16}$$

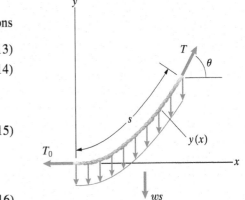

Figure 10.17
A cable subjected to a load distributed uniformly along its length.

The slope of the cable $dy/dx = \tan \theta$, so Eq. (10.15) can be written

$$\frac{dy}{dx} = as.$$

The derivative of this equation with respect to x is

$$\frac{d}{dx}\left(\frac{dy}{dx}\right) = a\frac{ds}{dx}. \tag{10.17}$$

By using the relation

$$ds^2 = dx^2 + dy^2,$$

we can write the derivative of s with respect to x as

$$\frac{ds}{dx} = \sqrt{1 + \left(\frac{dy}{dx}\right)^2} = \sqrt{1 + \sigma^2}, \tag{10.18}$$

where

$$\sigma = \frac{dy}{dx} = \tan \theta$$

is the slope. Now, with Eq. (10.18), we can write Eq. (10.17) as

$$\frac{d\sigma}{\sqrt{1 + \sigma^2}} = a\, dx.$$

The slope $\sigma = 0$ at $x = 0$. Integrating this equation yields

$$\int_0^{\sigma} \frac{d\sigma}{\sqrt{1 + \sigma^2}} = \int_0^x a\, dx,$$

and we obtain the slope as a function of x:

$$\sigma = \frac{dy}{dx} = \frac{1}{2}(e^{ax} - e^{-ax}) = \sinh ax. \tag{10.19}$$

Then, integrating this equation with respect to x yields the curve described by the cable, which is called a *catenary*:

$$y = \frac{1}{2a}(e^{ax} + e^{-ax} - 2) = \frac{1}{a}(\cosh ax - 1). \tag{10.20}$$

Tension of the Cable

Using Eq. (10.14) and the relation $dx = \cos\theta\, ds$, we obtain

$$T = \frac{T_0}{\cos\theta} = T_0 \frac{ds}{dx}.$$

Substituting Eq. (10.18) into this expression and using Eq. (10.19) yields the tension in the cable as a function of x:

$$T = T_0 \sqrt{1 + \frac{1}{4}(e^{ax} - e^{-ax})^2} = T_0 \cosh ax. \qquad (10.21)$$

Length of the Cable

From Eq. (10.15), the length s of the cable from the origin to the point at which the angle between the cable and the x axis equals θ is

$$s = \frac{1}{a}\tan\theta = \frac{\sigma}{a}.$$

Substituting Eq. (10.19) into this equation, we obtain an expression for the length s of the cable in the horizontal interval from its lowest point to x:

$$s = \frac{1}{2a}(e^{ax} - e^{-ax}) = \frac{\sinh ax}{a}. \qquad (10.22)$$

RESULTS

A suspended cable is subjected to a vertical load uniformly distributed *along the length of the cable*. The origin of the coordinate system is at the cable's lowest point. The curve described by the cable is the *catenary*

$$y = \frac{1}{2a}(e^{ax} + e^{-ax} - 2) = \frac{1}{a}(\cosh ax - 1). \quad (10.20)$$

The parameter $a = w/T_0$, where w is the magnitude of the distributed load and T_0 is the tension in the cable at its lowest point.

$$T = T_0\sqrt{1 + \frac{1}{4}(e^{ax} - e^{-ax})^2} = T_0 \cosh ax. \qquad (10.21)$$

Tension
The tension in the cable in terms of the tension at the lowest point and the horizontal coordinate x relative to the cable's lowest point.

$$s = \frac{1}{2a}(e^{ax} - e^{-ax}) = \frac{\sinh ax}{a}. \qquad (10.22)$$

Length
The length of the cable measured from the lowest point $x = 0$ to the point with horizontal coordinate x.

Active Example 10.8   Cable Loaded by Its Own Weight (▶ *Related Problem 10.59*)

The mass per unit length of the cable is 1 kg/m. The tension at its lowest point is 50 N. Determine the height h of its attachment points relative to the lowest point.

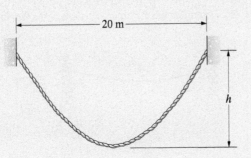

Strategy

The cable is subjected to a load $w = (9.81 \text{ m/s}^2)(1 \text{ kg/m}) = 9.81 \text{ N/m}$ distributed uniformly along its length. Because w and T_0 are known, we can determine $a = w/T_0$. Then we can use Eq. (10.20) to determine h.

Solution

$$a = \frac{w}{T_0} = \frac{9.81 \text{ N/m}}{50 \text{ N}} = 0.196 \text{ m}^{-1}.$$ ◀─── Determine the parameter a.

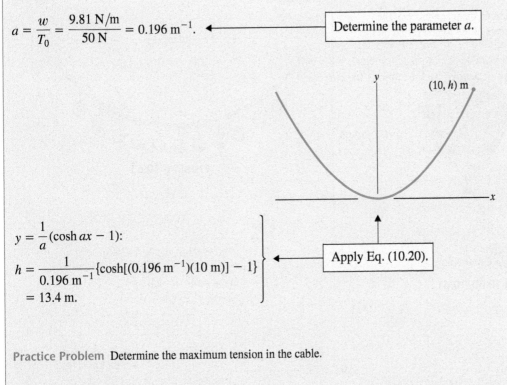

$$y = \frac{1}{a}(\cosh ax - 1):$$

$$h = \frac{1}{0.196 \text{ m}^{-1}}\{\cosh[(0.196 \text{ m}^{-1})(10 \text{ m})] - 1\}$$

$$= 13.4 \text{ m}.$$

◀─── Apply Eq. (10.20).

Practice Problem Determine the maximum tension in the cable.

Answer: 181 N.

Problems

▶ **10.59** The mass of the rope per unit length is 0.1 kg/m. The tension at its lowest point is 4.6 N. Using the approach described in Active Example 10.8, determine (a) the maximum tension in the rope and (b) the rope's length.

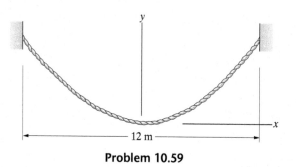

Problem 10.59

10.60 The stationary balloon's tether is horizontal at point O where it is attached to the truck. The mass per unit length of the tether is 0.45 kg/m. The tether exerts a 50-N horizontal force on the truck. The horizontal distance from point O to point A where the tether is attached to the balloon is 20 m. What is the height of point A relative to point O?

10.61 In Problem 10.60, determine the magnitudes of the horizontal and vertical components of the force exerted on the balloon at A by the tether.

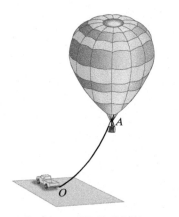

Problems 10.60/10.61

10.62 The mass per unit length of lines AB and BC is 2 kg/m. The tension at the lowest point of cable AB is 1.8 kN. The two lines exert equal horizontal forces at B.

(a) Determine the sags h_1 and h_2.

(b) Determine the maximum tensions in the two lines.

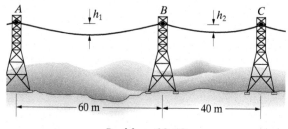

Problem 10.62

10.63 The rope is loaded by 2-kg masses suspended at 1-m intervals along its length. The mass of the rope itself is negligible. The tension in the rope at its lowest point is 100 N. Determine h and the maximum tension in the rope.

Strategy: Obtain an approximate answer by modeling the discrete loads on the rope as a load uniformly distributed along its length.

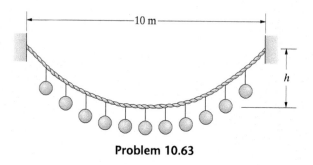

Problem 10.63

10.6 Discrete Loads

Our first applications of equilibrium in Chapter 3 involved determining the tensions in cables supporting suspended objects. In this section we consider the case of an arbitrary number N of objects suspended from a cable (Fig. 10.18a). We assume that the weight of the cable can be neglected in comparison to the suspended weights and that the cable is sufficiently flexible that we can approximate its shape by a series of straight segments.

Determining the Configuration and Tensions

Suppose that the horizontal distances $b_1, b_2, \ldots, b_{N+1}$ are known and that the vertical distance h_{N+1} specifying the position of the cable's right attachment point is known. We have two objectives: (1) to determine the configuration (shape) of the cable by solving for the vertical distances $h_1, h_2, \ldots, h_N$ specifying the positions of the attachment points of the weights and (2) to determine the tensions in the segments $1, 2, \ldots, N + 1$ of the cable.

We begin by drawing a free-body diagram, cutting the cable at its left attachment point and just to the right of the weight W_1 (Fig. 10.18b). We resolve the tension in the cable at the left attachment point into its horizontal and vertical components T_h and T_v. Summing moments about the attachment point A_1, we obtain the equation

$$\Sigma M_{\text{point } A_1} = h_1 T_h - b_1 T_v = 0.$$

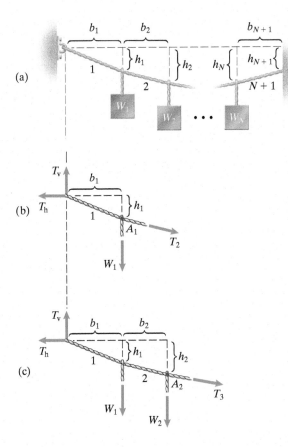

Figure 10.18
(a) N weights suspended from a cable.
(b) The first free-body diagram.
(c) The second free-body diagram.

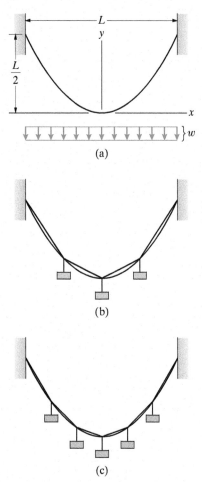

(a)

(b)

(c)

Figure 10.19
(a) Cable subjected to a continuous load.
(b) Cable with three discrete loads.
(c) Cable with five discrete loads.

Our next step is to obtain a free-body diagram by cutting the cable at its left attachment point and just to the right of the weight W_2 (Fig. 10.18c). Summing moments about A_2, we obtain

$$\Sigma M_{\text{point } A_2} = h_2 T_{\text{h}} - (b_1 + b_2)T_{\text{v}} + b_2 W_1 = 0.$$

Proceeding in this way, cutting the cable just to the right of each of the N weights, we obtain N equations. We can also draw a free-body diagram by cutting the cable at its left and right attachment points and sum moments about the right attachment point. In this way, we obtain $N + 1$ equations in terms of $N + 2$ unknowns: the two components of the tension T_{h} and T_{v} and the vertical positions of the attachment points $h_1, h_2, \ldots, h_N$. If the vertical position of just one attachment point is also specified, we can solve the system of equations for the vertical positions of the other attachment points, determining the configuration of the cable.

Once we know the configuration of the cable and the force T_{h}, the tension in any segment can be determined by cutting the cable at the left attachment point and within the segment and summing forces in the horizontal direction.

Comments on Continuous and Discrete Models

By comparing cables subjected to distributed and discrete loads, we can make some observations about how continuous and discrete systems are modeled in engineering. Consider a cable subjected to a horizontally distributed load w (Fig. 10.19a). The total force exerted on it is wL. Since the cable passes through the point $x = L/2$, $y = L/2$, we find from Eq. (10.10) that $a = 4/L$, so the equation for the curve described by the cable is $y = (2/L)x^2$.

In Fig. 10.19b, we compare the shape of the cable with the distributed load to that of a cable of negligible weight subjected to three discrete loads $W = wL/3$ with equal horizontal spacing. (We chose the dimensions of the cable with discrete loads so that the heights of the two cables would be equal at their midpoints.) In Fig. 10.19c, we compare the shape of the cable with the distributed load to that of a cable subjected to five discrete loads $W = wL/5$ with equal horizontal spacing. In Figs. 10.20a and 10.20b, we compare the tension in the cable subjected to the distributed load to those in the cables subjected to three and five discrete loads.

The shape and the tension in the cable with a distributed load are approximated by the shapes and tensions in the cables with discrete loads. Although the approximation of the tension is less impressive than the approximation of the

Figure 10.20
(a) The tension in a cable with a continuous load compared to the cable with three discrete loads.
(b) The tension in a cable with a continuous load compared to the cable with five discrete loads.

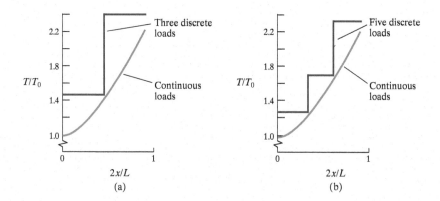

shape, it is clear that the former can be improved by increasing the number of discrete loads.

This approach, approximating a continuous distribution by a discrete model, is very important in engineering. It is the starting point of the finite difference and finite element methods. The opposite approach, modeling discrete systems by continuous models, is also widely used, for example when the forces exerted on a bridge by traffic are modeled as a distributed load.

RESULTS

The weights, the horizontal distances $b_1, b_2, ..., b_{N+1}$, and the vertical distance h_{N+1} are known. The objective is to determine the vertical distances $h_1, h_2, ..., h_N$ and the tensions in the segments of the cable.

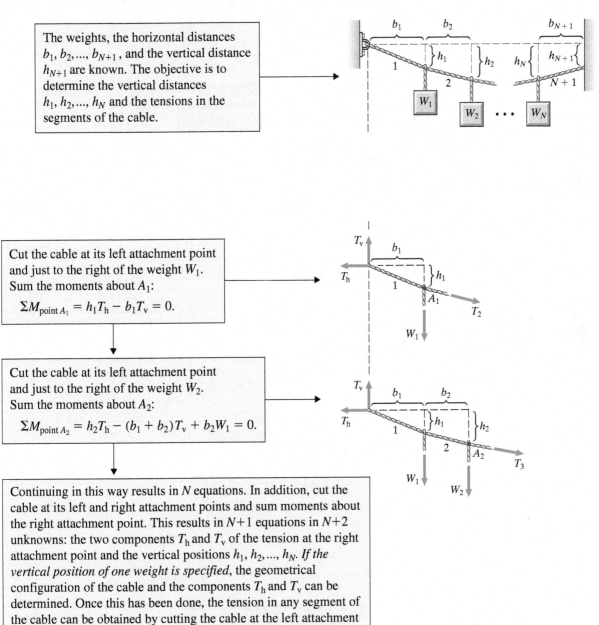

Cut the cable at its left attachment point and just to the right of the weight W_1. Sum the moments about A_1:

$$\Sigma M_{\text{point } A_1} = h_1 T_{\text{h}} - b_1 T_{\text{v}} = 0.$$

Cut the cable at its left attachment point and just to the right of the weight W_2. Sum the moments about A_2:

$$\Sigma M_{\text{point } A_2} = h_2 T_{\text{h}} - (b_1 + b_2) T_{\text{v}} + b_2 W_1 = 0.$$

Continuing in this way results in N equations. In addition, cut the cable at its left and right attachment points and sum moments about the right attachment point. This results in $N+1$ equations in $N+2$ unknowns: the two components T_{h} and T_{v} of the tension at the right attachment point and the vertical positions $h_1, h_2, ..., h_N$. *If the vertical position of one weight is specified*, the geometrical configuration of the cable and the components T_{h} and T_{v} can be determined. Once this has been done, the tension in any segment of the cable can be obtained by cutting the cable at the left attachment point and within the segment and summing forces in the horizontal direction.

Cable Subjected to Discrete Loads (▶ *Related Problem 10.64*)

The cable supports two masses $m_1 = 10$ kg and $m_2 = 20$ kg. Determine the vertical distance h_2.

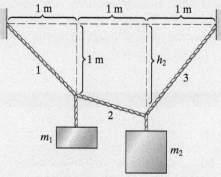

Strategy

By following the procedure described in Results, we can obtain three equations in terms of the horizontal and vertical components of the tension at the right attachment point and the vertical distance h_2.

Solution

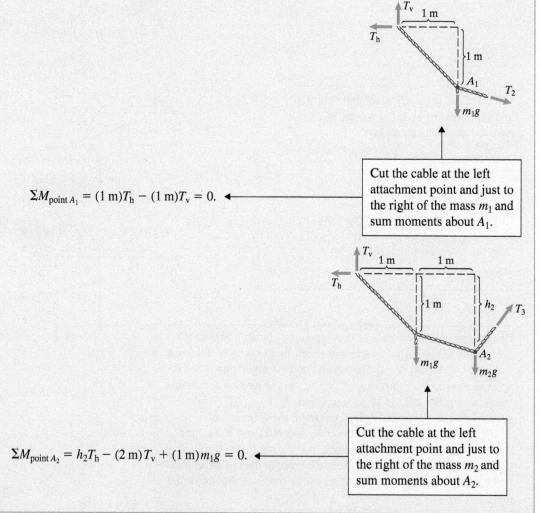

$$\Sigma M_{\text{point } A_1} = (1 \text{ m})T_h - (1 \text{ m})T_v = 0.$$

Cut the cable at the left attachment point and just to the right of the mass m_1 and sum moments about A_1.

$$\Sigma M_{\text{point } A_2} = h_2 T_h - (2 \text{ m})T_v + (1 \text{ m})m_1 g = 0.$$

Cut the cable at the left attachment point and just to the right of the mass m_2 and sum moments about A_2.

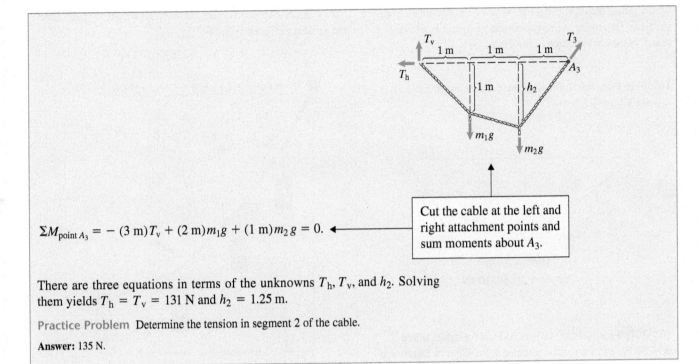

$\Sigma M_{\text{point } A_3} = -(3\text{ m})T_v + (2\text{ m})m_1 g + (1\text{ m})m_2 g = 0.$ ◀—

Cut the cable at the left and right attachment points and sum moments about A_3.

There are three equations in terms of the unknowns T_h, T_v, and h_2. Solving them yields $T_h = T_v = 131$ N and $h_2 = 1.25$ m.

Practice Problem Determine the tension in segment 2 of the cable.

Answer: 135 N.

Problems

▶ **10.64** In Active Example 10.9, what are the tensions in cable segments 1 and 3?

10.65 Each lamp weighs 12 lb.
(a) What is the length of the wire $ABCD$ needed to suspend the lamps as shown?
(b) What is the maximum tension in the wire?

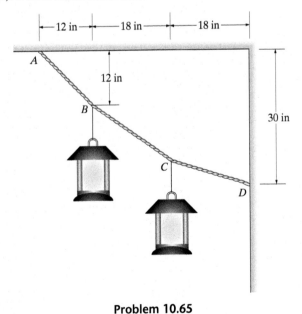

Problem 10.65

10.66 Two weights, $W_1 = W_2 = 50$ lb, are suspended from a cable. The vertical distance $h_1 = 4$ ft.
(a) Determine the vertical distance h_2.
(b) What is the maximum tension in the cable?

10.67 The weights are $W_1 = 50$ lb and $W_2 = 100$ lb, and the vertical distance $h_1 = 4$ ft.
(a) Determine the vertical distance h_2.
(b) What is the maximum tension in the cable?

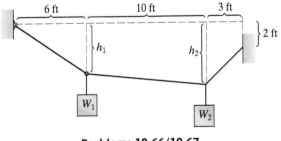

Problems 10.66/10.67

10.68 Three identical masses $m = 10$ kg are suspended from the cable. Determine the vertical distances h_1 and h_3 and draw a sketch of the configuration of the cable.

10.69 In Problem 10.68, what are the tensions in cable segments 1 and 2?

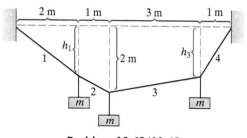

Problem 10.68/10.69

10.70 Three masses are suspended from the cable, where $m = 30$ kg, and the vertical distance $h_1 = 400$ mm. Determine the vertical distances h_2 and h_3.

10.71 In Problem 10.70, what is the maximum tension in the cable, and where does it occur?

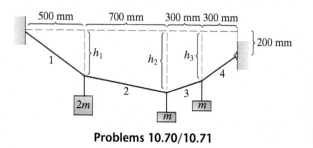

Problems 10.70/10.71

10.72 Each suspended object has the same weight W. Determine the vertical distances h_2 and h_3.

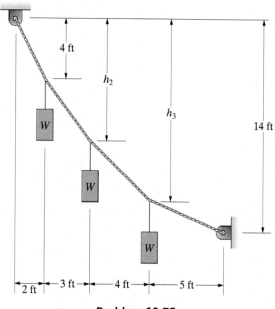

Problem 10.72

LIQUIDS AND GASES

10.7 Pressure and the Center of Pressure

BACKGROUND

Wind forces on buildings and aerodynamic forces on cars and airplanes are examples of forces that are distributed over areas. The downward force exerted on the bed of a dump truck by a load of gravel is distributed over the area of the bed. The upward force that supports a building is distributed over the area of its foundation. Loads distributed over the roofs of buildings by accumulated snow can be hazardous. Many forces of concern in engineering are distributed over areas. In this section we analyze the most familiar example, the force exerted by the pressure of a gas or liquid.

A surface immersed in a gas or liquid is subjected to forces exerted by molecular impacts. If the gas or liquid is stationary, the load can be described by a function p, the *pressure*, defined such that the normal force exerted on a differential element dA of the surface is $p\, dA$ (Figs. 10.21a and b). (Notice the parallel between the pressure and a load w distributed along a line, which is defined such that the force on a differential element dx of the line is $w\, dx$.)

The dimensions of p are (force)/(area). In U.S. Customary units, pressure can be expressed in pounds per square foot or pounds per square inch (psi). In SI units, pressure can be expressed in newtons per square meter, which are called pascals (Pa).

In some applications, it is convenient to use the *gage pressure*

$$p_g = p - p_{atm}, \qquad\qquad (10.23)$$

where p_{atm} is the pressure of the atmosphere. Atmospheric pressure varies with location and climatic conditions. Its value at sea level is approximately 1×10^5 Pa in SI units and 14.7 psi or 2120 lb/ft^2 in U.S. Customary units.

Center of Pressure

If the distributed force due to pressure on a surface is represented by an equivalent force, the point at which the line of action of the force intersects the surface

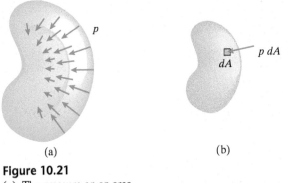

(a)   (b)

Figure 10.21
(a) The pressure on an area.
(b) The force on an element dA is $p\, dA$.

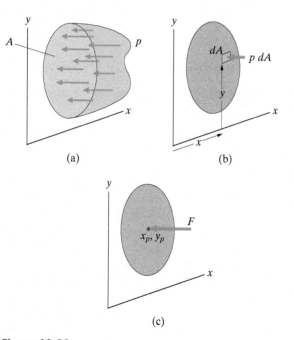

Figure 10.22
(a) A plane area subjected to pressure.
(b) The force on a differential element dA.
(c) The total force acting at the center of pressure.

is called the *center of pressure*. Consider a plane area A subjected to a pressure p and introduce a coordinate system such that the area lies in the x–y plane (Fig. 10.22a). The normal force on each differential element of area dA is $p\,dA$ (Fig. 10.22b), so the total normal force on A is

$$F = \int_A p\,dA. \tag{10.24}$$

Now we will determine the coordinates (x_p, y_p) of the center of pressure (Fig. 10.22c). Equating the moment of F about the origin to the total moment due to the pressure about the origin gives

$$(x_p\mathbf{i} + y_p\mathbf{j}) \times (-F\mathbf{k}) = \int_A (x\mathbf{i} + y\mathbf{j}) \times (-p\,dA\,\mathbf{k}),$$

and using Eq. (10.24), we obtain

$$x_p = \frac{\displaystyle\int_A xp\,dA}{\displaystyle\int_A p\,dA}, \quad y_p = \frac{\displaystyle\int_A yp\,dA}{\displaystyle\int_A p\,dA}. \tag{10.25}$$

These equations determine the position of the center of pressure when the pressure p is known. If the pressure p is uniform, the total normal force

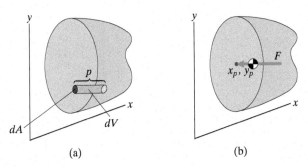

Figure 10.23
(a) The differential element $dV = p\,dA$.
(b) The line of action of F passes through the centroid of V.

is $F = pA$ and Eqs. (10.25) indicate that the center of pressure is the centroid of A.

In Chapter 7 it was shown that if we calculate the "area" defined by a load distributed along a line and place the resulting force at its centroid, the force is equivalent to the distributed load. A similar result holds for a pressure distributed on a plane area. The term $p\,dA$ in Eq. (10.24) is equal to a differential element dV of the "volume" between the surface defined by the pressure distribution and the area A (Fig. 10.23a). The total force exerted by the pressure is therefore equal to this "volume":

$$F = \int_V dV = V.$$

Substituting $p\,dA = dV$ into Eqs. (10.25), we obtain

$$x_p = \frac{\displaystyle\int_V x\,dV}{\displaystyle\int_V dV}, \quad y_p = \frac{\displaystyle\int_V y\,dV}{\displaystyle\int_V dV}.$$

The center of pressure coincides with the x and y coordinates of the centroid of the "volume" (Fig. 10.23).

Pressure in a Stationary Liquid

Designers of pressure vessels and piping, ships, dams, and other submerged structures must be concerned with forces and moments exerted by water pressure. The pressure in a liquid at rest increases with depth, which you can confirm by descending to the bottom of a swimming pool and noting the effect of the pressure on your ears. If we restrict ourselves to changes in depth for which changes in the density of the liquid can be neglected, we can determine the dependence of the pressure on depth by using a simple free-body diagram.

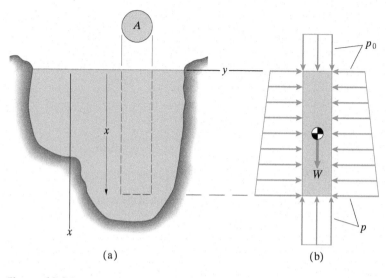

Figure 10.24
(a) A cylindrical volume that extends to a depth x
   in a body of stationary liquid.
(b) Free-body diagram of the cylinder.

Introducing a coordinate system with its origin at the surface of the liquid and the positive x axis downward (Fig. 10.24a), we draw a free-body diagram of a cylinder of liquid that extends from the surface to a depth x (Fig. 10.24b). The top of the cylinder is subjected to the pressure at the surface, which we call p_0. The sides and bottom of the cylinder are subjected to pressure by the surrounding liquid, which increases from p_0 at the surface to a value p at the depth x. The volume of the cylinder is Ax, where A is its cross-sectional area. Therefore, its weight is $W = \gamma Ax$, where γ is the weight density of the liquid. (Recall that the weight and mass densities are related by $\gamma = \rho g$.) Since the liquid is stationary, the cylinder is in equilibrium. From the equilibrium equation

$$\Sigma F_x = p_0 A - pA + \gamma Ax = 0,$$

we obtain a simple expression for the pressure p of the liquid at depth x:

$$p = p_0 + \gamma x. \tag{10.26}$$

Thus, the pressure increases linearly with depth, and the derivation we have used illustrates why: The pressure at a given depth literally holds up the liquid above that depth. If the surface of the liquid is open to the atmosphere, $p_0 = p_{\text{atm}}$, and we can write Eq. (10.26) in terms of the gage pressure $p_g = p - p_{\text{atm}}$ as

$$p_g = \gamma x. \tag{10.27}$$

In SI units, the density of water at sea level conditions is $\rho = 1000 \text{ kg/m}^3$, so its weight density is approximately $\gamma = \rho g = 9.81 \text{ kN/m}^3$. In U.S. Customary units, the weight density of water is approximately 62.4 lb/ft^3.

Definition of the Pressure

The pressure p of a liquid or gas is defined such that the normal force exerted on an element of area dA of a surface is $p\,dA$. The force and moment due to a distribution of pressure on a surface can be determined by integration.

The *gage pressure* is defined by

$$p_g = p - p_{atm}, \qquad (10.23)$$

where p_{atm} is atmospheric pressure. Atmospheric pressure at sea level is approximately 1×10^5 Pa in SI units and 14.7 psi or 2120 lb/ft^2 in U.S. Customary units.

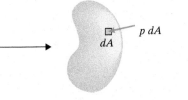

Center of Pressure

If the force exerted by a distribution of pressure on an area A is represented by an equivalent force vector, the point where the line of action of the force vector intersects A is called the *center of pressure*.

Volume Analogy

The total force F exerted on a plane area A by a distribution of pressure p is equal to the "volume" between A and the function p. If F is represented by a force vector acting at the centroid of the "volume," the force vector is equivalent to the distribution of pressure—its line of action intersects A at the center of pressure.

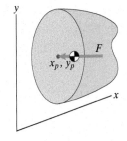

Pressure in a Stationary Liquid

The pressure at a depth x in a stationary liquid is

$$p = p_0 + \gamma x, \qquad (10.26)$$

where p_0 is the pressure at the surface and $\gamma = \rho g$ is the weight density of the liquid. The weight density of water is 9.81 kN/m^3 in SI units and 62.4 lb/ft^3 in U.S. Customary units. If $p_0 = p_{atm}$, Eq. (10.26) can be expressed in terms of the gage pressure as

$$p_g = \gamma x. \qquad (10.27)$$

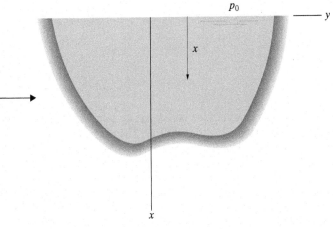

Active Example 10.10 Gate Loaded by a Pressure Distribution (▶ *Related Problem 10.78*)

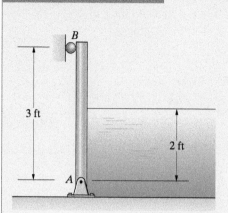

A stationary body of water exerts pressure on the right side of the gate AB. The width of the gate (its dimension into the page) is 3 ft, and the gate weighs 100 lb. The weight density of the water is 62.4 lb/ft³. Determine the reactions on the gate at the supports A and B.

Strategy

We will use integration to determine the force and moment exerted on the gate by the pressure of the water. We can then apply equilibrium to the free-body diagram of the gate to determine the reactions at A and B.

Solution

The left face of the gate and the right face above the surface of the water are exposed to atmospheric pressure. From Eqs. (10.23) and (10.26), the pressure in the water is the sum of atmospheric pressure and the gage pressure $p_g = \gamma x$, where x is measured downward from the surface of the water. The effects of atmospheric pressure on the gate cancel, so *only the forces and moments exerted on the gate by the gage pressure must be considered.*

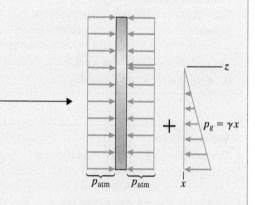

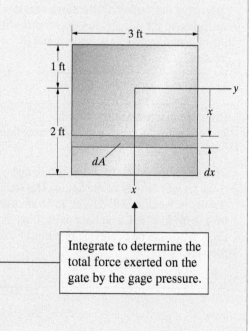

Integrate to determine the total force exerted on the gate by the gage pressure.

The origin of the coordinate system is at the surface of the water. The element of area of the gate $dA = (3 \text{ ft}) dx$. The total force exerted by the gage pressure is

$$F = \int_A p_g \, dA = \int_0^2 (\gamma x)(3 \text{ ft}) dx = 374 \text{ lb}.$$

The total moment about the y axis exerted by the gage pressure is

$$M = \int_A x p_g \, dA = \int_0^2 x(\gamma x)(3 \text{ ft}) \, dx = 499 \text{ ft-lb}.$$

Integrate to determine the total moment exerted on the gate by the gage pressure.

$$x_p = \frac{M}{F} = \frac{499 \text{ ft-lb}}{374 \text{ lb}} = 1.33 \text{ ft}.$$

Determine the position of the center of pressure from the condition for equivalence $M = x_p F$.

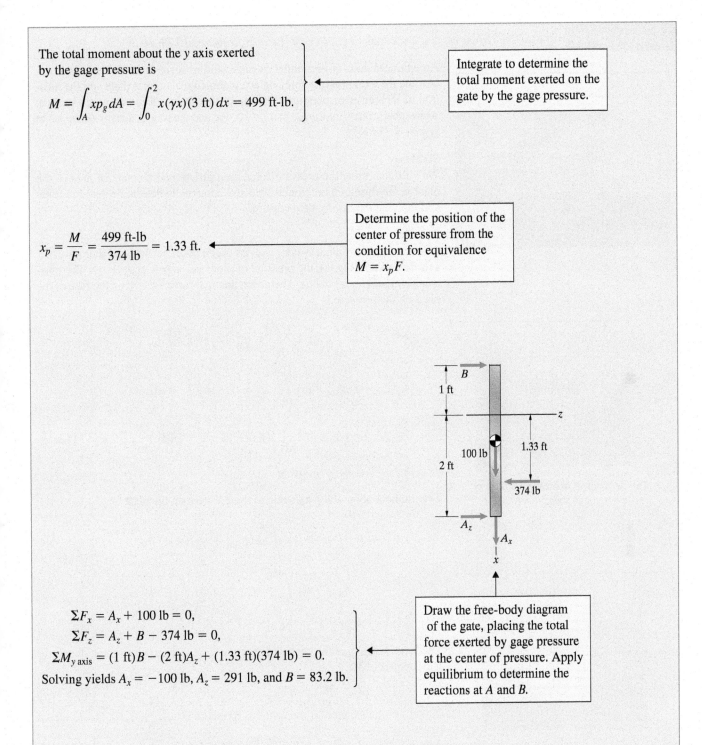

$$\Sigma F_x = A_x + 100 \text{ lb} = 0,$$
$$\Sigma F_z = A_z + B - 374 \text{ lb} = 0,$$
$$\Sigma M_{y \text{ axis}} = (1 \text{ ft})B - (2 \text{ ft})A_z + (1.33 \text{ ft})(374 \text{ lb}) = 0.$$
Solving yields $A_x = -100$ lb, $A_z = 291$ lb, and $B = 83.2$ lb.

Draw the free-body diagram of the gate, placing the total force exerted by gage pressure at the center of pressure. Apply equilibrium to determine the reactions at A and B.

Practice Problem Determine the reactions on the gate at the supports A and B. To do so, use the volume analogy to determine the total force exerted by gage pressure on the gate and the location of the center of pressure.

Answer: $A_x = -100$ lb, $A_z = 291$ lb, $B = 83.2$ lb.

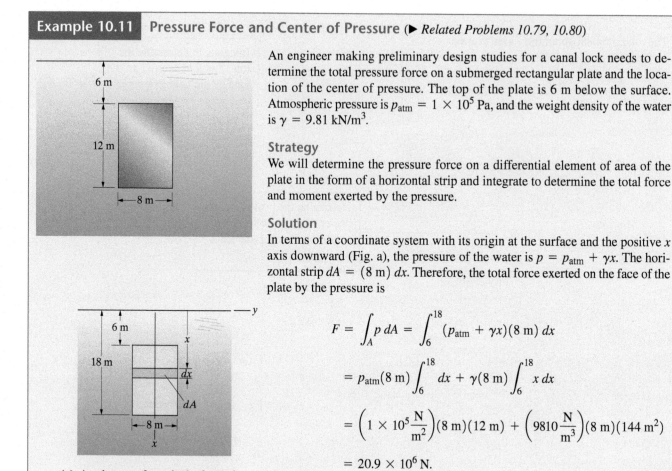

| **Example 10.11** | **Pressure Force and Center of Pressure** (▶ *Related Problems 10.79, 10.80*) |

An engineer making preliminary design studies for a canal lock needs to determine the total pressure force on a submerged rectangular plate and the location of the center of pressure. The top of the plate is 6 m below the surface. Atmospheric pressure is $p_{atm} = 1 \times 10^5$ Pa, and the weight density of the water is $\gamma = 9.81$ kN/m³.

Strategy
We will determine the pressure force on a differential element of area of the plate in the form of a horizontal strip and integrate to determine the total force and moment exerted by the pressure.

Solution
In terms of a coordinate system with its origin at the surface and the positive x axis downward (Fig. a), the pressure of the water is $p = p_{atm} + \gamma x$. The horizontal strip $dA = (8 \text{ m}) \, dx$. Therefore, the total force exerted on the face of the plate by the pressure is

$$F = \int_A p \, dA = \int_6^{18} (p_{atm} + \gamma x)(8 \text{ m}) \, dx$$

$$= p_{atm}(8 \text{ m}) \int_6^{18} dx + \gamma(8 \text{ m}) \int_6^{18} x \, dx$$

$$= \left(1 \times 10^5 \frac{\text{N}}{\text{m}^2}\right)(8 \text{ m})(12 \text{ m}) + \left(9810 \frac{\text{N}}{\text{m}^3}\right)(8 \text{ m})(144 \text{ m}^2)$$

$$= 20.9 \times 10^6 \text{ N}.$$

(a) An element of area in the form of a horizontal strip.

The moment about the y axis due to the pressure on the plate is

$$M = \int_A xp \, dA = \int_6^{18} x(p_{atm} + \gamma x)(8 \text{ m}) \, dx$$

$$= p_{atm}(8 \text{ m}) \int_6^{18} x \, dx + \gamma(8 \text{ m}) \int_6^{18} x^2 \, dx$$

$$= 262 \times 10^6 \text{ N-m}.$$

The force F acting at the center of pressure (Fig. b) exerts a moment about the y axis equal to M:

$$x_p F = M.$$

Therefore, the location of the center of pressure is

$$x_p = \frac{M}{F} = \frac{262 \text{ MN-m}}{20.9 \text{ MN}} = 12.5 \text{ m}.$$

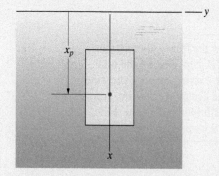

(b) The center of pressure.

Critical Thinking
Notice that the center of pressure does not coincide with the centroid of the area. The center of pressure of a plane area generally coincides with the centroid of the area only when the pressure is uniformly distributed. In this example, the pressure increases with depth, and as a result, the center of pressure is below the centroid.

Example 10.12 Determination of a Pressure Force (▶ *Related Problem 10.91*)

The container is filled with a liquid with weight density γ. Determine the force exerted by the pressure of the liquid on the cylindrical wall *AB*.

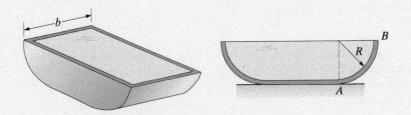

Strategy
The pressure of the liquid on the cylindrical wall varies with depth (Fig. a). It is the force exerted by this pressure distribution we want to determine. We could determine it by integrating over the cylindrical surface, but we can avoid that by drawing a free-body diagram of the quarter-cylinder of liquid to the right of *A*.

Solution
We draw the free-body diagram of the quarter-cylinder of liquid in Fig. b. The pressure distribution on the cylindrical surface of the liquid is the same one that acts on the cylindrical wall. If we denote the force exerted on the liquid by this pressure distribution by $\mathbf{F}_p$, the force exerted by the liquid on the cylindrical wall is $-\mathbf{F}_p$.

The other forces parallel to the *x*–*y* plane that act on the quarter-cylinder of liquid are its weight, atmospheric pressure at the upper surface, and the pressure distribution of the liquid on the left side. The volume of liquid is $(\frac{1}{4}\pi R^2)b$, so the force exerted on the free-body diagram by the weight of the liquid is $\frac{1}{4}\gamma\pi R^2 b\mathbf{i}$. The force exerted on the upper surface by atmospheric pressure is $Rbp_{\text{atm}}\mathbf{i}$.

We can integrate to determine the force exerted by the pressure on the left side of the free-body diagram. Its magnitude is

$$\int_A p\, dA = \int_0^R (p_{\text{atm}} + \gamma x)b\, dx = Rb\left(p_{\text{atm}} + \frac{1}{2}\gamma R\right).$$

From the equilibrium equation

$$\Sigma\mathbf{F} = \frac{1}{4}\gamma\pi R^2 b\mathbf{i} + Rbp_{\text{atm}}\mathbf{i} + Rb\left(p_{\text{atm}} + \frac{1}{2}\gamma R\right)\mathbf{j} + \mathbf{F}_p = 0,$$

we obtain the force exerted on the wall *AB* by the pressure of the liquid:

$$-\mathbf{F}_p = Rb\left(p_{\text{atm}} + \frac{\pi}{4}\gamma R\right)\mathbf{i} + Rb\left(p_{\text{atm}} + \frac{1}{2}\gamma R\right)\mathbf{j}.$$

Critical Thinking
The need to integrate over a curved surface to calculate a pressure force can often be avoided by choosing a suitable free-body diagram as we have done in this example.

(a) The pressure of the liquid on the wall *AB*.

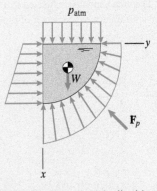

(b) Free-body diagram of the liquid to the right of *A*.

Problems

10.73 An engineer planning a water system for a new community estimates that at maximum expected usage, the pressure drop between the central system and the farthest planned fire hydrant will be 25 psi. Firefighting personnel indicate that a gage pressure of 40 psi at the fire hydrant is required. The weight density of the water is $\gamma = 62.4$ lb/ft^3. How tall would a water tower at the central system have to be to provide the needed pressure?

10.74 A cube of material is suspended below the surface of a liquid of weight density γ. By calculating the forces exerted on the faces of the cube by pressure, show that their sum is an upward force of magnitude γb^3.

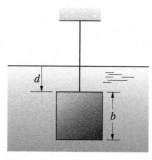

Problem 10.74

10.75 The area shown is subjected to a *uniform* pressure $p_{atm} = 1 \times 10^5$ Pa.

(a) What is the total force exerted on the area by the pressure?

(b) What is the moment about the *y*-axis due to the pressure on the area?

10.76 The area shown is subjected to a *uniform* pressure. Determine the coordinates of the center of pressure.

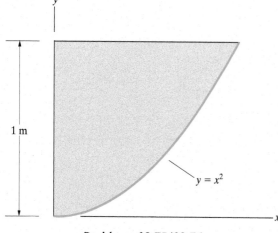

Problems 10.75/10.76

10.77 The area shown is subjected to a *uniform* pressure $p_{atm} = 14.7$ psi.

(a) What is the total force exerted on the area by the pressure?

(b) What is the moment about the *y* axis due to the pressure on the area?

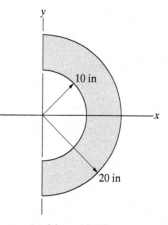

Problem 10.77

▶ **10.78** In Active Example 10.10, suppose that the water depth relative to point *A* is increased from 2 ft to 3 ft. Determine the reactions on the gate at the supports *A* and *B*.

▶ **10.79** The top of the rectangular plate is 2 m below the surface of a lake. Atmospheric pressure is $p_{atm} = 1 \times 10^5$ Pa and the mass density of the water is $\rho = 1000$ kg/m^3.

(a) What is the maximum pressure exerted on the plate by the water?

(b) Determine the force exerted on a face of the plate by the pressure of the water. (See Example 10.11.)

▶ **10.80** In Problem 10.79, how far below the top of the plate is the center of pressure located? (See Example 10.11.)

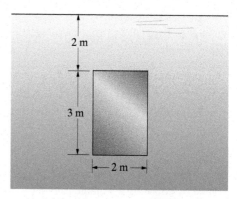

Problems 10.79/10.80

10.81 The width of the dam (the dimension into the page) is 100 m. The mass density of the water is $\rho = 1000 \text{ kg/m}^3$. Determine the force exerted on the dam by the gage pressure of the water

(a) by integration;

(b) by calculating the "volume" of the pressure distribution.

10.82 In Problem 10.81, how far down from the surface of the water is the center of pressure due to the gage pressure of the water on the dam?

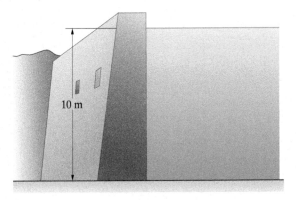

10 m

Problems 10.81/10.82

10.83 The width of the gate (the dimension into the page) is 3 m. Atmospheric pressure is $p_{atm} = 1 \times 10^5$ Pa and the mass density of the water is $\rho = 1000 \text{ kg/m}^3$. Determine the horizontal force and couple exerted on the gate by its built-in support A.

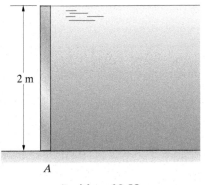

2 m

A

Problem 10.83

10.84 The homogeneous gate weighs 100 lb, and its width (the dimension into the page) is 3 ft. The weight density of the water is $\gamma = 62.4 \text{ lb/ft}^3$, and atmospheric pressure is $p_{atm} = 2120 \text{ lb/ft}^2$. Determine the reactions at A and B.

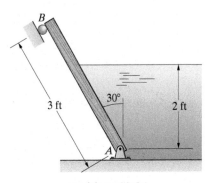

B

3 ft   30°   2 ft

A

Problem 10.84

10.85 The width of the gate (the dimension into the page) is 2 m and there is water of depth $d = 1$ m on the right side. Atmospheric pressure is $p_{atm} = 1 \times 10^5$ Pa and the mass density of the water is $\rho = 1000 \text{ kg/m}^3$. Determine the horizontal forces exerted on the gate at A and B.

10.86 The gate in Problem 10.85 is designed to rotate and release the water when the depth d exceeds a certain value. What is that depth?

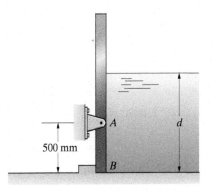

A d

500 mm

B

Problems 10.85/10.86

10.87* The dam has water of depth 4 ft on one side. The width of the dam (the dimension into the page) is 8 ft. The weight density of the water is $\gamma = 62.4$ lb/ft^3, and atmospheric pressure $p_{atm} = 2120$ lb/ft^2. If you neglect the weight of the dam, what are the reactions at A and B?

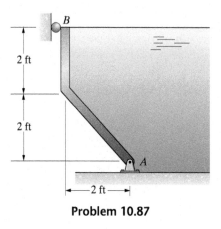

Problem 10.87

10.88* The dam has water of depth 4 ft on one side. The width of the dam (the dimension into the page) is 8 ft. The weight density of the water is $\gamma = 62.4$ lb/ft^3, and atmospheric pressure is $p_{atm} = 2120$ lb/ft^2. If you neglect the weight of the dam, what are the reactions at A and B?

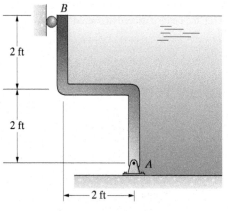

Problem 10.88

10.89 Consider a plane, vertical area A below the surface of a liquid. Let p_0 be the pressure at the surface.

(a) Show that the force exerted by pressure on the area is $F = \bar{p}A$, where $\bar{p} = p_0 + \gamma\bar{x}$ is the pressure of the liquid at the centroid of the area.

(b) Show that the x coordinate of the center of pressure is

$$x_p = \bar{x} + \frac{\gamma I_{y'}}{\bar{p}A},$$

where $I_{y'}$ is the moment of inertia of the area about the y' axis through its centroid.

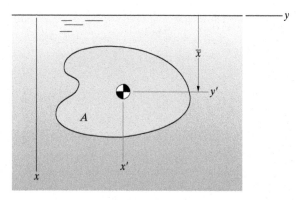

Problem 10.89

10.90 A circular plate of 1-m radius is below the surface of a stationary pool of water. Atmospheric pressure is $p_{atm} = 1 \times 10^5$ Pa, and the mass density of the water is $\rho = 1000$ kg/m^3. Determine (a) the force exerted on a face of the plate by the pressure of the water; (b) the x coordinate of the center of pressure. (See Problem 10.89.)

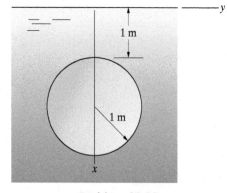

Problem 10.90

▶ 10.91* The tank consists of a cylinder with hemispherical ends. It is filled with water ($\rho = 1000 \text{ kg/m}^3$). The pressure of the water at the top of the tank is 140 kPa. Determine the magnitude of the force exerted by the pressure of the water on each hemispherical end of the tank. (See Example 10.12.)

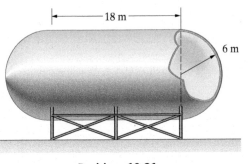

18 m

6 m

Problem 10.91

10.92 An object of volume V and weight W is suspended below the surface of a stationary liquid of weight density γ (Fig. a). Show that the tension in the cord is $W - V\gamma$. In other words, show that the pressure distribution on the surface of the object exerts an upward force equal to the product of the object's volume and the weight density of the water. This result is due to Archimedes (287–212 B.C.).

Strategy: Draw the free-body diagram of a volume of liquid that has the same shape and position as the object (Fig. b).

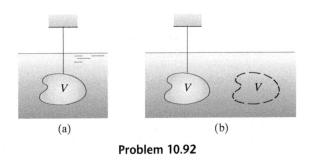

(a)   (b)

Problem 10.92

Review Problems

10.93 Determine the internal forces and moment at B (a) if $x = 250$ mm; (b) if $x = 750$ mm.

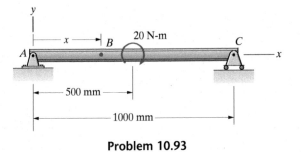

Problem 10.93

10.94 Determine the internal forces and moment (a) at B; (b) at C.

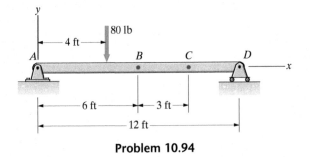

Problem 10.94

10.95 (a) Determine the maximum bending moment in the beam and the value of x where it occurs.

(b) Show that the equations for V and M as functions of x satisfy the equation $V = dM/dx$.

10.96 Draw the shear force and bending moment diagrams for the beam in Problem 10.95.

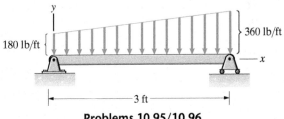

Problems 10.95/10.96

10.97 Determine the shear force and bending moment diagrams for the beam.

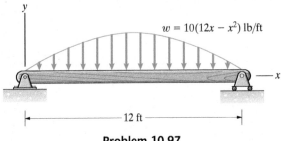

Problem 10.97

10.98 Determine V and M as functions of x for the beam ABC.

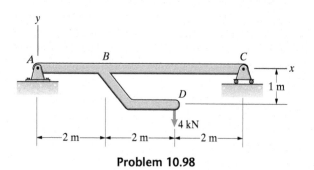

Problem 10.98

10.99 Draw the shear force and bending moment diagrams for beam ABC.

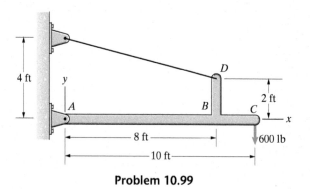

Problem 10.99

10.100 Determine the internal forces and moments at A.

10.101 Draw the shear force and bending moment diagrams of beam BC.

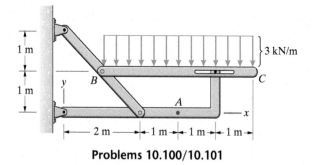

Problems 10.100/10.101

10.102 Determine the internal forces and moment at B
(a) if $x = 250$ mm;
(b) if $x = 750$ mm.

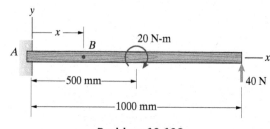

Problem 10.102

10.103 Draw the shear force and bending moment diagrams.

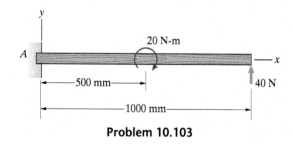

Problem 10.103

10.104 The homogeneous beam weighs 1000 lb. What are the internal forces and bending moment at its midpoint?

10.105 The homogeneous beam weighs 1000 lb. Draw the shear force and bending moment diagrams.

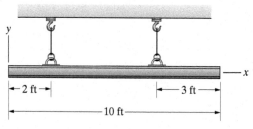

Problems 10.104/10.105

10.106 At A the main cable of the suspension bridge is horizontal and its tension is 1×10^8 lb.

(a) Determine the distributed load acting on the cable.

(b) What is the tension at B?

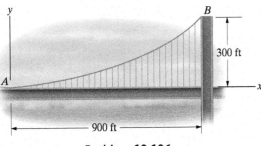

Problem 10.106

10.107 The power line has a mass of 1.4 kg/m. If the line will safely support a tension of 5 kN, determine whether it will safely support an ice accumulation of 0.4 kg/m.

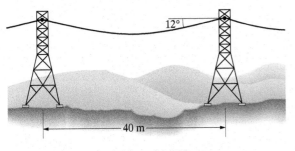

Problem 10.107

10.108 The water depth at the center of the elliptical window is 20 ft. Determine the magnitude of the net force exerted on the window by the pressure of the seawater ($\gamma = 64$ lb/ft^3) and the atmospheric pressure of the air on the opposite side. (See Problem 10.89.)

10.109 The water depth at the center of the elliptical window is 20 ft. Determine the magnitude of the net moment exerted on the window about the horizontal axis L by the pressure of the seawater ($\gamma = 64$ lb/ft^3) and the atmospheric pressure of the air on the opposite side. (See Problem 10.89.)

Problems 10.108/10.109

10.110* The gate has water of 2-m depth on one side. The width of the gate (the dimension into the page) is 4 m, and its mass is 160 kg. The mass density of the water is $\rho = 1000$ kg/m^3, and atmospheric pressure is $p_{\text{atm}} = 1 \times 10^5$ Pa. Determine the reactions on the gate at A and B. (The support at B exerts only a horizontal reaction on the gate.)

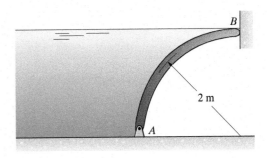

Problem 10.110

10.111 A spherical tank of 400-mm inner radius is full of water ($\rho = 1000$ kg/m^3). The pressure of the water at the top of the tank is 4×10^5 Pa.

(a) What is the pressure of the water at the bottom of the tank?

(b) What is the total force exerted on the inner surface of the tank by the pressure of the water?

Strategy: For (b), draw a free-body diagram of the sphere of water in the tank.

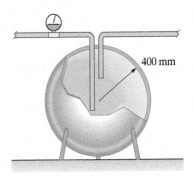

Problem 10.111

CHAPTER
11

Virtual Work and Potential Energy

When a spring is stretched, the work performed is stored in the spring as potential energy. Raising a load with a crane increases its gravitational potential energy. In this chapter we define work and potential energy and introduce a general and powerful result called the principle of virtual work.

◀ The torsional spring stores potential energy that drives the clock mechanism. In this chapter we use the concepts of virtual work and potential energy to analyze objects in equilibrium.

11.1 Virtual Work

BACKGROUND

The principle of virtual work is a statement about work done by forces and couples when an object or structure is subjected to various hypothetical motions. Before we can introduce this principle, we must define work.

Work

Consider a force acting on an object at a point P (Fig. 11.1a). Suppose that the object undergoes an infinitesimal motion, so that P has a differential displacement $d\mathbf{r}$ (Fig. 11.1b). The *work dU* done by $\mathbf{F}$ as a result of the displacement $d\mathbf{r}$ is defined to be

$$dU = \mathbf{F} \cdot d\mathbf{r}. \tag{11.1}$$

From the definition of the dot product, $dU = (|\mathbf{F}| \cos \theta)|d\mathbf{r}|$, where θ is the angle between $\mathbf{F}$ and $d\mathbf{r}$ (Fig. 11.1c). The work is equal to the product of the component of $\mathbf{F}$ in the direction of $d\mathbf{r}$ and the magnitude of $d\mathbf{r}$. Notice that if the component of $\mathbf{F}$ parallel to $d\mathbf{r}$ points in the direction opposite to $d\mathbf{r}$, the work is negative. Also notice that if $\mathbf{F}$ is perpendicular to $d\mathbf{r}$, the work is zero. The dimensions of work are (force) $\times$ (length).

Now consider a couple acting on an object (Fig. 11.2a). The moment due to the couple is $M = Fh$ in the counterclockwise direction. If the object rotates through an infinitesimal counterclockwise angle $d\alpha$ (Fig. 11.2b), the points of application of the forces are displaced through differential distances $\frac{1}{2}h\, d\alpha$. Consequently, the total work done is $dU = F(\frac{1}{2}h\, d\alpha) + F(\frac{1}{2}h\, d\alpha) = M\, d\alpha$.

We see that when an object acted on by a couple M is rotated through an angle $d\alpha$ in the same direction as the couple (Fig. 11.2c), the resulting work is

$$dU = M\, d\alpha. \tag{11.2}$$

If the direction of the couple is opposite to the direction of $d\alpha$, the work is negative.

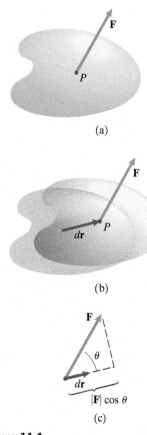

Figure 11.1
(a) A force $\mathbf{F}$ acting on an object.
(b) A displacement $d\mathbf{r}$ of P.
(c) The work $dU = (|\mathbf{F}| \cos \theta)|d\mathbf{r}|$.

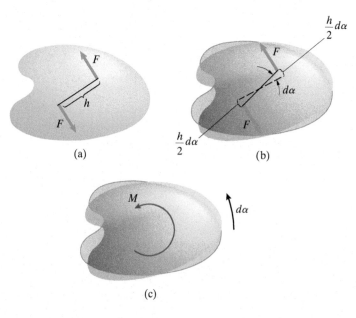

Figure 11.2
(a) A couple acting on an object.
(b) An infinitesimal rotation of the object.
(c) An object acted on by a couple M rotating through an angle $d\alpha$.

Principle of Virtual Work

Now that we have defined the work done by forces and couples, we can introduce the principle of virtual work. Before stating it, we first discuss an example to give you context for understanding the principle.

The homogeneous bar in Fig. 11.3a is supported by the wall and by the pin support at A and is loaded by a couple M. The free-body diagram of the bar is shown in Fig. 11.3b. The equilibrium equations are

$$\Sigma F_x = A_x - N = 0, \tag{11.3}$$

$$\Sigma F_y = A_y - W = 0, \tag{11.4}$$

$$\Sigma M_{\text{point } A} = NL \sin \alpha - W \frac{1}{2} L \cos \alpha - M = 0. \tag{11.5}$$

We can solve these three equations for the reactions A_x, A_y, and N. However, we have a different objective.

Consider the following question: If the bar is acted on by the forces and couple in Fig. 11.3b and we subject it to a hypothetical infinitesimal translation in the x direction, as shown in Fig. 11.4, what work is done? The hypothetical displacement δx is called a *virtual displacement* of the bar, and the resulting work δU is called the *virtual work*. The pin support and the wall prevent the bar from actually moving in the x direction: the virtual displacement is a theoretical artifice. Our objective is to calculate the resulting virtual work:

$$\delta U = A_x \delta_x + (-N) \delta_x = (A_x - N) \delta_x. \tag{11.6}$$

The forces A_y and W do no work because they are perpendicular to the displacements of their points of application. The couple M also does no work, because the bar does not rotate. Comparing this equation with Eq. (11.3), we find that the virtual work equals zero.

Next, we give the bar a virtual translation in the y direction (Fig. 11.5). The resulting virtual work is

$$\delta U = A_y \delta y + (-W) \delta y = (A_y - W) \delta y. \tag{11.7}$$

From Eq. (11.4), the virtual work again equals zero.

Finally, we give the bar a virtual rotation while holding point A fixed (Fig. 11.6a). The forces A_x and A_y do no work because their point of application does not move. The work done by the couple M is $-M \, \delta \alpha$, because its direction is opposite to that of the rotation. The displacements of the points of application of the forces N and W are shown in Fig. 11.6b, and the components of the forces in the direction of the displacements are shown in Fig. 11.6c. The work done by N is $(N \sin \alpha)(L \, \delta \alpha)$, and the work done by W is $(-W \cos \alpha)(\frac{1}{2} L \, \delta \alpha)$. The total work is

$$\delta U = (N \sin \alpha)(L \delta \alpha) + (-W \cos \alpha)\left(\frac{1}{2} L \delta \alpha\right) - M \, \delta \alpha$$

$$= \left(NL \sin \alpha - W \frac{1}{2} L \cos \alpha - M\right) \delta \alpha. \tag{11.8}$$

From Eq. (11.5), the virtual work resulting from the virtual rotation is also zero.

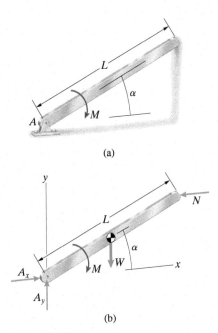

Figure 11.3
(a) A bar subjected to a couple M.
(b) Free-body diagram of the bar.

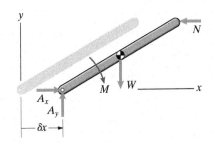

Figure 11.4
A virtual displacement δx.

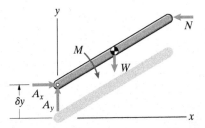

Figure 11.5
A virtual displacement δy.

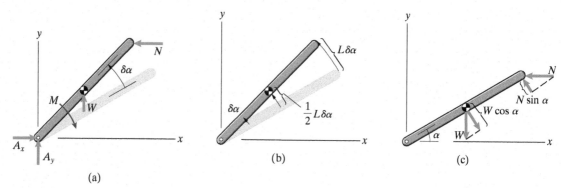

Figure 11.6
(a) A virtual rotation $\delta\alpha$.
(b) Displacements of the points of application of N and W.
(c) Components of N and W in the direction of the displacements.

We have shown that for three virtual motions of the bar, the virtual work is zero. These results are examples of a form of the principle of virtual work:

If an object is in equilibrium, the virtual work done by the external forces and couples acting on it is zero for any virtual translation or rotation:

$$\delta U = 0. \tag{11.9}$$

As our example illustrates, this principle can be used to derive the equilibrium equations for an object. Subjecting the bar to virtual translations δx and δy and a virtual rotation $\delta\alpha$ results in Eqs. (11.6)–(11.8). Because the virtual work must be zero in each case, we obtain Eqs. (11.3)–(11.5). But there is no advantage to this approach compared to simply drawing the free-body diagram of the object and writing the equations of equilibrium in the usual way. The advantages of the principle of virtual work become evident when we consider structures.

Application to Structures

The principle of virtual work stated in the preceding section applies to each member of a structure. By subjecting certain types of structures in equilibrium to virtual motions and calculating the total virtual work, we can determine unknown reactions at their supports as well as internal forces in their members. The procedure involves finding virtual motions that result in virtual work being done both by known loads and by unknown forces and couples.

Suppose that we want to determine the axial load in member BD of the truss in Fig. 11.7a. The other members of the truss are subjected to the 4-kN load and the forces exerted on them by member BD (Fig. 11.7b). If we give the structure a virtual rotation $\delta\alpha$ as shown in Fig. 11.7c, virtual work is done by the force T_{BD} acting at B and by the 4-kN load at C. Furthermore, the virtual work done by these two forces is the total virtual work done on the members of the structure, because the virtual work done by the internal forces they exert on each other cancels out. For example, consider joint C (Fig. 11.7d). The force T_{BC} is the axial load in member BC. The virtual work done at C on member BC is $T_{BC}(1.4\text{ m})\,\delta\alpha$, and the work done at C on member CD is $(4\text{ kN} - T_{BC})(1.4\text{ m})\,\delta\alpha$. When we add up the virtual work done on the members to obtain the total virtual work on the structure, the virtual work due to the internal force T_{BC} cancels out. (If the members exerted

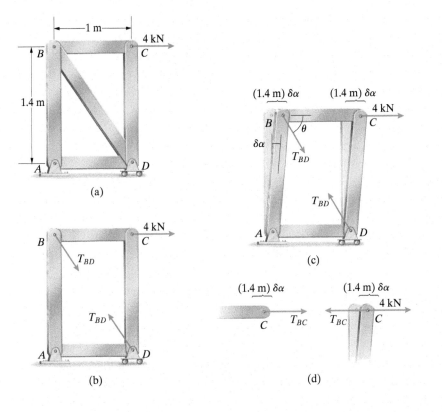

(a)

(b)

(c)

(d)

Figure 11.7
(a) A truss with a 4-kN load.
(b) Forces exerted by member BD.
(c) A virtual motion of the structure.
(d) Calculating the virtual work on members BC and CD at the joint C.

an internal *couple* on each other at C—for example, as a result of friction in the pin support—the virtual work would not cancel out.) Therefore, we can ignore internal forces in calculating the total virtual work on the structure:

$$\delta U = (T_{BD} \cos \theta)(1.4 \text{ m}) \, \delta\alpha + (4 \text{ kN})(1.4 \text{ m}) \, \delta\alpha = 0.$$

The angle $\theta = \arctan(1.4/1) = 54.5°$. Solving this equation, we obtain $T_{BD} = -6.88$ kN.

RESULTS

Work

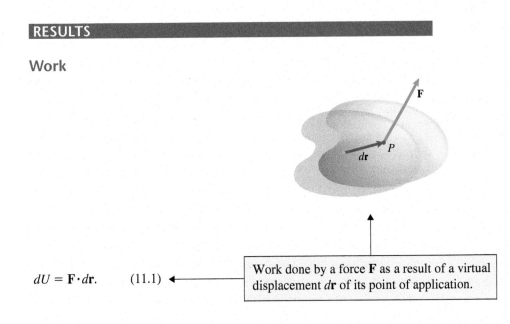

$$dU = \mathbf{F} \cdot d\mathbf{r}. \qquad (11.1)$$

Work done by a force $\mathbf{F}$ as a result of a virtual displacement $d\mathbf{r}$ of its point of application.

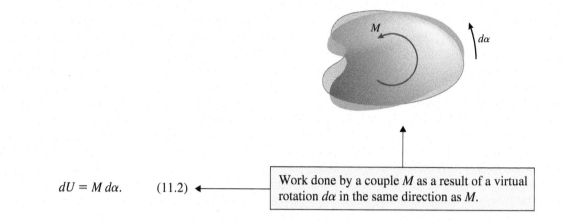

$$dU = M \, d\alpha. \qquad (11.2)$$

Work done by a couple M as a result of a virtual rotation $d\alpha$ in the same direction as M.

Principle of Virtual Work

If an object is in equilibrium, the virtual work done by the external forces and couples acting on it is zero for any virtual translation or rotation:

$$\delta U = 0. \qquad (11.9)$$

A virtual displacement or rotation is a *hypothetical* infinitesimal displacement or rotation.

The principle of virtual work can be applied to structures if no net work is done by internal forces and couples the members exert on each other. This involves two steps:

1. Choose a virtual motion—Identify a virtual motion of the structure that results in virtual work being done by known loads and by an unknown force or couple that is to be determined.
2. Determine the virtual work—Calculate the total virtual work resulting from the virtual motion to obtain an equation for the unknown force or couple.

Active Example 11.1 Applying Virtual Work to a Structure (▶ *Related Problems 11.12–11.16*)

Use the principle of virtual work to determine the horizontal reaction on the structure at C.

Strategy

Even though the structure is fixed at A and C, it can be subjected to *hypothetical* virtual motions. We must choose a virtual motion for which the horizontal reaction at C and the known external loads on the structure do work. By calculating the resulting virtual work, we can determine the horizontal reaction at C.

Solution

Free-body diagram of the structure. The objective is to determine C_x.

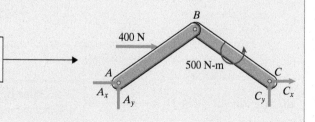

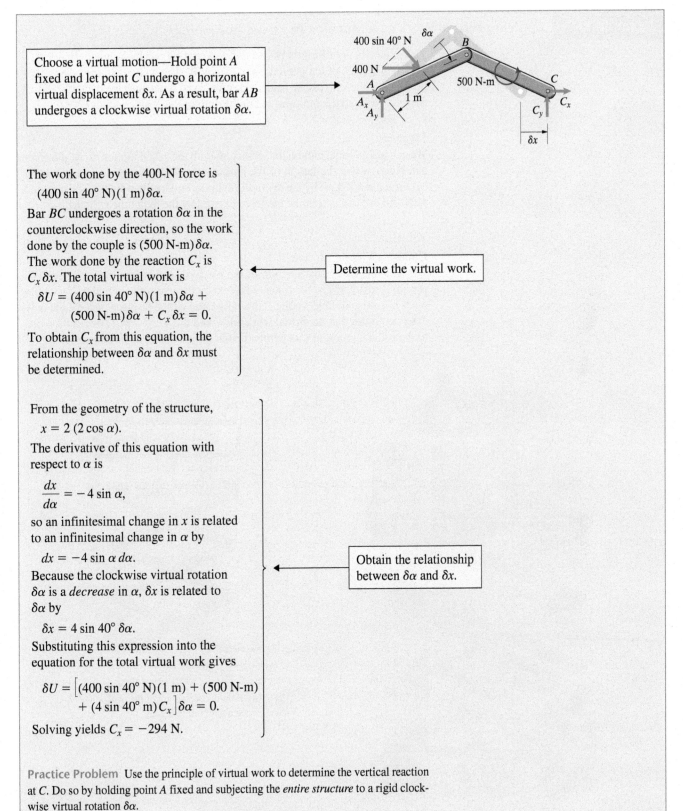

Choose a virtual motion—Hold point A fixed and let point C undergo a horizontal virtual displacement δx. As a result, bar AB undergoes a clockwise virtual rotation $\delta \alpha$.

The work done by the 400-N force is $(400 \sin 40° \text{ N})(1 \text{ m}) \delta \alpha$.

Bar BC undergoes a rotation $\delta \alpha$ in the counterclockwise direction, so the work done by the couple is $(500 \text{ N-m}) \delta \alpha$. The work done by the reaction C_x is $C_x \delta x$. The total virtual work is

$$\delta U = (400 \sin 40° \text{ N})(1 \text{ m}) \delta \alpha +$$
$$(500 \text{ N-m}) \delta \alpha + C_x \delta x = 0.$$

To obtain C_x from this equation, the relationship between $\delta \alpha$ and δx must be determined.

> Determine the virtual work.

From the geometry of the structure,

$$x = 2 (2 \cos \alpha).$$

The derivative of this equation with respect to α is

$$\frac{dx}{d\alpha} = -4 \sin \alpha,$$

so an infinitesimal change in x is related to an infinitesimal change in α by

$$dx = -4 \sin \alpha \, d\alpha.$$

Because the clockwise virtual rotation $\delta \alpha$ is a *decrease* in α, δx is related to $\delta \alpha$ by

$$\delta x = 4 \sin 40° \, \delta \alpha.$$

Substituting this expression into the equation for the total virtual work gives

$$\delta U = \left[(400 \sin 40° \text{ N})(1 \text{ m}) + (500 \text{ N-m}) \right.$$
$$\left. + (4 \sin 40° \text{ m}) C_x \right] \delta \alpha = 0.$$

Solving yields $C_x = -294$ N.

> Obtain the relationship between $\delta \alpha$ and δx.

Practice Problem Use the principle of virtual work to determine the vertical reaction at C. Do so by holding point A fixed and subjecting the *entire structure* to a rigid clockwise virtual rotation $\delta \alpha$.

Answer: $C_y = -79.3$ N.

| Example 11.2 | Applying Virtual Work to a Machine (▶ *Related Problem 11.21*) |

The extensible platform is raised and lowered by the hydraulic cylinder *BC*. The total weight of the platform and men is *W*. The weights of the beams supporting the platform can be neglected. What axial force must the hydraulic cylinder exert to hold the platform in equilibrium in the position shown?

Strategy
We can use a virtual motion that coincides with the actual motion of the platform and beams when the length of the hydraulic cylinder changes. By calculating the virtual work done by the hydraulic cylinder and by the weight of the men and platform, we can determine the force exerted by the hydraulic cylinder.

Solution

Choose a Virtual Motion
We draw the free-body diagram of the platform and beams in Fig. a. Our objective is to determine the force *F* exerted by the hydraulic cylinder. If we hold point *A* fixed and subject point *C* to a horizontal virtual displacement δx, the only external forces that do virtual work are *F* and the weight *W*. (The reaction due to the roller support at *C* is perpendicular to the virtual displacement.)

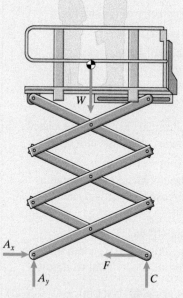

(a) Free-body diagram of the platform and supporting beams.

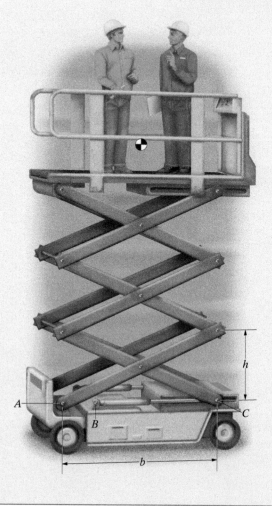

Determine the Virtual Work   The virtual work done by the force F as point C undergoes a virtual displacement δx to the right (Fig. b) is $-F\,\delta x$. To determine the virtual work done by the weight W, we must determine the vertical displacement of point D in Fig. b when point C moves to the right a distance δx. The dimensions b and h are related by

$$b^2 + h^2 = L^2,$$

where L is the length of the beam AD. Taking the derivative of this equation with respect to b, we obtain

$$2b + 2h\frac{dh}{db} = 0,$$

which we can solve for dh in terms of db:

$$dh = -\frac{b}{h}db.$$

Thus, when b increases an amount δx, the dimension h *decreases* an amount $(b/h)\,\delta x$. Because there are three pairs of beams, the platform moves downward a distance $(3b/h)\,\delta x$, and the virtual work done by the weight is $(3b/h)W\,\delta x$. The total virtual work is

$$\delta U = \left[-F + \left(\frac{3b}{h}\right)W\right]\delta x = 0,$$

and we obtain $F = (3b/h)W$.

Critical Thinking
We designed this example to demonstrate how advantageous the method of virtual work can be for certain types of problems. You can see that it would be very tedious to draw the free-body diagrams of the individual members of the frame supporting the platform and solve the equilibrium equations to determine the force exerted by the hydraulic cylinder. In contrast, it was relatively simple to determine the virtual work done by the external forces acting on the frame.

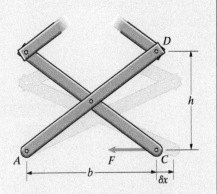

(b) A virtual displacement in which A remains fixed and C moves horizontally.

Problems

The following problems are to be solved using the principle of virtual work.

11.1   Determine the reactions at A.

Strategy: Subject the beam to three virtual motions: **1.** a horizontal displacement δx; **2.** a vertical displacement δy; and **3.** a rotation $\delta\theta$ about A.

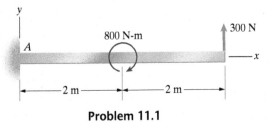

Problem 11.1

11.2   (a) Determine the virtual work done by the 2-kN force and the 2.4 kN-m couple when the beam is rotated through a counterclockwise angle $\delta\theta$ about point A.

(b) Use the result of (a) to determine the reaction at B.

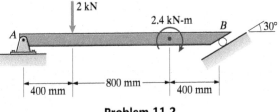

Problem 11.2

11.3 Determine the tension in the cable.

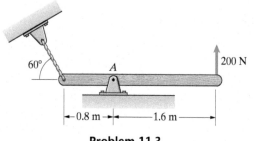

Problem 11.3

11.4 The L-shaped bar is in equilibrium. Determine F.

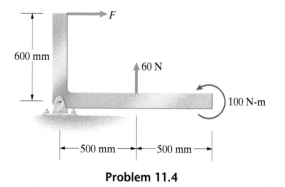

Problem 11.4

11.5 The dimension $L = 4$ ft and $w_0 = 300$ lb/ft. Determine the reactions at A and B.

Strategy: To determine the virtual work done by the distributed load, represent it by an equivalent force.

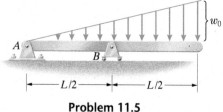

Problem 11.5

11.6 Determine the reactions at A and B.

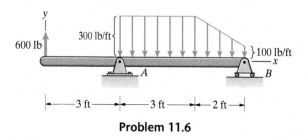

Problem 11.6

11.7 The mechanism is in equilibrium. Determine the force R in terms of F.

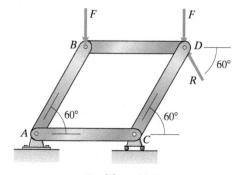

Problem 11.7

11.8 Determine the reaction at the roller support.

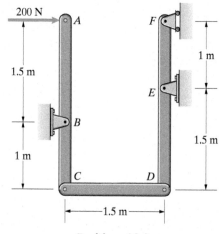

Problem 11.8

11.9 Determine the couple M necessary for the mechanism to be in equilibrium.

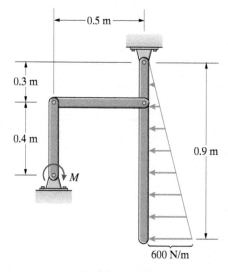

Problem 11.9

11.10 The system is in equilibrium. The total mass of the suspended load and assembly A is 120 kg.

(a) By using equilibrium, determine the force F.

(b) Using the result of (a) and the principle of virtual work, determine the distance the suspended load rises if the cable is pulled upward 300 mm at B.

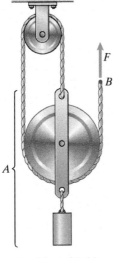

Problem 11.10

11.11 Determine the force P necessary for the mechanism to be in equilibrium.

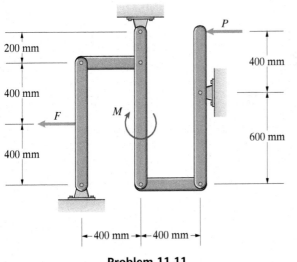

Problem 11.11

▶ **11.12*** Show that δx is related to $\delta \alpha$ by

$$\delta x = (L_1 \tan \beta) \, \delta \alpha.$$

(See Active Example 11.1.)

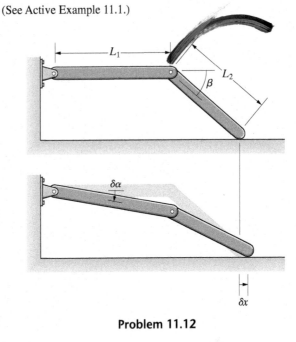

Problem 11.12

▶ **11.13** The horizontal surface is smooth. Determine the horizontal force F necessary for the system to be in equilibrium. (See Active Example 11.1.)

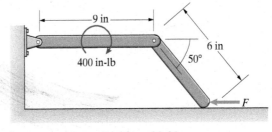

Problem 11.13

▶ 11.14* Show that δx is related to $\delta \alpha$ by

$$\delta x = \frac{L_1 x \sin \alpha}{x - L_1 \cos \alpha} \delta \alpha.$$

Strategy: Write the law of cosines in terms of α and take the derivative of the resulting equation with respect to α. (See Active Example 11.1.)

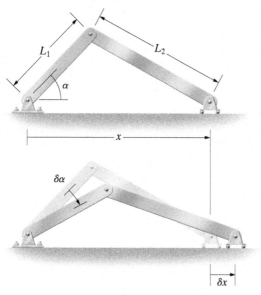

Problem 11.14

▶ 11.15 The linkage is in equilibrium. What is the force F? (See Active Example 11.1.)

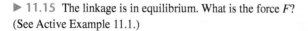
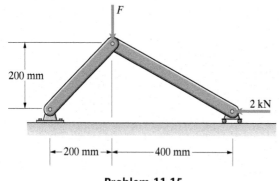

Problem 11.15

▶ 11.16 The linkage is in equilibrium. What is the force F? (See Active Example 11.1.)

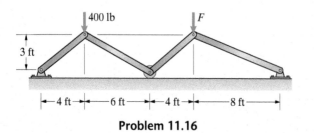

Problem 11.16

11.17 Bar AC is connected to bar BD by a pin that fits in the smooth vertical slot. The masses of the bars are negligible. If $M_A = 30$ N-m, what couple M_B is necessary for the system to be in equilibrium?

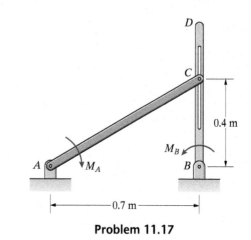

Problem 11.17

11.18 The angle $\alpha = 20°$, and the force exerted on the stationary piston by pressure is 4 kN toward the left. What couple M is necessary to keep the system in equilibrium?

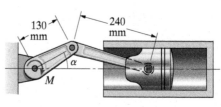

Problem 11.18

11.19 The structure is subjected to a 400-N load and is held in place by a horizontal cable. Determine the tension in the cable.

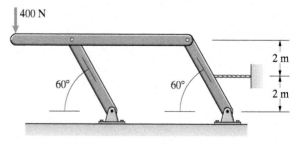

Problem 11.19

11.20 If the load on the car jack is $L = 6.5$ kN, what is the tension in the threaded shaft between A and B?

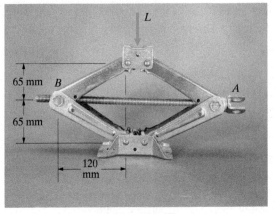

Problem 11.20

▶ **11.21** Determine the reactions at A and B. (Use the equilibrium equations to determine the horizontal components of the reactions, and use the procedure described in Example 11.2 to determine the vertical components.)

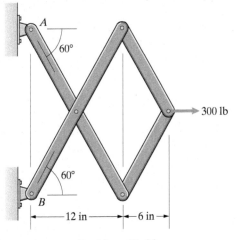

Problem 11.21

11.22 This device raises a load W by extending the hydraulic actuator DE. The bars AD and BC are each 2 m long, and the distances $b = 1.4$ m and $h = 0.8$ m. If $W = 4$ kN, what force must the actuator exert to hold the load in equilibrium?

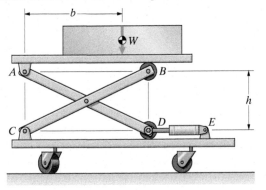

Problem 11.22

11.23 Determine the force P necessary for the mechanism to be in equilibrium.

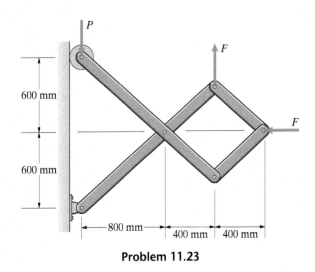

Problem 11.23

11.24 The collar A slides on the smooth vertical bar. The masses are $m_A = 20$ kg and $m_B = 10$ kg.

(a) If the collar A is given an upward virtual displacement δy, what is the resulting downward displacement of the mass B?

(b) Use virtual work to determine the tension in the spring.

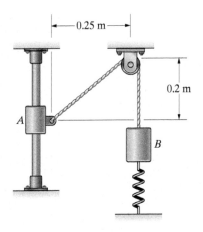

Problem 11.24

11.2 Potential Energy

BACKGROUND

The work of a force $\mathbf{F}$ due to a differential displacement of its point of application is

$$dU = \mathbf{F} \cdot d\mathbf{r}.$$

If a function of position V exists such that for any $d\mathbf{r}$,

$$dU = \mathbf{F} \cdot d\mathbf{r} = -dV, \tag{11.10}$$

the function V is called the *potential energy* associated with the force $\mathbf{F}$, and $\mathbf{F}$ is said to be *conservative*. (The negative sign in this equation is in keeping with the interpretation of V as "potential" energy. Positive work results from a decrease in V.) If the forces that do work on a system are conservative, we will show that the total potential energy of the system can be used to determine its equilibrium positions.

Examples of Conservative Forces

Weights of objects and the forces exerted by linear springs are conservative. In the following sections, we derive the potential energies associated with these forces.

Weight In terms of a coordinate system with its y axis upward, the force exerted by the weight of an object is $\mathbf{F} = -W\mathbf{j}$ (Fig. 11.8a). If we give the object an arbitrary displacement $d\mathbf{r} = dx\mathbf{i} + dy\mathbf{j} + dz\mathbf{k}$ (Fig. 11.8b), the work done by its weight is

$$dU = \mathbf{F} \cdot d\mathbf{r} = (-W\mathbf{j}) \cdot (dx\mathbf{i} + dy\mathbf{j} + dz\mathbf{k}) = -W\,dy.$$

We seek a potential energy V such that

$$dU = -W\,dy = -dV \tag{11.11}$$

or

$$\frac{dV}{dy} = W.$$

If we neglect the variation in the weight with height and integrate, we obtain

$$V = Wy + C.$$

The constant C is arbitrary. Since this function satisfies Eq. (11.11) for any value of C, we will let $C = 0$. The position of the origin of the coordinate system can also be chosen arbitrarily. Thus, the potential energy associated with the weight of an object is

$$V = Wy, \tag{11.12}$$

where y is the height of the object above some chosen reference level, or *datum*.

Springs Consider a linear spring connecting an object to a fixed support (Fig. 11.9a). In terms of the stretch $S = r - r_0$, where r is the length of the spring and r_0 is its unstretched length, the force exerted on the object is kS (Fig. 11.9b). If the point at which the spring is attached to the object undergoes a differential displacement $d\mathbf{r}$ (Fig. 11.9c), the work done by the force on the object is

$$dU = -kS\,dS,$$

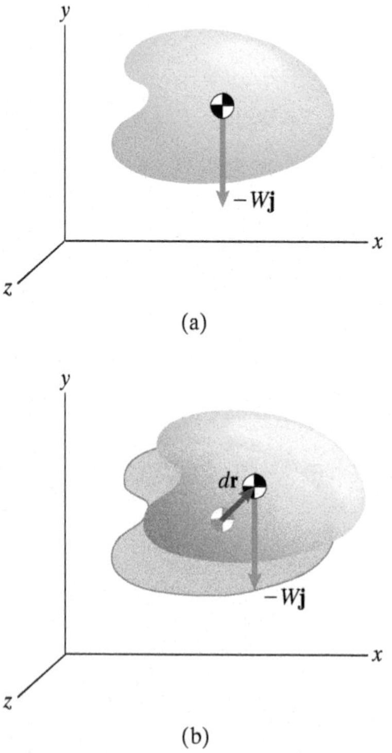

Figure 11.8
(a) Force exerted by the weight of an object.
(b) A differential displacement.

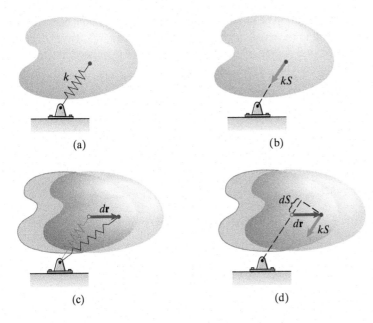

Figure 11.9
(a) A spring connected to an object.
(b) The force exerted on the object.
(c) A differential displacement of the object.
(d) The work done by the force is
$dU = -kS\,dS$.

where dS is the increase in the stretch of the spring resulting from the displacement (Fig. 11.9d). We seek a potential energy V such that

$$dU = -kS\,dS = -dV \qquad (11.13)$$

or

$$\frac{dV}{dS} = kS.$$

Integrating this equation and letting the integration constant be zero, we obtain the potential energy associated with the force exerted by a linear spring:

$$V = \frac{1}{2}kS^2. \qquad (11.14)$$

Notice that V is positive if the spring is either stretched (S is positive) or compressed (S is negative). Potential energy (the potential to do work) is stored in a spring by either stretching or compressing it.

Principle of Virtual Work for Conservative Forces

Because the work done by a conservative force is expressed in terms of its potential energy through Eq. (11.10), we can give an alternative statement of the principle of virtual work when an object is subjected to conservative forces:

Let an object be in equilibrium. If the forces and couples that do work on the object as the result of a virtual translation or rotation are conservative, the change in the total potential energy is zero:

$$\delta V = 0. \qquad (11.15)$$

We emphasize that it is not necessary that all of the forces and couples acting on the object be conservative for this result to hold; it is necessary only that the forces and couples that do work be conservative. This principle also applies to a system of interconnected objects if the external forces that do work are conservative and the internal forces at the connections between objects either do no work or are conservative. Such a system is called a *conservative system*.

If the position of a system can be specified by a single coordinate q, the system is said to have one *degree of freedom*. The total potential energy of a conservative, one-degree-of-freedom system can be expressed in terms of q, and we can write Eq. (11.15) as

$$\delta V = \frac{dV}{dq}\delta q = 0.$$

Thus, when the object or system is in equilibrium, the derivative of its total potential energy with respect to q is zero:

$$\frac{dV}{dq} = 0. \tag{11.16}$$

We can use this equation to determine the values of q at which the system is in equilibrium.

Stability of Equilibrium

Suppose that a homogeneous bar of weight W and length L is suspended from a pin support at one end. In terms of the angle α shown in Fig. 11.10a, the height of the center of mass relative to the pinned end is $-\frac{1}{2}L\cos\alpha$. Choosing the level of the pin support as the datum, we can therefore express the potential energy associated with the weight of the bar as

$$V = -\frac{1}{2}WL\cos\alpha.$$

When the bar is in equilibrium,

$$\frac{dV}{d\alpha} = \frac{1}{2}WL\sin\alpha = 0.$$

This condition is satisfied when $\alpha = 0$ (Fig. 11.10b) and also when $\alpha = 180°$ (Fig. 11.10c).

There is a fundamental difference between the two equilibrium positions of the bar. In the position shown in Fig. 11.10b, if we displace the bar slightly from its equilibrium position and release it, the bar will remain near the equilibrium position. We say that this equilibrium position is *stable*. When the bar is in the position shown in Fig. 11.10c, if we displace it slightly and release it, the bar will move away from the equilibrium position. This equilibrium position is *unstable*.

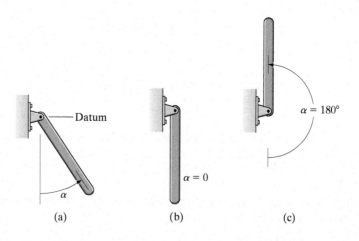

Figure 11.10
(a) A bar suspended from one end.
(b) The equilibrium position $\alpha = 0$.
(c) The equilibrium position $\alpha = 180°$.

The graph of the bar's potential energy V as a function of α is shown in Fig. 11.11a. The potential energy is a minimum at the stable equilibrium position $\alpha = 0$ and a maximum at the unstable equilibrium position $\alpha = 180°$. The derivative of V (Fig. 11.11b) equals zero at both equilibrium positions. The second derivative of V (Fig. 11.11c) is positive at the stable equilibrium position $\alpha = 0$ and negative at the unstable equilibrium position $\alpha = 180°$.

If a conservative, one-degree-of-freedom system is in equilibrium and the second derivative of V evaluated at the equilibrium position is positive, the equilibrium position is stable. If the second derivative of V is negative, it is unstable (Fig. 11.12).

$$\frac{dV}{dq} = 0, \qquad \frac{d^2V}{dq^2} > 0: \qquad \text{Stable equilibrium}$$

$$\frac{dV}{dq} = 0, \qquad \frac{d^2V}{dq^2} < 0: \qquad \text{Unstable equilibrium}$$

Proving these results requires analyzing the motion of the system near an equilibrium position.

Using potential energy to analyze the equilibrium of one-degree-of-freedom systems typically involves three steps:

1. **Determine the potential energy**—Express the total potential energy in terms of a single coordinate that specifies the position of the system.
2. **Find the equilibrium positions**—By calculating the first derivative of the potential energy, determine the equilibrium position or positions.
3. **Examine the stability**—Use the sign of the second derivative of the potential energy to determine whether the equilibrium positions are stable.

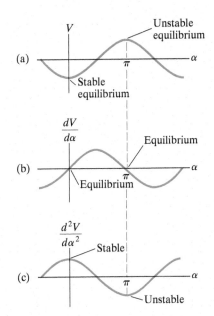

Figure 11.11
Graphs of V, $dV/d\alpha$, and $d^2V/d\alpha^2$.

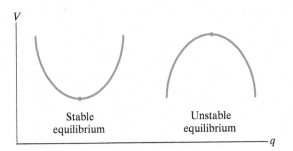

Figure 11.12
Graphs of the potential energy V as a function of the coordinate q that exhibit stable and unstable equilibrium positions.

RESULTS

Potential Energy

If a function of position V exists such that, for any infinitesimal displacement $d\mathbf{r}$, the work done by a force $\mathbf{F}$ is

$$dU = \mathbf{F} \cdot d\mathbf{r} = -dV,$$

then V is called the *potential energy* associated with the force and $\mathbf{F}$ is said to be *conservative*.

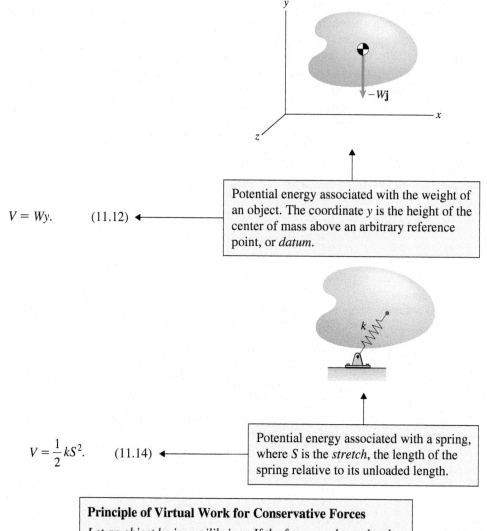

$$V = Wy. \qquad (11.12)$$

Potential energy associated with the weight of an object. The coordinate y is the height of the center of mass above an arbitrary reference point, or *datum*.

$$V = \frac{1}{2}kS^2. \qquad (11.14)$$

Potential energy associated with a spring, where S is the *stretch*, the length of the spring relative to its unloaded length.

Principle of Virtual Work for Conservative Forces

Let an object be in equilibrium. If the forces and couples that do work on the object as the result of a virtual translation or rotation are conservative, the change in the total potential energy is zero:

$$\delta V = 0. \qquad (11.15)$$

This principle also applies to a system of interconnected objects if the external forces that do work are conservative and the internal forces between objects either do no work or are conservative. Such a system is called a *conservative system*.

If the position of a system can be specified by a single coordinate q, the system is said to have one *degree of freedom*. When a conservative, one-degree-of-freedom system is in equilibrium,

$$\frac{dV}{dq} = 0. \qquad (11.16)$$

If the second derivative of V with respect to q is positive, the equilibrium position is stable, and if the second derivative is negative, the equilibrium position is unstable.

Active Example 11.3 Stability of a Conservative System (▶ *Related Problems 11.27–11.29*)

A crate of weight W is suspended from the ceiling by a spring. The coordinate x measures the vertical position of the center of mass of the crate relative to its position when the spring is unstretched. Determine the equilibrium position of the crate relative to its position when the spring is unstretched.

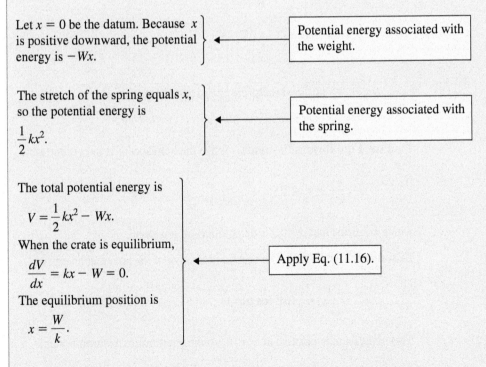

Strategy

The forces acting on the crate—its weight and the force exerted by the spring— are conservative. We can express the total potential energy in terms of the coordinate x and use Eq. (11.16) to determine the equilibrium position.

Solution

Let $x = 0$ be the datum. Because x is positive downward, the potential energy is $-Wx$.

> Potential energy associated with the weight.

The stretch of the spring equals x, so the potential energy is

$$\frac{1}{2}kx^2.$$

> Potential energy associated with the spring.

The total potential energy is

$$V = \frac{1}{2}kx^2 - Wx.$$

When the crate is equilibrium,

$$\frac{dV}{dx} = kx - W = 0.$$

> Apply Eq. (11.16).

The equilibrium position is

$$x = \frac{W}{k}.$$

Practice Problem Determine whether the equilibrium position of the crate is stable.

Answer: Yes.

Example 11.4 Stability of an Equilibrium Position (▶ *Related Problems 11.31, 11.32*)

The homogeneous hemisphere is at rest on the plane surface. Show that it is in equilibrium in the position shown. Is the equilibrium position stable?

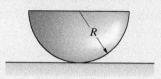

Strategy

To determine whether the hemisphere is in equilibrium and whether its equilibrium is stable, we must introduce a coordinate that specifies its orientation and express its potential energy in terms of that coordinate. We can use as the coordinate the angle of rotation of the hemisphere relative to the position shown.

Solution

Determine the Potential Energy Suppose that the hemisphere is rotated through an angle α relative to its original position (Fig. a). Then, from the datum shown, the potential energy associated with the weight W of the hemisphere is

$$V = -\frac{3}{8}RW \cos \alpha.$$

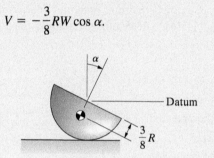

(a) The hemisphere rotated through an angle α.

Find the Equilibrium Positions When the hemisphere is in equilibrium,

$$\frac{dV}{d\alpha} = \frac{3}{8}RW \sin \alpha = 0,$$

which confirms that $\alpha = 0$ is an equilibrium position.

Examine the Stability The second derivative of the potential energy is

$$\frac{d^2V}{d\alpha^2} = \frac{3}{8}RW \cos \alpha.$$

This expression is positive at $\alpha = 0$, so the equilibrium position is stable.

Critical Thinking

Notice that we ignored the normal force exerted on the hemisphere by the plane surface. This force does no work and so does not affect the potential energy.

Example 11.5 | **Stability of an Equilibrium Position** (▶ *Related Problems 11.41, 11.42*)

The pinned bars are held in place by the linear spring. Each bar has weight W and length L. The spring is unstretched when $\alpha = 0$, and the bars are in equilibrium when $\alpha = 60°$. Determine the spring constant k, and determine whether the equilibrium position is stable or unstable.

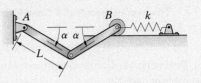

Strategy

The only forces that do work on the bars are their weights and the force exerted by the spring. By expressing the total potential energy in terms of α and using Eq. (11.16), we will obtain an equation we can solve for the spring constant k.

Solution

Determine the Potential Energy   If we use the datum shown in Fig. a, the potential energy associated with the weights of the two bars is

$$W\left(-\frac{1}{2}L \sin \alpha\right) + W\left(-\frac{1}{2}L \sin \alpha\right) = -WL \sin \alpha.$$

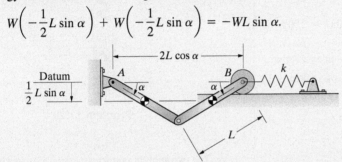

(a) Determining the total potential energy.

The spring is unstretched when $\alpha = 0$ and the distance between points A and B is $2L \cos \alpha$ (Fig. a), so the stretch of the spring is $2L - 2L \cos \alpha$. Therefore, the potential energy associated with the spring is $\frac{1}{2}k(2L - 2L \cos \alpha)^2$, and the total potential energy is

$$V = -WL \sin \alpha + 2kL^2(1 - \cos \alpha)^2.$$

When the system is in equilibrium,

$$\frac{dV}{d\alpha} = -WL \cos \alpha + 4kL^2 (\sin \alpha)(1 - \cos \alpha) = 0.$$

Because the system is in equilibrium when $\alpha = 60°$, we can solve this equation for the spring constant in terms of W and L:

$$k = \frac{W \cos \alpha}{4L (\sin \alpha)(1 - \cos \alpha)} = \frac{W \cos 60°}{4L (\sin 60°)(1 - \cos 60°)} = \frac{0.289W}{L}$$

Examine the Stability   The second derivative of the potential energy is

$$\frac{d^2V}{d\alpha^2} = WL \sin \alpha + 4kL^2 (\cos \alpha - \cos^2 \alpha + \sin^2 \alpha)$$

$$= WL \sin 60° + 4kL^2 (\cos 60° - \cos^2 60° + \sin^2 60°)$$

$$= 0.866WL + 4kL^2.$$

This is a positive number, so the equilibrium position is stable.

Critical Thinking

How do you know when you can apply the principle of virtual work for conservative forces to a system? The system must be conservative, which means that the forces and couples that do work when the system undergoes a virtual motion are conservative. Conservative forces are forces for which a potential energy exists. In this example, work is done by the weights of the bars and the force exerted by the spring, which are conservative forces.

Problems

11.25 The potential energy of a conservative system is given by $V = 2x^3 + 3x^2 - 12x$.

(a) For what values of x is the system in equilibrium?

(b) Determine whether the equilibrium positions you found in (a) are stable or unstable.

11.26 The potential energy of a conservative system is given by $V = 2q^3 - 21q^2 + 72q$.

(a) For what values of q is the system in equilibrium?

(b) Determine whether the equilibrium positions you found in (a) are stable or unstable.

▶ **11.27** The mass $m = 2$ kg and the spring constant $k = 100$ N/m. The spring is unstretched when $x = 0$.

(a) Determine the value of x for which the mass is in equilibrium.

(b) Is the equilibrium position stable or unstable? (See Example 11.3.)

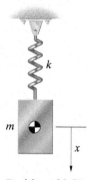

Problem 11.27

▶ **11.28** The *nonlinear* spring exerts a force $-kx + \varepsilon x^3$ on the mass, where k and ε are constants. Determine the potential energy V associated with the force exerted on the mass by the spring. (See Example 11.3.)

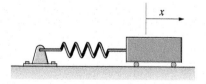

Problem 11.28

▶ **11.29** The 1-kg mass is suspended from the nonlinear spring described in Problem 11.28. The constants $k = 10$ and $\varepsilon = 1$, where x is in meters.

(a) Show that the mass is in equilibrium when $x = 1.12$ m and when $x = 2.45$ m.

(b) Determine whether the equilibrium positions are stable or unstable. (See Example 11.3.)

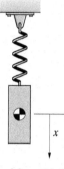

Problem 11.29

11.30 The two straight segments of the bar are each of weight W and length L. Determine whether the equilibrium position shown is stable if (a) $0 < \alpha_0 < 90°$; (b) $90° < \alpha_0 < 180°$.

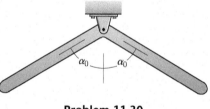

Problem 11.30

▶ **11.31** The homogeneous composite object consists of a hemisphere and a cylinder. It is at rest on the plane surface. Show that this equilibrium position is stable only if $L < R/\sqrt{2}$. (See Example 11.4.)

Problem 11.31

▶ 11.32 The homogeneous composite object consists of a half-cylinder and a triangular prism. It is at rest on the plane surface. Show that this equilibrium position is stable only if $h < \sqrt{2}\,R$. (See Example 11.4.)

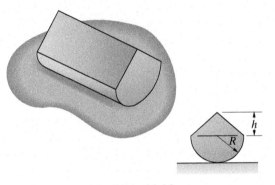

Problem 11.32

11.33 The homogeneous bar has weight W, and the spring is unstretched when the bar is vertical ($\alpha = 0$).

(a) Use potential energy to show that the bar is in equilibrium when $\alpha = 0$.

(b) Show that the equilibrium position $\alpha = 0$ is stable only if $2kL > W$.

11.34 Suppose that the bar in Problem 11.33 is in equilibrium when $\alpha = 20°$.

(a) Show that the spring constant $k = 0.490\ W/L$.

(b) Determine whether the equilibrium position is stable.

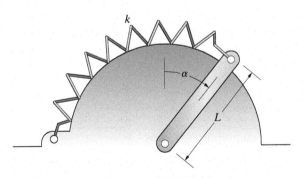

Problems 11.33/11.34

11.35 The bar AB has mass m and length L. The spring is unstretched when the bar is vertical ($\alpha = 0$). The light collar C slides on the smooth vertical bar so that the spring remains horizontal. Show that the equilibrium position $\alpha = 0$ is stable only if $2kL > mg$.

11.36 The bar AB in Problem 11.35 has mass $m = 4$ kg, length 2 m, and the spring constant is $k = 12$ N/m.

(a) Determine the value of α in the range $0 < \alpha < 90°$ for which the bar is in equilibrium.

(b) Is the equilibrium position determined in part (a) stable?

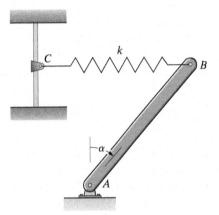

Problems 11.35/11.36

11.37 The bar AB has weight W and length L. The spring is unstretched when the bar is vertical ($\alpha = 0$). The light collar C slides on the smooth horizontal bar so that the spring remains vertical. Show that the equilibrium position $\alpha = 0$ is unstable.

11.38 The bar AB described in Problem 11.37 has a mass of 2 kg, and the spring constant is $k = 80$ N/m.

(a) Determine the value of α in the range $0 < \alpha < 90°$ for which the bar is in equilibrium.

(b) Is the equilibrium position determined in (a) stable?

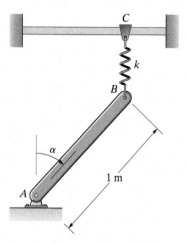

Problems 11.37/11.38

11.39 Each homogeneous bar is of mass m and length L. The spring is unstretched when $\alpha = 0$. If $mg = kL$, determine the value of α in the range $0 < \alpha < 90°$ for which the system is in equilibrium.

11.40 Determine whether the equilibrium position found in Problem 11.39 is stable or unstable.

Problems 11.39/11.40

▶ **11.41** The pinned bars are held in place by the linear spring. Each bar has weight W and length L. The spring is unstretched when $\alpha = 90°$. Determine the value of α in the range $0 < \alpha < 90°$ for which the system is in equilibrium. (See Example 11.5.)

▶ **11.42** Determine whether the equilibrium position found in Problem 11.41 is stable or unstable. (See Example 11.5.)

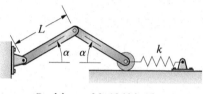

Problems 11.41/11.42

11.43 The bar weighs 15 lb. The spring is unstretched when $\alpha = 0$. The bar is in equilibrium when $\alpha = 30°$. Determine the spring constant k.

11.44 Determine whether the equilibrium positions of the bar in Problem 11.43 are stable or unstable.

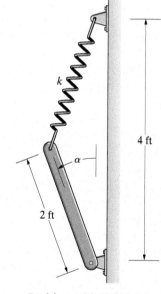

Problems 11.43/11.44

Review Problems

11.45 (a) Determine the couple exerted on the beam at A.

(b) Determine the vertical force exerted on the beam at A.

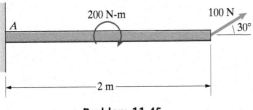

Problem 11.45

11.46 The structure is subjected to a 20 kN-m couple. Determine the horizontal reaction at C.

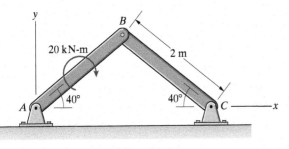

Problem 11.46

11.47 The "rack and pinion" mechanism is used to exert a vertical force on a sample at A for a stamping operation. If a force $F = 30$ lb is exerted on the handle, use the principle of virtual work to determine the force exerted on the sample.

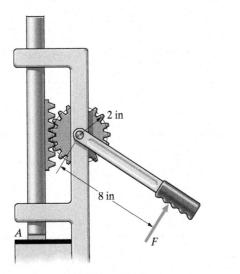

Problem 11.47

11.48 If you were assigned to calculate the force exerted on the bolt by the pliers when the grips are subjected to forces F as shown in Fig. a, you could carefully measure the dimensions, draw free-body diagrams, and use the equilibrium equations. But another approach would be to measure the change in the distance between the jaws when the distance between the handles is changed by a small amount. If your measurements indicate that the distance d in Fig. b decreases by 1 mm when D is decreased 8 mm, what is the approximate value of the force exerted on the bolt by each jaw when the forces F are applied?

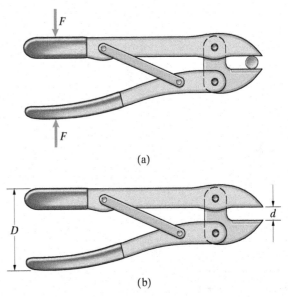

Problem 11.48

11.49 The system is in equilibrium. The total weight of the suspended load and assembly A is 300 lb.

(a) By using equilibrium, determine the force F.

(b) Using the result of (a) and the principle of virtual work, determine the distance the suspended load rises if the cable is pulled downward 1 ft at B.

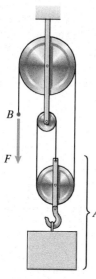

Problem 11.49

11.50 The system is in equilibrium.

(a) By drawing free-body diagrams and using equilibrium equations, determine the couple M.

(b) Using the result of (a) and the principle of virtual work, determine the angle through which pulley B rotates if pulley A rotates through an angle α.

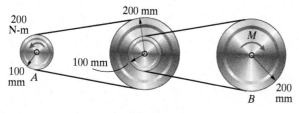

Problem 11.50

11.51 The mechanism is in equilibrium. Neglect friction between the horizontal bar and the collar. Determine M in terms of F, α, and L.

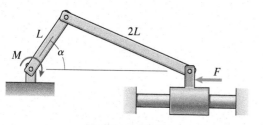

Problem 11.51

11.52 In an injection casting machine, a couple M applied to arm AB exerts a force on the injection piston at C. Given that the horizontal component of the force exerted at C is 4 kN, use the principle of virtual work to determine M.

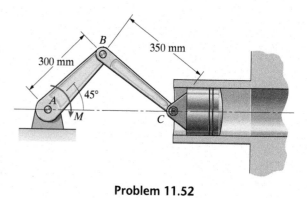

Problem 11.52

11.53 Show that if bar AB is subjected to a clockwise virtual rotation $\delta\alpha$, bar CD undergoes a counterclockwise virtual rotation $(b/a)\,\delta\alpha$.

11.54 The system is in equilibrium, $a = 800$ mm, and $b = 400$ mm. Use the principle of virtual work to determine the force F.

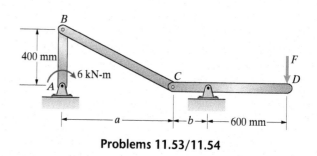

Problems 11.53/11.54

11.55 Show that if bar *AB* is subjected to a clockwise virtual rotation $\delta\alpha$, bar *CD* undergoes a clockwise virtual rotation $[ad/(ac + bc - bd)]\,\delta\alpha$.

11.56 The system is in equilibrium, $a = 300$ mm, $b = 350$ mm, $c = 350$ mm, and $d = 200$ mm. Use the principle of virtual work to determine the couple *M*.

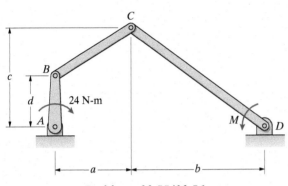

Problems 11.55/11.56

11.57 The mass of the bar is 10 kg, and it is 1 m in length. Neglect the masses of the two collars. The spring is unstretched when the bar is vertical ($\alpha = 0$), and the spring constant is $k = 100$ N/m. Determine the values of α at which the bar is in equilibrium.

11.58 Determine whether the equilibrium positions of the bar in Problem 11.57 are stable or unstable.

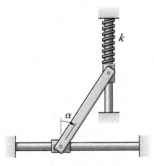

Problems 11.57/11.58

11.59 The spring is unstretched when $\alpha = 90°$. Determine the value of α in the range $0 < \alpha < 90°$ for which the system is in equilibrium.

11.60 Determine whether the equilibrium position found in Problem 11.59 is stable or unstable.

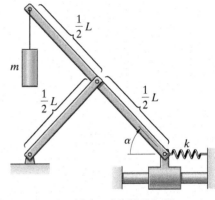

Problems 11.59/11.60

11.61 The hydraulic cylinder *C* exerts a horizontal force at *A*, raising the weight *W*. Determine the magnitude of the force the hydraulic cylinder must exert to support the weight in terms of *W* and α.

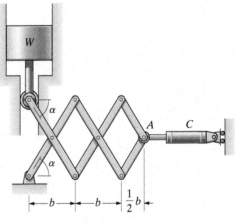

Problem 11.61

11.62 The homogeneous composite object consists of a hemisphere and a cone. It is at rest on the plane surface. Show that this equilibrium position is stable only if $h < \sqrt{3}R$.

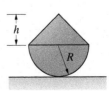

Problem 11.62

APPENDIX

A

Review of Mathematics

A.1 Algebra

Quadratic Equations

The solutions of the quadratic equation

$$ax^2 + bx + c = 0$$

are

$$x = \frac{-b \pm \sqrt{b^2 - 4ac}}{2a}.$$

Natural Logarithms

The natural logarithm of a positive real number x is denoted by $\ln x$. It is defined to be the number such that

$$e^{\ln x} = x,$$

where $e = 2.7182\ldots$ is the base of natural logarithms.

Logarithms have the following properties:

$$\ln(xy) = \ln x + \ln y,$$
$$\ln(x/y) = \ln x - \ln y,$$
$$\ln y^x = x \ln y.$$

A.2 Trigonometry

The trigonometric functions for a right triangle are

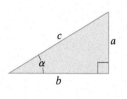

$$\sin \alpha = \frac{1}{\csc \alpha} = \frac{a}{c}, \quad \cos \alpha = \frac{1}{\sec \alpha} = \frac{b}{c}, \quad \tan \alpha = \frac{1}{\cot \alpha} = \frac{a}{b}.$$

The sine and cosine satisfy the relation

$$\sin^2 \alpha + \cos^2 \alpha = 1,$$

and the sine and cosine of the sum and difference of two angles satisfy

$$\sin(\alpha + \beta) = \sin \alpha \cos \beta + \cos \alpha \sin \beta,$$
$$\sin(\alpha - \beta) = \sin \alpha \cos \beta - \cos \alpha \sin \beta,$$
$$\cos(\alpha + \beta) = \cos \alpha \cos \beta - \sin \alpha \sin \beta,$$
$$\cos(\alpha - \beta) = \cos \alpha \cos \beta + \sin \alpha \sin \beta.$$

The **law of cosines** for an arbitrary triangle is

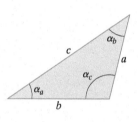

$$c^2 = a^2 + b^2 - 2ab \cos \alpha_c,$$

and the **law of sines** is

$$\frac{\sin \alpha_a}{a} = \frac{\sin \alpha_b}{b} = \frac{\sin \alpha_c}{c}.$$

A.3 Derivatives

$$\frac{d}{dx}x^n = nx^{n-1} \qquad\qquad \frac{d}{dx}\sin x = \cos x \qquad\qquad \frac{d}{dx}\sinh x = \cosh x$$

$$\frac{d}{dx}e^x = e^x \qquad\qquad \frac{d}{dx}\cos x = -\sin x \qquad\qquad \frac{d}{dx}\cosh x = \sinh x$$

$$\frac{d}{dx}\ln x = \frac{1}{x} \qquad\qquad \frac{d}{dx}\tan x = \frac{1}{\cos^2 x} \qquad\qquad \frac{d}{dx}\tanh x = \frac{1}{\cosh^2 x}$$

A.4 Integrals

$$\int x^n \, dx = \frac{x^{n+1}}{n+1} \quad (n \neq -1)$$

$$\int x^{-1} \, dx = \ln x$$

$$\int (a + bx)^{1/2} \, dx = \frac{2}{3b}(a + bx)^{3/2}$$

$$\int x(a + bx)^{1/2} \, dx = -\frac{2(2a - 3bx)(a + bx)^{3/2}}{15b^2}$$

$$\int (1 + a^2x^2)^{1/2} \, dx = \frac{1}{2}\left\{ x(1 + a^2x^2)^{1/2} \right.$$

$$\left. + \frac{1}{a}\ln\left[x + \left(\frac{1}{a^2} + x^2 \right)^{1/2} \right] \right\}$$

$$\int x(1 + a^2x^2)^{1/2} \, dx = \frac{a}{3}\left(\frac{1}{a^2} + x^2 \right)^{3/2}$$

$$\int x^2(1 + a^2x^2)^{1/2} \, dx = \frac{1}{4}ax\left(\frac{1}{a^2} + x^2 \right)^{3/2}$$

$$-\frac{1}{8a^2}x(1 + a^2x^2)^{1/2} - \frac{1}{8a^3}\ln\left[x + \left(\frac{1}{a^2} + x^2 \right)^{1/2} \right]$$

$$\int (1 - a^2x^2)^{1/2} \, dx = \frac{1}{2}\left[x(1 - a^2x^2)^{1/2} + \frac{1}{a}\arcsin ax \right]$$

$$\int x(1 - a^2x^2)^{1/2} \, dx = -\frac{a}{3}\left(\frac{1}{a^2} - x^2 \right)^{3/2}$$

$$\int x^2(a^2 - x^2)^{1/2} \, dx = -\frac{1}{4}x(a^2 - x^2)^{3/2}$$

$$+\frac{1}{8}a^2\left[x(a^2 - x^2)^{1/2} + a^2 \arcsin\frac{x}{a} \right]$$

$$\int \frac{dx}{(1 + a^2x^2)^{1/2}} = \frac{1}{a}\ln\left[x + \left(\frac{1}{a^2} + x^2 \right)^{1/2} \right]$$

$$\int \frac{dx}{(1 - a^2x^2)^{1/2}} = \frac{1}{a}\arcsin ax \quad \text{or} \quad -\frac{1}{a}\arccos ax$$

$$\int \sin x \, dx = -\cos x$$

$$\int \cos x \, dx = \sin x$$

$$\int \sin^2 x \, dx = -\frac{1}{2}\sin x \cos x + \frac{1}{2}x$$

$$\int \cos^2 x \, dx = \frac{1}{2}\sin x \cos x + \frac{1}{2}x$$

$$\int \sin^3 x \, dx = -\frac{1}{3}\cos x(\sin^2 x + 2)$$

$$\int \cos^3 x \, dx = \frac{1}{3}\sin x(\cos^2 x + 2)$$

$$\int \cos^4 x \, dx = \frac{3}{8}x + \frac{1}{4}\sin 2x + \frac{1}{32}\sin 4x$$

$$\int \sin^n x \cos x \, dx = \frac{(\sin x)^{n+1}}{n+1} \quad (n \neq -1)$$

$$\int \sinh x \, dx = \cosh x$$

$$\int \cosh x \, dx = \sinh x$$

$$\int \tanh x \, dx = \ln \cosh x$$

$$\int e^{ax} \, dx = \frac{e^{ax}}{a}$$

$$\int xe^{ax} \, dx = \frac{e^{ax}}{a^2}(ax - 1)$$

A.5 Taylor Series

The Taylor series of a function $f(x)$ is

$$f(a + x) = f(a) + f'(a)x + \frac{1}{2!}f''(a)x^2 + \frac{1}{3!}f'''(a)x^3 + \cdots,$$

where the primes indicate derivatives.

Some useful Taylor series are

$$e^x = 1 + x + \frac{x^2}{2!} + \frac{x^3}{3!} + \cdots,$$

$$\sin(a + x) = \sin a + (\cos a)x - \frac{1}{2}(\sin a)x^2 - \frac{1}{6}(\cos a)x^3 + \cdots,$$

$$\cos(a + x) = \cos a - (\sin a)x - \frac{1}{2}(\cos a)x^2 + \frac{1}{6}(\sin a)x^3 + \cdots,$$

$$\tan(a + x) = \tan a + \left(\frac{1}{\cos^2 a}\right)x + \left(\frac{\sin a}{\cos^3 a}\right)x^2$$

$$+ \left(\frac{\sin^2 a}{\cos^4 a} + \frac{1}{3\cos^2 a}\right)x^3 + \cdots.$$

APPENDIX

B

Properties of Areas and Lines

B.1 Areas

The coordinates of the centroid of the area A are

$$\bar{x} = \frac{\int_A x \, dA}{\int_A dA}, \qquad \bar{y} = \frac{\int_A y \, dA}{\int_A dA}.$$

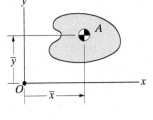

The moment of inertia about the x axis I_x, the moment of inertia about the y axis I_y, and the product of inertia I_{xy} are

$$I_x = \int_A y^2 \, dA, \qquad I_y = \int_A x^2 \, dA, \qquad I_{xy} = \int_A xy \, dA.$$

The polar moment of inertia about O is

$$J_O = \int_A r^2 \, dA = \int_A (x^2 + y^2) \, dA = I_x + I_y.$$

Area $= bh$

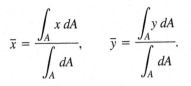

$$I_x = \frac{1}{3}bh^3, \qquad I_y = \frac{1}{3}hb^3, \qquad I_{xy} = \frac{1}{4}b^2h^2$$

$$I_{x'} = \frac{1}{12}bh^3, \qquad I_{y'} = \frac{1}{12}hb^3, \qquad I_{x'y'} = 0$$

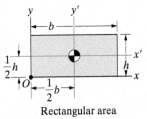

Rectangular area

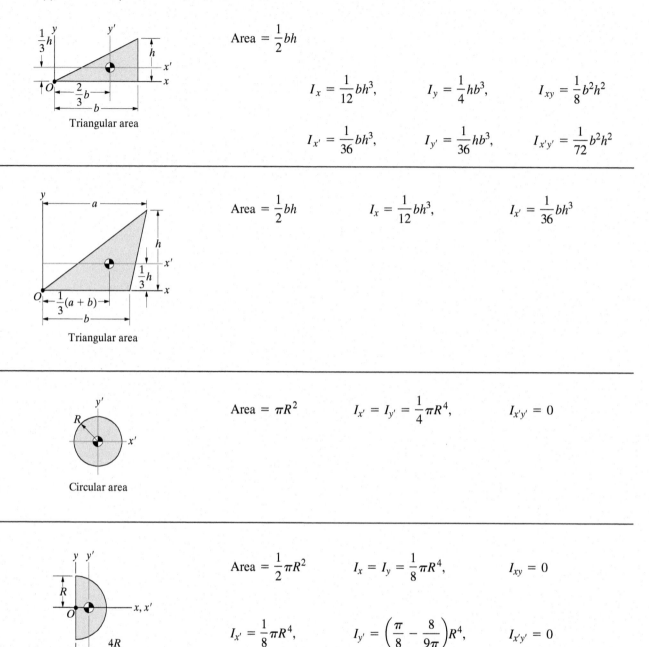

Triangular area

$$\text{Area} = \frac{1}{2}bh$$

$$I_x = \frac{1}{12}bh^3, \qquad I_y = \frac{1}{4}hb^3, \qquad I_{xy} = \frac{1}{8}b^2h^2$$

$$I_{x'} = \frac{1}{36}bh^3, \qquad I_{y'} = \frac{1}{36}hb^3, \qquad I_{x'y'} = \frac{1}{72}b^2h^2$$

Triangular area

$$\text{Area} = \frac{1}{2}bh \qquad I_x = \frac{1}{12}bh^3, \qquad I_{x'} = \frac{1}{36}bh^3$$

Circular area

$$\text{Area} = \pi R^2 \qquad I_{x'} = I_{y'} = \frac{1}{4}\pi R^4, \qquad I_{x'y'} = 0$$

Semicircular area

$$\text{Area} = \frac{1}{2}\pi R^2 \qquad I_x = I_y = \frac{1}{8}\pi R^4, \qquad I_{xy} = 0$$

$$I_{x'} = \frac{1}{8}\pi R^4, \qquad I_{y'} = \left(\frac{\pi}{8} - \frac{8}{9\pi}\right)R^4, \qquad I_{x'y'} = 0$$

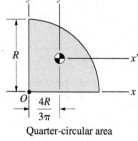

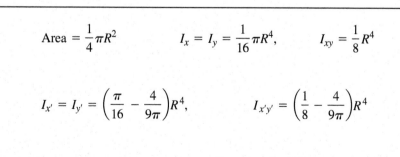

Quarter-circular area

$$\text{Area} = \frac{1}{4}\pi R^2 \qquad I_x = I_y = \frac{1}{16}\pi R^4, \qquad I_{xy} = \frac{1}{8}R^4$$

$$I_{x'} = I_{y'} = \left(\frac{\pi}{16} - \frac{4}{9\pi}\right)R^4, \qquad I_{x'y'} = \left(\frac{1}{8} - \frac{4}{9\pi}\right)R^4$$

Area $= \alpha R^2$

$$I_x = \frac{1}{4}R^4\left(\alpha - \frac{1}{2}\sin 2\alpha\right), \qquad I_y = \frac{1}{4}R^4\left(\alpha + \frac{1}{2}\sin 2\alpha\right),$$

$$I_{xy} = 0$$

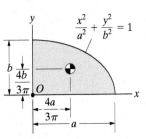

Circular sector

Area $= \frac{1}{4}\pi ab$

$$I_x = \frac{1}{16}\pi ab^3, \qquad I_y = \frac{1}{16}\pi a^3 b, \qquad I_{xy} = \frac{1}{8}a^2 b^2$$

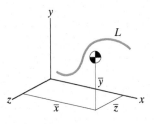

Quarter-elliptical area

Area $= \dfrac{cb^{n+1}}{n+1}$

$$I_x = \frac{c^3 b^{3n+1}}{9n+3}, \qquad I_y = \frac{cb^{n+3}}{n+3}, \qquad I_{xy} = \frac{c^2 b^{2n+2}}{4n+4}$$

Spandrel

B.2 Lines

The coordinates of the centroid of the line L are

$$\bar{x} = \frac{\displaystyle\int_L x\, dL}{\displaystyle\int_L dL}, \qquad \bar{y} = \frac{\displaystyle\int_L y\, dL}{\displaystyle\int_L dL}, \qquad \bar{z} = \frac{\displaystyle\int_L z\, dL}{\displaystyle\int_L dL}.$$

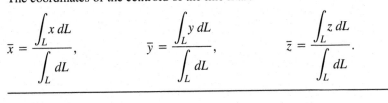

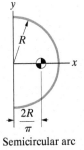

Semicircular arc

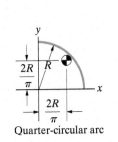

Quarter-circular arc

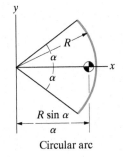

Circular arc

APPENDIX

C

Properties of Volumes and Homogeneous Objects

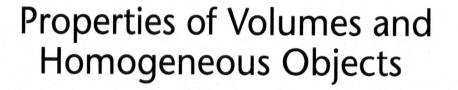

The coordinates of the centroid of the volume V are

$$\bar{x} = \frac{\int_V x\, dV}{\int_V dV}, \qquad \bar{y} = \frac{\int_V y\, dV}{\int_V dV}, \qquad \bar{z} = \frac{\int_V z\, dV}{\int_V dV}.$$

The center of mass of a homogeneous object coincides with the centroid of its volume.

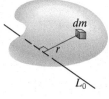

The moment of inertia of the object about the axis L_0 is

$$I_0 = \int_m r^2\, dm.$$

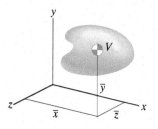

Slender bar

$$I_{x\,\text{axis}} = 0, \qquad I_{y\,\text{axis}} = I_{z\,\text{axis}} = \frac{1}{3}ml^2$$

$$I_{x'\,\text{axis}} = 0, \qquad I_{y'\,\text{axis}} = I_{z'\,\text{axis}} = \frac{1}{12}ml^2$$

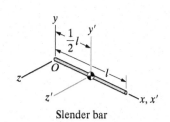

Thin circular plate

$$I_{x'\,\text{axis}} = I_{y'\,\text{axis}} = \frac{1}{4}mR^2, \qquad I_{z'\,\text{axis}} = \frac{1}{2}mR^2$$

$$I_{x \text{ axis}} = \frac{1}{3}mh^2, \qquad I_{y \text{ axis}} = \frac{1}{3}mb^2, \qquad I_{z \text{ axis}} = \frac{1}{3}m(b^2 + h^2)$$

$$I_{x' \text{ axis}} = \frac{1}{12}mh^2, \qquad I_{y' \text{ axis}} = \frac{1}{12}mb^2, \qquad I_{z' \text{ axis}} = \frac{1}{12}m(b^2 + h^2)$$

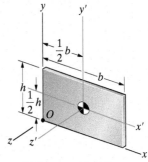

Thin rectangular plate

$$I_{x \text{ axis}} = \frac{m}{A}I_x, \qquad I_{y \text{ axis}} = \frac{m}{A}I_y, \qquad I_{z \text{ axis}} = I_{x \text{ axis}} + I_{y \text{ axis}}$$

The terms I_x and I_y are the moments of inertia of the plate's cross-sectional area A about the x and y axes.

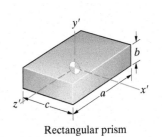

Thin plate

$$\text{Volume} = abc$$

$$I_{x' \text{ axis}} = \frac{1}{12}m(a^2 + b^2), \qquad I_{y' \text{ axis}} = \frac{1}{12}m(a^2 + c^2),$$

$$I_{z' \text{ axis}} = \frac{1}{12}m(b^2 + c^2)$$

Rectangular prism

$$\text{Volume} = \pi R^2 l$$

$$I_{x \text{ axis}} = I_{y \text{ axis}} = m\left(\frac{1}{3}l^2 + \frac{1}{4}R^2\right), \qquad I_{z \text{ axis}} = \frac{1}{2}mR^2$$

$$I_{x' \text{ axis}} = I_{y' \text{ axis}} = m\left(\frac{1}{12}l^2 + \frac{1}{4}R^2\right), \qquad I_{z' \text{ axis}} = \frac{1}{2}mR^2$$

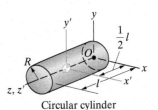

Circular cylinder

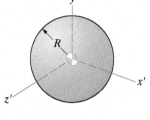

Circular cone

$$\text{Volume} = \frac{1}{3}\pi R^2 h$$

$$I_{x\text{ axis}} = I_{y\text{ axis}} = m\left(\frac{3}{5}h^2 + \frac{3}{20}R^2\right), \qquad I_{z\text{ axis}} = \frac{3}{10}mR^2$$

$$I_{x'\text{ axis}} = I_{y'\text{ axis}} = m\left(\frac{3}{80}h^2 + \frac{3}{20}R^2\right), \qquad I_{z'\text{ axis}} = \frac{3}{10}mR^2$$

Sphere

$$\text{Volume} = \frac{4}{3}\pi R^3$$

$$I_{x'\text{ axis}} = I_{y'\text{ axis}} = I_{z'\text{ axis}} = \frac{2}{5}mR^2$$

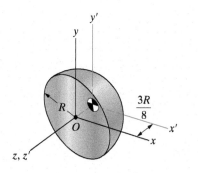

Hemisphere

$$\text{Volume} = \frac{2}{3}\pi R^3$$

$$I_{x\text{ axis}} = I_{y\text{ axis}} = I_{z\text{ axis}} = \frac{2}{5}mR^2$$

$$I_{x'\text{ axis}} = I_{y'\text{ axis}} = \frac{83}{320}mR^2, \quad I_{z'\text{ axis}} = \frac{2}{5}mR^2$$

Solutions to Practice Problems

Active Example 1.1

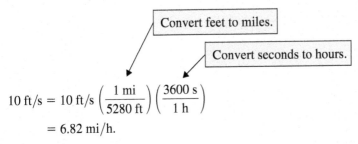

Convert feet to miles.

Convert seconds to hours.

$$10 \text{ ft/s} = 10 \text{ ft/s} \left(\frac{1 \text{ mi}}{5280 \text{ ft}} \right) \left(\frac{3600 \text{ s}}{1 \text{ h}} \right)$$

$$= 6.82 \text{ mi/h}.$$

Active Example 1.4

$$W = mg = (0.397 \text{ kg})(9.81 \text{ m/s}^2) = 3.89 \text{ N.} \longleftarrow$$

Use Eq. (1.6) to calculate the weight in newtons.

Active Example 2.1

Drawing the vectors **U** and **2V** to scale, place them head to tail.

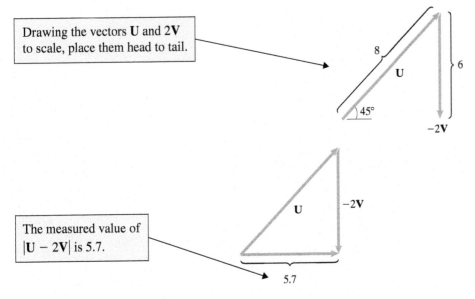

The measured value of $|\mathbf{U} - 2\mathbf{V}|$ is 5.7.

Active Example 2.3

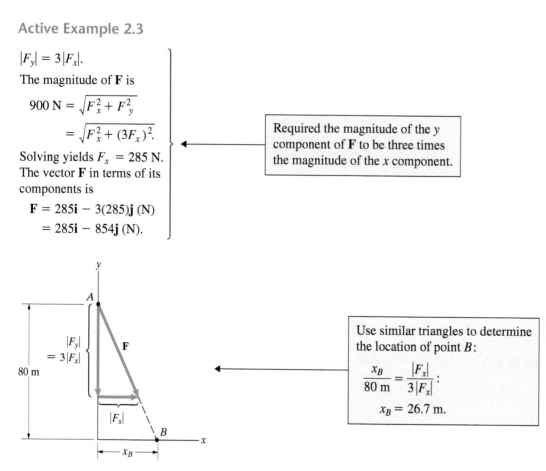

$|F_y| = 3|F_x|.$

The magnitude of **F** is

$900 \text{ N} = \sqrt{F_x^2 + F_y^2}$

$= \sqrt{F_x^2 + (3F_x)^2}.$

Solving yields $F_x = 285$ N. The vector **F** in terms of its components is

$\mathbf{F} = 285\mathbf{i} - 3(285)\mathbf{j} \text{ (N)}$

$= 285\mathbf{i} - 854\mathbf{j} \text{ (N)}.$

> Required the magnitude of the y component of **F** to be three times the magnitude of the x component.

> Use similar triangles to determine the location of point B:
>
> $\dfrac{x_B}{80 \text{ m}} = \dfrac{|F_x|}{3|F_x|}:$
>
> $x_B = 26.7$ m.

Active Example 2.6

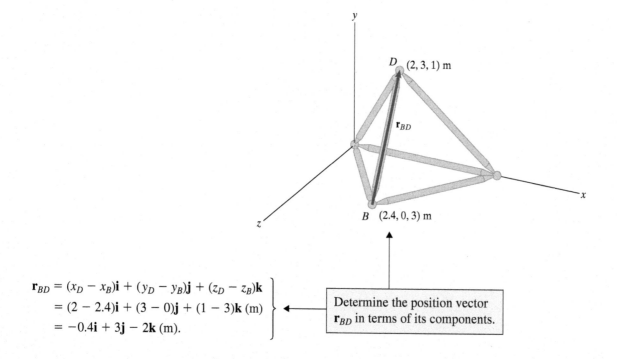

$\mathbf{r}_{BD} = (x_D - x_B)\mathbf{i} + (y_D - y_B)\mathbf{j} + (z_D - z_B)\mathbf{k}$

$= (2 - 2.4)\mathbf{i} + (3 - 0)\mathbf{j} + (1 - 3)\mathbf{k} \text{ (m)}$

$= -0.4\mathbf{i} + 3\mathbf{j} - 2\mathbf{k} \text{ (m)}.$

> Determine the position vector $\mathbf{r}_{BD}$ in terms of its components.

$$|\mathbf{r}_{BD}| = \sqrt{r_{BDx}^2 + r_{BDy}^2 + r_{BDz}^2}$$

$$= \sqrt{(-0.4 \text{ m})^2 + (3 \text{ m})^2 + (-2 \text{ m})^2}$$

$$= 3.63 \text{ m}.$$

Calculate the magnitude of $\mathbf{r}_{BD}$.

$$\mathbf{e}_{BD} = \frac{\mathbf{r}_{BD}}{|\mathbf{r}_{BD}|}$$

$$= \frac{-0.4\mathbf{i} + 3\mathbf{j} - 2\mathbf{k} \text{ (m)}}{3.63 \text{ (m)}}$$

$$= -0.110\mathbf{i} + 0.827\mathbf{j} - 0.551\mathbf{k}.$$

Divide $\mathbf{r}_{BD}$ by its magnitude to obtain $\mathbf{e}_{BD}$ in terms of its components.

Active Example 2.11

The vectors $\mathbf{U}$ and $\mathbf{V}$ are perpendicular if $\mathbf{U} \cdot \mathbf{V} = 0$. Use this condition to determine V_x.

$$\mathbf{U} \cdot \mathbf{V} = U_x V_x + U_y V_y + U_z V_z$$

$$= (6)V_x + (-5)(2) + (-3)(2)$$

$$= 6V_x - 16.$$

Calculate $\mathbf{U} \cdot \mathbf{V}$ in terms of the components of the vectors.

$$\mathbf{U} \cdot \mathbf{V} = 6V_x - 16 = 0,$$

$$V_x = 2.67.$$

Equate $\mathbf{U} \cdot \mathbf{V}$ to zero and solve for V_x.

Active Example 2.14

The cross product $\mathbf{U} \times \mathbf{V}$ is perpendicular to $\mathbf{U}$ and perpendicular to $\mathbf{V}$. By determining the vector $\mathbf{U} \times \mathbf{V}$ in terms of its components and dividing it by its magnitude $|\mathbf{U} \times \mathbf{V}|$, we can obtain the components of a unit vector that is perpendicular to $\mathbf{U}$ and perpendicular to $\mathbf{V}$.

$$\mathbf{U} \times \mathbf{V} = (U_y V_z - U_z V_y)\mathbf{i} - (U_x V_z - U_z V_x)\mathbf{j}$$

$$+ (U_x V_y - U_y V_x)\mathbf{k}$$

$$= [(2)(-4) - (-1)(-3)]\mathbf{i} - [(3)(-4) - (-1)(5)]\mathbf{j}$$

$$+ [(3)(-3) - (2)(5)]\mathbf{k}$$

$$= -11\mathbf{i} + 7\mathbf{j} - 19\mathbf{k}.$$

Calculate $\mathbf{U} \times \mathbf{V}$ in terms of the components of the vectors.

$$|\mathbf{U} \times \mathbf{V}| = \sqrt{(-11)^2 + (7)^2 + (-19)^2}$$

$$= 23.0.$$

$$\frac{\mathbf{U} \times \mathbf{V}}{|\mathbf{U} \times \mathbf{V}|} = \frac{-11\mathbf{i} + 7\mathbf{j} - 19\mathbf{k}}{23.0}$$

$$= -0.477\mathbf{i} + 0.304\mathbf{j} - 0.825\mathbf{k}.$$

Divide the vector $\mathbf{U} \times \mathbf{V}$ by its magnitude.

Active Example 3.1

Draw the Free-Body Diagram of the Car

Draw a sketch of the isolated car.

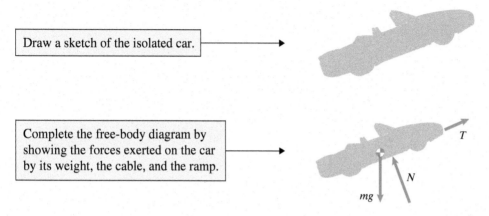

Complete the free-body diagram by showing the forces exerted on the car by its weight, the cable, and the ramp.

Apply the Equilibrium Equations

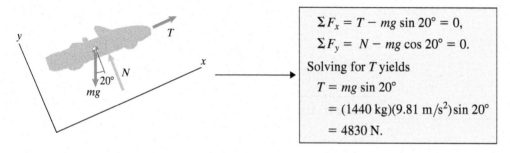

$$\Sigma F_x = T - mg \sin 20° = 0,$$
$$\Sigma F_y = N - mg \cos 20° = 0.$$

Solving for T yields

$$T = mg \sin 20°$$
$$= (1440 \text{ kg})(9.81 \text{ m/s}^2) \sin 20°$$
$$= 4830 \text{ N}.$$

Active Example 3.5

Draw the Free-Body Diagram and Apply Equilibrium

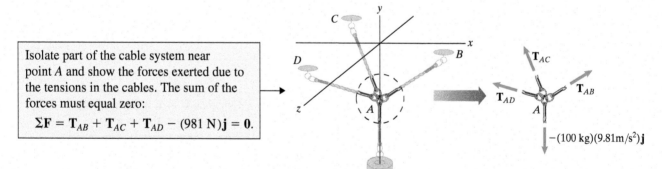

Isolate part of the cable system near point A and show the forces exerted due to the tensions in the cables. The sum of the forces must equal zero:

$$\Sigma \mathbf{F} = \mathbf{T}_{AB} + \mathbf{T}_{AC} + \mathbf{T}_{AD} - (981 \text{ N})\mathbf{j} = \mathbf{0}.$$

Write the Forces in Terms of Their Components

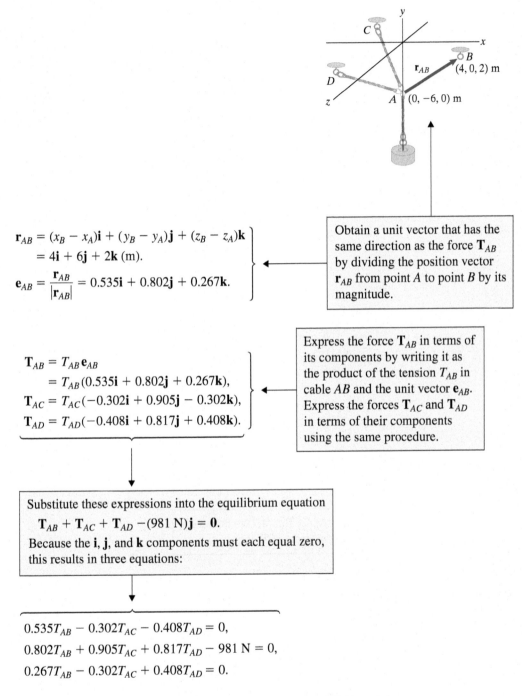

$$\mathbf{r}_{AB} = (x_B - x_A)\mathbf{i} + (y_B - y_A)\mathbf{j} + (z_B - z_A)\mathbf{k}$$
$$= 4\mathbf{i} + 6\mathbf{j} + 2\mathbf{k} \text{ (m)}.$$

$$\mathbf{e}_{AB} = \frac{\mathbf{r}_{AB}}{|\mathbf{r}_{AB}|} = 0.535\mathbf{i} + 0.802\mathbf{j} + 0.267\mathbf{k}.$$

> Obtain a unit vector that has the same direction as the force $\mathbf{T}_{AB}$ by dividing the position vector $\mathbf{r}_{AB}$ from point A to point B by its magnitude.

$$\mathbf{T}_{AB} = T_{AB}\,\mathbf{e}_{AB}$$
$$= T_{AB}(0.535\mathbf{i} + 0.802\mathbf{j} + 0.267\mathbf{k}),$$
$$\mathbf{T}_{AC} = T_{AC}(-0.302\mathbf{i} + 0.905\mathbf{j} - 0.302\mathbf{k}),$$
$$\mathbf{T}_{AD} = T_{AD}(-0.408\mathbf{i} + 0.817\mathbf{j} + 0.408\mathbf{k}).$$

> Express the force $\mathbf{T}_{AB}$ in terms of its components by writing it as the product of the tension T_{AB} in cable AB and the unit vector $\mathbf{e}_{AB}$. Express the forces $\mathbf{T}_{AC}$ and $\mathbf{T}_{AD}$ in terms of their components using the same procedure.

> Substitute these expressions into the equilibrium equation
> $$\mathbf{T}_{AB} + \mathbf{T}_{AC} + \mathbf{T}_{AD} - (981 \text{ N})\mathbf{j} = 0.$$
> Because the $\mathbf{i}$, $\mathbf{j}$, and $\mathbf{k}$ components must each equal zero, this results in three equations:

$$0.535T_{AB} - 0.302T_{AC} - 0.408T_{AD} = 0,$$
$$0.802T_{AB} + 0.905T_{AC} + 0.817T_{AD} - 981 \text{ N} = 0,$$
$$0.267T_{AB} - 0.302T_{AC} + 0.408T_{AD} = 0.$$

Solving these three equations yields $T_{AB} = 432$ N, $T_{AC} = 574$ N, and $T_{AD} = 141$ N.

Active Example 4.1

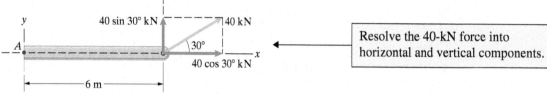

The magnitude of the moment of the horizontal component about A is zero. The magnitude of the moment of the vertical component is $(6\text{ m})(40\sin 30° \text{ N}) = 120$ kN-m. Its direction is counterclockwise, so the sum of the moments is $M_A = 120$ kN-m.

Resolve the 40-kN force into horizontal and vertical components.

Calculate the sum of the moments of the components about A.

Active Example 4.4

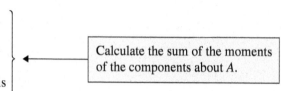

$$\mathbf{r}_{AC} = (x_C - x_A)\mathbf{i} + (y_C - y_A)\mathbf{j} + (z_C - z_A)\mathbf{k}$$
$$= 7\mathbf{i} + \mathbf{j} - 5\mathbf{k} \text{ (ft).}$$
$$\mathbf{M}_A = \mathbf{r}_{AC} \times \mathbf{F}$$

$$= \begin{vmatrix} \mathbf{i} & \mathbf{j} & \mathbf{k} \\ 7 & 1 & -5 \\ -40 & 70 & -40 \end{vmatrix}$$

$$= 310\mathbf{i} + 480\mathbf{j} + 530\mathbf{k} \text{ (ft-lb).}$$

(a) Apply Eq. (4.2) to determine the moment of $\mathbf{F}$ about point A.

$$D = \frac{|\mathbf{M}_A|}{|\mathbf{F}|}$$

$$= \frac{\sqrt{(310)^2 + (480)^2 + (530)^2} \text{ ft-lb}}{90 \text{ lb}}$$

$$= 8.66 \text{ ft.}$$

(b) Use the relation $|\mathbf{M}_A| = D|\mathbf{F}|$, where D is the perpendicular distance from A to the line of action of $\mathbf{F}$.

Active Example 4.6

$$\mathbf{r} = (x_A - x_C)\mathbf{i} + (y_A - y_C)\mathbf{j} + (z_A - z_C)\mathbf{k}$$
$$= 4\mathbf{i} - 2\mathbf{j} + 2\mathbf{k} \text{ (m)}.$$

Determine the components of the vector from point C to the point of application of $\mathbf{F}$.

$$\mathbf{M}_C = \mathbf{r} \times \mathbf{F}$$

$$= \begin{vmatrix} \mathbf{i} & \mathbf{j} & \mathbf{k} \\ 4 & -2 & 2 \\ -2 & 6 & 3 \end{vmatrix}$$

$$= -18\mathbf{i} - 16\mathbf{j} + 20\mathbf{k} \text{ (kN-m)}.$$

Calculate the moment of $\mathbf{F}$ about point C.

$$\mathbf{M}_{BC} = (\mathbf{e}_{BC} \cdot \mathbf{M}_C)\mathbf{e}_{BC}$$
$$= [(0)(-18) + (0.8)(-16) + (-0.6)(20)]\mathbf{e}_{BC}$$
$$= -24.8\mathbf{e}_{BC} \text{ (kN-m)}.$$

Apply Eq. (4.4) to determine the moment of $\mathbf{F}$ about the axis BC. Although the moment of $\mathbf{F}$ about point C is not the same as the moment of $\mathbf{F}$ about point B, their components parallel to the axis BC are the same.

Active Example 4.9

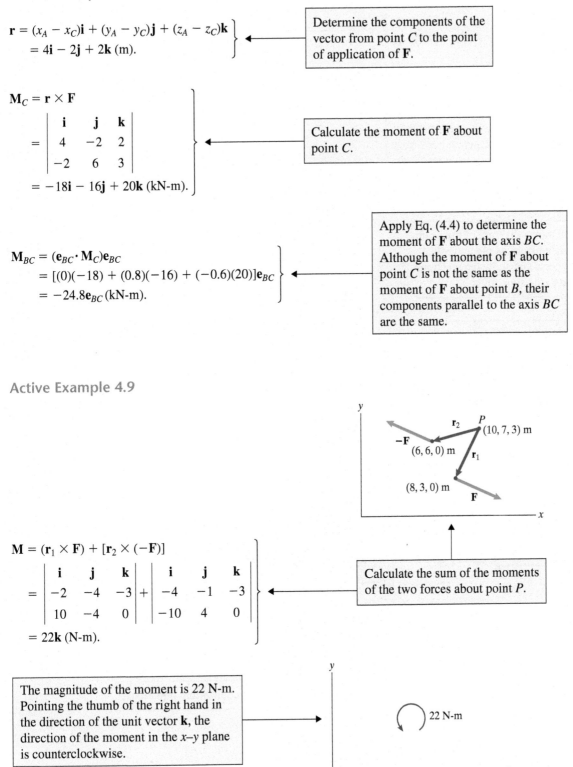

$$\mathbf{M} = (\mathbf{r}_1 \times \mathbf{F}) + [\mathbf{r}_2 \times (-\mathbf{F})]$$

$$= \begin{vmatrix} \mathbf{i} & \mathbf{j} & \mathbf{k} \\ -2 & -4 & -3 \\ 10 & -4 & 0 \end{vmatrix} + \begin{vmatrix} \mathbf{i} & \mathbf{j} & \mathbf{k} \\ -4 & -1 & -3 \\ -10 & 4 & 0 \end{vmatrix}$$

$$= 22\mathbf{k} \text{ (N-m)}.$$

Calculate the sum of the moments of the two forces about point P.

The magnitude of the moment is 22 N-m. Pointing the thumb of the right hand in the direction of the unit vector $\mathbf{k}$, the direction of the moment in the x–y plane is counterclockwise.

22 N-m

Active Example 4.12

$$F' = F$$
$$= 20\mathbf{i} + 15\mathbf{j} - 5\mathbf{k} \text{ (kN)}.$$

> The force $\mathbf{F}'$ must equal the sum of the forces in system 2.

$$M' = \begin{vmatrix} \mathbf{i} & \mathbf{j} & \mathbf{k} \\ 4 & 3 & -2 \\ 20 & 15 & -5 \end{vmatrix} + (-105\mathbf{i} + 110\mathbf{j} + 90\mathbf{k})$$
$$= -90\mathbf{i} + 90\mathbf{j} + 90\mathbf{k} \text{ (kN-m)}.$$

> The couple $\mathbf{M}'$ must equal the sum of the moments about the origin due to the forces and moments in system 2.

Active Example 5.1

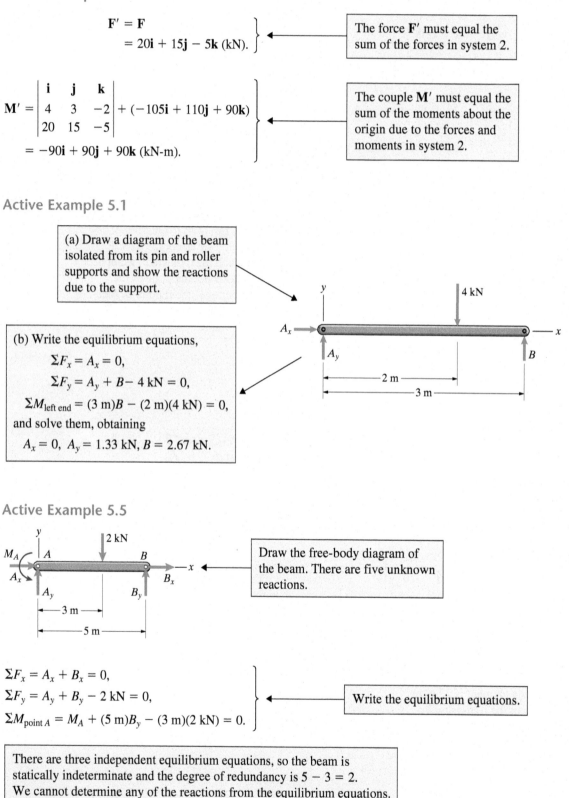

> (a) Draw a diagram of the beam isolated from its pin and roller supports and show the reactions due to the support.

> (b) Write the equilibrium equations,
> $$\Sigma F_x = A_x = 0,$$
> $$\Sigma F_y = A_y + B - 4 \text{ kN} = 0,$$
> $$\Sigma M_{\text{left end}} = (3 \text{ m})B - (2 \text{ m})(4 \text{ kN}) = 0,$$
> and solve them, obtaining
> $$A_x = 0, \ A_y = 1.33 \text{ kN}, B = 2.67 \text{ kN}.$$

Active Example 5.5

> Draw the free-body diagram of the beam. There are five unknown reactions.

$$\Sigma F_x = A_x + B_x = 0,$$
$$\Sigma F_y = A_y + B_y - 2 \text{ kN} = 0,$$
$$\Sigma M_{\text{point } A} = M_A + (5 \text{ m})B_y - (3 \text{ m})(2 \text{ kN}) = 0.$$

> Write the equilibrium equations.

> There are three independent equilibrium equations, so the beam is statically indeterminate and the degree of redundancy is $5 - 3 = 2$. We cannot determine any of the reactions from the equilibrium equations.

Active Example 5.7

Draw the Free-Body Diagram of the Bar

Isolate the bar and show the reactions exerted by the cables and the ball and socket support.

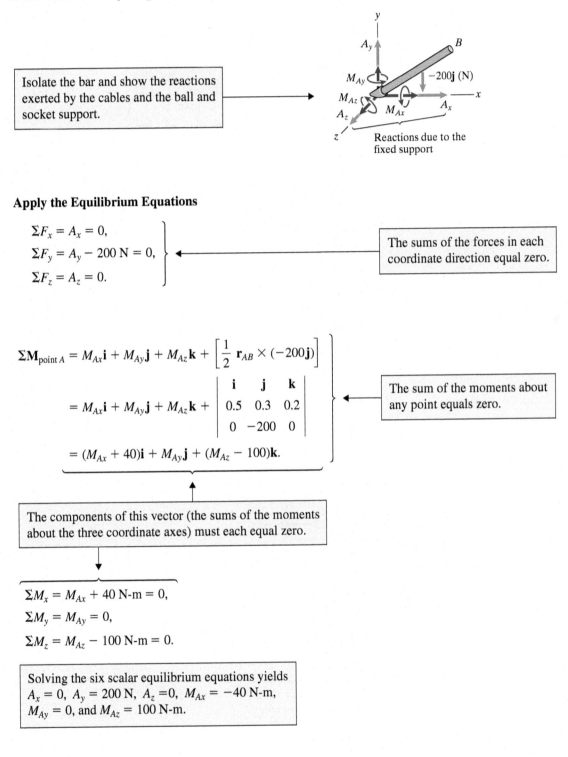

Reactions due to the fixed support

Apply the Equilibrium Equations

$$\Sigma F_x = A_x = 0,$$
$$\Sigma F_y = A_y - 200\ \text{N} = 0,$$
$$\Sigma F_z = A_z = 0.$$

The sums of the forces in each coordinate direction equal zero.

$$\Sigma \mathbf{M}_{\text{point }A} = M_{Ax}\mathbf{i} + M_{Ay}\mathbf{j} + M_{Az}\mathbf{k} + \left[\frac{1}{2}\ \mathbf{r}_{AB} \times (-200\mathbf{j}) \right]$$

$$= M_{Ax}\mathbf{i} + M_{Ay}\mathbf{j} + M_{Az}\mathbf{k} + \begin{vmatrix} \mathbf{i} & \mathbf{j} & \mathbf{k} \\ 0.5 & 0.3 & 0.2 \\ 0 & -200 & 0 \end{vmatrix}$$

$$= (M_{Ax} + 40)\mathbf{i} + M_{Ay}\mathbf{j} + (M_{Az} - 100)\mathbf{k}.$$

The sum of the moments about any point equals zero.

The components of this vector (the sums of the moments about the three coordinate axes) must each equal zero.

$$\Sigma M_x = M_{Ax} + 40\ \text{N-m} = 0,$$
$$\Sigma M_y = M_{Ay} = 0,$$
$$\Sigma M_z = M_{Az} - 100\ \text{N-m} = 0.$$

Solving the six scalar equilibrium equations yields $A_x = 0$, $A_y = 200$ N, $A_z = 0$, $M_{Ax} = -40$ N-m, $M_{Ay} = 0$, and $M_{Az} = 100$ N-m.

Active Example 5.10

The force exerted on the plate by the bar AB must be directed along the line between A and B, and the line of action of the weight of the plate is vertical, so the three forces on the plate are not parallel. Therefore they must be concurrent.

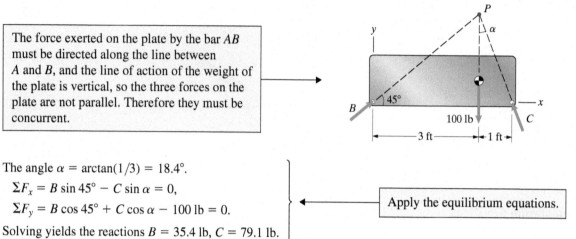

The angle $\alpha = \arctan(1/3) = 18.4°$.

$\Sigma F_x = B \sin 45° - C \sin \alpha = 0$,

$\Sigma F_y = B \cos 45° + C \cos \alpha - 100 \text{ lb} = 0$.

Solving yields the reactions $B = 35.4$ lb, $C = 79.1$ lb.

Apply the equilibrium equations.

Active Example 6.1

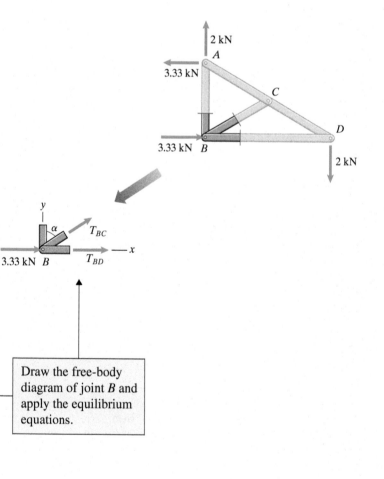

The angle $\alpha = \arctan(5/3) = 59.0°$.

$\Sigma F_x = T_{BC} \sin \alpha + T_{BD} + 3.33 \text{ kN} = 0$,

$\Sigma F_y = T_{BC} \cos \alpha = 0$.

Solving yields $T_{BC} = 0$ and $T_{BD} = -3.33$ kN. The axial force in member BC is zero and the axial force in member BD is 3.33 kN in compression, or

   BC: zero, BD: 3.33 kN (C).

(Notice that joint C is one of the "special joints" we discussed. We could have determined by observation that $T_{BC} = 0$.)

Draw the free-body diagram of joint B and apply the equilibrium equations.

Active Example 6.3

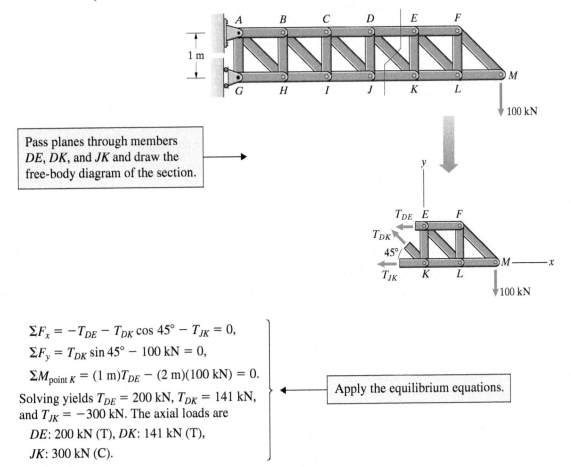

Pass planes through members *DE*, *DK*, and *JK* and draw the free-body diagram of the section.

Apply the equilibrium equations.

$\Sigma F_x = -T_{DE} - T_{DK} \cos 45° - T_{JK} = 0,$

$\Sigma F_y = T_{DK} \sin 45° - 100 \text{ kN} = 0,$

$\Sigma M_{\text{point } K} = (1 \text{ m})T_{DE} - (2 \text{ m})(100 \text{ kN}) = 0.$

Solving yields $T_{DE} = 200$ kN, $T_{DK} = 141$ kN, and $T_{JK} = -300$ kN. The axial loads are

DE: 200 kN (T), *DK*: 141 kN (T),

JK: 300 kN (C).

Active Example 6.5

We can determine the axial forces in members AB and AC by analyzing joint A.

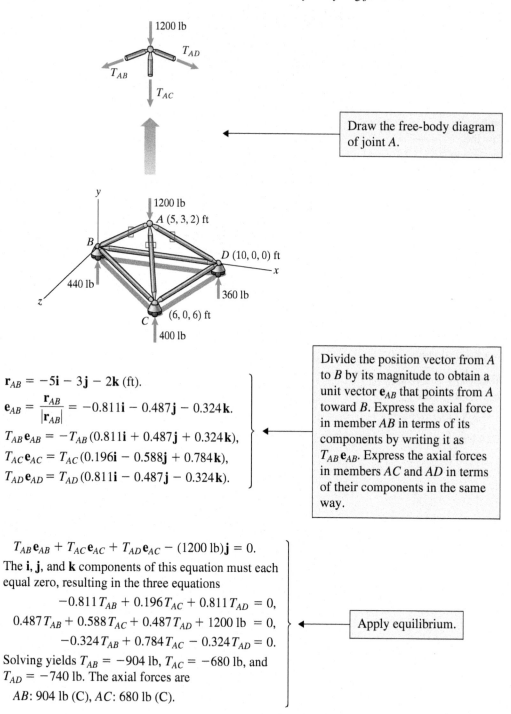

Draw the free-body diagram of joint A.

Divide the position vector from A to B by its magnitude to obtain a unit vector $\mathbf{e}_{AB}$ that points from A toward B. Express the axial force in member AB in terms of its components by writing it as $T_{AB}\mathbf{e}_{AB}$. Express the axial forces in members AC and AD in terms of their components in the same way.

$$\mathbf{r}_{AB} = -5\mathbf{i} - 3\mathbf{j} - 2\mathbf{k} \ (\text{ft}).$$

$$\mathbf{e}_{AB} = \frac{\mathbf{r}_{AB}}{|\mathbf{r}_{AB}|} = -0.811\mathbf{i} - 0.487\mathbf{j} - 0.324\mathbf{k}.$$

$$T_{AB}\mathbf{e}_{AB} = -T_{AB}(0.811\mathbf{i} + 0.487\mathbf{j} + 0.324\mathbf{k}),$$

$$T_{AC}\mathbf{e}_{AC} = T_{AC}(0.196\mathbf{i} - 0.588\mathbf{j} + 0.784\mathbf{k}),$$

$$T_{AD}\mathbf{e}_{AD} = T_{AD}(0.811\mathbf{i} - 0.487\mathbf{j} - 0.324\mathbf{k}).$$

Apply equilibrium.

$$T_{AB}\mathbf{e}_{AB} + T_{AC}\mathbf{e}_{AC} + T_{AD}\mathbf{e}_{AC} - (1200 \text{ lb})\mathbf{j} = 0.$$

The $\mathbf{i}$, $\mathbf{j}$, and $\mathbf{k}$ components of this equation must each equal zero, resulting in the three equations

$$-0.811\,T_{AB} + 0.196\,T_{AC} + 0.811\,T_{AD} = 0,$$

$$0.487\,T_{AB} + 0.588\,T_{AC} + 0.487\,T_{AD} + 1200 \text{ lb} = 0,$$

$$-0.324\,T_{AB} + 0.784\,T_{AC} - 0.324\,T_{AD} = 0.$$

Solving yields $T_{AB} = -904$ lb, $T_{AC} = -680$ lb, and $T_{AD} = -740$ lb. The axial forces are

AB: 904 lb (C), AC: 680 lb (C).

Active Example 6.6

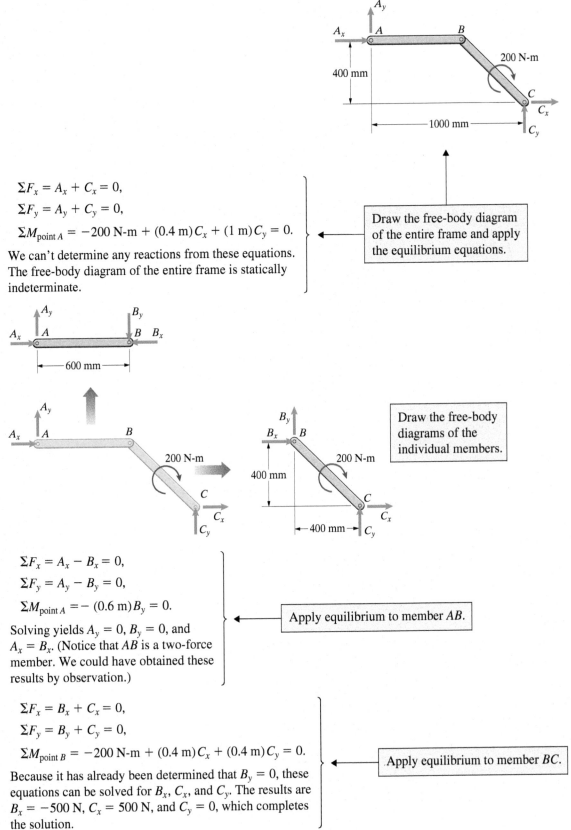

$\Sigma F_x = A_x + C_x = 0$,

$\Sigma F_y = A_y + C_y = 0$,

$\Sigma M_{\text{point } A} = -200 \text{ N-m} + (0.4 \text{ m})C_x + (1 \text{ m})C_y = 0$.

We can't determine any reactions from these equations. The free-body diagram of the entire frame is statically indeterminate.

Draw the free-body diagram of the entire frame and apply the equilibrium equations.

Draw the free-body diagrams of the individual members.

$\Sigma F_x = A_x - B_x = 0$,

$\Sigma F_y = A_y - B_y = 0$,

$\Sigma M_{\text{point } A} = -(0.6 \text{ m})B_y = 0$.

Solving yields $A_y = 0$, $B_y = 0$, and $A_x = B_x$. (Notice that AB is a two-force member. We could have obtained these results by observation.)

Apply equilibrium to member AB.

$\Sigma F_x = B_x + C_x = 0$,

$\Sigma F_y = B_y + C_y = 0$,

$\Sigma M_{\text{point } B} = -200 \text{ N-m} + (0.4 \text{ m})C_x + (0.4 \text{ m})C_y = 0$.

Because it has already been determined that $B_y = 0$, these equations can be solved for B_x, C_x, and C_y. The results are $B_x = -500$ N, $C_x = 500$ N, and $C_y = 0$, which completes the solution.

Apply equilibrium to member BC.

Active Example 7.1

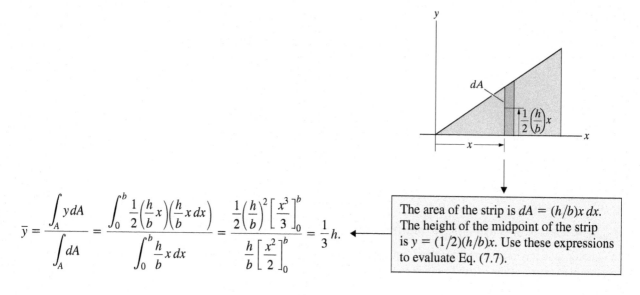

$$\bar{y} = \frac{\int_A y\,dA}{\int_A dA} = \frac{\int_0^b \frac{1}{2}\left(\frac{h}{b}x\right)\left(\frac{h}{b}x\,dx\right)}{\int_0^b \frac{h}{b}x\,dx} = \frac{\frac{1}{2}\left(\frac{h}{b}\right)^2\left[\frac{x^3}{3}\right]_0^b}{\frac{h}{b}\left[\frac{x^2}{2}\right]_0^b} = \frac{1}{3}h.$$

The area of the strip is $dA = (h/b)x\,dx$. The height of the midpoint of the strip is $y = (1/2)(h/b)x$. Use these expressions to evaluate Eq. (7.7).

Active Example 7.3

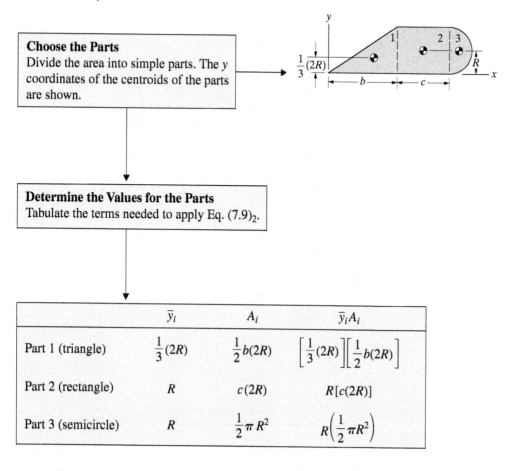

Choose the Parts
Divide the area into simple parts. The y coordinates of the centroids of the parts are shown.

Determine the Values for the Parts
Tabulate the terms needed to apply Eq. (7.9)$_2$.

	$\bar{y}_i$	A_i	$\bar{y}_i A_i$
Part 1 (triangle)	$\frac{1}{3}(2R)$	$\frac{1}{2}b(2R)$	$\left[\frac{1}{3}(2R)\right]\left[\frac{1}{2}b(2R)\right]$
Part 2 (rectangle)	R	$c(2R)$	$R[c(2R)]$
Part 3 (semicircle)	R	$\frac{1}{2}\pi R^2$	$R\left(\frac{1}{2}\pi R^2\right)$

$$\bar{y} = \frac{\bar{y}_1 A_1 + \bar{y}_2 A_2 + \bar{y}_3 A_3}{A_1 + A_2 + A_3}$$

$$= \frac{\left[\frac{1}{3}(2R)\right]\left[\frac{1}{2}b(2R)\right] + R[c(2R)] + R\left(\frac{1}{2}\pi R^2\right)}{\frac{1}{2}b(2R) + c(2R) + \frac{1}{2}\pi R^2}.$$

Calculate the Centroid
Use Eq. $(7.9)_2$ to determine the y component of the centroid.

Active Example 7.5

(a)

$$w = ax + b.$$

Write w as an arbitrary *linear* function of x.

$$0 = a(0) + b,$$
$$100 \text{ N/m} = a(12 \text{ m}) + b.$$

Solving yields $a = (100/12) \text{ N/m}^2$ and $b = 0$. Therefore

$$w = \frac{100}{12} x \text{ N/m}.$$

Use the known values of w at $x = 0$ and at $x = 12$ m to determine the constants a and b.

(b)

$$F = \int_L w\, dx$$
$$= \int_0^{12} \frac{100}{12} x\, dx$$
$$= 600 \text{ N}.$$

Apply Eq. (7.10) to determine the downward force exerted by the distributed load.

$$M = \int_L xw\, dx$$
$$= \int_0^{12} \frac{100}{12} x^2\, dx$$
$$= 4800 \text{ N-m}.$$

Apply Eq. (7.11) to determine the clockwise moment about the origin exerted by the distributed load.

Active Example 7.8

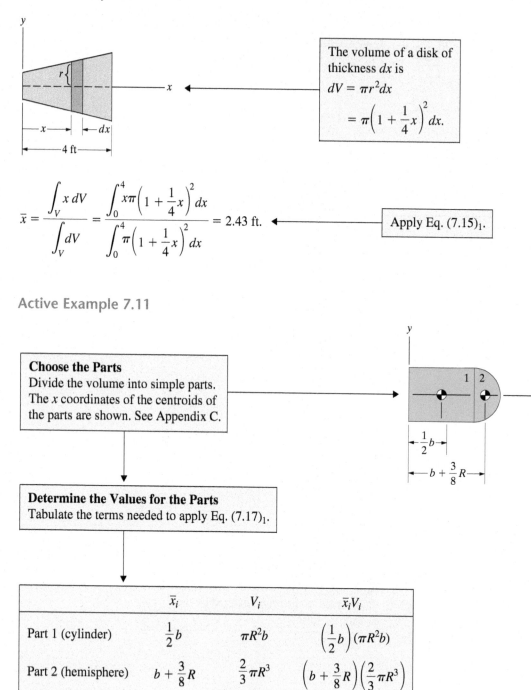

The volume of a disk of thickness dx is
$$dV = \pi r^2 dx$$
$$= \pi\left(1 + \frac{1}{4}x\right)^2 dx.$$

$$\bar{x} = \frac{\int_V x\, dV}{\int_V dV} = \frac{\int_0^4 x\pi\left(1 + \frac{1}{4}x\right)^2 dx}{\int_0^4 \pi\left(1 + \frac{1}{4}x\right)^2 dx} = 2.43 \text{ ft.}$$

Apply Eq. (7.15)$_1$.

Active Example 7.11

Choose the Parts
Divide the volume into simple parts. The x coordinates of the centroids of the parts are shown. See Appendix C.

Determine the Values for the Parts
Tabulate the terms needed to apply Eq. (7.17)$_1$.

	$\bar{x}_i$	V_i	$\bar{x}_i V_i$
Part 1 (cylinder)	$\frac{1}{2}b$	$\pi R^2 b$	$\left(\frac{1}{2}b\right)(\pi R^2 b)$
Part 2 (hemisphere)	$b + \frac{3}{8}R$	$\frac{2}{3}\pi R^3$	$\left(b + \frac{3}{8}R\right)\left(\frac{2}{3}\pi R^3\right)$

$$\bar{x} = \frac{\bar{x}_1 V_1 + \bar{x}_2 V_2}{V_1 + V_2}$$

$$= \frac{\left(\frac{1}{2}b\right)\left(\pi R^2 b\right) + \left(b + \frac{3}{8}R\right)\left(\frac{2}{3}\pi R^3\right)}{\pi R^2 b + \frac{2}{3}\pi R^3}.$$

Calculate the Centroid
Use Eq. $(7.17)_1$ to determine the x component of the centroid.

Active Example 7.14

Revolving this triangle area about the x axis generates the volume of the cone. The y coordinate of the centroid of the area is shown. The area of the triangle is $A = \frac{1}{2}hR$. The volume of the cone is

$$V = 2\pi \bar{y}_T A = \frac{1}{3}\pi h R^2.$$

$$\bar{y}_T = \frac{1}{3}R$$

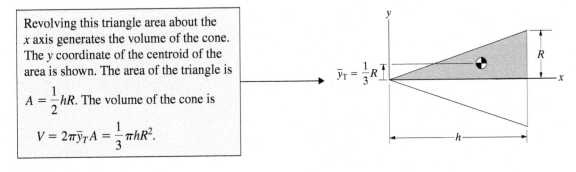

Active Example 7.16

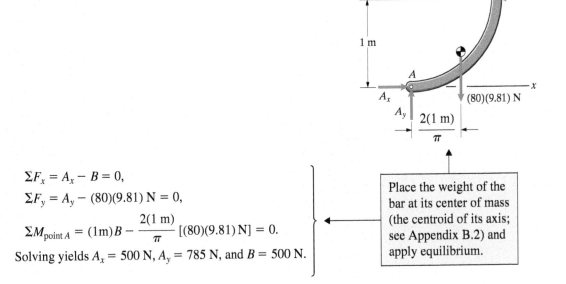

$$\Sigma F_x = A_x - B = 0,$$

$$\Sigma F_y = A_y - (80)(9.81)\text{ N} = 0,$$

$$\Sigma M_{\text{point }A} = (1\text{m})B - \frac{2(1\text{ m})}{\pi}\left[(80)(9.81)\text{ N}\right] = 0.$$

Solving yields $A_x = 500$ N, $A_y = 785$ N, and $B = 500$ N.

Place the weight of the bar at its center of mass (the centroid of its axis; see Appendix B.2) and apply equilibrium.

Active Example 7.18

The center of mass coincides with the centroid of the volume of the bar, so

$$\bar{y}_1 = \frac{1}{2}(240 \text{ mm}) = 120 \text{ mm}.$$

→ Center of mass of bar 1.

The y coordinate of the centroid of the volume is

$$\bar{y}_2 = \frac{1}{2}(80 \text{ mm}) = 40 \text{ mm}.$$

→ Center of mass of bar 2.

$$\bar{y} = \frac{\bar{y}_1 m_1 + \bar{y}_2 m_2}{m_1 + m_2}$$

$$= \frac{(120 \text{ mm})(10.8 \text{ kg}) + (40 \text{ mm})(5.99 \text{ kg})}{10.8 \text{ kg} + 5.99 \text{ kg}}$$

$$= 91.4 \text{ mm}.$$

→ Apply Eq. (7.27)₂.

Active Example 8.1

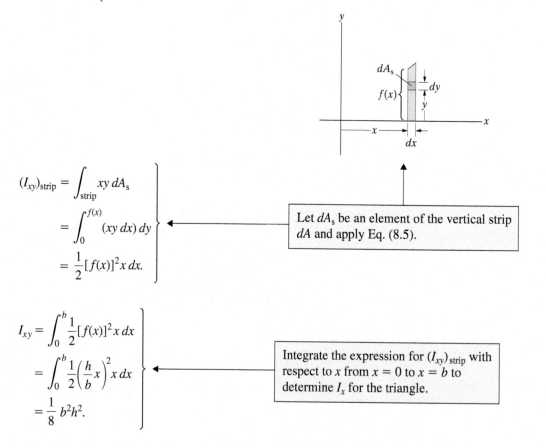

$$(I_{xy})_{\text{strip}} = \int_{\text{strip}} xy \, dA_s$$

$$= \int_0^{f(x)} (xy \, dx) \, dy$$

$$= \frac{1}{2}[f(x)]^2 x \, dx.$$

Let dA_s be an element of the vertical strip dA and apply Eq. (8.5).

$$I_{xy} = \int_0^b \frac{1}{2}[f(x)]^2 x \, dx$$

$$= \int_0^b \frac{1}{2}\left(\frac{h}{b}x\right)^2 x \, dx$$

$$= \frac{1}{8} b^2 h^2.$$

Integrate the expression for $(I_{xy})_{\text{strip}}$ with respect to x from $x = 0$ to $x = b$ to determine I_x for the triangle.

Active Example 8.3

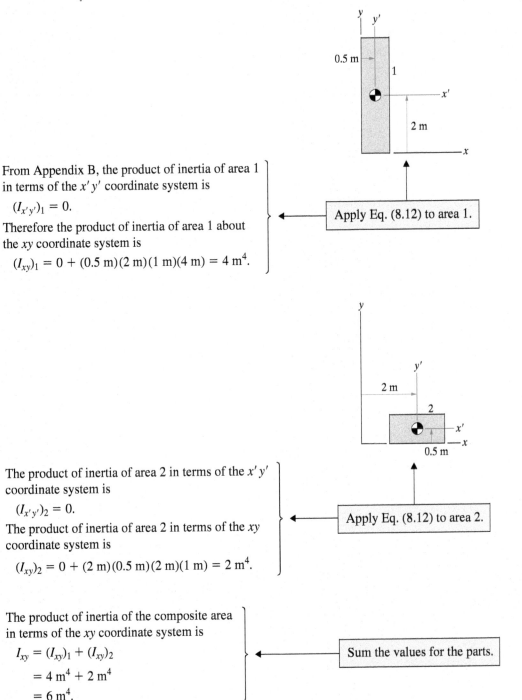

From Appendix B, the product of inertia of area 1 in terms of the $x'y'$ coordinate system is

$(I_{x'y'})_1 = 0.$

Therefore the product of inertia of area 1 about the xy coordinate system is

$(I_{xy})_1 = 0 + (0.5\ \text{m})(2\ \text{m})(1\ \text{m})(4\ \text{m}) = 4\ \text{m}^4.$

Apply Eq. (8.12) to area 1.

The product of inertia of area 2 in terms of the $x'y'$ coordinate system is

$(I_{x'y'})_2 = 0.$

The product of inertia of area 2 in terms of the xy coordinate system is

$(I_{xy})_2 = 0 + (2\ \text{m})(0.5\ \text{m})(2\ \text{m})(1\ \text{m}) = 2\ \text{m}^4.$

Apply Eq. (8.12) to area 2.

The product of inertia of the composite area in terms of the xy coordinate system is

$I_{xy} = (I_{xy})_1 + (I_{xy})_2$

$= 4\ \text{m}^4 + 2\ \text{m}^4$

$= 6\ \text{m}^4.$

Sum the values for the parts.

Active Example 8.6

$$\tan 2\theta_p = \frac{2I_{xy}}{I_y - I_x} = \frac{2(6)}{16 - 9} = 1.71.$$

This yields $\theta_p = 29.9°$.

Determine θ_p from Eq. (8.26).

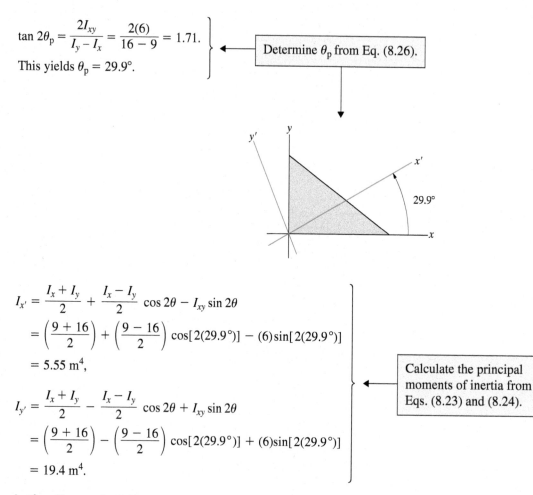

$$I_{x'} = \frac{I_x + I_y}{2} + \frac{I_x - I_y}{2} \cos 2\theta - I_{xy} \sin 2\theta$$

$$= \left(\frac{9 + 16}{2}\right) + \left(\frac{9 - 16}{2}\right) \cos[2(29.9°)] - (6)\sin[2(29.9°)]$$

$$= 5.55 \text{ m}^4,$$

$$I_{y'} = \frac{I_x + I_y}{2} - \frac{I_x - I_y}{2} \cos 2\theta + I_{xy} \sin 2\theta$$

$$= \left(\frac{9 + 16}{2}\right) - \left(\frac{9 - 16}{2}\right) \cos[2(29.9°)] + (6)\sin[2(29.9°)]$$

$$= 19.4 \text{ m}^4.$$

Calculate the principal moments of inertia from Eqs. (8.23) and (8.24).

Active Example 8.8

Place the point $1'$ at one of the points where the Mohr's circle intersects the horizontal axis. The principal moments of inertia are $I_{x'} = 7.5 \text{ ft}^4$, $I_{y'} = 24.5 \text{ ft}^4$. The angle measured counterclockwise from point 1 to point $1'$ is $2\theta_p = 135°$, so $\theta_p = 67.5°$.

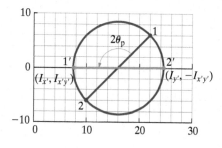

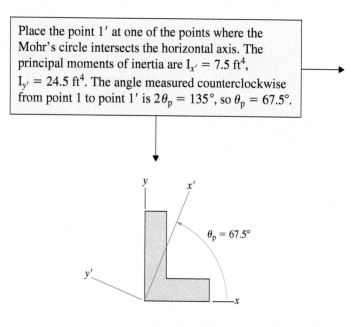

Active Example 8.9

From Appendix B,

$$I_y = \frac{1}{4} hb^3.$$

> Determine the moment of inertia of the area of the plate about the y axis.

The moment of inertia of the plate about the y axis is

$$I_{y\,\text{axis}} = \frac{m}{A} I_y$$

$$= \frac{m}{\frac{1}{2} bh} \left(\frac{1}{4} hb^3 \right)$$

$$= \frac{1}{2} mb^2.$$

> Apply Eq. (8.30).

Active Example 8.11

> Treat the object as a composite object made up of the bars 1 and 2. The distance between the axis L_O and parallel axes through the centers of mass of the two bars are shown.

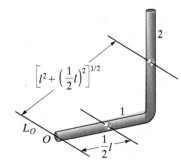

$$(I_O)_1 = I + d^2 m$$

$$= \frac{1}{12} ml^2 + \left(\frac{1}{2} l \right)^2 m$$

$$= \frac{1}{3} ml^2.$$

> Apply the parallel-axis theorem to bar 1.

$$(I_O)_2 = I + d^2 m$$

$$= \frac{1}{12} ml^2 + \left[l^2 + \left(\frac{1}{2} l \right)^2 \right] m$$

$$= \frac{4}{3} ml^2.$$

> Apply the parallel-axis theorem to bar 2.

$$I_O = (I_O)_1 + (I_O)_2$$

$$= \frac{1}{3} ml^2 + \frac{4}{3} ml^2$$

$$= \frac{5}{3} ml^2.$$

> Sum the results.

Active Example 9.1

Draw the free-body diagram of the crate. It is assumed that slip of the crate *up* the ramp is impending, so the direction of the friction force on the crate is *down* the ramp and its magnitude is $\mu_s N$.

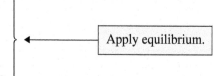

$$\Sigma F_x = T - N \sin 20° - \mu_s N \cos 20° = 0,$$
$$\Sigma F_y = N \cos 20° - \mu_s N \sin 20° - W = 0.$$

Solving these equations yields $T = 161$ lb.

Apply equilibrium.

Active Example 9.4

Draw the free-body diagram of the wedge assuming that $F = 0$ and that slip of the wedge out of the log is impending.

The sum of the forces in the vertical direction is

$$2N \sin\left(\frac{\alpha}{2}\right) - 2\mu_s N \cos\left(\frac{\alpha}{2}\right) = 0.$$

The wedge is in equilibrium if

$$\mu_s = \tan\left(\frac{\alpha}{2}\right) = \tan\left(\frac{10°}{2}\right) = 0.0875.$$

This is the *minimum* static coefficient of friction necessary for the wedge to remain in place in the log, so it will not slip out.

Apply equilibrium.

Active Example 9.5

The force $F = 200$ lb, the slope of the thread is $\alpha = 1.14°$, and the angle of friction is

$$\theta_k = \arctan \mu_k = \arctan (0.22) = 12.4°.$$

Substituting these values into Eq. (9.11),

$$M = rF \tan(\theta_k - \alpha)$$
$$= (1.6 \text{ in})(200 \text{ lb}) \tan(12.4° - 1.14°)$$
$$= 63.8 \text{ in-lb}.$$

Apply Eq. (9.11).

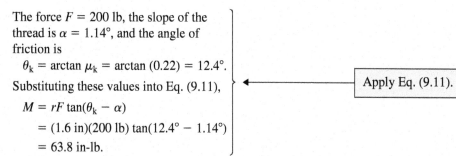

Active Example 9.6

$M = (6 \text{ in})(W - T)$.

> The pulley moves in the counterclockwise direction. Express the counterclockwise couple on the pulley in terms of T and W.

The angle of kinetic friction is

$\theta_k = \arctan \mu_k = \arctan(0.2) = 11.3°$.

Equation (9.12) is

$M = rF \sin \theta_k$:

$(6 \text{ in})(W - T) = (0.5 \text{ in})\sqrt{(W + T \sin 45°)^2 + (T \cos 45°)^2} \sin 11.3°$.

Setting $W = 1000$ lb and solving yields

$T = 970$ lb.

> Apply Eq. (9.12).

Active Example 9.7

The radii $r_o = 1.75$ in and $r_i = 0.5$ in.

$\alpha = \arctan[b/(r_o - r_i)] = \arctan[5/(1.75 - 0.5)] = 76.0°$.

> Determine the angle α.

$M = \dfrac{2\mu_k F}{3 \cos \alpha} \dfrac{r_o^3 - r_i^3}{r_o^2 - r_i^2}$

$= \dfrac{2(0.18)(200 \text{ lb})}{3 \cos 76.0°}\left[\dfrac{(1.75 \text{ in})^3 - (0.5 \text{ in})^3}{(1.75 \text{ in})^2 - (0.5 \text{ in})^2}\right]$

$= 184$ in-lb.

> Apply Eq. (9.13).

Active Example 9.9

$T = We^{\mu_s \beta} = (100 \text{ lb})e^{(0.2)(\pi/2)} = 137$ lb.

> Apply Eq. (9.17) to the left cylinder. Assume that slip of the rope in the direction of the force T is impending.

$F = Te^{(0.4)(\pi/2)} = (137 \text{ lb})e^{(0.4)(\pi/2)} = 257$ lb.

> Apply Eq. (9.17) to the right cylinder. Assume that slip of the rope in the direction of the force F is impending.

Active Example 10.1

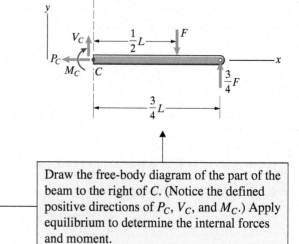

$$\Sigma F_x = -P_C = 0,$$

$$\Sigma F_y = V_C - F + \frac{3}{4}F = 0,$$

$$\Sigma M_{\text{point }C} = -M_C - \left(\frac{1}{2}L\right)F + \left(\frac{3}{4}L\right)\left(\frac{3}{4}F\right) = 0.$$

Solving yields $P_C = 0$, $V_C = \frac{1}{4}F$, and $M_C = \frac{1}{16}LF$.

Draw the free-body diagram of the part of the beam to the right of C. (Notice the defined positive directions of P_C, V_C, and M_C.) Apply equilibrium to determine the internal forces and moment.

Active Example 10.3

(a) Pass a plane through the beam at an arbitrary position x between B and C. The simplest free-body diagram is obtained by isolating the part of the beam to the right of the plane.

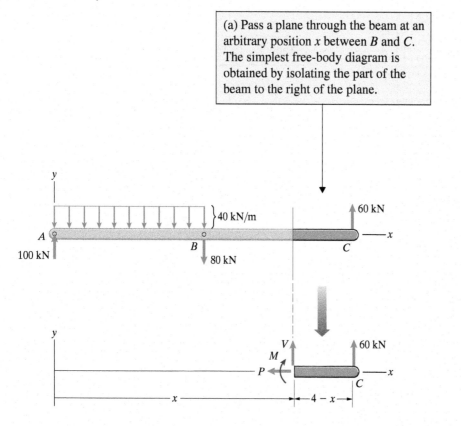

$$\Sigma F_x = -P = 0,$$

$$\Sigma F_y = V + 60 = 0,$$

$$\Sigma M_{\text{left end}} = -M + 60(4 - x) = 0.$$

Apply equilibrium to determine V and M.

Solving yields

$$\left.\begin{array}{l} V = -60 \text{ kN} \\ M = 60(4 - x) \text{ kN-m} \end{array}\right\} \ 2 < x < 4 \text{ m.}$$

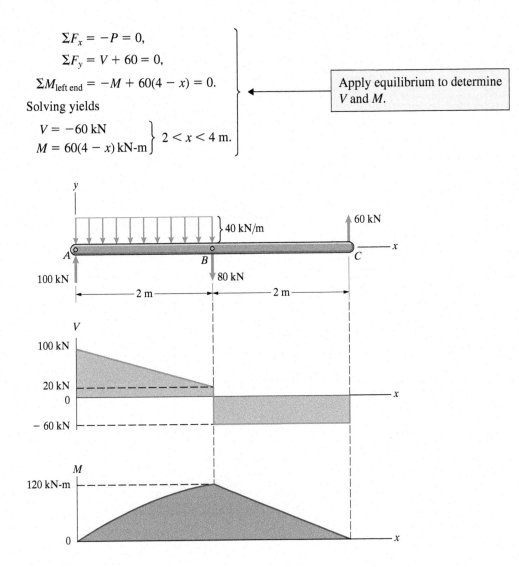

Active Example 10.4

No couple is applied to the beam at A, so the bending moment at A is zero. The shear force between A and B is $V = 100 - 40x$ kN. With this expression, Eq. (10.6) can be integrated to determine the bending moment between A and B:

$$\int_0^M dM = \int_0^x (100 - 40x)\,dx :$$

$$M = 100x - 20x^2 \text{ kN-m.}$$

The value of M at B is

$$100(2) - 20(2)^2 = 120 \text{ kN-m.}$$

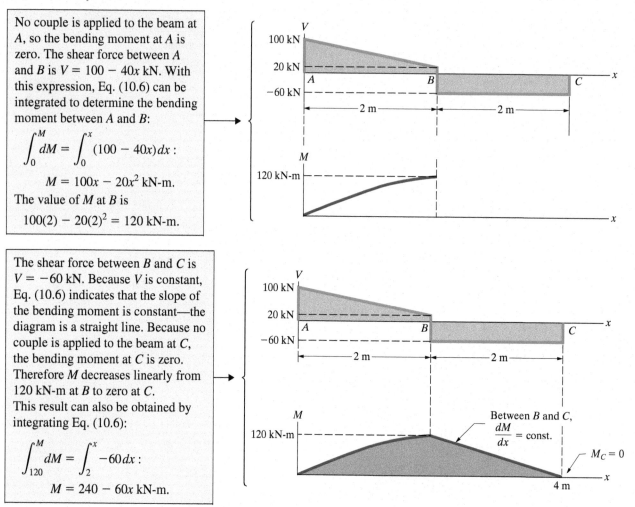

The shear force between B and C is $V = -60$ kN. Because V is constant, Eq. (10.6) indicates that the slope of the bending moment is constant—the diagram is a straight line. Because no couple is applied to the beam at C, the bending moment at C is zero. Therefore M decreases linearly from 120 kN-m at B to zero at C. This result can also be obtained by integrating Eq. (10.6):

$$\int_{120}^M dM = \int_2^x -60\,dx :$$

$$M = 240 - 60x \text{ kN-m.}$$

Active Example 10.6

The tension is given by Eq. (10.11) in terms of the tension at the lowest point and the horizontal coordinate relative to the lowest point. From Eq. (10.11), the maximum tension clearly occurs where the horizontal distance from the lowest point is the greatest, which in this example is the left attachment point. The maximum tension is

$$T = T_0\sqrt{1 + a^2 x_L^2}$$

$$= (686 \text{ lb})\sqrt{1 + (0.146 \text{ ft}^{-1})^2 (-23.4 \text{ ft})^2}$$

$$= 2440 \text{ lb.}$$

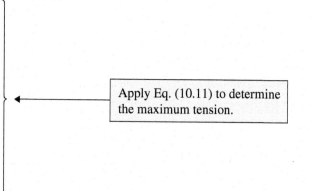

Apply Eq. (10.11) to determine the maximum tension.

Active Example 10.8

The maximum tension occurs where the horizontal distance from the lowest point is the greatest, at $x = 10$ m:

$$T = T_0 \cosh ax$$

$$= (50 \text{ N}) \cosh[(0.196 \text{ m}^{-1})(10 \text{ m})]$$

$$= 181 \text{ N}.$$

Apply Eq. (10.21).

Active Example 10.9

The angle α is

$$\alpha = \arctan\left(\frac{h_2 - 1 \text{ m}}{1 \text{ m}}\right)$$

$$= \arctan\left(\frac{1.25 \text{ m} - 1 \text{ m}}{1 \text{ m}}\right)$$

$$= 14.0°.$$

The sum of the horizontal forces is

$$T_2 \cos \alpha - T_h = 0,$$

yielding

$$T_2 = \frac{T_h}{\cos \alpha}$$

$$= \frac{131 \text{ N}}{\cos 14.0°}$$

$$= 135 \text{ N}.$$

Cut the cable at the left attachment point and within segment 2 and sum forces in the horizontal direction.

Active Example 10.10

The gage pressure $p_g = \gamma x$ increases linearly from $p_g = 0$ at the surface of the water to $p_g = (2\ \text{ft})\gamma$ at the bottom of the gate. The centroid of the distribution is shown.

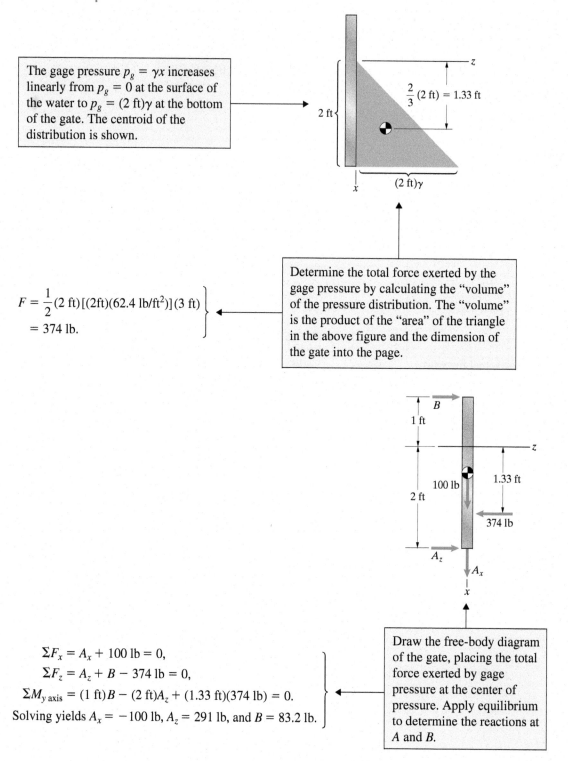

$$F = \frac{1}{2}(2\ \text{ft})\,[(2\text{ft})(62.4\ \text{lb/ft}^2)]\,(3\ \text{ft})$$
$$= 374\ \text{lb}.$$

Determine the total force exerted by the gage pressure by calculating the "volume" of the pressure distribution. The "volume" is the product of the "area" of the triangle in the above figure and the dimension of the gate into the page.

$$\Sigma F_x = A_x + 100\ \text{lb} = 0,$$
$$\Sigma F_z = A_z + B - 374\ \text{lb} = 0,$$
$$\Sigma M_{y\ \text{axis}} = (1\ \text{ft})B - (2\ \text{ft})A_z + (1.33\ \text{ft})(374\ \text{lb}) = 0.$$

Solving yields $A_x = -100\ \text{lb}$, $A_z = 291\ \text{lb}$, and $B = 83.2\ \text{lb}$.

Draw the free-body diagram of the gate, placing the total force exerted by gage pressure at the center of pressure. Apply equilibrium to determine the reactions at A and B.

Active Example 11.1

The work done by the 400-N force is $(400 \sin 40° \text{ N})(1 \text{ m})\delta\alpha$.
Bar BC undergoes a rotation $\delta\alpha$ in the clockwise direction, so
the work done by the couple is $-(500 \text{ N-m})\delta\alpha$. The work done
by the reaction C_y is $-C_y 2(2\cos 40°)\delta\alpha$. The total virtual work is

$$\delta U = (400 \sin 40° \text{ N})(1 \text{ m})\delta\alpha - (500 \text{ N-m})\delta\alpha - C_y 2(2\cos 40°)\delta\alpha = 0.$$

Solving yields $C_y = -79.3 \text{ N}$.

Determine the virtual work.

Active Example 11.3

The derivative of the potential energy
with respect to the coordinate x is

$$\frac{dV}{dx} = kx - W.$$

The second derivative is

$$\frac{d^2V}{dx^2} = k,$$

which is positive. The equilibrium
position is stable.

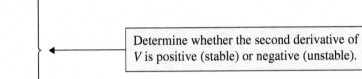

Determine whether the second derivative of
V is positive (stable) or negative (unstable).

Answers to Even-Numbered Problems

Chapter 1

1.2 (a) $e = 2.7183$; (b) $e^2 = 7.3891$; (c) $e^2 = 7.3892$.

1.4 17.8 m^2.

1.6 The 1-in wrench fits the 25-mm nut.

1.8 (a) 267 mi/h; (b) 392 ft/s.

1.10 310 N-m.

1.12 $g = 32.2 \text{ ft/s}^2$.

1.14 (a) 0.0208 m^2; (b) 32.2 in^2.

1.16 $2.07 \times 10^6 \text{ Pa}$.

1.18 27.4 lb/ft.

1.20 (a) kg-m/s; (b) 2.70 slug-ft/s.

1.22 (a) 0.397 kg; (b) 0.643 N.

1.24 (a) 4.60×10^{19} slugs; (b) 6.71×10^{20} kg.

1.26 163 lb.

1.28 32.1 km.

1.30 345,000 km.

Chapter 2

2.2 $|\mathbf{F}_{AB} + \mathbf{F}_{AC}| = 146 \text{ kN}$, direction is $32°$ above the horizontal.

2.4 $|\mathbf{F}_A + \mathbf{F}_B + \mathbf{F}_C| = 83 \text{ N}$.

2.6 $|\mathbf{r}_{AC}| = 181 \text{ mm}$.

2.8 $|\mathbf{F}_B| = 86.6 \text{ N}$, $|\mathbf{F}_C| = 50.0 \text{ N}$.

2.10 $|\mathbf{L}| = 453 \text{ lb}$, $|\mathbf{D}| = 211 \text{ lb}$.

2.12 $|\mathbf{F}_{BA}| = 174 \text{ lb}$.

2.14 $|\mathbf{r}_{BC}| = 390 \text{ m}$, $\alpha = 21.2°$.

2.18 $F_y = -102 \text{ MN}$.

2.20 $|\mathbf{F}| = 447 \text{ kip}$.

2.22 $V_x = 16$, $V_y = 12$ or $V_x = -16$, $V_y = -12$.

2.24 (a) $\mathbf{F} = 56.4\mathbf{i} + 20.5\mathbf{j}$ (lb); (b) 97.4 lb.

2.26 $\mathbf{r}_{AD} = -1.8\mathbf{i} - 0.3\mathbf{j}$ (m), $|\mathbf{r}_{AD}| = 1.825 \text{ m}$.

2.28 $\mathbf{r}_{AB} - \mathbf{r}_{BC} = \mathbf{i} - 1.73\mathbf{j}$ (m).

2.30 (a) $\mathbf{r}_{AB} = 48\mathbf{i} + 15\mathbf{j}$ (in);
(b) $\mathbf{r}_{BC} = -53\mathbf{i} + 5\mathbf{j}$ (in);
(c) $|\mathbf{r}_{AB} + \mathbf{r}_{BC}| = 20.6 \text{ in}$.

2.32 (a) $\mathbf{r}_{AB} = 52.0\mathbf{i} + 30\mathbf{j}$ (mm);
(b) $\mathbf{r}_{AB} = -42.4\mathbf{i} - 42.4\mathbf{j}$ (mm).

2.34 $x_B = 785 \text{ m}$, $y_B = 907 \text{ m}$ or $x_B = 255 \text{ m}$,
$y_B = 1173 \text{ m}$.

2.36 $\mathbf{e}_{CA} = 0.458\mathbf{i} - 0.889\mathbf{j}$.

2.38 $\mathbf{e} = 0.806\mathbf{i} + 0.593\mathbf{j}$.

2.40 $\mathbf{F} = -937\mathbf{i} + 750\mathbf{j}$ (N).

2.42 14,500 lb.

2.44 $|\mathbf{F}_{BA}| = 802 \text{ N}$.

2.46 $|\mathbf{F}_A| = 1720 \text{ lb}$, $\alpha = 33.3°$.

2.48 $57.9° \le \alpha \le 90°$.

2.50 $|\mathbf{F}_A| = 10 \text{ kN}$, $|\mathbf{F}_D| = 8.66 \text{ kN}$.

2.52 $|\mathbf{L}| = 214 \text{ lb}$, $|\mathbf{D}| = 85.4 \text{ lb}$.

2.54 $|\mathbf{F}_A| = 68.2 \text{ kN}$.

2.56 $|\mathbf{F}_{AC}| = 2.11 \text{ kN}$, $|\mathbf{F}_{AD}| = 2.76 \text{ kN}$.

2.58 $x = 75 - 0.880s$, $y = 12 + 0.476s$.

2.60 $\mathbf{r} = (0.814s - 6)\mathbf{i} + (0.581s + 1)\mathbf{j}$ (m).

2.62 $e_z = \dfrac{2}{3}$ or $e_z = -\dfrac{2}{3}$.

2.64 $U_x = 3.61$, $U_y = -7.22$, $U_z = -28.89$
or $U_x = -3.61$, $U_y = 7.22$, $U_z = 28.89$.

2.66 (a) $|\mathbf{U}| = 7$, $|\mathbf{V}| = 13$;
(b) $|3\mathbf{U} + 2\mathbf{V}| = 27.5$.

2.68 (a) $\cos\theta_x = 0.333$, $\cos\theta_y = -0.667$,
$\cos\theta_z = -0.667$;
(b) $\mathbf{e} = 0.333\mathbf{i} - 0.667\mathbf{j} - 0.667\mathbf{k}$.

2.70 $\mathbf{F} = -0.5\mathbf{i} + 0.2\mathbf{j} + 0.843\mathbf{k}$.

2.72 $\mathbf{r}_{BD} = -\mathbf{i} + 3\mathbf{j} - 2\mathbf{k}$ (m), $|\mathbf{r}_{BD}| = 3.74 \text{ m}$.

2.74 $\mathbf{e}_{CD} = -0.535\mathbf{i} + 0.802\mathbf{j} + 0.267\mathbf{k}$.

2.76 $\mathbf{F} = 300\mathbf{i} + 477\mathbf{j} + 205\mathbf{k}$ (lb).

2.78 (a) $|\mathbf{r}_{AB}| = 16.2 \text{ m}$;
(b) $\cos\theta_x = 0.615$, $\cos\theta_y = -0.492$,
$\cos\theta_z = -0.615$.

2.80 $\mathbf{r}_{AR}$: $\cos\theta_x = 0.667$, $\cos\theta_y = 0.667$,
$\cos\theta_z = 0.333$. $\mathbf{r}_{BR}$: $\cos\theta_x = -0.242$,
$\cos\theta_y = 0.970$, $\cos\theta_z = 0$.

2.82 $h = 8848 \text{ m}$ (29,030 ft).

2.84 $|\mathbf{F}_A + \mathbf{F}_B| = 217 \text{ lb}$.

2.86 $\mathbf{F} = 474\mathbf{i} + 516\mathbf{j} + 565\mathbf{k}$ (N).

2.88 (a) $\mathbf{e}_{BC} = -0.286\mathbf{i} - 0.857\mathbf{j} + 0.429\mathbf{k}$;
(b) $\mathbf{F} = -2.29\mathbf{i} - 6.86\mathbf{j} + 3.43\mathbf{k}$ (kN).

2.90 $|\mathbf{F}| = 424 \text{ lb}$.

2.92 259 lb.

2.94 $|\mathbf{F}_{AC}| = 1116 \text{ N}$, $|\mathbf{F}_{AD}| = 910 \text{ N}$.

2.96 $\mathbf{T} = -15.4\mathbf{i} + 27.0\mathbf{j} + 7.7\mathbf{k}$ (lb).

2.98 $\mathbf{T} = -41.1\mathbf{i} + 28.8\mathbf{j} + 32.8\mathbf{k}$ (N).

2.100 $32.4°$.

2.102 Either $|\mathbf{V}| = 0$ or $\mathbf{V}$ is perpendicular to $\mathbf{U}$.

2.104 $U_x = 2.857$, $V_y = 0.857$, $W_z = -3.143$.

2.108 $\theta = 62.3°$.

2.110 $\theta = 53.5°$.

2.112 $14.0\mathbf{i} + 11.2\mathbf{j} - 8.40\mathbf{k}$ (N).

2.114 (a) $42.5°$; (b) $-423\mathbf{j} + 604\mathbf{k}$ (lb).

2.116 $\mathbf{F}_p = 5.54\mathbf{j} + 3.69\mathbf{k}$ (N),
$\mathbf{F}_n = 10\mathbf{i} + 6.46\mathbf{j} - 9.69\mathbf{k}$ (N).

2.118 $\mathbf{T}_n = -37.1\mathbf{i} + 31.6\mathbf{j} + 8.2\mathbf{k}$ (N).

2.120 $\mathbf{F}_p = -0.1231\mathbf{i} + 0.0304\mathbf{j} - 0.1216\mathbf{k}$ (lb).

2.122 $\mathbf{v}_p = -1.30\mathbf{i} - 1.68\mathbf{j} - 3.36\mathbf{k}$ (m/s).

2.124 (a) $\mathbf{U} \times \mathbf{V} = 44\mathbf{i} + 56\mathbf{j} - 16\mathbf{k}$.

2.126 $2180\mathbf{i} + 1530\mathbf{j} - 1750\mathbf{k}$ (ft-lb).

2.128 Either $|\mathbf{V}| = 0$ or $\mathbf{V}$ is parallel to $\mathbf{U}$.

2.130 (a), (c) $\mathbf{U} \times \mathbf{V} = -51.8\mathbf{k}$; (b), (d) $\mathbf{V} \times \mathbf{U} = 51.8\mathbf{k}$.

2.134 (a) $\mathbf{r}_{OA} \times \mathbf{r}_{OB} = -4\mathbf{i} + 36\mathbf{j} + 32\mathbf{k}$ (m^2);
(b) $-0.083\mathbf{i} + 0.745\mathbf{j} + 0.662\mathbf{k}$
or $0.083\mathbf{i} - 0.745\mathbf{j} - 0.662\mathbf{k}$.

2.136 $\mathbf{r}_{AB} \times \mathbf{F} = -2400\mathbf{i} + 9600\mathbf{j} + 7200\mathbf{k}$ (ft-lb).

2.138 $\mathbf{r}_{CA} \times \mathbf{T} = -4.72\mathbf{i} - 3.48\mathbf{j} - 7.96\mathbf{k}$ (N-m).

2.140 $x_B = 2.81$ m, $y_B = 6.75$ m, $z_B = 3.75$ m.

2.144 1.8×10^6 mm^2.

2.146 $U_y = -2$.

2.148 $|\mathbf{A}| = 1110$ lb, $\alpha = 29.7°$.

2.150 $|\mathbf{E}| = 313$ lb, $|\mathbf{F}| = 140$ lb.

2.152 $\mathbf{e}_{AB} = 0.625\mathbf{i} - 0.469\mathbf{j} - 0.625\mathbf{k}$.

2.154 $\mathbf{F}_p = 8.78\mathbf{i} - 6.59\mathbf{j} - 8.78\mathbf{k}$ (lb).

2.156 $\mathbf{r}_{BA} \times \mathbf{F} = -70\mathbf{i} + 40\mathbf{j} - 100\mathbf{k}$ (ft-lb).

2.158 (a), (b) $686\mathbf{i} - 486\mathbf{j} - 514\mathbf{k}$ (ft-lb).

2.160 (a) $\mathbf{F} = 139\mathbf{i} + 58.2\mathbf{j} + 80\mathbf{k}$ (lb); (b) $\theta_x = 35.5°$, $\theta_y = 70°$, $\theta_z = 62.0°$.

2.162 $\mathbf{F}_p = 1.29\mathbf{i} - 3.86\mathbf{j} + 2.57\mathbf{k}$ (kN), $\mathbf{F}_n = -1.29\mathbf{i} - 2.14\mathbf{j} - 2.57\mathbf{k}$ (kN).

2.164 $\mathbf{r}_{AG} \times \mathbf{W} = -16.4\mathbf{i} - 82.4\mathbf{k}$ (N-m).

2.166 $\mathbf{r}_{BC} \times \mathbf{T} = 33.3\mathbf{i} - 125\mathbf{j} - 183\mathbf{k}$ (N-m).

Chapter 3

3.2 $F_2 = 4.77$ lb, $\alpha = 35.2°$.

3.4 $T_{AB} = T_{AC} = 1.53$ kN.

3.6 $T = 785$ N, $P = 823$ N.

3.8 $k = 1960$ N/m, $m_A = 4$ kg, $m_B = 6$ kg.

3.10 (a) $|N_{\text{crane}}| = 197$ kN, $|f_{\text{crane}}| = 0.707$ kN;
(b) $|N_{\text{caisson}}| = 3.22$ kN, $|f_{\text{caisson}}| = 0.707$ kN.

3.12 (a) $|N| = 11.06$ kN, $|f| = 4.03$ kN;
(b) $\alpha = 31.0°$.

3.14 (a) 254 lb; (b) 41.8°.

3.16 5.91 kN.

3.18 (a) 128 N; (b) 98.1 N.

3.20 $T_{\text{left}} = 299$ lb, $T_{\text{right}} = 300$ lb.

3.22 188 lb.

3.24 (a) 66.1 lb; (b) 12.3 lb.

3.26 $T_{AB} = 2.75$ kN, $T_{BC} = 2.06$ kN.

3.28 Upper cable tension is $0.828W$, lower cable tension is $0.132W$.

3.30 $T_{AB} = 1.21$ N, $T_{AD} = 2.76$ N.

3.32 $m = 12.2$ kg.

3.34 $F_B = 3680$ lb, $F_C = 2330$ lb.

3.36 $h = b$.

3.38 $T_{AB} = 688$ lb.

3.40 $T_{AB} = 64.0$ kN, $T_{BC} = 61.0$ kN.

3.44 $\alpha = 79.7°$, $T_{AB} = 120$ N, $T_{BC} = 21.4$ N, $T_{CD} = 62.6$ N.

3.46 $W_1 = 133$ lb.

3.48 (b) Left surface: 36.6 lb; right surface: 25.9 lb.

3.50 $k = 1420$ N/m.

3.52 $T = mgL/(h + R)$.

3.56 $m_2 = 12.5$ kg.

3.58 (a) $T = W/2$; (b) $T = W/4$; (c) $T = W/8$.

3.60 $L = 131.1$ kN, $D = 36.0$ kN.

3.62 (a) $\gamma = -14.0°$; (b) 4 km.

3.64 $T_{AB} = 405$ lb, $T_{AC} = 395$ lb, $T_{AD} = 103$ lb.

3.66 $T_{AB} = 1.54$ lb, $T_{AC} = 1.85$ lb.

3.68 Two at B, three at C, and three at D.

3.70 $T_{AB} = 9390$ lb, $T_{AC} = 5390$ lb, $T_{AD} = 10,980$ lb.

3.72 $D = 1176$ N, $T_{OA} = 6774$ N.

3.74 $T_{BC} = 1.61$ kN, $T_{BD} = 1.01$ kN.

3.76 $T_{EF} = T_{EG} = 738$ kN.

3.78 (a) The tension = 2.70 kN;
(b) The force exerted by the bar = $1.31\mathbf{i} - 1.31\mathbf{k}$ (kN).

3.80 $T_{AB} = 357$ N.

3.82 $F = 36.6$ N.

3.84 $W = 25.0$ lb.

3.86 (a) 83.9 lb; (b) 230.5 lb.

3.88 $T = mg/26$.

3.90 $F = 162.0$ N.

3.92 $T_{AB} = 420$ N, $T_{AC} = 533$ N, $|\mathbf{F}_S| = 969$ N.

3.94 $N = 2580$ lb, $f = 995$ lb.

3.96 $T_{AC} = 16.7$ lb, $T_{AD} = 17.2$ lb, $T_{AE} = 9.21$ lb.

3.98 Normal force = 12.15 kN, friction force = 4.03 kN.

Chapter 4

4.2 134 N-m.

4.4 $F = 36.2$ N.

4.6 25.0 kN-m clockwise.

4.8 $L = 2.4$ m.

4.10 $15.8° \le \alpha \le 37.3°$.

4.12 0.961 kN-m counterclockwise.

4.14 $M_S = 611$ in-lb.

4.16 $M_P = 298$ N-m.

4.18 410 N-m counterclockwise.

4.20 (a) $F_B = 37.5$ lb, $F_C = 22.5$ lb, $F_D = 26.0$ lb;
(b) Zero.

4.22 (a) $A = 56.6$ lb, $B = 24.4$ lb, $C = 12.2$ lb;
(b) Zero.

4.24 640 lb.

4.26 $M = 2.39$ kN-m.

4.28 (a) $A_x = 18.1$ kN, $A_y = -29.8$ kN, $B = -20.4$ kN;
(b) Zero.

4.30 (a) $A_x = 300$ lb, $A_y = 240$ lb, $B = 280$ lb;
(b) Zero.

4.32 60.4 ft-lb.

4.34 -22.3 ft-lb.

4.36 $M = -2340$ N-m.

4.38 $T_{AB} = T_{AC} = 223$ kN.

4.40 617 N-m.

4.42 $M_A = -3.00$ kN-m, $M_D = 7.50$ kN-m.

4.44 796 N.

4.46 (a), (b) $480\mathbf{k}$ (N-m).

4.48 (a) $800\mathbf{k}$ (kN-m);
(b) $-400\mathbf{k}$ (kN-m).

4.50 $\mathbf{F} = 20\mathbf{i} + 40\mathbf{j}$ (N).

4.52 $\mathbf{M}_O = -5600\mathbf{k}$ (ft-lb).

4.54 (a), (b) 1270 N-m.

4.56 128 ft-lb.

4.58 985 ft-lb.

4.60 58.0 kN.

4.62 (a) $|\mathbf{F}| = 1586$ N;
(b) $|\mathbf{F}| = 1584$ N.

4.64 $-16.4\mathbf{i} - 111.9\mathbf{k}$ (N-m).

4.66 $\mathbf{F} = 4\mathbf{i} - 4\mathbf{j} + 2\mathbf{k}$ (kN) or
$\mathbf{F} = 4\mathbf{i} - 3.38\mathbf{j} + 2.92\mathbf{k}$ (kN).

4.68 $\mathbf{M}_D = 1.25\mathbf{i} + 1.25\mathbf{j} - 6.25\mathbf{k}$ (kN-m).

4.70 $T_{AC} = 2.23$ kN, $T_{AD} = 2.43$ kN.

4.72 $T_{AB} = 1.60$ kN, $T_{AC} = 1.17$ kN.

4.74 $T_{BC} = 886$ N, $T_{BD} = 555$ N.

4.76 $\mathbf{M} = 482\mathbf{k}$ (kN-m).

4.78 (a) $\mathbf{M}_{x\text{ axis}} = 80\mathbf{i}$ (N-m);
(b) $\mathbf{M}_{y\text{ axis}} = -140\mathbf{j}$ (N-m);
(c) $\mathbf{M}_{z\text{ axis}} = \mathbf{0}$.

4.80 (a) Zero; (b) $2.7\mathbf{k}$ (kN-m).

4.82 (a) $\mathbf{M}_{x\text{ axis}} = -16\mathbf{i}$ (kN-m);
(b) $\mathbf{M}_{z\text{ axis}} = 15\mathbf{k}$ (kN-m).

4.84 $\mathbf{F} = 80\mathbf{i} + 80\mathbf{j} + 40\mathbf{k}$ (lb).

4.86 $-16.4\mathbf{i}$ (N-m).

4.88 (a), (b) $\mathbf{M}_{AB} = -76.1\mathbf{i} - 95.1\mathbf{j}$ (N-m).

4.90 $\mathbf{M}_{AO} = 119.1\mathbf{j} + 79.4\mathbf{k}$ (N-m).

4.92 $\mathbf{M}_{AB} = 77.1\mathbf{j} - 211.9\mathbf{k}$ (ft-lb).

4.94 $\mathbf{M}_{y\text{ axis}} = 215\mathbf{j}$ (N-m).

4.96 $\mathbf{M}_{x\text{ axis}} = 44\mathbf{i}$ (N-m).

4.98 $-338\mathbf{j}$ (ft-lb).

4.100 $|\mathbf{F}| = 13$ lb.

4.102 $\mathbf{M}_{\text{axis}} = -478\mathbf{i} - 174\mathbf{k}$ (N-m).

4.104 1 N-m.

4.106 $124\mathbf{k}$ (ft-lb).

4.108 28 N-m clockwise.

4.110 $\alpha = 30.9°$ or $\alpha = 71.8°$.

4.112 (b) $FL\cos 30°$.

4.114 40 ft-lb clockwise, or $-40\mathbf{k}$ (ft-lb).

4.116 2200 ft-lb clockwise.

4.118 (a) $C = 26$ kN-m; (b) Zero.

4.120 (a) $\mathbf{M} = -14\mathbf{i} - 10\mathbf{j} - 8\mathbf{k}$ (kN-m); (b) $D = 6.32$ m.

4.122 356 ft-lb.

4.124 $|\mathbf{M}| = 6.13$ kN-m.

4.126 $M_{Cy} = 7$ kN-m, $M_{Cz} = -2$ kN-m.

4.128 Yes.

4.130 Systems 1, 2, and 4 are equivalent.

4.134 $F = 265$ N.

4.136 $F = 70$ lb, $M = 130$ in-lb.

4.138 (a) $\mathbf{F} = -10\mathbf{j}$ (lb), $M = -10$ ft-lb; (b) $D = 1$ ft.

4.140 $\mathbf{F} = 200\mathbf{i} + 180\mathbf{j}$ (N), $d = 0.317$ m.

4.142 (a) $A_x = 12$ kip, $A_y = 10$ kip, $B = -10$ kip;
(b) $\mathbf{F} = -12\mathbf{i}$ (kip), intersects at $y = 5$ ft;
(c) They are both zero.

4.144 $\mathbf{F} = 104\mathbf{j}$ (kN), $M = 13.2$ kN-m counterclockwise.

4.146 $\mathbf{F} = 100\mathbf{j}$ (lb), $\mathbf{M} = \mathbf{0}$.

4.148 (a) $\mathbf{F} = 920\mathbf{i} - 390\mathbf{j}$ (N), $M = -419$ N-m;
(b) intersects at $y = 456$ mm.

4.150 $\mathbf{F} = 800\mathbf{j}$ (lb), intersects at $x = 7.5$ in.

4.152 (a) $-360\mathbf{k}$ (in-lb);
(b) $-36\mathbf{j}$ (in-lb);
(c) $\mathbf{F} = 10\mathbf{i} - 30\mathbf{j} + 3\mathbf{k}$ (lb),
$\mathbf{M} = -36\mathbf{j} - 360\mathbf{k}$ (in-lb).

4.154 $x = 2.00$ ft, $z = -0.857$ ft.

4.156 $\mathbf{F} = 100\mathbf{j} + 80\mathbf{k}$ (N), $\mathbf{M} = 240\mathbf{j} - 300\mathbf{k}$ (N-m).

4.158 (a) $\mathbf{F} = \mathbf{0}$, $\mathbf{M} = rA\mathbf{i}$;
(b) $\mathbf{F}' = \mathbf{0}$, $\mathbf{M}' = rA\mathbf{i}$.

4.160 (a) $\mathbf{F} = \mathbf{0}$, $\mathbf{M} = 4.60\mathbf{i} + 1.86\mathbf{j} - 3.46\mathbf{k}$ (kN-m);
(b) 6.05 kN-m.

4.162 $\mathbf{F} = -20\mathbf{i} + 20\mathbf{j} + 10\mathbf{k}$ (lb),
$\mathbf{M} = 50\mathbf{i} + 250\mathbf{j} + 100\mathbf{k}$ (in-lb).

4.164 (a) $\mathbf{F} = 28\mathbf{k}$ (kip), $\mathbf{M} = 96\mathbf{i} - 192\mathbf{j}$ (ft-kip);
(b) $x = 6.86$ ft, $y = 3.43$ ft.

4.166 $\mathbf{F} = 100\mathbf{i} + 20\mathbf{j} - 20\mathbf{k}$ (N),
$\mathbf{M} = -143\mathbf{i} + 406\mathbf{j} - 280\mathbf{k}$ (N-m).

4.168 $\mathbf{M}_p = 0$, line of action intersects at $y = 0$, $z = 2$ ft.

4.170 $x = 2.41$ m, $y = 3.80$ m.

4.172 $\mathbf{F} = 40.8\mathbf{i} + 40.8\mathbf{j} + 81.6\mathbf{k}$ (N),
$\mathbf{M} = -179.6\mathbf{i} + 391.9\mathbf{j} - 32.7\mathbf{k}$ (N-m).

4.174 (a) $320\mathbf{i}$ (in-lb);
(b) $\mathbf{F} = -20\mathbf{k}$ (lb), $\mathbf{M} = 320\mathbf{i} + 660\mathbf{j}$ (in-lb);
(c) $M_t = 0$, $x = 33$ in, $y = -16$ in.

4.176 $|\mathbf{M}_P| = 244$ N-m.

4.178 (a) -76.2 N-m;
(b) -66.3 N-m.

4.180 $|\mathbf{F}| = 224$ lb, $|\mathbf{M}| = 1600$ ft-lb.

4.182 501 lb.

4.184 $-228.1\mathbf{i} - 68.4\mathbf{k}$ (N-m).

4.186 $\mathbf{M}_{x\text{ axis}} = -153\mathbf{i}$ (ft-lb).

4.188 $\mathbf{M}_{CD} = -173\mathbf{i} + 1038\mathbf{k}$ (ft-lb).

4.190 (a) $\mathbf{T}_{AB} = \mathbf{T}_{CD} = 173.2$ lb;
(b) $\mathbf{F} = 300\mathbf{j}$ (lb) at $x = 4$ ft.

4.192 $\mathbf{F} = -20\mathbf{i} + 70\mathbf{j}$ (N), $M = 22$ N-m.

4.194 $\mathbf{F}' = -100\mathbf{i} + 40\mathbf{j} + 30\mathbf{k}$ (lb),
$\mathbf{M} = -80\mathbf{i} + 200\mathbf{k}$ (in-lb).

4.196 $\mathbf{F} = 1166\mathbf{i} + 566\mathbf{j}$ (N), $y = 13.9$ m.

4.198 $\mathbf{F} = 190\mathbf{j}$ (N), $\mathbf{M} = -98\mathbf{i} + 184\mathbf{k}$ (N-m).

4.200 $\mathbf{F} = -0.364\mathbf{i} + 4.908\mathbf{j} + 1.090\mathbf{k}$ (kN),
$\mathbf{M} = -0.131\mathbf{i} - 0.044\mathbf{j} + 1.112\mathbf{k}$ (kN-m).

Chapter 5

5.2 $A_x = -1$ kN, $A_y = -5.73$ kN,
$M_A = -22.9$ kN-m.

5.4 Tension is 386 lb, $B_x = 493$ lb, $B_y = 186$ lb.

5.6 (b) $A_x = 0$, $A_y = -1.85$ kN, $B_y = 2.74$ kN.

5.8 (b) $A_x = 0$, $A_y = -5$ kN, $B_y = 15$ kN.

5.10 (b) $A = 100$ lb, $B = 200$ lb.

5.12 (b) $A_x = 502$ N, $A_y = 870$ N.

5.14 (b) $A_x = 4$ kN, $A_y = -2.8$ kN, $B_y = 2.8$ kN.

5.16 On each hand, 66.3 lb. On each foot, 23.7 lb.

5.18 $A_x = -100$ lb, $A_y = -225$ lb, $E = 625$ lb.

5.20 $k = 3380$ N/m, $B_x = -188.0$ N, $B_y = 98.7$ N.

5.22 5.93 kN.

5.24 $R = 12.5$ lb, $B_x = 11.3$ lb, $B_y = 15.3$ lb.

5.26 (a) 21.2 lb; (b) 30 lb.

5.28 $W_L = 1125$ lb.

5.30 6.23 lb.

5.32 $T = 3.68$ lb.

5.34 $T_{AE} = 28.6$ lb, $D_x = -26.9$ lb, $D_y = 32.2$ lb.

5.36 $A_x = -1.83$ kN, $A_y = 2.10$ kN, $B_y = 2.46$ kN.

5.38 $A_x = -200$ lb, $A_y = -100$ lb, $M_A = 1600$ ft-lb.

5.40 $k = 3.21$ lb/ft.

5.42 $A_x = 3.46$ kN, $A_y = -2$ kN,
$B_x = -3.46$ kN, $B_y = 2$ kN.

5.44 $\mathbf{F} = 28.3\mathbf{i} + 58.3\mathbf{j}$ (lb), $D = 7.03$ ft, $A_x = -28.3$ lb,
$A_y = -58.3$ lb, $M_A = -410$ ft-lb.

5.46 $A_x = -1.57$ kN, $A_y = 1.57$ kN, $E_x = 1.57$ kN.

5.48 $A_x = 0$, $A_y = 200$ lb, $M_A = 900$ ft-lb.

5.50 $A_x = 57.7$ lb, $A_y = -13.3$ lb, $B = 15.3$ lb.

5.52 $W = 15$ kN.

5.54 (b) $C_x = 500$ N, $C_y = -200$ N.

5.56 $T_{BC} = 5.45$ lb, $A_x = 5.03$ lb, $A_y = 7.90$ lb.

5.58 20.3 kN.

5.60 $W_2 = 2484$ lb, $A_x = -2034$ lb, $A_y = 2425$ lb.

5.62 $W = 46.2$ N, $A_x = 22.3$ N, $A_y = 61.7$ N.

5.64 $F = 44.5$ lb, $A_x = 25.3$ lb, $A_y = -1.9$ lb.

5.66 $W = 132$ lb.

5.68

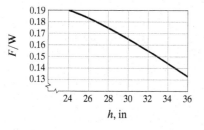

5.76 (1) and (2) are improperly supported. For (3), reactions
are $A = F/2$, $B = F/2$, $C = F$.

5.78 (b) $A_x = -6.53$ kN, $A_y = -3.27$ kN,
$A_z = 3.27$ kN, $M_{Ax} = 0$, $M_{Ay} = -6.53$ kN-m,
$M_{Az} = -6.53$ kN-m.

5.80 374 lb.

5.82 $C_x = -349$ lb, $C_y = 698$ lb,
$C_z = 175$ lb, $M_{Cx} = -3490$ ft-lb,
$M_{Cy} = -2440$ ft-lb, $M_{Cz} = 2790$ ft-lb.

5.84 (a) $-17.8\mathbf{i} - 62.8\mathbf{k}$ (N-m);
(b) $A_x = 0$, $A_y = 360$ N, $A_z = 0$,
$M_{Ax} = 17.8$ N-m, $M_{Ay} = 0$, $M_{Az} = 62.8$ N-m.

5.86 $A_x = 166.7$ N, $A_y = 200$ N, $A_z = 66.7$ N,
$T_{BC} = 100$ N, $T_{BD} = 170$ N.

5.88 $|\mathbf{F}| = 10.9$ kN.

5.90 $T_{AB} = 553$ lb, $T_{AC} = 289$ lb,
$O_x = 632$ lb, $O_y = 574$ lb, $O_z = 0$.

5.92 $x = 0.1$ m, $z = 0.133$ m.

5.94 $T_{BD} = 50.2$ lb, $A_x = -34.4$ lb,
$A_y = 17.5$ lb, $A_z = -24.1$ lb,
$M_{Ax} = 0$, $M_{Ay} = 192.5$ in-lb.

5.96 $\mathbf{F} = 4\mathbf{j}$ (kN) at $x = 0$, $z = 0.15$ m.

5.98 (b) $A_x = -0.74$ kN, $A_y = 1$ kN, $A_z = -0.64$ kN,
$B_x = 0.74$ kN, $B_z = 0.64$ kN.

5.100 $F_y = 34.5$ lb.

5.102 $T_{BD} = 1.47$ kN, $T_{BE} = 1.87$ kN,
$A_x = 0$, $A_y = 4.24$ kN, $A_z = 0$.

5.104 $T = 139$ lb, $A_x = 46.4$ lb, $A_y = -26.8$ lb,
$A_z = 31.7$ lb, $M_{Ax} = -63.4$ ft-lb,
$M_{Ay} = -110$ ft-lb.

5.106 Tension is 60 N, $B_x = -10$ N, $B_y = 90$ N,
$B_z = 10$ N, $M_{By} = 1$ N-m, $M_{Bz} = -3$ N-m.

5.108 Tension is 60 N, $B_x = -10$ N, $B_y = 75$ N,
$B_z = 15$ N, $C_y = 15$ N, $C_z = -5$ N.

5.110 $A_x = -2.86$ kip, $A_y = 17.86$ kip, $A_z = -8.10$ kip,
$B_y = 3.57$ kip, $B_z = 12.38$ kip.

5.112 $A_x = 0$, $A_y = 400$ N, $B_x = 1000$ N,
$B_y = -400$ N, $B_z = 0$, $T = 1080$ N.

5.114 $|\mathbf{A}| = 8.54$ kN, $|\mathbf{B}| = 10.75$ kN.

5.116 $A_x = 3.62$ kN, $A_y = 5.89$ kN, $A_z = 5.43$ kN,
$C_x = 8.15$ kN, $C_y = 0$, $C_z = 0.453$ kN.

5.118 $T_{AB} = 488$ lb, $T_{CD} = 373$ lb, reaction is
$31\mathbf{i} + 823\mathbf{j} - 87\mathbf{k}$ (lb).

5.120 $A_x = -76.7$ N, $A_y = 97.0$ N, $A_z = -54.3$ N,
$M_{Ax} = -2.67$ N-m, $M_{Ay} = 6.39$ N-m,
$M_{Az} = 2.13$ N-m.

5.122 (a) 60 lb;
(b) $A_x = 38.1$ lb, $A_y = 46.3$ lb or $A_x = -38.1$ lb,
$A_y = -46.3$ lb.

5.124 Tension is 33.3 lb; magnitude of reaction is 44.1 lb.

5.126 $\alpha = 10.9°$, $F_A = 1.96$ kN, $F_B = 2.27$ kN.

5.128 (a) No, because of the 3 kN-m couple; (b) magnitude at
A is 7.88 kN; magnitude at B is 6.66 kN; (c) no.

5.130 (b) $A_x = -8$ kN, $A_y = 2$ kN, $C_x = 8$ kN.

5.134 (b) $T_A = 7.79$ lb, $T_B = 10.28$ lb; (c) 6.61 lb.

5.136 (a) There are four unknown reactions and three equilibrium equations; (b) $A_x = -50$ lb, $B_x = 50$ lb.

5.138 (b) Force on nail = 55 lb, normal force = 50.77 lb,
friction force = 9.06 lb.

5.140 $k = 13,500$ N/m.

5.142 $A_y = 727$ lb, $H_x = 225$ lb, $H_y = 113$ lb.

5.144 $\alpha = 0$ and $\alpha = 59.4°$.

5.146 The force is 800 N upward; its line of action passes
through the midpoint of the plate.

5.148 $m = 67.2$ kg.

5.150 $\alpha = 90°$, $T_{BC} = W/2$, $A = W/2$.

Chapter 6

6.2 AB: 915 N (C); AC: 600 N (C); BC: 521 N (T).

6.4 BC: 800 lb (T); CD: 600 lb (C).

6.6 (a) Tension: 2.43 kN in AB and BD.
Compression: 2.88 kN in CD.
(b) Tension: 1.74 kN in BD.
Compression: 1.60 kN in CD.

6.8 Tension, 31.9 kip in AC, CE, EG, and GH. Compression, 42.5 kip in BD and DF.

6.10 BD: zero; CD: 10 kN (T); CE: 16 kN (C).

6.12 (a) Tension: 5540 lb in BD. Compression: 7910 lb in CE.
(b) Tension: 2770 lb in BD. Compression: 3760 lb in CE.

6.14 $F = 8.33$ kN.

6.16 DE: 3.66 kN (C); DF: 1.45 kN (C); DG: 3.36 kN (T).

6.18 AB: 10.56 kN (T); AC: 17.58 kN (C); BC: 6.76 kN (T);
BD: 1.81 kN (T); CD: 16.23 kN (C).

6.20 AB: 375 lb (C); AC: 625 lb (T); BC: 300 lb (T).
6.22 BC: 90.1 kN (T); CD: 90.1 kN (C); CE: 300 kN (T).
6.24 BC: 1200 kN (C); BI: 300 kN (T); BJ: 636 kN (T).
6.26 AB: 2520 lb (C); BC: 2160 lb (C); CD: 1680 lb (C).
6.32 BC: 400 kN (T), BI: 141 kN (T), HI: 500 kN (C).
6.34   (a), (b) 141 kN (C).
6.36 AB: 1.33F (C); BC: 1.33F (C); CE: 1.33F (T).
6.38 BD: 95.6 kip (C); BE: 41.1 kip (T); CE: 58.4 kip (T).
6.40 DF: 69.1 kip (C); DG: 29.4 kip (C); EG: 95.6 kip (T).
6.42 EK: 240 kN (T).
6.44 AC: 2000 lb (C); BC: 800 lb (T); BD: 1000 lb (T).
6.46 DF: 16 kN (T); DG: 6.67 kN (C); EG: 26.7 kN (C).
6.48   2.50 kN (C).
6.50 CE: 680 kN (T); CF: 374 kN (C); DF: 375 kN (C).
6.52   (a) 1160 lb (C).
6.54 IL: 16 kN (C); KM: 24 kN (T).
6.58 AD: 4.72 kN (C); BD: 4.16 kN CD (C);
       CD: 4.85 kN (C).
6.60 AB, AC, AD: 0.408F (C).
6.62 AB: 379 lb (C); AC: 665 lb (C); AD: 160 lb (C).
6.64 BC: 32.7 kN (T); BD: 45.2 kN (T); BE: 112.1 kN (C).
6.66 $P_3 = -315$ kN.
6.68   5.59 kN (C) in each member.
6.70 $A_x = 400$ N, $A_y = -900$ N, $B_x = -400$ N,
       $B_y = 900$ N, $M_A = -540$ N-m.
6.72 $C_x = 736$ N, $C_y = 2450$ N, $E_x = 245$ N,
       $E_y = -1720$ N.
6.74 $C_x = 66.7$ lb, $C_y = 24$ lb.
6.76 $A_x = 0$, $A_y = -400$ N, $C_x = -600$ N,
       $C_y = -300$ N, $D_x = 0$, $D_y = 1000$ N.
6.78 $D_x = -1475$ N, $D_y = -516$ N, $E_x = 0$,
       $E_y = -516$ N, $M_E = 619$ N-m.
6.80 $A_x = -2.35$ kN, $A_y = 2.35$ kN,
       $B_x = 0$, $B_y = -4.71$ kN,
       $C_x = 2.35$ kN, $C_y = 2.35$ kN.
6.82   Tension = 62.5 lb, $F_x = -75$ lb, $F_y = 25$ lb.
6.84 $B_x = -400$ lb, $B_y = -300$ lb, $C_x = 400$ lb,
       $C_y = 200$ lb, $D_x = 0$, $D_y = 100$ lb.
6.86 $A_x = -150$ lb, $A_y = 120$ lb, $B_x = 180$ lb,
       $B_y = -30$ lb, $D_x = -30$ lb, $D_y = -90$ lb.
6.88 $A_x = -310$ lb, $A_y = -35$ lb, $B_x = 80$ lb,
       $B_y = -80$ lb, $C_x = 310$ lb, $C_y = 195$ lb,
       $D_x = -80$ lb, $D_y = -80$ lb.
6.90 $A_x = 170$ lb, $A_y = 129$ lb,
       $B_x = -170$ lb, $B_y = -209$ lb.
6.94 $A_x = -22$ lb, $A_y = 15$ lb,
       $C_x = -14$ lb, $C_y = 3$ lb.
6.96   300 lb (C).
6.98 B: 73.5 N; C: 88.8 N.
6.100 $T_{BC} = 1410$ N, $T_{DF} = 625$ N.
6.102 $A_x = 2$ kN, $A_y = -1.52$ kN,
       $B_x = -2$ kN, $B_y = 1.52$ kN.
6.104 $E_x = 604$ lb, $E_y = 179$ lb, axial force is 616 lb.
6.106 100 N.
6.108 At B: 1750 N. DE: 1320 N (C).
6.110 742 lb.
6.112 1150 lb.

6.114 $K_x = 847$ N, $K_y = 363$ N.
6.116 $T_{AB} = 7.14$ kN (C), $T_{AC} = 5.71$ kN (T),
       $T_{BC} = 10$ kN (T).
6.118 BC: 120 kN (C); BG: 42.4 kN (T); FG: 90 kN (T).
6.120 AB: 125 lb (C); AC: zero; BC: 188 lb (T);
       BD: 225 lb (C); CD: 125 lb (C); CE: 225 lb (T).
6.122 $T_{BD} = 13.3$ kN (T), $T_{CD} = 11.7$ kN (T),
       $T_{CE} = 28.3$ kN (C).
6.124 AC: 480 N (T); CD: 240 N (C); CF: 300 N (T).
6.126 Tension: member AC, 480 lb (T);
       Compression: member BD, 633 lb (C).
6.128 CD: 11.42 kN (C); CJ: 4.17 kN (C); IJ: 12.00 kN
       (T).
6.130 AB: 7.20 kN (C); AC: 4.56 kN (C).
6.132 $A_x = -1.57$ kN, $A_y = 1.18$ kN,
       $B_x = 0$, $B_y = -2.35$ kN, $C_x = 1.57$ kN,
       $C_y = 1.18$ kN.
6.134 $B_x = 3820$ lb, $B_y = 6690$ lb, $C = 9020$ lb,
       $D_x = -1390$ lb, $D_y = -1930$ lb.
6.136 973 N.
6.138 $A_x = -52.33$ kN, $A_y = -43.09$ kN,
       $E_x = 0.81$ kN, $E_y = -14.86$ kN.

Chapter 7

7.2 $\bar{x} = 3/8$.
7.4 $\bar{x} = 1.25$, $\bar{y} = 0.825$.
7.8 $\bar{x} = 0.711$ ft, $y = 0.584$ ft.
7.10 $\bar{x} = 0$, $y = 1.6$ ft.
7.12 $\bar{x} = 8$, $\bar{y} = 3.6$.
7.14 $\bar{x} = 0.533$.
7.16 $\bar{x} = 1$.
7.18 $\bar{y} = -7.6$.
7.20 $\bar{y} = 2.53$.
7.22 $a = 0.656$, $b = 6.56 \times 10^{-5}$ m^{-2}.
7.24 $\bar{x} = \bar{y} = 4R/3\pi$.
7.26 $\bar{x} = 3.31$.
7.28 $\bar{x} = 116$ mm.
7.30 $\bar{x} = 9.90$ in, $\bar{y} = 0$.
7.32 $\bar{x} = 23.9$ in, $\bar{y} = 33.3$ in.
7.34 $\bar{x} = 2.88$ ft, $\bar{y} = 3.20$ ft.
7.36 $\bar{x} = 3.67$ mm, $\bar{y} = 21.52$ mm.
7.38 $b = 39.6$ mm, $h = 18.2$ mm.
7.40 $\bar{x} = 9.64$ m, $\bar{y} = 4.60$ m.
7.44 $\bar{x} = 6.47$ ft, $\bar{y} = 10.60$ ft.
7.46 $A_x = 0$, $A_y = 160$ N, $B = 200$ N.
7.48 $A_x = -1200$ N, $A_y = 800$ N, $B = 2200$ N.
7.50 $A_x = 0$, $A_y = 10$ kN, $M_A = -31.3$ kN-m.
7.52 $A_x = 0$, $A_y = 4.17$ kN, $B_y = 8.83$ kN.
7.54 $A_y = 3267$ lb, $B_x = -800$ lb, $B_y = -1267$ lb.
7.56 BD: 21.3 kN (C); CD: 3.77 kN (C); CE: 24 kN (T).
7.58 $A_x = -18$ kN, $A_y = 20$ kN,
       $B_x = 0$, $B_y = -4$ kN,
       $C_x = 18$ kN, $C_y = -16$ kN.
7.60 $V = 275$ m^3,   height = 2.33 m.
7.62 $V = 4.16$ m^3, $\bar{x} = 1.41$ m.
7.64 $\bar{x} = 0.675R$, $\bar{y} = 0$, $\bar{z} = 0$.

7.66 $\bar{y} = 0.410$.

7.68 $\bar{x} = 3.24$.

7.70 $\bar{x} = R\sin\alpha/\alpha, \quad \bar{y} = R(1 - \cos\alpha)/\alpha$.

7.72 $\bar{x} = 38.3$ mm.

7.74 $\bar{x} = -128$ mm, $\bar{y} = \bar{z} = 0$.

7.76 $\bar{x} = 0, \ \bar{y} = 43.7$ mm, $\bar{z} = 38.2$ mm.

7.78 $\bar{x} = 229.5$ mm, $\bar{y} = \bar{z} = 0$.

7.80 $\bar{x} = 23.65$ mm, $\bar{y} = 36.63$ mm, $\bar{z} = 3.52$ mm.

7.82 $\bar{x} = 6$ m, $\bar{y} = 1.83$ m.

7.84 $\bar{x} = 65.9$ mm, $\bar{y} = 21.7$ mm, $\bar{z} = 68.0$ mm.

7.86 $A = \frac{3}{4}\pi R\sqrt{h^2 + R^2}$.

7.88 $\bar{y}_s = 4R/3\pi$.

7.90 $\bar{y} = 0.410$.

7.92 $A = 138$ ft^2.

7.94 $V = 0.0377$ m^3.

7.96 $V = 2.48 \times 10^6$ mm^3.

7.98   Volume $= 0.0266$ m^3.

7.100 $A_x = 0, \ A_y = 294$ N, $B_y = 196$ N.

7.102 $A_x = 0, \ A_y = 316$ N, $B = 469$ N.

7.104 $\bar{x} = 6.59$ in, $\bar{y} = 2.17$ in, $\bar{z} = 6.80$ in.

7.106 $A_x = 0, \ A_y = 3.16$ kN, $M_A = 1.94$ kN-m.

7.108 $\bar{x} = 121$ mm, $\bar{y} = 0, \ \bar{z} = 0$.

7.110 $\bar{x}_3 = 82$ mm, $\bar{y}_3 = 122$ mm, $\bar{z}_3 = 16$ mm.

7.112 (a) $\bar{x} = 5.17$ m; (b) $A_x = -50$ kN, $A_y = -25.0$ kN, $G = 33.8$ kN.

7.114 Mass $= 408$ kg, $\bar{x} = 2.5$ m, $\bar{y} = -1.5$ m.

7.116 $\bar{x} = 20.10$ in, $\bar{y} = 8.03$ in, $\bar{z} = 15.35$ in.

7.118 $\bar{x} = 3/8, \ \bar{y} = 3/5$.

7.120 $\bar{x} = 87.3$ mm, $\bar{y} = 55.3$ mm.

7.122 917 N (T).

7.124 $A_x = 7$ kN, $A_y = -6$ kN, $D_x = 4$ kN, $D_y = 0$.

7.126 $\bar{x} = 1.87$ m.

7.128 $A = 682$ in^2.

7.130 $\bar{x} = 110$ mm.

7.132 $\bar{x} = 1.70$ m.

7.134 $\bar{x} = 25.24$ mm, $\bar{y} = 8.02$ mm, $\bar{z} = 27.99$ mm.

7.136 (a) $\bar{x} = 1.511$ m; (b) $\bar{x} = 1.611$ m.

7.138 $A = 80.7$ kN, $B = 171.6$ kN.

Chapter 8

8.2 $I_x = 0.0288$ m^4, $k_x = 0.346$ m.

8.4   (a) $I_y = 12.8 \times 10^5$ mm^4; (b) $I_{y'} = 3.2 \times 10^5$ mm^4.

8.6 $I_y = 0.175$ m^4, $k_y = 0.624$ m.

8.8 $I_{xy} = 0.0638$ m^4.

8.10 $I_x = 1.69$.

8.12 $I_{xy} = 0.583$.

8.14 $I_x = 1330, \ k_x = 4.30$.

8.16 $I_{xy} = 2070$.

8.18 $I_x = 953, \ k_x = 6.68$.

8.20 (a) $I_x = \frac{1}{8}\pi R^4, \ k_x = \frac{1}{2}R$.

8.22 $I_y = 49.09$ m^4, $k_y = 2.50$ m.

8.24 $I_y = 522, \ k_y = 2.07$.

8.28 $I_y = 10$ m^4, $k_y = 1.29$ m.

8.30 $I_x = 6.00 \times 10^6$ mm^4, $k_x = 23.5$ mm.

8.32 $I_y = 0.0125$ m^4, $k_y = 0.177$ m.

8.34 $I_y = 3.6 \times 10^5$ mm^4, $J_O = 1 \times 10^6$ mm^4.

8.36 $I_x = 2.65 \times 10^8$ mm^4, $k_x = 129$ mm.

8.38 $I_x = 7.79 \times 10^7$ mm^4, $k_x = 69.8$ mm.

8.40 $I_{xy} = 1.08 \times 10^7$ mm^4.

8.42 $J_O = 363$ ft^4, $k_O = 4.92$ ft.

8.44 $I_x = 10.7$ ft^4, $k_x = 0.843$ ft.

8.46 $I_{xy} = 7.1$ ft^4.

8.48 $J_O = 5.63 \times 10^7$ mm^4, $k_O = 82.1$ mm.

8.50 $I_x = 1.08 \times 10^7$ mm^4, $k_x = 36.0$ mm.

8.52 $J_O = 1.58 \times 10^7$ mm^4, $k_O = 43.5$ mm.

8.54 $J_O = 2.35 \times 10^5$ in^4, $k_O = 15.1$ in.

8.56 $I_x = 49.7$ m^4, $k_x = 2.29$ m.

8.58 $I_y = 2.55 \times 10^6$ in^4, $k_y = 27.8$ in.

8.60 $I_{xy} = 2.54 \times 10^6$ in^4.

8.62 $I_x = 1.26 \times 10^6$ in^4, $k_x = 19.5$ in.

8.64 $I_y = 4.34 \times 10^4$ in^4, $k_y = 10.5$ in.

8.66 $I_{xy} = 4.83 \times 10^4$ in^4.

8.68 $J_O = 4.01 \times 10^4$ in^4, $k_O = 14.6$ in.

8.70 $I_x = 8.89 \times 10^3$ in^4, $k_x = 7.18$ in.

8.72 $I_y = 3.52 \times 10^3$ in^4, $k_y = 4.52$ in.

8.74 $I_{xy} = 995$ in^4.

8.76 $J_O = 5.80 \times 10^6$ mm^4, $k_O = 37.5$ mm.

8.78 $I_x = 1470$ in^4, $I_y = 3120$ in^4.

8.80 $I_x = 4020$ in^4, $I_y = 6980$ in^4, or $I_x = 6820$ in^4, $I_y = 4180$ in^4.

8.82 $I_x = 4.01 \times 10^6$ mm^4.

8.86 $I_x = 59.8 \times 10^6$ mm^4, $I_y = 18.0 \times 10^6$ mm^4.

8.88 $I_{x'} = 7.80$ ft^4, $I_{y'} = 24.2$ ft^4, $I_{x'y'} = -2.20$ ft^4.

8.90 $I_{x'} = 1.20 \times 10^6$ in^4, $I_{y'} = 7.18 \times 10^5$ in^4, $I_{x'y'} = 2.11 \times 10^5$ in^4.

8.92 $\theta_p = -12.1°$, principal moments of inertia are 80.2×10^{-6} m^4 and 27.7×10^{-6} m^4.

8.94 $I_{x'} = 7.80$ ft^4 $I_{y'} = 24.2$ ft^4 $I_{x'y'} = -2.20$ ft^4.

8.96 $I_{x'} = 1.20 \times 10^6$ in^4, $I_{y'} = 7.18 \times 10^5$ in^4, $I_{x'y'} = 2.11 \times 10^5$ in^4.

8.98 $\theta_p = -12.1°$, principal moments of inertia are 80.2×10^{-6} m^4 and 27.7×10^{-6} m^4.

8.100 $I_O = 14$ kg-m^2.

8.102 $I_{z\,\text{axis}} = 15.1$ kg-m^2.

8.104 $I_{x\,\text{axis}} = 0.667$ kg-m^2, $I_{y\,\text{axis}} = 2.67$ kg-m^2.

8.106 $I_{y\,\text{axis}} = 1.99$ slug-ft^2.

8.108 20.8 kg-m^2.

8.110 $I_O = \frac{17}{12}ml^2$.

8.112 $I_{z\,\text{axis}} = 47.0$ kg-m^2.

8.114 $I_{z\,\text{axis}} = 0.0803$ slug-ft^2.

8.116   3810 slug-ft^2.

8.118 $I_{z \text{ axis}} = 9.00$ kg-m^2.

8.120 $I_{y \text{ axis}} = 0.0881$ slug-ft^2.

8.122 $I_{x \text{ axis}} = m\left(\frac{1}{3}l^2 + \frac{1}{4}R^2\right)$.

8.124 $I_{x \text{ axis}} = I_{y \text{ axis}} = m\left(\frac{3}{20}R^2 + \frac{3}{5}h^2\right)$.

8.126 $I_{x \text{ axis}} = \frac{1}{6}mh^2 + \frac{1}{3}ma^2$.

8.128 $I_{x \text{ axis}} = 0.221$ kg-m^2.

8.130 $I_{x'} = 0.995$ kg-m^2, $I_{y'} = 20.1$ kg-m^2.

8.132 $I_{z \text{ axis}} = 0.00911$ kg-m^2.

8.134 $I_O = 0.00367$ kg-m^2.

8.136 $I_{z \text{ axis}} = 0.714$ slug-ft^2.

8.138 $I_y = \frac{1}{5}, \quad k_y = \sqrt{\frac{3}{5}}$.

8.140 $J_O = \frac{26}{105}, \quad k_O = \sqrt{\frac{26}{35}}$.

8.142 $I_y = 12.8, \quad k_y = 2.19$.

8.144 $I_{xy} = 2.13$.

8.146 $I_{x'} = 0.183, \quad k_{x'} = 0.262$.

8.148 $I_y = 2.75 \times 10^7$ mm^4, $k_y = 43.7$ mm.

8.150 $I_x = 5.03 \times 10^7$ mm^4, $k_x = 59.1$ mm.

8.152 $I_y = 94.2$ ft^4, $k_y = 2.24$ ft.

8.154 $I_x = 396$ ft^4, $k_x = 3.63$ ft.

8.156 $\theta_p = 19.5°, \quad 20.3$ m^4, 161 m^4.

8.158 $I_{y \text{ axis}} = 0.0702$ kg-m^2.

8.160 $I_{z \text{ axis}} = \frac{1}{10}mw^2$.

8.162 $I_{x \text{ axis}} = 3.83$ slug-ft^2.

8.164   0.537 kg-m^2.

Chapter 9

9.2   1.04 lb.

9.4   (a) $\alpha = 38.7°$;   (b) $\alpha = 11.3°$.

9.6   (a) No;   (b) 20.4 lb.

9.8   177 N.

9.10   20 lb.

9.12 $\alpha = 14.0°$.

9.14   (a) $T = 56.5$ N.

9.16   (a) Yes. The force is $\mu_s W$;   (b) $3\mu_s W$.

9.18 $89.6 \leq T \leq 110.4$ lb.

9.20 $F = 267$ N.

9.22 $M = hrF\mu_k / [2(h + b\mu_k)]$.

9.24   9.40 ft-lb.

9.26 $\alpha = 33.4°$.

9.28 $\alpha = 28.3°$.

9.30   (a) $M = 162$ in-lb;   (b) $M = 135$ in-lb.

9.32 $M = \mu_s RW[\sin\alpha + \mu_s(1 - \cos\alpha)]/[(1 + \mu_s^2)\sin\alpha]$.

9.34 $\alpha = 39.6°$.

9.36   (a) $T = 9.42$ lb;   (b) $T = 33.3$ lb.

9.40 $y = 234$ mm.

9.42 $\alpha = 9.27°$.

9.44 $F = 44$ lb.

9.48 $\alpha = 1.54°, \quad P = 202$ N.

9.50   (a) $F = \mu_s W$;
      (b) $F = (W/2)(\mu_{sA} + \mu_{sB})/[1 + (h/b)(\mu_{sA} - \mu_{sB})]$.

9.52 $F/2$.

9.54   333 N.

9.56 $F = 74.3$ lb.

9.58   (a) $f = 24.5$ N;   (b) $\mu_s = 0.503$.

9.60   (a) $f = 8$ kN;   (b) $\mu_s = 0.533$.

9.62 $\mu_s = 0.432$.

9.64 $\mu_s = 0.901$.

9.66 $F = 139$ lb.

9.68 $F = 102$ lb.

9.70 $F = 1360$ lb.

9.72 $F = 156$ N.

9.74   343 kg.

9.76   No. The minimum value of μ_s required is 0.176.

9.78 $F = 1160$ N.

9.80   1.84 N-m.

9.82   (a) 0.967 in-lb;   (b) 0.566 in-lb.

9.84   (a) 2.39 ft-lb;   (b) 1.20 ft-lb.

9.86   11.8 ft-lb.

9.88   108 in-lb.

9.90   27.4 in-lb.

9.92   4.18 N-m.

9.94   4.88 N-m.

9.96   17.4 N-m.

9.98 $W = 1.55$ lb.

9.100   106 N.

9.102   51.9 lb.

9.104 $T = 40.9$ N.

9.106 $F_B = 207$ N.

9.108 $M = 1.92$ ft-lb.

9.110 $T = 346$ N.

9.112 $M = 160$ in-lb.

9.114 $M = 12.7$ N-m.

9.116 $M = 7.81$ N-m.

9.118 $M = 5.20$ N-m.

9.120   (a) $M = 93.5$ N-m;   (b) 8.17 percent.

9.122   9.51 ft-lb.

9.124   80.1 lb.

9.126 $T_C = 107$ N.

9.128 $M = rW(e^{\pi\mu_k} - 1)$.

9.130   (a) 14.2 lb;   (b) 128.3 lb.

9.132   13.1 lb.

9.134 $M_A = 65.2$ N-m, $M_B = 32.6$ N-m.

9.136   (a) $f = 10.3$ lb.

9.138 $F = 290$ lb.

9.140 $\alpha = 65.7°$.

9.142 $\alpha = 24.2°$.

9.144 $b = (h/\mu_s - t)/2$.

9.146 $h = 5.82$ in.

9.148   286 lb.

9.150   1130 kg, torque $= 2.67$ kN-m.

9.152 $f = 2.63$ N.

9.154 $\mu_s = 0.272$.

9.156 $M = 1.13$ N-m.

9.158 $P = 43.5$ N.

9.160   146 lb.

9.162   (a) $W = 106$ lb;   (b) $W = 273$ lb.

Chapter 10

10.2 $P_A = 0$, $V_A = 100$ N, $M_A = 40$ N-m.

10.4 $P_A = 0$, $V_A = 400$ lb, $M_A = -1900$ ft-lb.

10.6 (a) $P_A = 0$, $V_A = 4$ kN, $M_A = 4$ kN-m;
(b) $P_A = 0$, $V_A = 2$ kN, $M_A = 3$ kN-m.

10.8 $P_B = 0$, $V_B = 40$ N, $M_B = 373$ N-m.

10.10 $P_A = 0$, $V_A = -400$ lb, $M_A = 267$ ft-lb.

10.12 (a) $P_B = 0$, $V_B = -31$ lb, $M_B = 572$ ft-lb;
(b) $P_B = 0$, $V_B = 24$ lb, $M_B = 600$ ft-lb.

10.14 $P_A = 0$, $V_A = -2$ kN, $M_A = 6$ kN-m.

10.16 $P_A = 300$ N, $V_A = -150$ N, $M_A = 330$ N-m.

10.18 $P_A = 4$ kN, $V_A = 6$ kN, $M_A = 4.8$ kN-m.

10.20 $P_A = 0$, $V_A = -6$ kN, $M_A = 6$ kN-m.

10.22 $V = 400$ lb, $M = 400x$ ft-lb.

10.24 (a) $V = (5/2)(12 - x)^2$ lb,
$M = -(5/6)(12 - x)^3$ ft-lb.

10.26 $V = -600$ N, $M = -600x$ N-m.

10.28 (a) $0 < x < 6$ ft, $P = 0$, $V = 50$ lb,
$M = 50x$ ft-lb; $6 < x < 12$ ft,
$P = 0$, $V = 50 - (25/3)(x - 6)^2$ lb,
$M = 50x - (25/9)(x - 6)^3$ ft-lb;
(b)

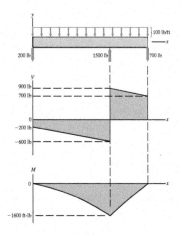

10.30 No. The maximum bending moment magnitude is 8 kN-m.

10.32 $M = 54.2$ N-m at $x = 233$ mm.

10.34

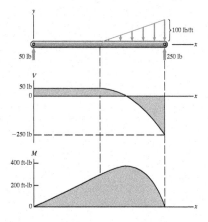

10.36 $V = -100x^2$ lb, $M = -33.3x^3 + 1800$ ft-lb.

10.38

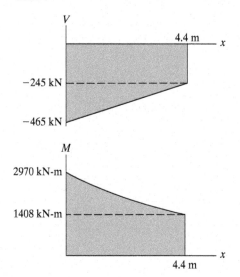

10.40

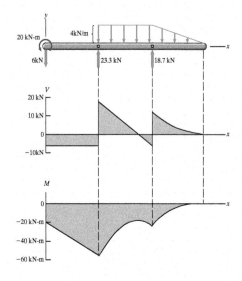

10.42

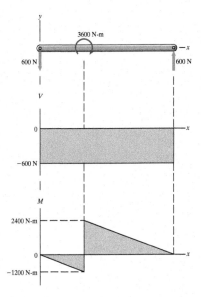

10.44

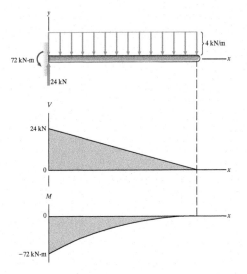

10.46

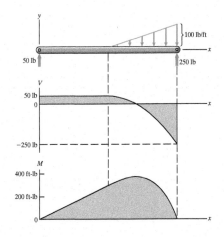

10.48

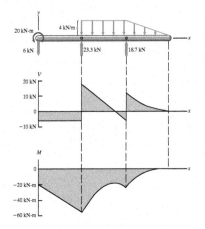

10.50 759 kip.
10.52 (a) $T_{max} = 86.2$ kN; (b) 36.14 m.
10.54 AC: 1061 N (T), BC: 1200 N (C).
10.56 Length = 108.3 m, $h = 37.2$ m.
10.58

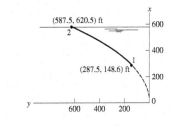

10.60 22.8 m.
10.62 (a) $h_1 = 4.95$ m, $h_2 = 2.19$ m;
  (b) $T_{AB} = 1.90$ kN, $T_{BC} = 1.84$ kN.
10.64 $T_1 = 185$ N, $T_3 = 209$ N.
10.66 (a) $h_2 = 4$ ft; (b) 90.1 lb.
10.68 $h_1 = 1.739$ m, $h_3 = 0.957$ m.
10.70 $h_2 = 464$ mm, $h_3 = 385$ mm.
10.72 $h_2 = 8.38$ ft, $h_3 = 12.08$ ft.
10.76 $x_p = 3/8$ m, $y_p = 3/5$ m.
10.78 $A_x = -100$ lb, $A_z = 562$ lb, $B = 281$ lb.
10.80 1.55 m.
10.82 6.67 m.
10.84 A: 257 lb to the right, 248 lb upward; B: 136 lb.
10.86 $d = 1.5$ m.
10.88 $A_x = 2160$ lb, $A_y = 2000$ lb, $B_x = 1830$ lb.
10.90 (a) 376 kN; (b) $x_p = 2.02$ m.
10.94 (a) $P_B = 0$, $V_B = -26.7$ lb, $M_B = 160$ ft-lb;
  (b) $P_C = 0$, $V_C = -26.7$ lb, $M_C = 80$ ft-lb.

10.96

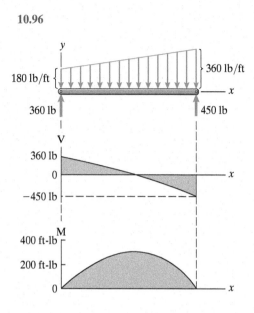

10.98 $0 < x < 2$ m, $P = 0$, $V = 1.33$ kN,
$M = 1.33x$ kN-m;
$2 < x < 6$ m, $P = 0$, $V = -2.67$ kN,
$M = 2.67(6 - x)$ kN-m.
10.100 $P_A = 0$, $V_A = 8$ kN, $M_A = -8$ kN-m.
10.102 (a) $P_B = 0$, $V_B = -40$ N, $M_B = 10$ N-m;
(b) $P_B = 0$, $V_B = -40$ N, $M_B = 10$ N-m.
10.104 $P = 0$, $V = -100$ lb, $M = -50$ ft-lb.
10.106 (a) $w = 74{,}100$ lb/ft; (b) 1.20×10^8 lb.
10.108 84.4 kip.
10.110 A: 44.2 kN to the left, 35.3 kN upward; B: 34.3 kN.

Chapter 11

11.2 (a) Work $= -3.20 \, \delta\theta$ kN-m; (b) $B = 2.31$ kN.
11.4 $F = 217$ N.
11.6 $A_x = 0$, $A_y = -237$ lb, $B_y = 937$ lb.
11.8 $F = 450$ N.
11.10 (a) $F = 392$ N; (b) 100 mm.
11.16 $F = 360$ lb.
11.18 $M = 270$ N-m.
11.20 12 kN.
11.22 9.17 kN.
11.24 (a) $0.625 \, \delta y$; (b) 216 N.
11.26 (a) $q = 3$, $q = 4$;
(b) $q = 3$ is unstable and $q = 4$ is stable.
11.28 $V = \frac{1}{2}kx^2 - \frac{1}{4}\varepsilon x^4$.
11.30 (a) Stable; (b) Unstable.
11.34 (b) It is stable.
11.36 (a) $\alpha = 35.2°$; (b) No.
11.38 (a) $\alpha = 28.7°$; (b) Yes.
11.40 Stable.
11.42 Unstable.
11.44 $\alpha = 0$ is unstable and $\alpha = 30°$ is stable.
11.46 $C_x = -7.78$ kN.
11.48 $8F$.
11.50 (a) $M = 800$ N-m; (b) $\alpha/4$.
11.52 $M = 1.50$ kN-m.
11.54 $F = 5$ kN.
11.56 $M = 63$ N-m.
11.58 $\alpha = 0$ is unstable and $\alpha = 59.4°$ is stable.
11.60 Unstable.
11.62 $\alpha = 30°$.

Index

Engineering Mechanics
DYNAMICS

CHAPTER

12

Introduction

How do engineers design and construct the devices we use, from simple objects such as chairs and pencil sharpeners to complicated ones such as dams, cars, airplanes, and spacecraft? They must have a deep understanding of the physics underlying the design of such devices and must be able to use mathematical models to predict their behavior. Students of engineering begin to learn how to analyze and predict the behaviors of physical systems by studying mechanics.

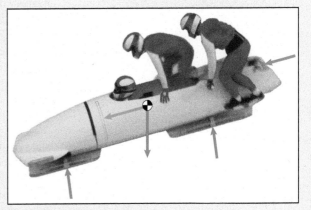

◄ The motions of the bobsled and its crew—their positions, velocities, and accelerations—can be analyzed using the equations of dynamics. Engineers use dynamics to predict the motions of objects.

12.1 Engineering and Mechanics

BACKGROUND

How can engineers design complex systems and predict their characteristics before they are constructed? Engineers have always relied on their knowledge of previous designs, experiments, ingenuity, and creativity to develop new designs. Modern engineers add a powerful technique: They develop mathematical equations based on the physical characteristics of the devices they design. With these mathematical models, engineers predict the behavior of their designs, modify them, and test them prior to their actual construction. Aerospace engineers use mathematical models to predict the paths the space shuttle will follow in flight. Civil engineers use mathematical models to analyze the effects of loads on buildings and foundations.

At its most basic level, mechanics is the study of forces and their effects. Elementary mechanics is divided into *statics*, the study of objects in equilibrium, and *dynamics*, the study of objects in motion. The results obtained in elementary mechanics apply directly to many fields of engineering. Mechanical and civil engineers designing structures use the equilibrium equations derived in statics. Civil engineers analyzing the responses of buildings to earthquakes and aerospace engineers determining the trajectories of satellites use the equations of motion derived in dynamics.

Mechanics was the first analytical science. As a result, fundamental concepts, analytical methods, and analogies from mechanics are found in virtually every field of engineering. Students of chemical and electrical engineering gain a deeper appreciation for basic concepts in their fields, such as equilibrium, energy, and stability, by learning them in their original mechanical contexts. By studying mechanics, they retrace the historical development of these ideas.

Mechanics consists of broad principles that govern the behavior of objects. In this book we describe these principles and provide examples that demonstrate some of their applications. Although it is essential that you practice working problems similar to these examples, and we include many problems of this kind, our objective is to help you understand the principles well enough to apply them to situations that are new to you. Each generation of engineers confronts new problems.

Problem Solving

In the study of mechanics, you learn problem-solving procedures that you will use in succeeding courses and throughout your career. Although different types of problems require different approaches, the following steps apply to many of them:

- Identify the information that is given and the information, or answer, you must determine. It's often helpful to restate the problem in your own words. When appropriate, make sure you understand the physical system or model involved.

- Develop a *strategy* for the problem. This means identifying the principles and equations that apply and deciding how you will use them to solve the problem. Whenever possible, draw diagrams to help visualize and solve the problem.

- Whenever you can, try to predict the answer. This will develop your intuition and will often help you recognize an incorrect answer.

- Solve the equations and, whenever possible, interpret your results and compare them with your prediction. This last step is a *reality check*. Is your answer reasonable?

Numbers

Engineering measurements, calculations, and results are expressed in numbers. You need to know how we express numbers in the examples and problems and how to express the results of your own calculations.

Significant Digits   This term refers to the number of meaningful (that is, accurate) digits in a number, counting to the right starting with the first non-zero digit. The two numbers 7.630 and 0.007630 are each stated to four significant digits. If only the first four digits in the number 7,630,000 are known to be accurate, this can be indicated by writing the number in scientific notation as 7.630×10^6.

If a number is the result of a measurement, the significant digits it contains are limited by the accuracy of the measurement. If the result of a measurement is stated to be 2.43, this means that the actual value is believed to be closer to 2.43 than to 2.42 or 2.44.

Numbers may be rounded off to a certain number of significant digits. For example, we can express the value of π to three significant digits, 3.14, or we can express it to six significant digits, 3.14159. When you use a calculator or computer, the number of significant digits is limited by the number of digits the machine is designed to carry.

Use of Numbers in This Book   You should treat numbers given in problems as exact values and not be concerned about how many significant digits they contain. If a problem states that a quantity equals 32.2, you can assume its value is 32.200. . . . We generally express intermediate results and answers in the examples and the answers to the problems to at least three significant digits. If you use a calculator, your results should be that accurate. Be sure to avoid round-off errors that occur if you round off intermediate results when making a series of calculations. Instead, carry through your calculations with as much accuracy as you can by retaining values in your calculator.

Space and Time

Space simply refers to the three-dimensional universe in which we live. Our daily experiences give us an intuitive notion of space and the locations, or positions, of points in space. The distance between two points in space is the length of the straight line joining them.

Measuring the distance between points in space requires a unit of length. We use both the International System of units, or SI units, and U.S. Customary units. In SI units, the unit of length is the meter (m). In U.S. Customary units, the unit of length is the foot (ft).

Time is, of course, familiar—our lives are measured by it. The daily cycles of light and darkness and the hours, minutes, and seconds measured by our clocks and watches give us an intuitive notion of time. Time is measured by the intervals between repeatable events, such as the swings of a clock pendulum or the vibrations of a quartz crystal in a watch. In both SI units and U.S. Customary units, the unit of time is the second (s). The minute (min), hour (h), and day are also frequently used.

If the position of a point in space relative to some reference point changes with time, the rate of change of its position is called its *velocity*, and the rate of change of its velocity is called its *acceleration*. In SI units, the velocity is expressed in meters per second (m/s) and the acceleration is expressed in meters per second per second, or meters per second squared (m/s^2). In U.S.

Customary units, the velocity is expressed in feet per second (ft/s) and the acceleration is expressed in feet per second squared (ft/s^2).

Newton's Laws

Elementary mechanics was established on a firm basis with the publication in 1687 of *Philosophiae Naturalis Principia Mathematica*, by Isaac Newton. Although highly original, it built on fundamental concepts developed by many others during a long and difficult struggle toward understanding (Fig. 12.1).

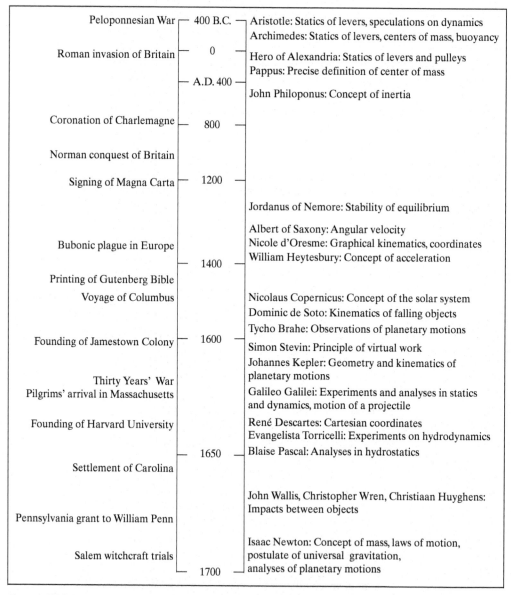

Figure 12.1

Chronology of developments in mechanics up to the publication of Newton's *Principia* in relation to other events in history.

Newton stated three "laws" of motion, which we express in modern terms:

1. *When the sum of the forces acting on a particle is zero, its velocity is constant. In particular, if the particle is initially stationary, it will remain stationary.*

2. *When the sum of the forces acting on a particle is not zero, the sum of the forces is equal to the rate of change of the linear momentum of the particle. If the mass is constant, the sum of the forces is equal to the product of the mass of the particle and its acceleration.*

3. *The forces exerted by two particles on each other are equal in magnitude and opposite in direction.*

Notice that we did not define force and mass before stating Newton's laws. The modern view is that these terms are defined by the second law. To demonstrate, suppose that we choose an arbitrary object and define it to have unit mass. Then we define a unit of force to be the force that gives our unit mass an acceleration of unit magnitude. In principle, we can then determine the mass of any object: We apply a unit force to it, measure the resulting acceleration, and use the second law to determine the mass. We can also determine the magnitude of any force: We apply it to our unit mass, measure the resulting acceleration, and use the second law to determine the force.

Thus Newton's second law gives precise meanings to the terms *mass* and *force*. In SI units, the unit of mass is the kilogram (kg). The unit of force is the newton (N), which is the force required to give a mass of one kilogram an acceleration of one meter per second squared. In U.S. Customary units, the unit of force is the pound (lb). The unit of mass is the slug, which is the amount of mass accelerated at one foot per second squared by a force of one pound.

Although the results we discuss in this book are applicable to many of the problems met in engineering practice, there are limits to the validity of Newton's laws. For example, they don't give accurate results if a problem involves velocities that are not small compared to the velocity of light (3×10^8 m/s). Einstein's special theory of relativity applies to such problems. Elementary mechanics also fails in problems involving dimensions that are not large compared to atomic dimensions. Quantum mechanics must be used to describe phenomena on the atomic scale.

International System of Units

In SI units, length is measured in meters (m) and mass in kilograms (kg). Time is measured in seconds (s), although other familiar measures such as minutes (min), hours (h), and days are also used when convenient. Meters, kilograms, and seconds are called the *base units* of the SI system. Force is measured in newtons (N). Recall that these units are related by Newton's second law: One newton is the force required to give an object of one kilogram mass an acceleration of one meter per second squared:

$$1 \text{ N} = (1 \text{ kg})(1 \text{ m/s}^2) = 1 \text{ kg-m/s}^2.$$

Because the newton can be expressed in terms of the base units, it is called a *derived unit*.

To express quantities by numbers of convenient size, multiples of units are indicated by prefixes. The most common prefixes, their abbreviations, and the multiples they represent are shown in Table 12.1. For example, 1 km is 1 kilometer, which is 1000 m, and 1 Mg is 1 megagram, which is 10^6 g, or 1000 kg. We frequently use kilonewtons (kN).

Table 12.1 The common prefixes used in SI units and the multiples they represent.

Prefix	Abbreviation	Multiple
nano-	n	10^{-9}
micro-	μ	10^{-6}
milli-	m	10^{-3}
kilo-	k	10^3
mega-	M	10^6
giga-	G	10^9

U.S. Customary Units

In U.S. Customary units, length is measured in feet (ft) and force is measured in pounds (lb). Time is measured in seconds (s). These are the base units of the U.S. Customary system. In this system of units, mass is a derived unit. The unit of mass is the slug, which is the mass of material accelerated at one foot per second squared by a force of one pound. Newton's second law states that

$$1 \text{ lb} = (1 \text{ slug})(1 \text{ ft/s}^2).$$

From this expression we obtain

$$1 \text{ slug} = 1 \text{ lb-s}^2/\text{ft}.$$

We use other U.S. Customary units such as the mile (1 mi = 5280 ft) and the inch (1 ft = 12 in). We also use the kilopound (kip), which is 1000 lb.

Angular Units

In both SI and U.S. Customary units, angles are normally expressed in radians (rad). We show the value of an angle θ in radians in Fig. 12.2. It is defined to be the ratio of the part of the circumference subtended by θ to the radius of the circle. Angles are also expressed in degrees. Since there are 360 degrees (360°) in a complete circle, and the complete circumference of the circle is $2\pi R$, 360° equals 2π rad.

Equations containing angles are nearly always derived under the assumption that angles are expressed in radians. Therefore, when you want to substitute the value of an angle expressed in degrees into an equation, you should first convert it into radians. A notable exception to this rule is that many calculators are designed to accept angles expressed in either degrees or radians when you use them to evaluate functions such as sin θ.

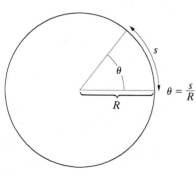

Figure 12.2
Definition of an angle in radians.

Conversion of Units

Many situations arise in engineering practice that require values expressed in one kind of unit to be converted into values in other units. For example, if some of the data to be used in an equation are given in SI units and some are given in U.S. Customary units, they must all be expressed in terms of one system of units before they are substituted into the equation. Converting units is straightforward, although it must be done with care.

Suppose that we want to express 1 mile per hour (mi/h) in terms of feet per second (ft/s). Because 1 mile equals 5280 feet and 1 hour equals 3600 seconds, we can treat the expressions

$$\left(\frac{5280 \text{ ft}}{1 \text{ mi}}\right) \quad \text{and} \quad \left(\frac{1 \text{ h}}{3600 \text{ s}}\right)$$

as ratios whose values are 1. In this way, we obtain

$$1 \text{ mi/h} = (1 \text{ mi/h})\left(\frac{5280 \text{ ft}}{1 \text{ mi}}\right)\left(\frac{1 \text{ h}}{3600 \text{ s}}\right) = 1.47 \text{ ft/s}.$$

Some useful unit conversions are given in Table 12.2.

Table 12.2 Unit conversions.

Time	1 minute	=	60 seconds
	1 hour	=	60 minutes
	1 day	=	24 hours
Length	1 foot	=	12 inches
	1 mile	=	5280 feet
	1 inch	=	25.4 millimeters
	1 foot	=	0.3048 meters
Angle	2π radians	=	360 degrees
Mass	1 slug	=	14.59 kilograms
Force	1 pound	=	4.448 newtons

RESULTS

- Identify the given information and the answer that must be determined.
- Develop a strategy; identify principles and equations that apply and how they will be used.
- Try to predict the answer whenever possible.
- Obtain the answer and, whenever possible, interpret it and compare it with the prediction.

Problem Solving: These steps apply to many types of problems.

SI Units—The *base units* are time in seconds (s), length in meters (m), and mass in kilograms (kg). The unit of force is the newton (N), which is the force required to accelerate a mass of one kilogram at one meter per second squared.

U.S. Customary Units—The base units are time in seconds (s), length in feet (ft), and force in pounds (lb). The unit of mass is the slug, which is the mass accelerated at one foot per second squared by a force of one pound.

Systems of units.

$$\theta = \frac{s}{R}$$

Definition of an angle in radians.

Equivalent quantities, such as 1 hour = 60 minutes, can be written as ratios whose values are 1:

$$\left(\frac{1\,\text{h}}{60\,\text{min}}\right) = 1,$$

and used to convert units. For example,

$$15\,\text{min} = 15\,\text{min}\left(\frac{1\,\text{h}}{60\,\text{min}}\right) = 0.25\,\text{h}.$$

Conversion of units.

A comprehensive resource on units has been compiled by Russ Rowlett of the University of North Carolina at Chapel Hill and made available online at www.unc.edu/~rowlett/units.

Active Example 12.1 | Converting Units (▶ *Related Problem 12.11*)

A man is riding a bicycle at a speed of 6 meters per second (m/s). How fast is he going in kilometers per hour (km/h)?

Strategy
One kilometer is 1000 meters and one hour is 60 minutes × 60 seconds = 3600 seconds. We can use these unit conversions to determine his speed in km/h.

Solution

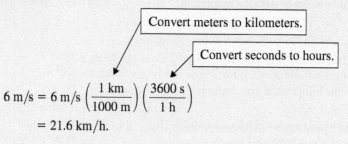

Convert meters to kilometers.

Convert seconds to hours.

$$6 \text{ m/s} = 6 \text{ m/s} \left(\frac{1 \text{ km}}{1000 \text{ m}} \right) \left(\frac{3600 \text{ s}}{1 \text{ h}} \right)$$

$$= 21.6 \text{ km/h}.$$

Practice Problem A man is riding a bicycle at a speed of 10 feet per second (ft/s). How fast is he going in miles per hour (mi/h)?
Answer: 6.82 mi/h.

Example 12.2 | Converting Units of Pressure (▶ *Related Problem 12.16*)

Deep Submersible Vehicle.

The pressure exerted at a point of the hull of the deep submersible vehicle is 3.00×10^6 Pa (pascals). A pascal is 1 newton per square meter. Determine the pressure in pounds per square foot.

Strategy
From Table 12.2, 1 pound = 4.448 newtons and 1 foot = 0.3048 meters. With these unit conversions we can calculate the pressure in pounds per square foot.

Solution
The pressure (to three significant digits) is

$$3.00 \times 10^6 \text{ N/m}^2 = (3.00 \times 10^6 \text{ N/m}^2) \left(\frac{1 \text{ lb}}{4.448 \text{ N}} \right) \left(\frac{0.3048 \text{ m}}{1 \text{ ft}} \right)^2$$

$$= 62{,}700 \text{ lb/ft}^2.$$

Critical Thinking
How could we have obtained this result in a more direct way? Notice from the table of unit conversions in the inside front cover that 1 Pa = 0.0209 lb/ft². Therefore,

$$3.00 \times 10^6 \text{ N/m}^2 = (3.00 \times 10^6 \text{ N/m}^2) \left(\frac{0.0209 \text{ lb/ft}^2}{1 \text{ N/m}^2} \right)$$

$$= 62{,}700 \text{ lb/ft}^2.$$

| Example 12.3 | Determining Units from an Equation (▶ *Related Problem 12.20*) |

Suppose that in Einstein's equation

$$E = mc^2,$$

the mass m is in kilograms and the velocity of light c is in meters per second.

(a) What are the SI units of E?
(b) If the value of E in SI units is 20, what is its value in U.S. Customary base units?

Strategy

(a) Since we know the units of the terms m and c, we can deduce the units of E from the given equation.
(b) We can use the unit conversions for mass and length from Table 12.2 to convert E from SI units to U.S. Customary units.

Solution

(a) From the equation for E,

$$E = (m \text{ kg})(c \text{ m/s})^2,$$

the SI units of E are kg-m^2/s^2.
(b) From Table 12.2, 1 slug = 14.59 kg and 1 ft = 0.3048 m. Therefore,

$$1 \text{ kg-m}^2/\text{s}^2 = (1 \text{ kg-m}^2/\text{s}^2)\left(\frac{1 \text{ slug}}{14.59 \text{ kg}}\right)\left(\frac{1 \text{ ft}}{0.3048 \text{ m}}\right)^2$$

$$= 0.738 \text{ slug-ft}^2/\text{s}^2.$$

The value of E in U.S. Customary units is

$$E = (20)(0.738) = 14.8 \text{ slug-ft}^2/\text{s}^2.$$

Critical Thinking

In part (a), how did we know that we could determine the units of E by determining the units of mc^2? The dimensions, or units, of each term in an equation must be the same. For example, in the equation $a + b = c$, the dimensions of each of the terms a, b, and c must be the same. The equation is said to be *dimensionally homogeneous*. This requirement is expressed by the colloquial phrase "Don't compare apples and oranges."

Problems

12.1 The value of π is 3.14159265.... If C is the circumference of a circle and r is its radius, determine the value of r/C to four significant digits.

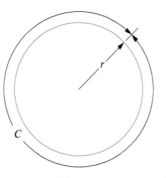

Problem 12.1

12.2 The base of natural logarithms is $e = 2.718281828\ldots$.

(a) Express e to five significant digits.

(b) Determine the value of e^2 to five significant digits.

(c) Use the value of e you obtained in part (a) to determine the value of e^2 to five significant digits.

[Part (c) demonstrates the hazard of using rounded-off values in calculations.]

12.3 A machinist drills a circular hole in a panel with a nominal radius $r = 5$ mm. The actual radius of the hole is in the range $r = 5 \pm 0.01$ mm.

(a) To what number of significant digits can you express the radius?

(b) To what number of significant digits can you express the area of the hole?

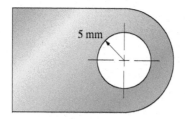

Problem 12.3

12.4 The opening in the soccer goal is 24 ft wide and 8 ft high, so its area is 24 ft $\times$ 8 ft $= 192$ ft^2. What is its area in m^2 to three significant digits?

Problem 12.4

12.5 The Burj Dubai, scheduled for completion in 2008, will be the world's tallest building with a height of 705 m. The area of its ground footprint will be 8000 m^2. Convert its height and footprint area to U.S. Customary units to three significant digits.

Problem 12.5

12.6 Suppose that you have just purchased a Ferrari F355 coupe and you want to know whether you can use your set of SAE (U.S. Customary unit) wrenches to work on it. You have wrenches with widths w = 1/4 in, 1/2 in, 3/4 in, and 1 in, and the car has nuts with dimensions n = 5 mm, 10 mm, 15 mm, 20 mm, and 25 mm. Defining a wrench to fit if w is no more than 2% larger than n, which of your wrenches can you use?

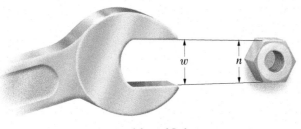

Problem 12.6

12.7 Suppose that the height of Mt. Everest is known to be between 29,032 ft and 29,034 ft. Based on this information, to how many significant digits can you express the height (a) in feet? (b) in meters?

12.8 The maglev (magnetic levitation) train from Shanghai to the airport at Pudong reaches a speed of 430 km/h. Determine its speed (a) in mi/h; (b) in ft/s.

Problem 12.8

12.9 In the 2006 Winter Olympics, the men's 15-km cross-country skiing race was won by Andrus Veerpalu of Estonia in a time of 38 minutes, 1.3 seconds. Determine his average speed (the distance traveled divided by the time required) to three significant digits (a) in km/h; (b) in mi/h.

12.10 The Porsche's engine exerts 229 ft-lb (foot-pounds) of torque at 4600 rpm. Determine the value of the torque in N-m (newton-meters).

Problem 12.10

▶ **12.11** The *kinetic energy* of the man in Active Example 12.1 is defined by $\frac{1}{2}mv^2$, where m is his mass and v is his velocity. The man's mass is 68 kg and he is moving at 6 m/s, so his kinetic energy is $\frac{1}{2}(68 \text{ kg})(6 \text{ m/s})^2 = 1224 \text{ kg-m}^2/\text{s}^2$. What is his kinetic energy in U.S. Customary units?

12.12 The acceleration due to gravity at sea level in SI units is g = 9.81 m/s^2. By converting units, use this value to determine the acceleration due to gravity at sea level in U.S. Customary units.

12.13 A *furlong per fortnight* is a facetious unit of velocity, perhaps made up by a student as a satirical comment on the bewildering variety of units engineers must deal with. A furlong is 660 ft (1/8 mile). A fortnight is 2 weeks (14 nights). If you walk to class at 2 m/s, what is your speed in furlongs per fortnight to three significant digits?

12.14 Determine the cross-sectional area of the beam (a) in m^2; (b) in in^2.

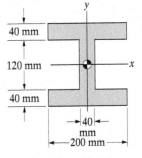

Problem 12.14

12.15 The cross-sectional area of the C12×30 American Standard Channel steel beam is $A = 8.81$ in^2. What is its cross-sectional area in mm^2?

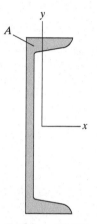

Problem 12.15

▶ **12.16** A pressure transducer measures a value of 300 lb/in^2. Determine the value of the pressure in pascals. A pascal (Pa) is one newton per square meter.

12.17 A horsepower is 550 ft-lb/s. A watt is 1 N-m/s. Determine how many watts are generated by the engines of the passenger jet if they are producing 7000 horsepower.

Problem 12.17

12.18 Distributed loads on beams are expressed in units of force per unit length. If the value of a distributed load is 400 N/m, what is its value in lb/ft?

12.19 The moment of inertia of the rectangular area about the x axis is given by the equation

$$I = \tfrac{1}{3}bh^3.$$

The dimensions of the area are $b = 200$ mm and $h = 100$ mm. Determine the value of I to four significant digits in terms of (a) mm^4, (b) m^4, and (c) in^4.

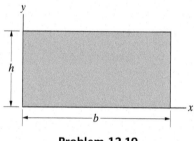

Problem 12.19

▶ **12.20** In Example 12.3, instead of Einstein's equation consider the equation $L = mc$, where the mass m is in kilograms and the velocity of light c is in meters per second. (a) What are the SI units of L? (b) If the value of L in SI units is 12, what is its value in U.S. Customary base units?

12.21 The equation

$$\sigma = \frac{My}{I}$$

is used in the mechanics of materials to determine normal stresses in beams.

(a) When this equation is expressed in terms of SI base units, M is in newton-meters (N-m), y is in meters (m), and I is in meters to the fourth power (m^4). What are the SI units of σ?

(b) If $M = 2000$ N-m, $y = 0.1$ m, and $I = 7 \times 10^{-5}$ m^4, what is the value of σ in U.S. Customary base units?

12.2 Newtonian Gravitation

BACKGROUND

Newton postulated that the gravitational force between two particles of mass m_1 and m_2 that are separated by a distance r (Fig. 12.3) is

$$F = \frac{Gm_1m_2}{r^2}, \tag{12.1}$$

where G is called the universal gravitational constant. The value of G in SI units is 6.67×10^{-11} N-m^2/kg^2. Based on this postulate, he calculated the gravitational force between a particle of mass m_1 and a homogeneous sphere of mass m_2 and found that it is also given by Eq. (12.1), with r denoting the distance from the particle to the center of the sphere. Although the earth is not a homogeneous sphere, we can use this result to approximate the weight of an object of mass m due to the gravitational attraction of the earth. We have

$$W = \frac{Gmm_E}{r^2}, \tag{12.2}$$

where m_E is the mass of the earth and r is the distance from the center of the earth to the object. Notice that the weight of an object depends on its location relative to the center of the earth, whereas the mass of the object is a measure of the amount of matter it contains and doesn't depend on its position.

When an object's weight is the only force acting on it, the resulting acceleration is called the acceleration due to gravity. In this case, Newton's second law states that $W = ma$, and from Eq. (12.2) we see that the acceleration due to gravity is

$$a = \frac{Gm_E}{r^2}. \tag{12.3}$$

The *acceleration due to gravity at sea level* is denoted by g. Denoting the radius of the earth by R_E, we see from Eq. (12.3) that $Gm_E = gR_E^2$. Substituting this result into Eq. (12.3), we obtain an expression for the acceleration due to gravity at a distance r from the center of the earth in terms of the acceleration due to gravity at sea level:

$$a = g\frac{R_E^2}{r^2}. \tag{12.4}$$

Since the weight of the object $W = ma$, the weight of an object at a distance r from the center of the earth is

$$W = mg\frac{R_E^2}{r^2}. \tag{12.5}$$

At sea level $(r = R_E)$, the weight of an object is given in terms of its mass by the simple relation

$$W = mg. \tag{12.6}$$

The value of g varies from location to location on the surface of the earth. The values we use in examples and problems are $g = 9.81$ m/s^2 in SI units and $g = 32.2$ ft/s^2 in U.S. Customary units.

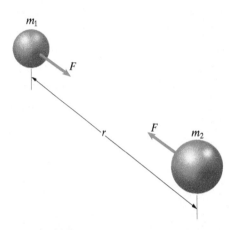

Figure 12.3
The gravitational forces between two particles are equal in magnitude and directed along the line between them.

RESULTS

The gravitational force between two particles of mass m_1 and m_2 that are separated by a distance r is

$$F = \frac{Gm_1m_2}{r^2}, \tag{12.1}$$

where G is the universal gravitational constant. The value of G in SI units is

$$6.67 \times 10^{-11} \text{ N-m}^2/\text{kg}^2.$$

> Newtonian gravitation.

When the earth is modeled as a homogeneous sphere of radius R_E, the acceleration due to gravity at a distance r from the center is

$$a = g\frac{R_E^2}{r^2}, \tag{12.4}$$

where g is the acceleration due to gravity at sea level.

> Acceleration due to gravity of the earth.

$$W = mg, \tag{12.6}$$

where m is the mass of the object and g is the acceleration due to gravity at sea level.

> Weight of an object at sea level.

Active Example 12.4   Weight and Mass (▶ *Related Problem 12.22*)

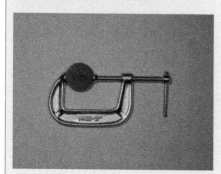

The C-clamp weighs 14 oz at sea level. [16 oz (ounces) = 1 lb.] The acceleration due to gravity at sea level is $g = 32.2 \text{ ft/s}^2$. What is the mass of the C-clamp in slugs?

Strategy
We must first determine the weight of the C-clamp in pounds. Then we can use Eq. (12.6) to determine the mass in slugs.

Solution

$$14 \text{ oz} = 14 \text{ oz} \left(\frac{1 \text{ lb}}{16 \text{ oz}}\right) = 0.875 \text{ lb}.$$

> Convert the weight from ounces to pounds.

$$m = \frac{W}{g} = \frac{0.875 \text{ lb}}{32.2 \text{ ft/s}^2} = 0.0272 \text{ slug}.$$

> Use Eq. (12.6) to calculate the mass in slugs.

Practice Problem The mass of the C-clamp is 0.397 kg. The acceleration due to gravity at sea level is $g = 9.81 \text{ m/s}^2$. What is the weight of the C-clamp at sea level in newtons?

Answer: 3.89 N.

Example 12.5   Determining an Object's Weight (▶ *Related Problem 12.27*)

When the Mars Exploration Rover was fully assembled, its mass was 180 kg. The acceleration due to gravity at the surface of Mars is 3.68 m/s^2 and the radius of Mars is 3390 km.

(a) What was the rover's weight when it was at sea level on Earth?

(b) What is the rover's weight on the surface of Mars?

(c) The entry phase began when the spacecraft reached the Mars atmospheric entry interface point at 3522 km from the center of Mars. What was the rover's weight at that point?

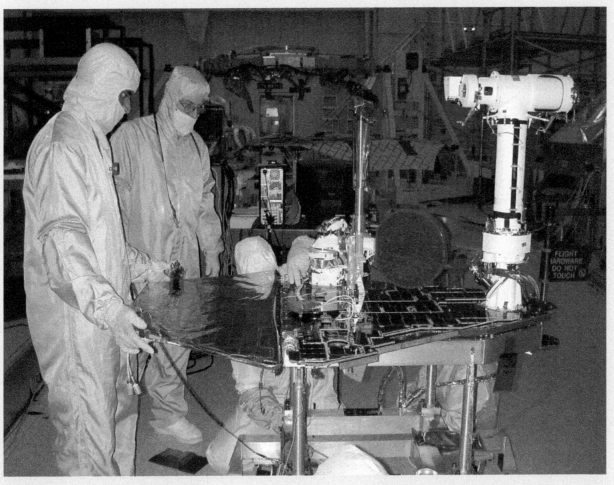

Mars Exploration Rover being assembled.

Strategy

The rover's weight at sea level on Earth is given by Eq. (12.6) with $g = 9.81$ m/s^2.

We can determine the weight on the surface of Mars by using Eq. (12.6) with the acceleration due to gravity equal to 3.68 m/s^2.

To determine the rover's weight as it began the entry phase, we can write an equation for Mars equivalent to Eq. (12.5).

Solution

(a) The weight at sea level on Earth is

$$W = mg$$
$$= (180 \text{ kg})(9.81 \text{ m/s}^2)$$
$$= 1770 \text{ N } (397 \text{ lb}).$$

(b) Let $g_M = 3.68$ m/s^2 be the acceleration due to gravity at the surface of Mars. Then the weight of the rover on the surface of Mars is

$$W = mg_M$$
$$= (180 \text{ kg})(3.68 \text{ m/s}^2)$$
$$= 662 \text{ N } (149 \text{ lb}).$$

(c) Let $R_M = 3390$ km be the radius of Mars. From Eq. (12.5), the rover's weight when it is 3522 km above the center of Mars is

$$W = mg_M \frac{R_M^2}{r^2}$$
$$= (180 \text{ kg})(3.68 \text{ m/s}^2) \frac{(3{,}390{,}000 \text{ m})^2}{(3{,}522{,}000 \text{ m})^2}$$
$$= 614 \text{ N } (138 \text{ lb}).$$

Critical Thinking

In part (c), how did we know that we could apply Eq. (12.5) to Mars? Equation (12.5) is applied to Earth based on modeling it as a homogeneous sphere. It can be applied to other celestial objects under the same assumption. The accuracy of the results depends on how aspherical and inhomogeneous the object is.

Problems

▶ **12.22** The acceleration due to gravity on the surface of the moon is 1.62 m/s². (a) What would the mass of the C-clamp in Active Example 12.4 be on the surface of the moon? (b) What would the weight of the C-clamp in newtons be on the surface of the moon?

12.23 The 1 ft × 1 ft × 1 ft cube of iron weighs 490 lb at sea level. Determine the weight in newtons of a 1 m × 1 m × 1 m cube of the same material at sea level.

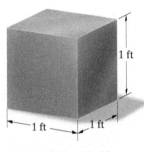

Problem 12.23

12.24 The area of the Pacific Ocean is 64,186,000 square miles and its average depth is 12,925 ft. Assume that the weight per unit volume of ocean water is 64 lb/ft³. Determine the mass of the Pacific Ocean (a) in slugs; (b) in kilograms.

12.25 The acceleration due to gravity at sea level is $g = 9.81$ m/s². The radius of the earth is 6370 km. The universal gravitational constant is $G = 6.67 \times 10^{-11}$ N-m²/kg². Use this information to determine the mass of the earth.

12.26 A person weighs 180 lb at sea level. The radius of the earth is 3960 mi. What force is exerted on the person by the gravitational attraction of the earth if he is in a space station in orbit 200 mi above the surface of the earth?

▶ **12.27** The acceleration due to gravity on the surface of the moon is 1.62 m/s². The moon's radius is $R_M = 1738$ km. (See Example 12.5.)

(a) What is the weight in newtons on the surface of the moon of an object that has a mass of 10 kg?

(b) Using the approach described in Example 12.5, determine the force exerted on the object by the gravity of the moon if the object is located 1738 km above the moon's surface.

12.28 If an object is near the surface of the earth, the variation of its weight with distance from the center of the earth can often be neglected. The acceleration due to gravity at sea level is $g = 9.81$ m/s². The radius of the earth is 6370 km. The weight of an object at sea level is mg, where m is its mass. At what height above the surface of the earth does the weight of the object decrease to $0.99mg$?

12.29 The planet Neptune has an equatorial diameter of 49,532 km and its mass is 1.0247×10^{26} kg. If the planet is modeled as a homogeneous sphere, what is the acceleration due to gravity at its surface? (The universal gravitational constant is $G = 6.67 \times 10^{-11}$ N-m²/kg².)

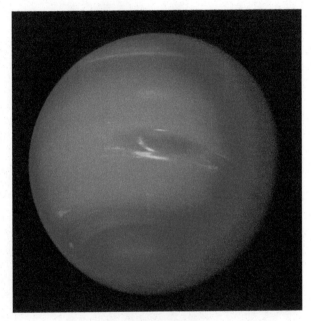

Problem 12.29

12.30 At a point between the earth and the moon, the magnitude of the force exerted on an object by the earth's gravity equals the magnitude of the force exerted on the object by the moon's gravity. What is the distance from the center of the earth to that point to three significant digits? The distance from the center of the earth to the center of the moon is 383,000 km, and the radius of the earth is 6370 km. The radius of the moon is 1738 km, and the acceleration due to gravity at its surface is 1.62 m/s².

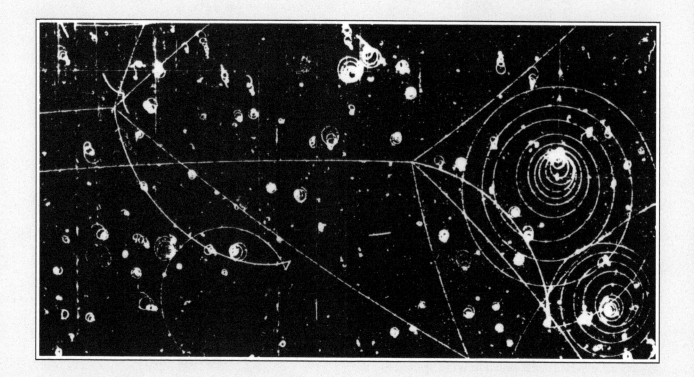

Motion of a Point

In this chapter we begin the study of motion. We are not yet concerned with the properties of objects or the causes of their motions—our objective is simply to describe and analyze the motion of a point in space. After defining the position, velocity, and acceleration of a point, we consider the simplest case, motion along a straight line. We then show how motion of a point along an arbitrary path, or *trajectory*, is expressed and analyzed using various coordinate systems.

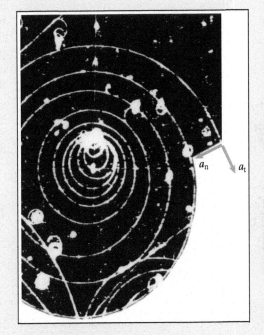

◀ The lines show the paths followed by subatomic particles moving in a magnetic field. The particles with curved paths have both tangential and normal components of acceleration.

13.1 Position, Velocity, and Acceleration

BACKGROUND

If you observe people in a room, such as a group at a party, you perceive their positions relative to the room. For example, some people may be in the back of the room, some in the middle of the room, and so forth. The room is your "frame of reference." To make this idea precise, we can introduce a cartesian coordinate system with its axes aligned with the walls of the room as in Fig. 13.1a and specify the position of a person (actually, the position of some point of the person, such as his or her center of mass) by specifying the components of the position vector **r** relative to the origin of the coordinate system. This coordinate system is a convenient reference frame for objects in the room. If you are sitting in an airplane, you perceive the positions of objects within the airplane relative to the airplane. In this case, the interior of the airplane is your frame of reference. To precisely specify the position of a person within the airplane, we can introduce a cartesian coordinate system that is fixed relative to the airplane and measure the position of the person's center of mass by specifying the components of the position vector **r** relative to the origin (Fig. 13.1b). A *reference frame* is simply a coordinate system that is suitable for specifying positions of points. You may be familiar only with cartesian coordinates. We discuss other examples in this chapter and continue our discussion of reference frames throughout the book.

We can describe the position of a point P relative to a given reference frame with origin O by the *position vector* **r** from O to P (Fig. 13.2a). Suppose that P is in motion relative to the chosen reference frame, so that **r** is a function of time t (Fig. 13.2b). We express this by the notation

$$\mathbf{r} = \mathbf{r}(t).$$

The *velocity* of P relative to the given reference frame at time t is defined by

$$\mathbf{v} = \frac{d\mathbf{r}}{dt} = \lim_{\Delta t \to 0} \frac{\mathbf{r}(t + \Delta t) - \mathbf{r}(t)}{\Delta t}, \tag{13.1}$$

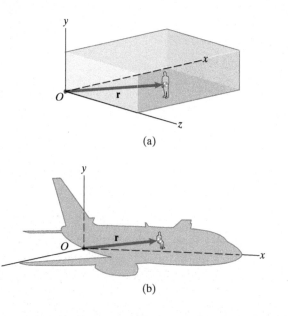

(a)

(b)

Figure 13.1
Convenient reference frames for specifying positions of objects
(a) in a room;
(b) in an airplane.

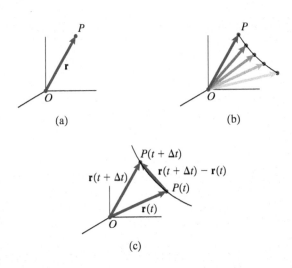

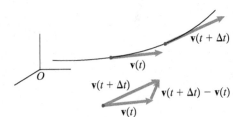

Figure 13.2
(a) The position vector $\mathbf{r}$ of P relative to O.
(b) Motion of P relative to the reference frame.
(c) Change in position of P from t to $t + \Delta t$.

where the vector $\mathbf{r}(t + \Delta t) - \mathbf{r}(t)$ is the change in position, or *displacement*, of P during the interval of time Δt (Fig. 13.2c). Thus, the velocity is the rate of change of the position of P.

The dimensions of a derivative are determined just as if it is a ratio, so the dimensions of $\mathbf{v}$ are (distance)/(time). The reference frame being used is often obvious, and we simply call $\mathbf{v}$ the velocity of P. However, remember that the position and velocity of a point can be specified only relative to some reference frame.

Notice in Eq. (13.1) that the derivative of a vector with respect to time is defined in exactly the same way as is the derivative of a scalar function. As a result, the derivative of a vector shares some of the properties of the derivative of a scalar function. We will use two of these properties: the derivative with respect to time, or time derivative, of the sum of two vector functions $\mathbf{u}$ and $\mathbf{w}$ is

$$\frac{d}{dt}(\mathbf{u} + \mathbf{w}) = \frac{d\mathbf{u}}{dt} + \frac{d\mathbf{w}}{dt},$$

and the time derivative of the product of a scalar function f and a vector function $\mathbf{u}$ is

$$\frac{d(f\mathbf{u})}{dt} = \frac{df}{dt}\mathbf{u} + f\frac{d\mathbf{u}}{dt}.$$

The *acceleration* of P relative to the given reference frame at time t is defined by

$$\mathbf{a} = \frac{d\mathbf{v}}{dt} = \lim_{\Delta t \to 0} \frac{\mathbf{v}(t + \Delta t) - \mathbf{v}(t)}{\Delta t}, \tag{13.2}$$

where $\mathbf{v}(t + \Delta t) - \mathbf{v}(t)$ is the change in the velocity of P during the interval of time Δt (Fig. 13.3). The acceleration is the rate of change of the velocity of P at time t (the second time derivative of the displacement), and its dimensions are (distance)/(time)^2.

We have defined the velocity and acceleration of P relative to the origin O of the reference frame. We can show that *a point has the same velocity and acceleration relative to any fixed point in a given reference frame*. Let O' be an arbitrary fixed point, and let $\mathbf{r}'$ be the position vector from O' to P (Fig. 13.4a). The velocity of P relative to O' is $\mathbf{v}' = d\mathbf{r}'/dt$. The velocity of P relative to the

Figure 13.3
Change in the velocity of P from t to $t + \Delta t$.

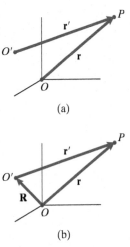

Figure 13.4
(a) Position vectors of P relative to O and O'.
(b) Position vector of O' relative to O.

origin O is $\mathbf{v} = d\mathbf{r}/dt$. We wish to show that $\mathbf{v}' = \mathbf{v}$. Let $\mathbf{R}$ be the vector from O to O' (Fig. 13.4b), so that

$$\mathbf{r}' = \mathbf{r} - \mathbf{R}.$$

Since the vector $\mathbf{R}$ is constant, the velocity of P relative to O' is

$$\mathbf{v}' = \frac{d\mathbf{r}'}{dt} = \frac{d\mathbf{r}}{dt} - \frac{d\mathbf{R}}{dt} = \frac{d\mathbf{r}}{dt} = \mathbf{v}.$$

The acceleration of P relative to O' is $\mathbf{a}' = d\mathbf{v}'/dt$, and the acceleration of P relative to O is $\mathbf{a} = d\mathbf{v}/dt$. Since $\mathbf{v}' = \mathbf{v}$, $\mathbf{a}' = \mathbf{a}$. Thus, the velocity and acceleration of a point P relative to a given reference frame do not depend on the location of the fixed reference point used to specify the position of P.

RESULTS

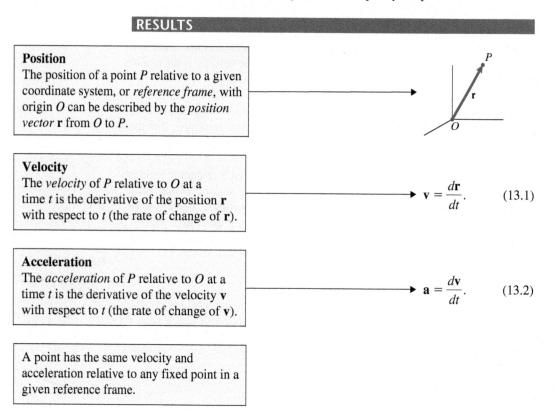

Position
The position of a point P relative to a given coordinate system, or *reference frame*, with origin O can be described by the *position vector* $\mathbf{r}$ from O to P.

Velocity
The *velocity* of P relative to O at a time t is the derivative of the position $\mathbf{r}$ with respect to t (the rate of change of $\mathbf{r}$).

$$\mathbf{v} = \frac{d\mathbf{r}}{dt}. \qquad (13.1)$$

Acceleration
The *acceleration* of P relative to O at a time t is the derivative of the velocity $\mathbf{v}$ with respect to t (the rate of change of $\mathbf{v}$).

$$\mathbf{a} = \frac{d\mathbf{v}}{dt}. \qquad (13.2)$$

A point has the same velocity and acceleration relative to any fixed point in a given reference frame.

13.2 Straight-Line Motion

BACKGROUND

We discuss this simple type of motion primarily so that you can gain experience and insight before proceeding to the general case of the motion of a point. But engineers must analyze straight-line motions in many practical situations, such as the motion of a vehicle on a straight road or the motion of a piston in an internal combustion engine.

Description of the Motion

Consider a straight line through the origin O of a given reference frame. We assume that the direction of the line relative to the reference frame is fixed.

(For example, the x axis of a cartesian coordinate system passes through the origin and has fixed direction relative to the reference frame.) We can specify the position of a point P on such a line relative to O by a coordinate s measured along the line from O to P. In Fig. 13.5a we define s to be positive to the right, so s is positive when P is to the right of O and negative when P is to the left of O. The *displacement* of P during an interval of time from t_0 to t is the change in the position $s(t) - s(t_0)$, where $s(t)$ denotes the position at time t.

By introducing a unit vector $\mathbf{e}$ that is parallel to the line and points in the positive s direction (Fig. 13.5b), we can write the position vector of P relative to O as

$$\mathbf{r} = s\mathbf{e}.$$

Because the magnitude and direction of $\mathbf{e}$ are constant, $d\mathbf{e}/dt = \mathbf{0}$, and so the velocity of P relative to O is

$$\mathbf{v} = \frac{d\mathbf{r}}{dt} = \frac{ds}{dt}\mathbf{e}.$$

We can write the velocity vector as $\mathbf{v} = v\mathbf{e}$, obtaining the scalar equation

$$v = \frac{ds}{dt}.$$

The velocity v of point P along the straight line is the rate of change of the position s. Notice that v is equal to the slope at time t of the line tangent to the graph of s as a function of time (Fig. 13.6).

The acceleration of P relative to O is

$$\mathbf{a} = \frac{d\mathbf{v}}{dt} = \frac{d}{dt}(v\mathbf{e}) = \frac{dv}{dt}\mathbf{e}.$$

Writing the acceleration vector as $\mathbf{a} = a\mathbf{e}$, we obtain the scalar equation

$$a = \frac{dv}{dt} = \frac{d^2s}{dt^2}.$$

The acceleration a is equal to the slope at time t of the line tangent to the graph of v as a function of time (Fig. 13.7).

By introducing the unit vector $\mathbf{e}$, we have obtained scalar equations describing the motion of P. The position is specified by the coordinate s, and the velocity and acceleration are governed by the equations

$$v = \frac{ds}{dt} \tag{13.3}$$

and

$$a = \frac{dv}{dt}. \tag{13.4}$$

Applying the *chain rule* of differential calculus, we can write the derivative of the velocity with respect to time as

$$\frac{dv}{dt} = \frac{dv}{ds}\frac{ds}{dt},$$

obtaining an alternative expression for the acceleration that is often useful:

$$a = \frac{dv}{ds}v. \tag{13.5}$$

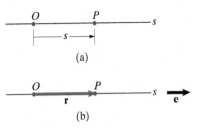

Figure 13.5
(a) The coordinate s from O to P.
(b) The unit vector $\mathbf{e}$ and position vector $\mathbf{r}$.

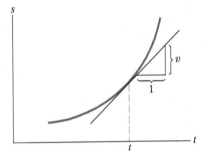

Figure 13.6
The slope of the straight line tangent to the graph of s versus t is the velocity at time t.

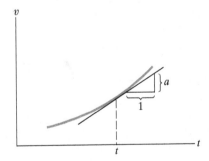

Figure 13.7
The slope of the straight line tangent to the graph of v versus t is the acceleration at time t.

Figure 13.8
The coordinate s measures the position of the center of mass of the truck relative to a reference point.

Analysis of the Motion

In some situations, the position s of a point of an object is known as a function of time. Engineers use methods such as radar and laser-Doppler interferometry to measure positions as functions of time. In this case, we can obtain the velocity and acceleration as functions of time from Eqs. (13.3) and (13.4) by differentiation. For example, if the position of the truck in Fig. 13.8 during the interval of time from $t = 2$ s to $t = 4$ s is given by the equation

$$s = 6 + \frac{1}{3}t^3 \text{ m,}$$

then the velocity and acceleration of the truck during that interval of time are

$$v = \frac{ds}{dt} = t^2 \text{ m/s}$$

and

$$a = \frac{dv}{dt} = 2t \text{ m/s}^2.$$

However, it is more common to know an object's acceleration than to know its position, because the acceleration of an object can be determined by Newton's second law when the forces acting on it are known. When the acceleration is known, we can determine the velocity and position from Eqs. (13.3)–(13.5) by integration.

Acceleration Specified as a Function of Time   If the acceleration is a known function of time, we can integrate the relation

$$\frac{dv}{dt} = a \tag{13.6}$$

with respect to time to determine the velocity as a function of time. We obtain

$$v = \int a\, dt + A,$$

where A is an integration constant. Then we can integrate the relation

$$\frac{ds}{dt} = v \tag{13.7}$$

to determine the position as a function of time,

$$s = \int v\, dt + B,$$

where B is another integration constant. We would need additional information about the motion, such as the values of v and s at a given time, to determine the constants A and B.

Instead of using indefinite integrals, we can write Eq. (13.6) as

$$dv = a\, dt$$

and integrate in terms of definite integrals:

$$\int_{v_0}^{v} dv = \int_{t_0}^{t} a \, dt. \qquad (13.8)$$

The lower limit v_0 is the velocity at time t_0, and the upper limit v is the velocity at an arbitrary time t. Evaluating the integral on the left side of Eq. (13.8), we obtain an expression for the velocity as a function of time:

$$v = v_0 + \int_{t_0}^{t} a \, dt. \qquad (13.9)$$

We can then write Eq. (13.7) as

$$ds = v \, dt$$

and integrate in terms of definite integrals to obtain

$$\int_{s_0}^{s} ds = \int_{t_0}^{t} v \, dt,$$

where the lower limit s_0 is the position at time t_0 and the upper limit s is the position at an arbitrary time t. Evaluating the integral on the left side, we obtain the position as a function of time:

$$s = s_0 + \int_{t_0}^{t} v \, dt. \qquad (13.10)$$

Although we have shown how to determine the velocity and position when the acceleration is known as a function of time, don't try to remember results such as Eqs. (13.9) and (13.10). As we will demonstrate in the examples, we recommend that straight-line motion problems be solved by using Eqs. (13.3)–(13.5).

We can make some useful observations from Eqs. (13.9) and (13.10):

- The area defined by the graph of the acceleration of P as a function of time from t_0 to t is equal to the change in the velocity from t_0 to t (Fig. 13.9a).
- The area defined by the graph of the velocity of P as a function of time from t_0 to t is equal to the change in position from t_0 to t (Fig. 13.9b).

These relationships can often be used to obtain a qualitative understanding of an object's motion, and in some cases can even be used to determine the object's motion quantitatively.

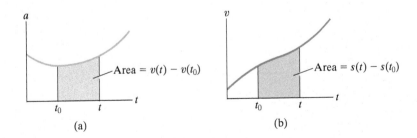

(a)   (b)

Figure 13.9
Relations between areas defined by the graphs of the acceleration and velocity of P and changes in its velocity and position.

Constant Acceleration In some situations, the acceleration of an object is constant or nearly constant. For example, if a dense object such as a golf ball or a rock is dropped and doesn't fall too far, the object's acceleration is approximately equal to the acceleration due to gravity at sea level.

Let the acceleration be a known constant a_0. From Eqs. (13.9) and (13.10), the velocity and position as functions of time are

$$v = v_0 + a_0(t - t_0) \tag{13.11}$$

and

$$s = s_0 + v_0(t - t_0) + \frac{1}{2}a_0(t - t_0)^2, \tag{13.12}$$

where s_0 and v_0 are the position and velocity, respectively, at time t_0. Notice that *if the acceleration is constant, the velocity is a linear function of time.*

From Eq. (13.5), we can write the acceleration as

$$a_0 = \frac{dv}{ds}v.$$

Writing this expression as $v\,dv = a_0\,ds$ and integrating,

$$\int_{v_0}^{v} v\,dv = \int_{s_0}^{s} a_0\,ds,$$

we obtain an equation for the velocity as a function of position:

$$v^2 = v_0^2 + 2a_0(s - s_0). \tag{13.13}$$

Although Eqs. (13.11)–(13.13) can be useful *when the acceleration is constant,* they must not be used otherwise.

RESULTS

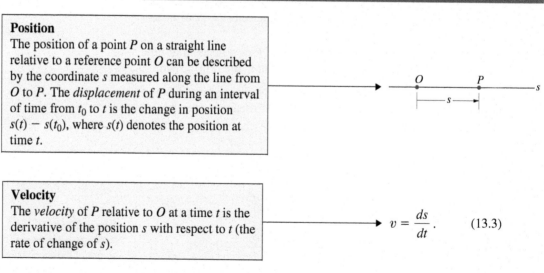

Position
The position of a point P on a straight line relative to a reference point O can be described by the coordinate s measured along the line from O to P. The *displacement* of P during an interval of time from t_0 to t is the change in position $s(t) - s(t_0)$, where $s(t)$ denotes the position at time t.

Velocity
The *velocity* of P relative to O at a time t is the derivative of the position s with respect to t (the rate of change of s).

$$v = \frac{ds}{dt}. \tag{13.3}$$

Acceleration
The *acceleration* of P relative to O at a time t is the derivative of the velocity v with respect to t (the rate of change of v).

$$a = \frac{dv}{dt}. \qquad (13.4)$$

Applying the chain rule

$$a = \frac{dv}{dt} = \frac{dv}{ds}\frac{ds}{dt}$$

results in an alternative expression for the acceleration that is often useful.

$$a = \frac{dv}{ds}v. \qquad (13.5)$$

When the Acceleration is Known as a Function of Time

The acceleration can be integrated with respect to time to determine the velocity as a function of time. A is an integration constant.

$$\begin{cases} \dfrac{dv}{dt} = a, \\[2mm] v = \displaystyle\int a\,dt + A. \end{cases}$$

Alternatively, definite integrals can be used to determine the velocity. Here v_0 is the velocity at time t_0, and v is the velocity at time t. This result shows that the change in the velocity from time t_0 to time t is equal to the area defined by the graph of the acceleration from time t_0 to time t.

$$\begin{cases} \displaystyle\int_{v_0}^{v} dv = \int_{t_0}^{t} a\,dt, \\[2mm] v = v_0 + \displaystyle\int_{t_0}^{t} a\,dt. \end{cases}$$

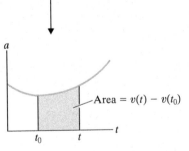

Area $= v(t) - v(t_0)$

When the Velocity is Known as a Function of Time

The velocity can be integrated with respect to time to determine the position as a function of time. B is an integration constant.

$$\begin{cases} \dfrac{ds}{dt} = v, \\[2mm] s = \displaystyle\int v\,dt + B. \end{cases}$$

Definite integrals can be used to determine the position. Here s_0 is the velocity at time t_0, and s is the velocity at time t. This result shows that the change in the position from time t_0 to time t is equal to the area defined by the graph of the velocity from time t_0 to time t.

$$\begin{cases} \int_{s_0}^{s} dv = \int_{t_0}^{t} v\, dt, \\[2mm] s = s_0 + \int_{t_0}^{t} v\, dt. \end{cases}$$

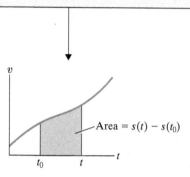

Area $= s(t) - s(t_0)$

When the Acceleration is Constant

Suppose that *the acceleration is a constant* $a = a_0$. Equations (13.3)–(13.5) can be integrated to obtain these convenient results for the velocity v and position s at time t. Here v_0 is the velocity at time t_0, and s_0 is the position at time t_0.

$$\begin{cases} v = v_0 + a_0(t - t_0), & (13.11) \\[2mm] s = s_0 + v_0(t - t_0) + \dfrac{1}{2} a_0(t - t_0)^2, & (13.12) \\[2mm] v^2 = v_0^2 + 2a_0(s - s_0). & (13.13) \end{cases}$$

Active Example 13.1   Acceleration that is a Function of Time (▶ *Related Problem 13.12*)

The acceleration (in m/s²) of point P relative to point O is given as a function of time by $a = 3t^2$, where t is in seconds. At $t = 1$ s, the position of P is $s = 3$ m, and at $t = 2$ s, the position of P is $s = 7.5$ m. What is the position of P at $t = 3$ s?

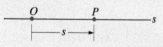

Strategy
Because the acceleration is given as a function of time, we can integrate it to obtain an equation for the velocity as a function of time. Then we can integrate the velocity to obtain an equation for the position as a function of time. The resulting equations will contain two unknown integration constants. We can evaluate them by using the given values of the position at $t = 1$ s and $t = 2$ s.

Solution

| Integrate the acceleration to determine the velocity as a function of time. A is an integration constant. | $\begin{cases} a = \dfrac{dv}{dt} = 3t^2, \\[2mm] v = t^3 + A. \end{cases}$ |

| Integrate the velocity to determine the position as a function of time. B is an integration constant. | $\begin{cases} v = \dfrac{ds}{dt} = t^3 + A, \\[2mm] s = \dfrac{1}{4}t^4 + At + B. \end{cases}$ |

| Use the known positions at $t = 1$ s and at $t = 2$ s to determine A and B, obtaining $A = 0.75$ and $B = 2$. | $\begin{cases} s\big|_{t=1\,s} = 3 = \dfrac{1}{4}(1)^4 + A(1) + B, \\[2mm] s\big|_{t=2\,s} = 7.5 = \dfrac{1}{4}(2)^4 + A(2) + B. \end{cases}$ |

| Determine the position at $t = 3$ s. | $\begin{cases} s = \dfrac{1}{4}t^4 + 0.75t + 2 : \\[2mm] s\big|_{t=3\,s} = \dfrac{1}{4}(3)^4 + 0.75(3) + 2 = 24.5 \text{ m.} \end{cases}$ |

Practice Problem The acceleration (in ft/s^2) of point P relative to point O is given as a function of time by $a = 2t$, where t is in seconds. At $t = 3$ s, the position and velocity of P are $s = 30$ ft and $v = 14$ ft/s. What are the position and velocity of P at $t = 10$ s?

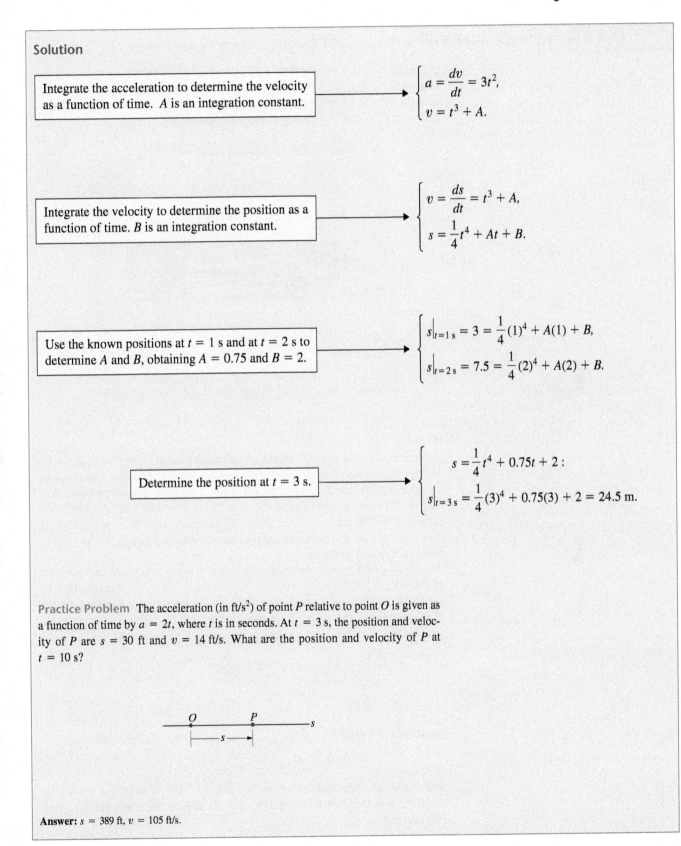

Answer: $s = 389$ ft, $v = 105$ ft/s.

Example 13.2 Straight-Line Motion with Constant Acceleration (▶ *Related Problem 13.1*)

Engineers testing a vehicle that will be dropped by parachute estimate that the vertical velocity of the vehicle when it reaches the ground will be 6 m/s. If they drop the vehicle from the test rig shown, from what height h should they drop it to match the impact velocity of the parachute drop?

Strategy

If the only significant force acting on an object near the earth's surface is its weight, the acceleration of the object is approximately constant and equal to the acceleration due to gravity at sea level. Therefore, we can assume that the vehicle's acceleration during its short fall is $g = 9.81$ m/s^2. We can integrate Eqs. (13.3) and (13.4) to obtain the vehicle's velocity and position as functions of time and then use them to determine the position of the vehicle when its velocity is 6 m/s.

Solution

Let $t = 0$ be the time at which the vehicle is dropped, and let s be the position of the bottom of the cushioning material beneath the vehicle relative to its position at $t = 0$ (Fig. a). The vehicle's acceleration is $a = 9.81$ m/s^2.

From Eq. (13.4),

$$\frac{dv}{dt} = a = 9.81 \text{ m/s}^2.$$

Integrating, we obtain

$$v = 9.81t + A,$$

where A is an integration constant. Because the vehicle is at rest when it is released, $v = 0$ at $t = 0$. Therefore, $A = 0$, and the vehicle's velocity as a function of time is

$$v = 9.81t \text{ m/s}.$$

(a) The coordinate s measures the position of the bottom of the platform relative to its initial position.

We substitute this result into Eq. (13.3) to get

$$\frac{ds}{dt} = v = 9.81t$$

and integrate, obtaining

$$s = 4.91t^2 + B.$$

The position $s = 0$ when $t = 0$, so the integration constant $B = 0$, and the position as a function of time is

$$s = 4.91t^2.$$

From our equation for the velocity as a function of time, the time necessary for the vehicle to reach 6 m/s as it falls is

$$t = \frac{v}{9.81 \text{ m/s}^2} = \frac{6 \text{ m/s}}{9.81 \text{ m/s}^2} = 0.612 \text{ s}.$$

Substituting this time into our equation for the position as a function of time yields the required height h:

$$h = 4.91t^2 = 4.91(0.612)^2 = 1.83 \text{ m}.$$

Critical Thinking

Notice that we could have determined the height h from which the vehicle should be dropped in a simpler way by using Eq. (13.13), which relates the velocity to the position.

$$v^2 = v_0^2 + 2a_0(s - s_0):$$

$$(6 \text{ m/s})^2 = 0 + 2(9.81 \text{ m/s}^2)(h - 0).$$

Solving, we obtain $h = 1.83$ m. But it is essential to remember that Eqs. (13.11)–(13.13) apply *only when the acceleration is constant*, as it is in this example.

Example 13.3   Graphical Solution of Straight-Line Motion (▶ *Related Problem 13.26*)

The cheetah, *Acinonyx jubatus*, can run as fast as 75 mi/h. If you assume that the animal's acceleration is constant and that it reaches top speed in 4 s, what distance can the cheetah cover in 10 s?

Strategy
The acceleration has a constant value for the first 4 s and is then zero. We can determine the distance traveled during each of these "phases" of the motion and sum them to obtain the total distance covered. We do so both analytically and graphically.

Solution
The top speed in terms of feet per second is

$$75 \text{ mi/h} = (75 \text{ mi/h})\left(\frac{5280 \text{ ft}}{1 \text{ mi}}\right)\left(\frac{1 \text{ h}}{3600 \text{ s}}\right) = 110 \text{ ft/s}.$$

First Method   Let a_0 be the acceleration during the first 4 s. We integrate Eq. (13.4) to get

$$\int_0^v dv = \int_0^t a_0 \, dt,$$

$$\left[v\right]_0^v = a_0\left[t\right]_0^t,$$

$$v - 0 = a_0(t - 0),$$

obtaining the velocity as a function of time during the first 4 s:

$$v = a_0 t \text{ ft/s}.$$

When $t = 4$ s, $v = 110$ ft/s; so $a_0 = 110/4 = 27.5$ ft/s^2. Therefore, the velocity during the first 4 s is $v = 27.5t$ ft/s. Now we integrate Eq. (13.3),

$$\int_0^s ds = \int_0^t 27.5t \, dt,$$

$$\left[s\right]_0^s = 27.5\left[\frac{t^2}{2}\right]_0^t,$$

$$s - 0 = 27.5\left(\frac{t^2}{2} - 0\right),$$

obtaining the position as a function of time during the first 4 s:

$$s = 13.75t^2 \text{ ft}.$$

At $t = 4$ s, the position is $s = 13.75(4)^2 = 220$ ft.

From $t = 4$ s to $t = 10$ s, the velocity $v = 110$ ft/s. We write Eq. (13.3) as

$$ds = v\, dt = 110\, dt$$

and integrate to determine the distance traveled during the second phase of the motion,

$$\int_0^s ds = \int_4^{10} 110\, dt,$$

$$\left[s \right]_0^s = 110 \left[t \right]_4^{10},$$

$$s - 0 = 110(10 - 4),$$

obtaining $s = 660$ ft. The total distance the cheetah travels is $220 \text{ ft} + 660 \text{ ft} = 880$ ft, or 293 yd, in 10 s.

Second Method   We draw a graph of the cheetah's velocity as a function of time in Fig. a. The acceleration is constant during the first 4 s of motion, so the velocity is a linear function of time from $v = 0$ at $t = 0$ to $v = 110$ ft/s at $t = 4$ s. The velocity is constant during the last 6 s. The total distance covered is the sum of the areas during the two phases of motion:

$$\tfrac{1}{2}(4 \text{ s})(110 \text{ ft/s}) + (6 \text{ s})(110 \text{ ft/s}) = 220 \text{ ft} + 660 \text{ ft} = 880 \text{ ft}.$$

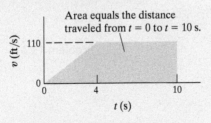

(a) The cheetah's velocity as a function of time.

Critical Thinking

Notice that in the first method we used definite, rather than indefinite, integrals to determine the cheetah's velocity and position as functions of time. You should rework the example using indefinite integrals and compare your results with ours. Whether to use definite or indefinite integrals is primarily a matter of taste, but you need to be familiar with both procedures.

Problems

The problems that follow involve straight-line motion. The time t is in seconds unless otherwise stated.

▶ 13.1 In Example 13.2, suppose that the vehicle is dropped from a height $h = 6$ m. (a) What is its downward velocity 1 s after it is released? (b) What is its downward velocity just before it reaches the ground?

13.2 The milling machine is programmed so that during the interval of time from $t = 0$ to $t = 2$ s, the position of its head (in inches) is given as a function of time by $s = 4t - 2t^2$. What are the velocity (in in/s) and acceleration (in in/s^2) of the head at $t = 1$ s?

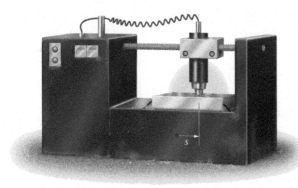

Problem 13.2

13.3 In an experiment to estimate the acceleration due to gravity, a student drops a ball at a distance of 1 m above the floor. His lab partner measures the time it takes to fall and obtains an estimate of 0.46 s.

(a) What do they estimate the acceleration due to gravity to be?

(b) Let s be the ball's position relative to the floor. Using the value of the acceleration due to gravity that they obtained, and assuming that the ball is released at $t = 0$, determine s (in m) as a function of time.

Problem 13.3

13.4 The boat's position during the interval of time from $t = 2$ s to $t = 10$ s is given by $s = 4t + 1.6t^2 - 0.08t^3$ m.

(a) Determine the boat's velocity and acceleration at $t = 4$ s.

(b) What is the boat's maximum velocity during this interval of time, and when does it occur?

Problem 13.4

13.5 The rocket starts from rest at $t = 0$ and travels straight up. Its height above the ground as a function of time can be approximated by $s = bt^2 + ct^3$, where b and c are constants. At $t = 10$ s, the rocket's velocity and acceleration are $v = 229$ m/s and $a = 28.2$ m/s². Determine the time at which the rocket reaches supersonic speed (325 m/s). What is its altitude when that occurs?

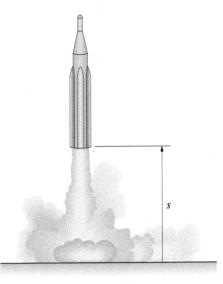

Problem 13.5

13.6 The position of a point during the interval of time from $t = 0$ to $t = 6$ s is given by $s = -\frac{1}{2}t^3 + 6t^2 + 4t$ m.

(a) What is the maximum velocity during this interval of time, and at what time does it occur?

(b) What is the acceleration when the velocity is a maximum?

13.7 The position of a point during the interval of time from $t = 0$ to $t = 3$ s is $s = 12 + 5t^2 - t^3$ ft.

(a) What is the maximum velocity during this interval of time, and at what time does it occur?

(b) What is the acceleration when the velocity is a maximum?

13.8 The rotating crank causes the position of point P as a function of time to be $s = 0.4 \sin(2\pi t)$ m.

(a) Determine the velocity and acceleration of P at $t = 0.375$ s.

(b) What is the maximum magnitude of the velocity of P?

(c) When the magnitude of the velocity of P is a maximum, what is the acceleration of P?

13.9 For the mechanism in Problem 13.8, draw graphs of the position s, velocity v, and acceleration a of point P as functions of time for $0 \le t \le 2$ s. Using your graphs, confirm that the slope of the graph of s is zero at times for which v is zero and that the slope of the graph of v is zero at times for which a is zero.

Problems 13.8/13.9

13.10 A seismograph measures the horizontal motion of the ground during an earthquake. An engineer analyzing the data determines that for a 10-s interval of time beginning at $t = 0$ the position is approximated by $s = 100 \cos(2\pi t)$ mm. What are (a) the maximum velocity and (b) the maximum acceleration of the ground during the 10-s interval?

13.11 In an assembly operation, the robot's arm moves along a straight horizontal line. During an interval of time from $t = 0$ to $t = 1$ s, the position of the arm is given by $s = 30t^2 - 20t^3$ mm.

(a) Determine the maximum velocity during this interval of time.

(b) What are the position and acceleration when the velocity is a maximum?

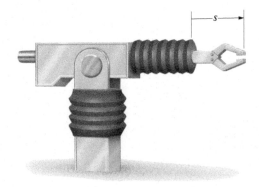

Problem 13.11

▶ **13.12** In Active Example 13.1, the acceleration (in m/s^2) of point P relative to point O is given as a function of time by $a = 3t^2$. Suppose that at $t = 0$ the position and velocity of P are $s = 5$ m and $v = 2$ m/s. Determine the position and velocity of P at $t = 4$ s.

13.13 The Porsche starts from rest at time $t = 0$. During the first 10 seconds of its motion, its velocity in km/h is given as a function of time by $v = 22.8t - 0.88t^2$, where t is in seconds. (a) What is the car's maximum acceleration in m/s^2, and when does it occur? (b) What distance in km does the car travel during the 10 seconds?

Problem 13.13

13.14 The acceleration of a point is $a = 20t$ m/s^2. When $t = 0$, $s = 40$ m and $v = -10$ m/s. What are the position and velocity at $t = 3$ s?

13.15 The acceleration of a point is $a = 60t - 36t^2$ ft/s^2. When $t = 0$, $s = 0$ and $v = 20$ ft/s. What are the position and velocity as functions of time?

13.16 As a first approximation, a bioengineer studying the mechanics of bird flight assumes that the snow petrel takes off with constant acceleration. Video measurements indicate that a bird requires a distance of 4.3 m to take off and is moving at 6.1 m/s when it does. What is its acceleration?

13.17 Progressively developing a more realistic model, the bioengineer next models the acceleration of the snow petrel by an equation of the form $a = C(1 + \sin \omega t)$, where C and ω are constants. From video measurements of a bird taking off, he estimates that $\omega = 18$ and determines that the bird requires 1.42 s to take off and is moving at 6.1 m/s when it does. What is the constant C?

Problems 13.16/13.17

13.18 Missiles designed for defense against ballistic missiles have attained accelerations in excess of 100 g's, or 100 times the acceleration due to gravity. Suppose that the missile shown lifts off from the ground and has a constant acceleration of 100 g's. How long does it take to reach an altitude of 3000 m? How fast is it going when it reaches that altitude?

13.19 Suppose that the missile shown lifts off from the ground and, because it becomes lighter as its fuel is expended, its acceleration (in g's) is given as a function of time in seconds by

$$a = \frac{100}{1 - 0.2t}.$$

What is the missile's velocity in miles per hour 1 s after liftoff?

Problems 13.18/13.19

13.20 The airplane releases its drag parachute at time $t = 0$. Its velocity is given as a function of time by

$$v = \frac{80}{1 + 0.32t} \text{ m/s.}$$

What is the airplane's acceleration at $t = 3$ s?

13.21 How far does the airplane in Problem 13.20 travel during the interval of time from $t = 0$ to $t = 10$ s?

Problems 13.20/13.21

13.22 The velocity of a bobsled is $v = 10t$ ft/s. When $t = 2$ s, the position of the sled is $s = 25$ ft. What is its position when $t = 10$ s?

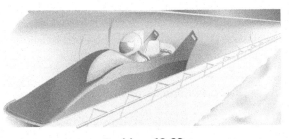

Problem 13.22

13.23 In September 2003, Tony Schumacher started from rest and drove a quarter mile (1320 ft) in 4.498 s in a National Hot Rod Association race. His speed as he crossed the finish line was 328.54 mi/h. Assume that the car's acceleration can be expressed by a linear function of time $a = b + ct$.

(a) Determine the constants b and c.

(b) What was the car's speed 2 s after the start of the race?

13.24 The velocity of an object is $v = 200 - 2t^2$ m/s. When $t = 3$ s, the position of the object is $s = 600$ m. What are the position and acceleration of the object at $t = 6$ s?

13.25 An inertial navigation system measures the acceleration of a vehicle from $t = 0$ to $t = 6$ s and determines it to be $a = 2 + 0.1t$ m/s². At $t = 0$, the vehicle's position and velocity are $s = 240$ m and $v = 42$ m/s, respectively. What are the vehicle's position and velocity at $t = 6$ s?

▶ **13.26** In Example 13.3, suppose that the cheetah's acceleration is constant and it reaches its top speed of 75 mi/h in 5 s. What distance can it cover in 10 s?

13.27 The graph shows the airplane's acceleration during take-off. What is the airplane's velocity when it rotates (lifts off) at $t = 30$ s?

13.28 The graph shows the airplane's acceleration during take-off. What distance has the airplane traveled when it lifts off at $t = 30$ s?

Problems 13.27/13.28

13.29 The car is traveling at 30 mi/h when the traffic light 295 ft ahead turns yellow. The driver takes 1 s to react before he applies the brakes.

(a) After he applies the brakes, what constant rate of deceleration will cause the car to come to a stop just as it reaches the light?

(b) How long does it take the car to travel the 295 ft to the light?

13.30 The car is traveling at 30 mi/h when the traffic light 295 ft ahead turns yellow. The driver takes 1 s to react before he applies the accelerator. If the car has a constant acceleration of 5 ft/s² and the light remains yellow for 5 s, will the car reach the light before it turns red? How fast is the car moving when it reaches the light?

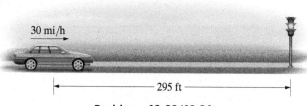

Problems 13.29/13.30

13.31 A high-speed rail transportation system has a top speed of 100 m/s. For the comfort of the passengers, the magnitude of the acceleration and deceleration is limited to 2 m/s². Determine the minimum time required for a trip of 100 km.

Strategy: A graphical approach can help you solve this problem. Recall that the change in the position from an initial time t_0 to a time t is equal to the area defined by the graph of the velocity as a function of time from t_0 to t.

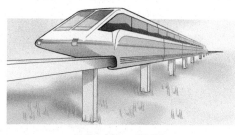

Problem 13.31

13.32 The nearest star, Proxima Centauri, is 4.22 light years from the Earth. Ignoring relative motion between the solar system and Proxima Centauri, suppose that a spacecraft accelerates from the vicinity of the Earth at $0.01g$ (0.01 times the acceleration due to gravity at sea level) until it reaches one-tenth the speed of light, coasts until it is time to decelerate, and then decelerates at $0.01g$ until it comes to rest in the vicinity of Proxima Centauri. How long does the trip take? (Light travels at 3×10^8 m/s.)

13.33 A race car starts from rest and accelerates at $a = 5 + 2t$ ft/s² for 10 s. The brakes are then applied, and the car has a constant acceleration $a = -30$ ft/s² until it comes to rest. Determine (a) the maximum velocity, (b) the total distance traveled, and (c) the total time of travel.

13.34 When $t = 0$, the position of a point is $s = 6$ m, and its velocity is $v = 2$ m/s. From $t = 0$ to $t = 6$ s, the acceleration of the point is $a = 2 + 2t^2$ m/s². From $t = 6$ s until it comes to rest, its acceleration is $a = -4$ m/s².

(a) What is the total time of travel?

(b) What total distance does the point move?

13.35 Zoologists studying the ecology of the Serengeti Plain estimate that the average adult cheetah can run 100 km/h and the average springbok can run 65 km/h. If the animals run along the same straight line, start at the same time, are each assumed to have constant acceleration, and reach top speed in 4 s, how close must a cheetah be when the chase begins to catch a springbok in 15 s?

13.36 Suppose that a person unwisely drives 75 mi/h in a 55 mi/h zone and passes a police car going 55 mi/h in the same direction. If the police officers begin constant acceleration at the instant they are passed and increase their velocity to 80 mi/h in 4 s, how long does it take them to be even with the pursued car?

13.37 If $\theta = 1$ rad and $d\theta/dt = 1$ rad/s, what is the velocity of P relative to O?

> *Strategy:* You can write the position of P relative to O as
>
> $$s = (2 \text{ ft}) \cos \theta + (2 \text{ ft}) \cos \theta$$
>
> and then take the derivative of this expression with respect to time to determine the velocity.

13.38 If $\theta = 1$ rad, $d\theta/dt = -2$ rad/s, and $d^2\theta/dt^2 = 0$, what are the velocity and acceleration of P relative to O?

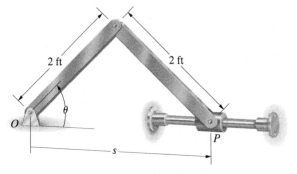

Problems 13.37/13.38

13.39* If $\theta = 1$ rad and $d\theta/dt = 1$ rad/s, what is the velocity of P relative to O?

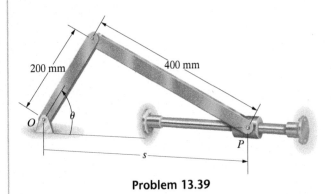

Problem 13.39

13.3 Straight-Line Motion When the Acceleration Depends on Velocity or Position

BACKGROUND

Acceleration Specified as a Function of Velocity   Aerodynamic and hydrodynamic forces can cause an object's acceleration to depend on its velocity (Fig. 13.10). Suppose that the acceleration is a known function of velocity—that is,

$$\frac{dv}{dt} = a(v). \tag{13.14}$$

We cannot integrate this equation with respect to time to determine the velocity, because $a(v)$ is not known as a function of time. But we can *separate variables*, putting terms involving v on one side of the equation and terms involving t on the other side:

$$\frac{dv}{a(v)} = dt. \tag{13.15}$$

We can now integrate, obtaining

$$\int_{v_0}^{v} \frac{dv}{a(v)} = \int_{t_0}^{t} dt, \tag{13.16}$$

Figure 13.10
Aerodynamic and hydrodynamic forces depend on an object's velocity. As the bullet slows, the aerodynamic drag force resisting its motion decreases.
© 1973 Kim Vandiver & Harold Edgerton, Courtesy of Palm Press, Inc.

where v_0 is the velocity at time t_0. In principle, we can solve this equation for the velocity as a function of time and then integrate the relation

$$\frac{ds}{dt} = v$$

to determine the position as a function of time.

By using the chain rule, we can also determine the velocity as a function of the position. Writing the acceleration as

$$\frac{dv}{dt} = \frac{dv}{ds}\frac{ds}{dt} = \frac{dv}{ds}v$$

and substituting it into Eq. (13.14), we obtain

$$\frac{dv}{ds}v = a(v).$$

Separating variables yields

$$\frac{v\,dv}{a(v)} = ds.$$

Integrating,

$$\int_{v_0}^{v} \frac{v\,dv}{a(v)} = \int_{s_0}^{s} ds,$$

we can obtain a relation between the velocity and the position.

Acceleration Specified as a Function of Position   Gravitational forces and forces exerted by springs can cause an object's acceleration to depend on its position. If the acceleration is a known function of position—that is,

$$\frac{dv}{dt} = a(s), \tag{13.17}$$

we cannot integrate with respect to time to determine the velocity, because $a(s)$ is not known as a function of time. Moreover, we cannot separate variables, because the equation contains three variables: v, t, and s. However, by using the chain rule

$$\frac{dv}{dt} = \frac{dv}{ds}\frac{ds}{dt} = \frac{dv}{ds}v,$$

we can write Eq. (13.17) as

$$\frac{dv}{ds}v = a(s).$$

Now we can separate variables,

$$v\,dv = a(s)\,ds, \tag{13.18}$$

and we integrate:

$$\int_{v_0}^{v} v\, dv = \int_{s_0}^{s} a(s)\, ds. \qquad (13.19)$$

In principle, we can solve this equation for the velocity as a function of the position:

$$v = \frac{ds}{dt} = v(s). \qquad (13.20)$$

Then we can separate variables in this equation and integrate to determine the position as a function of time:

$$\int_{s_0}^{s} \frac{ds}{v(s)} = \int_{t_0}^{t} dt.$$

RESULTS

| When the acceleration is known as a function of velocity, $a = a(v)$. | Separate variables, $\dfrac{dv}{dt} = a(v)$: $$\dfrac{dv}{a(v)} = dt,$$ and integrate to determine the velocity as a function of time. Or, first apply the chain rule, $$\dfrac{dv}{dt} = \dfrac{dv}{ds}\dfrac{ds}{dt} = \dfrac{dv}{ds}v = a(v),$$ then separate variables, $$\dfrac{v\,dv}{a(v)} = ds,$$ and integrate to determine the velocity as a function of position. |
| When the acceleration is known as a function of position, $a = a(s)$. | Apply the chain rule, $$\dfrac{dv}{dt} = \dfrac{dv}{ds}\dfrac{ds}{dt} = \dfrac{dv}{ds}v = a(s),$$ then separate variables, $$v\,dv = a(s)\,ds,$$ and integrate to determine the velocity as a function of position. |

Active Example 13.4 Acceleration that is a Function of Velocity (▶ *Related Problem 13.40*)

After deploying its drag parachute, the airplane's acceleration (in m/s²) is $a = -0.004v^2$, where v is the velocity in m/s. Determine the time required for the plane's velocity to decrease from 80 m/s to 10 m/s.

Strategy

The airplane's acceleration is known as a function of its velocity. Writing the acceleration as $a = dv/dt$, we can separate variables and integrate to determine the velocity as a function of time.

Solution

Separate variables. ⟶
$$\begin{cases} \dfrac{dv}{dt} = -0.004v^2 : \\ \dfrac{dv}{v^2} = -0.004\,dt. \end{cases}$$

Integrate, defining $t = 0$ to be the time at which the velocity is 80 m/s. Here v is the velocity at time t. ⟶
$$\begin{cases} \displaystyle\int_{80}^{v} \dfrac{dv}{v^2} = -0.004 \int_{0}^{t} dt, \\ \left[-\dfrac{1}{v}\right]_{80}^{v} = -0.004\left[t\right]_{0}^{t}, \\ -\dfrac{1}{v} + \dfrac{1}{80} = -0.004t. \end{cases}$$

Solve for t in terms of the velocity. From this equation, we find that the time required for the volocity to decrease to 10 m/s is 21.9 s. The graph shows the airplane's velocity as a function of time. ⟶ $$t = 250\left(\dfrac{1}{v} - \dfrac{1}{80}\right).$$

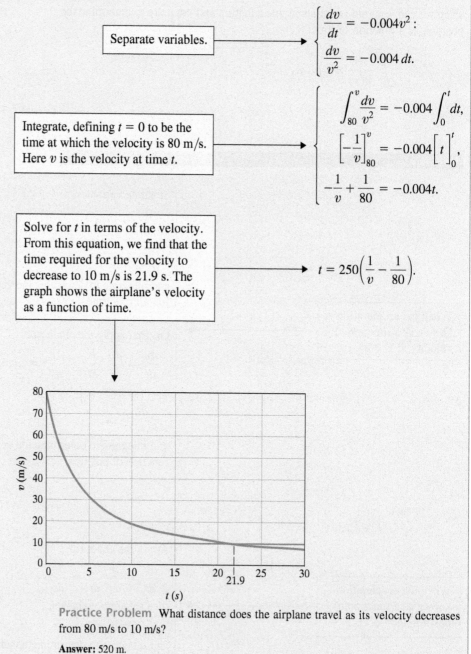

Practice Problem What distance does the airplane travel as its velocity decreases from 80 m/s to 10 m/s?

Answer: 520 m.

| Example 13.5 | Gravitational (Position-Dependent) Acceleration (▶ *Related Problem 13.62*) |

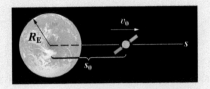

In terms of the distance s from the center of the earth, the magnitude of the acceleration due to gravity is gR_E^2/s^2, where R_E is the radius of the earth. (See the discussion of gravity in Section 12.2.) If a spacecraft is a distance s_0 from the center of the earth, what outward velocity v_0 must it be given to reach a specified distance h from the earth's center?

Strategy
The acceleration is known as a function of the position s. We can apply the chain rule and separate variables, then integrate to determine the velocity as a function of s.

Solution
The acceleration due to gravity is *toward* the center of the earth:

$$a = -\frac{gR_E^2}{s^2}.$$

Applying the chain rule results in

$$a = \frac{dv}{dt} = \frac{dv}{ds}\frac{ds}{dt} = \frac{dv}{ds}v = -\frac{gR_E^2}{s^2}.$$

Separating variables, we obtain

$$v\,dv = -\frac{gR_E^2}{s^2}ds.$$

We integrate this equation using the initial condition ($v = v_0$ when $s = s_0$) as the lower limits and the final condition ($v = 0$ when $s = h$) as the upper limits:

$$\int_{v_0}^{0} v\,dv = -\int_{s_0}^{h} \frac{gR_E^2}{s^2}ds,$$

$$\left[\frac{v^2}{2}\right]_{v_0}^{0} = gR_E^2\left[\frac{1}{s}\right]_{s_0}^{h},$$

$$0 - \frac{v_0^2}{2} = gR_E^2\left(\frac{1}{h} - \frac{1}{s_0}\right).$$

Solving for v_0, we obtain the initial velocity necessary for the spacecraft to reach a distance h:

$$v_0 = \sqrt{2gR_E^2\left(\frac{1}{s_0} - \frac{1}{h}\right)}.$$

Critical Thinking
We can make an interesting and important observation from the result of this example. Notice that as the distance h increases, the necessary initial velocity v_0 approaches a finite limit. This limit,

$$v_{esc} = \lim_{h\to\infty} v_0 = \sqrt{\frac{2gR_E^2}{s_0}},$$

is called the *escape velocity*. In the absence of other effects, an object with this initial velocity will continue moving outward indefinitely. The existence of an escape velocity makes it feasible to send spacecraft to other planets. Once escape velocity is attained, it isn't necessary to expend additional fuel to keep going.

Problems

▶ 13.40 In Active Example 13.4, determine the time required for the plane's velocity to decrease from 50 m/s to 10 m/s.

13.41 An engineer designing a system to control a router for a machining process models the system so that the router's acceleration (in in/s^2) during an interval of time is given by $a = -0.4v$, where v is the velocity of the router in in/s. When $t = 0$, the position is $s = 0$ and the velocity is $v = 2$ in/s. What is the position at $t = 3$ s?

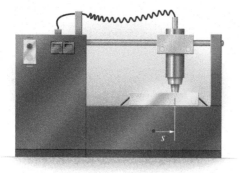

Problem 13.41

13.42 The boat is moving at 10 m/s when its engine is shut down. Due to hydrodynamic drag, its subsequent acceleration is $a = -0.05v^2$ m/s^2, where v is the velocity of the boat in m/s. What is the boat's velocity 4 s after the engine is shut down?

13.43 In Problem 13.42, what distance does the boat move in the 4 s following the shutdown of its engine?

Problems 13.42/13.43

13.44 A steel ball is released from rest in a container of oil. Its downward acceleration is $a = 2.4 - 0.6v$ in/s^2, where v is the ball's velocity in in/s. What is the ball's downward velocity 2 s after it is released?

13.45 In Problem 13.44, what distance does the ball fall in the first 2 s after its release?

Problems 13.44/13.45

13.46 The greatest ocean depth yet discovered is the Marianas Trench in the western Pacific Ocean. A steel ball released at the surface requires 64 min to reach the bottom. The ball's downward acceleration is $a = 0.9g - cv$, where $g = 9.81$ m/s^2 and the constant $c = 3.02$ s^{-1}. What is the depth of the Marianas Trench in kilometers?

13.47 The acceleration of a regional airliner during its takeoff run is $a = 14 - 0.0003v^2$ ft/s^2, where v is its velocity in ft/s. How long does it take the airliner to reach its takeoff speed of 200 ft/s?

13.48 In Problem 13.47, what distance does the airliner require to take off?

13.49 A sky diver jumps from a helicopter and is falling straight down at 30 m/s when her parachute opens. From then on, her downward acceleration is approximately $a = g - cv^2$, where $g = 9.81$ m/s^2 and c is a constant. After an initial "transient" period, she descends at a nearly constant velocity of 5 m/s.

(a) What is the value of c, and what are its SI units?

(b) What maximum deceleration is the sky diver subjected to?

(c) What is her downward velocity when she has fallen 2 m from the point at which her parachute opens?

Problem 13.49

13.50 The rocket sled starts from rest and accelerates at $a = 30 + 2t$ m/s^2 until its velocity is 400 m/s. It then hits a water brake and its acceleration is $a = -0.003v^2$ m/s^2 until its velocity decreases to 100 m/s. What total distance does the sled travel?

13.51 In Problem 13.50, what is the sled's total time of travel?

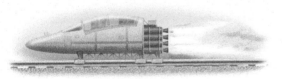

Problems 13.50/13.51

13.52 A car's acceleration is related to its position by $a = 0.01s$ m/s^2. When $s = 100$ m, the car is moving at 12 m/s. How fast is the car moving when $s = 420$ m?

13.53 Engineers analyzing the motion of a linkage determine that the velocity of an attachment point is given by $v = A + 4s^2$ ft/s, where A is a constant. When $s = 2$ ft, the acceleration of the point is measured and determined to be $a = 320$ ft/s^2. What is the velocity of the point when $s = 2$ ft?

13.54 The acceleration of an object is given as a function of its position in feet by $a = 2s^2$ ft/s^2. When $s = 0$, its velocity is $v = 1$ ft/s. What is the velocity of the object when $s = 2$ ft?

13.55 Gas guns are used to investigate the properties of materials subjected to high-velocity impacts. A projectile is accelerated through the barrel of the gun by gas at high pressure. Assume that the acceleration of the projectile in m/s^2 is given by $a = c/s$, where s is the position of the projectile in the barrel in meters and c is a constant that depends on the initial gas pressure behind the projectile. The projectile starts from rest at $s = 1.5$ m and accelerates until it reaches the end of the barrel at $s = 3$ m. Determine the value of the constant c necessary for the projectile to leave the barrel with a velocity of 200 m/s.

13.56 If the propelling gas in the gas gun described in Problem 13.55 is air, a more accurate modeling of the acceleration of the projectile is obtained by assuming that the acceleration of the projectile is given by $a = c/s^\gamma$, where $\gamma = 1.4$ is the ratio of specific heat for air. (This means that an isentropic expansion process is assumed instead of the isothermal process assumed in Problem 13.55.) Determine the value of the constant c necessary for the projectile to leave the barrel with a velocity of 200 m/s.

Problems 13.55/13.56

13.57 A spring–mass oscillator consists of a mass and a spring connected as shown. The coordinate s measures the displacement of the mass relative to its position when the spring is unstretched. If the spring is linear, the mass is subjected to a deceleration proportional to s. Suppose that $a = -4s$ m/s^2 and that you give the mass a velocity $v = 1$ m/s in the position $s = 0$.

(a) How far will the mass move to the right before the spring brings it to a stop?

(b) What will be the velocity of the mass when it has returned to the position $s = 0$?

13.58 In Problem 13.57, suppose that at $t = 0$ you release the mass from rest in the position $s = 1$ m. Determine the velocity of the mass as a function of s as it moves from the initial position to $s = 0$.

13.59 A spring–mass oscillator consists of a mass and a spring connected as shown. The coordinate s measures the displacement of the mass relative to its position when the spring is unstretched. Suppose that the nonlinear spring subjects the mass to an acceleration $a = -4s - 2s^3$ m/s^2 and that you give the mass a velocity $v = 1$ m/s in the position $s = 0$.

(a) How far will the mass move to the right before the spring brings it to a stop?

(b) What will be the velocity of the mass when it has returned to the position $s = 0$?

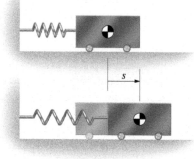

Problems 13.57–13.59

13.60 The mass is released from rest with the springs unstretched. Its downward acceleration is $a = 32.2 - 50s$ ft/s^2, where s is the position of the mass measured from the position in which it is released.

(a) How far does the mass fall?

(b) What is the maximum velocity of the mass as it falls?

13.61 Suppose that the mass in Problem 13.60 is in the position $s = 0$ and is given a downward velocity of 10 ft/s.

(a) How far does the mass fall?

(b) What is the maximum velocity of the mass as it falls?

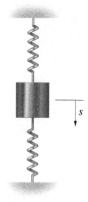

Problems 13.60/13.61

▶ **13.62** If a spacecraft is 100 mi above the surface of the earth, what initial velocity v_0 straight away from the earth would be required for the vehicle to reach the moon's orbit, 238,000 mi from the center of the earth? The radius of the earth is 3960 mi. Neglect the effect of the moon's gravity. (See Example 13.5.)

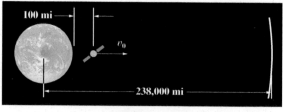

Problem 13.62

13.63 The moon's radius is 1738 km. The magnitude of the acceleration due to gravity of the moon at a distance s from the center of the moon is

$$\frac{4.89 \times 10^{12}}{s^2} \text{ m/s}^2.$$

Suppose that a spacecraft is launched straight up from the moon's surface with a velocity of 2000 m/s.

(a) What will the magnitude of its velocity be when it is 1000 km above the surface of the moon?

(b) What maximum height above the moon's surface will it reach?

13.64* The velocity of an object subjected only to the earth's gravitational field is

$$v = \left[v_0^2 + 2gR_E^2 \left(\frac{1}{s} - \frac{1}{s_0} \right) \right]^{1/2},$$

where s is the object's position relative to the center of the earth, v_0 is the object's velocity at position s_0, and R_E is the earth's radius. Using this equation, show that the object's acceleration is given as a function of s by $a = -gR_E^2/s^2$.

13.65 Suppose that a tunnel could be drilled straight through the earth from the North Pole to the South Pole and the air was evacuated. An object dropped from the surface would fall with acceleration $a = -gs/R_E$, where g is the acceleration of gravity at sea level, R_E is the radius of the earth, and s is the distance of the object from the center of the earth. (The acceleration due to gravity is equal to zero at the center of the earth and increases linearly with distance from the center.) What is the magnitude of the velocity of the dropped object when it reaches the center of the earth?

13.66* Determine the time in seconds required for the object in Problem 13.65 to fall from the surface of the earth to the center. The earth's radius is 6370 km.

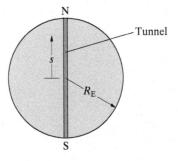

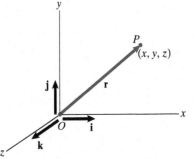

Problems 13.65/13.66

13.4 Curvilinear Motion—Cartesian Coordinates

BACKGROUND

The motion of a point along a straight line can be described by the scalars s, v, and a. But if a point describes a *curvilinear* path relative to some reference frame, we must specify its motion in terms of its position, velocity, and acceleration vectors. In many cases, the motion of the point can be analyzed conveniently by expressing the vectors in terms of cartesian coordinates.

Let $\mathbf{r}$ be the position vector of a point P relative to the origin O of a cartesian reference frame (Fig. 13.11). The components of $\mathbf{r}$ are the x, y, and z coordinates of P:

$$\mathbf{r} = x\mathbf{i} + y\mathbf{j} + z\mathbf{k}.$$

The unit vectors $\mathbf{i}$, $\mathbf{j}$, and $\mathbf{k}$ each have constant magnitude and constant direction relative to the reference frame, so the velocity of P relative to the reference frame is

$$\mathbf{v} = \frac{d\mathbf{r}}{dt} = \frac{dx}{dt}\mathbf{i} + \frac{dy}{dt}\mathbf{j} + \frac{dz}{dt}\mathbf{k}. \tag{13.21}$$

Expressing the velocity in terms of scalar components yields

$$\mathbf{v} = v_x\mathbf{i} + v_y\mathbf{j} + v_z\mathbf{k}, \tag{13.22}$$

Figure 13.11
A cartesian coordinate system with origin O.

from which we obtain scalar equations relating the components of the velocity to the coordinates of P:

$$v_x = \frac{dx}{dt}, \quad v_y = \frac{dy}{dt}, \quad v_z = \frac{dz}{dt}. \tag{13.23}$$

The acceleration of P is

$$\mathbf{a} = \frac{d\mathbf{v}}{dt} = \frac{dv_x}{dt}\mathbf{i} + \frac{dv_y}{dt}\mathbf{j} + \frac{dv_z}{dt}\mathbf{k}.$$

By expressing the acceleration in terms of scalar components as

$$\mathbf{a} = a_x\mathbf{i} + a_y\mathbf{j} + a_z\mathbf{k}, \tag{13.24}$$

we obtain the scalar equations

$$a_x = \frac{dv_x}{dt}, \quad a_y = \frac{dv_y}{dt}, \quad a_z = \frac{dv_z}{dt}. \tag{13.25}$$

Equations (13.23) and (13.25) describe the motion of a point relative to a cartesian coordinate system. Notice that the equations describing the motion in each coordinate direction are identical in form to the equations that describe the motion of a point along a straight line. As a consequence, the motion in each coordinate direction can often be analyzed using the methods we applied to straight-line motion.

The *projectile problem* is the classic example of this kind. If an object is thrown through the air and aerodynamic drag is negligible, the object accelerates downward with the acceleration due to gravity. In terms of a fixed cartesian coordinate system with its y axis upward, the acceleration is given by $a_x = 0$, $a_y = -g$, and $a_z = 0$. Suppose that at $t = 0$ the projectile is located at the origin and has velocity v_0 in the x–y plane at an angle θ_0 above the horizontal (Fig. 13.12). At $t = 0$, $x = 0$ and $v_x = v_0 \cos \theta_0$. The acceleration in the x direction is zero—that is,

$$a_x = \frac{dv_x}{dt} = 0.$$

Therefore v_x is constant and remains equal to its initial value:

$$v_x = \frac{dx}{dt} = v_0 \cos \theta_0. \tag{13.26}$$

(This result may seem unrealistic. The reason is that your intuition, based upon everyday experience, accounts for drag, whereas the analysis presented here does not.) Integrating Eq. (13.26) yields

$$\int_0^x dx = \int_0^t v_0 \cos \theta_0 \, dt,$$

whereupon we obtain the x coordinate of the object as a function of time:

$$x = (v_0 \cos \theta_0)t. \tag{13.27}$$

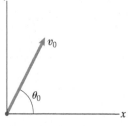

Figure 13.12
Initial conditions for a projectile problem.

Thus we have determined the position and velocity of the projectile in the x direction as functions of time without considering the projectile's motion in the y or z direction.

At $t = 0$, $y = 0$ and $v_y = v_0 \sin \theta_0$. The acceleration in the y direction is

$$a_y = \frac{dv_y}{dt} = -g.$$

Integrating, we obtain

$$\int_{v_0 \sin \theta_0}^{v_y} dv_y = \int_0^t -g \, dt,$$

from which it follows that

$$v_y = \frac{dy}{dt} = v_0 \sin \theta_0 - gt. \tag{13.28}$$

Integrating this equation yields

$$\int_0^y dy = \int_0^t (v_0 \sin \theta_0 - gt) \, dt,$$

and we find that the y coordinate as a function of time is

$$y = (v_0 \sin \theta_0)t - \tfrac{1}{2}gt^2. \tag{13.29}$$

Notice from this analysis that the same vertical velocity and position are obtained by throwing the projectile straight up with initial velocity $v_0 \sin \theta_0$ (Figs. 13.13a, b). The vertical motion is completely independent of the horizontal motion.

By solving Eq. (13.27) for t and substituting the result into Eq. (13.29), we obtain an equation describing the parabolic trajectory of the projectile:

$$y = (\tan \theta_0)x - \frac{g}{2v_0^2 \cos^2 \theta_0} x^2. \tag{13.30}$$

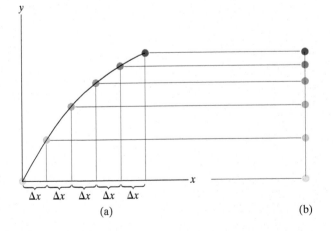

(a)

(b)

Figure 13.13
(a) Positions of the projectile at equal time intervals Δt. The distance $\Delta x = v_0(\cos \theta_0) \, \Delta t$.
(b) Positions at equal time intervals Δt of a projectile given an initial vertical velocity equal to $v_0 \sin \theta_0$.

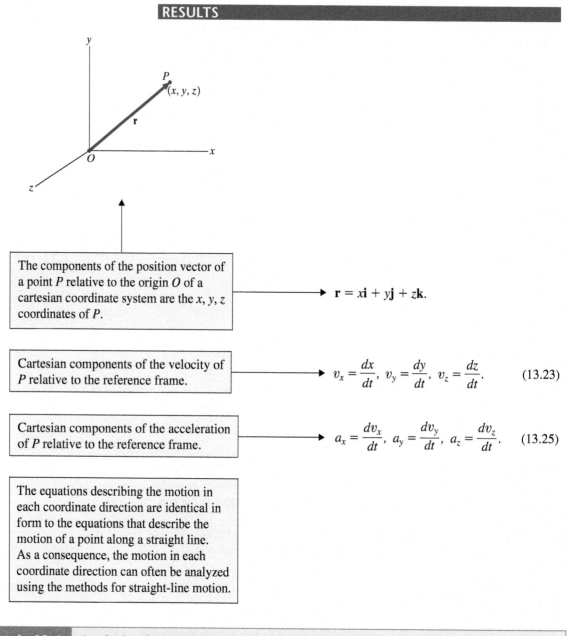

The components of the position vector of a point P relative to the origin O of a cartesian coordinate system are the x, y, z coordinates of P. $\longrightarrow$ $\mathbf{r} = x\mathbf{i} + y\mathbf{j} + z\mathbf{k}.$

Cartesian components of the velocity of P relative to the reference frame. $\longrightarrow$ $v_x = \dfrac{dx}{dt}, \ v_y = \dfrac{dy}{dt}, \ v_z = \dfrac{dz}{dt}.$ (13.23)

Cartesian components of the acceleration of P relative to the reference frame. $\longrightarrow$ $a_x = \dfrac{dv_x}{dt}, \ a_y = \dfrac{dv_y}{dt}, \ a_z = \dfrac{dv_z}{dt}.$ (13.25)

The equations describing the motion in each coordinate direction are identical in form to the equations that describe the motion of a point along a straight line. As a consequence, the motion in each coordinate direction can often be analyzed using the methods for straight-line motion.

Active Example 13.6 Analysis of Motion in Terms of Cartesian Components
(▶ *Related Problem 13.67*)

During a test flight in which a helicopter starts from rest at $t = 0$ at the origin of the coordinate system shown and moves in the x–y plane, onboard accelerometers indicate that its components of acceleration (in m/s^2) during the interval of time from $t = 0$ to $t = 10$ s are

$$a_x = 0.6t,$$

$$a_y = 1.8 - 0.36t,$$

What is the magnitude of the helicopter's velocity at $t = 6$ s?

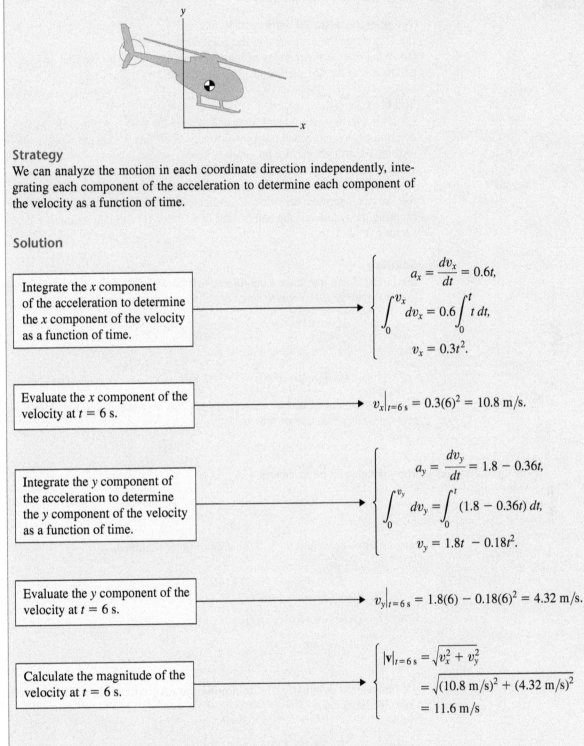

Strategy

We can analyze the motion in each coordinate direction independently, integrating each component of the acceleration to determine each component of the velocity as a function of time.

Solution

Integrate the x component of the acceleration to determine the x component of the velocity as a function of time.	$$\begin{cases} a_x = \dfrac{dv_x}{dt} = 0.6t, \\ \displaystyle\int_0^{v_x} dv_x = 0.6 \int_0^t t\, dt, \\ v_x = 0.3t^2. \end{cases}$$
Evaluate the x component of the velocity at $t = 6$ s.	$v_x\big|_{t=6\,\text{s}} = 0.3(6)^2 = 10.8 \text{ m/s}.$
Integrate the y component of the acceleration to determine the y component of the velocity as a function of time.	$$\begin{cases} a_y = \dfrac{dv_y}{dt} = 1.8 - 0.36t, \\ \displaystyle\int_0^{v_y} dv_y = \int_0^t (1.8 - 0.36t)\, dt, \\ v_y = 1.8t - 0.18t^2. \end{cases}$$
Evaluate the y component of the velocity at $t = 6$ s.	$v_y\big|_{t=6\,\text{s}} = 1.8(6) - 0.18(6)^2 = 4.32 \text{ m/s}.$
Calculate the magnitude of the velocity at $t = 6$ s.	$$\begin{cases} |\mathbf{v}|_{t=6\,\text{s}} = \sqrt{v_x^2 + v_y^2} \\ \qquad = \sqrt{(10.8 \text{ m/s})^2 + (4.32 \text{ m/s})^2} \\ \qquad = 11.6 \text{ m/s} \end{cases}$$

Practice Problem Determine the position vector of the helicopter at $t = 6$ s relative to its position at $t = 0$.

Answer: $\mathbf{r}\big|_{t=6\,\text{s}} = 21.6\mathbf{i} + 19.4\mathbf{j}$ (m).

Example 13.7 | **A Projectile Problem** (▶ *Related Problem 13.69*)

The skier leaves the 20° surface at 10 m/s.
(a) Determine the distance d to the point where he lands.
(b) What are the magnitudes of his components of velocity parallel and perpendicular to the 45° surface just before he lands?

Strategy

(a) By neglecting aerodynamic drag and treating the skier as a projectile, we can determine his velocity and position as functions of time. Using the equation describing the straight surface on which he lands, we can relate his horizontal and vertical coordinates at impact and thereby obtain an equation for the time at which he lands. Knowing the time, we can determine his position and velocity.
(b) We can determine his velocity parallel and perpendicular to the 45° surface by using the result that the component of a vector $\mathbf{U}$ in the direction of a unit vector $\mathbf{e}$ is $(\mathbf{e} \cdot \mathbf{U})\mathbf{e}$.

Solution

(a) In Fig. a, we introduce a coordinate system with its origin where the skier leaves the surface. His components of velocity at that instant ($t = 0$) are

$$v_x = 10 \cos 20° = 9.40 \text{ m/s}$$

and

$$v_y = -10 \sin 20° = -3.42 \text{ m/s}.$$

The x component of acceleration is zero, so v_x is constant and the skier's x coordinate as a function of time is

$$x = 9.40t \text{ m}.$$

The y component of acceleration is

$$a_y = \frac{dv_y}{dt} = -9.81 \text{ m/s}^2.$$

Integrating to determine v_y as a function of time, we obtain

$$\int_{-3.42}^{v_y} dv_y = \int_0^t -9.81 \, dt,$$

from which it follows that

$$v_y = \frac{dy}{dt} = -3.42 - 9.81t \text{ m/s}.$$

We integrate this equation to determine the y coordinate as a function of time. We have

$$\int_0^y dy = \int_0^t (-3.42 - 9.81t) \, dt,$$

yielding

$$y = -3.42t - 4.905t^2 \text{ m}.$$

(a)

The slope of the surface on which the skier lands is -1, so the linear equation describing it is $y = (-1)x + A$, where A is a constant. At $x = 0$, the y coordinate of the surface is -3 m, so $A = -3$ m and the equation describing the 45° surface is

$$y = -x - 3 \text{ m}.$$

Substituting our equations for x and y as functions of time into this equation, we obtain an equation for the time at which the skier lands:

$$-3.42t - 4.905t^2 = -9.40t - 3.$$

Solving for t, we get $t = 1.60$ s. Therefore, his coordinates when he lands are

$$x = 9.40(1.60) = 15.0 \text{ m}$$

and

$$y = -3.42(1.60) - 4.905(1.60)^2 = -18.0 \text{ m},$$

and the distance d is

$$d = \sqrt{(15.0)^2 + (18.0 - 3)^2} = 21.3 \text{ m}.$$

(b) The components of the skier's velocity just before he lands are

$$v_x = 9.40 \text{ m/s}$$

and

$$v_y = -3.42 - 9.81(1.60) = -19.1 \text{ m/s},$$

and the magnitude of his velocity is $|\mathbf{v}| = \sqrt{(9.40)^2 + (-19.1)^2} = 21.3$ m/s. Let $\mathbf{e}$ be a unit vector that is parallel to the slope on which he lands (Fig. a):

$$\mathbf{e} = \cos 45°\mathbf{i} - \sin 45°\mathbf{j}.$$

The component of the velocity parallel to the surface is

$$(\mathbf{e} \cdot \mathbf{v})\mathbf{e} = [(\cos 45°\mathbf{i} - \sin 45°\mathbf{j}) \cdot (9.40\mathbf{i} - 19.1\mathbf{j})]\mathbf{e}$$

$$= 20.2\mathbf{e} \text{ (m/s)}.$$

The magnitude of the skier's velocity parallel to the surface is 20.2 m/s. Therefore, the magnitude of his velocity perpendicular to the surface is

$$\sqrt{|\mathbf{v}|^2 - (20.2)^2} = 6.88 \text{ m/s}.$$

Critical Thinking

The key to solving this problem was that we knew the skier's acceleration. Knowing the acceleration, we were able to determine the components of his velocity and position as functions of time. Notice how we determined the position at which he landed on the slope. We knew that at the instant he landed, his x and y coordinates specified a point on the straight line defining the surface of the slope. By substituting his x and y coordinates as functions of time into the equation for the straight line defining the slope, we were able to solve for the time at which he landed. Knowing the time, we could determine his velocity and position at that instant.

Problems

▶ **13.67** In a second test, the coordinates of the position (in m) of the helicopter in Active Example 13.6 are given as functions of time by

$$x = 4 + 2t,$$
$$y = 4 + 4t + t^2.$$

(a) What is the magnitude of the helicopter's velocity at $t = 3$ s?

(b) What is the magnitude of the helicopter's acceleration at $t = 3$ s?

13.68 In terms of a particular reference frame, the position of the center of mass of the F-14 at the time shown ($t = 0$) is $\mathbf{r} = 10\mathbf{i} + 6\mathbf{j} + 22\mathbf{k}$ (m). The velocity from $t = 0$ to $t = 4$ s is $\mathbf{v} = (52 + 6t)\mathbf{i} + (12 + t^2)\mathbf{j} - (4 + 2t^2)\mathbf{k}$ (m/s). What is the position of the center of mass of the plane at $t = 4$ s?

Problem 13.68

▶ **13.69** In Example 13.7, suppose that the angle between the horizontal and the slope on which the skier lands is 30° instead of 45°. Determine the distance d to the point where he lands.

13.70 A projectile is launched from ground level with initial velocity $v_0 = 20$ m/s. Determine its range R if (a) $\theta_0 = 30°$, (b) $\theta_0 = 45°$, and (c) $\theta_0 = 60°$.

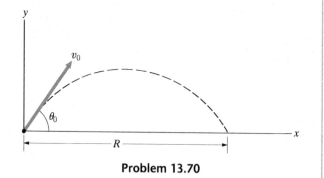

Problem 13.70

13.71 Immediately after the bouncing golf ball leaves the floor, its components of velocity are $v_x = 0.662$ m/s and $v_y = 3.66$ m/s.

(a) Determine the horizontal distance from the point where the ball left the floor to the point where it hits the floor again.

(b) The ball leaves the floor at $x = 0$, $y = 0$. Determine the ball's y coordinate as a function of x. (The parabolic function you obtain is shown superimposed on the photograph of the ball.)

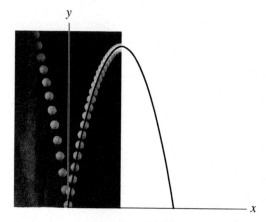

Problem 13.71

13.72 Suppose that you are designing a mortar to launch a rescue line from coast guard vessels to ships in distress. The light line is attached to a weight fired by the mortar. Neglect aerodynamic drag and the weight of the line for your preliminary analysis. If you want the line to be able to reach a ship 300 ft away when the mortar is fired at 45° above the horizontal, what muzzle velocity is required?

13.73 In Problem 13.72, what maximum height above the point from which it was fired is reached by the weight?

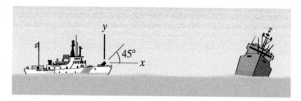

Problems 13.72/13.73

13.74 When the athlete releases the shot, it is 1.82 m above the ground and its initial velocity is $v_0 = 13.6$ m/s. Determine the horizontal distance the shot travels from the point of release to the point where it hits the ground.

Problem 13.74

13.75 A pilot wants to drop survey markers at remote locations in the Australian outback. If he flies at a constant velocity $v_0 = 40$ m/s at altitude $h = 30$ m and the marker is released with zero velocity relative to the plane, at what horizontal distance d from the desired impact point should the marker be released?

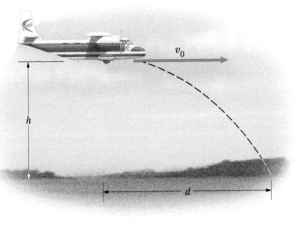

Problem 13.75

13.76 If the pitching wedge the golfer is using gives the ball an initial angle $\theta_0 = 50°$, what range of velocities v_0 will cause the ball to land within 3 ft of the hole? (Assume that the hole lies in the plane of the ball's trajectory.)

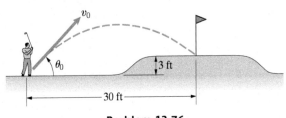

Problem 13.76

13.77 A batter strikes a baseball 3 ft above home plate and pops it up. The second baseman catches it 6 ft above second base 3.68 s after it was hit. What was the ball's initial velocity, and what was the angle between the ball's initial velocity vector and the horizontal?

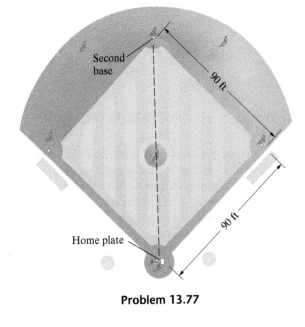

Problem 13.77

13.78 A baseball pitcher releases a fastball with an initial velocity $v_0 = 90$ mi/h. Let θ be the initial angle of the ball's velocity vector above the horizontal. When it is released, the ball is 6 ft above the ground and 58 ft from the batter's plate. The batter's strike zone extends from 1 ft 10 in above the ground to 4 ft 6 in above the ground. Neglecting aerodynamic effects, determine whether the ball will hit the strike zone (a) if $\theta = 1°$ and (b) if $\theta = 2°$.

13.79 In Problem 13.78, assume that the pitcher releases the ball at an angle $\theta = 1°$ above the horizontal, and determine the range of velocities v_0 (in ft/s) within which he must release the ball to hit the strike zone.

Problems 13.78/13.79

13.80 A zoology graduate student is provided with a bow and an arrow tipped with a syringe of sedative and is assigned to measure the temperature of a black rhinoceros (*Diceros bicornis*). The range of his bow when it is fully drawn and aimed 45° above the horizontal is 100 m. A truculent rhino suddenly charges straight toward him at 30 km/h. If he fully draws his bow and aims 20° above the horizontal, how far away should the rhino be when the student releases the arrow?

Problem 13.80

13.81 The crossbar of the goalposts in American football is $y_c = 10$ ft above the ground. To kick a field goal, the kicker must make the ball go between the two uprights supporting the crossbar, and the ball must be above the crossbar when it does so. Suppose that the kicker attempts a 40-yd field goal ($x_c = 120$ ft) and kicks the ball with initial velocity $v_0 = 70$ ft/s and angle $\theta_0 = 40°$. By what vertical distance does the ball clear the crossbar?

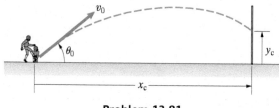

Problem 13.81

13.82* An American football quarterback stands at A. At the instant the quarterback throws the football, the receiver is at B running at 20 ft/s toward C, where he catches the ball. The ball is thrown at an angle of 45° above the horizontal, and it is thrown and caught at the same height above the ground. Determine the magnitude of the ball's initial velocity and the length of time it is in the air.

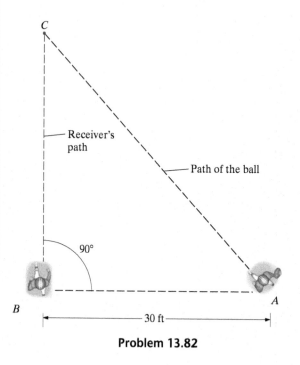

Problem 13.82

13.83 The cliff divers of Acapulco, Mexico, must time their dives so that they enter the water at the crest (high point) of a wave. The crests of the waves are 1 m above the mean water depth $h = 4$ m. The horizontal velocity of the waves is equal to $\sqrt{gh}$. The diver's aiming point is 2 m out from the base of the cliff. Assume that his velocity is horizontal when he begins the dive.

(a) What is the magnitude of the diver's velocity when he enters the water?

(b) How far from his aiming point must a wave crest be when he dives in order for him to enter the water at the crest?

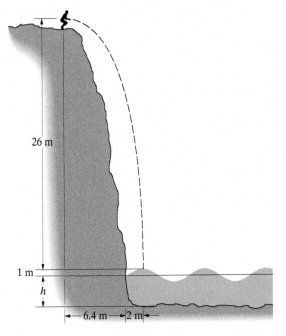
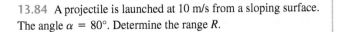

Problem 13.83

13.84 A projectile is launched at 10 m/s from a sloping surface. The angle $\alpha = 80°$. Determine the range R.

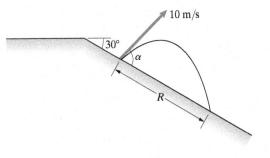

Problem 13.84

13.85 A projectile is launched at 100 ft/s at 60° above the horizontal. The surface on which it lands is described by the equation shown. Determine the x coordinate of the point of impact.

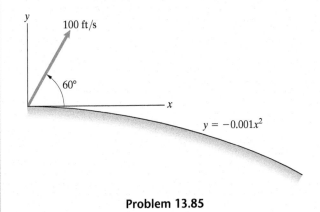

Problem 13.85

13.86 At $t = 0$, a steel ball in a tank of oil is given a horizontal velocity $\mathbf{v} = 2\mathbf{i}$ (m/s). The components of the ball's acceleration, in m/s^2, are $a_x = -1.2v_x$, $a_y = -8 - 1.2v_y$, and $a_z = -1.2v_z$. What is the velocity of the ball at $t = 1$ s?

13.87 In Problem 13.86, what is the position of the ball at $t = 1$ s relative to its position at $t = 0$?

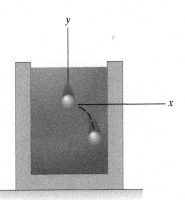

Problems 13.86/13.87

13.88 The point P moves along a circular path with radius R. Show that the magnitude of its velocity is $|\mathbf{v}| = R|d\theta/dt|$.

Strategy: Use Eqs. (13.23).

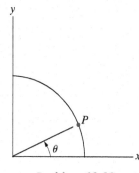

Problem 13.88

13.89 If $y = 150$ mm, $dy/dt = 300$ mm/s, and $d^2y/dt^2 = 0$, what are the magnitudes of the velocity and acceleration of point P?

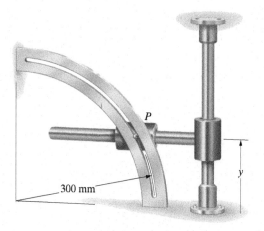

300 mm

Problem 13.89

13.90* A car travels at a constant speed of 100 km/h on a straight road of increasing grade whose vertical profile can be approximated by the equation shown. When the car's horizontal coordinate is $x = 400$ m, what is the car's acceleration?

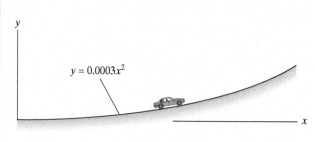

$y = 0.0003x^2$

Problem 13.90

13.91* Suppose that a projectile has the initial conditions shown in Fig. 13.12. Show that in terms of the $x'y'$ coordinate system with its origin at the highest point of the trajectory, the equation describing the trajectory is

$$y' = -\frac{g}{2v_0^2 \cos^2 \theta_0}(x')^2.$$

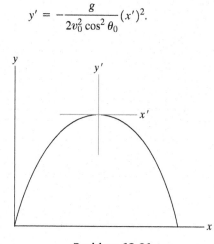

Problem 13.91

13.92* The acceleration components of a point are $a_x = -4 \cos 2t$, $a_y = -4 \sin 2t$, and $a_z = 0$. At $t = 0$, the position and velocity of the point are $\mathbf{r} = \mathbf{i}$ and $\mathbf{v} = 2\mathbf{j}$. Show that (a) the magnitude of the velocity is constant, (b) the velocity and acceleration vectors are perpendicular, (c) the magnitude of the acceleration is constant and points toward the origin, and (d) the trajectory of the point is a circle with its center at the origin.

13.5 Angular Motion

We have seen that in some cases the curvilinear motion of a point can be analyzed by using cartesian coordinates. In the sections that follow, we describe problems that can be analyzed more simply in terms of other coordinate systems. To help you understand our discussion of these alternative coordinate systems, we introduce two preliminary topics in this section: the angular motion of a line in a plane and the time derivative of a unit vector rotating in a plane.

Angular Motion of a Line

We can specify the angular position of a line L in a particular plane relative to a reference line L_0 in the plane by an angle θ (Fig. 13.14). The *angular velocity* of L relative to L_0 is defined by

$$\omega = \frac{d\theta}{dt}, \tag{13.31}$$

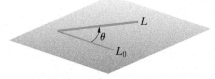

Figure 13.14
A line L and a reference line L_0 in a plane.

and the *angular acceleration* of L relative to L_0 is defined by

$$\alpha = \frac{d\omega}{dt} = \frac{d^2\theta}{dt^2}. \tag{13.32}$$

The dimensions of the angular position, angular velocity, and angular acceleration are rad, rad/s, and rad/s^2, respectively. Although these quantities are often expressed in terms of degrees or revolutions instead of radians, convert them into radians before using them in calculations.

Rotating Unit Vector

The directions of the unit vectors $\mathbf{i}$, $\mathbf{j}$, and $\mathbf{k}$ relative to the cartesian reference frame are constant. However, in other coordinate systems, the unit vectors used to describe the motion of a point rotate as the point moves. To obtain expressions for the velocity and acceleration in such coordinate systems, we must know the time derivative of a rotating unit vector.

We can describe the angular motion of a unit vector $\mathbf{e}$ in a plane just as we described the angular motion of a line. The direction of $\mathbf{e}$ relative to a reference line L_0 is specified by the angle θ in Fig. 13.15a, and the rate of rotation of $\mathbf{e}$ relative to L_0 is specified by the angular velocity

$$\omega = \frac{d\theta}{dt}.$$

The time derivative of $\mathbf{e}$ is defined by

$$\frac{d\mathbf{e}}{dt} = \lim_{\Delta t \to 0} \frac{\mathbf{e}(t + \Delta t) - \mathbf{e}(t)}{\Delta t}.$$

Figure 13.15b shows the vector $\mathbf{e}$ at time t and at time $t + \Delta t$. The change in $\mathbf{e}$ during this interval is $\Delta\mathbf{e} = \mathbf{e}(t + \Delta t) - \mathbf{e}(t)$, and the angle through which $\mathbf{e}$ rotates is $\Delta\theta = \theta(t + \Delta t) - \theta(t)$. The triangle in Fig. 13.15b is isosceles, so

$$|\Delta\mathbf{e}| = 2|\mathbf{e}| \sin(\Delta\theta/2) = 2 \sin(\Delta\theta/2).$$

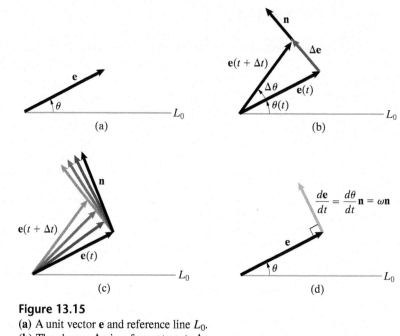

Figure 13.15
(a) A unit vector **e** and reference line L_0.
(b) The change $\Delta\mathbf{e}$ in **e** from t to $t + \Delta t$.
(c) As Δt goes to zero, **n** becomes perpendicular to $\mathbf{e}(t)$.
(d) The time derivative of **e**.

To write the vector $\Delta\mathbf{e}$ in terms of this expression, we introduce a unit vector **n** that points in the direction of $\Delta\mathbf{e}$ (Fig. 13.15b):

$$\Delta\mathbf{e} = |\Delta\mathbf{e}|\mathbf{n} = 2\sin(\Delta\theta/2)\mathbf{n}.$$

In terms of this expression, the time derivative of **e** is

$$\frac{d\mathbf{e}}{dt} = \lim_{\Delta t \to 0} \frac{\Delta\mathbf{e}}{\Delta t} = \lim_{\Delta t \to 0} \frac{2\sin(\Delta\theta/2)\mathbf{n}}{\Delta t}.$$

To evaluate the limit, we write it in the form

$$\frac{d\mathbf{e}}{dt} = \lim_{\Delta t \to 0} \frac{\sin(\Delta\theta/2)}{\Delta\theta/2} \frac{\Delta\theta}{\Delta t}\mathbf{n}.$$

In the limit as Δt approaches zero, $\sin(\Delta\theta/2)/(\Delta\theta/2) = 1$, $\Delta\theta/\Delta t = d\theta/dt$, and the unit vector **n** is perpendicular to $\mathbf{e}(t)$ (Fig. 13.15c). Therefore, the time derivative of **e** is

$$\frac{d\mathbf{e}}{dt} = \frac{d\theta}{dt}\mathbf{n} = \omega\mathbf{n}, \tag{13.33}$$

where **n** is a unit vector that is perpendicular to **e** and points in the positive θ direction (Fig. 13.15d). In the sections that follow, we use this result in deriving expressions for the velocity and acceleration of a point in different coordinate systems.

Angular Motion of a Line

> **Angular Position**
> The angular position of a line L in a plane relative to a reference line L_0 in the plane can be described by an angle θ.

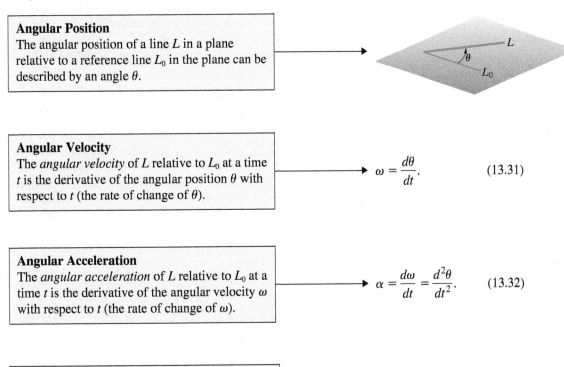

> **Angular Velocity**
> The *angular velocity* of L relative to L_0 at a time t is the derivative of the angular position θ with respect to t (the rate of change of θ).

$$\omega = \frac{d\theta}{dt}. \qquad (13.31)$$

> **Angular Acceleration**
> The *angular acceleration* of L relative to L_0 at a time t is the derivative of the angular velocity ω with respect to t (the rate of change of ω).

$$\alpha = \frac{d\omega}{dt} = \frac{d^2\theta}{dt^2}. \qquad (13.32)$$

> The equations relating the angular position θ, the angular velocity ω, and the angular acceleration α are identical in form to the equations that relate the position s, the velocity v, and the acceleration a in the motion of a point along a straight line. As a consequence, problems involving angular motion can be solved using the same methods that were applied to straight-line motion.

Straight-Line Motion	Angular Motion
$v = \dfrac{ds}{dt}$	$\omega = \dfrac{d\theta}{dt}$
$a = \dfrac{dv}{dt} = \dfrac{d^2s}{dt^2}$	$\alpha = \dfrac{d\omega}{dt} = \dfrac{d^2\theta}{dt^2}$

Rotating Unit Vector

> Let $\mathbf{e}$ be a unit vector that rotates in a plane relative to a reference line L_0 in the plane.

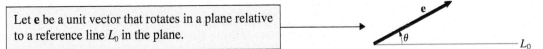

> The derivative of $\mathbf{e}$ with respect to time is
> $$\frac{d\mathbf{e}}{dt} = \frac{d\theta}{dt}\mathbf{n} = \omega\mathbf{n}, \quad (13.33)$$
> where $\mathbf{n}$ is a unit vector that is perpendicular to $\mathbf{e}$ and points in the positive θ direction.

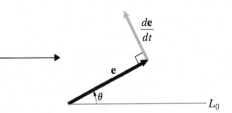

Active Example 13.8 Analysis of Angular Motion (▶ *Related Problem 13.96*)

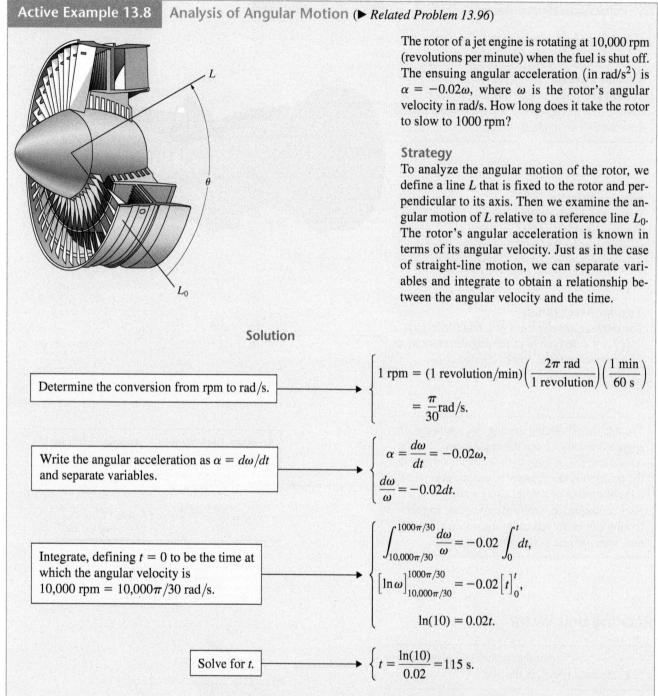

The rotor of a jet engine is rotating at 10,000 rpm (revolutions per minute) when the fuel is shut off. The ensuing angular acceleration (in rad/s²) is $\alpha = -0.02\omega$, where ω is the rotor's angular velocity in rad/s. How long does it take the rotor to slow to 1000 rpm?

Strategy

To analyze the angular motion of the rotor, we define a line L that is fixed to the rotor and perpendicular to its axis. Then we examine the angular motion of L relative to a reference line L_0. The rotor's angular acceleration is known in terms of its angular velocity. Just as in the case of straight-line motion, we can separate variables and integrate to obtain a relationship between the angular velocity and the time.

Solution

Determine the conversion from rpm to rad/s.

$$1 \text{ rpm} = (1 \text{ revolution/min})\left(\frac{2\pi \text{ rad}}{1 \text{ revolution}}\right)\left(\frac{1 \text{ min}}{60 \text{ s}}\right)$$

$$= \frac{\pi}{30}\text{rad/s}.$$

Write the angular acceleration as $\alpha = d\omega/dt$ and separate variables.

$$\alpha = \frac{d\omega}{dt} = -0.02\omega,$$

$$\frac{d\omega}{\omega} = -0.02dt.$$

Integrate, defining $t = 0$ to be the time at which the angular velocity is 10,000 rpm = $10,000\pi/30$ rad/s.

$$\int_{10,000\pi/30}^{1000\pi/30} \frac{d\omega}{\omega} = -0.02 \int_0^t dt,$$

$$\left[\ln\omega\right]_{10,000\pi/30}^{1000\pi/30} = -0.02\left[t\right]_0^t,$$

$$\ln(10) = 0.02t.$$

Solve for t.

$$t = \frac{\ln(10)}{0.02} = 115 \text{ s}.$$

Practice Problem Determine the number of revolutions the rotor turns as it decelerates from 10,000 rpm to 1000 rpm. Begin by applying the chain rule to the angular acceleration:

$$\alpha = \frac{d\omega}{dt} = \frac{d\omega}{d\theta}\frac{d\theta}{dt} = \frac{d\omega}{d\theta}\omega.$$

Answer: 7500 revolutions.

Problems

13.93 When an airplane touches down at $t = 0$, a stationary wheel is subjected to a constant angular acceleration $\alpha = 110$ rad/s^2 until $t = 1$ s.

(a) What is the wheel's angular velocity at $t = 1$ s?

(b) At $t = 0$, the angle $\theta = 0$. Determine θ in radians and in revolutions at $t = 1$ s.

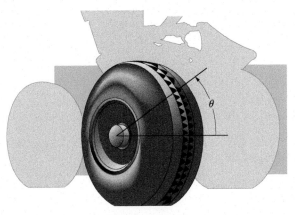

Problem 13.93

13.94 Let L be a line from the center of the earth to a fixed point on the equator, and let L_0 be a fixed reference direction. The figure views the earth from above the north pole.

(a) Is $d\theta/dt$ positive or negative? (Remember that the sun rises in the east.)

(b) Determine the approximate value of $d\theta/dt$ in rad/s and use it to calculate the angle through which the earth rotates in one hour.

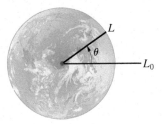

Problem 13.94

13.95 The angular acceleration of the line L relative to the line L_0 is given as a function of time by $\alpha = 2.5 - 1.2t$ rad/s^2. At $t = 0$, $\theta = 0$ and the angular velocity of L relative to L_0 is $\omega = 5$ rad/s. Determine θ and ω at $t = 3$ s.

Problem 13.95

▶ **13.96** In Active Example 13.8, suppose that the angular acceleration of the rotor is $\alpha = -0.00002\omega^2$, where ω is the angular velocity of the rotor in rad/s. How long does it take the rotor to slow from 10,000 rpm to 1000 rpm?

13.97 The astronaut is not rotating. He has an orientation control system that can subject him to a constant angular acceleration of 0.1 rad/s^2 about the vertical axis in either direction. If he wants to rotate 180° about the vertical axis (that is, rotate so that he is facing toward the left) and not be rotating in his new orientation, what is the minimum time in which he could achieve the new orientation?

13.98 The astronaut is not rotating. He has an orientation control system that can subject him to a constant angular acceleration of 0.1 rad/s^2 about the vertical axis in either direction. Refer to Problem 13.97. For safety, the control system will not allow his angular velocity to exceed 15° per second. If he wants to rotate 180° about the vertical axis (that is, rotate so that he is facing toward the left) and not be rotating in his new orientation, what is the minimum time in which he could achieve the new orientation?

Problems 13.97/13.98

13.99 The rotor of an electric generator is rotating at 200 rpm when the motor is turned off. Due to frictional effects, the angular acceleration of the rotor after the motor is turned off is $\alpha = -0.01\omega$ rad/s^2, where ω is the angular velocity in rad/s.

(a) What is the rotor's angular velocity one minute after the motor is turned off?

(b) After the motor is turned off, how many revolutions does the rotor turn before it comes to rest?

　　Strategy: To do part (b), use the chain rule to write the angular acceleration as

$$\alpha = \frac{d\omega}{dt} = \frac{d\omega}{d\theta}\frac{d\theta}{dt} = \frac{d\omega}{d\theta}\omega.$$

13.100 The needle of a measuring instrument is connected to a *torsional spring* that gives it an angular acceleration $\alpha = -4\theta$ rad/s^2, where θ is the needle's angular position in radians relative to a reference direction. The needle is given an angular velocity $\omega = 2$ rad/s in the position $\theta = 0$.

(a) What is the magnitude of the needle's angular velocity when $\theta = 30°$?

(b) What maximum angle θ does the needle reach before it rebounds?

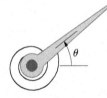

Problem 13.100

13.101 The angle θ measures the direction of the unit vector **e** relative to the x axis. The angular velocity of **e** is $\omega = d\theta/dt = 2$ rad/s, a constant. Determine the derivative $d\mathbf{e}/dt$ when $\theta = 90°$ in two ways:

(a) Use Eq. (13.33).

(b) Express the vector **e** in terms of its x and y components and take the time derivative of **e**.

13.102 The angle θ measures the direction of the unit vector **e** relative to the x axis. The angle θ is given as a function of time by $\theta = 2t^2$ rad. What is the vector $d\mathbf{e}/dt$ at $t = 4$ s?

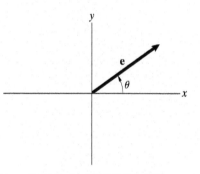

Problems 13.101/13.102

13.103 The line OP is of constant length R. The angle $\theta = \omega_0 t$, where ω_0 is a constant.

(a) Use the relations

$$v_x = \frac{dx}{dt} \quad \text{and} \quad v_y = \frac{dy}{dt}$$

to determine the velocity of point P relative to O.

(b) Use Eq. (13.33) to determine the velocity of point P relative to O, and confirm that your result agrees with the result of part (a).

　　Strategy: In part (b), write the position vector of P relative to O as $\mathbf{r} = R\mathbf{e}$, where **e** is a unit vector that points from O toward P.

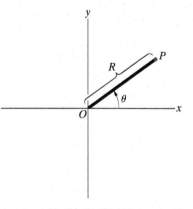

Problem 13.103

13.6 Curvilinear Motion—Normal and Tangential Components

In this method of describing curvilinear motion, we specify the position of a point by a coordinate measured *along its path* and express the velocity and acceleration in terms of their components tangential and normal (perpendicular) to the path. Normal and tangential components are particularly useful when a point moves along a circular path. Furthermore, they give us unique insight into the character of the velocity and acceleration in curvilinear motion. We first discuss motion in a planar path because of its conceptual simplicity.

Planar Motion

Consider a point P moving along a plane curvilinear path relative to some reference frame (Fig. 13.16a). The position vector $\mathbf{r}$ specifies the position of P relative to the reference point O, and the coordinate s measures P's position along the path relative to a point O' on the path. The velocity of P relative to O is

$$\mathbf{v} = \frac{d\mathbf{r}}{dt} = \lim_{\Delta t \to 0} \frac{\mathbf{r}(t + \Delta t) - \mathbf{r}(t)}{\Delta t} = \lim_{\Delta t \to 0} \frac{\Delta \mathbf{r}}{\Delta t}, \qquad (13.34)$$

where $\Delta \mathbf{r} = \mathbf{r}(t + \Delta t) - \mathbf{r}(t)$ (Fig. 13.16b). We denote the distance traveled along the path from t to $t + \Delta t$ by Δs. By introducing a unit vector $\mathbf{e}$ defined to point in the direction of $\Delta \mathbf{r}$, we can write Eq. (13.34) as

$$\mathbf{v} = \lim_{\Delta t \to 0} \frac{\Delta s}{\Delta t} \mathbf{e}.$$

As Δt approaches zero, $\Delta s / \Delta t$ becomes ds/dt and $\mathbf{e}$ becomes a unit vector tangent to the path at the position of P at time t, which we denote by $\mathbf{e}_t$ (Fig. 13.16c):

$$\mathbf{v} = v\mathbf{e}_t = \frac{ds}{dt} \mathbf{e}_t. \qquad (13.35)$$

The velocity of a point in curvilinear motion is a vector whose magnitude equals the rate of change of distance traveled along the path and whose direction is tangent to the path.

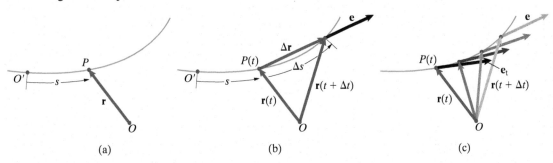

(a) (b) (c)

Figure 13.16
(a) The position of P along its path is specified by the coordinate s.
(b) Position of P at time t and at time $t + \Delta t$.
(c) The limit of $\mathbf{e}$ as $\Delta t \to 0$ is a unit vector tangent to the path.

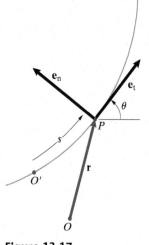

Figure 13.17
The path angle θ.

To determine the acceleration of P, we take the time derivative of Eq. (13.35):

$$\mathbf{a} = \frac{d\mathbf{v}}{dt} = \frac{dv}{dt}\mathbf{e}_t + v\frac{d\mathbf{e}_t}{dt}. \tag{13.36}$$

If the path is not a straight line, the unit vector $\mathbf{e}_t$ rotates as P moves. As a consequence, the time derivative of $\mathbf{e}_t$ is not zero. In the previous section, we derived an expression for the time derivative of a rotating unit vector in terms of the unit vector's angular velocity [Eq. (13.33)]. To use that result, we define the *path angle* θ specifying the direction of $\mathbf{e}_t$ relative to a reference line (Fig. 13.17). Then, from Eq. (13.33), the time derivative of $\mathbf{e}_t$ is

$$\frac{d\mathbf{e}_t}{dt} = \frac{d\theta}{dt}\mathbf{e}_n,$$

where $\mathbf{e}_n$ is a unit vector that is normal to $\mathbf{e}_t$ and points in the positive θ direction if $d\theta/dt$ is positive. Substituting this expression into Eq. (13.36), we obtain the acceleration of P:

$$\mathbf{a} = \frac{dv}{dt}\mathbf{e}_t + v\frac{d\theta}{dt}\mathbf{e}_n. \tag{13.37}$$

We can derive this result in another way that is less rigorous, but that gives additional insight into the meanings of the tangential and normal components of the acceleration. Figure 13.18a shows the velocity of P at times t and $t + \Delta t$. In Fig. 13.18b, you can see that the change in the velocity, $\mathbf{v}(t + \Delta t) - \mathbf{v}(t)$, consists of two components. The component Δv, which is tangent to the path at time t, is due to the change in the magnitude of the velocity. The component $v\Delta\theta$, which is perpendicular to the path at time t, is due to the change in the direction of the velocity vector. Thus, the change in the velocity is (approximately)

$$\mathbf{v}(t + \Delta t) - \mathbf{v}(t) = \Delta v\,\mathbf{e}_t + v\Delta\theta\,\mathbf{e}_n.$$

To obtain the acceleration, we divide this expression by Δt and take the limit as $\Delta t \to 0$:

$$\mathbf{a} = \lim_{\Delta t \to 0}\frac{\Delta\mathbf{v}}{\Delta t} = \lim_{\Delta t \to 0}\left(\frac{\Delta v}{\Delta t}\mathbf{e}_t + v\frac{\Delta\theta}{\Delta t}\mathbf{e}_n\right)$$

$$= \frac{dv}{dt}\mathbf{e}_t + v\frac{d\theta}{dt}\mathbf{e}_n.$$

Figure 13.18
(a) Velocity of P at t and at $t + \Delta t$.
(b) The tangential and normal components of the change in the velocity.

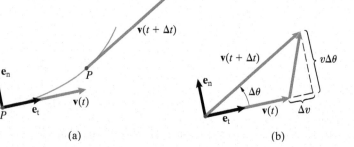

(a)

(b)

Thus, we again obtain Eq. (13.37). However, this derivation clearly points out that the tangential component of the acceleration arises from the rate of change of the magnitude of the velocity, whereas the normal component arises from the rate of change in the direction of the velocity vector. Notice that if the path is a straight line at time t, the normal component of the acceleration equals zero, because in that case $d\theta/dt$ is zero.

We can express the acceleration in another form that often is more convenient to use. Figure 13.19 shows the positions on the path reached by P at times t and $t + dt$. If the path is curved, straight lines extended from these points perpendicular to the path will intersect as shown. The distance ρ from the path to the point where these two lines intersect is called the *instantaneous radius of curvature* of the path. (If the path is circular, ρ is simply the radius of the path.) The angle $d\theta$ is the change in the path angle, and ds is the distance traveled from t to $t + \Delta t$. You can see from the figure that ρ is related to ds by

$$ds = \rho \, d\theta.$$

Dividing by dt, we obtain

$$\frac{ds}{dt} = v = \rho \frac{d\theta}{dt}.$$

Using this relation, we can write Eq. (13.37) as

$$\mathbf{a} = \frac{dv}{dt}\mathbf{e}_t + \frac{v^2}{\rho}\mathbf{e}_n.$$

For a given value of v, the normal component of the acceleration depends on the instantaneous radius of curvature. The greater the curvature of the path, the greater is the normal component of acceleration. When the acceleration is expressed in this way, the unit vector $\mathbf{e}_n$ must be defined to point toward the *concave* side of the path (Fig. 13.20).

Thus, the velocity and acceleration in terms of normal and tangential components are (Fig. 13.21)

$$\mathbf{v} = v\mathbf{e}_t = \frac{ds}{dt}\mathbf{e}_t \tag{13.38}$$

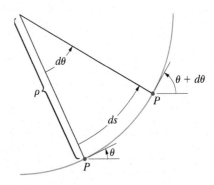

Figure 13.19
The instantaneous radius of curvature, ρ.

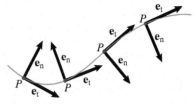

Figure 13.20
The unit vector normal to the path points toward the concave side.

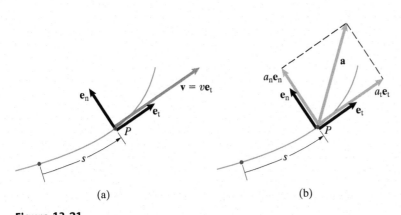

(a)   (b)

Figure 13.21
Normal and tangential components of the velocity (a) and acceleration (b).

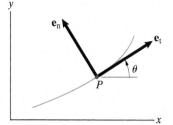

Figure 13.22
A point P moving in the x–y plane.

and

$$\mathbf{a} = a_t\mathbf{e}_t + a_n\mathbf{e}_n, \tag{13.39}$$

where

$$a_t = \frac{dv}{dt} \quad \text{and} \quad a_n = v\frac{d\theta}{dt} = \frac{v^2}{\rho}. \tag{13.40}$$

If the motion occurs in the x–y plane of a cartesian reference frame (Fig. 13.22) and θ is the angle between the x axis and the unit vector $\mathbf{e}_t$, the unit vectors $\mathbf{e}_t$ and $\mathbf{e}_n$ are related to the cartesian unit vectors by

$$\mathbf{e}_t = \cos\theta\,\mathbf{i} + \sin\theta\,\mathbf{j}$$

and

$$\mathbf{e}_n = -\sin\theta\,\mathbf{i} + \cos\theta\,\mathbf{j}. \tag{13.41}$$

If the path in the x–y plane is described by a function $y = y(x)$, it can be shown that the instantaneous radius of curvature is given by

$$\rho = \frac{\left[1 + \left(\dfrac{dy}{dx}\right)^2\right]^{3/2}}{\left|\dfrac{d^2y}{dx^2}\right|}. \tag{13.42}$$

Circular Motion

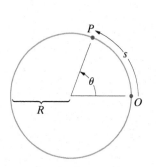

Figure 13.23
A point moving in a circular path.

If a point P moves in a plane circular path of radius R (Fig. 13.23), the distance s is related to the angle θ by

$$s = R\theta \quad \text{(circular path)}.$$

Using this relation, then, we can specify the position of P along the circular path by either s or θ. Taking the time derivative of the equation, we obtain a relation between $v = ds/dt$ and the angular velocity of the line from the center of the path to P:

$$v = R\frac{d\theta}{dt} = R\omega \quad \text{(circular path)}. \tag{13.43}$$

Taking another time derivative, we obtain a relation between the tangential component of the acceleration $a_t = dv/dt$ and the angular acceleration:

$$a_t = R\frac{d\omega}{dt} = R\alpha \quad \text{(circular path)}. \tag{13.44}$$

For this circular path, the instantaneous radius of curvature $\rho = R$, so the normal component of the acceleration is

$$a_n = \frac{v^2}{R} = R\omega^2 \quad \text{(circular path)}. \tag{13.45}$$

Because problems involving circular motion of a point are so common, these relations are worth remembering. But you must be careful to use them *only* when the path is circular.

Three-Dimensional Motion

Although most applications of normal and tangential components involve the motion of a point in a plane, we briefly discuss three-dimensional motion for the insight it provides into the nature of the velocity and acceleration. If we consider the motion of a point along a three-dimensional path relative to some reference frame, the steps leading to Eq. (13.38) are unaltered. The velocity is

$$\mathbf{v} = v\mathbf{e}_t = \frac{ds}{dt}\mathbf{e}_t, \tag{13.46}$$

where $v = ds/dt$ is the rate of change of distance along the path and the unit vector $\mathbf{e}_t$ is tangent to the path and points in the direction of motion. We take the time derivative of this equation to obtain the acceleration:

$$\mathbf{a} = \frac{d\mathbf{v}}{dt} = \frac{dv}{dt}\mathbf{e}_t + v\frac{d\mathbf{e}_t}{dt}.$$

As the point moves along its three-dimensional path, the direction of the unit vector $\mathbf{e}_t$ changes. In the case of motion of a point in a plane, this unit vector rotates in the plane, but in three-dimensional motion, the picture is more complicated. Figure 13.24a shows the path seen from a viewpoint perpendicular to the plane containing the vector $\mathbf{e}_t$ at times t and $t + dt$. This plane is called the *osculating plane*. It can be thought of as the instantaneous plane of rotation of the unit vector $\mathbf{e}_t$, and its orientation will generally change as P moves along its path. Since $\mathbf{e}_t$ is rotating in the osculating plane at time t, its time derivative is

$$\frac{d\mathbf{e}_t}{dt} = \frac{d\theta}{dt}\mathbf{e}_n, \tag{13.47}$$

where $d\theta/dt$ is the angular velocity of $\mathbf{e}_t$ in the osculating plane and the unit vector $\mathbf{e}_n$ is defined as shown in Fig. 13.24b. The vector $\mathbf{e}_n$ is perpendicular to $\mathbf{e}_t$, parallel to the osculating plane, and directed toward the concave side of the path. Therefore the acceleration is

$$\mathbf{a} = \frac{dv}{dt}\mathbf{e}_t + v\frac{d\theta}{dt}\mathbf{e}_n. \tag{13.48}$$

In the same way as in the case of motion in a plane, we can also express the acceleration in terms of the instantaneous radius of curvature of the path (Fig. 13.24c):

$$\mathbf{a} = \frac{dv}{dt}\mathbf{e}_t + \frac{v^2}{\rho}\mathbf{e}_n. \tag{13.49}$$

We see that the expressions for the velocity and acceleration in normal and tangential components for three-dimensional motion are identical in form to the expressions for planar motion. The velocity is a vector whose magnitude equals the rate of change of distance traveled along the path and whose direction is tangent to the path. The acceleration has a component tangential to the path equal to the rate of change of the magnitude of the velocity and a component perpendicular to the path that depends on the magnitude of the velocity and the instantaneous radius of curvature of the path. In planar motion, the unit vector $\mathbf{e}_n$ is parallel to the plane of the motion. In three-dimensional motion,

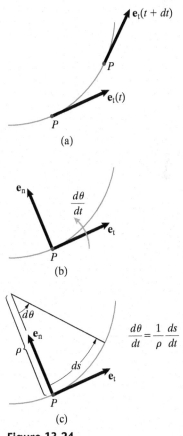

Figure 13.24
(a) Defining the osculating plane.
(b) Definition of the unit vector $\mathbf{e}_n$.
(c) The instantaneous radius of curvature.

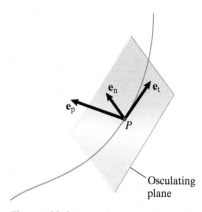

Figure 13.25
Defining the third unit vector $\mathbf{e}_p$.

$\mathbf{e}_n$ is parallel to the osculating plane, whose orientation depends on the nature of the path. Notice from Eq. (13.47) that $\mathbf{e}_n$ can be expressed in terms of $\mathbf{e}_t$ by

$$\mathbf{e}_n = \frac{\dfrac{d\mathbf{e}_t}{dt}}{\left|\dfrac{d\mathbf{e}_t}{dt}\right|}. \tag{13.50}$$

As the final step necessary to establish a three-dimensional coordinate system, we introduce a third unit vector that is perpendicular to both $\mathbf{e}_t$ and $\mathbf{e}_n$ by the definition

$$\mathbf{e}_p = \mathbf{e}_t \times \mathbf{e}_n. \tag{13.51}$$

The unit vector $\mathbf{e}_p$ is perpendicular to the osculating plane (Fig. 13.25).

RESULTS

Normal and Tangential Components in Planar Motion

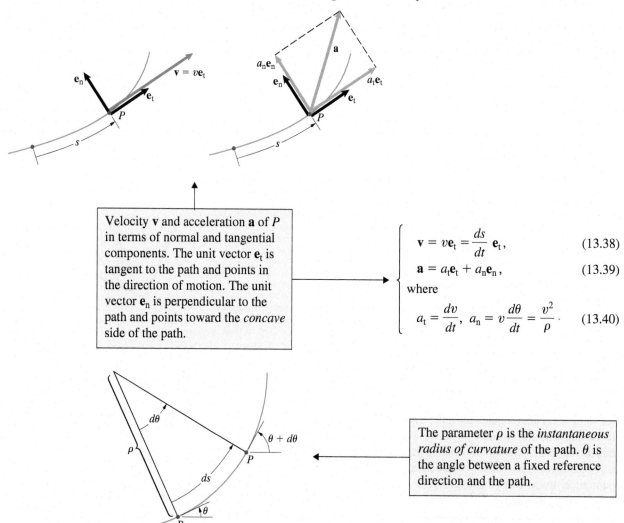

Velocity $\mathbf{v}$ and acceleration $\mathbf{a}$ of P in terms of normal and tangential components. The unit vector $\mathbf{e}_t$ is tangent to the path and points in the direction of motion. The unit vector $\mathbf{e}_n$ is perpendicular to the path and points toward the *concave* side of the path.

$$\mathbf{v} = v\mathbf{e}_t = \frac{ds}{dt}\,\mathbf{e}_t, \tag{13.38}$$

$$\mathbf{a} = a_t\mathbf{e}_t + a_n\mathbf{e}_n, \tag{13.39}$$

where

$$a_t = \frac{dv}{dt}, \quad a_n = v\frac{d\theta}{dt} = \frac{v^2}{\rho}. \tag{13.40}$$

The parameter ρ is the *instantaneous radius of curvature* of the path. θ is the angle between a fixed reference direction and the path.

Motion in the *x–y* Plane of a Cartesian Reference Frame

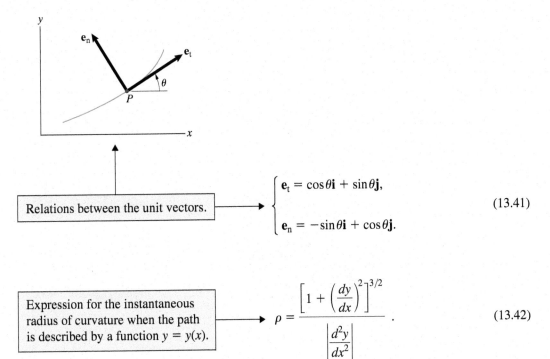

| Relations between the unit vectors. | $\begin{cases} \mathbf{e}_t = \cos\theta\mathbf{i} + \sin\theta\mathbf{j}, \\[2mm] \mathbf{e}_n = -\sin\theta\mathbf{i} + \cos\theta\mathbf{j}. \end{cases}$ | (13.41) |

| Expression for the instantaneous radius of curvature when the path is described by a function $y = y(x)$. | $\rho = \dfrac{\left[1 + \left(\dfrac{dy}{dx}\right)^2\right]^{3/2}}{\left|\dfrac{d^2y}{dx^2}\right|}.$ | (13.42) |

Motion in a Circular Path

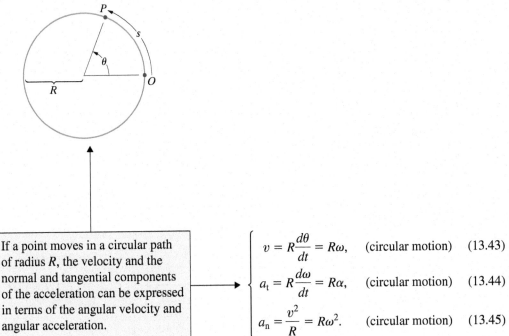

| If a point moves in a circular path of radius R, the velocity and the normal and tangential components of the acceleration can be expressed in terms of the angular velocity and angular acceleration. | $\begin{cases} v = R\dfrac{d\theta}{dt} = R\omega, & \text{(circular motion)} \quad (13.43) \\[3mm] a_t = R\dfrac{d\omega}{dt} = R\alpha, & \text{(circular motion)} \quad (13.44) \\[3mm] a_n = \dfrac{v^2}{R} = R\omega^2. & \text{(circular motion)} \quad (13.45) \end{cases}$ |

Active Example 13.9 | **Motion in Terms of Normal and Tangential Components**
(▶ *Related Problem 13.104*)

The motorcycle starts from rest at $t = 0$ on a circular track with a 400-m radius. The tangential component of the motorcycle's acceleration (in m/s^2) is given as a function of time by $a_t = 2 + 0.2t$. What is the motorcycle's velocity in terms of normal and tangential components at $t = 10$ s? What distance s has the motorcycle moved along the track at $t = 10$ s?

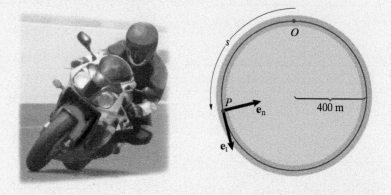

Strategy

Let s be the distance along the track from the initial position O of the motorcycle to its position at time t. Knowing the tangential acceleration as a function of time, we can integrate to determine the velocity v and position s as functions of time.

Solution

Integrate the tangential acceleration to determine the velocity as a function of time.	$\begin{cases} a_t = \dfrac{dv}{dt} = 2 + 0.2t, \\[2mm] \displaystyle\int_0^v dv = \int_0^t (2 + 0.2t)dt, \\[2mm] v = 2t + 0.1t^2 \, \text{m/s}. \end{cases}$
Evaluate the velocity at $t = 10$ s.	$\begin{cases} v\big|_{t=10\text{ s}} = 2(10) + 0.1(10)^2 \\[2mm] \qquad = 30 \text{ m/s}. \end{cases}$
Express the velocity at $t = 10$ s as a vector in terms of normal and tangential components.	$\begin{cases} \mathbf{v} = v\mathbf{e}_t \\[2mm] \quad = 30\mathbf{e}_t \, \text{(m/s)}. \end{cases}$
Integrate $v = ds/dt$ to determine the position as a function of time.	$\begin{cases} v = \dfrac{ds}{dt} = 2t + 0.1t^2, \\[2mm] \displaystyle\int_0^s ds = \int_0^t (2t + 0.1t^2)dt, \\[2mm] s = t^2 + \dfrac{0.1}{3}t^3 \, \text{m}. \end{cases}$

| Evaluate the position at $t = 10$ s. | $\rightarrow$ | $\begin{cases} s\vert_{t=10 \text{ s}} = (10)^2 + \dfrac{0.1}{3}(10)^3 \\ \qquad = 133 \text{ m.} \end{cases}$ |

Practice Problem Determine the motorcycle's acceleration in terms of normal and tangential components at $t = 10$ s.

Answer: $\mathbf{a} = 4\mathbf{e}_t + 2.25\mathbf{e}_n \ (\text{m/s}^2)$.

Example 13.10 The Circular-Orbit Problem (▶ *Related Problem 13.114*)

A satellite is in a circular orbit of radius R around the earth. What is its velocity?

Strategy
The acceleration due to gravity at a distance R from the center of the earth is gR_E^2/R^2, where R_E is the radius of the earth. (See Eq. 12.4.) By using this expression together with the equation for the acceleration in terms of normal and tangential components, we can obtain an equation for the satellite's velocity.

Solution
In terms of normal and tangential components (Fig. a), the acceleration of the satellite is

$$\mathbf{a} = \frac{dv}{dt}\mathbf{e}_t + \frac{v^2}{R}\mathbf{e}_n.$$

This expression must equal the acceleration due to gravity toward the center of the earth:

$$\frac{dv}{dt}\mathbf{e}_t + \frac{v^2}{R}\mathbf{e}_n = \frac{gR_E^2}{R^2}\mathbf{e}_n. \qquad (1)$$

Because there is no $\mathbf{e}_t$ component on the right side of Eq. (1), we conclude that the magnitude of the satellite's velocity is constant:

$$\frac{dv}{dt} = 0.$$

Equating the $\mathbf{e}_n$ components in Eq. (1) and solving for v, we obtain

$$v = \sqrt{\frac{gR_E^2}{R}}.$$

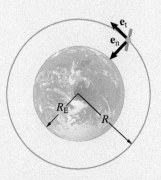

(a) Describing the satellite's motion in terms of normal and tangential components.

Critical Thinking
In Example 13.5 we determined the escape velocity of an object traveling straight away from the earth in terms of its initial distance from the center of the earth. The escape velocity for an object at a distance R from the center of the earth, $v_{esc} = \sqrt{2gR_E^2/R}$, is only $\sqrt{2}$ times the velocity of an object in a circular orbit of radius R. This explains why it was possible to begin launching probes to other planets not long after the first satellites were placed in earth orbit.

Example 13.11	Relating Cartesian Components to Normal and Tangential Components

(▶ *Related Problem 13.122*)

During a flight in which a helicopter starts from rest at $t = 0$, the cartesian components of its acceleration are

$$a_x = 0.6t \text{ m/s}^2$$

and

$$a_y = 1.8 - 0.36t \text{ m/s}^2.$$

What are the normal and tangential components of the helicopter's acceleration and the instantaneous radius of curvature of its path at $t = 4$ s?

Strategy

We can integrate the cartesian components of acceleration to determine the cartesian components of the velocity at $t = 4$ s, and then we can determine the components of the tangential unit vector $\mathbf{e}_t$ by dividing the velocity vector by its magnitude: $\mathbf{e}_t = \mathbf{v}/|\mathbf{v}|$. Next, we can determine the tangential component of the acceleration by evaluating the dot product of the acceleration vector with $\mathbf{e}_t$. Knowing the tangential component of the acceleration, we can then evaluate the normal component and determine the radius of curvature of the path from the relation $a_n = v^2/\rho$.

Solution

Integrating the components of acceleration with respect to time (see Active Example 13.6), we find that the cartesian components of the velocity are

$$v_x = 0.3t^2 \text{ m/s}$$

and

$$v_y = 1.8t - 0.18t^2 \text{ m/s}.$$

At $t = 4$ s, $v_x = 4.80$ m/s and $v_y = 4.32$ m/s. The tangential unit vector $\mathbf{e}_t$ at $t = 4$ s is (Fig. a)

$$\mathbf{e}_t = \frac{\mathbf{v}}{|\mathbf{v}|} = \frac{4.80\mathbf{i} + 4.32\mathbf{j}}{\sqrt{(4.80)^2 + (4.32)^2}} = 0.743\mathbf{i} + 0.669\mathbf{j}.$$

The components of the acceleration at $t = 4$ s are

$$a_x = 0.6(4) = 2.4 \text{ m/s}^2$$

and

$$a_y = 1.8 - 0.36(4) = 0.36 \text{ m/s}^2,$$

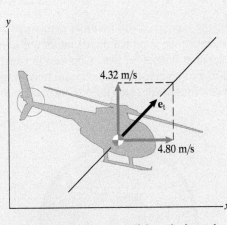

(a) Cartesian components of the velocity and the vector $\mathbf{e}_t$.

so the tangential component of the acceleration at $t = 4$ s is

$$a_t = \mathbf{e}_t \cdot \mathbf{a}$$

$$= (0.743\mathbf{i} + 0.669\mathbf{j}) \cdot (2.4\mathbf{i} + 0.36\mathbf{j})$$

$$= 2.02 \text{ m/s}^2.$$

The magnitude of the acceleration is $\sqrt{(2.4)^2 + (0.36)^2} = 2.43$ m/s^2, so the magnitude of the normal component of the acceleration is

$$a_n = \sqrt{|\mathbf{a}|^2 - a_t^2} = \sqrt{(2.43)^2 - (2.02)^2} = 1.34 \text{ m/s}^2.$$

The radius of curvature of the path is thus

$$\rho = \frac{|\mathbf{v}|^2}{a_n} = \frac{(4.80)^2 + (4.32)^2}{1.34} = 31.2 \text{ m}.$$

Critical Thinking

The cartesian components of a vector are parallel to the axes of the cartesian coordinate system, whereas the normal and tangential components are normal and tangential to the *path*. In this example, the cartesian components of the acceleration of the helicopter were given as functions of time. How could we determine the normal and tangential components of the acceleration at $t = 4$ s without knowing the path? Notice that we used the fact that *the velocity vector is tangent to the path*. We integrated the cartesian components of the acceleration to determine the cartesian components of the velocity at $t = 4$ s. That told us the direction of the path. By dividing the velocity vector by its magnitude, we obtained a unit vector tangent to the path that pointed in the direction of the motion, which is the vector $\mathbf{e}_t$.

Example 13.12 | Centrifuge (▶ *Related Problem 13.128*)

The distance from the center of the medical centrifuge to its samples is 300 mm. When the centrifuge is turned on, its motor and control system give it an angular acceleration $\alpha = A - B\omega^2$. Choose the constants A and B so that the samples will be subjected to a maximum horizontal acceleration of 12,000 g's and the centrifuge will reach 90% of its maximum operating speed in 2 min.

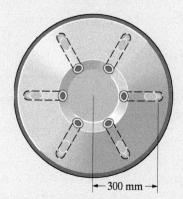

300 mm

Strategy

Since we know both the radius of the circular path in which the samples move and the horizontal acceleration to which they are to be subjected, we can solve for the operating angular velocity of the centrifuge. We will use the given angular acceleration to determine the centrifuge's angular velocity as a function of time in terms of the constants A and B. We can then use the operating angular velocity and the condition that the centrifuge reach 90% of the operating angular velocity in 2 min to determine the constants A and B.

Solution

From Eq. (13.45), the samples are subjected to a normal acceleration

$$a_n = R\omega^2.$$

Setting $a_n = (12,000)(9.81)$ m/s^2 and $R = 0.3$ m and solving for the angular velocity, we find that the desired maximum operating speed is $\omega_{max} = 626$ rad/s.

The angular acceleration is

$$\alpha = \frac{d\omega}{dt} = A - B\omega^2.$$

We separate variables to get

$$\frac{d\omega}{A - B\omega^2} = dt.$$

Then we integrate to determine ω as a function of time, assuming that the centrifuge starts from rest at $t = 0$:

$$\int_0^\omega \frac{d\omega}{A - B\omega^2} = \int_0^t dt.$$

Evaluating the integrals, we obtain

$$\frac{1}{2\sqrt{AB}}\ln\left(\frac{A + \sqrt{AB}\,\omega}{A - \sqrt{AB}\,\omega}\right) = t.$$

The solution of this equation for ω is

$$\omega = \sqrt{\frac{A}{B}}\left(\frac{e^{2\sqrt{AB}\,t} - 1}{e^{2\sqrt{AB}\,t} + 1}\right).$$

As t becomes large, ω approaches $\sqrt{A/B}$, so we have the condition that

$$\sqrt{\frac{A}{B}} = \omega_{max} = 626 \text{ rad/s}, \tag{1}$$

and we can write the equation for ω as

$$\omega = \omega_{max}\left(\frac{e^{2\sqrt{AB}\,t} - 1}{e^{2\sqrt{AB}\,t} + 1}\right). \tag{2}$$

We also have the condition that $\omega = 0.9\omega_{max}$ after 2 min. Setting $\omega = 0.9\omega_{max}$ and $t = 120$ s in Eq. (2) and solving for $\sqrt{AB}$, we obtain

$$\sqrt{AB} = \frac{\ln(19)}{240}.$$

We solve this equation together with Eq. (1), obtaining $A = 7.69 \text{ rad/s}^2$ and $B = 1.96 \times 10^{-5} \text{ rad}^{-1}$. The graph shows the angular velocity of the centrifuge as a function of time.

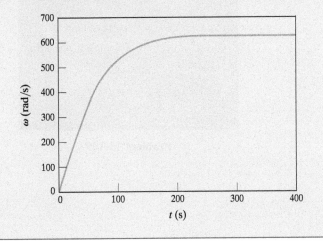

Problems

▶ **13.104** In Active Example 13.9, determine the motorcycle's velocity and acceleration in terms of normal and tangential components at $t = 5$ s.

13.105 The armature starts from rest at $t = 0$ and has constant angular acceleration $\alpha = 2$ rad/s². At $t = 4$ s, what are the velocity and acceleration of point P relative to point O in terms of normal and tangential components?

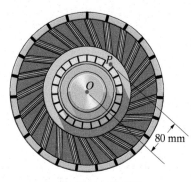

Problem 13.105

13.106 Suppose that you want to design a medical centrifuge to subject samples to normal accelerations of 1000 g's.

(a) If the distance from the center of the centrifuge to the sample is 300 mm, what speed of rotation in rpm is necessary?

(b) If you want the centrifuge to reach its design rpm in 1 min, what constant angular acceleration is necessary?

13.107 The medical centrifuge starts from rest at $t = 0$ and is subjected to a constant angular acceleration $\alpha = 3$ rad/s². What is the magnitude of the total acceleration to which the samples are subjected at $t = 1$ s?

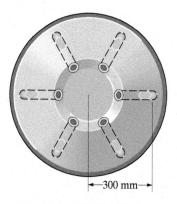

Problems 13.106/13.107

13.108 A centrifuge used to subject engineering components to high acceleration has a radius of 8 m. It starts from rest at $t = 0$, and during its two-minute acceleration phase it is programmed so that its angular acceleration is given as a function of time in seconds by $\alpha = 0.192 - 0.0016t$ rad/s². At $t = 120$ s, what is the magnitude of the acceleration a component is subjected to?

Problem 13.108

13.109 A powerboat being tested for maneuverability is started from rest at $t = 0$ and driven in a circular path 12 m in radius. The tangential component of the boat's acceleration as a function of time is $a_t = 0.4t$ m/s².

(a) What are the boat's velocity and acceleration in terms of normal and tangential components at $t = 4$ s?

(b) What distance does the boat move along its circular path from $t = 0$ to $t = 4$ s?

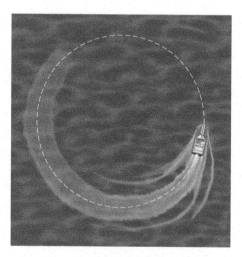

Problem 13.109

13.110 The angle $\theta = 2t^2$ rad.

(a) What are the velocity and acceleration of point P in terms of normal and tangential components at $t = 1$ s?

(b) What distance along the circular path does point P move from $t = 0$ to $t = 1$ s?

13.111 The angle $\theta = 2t^2$ rad. What are the velocity and acceleration of point P in terms of normal and tangential components when P has gone one revolution around the circular path starting at $t = 0$?

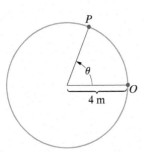

Problems 13.110/13.111

13.112 At the instant shown, the crank AB is rotating with a counterclockwise angular velocity of 5000 rpm. Determine the velocity of point B (a) in terms of normal and tangential components and (b) in terms of cartesian components.

13.113 The crank AB is rotating with a constant counterclockwise angular velocity of 5000 rpm. Determine the acceleration of point B (a) in terms of normal and tangential components and (b) in terms of cartesian components.

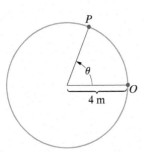

Problems 13.112/13.113

▶ **13.114** Suppose that a circular tunnel of radius R could be dug beneath the equator. In principle, a satellite could be placed in orbit about the center of the earth within the tunnel. The acceleration due to gravity in the tunnel would be gR/R_E, where g is the acceleration due to gravity at sea level and R_E is the earth's radius. Determine the velocity of the satellite and show that the time required to complete one orbit is independent of the radius R. (See Example 13.10.)

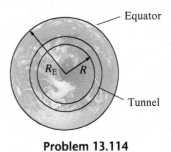

Problem 13.114

13.115 At the instant shown, the magnitude of the airplane's velocity is 130 m/s, its tangential component of acceleration is $a_t = -4$ m/s^2, and the rate of change of its path angle is $d\theta/dt = 5°/$s.

(a) What are the airplane's velocity and acceleration in terms of normal and tangential components?

(b) What is the instantaneous radius of curvature of the airplane's path?

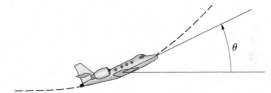

Problem 13.115

13.116 In the preliminary design of a sun-powered car, a group of engineering students estimates that the car's acceleration will be 0.6 m/s². Suppose that the car starts from rest at A, and the tangential component of its acceleration is $a_t = 0.6$ m/s². What are the car's velocity and acceleration in terms of normal and tangential components when it reaches B?

13.117 After subjecting a car design to wind-tunnel testing, the students estimate that the tangential component of the car's acceleration will be $a_t = 0.6 - 0.002v^2$ m/s², where v is the car's velocity in m/s. If the car starts from rest at A, what are its velocity and acceleration in terms of normal and tangential components when it reaches B?

13.118 Suppose that the tangential component of acceleration of a car is given in terms of the car's position by $a_t = 0.4 - 0.001s$ m/s², where s is the distance the car travels along the track from point A. What are the car's velocity and acceleration in terms of normal and tangential components at point B?

Problems 13.116–13.118

13.119 The car increases its speed at a constant rate from 40 mi/h at A to 60 mi/h at B. What is the magnitude of its acceleration 2 s after it passes point A?

13.120 The car increases its speed at a constant rate from 40 mi/h at A to 60 mi/h at B. Determine the magnitude of its acceleration when it has traveled along the road a distance (a) 120 ft from A and (b) 160 ft from A.

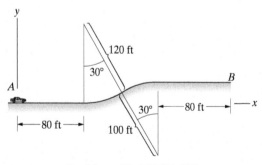

Problems 13.119/13.120

13.121 Astronaut candidates are to be tested in a centrifuge with 10-m radius that rotates in the horizontal plane. Test engineers want to subject the candidates to an acceleration of 5 g's, or five times the acceleration due to gravity. Earth's gravity effectively exerts an acceleration of 1 g in the vertical direction. Determine the angular velocity of the centrifuge in revolutions per second so that the magnitude of the total acceleration is 5 g's.

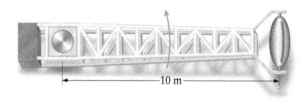

Problem 13.121

▶ **13.122** In Example 13.11, what is the helicopter's velocity in terms of normal and tangential components at $t = 4$ s?

13.123 The athlete releases the shot with velocity $v = 16$ m/s.

(a) What are the velocity and acceleration of the shot in terms of normal and tangential components when it is at the highest point of its trajectory?

(b) What is the instantaneous radius of curvature of the shot's path when it is at the highest point of its trajectory?

13.124 At $t = 0$, the athlete releases the shot with velocity $v = 16$ m/s.

(a) What are the velocity and acceleration of the shot in terms of normal and tangential components at $t = 0.3$ s?

(b) Use the relation $a_n = v^2/\rho$ to determine the instantaneous radius of curvature of the shot's path at $t = 0.3$ s.

13.125 At $t = 0$, the athlete releases the shot with velocity $v = 16$ m/s. Use Eq. (13.42) to determine the instantaneous radius of curvature of the shot's path at $t = 0.3$ s.

Problems 13.123–13.125

13.126 The cartesian coordinates of a point moving in the x–y plane are

$$x = 20 + 4t^2 \text{ m} \quad \text{and} \quad y = 10 - t^3 \text{ m}.$$

What is the instantaneous radius of curvature of the path of the point at $t = 3$ s?

13.127 The helicopter starts from rest at $t = 0$. The cartesian components of its acceleration are $a_x = 0.6t$ m/s^2 and $a_y = 1.8 - 0.36t$ m/s^2. Determine the tangential and normal components of the acceleration at $t = 6$ s.

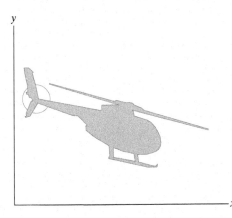

Problem 13.127

▶ **13.128** Suppose that when the centrifuge in Example 13.12 is turned on, its motor and control system give it an angular acceleration (in rad/s^2) $\alpha = 12 - 0.02\omega$, where ω is the centrifuge's angular velocity. Determine the tangential and normal components of the acceleration of the samples at $t = 0.2$ s.

13.129* For astronaut training, the airplane shown is to achieve "weightlessness" for a short period of time by flying along a path such that its acceleration is $a_x = 0$ and $a_y = -g$. If the velocity of the plane at O at time $t = 0$ is $\mathbf{v} = v_0\mathbf{i}$, show that the autopilot must fly the airplane so that its tangential component of acceleration as a function of time is

$$a_t = g \frac{gt/v_0}{\sqrt{1 + (gt/v_0)^2}}.$$

13.130* In Problem 13.129, what is the airplane's normal component of acceleration as a function of time?

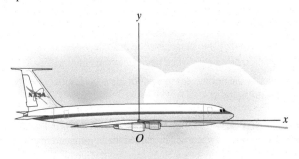

Problems 13.129/13.130

13.131 If $y = 100$ mm, $dy/dt = 200$ mm/s, and $d^2y/dt^2 = 0$, what are the velocity and acceleration of P in terms of normal and tangential components?

13.132* Suppose that the point P moves upward in the slot with velocity $\mathbf{v} = 300\mathbf{e}_t$ (mm/s). When $y = 150$ mm, what are dy/dt and d^2y/dt^2?

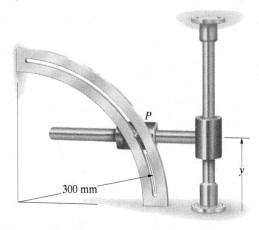

Problems 13.131/13.132

13.133* A car travels at 100 km/h on a straight road of increasing grade whose vertical profile can be approximated by the equation shown. When the car's horizontal coordinate is $x = 400$ m, what are the tangential and normal components of the car's acceleration?

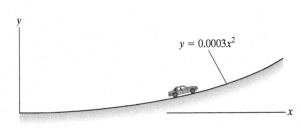

Problem 13.133

13.134 A boy rides a skateboard on the concrete surface of an empty drainage canal described by the equation shown. He starts at $y = 20$ ft, and the magnitude of his velocity is approximated by $v = \sqrt{2(32.2)(20 - y)}$ ft/s.

(a) Use Eq. (13.42) to determine the instantaneous radius of curvature of the boy's path when he reaches the bottom.

(b) What is the normal component of his acceleration when he reaches the bottom?

13.135 In Problem 13.134, what is the normal component of the boy's acceleration when he has passed the bottom and reached $y = 10$ ft?

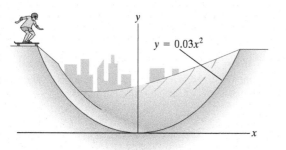

Problems 13.134/13.135

13.136* By using Eqs. (13.41): (a) show that the relations between the cartesian unit vectors and the unit vectors $\mathbf{e}_t$ and $\mathbf{e}_n$ are

$$\mathbf{i} = \cos\theta\,\mathbf{e}_t - \sin\theta\,\mathbf{e}_n$$

$$\mathbf{j} = \sin\theta\,\mathbf{e}_t + \cos\theta\,\mathbf{e}_n.$$

(b) Show that

$$\frac{d\mathbf{e}_t}{dt} = \frac{d\theta}{dt}\mathbf{e}_n \quad \text{and} \quad \frac{d\mathbf{e}_n}{dt} = -\frac{d\theta}{dt}\mathbf{e}_t.$$

13.7 Curvilinear Motion—Polar and Cylindrical Coordinates

BACKGROUND

Polar coordinates are often used to describe the curvilinear motion of a point. Circular motion, certain orbit problems, and, more generally, *central-force* problems, in which the acceleration of a point is directed toward a given point, can be expressed conveniently in polar coordinates.

Consider a point P in the x–y plane of a cartesian coordinate system. We can specify the position of P relative to the origin O either by its cartesian coordinates x, y or by its polar coordinates r, θ (Fig. 13.26a). To express vectors in terms of polar coordinates, we define a unit vector $\mathbf{e}_r$ that points in the direction of the radial line from the origin to P and a unit vector $\mathbf{e}_\theta$ that is perpendicular to $\mathbf{e}_r$ and points in the direction of increasing θ (Fig. 13.26b). In terms of these vectors, the position vector $\mathbf{r}$ from O to P is

$$\mathbf{r} = r\mathbf{e}_r. \tag{13.52}$$

(Notice that $\mathbf{r}$ has no component in the direction of $\mathbf{e}_\theta$.)

We can determine the velocity of P in terms of polar coordinates by taking the time derivative of Eq. (13.52):

$$\mathbf{v} = \frac{d\mathbf{r}}{dt} = \frac{dr}{dt}\mathbf{e}_r + r\frac{d\mathbf{e}_r}{dt}. \tag{13.53}$$

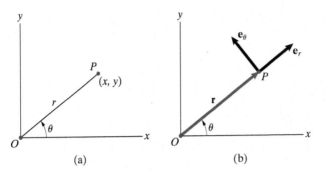

Figure 13.26
(a) The polar coordinates of P.
(b) The unit vectors $\mathbf{e}_r$ and $\mathbf{e}_\theta$ and the position vector $\mathbf{r}$.

As P moves along a curvilinear path, the unit vector $\mathbf{e}_r$ rotates with angular velocity $\omega = d\theta/dt$. Therefore, from Eq. (13.33), we can express the time derivative of $\mathbf{e}_r$ in terms of $\mathbf{e}_\theta$ as

$$\frac{d\mathbf{e}_r}{dt} = \frac{d\theta}{dt}\mathbf{e}_\theta. \qquad (13.54)$$

Substituting this result into Eq. (13.53), we obtain the velocity of P:

$$\mathbf{v} = \frac{dr}{dt}\mathbf{e}_r + r\frac{d\theta}{dt}\mathbf{e}_\theta = \frac{dr}{dt}\mathbf{e}_r + r\omega\mathbf{e}_\theta. \qquad (13.55)$$

We can get this result in another way that is less rigorous, but more direct and intuitive. Figure 13.27 shows the position vector of P at times t and $t + \Delta t$. The change in the position vector, $\mathbf{r}(t + \Delta t) - \mathbf{r}(t)$, consists of two components. The component Δr is due to the change in the radial position r and is in the $\mathbf{e}_r$ direction. The component $r\Delta\theta$ is due to the change in θ and is in the $\mathbf{e}_\theta$ direction. Thus, the change in the position of P is (approximately)

$$\mathbf{r}(t + \Delta t) - \mathbf{r}(t) = \Delta r\mathbf{e}_r + r\Delta\theta\mathbf{e}_\theta.$$

Dividing this expression by Δt and taking the limit as $\Delta t \to 0$, we obtain the velocity of P:

$$\mathbf{v} = \lim_{\Delta t \to 0}\left(\frac{\Delta r}{\Delta t}\mathbf{e}_r + r\frac{\Delta\theta}{\Delta t}\mathbf{e}_\theta\right)$$

$$= \frac{dr}{dt}\mathbf{e}_r + r\omega\mathbf{e}_\theta.$$

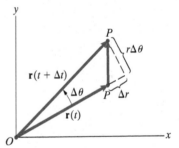

Figure 13.27
The position vector of P at t and $t + \Delta t$.

One component of the velocity is in the radial direction and is equal to the rate of change of the radial position r. The other component is normal, or *transverse*, to the radial direction and is proportional to the radial distance and to the rate of change of θ.

We obtain the acceleration of P by taking the time derivative of Eq. (13.55):

$$\mathbf{a} = \frac{d\mathbf{v}}{dt} = \frac{d^2r}{dt^2}\mathbf{e}_r + \frac{dr}{dt}\frac{d\mathbf{e}_r}{dt} + \frac{dr}{dt}\frac{d\theta}{dt}\mathbf{e}_\theta + r\frac{d^2\theta}{dt^2}\mathbf{e}_\theta + r\frac{d\theta}{dt}\frac{d\mathbf{e}_\theta}{dt}. \qquad (13.56)$$

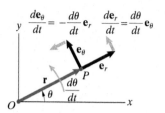

$$\frac{d\mathbf{e}_\theta}{dt} = -\frac{d\theta}{dt}\mathbf{e}_r \qquad \frac{d\mathbf{e}_r}{dt} = \frac{d\theta}{dt}\mathbf{e}_\theta$$

Figure 13.28
Time derivatives of $\mathbf{e}_r$ and $\mathbf{e}_\theta$.

The time derivative of the unit vector $\mathbf{e}_r$ due to the rate of change of θ is given by Eq. (13.54). As P moves, $\mathbf{e}_\theta$ also rotates with angular velocity $d\theta/dt$ (Fig. 13.28). You can see from this figure that the time derivative of $\mathbf{e}_\theta$ is in the $-\mathbf{e}_r$ direction if $d\theta/dt$ is positive:

$$\frac{d\mathbf{e}_\theta}{dt} = -\frac{d\theta}{dt}\mathbf{e}_r.$$

Substituting this expression and Eq. (13.54) into Eq. (13.56), we obtain the acceleration of P:

$$\mathbf{a} = \left[\frac{d^2r}{dt^2} - r\left(\frac{d\theta}{dt}\right)^2\right]\mathbf{e}_r + \left[r\frac{d^2\theta}{dt^2} + 2\frac{dr}{dt}\frac{d\theta}{dt}\right]\mathbf{e}_\theta.$$

Thus, the velocity and acceleration are respectively (Fig. 13.29)

$$\mathbf{v} = v_r\mathbf{e}_r + v_\theta\mathbf{e}_\theta = \frac{dr}{dt}\mathbf{e}_r + r\omega\mathbf{e}_\theta \tag{13.57}$$

and

$$\mathbf{a} = a_r\mathbf{e}_r + a_\theta\mathbf{e}_\theta, \tag{13.58}$$

where

$$a_r = \frac{d^2r}{dt^2} - r\left(\frac{d\theta}{dt}\right)^2 = \frac{d^2r}{dt^2} - r\omega^2$$

$$\tag{13.59}$$

$$a_\theta = r\frac{d^2\theta}{dt^2} + 2\frac{dr}{dt}\frac{d\theta}{dt} = r\alpha + 2\frac{dr}{dt}\omega.$$

The term $-r\omega^2$ in the radial component of the acceleration is called the *centripetal acceleration*, and the term $2(dr/dt)\omega$ in the transverse component is called the *Coriolis acceleration*.

The unit vectors $\mathbf{e}_r$ and $\mathbf{e}_\theta$ are related to the cartesian unit vectors by

$$\mathbf{e}_r = \cos\theta\mathbf{i} + \sin\theta\mathbf{j}$$

and

$$\tag{13.60}$$

$$\mathbf{e}_\theta = -\sin\theta\mathbf{i} + \cos\theta\mathbf{j}.$$

Figure 13.29
Radial and transverse components of the velocity (a) and acceleration (b).

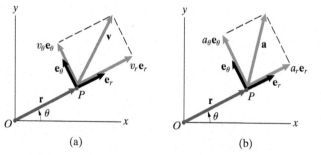

Circular Motion   Circular motion can be conveniently described using either radial and transverse or normal and tangential components. Let us compare these two methods of expressing the velocity and acceleration of a point P moving in a circular path of radius R (Fig. 13.30). Because the polar coordinate $r = R$ is constant, Eq. (13.57) for the velocity reduces to

$$\mathbf{v} = R\omega\mathbf{e}_\theta.$$

In terms of normal and tangential components, the velocity is

$$\mathbf{v} = v\mathbf{e}_t.$$

Notice in Fig. 13.30 that $\mathbf{e}_\theta = \mathbf{e}_t$. Comparing these two expressions for the velocity, we obtain the relation between the velocity and the angular velocity in circular motion:

$$v = R\omega.$$

From Eqs. (13.58) and (13.59), the acceleration for a circular path of radius R in terms of polar coordinates is

$$\mathbf{a} = -R\omega^2\mathbf{e}_r + R\alpha\mathbf{e}_\theta,$$

and the acceleration in terms of normal and tangential components is

$$\mathbf{a} = \frac{dv}{dt}\mathbf{e}_t + \frac{v^2}{R}\mathbf{e}_n.$$

The unit vector $\mathbf{e}_r = -\mathbf{e}_n$. Because of the relation $v = R\omega$, the normal components of acceleration are equal: $v^2/R = R\omega^2$. Equating the transverse and tangential components, we obtain the relation

$$\frac{dv}{dt} = a_t = R\alpha.$$

Cylindrical Coordinates   Polar coordinates describe the motion of a point P in the x–y plane. We can describe three-dimensional motion by using *cylindrical coordinates* r, θ, and z (Fig. 13.31). The cylindrical coordinates r and θ are the polar coordinates of P, measured in the plane parallel to the x–y plane, and the

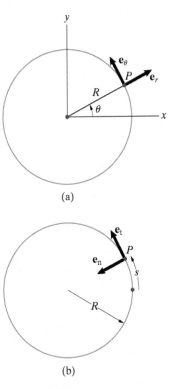

(a)

(b)

Figure 13.30
A point P moving in a circular path.
(a) Polar coordinates.
(b) Normal and tangential components.

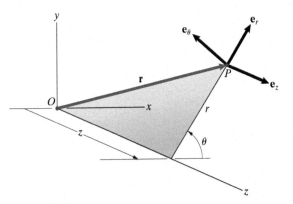

Figure 13.31
Cylindrical coordinates r, θ, and z of point P and the unit vectors $\mathbf{e}_r$, $\mathbf{e}_\theta$, and $\mathbf{e}_z$.

definitions of the unit vectors $\mathbf{e}_r$ and $\mathbf{e}_\theta$ are unchanged. The position of P perpendicular to the x–y plane is measured by the coordinate z, and the unit vector $\mathbf{e}_z$ points in the positive z axis direction.

In terms of cylindrical coordinates, the position vector $\mathbf{r}$ is the sum of the expression for the position vector in polar coordinates and the z component:

$$\mathbf{r} = r\mathbf{e}_r + z\mathbf{e}_z. \tag{13.61}$$

(The polar coordinate r is not the magnitude of $\mathbf{r}$, except when P lies in the x–y plane.) By taking time derivatives, we obtain the velocity

$$\mathbf{v} = \frac{d\mathbf{r}}{dt} = v_r\mathbf{e}_r + v_\theta\mathbf{e}_\theta + v_z\mathbf{e}_z$$

$$= \frac{dr}{dt}\mathbf{e}_r + r\omega\,\mathbf{e}_\theta + \frac{dz}{dt}\mathbf{e}_z \tag{13.62}$$

and acceleration

$$\mathbf{a} = \frac{d\mathbf{v}}{dt} = a_r\mathbf{e}_r + a_\theta\mathbf{e}_\theta + a_z\mathbf{e}_z, \tag{13.63}$$

where

$$a_r = \frac{d^2r}{dt^2} - r\omega^2, \quad a_\theta = r\alpha + 2\frac{dr}{dt}\omega, \quad \text{and} \quad a_z = \frac{d^2z}{dt^2}. \tag{13.64}$$

Notice that Eqs. (13.62) and (13.63) reduce to the polar coordinate expressions for the velocity and acceleration, Eqs. (13.57) and (13.58), when P moves along a path in the x–y plane.

RESULTS

Polar Coordinates

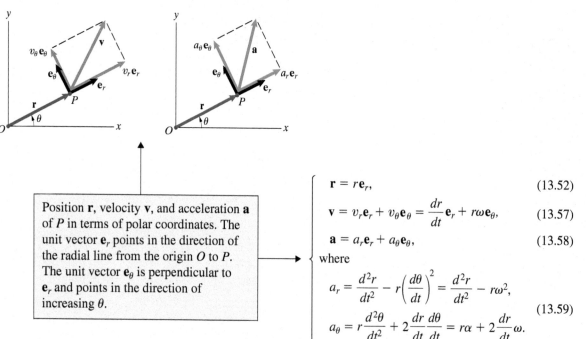

Position $\mathbf{r}$, velocity $\mathbf{v}$, and acceleration $\mathbf{a}$ of P in terms of polar coordinates. The unit vector $\mathbf{e}_r$ points in the direction of the radial line from the origin O to P. The unit vector $\mathbf{e}_\theta$ is perpendicular to $\mathbf{e}_r$ and points in the direction of increasing θ.

$$\mathbf{r} = r\mathbf{e}_r, \tag{13.52}$$

$$\mathbf{v} = v_r\mathbf{e}_r + v_\theta\mathbf{e}_\theta = \frac{dr}{dt}\mathbf{e}_r + r\omega\mathbf{e}_\theta, \tag{13.57}$$

$$\mathbf{a} = a_r\mathbf{e}_r + a_\theta\mathbf{e}_\theta, \tag{13.58}$$

where

$$a_r = \frac{d^2r}{dt^2} - r\left(\frac{d\theta}{dt}\right)^2 = \frac{d^2r}{dt^2} - r\omega^2,$$

$$a_\theta = r\frac{d^2\theta}{dt^2} + 2\frac{dr}{dt}\frac{d\theta}{dt} = r\alpha + 2\frac{dr}{dt}\omega. \tag{13.59}$$

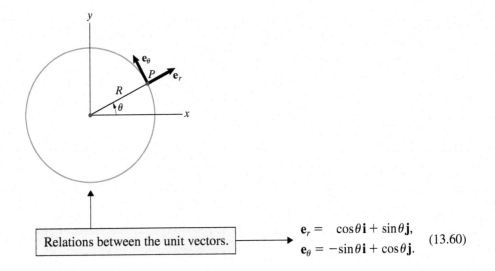

Relations between the unit vectors.

$$\begin{aligned} \mathbf{e}_r &= \cos\theta\,\mathbf{i} + \sin\theta\,\mathbf{j}, \\ \mathbf{e}_\theta &= -\sin\theta\,\mathbf{i} + \cos\theta\,\mathbf{j}. \end{aligned} \quad (13.60)$$

Cylindrical Coordinates

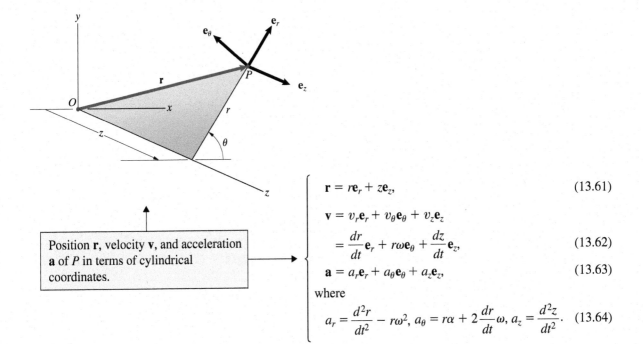

Position $\mathbf{r}$, velocity $\mathbf{v}$, and acceleration $\mathbf{a}$ of P in terms of cylindrical coordinates.

$$\mathbf{r} = r\mathbf{e}_r + z\mathbf{e}_z, \qquad (13.61)$$

$$\begin{aligned} \mathbf{v} &= v_r\mathbf{e}_r + v_\theta\mathbf{e}_\theta + v_z\mathbf{e}_z \\ &= \frac{dr}{dt}\mathbf{e}_r + r\omega\mathbf{e}_\theta + \frac{dz}{dt}\mathbf{e}_z, \end{aligned} \qquad (13.62)$$

$$\mathbf{a} = a_r\mathbf{e}_r + a_\theta\mathbf{e}_\theta + a_z\mathbf{e}_z, \qquad (13.63)$$

where

$$a_r = \frac{d^2r}{dt^2} - r\omega^2, \; a_\theta = r\alpha + 2\frac{dr}{dt}\omega, \; a_z = \frac{d^2z}{dt^2}. \quad (13.64)$$

Active Example 13.13   Analyzing Motion in Terms of Polar Coordinates (▶ *Related Problem 13.138*)

The robot arm is programmed so that the point P traverses the path described by

$$r = 1 - 0.5 \cos 2\pi t \text{ m},$$
$$\theta = 0.5 - 0.2 \sin 2\pi t \text{ rad}.$$

What is the velocity of P in terms of polar coordinates at $t = 0.8$ s?

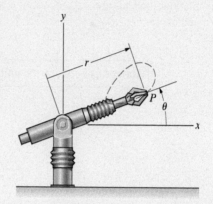

Strategy
The polar coordinates r and θ of P are known as functions of time, so we can determine the derivatives in the expression for the velocity in terms of polar coordinates and evaluate the velocity at $t = 0.8$ s.

Solution

Determine the derivatives in the expression for the velocity.	$\begin{cases} \dfrac{dr}{dt} = \pi \sin 2\pi t, \\[2mm] \dfrac{d\theta}{dt} = -0.4\pi \cos 2\pi t. \end{cases}$

Determine the velocity as a function of time.	$\begin{cases} \mathbf{v} = \dfrac{dr}{dt}\mathbf{e}_r + r\dfrac{d\theta}{dt}\mathbf{e}_\theta \\[2mm] \qquad = \pi \sin 2\pi t\, \mathbf{e}_r + (1 - 0.5\cos 2\pi t)(-0.4\pi \cos 2\pi t)\mathbf{e}_\theta. \end{cases}$

Evaluate the velocity at $t = 0.8$ s.	$\mathbf{v} = -2.99\mathbf{e}_r - 0.328\mathbf{e}_\theta$ (m/s).

Practice Problem What is the acceleration of P in terms of polar coordinates at $t = 0.8$ s?

Answer: $\mathbf{a} = 5.97\mathbf{e}_r - 4.03\mathbf{e}_\theta$ (m/s²).

| Example 13.14 | **Expressing Motion in Terms of Polar Coordinates** (▶ *Related Problem 13.141*) |

Suppose that you are standing on a large disk (say, a merry-go-round) rotating with constant angular velocity ω_0 and you start walking at constant speed v_0 along a straight radial line painted on the disk. What are your velocity and acceleration when you are a distance r from the center of the disk?

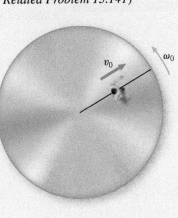

Strategy

We can describe your motion in terms of polar coordinates (Fig. a). By using the information given about your motion and the motion of the disk, we can evaluate the terms in the expressions for the velocity and acceleration in terms of polar coordinates.

Solution

The speed with which you walk along the radial line is the rate of change of r, $dr/dt = v_0$, and the angular velocity of the disk is the rate of change of θ, $\omega = \omega_0$. Your velocity is

$$\mathbf{v} = \frac{dr}{dt}\mathbf{e}_r + r\omega\mathbf{e}_\theta = v_0\mathbf{e}_r + r\omega_0\mathbf{e}_\theta.$$

Your velocity consists of two components: a radial component due to the speed at which you are walking and a transverse component due to the disk's rate of rotation. The transverse component increases as your distance from the center of the disk increases.

Your walking speed $v_0 = dr/dt$ is constant, so $d^2r/dt^2 = 0$. Also, the disk's angular velocity $\omega_0 = d\theta/dt$ is constant, so $d^2\theta/dt^2 = 0$. The radial component of your acceleration is

$$a_r = \frac{d^2r}{dt^2} - r\omega^2 = -r\omega_0^2,$$

and the transverse component is

$$a_\theta = r\alpha + 2\frac{dr}{dt}\omega = 2v_0\omega_0.$$

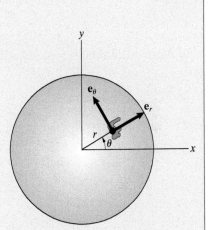

(a) Your position in terms of polar coordinates.

Critical Thinking

Why didn't we use normal and tangential components to determine your velocity and acceleration? The reason they would not be convenient in this example is that the path is not known, and normal and tangential components are defined in terms of the path.

If you have ever tried walking on a merry-go-round, you know that it is a difficult proposition. This example indicates why. Subjectively, you are walking along a straight line with constant velocity, but you are actually experiencing the centripetal acceleration a_r and the Coriolis acceleration a_θ due to the disk's rotation.

| Example 13.15 | Velocity in Terms of Polar and Cartesian Components |

(▶ *Related Problems 13.155 and 13.156*)

In the cam–follower mechanism shown, the slotted bar rotates with constant angular velocity $\omega = 4$ rad/s, and the radial position of the follower is determined by the elliptic profile of the stationary cam. The path of the follower is described by the polar equation

$$r = \frac{0.15}{1 + 0.5 \cos \theta} \text{ m.}$$

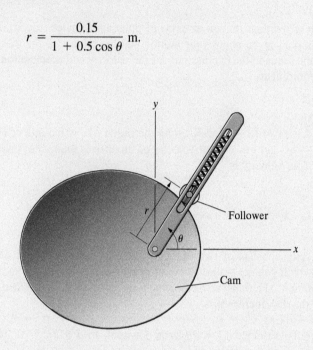

Determine the velocity of the follower when $\theta = 45°$ in terms of (a) polar coordinates and (b) cartesian coordinates.

Strategy

By taking the time derivative of the polar equation for the profile of the cam, we can obtain a relation between the known angular velocity and the radial component of velocity that permits us to evaluate the velocity in terms of polar coordinates. Then, by using Eqs. (13.60), we can obtain the velocity in terms of cartesian coordinates.

Solution

(a) The polar equation for the cam profile is of the form $r = r(\theta)$. Taking its derivative with respect to time, we obtain

$$\frac{dr}{dt} = \frac{dr(\theta)}{d\theta} \frac{d\theta}{dt}$$

$$= \frac{d}{d\theta} \left(\frac{0.15}{1 + 0.5 \cos \theta} \right) \frac{d\theta}{dt}$$

$$= \left[\frac{0.075 \sin \theta}{(1 + 0.5 \cos \theta)^2} \right] \frac{d\theta}{dt}.$$

The velocity of the follower in polar coordinates is therefore

$$\mathbf{v} = \frac{dr}{dt}\mathbf{e}_r + r\frac{d\theta}{dt}\mathbf{e}_\theta$$

$$= \left[\frac{0.075\sin\theta}{(1 + 0.5\cos\theta)^2}\right]\frac{d\theta}{dt}\mathbf{e}_r + \left(\frac{0.15}{1 + 0.5\cos\theta}\right)\frac{d\theta}{dt}\mathbf{e}_\theta.$$

The angular velocity $\omega = d\theta/dt = 4$ rad/s, so we can evaluate the polar components of the velocity when $\theta = 45°$, obtaining

$$\mathbf{v} = 0.116\mathbf{e}_r + 0.443\mathbf{e}_\theta \text{ (m/s)}.$$

(b) Substituting Eqs. (13.60) with $\theta = 45°$ into the polar coordinate expression for the velocity, we obtain the velocity in terms of cartesian coordinates:

$$\mathbf{v} = 0.116\mathbf{e}_r + 0.443\mathbf{e}_\theta$$

$$= 0.116(\cos 45°\mathbf{i} + \sin 45°\mathbf{j}) + 0.443(-\sin 45°\mathbf{i} + \cos 45°\mathbf{j})$$

$$= -0.232\mathbf{i} + 0.395\mathbf{j} \text{ (m/s)}.$$

Critical Thinking

Notice that, in determining the velocity of the follower, we made the tacit assumption that it stays in contact with the surface of the cam as the bar rotates. Designers of cam mechanisms must insure that the spring is sufficiently strong so that the follower does not lose contact with the surface. In Chapter 14 we introduce the concepts needed to analyze such problems.

Problems

13.137 The polar coordinates of the collar A are given as functions of time in seconds by $r = 1 + 0.2t^2$ ft and $\theta = 2t$ rad. What are the magnitudes of the velocity and acceleration of the collar at $t = 2$ s?

▶ **13.138** In Active Example 13.13, suppose that the robot arm is reprogrammed so that the point P traverses the path described by

$$r = 1 - 0.5\sin 2\pi t \text{ m},$$

$$\theta = 0.5 - 0.2\cos 2\pi t \text{ rad}.$$

What is the velocity of P in terms of polar coordinates at $t = 0.8$ s?

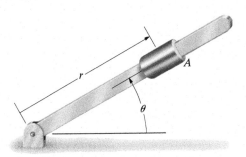

Problem 13.137

13.139 At the instant shown, $r = 3$ m and $\theta = 30°$. The cartesian components of the velocity of point A are $v_x = 2$ m/s and $v_y = 8$ m/s.

(a) Determine the velocity of point A in terms of polar coordinates.

(b) What is the angular velocity $d\theta/dt$ of the crane at the instant shown?

13.140 The polar coordinates of point A of the crane are given as functions of time in seconds by $r = 3 + 0.2t^2$ m and $\theta = 0.02t^2$ rad. Determine the acceleration of point A in terms of polar coordinates at $t = 3$ s.

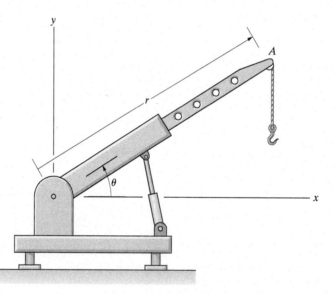

Problems 13.139/13.140

▶ **13.141** The radial line rotates with a constant angular velocity of 2 rad/s. Point P moves along the line at a constant speed of 4 m/s. Determine the magnitudes of the velocity and acceleration of P when $r = 2$ m. (See Example 13.14.)

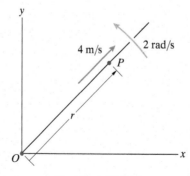

Problem 13.141

13.142 At the instant shown, the coordinates of the collar A are $x = 2.3$ ft, $y = 1.9$ ft. The collar is sliding on the bar from B toward C at a constant speed of 4 ft/s.

(a) What is the velocity of the collar in terms of polar coordinates?

(b) Use the answer to part (a) to determine the angular velocity of the radial line from the origin to the collar A at the instant shown.

13.143 At the instant shown, the coordinates of the collar A are $x = 2.3$ ft, $y = 1.9$ ft. The collar is sliding on the bar from B toward C at a constant speed of 4 ft/s.

(a) What is the acceleration of the collar in terms of polar coordinates?

(b) Use the answer to part (a) to determine the angular acceleration of the radial line from the origin to the collar A at the instant shown.

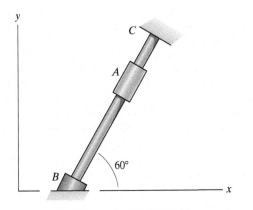

Problems 13.142/13.143

13.144* A boat searching for underwater archaeological sites in the Aegean Sea moves at 4 knots and follows the path $r = 10\theta$ m, where θ is in radians. (A knot is one nautical mile, or 1852 meters, per hour.) When $\theta = 2\pi$ rad, determine the boat's velocity (a) in terms of polar coordinates and (b) in terms of cartesian coordinates.

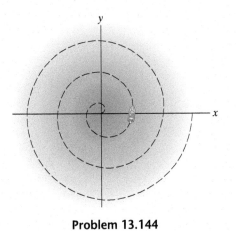

Problem 13.144

13.145 The collar A slides on the circular bar. The radial position of A (in meters) is given as a function of θ by $r = 2\cos\theta$. At the instant shown, $\theta = 25°$ and $d\theta/dt = 4$ rad/s. Determine the velocity of A in terms of polar coordinates.

13.146 In Problem 13.145, $d^2\theta/dt^2 = 0$ at the instant shown. Determine the acceleration of A in terms of polar coordinates.

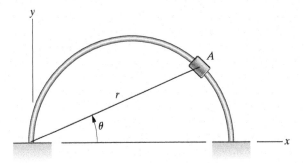

Problems 13.145/13.146

13.147 The radial coordinate of the earth satellite is related to its angular position θ by

$$r = \frac{1.91 \times 10^7}{1 + 0.5\cos\theta} \text{ m.}$$

The product of the radial position and the transverse component of the velocity is

$$rv_\theta = 8.72 \times 10^{10} \text{ m}^2/\text{s.}$$

What is the satellite's velocity in terms of polar coordinates when $\theta = 90°$?

13.148* In Problem 13.147, what is the satellite's acceleration in terms of polar coordinates when $\theta = 90°$?

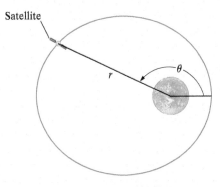

Problems 13.147/13.148

13.149 A bead slides along a wire that rotates in the x–y plane with constant angular velocity ω_0. The radial component of the bead's acceleration is zero. The radial component of its velocity is v_0 when $r = r_0$. Determine the polar components of the bead's velocity as a function of r.

Strategy: The radial component of the bead's velocity is

$$v_r = \frac{dr}{dt},$$

and the radial component of its acceleration is

$$a_r = \frac{d^2 r}{dt^2} - r\left(\frac{d\theta}{dt}\right)^2 = \frac{dv_r}{dt} - r\omega_0^2.$$

By using the chain rule,

$$\frac{dv_r}{dt} = \frac{dv_r}{dr}\frac{dr}{dt} = \frac{dv_r}{dr}v_r,$$

you can express the radial component of the acceleration in the form

$$a_r = \frac{dv_r}{dr}v_r - r\omega_0^2.$$

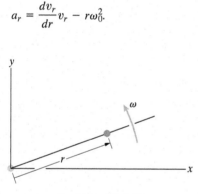

Problem 13.149

13.150 If the motion of a point in the x–y plane is such that its transverse component of acceleration a_θ is zero, show that the product of its radial position and its transverse velocity is constant: $r v_\theta = $ constant.

13.151* From astronomical data, Johannes Kepler deduced that the line from the sun to a planet traces out equal areas in equal times (Fig. a). Show that this result follows from the fact that the transverse component a_θ of the planet's acceleration is zero. [When r changes by an amount dr and θ changes by an amount $d\theta$ (Fig. b), the resulting differential element of area is $dA = \frac{1}{2}r(r\,d\theta)$.]

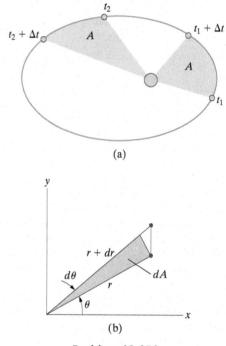

(a)

(b)

Problem 13.151

13.152 The bar rotates in the x–y plane with constant angular velocity $\omega_0 = 12$ rad/s. The radial component of acceleration of the collar C (in m/s^2) is given as a function of the radial position in meters by $a_r = -8r$. When $r = 1$ m, the radial component of velocity of C is $v_r = 2$ m/s. Determine the velocity of C in terms of polar coordinates when $r = 1.5$ m.

Strategy: Use the chain rule to write the first term in the radial component of the acceleration as

$$\frac{d^2r}{dt^2} = \frac{dv_r}{dt} = \frac{dv_r}{dr}\frac{dr}{dt} = \frac{dv_r}{dr}v_r.$$

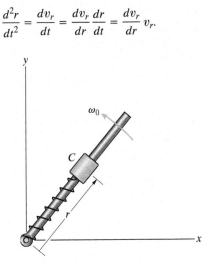

Problem 13.152

13.153 The hydraulic actuator moves the pin P upward with velocity $\mathbf{v} = 2\mathbf{j}$ (m/s). Determine the velocity of the pin in terms of polar coordinates and the angular velocity of the slotted bar when $\theta = 35°$.

13.154 The hydraulic actuator moves the pin P upward with constant velocity $\mathbf{v} = 2\mathbf{j}$ (m/s). Determine the acceleration of the pin in terms of polar coordinates and the angular acceleration of the slotted bar when $\theta = 35°$.

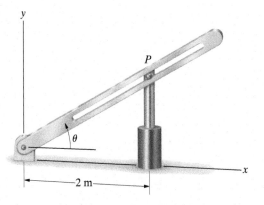

Problems 13.153/13.154

▶ 13.155 In Example 13.15, determine the velocity of the cam follower when $\theta = 135°$ (a) in terms of polar coordinates and (b) in terms of cartesian coordinates.

▶ 13.156* In Example 13.15, determine the acceleration of the cam follower when $\theta = 135°$ (a) in terms of polar coordinates and (b) in terms of cartesian coordinates.

13.157 In the cam–follower mechanism, the slotted bar rotates with constant angular velocity $\omega = 10$ rad/s and the radial position of the follower A is determined by the profile of the stationary cam. The path of the follower is described by the polar equation

$$r = 1 + 0.5 \cos 2\theta \text{ ft.}$$

Determine the velocity of the cam follower when $\theta = 30°$ (a) in terms of polar coordinates and (b) in terms of cartesian coordinates.

13.158* In Problem 13.157, determine the acceleration of the cam follower when $\theta = 30°$ (a) in terms of polar coordinates and (b) in terms of cartesian coordinates.

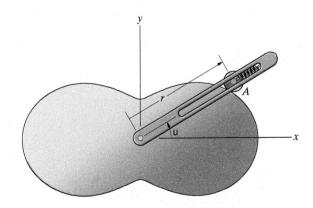

Problems 13.157/13.158

13.159* The cartesian coordinates of a point P in the x–y plane are related to the polar coordinates of the point by the equations $x = r \cos \theta$ and $y = r \sin \theta$.

(a) Show that the unit vectors $\mathbf{i}$ and $\mathbf{j}$ are related to the unit vectors $\mathbf{e}_r$ and $\mathbf{e}_\theta$ by

$$\mathbf{i} = \cos \theta \, \mathbf{e}_r - \sin \theta \, \mathbf{e}_\theta$$

and

$$\mathbf{j} = \sin \theta \, \mathbf{e}_r + \cos \theta \, \mathbf{e}_\theta.$$

(b) Beginning with the expression for the position vector of P in terms of cartesian coordinates, $\mathbf{r} = x\mathbf{i} + y\mathbf{j}$, derive Eq. (13.52) for the position vector in terms of polar coordinates.

(c) By taking the time derivative of the position vector of point P expressed in terms of cartesian coordinates, derive Eq. (13.55) for the velocity in terms of polar coordinates.

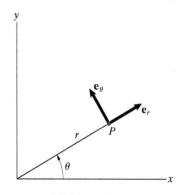

Problem 13.159

13.160 The airplane flies in a straight line at 400 mi/h. The radius of its propeller is 5 ft, and the propeller turns at 2000 rpm in the counterclockwise direction when seen from the front of the airplane. Determine the velocity and acceleration of a point on the tip of the propeller in terms of cylindrical coordinates. (Let the z axis be oriented as shown in the figure.)

Problem 13.160

13.161 A charged particle P in a magnetic field moves along the spiral path described by $r = 1$ m, $\theta = 2z$ rad, where z is in meters. The particle moves along the path in the direction shown with constant speed $|\mathbf{v}| = 1$ km/s. What is the velocity of the particle in terms of cylindrical coordinates?

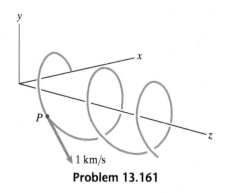

Problem 13.161

13.8 Relative Motion

BACKGROUND

We have discussed the curvilinear motion of a point relative to a given reference frame. In many applications, it is necessary to analyze the motions of two or more points relative to a reference frame and also their motions relative to each other. As a simple example, consider a passenger on a moving bus. If he walks down the aisle, the position and velocity that are important to him are his position in the bus and how fast he is moving down the aisle. His subjective motion is relative to the bus. But he also has a position and velocity relative to the earth. It would be convenient to have a framework for analyzing the bus's motion relative to the earth, the passenger's motion relative to the bus, and his motion relative to the earth. We develop such a framework in this section, introducing concepts and terminology that will be used in many contexts throughout the book.

Let A and B be two points whose motions we want to describe relative to a reference frame with origin O. We denote the positions of A and B relative to O by $\mathbf{r}_A$ and $\mathbf{r}_B$ (Fig. 13.32). We also want to describe the motion of point A relative to point B, and denote the position of A relative to B by $\mathbf{r}_{A/B}$. These vectors are related by

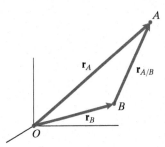

Figure 13.32

$$\mathbf{r}_A = \mathbf{r}_B + \mathbf{r}_{A/B}. \tag{13.65}$$

Stated in words, *the position of A is equal to the position of B plus the position of A relative to B*. Notice that when we simply say the "position of A" or "position of B," we mean their positions relative to O. The derivative of Eq. (13.65) with respect to time is

$$\frac{d\mathbf{r}_A}{dt} = \frac{d\mathbf{r}_B}{dt} + \frac{d\mathbf{r}_{A/B}}{dt}.$$

We write this equation as

$$\mathbf{v}_A = \mathbf{v}_B + \mathbf{v}_{A/B}, \tag{13.66}$$

where $\mathbf{v}_A$ is the velocity of A relative to O, $\mathbf{v}_B$ is the velocity of B relative to O, and $\mathbf{v}_{A/B} = d\mathbf{r}_{A/B}/dt$ is the velocity of A relative to B. *The velocity of A is equal to the velocity of B plus the velocity of A relative to B*. We now take the derivative of Eq. (13.66) with respect to time,

$$\frac{d\mathbf{v}_A}{dt} = \frac{d\mathbf{v}_B}{dt} + \frac{d\mathbf{v}_{A/B}}{dt},$$

and write this equation as

$$\mathbf{a}_A = \mathbf{a}_B + \mathbf{a}_{A/B}. \tag{13.67}$$

The term $\mathbf{a}_A$ is the acceleration of A relative to O, $\mathbf{a}_B$ is the acceleration of B relative to O, and $\mathbf{a}_{A/B} = d\mathbf{v}_{A/B}/dt$ is the acceleration of A relative to B. *The acceleration of A is equal to the acceleration of B plus the acceleration of A relative to B*.

Although they are simple in form, Eqs. (13.65)–(13.67) and the underlying concepts are extremely useful, and we apply them in a variety of contexts throughout the book.

RESULTS

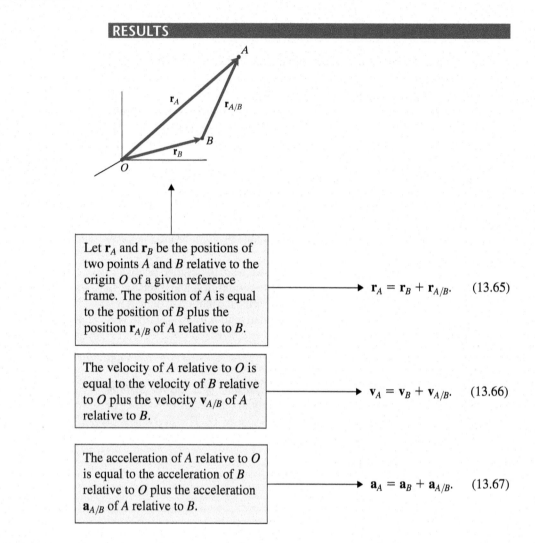

Let $\mathbf{r}_A$ and $\mathbf{r}_B$ be the positions of two points A and B relative to the origin O of a given reference frame. The position of A is equal to the position of B plus the position $\mathbf{r}_{A/B}$ of A relative to B.

$$\mathbf{r}_A = \mathbf{r}_B + \mathbf{r}_{A/B}. \quad (13.65)$$

The velocity of A relative to O is equal to the velocity of B relative to O plus the velocity $\mathbf{v}_{A/B}$ of A relative to B.

$$\mathbf{v}_A = \mathbf{v}_B + \mathbf{v}_{A/B}. \quad (13.66)$$

The acceleration of A relative to O is equal to the acceleration of B relative to O plus the acceleration $\mathbf{a}_{A/B}$ of A relative to B.

$$\mathbf{a}_A = \mathbf{a}_B + \mathbf{a}_{A/B}. \quad (13.67)$$

Active Example 13.16 Motion of a Ship in a Current (▶ *Related Problem 13.167*)

A ship moving at 5 knots (nautical miles per hour) relative to the water is in a uniform current flowing east at 2 knots. If the helmsman wants to travel northwest relative to the earth, what direction must he point the ship? What is the resulting magnitude of the ship's velocity relative to the earth?

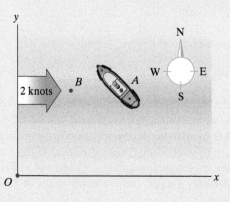

Strategy

Let the reference frame shown be stationary with respect to the earth. We denote the ship by A and define B to be a point that is stationary *relative to the water.* That is, point B is moving toward the east at 2 knots. By applying Eq. (13.66), we can determine the direction of the ship's motion relative to the water and the magnitude of the ship's velocity relative to the earth.

Solution

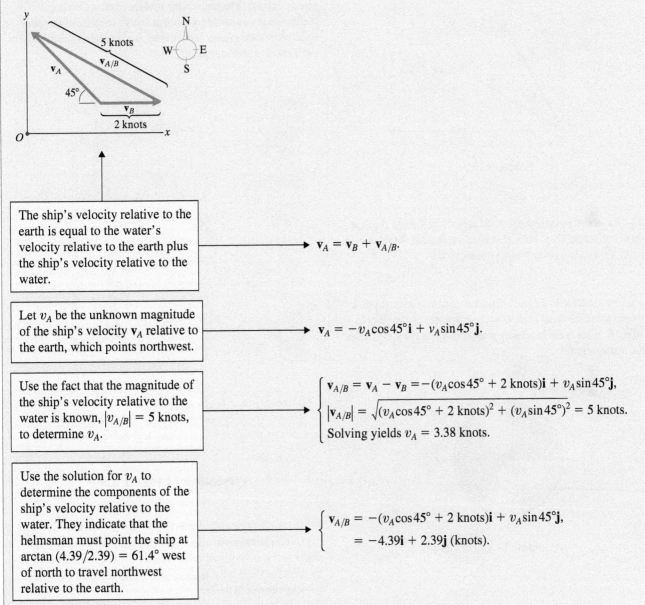

The ship's velocity relative to the earth is equal to the water's velocity relative to the earth plus the ship's velocity relative to the water.

$$\mathbf{v}_A = \mathbf{v}_B + \mathbf{v}_{A/B}.$$

Let v_A be the unknown magnitude of the ship's velocity $\mathbf{v}_A$ relative to the earth, which points northwest.

$$\mathbf{v}_A = -v_A\cos 45°\mathbf{i} + v_A\sin 45°\mathbf{j}.$$

Use the fact that the magnitude of the ship's velocity relative to the water is known, $|v_{A/B}| = 5$ knots, to determine v_A.

$$\begin{cases} \mathbf{v}_{A/B} = \mathbf{v}_A - \mathbf{v}_B = -(v_A\cos 45° + 2\ \text{knots})\mathbf{i} + v_A\sin 45°\mathbf{j}, \\ |\mathbf{v}_{A/B}| = \sqrt{(v_A\cos 45° + 2\ \text{knots})^2 + (v_A\sin 45°)^2} = 5\ \text{knots}. \\ \text{Solving yields } v_A = 3.38\ \text{knots}. \end{cases}$$

Use the solution for v_A to determine the components of the ship's velocity relative to the water. They indicate that the helmsman must point the ship at $\arctan(4.39/2.39) = 61.4°$ west of north to travel northwest relative to the earth.

$$\begin{cases} \mathbf{v}_{A/B} = -(v_A\cos 45° + 2\ \text{knots})\mathbf{i} + v_A\sin 45°\mathbf{j}, \\ \qquad = -4.39\mathbf{i} + 2.39\mathbf{j}\ (\text{knots}). \end{cases}$$

Practice Problem If the helmsman wants to travel due north relative to the earth, what direction must he point the ship? What is the resulting magnitude of the ship's velocity relative to the earth?

Answer: 23.6° west of north, 4.58 knots.

Problems

13.162 At $t = 0$, two projectiles A and B are simultaneously launched from O with the initial velocities and elevation angles shown. Determine the velocity of projectile A relative to projectile B (a) at $t = 0.5$ s and (b) at $t = 1$ s.

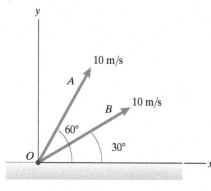

Problem 13.162

13.163 Relative to the earth-fixed coordinate system, the disk rotates about the fixed point O at 10 rad/s. What is the velocity of point A relative to point B at the instant shown?

13.164 Relative to the earth-fixed coordinate system, the disk rotates about the fixed point O with a constant angular velocity of 10 rad/s. What is the acceleration of point A relative to point B at the instant shown?

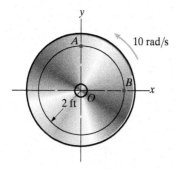

Problems 13.163/13.164

13.165 The train on the circular track is traveling at 50 ft/s. The train on the straight track is traveling at 20 ft/s. In terms of the earth-fixed coordinate system shown, what is the velocity of passenger A relative to passenger B?

13.166 The train on the circular track is traveling at a constant speed of 50 ft/s. The train on the straight track is traveling at 20 ft/s and is increasing its speed at 2 ft/s². In terms of the earth-fixed coordinate system shown, what is the acceleration of passenger A relative to passenger B?

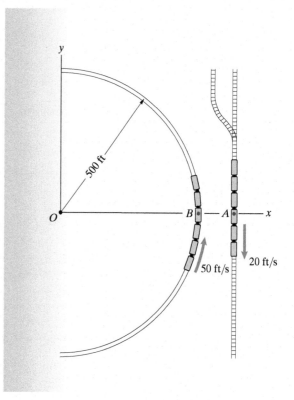

Problems 13.165/13.166

▶ **13.167** In Active Example 13.16, suppose that the velocity of the current increases to 3 knots flowing east. If the helmsman wants to travel northwest relative to the earth, what direction must he point the ship? What is the resulting magnitude of the ship's velocity relative to the earth?

13.168 A private pilot wishes to fly from a city P to a city Q that is 200 km directly north of city P. The airplane will fly with an airspeed of 290 km/h. At the altitude at which the airplane will be flying, there is an east wind (that is, the wind's direction is west) with a speed of 50 km/h. What direction should the pilot point the airplane to fly directly from city P to city Q? How long will the trip take?

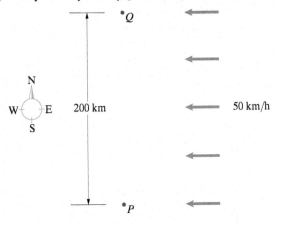

Problem 13.168

13.169 The river flows north at 3 m/s. (Assume that the current is uniform.) If you want to travel in a straight line from point C to point D in a boat that moves at a constant speed of 10 m/s relative to the water, in what direction should you point the boat? How long does it take to make the crossing?

13.170 The river flows north at 3 m/s. (Assume that the current is uniform.) What minimum speed must a boat have relative to the water in order to travel in a straight line from point C to point D? How long does it take to make the crossing?

Strategy: Draw a vector diagram showing the relationships of the velocity of the river relative to the earth, the velocity of the boat relative to the river, and the velocity of the boat relative to the earth. See which direction of the velocity of the boat relative to the river causes its magnitude to be a minimum.

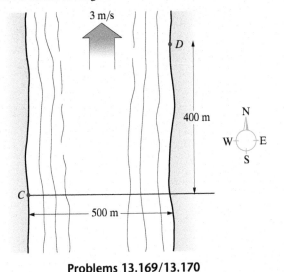

Problems 13.169/13.170

13.171* Relative to the earth, the sailboat sails north with speed $v_0 = 6$ knots (nautical miles per hour) and then sails east at the same speed. The telltale indicates the direction of the wind *relative to the boat*. Determine the direction and magnitude of the wind's velocity (in knots) relative to the earth.

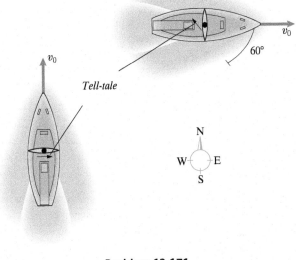

Problem 13.171

Review Problems

13.172 Suppose that you throw a ball straight up at 10 m/s and release it at 2 m above the ground.

(a) What maximum height above the ground does the ball reach?

(b) How long after you release it does the ball hit the ground?

(c) What is the magnitude of its velocity just before it hits the ground?

13.173 Suppose that you must determine the duration of the yellow light at a highway intersection. Assume that cars will be approaching the intersection traveling as fast as 65 mi/h, that drivers' reaction times are as long as 0.5 s, and that cars can safely achieve a deceleration of at least 0.4 g.

(a) How long must the light remain yellow to allow drivers to come to a stop safely before the light turns red?

(b) What is the minimum distance cars must be from the intersection when the light turns yellow to come to a stop safely at the intersection?

13.174 The acceleration of a point moving along a straight line is $a = 4t + 2$ m/s^2. When $t = 2$ s, the position of the point is $s = 36$ m, and when $t = 4$ s, its position is $s = 90$ m. What is the velocity of the point when $t = 4$ s?

13.175 A model rocket takes off straight up. Its acceleration during the 2 s its motor burns is 25 m/s^2. Neglect aerodynamic drag, and determine

(a) the maximum velocity of the rocket during the flight and

(b) the maximum altitude the rocket reaches.

13.176 In Problem 13.175, if the rocket's parachute fails to open, what is the total time of flight from takeoff until the rocket hits the ground?

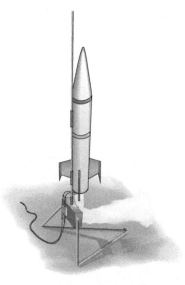

Problems 13.175/13.176

13.177 The acceleration of a point moving along a straight line is $a = -cv^3$, where c is a constant. If the velocity of the point is v_0, what distance does the point move before its velocity decreases to $v_0/2$?

13.178 Water leaves the nozzle at 20° above the horizontal and strikes the wall at the point indicated. What is the velocity of the water as it leaves the nozzle?

Strategy: Determine the motion of the water by treating each particle of water as a projectile.

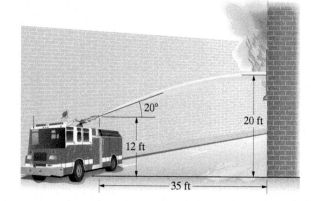

Problem 13.178

13.179 In practice, the quarterback throws the football with velocity v_0 at 45° above the horizontal. At the same instant, the receiver standing 20 ft in front of him starts running straight downfield at 10 ft/s and catches the ball. Assume that the ball is thrown and caught at the same height above the ground. What is the velocity v_0?

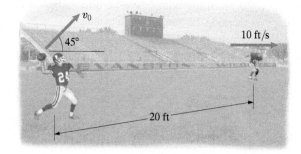

Problem 13.179

13.180 The constant velocity $v = 2$ m/s. What are the magnitudes of the velocity and acceleration of point P when $x = 0.25$ m?

13.181 The constant velocity $v = 2$ m/s. What is the acceleration of point P in terms of normal and tangential components when $x = 0.25$ m?

13.182 The constant velocity $v = 2$ m/s. What is the acceleration of point P in terms of polar coordinates when $x = 0.25$ m?

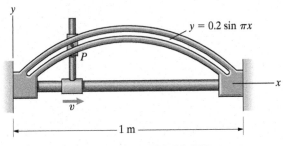

Problems 13.180–13.182

13.183 A point P moves along the spiral path $r = (0.1)\theta$ ft, where θ is in radians. The angular position $\theta = 2t$ rad, where t is in seconds, and $r = 0$ at $t = 0$. Determine the magnitudes of the velocity and acceleration of P at $t = 1$ s.

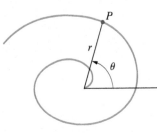

Problem 13.183

13.184 In the cam–follower mechanism, the slotted bar rotates with constant angular velocity $\omega = 12$ rad/s, and the radial position of the follower A is determined by the profile of the stationary cam. The slotted bar is pinned a distance $h = 0.2$ m to the left of the center of the circular cam. The follower moves in a circular path 0.42 m in radius. Determine the velocity of the follower when $\theta = 40°$ (a) in terms of polar coordinates and (b) in terms of cartesian coordinates.

13.185* In Problem 13.184, determine the acceleration of the follower when $\theta = 40°$ (a) in terms of polar coordinates and (b) in terms of cartesian coordinates.

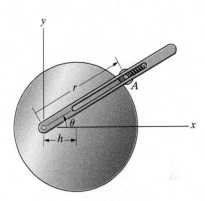

Problems 13.184/13.185

Design Project

Design and carry out experiments to measure the acceleration due to gravity. Galileo (1564–1642) did so by measuring the motions of falling objects. Use his method, but also try to devise other approaches that may result in improved accuracy. Investigate the repeatability of your measurements. Write a brief report describing your experiments, discussing possible sources of error, and presenting your results.

CHAPTER
14

Force, Mass, and Acceleration

Until now, we have analyzed motions of objects without considering the forces that cause them. In this chapter we relate cause and effect: By drawing the free-body diagram of an object to identify the forces acting on it, we can use Newton's second law to determine the acceleration of the object. Alternatively, when we know an object's acceleration, we can use Newton's second law to obtain information about the forces acting on it.

◀ The normal force exerted on his skies by the snow gives the skier a normal component of acceleration, resulting in his curved path.

14.1 Newton's Second Law

BACKGROUND

Newton stated that the total force on a particle is equal to the rate of change of its *linear momentum*, which is the product of its mass and velocity:

$$\mathbf{f} = \frac{d}{dt}(m\mathbf{v}).$$

If the particle's mass is constant, the total force equals the product of its mass and acceleration:

$$\mathbf{f} = m\frac{d\mathbf{v}}{dt} = m\mathbf{a}. \tag{14.1}$$

We pointed out in Chapter 12 that the second law gives precise meanings to the terms *force* and *mass*. Once a unit of mass is chosen, a unit of force is defined to be the force necessary to give one unit of mass an acceleration of unit magnitude. For example, the unit of force in SI units, the newton, is the force necessary to give a mass of one kilogram an acceleration of one meter per second squared. In principle, the second law then gives the value of any force and the mass of any object. By subjecting a one-kilogram mass to an arbitrary force and measuring the acceleration of the mass, we can solve the second law for the direction of the force and its magnitude in newtons. By subjecting an arbitrary mass to a one-newton force and again measuring the acceleration, we can solve the second law for the value of the mass in kilograms.

If the mass of a particle and the total force acting on it are known, Newton's second law determines its acceleration. In Chapter 13, we described how to determine the velocity, position, and trajectory of a point whose acceleration is known. Therefore, with the second law, a particle's motion can be determined when the total force acting on it is known, or the total force can be determined when the motion is known.

Equation of Motion for the Center of Mass

Newton's second law is postulated for a particle, or small element of matter, but an equation of precisely the same form describes the motion of the center of mass of an arbitrary object. We can show that the total external force on an arbitrary object is equal to the product of its mass and the acceleration of its center of mass.

To do so, we consider an arbitrary system of N particles. Let m_i be the mass of the ith particle, and let $\mathbf{r}_i$ be its position vector (Fig. 14.1a). Let m be the total mass of the particles; that is,

$$m = \sum_i m_i,$$

where the summation sign with subscript i means "the sum over i from 1 to N." The position of the center of mass of the system is

$$\mathbf{r} = \frac{\sum_i m_i \mathbf{r}_i}{m}.$$

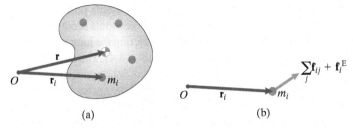

Figure 14.1
(a) Dividing an object into particles. The vector $\mathbf{r}_i$ is
the position vector of the *i*th particle, and $\mathbf{r}$ is the
position vector of the object's center of mass.
(b) Forces on the *i*th particle.

By taking two time derivatives of this expression, we obtain

$$\sum_i m_i \frac{d^2\mathbf{r}_i}{dt^2} = m\frac{d^2\mathbf{r}}{dt^2} = m\mathbf{a}, \tag{14.2}$$

where $\mathbf{a}$ is the acceleration of the center of mass.

The *i*th particle of the system may be acted upon by forces exerted by the other particles. Let $\mathbf{f}_{ij}$ be the force exerted on the *i*th particle by the *j*th particle. Then Newton's third law states that the *i*th particle exerts a force on the *j*th particle of equal magnitude and opposite direction: $\mathbf{f}_{ji} = -\mathbf{f}_{ij}$. If the external force on the *i*th particle (i.e., the total force exerted on the *i*th particle by objects other than the object we are considering) is denoted by $\mathbf{f}_i^E$, Newton's second law for the *i*th particle is (Fig. 14.1b)

$$\sum_j \mathbf{f}_{ij} + \mathbf{f}_i^E = m_i \frac{d^2\mathbf{r}_i}{dt^2}.$$

We can write this equation for each particle of the system. Summing the resulting equations from $i = 1$ to N, we obtain

$$\sum_i \sum_j \mathbf{f}_{ij} + \sum_i \mathbf{f}_i^E = \sum_i m_i \frac{d^2\mathbf{r}_i}{dt^2}. \tag{14.3}$$

The first term on the left side, the sum of the internal forces on the system, is zero due to Newton's third law:

$$\sum_i \sum_j \mathbf{f}_{ij} = \mathbf{f}_{12} + \mathbf{f}_{21} + \mathbf{f}_{13} + \mathbf{f}_{31} + \cdots = \mathbf{0}.$$

The second term on the left side of Eq. (14.3) is the sum of the external forces on the system. Denoting this sum by $\Sigma\mathbf{F}$ and using Eq. (14.2), we conclude that *the sum of the external forces equals the product of the total mass and the acceleration of the center of mass:*

$$\Sigma\mathbf{F} = m\mathbf{a}. \tag{14.4}$$

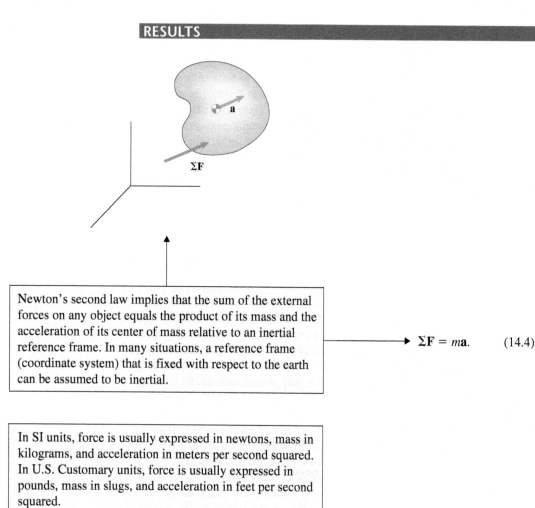

RESULTS

Newton's second law implies that the sum of the external forces on any object equals the product of its mass and the acceleration of its center of mass relative to an inertial reference frame. In many situations, a reference frame (coordinate system) that is fixed with respect to the earth can be assumed to be inertial.

$$\Sigma \mathbf{F} = m\mathbf{a}. \qquad (14.4)$$

In SI units, force is usually expressed in newtons, mass in kilograms, and acceleration in meters per second squared. In U.S. Customary units, force is usually expressed in pounds, mass in slugs, and acceleration in feet per second squared.

14.2 Applications—Cartesian Coordinates and Straight-Line Motion

By drawing the free-body diagram of an object, the external forces acting on it can be identified and Newton's second law used to determine the object's acceleration. Conversely, if the motion of an object is known, Newton's second law can be used to determine the total external force on the object. In particular, if an object's acceleration in a particular direction is known to be zero, the sum of the external forces in that direction must equal zero.

If we express the sum of the forces acting on an object of mass m and the acceleration of its center of mass in terms of their components in a cartesian reference frame (Fig. 14.3), Newton's second law states that

$$\Sigma \mathbf{F} = m\mathbf{a},$$

or

$$(\Sigma F_x \mathbf{i} + \Sigma F_y \mathbf{j} + \Sigma F_z \mathbf{k}) = m(a_x \mathbf{i} + a_y \mathbf{j} + a_z \mathbf{k}).$$

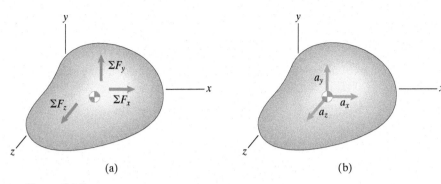

Figure 14.3
(a) Cartesian components of the sum of the forces on an object.
(b) Components of the acceleration of the center of mass of the object.

Equating x, y, and z components, we obtain three scalar equations of motion:

$$\Sigma F_x = ma_x, \qquad \Sigma F_y = ma_y, \qquad \Sigma F_z = ma_z. \qquad (14.5)$$

The total force in each coordinate direction equals the product of the mass and the component of the acceleration in that direction.

If an object's motion is confined to the x–y plane, $a_z = 0$, so the sum of the forces in the z direction is zero. Thus, when the motion is confined to a fixed plane, the component of the total force normal to that plane equals zero. For straight-line motion along the x axis (Fig. 14.4a), Eqs. (14.5) are

$$\Sigma F_x = ma_x, \qquad \Sigma F_y = 0, \quad \text{and} \quad \Sigma F_z = 0.$$

We see that in straight-line motion, the components of the total force perpendicular to the line equal zero, and the component of the total force tangent to the line equals the product of the mass and the acceleration along the line (Fig. 14.4b).

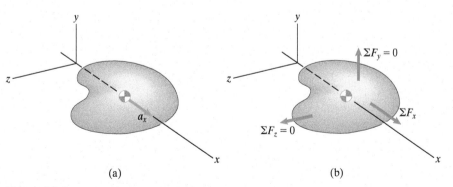

Figure 14.4
(a) Acceleration of an object in straight-line motion along the x axis.
(b) The y and z components of the total force acting on the object equal zero.

Straight-Line Motion (▶ *Related Problem 14.1*)

The 100-lb crate is released from rest on the inclined surface at time $t = 0$. The coefficients of friction between the crate and the inclined surface are $\mu_s = 0.2$ and $\mu_k = 0.15$. How fast is the crate moving at $t = 1$ s?

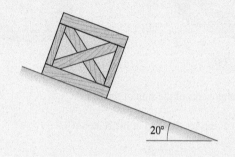

Strategy

We must first determine whether the crate slips when it is released. By assuming that it remains stationary, we can solve for the friction force necessary to keep the crate in equilibrium and see if it exceeds the maximum static friction force the surfaces will support. If the crate slips, we can use Newton's second law to determine its acceleration down the inclined surface. Once the acceleration is known, we can integrate it to determine the velocity of the crate as a function of time.

Solution

Draw the free-body diagram of the crate. The external forces are the weight of the crate and the normal and friction forces exerted by the inclined surface.

Assuming that the crate is stationary, use the equilibrium equations to determine the friction force necessary for equilibrium and the normal force.

$$\Sigma F_x = W \sin 20° - f = 0,$$
$$\Sigma F_y = N - W \cos 20° = 0.$$

Solving yields

$$f = W \sin 20° = (100\ \text{lb}) \sin 20° = 34.2\ \text{lb},$$
$$N = W \cos 20° = (100\ \text{lb}) \cos 20° = 94.0\ \text{lb}.$$

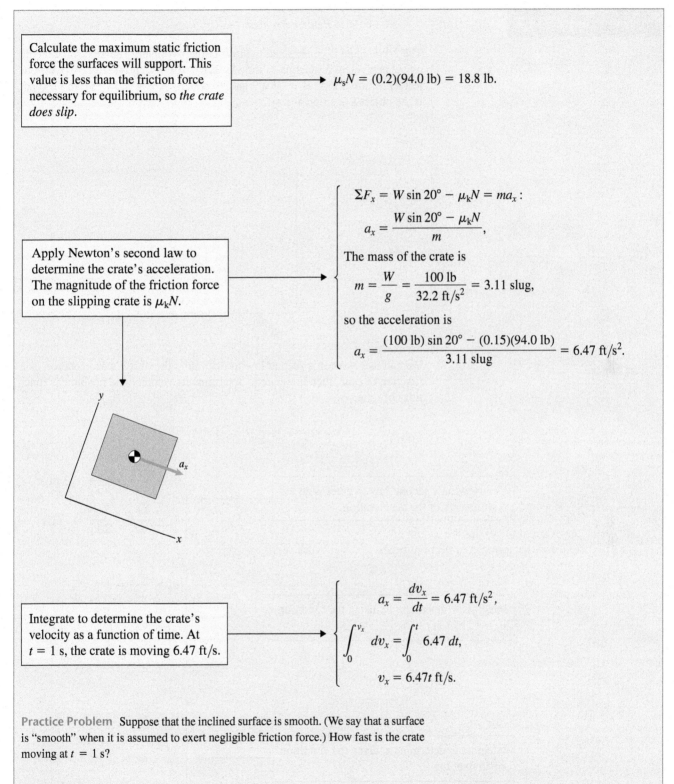

Calculate the maximum static friction force the surfaces will support. This value is less than the friction force necessary for equilibrium, so *the crate does slip*.

$\mu_s N = (0.2)(94.0 \text{ lb}) = 18.8 \text{ lb}.$

Apply Newton's second law to determine the crate's acceleration. The magnitude of the friction force on the slipping crate is $\mu_k N$.

$$\Sigma F_x = W \sin 20° - \mu_k N = m a_x :$$

$$a_x = \frac{W \sin 20° - \mu_k N}{m},$$

The mass of the crate is

$$m = \frac{W}{g} = \frac{100 \text{ lb}}{32.2 \text{ ft/s}^2} = 3.11 \text{ slug},$$

so the acceleration is

$$a_x = \frac{(100 \text{ lb}) \sin 20° - (0.15)(94.0 \text{ lb})}{3.11 \text{ slug}} = 6.47 \text{ ft/s}^2.$$

Integrate to determine the crate's velocity as a function of time. At $t = 1$ s, the crate is moving 6.47 ft/s.

$$a_x = \frac{dv_x}{dt} = 6.47 \text{ ft/s}^2,$$

$$\int_0^{v_x} dv_x = \int_0^t 6.47 \, dt,$$

$$v_x = 6.47t \text{ ft/s}.$$

Practice Problem Suppose that the inclined surface is smooth. (We say that a surface is "smooth" when it is assumed to exert negligible friction force.) How fast is the crate moving at $t = 1$ s?

Answer: 11.0 ft/s.

| Active Example 14.2 | Cartesian Coordinates (▶ *Related Problem 14.10*) |

The 2-kg object is constrained to move in the *x*–*y* plane. The total force on the object is given as a function of time by $\Sigma\mathbf{F} = 6\mathbf{i} + 2t\mathbf{j}$ (N). At $t = 0$, the object's position is $\mathbf{r} = 5\mathbf{i} + 3\mathbf{j}$ (m) and its velocity is $\mathbf{v} = 12\mathbf{i} + 5\mathbf{j}$ (m/s). What is the object's position at $t = 3$ s?

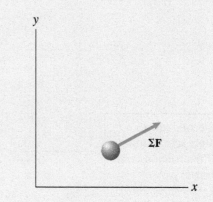

Strategy

We can use Newton's second law to determine the object's acceleration as a function of time, then integrate to determine its velocity and position as functions of time.

Solution

Use Newton's second law to determine the components of the acceleration.

$$\begin{cases} a_x = \dfrac{\Sigma F_x}{m} = \dfrac{6\ \text{N}}{2\ \text{kg}} = 3\ \text{m/s}^2, \\[2mm] a_y = \dfrac{\Sigma F_y}{m} = \dfrac{2t\ \text{N}}{2\ \text{kg}} = t\ \text{m/s}^2. \end{cases}$$

Integrate to determine v_x, using the condition $v_x = 12$ m/s at $t = 0$.

$$\begin{cases} a_x = \dfrac{dv_x}{dt} = 3\ \text{m/s}^2, \\[2mm] \displaystyle\int_{12}^{v_x} dv_x = \int_{0}^{t} 3\ dt, \\[2mm] v_x = 12 + 3t\ \text{m/s}. \end{cases}$$

Integrate to determine x, using the condition $x = 5$ m at $t = 0$.

$$\begin{cases} v_x = \dfrac{dx}{dt} = 12 + 3t\ \text{m/s}, \\[2mm] \displaystyle\int_{5}^{x} dx = \int_{0}^{t} (12 + 3t)\ dt, \\[2mm] x = 5 + 12t + \dfrac{3}{2}t^2\ \text{m}. \end{cases}$$

Integrate to determine v_y, using the condition $v_y = 5$ m/s at $t = 0$.

$$\begin{cases} a_y = \dfrac{dv_y}{dt} = t \text{ m/s}^2. \\[2mm] \displaystyle\int_5^{v_y} dv_y = \int_0^t t \, dt, \\[2mm] v_y = 5 + \tfrac{1}{2} t^2 \text{ m/s}. \end{cases}$$

Integrate to determine y, using the condition $y = 3$ m at $t = 0$.

$$\begin{cases} v_y = \dfrac{dy}{dt} = 5 + \tfrac{1}{2}t^2 \text{ m/s}. \\[2mm] \displaystyle\int_3^y dy = \int_0^t \left(5 + \tfrac{1}{2}t^2\right) dt, \\[2mm] y = 3 + 5t + \tfrac{1}{6} t^3 \text{ m}. \end{cases}$$

Determine the position at $t = 3$ s.

$$\begin{cases} x|_{t=3\text{ s}} = 5 + 12(3) + \tfrac{3}{2}(3)^2 = 54.5 \text{ m}, \\[2mm] y|_{t=3\text{ s}} = 3 + 5(3) + \tfrac{1}{6}(3)^3 = 22.5 \text{ m}. \end{cases}$$

Practice Problem The 10-lb object is constrained to move in the x–y plane. The object's position is given as a function of time by $\mathbf{r} = 8t^2\mathbf{i} + t^3\mathbf{j}$ (ft). What total force acts on the object at $t = 4$ s?

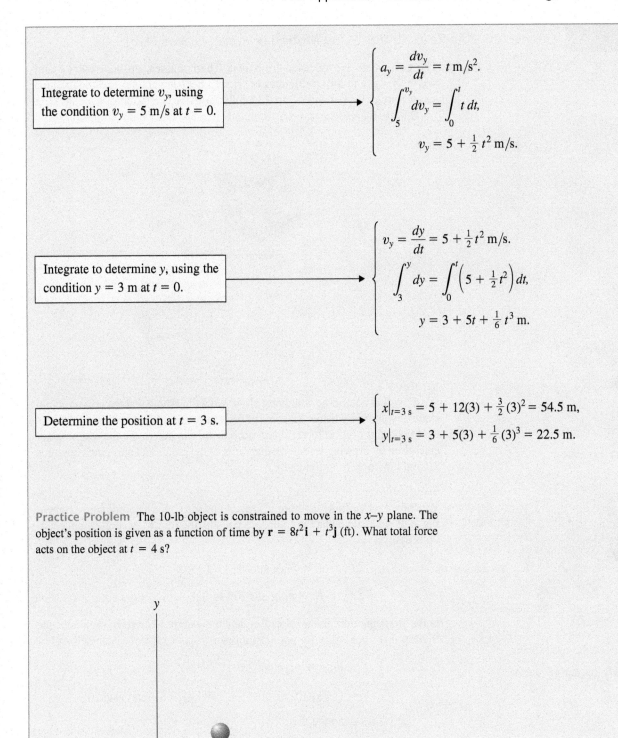

Answer: $\Sigma\mathbf{F} = 4.97\mathbf{i} + 7.45\mathbf{j}$ (lb).

Example 14.3	Connected Objects in Straight-Line Motion (▶ *Related Problem 14.28*)

The two crates are released from rest. Their masses are $m_A = 40$ kg and $m_B = 30$ kg, and the coefficients of friction between crate A and the inclined surface are $\mu_s = 0.2$ and $\mu_k = 0.15$. What is the acceleration of the crates?

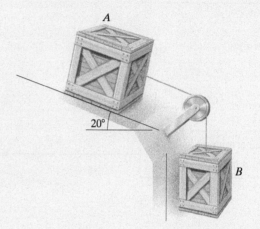

Strategy

We must first determine whether A slips. We will assume that the crates remain stationary and see whether the force of friction necessary for equilibrium exceeds the maximum friction force. If slip occurs, we can determine the resulting acceleration by drawing free-body diagrams of the crates and applying Newton's second law to them individually.

Solution

We draw the free-body diagram of crate A and introduce a coordinate system in Fig. a. If we assume that the crate does not slip, the following equilibrium equations apply:

$$\Sigma F_x = T + m_A g \sin 20° - f = 0;$$

$$\Sigma F_y = N - m_A g \cos 20° = 0.$$

In the first equation, the tension T equals the weight of crate B; therefore, the friction force necessary for equilibrium is

$$f = m_B g + m_A g \sin 20°$$

$$= (30\ \text{kg})(9.81\ \text{m/s}^2) + (40\ \text{kg})(9.81\ \text{m/s}^2) \sin 20°$$

$$= 429\ \text{N}.$$

The normal force $N = m_A g \cos 20°$, so the maximum friction force the surface will support is

$$f_{max} = \mu_s N$$

$$= (0.2)[(40\ \text{kg})(9.81\ \text{m/s}^2) \cos 20°]$$

$$= 73.7\ \text{N}.$$

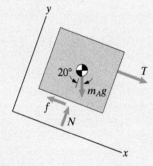

(a) Free-body diagram of crate A.

Crate A will therefore slip, and the friction force is $f = \mu_k N$. We show the crate's acceleration down the plane in Fig. b. Its acceleration perpendicular to the plane is zero (i.e., $a_y = 0$). Applying Newton's second law yields

$$\Sigma F_x = T + m_A g \sin 20° - \mu_k N = m_A a_x$$

$$\Sigma F_y = N - m_A g \cos 20° = 0.$$

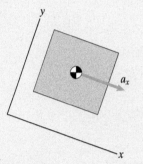

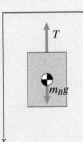

(b) The crate's acceleration.

(c) Free-body diagram of crate B.

In this case, *we do not know* the tension T, because crate B is not in equilibrium. We show the free-body diagram of crate B and its vertical acceleration in Figs. c and d. The equation of motion is

$$\Sigma F_x = m_B g - T = m_B a_x.$$

(In terms of the two coordinate systems we use, the two crates have the same acceleration a_x.) Thus, by applying Newton's second law to both crates, we have obtained three equations in terms of the unknowns T, N, and a_x. Solving for a_x, we obtain $a_x = 5.33$ m/s^2.

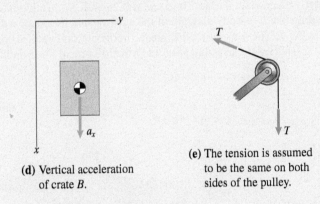

(d) Vertical acceleration of crate B.

(e) The tension is assumed to be the same on both sides of the pulley.

Critical Thinking

Notice that we assumed the tension in the cable to be the same on each side of the pulley (Fig. e). In fact, however, the tensions must be different, because a moment is necessary to cause angular acceleration of the pulley. For now, our only recourse is to assume that the pulley is light enough that the moment necessary to accelerate it is negligible. In Chapter 18 we include the analysis of the angular motion of the pulley in problems of this type and obtain more realistic solutions.

Example 14.4 | **Application to Straight-Line Motion** (▶ *Related Problem 14.45*)

The airplane touches down on the aircraft carrier with a horizontal velocity of 50 m/s relative to the carrier. The arresting gear exerts a horizontal force of magnitude $T_x = 10,000v$ newtons (N), where v is the plane's velocity in meters per second. The plane's mass is 6500 kg.

(a) What maximum horizontal force does the arresting gear exert on the plane?

(b) If other horizontal forces can be neglected, what distance does the plane travel before coming to rest?

Strategy

(a) Since the plane begins to decelerate when it contacts the arresting gear, the maximum force occurs at first contact when $v = 50$ m/s.

(b) The horizontal force exerted by the arresting gear equals the product of the plane's mass and its acceleration. Once we know the acceleration, we can integrate to determine the distance required for the plane to come to rest.

Solution

(a) We draw the free-body diagram of the airplane and introduce a coordinate system in Fig. a. The forces T_x and T_y are the horizontal and vertical components of force exerted by the arresting gear, and N is the vertical force on the landing

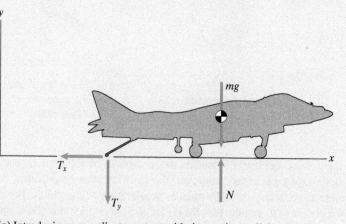

(a) Introducing a coordinate system with the x axis parallel to the horizontal force.

gear. The horizontal force on the plane is $\Sigma F_x = -T_x = -10{,}000v$ N. The magnitude of the maximum force is

$$10{,}000v = (10{,}000)(50) = 500{,}000 \text{ N},$$

or 112,400 lb.

(b) In terms of the plane's horizontal component of acceleration (Fig. b), we obtain the equation of motion

$$\Sigma F_x = ma_x:$$

$$-10{,}000v_x = ma_x.$$

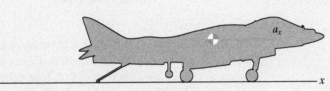

(b) The airplane's horizontal acceleration.

The airplane's acceleration is a function of its velocity. We use the chain rule to express the acceleration in terms of a derivative with respect to x:

$$ma_x = m\frac{dv_x}{dt} = m\frac{dv_x}{dx}\frac{dx}{dt} = m\frac{dv_x}{dx}v_x = -10{,}000v_x.$$

Now we separate variables and integrate, defining $x = 0$ to be the position at which the plane contacts the arresting gear:

$$\int_{50}^{0} m\,dv_x = -\int_{0}^{x} 10{,}000\,dx.$$

Evaluating the integrals and solving for x, we obtain

$$x = \frac{50m}{10{,}000} = \frac{(50)(6500)}{10{,}000} = 32.5 \text{ m}.$$

Critical Thinking

The force exerted by the arresting gear depended on the airplane's velocity, which resulted in an acceleration that depended on velocity. Our use of the chain rule to determine the velocity as a function of position when the acceleration is a function of the velocity is discussed in Section 13.3.

Problems

▶ **14.1** In Active Example 14.1, suppose that the coefficient of kinetic friction between the crate and the inclined surface is $\mu_k = 0.12$. Determine the distance the crate has moved down the inclined surface at $t = 1$ s.

14.2 The mass of the Sikorsky UH-60A helicopter is 9300 kg. It takes off vertically with its rotor exerting a constant upward thrust of 112 kN.

(a) How fast is the helicopter rising 3 s after it takes off?

(b) How high has it risen 3 s after it takes off?

   Strategy: Be sure to draw the free-body diagram of the helicopter.

14.3 The mass of the Sikorsky UH-60A helicopter is 9300 kg. It takes off vertically at $t = 0$. The pilot advances the throttle so that the upward thrust of its engine (in kN) is given as a function of time in seconds by $T = 100 + 2t^2$.

(a) How fast is the helicopter rising 3 s after it takes off?

(b) How high has it risen 3 s after it takes off?

Problems 14.2/14.3

14.4 The horizontal surface is smooth. The 30-lb box is at rest when the constant force F is applied. Two seconds later, the box is moving to the right at 20 ft/s. Determine F.

14.5 The coefficient of kinetic friction between the 30-lb box and the horizontal surface is $\mu_k = 0.1$. The box is at rest when the constant force F is applied. Two seconds later, the box is moving to the right at 20 ft/s. Determine F.

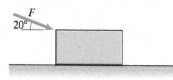

Problems 14.4/14.5

14.6 The inclined surface is smooth. The velocity of the 14-kg box is zero when it is subjected to a constant horizontal force $F = 20$ N. What is the velocity of the box two seconds later?

14.7 The coefficient of kinetic friction between the 14-kg box and the inclined surface is $\mu_k = 0.1$. The velocity of the box is zero when it is subjected to a constant horizontal force $F = 20$ N. What is the velocity of the box two seconds later?

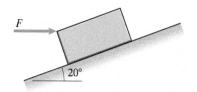

Problems 14.6/14.7

14.8 The 170-lb skier is schussing on a 25° slope. At the instant shown, he is moving at 40 ft/s. The kinetic coefficient of friction between his skis and the snow is $\mu_k = 0.08$. If he makes no attempt to check his speed, how long does it take for it to increase to 60 ft/s?

14.9 The 170-lb skier is schussing on a 25° slope. At the instant shown, he is moving at 40 ft/s. The kinetic coefficient of friction between his skis and the snow is $\mu_k = 0.08$. Aerodynamic drag exerts a resisting force on him of magnitude $0.015v^2$, where v is the magnitude of his velocity. If he makes no attempt to check his speed, how long does it take for it to increase to 60 ft/s?

Problems 14.8/14.9

▶ **14.10** The total external force on the 10-kg object is constant and equal to $\Sigma \mathbf{F} = 90\mathbf{i} - 60\mathbf{j} + 20\mathbf{k}$ (N). At time $t = 0$, its velocity is $\mathbf{v} = -14\mathbf{i} + 26\mathbf{j} + 32\mathbf{k}$ (m/s). What is its velocity at $t = 4$ s? (See Active Example 14.2.)

14.11 The total external force on the 10-kg object shown in Problem 14.10 is given as a function of time by $\Sigma\mathbf{F} = (-20t + 90)\mathbf{i} - 60\mathbf{j} + (10t + 40)\mathbf{k}$ (N). At time $t = 0$, its position is $\mathbf{r} = 40\mathbf{i} + 30\mathbf{j} - 360\mathbf{k}$ (m) and its velocity is $\mathbf{v} = -14\mathbf{i} + 26\mathbf{j} + 32\mathbf{k}$ (m/s). What is its position at $t = 4$ s?

14.12 The position of the 10-kg object is given as a function of time by $\mathbf{r} = (20t^3 - 300)\mathbf{i} + 60t^2\mathbf{j} + (6t^4 - 40t^2)\mathbf{k}$ (m). What is the total external force on the object at $t = 2$ s?

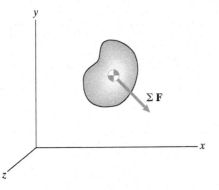

Problems 14.10–14.12

14.13 The total force exerted on the 80,000-lb launch vehicle by the thrust of its engine, its weight, and aerodynamic forces during the interval of time from $t = 2$ s to $t = 4$ s is given as a function of time by $\Sigma\mathbf{F} = (2000 - 400t^2)\mathbf{i} + (5200 + 440t)\mathbf{j} + (800 + 60t^2)\mathbf{k}$ (lb). At $t = 2$ s, its velocity is $\mathbf{v} = 12\mathbf{i} + 220\mathbf{j} - 30\mathbf{k}$ (ft/s). What is its velocity at $t = 4$ s?

Problem 14.13

14.14 At the instant shown, the horizontal component of acceleration of the 26,000-lb airplane due to the sum of the external forces acting on it is 14 ft/s². If the pilot suddenly increases the magnitude of the thrust T by 4000 lb, what is the horizontal component of the plane's acceleration immediately afterward?

Problem 14.14

14.15 At the instant shown, the rocket is traveling straight up at 100 m/s. Its mass is 90,000 kg and the thrust of its engine is 2400 kN. Aerodynamic drag exerts a resisting force (in newtons) of magnitude $0.8v^2$, where v is the magnitude of the velocity. How long does it take for the rocket's velocity to increase to 200 m/s?

Problem 14.15

14.16 A 2-kg cart containing 8 kg of water is initially stationary (Fig. P14.16a). The center of mass of the "object" consisting of the cart and water is at $x = 0$. The cart is subjected to the time-dependent force shown in Fig. P14.16b, where $F_0 = 5$ N and $t_0 = 2$ s. Assume that no water spills out of the cart and that the horizontal forces exerted on the wheels by the floor are negligible.

(a) Do you know the acceleration of the cart during the period $0 < t < t_0$?

(b) Do you know the acceleration of the center of mass of the "object" consisting of the cart and water during the period $0 < t < t_0$?

(c) What is the x coordinate of the center of mass of the "object" when $t > 2t_0$?

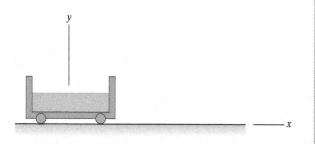

(a)

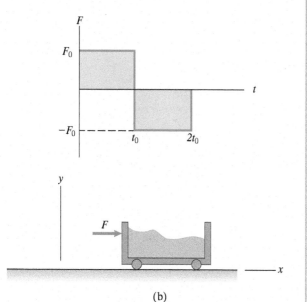

(b)

Problem 14.16

14.17 The combined weight of the motorcycle and rider is 360 lb. The coefficient of kinetic friction between the tires and the road is $\mu_k = 0.8$. The rider starts from rest, spinning the rear wheel. Neglect the horizontal force exerted on the front wheel by the road. In two seconds, the motorcycle moves 35 ft. What was the normal force between the rear wheel and the road?

Problem 14.17

14.18 The mass of the bucket B is 180 kg. From $t = 0$ to $t = 2$ s, the x and y coordinates of the center of mass of the bucket are

$$x = -0.2t^3 + 0.05t^2 + 10 \text{ m,}$$
$$y = 0.1t^2 + 0.4t + 6 \text{ m.}$$

Determine the x and y components of the force exerted on the bucket by its supports at $t = 1$ s.

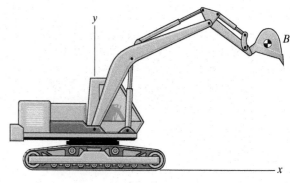

Problem 14.18

14.19 During a test flight in which a 9000-kg helicopter starts from rest at $t = 0$, the acceleration of its center of mass from $t = 0$ to $t = 10$ s is

$$\mathbf{a} = 0.6t\mathbf{i} + (1.8 - 0.36t)\mathbf{j} \ (\text{m/s}^2).$$

What is the magnitude of the total external force on the helicopter (including its weight) at $t = 6$ s?

14.20 The engineers conducting the test described in Problem 14.19 want to express the total force on the helicopter at $t = 6$ s in terms of three forces: the weight W, a component T tangent to the path, and a component L normal to the path. What are the values of W, T, and L?

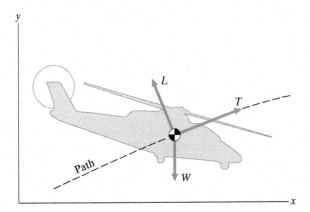

Problems 14.19/14.20

14.21 At the instant shown, the 11,000-kg airplane's velocity is $\mathbf{v} = 270\mathbf{i}$ (m/s). The forces acting on the plane are its weight, the thrust $T = 110$ kN, the lift $L = 260$ kN, and the drag $D = 34$ kN. (The x axis is parallel to the airplane's path.) Determine the magnitude of the airplane's acceleration.

14.22 At the instant shown, the 11,000-kg airplane's velocity is $\mathbf{v} = 300\mathbf{i}$ (m/s). The rate of change of the magnitude of the velocity is $dv/dt = 5$ m/s^2. The radius of curvature of the airplane's path is 4500 m, and the y axis points toward the concave side of the path. The thrust is $T = 120,000$ N. Determine the lift L and drag D.

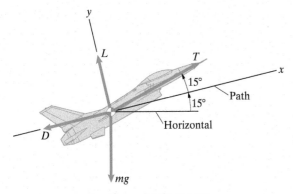

Problems 14.21/14.22

14.23 The coordinates in meters of the 360-kg sport plane's center of mass relative to an earth-fixed reference frame during an interval of time are

$$x = 20t - 1.63t^2,$$
$$y = 35t - 0.15t^3,$$

and

$$z = -20t - 1.38t^2,$$

where t is the time in seconds. The y axis points upward. The forces exerted on the plane are its weight, the thrust vector $\mathbf{T}$ exerted by its engine, the lift force vector $\mathbf{L}$, and the drag force vector $\mathbf{D}$. At $t = 4$ s, determine $\mathbf{T} + \mathbf{L} + \mathbf{D}$.

14.24 The force in newtons exerted on the 360-kg sport plane in Problem 14.23 by its engine, the lift force, and the drag force during an interval of time is

$$\mathbf{T} + \mathbf{L} + \mathbf{D} = (-1000 + 280t)\mathbf{i} + (4000 - 430t)\mathbf{j}$$
$$+ (720 + 200t)\mathbf{k},$$

where t is the time in seconds. If the coordinates of the plane's center of mass are $(0, 0, 0)$ and its velocity is $20\mathbf{i} + 35\mathbf{j} - 20\mathbf{k}$ (m/s) at $t = 0$, what are the coordinates of the center of mass at $t = 4$ s?

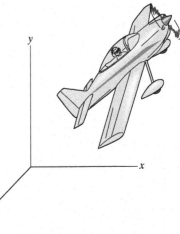

Problems 14.23/14.24

14.25 The robot manipulator is programmed so that $x = 40 + 24t^2$ mm, $y = 4t^3$ mm, and $z = 0$ during the interval of time from $t = 0$ to $t = 4$ s. The y axis points upward. What are the x and y components of the total force exerted by the jaws of the manipulator on the 2-kg widget A at $t = 3$ s?

14.26 The robot manipulator is programmed so that it is stationary at $t = 0$ and the components of the acceleration of A are $a_x = 400 - 0.8v_x$ mm/s^2 and $a_y = 200 - 0.4v_y$ mm/s^2 from $t = 0$ to $t = 2$ s, where v_x and v_y are the components of the velocity in mm/s. The y axis points upward. What are the x and y components of the total force exerted by the jaws of the manipulator on the 2-kg widget A at $t = 1$ s?

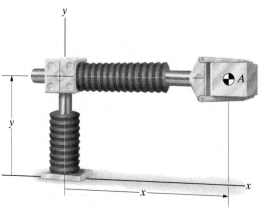

Problems 14.25/14.26

14.27 In the sport of curling, the object is to slide a "stone" weighing 44 lb onto the center of a target located 31 yards from the point of release. In terms of the coordinate system shown, the point of release is at $x = 0$, $y = 0$. Suppose that a shot comes to rest at $x = 31.0$ yards, $y = 1$ yard. Assume that the coefficient of kinetic friction is constant and equal to $\mu_k = 0.01$. What were the x and y components of the stone's velocity at release?

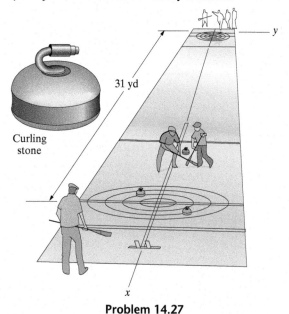

Problem 14.27

▶ **14.28** The two masses are released from rest. How fast are they moving at $t = 0.5$ s? (See Example 14.3.)

2 kg   5 kg

Problem 14.28

14.29 The two weights are released from rest. The horizontal surface is smooth. (a) What is the tension in the cable after the weights are released? (b) How fast are the weights moving one second after they are released?

14.30 The two weights are released from rest. The coefficient of kinetic friction between the horizontal surface and the 5-lb weight is $\mu_k = 0.18$. (a) What is the tension in the cable after the weights are released? (b) How fast are the weights moving one second after they are released?

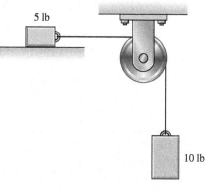

5 lb

10 lb

Problems 14.29/14.30

14.31 The mass of each box is 14 kg. One second after they are released from rest, they have moved 0.3 m from their initial positions. What is the coefficient of kinetic friction between the boxes and the surface?

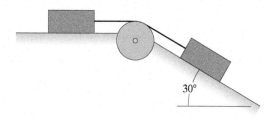

30°

Problem 14.31

14.32 The masses $m_A = 15$ kg and $m_B = 30$ kg, and the coefficients of friction between all of the surfaces are $\mu_s = 0.4$ and $\mu_k = 0.35$. The blocks are stationary when the constant force F is applied. Determine the resulting acceleration of block B if (a) $F = 200$ N; (b) $F = 400$ N.

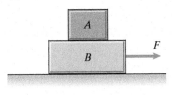

Problem 14.32

14.33 The crane's trolley at A moves to the right with constant acceleration, and the 800-kg load moves without swinging.

(a) What is the acceleration of the trolley and load?

(b) What is the sum of the tensions in the parallel cables supporting the load?

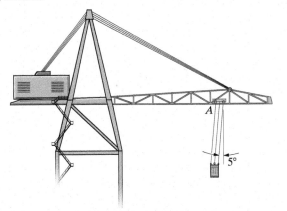

Problem 14.33

14.34 The mass of A is 30 kg and the mass of B is 5 kg. The horizontal surface is smooth. The constant force F causes the system to accelerate. The angle $\theta = 20°$ is constant. Determine F.

14.35 The mass of A is 30 kg and the mass of B is 5 kg. The coefficient of kinetic friction between A and the horizontal surface is $\mu_k = 0.2$. The constant force F causes the system to accelerate. The angle $\theta = 20°$ is constant. Determine F.

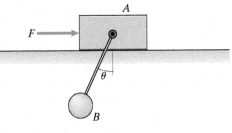

Problems 14.34/14.35

14.36 The 100-lb crate is initially stationary. The coefficients of friction between the crate and the inclined surface are $\mu_s = 0.2$ and $\mu_k = 0.16$. Determine how far the crate moves from its initial position in 2 s if the horizontal force $F = 90$ lb.

14.37 In Problem 14.36, determine how far the crate moves from its initial position in 2 s if the horizontal force $F = 30$ lb.

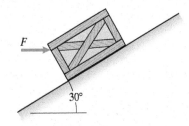

Problems 14.36/14.37

14.38 The crate has a mass of 120 kg, and the coefficients of friction between it and the sloping dock are $\mu_s = 0.6$ and $\mu_k = 0.5$.

(a) What tension must the winch exert on the cable to start the stationary crate sliding up the dock?

(b) If the tension is maintained at the value determined in part (a), what is the magnitude of the crate's velocity when it has moved 2 m up the dock?

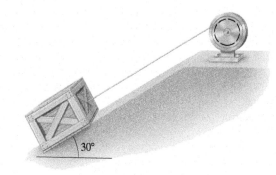

Problem 14.38

14.39 The coefficients of friction between the load A and the bed of the utility vehicle are $\mu_s = 0.4$ and $\mu_k = 0.36$. If the floor is level ($\theta = 0$), what is the largest acceleration (in m/s^2) of the vehicle for which the load will not slide on the bed?

14.40 The coefficients of friction between the load A and the bed of the utility vehicle are $\mu_s = 0.4$ and $\mu_k = 0.36$. The angle $\theta = 20°$. Determine the largest forward and rearward accelerations of the vehicle for which the load will not slide on the bed.

Problems 14.39/14.40

14.41 The package starts from rest and slides down the smooth ramp. The hydraulic device B exerts a constant 2000-N force and brings the package to rest in a distance of 100 mm from the point at which it makes contact. What is the mass of the package?

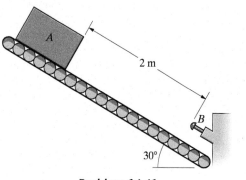

Problem 14.41

14.42 The force exerted on the 10-kg mass by the linear spring is $F = -ks$, where k is the spring constant and s is the displacement of the mass relative to its position when the spring is unstretched. The value of k is 40 N/m. The mass is in the position $s = 0$ and is given an initial velocity of 4 m/s toward the right. Determine the velocity of the mass as a function of s.

Strategy: Use the chain rule to write the acceleration as

$$\frac{dv}{dt} = \frac{dv}{ds}\frac{ds}{dt} = \frac{dv}{ds}v.$$

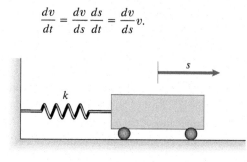

Problem 14.42

14.43 The 450-kg boat is moving at 10 m/s when its engine is shut down. The magnitude of the hydrodynamic drag force (in newtons) is $40v^2$, where v is the magnitude of the velocity in m/s. When the boat's velocity has decreased to 1 m/s, what distance has it moved from its position when the engine was shut down?

Problem 14.43

14.44 A sky diver and his parachute weigh 200 lb. He is falling vertically at 100 ft/s when his parachute opens. With the parachute open, the magnitude of the drag force (in pounds) is $0.5 v^2$.

(a) What is the magnitude of the sky diver's acceleration at the instant the parachute opens?

(b) What is the magnitude of his velocity when he has descended 20 ft from the point where his parachute opens?

Problem 14.44

▶ **14.45** The Panavia Tornado with a mass of 18,000 kg lands at a speed of 213 km/h. The decelerating force (in newtons) exerted on it by its thrust reversers and aerodynamic drag is $80,000 + 2.5v^2$, where v is the airplane's velocity in m/s. What is the length of the airplane's landing roll? (See Example 14.4.)

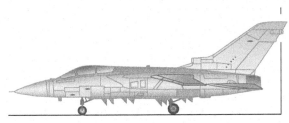

Problem 14.45

14.46 A 200-lb bungee jumper jumps from a bridge 130 ft above a river. The bungee cord has an unstretched length of 60 ft and has a spring constant $k = 14$ lb/ft.

(a) How far above the river is the jumper when the cord brings him to a stop?

(b) What maximum force does the cord exert on him?

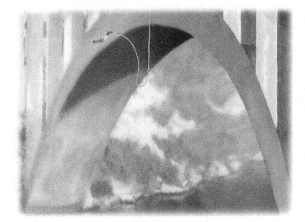

Problem 14.46

14.47 A helicopter weighs 20,500 lb. It takes off vertically from sea level, and its upward velocity in ft/s is given as a function of its altitude h in feet by $v = 66 - 0.01h$.

(a) How long does it take the helicopter to climb to an altitude of 4000 ft?

(b) What is the sum of the vertical forces on the helicopter when its altitude is 2000 ft?

14.48 In a cathode-ray tube, an electron (mass $= 9.11 \times 10^{-31}$ kg) is projected at O with velocity $\mathbf{v} = (2.2 \times 10^7)\mathbf{i}$ (m/s). While the electron is between the charged plates, the electric field generated by the plates subjects it to a force $\mathbf{F} = -eE\mathbf{j}$, where the charge of the electron $e = 1.6 \times 10^{-19}$ C (coulombs) and the electric field strength $E = 15$ kN/C. External forces on the electron are negligible when it is not between the plates. Where does the electron strike the screen?

14.49 In Problem 14.48, determine where the electron strikes the screen if the electric field strength is $E = 15\sin(\omega t)$ kN/C, where the frequency $\omega = 2 \times 10^9$ s^{-1}.

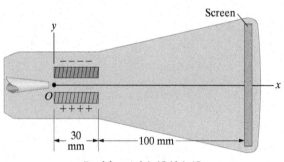

Problems 14.48/14.49

14.50 An astronaut wants to travel from a space station to a satellite S that needs repair. She departs the space station at O. A spring-loaded launching device gives her maneuvering unit an initial velocity of 1 m/s (relative to the space station) in the y direction. At that instant. the position of the satellite is $x = 70$ m, $y = 50$ m, $z = 0$, and it is drifting at 2 m/s (relative to the station) in the x direction. The astronaut intercepts the satellite by applying a constant thrust parallel to the x axis. The total mass of the astronaut and her maneuvering unit is 300 kg.

(a) How long does it take the astronaut to reach the satellite?

(b) What is the magnitude of the thrust she must apply to make the intercept?

(c) What is the astronaut's velocity *relative to the satellite* when she reaches it?

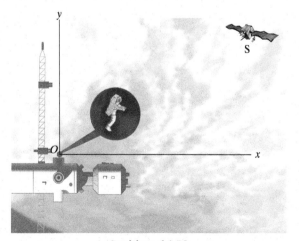

Problem 14.50

14.51 What is the acceleration of the 8-kg collar A relative to the smooth bar?

14.52 Determine the acceleration of the 8-kg collar A relative to the bar if the coefficient of kinetic friction between the collar and the bar is $\mu_k = 0.1$.

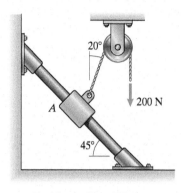

Problems 14.51/14.52

14.53 The force $F = 50$ lb. What is the magnitude of the acceleration of the 20-lb collar A along the smooth bar at the instant shown?

14.54* In Problem 14.53, determine the magnitude of the acceleration of the 20-lb collar A along the bar at the instant shown if the coefficient of static friction between the collar and the bar is $\mu_k = 0.2$.

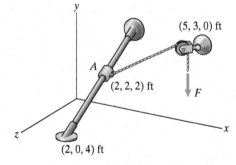

Problems 14.53/14.54

14.55 The 6-kg collar starts from rest at position A, where the coordinates of its center of mass are (400, 200, 200) mm, and slides up the smooth bar to position B, where the coordinates of its center of mass are (500, 400, 0) mm, under the action of a constant force $\mathbf{F} = -40\mathbf{i} + 70\mathbf{j} - 40\mathbf{k}$ (N). How long does the collar take to go from A to B?

14.56* In Problem 14.55, how long does the collar take to go from A to B if the coefficient of kinetic friction between the collar and the bar is $\mu_k = 0.2$?

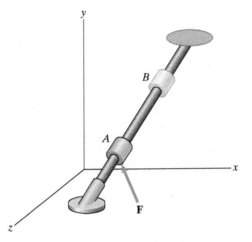

Problems 14.55/14.56

14.57 The crate is drawn across the floor by a winch that retracts the cable at a constant rate of 0.2 m/s. The crate's mass is 120 kg, and the coefficient of kinetic friction between the crate and the floor is $\mu_k = 0.24$.

(a) At the instant shown, what is the tension in the cable?

(b) Obtain a "quasi-static" solution for the tension in the cable by ignoring the crate's acceleration. Compare this solution with your result in part (a).

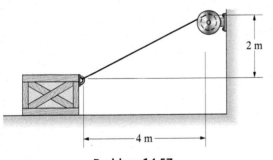

Problem 14.57

14.58 If $y = 100$ mm, $dy/dt = 600$ mm/s, and $d^2y/dt^2 = -200$ mm/s^2, what horizontal force is exerted on the 0.4-kg slider A by the smooth circular slot?

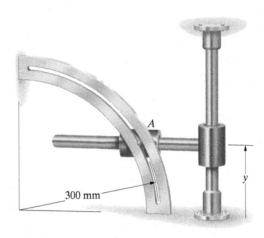

Problem 14.58

14.59 The 1-kg collar P slides on the vertical bar and has a pin that slides in the curved slot. The vertical bar moves with constant velocity $v = 2$ m/s. The y axis points upward. What are the x and y components of the total force exerted on the collar by the vertical bar and the slotted bar when $x = 0.25$ m?

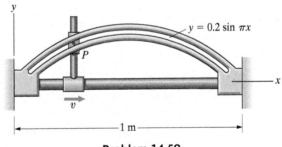

Problem 14.59

14.60* The 1360-kg car travels along a straight road of increasing grade whose vertical profile is given by the equation shown. The magnitude of the car's velocity is a constant 100 km/h. When $x = 200$ m, what are the x and y components of the total force acting on the car (including its weight)?

   Strategy: You know that the tangential component of the car's acceleration is zero. You can use this condition together with the equation for the profile of the road to determine the x and y components of the car's acceleration.

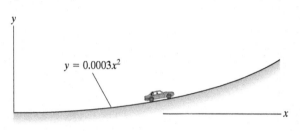

$$y = 0.0003x^2$$

Problem 14.60

14.61* The two 100-lb blocks are released from rest. Determine the magnitudes of their accelerations if friction at all the contacting surfaces is negligible.

   Strategy: Use the fact that the components of the accelerations of the blocks perpendicular to their mutual interface must be equal.

14.62* The two 100-lb blocks are released from rest. The coefficient of kinetic friction between all contacting surfaces is $\mu_k = 0.1$. How long does it take block A to fall 1 ft?

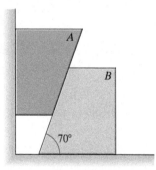

Problems 14.61/14.62

14.63 The 3000-lb vehicle has left the ground after driving over a rise. At the instant shown, it is moving horizontally at 30 mi/h and the bottoms of its tires are 24 in above the (approximately) level ground. The earth-fixed coordinate system is placed with its origin 30 in above the ground, at the height of the vehicle's center of mass when the tires first contact the ground. (Assume that the vehicle remains horizontal.) When that occurs, the vehicle's center of mass initially continues moving downward and then rebounds upward due to the flexure of the suspension system. While the tires are in contact with the ground, the force exerted on them by the ground is $-2400\mathbf{i} - 18,000y\mathbf{j}$ (lb), where y is the vertical position of the center of mass in feet. When the vehicle rebounds, what is the vertical component of the velocity of the center of mass at the instant the wheels leave the ground? (The wheels leave the ground when the center of mass is at $y = 0$.)

24 in

30 in

30 in

24 in

Problem 14.63

14.64* A steel sphere in a tank of oil is given an initial velocity $\mathbf{v} = 2\mathbf{i}$ (m/s) at the origin of the coordinate system shown. The radius of the sphere is 15 mm. The density of the steel is 8000 kg/m^3 and the density of the oil is 980 kg/m^3. If V is the sphere's volume, the (upward) buoyancy force on the sphere is equal to the weight of a volume V of oil. The magnitude of the hydrodynamic drag force $\mathbf{D}$ on the sphere as it falls is $|\mathbf{D}| = 1.6|\mathbf{v}|$ N, where $|\mathbf{v}|$ is the magnitude of the sphere's velocity in m/s. What are the x and y components of the sphere's velocity at $t = 0.1$ s?

14.65* In Problem 14.64, what are the x and y coordinates of the sphere at $t = 0.1$ s?

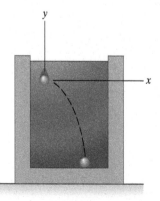

Problems 14.64/14.65

14.3 Applications—Normal and Tangential Components

When an object moves in a curved path, we can resolve the sum of the forces acting on it into normal and tangential components (Fig. 14.5a). We can also express the object's acceleration in terms of normal and tangential components (Fig. 14.5b) and write Newton's second law, $\Sigma \mathbf{F} = m\mathbf{a}$, in the form

$$\Sigma F_t \mathbf{e}_t + \Sigma F_n \mathbf{e}_n = m(a_t \mathbf{e}_t + a_n \mathbf{e}_n), \qquad (14.6)$$

where

$$a_t = \frac{dv}{dt} \quad \text{and} \quad a_n = \frac{v^2}{\rho}.$$

Equating the normal and tangential components in Eq. (14.6), we obtain two scalar equations of motion:

$$\Sigma F_t = ma_t = m\frac{dv}{dt}, \qquad \Sigma F_n = ma_n = m\frac{v^2}{\rho}. \qquad (14.7)$$

The sum of the forces in the tangential direction equals the product of the mass and the rate of change of the magnitude of the velocity, and the sum of the forces in the normal direction equals the product of the mass and the normal component of acceleration. If the path of the object's center of mass lies in a plane, the acceleration of the center of mass perpendicular to the plane is zero, so the sum of the forces perpendicular to the plane is zero.

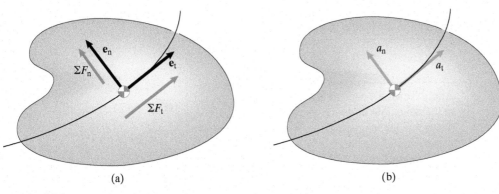

(a)   (b)

Figure 14.5
(a) Normal and tangential components of the sum of the forces on an object.
(b) Normal and tangential components of the acceleration of the center of mass of the object.

Tangential and Normal Components (▶ *Related Problem 14.66*)

The boat, which with its passengers weighs 1200 lb, is moving at 20 ft/s in a circular path with radius $R = 40$ ft. At $t = 0$, the driver advances the throttle so that the tangential component of the total force acting on the boat increases to 100 lb and remains constant. He continues following the same circular path. At $t = 2$ s, determine the magnitude of the boat's velocity and the total force acting on the boat in the direction perpendicular to its path.

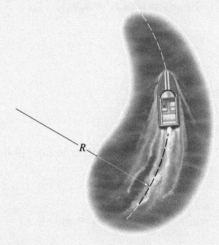

Strategy

We can apply Newton's second law in the tangential direction to determine the tangential component of the boat's acceleration and integrate the acceleration to obtain the velocity as a function of time. Once the velocity at $t = 2$ s has been determined, we can apply Newton's second law in the direction perpendicular to the boat's path to determine the total normal force at $t = 2$ s.

Solution

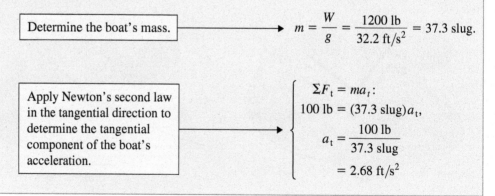

Determine the boat's mass.

$$m = \frac{W}{g} = \frac{1200 \text{ lb}}{32.2 \text{ ft/s}^2} = 37.3 \text{ slug}.$$

Apply Newton's second law in the tangential direction to determine the tangential component of the boat's acceleration.

$$\Sigma F_t = ma_t:$$
$$100 \text{ lb} = (37.3 \text{ slug})a_t,$$
$$a_t = \frac{100 \text{ lb}}{37.3 \text{ slug}}$$
$$= 2.68 \text{ ft/s}^2$$

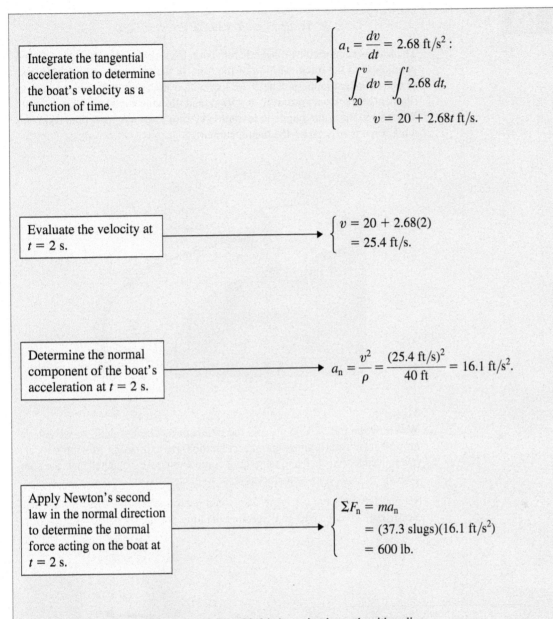

Integrate the tangential acceleration to determine the boat's velocity as a function of time.

$$\begin{cases} a_t = \dfrac{dv}{dt} = 2.68 \text{ ft/s}^2 : \\ \displaystyle\int_{20}^{v} dv = \int_{0}^{t} 2.68 \, dt, \\ v = 20 + 2.68t \text{ ft/s.} \end{cases}$$

Evaluate the velocity at $t = 2$ s.

$$\begin{cases} v = 20 + 2.68(2) \\ = 25.4 \text{ ft/s.} \end{cases}$$

Determine the normal component of the boat's acceleration at $t = 2$ s.

$$a_n = \frac{v^2}{\rho} = \frac{(25.4 \text{ ft/s})^2}{40 \text{ ft}} = 16.1 \text{ ft/s}^2.$$

Apply Newton's second law in the normal direction to determine the normal force acting on the boat at $t = 2$ s.

$$\begin{cases} \Sigma F_n = ma_n \\ = (37.3 \text{ slugs})(16.1 \text{ ft/s}^2) \\ = 600 \text{ lb.} \end{cases}$$

Practice Problem The boat is moving at 20 ft/s in a circular path with radius $R = 40$ ft. Suppose that at $t = 0$ the driver advances the throttle so that the tangential component of the total force (in pounds) acting on the boat is given as a function of time by $\Sigma F_t = 200t$. He continues following the same circular path. At $t = 2$ s, determine the magnitude of the boat's velocity and the total force acting on the boat in the direction perpendicular to its path.

Answer: Velocity is 30.7 ft/s, normal force is 880 lb.

Active Example 14.6   Train on a Banked Track (▶ *Related Problem 14.79*)

The train is supported by magnetic repulsion forces exerted in the direction perpendicular to the track. Motion of the train in the transverse direction is prevented by lateral supports. The 20,000-kg train is traveling at 30 m/s on a circular segment of track with radius $R = 150$ m, and the bank angle of the track is 40°. What force must the magnetic levitation system exert to support the train, and what force is exerted by the lateral supports?

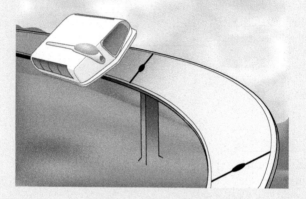

Strategy

We know the train's velocity and the radius of its circular path, so we can determine its normal component of acceleration. By expressing Newton's second law in terms of normal and tangential components, we can determine the components of force normal and transverse to the track.

Solution

View of the train from above showing the normal and tangential unit vectors. The vector $\mathbf{e}_t$ is tangential to the train's path and the vector $\mathbf{e}_n$ points toward the center of its circular path.

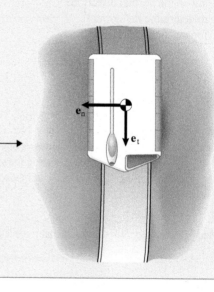

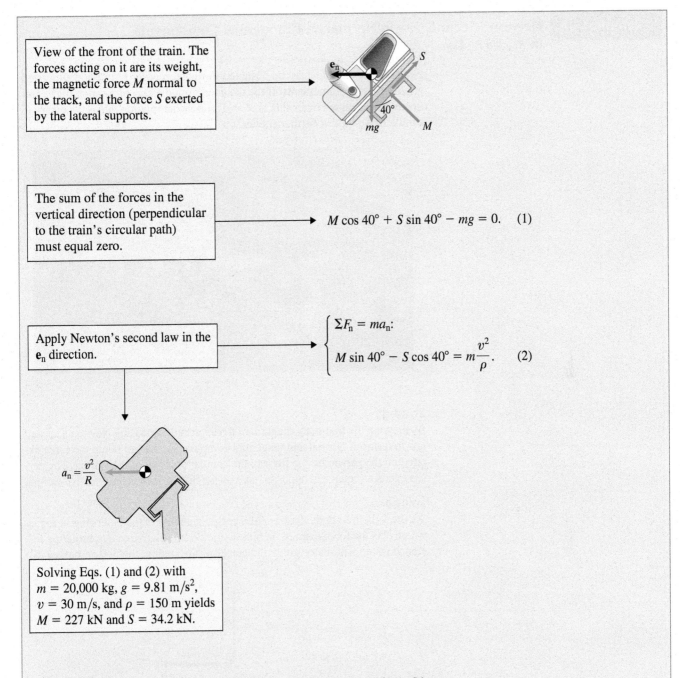

View of the front of the train. The forces acting on it are its weight, the magnetic force M normal to the track, and the force S exerted by the lateral supports.

The sum of the forces in the vertical direction (perpendicular to the train's circular path) must equal zero.

$$M \cos 40° + S \sin 40° - mg = 0. \quad (1)$$

Apply Newton's second law in the e_n direction.

$$\begin{cases} \Sigma F_n = ma_n: \\ M \sin 40° - S \cos 40° = m\dfrac{v^2}{\rho}. \quad (2) \end{cases}$$

$$a_n = \frac{v^2}{R}$$

Solving Eqs. (1) and (2) with $m = 20{,}000$ kg, $g = 9.81$ m/s^2, $v = 30$ m/s, and $\rho = 150$ m yields $M = 227$ kN and $S = 34.2$ kN.

Practice Problem For what speed v of the train would the lateral force S be zero? (This is the optimum speed for the train to travel on the banked track. If you were a passenger, you would not need to exert any lateral force to remain in place in your seat.)

Answer: 35.1 m/s.

| Example 14.7 | Newton's Second Law in Normal and Tangential Components |

(▶ *Related Problem 14.73*)

Future space stations may be designed to rotate in order to provide simulated gravity for their inhabitants. If the distance from the axis of rotation of the station to the occupied outer ring is $R = 100$ m, what rotation rate is necessary to simulate one-half of earth's gravity?

Strategy

By drawing the free-body diagram of a person and expressing Newton's second law in terms of normal and tangential components, we can relate the force exerted on the person by the floor to the angular velocity of the station. The person exerts an equal and opposite force on the floor, which is his effective weight.

Solution

We draw the free-body diagram of a person standing in the outer ring in Fig. a, where N is the force exerted on him by the floor. Relative to a nonrotating reference frame with its origin at the center of the station, the person moves in a

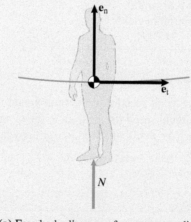

(a) Free-body diagram of a person standing in the occupied ring.

circular path of radius R. His normal and tangential components of acceleration are shown in Fig. b. Applying Eqs. (14.7), we obtain

$$\Sigma F_t = 0 = m\frac{dv}{dt}$$

and

$$\Sigma F_n = N = m\frac{v^2}{R}.$$

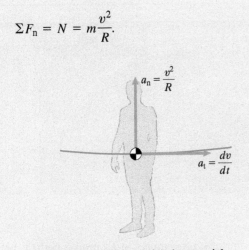

(b) The person's normal and tangential components of acceleration.

The first equation simply indicates that the magnitude of the person's velocity is constant. The second equation tells us the force N. The magnitude of his velocity is $v = R\omega$, where ω is the angular velocity of the station. If one-half of earth's gravity is simulated, $N = \frac{1}{2}mg$. Therefore

$$N = \frac{1}{2}mg = m\frac{(R\omega)^2}{R}.$$

Solving for ω, we obtain the necessary angular velocity of the station:

$$\omega = \sqrt{\frac{g}{2R}} = \sqrt{\frac{9.81 \text{ m/s}^2}{2(100 \text{ m})}} = 0.221 \text{ rad/s}.$$

This is one revolution every 28.4 s.

Critical Thinking

When you are standing in a room, the floor pushes upward on you with a force N equal to your weight. The effect of gravity on your body is indistinguishable from the effect of a force of magnitude N pushing on your feet and accelerating you upward with acceleration g in the absence of gravity. (This observation was one of Einstein's starting points in developing his general theory of relativity.) This is the basis of simulating gravity by using rotation, and it explains why we set $N = mg/2$ in this example to simulate one-half of earth's gravity.

| **Example 14.8** | **Motor Vehicle Dynamics** (▶ *Related Problems 14.89, 14.90*) |

A civil engineer's preliminary design for a freeway off-ramp is circular with radius $R = 60$ m. If she assumes that the coefficient of static friction between tires and road is at least $\mu_s = 0.4$, what is the maximum speed at which vehicles can enter the ramp without losing traction?

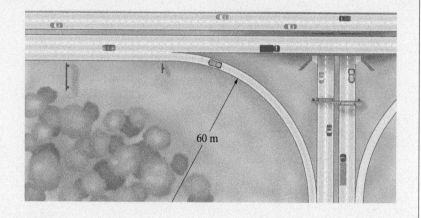

60 m

Strategy

Since a vehicle on the off-ramp moves in a circular path, it has a normal component of acceleration that depends on its velocity. The necessary normal component of force is exerted by friction between the tires and the road, and the friction force cannot be greater than the product of μ_s and the normal force. By assuming that the friction force is equal to this value, we can determine the maximum velocity for which slipping will not occur.

Solution

We view the free-body diagram of a car on the off-ramp from above the car in Fig. a and from the front of the car in Fig. b. In Fig. c, we show the car's acceleration, which is perpendicular to the circular path of the car and toward the center of the path. The sum of the forces in the $\mathbf{e}_n$ direction equals the product of the mass and the normal component of the acceleration; that is,

$$\Sigma F_n = ma_n = m\frac{v^2}{R},$$

or

$$f = m\frac{v^2}{R}.$$

The required friction force increases as v increases. The maximum friction force the surfaces will support is $f_{max} = \mu_s N = \mu_s mg$. Therefore, the maximum velocity for which slipping does not occur is

$$v = \sqrt{\mu_s gR} = \sqrt{0.4(9.81 \text{ m/s}^2)(60 \text{ m})} = 15.3 \text{ m/s},$$

or 55.2 km/h (34.3 mi/h).

(a) Top view of the free-body diagram.

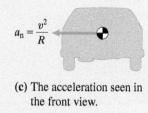

(b) Front view of the free-body diagram.

$a_n = \dfrac{v^2}{R}$

(c) The acceleration seen in the front view.

Problems

▶ **14.66** The boat in Active Example 14.5 weighs 1200 lb with its passengers. Suppose that the boat is moving at a constant speed of 20 ft/s in a circular path with radius $R = 40$ ft. Determine the tangential and normal components of force acting on the boat.

14.67 In preliminary design studies for a sun-powered car, it is estimated that the mass of the car and driver will be 100 kg and the torque produced by the engine will result in a 60-N tangential force on the car. Suppose that the car starts from rest on the track at A and is subjected to a constant 60-N tangential force. Determine the magnitude of the car's velocity and the normal component of force on the car when it reaches B.

14.68 In a test of a sun-powered car, the mass of the car and driver is 100 kg. The car starts from rest on the track at A, moving toward the right. The tangential force exerted on the car (in newtons) is given as a function of time by $\Sigma F_t = 20 + 1.2t$. Determine the magnitude of the car's velocity and the normal component of force on the car at $t = 40$ s.

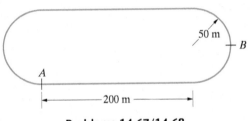

Problems 14.67/14.68

14.69 An astronaut candidate with a mass of 72 kg is tested in a centrifuge with a radius of 10 m. The centrifuge rotates in the horizontal plane. It starts from rest at time $t = 0$ and has a constant angular acceleration of 0.2 rad/s^2. Determine the magnitude of the horizontal force exerted on him by the centrifuge (a) at $t = 0$; (b) at $t = 10$ s.

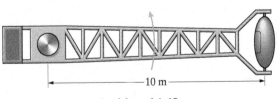

Problem 14.69

14.70 The circular disk lies *in the horizontal plane*. At the instant shown, the disk rotates with a counterclockwise angular velocity of 4 rad/s and a counterclockwise angular acceleration of 2 rad/s^2. The 0.5-kg slider A is supported horizontally by the smooth slot and the string attached at B. Determine the tension in the string and the magnitude of the horizontal force exerted on the slider by the slot.

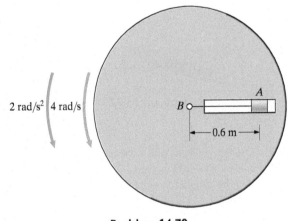

Problem 14.70

14.71 The circular disk lies *in the horizontal plane* and rotates with a constant counterclockwise angular velocity of 4 rad/s. The 0.5-kg slider A is supported horizontally by the smooth slot and the string attached at B. Determine the tension in the string and the magnitude of the horizontal force exerted on the slider by the slot.

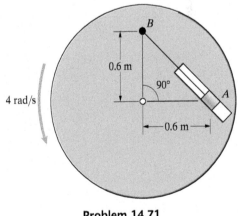

Problem 14.71

14.72 The 32,000-lb airplane is flying in the vertical plane at 420 ft/s. At the instant shown, the angle $\theta = 30°$, and the cartesian components of the plane's acceleration are
$a_x = -6$ ft/s^2, $a_y = 30$ ft/s^2.

(a) What are the tangential and normal components of the total force acting on the airplane (including its weight)?

(b) What is $d\theta/dt$ in degrees per second?

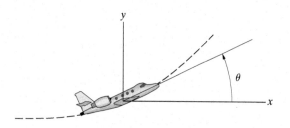

Problem 14.72

▶ **14.73** Consider a person with a mass of 72 kg who is in the space station described in Example 14.7. When he is in the occupied outer ring, his simulated weight in newtons is $\frac{1}{2}(72 \text{ kg})(9.81 \text{ m/s}^2) = 353$ N. Suppose that he climbs to a position in one of the radial tunnels that leads to the center of the station. Let r be his distance in meters from the center of the station. (a) Determine his simulated weight in his new position in terms of r. (b) What would his simulated weight be when he reaches the center of the station?

14.74 Small parts on a conveyer belt moving with constant velocity v are allowed to drop into a bin. Show that the angle θ at which the parts start sliding on the belt satisfies the equation

$$\cos \theta - \frac{1}{\mu_s} \sin \theta = \frac{v^2}{gR},$$

where μ_s is the coefficient of static friction between the parts and the belt.

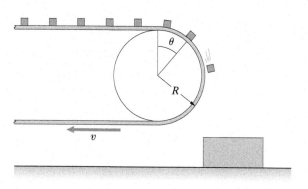

Problem 14.74

14.75 The 1-slug mass m rotates around the vertical pole in a horizontal circular path. The angle $\theta = 30°$ and the length of the string is $L = 4$ ft. What is the magnitude of the velocity of the mass?

Strategy: Notice that the vertical acceleration of the mass is zero. Draw the free-body diagram of the mass and write Newton's second law in terms of tangential and normal components.

14.76 The 1-slug mass m rotates around the vertical pole in a horizontal circular path. The length of the string is $L = 4$ ft. Determine the magnitude of the velocity of the mass and the angle θ if the tension in the string is 50 lb.

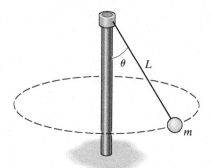

Problems 14.75/14.76

14.77 The 10-kg mass m rotates around the vertical pole in a horizontal circular path of radius $R = 1$ m. If the magnitude of the velocity of the mass is $v = 3$ m/s, what are the tensions in the strings A and B?

14.78 The 10-kg mass m rotates around the vertical pole in a horizontal circular path of radius $R = 1$ m. For what range of values of the velocity v of the mass will the mass remain in the circular path described?

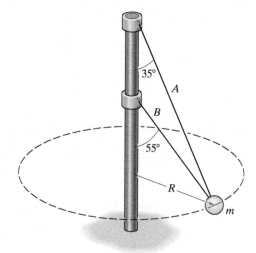

Problems 14.77/14.78

▶ 14.79 Suppose you are designing a monorail transportation system that will travel at 50 m/s, and you decide that the angle θ that the cars swing out from the vertical when they go through a turn must not be larger than 20°. If the turns in the track consist of circular arcs of constant radius R, what is the minimum allowable value of R? (See Active Example 14.6.)

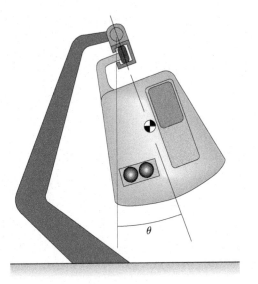

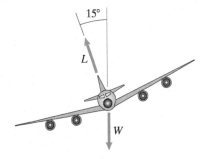

Problem 14.79

14.80 An airplane of weight $W = 200{,}000$ lb makes a turn at constant altitude and at constant velocity $v = 600$ ft/s. The bank angle is 15°.

(a) Determine the lift force L.

(b) What is the radius of curvature of the plane's path?

Problem 14.80

14.81 The suspended 2-kg mass m is stationary.

(a) What are the tensions in the strings A and B?

(b) If string A is cut, what is the tension in string B immediately afterward?

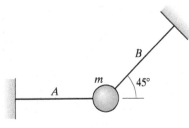

Problem 14.81

14.82 The airplane flies with constant velocity v along a circular path in the vertical plane. The radius of the airplane's circular path is 2000 m. The mass of the pilot is 72 kg.

(a) The pilot will experience "weightlessness" at the top of the circular path if the airplane exerts no net force on him at that point. Draw a free-body diagram of the pilot and use Newton's second law to determine the velocity v necessary to achieve this condition.

(b) Suppose that you don't want the force exerted on the pilot by the airplane to exceed four times his weight. If he performs this maneuver at $v = 200$ m/s, what is the minimum acceptable radius of the circular path?

Problem 14.82

14.83 The smooth circular bar rotates with constant angular velocity ω_0 about the vertical axis AB. The radius $R = 0.5$ m. The mass m remains stationary relative to the circular bar at $\beta = 40°$. Determine ω_0.

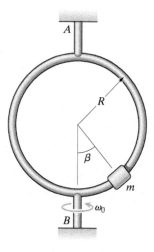

Problem 14.83

14.84 The force exerted on a charged particle by a magnetic field is

$$\mathbf{F} = q\mathbf{v} \times \mathbf{B},$$

where q and $\mathbf{v}$ are the charge and velocity vector of the particle and $\mathbf{B}$ is the magnetic field vector. A particle of mass m and positive charge q is projected at O with velocity $\mathbf{v} = v_0\mathbf{i}$ into a uniform magnetic field $\mathbf{B} = B_0\mathbf{k}$. Using normal and tangential components, show that (a) the magnitude of the particle's velocity is constant and (b) the particle's path is a circle with radius mv_0/qB_0.

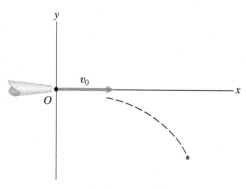

Problem 14.84

14.85 The mass m is attached to a string that is wrapped around the fixed post of radius R. At $t = 0$, the mass is given a velocity v_0 as shown. Neglect external forces on m other than the force exerted by the string. Determine the tension in the string as a function of the angle θ.

Strategy: The velocity vector of the mass is perpendicular to the string. Express Newton's second law in terms of normal and tangential components.

14.86 The mass m is attached to a string that is wrapped around the fixed post of radius R. At $t = 0$, the mass is given a velocity v_0 as shown. Neglect external forces on m other than the force exerted by the string. Determine the angle θ as a function of time.

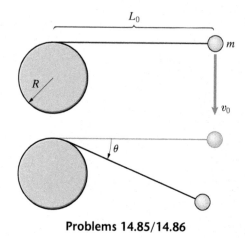

Problems 14.85/14.86

14.87 The sum of the forces in newtons exerted on the 360-kg sport plane (including its weight) during an interval of time is

$$(-1000 + 280t)\mathbf{i} + (480 - 430t)\mathbf{j} + (720 + 200t)\mathbf{k},$$

where t is the time in seconds. At $t = 0$, the velocity of the plane's center of mass relative to the earth-fixed reference frame is $20\mathbf{i} + 35\mathbf{j} - 20\mathbf{k}$ (m/s). If you resolve the sum of the forces on the plane into components tangent and normal to the plane's path at $t = 2$ s, what are the values of ΣF_t and ΣF_n?

14.88 In Problem 14.87, what is the instantaneous radius of curvature of the plane's path at $t = 2$ s? The vector components of the sum of the forces in the directions tangential and normal to the path lie in the osculating plane. Determine the components of a unit vector perpendicular to the osculating plane at $t = 2$ s.

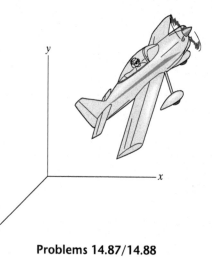

Problems 14.87/14.88

▶ **14.89** The freeway off-ramp is circular with 60-m radius (Fig. a). The off-ramp has a slope $\beta = 15°$ (Fig. b). If the coefficient of static friction between the tires of a car and the road is $\mu_s = 0.4$, what is the maximum speed at which it can enter the ramp without losing traction? (See Example 14.8.)

▶ **14.90*** The freeway off-ramp is circular with 60-m radius (Fig. a). The off-ramp has a slope β (Fig. b). If the coefficient of static friction between the tires of a car and the road is $\mu_s = 0.4$ what minimum slope β is needed so that the car could (in theory) enter the off-ramp at any speed without losing traction? (See Example 14.8.)

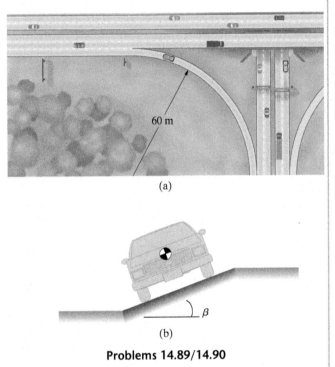

60 m

(a)

β

(b)

Problems 14.89/14.90

14.91 A car traveling at 30 m/s is at the top of a hill. The coefficient of kinetic friction between the tires and the road is $\mu_k = 0.8$. The instantaneous radius of curvature of the car's path is 200 m. If the driver applies the brakes and the car's wheels lock, what is the resulting deceleration of the car in the direction tangent to its path?

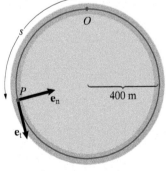

Problem 14.91

14.92 A car traveling at 30 m/s is at the bottom of a depression. The coefficient of kinetic friction between the tires and the road is $\mu_k = 0.8$. The instantaneous radius of curvature of the car's path is 200 m. If the driver applies the brakes and the car's wheel's lock, what is the resulting deceleration of the car in the direction tangential to its path? Compare your answer to that of Problem 14.91.

Problem 14.92

14.93 The combined mass of the motorcycle and rider is 160 kg. The motorcycle starts from rest at $t = 0$ and moves along a circular track with a 400-m radius. The tangential component of acceleration of the motorcycle as a function of time is $a_t = 2 + 0.2t$ m/s^2. The coefficient of static friction between the tires and the track is $\mu_s = 0.8$. How long after it starts does the motorcycle reach the limit of adhesion, which means that its tires are on the verge of slipping? How fast is the motorcycle moving when that occurs?

Strategy: Draw a free-body diagram showing the tangential and normal components of force acting on the motorcycle.

O

s

P

$\mathbf{e}_n$

$\mathbf{e}_t$

400 m

Problem 14.93

14.4 Applications—Polar and Cylindrical Coordinates

When an object moves in a planar curved path, we can describe the motion of the center of mass of the object in terms of polar coordinates. Resolving the sum of the forces parallel to the plane into polar components (Fig. 14.6a) and expressing the acceleration of the center of mass in terms of polar components (Fig. 14.6b), we can write Newton's second law, $\Sigma\mathbf{F} = m\mathbf{a}$, in the form

$$\Sigma F_r \mathbf{e}_r + \Sigma F_\theta \mathbf{e}_\theta = m(a_r \mathbf{e}_r + a_\theta \mathbf{e}_\theta), \tag{14.8}$$

where

$$a_r = \frac{d^2 r}{dt^2} - r\left(\frac{d\theta}{dt}\right)^2 = \frac{d^2 r}{dt^2} - r\omega^2$$

and

$$a_\theta = r\frac{d^2\theta}{dt^2} + 2\frac{dr}{dt}\frac{d\theta}{dt} = r\alpha + 2\frac{dr}{dt}\omega.$$

Equating the $\mathbf{e}_r$ and $\mathbf{e}_\theta$ components in Eq. (14.8), we obtain the scalar equations

$$\Sigma F_r = ma_r = m\left(\frac{d^2 r}{dt^2} - r\omega^2\right) \tag{14.9}$$

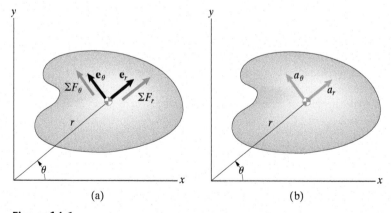

(a)   (b)

Figure 14.6
Polar components of (a) the sum of the forces and (b) the acceleration of the center of mass.

and

$$\Sigma F_\theta = ma_\theta = m\left(r\alpha + 2\frac{dr}{dt}\omega \right).$$ (14.10)

We can describe the three-dimensional motion of an object using cylindrical coordinates, in which the position of the center of mass perpendicular to the x–y plane is measured by the coordinate z and the unit vector $\mathbf{e}_z$ points in the positive z direction. We resolve the sum of the forces into radial, transverse, and z components (Fig. 14.7a) and express the acceleration of the center of mass in terms of radial, transverse, and z components (Fig. 14.7b). The three scalar equations of motion are the polar equations (14.9) and (14.10) and the equation of motion in the z direction,

$$\Sigma F_z = ma_z = m\frac{dv_z}{dt} = m\frac{d^2z}{dt^2}.$$ (14.11)

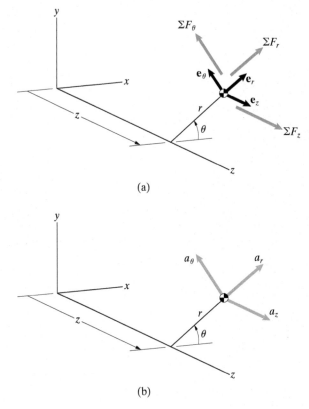

(a)

(b)

Figure 14.7
(a) Components of the sum of the forces on an object in
    cylindrical coordinates.
(b) Components of the acceleration of the center of mass.

Active Example 14.9 | **Polar Coordinates** (▶ *Related Problems 14.98, 14.99*)

The smooth bar rotates *in the horizontal plane* with constant angular velocity ω_0. The unstretched length of the linear spring is r_0. The collar A has mass m and is released at $r = r_0$ with no radial velocity. Determine the radial velocity of the collar as a function of r.

Strategy

The only force on the collar in the radial direction is the force of the spring, which we can express in polar coordinates in terms of r. By integrating Eq. (14.9), we can determine the radial component of the velocity $v_r = dr/dt$ as a function of r.

Solution

Free-body diagram of the collar A with the force exerted by the spring expressed in terms of r. The bar exerts a transverse force N on the collar.

Apply Newton's second law in the radial direction. This results in an equation for the radial component of the acceleration as a function of r.

$$\Sigma F_r = ma_r:$$

$$-k(r - r_0) = m\left(\frac{d^2r}{dt^2} - r\omega^2\right) = m\left(\frac{dv_r}{dt} - r\omega_0^2\right),$$

which we can write as

$$\frac{dv_r}{dt} = r\omega_0^2 - \frac{k}{m}(r - r_0).$$

Use the chain rule to express the radial acceleration in terms of r instead of t, separate variables, and integrate.

$$\frac{dv_r}{dt} = \frac{dv_r}{dr}\frac{dr}{dt} = \frac{dv_r}{dr}v_r = r\omega_0^2 - \frac{k}{m}(r - r_0),$$

$$\int_0^{v_r} v_r\,dv_r = \int_{r_0}^{r}\left[\left(\omega_0^2 - \frac{k}{m}\right)r + \frac{k}{m}r_0\right]dr,$$

$$\frac{1}{2}v_r^2 = \frac{1}{2}\left(\omega_0^2 - \frac{k}{m}\right)(r^2 - r_0^2) + \frac{k}{m}r_0(r - r_0).$$

Solve to obtain the radial velocity as a function of r.

$$v_r = \sqrt{\left(\omega_0^2 - \frac{k}{m}\right)(r^2 - r_0^2) + \frac{2k}{m}r_0(r - r_0)}.$$

Practice Problem Determine the transverse force N exerted on the collar by the bar as a function of r.

Answer: $N = 2m\omega_0\sqrt{\left(\omega_0^2 - \frac{k}{m}\right)(r^2 - r_0^2) + \frac{2k}{m}r_0(r - r_0)}.$

Problems

14.94 The center of mass of the 12-kg object moves in the x–y plane. Its polar coordinates are given as functions of time by $r = 12 - 0.4t^2$ m, $\theta = 0.02t^3$ rad. Determine the polar components of the total force acting on the object at $t = 2$ s.

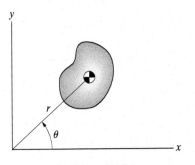

Problem 14.94

14.95 A 100-lb person walks on a large disk that rotates with constant angular velocity $\omega_0 = 0.3$ rad/s. He walks at a constant speed $v_0 = 5$ ft/s along a straight radial line painted on the disk. Determine the polar components of the horizontal force exerted on him when he is 6 ft from the center of the disk. (How are these forces exerted on him?)

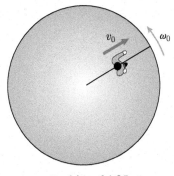

Problem 14.95

14.96 The robot is programmed so that the 0.4-kg part A describes the path

$$r = 1 - 0.5 \cos 2\pi t \text{ m},$$
$$\theta = 0.5 - 0.2 \sin 2\pi t \text{ rad}.$$

Determine the polar components of force exerted on A by the robot's jaws at $t = 2$ s.

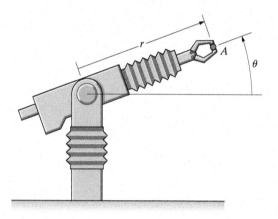

Problem 14.96

14.97 A 50-lb object P moves along the spiral path $r = (0.1)\theta$ ft, where θ is in radians. Its angular position is given as a function of time by $\theta = 2t$ rad, and $r = 0$ at $t = 0$. Determine the polar components of the total force acting on the object at $t = 4$ s.

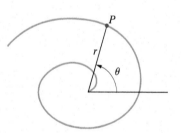

Problem 14.97

▶ **14.98** The smooth bar rotates *in the horizontal plane* with constant angular velocity $\omega_0 = 60$ rpm. If the radial velocity of the 1-kg collar A is $v_r = 10$ m/s when its radial position is $r = 1$ m, what is its radial velocity when $r = 2$ m? (See Active Example 14.9.)

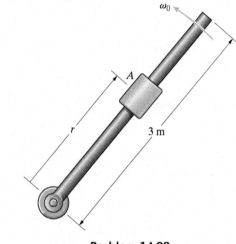

Problem 14.98

▶ **14.99** The smooth bar rotates *in the horizontal plane* with constant angular velocity $\omega_0 = 60$ rpm. The spring constant is $k = 20$ N/m and the unstretched length of the spring is 3 m. If the radial velocity of the 1-kg collar A is $v_r = 10$ m/s when its radial position is $r = 1$ m, what is its radial velocity when $r = 2$ m? (See Active Example 14.9.)

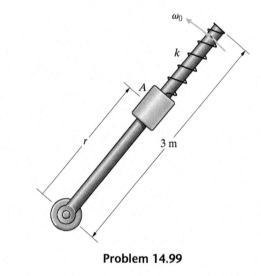

Problem 14.99

14.100 The 2-kg mass m is released from rest with the string horizontal. The length of the string is $L = 0.6$ m. By using Newton's second law in terms of polar coordinates, determine the magnitude of the velocity of the mass and the tension in the string when $\theta = 45°$.

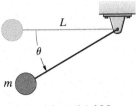

Problem 14.100

14.101 The 1-lb block A is given an initial velocity $v_0 = 14$ ft/s to the right when it is in the position $\theta = 0$, causing it to slide up the smooth circular surface. By using Newton's second law in terms of polar coordinates, determine the magnitude of the velocity of the block when $\theta = 60°$.

14.102 The 1-lb block is given an initial velocity $v_0 = 14$ ft/s to the right when it is in the position $\theta = 0$, causing it to slide up the smooth circular surface. Determine the normal force exerted on the block by the surface when $\theta = 60°$.

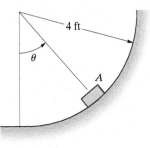

Problems 14.101/14.102

14.103 The skier passes point A going 17 m/s. From A to B, the radius of his circular path is 6 m. By using Newton's second law in terms of polar coordinates, determine the magnitude of the skier's velocity as he leaves the jump at B. Neglect tangential forces other than the tangential component of his weight.

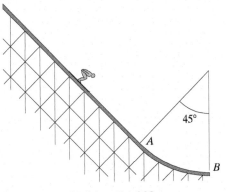

Problem 14.103

14.104* A 2-kg mass rests on a flat horizontal bar. The bar begins rotating *in the vertical plane* about O with a constant angular acceleration of 1 rad/s². The mass is observed to slip relative to the bar when the bar is 30° above the horizontal. What is the static coefficient of friction between the mass and the bar? Does the mass slip toward or away from O?

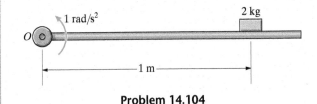

Problem 14.104

14.105* The 1/4-lb slider A is pushed along the circular bar by the slotted bar. The circular bar lies *in the horizontal plane*. The angular position of the slotted bar is $\theta = 10t^2$ rad. Determine the polar components of the total external force exerted on the slider at $t = 0.2$ s.

14.106* The 1/4-lb slider A is pushed along the circular bar by the slotted bar. The circular bar lies *in the vertical plane*. The angular position of the slotted bar is $\theta = 10t^2$ rad. Determine the polar components of the total force exerted on the slider by the circular and slotted bars at $t = 0.25$ s.

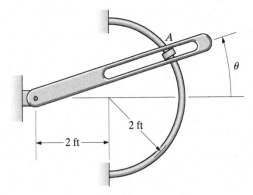

Problems 14.105/14.106

14.107* The slotted bar rotates *in the horizontal plane* with constant angular velocity ω_0. The mass m has a pin that fits in the slot of the bar. A spring holds the pin against the surface of the fixed cam. The surface of the cam is described by $r = r_0(2 - \cos \theta)$. Determine the polar components of the total external force exerted on the pin as functions of θ.

14.108* In Problem 14.107, suppose that the unstretched length of the spring is r_0. Determine the smallest value of the spring constant k for which the pin will remain on the surface of the cam.

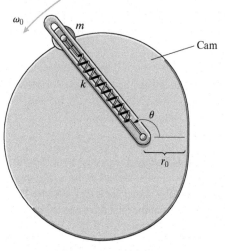

Problems 14.107/14.108

14.109 A charged particle P in a magnetic field moves along the spiral path described by $r = 1$ m, $\theta = 2z$ rad, where z is in meters. The particle moves along the path in the direction shown with constant speed $|\mathbf{v}| = 1$ km/s. The mass of the particle is 1.67×10^{-27} kg. Determine the sum of the forces on the particle in terms of cylindrical coordinates.

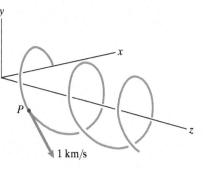

Problem 14.109

14.110 At the instant shown, the cylindrical coordinates of the 4-kg part A held by the robotic manipulator are $r = 0.6$ m, $\theta = 25°$, and $z = 0.8$ m. (The coordinate system is fixed with respect to the earth, and the y axis points upward.) A's radial position is increasing at $dr/dt = 0.2$ m/s, and $d^2r/dt^2 = -0.4$ m/s². The angle θ is increasing when $d\theta/dt = 1.2$ rad/s, and $d^2\theta/dt^2 = 2.8$ rad/s². The base of the manipulator arm is accelerating in the z direction at $d^2z/dt^2 = 2.5$ m/s². Determine the force vector exerted on A by the manipulator in terms of cylindrical coordinates.

14.111 Suppose that the robotic manipulator is used in a space station to investigate zero-g manufacturing techniques. During an interval of time, the manipulator is programmed so that the cylindrical coordinates of the 4-kg part A are $\theta = 0.15t^2$ rad, $r = 0.5(1 + \sin \theta)$ m, and $z = 0.8(1 + \theta)$ m. Determine the force vector exerted on A by the manipulator at $t = 2$ s in terms of cylindrical coordinates.

14.112* In Problem 14.111, draw a graph of the magnitude of the force exerted on part A by the manipulator as a function of time from $t = 0$ to $t = 5$ s, and use the graph to estimate the maximum force during that interval of time.

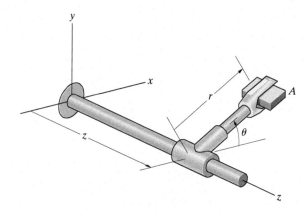

Problems 14.110–14.112

14.5 Orbital Mechanics

BACKGROUND

It is appropriate to include a discussion of orbital mechanics in our chapter on applications of Newton's second law. Newton's analytical determination of the elliptical orbits of the planets, which had been deduced from observational data by Johannes Kepler, was a triumph for Newtonian mechanics and confirmation of the inverse-square relation for gravitational acceleration.

We can use Newton's second law expressed in polar coordinates to determine the orbit of an earth satellite or a planet. Suppose that at $t = 0$ a satellite has an initial velocity v_0 at a distance r_0 from the center of the earth (Fig. 14.8a). We assume that the initial velocity is perpendicular to the line from the center of the earth to the satellite. The satellite's position during its subsequent motion is specified by its polar coordinates (r, θ), where θ is measured from the satellite's position at $t = 0$ (Fig. 14.8b). Our objective is to determine r as a function of θ.

Determination of the Orbit

If we model the earth as a homogeneous sphere, the force exerted on the satellite by gravity at a distance r from the center of the earth is mgR_E^2/r^2, where R_E is the earth's radius. (See Eq. 12.5.) From Eq. (14.9), the equation of motion in the radial direction is

$$\Sigma F_r = ma_r:$$

$$-\frac{mgR_E^2}{r^2} = m\left[\frac{d^2r}{dt^2} - r\left(\frac{d\theta}{dt}\right)^2\right].$$

From Eq. (14.10), the equation of motion in the transverse direction is

$$\Sigma F_\theta = ma_\theta:$$

$$0 = m\left(r\frac{d^2\theta}{dt^2} + 2\frac{dr}{dt}\frac{d\theta}{dt}\right).$$

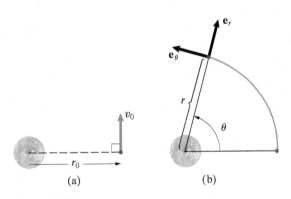

Figure 14.8
(a) Initial position and velocity of an earth satellite.
(b) Specifying the subsequent path in terms of polar coordinates.

We therefore obtain the two equations

$$\frac{d^2r}{dt^2} - r\left(\frac{d\theta}{dt}\right)^2 = -\frac{gR_E^2}{r^2} \tag{14.12}$$

and

$$r\frac{d^2\theta}{dt^2} + 2\frac{dr}{dt}\frac{d\theta}{dt} = 0. \tag{14.13}$$

We can write Eq. (14.13) in the form

$$\frac{1}{r}\frac{d}{dt}\left(r^2\frac{d\theta}{dt}\right) = 0,$$

which indicates that

$$r^2\frac{d\theta}{dt} = rv_\theta = \text{constant}. \tag{14.14}$$

At $t = 0$, the components of the velocity are $v_r = 0$ and $v_\theta = v_0$, and the radial position is $r = r_0$. We can therefore write the constant in Eq. (14.14) in terms of the initial conditions:

$$r^2\frac{d\theta}{dt} = rv_\theta = r_0v_0. \tag{14.15}$$

Using this equation to eliminate $d\theta/dt$ from Eq. (14.12), we obtain

$$\frac{d^2r}{dt^2} - \frac{r_0^2v_0^2}{r^3} = -\frac{gR_E^2}{r^2}. \tag{14.16}$$

We can solve this differential equation by introducing the change of variable

$$u = \frac{1}{r}. \tag{14.17}$$

In doing so, we will also change the independent variable from t to θ, because we want to determine r as a function of the angle θ instead of time. To express Eq. (14.16) in terms of u, we must determine d^2r/dt^2 in terms of u. Using the chain rule, we write the derivative of r with respect to time as

$$\frac{dr}{dt} = \frac{d}{dt}\left(\frac{1}{u}\right) = -\frac{1}{u^2}\frac{du}{dt} = -\frac{1}{u^2}\frac{du}{d\theta}\frac{d\theta}{dt}. \tag{14.18}$$

Notice from Eq. (14.15) that

$$\frac{d\theta}{dt} = \frac{r_0v_0}{r^2} = r_0v_0u^2. \tag{14.19}$$

Substituting this expression into Eq. (14.18), we obtain

$$\frac{dr}{dt} = -r_0 v_0 \frac{du}{d\theta}. \tag{14.20}$$

We differentiate Eq. (14.20) with respect to time and apply the chain rule again:

$$\frac{d^2r}{dt^2} = \frac{d}{dt}\left(-r_0 v_0 \frac{du}{d\theta}\right) = -r_0 v_0 \frac{d\theta}{dt}\frac{d}{d\theta}\left(\frac{du}{d\theta}\right) = -r_0 v_0 \frac{d\theta}{dt}\frac{d^2u}{d\theta^2}.$$

Using Eq. (14.19) to eliminate $d\theta/dt$ from this expression, we obtain the second time derivative of r in terms of u:

$$\frac{d^2r}{dt^2} = -r_0^2 v_0^2 u^2 \frac{d^2u}{d\theta^2}.$$

Substituting this result into Eq. (14.16) yields a linear differential equation for u as a function of θ:

$$\frac{d^2u}{d\theta^2} + u = \frac{gR_E^2}{r_0^2 v_0^2}.$$

The general solution of this equation is

$$u = A \sin\theta + B \cos\theta + \frac{gR_E^2}{r_0^2 v_0^2}, \tag{14.21}$$

where A and B are constants. We can use the initial conditions to determine A and B. When $\theta = 0$, $u = 1/r_0$. Also, when $\theta = 0$ the radial component of velocity $v_r = dr/dt = 0$, so from Eq. (14.20) we see that $du/d\theta = 0$. From these two conditions, we obtain

$$A = 0 \quad \text{and} \quad B = \frac{1}{r_0} - \frac{gR_E^2}{r_0^2 v_0^2}.$$

Substituting these results into Eq. (14.21), we can write the resulting solution for $r = 1/u$ as

$$\frac{r}{r_0} = \frac{1 + \varepsilon}{1 + \varepsilon \cos\theta}, \tag{14.22}$$

where

$$\varepsilon = \frac{r_0 v_0^2}{gR_E^2} - 1. \tag{14.23}$$

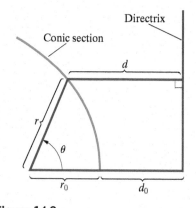

Figure 14.9
If the ratio r/d is constant, the curve describes a conic section.

Types of Orbits

The curve called a *conic section* (Fig. 14.9) has the property that the ratio of r to the perpendicular distance d to a straight line called the *directrix* is constant. This ratio, $r/d = r_0/d_0$, is called the *eccentricity* of the curve. From the figure, we see that

$$r \cos \theta + d = r_0 + d_0,$$

which we can write as

$$\frac{r}{r_0} = \frac{1 + (r_0/d_0)}{1 + (r_0/d_0) \cos \theta}.$$

Comparing this expression with Eq. (14.22), we see that *the satellite's orbit describes a conic section with eccentricity* ε. The value of the eccentricity determines the character of the orbit.

Circular Orbit   If the initial velocity v_0 is chosen so that $\varepsilon = 0$, Eq. (14.22) reduces to $r = r_0$ and the orbit is circular (Fig. 14.10). Setting $\varepsilon = 0$ in Eq. (14.23) and solving for v_0, we obtain

$$v_0 = \sqrt{\frac{gR_{\text{E}}^2}{r_0}}, \tag{14.24}$$

which agrees with the velocity for a circular orbit we obtained by a different method in Example 13.5.

Elliptic Orbit   If $0 < \varepsilon < 1$, the orbit is an ellipse (Fig. 14.10). The maximum radius of the ellipse occurs when $\theta = 180°$. Setting θ equal to $180°$ in Eq. (14.22), we obtain an expression for the maximum radius of the ellipse in terms of the initial radius and ε:

$$r_{\text{max}} = r_0 \left(\frac{1 + \varepsilon}{1 - \varepsilon} \right). \tag{14.25}$$

Parabolic Orbit   Notice from Eq. (14.25) that the maximum radius of the elliptic orbit increases without limit as $\varepsilon \to 1$. When $\varepsilon = 1$, the orbit is a parabola (Fig. 14.10). The corresponding velocity v_0 is the minimum initial velocity for which the radius r increases without limit, which is the escape velocity. Setting $\varepsilon = 1$ in Eq. (14.23) and solving for v_0, we obtain

$$v_0 = \sqrt{\frac{2gR_{\text{E}}^2}{r_0}}.$$

This is the same value for the escape velocity we obtained in Example 13.5 for the case of an object moving in a straight path directly away from the center of the earth.

Hyperbolic Orbit   If $\varepsilon > 1$, the orbit is a hyperbola (Fig. 14.10).

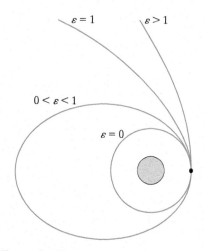

Figure 14.10
Orbits for different eccentricities.

The solution we have presented, based on the assumption that the earth is a homogeneous sphere, approximates the orbit of an earth satellite. Determining the orbit accurately requires taking into account the variations in the earth's gravitational field due to its actual mass distribution. Similarly, depending on the accuracy required, determining the orbit of a planet around the sun may require accounting for perturbations due to the gravitational attractions of the other planets.

RESULTS

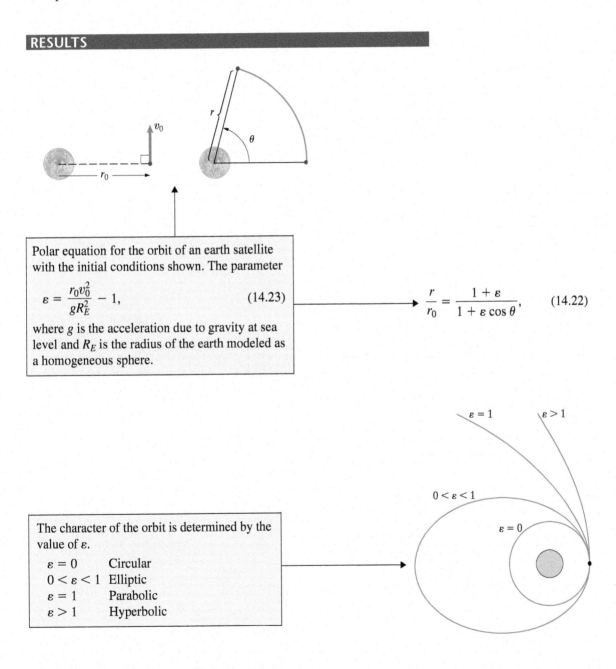

Polar equation for the orbit of an earth satellite with the initial conditions shown. The parameter

$$\varepsilon = \frac{r_0 v_0^2}{g R_E^2} - 1, \qquad (14.23)$$

where g is the acceleration due to gravity at sea level and R_E is the radius of the earth modeled as a homogeneous sphere.

$$\frac{r}{r_0} = \frac{1 + \varepsilon}{1 + \varepsilon \cos \theta}, \qquad (14.22)$$

The character of the orbit is determined by the value of ε.

$\varepsilon = 0$ Circular
$0 < \varepsilon < 1$ Elliptic
$\varepsilon = 1$ Parabolic
$\varepsilon > 1$ Hyperbolic

The radial position of the satellite and its transverse component of velocity satisfy the relation $r v_\theta = $ constant.   (14.14)

Active Example 14.10   Orbit of an Earth Satellite (▶ *Related Problem 14.115*)

An earth satellite is placed into orbit with an initial velocity $v_0 = 9240$ m/s. The satellite's initial position is 6600 km from the center of the earth. Show that the resulting orbit is elliptic and determine its maximum radius. The earth's radius is 6370 km.

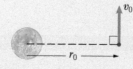

Strategy

We must calculate the value of ε from Eq. (14.23) to determine the type of orbit. We can use the polar equation for the orbit, Eq. (14.22), to obtain the maximum radius.

Solution

Calculate the value of ε. The orbit is elliptic.

$$\begin{cases} \varepsilon = \dfrac{r_0 v_0^2}{gR_E^2} - 1 \\[2mm] = \dfrac{(6600 \times 10^3 \text{ m})(9240 \text{ m/s})^2}{(9.81 \text{ m/s}^2)(6370 \times 10^3 \text{ m})^2} - 1 \\[2mm] = 0.416. \end{cases}$$

Determine the maximum radius from the polar equation for the orbit with $\theta = 180°$. The graph of the elliptic orbit is shown.

$$\begin{cases} r_{max} = r_0\left(\dfrac{1 + \varepsilon}{1 + \varepsilon \cos 180°}\right) \\[2mm] = r_0\left(\dfrac{1 + \varepsilon}{1 - \varepsilon}\right) \\[2mm] = (6600 \text{ km})\left(\dfrac{1 + 0.416}{1 - 0.416}\right) \\[2mm] = 16{,}000 \text{ km}. \end{cases}$$

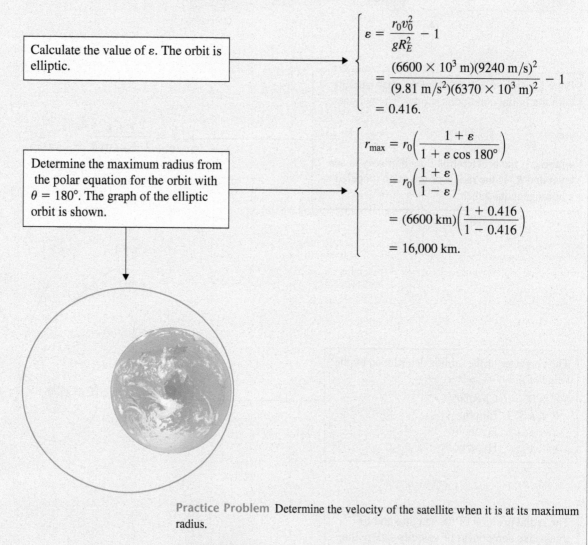

Practice Problem Determine the velocity of the satellite when it is at its maximum radius.

Answers: 3810 m/s.

Problems

Use the values $R_E = 6370$ km $= 3960$ mi for the radius of the earth.

14.113 The International Space Station is in a circular orbit 225 miles above the earth's surface.
(a) What is the magnitude of the velocity of the space station?
(b) How long does it take to complete one revolution?

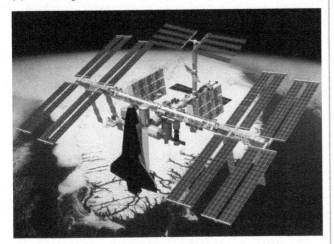

Problem 14.113

14.114 The moon is approximately 383,000 km from the earth. Assume that the moon's orbit around the earth is circular with velocity given by Eq. (14.24).
(a) What is the magnitude of the moon's velocity?
(b) How long does it take to complete one revolution around the earth?

▶ **14.115** Suppose that you place a satellite into an elliptic earth orbit with an initial radius $r_0 = 6700$ km and an initial velocity v_0 such that the maximum radius of the orbit is 13,400 km. (a) Determine v_0. (b) What is the magnitude of the satellite's velocity when it is at its maximum radius? (See Active Example 14.10.)

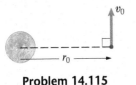

Problem 14.115

14.116 A satellite is given an initial velocity $v_0 = 6700$ m/s at a distance $r_0 = 2R_E$ from the center of the earth as shown in Fig. 14.8a. Draw a graph of the resulting orbit.

14.117 The time required for a satellite in a circular earth orbit to complete one revolution increases as the radius of the orbit increases. If you choose the radius properly, the satellite will complete one revolution in 24 hours. If a satellite is placed in such an orbit directly above the equator and moving from west to east, it will remain above the same point on the earth as the earth rotates beneath it. This type of orbit, conceived by Arthur C. Clarke, is called *geosynchronous*, and is used for communication and television broadcast satellites. Determine the radius of a geosynchronous orbit in km.

14.118* You can send a spacecraft from the earth to the moon in the following way: First, launch the spacecraft into a circular "parking" orbit of radius r_0 around the earth (Fig. P14.118a). Then, increase its velocity in the direction tangent to the circular orbit to a value v_0 such that it will follow an elliptic orbit whose maximum radius is equal to the radius r_M of the moon's orbit around the earth (Fig. P14.118b). The radius $r_M = 238{,}000$ mi. Let $r_0 = 4160$ mi. What velocity v_0 is necessary to send a spacecraft to the moon? (This description is simplified in that it disregards the effect of the moon's gravity.)

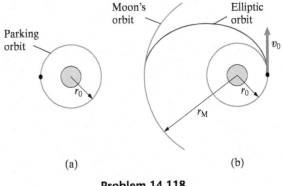

(a)   (b)

Problem 14.118

14.119* At $t = 0$, an earth satellite is a distance r_0 from the center of the earth and has an initial velocity v_0 in the direction shown. Show that the polar equation for the resulting orbit is

$$\frac{r}{r_0} = \frac{(\varepsilon + 1) \cos^2 \beta}{[(\varepsilon + 1) \cos^2 \beta - 1] \cos \theta - (\varepsilon + 1) \sin \beta \cos \beta \sin \theta + 1},$$

where $\varepsilon = (r_0 v_0^2 / g R_E^2) - 1$.

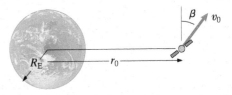

Problem 14.119

Review Problems

14.120 The Acura NSX can brake from 60 mi/h to a stop in a distance of 112 ft. The car weighs 3250 lb. (a) If you assume that the vehicle's deceleration is constant, what are its deceleration and the magnitude of the horizontal force its tires exert on the road? (b) If the car's tires are at the limit of adhesion (i.e., slip is impending), and the normal force exerted on the car by the road equals the car's weight, what is the coefficient of friction μ_s? (This analysis neglects the effects of horizontal and vertical aerodynamic forces.)

14.121 Using the coefficient of friction obtained in Problem 14.120, determine the highest constant speed at which the NSX could drive on a flat, circular track of 600-ft radius without skidding.

14.122 A "cog" engine hauls three cars of sightseers to a mountaintop in Bavaria. The mass of each car, including its passengers, is 10,000 kg, and the friction forces exerted by the wheels of the cars are negligible. Determine the forces in the couplings 1, 2, and 3 if (a) the engine is moving at constant velocity and (b) the engine is accelerating up the mountain at 1.2 m/s².

Problem 14.122

14.123 In a future mission, a spacecraft approaches the surface of an asteroid passing near the earth. Just before it touches down, the spacecraft is moving downward at a constant velocity relative to the surface of the asteroid and its downward thrust is 0.01 N. The computer decreases the downward thrust to 0.005 N, and an onboard laser interferometer determines that the acceleration of the spacecraft relative to the surface becomes 5×10^{-6} m/s² downward. What is the gravitational acceleration of the asteroid near its surface?

Problem 14.123

14.124 A car with a mass of 1470 kg, including its driver, is driven at 130 km/h over a slight rise in the road. At the top of the rise, the driver applies the brakes. The coefficient of static friction between the tires and the road is $\mu_s = 0.9$, and the radius of curvature of the rise is 160 m. Determine the car's deceleration at the instant the brakes are applied, and compare it with the deceleration on a level road.

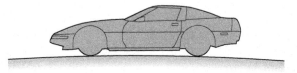

Problem 14.124

14.125 The car drives at constant velocity up the straight segment of road on the left. If the car's tires continue to exert the same tangential force on the road after the car has gone over the crest of the hill and is on the straight segment of road on the right, what will be the car's acceleration?

Problem 14.125

14.126 The aircraft carrier *Nimitz* weighs 91,000 tons. (A ton is 2000 lb.) Suppose that it is traveling at its top speed of approximately 30 knots (a knot is 6076 ft/h) when its engines are shut down. If the water exerts a drag force of magnitude $20{,}000v$ lb, where v is the carrier's velocity in feet per second, what distance does the carrier move before coming to rest?

14.127 If $m_A = 10$ kg, $m_B = 40$ kg, and the coefficient of kinetic friction between all surfaces is $\mu_k = 0.11$, what is the acceleration of B down the inclined surface?

14.128 If A weighs 20 lb, B weighs 100 lb, and the coefficient of kinetic friction between all surfaces is $\mu_k = 0.15$, what is the tension in the cord as B slides down the inclined surface?

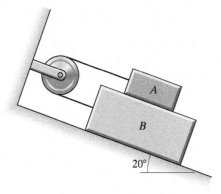

Problems 14.127/14.128

14.129 A gas gun is used to accelerate projectiles to high velocities for research on material properties. The projectile is held in place while gas is pumped into the tube to a high pressure p_0 on the left and the tube is evacuated on the right. The projectile is then released and is accelerated by the expanding gas. Assume that the pressure p of the gas is related to the volume V it occupies by $pV^\gamma = $ constant, where γ is a constant. If friction can be neglected, show that the velocity of the projectile at the position x is

$$ v = \sqrt{\frac{2p_0 A x_0^\gamma}{m(\gamma - 1)}\left(\frac{1}{x_0^{\gamma-1}} - \frac{1}{x^{\gamma-1}}\right)}, $$

where m is the mass of the projectile and A is the cross-sectional area of the tube.

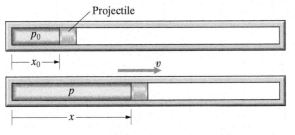

Problem 14.129

14.130 The weights of the blocks are $W_A = 120$ lb and $W_B = 20$ lb, and the surfaces are smooth. Determine the acceleration of block A and the tension in the cord.

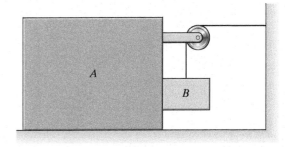

Problem 14.130

14.131 The 100-Mg space shuttle is in orbit when its engines are turned on, exerting a thrust force $\mathbf{T} = 10\mathbf{i} - 20\mathbf{j} + 10\mathbf{k}$ (kN) for 2 s. Neglect the resulting change in mass of the shuttle. At the end of the 2-s burn, fuel is still sloshing back and forth in the shuttle's tanks. What is the change in the velocity of the center of mass of the shuttle (including the fuel it contains) due to the 2-s burn?

14.132 The water skier contacts the ramp with a velocity of 25 mi/h parallel to the surface of the ramp. Neglecting friction and assuming that the tow rope exerts no force on him once he touches the ramp, estimate the horizontal length of the skier's jump from the end of the ramp.

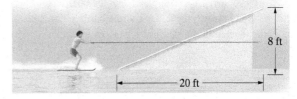

Problem 14.132

14.133 Suppose you are designing a roller-coaster track that will take the cars through a vertical loop of 40-ft radius. If you decide that, for safety, the downward force exerted on a passenger by his or her seat at the top of the loop should be at least one-half the passenger's weight, what is the minimum safe velocity of the cars at the top of the loop?

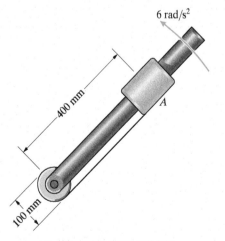

Problem 14.133

14.134 As the smooth bar rotates *in the horizontal plane*, the string winds up on the fixed cylinder and draws the 1-kg collar A inward. The bar starts from rest at $t = 0$ in the position shown and rotates with constant angular acceleration. What is the tension in the string at $t = 1$ s?

14.135 In Problem 14.134, suppose that the coefficient of kinetic friction between the collar and the bar is $\mu_k = 0.2$. What is the tension in the string at $t = 1$ s?

6 rad/s²

400 mm

A

100 mm

Problems 14.134/14.135

14.136 If you want to design the cars of a train to tilt as the train goes around curves in order to achieve maximum passenger comfort, what is the relationship between the desired tilt angle θ, the velocity v of the train, and the instantaneous radius of curvature, ρ, of the track?

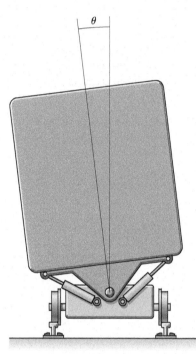

θ

Problem 14.136

14.137 To determine the coefficient of static friction between two materials, an engineer at the U.S. National Institute of Standards and Technology places a small sample of one material on a horizontal disk whose surface is made of the other material and then rotates the disk from rest with a constant angular acceleration of 0.4 rad/s². If she determines that the small sample slips on the disk after 9.903 s, what is the coefficient of friction?

200 mm

Problem 14.137

14.138* The 1-kg slider A is pushed along the curved bar by the slotted bar. The curved bar lies *in the horizontal plane*, and its profile is described by $r = 2(\theta/2\pi + 1)$ m, where θ is in radians. The angular position of the slotted bar is $\theta = 2t$ rad. Determine the polar components of the total external force exerted on the slider when $\theta = 120°$.

14.139* In Problem 14.138, suppose that the curved bar lies *in the vertical plane*. Determine the polar components of the total force exerted on A by the curved and slotted bars at $t = 0.5$ s.

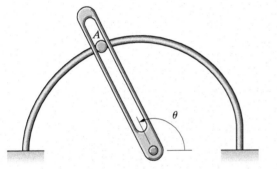

Problems 14.138/14.139

Design Project

The proposed design for an energy-absorbing bumper for a car exerts a decelerating force of magnitude $bs + cv$ on the car when it collides with a rigid obstacle, where s is the distance the car travels from the point where it contacts the obstacle and v is the car's velocity. Thus the force exerted on the car by the bumper is a function of the car's position and velocity.

(a) Suppose that at $t = 0$ the car contacts the obstacle with initial velocity v_0. Prove that the car's position is given as a function of time by

$$s = \frac{v_0}{2h}\left[e^{-(d-h)t} - e^{-(d+h)t}\right],$$

where $d = c/2m$, $h = \sqrt{d^2 - b/m}$, and m is the mass of the car. To do this, first show that this equation satisfies Newton's second law. Then confirm that it satisfies the initial conditions $s = 0$ and $v = v_0$ at $t = 0$.

(b) Investigate the effects of the car's mass, the initial velocity, and the constants b and c on the motion of the car when it strikes the obstacle. (Assume that $d^2 > b/m$.) Pay particular attention to how your choices for the constants b and c affect the maximum deceleration to which the occupants of the car would be subjected. Write a brief report presenting the results of your analysis and giving your conclusions concerning the design of energy-absorbing bumpers.

CHAPTER
15

Energy Methods

The concepts of energy and conservation of energy originated in large part from the study of classical mechanics. A simple transformation of Newton's second law results in an equation that motivates the definitions of work, kinetic energy (energy due to an object's motion), and potential energy (energy due to an object's position). This equation can greatly simplify the solution of problems involving certain forces that depend on an object's position, including gravitational forces and forces exerted by springs.

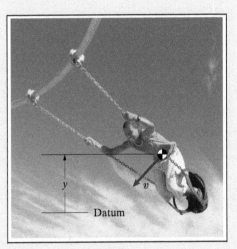

◀ A swing provides excitement by transforming the potential energy of height into the kinetic energy of motion. In this chapter we use energy methods to analyze motions of objects.

15.1 Work and Kinetic Energy

BACKGROUND

Principle of Work and Energy

We have used Newton's second law to relate the acceleration of an object's center of mass to external forces acting on it. We will now show how Newton's second law, which is a vector equation, can be transformed into a scalar equation that is extremely useful in particular circumstances. We begin with Newton's second law in the form

$$\Sigma \mathbf{F} = m\frac{d\mathbf{v}}{dt}, \tag{15.1}$$

and take the dot product of both sides with the velocity:

$$\Sigma \mathbf{F} \cdot \mathbf{v} = m\frac{d\mathbf{v}}{dt} \cdot \mathbf{v}. \tag{15.2}$$

We write the left side of this equation as

$$\Sigma \mathbf{F} \cdot \mathbf{v} = \Sigma \mathbf{F} \cdot \frac{d\mathbf{r}}{dt}$$

and write the right side as

$$m\frac{d\mathbf{v}}{dt} \cdot \mathbf{v} = \tfrac{1}{2}m\frac{d}{dt}(\mathbf{v} \cdot \mathbf{v}),$$

obtaining

$$\Sigma \mathbf{F} \cdot d\mathbf{r} = \tfrac{1}{2}m\, d(v^2), \tag{15.3}$$

where $v^2 = \mathbf{v} \cdot \mathbf{v}$ is the square of the magnitude of the velocity. The term on the left side of Eq. (15.3) is the *work* expressed in terms of the total external force on the object and an infinitesimal displacement $d\mathbf{r}$ of its center of mass. Integrating Eq. (15.3) yields

$$\int_{\mathbf{r}_1}^{\mathbf{r}_2} \Sigma \mathbf{F} \cdot d\mathbf{r} = \tfrac{1}{2}mv_2^2 - \tfrac{1}{2}mv_1^2, \tag{15.4}$$

where v_1 and v_2 are the magnitudes of the velocity of the center of mass of the object when it is at positions $\mathbf{r}_1$ and $\mathbf{r}_2$, respectively. The term $\tfrac{1}{2}mv^2$ is called the *kinetic energy* associated with the motion of the center of mass. Denoting the work done as the center of mass moves from $\mathbf{r}_1$ to $\mathbf{r}_2$ by

$$U_{12} = \int_{\mathbf{r}_1}^{\mathbf{r}_2} \Sigma \mathbf{F} \cdot d\mathbf{r}, \tag{15.5}$$

we obtain the principle of work and energy:

The work done on an object as it moves between two positions equals the change in its kinetic energy.

$$U_{12} = \tfrac{1}{2}mv_2^2 - \tfrac{1}{2}mv_1^2. \tag{15.6}$$

The dimensions of work, and therefore the dimensions of kinetic energy, are (force) × (length). In SI units, work is usually expressed in N-m or joules (J). In U.S. Customary units, work is usually expressed in ft-lb.

If the work done on an object as it moves between two positions can be evaluated, the principle of work and energy permits us to determine the change in the magnitude of the object's velocity. We can also apply this principle to a system of objects, equating the total work done by external forces to the change in the total kinetic energy of the system. But the principle must be applied with caution, because, as we demonstrate in Example 15.3, net work can be done on a system by internal forces.

Although the principle of work and energy relates a change in the position of an object to the change in its velocity, it is not convenient for obtaining other information about the motion of the object, such as the time it takes the object to move from one position to another. Furthermore, since the work is an integral with respect to position, we can usually evaluate it only when the forces doing work are known as functions of position. Despite these limitations, the principle is extremely useful for certain problems because the work can be determined very easily.

Evaluating the Work

Let us consider an object in curvilinear motion relative to an inertial reference frame (Fig. 15.1a) and specify its position by the coordinate s measured along its path from a reference point O. In terms of the tangential unit vector $\mathbf{e}_t$, the object's velocity is

$$\mathbf{v} = \frac{ds}{dt}\mathbf{e}_t.$$

Because $\mathbf{v} = d\mathbf{r}/dt$, we can multiply the velocity by dt to obtain an expression for the vector $d\mathbf{r}$ describing an infinitesimal displacement along the path (Fig. 15.1b):

$$d\mathbf{r} = \mathbf{v}\,dt = ds\,\mathbf{e}_t.$$

The work done by the external forces acting on the object as a result of the displacement $d\mathbf{r}$ is

$$\Sigma\mathbf{F}\cdot d\mathbf{r} = (\Sigma\mathbf{F}\cdot\mathbf{e}_t)\,ds = \Sigma F_t ds,$$

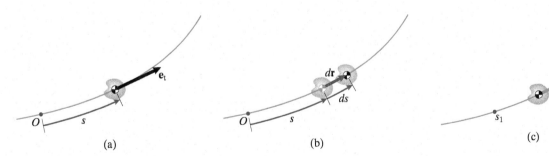

Figure 15.1
(a) The coordinate s and tangential unit vector.
(b) An infinitesimal displacement $d\mathbf{r}$.
(c) The work done from s_1 to s_2 is determined by the tangential component of the external forces.

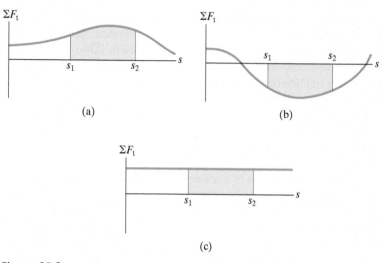

Figure 15.2
(a) The work equals the area defined by the graph of the tangential force as a function of the distance along the path.
(b) Negative work is done if the tangential force is opposite to the direction of the motion.
(c) The work done by a constant tangential force equals the product of the force and the distance.

where ΣF_t is the tangential component of the total force. Therefore, as the object moves from a position s_1 to a position s_2 (Fig. 15.1c), the work is

$$U_{12} = \int_{s_1}^{s_2} \Sigma F_t \, ds. \tag{15.7}$$

The work is equal to the integral of the tangential component of the total force with respect to distance along the path. Thus, the work done is equal to the area defined by the graph of the tangential force from s_1 to s_2 (Fig. 15.2a). *Components of force perpendicular to the path do no work.* Notice that if ΣF_t is opposite to the direction of motion over some part of the path, which means that the object is decelerating, the work is negative (Fig. 15.2b). If ΣF_t is constant between s_1 and s_2, the work is simply the product of the total tangential force and the displacement (Fig. 15.2c):

$$U_{12} = \Sigma F_t(s_2 - s_1). \qquad \text{Constant tangential force} \tag{15.8}$$

Power

Power is the rate at which work is done. The work done by the external forces acting on an object during an infinitesimal displacement $d\mathbf{r}$ is

$$\Sigma \mathbf{F} \cdot d\mathbf{r}.$$

We obtain the power P by dividing this expression by the interval of time dt during which the displacement takes place:

$$P = \Sigma \mathbf{F} \cdot \mathbf{v}. \tag{15.9}$$

This is the power transferred to or from the object, depending on whether P is positive or negative. In SI units, power is expressed in newton-meters per second, which is joules per second (J/s) or watts (W). In U.S. Customary units, power is expressed in foot-pounds per second or in the anachronistic horsepower (hp), which is 746 W or approximately 550 ft-lb/s.

Notice from Eq. (15.3) that the power equals the rate of change of the kinetic energy of the object:

$$P = \frac{d}{dt}\left(\tfrac{1}{2}mv^2\right).$$

Transferring power to and from an object causes its kinetic energy to increase and decrease, respectively. Using the preceding relation, we can write the average with respect to time of the power during an interval of time from t_1 to t_2 as

$$P_{av} = \frac{1}{t_2 - t_1}\int_{t_1}^{t_2} P \, dt = \frac{1}{t_2 - t_1}\int_{v_1^2}^{v_2^2} \tfrac{1}{2}m \, d(v^2).$$

Performing the integration, we find that the average power transferred to or from an object during an interval of time is equal to the change in its kinetic energy, or the work done, divided by the interval of time:

$$P_{av} = \frac{\tfrac{1}{2}mv_2^2 - \tfrac{1}{2}mv_1^2}{t_2 - t_1} = \frac{U_{12}}{t_2 - t_1}. \tag{15.10}$$

RESULTS

Principle of Work and Energy

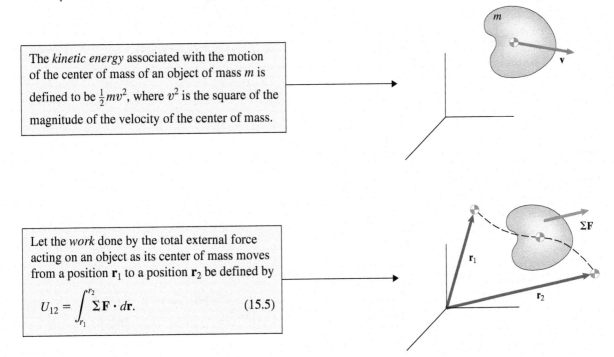

The *kinetic energy* associated with the motion of the center of mass of an object of mass m is defined to be $\tfrac{1}{2}mv^2$, where v^2 is the square of the magnitude of the velocity of the center of mass.

Let the *work* done by the total external force acting on an object as its center of mass moves from a position $\mathbf{r}_1$ to a position $\mathbf{r}_2$ be defined by

$$U_{12} = \int_{r_1}^{r_2} \Sigma\mathbf{F} \cdot d\mathbf{r}. \tag{15.5}$$

The *principle of work and energy* states that the work done on an object as it moves between two positions equals the change in its kinetic energy.

$$U_{12} = \frac{1}{2}mv_2^2 - \frac{1}{2}mv_1^2.$$ (15.6)

Evaluating the Work

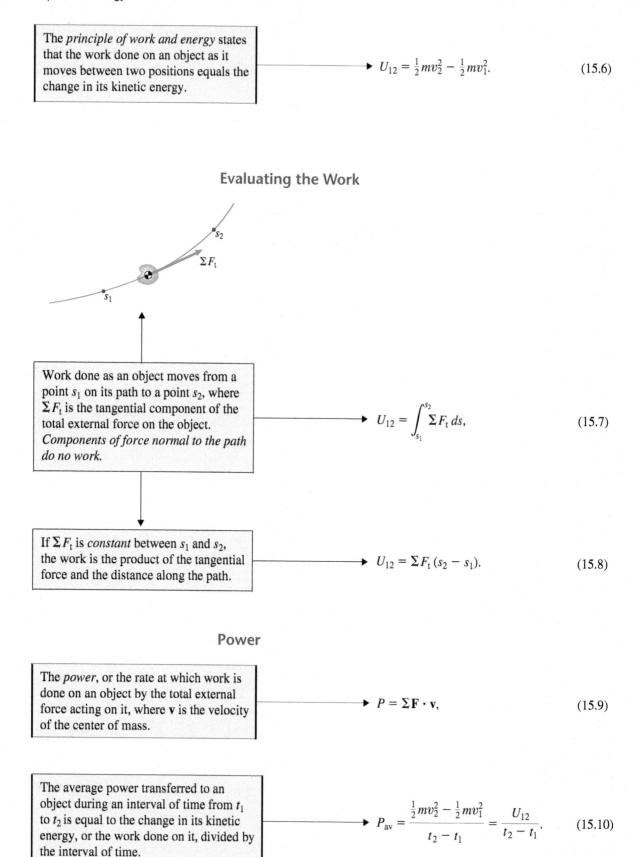

Work done as an object moves from a point s_1 on its path to a point s_2, where ΣF_t is the tangential component of the total external force on the object. *Components of force normal to the path do no work.*

$$U_{12} = \int_{s_1}^{s_2} \Sigma F_t \, ds,$$ (15.7)

If ΣF_t is *constant* between s_1 and s_2, the work is the product of the tangential force and the distance along the path.

$$U_{12} = \Sigma F_t \, (s_2 - s_1).$$ (15.8)

Power

The *power*, or the rate at which work is done on an object by the total external force acting on it, where **v** is the velocity of the center of mass.

$$P = \Sigma \mathbf{F} \cdot \mathbf{v},$$ (15.9)

The average power transferred to an object during an interval of time from t_1 to t_2 is equal to the change in its kinetic energy, or the work done on it, divided by the interval of time.

$$P_{av} = \frac{\frac{1}{2}mv_2^2 - \frac{1}{2}mv_1^2}{t_2 - t_1} = \frac{U_{12}}{t_2 - t_1}.$$ (15.10)

Active Example 15.1   **Work and Energy in Straight-Line Motion** (▶ *Related Problem 15.1*)

The 180-kg container A starts from rest in position $s = 0$. The horizontal force (in newtons) that is exerted on the container by the hydraulic piston is given as a function of the position s in meters by $F = 700 - 150s$. The coefficient of kinetic friction between the container and the floor is $\mu_k = 0.26$. What is the velocity of the container when it has reached the position $s = 1$ m?

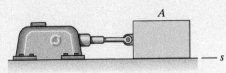

Strategy

The force acting on the container is given as a function of its position, so we can use Eq. (15.7) to determine the work done on it. By applying the principle of work and energy, we can determine the change in its velocity.

Solution

Draw the free-body diagram of the container and identify the forces that do work. The force exerted by the hydraulic cylinder and the friction force are tangent to the path. The normal force N is needed to calculate the friction force. The container has no acceleration in the vertical direction, so $N = (180 \text{ kg})(9.81 \text{ m/s}^2) = 1770$ N.

Evaluate the work done as the container moves from its initial position to $s = 1$ m.

$$U_{12} = \int_{s_1}^{s_2} \Sigma F_t \, ds$$

$$= \int_0^1 (F - \mu_k N) ds$$

$$= \int_0^1 [(700 - 150s) - (0.26)(1770)] ds$$

$$= 166 \text{ N-m.}$$

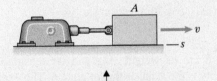

Apply the principle of work and energy to determine the container's velocity when it reaches $s = 1$ m. Solving yields $v_2 = 1.36$ m/s.

$$U_{12} = \tfrac{1}{2}mv_2^2 - \tfrac{1}{2}mv_1^2 :$$

$$166 \text{ N-m} = \tfrac{1}{2}(180 \text{ kg})v_2^2 - 0.$$

Practice Problem Suppose that the mass of the container A is 120 kg. What is its velocity when it has reached the position $s = 1$ m?

Answer: 2.31 m/s.

Example 15.2 **Applying Work and Energy to a System** (▶ *Related Problem 15.23*)

The two crates are released from rest. Their masses are $m_A = 40$ kg and $m_B = 30$ kg, and the kinetic coefficient of friction between crate A and the inclined surface is $\mu_k = 0.15$. What is the magnitude of the velocity of the crates when they have moved 400 mm?

Strategy

We will determine the velocity in two ways.

First Method By drawing free-body diagrams of each of the crates and applying the principle of work and energy to them individually, we can obtain two equations in terms of the magnitude of the velocity and the tension in the cable.

Second Method We can draw a single free-body diagram of the two crates, the cable, and the pulley and apply the principle of work and energy to the entire system.

Solution

First Method We draw the free-body diagram of crate A in Fig. a. The forces that do work as the crate moves down the plane are the forces tangential to its path: the tension T; the tangential component of the weight, $m_A g \sin 20°$; and the friction force $\mu_k N$. Because the acceleration of the crate normal to the surface is zero, $N = m_A g \cos 20°$. The magnitude v of the velocity at which A moves parallel to the surface equals the magnitude of the velocity at which B falls (Fig. b). Using Eq. (15.7) to determine the work, we equate the work done on A as it moves from $s_1 = 0$ to $s_2 = 0.4$ m to the change in the kinetic energy of A.

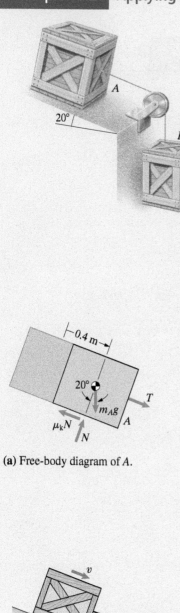

(a) Free-body diagram of A.

$$\int_{s_1}^{s_2} \Sigma F_t \, ds = \tfrac{1}{2} m v_2^2 - \tfrac{1}{2} m v_1^2:$$

$$\int_0^{0.4} [T + m_A g \sin 20° - \mu_k(m_A g \cos 20°)] \, ds = \tfrac{1}{2} m_A v_2^2 - 0. \quad (1)$$

The forces that do work on crate B are its weight $m_B g$ and the tension T (Fig. c). The magnitude of B's velocity is the same as that of crate A. The work done on B equals the change in its kinetic energy.

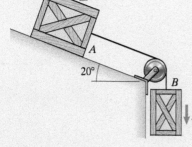

(b) The magnitude of the velocity of each crate is the same.

$$\int_{s_1}^{s_2} \Sigma F_t \, ds = \tfrac{1}{2} m v_2^2 - \tfrac{1}{2} m v_1^2:$$

$$\int_0^{0.4} (m_B g - T) \, ds = \tfrac{1}{2} m_B v_2^2 - 0. \quad (2)$$

By summing Eqs. (1) and (2), we eliminate T, obtaining

$$\int_0^{0.4} (m_A g \sin 20° - \mu_k m_A g \cos 20° + m_B g) \, ds = \tfrac{1}{2}(m_A + m_B)v_2^2:$$

$$[40 \sin 20° - (0.15)(40) \cos 20° + 30](9.81)(0.4) = \tfrac{1}{2}(40 + 30)v_2^2.$$

Solving for the velocity, we get $v_2 = 2.07$ m/s.

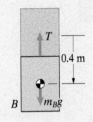

(c) Free-body diagram of B.

Second Method We draw the free-body diagram of the system consisting of the crates, cable, and pulley in Fig. d. Notice that the cable tension does not appear in this diagram. The reactions at the pin support of the pulley do no work, because the support does not move. The total work done by external forces on the system as the boxes move 400 mm is equal to the change in the total kinetic energy of the system.

$$\int_0^{0.4} \left[m_A g \sin 20° - \mu_k(m_A g \cos 20°) \right] ds + \int_0^{0.4} m_B g \, ds$$
$$= \tfrac{1}{2} m_A v_2^2 + \tfrac{1}{2} m_B v_2^2 - 0:$$

$$[40 \sin 20° - (0.15)(40) \cos 20° + 30](9.81)(0.4) = \tfrac{1}{2}(40 + 30)v_2^2.$$

This equation is identical to the one we obtained by applying the principle of work and energy to the individual crates.

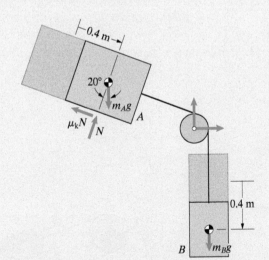

(d) Free-body diagram of the system.

Critical Thinking
You will often find it simpler to apply the principle of work and energy to an entire system instead of its separate parts. However, as we demonstrate in the next example, you need to be aware that internal forces in a system can do net work.

Example 15.3 | **Net Work by Internal Forces** (▶ *Related Problem 15.30*)

Crates A and B are released from rest. The coefficient of kinetic friction between A and B is μ_k, and friction between B and the inclined surface can be neglected. What is the velocity of the crates when they have moved a distance b?

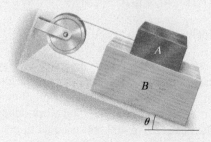

Strategy

By applying the principle of work and energy to each crate, we can obtain two equations in terms of the tension in the cable and the velocity.

Solution

We draw the free-body diagrams of the crates in Figs. a and b. The acceleration of A normal to the inclined surface is zero, so $N = m_A g \cos \theta$. The magnitudes of the velocities of A and B are equal (Fig. c). The work done on A equals the change in its kinetic energy.

$$U_{12} = \tfrac{1}{2}m_A v_2^2 - \tfrac{1}{2}m_A v_1^2:$$

$$\int_0^b (T - m_A g \sin \theta - \mu_k m_A g \cos \theta)\, ds = \tfrac{1}{2}m_A v_2^2. \tag{1}$$

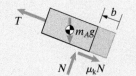

(a) Free-body diagram of A.

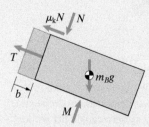

(b) Free-body diagram of B.

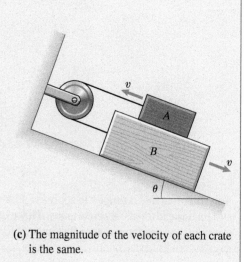

(c) The magnitude of the velocity of each crate is the same.

The work done on B equals the change in *its* kinetic energy.

$$U_{12} = \tfrac{1}{2}m_B v_2^2 - \tfrac{1}{2}m_B v_1^2:$$

$$\int_0^b (-T + m_B g \sin\theta - \mu_k m_A g \cos\theta)\, ds = \tfrac{1}{2}m_B v_2^2. \qquad (2)$$

Summing these equations to eliminate T and solving for v_2, we obtain

$$v_2 = \sqrt{2gb[(m_B - m_A)\sin\theta - 2\mu_k m_A \cos\theta]/(m_A + m_B)}.$$

Critical Thinking

If we attempt to solve this example by applying the principle of work and energy to the system consisting of the crates, the cable, and the pulley (Fig. d), we obtain an incorrect result. Equating the work done by external forces to the change in the total kinetic energy of the system, we obtain

$$\int_0^b m_B g \sin\theta\, ds - \int_0^b m_A g \sin\theta\, ds = \tfrac{1}{2}m_A v_2^2 + \tfrac{1}{2}m_B v_2^2:$$

$$(m_B g \sin\theta)b - (m_A g \sin\theta)b = \tfrac{1}{2}m_A v_2^2 + \tfrac{1}{2}m_B v_2^2.$$

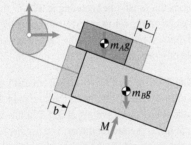

(d) Free-body diagram of the system.

But if we sum our work and energy equations for the individual crates—Eqs. (1) and (2)—we obtain the correct equation:

$$\underbrace{[(m_B g \sin\theta)b - (m_A g \sin\theta)b]}_{\substack{\text{Work done by}\\\text{external forces}}} + \underbrace{[-(2\mu_k m_A g \cos\theta)b]}_{\substack{\text{Work done by}\\\text{internal forces}}} = \tfrac{1}{2}m_A v_2^2 + \tfrac{1}{2}m_B v_2^2.$$

The internal frictional forces the crates exert on each other do net work on the system. We did not account for this work in applying the principle of work and energy to the free-body diagram of the entire system.

Problems

▶ **15.1** In Active Example 15.1, what is the velocity of the container when it has reached the position $s = 2$ m?

15.2 The mass of the Sikorsky UH-60A helicopter is 9300 kg. It takes off vertically with its rotor exerting a constant upward thrust of 112 kN. Use the principle of work and energy to determine how far it has risen when its velocity is 6 m/s.

   Strategy: Be sure to draw the free-body diagram of the helicopter.

Problem 15.2

15.3 The 20-lb box is at rest on the horizontal surface when the constant force $F = 5$ lb is applied. The coefficient of kinetic friction between the box and the surface is $\mu_k = 0.2$. Determine how fast the box is moving when it has moved 2 ft from its initial position (a) by applying Newton's second law; (b) by applying the principle of work and energy.

Problem 15.3

15.4 At the instant shown, the 30-lb box is moving up the smooth inclined surface at 2 ft/s. The constant force $F = 15$ lb. How fast will the box be moving when it has moved 1 ft up the surface from its present position?

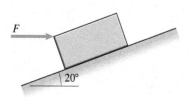

Problem 15.4

15.5 The 0.45-kg soccer ball is 1 m above the ground when it is kicked straight upward at 10 m/s. By using the principle of work and energy, determine: (a) how high above the ground the ball goes, (b) the magnitude of the ball's velocity when it falls back to a height of 1 m above the ground, (c) the magnitude of the ball's velocity immediately before it hits the ground.

15.6 Assume that the soccer ball is stationary the instant before it is kicked upward at 12 m/s. The duration of the kick is 0.02 s. What average power is transferred to the ball during the kick?

Problems 15.5/15.6

15.7 The 2000-lb drag racer starts from rest and travels a quarter-mile course. It completes the course in 4.524 seconds and crosses the finish line traveling at 325.77 mi/h. (a) How much work is done on the car as it travels the course? (b) Assume that the horizontal force exerted on the car is constant and use the principle of work and energy to determine it.

15.8 The 2000-lb drag racer starts from rest and travels a quarter-mile course. It completes the course in 4.524 seconds and crosses the finish line traveling at 325.77 mi/h. Assume that the horizontal force exerted on the car is constant. Determine (a) the maximum power and (b) the average power transferred to the car as it travels the quarter-mile course.

Problems 15.7/15.8

15.9 As the 32,000-lb airplane takes off, the tangential component of force exerted on it by its engines is $\Sigma F_t = 45,000$ lb. Neglecting other forces on the airplane, use the principle of work and energy to determine how much runway is required for its velocity to reach 200 mi/h.

15.10 As the 32,000-lb airplane takes off, the tangential component of force exerted on it by its engines is $\Sigma F_t = 45,000$ lb. Neglecting other forces on the airplane, determine (a) the maximum power and (b) the average power transferred to the airplane as its velocity increases from zero to 200 mi/h.

15.11 The 32,000-lb airplane takes off from rest in the position $s = 0$. The total tangential force exerted on it by its engines and aerodynamic drag (in pounds) is given as a function of its position s by $\Sigma F_t = 45,000 - 5.2s$. Use the principle of work and energy to determine how fast the airplane is traveling when its position is $s = 950$ ft.

Problems 15.9–15.11

15.12 The spring ($k = 20$ N/m) is unstretched when $s = 0$. The 5-kg cart is moved to the position $s = -1$ m and released from rest. What is the magnitude of its velocity when it is in the position $s = 0$?

15.13 The spring ($k = 20$ N/m) is unstretched when $s = 0$. The 5-kg cart is moved to the position $s = -1$ m and released from rest. What maximum distance down the sloped surface does the cart move relative to its initial position?

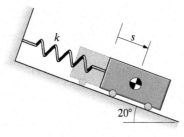

Problems 15.12/15.13

15.14 The force exerted on a car by a prototype crash barrier as the barrier crushes is $F = -(120s + 40s^3)$ lb, where s is the distance in feet from the initial contact. The effective length of the barrier is 18 ft. How fast can a 5000-lb car be moving and be brought to rest within the effective length of the barrier?

15.15 A 5000-lb car hits the crash barrier at 80 mi/h and is brought to rest in 0.11 seconds. What average power is transferred from the car during the impact?

Problems 15.14/15.15

15.16 A group of engineering students constructs a sun-powered car and tests it on a circular track with a 1000-ft radius. The car, with a weight of 460 lb including its occupant, starts from rest. The total tangential component of force on the car is

$$\Sigma F_t = 30 - 0.2s \text{ lb,}$$

where s is the distance (in ft) the car travels along the track from the position where it starts.

(a) Determine the work done on the car when it has gone a distance $s = 120$ ft.

(b) Determine the magnitude of the *total* horizontal force exerted on the car's tires by the road when it is at the position $s = 120$ ft.

15.17 At the instant shown, the 160-lb vaulter's center of mass is 8.5 ft above the ground, and the vertical component of his velocity is 4 ft/s. As his pole straightens, it exerts a vertical force on the vaulter of magnitude $180 + 2.8y^2$ lb, where y is the vertical position of his center of mass *relative to its position at the instant shown*. This force is exerted on him from $y = 0$ to $y = 4$ ft, when he releases the pole. What is the maximum height above the ground reached by the vaulter's center of mass?

Problem 15.17

15.18 The springs ($k = 25$ lb/ft) are unstretched when $s = 0$. The 50-lb weight is released from rest in the position $s = 0$.

(a) When the weight has fallen 1 ft, how much work has been done on it by each spring?

(b) What is the magnitude of the velocity of the weight when it has fallen 1 ft?

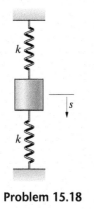

Problem 15.18

15.19 The coefficients of friction between the 160-kg crate and the ramp are $\mu_s = 0.3$ and $\mu_k = 0.28$.

(a) What tension T_0 must the winch exert to start the crate moving up the ramp?

(b) If the tension remains at the value T_0 after the crate starts sliding, what total work is done on the crate as it slides a distance $s = 3$ m up the ramp, and what is the resulting velocity of the crate?

15.20 In Problem 15.19, if the winch exerts a tension $T = T_0(1 + 0.1s)$ after the crate starts sliding, what total work is done on the crate as it slides a distance $s = 3$ m up the ramp, and what is the resulting velocity of the crate?

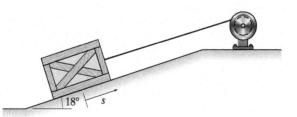

Problems 15.19/15.20

15.21 The 200-mm-diameter gas gun is evacuated on the right of the 8-kg projectile. On the left of the projectile, the tube contains gas with pressure $p_0 = 1 \times 10^5$ Pa (N/m^2). The force F is slowly increased, moving the projectile 0.5 m to the left from the position shown. The force is then removed, and the projectile accelerates to the right. If you neglect friction and assume that the pressure of the gas is related to its volume by $pV = $ constant, what is the velocity of the projectile when it has returned to its original position?

15.22 In Problem 15.21, if you assume that the pressure of the gas is related to its volume by $pV = $ constant while the gas is compressed (an isothermal process) and by $pV^{1.4} = $ constant while it is expanding (an isentropic process), what is the velocity of the projectile when it has returned to its original position?

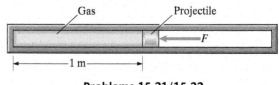

Problems 15.21/15.22

▶ 15.23 In Example 15.2, suppose that the angle between the inclined surface and the horizontal is increased from 20° to 30°. What is the magnitude of the velocity of the crates when they have moved 400 mm?

15.24 The system is released from rest. The 4-kg mass slides on the smooth horizontal surface. By using the principle of work and energy, determine the magnitude of the velocity of the masses when the 20-kg mass has fallen 1 m.

15.25 Solve Problem 15.24 if the coefficient of kinetic friction between the 4-kg mass and the horizontal surface is $\mu_k = 0.4$.

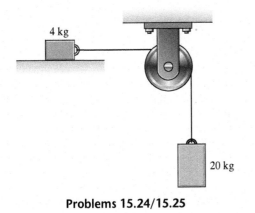

4 kg

20 kg

Problems 15.24/15.25

15.26 Each box weighs 50 lb and the inclined surfaces are smooth. The system is released from rest. Determine the magnitude of the velocities of the boxes when they have moved 1 ft.

15.27 Solve Problem 15.26 if the coefficient of kinetic friction between the boxes and the inclined surfaces is $\mu_k = 0.05$.

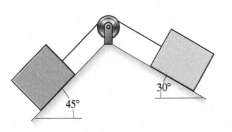

30°

45°

Problems 15.26/15.27

15.28 The masses of the three blocks are $m_A = 40$ kg, $m_B = 16$ kg, and $m_C = 12$ kg. Neglect the mass of the bar holding C in place. Friction is negligible. By applying the principle of work and energy to A and B individually, determine the magnitude of their velocity when they have moved 500 mm.

15.29 Solve Problem 15.28 by applying the principle of work and energy to the system consisting of A, B, the cable connecting them, and the pulley.

▶ 15.30 The masses of the three blocks are $m_A = 40$ kg, $m_B = 16$ kg, and $m_C = 12$ kg. The coefficient of kinetic friction between all surfaces is $\mu_k = 0.1$. Determine the magnitude of the velocity of blocks A and B when they have moved 500 mm. (See Example 15.3.)

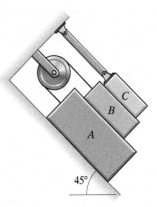

C

B

A

45°

Problems 15.28–15.30

15.2 Work Done by Particular Forces

BACKGROUND

We have seen that if the tangential component of the total external force on an object is known as a function of distance along the object's path, the principle of work and energy can be used to relate a change in the position of the object to the change in its velocity. For certain types of forces, however, not only can we determine the work without knowing the tangential component of the force as a function of distance along the path, but we don't even need to know the path. Two important examples are weight and the force exerted by a spring.

Weight

To evaluate the work done by an object's weight, we orient a cartesian coordinate system with the y axis upward and suppose that the object moves from position 1 with coordinates (x_1, y_1, z_1) to position 2 with coordinates (x_2, y_2, z_2) (Fig. 15.3a). The force exerted by the object's weight is $\mathbf{F} = -mg\mathbf{j}$. (Other forces may act on the object, but we are concerned only with the work done by its weight.) Because $\mathbf{v} = d\mathbf{r}/dt$, we can multiply the velocity, expressed in cartesian coordinates, by dt to obtain an expression for the vector $d\mathbf{r}$:

$$d\mathbf{r} = \left(\frac{dx}{dt}\mathbf{i} + \frac{dy}{dt}\mathbf{j} + \frac{dz}{dt}\mathbf{k}\right) dt = dx\,\mathbf{i} + dy\,\mathbf{j} + dz\,\mathbf{k}.$$

Taking the dot product of $\mathbf{F}$ and $d\mathbf{r}$ yields

$$\mathbf{F} \cdot d\mathbf{r} = (-mg\mathbf{j}) \cdot (dx\,\mathbf{i} + dy\,\mathbf{j} + dz\,\mathbf{k}) = -mg\,dy.$$

The work done as the object moves from position 1 to position 2 reduces to an integral with respect to y:

$$U_{12} = \int_{\mathbf{r}_1}^{\mathbf{r}_2} \mathbf{F} \cdot d\mathbf{r} = \int_{y_1}^{y_2} -mg\,dy.$$

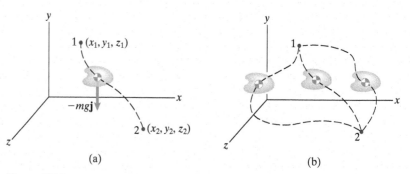

(a) (b)

Figure 15.3
(a) An object moving between two positions.
(b) The work done by the weight is the same for any path.

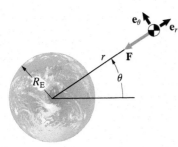

Figure 15.4
Expressing an object's weight
in polar coordinates.

Evaluating the integral, we obtain the work done by the weight of an object as
it moves between two positions:

$$U_{12} = -mg(y_2 - y_1). \tag{15.11}$$

The work is simply the product of the weight and the change in the object's
height. The work done is negative if the height increases and positive if it decreases. Notice that *the work done is independent of the path the object follows
from position* 1 *to position* 2 (Fig. 15.3b). Thus, we don't need to know the path
to determine the work done by an object's weight—we only need to know the
relative heights of the initial and final positions.

What work is done by an object's weight if we account for its variation
with distance from the center of the earth? In terms of polar coordinates, we
can write the weight of an object at a distance r from the center of the earth as
(Fig. 15.4)

$$\mathbf{F} = -\frac{mgR_E^2}{r^2}\mathbf{e}_r.$$

Using the expression for the velocity in polar coordinates, we obtain, for the
vector $d\mathbf{r} = \mathbf{v}\,dt$,

$$d\mathbf{r} = \left(\frac{dr}{dt}\mathbf{e}_r + r\frac{d\theta}{dt}\mathbf{e}_\theta\right)dt = dr\,\mathbf{e}_r + r\,d\theta\,\mathbf{e}_\theta. \tag{15.12}$$

The dot product of $\mathbf{F}$ and $d\mathbf{r}$ is

$$\mathbf{F}\cdot d\mathbf{r} = \left(-\frac{mgR_E^2}{r^2}\mathbf{e}_r\right)\cdot(dr\,\mathbf{e}_r + r\,d\theta\,\mathbf{e}_\theta) = -\frac{mgR_E^2}{r^2}dr,$$

so the work reduces to an integral with respect to r:

$$U_{12} = \int_{\mathbf{r}_1}^{\mathbf{r}_2} \mathbf{F}\cdot d\mathbf{r} = \int_{r_1}^{r_2} -\frac{mgR_E^2}{r^2}dr.$$

Evaluating the integral, we obtain the work done by an object's weight,
accounting for the variation of the weight with height:

$$U_{12} = mgR_E^2\left(\frac{1}{r_2} - \frac{1}{r_1}\right). \tag{15.13}$$

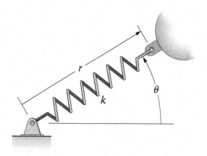
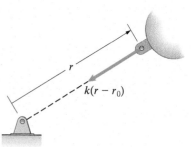

Figure 15.5
Expressing the force exerted by a linear spring in polar coordinates.

Again, the work is independent of the path from position 1 to position 2. To evaluate it, we only need to know the object's radial distance from the center of the earth at the two positions.

Springs

Suppose that a linear spring connects an object to a fixed support. In terms of polar coordinates (Fig. 15.5), the force exerted on the object is

$$\mathbf{F} = -k(r - r_0)\mathbf{e}_r,$$

where k is the spring constant and r_0 is the unstretched length of the spring. Using Eq. (15.12), we get the dot product of $\mathbf{F}$ and $d\mathbf{r}$:

$$\mathbf{F} \cdot d\mathbf{r} = [-k(r - r_0)\mathbf{e}_r] \cdot (dr\,\mathbf{e}_r + r\,d\theta\,\mathbf{e}_\theta) = -k(r - r_0)\,dr.$$

It is convenient to express the work done by a spring in terms of its *stretch*, defined by $S = r - r_0$. (Although the word *stretch* usually means an increase in length, we use the term more generally to denote the change in length of the spring. A negative stretch is a decrease in length.) In terms of this variable, $\mathbf{F} \cdot d\mathbf{r} = -kS\,dS$, and the work is

$$U_{12} = \int_{\mathbf{r}_1}^{\mathbf{r}_2} \mathbf{F} \cdot d\mathbf{r} = \int_{S_1}^{S_2} -kS\,dS.$$

The work done on an object by a spring attached to a fixed support is

$$U_{12} = -\tfrac{1}{2}k(S_2^2 - S_1^2), \tag{15.14}$$

where S_1 and S_2 are the values of the stretch at the initial and final positions. We don't need to know the object's path to determine the work done by the spring. Remember, however, that Eq. (15.14) applies only to a linear spring. In Fig. 15.6, we determine the work done in stretching a linear spring by calculating the area defined by the graph of the force as a function of S.

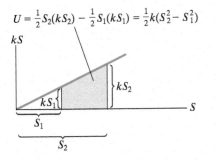

$$U = \tfrac{1}{2}S_2(kS_2) - \tfrac{1}{2}S_1(kS_1) = \tfrac{1}{2}k(S_2^2 - S_1^2)$$

Figure 15.6
Work done in stretching a linear spring from S_1 to S_2. (If $S_2 > S_1$, the work done *on* the spring is positive, so the work done *by* the spring is negative.)

RESULTS

<table>
<tr><td>

For some types of forces, the work done during a motion from a position 1 to a position 2 can be determined easily. Notice that *the work is independent of the path from 1 to 2.*

</td><td>

Weight
When the weight can be regarded as constant, the work is
$$U_{12} = -mg(y_2 - y_1), \quad (15.11)$$
where the positive y axis points upwards. *The work is the product of the weight and the change in height. It is negative if the height increases and positive if it decreases.*

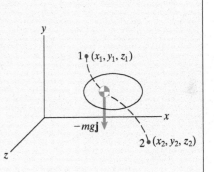

Variable Weight
when the variation of gravity with height must be considered, the work is
$$U_{12} = mgR_E^2\left(\frac{1}{r_2} - \frac{1}{r_1}\right), \quad (15.13)$$
where R_E is the radius of the earth.

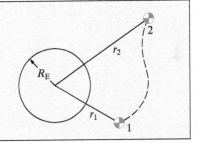

Springs
The work done on an object by a linear spring is
$$U_{12} = -\tfrac{1}{2}k(S_2^2 - S_1^2), \quad (15.14)$$
where S_1 and S_2 are the values of the stretch of the spring at the initial and final positions.

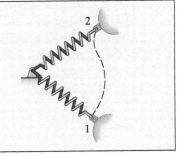

</td></tr>
</table>

Active Example 15.4 | Work Done by Weight and Springs (▶ *Related Problem 15.49*)

The 40-kg hammer is lifted to position 1 and released from rest. It falls and strikes a workpiece when it is in position 2. The spring constant is $k = 1500$ N/m, and the springs are unstretched when the hammer is in position 2. Neglect friction. What is the velocity of the hammer just before it strikes the workpiece?

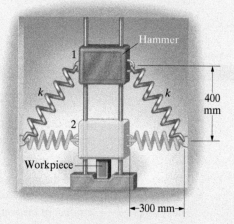

Strategy

Work is done on the hammer by its weight and by the forces exerted on it by the springs. We can apply the principle of work and energy to the motion of the hammer from position 1 to position 2 to determine the velocity at position 2.

Solution

Calculate the work done by the weight:
The hammer falls downward, so the work is positive, and its magnitude is the product of the weight and the change in height.

$$
\begin{cases}
U_{\text{weight}} = (\text{weight})(\text{change in height}) \\
\qquad = [(40 \text{ kg})(9.81 \text{ m/s}^2)](0.4 \text{ m}) \\
\qquad = 157 \text{ N-m.}
\end{cases}
$$

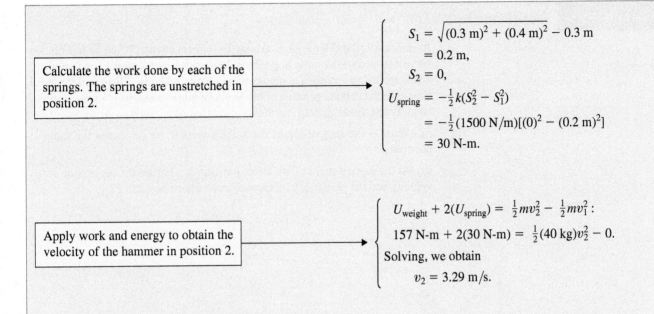

$$\begin{cases} S_1 = \sqrt{(0.3 \text{ m})^2 + (0.4 \text{ m})^2} - 0.3 \text{ m} \\ \qquad = 0.2 \text{ m}, \\ S_2 = 0, \\ U_{\text{spring}} = -\frac{1}{2} k(S_2^2 - S_1^2) \\ \qquad = -\frac{1}{2}(1500 \text{ N/m})[(0)^2 - (0.2 \text{ m})^2] \\ \qquad = 30 \text{ N-m}. \end{cases}$$

Calculate the work done by each of the springs. The springs are unstretched in position 2.

$$\begin{cases} U_{\text{weight}} + 2(U_{\text{spring}}) = \frac{1}{2} mv_2^2 - \frac{1}{2} mv_1^2 : \\ 157 \text{ N-m} + 2(30 \text{ N-m}) = \frac{1}{2}(40 \text{ kg})v_2^2 - 0. \\ \text{Solving, we obtain} \\ \qquad v_2 = 3.29 \text{ m/s}. \end{cases}$$

Apply work and energy to obtain the velocity of the hammer in position 2.

Practice Problem The 40-kg hammer is given a downward velocity of 2 m/s in position 1. It falls and strikes a workpiece when it is in position 2. The spring constant is $k = 1500$ N/m, and the springs are unstretched when the hammer is in position 1. Neglect friction. What is the velocity of the hammer just before it strikes the workpiece?

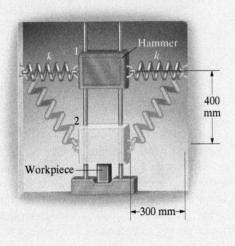

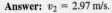

Answer: $v_2 = 2.97$ m/s.

Example 15.5   **Work Done by Weight** (▶ *Related Problem 15.31*)

At position 1, the skier is approaching his jump at 15 m/s. When he reaches the horizontal end of the ramp at position 2, 20 m below position 1, he jumps upward, achieving a vertical component of velocity of 3 m/s. (Disregard the small change in the vertical position of his center of mass due to his jumping motion.) Neglect aerodynamic drag and the frictional forces on his skis.

(a) What is the magnitude of the skier's velocity as he leaves the ramp at position 2?

(b) At the highest point of his jump, position 3, what are the magnitude of his velocity and the height of his center of mass above position 2?

Strategy

(a) If we neglect aerodynamic and frictional forces, the only force doing work from position 1 to position 2 is the skier's weight. The normal force exerted on his skis by the ramp does no work because it is perpendicular to his path. We need to know only the change in the skier's height from position 1 to position 2 to determine the work done by his weight, so we can apply the principle of work and energy to determine his velocity at position 2 before he jumps.

(b) From the time he leaves the ramp at position 2 until he reaches position 3, the only force acting on the skier is his weight, so the horizontal component of his velocity is constant. This means that we know the magnitude of his velocity at position 3, because he is moving horizontally at that point. Therefore, we can apply the principle of work and energy to his motion from position 2 to position 3 to determine his height above position 2.

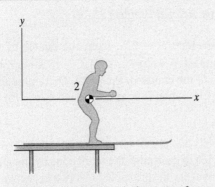

(a) The height of the skier's center of mass
is measured relative to position 2.

Solution

(a) We will use Eq. (15.11) to evaluate the work done by the skier's weight, measuring the height of his center of mass relative to position 2 (Fig. a). The principle of work and energy from position 1 to position 2 is

$$U_{12} = -mg(y_2 - y_1) = \tfrac{1}{2}mv_2^2 - \tfrac{1}{2}mv_1^2:$$
$$-m(9.81)(0 - 20) = \tfrac{1}{2}mv_2^2 - \tfrac{1}{2}m(15)^2.$$

Solving for v_2, we find that the skier's horizontal velocity at position 2 before he jumps upward is 24.8 m/s. After he jumps upward, the magnitude of his velocity at position 2 is $v_2' = \sqrt{(24.8)^2 + (3)^2} = 25.0$ m/s.

(b) The magnitude of the skier's velocity at position 3 is equal to the horizontal component of his velocity at position 2: $v_3 = v_2 = 24.8$ m/s. Applying work and energy to his motion from position 2 to position 3, we obtain

$$U_{23} = -mg(y_3 - y_2) \quad = \tfrac{1}{2}mv_3^2 - \tfrac{1}{2}m(v_2')^2:$$
$$-m(9.81)(y_3 - 0) = \tfrac{1}{2}m(24.8)^2 - \tfrac{1}{2}m(25.0)^2,$$

from which it follows that $y_3 = 0.459$ m.

Critical Thinking

Why didn't we need to include the effect of the normal force exerted on the skier by the ramp? The reason is that *it is perpendicular to his path and so does no work*. To obtain an accurate prediction of the skier's motion, we would need to account for the friction force exerted by the ramp and aerodynamic forces. Nevertheless, our approximate analysis in this example provides useful insight, showing how the work done by gravity as he descends increases his kinetic energy. Notice that the work done by gravity is determined by his change in height, not the length of his path.

| Example 15.6 | **Work Done by the Earth's Gravity** (▶ *Related Problem 15.74*) |

A spacecraft at a distance $r_1 = 2R_E$ from the center of the earth has a velocity of magnitude $v_1 = \sqrt{2gR_E/3}$ relative to a nonrotating reference frame with its origin at the center of the earth. Determine the magnitude of the spacecraft's velocity when it is at a distance $r_2 = 4R_E$ from the center of the earth.

Strategy

By applying Eq. (15.13) to determine the work done by the gravitational force on the spacecraft, we can use the principle of work and energy to determine the magnitude of the spacecraft's velocity.

Solution

From Eq. (15.13), the work done by gravity as the spacecraft moves from a distance r_1 from the center of the earth to a distance r_2 is

$$U_{12} = mgR_E^2\left(\frac{1}{r_2} - \frac{1}{r_1}\right).$$

Let v_2 be the magnitude of the velocity of the spacecraft when it is at a distance r_2 from the center of the earth. Applying the principle of work and energy yields

$$U_{12} = mgR_E^2\left(\frac{1}{r_2} - \frac{1}{r_1}\right) = \tfrac{1}{2}mv_2^2 - \tfrac{1}{2}mv_1^2.$$

We solve for v_2, obtaining

$$v_2 = \sqrt{v_1^2 + 2gR_E^2\left(\frac{1}{r_2} - \frac{1}{r_1}\right)}$$

$$= \sqrt{\left(\frac{2gR_E}{3}\right) + 2gR_E^2\left(\frac{1}{4R_E} - \frac{1}{2R_E}\right)}$$

$$= \sqrt{\frac{gR_E}{6}}.$$

The velocity $v_2 = v_1/2$.

Critical Thinking

Notice that we did not need to specify the direction of the spacecraft's initial velocity to determine the magnitude of its velocity at a different distance from the center of the earth. This illustrates the power of the principle of work and energy, as well as one of its limitations. Even if we know the direction of the initial velocity, the principle of work and energy tells us only the *magnitude* of the velocity at a different distance.

Problems

▶ **15.31** In Example 15.5, suppose that the skier is moving at 20 m/s when he is in position 1. Determine the horizontal component of his velocity when he reaches position 2, 20 m below position 1.

15.32 Suppose that you stand at the edge of a 200-ft cliff and throw rocks at 30 ft/s in the three directions shown. Neglecting aerodynamic drag, use the principle of work and energy to determine the magnitude of the velocity of the rock just before it hits the ground in each case.

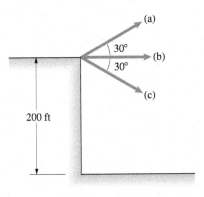

Problem 15.32

15.33 The 30-kg box is sliding down the smooth surface at 1 m/s when it is in position 1. Determine the magnitude of the box's velocity at position 2 in each case.

15.34 Solve Problem 15.33 if the coefficient of kinetic friction between the box and the inclined surface is $\mu_k = 0.2$.

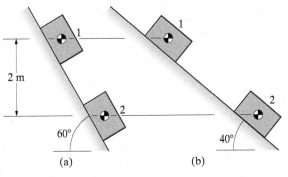

Problems 15.33/15.34

15.35 In case (a), a 5-lb ball is released from rest at position 1 and falls to position 2. In case (b), the ball is released from rest at position 1 and swings to position 2. For each case, use the principle of work and energy to determine the magnitude of the ball's velocity at position 2. [In case (b), notice that the force exerted on the ball by the string is perpendicular to the ball's path.]

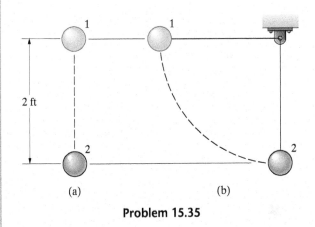

Problem 15.35

15.36 The 2-kg ball is released from rest in position 1 with the string horizontal. The length of the string is $L = 1$ m. What is the magnitude of the ball's velocity when it is in position 2?

15.37 The 2-kg ball is released from rest in position 1 with the string horizontal. The length of the string is $L = 1$ m. What is the tension in the string when the ball is in position 2?

Strategy: Draw the free-body diagram of the ball when it is in position 2 and write Newton's second law in terms of normal and tangential components.

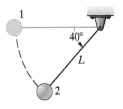

Problems 15.36/15.37

15.38 The 400-lb wrecker's ball swings at the end of a 25-ft cable. If the magnitude of the ball's velocity at position 1 is 4 ft/s, what is the magnitude of its velocity just before it hits the wall at position 2?

15.39 The 400-lb wrecker's ball swings at the end of a 25-ft cable. If the magnitude of the ball's velocity at position 1 is 4 ft/s, what is the maximum tension in the cable as the ball swings from position 1 to position 2?

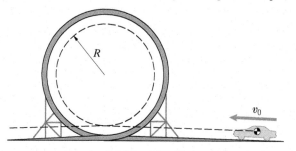

Problems 15.38/15.39

15.40 A stunt driver wants to drive a car through the circular loop of radius $R = 5$ m. Determine the minimum velocity v_0 at which the car can enter the loop and coast through without losing contact with the track. What is the car's velocity at the top of the loop?

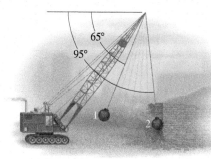

Problem 15.40

15.41 The 2-kg collar starts from rest at position 1 and slides down the smooth rigid wire. The y axis points upward. What is the magnitude of the velocity of the collar when it reaches position 2?

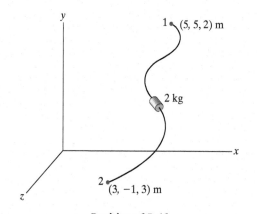

Problem 15.41

15.42 The 4-lb collar slides down the smooth rigid wire from position 1 to position 2. When it reaches position 2, the magnitude of its velocity is 24 ft/s. What was the magnitude of its velocity at position 1?

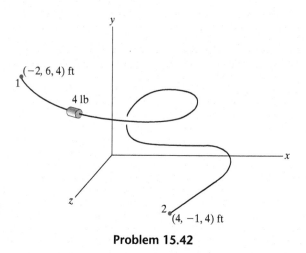

Problem 15.42

15.43 The forces acting on the 28,000-lb airplane are the thrust T and drag D, which are parallel to the airplane's path; the lift L, which is perpendicular to the path; and the weight W. The airplane climbs from an altitude of 3000 ft to an altitude of 10,000 ft. During the climb, the magnitude of its velocity decreases from 800 ft/s to 600 ft/s.

(a) What work is done on the airplane by its lift during the climb?

(b) What work is done by the thrust and drag combined?

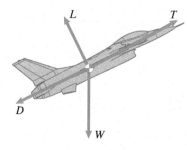

Problem 15.43

15.44 The 2400-lb car is traveling 40 mi/h at position 1. If the combined effect of the aerodynamic drag on the car and the tangential force exerted on its wheels by the road is that they exert no net tangential force on the car, what is the magnitude of the car's velocity at position 2?

15.45 The 2400-lb car is traveling 40 mi/h at position 1. If the combined effect of the aerodynamic drag on the car and the tangential force exerted on its wheels by the road is that they exert a constant 400-lb tangential force on the car in the direction of its motion, what is the magnitude of the car's velocity at position 2?

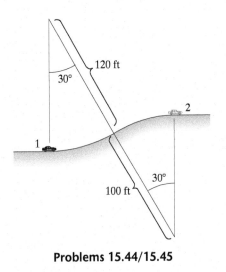

Problems 15.44/15.45

15.46 The mass of the rocket is 250 kg. Its engine has a constant thrust of 45 kN. The length of the launching ramp is 10 m. If the magnitude of the rocket's velocity when it reaches the end of the ramp is 52 m/s, how much work is done on the rocket by friction and aerodynamic drag?

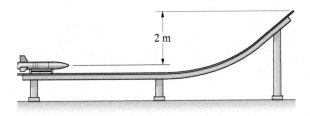

Problem 15.46

15.47 A bioengineer interested in the energy requirements of sports determines from videotape that when the athlete begins his motion to throw the 7.25-kg shot (Fig. P15.47a), the shot is stationary and 1.50 m above the ground. At the instant the athlete releases it (Fig. P15.47b), the shot is 2.10 m above the ground. The shot reaches a maximum height of 4.60 m above the ground and travels a horizontal distance of 18.66 m from the point where it was released. How much work does the athlete do on the shot from the beginning of his motion to the instant he releases it?

Problem 15.47

15.48 A small pellet of mass $m = 0.2$ kg starts from rest at position 1 and slides down the smooth surface of the cylinder to position 2, where $\theta = 30°$.

(a) What work is done on the pellet as it slides from position 1 to position 2?

(b) What is the magnitude of the pellet's velocity at position 2?

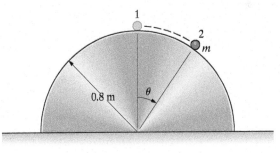

Problem 15.48

▶ **15.49** In Active Example 15.4, suppose that you want to increase the value of the spring constant k so that the velocity of the hammer just before it strikes the workpiece is 4 m/s. What is the required value of k?

15.50 Suppose that you want to design a bumper that will bring a 50-lb package moving at 10 ft/s to rest 6 in from the point of contact with the bumper. If friction is negligible, what is the necessary spring constant k?

15.51 In Problem 15.50, what spring constant is necessary if the coefficient of kinetic friction between the package and the floor is $\mu_k = 0.3$ and the package contacts the bumper moving at 10 ft/s?

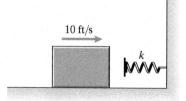

10 ft/s

k

Problems 15.50/15.51

15.52 The 50-lb package starts from rest, slides down the smooth ramp, and is stopped by the spring.

(a) If you want the package to be brought to rest 6 in from the point of contact, what is the necessary spring constant k?

(b) What maximum deceleration is the package subjected to?

15.53 The 50-lb package starts from rest, slides down the ramp, and is stopped by the spring. The coefficient of kinetic friction between the package and the ramp is $\mu_k = 0.12$. If you want the package to be brought to rest 6 in from the point of contact, what is the necessary spring constant k?

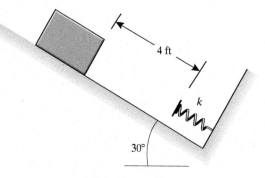

4 ft

k

30°

Problems 15.52/15.53

15.54 The system is released from rest with the spring un-stretched. The spring constant is $k = 200$ N/m. Determine the magnitude of the velocity of the masses when the right mass has fallen 1 m.

15.55 The system is released from rest with the spring unstretched. The spring constant is $k = 200$ N/m. What maximum downward velocity does the right mass attain as it falls?

4 kg

20 kg

k

Problems 15.54/15.55

15.56 The system is released from rest. The 4-kg mass slides on the smooth horizontal surface. The spring constant is $k = 100$ N/m, and the tension in the spring when the system is released is 50 N. By using the principle of work and energy, determine the magnitude of the velocity of the masses when the 20-kg mass has fallen 1 m.

15.57 Solve Problem 15.56 if the coefficient of kinetic friction between the 4-kg mass and the horizontal surface is $\mu_k = 0.4$.

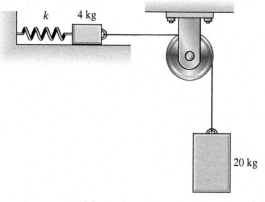

k 4 kg

20 kg

Problems 15.56/15.57

15.58 The 40-lb crate is released from rest on the smooth inclined surface with the spring unstretched. The spring constant is $k = 8$ lb/ft.

(a) How far down the inclined surface does the crate slide before it stops?

(b) What maximum velocity does the crate attain on its way down?

15.59 Solve Problem 15.58 if the coefficient of kinetic friction between the 4-kg mass and the horizontal surface is $\mu_k = 0.2$.

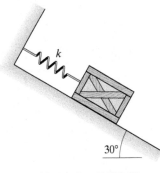

Problems 15.58/15.59

15.60 The 4-kg collar starts from rest in position 1 on the smooth bar with the spring unstretched. The spring constant is $k = 100$ N/m. How far does the collar fall relative to position 1?

15.61 In position 1 on the smooth bar, the 4-kg collar has a downward velocity of 1 m/s and the spring is unstretched. The spring constant is $k = 100$ N/m. What maximum downward velocity does the collar attain as it falls?

15.62 The 4-kg collar starts from rest in position 1 on the smooth bar. The tension in the spring in position 1 is 20 N. The spring constant is $k = 100$ N/m. How far does the collar fall relative to position 1?

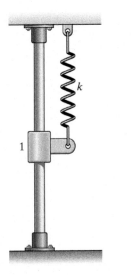

Problems 15.60–15.62

15.63 The 4-kg collar is released from rest in position 1 on the smooth bar. If the spring constant is $k = 6$ kN/m and the spring is unstretched in position 2, what is the velocity of the collar when it has fallen to position 2?

15.64 The 4-kg collar is released from rest in position 1 on the smooth bar. The spring constant is $k = 4$ kN/m. The tension in the spring in position 2 is 500 N. What is the velocity of the collar when it has fallen to position 2?

15.65 The 4-kg collar starts from rest in position 1 on the smooth bar. Its velocity when it has fallen to position 2 is 4 m/s. The spring is unstretched when the collar is in position 2. What is the spring constant k?

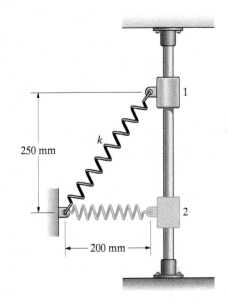

Problems 15.63–15.65

15.66 The 10-kg collar starts from rest at position 1 and slides along the smooth bar. The y axis points upward. The spring constant is $k = 100$ N/m, and the unstretched length of the spring is 2 m. What is the velocity of the collar when it reaches position 2?

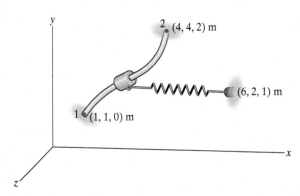

Problem 15.66

15.67 A spring-powered mortar is used to launch 10-lb packages of fireworks into the air. The package starts from rest with the spring compressed to a length of 6 in. The unstretched length of the spring is 30 in. If the spring constant is $k = 1300$ lb/ft, what is the magnitude of the velocity of the package as it leaves the mortar?

15.68 Suppose you want to design the mortar in Problem 15.67 to throw the package to a height of 150 ft above its initial position. Neglecting friction and drag, determine the necessary spring constant.

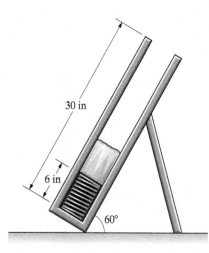

Problems 15.67/15.68

15.69 Suppose an object has a string or cable with *constant* tension T attached as shown. The force exerted on the object can be expressed in terms of polar coordinates as $\mathbf{F} = -T\mathbf{e}_r$. Show that the work done on the object as it moves along an *arbitrary* plane path from a radial position r_1 to a radial position r_2 is $U_{12} = -T(r_2 - r_1)$.

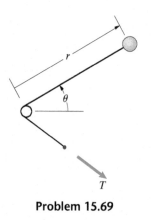

Problem 15.69

15.70 The 2-kg collar is initially at rest at position 1. A constant 100-N force is applied to the string, causing the collar to slide up the smooth vertical bar. What is the velocity of the collar when it reaches position 2? (See Problem 15.69.)

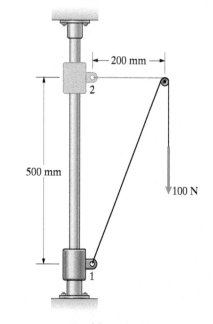

Problem 15.70

15.71 The 10-kg collar starts from rest at position 1. The tension in the string is 200 N, and the y axis points upward. If friction is negligible, what is the magnitude of the velocity of the collar when it reaches position 2? (See Problem 15.69.)

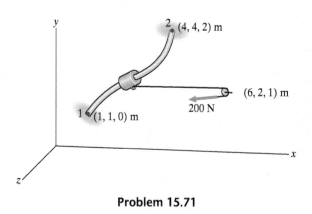

Problem 15.71

15.72 As the F/A-18 lands at 210 ft/s, the cable from A to B engages the airplane's arresting hook at C. The arresting mechanism maintains the tension in the cable at a constant value, bringing the 26,000-lb airplane to rest at a distance of 72 ft. What is the tension in the cable? (See Problem 15.69.)

15.73 If the airplane in Problem 15.72 lands at 240 ft/s, what distance does it roll before the arresting system brings it to rest?

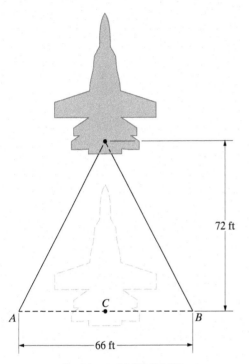

72 ft

A C B

66 ft

Problems 15.72/15.73

▶ **15.74** A spacecraft 320 km above the surface of the earth is moving at escape velocity $v_{esc} = 10,900$ m/s. What is its distance from the center of the earth when its velocity is 50 percent of its initial value? The radius of the earth is 6370 km. (See Example 15.6.)

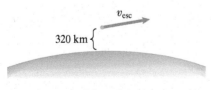

v_{esc}

320 km

Problem 15.74

15.75 A piece of ejecta is thrown up by the impact of a meteor on the moon. When it is 1000 km above the moon's surface, the magnitude of its velocity (relative to a nonrotating reference frame with its origin at the center of the moon) is 200 m/s. What is the magnitude of its velocity just before it strikes the moon's surface? The acceleration due to gravity at the surface of the moon is 1.62 m/s^2. The moon's radius is 1738 km.

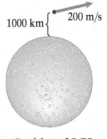

1000 km   200 m/s

Problem 15.75

15.76 A satellite in a circular orbit of radius r around the earth has velocity $v = \sqrt{gR_E^2/r}$, where $R_E = 6370$ km is the radius of the earth. Suppose you are designing a rocket to transfer a 900-kg communication satellite from a circular parking orbit with 6700-km radius to a circular geosynchronous orbit with 42,222-km radius. How much work must the rocket do on the satellite?

15.77 The force exerted on a charged particle by a magnetic field is

$$\mathbf{F} = q\mathbf{v} \times \mathbf{B},$$

where q and $\mathbf{v}$ are the charge and velocity of the particle and $\mathbf{B}$ is the magnetic field vector. Suppose that other forces on the particle are negligible. Use the principle of work and energy to show that the magnitude of the particle's velocity is constant.

15.3 Potential Energy and Conservative Forces

BACKGROUND

Potential Energy

The work done on an object by some forces can be expressed as the change of a function of the object's position called the potential energy. When all the forces that do work on a system have this property, we can state the principle of work and energy as a conservation law: The sum of the kinetic and potential energies is constant.

When we derived the principle of work and energy in Section 15.1 by integrating Newton's second law, we were able to evaluate the integral on one side of the equation, obtaining the change in the kinetic energy:

$$U_{12} = \int_{\mathbf{r}_1}^{\mathbf{r}_2} \Sigma \mathbf{F} \cdot d\mathbf{r} = \tfrac{1}{2}mv_2^2 - \tfrac{1}{2}mv_1^2. \tag{15.15}$$

Suppose we could determine a scalar function of position V such that

$$dV = -\Sigma \mathbf{F} \cdot d\mathbf{r}. \tag{15.16}$$

Then we could also evaluate the integral defining the work:

$$U_{12} = \int_{\mathbf{r}_1}^{\mathbf{r}_2} \Sigma \mathbf{F} \cdot d\mathbf{r} = \int_{V_1}^{V_2} - dV = -(V_2 - V_1), \tag{15.17}$$

where V_1 and V_2 are the values of V at the positions $\mathbf{r}_1$ and $\mathbf{r}_2$, respectively. Substituting this expression into Eq. (15.15), we obtain the principle of work and energy in the form

$$\tfrac{1}{2}mv_1^2 + V_1 = \tfrac{1}{2}mv_2^2 + V_2. \tag{15.18}$$

If the kinetic energy increases as the object moves from position 1 to position 2, the function V must decrease, and vice versa, as if V represents a reservoir of "potential" kinetic energy. For this reason, V is called the *potential energy*.

Equation (15.18) states that the sum of the kinetic and potential energies of an object has the same value at any two points. Energy is *conserved*. However, *there is an important restriction on the use of this result.* We arrived at Eq. (15.18) by assuming that a function V, the potential energy, exists that satisfies Eq. (15.16). This is true only for a limited class of forces, which are said to be *conservative*. We discuss conservative forces in the next section. If *all* of the forces that do work on an object are conservative, Eq. (15.18) can be applied, where V is the sum of the potential energies of the forces that do work on the object. Otherwise, Eq. (15.18) cannot be used. A system is said to be conservative if all of the forces that do work on the system are conservative. The sum of the kinetic and potential energies of a conservative system is conserved.

An object may be subjected to both conservative and nonconservative forces. When that is the case, it is often convenient to introduce the potential energies of the forces that are conservative into the statement of the principle of work and energy. To allow for this option, we write Eq. (15.15) as

$$\tfrac{1}{2}mv_1^2 + V_1 + U_{12} = \tfrac{1}{2}mv_2^2 + V_2. \tag{15.19}$$

When the principle of work and energy is written in this form, the term U_{12} includes the work done by all nonconservative forces acting on the object. If a conservative force does work on the object, there is a choice. The work can be calculated and included in U_{12}, *or* the force's potential energy can be included in V. This procedure can also be applied to a system that is subjected to both conservative and nonconservative forces. The sum of the kinetic and potential energies of a system in position 1 plus the work done as the system moves from position 1 to position 2 is equal to the total sum of the kinetic and potential energies in position 2.

Conservative Forces

We can apply conservation of energy only if the forces doing work on an object or system are conservative and we know (or can determine) their potential energies. In this section, we determine the potential energies of some conservative forces and use the results to demonstrate applications of conservation of energy. But before discussing forces that are conservative, we demonstrate with a simple example that *frictional forces are not conservative.*

The work done by a conservative force as an object moves from a position 1 to a position 2 is independent of the object's path. This result follows from Eq. (15.17), which states that the work depends only on the values of the potential energy at positions 1 and 2. Equation (15.17) also implies that if the object moves along a closed path, returning to position 1, the work done by a conservative force is zero. Suppose that a book of mass m rests on a table and you push it horizontally so that it slides along a path of length L. The magnitude of the force of friction is $\mu_k mg$, and the direction of the force is opposite to that of the book's motion (Fig. 15.7). The work done is

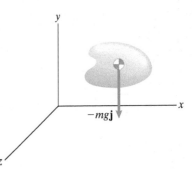

Figure 15.7
The book's path from position 1 to position 2. The force of friction points opposite to the direction of the motion.

$$U_{12} = \int_0^L -\mu_k mg \; ds = -\mu_k mgL.$$

The work is proportional to the length of the object's path and therefore is not independent of the path. As this simple example demonstrates, friction forces are not conservative.

The weight of an object and the force exerted by a spring attached to a fixed support are conservative forces. Using them as examples, we demonstrate how you can determine the potential energies of other conservative forces. We also use the potential energies of these forces in examples of the use of conservation of energy to analyze the motions of conservative systems.

Weight   To determine the potential energy associated with an object's weight, we use a cartesian coordinate system with its y axis pointing upward (Fig. 15.8). The weight is $\mathbf{F} = -mg\mathbf{j}$, and its dot product with the vector $d\mathbf{r}$ is

$$\mathbf{F} \cdot d\mathbf{r} = (-mg\mathbf{j}) \cdot (dx\,\mathbf{i} + dy\,\mathbf{j} + dz\,\mathbf{k}) = -mg\,dy.$$

From Eq. (15.16), the potential energy V must satisfy the relation

$$dV = -\mathbf{F} \cdot d\mathbf{r} = mg\,dy, \tag{15.20}$$

which we can write as

$$\frac{dV}{dy} = mg.$$

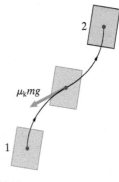

Figure 15.8
Weight of an object expressed in terms of a coordinate system with the y axis pointing upward.

Integrating this equation, we obtain

$$V = mgy + C,$$

where C, the constant of integration, is arbitrary. This expression satisfies Eq. (15.20) for any value of C. Another way of understanding why C is arbitrary is to notice in Eq. (15.18) that it is the difference in the potential energy between two positions that determines the change in the kinetic energy. We will let $C = 0$ and write the potential energy of the weight of an object as

$$V = mgy. \tag{15.21}$$

The potential energy is the product of the object's weight and height. The height can be measured from any convenient reference level, or *datum*. Since it is the difference in potential energy that determines the change in the kinetic energy, it is the difference in height that matters, not the level from which the height is measured.

The roller coaster (Fig. 15.9a) is a classic example of conservation of energy. If aerodynamic and frictional forces are neglected, the weight is the only force doing work, and the system is conservative. The potential energy of the roller coaster is proportional to the height of the track relative to a datum. In Fig. 15.9b,

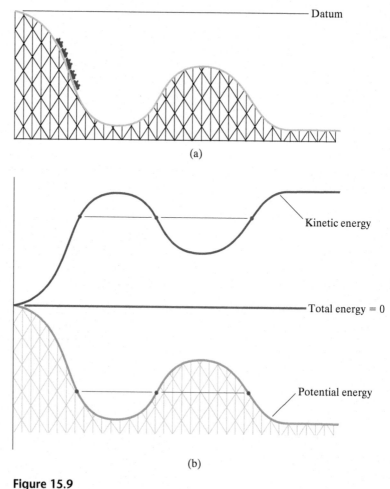

(a)

(b)

Figure 15.9
(a) Roller coaster and a reference level, or datum.
(b) The sum of the potential and kinetic energies is constant.

we assume that the roller coaster started from rest at the datum level. The sum of the kinetic and potential energies is constant, so the kinetic energy "mirrors" the potential energy. At points of the track that have equal heights, the magnitudes of the velocities are equal.

To account for the variation of weight with distance from the center of the earth, we can express the weight in polar coordinates as

$$\mathbf{F} = -\frac{mgR_E^2}{r^2}\mathbf{e}_r,$$

where r is the distance from the center of the earth (Fig. 15.10). From Eq. (15.12), the vector $d\mathbf{r}$ in terms of polar coordinates is

$$d\mathbf{r} = dr\,\mathbf{e}_r + r\,d\theta\,\mathbf{e}_\theta. \tag{15.22}$$

The potential energy must satisfy

$$dV = -\mathbf{F}\cdot d\mathbf{r} = \frac{mgR_E^2}{r^2}dr,$$

or

$$\frac{dV}{dr} = \frac{mgR_E^2}{r^2}.$$

We integrate this equation and let the constant of integration be zero, obtaining the potential energy

$$V = -\frac{mgR_E^2}{r}. \tag{15.23}$$

Compare this expression with the gravitational potential energy given by Eq. (15.21), in which the variation of the gravitational force with height is neglected. (See Problem 15.109.)

Springs In terms of polar coordinates, the force exerted on an object by a linear spring is

$$\mathbf{F} = -k(r - r_0)\mathbf{e}_r,$$

where r_0 is the unstretched length of the spring (Fig. 15.11). Using Eq. (15.22), we see that the potential energy must satisfy

$$dV = -\mathbf{F}\cdot d\mathbf{r} = k(r - r_0)\,dr.$$

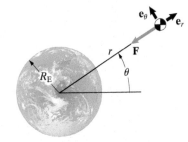

Figure 15.10
Expressing weight in terms of polar coordinates.

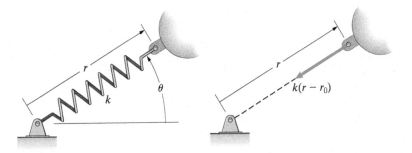

Figure 15.11
Expressing the force exerted by a linear spring in polar coordinates.

Expressed in terms of the stretch of the spring $S = r - r_0$, this equation is $dV = kS\, dS$, or

$$\frac{dV}{dS} = kS.$$

Integrating, we obtain the potential energy of a linear spring:

$$V = \tfrac{1}{2}kS^2. \tag{15.24}$$

RESULTS

Conservative Forces and Potential Energy

For a given force **F**, if there is a function of position V such that

$$dV = -\mathbf{F} \cdot d\mathbf{r},$$

then **F** is said to be *conservative*, and V is called the *potential energy* associated with **F**.

Conservation of Energy

If all of the forces that do work on an object are conservative, the sum of the kinetic energy and the total potential energy is the same at any two positions.

$$\tfrac{1}{2}mv_1^2 + V_1 = \tfrac{1}{2}mv_2^2 + V_2. \tag{15.18}$$

When both conservative and nonconservative forces do work on an object, the principle of work and energy can be expressed in terms of the potential energy V of the conservative forces and the work U_{12} done by nonconservative forces.

$$\tfrac{1}{2}mv_1^2 + V_1 + U_{12} = \tfrac{1}{2}mv_2^2 + V_2. \tag{15.19}$$

Applying conservation of energy typically involves three steps.

1. *Determine whether the forces are conservative.* Draw a free-body diagram to identify the forces that do work and confirm that they are conservative.
2. *Determine the potential energy.* Evaluate the potential energies of the forces that do work.
3. *Apply conservation of energy.* Equate the sum of the kinetic and potential energies at two positions. This results in an expression relating a change in position to the change in the kinetic energy.

Potential Energies Associated with Particular Forces

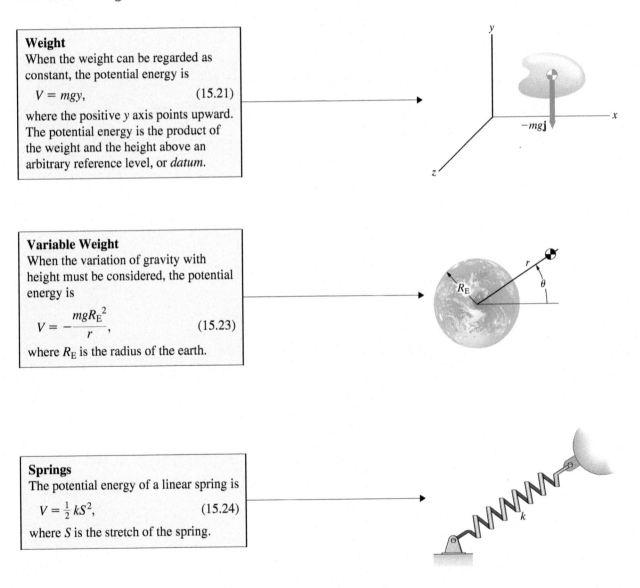

Weight
When the weight can be regarded as constant, the potential energy is

$$V = mgy, \qquad (15.21)$$

where the positive y axis points upward. The potential energy is the product of the weight and the height above an arbitrary reference level, or *datum*.

Variable Weight
When the variation of gravity with height must be considered, the potential energy is

$$V = -\frac{mgR_{\mathrm{E}}^2}{r}, \qquad (15.23)$$

where R_{E} is the radius of the earth.

Springs
The potential energy of a linear spring is

$$V = \tfrac{1}{2} kS^2, \qquad (15.24)$$

where S is the stretch of the spring.

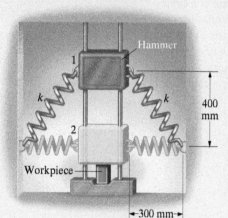

Active Example 15.7 Potential Energy of Weight and Springs (▶ *Related Problem 15.89*)

The 40-kg hammer is lifted to position 1 and released from rest. It falls and strikes a workpiece when it is in position 2. The spring constant is $k = 1500$ N/m, and the springs are unstretched when the hammer is in position 2. Neglect friction. Use conservation of energy to determine the hammer's velocity when it reaches position 2.

Strategy

We must confirm that the forces that do work on the hammer are conservative. If they are, we can determine the velocity of the hammer at position 2 by equating the sum of its kinetic and potential energies at position 1 to the sum of its kinetic and potential energies at position 2.

Solution

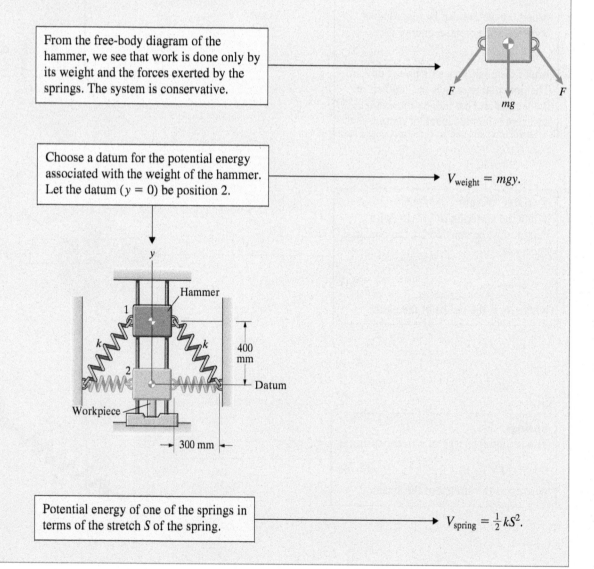

From the free-body diagram of the hammer, we see that work is done only by its weight and the forces exerted by the springs. The system is conservative.

Choose a datum for the potential energy associated with the weight of the hammer. Let the datum ($y = 0$) be position 2.

$V_{\text{weight}} = mgy.$

Potential energy of one of the springs in terms of the stretch S of the spring.

$V_{\text{spring}} = \frac{1}{2}kS^2.$

Calculate the stretch of one
of the springs at positions
1 and 2.

$$\begin{cases} S_1 = \sqrt{(0.3\ \text{m})^2 + (0.4\ \text{m})^2} - 0.3\ \text{m} \\ \qquad = 0.2\ \text{m}, \\ S_2 = 0. \end{cases}$$

Apply conservation of
energy to positions 1 and
2 to determine the velocity
at position 2.

$$\begin{cases} (V_{\text{weight}})_1 + 2(V_{\text{spring}})_1 + \tfrac{1}{2}mv_1^2 = (V_{\text{weight}})_2 + 2(V_{\text{spring}})_2 + \tfrac{1}{2}mv_2^2: \\[4pt] \qquad mgy_1 + 2\left(\tfrac{1}{2}kS_1^2\right) + \tfrac{1}{2}mv_1^2 = mgy_2 + 2\left(\tfrac{1}{2}kS_2^2\right) + \tfrac{1}{2}mv_2^2, \\[4pt] (40\ \text{kg})\left(9.81\ \text{m/s}^2\right)(0.4\ \text{m}) + 2\left[\tfrac{1}{2}(1500\ \text{N/m})(0.2\ \text{m})^2\right] + 0 \\[4pt] \qquad\qquad\qquad\qquad = 0 + 0 + \tfrac{1}{2}(40\ \text{kg})v_2^2. \\[6pt] \text{Solving, we obtain} \\ \qquad v_2 = 3.29\ \text{m/s}. \end{cases}$$

Practice Problem The 40-kg hammer is given a downward velocity of 2 m/s in position 1. It falls and strikes a workpiece when it is in position 2. The spring constant is $k = 1500$ N/m, and the springs are unstretched when the hammer is in position 1. Neglect friction. Use conservation of energy to determine the velocity of the hammer just before it strikes the workpiece.

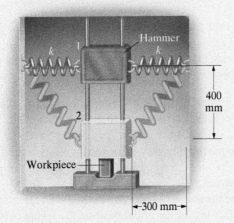

Answer: $v_2 = 2.97$ m/s.

Example 15.8 **Conservation of Energy of a System** (▶ *Related Problem 15.91*)

The spring ($k = 300$ N/m) is connected to the floor and to the 90-kg collar A. Collar A is at rest, supported by the spring, when the 135-kg box B is released from rest in the position shown. What are the velocities of A and B when B has fallen 1 m?

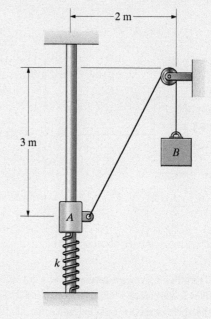

Strategy

If all of the forces that do work on the system are conservative, we can apply conservation of energy to obtain one equation in terms of the velocities of A and B when B has fallen 1 m. To complete the solution, we must also use kinematics to determine the relationship between the velocities of A and B.

Solution

Determine whether the System Is Conservative   We consider the collar A, box B, and pulley as a single system. From the free-body diagram of the system in Fig. a, we see that work is done only by the weights of the collar and box and the spring force F. The system is therefore conservative.

Determine the Potential Energy   Using the initial position of collar A as its datum, the potential energy associated with the weight of A when it has risen a distance x_A (Fig. b) is $V_A = m_A g x_A$. Using the initial position of box B as its datum, the potential energy associated with its weight when it has fallen a distance x_B is $V_B = -m_B g x_B$. (The minus sign is necessary because x_B is positive downward.)

To determine the potential energy associated with the spring force, we must account for the fact that in the initial position the spring is compressed by the weight of collar A. The spring is initially compressed a distance δ such that $m_A g = k\delta$ (Fig. c). When the collar has moved upward a distance x_A, the stretch of the spring is $S = x_A - \delta = x_A - m_A g/k$, so its potential energy is

$$V_S = \tfrac{1}{2}kS^2 = \tfrac{1}{2}k\left(x_A - \frac{m_A g}{k}\right)^2.$$

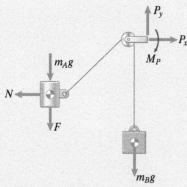

(a) Free-body diagram of the system.

The total potential energy of the system in terms of the displacements of the collar and box is

$$V = V_A + V_B + V_S$$

$$= m_A g x_A - m_B g x_B + \frac{1}{2} k \left(x_A - \frac{m_A g}{k} \right)^2.$$

Apply Conservation of Energy   The sum of the kinetic and potential energies of the system in its initial position and in the position shown in Fig. b must be equal. Denoting the total kinetic energy by T, we have

$$T_1 + V_1 = T_2 + V_2:$$

$$0 + \frac{1}{2} k \left(-\frac{m_A g}{k} \right)^2 = \frac{1}{2} m_A v_A^2 + \frac{1}{2} m_B v_B^2$$

$$+ m_A g x_A - m_B g x_B + \frac{1}{2} k \left(x_A - \frac{m_A g}{k} \right)^2. \tag{1}$$

We want to determine v_A and v_B when $x_B = 1$ m, but we have only one equation in terms of x_A, x_B, v_A, and v_B. To complete the solution, we must relate the displacement and velocity of the collar A to the displacement and velocity of the box B.

From Fig. b, the decrease in the length of the rope from A to the pulley as the collar rises must equal the distance the box falls:

$$\sqrt{(3 \text{ m})^2 + (2 \text{ m})^2} - \sqrt{(3 \text{ m} - x_A)^2 + (2 \text{ m})^2} = x_B.$$

Solving this equation for the value of x_A when $x_B = 1$ m, we obtain $x_A = 1.33$ m. By taking the derivative of this equation with respect to time, we also obtain a relation between v_A and v_B:

$$\left[\frac{3 \text{ m} - x_A}{\sqrt{(3 \text{ m} - x_A)^2 + (2 \text{ m})^2}} \right] v_A = v_B.$$

Setting $x_A = 1.33$ m, we determine from the preceding equation that

$$0.641 v_A = v_B.$$

We solve this equation together with Eq. (1) for the velocities of the collar and box when $x_A = 1.33$ m and $x_B = 1$ m, obtaining $v_A = 3.82$ m/s and $v_B = 2.45$ m/s.

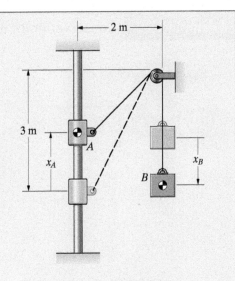

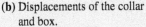

(b) Displacements of the collar and box.

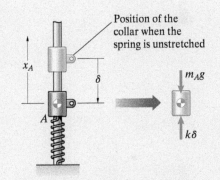

(c) Determining the initial compression of the spring.

Critical Thinking

Why didn't we have to consider the forces exerted on the collar and box by the rope? The reason is that they are internal forces when the collar, box, and pulley are regarded as a single system. This example clearly demonstrates the advantage of applying conservation of energy to an entire system whenever possible.

Example 15.9 **Conservation of Energy of a Spacecraft** (▶ *Related Problem 15.103*)

A spacecraft at a distance $r_0 = 2R_E$ from the center of the earth is moving outward with initial velocity $v_0 = \sqrt{2gR_E/3}$. Determine the velocity of the craft as a function of its distance from the center of the earth.

Strategy

The potential energy associated with the earth's gravity is given by Eq. (15.23). The initial radial position and velocity of the spacecraft are given, so we can use conservation of energy to determine its velocity as a function of its radial position.

Solution

Determine whether the System Is Conservative If work is done on the spacecraft by gravity alone, the system is conservative.

Determine the Potential Energy The potential energy associated with the weight of the spacecraft is given in terms of its distance r from the center of the earth by Eq. (15.23):

$$V = -\frac{mgR_E^2}{r}.$$

Apply Conservation of Energy Let v be the magnitude of the spacecraft's velocity at an arbitrary distance r. The sums of the potential and kinetic energies at r_0 and at r must be equal.

$$-\frac{mgR_E^2}{r_0} + \frac{1}{2}mv_0^2 = -\frac{mgR_E^2}{r} + \frac{1}{2}mv^2:$$

$$-\frac{mgR_E^2}{2R_E} + \frac{1}{2}m\left(\frac{2}{3}gR_E\right) = -\frac{mgR_E^2}{r} + \frac{1}{2}mv^2.$$

Solving for v, we find that the spacecraft's velocity as a function of r is

$$v = \sqrt{gR_E\left(\frac{2R_E}{r} - \frac{1}{3}\right)}.$$

Critical Thinking

The graph shows the kinetic energy, potential energy, and total energy as functions of r/R_E. The kinetic energy decreases and the potential energy increases as the spacecraft moves outward until its velocity decreases to zero at $r = 6R_E$.

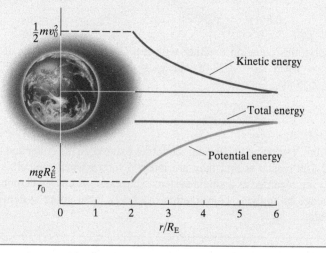

Problems

15.78 The 10-lb box is released from rest at position 1 and slides down the smooth inclined surface to position 2.

(a) If the datum is placed at the level of the floor as shown, what is the sum of the kinetic and potential energies of the box when it is in position 1?

(b) What is the sum of the kinetic and potential energies of the box when it is in position 2?

(c) Use conservation of energy to determine the magnitude of the box's velocity when it is in position 2.

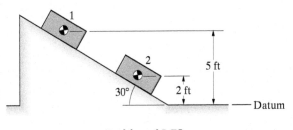

Problem 15.78

15.79 The 0.45-kg soccer ball is 1 m above the ground when it is kicked upward at 12 m/s. Use conservation of energy to determine the magnitude of the ball's velocity when it is 4 m above the ground. Obtain the answer by placing the datum (a) at the level of the ball's initial position and (b) at ground level.

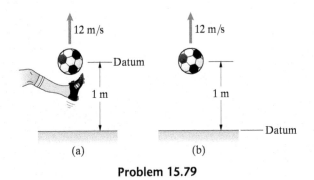

Problem 15.79

15.80 The Lunar Module used in the Apollo moon landings could make a safe landing if the magnitude of its vertical velocity at impact was no greater than 5 m/s. Use conservation of energy to determine the maximum height h at which the pilot could shut off the engine if the vertical velocity of the lander is (a) 2 m/s downward and (b) 2 m/s upward. The acceleration due to gravity at the moon's surface is 1.62 m/s^2.

Problem 15.80

15.81 The 0.4-kg collar starts from rest at position 1 and slides down the smooth rigid wire. The y axis points upward. Use conservation of energy to determine the magnitude of the velocity of the collar when it reaches point 2.

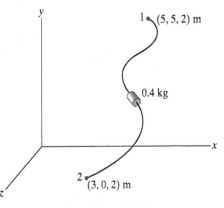

Problem 15.81

15.82 At the instant shown, the 20-kg mass is moving downward at 1.6 m/s. Let d be the downward displacement of the mass relative to its present position. Use conservation of energy to determine the magnitude of the velocity of the 20-kg mass when $d = 1$ m.

4 kg   20 kg

Problem 15.82

15.83 The mass of the ball is $m = 2$ kg, and the string's length is $L = 1$ m. The ball is released from rest in position 1 and swings to position 2, where $\theta = 40°$.

(a) Use conservation of energy to determine the magnitude of the ball's velocity at position 2.

(b) Draw graphs of the kinetic energy, the potential energy, and the total energy for values of θ from zero to 180°.

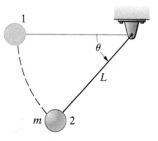

Problem 15.83

15.84 The mass of the ball is $m = 2$ kg and the string's length is $L = 1$ m. The ball is released from rest in position 1. When the string is vertical, it hits the fixed peg shown.

(a) Use conservation of energy to determine the minimum angle θ necessary for the ball to swing to position 2.

(b) If the ball is released at the minimum angle θ determined in part (a), what is the tension in the string just before and just after it hits the peg?

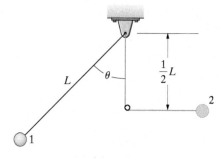

Problem 15.84

15.85 A small pellet of mass $m = 0.2$ kg starts from rest at position 1 and slides down the smooth surface of the cylinder to position 2. The radius $R = 0.8$ m. Use conservation of energy to determine the magnitude of the pellet's velocity at position 2 if $\theta = 45°$.

15.86 In Problem 15.85, what is the value of the angle θ at which the pellet loses contact with the surface of the cylinder?

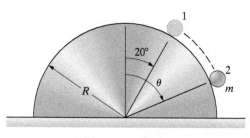

Problems 15.85/15.86

15.87 The bar is smooth. The 10-kg slider at *A* is given a downward velocity of 6.5 m/s.

(a) Use conservation of energy to determine whether the slider will reach point *C*. If it does, what is the magnitude of its velocity at point *C*?

(b) What is the magnitude of the normal force the bar exerts on the slider as it passes point *B*?

15.88 The bar is smooth. The 10-kg slider at *A* is given a downward velocity of 7.5 m/s.

(a) Use conservation of energy to determine whether the slider will reach point *D*. If it does, what is the magnitude of its velocity at point *D*?

(b) What is the magnitude of the normal force the bar exerts on the slider as it passes point *B*?

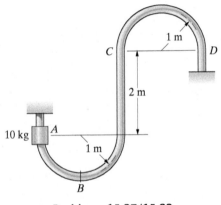

Problems 15.87/15.88

▶ **15.89** In Active Example 15.7, suppose that you want to increase the value of the spring constant *k* so that the velocity of the hammer just before it strikes the workpiece is 4 m/s. Use conservation of energy to determine the required value of *k*.

15.90 A rock climber of weight *W* has a rope attached a distance *h* below him for protection. Suppose that he falls, and assume that the rope behaves like a linear spring with unstretched length *h* and spring constant $k = C/h$, where *C* is a constant. Use conservation of energy to determine the maximum force exerted on the climber by the rope. (Notice that the maximum force is independent of *h*, which is a reassuring result for climbers: The maximum force resulting from a long fall is the same as that resulting from a short one.)

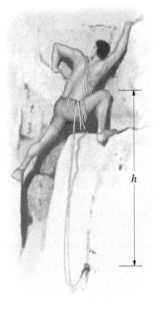

Problem 15.90

▶ **15.91** The collar *A* slides on the smooth horizontal bar. The spring constant $k = 40$ lb/ft. The weights are $W_A = 30$ lb and $W_B = 60$ lb. At the instant shown, the spring is unstretched and *B* is moving downward at 4 ft/s. Use conservation of energy to determine the velocity of *B* when it has moved downward 2 ft from its current position. (See Example 15.8.)

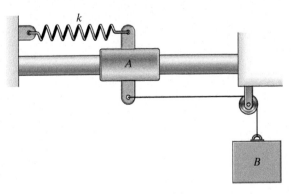

Problem 15.91

15.92 The spring constant $k = 700$ N/m. The masses $m_A = 14$ kg and $m_B = 18$ kg. The horizontal bar is smooth. At the instant shown, the spring is unstretched and the mass B is moving downward at 1 m/s. How fast is B moving when it has moved downward 0.2 m from its present position?

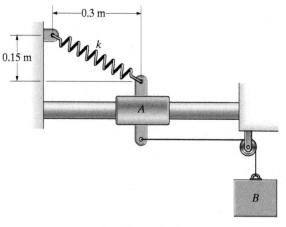

Problem 15.92

15.93 The semicircular bar is smooth. The unstretched length of the spring is 10 in. The 5-lb collar at A is given a downward velocity of 6 ft/s, and when it reaches B the magnitude of its velocity is 15 ft/s. Determine the spring constant k.

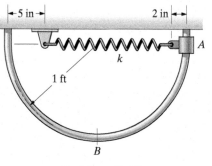

Problem 15.93

15.94 The mass $m = 1$ kg, the spring constant $k = 200$ N/m, and the unstretched length of the spring is 0.1 m. When the system is released from rest in the position shown, the spring contracts, pulling the mass to the right. Use conservation of energy to determine the magnitude of the velocity of the mass when the string and the spring are parallel.

15.95 In Problem 15.94, what is the tension in the string when the string and spring are parallel?

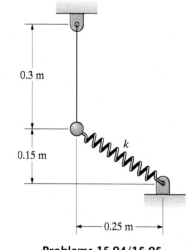

Problems 15.94/15.95

15.96 The force exerted on an object by a *nonlinear* spring is

$$\mathbf{F} = -[k(r - r_0) + q(r - r_0)^3]\mathbf{e}_r,$$

where k and q are constants and r_0 is the unstretched length of the spring. Determine the potential energy of the spring in terms of its stretch $S = r - r_0$.

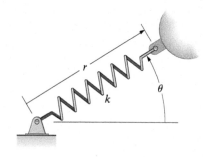

Problem 15.96

15.97 The 20-kg cylinder is released at the position shown and falls onto the linear spring ($k = 3000$ N/m). Use conservation of energy to determine how far down the cylinder moves after contacting the spring.

15.98 The 20-kg cylinder is released at the position shown and falls onto the *nonlinear* spring. In terms of the stretch S of the spring, its potential energy is $V = \frac{1}{2}kS^2 + \frac{1}{4}qS^4$, where $k = 3000$ N/m and $q = 4000$ N/m^3. What is the velocity of the cylinder when the spring has been compressed 0.5 m?

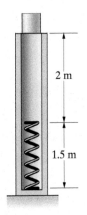

2 m

1.5 m

Problems 15.97/15.98

15.99 The string exerts a force of constant magnitude T on the object. Use polar coordinates to show that the potential energy associated with this force is $V = Tr$.

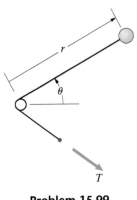

r

θ

T

Problem 15.99

15.100 The system is at rest in the position shown, with the 12-lb collar A resting on the spring ($k = 20$ lb/ft), when a constant 30-lb force is applied to the cable. What is the velocity of the collar when it has risen 1 ft? (See Problem 15.99.)

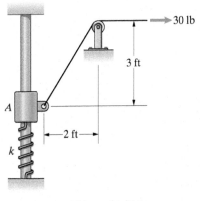

30 lb

3 ft

A

2 ft

k

Problem 15.100

15.101 A 1-kg disk slides on a smooth horizontal table and is attached to a string that passes through a hole in the table. A constant force $T = 10$ N is exerted on the string. At the instant shown, $r = 1$ m and the velocity of the disk in terms of polar coordinates is $\mathbf{v} = 6\mathbf{e}_\theta$ (m/s). Use conservation of energy to determine the magnitude of the velocity of the disk when $r = 2$ m. (See Problem 15.99.)

15.102 A 1-kg disk slides on a smooth horizontal table and is attached to a string that passes through a hole in the table. A constant force $T = 10$ N is exerted on the string. At the instant shown, $r = 1$ m and the velocity of the disk in terms of polar coordinates is $\mathbf{v} = 8\mathbf{e}_\theta$ (m/s). Because this is central-force motion, the product of the radial position r and the transverse component of velocity v_θ is constant. Use this fact and conservation of energy to determine the velocity of the disk in terms of polar coordinates when $r = 2$ m.

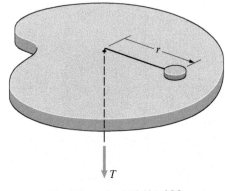

r

T

Problems 15.101/15.102

▶ **15.103** A satellite initially is inserted into orbit at a distance $r_0 = 8800$ km from the center of the earth. When it is at a distance $r = 18,000$ km from the center of the earth, the magnitude of its velocity is $v = 7000$ m/s. Use conservation of energy to determine its initial velocity v_0. The radius of the earth is 6370 km. (See Example 15.9.)

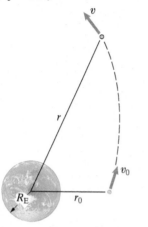

Problem 15.103

15.104 Astronomers detect an asteroid 100,000 km from the earth moving at 2 km/s relative to the center of the earth. Suppose the asteroid strikes the earth. Use conservation of energy to determine the magnitude of its velocity as it enters the atmosphere. (You can neglect the thickness of the atmosphere in comparison to the earth's 6370-km radius.)

15.105 A satellite is in the elliptic earth orbit shown. Its velocity in terms of polar coordinates when it is at the perigee A is $\mathbf{v} = 8640\mathbf{e}_\theta$ (m/s). Determine the velocity of the satellite in terms of polar coordinates when it is at point B.

15.106 Use conservation of energy to determine the magnitude of the velocity of the satellite in Problem 15.105 at the apogee C. Using your result, confirm numerically that the velocities at perigee and apogee satisfy the relation $r_A v_A = r_C v_C$.

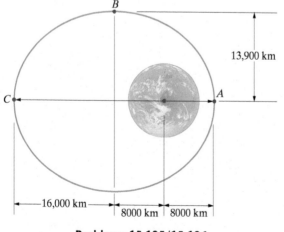

Problems 15.105/15.106

15.107 The *Voyager* and *Galileo* spacecraft observed volcanic plumes, believed to consist of condensed sulfur or sulfur dioxide gas, above the surface of the Jovian satellite Io. The plume observed above a volcano named Prometheus was estimated to extend 50 km above the surface. The acceleration due to gravity at the surface is 1.80 m/s². Using conservation of energy and neglecting the variation of gravity with height, determine the velocity at which a solid particle would have to be ejected to reach 50 km above Io's surface.

15.108 Solve Problem 15.107 using conservation of energy and accounting for the variation of gravity with height. The radius of Io is 1815 km.

Problems 15.107/15.108

15.109* What is the relationship between Eq. (15.21), which is the gravitational potential energy neglecting the variation of the gravitational force with height, and Eq. (15.23), which accounts for the variation? Express the distance from the center of the earth as $r = R_E + y$, where R_E is the earth's radius and y is the height above the surface, so that Eq. (15.23) can be written as

$$V = -\frac{mgR_E}{1 + \dfrac{y}{R_E}}.$$

By expanding this equation as a Taylor series in terms of y/R_E and assuming that $y/R_E \ll 1$, show that you obtain a potential energy equivalent to Eq. (15.21).

15.4 Relationships between Force and Potential Energy

BACKGROUND

Here we consider two questions: (1) Given a potential energy, how can we determine the corresponding force? (2) Given a force, how can we determine whether it is conservative? That is, how can we tell whether an associated potential energy exists?

The potential energy V of a force $\mathbf{F}$ is a function of position that satisfies the relation

$$dV = -\mathbf{F} \cdot d\mathbf{r}. \tag{15.25}$$

Let us express V in terms of a cartesian coordinate system:

$$V = V(x, y, z).$$

The differential of V is

$$dV = \frac{\partial V}{\partial x} dx + \frac{\partial V}{\partial y} dy + \frac{\partial V}{\partial z} dz. \tag{15.26}$$

Expressing $\mathbf{F}$ and $d\mathbf{r}$ in terms of cartesian components and taking their dot product yields

$$\mathbf{F} \cdot d\mathbf{r} = (F_x \mathbf{i} + F_y \mathbf{j} + F_z \mathbf{k}) \cdot (dx\mathbf{i} + dy\mathbf{j} + dz\mathbf{k})$$
$$= F_x\, dx + F_y\, dy + F_z\, dz.$$

Substituting this expression and Eq. (15.26) into Eq. (15.25), we obtain

$$\frac{\partial V}{\partial x} dx + \frac{\partial V}{\partial y} dy + \frac{\partial V}{\partial z} dz = -(F_x\, dx + F_y\, dy + F_z\, dz),$$

which implies that

$$F_x = -\frac{\partial V}{\partial x}, \qquad F_y = -\frac{\partial V}{\partial y}, \quad \text{and} \quad F_z = -\frac{\partial V}{\partial z}. \tag{15.27}$$

Given a potential energy V expressed in cartesian coordinates, we can use Eqs. (15.27) to determine the corresponding force. The force

$$\mathbf{F} = -\left(\frac{\partial V}{\partial x}\mathbf{i} + \frac{\partial V}{\partial y}\mathbf{j} + \frac{\partial V}{\partial z}\mathbf{k} \right) = -\nabla V, \tag{15.28}$$

where ∇V is the *gradient* of V. By using expressions for the gradient in terms of other coordinate systems, we can determine the force $\mathbf{F}$ when we know the potential energy in terms of those coordinate systems. For example, in terms of cylindrical coordinates,

$$\mathbf{F} = -\left(\frac{\partial V}{\partial r}\mathbf{e}_r + \frac{1}{r}\frac{\partial V}{\partial \theta}\mathbf{e}_\theta + \frac{\partial V}{\partial z}\mathbf{e}_z \right). \tag{15.29}$$

If a force $\mathbf{F}$ is conservative, its *curl* $\nabla \times \mathbf{F}$ is zero. The expression for the curl of $\mathbf{F}$ in cartesian coordinates is

$$\nabla \times \mathbf{F} = \begin{vmatrix} \mathbf{i} & \mathbf{j} & \mathbf{k} \\ \dfrac{\partial}{\partial x} & \dfrac{\partial}{\partial y} & \dfrac{\partial}{\partial z} \\ F_x & F_y & F_z \end{vmatrix}. \tag{15.30}$$

Substituting Eqs. (15.27) into this expression confirms that $\nabla \times \mathbf{F} = \mathbf{0}$ when $\mathbf{F}$ is conservative. The converse is also true. A force $\mathbf{F}$ is conservative if its curl is zero. We can use this condition to determine whether a given force is conservative. In terms of cylindrical coordinates, the curl of $\mathbf{F}$ is

$$\nabla \times \mathbf{F} = \frac{1}{r} \begin{vmatrix} \mathbf{e}_r & r\mathbf{e}_\theta & \mathbf{e}_z \\ \dfrac{\partial}{\partial r} & \dfrac{\partial}{\partial \theta} & \dfrac{\partial}{\partial z} \\ F_r & rF_\theta & F_z \end{vmatrix}. \tag{15.31}$$

RESULTS

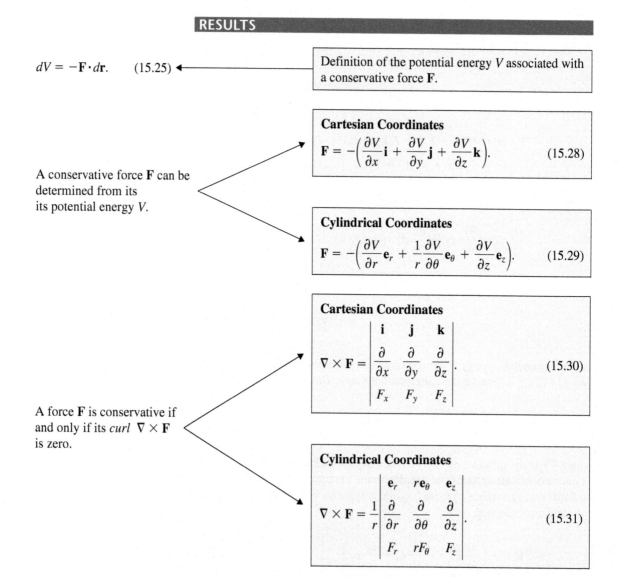

$dV = -\mathbf{F} \cdot d\mathbf{r}.$ (15.25) ← Definition of the potential energy V associated with a conservative force $\mathbf{F}$.

A conservative force $\mathbf{F}$ can be determined from its its potential energy V.

Cartesian Coordinates

$$\mathbf{F} = -\left(\frac{\partial V}{\partial x}\mathbf{i} + \frac{\partial V}{\partial y}\mathbf{j} + \frac{\partial V}{\partial z}\mathbf{k} \right). \tag{15.28}$$

Cylindrical Coordinates

$$\mathbf{F} = -\left(\frac{\partial V}{\partial r}\mathbf{e}_r + \frac{1}{r}\frac{\partial V}{\partial \theta}\mathbf{e}_\theta + \frac{\partial V}{\partial z}\mathbf{e}_z \right). \tag{15.29}$$

A force $\mathbf{F}$ is conservative if and only if its *curl* $\nabla \times \mathbf{F}$ is zero.

Cartesian Coordinates

$$\nabla \times \mathbf{F} = \begin{vmatrix} \mathbf{i} & \mathbf{j} & \mathbf{k} \\ \dfrac{\partial}{\partial x} & \dfrac{\partial}{\partial y} & \dfrac{\partial}{\partial z} \\ F_x & F_y & F_z \end{vmatrix}. \tag{15.30}$$

Cylindrical Coordinates

$$\nabla \times \mathbf{F} = \frac{1}{r} \begin{vmatrix} \mathbf{e}_r & r\mathbf{e}_\theta & \mathbf{e}_z \\ \dfrac{\partial}{\partial r} & \dfrac{\partial}{\partial \theta} & \dfrac{\partial}{\partial z} \\ F_r & rF_\theta & F_z \end{vmatrix}. \tag{15.31}$$

Active Example 15.10 Determining the Force from a Potential Energy

(▶ *Related Problems 15.112, 15.113*)

The potential energy associated with the weight of an object of mass m at a distance r from the center of the earth is (in cylindrical coordinates)

$$V = -\frac{mgR_E^2}{r},$$

where R_E is the radius of the earth. Use this expression to determine the force exerted on the object by its weight.

Strategy
The potential energy is expressed in cylindrical coordinates, so we can obtain the force from Eq. (15.29).

Solution

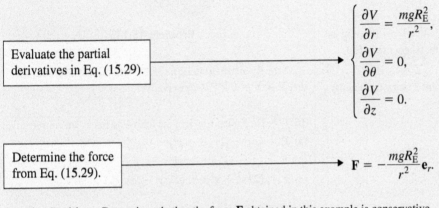

Evaluate the partial derivatives in Eq. (15.29).

$$\begin{cases} \dfrac{\partial V}{\partial r} = \dfrac{mgR_E^2}{r^2}, \\[2mm] \dfrac{\partial V}{\partial \theta} = 0, \\[2mm] \dfrac{\partial V}{\partial z} = 0. \end{cases}$$

Determine the force from Eq. (15.29).

$$\mathbf{F} = -\frac{mgR_E^2}{r^2}\mathbf{e}_r.$$

Practice Problem Determine whether the force $\mathbf{F}$ obtained in this example is conservative.

Answer: Yes

Problems

15.110 The potential energy associated with a force $\mathbf{F}$ acting on an object is $V = x^2 + y^3$ N-m, where x and y are in meters.

(a) Determine $\mathbf{F}$.

(b) Suppose that the object moves from position 1 to position 2 along path A, and then moves from position 1 to position 2 along path B. Determine the work done by $\mathbf{F}$ along each path.

15.111 An object is subjected to the force $\mathbf{F} = y\mathbf{i} - x\mathbf{j}$ (N), where x and y are in meters.

(a) Show that $\mathbf{F}$ is *not* conservative.

(b) Suppose the object moves from point 1 to point 2 along the paths A and B shown in Problem 15.110. Determine the work done by $\mathbf{F}$ along each path.

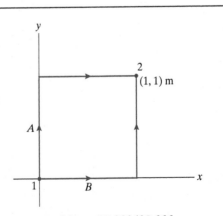

Problems 15.110/15.111

▶ 15.112 In terms of polar coordinates, the potential energy associated with the force **F** exerted on an object by a *nonlinear* spring is

$$V = \frac{1}{2}k(r - r_0)^2 + \frac{1}{4}q(r - r_0)^4,$$

where k and q are constants and r_0 is the unstretched length of the spring. Determine **F** in terms of polar coordinates. (See Active Example 15.10.)

▶ 15.113 In terms of polar coordinates, the force exerted on an object by a *nonlinear* spring is

$$\mathbf{F} = -[k(r - r_0) + q(r - r_0)^3]\mathbf{e}_r,$$

where k and q are constants and r_0 is the unstretched length of the spring. Use Eq. (15.31) to show that **F** is conservative. (See Active Example 15.10.)

15.114 The potential energy associated with a force **F** acting on an object is $V = -r\sin\theta + r^2\cos^2\theta$ ft-lb, where r is in feet.

(a) Determine **F**.

(b) If the object moves from point 1 to point 2 along the circular path, how much work is done by **F**?

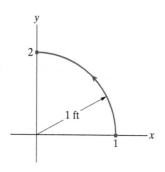

Problem 15.114

15.115 In terms of polar coordinates, the force exerted on an object of mass m by the gravity of a hypothetical two-dimensional planet is $\mathbf{F} = -(mg_T R_T/r)\mathbf{e}_r$, where g_T is the acceleration due to gravity at the surface, R_T is the radius of the planet, and r is the distance of the object from the center of the planet.

(a) Determine the potential energy associated with this gravitational force.

(b) If the object is given a velocity v_0 at a distance r_0, what is its velocity v as a function of r?

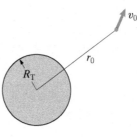

Problem 15.115

15.116 By substituting Eqs. (15.27) into Eq. (15.30), confirm that $\nabla \times \mathbf{F} = \mathbf{0}$ if **F** is conservative.

15.117 Determine which of the following forces are conservative:

(a) $\mathbf{F} = (3x^2 - 2xy)\mathbf{i} - x^2\mathbf{j}$;

(b) $\mathbf{F} = (x - xy^2)\mathbf{i} + x^2y\mathbf{j}$;

(c) $\mathbf{F} = (2xy^2 + y^3)\mathbf{i} + (2x^2y - 3xy^2)\mathbf{j}$.

Review Problems

15.118 The driver of a 3000-lb car moving at 40 mi/h applies an increasing force on the brake pedal. The magnitude of the resulting frictional force exerted on the car by the road is $f = 250 + 6s$ lb, where s is the car's horizontal position (in feet) relative to its position when the brakes were applied. Assuming that the car's tires do not slip; determine the distance required for the car to stop (a) by using Newton's second law and (b) by using the principle of work and energy.

15.119 Suppose that the car in Problem 15.118 is on wet pavement and the coefficients of friction between the tires and the road are $\mu_s = 0.4$ and $\mu_k = 0.35$. Determine the distance required for the car to stop.

Problems 15.118/15.119

15.120 An astronaut in a small rocket vehicle (combined mass $= 450$ kg) is hovering 100 m above the surface of the moon when he discovers that he is nearly out of fuel and can exert the thrust necessary to cause the vehicle to hover for only 5 more seconds. He quickly considers two strategies for getting to the surface: (a) Fall 20 m, turn on the thrust for 5 s, and then fall the rest of the way; (b) fall 40 m, turn on the thrust for 5 s, and then fall the rest of the way. Which strategy gives him the best chance of surviving? How much work is done by the engine's thrust in each case? ($g_{moon} = 1.62$ m/s^2.)

15.121 The coefficients of friction between the 20-kg crate and the inclined surface are $\mu_s = 0.24$ and $\mu_k = 0.22$. If the crate starts from rest and the horizontal force $F = 200$ N, what is the magnitude of the velocity of the crate when it has moved 2 m?

15.122 The coefficients of friction between the 20-kg crate and the inclined surface are $\mu_s = 0.24$ and $\mu_k = 0.22$. If the crate starts from rest and the horizontal force $F = 40$ N, what is the magnitude of the velocity of the crate when it has moved 2 m?

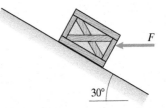

Problems 15.121/15.122

15.123 The Union Pacific Big Boy locomotive weighs 1.19 million lb, and the tractive effort (tangential force) of its drive wheels is 135,000 lb. If you neglect other tangential forces, what distance is required for the train to accelerate from zero to 60 mi/h?

15.124 In Problem 15.123, suppose that the acceleration of the locomotive as it accelerates from zero to 60 mi/h is $(F_0/m)(1 - v/88)$, where $F_0 = 135,000$ lb, m is the mass of the locomotive, and v is its velocity in feet per second.

(a) How much work is done in accelerating the train to 60 mi/h?

(b) Determine the locomotive's velocity as a function of time.

Problems 15.123/15.124

15.125 A car traveling 65 mi/h hits the crash barrier described in Problem 15.14. Determine the maximum deceleration to which the passengers are subjected if the car weighs (a) 2500 lb and (b) 5000 lb.

15.126 In a preliminary design for a mail-sorting machine, parcels moving at 2 ft/s slide down a smooth ramp and are brought to rest by a linear spring. What should the spring constant be if you don't want a 10-lb parcel to be subjected to a maximum deceleration greater than 10 *g*'s?

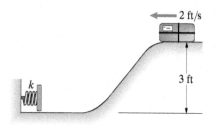

Problem 15.126

15.127 When the 1-kg collar is in position 1, the tension in the spring is 50 N, and the unstretched length of the spring is 260 mm. If the collar is pulled to position 2 and released from rest, what is its velocity when it returns to position 1?

15.128 When the 1-kg collar is in position 1, the tension in the spring is 100 N, and when the collar is in position 2, the tension in the spring is 400 N.

(a) What is the spring constant *k*?

(b) If the collar is given a velocity of 15 m/s at position 1, what is the magnitude of its velocity just before it reaches position 2?

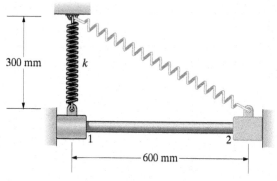

Problems 15.127/15.128

15.129 The 30-lb weight is released from rest with the two springs ($k_A = 30$ lb/ft, $k_B = 15$ lb/ft) unstretched.

(a) How far does the weight fall before rebounding?

(b) What maximum velocity does it attain?

Problem 15.129

15.130 The piston and the load it supports are accelerated upward by the gas in the cylinder. The total weight of the piston and load is 1000 lb. The cylinder wall exerts a constant 50-lb frictional force on the piston as it rises. The net force exerted on the piston by pressure is $(p_2 - p_{atm})A$, where p is the pressure of the gas, $p_{atm} = 2117$ lb/ft^2 is atmospheric pressure, and $A = 1$ ft^2 is the cross-sectional area of the piston. Assume that the product of p and the volume of the cylinder is constant. When $s = 1$ ft, the piston is stationary and $p = 5000$ lb/ft^2. What is the velocity of the piston when $s = 2$ ft?

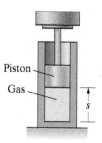

Problem 15.130

15.131 When a 22,000-kg rocket's engine burns out at an altitude of 2 km, the velocity of the rocket is 3 km/s and it is traveling at an angle of 60° relative to the horizontal. Neglect the variation in the gravitational force with altitude.

(a) If you neglect aerodynamic forces, what is the magnitude of the velocity of the rocket when it reaches an altitude of 6 km?

(b) If the actual velocity of the rocket when it reaches an altitude of 6 km is 2.8 km/s, how much work is done by aerodynamic forces as the rocket moves from 2 km to 6 km altitude?

15.132 The 12-kg collar A is at rest in the position shown at $t = 0$ and is subjected to the tangential force $F = 24 - 12t^2$ N for 1.5 s. Neglecting friction, what maximum height h does the collar reach?

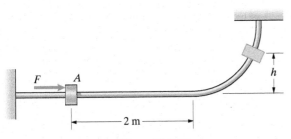

Problem 15.132

15.133 Suppose that, in designing a loop for a roller coaster's track, you establish as a safety criterion that at the top of the loop the normal force exerted on a passenger by the roller coaster should equal 10 percent of the passenger's weight. (That is, the passenger's "effective weight" pressing him down into his seat is 10 percent of his actual weight.) The roller coaster is moving at 62 ft/s when it enters the loop. What is the necessary instantaneous radius of curvature ρ of the track at the top of the loop?

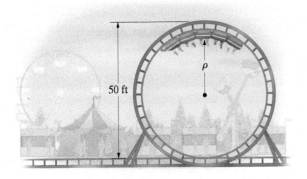

Problem 15.133

15.134 A 180-lb student runs at 15 ft/s, grabs a rope, and swings out over a lake. He releases the rope when his velocity is zero.

(a) What is the angle θ when he releases the rope?

(b) What is the tension in the rope just before he releases it?

(c) What is the maximum tension in the rope?

15.135 If the student in Problem 15.134 releases the rope when $\theta = 25°$, what maximum height does he reach relative to his position when he grabs the rope?

Problems 15.134/15.135

15.136 A boy takes a running start and jumps on his sled at position 1. He leaves the ground at position 2 and lands in deep snow at a distance $b = 25$ ft. How fast was he going at position 1?

15.137 In Problem 15.136, if the boy starts at position 1 going 15 ft/s, what distance b does he travel through the air?

Problems 15.136/15.137

15.138 The 1-kg collar A is attached to the linear spring $(k = 500 \text{ N/m})$ by a string. The collar starts from rest in the position shown, and the initial tension in the string is 100 N. What distance does the collar slide up the smooth bar?

Problem 15.138

15.139 The masses $m_A = 40$ kg and $m_B = 60$ kg. The collar A slides on the smooth horizontal bar. The system is released from rest. Use conservation of energy to determine the velocity of the collar A when it has moved 0.5 m to the right.

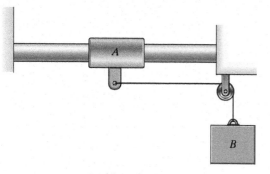

Problem 15.139

15.140 The spring constant is $k = 850$ N/m, $m_A = 40$ kg, and $m_B = 60$ kg. The collar A slides on the smooth horizontal bar. The system is released from rest in the position shown with the spring unstretched. Use conservation of energy to determine the velocity of the collar A when it has moved 0.5 m to the right.

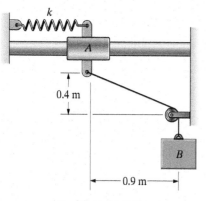

Problem 15.140

15.141 The y axis is vertical and the curved bar is smooth. If the magnitude of the velocity of the 4-lb slider is 6 ft/s at position 1, what is the magnitude of its velocity when it reaches position 2?

15.142 In Problem 15.141, determine the magnitude of the velocity of the slider when it reaches position 2 if it is subjected to the additional force $\mathbf{F} = 3x\mathbf{i} - 2\mathbf{j}$ (lb) during its motion.

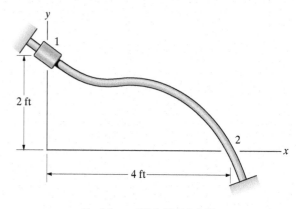

Problems 15.141/15.142

15.143 Suppose that an object of mass m is beneath the surface of the earth. In terms of a polar coordinate system with its origin at the earth's center, the gravitational force on the object is $-(mgr/R_E)\mathbf{e}_r$, where R_E is the radius of the earth. Show that the potential energy associated with the gravitational force is $V = mgr^2/2R_E$.

15.144 It has been pointed out that if tunnels could be drilled straight through the earth between points on the surface, trains could travel between those points using gravitational force for acceleration and deceleration. (The effects of friction and aerodynamic drag could be minimized by evacuating the tunnels and using magnetically levitated trains.) Suppose that such a train travels from the North Pole to a point on the equator. Disregard the earth's rotation. Determine the magnitude of the velocity of the train (a) when it arrives at the equator and (b) when it is halfway from the North Pole to the equator. The radius of the earth is $R_E = 3960$ mi. (See Problem 15.143.)

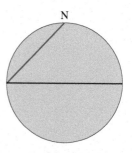

Problem 15.144

15.145 In Problem 15.123, what is the maximum power transferred to the locomotive during its acceleration?

15.146 Just before it lifts off, a 10,500-kg airplane is traveling at 60 ft/s. The total horizontal force exerted by the plane's engines is 189 kN, and the plane is accelerating at 15 m/s².

(a) How much power is being transferred to the plane by its engines?

(b) What is the total power being transferred to the plane?

Problem 15.146

15.147 The "Paris Gun" used by Germany in World War I had a range of 120 km, a 37.5-m barrel, and a muzzle velocity of 1550 m/s and fired a 120-kg shell.

(a) If you assume the shell's acceleration to be constant, what maximum power was transferred to the shell as it traveled along the barrel?

(b) What average power was transferred to the shell?

Problem 15.147

Design Project

Determine the specifications (unstretched length and spring constant k) for the elastic cord to be used at a bungee-jumping facility. Participants are to jump from a platform 150 ft above the ground. When they rebound, they must avoid an obstacle that extends 15 ft below the point at which they jumped. In determining the specifications for the cord, establish reasonable safety limits for the minimum distances by which participants must avoid the ground and obstacle. Account for the fact that participants will have different weights. If necessary, specify a maximum allowable weight for participants. Write a brief report presenting your analyses and making a design recommendation for the specifications of the cord.

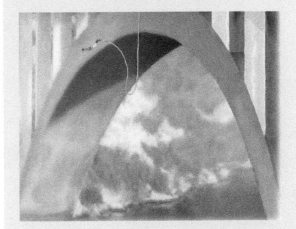

CHAPTER

16

Momentum Methods

Integrating Newton's second law with respect to time yields a relation between the time integral of the forces acting on an object and the change in the object's linear momentum. With this result, called the principle of impulse and momentum, we can determine the change in an object's velocity when the external forces are known as functions of time, analyze impacts between objects, and evaluate forces exerted by continuous flows of mass.

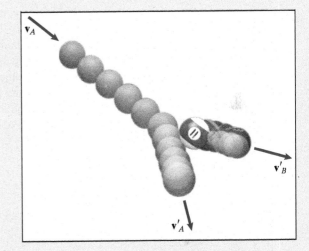

◄ The total linear momentum of the pool balls is approximately the same before and after their impact. In this chapter we use methods based on linear and angular momentum to analyze motions of objects.

16.1 Principle of Impulse and Momentum

BACKGROUND

The principle of work and energy is a very useful tool in mechanics. We can derive another useful tool for the analysis of motion by integrating Newton's second law with respect to time. We express Newton's second law in the form

$$\Sigma \mathbf{F} = m\frac{d\mathbf{v}}{dt}.$$

Then we integrate with respect to time to obtain

$$\int_{t_1}^{t_2} \Sigma \mathbf{F}\, dt = m\mathbf{v}_2 - m\mathbf{v}_1, \tag{16.1}$$

where $\mathbf{v}_1$ and $\mathbf{v}_2$ are the velocities of the center of mass of the object at the times t_1 and t_2. The term on the left is called the *linear impulse*, and $m\mathbf{v}$ is the *linear momentum*. Equation (16.1) is called the *principle of impulse and momentum*: The impulse applied to an object during an interval of time is equal to the change in the object's linear momentum (Fig. 16.1). The dimensions of the linear impulse and linear momentum are (mass) $\times$ (length)/(time).

The average with respect to time of the total force acting on an object from t_1 to t_2 is

$$\Sigma \mathbf{F}_{av} = \frac{1}{t_2 - t_1}\int_{t_1}^{t_2} \Sigma \mathbf{F}\, dt,$$

so we can write Eq. (16.1) as

$$(t_2 - t_1)\Sigma \mathbf{F}_{av} = m\mathbf{v}_2 - m\mathbf{v}_1. \tag{16.2}$$

With this equation, we can determine the average value of the total force acting on an object during a given interval of time if we know the change in the object's velocity.

A force that acts over a small interval of time but exerts a significant linear impulse is called an *impulsive force*. An impulsive force and its average with respect to time are shown in Fig. 16.2. Determining the time history of such a force is often impractical, but with Eq. (16.2) its average value can sometimes

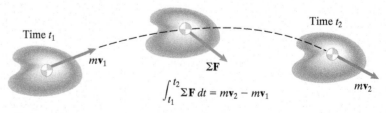

Time t_1

$m\mathbf{v}_1$

$\Sigma \mathbf{F}$

$\int_{t_1}^{t_2} \Sigma \mathbf{F}\, dt = m\mathbf{v}_2 - m\mathbf{v}_1$

Time t_2

$m\mathbf{v}_2$

Figure 16.1
Principle of impulse and momentum.

be determined. For example, a golf ball struck by a club is subjected to an impulsive force. By making high-speed motion pictures, the duration of the impact and the ball's velocity after the impact can be measured. Knowing the duration of the impact and the ball's change in linear momentum, we can determine the average force exerted by the club. (See Example 16.3.)

We can express Eqs. (16.1) and (16.2) in scalar forms that are often useful. The sum of the forces in the direction tangent to an object's path equals the product of the object's mass and the rate of change of its velocity along the path (see Eq. 14.7):

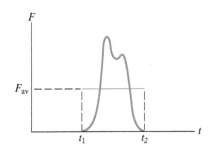

Figure 16.2
An impulsive force and its average value.

$$\Sigma F_{\mathrm{t}} = ma_{\mathrm{t}} = m\frac{dv}{dt}.$$

Integrating this equation with respect to time, we obtain

$$\int_{t_1}^{t_2} \Sigma F_{\mathrm{t}}\, dt = mv_2 - mv_1, \qquad (16.3)$$

where v_1 and v_2 are the velocities along the path at the times t_1 and t_2. The impulse applied to an object by the sum of the forces tangent to its path during an interval of time is equal to the change in the object's linear momentum along the path. In terms of the average with respect to time of the sum of the forces tangent to the path, or

$$\Sigma F_{\mathrm{t\,av}} = \frac{1}{t_2 - t_1}\int_{t_1}^{t_2} \Sigma F_{\mathrm{t}}\, dt,$$

we can write Eq. (16.3) as

$$(t_2 - t_1)\Sigma F_{\mathrm{t\,av}} = mv_2 - mv_1. \qquad (16.4)$$

This equation relates the average of the sum of the forces tangent to the path during an interval of time to the change in the velocity along the path.

Notice that Eq. (16.1) and the principle of work and energy, Eq. (15.6), are quite similar. They both relate an integral of the external forces to the change in an object's velocity. Equation (16.1) is a vector equation that determines the change in both the magnitude and direction of the velocity, whereas the principle of work and energy, a scalar equation, gives only the change in the magnitude of the velocity. But there is a greater difference between the two methods: In the case of impulse and momentum, there is no class of forces equivalent to the conservative forces that make the principle of work and energy so easy to apply.

When the external forces acting on an object are known as functions of time, the principle of impulse and momentum can be applied to determine the change in velocity of the object during an interval of time. Although this is an important result, it is not new. In Chapter 14, when we used Newton's second law to determine an object's acceleration and then integrated the acceleration with respect to time to determine the object's velocity, we were effectively applying the principle of impulse and momentum. However, in the rest of this chapter we show that this principle can be extended to new and interesting applications.

RESULTS

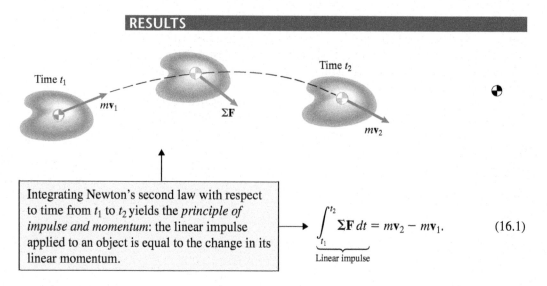

Integrating Newton's second law with respect to time from t_1 to t_2 yields the *principle of impulse and momentum*: the linear impulse applied to an object is equal to the change in its linear momentum.

$$\underbrace{\int_{t_1}^{t_2} \Sigma\mathbf{F}\,dt}_{\text{Linear impulse}} = m\mathbf{v}_2 - m\mathbf{v}_1. \qquad (16.1)$$

By introducing the average of the total force with respect to time from t_1 to t_2,

$$\Sigma\mathbf{F}_{av} = \frac{1}{t_2 - t_1}\int_{t_1}^{t_2} \Sigma\mathbf{F}\,dt,$$

the principle of impulse and momentum can be expressed in terms of the average force.

$$(t_2 - t_1)\Sigma\mathbf{F}_{av} = m\mathbf{v}_2 - m\mathbf{v}_1 \qquad (16.2)$$

A force that acts over a small interval of time but exerts significant linear impulse is called an *impulsive force*. The principle of impulse and moment often can be used to determine the average value of an impulsive force.

Alternative forms of the principle of impulse and momentum that are often useful. ΣF_t is the tangential component of the total force on an object and $\Sigma F_{t\,av}$ is the average of ΣF_t from t_1 to t_2. The terms v_1 and v_2 are the velocities along the path at t_1 and t_2.

$$\left\{ \begin{array}{ll} \displaystyle\int_{t_1}^{t_2} \Sigma F_t\,dt = mv_2 - mv_1, & (16.3) \\[2mm] (t_2 - t_1)\Sigma F_{t\,av} = mv_2 - mv_1. & (16.4) \end{array} \right.$$

Active Example 16.1   **Applying Impulse and Momentum** (▶ *Related Problem 16.7*)

A 1200-kg helicopter starts from rest at time $t = 0$. The components of the total force (in newtons) on the helicopter from $t = 0$ to $t = 10$ s are

$$\Sigma F_x = 720t,$$

$$\Sigma F_y = 2160 - 360t,$$

$$\Sigma F_z = 0.$$

Determine the helicopter's velocity at $t = 10$ s.

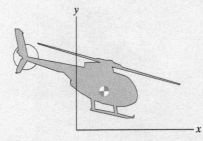

Strategy
We know the helicopter's velocity at $t = 0$ and the components of the total force acting on it as functions of time, so we can use Eq. (16.1) to determine its velocity at $t = 10$ s.

Solution

> Apply the principle of impulse and momentum from $t = 0$ to $t = 10$ s.

$$\int_{t_1}^{t_2} \Sigma \mathbf{F} \, dt = m\mathbf{v}_2 - m\mathbf{v}_1:$$

$$\int_0^{10} [720t\mathbf{i} + (2160 - 360t)\mathbf{j}]dt = (1200)\mathbf{v}_2 - (1200)(0),$$

$$\left[360t^2\mathbf{i} + (2160t - 180t^2)\mathbf{j}\right]_0^{10} = 1200\mathbf{v}_2,$$

$$36{,}000\mathbf{i} + 3600\mathbf{j} = 1200\mathbf{v}_2.$$

Solving for $\mathbf{v}_2$, the velocity at $t = 10$ s is $30\mathbf{i} + 3\mathbf{j}$ (m/s).

Practice Problem Suppose that the components of the total force acting on the helicopter from $t = 10$ s to $t = 20$ s are not known. But it is known that at $t = 20$ s, the helicopter's velocity is $36\mathbf{i} + 8\mathbf{j}$ (m/s). What is the average total force acting on it from $t = 10$ s to $t = 20$ s?

Answer: $720\mathbf{i} + 600\mathbf{j}$ (N).

Active Example 16.2   Impulse and Momentum Tangent to the Path (▶ *Related Problems 16.29, 16.30*)

The motorcycle starts from rest at time $t = 0$. The tangential component of the total force (in newtons) on it from $t = 0$ to $t = 30$ s is

$$\Sigma F_t = 300 - 9t.$$

The combined mass of the motorcycle and rider is 225 kg. What is the magnitude of the motorcycle's velocity at $t = 30$ s?

Strategy

The velocity at $t = 0$ is known and the tangential component of the total force is known as a function of time, so we can use Eq. (16.3) to determine the magnitude of the velocity at $t = 30$ s.

Solution

Apply Eq. (16.3) to the interval of time from $t = 0$ to $t = 30$ s.

$$\int_{t_1}^{t_2} \Sigma F_t \, dt = mv_2 - mv_1:$$

$$\int_0^{30} (300 - 9t)dt = 225v_2 - 225(0),$$

$$\left[300t - 4.5t^2 \right]_0^{30} = 225v_2,$$

$$4950 = 225v_2.$$

We obtain $v_2 = 22$ m/s.

Practice Problem   What is the average of the tangential component of the total force on the motorcycle from $t = 0$ to $t = 30$ s?

Answer: $\Sigma F_t = 165$ N.

| Example 16.3 | Determining an Impulsive Force (▶ *Related Problem 16.33*) |

A golf ball in flight is photographed at intervals of 0.001 s. The 1.62-oz ball is 1.68 in in diameter. If the club was in contact with the ball for 0.0006 s, estimate the average value of the impulsive force exerted by the club.

Strategy

By measuring the distance traveled by the ball in one of the 0.001-s intervals, we can estimate its velocity after being struck and then use Eq. (16.2) to determine the average total force on the ball.

Solution

By comparing the distance moved during one of the 0.001-s intervals with the known diameter of the ball, we estimate that the ball traveled 1.9 in and that its direction is 21° above the horizontal (Fig. a). The magnitude of the ball's velocity is

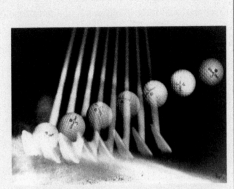

© Harold & Esther Edgerton Foundation, 2007, courtesy of Palm Press, Inc.

$$\frac{(1.9/12)\ \text{ft}}{0.001\ \text{s}} = 158\ \text{ft/s}.$$

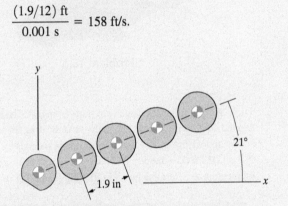

(a) Estimating the distance traveled during one 0.001-s interval.

The weight of the ball is $1.62/16 = 0.101$ lb, so its mass is $0.101/32.2 = 3.14 \times 10^{-3}$ slug. From Eq. (16.2), we obtain

$$(t_2 - t_1)\Sigma\mathbf{F}_{\text{av}} = m\mathbf{v}_2 - m\mathbf{v}_1:$$

$$(0.0006\ \text{s})\Sigma\mathbf{F}_{\text{av}} =$$

$$(3.14 \times 10^{-3}\ \text{slug})(158\ \text{ft/s})(\cos 21°\mathbf{i} + \sin 21°\mathbf{j}) - 0,$$

which yields

$$\Sigma\mathbf{F}_{\text{av}} = 775\mathbf{i} + 297\mathbf{j}\ (\text{lb}).$$

Critical Thinking

The average force during the time the club is in contact with the ball includes both the impulsive force exerted by the club and the ball's weight. In comparison with the large average impulsive force exerted by the club, the weight $(-0.101\mathbf{j}$ lb) is negligible.

Determining the time history of the force exerted on the ball by the club would require a complicated analysis accounting for the ball's deformation during the impact. In contrast, we were able to determine the *average* force exerted on the ball by a straightforward application of impulse and momentum.

Problems

16.1 The 20-kg crate is stationary at time $t = 0$. It is subjected to a horizontal force given as a function of time (in newtons) by $F = 10 + 2t^2$.

(a) Determine the magnitude of the linear impulse exerted on the crate from $t = 0$ to $t = 4$ s.

(b) Use the principle of impulse and momentum to determine how fast the crate is moving at $t = 4$ s.

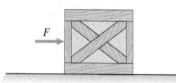

Problem 16.1

16.2 The 100-lb crate is released from rest on the inclined surface at time $t = 0$. The coefficient of kinetic friction between the crate and the surface is $\mu_k = 0.18$.

(a) Determine the magnitude of the linear impulse due to the forces acting on the crate from $t = 0$ to $t = 2$ s.

(b) Use the principle of impulse and momentum to determine how fast the crate is moving at $t = 2$ s.

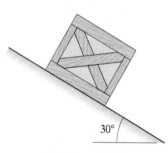

Problem 16.2

16.3 The mass of the helicopter is 9300 kg. It takes off vertically at time $t = 0$. The pilot advances the throttle so that the upward thrust of its engine (in kN) is given as a function of time in seconds by $T = 100 + 2t^2$.

(a) Determine the magnitude of the linear impulse due to the forces acting on the helicopter from $t = 0$ to $t = 3$ s.

(b) Use the principle of impulse and momentum to determine how fast the helicopter is moving at $t = 3$ s.

Problem 16.3

16.4 A 150 million-kg cargo ship starts from rest. The total force exerted on it by its engines and hydrodynamic drag (in newtons) can be approximated as a function of time in seconds by $\Sigma F_t = 937{,}500 - 0.65t^2$. Use the principle of impulse and momentum to determine how fast the ship is moving in 16 minutes.

Problem 16.4

16.5 The combined mass of the motorcycle and rider is 136 kg. The coefficient of kinetic friction between the motorcycle's tires and the road is $\mu_k = 0.6$. The rider starts from rest and spins the rear (drive) wheel. The normal force between the rear wheel and the road is 790 N.

(a) What impulse does the friction force on the rear wheel exert in 2 s?

(b) If you neglect other horizontal forces, what velocity is attained by the motorcycle in 2 s?

Problem 16.5

16.6 A bioengineer models the force generated by the wings of the 0.2-kg snow petrel by an equation of the form $F = F_0(1 + \sin \omega t)$, where F_0 and ω are constants. From video measurements of a bird taking off, he estimates that $\omega = 18$ and determines that the bird requires 1.42 s to take off and is moving at 6.1 m/s when it does. Use the principle of impulse and momentum to determine the constant F_0.

Problem 16.6

▶ **16.7** In Active Example 16.1, what is the average total force acting on the helicopter from $t = 0$ to $t = 10$ s?

16.8 At time $t = 0$, the velocity of the 15-kg object is $\mathbf{v} = 2\mathbf{i} + 3\mathbf{j} - 5\mathbf{k}$ (m/s). The total force acting on it from $t = 0$ to $t = 4$ s is

$$\Sigma\mathbf{F} = (2t^2 - 3t + 7)\mathbf{i} + 5t\mathbf{j} + (3t + 7)\mathbf{k} \text{ (N)}.$$

Use the principle of impulse and momentum to determine its velocity at $t = 4$ s.

16.9 At time $t = 0$, the velocity of the 15-kg object is $\mathbf{v} = 2\mathbf{i} + 3\mathbf{j} - 5\mathbf{k}$ (m/s). The total force acting on it from $t = 0$ to $t = 4$ s is

$$\Sigma\mathbf{F} = (2t^2 - 3t + 7)\mathbf{i} + 5t\mathbf{j} + (3t + 7)\mathbf{k} \text{ (N)}.$$

What is the average total force on the object during the interval of time from $t = 0$ to $t = 4$ s?

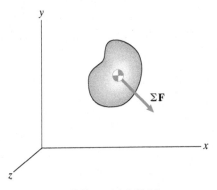

Problems 16.8/16.9

16.10 The 1-lb collar A is initially at rest in the position shown on the smooth horizontal bar. At $t = 0$, a force $\mathbf{F} = \frac{1}{20}t^2\mathbf{i} + \frac{1}{10}t\mathbf{j} - \frac{1}{30}t^3\mathbf{k}$ (lb) is applied to the collar, causing it to slide along the bar. What is the velocity of the collar at $t = 2$ s?

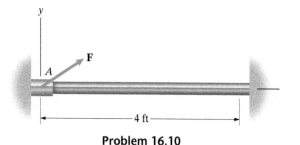

Problem 16.10

16.11 The y axis is vertical and the curved bar is smooth. The 4-lb slider is released from rest in position 1 and requires 1.2 s to slide to position 2. What is the magnitude of the average tangential force acting on the slider as it moves from position 1 to position 2?

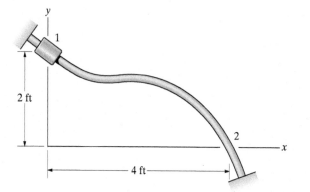

Problem 16.11

16.12 During the first 5 s of the 14,200-kg airplane's takeoff roll, the pilot increases the engine's thrust at a constant rate from 22 kN to its full thrust of 112 kN.

(a) What impulse does the thrust exert on the airplane during the 5 s?

(b) If you neglect other forces, what total time is required for the airplane to reach its takeoff speed of 46 m/s?

Problem 16.12

16.13 The 10-kg box starts from rest on the smooth surface and is subjected to the horizontal force described in the graph. Use the principle of impulse and momentum to determine how fast the box is moving at $t = 12$ s.

16.14 The 10-kg box starts from rest and is subjected to the horizontal force described in the graph. The coefficients of friction between the box and the surface are $\mu_s = \mu_k = 0.2$. Determine how fast the box is moving at $t = 12$ s.

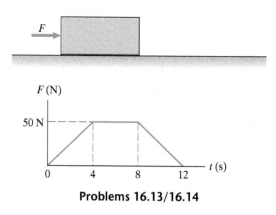

Problems 16.13/16.14

16.15 The crate has a mass of 120 kg, and the coefficients of friction between it and the sloping dock are $\mu_s = 0.6$ and $\mu_k = 0.5$. The crate starts from rest, and the winch exerts a tension $T = 1220$ N.

(a) What impulse is applied to the crate during the first second of motion?

(b) What is the crate's velocity after 1 s?

16.16 Solve Problem 16.15 if the crate starts from rest at $t = 0$ and the winch exerts a tension $T = 1220 + 200t$ N.

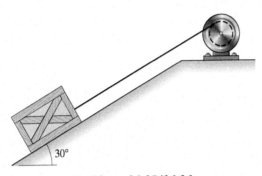

Problems 16.15/16.16

16.17 In an assembly-line process, the 20-kg package A starts from rest and slides down the smooth ramp. Suppose that you want to design the hydraulic device B to exert a constant force of magnitude F on the package and bring it to rest in 0.15 s. What is the required force F?

16.18 The 20-kg package A starts from rest and slides down the smooth ramp. If the hydraulic device B exerts a force of magnitude $F = 540(1 + 0.4t^2)$ N on the package, where t is in seconds measured from the time of first contact, what time is required to bring the package to rest?

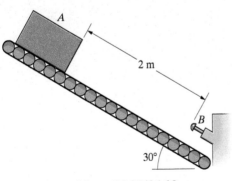

Problems 16.17/16.18

16.19 In a cathode-ray tube, an electron (mass $= 9.11 \times 10^{-31}$ kg) is projected at O with velocity $\mathbf{v} = (2.2 \times 10^7)\mathbf{i}$ (m/s). While the electron is between the charged plates, the electric field generated by the plates subjects it to a force $\mathbf{F} = -eE\mathbf{j}$. The charge on the electron is $e = 1.6 \times 10^{-19}$ C (coulombs), and the electric field strength is $E = 15 \sin(\omega t)$ kN/C, where the frequency $\omega = 2 \times 10^9$ s^{-1}.

(a) What impulse does the electric field exert on the electron while it is between the plates?

(b) What is the velocity of the electron as it leaves the region between the plates?

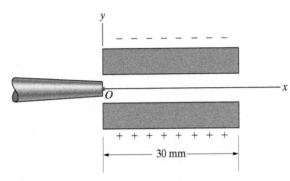

Problem 16.19

16.20 The two weights are released from rest at time $t = 0$. The coefficient of kinetic friction between the horizontal surface and the 5-lb weight is $\mu_k = 0.4$. Use the principle of impulse and momentum to determine the magnitude of the velocity of the 10-lb weight at $t = 1$ s.

Strategy: Apply the principle to each weight individually.

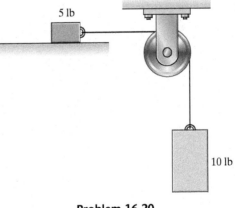

Problem 16.20

16.21 The two crates are released from rest. Their masses are $m_A = 40$ kg and $m_B = 30$ kg, and the coefficient of kinetic friction between crate A and the inclined surface is $\mu_k = 0.15$. What is the magnitude of the velocity of the crates after 1 s?

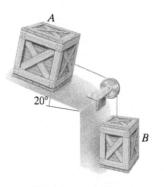

Problem 16.21

16.22 The two crates are released from rest. Their masses are $m_A = 20$ kg and $m_B = 80$ kg, and the surfaces are smooth. The angle $\theta = 20°$. What is the magnitude of the velocity of crate A after 1 s?

Strategy: Apply the principle of impulse and momentum to each crate individually.

16.23 The two crates are released from rest. Their masses are $m_A = 20$ kg and $m_B = 80$ kg. The coefficient of kinetic friction between the contacting surfaces is $\mu_k = 0.1$. The angle $\theta = 20°$. What is the magnitude of the velocity of crate A after 1 s?

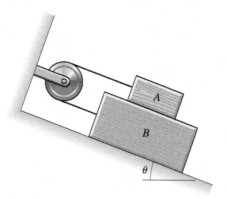

Problems 16.22/16.23

16.24 At $t = 0$, a 20-kg projectile is given an initial velocity $v_0 = 20$ m/s at $\theta_0 = 60°$ above the horizontal.

(a) By using Newton's second law to determine the acceleration of the projectile, determine its velocity at $t = 3$ s.

(b) What impulse is applied to the projectile by its weight from $t = 0$ to $t = 3$ s?

(c) Use the principle of impulse and momentum to determine the projectile's velocity at $t = 3$ s.

16.25 A soccer player kicks the stationary 0.45-kg ball to a teammate. The ball reaches a maximum height above the ground of 2 m at a horizontal distance of 5 m from the point where it was kicked. The duration of the kick was 0.04 seconds. Neglecting the effect of aerodynamic drag, determine the magnitude of the average force the player exerted on the ball.

Problem 16.25

16.26 An object of mass $m = 2$ kg slides with constant velocity $v_0 = 4$ m/s on a horizontal table (seen from above in the figure). The object is connected by a string of length $L = 1$ m to the fixed point O and is in the position shown, with the string parallel to the x axis, at $t = 0$.

(a) Determine the x and y components of the force exerted on the mass by the string as functions of time.

(b) Use your results from part (a) and the principle of impulse and momentum to determine the velocity vector of the mass at $t = 1$ s.

Strategy: To do part (a), write Newton's second law for the mass in terms of polar coordinates.

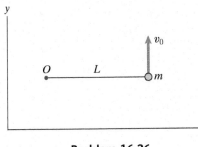

Problem 16.26

16.27 A rail gun, which uses an electromagnetic field to accelerate an object, accelerates a 30-g projectile from zero to 5 km/s in 0.0004 s. What is the magnitude of the average force exerted on the projectile?

16.28 The mass of the boat and its passenger is 420 kg. At time $t = 0$, the boat is moving at 14 m/s and its motor is turned off. The magnitude of the hydrodynamic drag force on the boat (in newtons) is given as a function of time by $830(1 - 0.08t)$. Determine how long it takes for the boat's velocity to decrease to 5 m/s.

Problem 16.28

▶ 16.29 The motorcycle starts from rest at $t = 0$ and travels along a circular track with 300-m radius. From $t = 0$ to $t = 10$ s, the component of the total force on the motorcycle tangential to its path is $\Sigma F_t = 600$ N. The combined mass of the motorcycle and rider is 150 kg. Use the principle of impulse and momentum to determine the magnitude of the motorcycle's velocity at $t = 10$ s. (See Active Example 16.2.)

▶ 16.30 The motorcycle starts from rest at $t = 0$ and travels along a circular track with 300-m radius. From $t = 0$ to $t = 10$ s, the component of the total tangential force on the motorcycle is given as a function of time by $\Sigma F_t = 460 + 3t^2$ N. The combined mass of the motorcycle and rider is 150 kg. Use the principle of impulse and momentum to determine the magnitude of the motorcycle's velocity at $t = 10$ s. (See Active Example 16.2.)

Problems 16.29/16.30

16.31 The titanium rotor of a Beckman Coulter ultracentrifuge used in biomedical research contains 2-gram samples at a distance of 41.9 mm from the axis of rotation. The rotor reaches its maximum speed of 130,000 rpm in 12 minutes.

(a) Determine the average tangential force exerted on a sample during the 12 minutes the rotor is accelerating.

(b) When the rotor is at its maximum speed, what normal acceleration are samples subjected to?

16.32 The angle θ between the horizontal and the airplane's path varies from $\theta = 0$ to $\theta = 30°$ at a constant rate of 5 degrees per second. During this maneuver, the airplane's thrust and aerodynamic drag are balanced, so that the only force exerted on the airplane in the direction tangent to its path is due to its weight. The magnitude of the airplane's velocity when $\theta = 0$ is 120 m/s. Use the principle of impulse and momentum to determine the magnitude of the velocity when $\theta = 30°$.

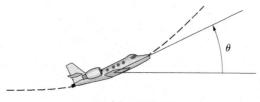

Problem 16.32

▶ 16.33 In Example 16.3, suppose that the mass of the golf ball is 0.046 kg and its diameter is 43 mm. The club is in contact with the ball for 0.0006 s, and the distance the ball travels in one 0.001-s interval is 50 mm. What is the magnitude of the average impulsive force exerted by the club?

16.34 In a test of an energy-absorbing bumper, a 2800-lb car is driven into a barrier at 5 mi/h. The duration of the impact is 0.4 seconds. When the car rebounds from the barrier, the magnitude of its velocity is 1.5 mi/h.

(a) What is the magnitude of the average horizontal force exerted on the car during the impact?

(b) What is the average deceleration of the car during the impact?

Problem 16.34

16.35 A bioengineer, using an instrumented dummy to test a protective mask for a hockey goalie, launches the 170-g puck so that it strikes the mask moving horizontally at 40 m/s. From photographs of the impact, she estimates its duration to be 0.02 s and observes that the puck rebounds at 5 m/s.

(a) What linear impulse does the puck exert?

(b) What is the average value of the impulsive force exerted on the mask by the puck?

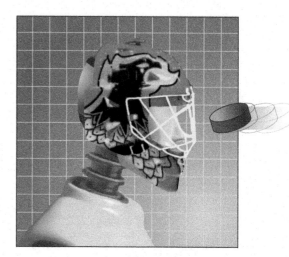

Problem 16.35

16.36 A fragile object dropped onto a hard surface breaks because it is subjected to a large impulsive force. If you drop a 2-oz watch from 4 ft above the floor, the duration of the impact is 0.001 s, and the watch bounces 2 in above the floor, what is the average value of the impulsive force?

16.37 The 0.45-kg soccer ball is given a kick with a 0.12-s duration that accelerates it from rest to a velocity of 12 m/s at 60° above the horizontal.

(a) What is the magnitude of the average total force exerted on the ball during the kick?

(b) What is the magnitude of the average force exerted on the ball by the player's foot during the kick?

Strategy: Use Eq. (16.2) to determine the average total force on the ball. To determine the average force exerted by the player's foot, you must subtract the ball's weight from the average total force.

Problem 16.37

16.38 An entomologist measures the motion of a 3-g locust during its jump and determines that the insect accelerates from rest to 3.4 m/s in 25 ms (milliseconds). The angle of takeoff is 55° above the horizontal. What are the horizontal and vertical components of the average impulsive force exerted by the locust's hind legs during the jump?

16.39 A 5-oz baseball is 3 ft above the ground when it is struck by a bat. The horizontal distance to the point where the ball strikes the ground is 180 ft. Photographic studies indicate that the ball was moving approximately horizontally at 100 ft/s before it was struck, the duration of the impact was 0.015 s, and the ball was traveling at 30° above the horizontal after it was struck. What was the magnitude of the average impulsive force exerted on the ball by the bat?

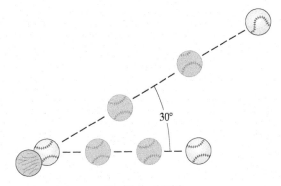

Problem 16.39

16.40 Paraphrasing the official rules of racquetball, a standard racquetball is $2\frac{1}{4}$ inches in diameter, weighs 1.4 ounces (16 ounces = 1 pound), and bounces between 68 and 72 inches from a 100-inch drop at a temperature between 70 and 74 degrees Fahrenheit. Suppose that a ball bounces 71 inches when it is dropped from a 100-inch height. If the duration of the impact is 0.08 s, what average force is exerted on the ball by the floor?

16.41 The masses $m_A = m_B$. The surface is smooth. At $t = 0$, A is stationary, the spring is unstretched, and B is given a velocity v_0 toward the right.

(a) In the subsequent motion, what is the velocity of the common center of mass of A and B?

(b) What are the velocities of A and B when the spring is unstretched?

Strategy: To do part (b), think about the motions of the masses relative to their common center of mass.

16.42 The masses $m_A = 40$ kg and $m_B = 30$ kg, and $k = 400$ N/m. The two masses are released from rest on the smooth surface with the spring stretched 1 m. What are the magnitudes of the velocities of the masses when the spring is unstretched?

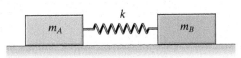

Problems 16.41/16.42

16.2 Conservation of Linear Momentum and Impacts

In this section, we consider the motions of several objects and show that if the effects of external forces can be neglected, the total linear momentum of the objects is conserved. (By *external forces*, we mean forces that are not exerted by the objects under consideration.) This result provides a powerful tool for analyzing interactions between objects, such as collisions, and also permits us to determine forces exerted on objects as a result of gaining or losing mass.

Conservation of Linear Momentum

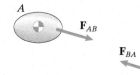
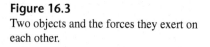

Figure 16.3
Two objects and the forces they exert on each other.

Consider the objects A and B in Fig. 16.3. $\mathbf{F}_{AB}$ is the force exerted on A by B, and $\mathbf{F}_{BA}$ is the force exerted on B by A. These forces could result from the two objects being in contact, for example, or could be exerted by a spring connecting them. As a consequence of Newton's third law, the two forces are equal and opposite, so that

$$\mathbf{F}_{AB} + \mathbf{F}_{BA} = \mathbf{0}. \tag{16.5}$$

Suppose that no external forces act on A and B or that the external forces are negligible in comparison with the forces that A and B exert on each other. Then we can apply the principle of impulse and momentum to each object for arbitrary times t_1 and t_2:

$$\int_{t_1}^{t_2} \mathbf{F}_{AB}\,dt = m_A \mathbf{v}_{A2} - m_A \mathbf{v}_{A1},$$

$$\int_{t_1}^{t_2} \mathbf{F}_{BA}\,dt = m_B \mathbf{v}_{B2} - m_B \mathbf{v}_{B1}.$$

If we sum these equations, the terms on the left cancel, and we obtain

$$m_A \mathbf{v}_{A1} + m_B \mathbf{v}_{B1} = m_A \mathbf{v}_{A2} + m_B \mathbf{v}_{B2},$$

which means that the *total linear momentum of A and B is conserved*:

$$m_A \mathbf{v}_A + m_B \mathbf{v}_B = \text{constant}. \tag{16.6}$$

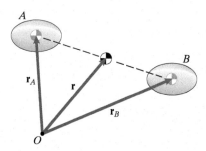

Figure 16.4
Position vector $\mathbf{r}$ of the common center of mass of A and B.

We can show that the velocity of the combined center of mass of the objects A and B (that is, the center of mass of A and B regarded as a single object) is also constant. Let $\mathbf{r}_A$ and $\mathbf{r}_B$ be the position vectors of their individual centers of mass (Fig. 16.4). The position of the combined center of mass is

$$\mathbf{r} = \frac{m_A \mathbf{r}_A + m_B \mathbf{r}_B}{m_A + m_B}.$$

By taking the derivative of this equation with respect to time and using Eq. (16.6), we obtain

$$\mathbf{v} = \frac{m_A \mathbf{v}_A + m_B \mathbf{v}_B}{m_A + m_B} = \text{constant}, \tag{16.7}$$

where $\mathbf{v} = d\mathbf{r}/dt$ is the velocity of the combined center of mass. Although the goal will usually be to determine the individual motions of the objects, knowing that the velocity of the combined center of mass is constant can contribute to our understanding of a problem, and in some instances the motion of the combined center of mass may be the only information that can be obtained.

Even when significant external forces act on A and B, if the external forces are negligible in a particular direction, Eqs. (16.6) and (16.7) apply in that direction. These equations also apply to an arbitrary number of objects: If the external forces acting on any collection of objects are negligible, the total linear momentum of the objects is conserved, and the velocity of their center of mass is constant.

Impacts

In machines that perform stamping or forging operations, dies impact against workpieces. Mechanical printers create images by impacting metal elements against the paper and platen. Vehicles impact each other intentionally, as when railroad cars are rolled against each other to couple them, and unintentionally in accidents. Impacts occur in many situations of concern in engineering. In this section, we consider a basic question: If we know the velocities of two objects before they collide, how can we determine their velocities afterward? In other words, what is the effect of the impact on the motions of the objects?

If colliding objects are not subjected to external forces, their total linear momentum must be the same before and after the impact. Even when they are subjected to external forces, the force of the impact is often so large, and its duration so brief, that the effect of external forces on the motions of the objects during the impact is negligible. Suppose that objects A and B with velocities $\mathbf{v}_A$ and $\mathbf{v}_B$ collide, and let $\mathbf{v}'_A$ and $\mathbf{v}'_B$ be their velocities after the impact (Fig. 16.5a). If the effects of external forces are negligible, then the total linear momentum of the system composed of A and B is conserved:

$$m_A\mathbf{v}_A + m_B\mathbf{v}_B = m_A\mathbf{v}'_A + m_B\mathbf{v}'_B. \tag{16.8}$$

Furthermore, the velocity $\mathbf{v}$ of the center of mass of A and B is the same before and after the impact. Thus, from Eq. (16.7),

$$\mathbf{v} = \frac{m_A\mathbf{v}_A + m_B\mathbf{v}_B}{m_A + m_B}. \tag{16.9}$$

If A and B adhere and remain together after they collide, they are said to undergo a *perfectly plastic impact*. Equation (16.9) gives the velocity of the center of mass of the object they form after the impact (Fig. 16.5b). A remarkable feature of this result is that we determine the velocity following the impact *without considering the physical nature of the impact*.

If A and B do not adhere, linear momentum conservation alone does not provide enough equations to determine their velocities after the impact. We first consider the case in which they travel along the same straight line before and after they collide.

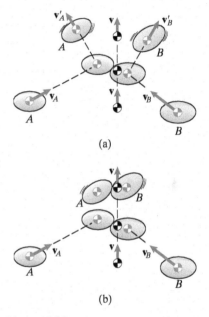

Figure 16.5
(a) Velocities of A and B before and after the impact, and the velocity $\mathbf{v}$ of their center of mass.

(b) A perfectly plastic impact.

Direct Central Impacts   Suppose that the centers of mass of A and B travel along the same straight line with velocities v_A and v_B before their impact (Fig. 16.6a). Let R be the magnitude of the force A and B exert on each other during the impact (Fig. 16.6b). We assume that the contacting surfaces are oriented so that R is parallel to the line along which the two objects travel and is directed toward their centers of mass. This condition, called *direct central impact*, means that A and B continue to travel along the same straight line after their impact (Fig. 16.6c). If the effects of external forces during the impact are negligible, the total linear momentum of the objects is conserved:

$$m_A v_A + m_B v_B = m_A v'_A + m_B v'_B. \tag{16.10}$$

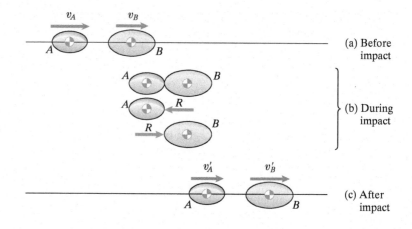

(a) Before impact

(b) During impact

(c) After impact

Figure 16.6
(a) Objects A and B traveling along the same straight line.
(b) During the impact, they exert a force R on each other.
(c) They travel along the same straight line after the central impact.

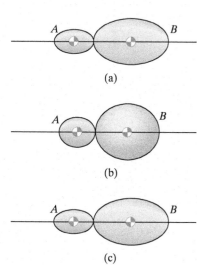

(a)

(b)

(c)

Figure 16.7
(a) First contact, $t = t_1$.
(b) Closest approach, $t = t_C$.
(c) End of contact, $t = t_2$.

However, we need another equation to determine the velocities v'_A and v'_B. To obtain it, we must consider the impact in more detail.

Let t_1 be the time at which A and B first come into contact (Fig. 16.7a). As a result of the impact, they will deform and their centers of mass will continue to approach each other. At a time t_C, their centers of mass will have reached their nearest proximity (Fig. 16.7b). At this time, the relative velocity of the two centers of mass is zero, so they have the same velocity. We denote it by v_C. The objects then begin to move apart and separate at a time t_2 (Fig. 16.7c). We apply the principle of impulse and momentum to A during the intervals of time from t_1 to the time of closest approach t_C and from t_C to t_2:

$$\int_{t_1}^{t_C} -R \, dt = m_A v_C - m_A v_A, \tag{16.11}$$

$$\int_{t_C}^{t_2} -R \, dt = m_A v'_A - m_A v_C. \tag{16.12}$$

Then we apply the principle to B for the same intervals of time:

$$\int_{t_1}^{t_C} R \, dt = m_B v_C - m_B v_B, \tag{16.13}$$

$$\int_{t_C}^{t_2} R \, dt = m_B v'_B - m_B v_C. \tag{16.14}$$

As a result of the impact, part of the objects' kinetic energy can be lost due to a variety of mechanisms, including permanent deformation and generation of heat and sound. As a consequence, the impulse they impart to each other during the "restitution" phase of the impact from t_C to t_2 is, in general, smaller than the impulse they impart from t_1 to t_C. The ratio of these impulses is called the *coefficient of restitution*:

$$e = \frac{\displaystyle\int_{t_C}^{t_2} R \, dt}{\displaystyle\int_{t_1}^{t_C} R \, dt}. \tag{16.15}$$

The value of e depends on the properties of the objects as well as their velocities and orientations when they collide, and it can be determined only by experiment or by a detailed analysis of the deformations of the objects during the impact.

If we divide Eq. (16.12) by Eq. (16.11) and divide Eq. (16.14) by Eq. (16.13), we can express the resulting equations in the forms

$$(v_C - v_A)e = v'_A - v_C$$

and

$$(v_C - v_B)e = v'_B - v_C.$$

Subtracting the first equation from the second, we obtain

$$e = \frac{v'_B - v'_A}{v_A - v_B}. \tag{16.16}$$

Thus, the coefficient of restitution is related in a simple way to the relative velocities of the objects before and after the impact. If e is known, Eq. (16.16) can be used together with the equation of conservation of linear momentum, Eq. (16.10), to determine v'_A and v'_B.

If $e = 0$, Eq. (16.16) indicates that $v'_B = v'_A$. The objects remain together after the impact, and the impact is perfectly plastic. If $e = 1$, it can be shown that the total kinetic energy is the same before and after the impact:

$$\tfrac{1}{2}m_A v_A^2 + \tfrac{1}{2}m_B v_B^2 = \tfrac{1}{2}m_A(v'_A)^2 + \tfrac{1}{2}m_B(v'_B)^2 \quad \text{(when } e = 1\text{).}$$

An impact in which kinetic energy is conserved is called *perfectly elastic*. Although this is sometimes a useful approximation, energy is lost in any impact in which material objects come into contact. If a collision can be heard, kinetic energy has been converted into sound. Permanent deformations of the colliding objects after the impact also represent losses of kinetic energy.

Oblique Central Impacts   We can extend the procedure used to analyze direct central impacts to the case in which the objects approach each other at an oblique angle. Suppose that A and B approach with arbitrary velocities $\mathbf{v}_A$ and $\mathbf{v}_B$ (Fig. 16.8) and that the forces they exert on each other during their impact are parallel to the x axis and point toward their centers of mass. No forces are exerted on A and B in the y or z directions, so their velocities in those directions are unchanged by the impact:

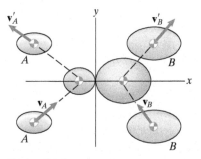

$$(\mathbf{v}'_A)_y = (\mathbf{v}_A)_y, \qquad (\mathbf{v}'_B)_y = (\mathbf{v}_B)_y,$$
$$(\mathbf{v}'_A)_z = (\mathbf{v}_A)_z, \qquad (\mathbf{v}'_B)_z = (\mathbf{v}_B)_z. \tag{16.17}$$

Figure 16.8
An oblique central impact.

In the x direction, linear momentum is conserved:

$$m_A(\mathbf{v}_A)_x + m_B(\mathbf{v}_B)_x = m_A(\mathbf{v}'_A)_x + m_B(\mathbf{v}'_B)_x. \tag{16.18}$$

By the same analysis we used to arrive at Eq. (16.16), the x components of velocity satisfy the relation

$$e = \frac{(\mathbf{v}'_B)_x - (\mathbf{v}'_A)_x}{(\mathbf{v}_A)_x - (\mathbf{v}_B)_x}. \tag{16.19}$$

We can analyze an oblique central impact in which an object A hits a stationary object B if friction is negligible. Suppose that B is constrained so that it cannot move relative to the inertial reference frame. For example, in Fig. 16.9, A strikes a wall B that is fixed relative to the earth. The y and z components of A's velocity are unchanged, because friction is neglected and the impact exerts no force in those directions. The x component of A's velocity after the impact is given by Eq. (16.19) with B's velocity equal to zero:

$$(\mathbf{v}'_A)_x = -e(\mathbf{v}_A)_x.$$

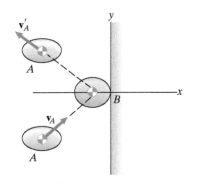

Figure 16.9
Impact with a stationary object.

Conservation of Linear Momentum

If the only forces acting on two objects A and B are the forces they exert on each other, *their total linear momentum is conserved* and *the velocity $\mathbf{v}$ of their common center of mass is constant.*

$$m_A\mathbf{v}_A + m_B\mathbf{v}_B = \text{constant.} \quad (16.6)$$

$$\mathbf{v} = \frac{m_A\mathbf{v}_A + m_B\mathbf{v}_B}{m_A + m_B} = \text{constant.} \quad (16.7)$$

Impacts

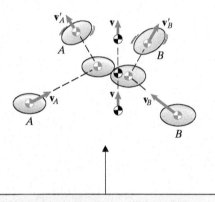

If two objects A and B collide and the effects of external forces are negligible, their total linear momentum and the velocity of their common center of mass are the same before and after the impact.

$$m_A\mathbf{v}_A + m_B\mathbf{v}_B = m_A\mathbf{v}_A' + m_B\mathbf{v}_B', \quad (16.8)$$

$$\mathbf{v} = \frac{m_A\mathbf{v}_A + m_B\mathbf{v}_B}{m_A + m_B}. \quad (16.9)$$

Perfectly Plastic Collision

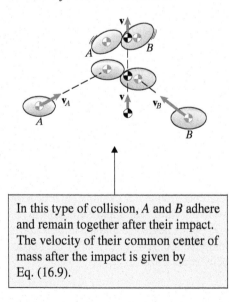

In this type of collision, A and B adhere and remain together after their impact. The velocity of their common center of mass after the impact is given by Eq. (16.9).

Direct Central Impact

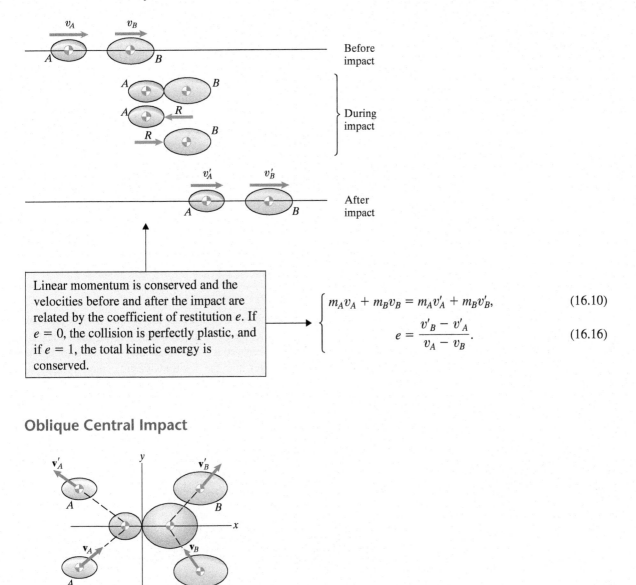

Linear momentum is conserved and the velocities before and after the impact are related by the coefficient of restitution e. If $e = 0$, the collision is perfectly plastic, and if $e = 1$, the total kinetic energy is conserved.

$$m_A v_A + m_B v_B = m_A v'_A + m_B v'_B, \tag{16.10}$$

$$e = \frac{v'_B - v'_A}{v_A - v_B}. \tag{16.16}$$

Oblique Central Impact

The components of velocity parallel to the plane of the impact are unchanged.

$$(\mathbf{v}'_A)_y = (\mathbf{v}_A)_y, \quad (\mathbf{v}'_B)_y = (\mathbf{v}_B)_y,$$

$$(\mathbf{v}'_A)_z = (\mathbf{v}_A)_z, \quad (\mathbf{v}'_B)_z = (\mathbf{v}_B)_z. \tag{16.17}$$

Linear momentum is conserved in the direction perpendicular to the plane of the impact and the components of velocity are related by the coefficient of restitution.

$$m_A (\mathbf{v}_A)_x + m_B (\mathbf{v}_B)_x = m_A (\mathbf{v}'_A)_x + m_B (\mathbf{v}'_B)_x. \tag{16.18}$$

$$e = \frac{(\mathbf{v}'_B)_x - (\mathbf{v}'_A)_x}{(\mathbf{v}_A)_x - (\mathbf{v}_B)_x}. \tag{16.19}$$

Active Example 16.4 | **Conservation of Linear Momentum** (▶ *Related Problem 16.43*)

A person of mass m_P stands at the center of a stationary barge of mass m_B. Suppose that the person runs to the right-hand end of the barge and stops. What are her position and the barge's position relative to their original positions? Neglect horizontal forces exerted on the barge by the water.

Strategy

The only horizontal forces on the person and barge are the forces they exert on each other, so the horizontal velocity of their common center of mass must be constant. Its velocity is initially zero, so it must remain zero—it does not move. We can use this condition to determine the positions of the centers of mass of the person and the barge when the person is at the right end of the barge.

Solution

Place the origin of the coordinate system at the initial position of the common center of mass of the person and barge. Let x_P be the person's position when she has arrived at the right end of the barge, and let x_B be the barge's position *to the left of the origin*.

The position of the common center of mass of the person and the barge must remain $\bar{x} = 0$.

$$\bar{x} = \frac{x_P m_P + (-x_B)m_B}{m_P + m_B} = 0.$$

Solving this equation together with the relation $x_P + x_B = L/2$ yields the two positions.

$$x_P = \frac{m_B L}{2(m_P + m_B)}, \qquad x_B = \frac{m_P L}{2(m_P + m_B)}.$$

Practice Problem If the person leaves her initial position at the center of the stationary barge and starts running with velocity v_P toward the right, what is the resulting velocity of the barge?

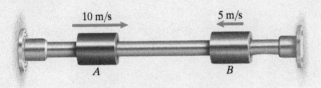

Answer: $v_B = (m_P/m_B)v_P$ toward the left.

Active Example 16.5   Analyzing an Impact (▶ *Related Problem 16.60*)

The 4-kg masses A and B slide on the smooth horizontal bar with the velocities shown. Determine their velocities after they collide if their coefficient of restitution is $e = 0.8$.

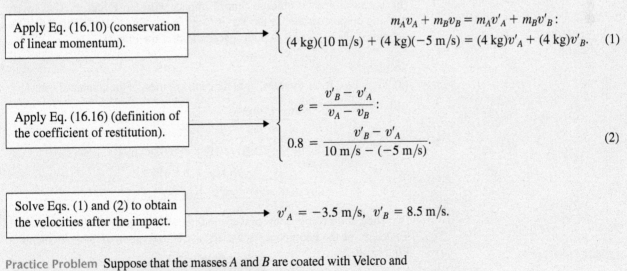

Strategy

Knowing the masses, the velocities before the collision, and the coefficient of restitution, we can use Eqs. (16.10) and (16.16) to determine the velocities of the two masses after the collision.

Apply Eq. (16.10) (conservation of linear momentum).	$\left\{ \begin{array}{l} m_A v_A + m_B v_B = m_A v'_A + m_B v'_B : \\ (4\ \text{kg})(10\ \text{m/s}) + (4\ \text{kg})(-5\ \text{m/s}) = (4\ \text{kg})v'_A + (4\ \text{kg})v'_B. \quad (1) \end{array} \right.$
Apply Eq. (16.16) (definition of the coefficient of restitution).	$\left\{ \begin{array}{l} e = \dfrac{v'_B - v'_A}{v_A - v_B} : \\ 0.8 = \dfrac{v'_B - v'_A}{10\ \text{m/s} - (-5\ \text{m/s})}. \quad (2) \end{array} \right.$
Solve Eqs. (1) and (2) to obtain the velocities after the impact.	$v'_A = -3.5\ \text{m/s}, \quad v'_B = 8.5\ \text{m/s}.$

Practice Problem Suppose that the masses A and B are coated with Velcro and stick together when they collide. What is their velocity after the impact?

Answer: 2.5 m/s.

Example 16.6	Applying Momentum Methods to Spacecraft Docking (▶ *Related Problem 16.77*)

The *Apollo* command-service module (*A*) attempts to dock with the *Soyuz* capsule (*B*), July 15, 1975. Their masses are $m_A = 18$ Mg and $m_B = 6.6$ Mg. The *Soyuz* is stationary relative to the reference frame shown, and the command-service module approaches with velocity $\mathbf{v}_A = 0.2\mathbf{i} + 0.03\mathbf{j} - 0.02\mathbf{k}$ (m/s).
(a) If the first attempt at docking is successful, what is the velocity of the center of mass of the combined vehicles afterward?
(b) If the first attempt is unsuccessful and the coefficient of restitution of the resulting impact is $e = 0.95$, what are the velocities of the two spacecraft after the impact?

Strategy
(a) If the docking is successful, the impact is perfectly plastic, and we can use Eq. (16.9) to determine the velocity of the center of mass of the combined object after the impact.
(b) By assuming an oblique central impact with the forces exerted by the docking collars parallel to the *x* axis, we can use Eqs. (16.18) and (16.19) to determine the velocities of both spacecraft after the impact.

Solution
(a) From Eq. (16.9), the velocity of the center of mass of the combined vehicles is

$$\mathbf{v} = \frac{m_A\mathbf{v}_A + m_B\mathbf{v}_B}{m_A + m_B}$$

$$= \frac{(18 \text{ Mg})[0.2\mathbf{i} + 0.03\mathbf{j} - 0.02\mathbf{k} \text{ (m/s)}] + \mathbf{0}}{18 \text{ Mg} + 6.6 \text{ Mg}}$$

$$= 0.146\mathbf{i} + 0.0220\mathbf{j} - 0.0146\mathbf{k} \text{ (m/s)}.$$

(b) The *y* and *z* components of the velocities of both spacecraft are unchanged. To determine the *x* components, we first use conservation of linear momentum, Eq. (16.18).

$$m_A(\mathbf{v}_A)_x + m_B(\mathbf{v}_B)_x = m_A(\mathbf{v}_A')_x + m_B(\mathbf{v}_B')_x:$$

$$(18 \text{ Mg})(0.2 \text{ m/s}) + 0 = (18 \text{ Mg})(\mathbf{v}_A')_x + (6.6 \text{ Mg})(\mathbf{v}_B')_x.$$

We then use the coefficient of restitution, Eq. (16.19), to obtain

$$e = \frac{(v'_B)_x - (v'_A)_x}{(v_A)_x - (v_B)_x}:$$

$$0.95 = \frac{(v'_B)_x - (v'_A)_x}{0.2 \text{ m/s} - 0}.$$

Solving these two equations yields $(v'_A)_x = 0.0954$ m/s and $(v'_B)_x = 0.285$ m/s, so the velocities of the spacecraft after the impact are

$$v'_A = 0.0954i + 0.03j - 0.02k \text{ (m/s)},$$

$$v'_B = 0.285i \text{ (m/s)}.$$

Critical Thinking

Why are calculations of this kind useful? Analytical simulations of the impact of the two spacecraft as they docked were used in designing the docking mechanisms and also in training the astronauts who performed the docking maneuver.

Problems

▶ **16.43** A girl weighing 80 lb stands at rest on a 325-lb floating platform. She starts running at 10 ft/s *relative to the platform* and runs off the end. Neglect the horizontal force exerted on the platform by the water.

(a) After she starts running, what is her velocity relative to the water?

(b) While she is running, what is the velocity of the common center of mass of the girl and the platform relative to the water? (See Active Example 16.4.)

16.44 Two railroad cars with weights $W_A = 120{,}000$ lb and $W_B = 70{,}000$ lb collide and become coupled together. Car A is full, and car B is half full, of carbolic acid. When the cars collide, the acid in B sloshes back and forth violently.

(a) Immediately after the impact, what is the velocity of the common center of mass of the two cars?

(b) When the sloshing in B has subsided, what is the velocity of the two cars?

16.45 The weights of the railroad cars are $W_A = 120{,}000$ lb and $W_B = 70{,}000$ lb. The railroad track has a constant slope of 0.2 degrees upward toward the right. If the cars are 6 ft apart at the instant shown, what is the velocity of their common center of mass immediately after they become coupled together?

Problem 16.43

A ⟶ 2 ft/s   B ⟶ 1 ft/s

Problems 16.44/16.45

16.46 The 400-kg satellite S traveling at 7 km/s is hit by a 1-kg meteor M traveling at 12 km/s. The meteor is embedded in the satellite by the impact. Determine the magnitude of the velocity of their common center of mass after the impact and the angle β between the path of the center of mass and the original path of the satellite.

16.47 The 400-kg satellite S traveling at 7 km/s is hit by a 1-kg meteor M. The meteor is embedded in the satellite by the impact. What would the magnitude of the velocity of the meteor need to be to cause the angle β between the original path of the satellite and the path of the center of mass of the combined satellite and meteor after the impact to be 0.5°? What is the magnitude of the velocity of the center of mass after the impact?

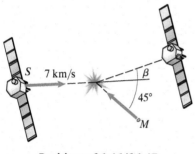

Problems 16.46/16.47

16.48 A 68-kg astronaut is initially stationary at the left side of an experiment module within an orbiting space shuttle. The 105,000-kg shuttle's center of mass is 4 m to the astronaut's right. He launches himself toward the center of mass at 1 m/s *relative to the shuttle*. He travels 8 m relative to the shuttle before bringing himself to rest at the opposite wall of the experiment module.

(a) What is the change in the magnitude of the shuttle's velocity relative to its original velocity while the astronaut is in motion?

(b) What is the change in the magnitude of the shuttle's velocity relative to its original velocity after his "flight"?

(c) Where is the shuttle's center of mass relative to the astronaut after his "flight"?

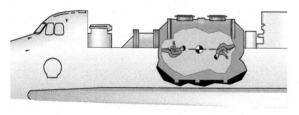

Problem 16.48

16.49 An 80-lb boy sitting in a stationary 20-lb wagon wants to simulate rocket propulsion by throwing bricks out of the wagon. Neglect horizontal forces on the wagon's wheels. If the boy has three bricks weighing 10 lb each and throws them with a horizontal velocity of 10 ft/s relative to the wagon, determine the velocity he attains (a) if he throws the bricks one at a time and (b) if he throws them all at once.

Problem 16.49

16.50 The catapult, designed to throw a line to ships in distress, throws a 2-kg projectile. The mass of the catapult is 36 kg, and it rests on a smooth surface. If the velocity of the projectile *relative to the earth* as it leaves the tube is 50 m/s at $\theta_0 = 30°$ relative to the horizontal, what is the resulting velocity of the catapult toward the left?

16.51 The catapult, which has a mass of 36 kg and throws a 2-kg projectile, rests on a smooth surface. The velocity of the projectile *relative to the catapult* as it leaves the tube is 50 m/s at $\theta_0 = 30°$ relative to the horizontal. What is the resulting velocity of the catapult toward the left?

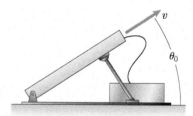

Problems 16.50/16.51

16.52 A bullet with a mass of 3.6 grams is moving horizontally with velocity v and strikes a 5-kg block of wood, becoming embedded in it. After the impact, the bullet and block slide 24 mm across the floor. The coefficient of kinetic friction between the block and the floor is $\mu_k = 0.4$. Determine the velocity v.

Problem 16.52

16.53 A 0.12-ounce bullet hits a suspended 15-lb block of wood and becomes embedded in it. The angle through which the wires supporting the block rotate as a result of the impact is measured and determined to be 7°. What was the bullet's velocity?

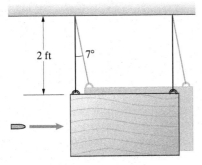

Problem 16.53

16.54 The overhead conveyor drops the 12-kg package A into the 1.6-kg carton B. The package is tacky and sticks to the bottom of the carton. If the coefficient of friction between the carton and the horizontal conveyor is $\mu_k = 0.2$, what distance does the carton slide after the impact?

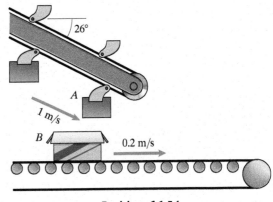

Problem 16.54

16.55 A 12,000-lb bus collides with a 2800-lb car. The velocity of the bus before the collision is $\mathbf{v}_B = 18\mathbf{i}$ (ft/s) and the velocity of the car is $\mathbf{v}_C = 33\mathbf{j}$ (ft/s). The two vehicles become entangled and remain together after the collision. The coefficient of kinetic friction between the vehicles' tires and the road is $\mu_k = 0.6$.

(a) What is the velocity of the common center of mass of the two vehicles immediately after the collision?

(b) Determine the approximate final position of the common center of mass of the vehicles relative to its position when the collision occurred. (Assume that the tires skid, not roll, on the road.)

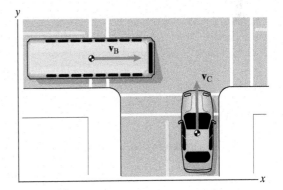

Problem 16.55

16.56 The velocity of the 200-kg astronaut A relative to the space station is $40\mathbf{i} + 30\mathbf{j}$ (mm/s). The velocity of the 300-kg structural member B relative to the station is $-20\mathbf{i} + 30\mathbf{j}$ (mm/s). When they approach each other, the astronaut grasps and clings to the structural member.

(a) What is the velocity of their common center of mass when they arrive at the station?

(b) Determine the approximate position at which they contact the station.

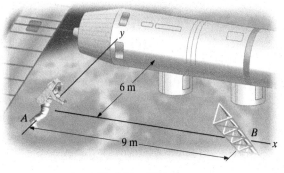

Problem 16.56

16.57 The weights of the two objects are $W_A = 5$ lb and $W_B = 8$ lb. Object A is moving at $v_A = 2$ ft/s and undergoes a perfectly elastic impact with the stationary object B. Determine the velocities of the objects after the impact.

16.58 The weights of the two objects are $W_A = 5$ lb and $W_B = 8$ lb. Object A is moving at $v_A = 2$ ft/s and undergoes a direct central impact with the stationary object B. The coefficient of restitution is $e = 0.8$. Determine the velocities of the objects after the impact.

Problems 16.57/16.58

16.59 The objects A and B with velocities $v_A = 20$ m/s and $v_B = 4$ m/s undergo a direct central impact. Their masses are $m_A = 8$ kg and $m_B = 12$ kg. After the impact, the object B is moving to the right at 16 m/s. What is the coefficient of restitution?

Problem 16.59

▶ **16.60** The 8-kg mass A and the 12-kg mass B slide on the smooth horizontal bar with the velocities shown. The coefficient of restitution is $e = 0.2$. Determine the velocities of the masses after they collide. (See Active Example 16.5.)

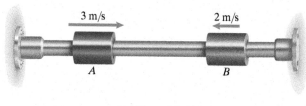

Problem 16.60

16.61 In a study of the effects of an accident on simulated occupants, the 1900-lb car with velocity $v_A = 30$ mi/h collides with the 2800-lb car with velocity $v_B = 20$ mi/h. The coefficient of restitution of the impact is $e = 0.15$. What are the velocities of the cars immediately after the collision?

16.62 In a study of the effects of an accident on simulated occupants, the 1900-lb car with velocity $v_A = 30$ mi/h collides with the 2800-lb car with velocity $v_B = 20$ mi/h. The coefficient of restitution of the impact is $e = 0.15$. The duration of the collision is 0.22 s. Determine the magnitude of the average acceleration to which the occupants of each car are subjected.

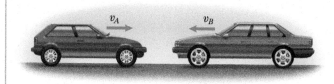

Problems 16.61/16.62

16.63 The balls are of equal mass m. Balls B and C are connected by an unstretched spring and are stationary. Ball A moves toward ball B with velocity v_A. The impact of A with B is perfectly elastic ($e = 1$).

(a) What is the velocity of the common center of mass of balls B and C immediately after the impact?

(b) What is the velocity of the common center of mass of B and C at time t after the impact?

16.64 In Problem 16.63, what is the maximum compressive force in the spring as a result of the impact?

16.65* The balls are of equal mass m. Balls B and C are connected by an unstretched spring and are stationary. Ball A moves toward ball B with velocity v_A. The impact of A with B is perfectly elastic ($e = 1$). Suppose that you interpret this as an impact between ball A and an "object" D consisting of the connected balls B and C.

(a) What is the coefficient of restitution of the impact between A and D?

(b) If you consider the total energy after the impact to be the sum of the kinetic energies $\frac{1}{2}m(v'_A)^2 + \frac{1}{2}(2m)(v'_D)^2$, where v'_D is the velocity of the center of mass of D after the impact, how much energy is "lost" as a result of the impact?

(c) How much energy is actually lost as a result of the impact? (This problem is an interesting model for one of the mechanisms of energy loss in impacts between objects. The energy "loss" calculated in part (b) is transformed into "internal energy"—the vibrational motions of B and C relative to their common center of mass.)

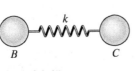

Problems 16.63–16.65

16.66 Suppose that you investigate an accident in which a 3400-lb car A struck a parked 1960-lb car B. All four of car B's wheels were locked, and skid marks indicate that it slid 20 ft after the impact. If you estimate the coefficient of kinetic friction between B's tires and the road to be $\mu_k = 0.8$ and the coefficient of restitution of the impact to be $e = 0.2$, what was A's velocity v_A just before the impact? (Assume that only one impact occurred.)

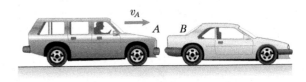

Problem 16.66

16.67 When the player releases the ball from rest at a height of 5 ft above the floor, it bounces to a height of 3.5 ft. If he throws the ball downward, releasing it at 3 ft above the floor, how fast would he need to throw it so that it would bounce to a height of 12 ft?

Problem 16.67

16.68 The 0.45-kg soccer ball is 1 m above the ground when it is kicked upward at 12 m/s. If the coefficient of restitution between the ball and the ground is $e = 0.6$, what maximum height above the ground does the ball reach on its first bounce?

16.69 The 0.45-kg soccer ball is stationary just before it is kicked upward at 12 m/s. If the impact lasts 0.02 s, what average force is exerted on the ball by the player's foot?

12 m/s

1 m

Problems 16.68/16.69

16.70 By making measurements directly from the photograph of the bouncing golf ball, estimate the coefficient of restitution.

16.71 If you throw the golf ball in Problem 16.70 horizontally at 2 ft/s and release it 4 ft above the surface, what is the distance between the first two bounces?

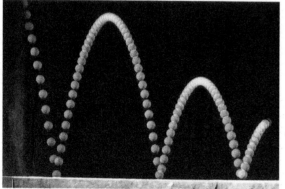

© Harold & Esther Edgerton Foundation, 2007, courtesy of Palm Press, Inc.

Problems 16.70/16.71

16.72 In a forging operation, the 100-lb weight is lifted into position 1 and released from rest. It falls and strikes a workpiece in position 2. If the weight is moving at 15 ft/s immediately before the impact and the coefficient of restitution is $e = 0.3$, what is the velocity of the weight immediately after the impact?

16.73 The 100-lb weight is released from rest in position 1. The spring constant is $k = 120$ lb/ft, and the springs are unstretched in position 2. If the coefficient of restitution of the impact of the weight with the workpiece in position 2 is $e = 0.6$, what is the magnitude of the velocity of the weight immediately after the impact?

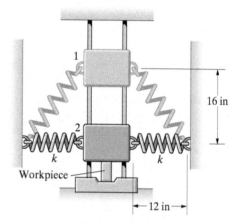

16 in

k k

Workpiece

12 in

Problems 16.72/16.73

16.74* A bioengineer studying helmet design uses an experimental apparatus that launches a 2.4-kg helmet containing a 2-kg model of the human head against a rigid surface at 6 m/s. The head suspended within the helmet is not immediately affected by the impact of the helmet with the surface and continues to move to the right at 6 m/s, so the head then undergoes an impact with the helmet. If the coefficient of restitution of the helmet's impact with the surface is 0.85 and the coefficient of restitution of the subsequent impact of the head with the helmet is 0.15, what is the velocity of the head after its initial impact with the helmet?

16.75* (a) If the duration of the impact of the head with the helmet in Problem 16.74 is 0.004 s, what is the magnitude of the average force exerted on the head by the impact?

(b) Suppose that the simulated head alone strikes the rigid surface at 6 m/s, the coefficient of restitution is 0.5, and the duration of the impact is 0.0002 s. What is the magnitude of the average force exerted on the head by the impact?

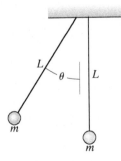

Problems 16.74/16.75

16.76 Two small balls, each of 1-lb weight, hang from strings of length $L = 3$ ft. The left ball is released from rest with $\theta = 35°$. The coefficient of restitution of the impact is $e = 0.9$. Through what maximum angle does the right ball swing?

Problem 16.76

▶ 16.77 In Example 16.6, if the *Apollo* command-service module approaches the *Soyuz* spacecraft with velocity $0.25\mathbf{i} + 0.04\mathbf{j} + 0.01\mathbf{k}$ (m/s) and the docking is successful, what is the velocity of the center of mass of the combined vehicles afterward?

16.78 The 3-kg object A and 8-kg object B undergo an oblique central impact. The coefficient of restitution is $e = 0.8$. Before the impact, $\mathbf{v}_A = 10\mathbf{i} + 4\mathbf{j} + 8\mathbf{k}$ (m/s) and $\mathbf{v}_B = -2\mathbf{i} - 6\mathbf{j} + 5\mathbf{k}$ (m/s). What are the velocities of A and B after the impact?

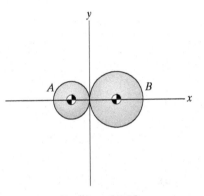

Problem 16.78

16.79 A baseball bat (shown with the bat's axis perpendicular to the page) strikes a thrown baseball. Before their impact, the velocity of the baseball is $\mathbf{v}_b = 132(\cos 45°\mathbf{i} + \cos 45°\mathbf{j})$ (ft/s) and the velocity of the bat is $\mathbf{v}_B = 60(-\cos 45°\mathbf{i} - \cos 45°\mathbf{j})$ (ft/s). Neglect the change in the velocity of the bat due to the direct central impact. The coefficient of restitution is $e = 0.2$. What is the ball's velocity after the impact? Assume that the baseball and the bat are moving horizontally. Does the batter achieve a potential hit or a foul ball?

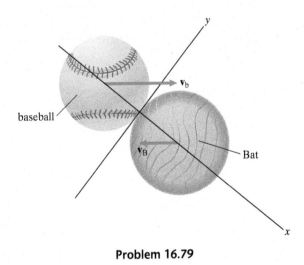

Problem 16.79

16.80 The cue gives the cue ball A a velocity parallel to the y axis. The cue ball hits the eight ball B and knocks it straight into the corner pocket. If the magnitude of the velocity of the cue ball just before the impact is 2 m/s and the coefficient of restitution is $e = 1$, what are the velocity vectors of the two balls just after the impact? (The balls are of equal mass.)

16.81 In Problem 16.80, what are the velocity vectors of the two balls just after the impact if the coefficient of restitution is $e = 0.9$?

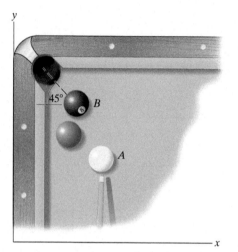

Problems 16.80/16.81

16.82 If the coefficient of restitution is the same for both impacts, show that the cue ball's path after two banks is parallel to its original path.

Problem 16.82

16.83 The velocity of the 170-g hockey puck is $\mathbf{v}_P = 10\mathbf{i} - 4\mathbf{j}$ (m/s). If you neglect the change in the velocity $\mathbf{v}_S = v_S\mathbf{j}$ of the stick resulting from the impact, and if the coefficient of restitution is $e = 0.6$, what should v_S be to send the puck toward the goal?

16.84 In Problem 16.83, if the stick responds to the impact the way an object with the same mass as the puck would and the coefficient of restitution is $e = 0.6$, what should v_S be to send the puck toward the goal?

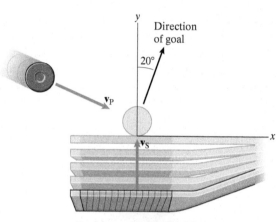

Problems 16.83/16.84

16.3 Angular Momentum

BACKGROUND

In this section we derive a result, analogous to the principle of impulse and momentum, that relates the integral of a moment with respect to time to the change in a quantity called the angular momentum.

Principle of Angular Impulse and Momentum

We describe the position of an object relative to an inertial reference frame with origin O by the position vector $\mathbf{r}$ from O to the object's center of mass (Fig. 16.10a). Recall that we obtained the very useful principle of work and energy by taking the dot product of Newton's second law with the velocity. Here we obtain another useful result by taking the cross product of Newton's second law with the position vector. This procedure gives us a relation between the moment of the external forces about O and the object's motion.

We take the cross product of Newton's second law with $\mathbf{r}$:

$$\mathbf{r} \times \Sigma\mathbf{F} = \mathbf{r} \times m\mathbf{a} = \mathbf{r} \times m\frac{d\mathbf{v}}{dt}. \tag{16.20}$$

Notice that the derivative of the quantity $\mathbf{r} \times m\mathbf{v}$ with respect to time is

$$\frac{d}{dt}(\mathbf{r} \times m\mathbf{v}) = \underbrace{\left(\frac{d\mathbf{r}}{dt} \times m\mathbf{v}\right)}_{= \mathbf{0}} + \left(\mathbf{r} \times m\frac{d\mathbf{v}}{dt}\right).$$

(The first term on the right side is zero because $d\mathbf{r}/dt = \mathbf{v}$ and the cross product of parallel vectors is zero.) Using this result, we can write Eq. (16.20) as

$$\mathbf{r} \times \Sigma\mathbf{F} = \frac{d\mathbf{H}_O}{dt}, \tag{16.21}$$

where the vector

$$\mathbf{H}_O = \mathbf{r} \times m\mathbf{v} \tag{16.22}$$

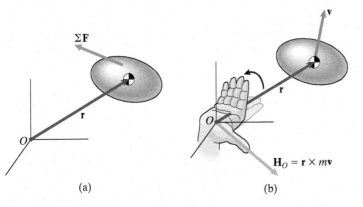

(a)   (b)

Figure 16.10
(a) The position vector and the total external force on an object.
(b) The angular momentum vector and the right-hand rule for determining its direction.

is called the *angular momentum* about O (Fig. 16.10b). If we interpret the angular momentum as the moment of the linear momentum of the object about point O, Eq. (16.21) states that the moment $\mathbf{r} \times \Sigma\mathbf{F}$ equals the rate of change of the moment of momentum about point O. *If the moment is zero during an interval of time, $\mathbf{H}_O$ is constant.*

Integrating Eq. (16.21) with respect to time, we obtain

$$\int_{t_1}^{t_2} (\mathbf{r} \times \Sigma\mathbf{F}) \, dt = (\mathbf{H}_O)_2 - (\mathbf{H}_O)_1. \qquad (16.23)$$

The integral on the left is called the *angular impulse*, and the equation itself is called the *principle of angular impulse and momentum*: The angular impulse applied to an object during an interval of time is equal to the change in the object's angular momentum. If we know the moment $\mathbf{r} \times \Sigma\mathbf{F}$ as a function of time, we can determine the change in the angular momentum. The dimensions of the angular impulse and angular momentum are (mass) $\times$ (length)^2/(time).

Central-Force Motion

If the total force acting on an object remains directed toward a point that is fixed relative to an inertial reference frame, the object is said to be in *central-force motion*. The fixed point is called the *center* of the motion. Orbits are the most familiar instances of central-force motion. For example, the gravitational force on an earth satellite remains directed toward the center of the earth.

If we place the reference point O at the center of the motion (Fig. 16.11a), the position vector $\mathbf{r}$ is parallel to the total force, so $\mathbf{r} \times \Sigma\mathbf{F}$ equals zero. Therefore, Eq. (16.23) indicates that in central-force motion, an object's angular momentum is conserved:

$$\mathbf{H}_O = \text{constant}. \qquad (16.24)$$

In plane central-force motion, we can express $\mathbf{r}$ and $\mathbf{v}$ in cylindrical coordinates (Fig. 16.11b):

$$\mathbf{r} = r\mathbf{e}_r, \qquad \mathbf{v} = v_r\mathbf{e}_r + v_\theta\mathbf{e}_\theta.$$

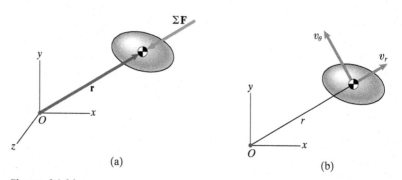

(a)   (b)

Figure 16.11
(a) Central-force motion.
(b) Expressing the position and velocity in cylindrical coordinates.

Substituting these expressions into Eq. (16.22), we obtain the angular momentum:

$$\mathbf{H}_O = (r\mathbf{e}_r) \times m(v_r\mathbf{e}_r + v_\theta\mathbf{e}_\theta) = mrv_\theta\mathbf{e}_z.$$

From this expression we see that

$$rv_\theta = \text{constant}. \tag{16.25}$$

RESULTS

Angular Momentum

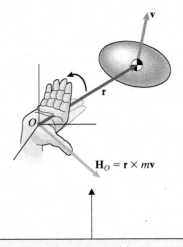

$$\mathbf{H}_O = \mathbf{r} \times m\mathbf{v}$$

The cross product of the position vector of the center of mass with the linear momentum is called the *angular momentum* associated with the motion of the center of mass.

$$\mathbf{H}_O = \mathbf{r} \times m\mathbf{v}. \tag{16.22}$$

Principle of Angular Impulse and Momentum

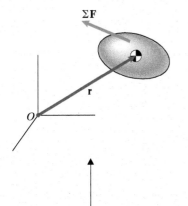

The integral with respect to time of the moment exerted by the external forces on an object, the *angular impulse*, is equal to the change in its linear momentum. If the moment is zero during an interval of time, the angular momentum is constant.

$$\underbrace{\int_{t_1}^{t_2} (\mathbf{r} \times \Sigma\mathbf{F})\,dt}_{\text{Angular impulse}} = (\mathbf{H}_O)_2 - (\mathbf{H}_O)_1. \tag{16.23}$$

Central-Force Motion

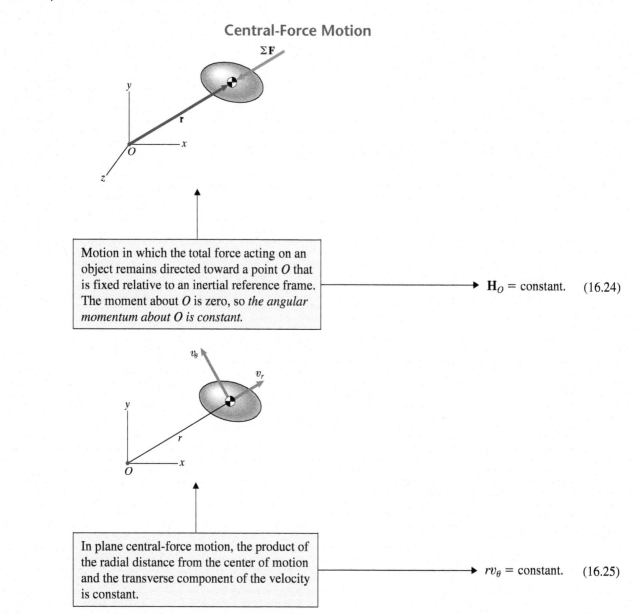

Motion in which the total force acting on an object remains directed toward a point O that is fixed relative to an inertial reference frame. The moment about O is zero, so *the angular momentum about O is constant.*

$\mathbf{H}_O = \text{constant.}$ (16.24)

In plane central-force motion, the product of the radial distance from the center of motion and the transverse component of the velocity is constant.

$rv_\theta = \text{constant.}$ (16.25)

Active Example 16.7 Angular Impulse and Momentum (▶ *Related Problem 16.91*)

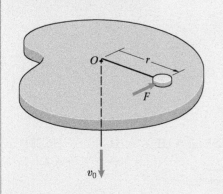

The disk of mass m slides on the smooth horizontal table. It is attached to a string that is drawn through a hole in the table at O at constant velocity v_0. At $t = 0$, the radial position is $r = r_0$ and the transverse velocity of the disk is zero. The disk is subjected to a force that has constant magnitude F and is perpendicular to the string. Determine the disk's velocity as a function of time.

Strategy

By expressing r as a function of time, we can determine the moment of the force on the disk about O as a function of time. The disk's angular momentum depends on its velocity, so we can apply the principle of angular impulse and momentum to obtain information about the velocity as a function of time.

Solution

Express r in terms of the initial radius r_0 and the constant velocity v_0.	$r = r_0 - v_0 t.$

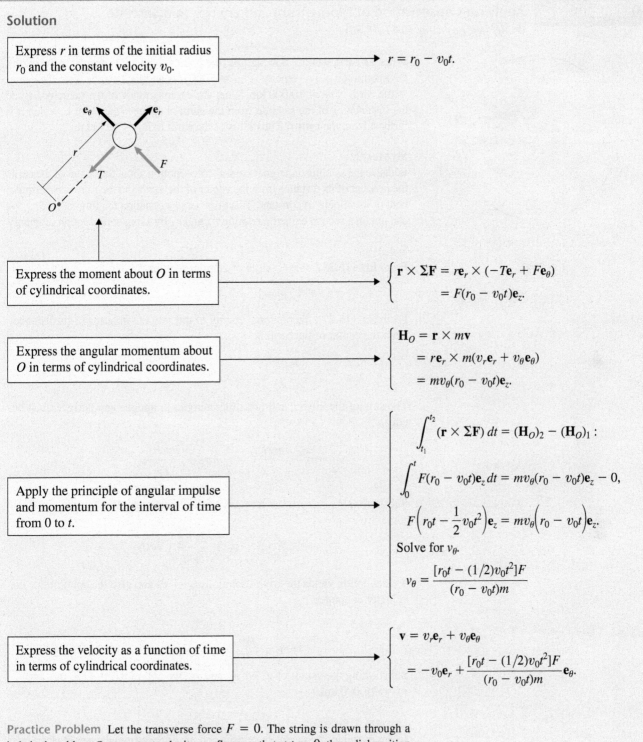

Express the moment about O in terms of cylindrical coordinates.	$\begin{cases} \mathbf{r} \times \Sigma \mathbf{F} = r\mathbf{e}_r \times (-T\mathbf{e}_r + F\mathbf{e}_\theta) \\ \qquad = F(r_0 - v_0 t)\mathbf{e}_z. \end{cases}$

Express the angular momentum about O in terms of cylindrical coordinates.	$\begin{cases} \mathbf{H}_O = \mathbf{r} \times m\mathbf{v} \\ \qquad = r\mathbf{e}_r \times m(v_r\mathbf{e}_r + v_\theta\mathbf{e}_\theta) \\ \qquad = mv_\theta(r_0 - v_0 t)\mathbf{e}_z. \end{cases}$

Apply the principle of angular impulse and momentum for the interval of time from 0 to t.	$\begin{cases} \displaystyle\int_{t_1}^{t_2} (\mathbf{r} \times \Sigma \mathbf{F})\, dt = (\mathbf{H}_O)_2 - (\mathbf{H}_O)_1 : \\[2mm] \displaystyle\int_0^t F(r_0 - v_0 t)\mathbf{e}_z\, dt = mv_\theta(r_0 - v_0 t)\mathbf{e}_z - 0, \\[2mm] F\!\left(r_0 t - \dfrac{1}{2}v_0 t^2\right)\mathbf{e}_z = mv_\theta\!\left(r_0 - v_0 t\right)\mathbf{e}_z. \\ \text{Solve for } v_\theta. \\[2mm] v_\theta = \dfrac{[r_0 t - (1/2)v_0 t^2]F}{(r_0 - v_0 t)m} \end{cases}$

Express the velocity as a function of time in terms of cylindrical coordinates.	$\begin{cases} \mathbf{v} = v_r\mathbf{e}_r + v_\theta\mathbf{e}_\theta \\[2mm] \quad = -v_0\mathbf{e}_r + \dfrac{[r_0 t - (1/2)v_0 t^2]F}{(r_0 - v_0 t)m}\mathbf{e}_\theta. \end{cases}$

Practice Problem Let the transverse force $F = 0$. The string is drawn through a hole in the table at O at constant velocity v_0. Suppose that at $t = 0$, the radial position is $r = r_0$ and the transverse velocity of the disk is v_0. Use the fact that the disk is in central-force motion to determine its velocity as a function of time.

Answer: $\mathbf{v} = -v_0\mathbf{e}_r + \dfrac{r_0 v_0}{r_0 - v_0 t}\mathbf{e}_\theta.$

| Example 16.8 | Applying Conservation of Momentum and Energy to a Satellite |

(▶ *Related Problems 16.87, 16.88*)

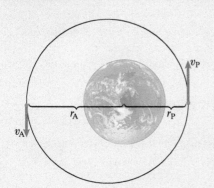

When an earth satellite is at perigee (the point at which it is nearest to the earth), the magnitude of its velocity is $v_P = 7000$ m/s and its distance from the center of the earth is $r_P = 10{,}000$ km. What are the magnitude of the velocity v_A and the distance r_A of the satellite from the earth at apogee (the point at which it is farthest from the earth)? The radius of the earth is $R_E = 6370$ km.

Strategy

Because the satellite undergoes central-force motion about the center of the earth, the product of its distance from the center of the earth and the transverse component of its velocity is constant. This gives us one equation relating v_A and r_A. We can obtain a second equation relating v_A and r_A by using conservation of energy.

Solution

From Eq. (16.25), conservation of angular momentum requires that

$$r_A v_A = r_P v_P.$$

From Eq. (15.27), the potential energy of the satellite in terms of its distance from the center of the earth is

$$V = -\frac{mgR_E^2}{r}.$$

The sum of the kinetic and potential energies at apogee and perigee must be equal:

$$\frac{1}{2}mv_A^2 - \frac{mgR_E^2}{r_A} = \frac{1}{2}mv_P^2 - \frac{mgR_E^2}{r_P}.$$

Substituting $r_A = r_P v_P / v_A$ into this equation and rearranging terms, we obtain

$$(v_A - v_P)\left(v_A + v_P - \frac{2gR_E^2}{r_P v_P}\right) = 0.$$

This equation yields the trivial solution $v_A = v_P$ and also the solution for the velocity at apogee:

$$v_A = \frac{2gR_E^2}{r_P v_P} - v_P.$$

Substituting the values of g, R_E, r_P, and v_P, we obtain $v_A = 4370$ m/s and $r_A = 16{,}000$ km.

Critical Thinking

In this example, the satellite's velocity and its distance from the center of the earth at perigee were known. Notice that with this information, you can use Eq. (16.25) to determine the transverse component of the satellite's velocity v_θ at *any* given radial position. You can use conservation of energy to determine the magnitude of the satellite's velocity at the same radial position, which means that you can also determine the radial component of the satellite's velocity v_r.

Problems

16.85 At the instant shown ($t_1 = 0$), the position of the 2-kg object's center of mass is $\mathbf{r} = 6\mathbf{i} + 4\mathbf{j} + 2\mathbf{k}$ (m) and its velocity is $\mathbf{v} = -16\mathbf{i} + 8\mathbf{j} - 12\mathbf{k}$ (m/s). No external forces act on the object. What is the object's angular momentum about the origin O at $t_2 = 1$ s?

16.86 The total external force on the 2-kg object is given as a function of time by $\Sigma\mathbf{F} = 2t\mathbf{i} + 4\mathbf{j}$ (N). At time $t_1 = 0$, the object's position and velocity are $\mathbf{r} = \mathbf{0}$ and $\mathbf{v} = \mathbf{0}$.

(a) Use Newton's second law to determine the object's velocity $\mathbf{v}$ and position $\mathbf{r}$ as functions of time.

(b) By integrating $\mathbf{r} \times \Sigma\mathbf{F}$ with respect to time from $t_1 = 0$ to $t_2 = 6$ s, determine the angular impulse about O exerted on the object during this interval of time.

(c) Use the results of part (a) to determine the change in the object's angular momentum from $t_1 = 0$ to $t_2 = 6$ s.

▶ **16.87** A satellite is in the elliptic earth orbit shown. Its velocity at perigee A is 8640 m/s. The radius of the earth is 6370 km.

(a) Use conservation of angular momentum to determine the magnitude of the satellite's velocity at apogee C.

(b) Use conservation of energy to determine the magnitude of the velocity at C.

(See Example 16.8.)

▶ **16.88** For the satellite in Problem 16.87, determine the magnitudes of the radial velocity v_r and transverse velocity v_θ at B.

(See Example 16.8.)

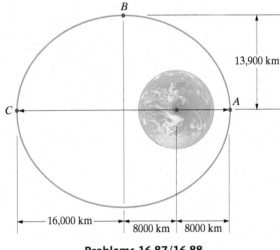

Problems 16.87/16.88

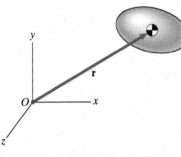

Problems 16.85/16.86

16.89 The bar rotates *in the horizontal plane* about a smooth pin at the origin. The 2-kg sleeve A slides on the smooth bar, and the mass of the bar is negligible in comparison to the mass of the sleeve. The spring constant is $k = 40$ N/m, and the spring is unstretched when $r = 0$. At $t = 0$, the radial position of the sleeve is $r = 0.2$ m and the angular velocity of the bar is $\omega_0 = 6$ rad/s. What is the angular velocity of the bar when $r = 0.25$ m?

16.90 The bar rotates *in the horizontal plane* about a smooth pin at the origin. The 2-kg sleeve A slides on the smooth bar, and the mass of the bar is negligible in comparison to the mass of the sleeve. The spring constant is $k = 40$ N/m, and the spring is unstretched when $r = 0$. At $t = 0$, the radial position of the sleeve is $r = 0.2$ m, its radial velocity is $v_r = 0$, and the angular velocity of the bar is $\omega_0 = 6$ rad/s. What are the angular velocity of the bar and the radial velocity of the sleeve when $r = 0.25$ m?

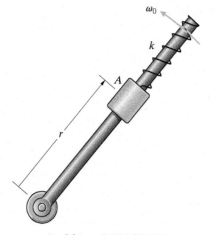

Problems 16.89/16.90

▶ **16.91** A 2-kg disk slides on a smooth horizontal table and is connected to an elastic cord whose tension is $T = 6r$ N, where r is the radial position of the disk in meters. If the disk is at $r = 1$ m and is given an initial velocity of 4 m/s in the transverse direction, what are the magnitudes of the radial and transverse components of its velocity when $r = 2$ m? (See Active Example 16.7.)

16.92 In Problem 16.91, determine the maximum value of r reached by the disk.

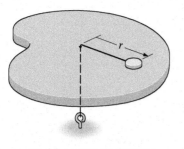

Problems 16.91/16.92

16.93 A 1-kg disk slides on a smooth horizontal table and is attached to a string that passes through a hole in the table.

(a) If the mass moves in a circular path of constant radius $r = 1$ m with a velocity of 2 m/s, what is the tension T?

(b) Starting from the initial condition described in part (a), the tension T is increased in such a way that the string is pulled through the hole at a constant rate until $r = 0.5$ m. Determine the value of T as a function of r while this is taking place.

16.94 In Problem 16.93, how much work is done on the mass in pulling the string through the hole as described in part (b)?

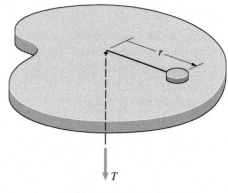

Problems 16.93/16.94

16.95 Two gravity research satellites ($m_A = 250$ kg, $m_B = 50$ kg) are tethered by a cable. The satellites and cable rotate with angular velocity $\omega_0 = 0.25$ revolution per minute. Ground controllers order satellite A to slowly unreel 6 m of additional cable. What is the angular velocity afterward?

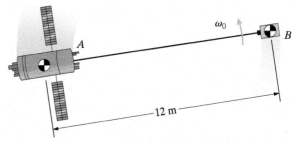

Problem 16.95

16.96 The astronaut moves in the x–y plane at the end of a 10-m tether attached to a large space station at O. The total mass of the astronaut and his equipment is 120 kg.

(a) What is the astronaut's angular momentum about O before the tether becomes taut?

(b) What is the magnitude of the component of his velocity perpendicular to the tether immediately after the tether becomes taut?

16.97 The astronaut moves in the x–y plane at the end of a 10-m tether attached to a large space station at O. The total mass of the astronaut and his equipment is 120 kg. The coefficient of restitution of the "impact" that occurs when he comes to the end of the tether is $e = 0.8$. What are the x and y components of his velocity immediately after the tether becomes taut?

16.98 A ball suspended from a string that goes through a hole in the ceiling at O moves with velocity v_A in a horizontal circular path of radius r_A. The string is then drawn through the hole until the ball moves with velocity v_B in a horizontal circular path of radius r_B. Use the principle of angular impulse and momentum to show that $r_A v_A = r_B v_B$.

Strategy: Let $\mathbf{e}$ be a unit vector that is perpendicular to the ceiling. Although this is not a central-force problem—the ball's weight does not point toward O—you can show that $\mathbf{e} \cdot (\mathbf{r} \times \Sigma \mathbf{F}) = 0$, so that $\mathbf{e} \cdot \mathbf{H}_O$ is conserved.

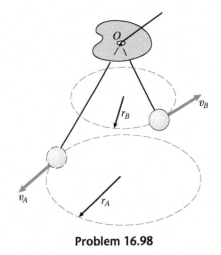

Problem 16.98

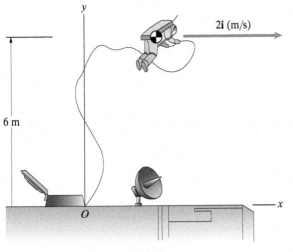

Problems 16.96/16.97

16.4 Mass Flows

BACKGROUND

In this section, we use conservation of linear momentum to determine the force exerted on an object as a result of emitting or absorbing a continuous flow of mass. The resulting equation applies to a variety of situations, including determining the thrust of a rocket and calculating the forces exerted on objects by flows of liquids or granular materials.

Suppose that an object of mass m and velocity $\mathbf{v}$ is subjected to no external forces (Fig. 16.12a) and emits an element of mass Δm_f with velocity $\mathbf{v}_f$ *relative to the object* (Fig. 16.12b). We denote the new velocity of the object by $\mathbf{v} + \Delta \mathbf{v}$. The linear momentum of the object before the element of mass is emitted equals the total linear momentum of the object and the element afterward:

$$m\mathbf{v} = (m - \Delta m_f)(\mathbf{v} + \Delta \mathbf{v}) + \Delta m_f(\mathbf{v} + \mathbf{v}_f).$$

Evaluating the products and simplifying, we obtain

$$m\Delta \mathbf{v} + \Delta m_f \mathbf{v}_f - \Delta m_f \Delta \mathbf{v} = 0. \tag{16.26}$$

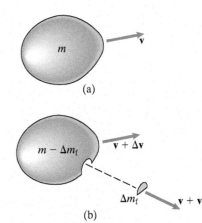

Figure 16.12
An object's mass and velocity (a) before and (b) after emitting an element of mass.

Now we assume that instead of shedding a discrete element of mass, the object emits a continuous flow of mass and that Δm_f is the amount emitted in an interval of time Δt. We divide Eq. (16.26) by Δt and write the resulting equation as

$$m\frac{\Delta \mathbf{v}}{\Delta t} + \frac{\Delta m_f}{\Delta t}\mathbf{v}_f - \frac{\Delta m_f}{\Delta t}\frac{\Delta \mathbf{v}}{\Delta t}\Delta t = 0.$$

Taking the limit of this equation as $\Delta t \rightarrow 0$, we obtain

$$-\frac{dm_f}{dt}\mathbf{v}_f = m\mathbf{a},$$

where $\mathbf{a}$ is the acceleration of the object's center of mass and the term dm_f/dt is the *mass flow rate*—the rate at which mass flows from the object. Comparing this equation with Newton's second law, we conclude that a flow of mass *from* an object exerts a force

$$\mathbf{F}_f = -\frac{dm_f}{dt}\mathbf{v}_f \tag{16.27}$$

on the object. The force is proportional to the mass flow rate and to the magnitude of the relative velocity of the flow, and its direction is *opposite* to the direction of the relative velocity. Conversely, a flow of mass *to* an object exerts a force in the *same* direction as the relative velocity.

RESULTS

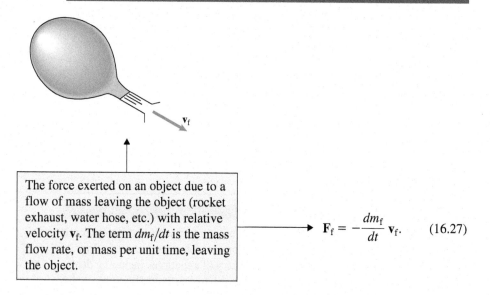

The force exerted on an object due to a flow of mass leaving the object (rocket exhaust, water hose, etc.) with relative velocity $\mathbf{v}_f$. The term dm_f/dt is the mass flow rate, or mass per unit time, leaving the object.

$$\mathbf{F}_f = -\frac{dm_f}{dt}\mathbf{v}_f. \tag{16.27}$$

Active Example 16.9 **Force Resulting from a Mass Flow** (▶ *Related Problem 16.113*)

The rocket sled is slowed by a water brake after its motor has burned out. A tube extends downward into a trough of water between the tracks. The open end of the tube points forward so that water enters the tube in the direction parallel to the x axis as the sled moves forward. The other end of the tube points upward, so that the water flows out in the direction parallel to the y axis. If v is the sled's velocity, the water enters the sled with velocity v relative to the sled and flows out with the same velocity. The mass flow rate through the tube is $\rho v A$, where

$\rho = 1000$ kg/m^3 is the density of the water and $A = 0.01$ m^2 is the cross-sectional area of the tube. At an instant when $v = 300$ m/s, what force is exerted on the sled by the flow of water entering it?

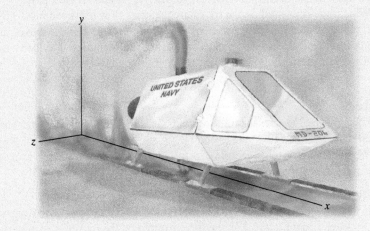

Strategy

We can use Eq. (16.27) to determine the force exerted by the water entering the sled.

Solution

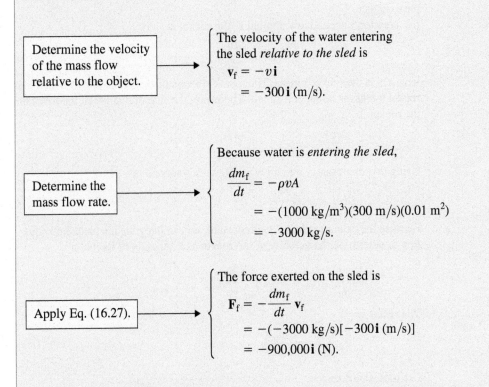

| Determine the velocity of the mass flow relative to the object. | → | The velocity of the water entering the sled *relative to the sled* is $$\mathbf{v}_f = -v\mathbf{i}$$ $$= -300\,\mathbf{i}\ (\text{m/s}).$$ |

| Determine the mass flow rate. | → | Because water is *entering the sled*, $$\frac{dm_f}{dt} = -\rho v A$$ $$= -(1000\ \text{kg/m}^3)(300\ \text{m/s})(0.01\ \text{m}^2)$$ $$= -3000\ \text{kg/s}.$$ |

| Apply Eq. (16.27). | → | The force exerted on the sled is $$\mathbf{F}_f = -\frac{dm_f}{dt}\mathbf{v}_f$$ $$= -(-3000\ \text{kg/s})[-300\mathbf{i}\ (\text{m/s})]$$ $$= -900{,}000\mathbf{i}\ (\text{N}).$$ |

Practice Problem Use Eq. (16.27) to determine the force exerted on the sled by the flow of water leaving it.

Answer: $\mathbf{F}_f = -900{,}000\mathbf{j}$ (N).

Example 16.10   **Thrust of a Rocket** (▶ *Related Problem 16.110*)

The classic example of a force created by a mass flow is the rocket. The one shown has a uniform, constant exhaust velocity v_f parallel to the x axis.
(a) What force is exerted on the rocket by the mass flow of its exhaust?
(b) If the force determined in part (a) is the only force acting on the rocket, and it starts from rest with an initial mass m_0, determine the rocket's velocity as a function of its mass m.

Strategy

(a) Equation (16.27) gives the force exerted on the rocket in terms of the exhaust velocity and the mass flow rate of fuel.
(b) We can use Newton's second law to obtain an equation for the rocket's velocity as a function of its mass.

Solution

(a) In terms of the coordinate system shown, the velocity vector of the exhaust is $\mathbf{v}_f = -v_f \mathbf{i}$. From Eq. (16.27), the force exerted on the rocket is

$$\mathbf{F}_f = -\frac{dm_f}{dt}\mathbf{v}_f = \frac{dm_f}{dt}v_f\mathbf{i},$$

where dm_f/dt is the mass flow rate of the rocket's fuel. The force exerted on the rocket by its exhaust is toward the right, opposite to the direction of the flow of the exhaust.

(b) Newton's second law applied to the rocket is

$$\Sigma F_x = \frac{dm_f}{dt}v_f = m\frac{dv_x}{dt},$$

where m is the rocket's mass. The mass flow rate of fuel is the rate at which the rocket's mass is being consumed. Therefore, the rate of change of the mass of the rocket is

$$\frac{dm}{dt} = -\frac{dm_f}{dt}.$$

Using this expression, we can write Newton's second law as

$$dv_x = -v_f\frac{dm}{m}.$$

Because the exhaust velocity v_f is constant, we can integrate the preceding equation to determine the velocity of the rocket as a function of its mass:

$$\int_0^{v_x} dv_x = -v_f\int_{m_0}^m \frac{dm}{m}.$$

The result is

$$v_x = v_f\ln\left(\frac{m_0}{m}\right).$$

Critical Thinking

The velocity attained by the rocket is determined by the exhaust velocity and the amount of mass expended. Thus, a rocket can gain more velocity by expending more of its mass. However, notice that increasing the ratio m_0/m from 10 to 100 increases the velocity attained by only a factor of two. In contrast, increasing the exhaust velocity results in a proportional increase in the rocket's velocity. Rocket engineers use fuels such as liquid oxygen and liquid hydrogen because they produce a relatively large exhaust velocity. This objective has also led to research on rocket engines that use electromagnetic fields to accelerate charged particles of fuel to large velocities.

Example 16.11   Force Resulting from a Mass Flow (▶ *Related Problem 16.104*)

A horizontal stream of water with velocity v_0 and mass flow rate dm_f/dt hits a plate that deflects the water in the horizontal plane through an angle θ. Assume that the magnitude of the velocity of the water when it leaves the plate is approximately equal to v_0. What force is exerted on the plate by the water?

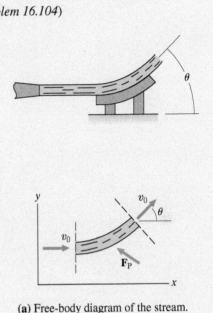

Strategy
We can determine the force exerted on the plate by treating the part of the stream in contact with the plate as an object with mass flows entering and leaving it.

Solution
In Fig. a, we draw the free-body diagram of the part of the stream in contact with the plate. Streams of mass with velocity v_0 enter and leave this "object," and $\mathbf{F_P}$ is the force exerted on the stream by the plate. We wish to determine the force $-\mathbf{F_P}$ exerted on the plate by the stream. First we consider the departing stream of water. The mass flow rate of water leaving the free-body diagram must be equal to the mass flow rate entering. In terms of the coordinate system shown, the velocity of the departing stream is

$$\mathbf{v_f} = v_0 \cos\theta\,\mathbf{i} + v_0 \sin\theta\,\mathbf{j}.$$

Let $\mathbf{F_D}$ be the force exerted on the object by the departing stream. From Eq. (16.27),

$$\mathbf{F_D} = -\frac{dm_f}{dt}\mathbf{v_f} = -\frac{dm_f}{dt}(v_0 \cos\theta\,\mathbf{i} + v_0 \sin\theta\,\mathbf{j}).$$

The velocity of the entering stream is $\mathbf{v_f} = v_0\mathbf{i}$. Since this flow is entering the object rather than leaving it, the resulting force $\mathbf{F_E}$ is in the same direction as the relative velocity:

$$\mathbf{F_E} = \frac{dm_f}{dt}\mathbf{v_f} = \frac{dm_f}{dt}v_0\mathbf{i}.$$

The sum of the forces on the free-body diagram must equal zero:

$$\mathbf{F_D} + \mathbf{F_E} + \mathbf{F_P} = \mathbf{0}.$$

Hence, the force exerted on the plate by the water is (Fig. b)

$$-\mathbf{F_P} = \mathbf{F_D} + \mathbf{F_E} = \frac{dm_f}{dt}v_0[(1 - \cos\theta)\mathbf{i} - \sin\theta\,\mathbf{j}].$$

(a) Free-body diagram of the stream.

(b) Force exerted on the plate.

Critical Thinking
This simple example affords an insight into how turbine blades and airplane wings can create forces by deflecting streams of liquid or gas (Fig. c).

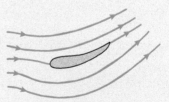

(c) Pattern of moving fluid around an airplane wing.

Example 16.12 | **Jet Engine** (▶ *Related Problems 16.118–16.120*)

In the turbojet engine, a mass flow rate dm_c/dt of inlet air enters the compressor with velocity v_i. The air is mixed with fuel and ignited in the combustion chamber. The mixture then flows through the turbine, which powers the compressor. The exhaust, with a mass flow rate equal to that of the air plus the mass flow rate of the fuel $(dm_c/dt + dm_f/dt)$, exits at a high exhaust velocity v_e, exerting a large force on the engine. Suppose that $dm_c/dt = 13.5$ kg/s and $dm_f/dt = 0.130$ kg/s. The inlet air velocity is $v_i = 120$ m/s and the exhaust velocity is $v_e = 480$ m/s. What is the engine's thrust?

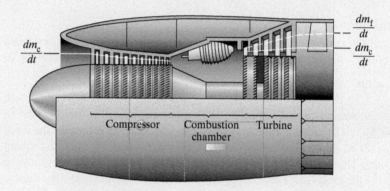

Compressor   Combustion   Turbine
              chamber

Strategy

We can determine the engine's thrust by using Eq. (16.27). We must include both the force exerted by the engine's exhaust and the force exerted by the mass flow of air entering the compressor to determine the net thrust.

Solution

The engine's exhaust exerts a force to the left equal to the product of the mass flow rate of the fuel–air mixture and the exhaust velocity. The inlet air exerts a force to the right equal to the product of the mass flow rate of the inlet air and the inlet velocity. The engine's thrust (the net force to the left) is

$$T = \left(\frac{dm_c}{dt} + \frac{dm_f}{dt}\right)v_e - \frac{dm_c}{dt}v_i$$

$$= (13.5 + 0.130)(480) - (13.5)(120)$$

$$= 4920 \text{ N}.$$

Problems

16.99 The Cheverton fire-fighting and rescue boat can pump 3.8 kg/s of water from each of its two pumps at a velocity of 44 m/s. If both pumps point in the same direction, what total force do they exert on the boat?

Problem 16.99

16.100 The mass flow rate of water through the nozzle is 1.6 slugs/s. Determine the magnitude of the horizontal force exerted on the truck by the flow of water.

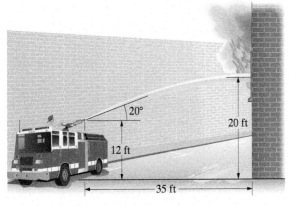

Problem 16.100

16.101 The front-end loader moves at a constant speed of 2 mi/h scooping up iron ore. The constant horizontal force exerted on the loader by the road is 400 lb. What weight of iron ore is scooped up in 3 s?

Problem 16.101

16.102 The snowblower moves at 1 m/s and scoops up 750 kg/s of snow. Determine the force exerted by the entering flow of snow.

16.103 The snowblower scoops up 750 kg/s of snow. It blows the snow out the side at 45° above the horizontal from a port 2 m above the ground and the snow lands 20 m away. What horizontal force is exerted on the blower by the departing flow of snow?

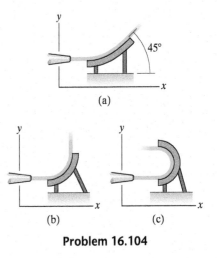

Problems 16.102/16.103

▶ **16.104** A nozzle ejects a stream of water horizontally at 40 m/s with a mass flow rate of 30 kg/s, and the stream is deflected in the horizontal plane by a plate. Determine the force exerted on the plate by the stream in cases (a), (b), and (c). (See Example 16.11.)

Problem 16.104

16.105* A stream of water with velocity 80**i** (m/s) and a mass flow rate of 6 kg/s strikes a turbine blade moving with constant velocity 20**i** (m/s).

(a) What force is exerted on the blade by the water?

(b) What is the magnitude of the velocity of the water as it leaves the blade?

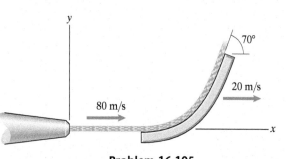

Problem 16.105

16.106 At the instant shown, the nozzle A of the lawn sprinkler is located at (0.1, 0, 0) m. Water exits each nozzle at 8 m/s relative to the nozzle with a mass flow rate of 0.22 kg/s. At the instant shown, the flow relative to the nozzle at A is in the direction of the unit vector

$$\mathbf{e} = \frac{1}{\sqrt{3}}\mathbf{i} - \frac{1}{\sqrt{3}}\mathbf{j} + \frac{1}{\sqrt{3}}\mathbf{k}.$$

Determine the total moment about the z axis exerted on the sprinkler by the flows from all four nozzles.

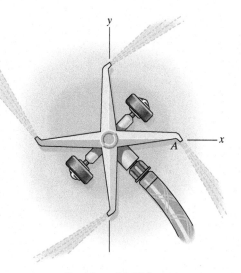

Problem 16.106

16.107 A 45-kg/s flow of gravel exits the chute at 2 m/s and falls onto a conveyer moving at 0.3 m/s. Determine the components of the force exerted on the conveyer by the flow of gravel if $\theta = 0$.

16.108 Solve Problem 16.107 if $\theta = 30°$.

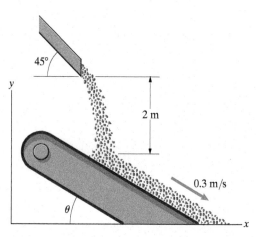

Problems 16.107/16.108

16.109 Suppose that you are designing a toy car that will be propelled by water that squirts from an internal tank at 10 ft/s relative to the car. The total weight of the car and its water "fuel" is to be 2 lb. If you want the car to achieve a maximum speed of 12 ft/s, what part of the total weight must be water?

Problem 16.109

▶ **16.110** The rocket consists of a 1000-kg payload and a 9000-kg booster. Eighty percent of the booster's mass is fuel, and its exhaust velocity is 1200 m/s. If the rocket starts from rest and external forces are neglected, what velocity will it attain? (See Example 16.10.)

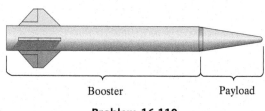

Booster   Payload

Problem 16.110

16.111* The rocket consists of a 1000-kg payload and a booster. The booster has two stages whose total mass is 9000 kg. Eighty percent of the mass of each stage is fuel, and the exhaust velocity of each stage is 1200 m/s. When the fuel of stage 1 is expended, it is discarded and the motor of stage 2 is ignited. Assume that the rocket starts from rest and neglect external forces. Determine the velocity attained by the rocket if the masses of the stages are $m_1 = 6000$ kg and $m_2 = 3000$ kg. Compare your result to the answer to Problem 16.110.

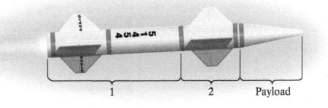

Problem 16.111

16.112 A rocket of initial mass m_0 takes off straight up. Its exhaust velocity v_f and the mass flow rate of its engine $\dot{m}_f = dm_f/dt$ are constant. Show that, during the initial part of the flight, when aerodynamic drag is negligible, the rocket's upward velocity as a function of time is

$$v = v_f \ln\left(\frac{m_0}{m_0 - \dot{m}_f t}\right) - gt.$$

Problem 16.112

▶ **16.113** The mass of the rocket sled in Active Example 16.9 is 440 kg. Assuming that the only significant force acting on the sled in the direction of its motion is the force exerted by the flow of water entering it, what distance is required for the sled to decelerate from 300 m/s to 100 m/s?

16.114* Suppose that you grasp the end of a chain that weighs 3 lb/ft and lift it straight up off the floor at a constant speed of 2 ft/s.

(a) Determine the upward force F you must exert as a function of the height s.

(b) How much work do you do in lifting the top of the chain to $s = 4$ ft?

   Strategy: Treat the part of the chain you have lifted as an object that is gaining mass.

16.115* Solve Problem 16.114, assuming that you lift the end of the chain straight up off the floor with a constant acceleration of 2 ft/s^2.

Problems 16.114/16.115

16.116* It has been suggested that a heavy chain could be used to gradually stop an airplane that rolls past the end of the runway. A hook attached to the end of the chain engages the plane's nose wheel, and the plane drags an increasing length of the chain as it rolls. Let m be the airplane's mass and v_0 its initial velocity, and let ρ_L be the mass per unit length of the chain. Neglecting friction and aerodynamic drag, what is the airplane's velocity as a function of s?

16.117* In Problem 16.116, the frictional force exerted on the chain by the ground would actually dominate other forces as the distance s increases. If the coefficient of kinetic friction between the chain and the ground is μ_k and you neglect all forces except the frictional force, what is the airplane's velocity as a function of s?

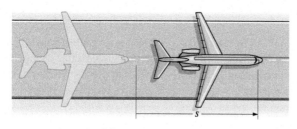

Problems 16.116/16.117

▶ **16.118** A turbojet engine is being operated on a test stand. The mass flow rate of air entering the compressor is 13.5 kg/s and the mass flow rate of fuel is 0.13 kg/s. The effective velocity of the air entering the compressor is zero, and the exhaust velocity is 500 m/s. What is the thrust of the engine? (See Example 16.12.)

▶ **16.119** A turbojet engine is in an airplane flying at 400 km/h. The mass flow rate of air entering the compressor is 13.5 kg/s and the mass flow rate of fuel is 0.13 kg/s. The effective velocity of the air entering the inlet is equal to the airplane's velocity, and the exhaust velocity (relative to the airplane) is 500 m/s. What is the thrust of the engine? (See Example 16.12.)

▶ **16.120** A turbojet engine's thrust reverser causes the exhaust to exit the engine at 20° from the engine centerline. The mass flow rate of air entering the compressor is 44 kg/s, and the air enters at 60 m/s. The mass flow rate of fuel is 1.5 kg/s, and the exhaust velocity is 370 m/s. What braking force does the engine exert on the airplane? (See Example 16.12.)

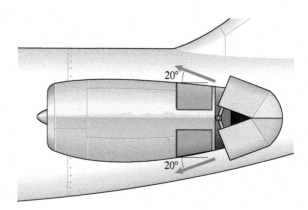

Problem 16.120

Review Problems

16.121 The total external force on a 10-kg object is constant and equal to $90\mathbf{i} - 60\mathbf{j} + 20\mathbf{k}$ (N). At $t = 2$ s, the object's velocity is $-8\mathbf{i} + 6\mathbf{j}$ (m/s).

(a) What impulse is applied to the object from $t = 2$ s to $t = 4$ s?

(b) What is the object's velocity at $t = 4$ s?

16.122 The total external force on an object is $\mathbf{F} = 10t\mathbf{i} + 60\mathbf{j}$ (lb). At $t = 0$, the object's velocity is $\mathbf{v} = 20\mathbf{j}$ (ft/s). At $t = 12$ s, the x component of its velocity is 48 ft/s.

(a) What impulse is applied to the object from $t = 0$ to $t = 6$ s?

(b) What is the object's velocity at $t = 6$ s?

16.123 An aircraft arresting system is used to stop airplanes whose braking systems fail. The system stops a 47,500-kg airplane moving at 80 m/s in 9.15 s.

(a) What impulse is applied to the airplane during the 9.15 s?

(b) What is the average deceleration to which the passengers are subjected?

Problem 16.123

16.124 The 1895 Austrian 150-mm howitzer had a 1.94-m barrel, possessed a muzzle velocity of 300 m/s, and fired a 38-kg shell. If the shell took 0.013 s to travel the length of the barrel, what average force was exerted on the shell?

16.125 An athlete throws a shot weighing 16 lb. When he releases it, the shot is 7 ft above the ground and its components of velocity are $v_x = 31$ ft/s and $v_y = 26$ ft/s.

(a) Suppose the athlete accelerates the shot from rest in 0.8 s, and assume as a first approximation that the force **F** he exerts on the shot is constant. Use the principle of impulse and momentum to determine the x and y components of **F**.

(b) What is the horizontal distance from the point where he releases the shot to the point where it strikes the ground?

Problem 16.125

16.126 The 6000-lb pickup truck A moving at 40 ft/s collides with the 4000-lb car B moving at 30 ft/s.

(a) What is the magnitude of the velocity of their common center of mass after the impact?

(b) Treat the collision as a perfectly plastic impact. How much kinetic energy is lost?

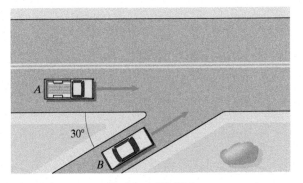

Problem 16.126

16.127 Two hockey players ($m_A = 80$ kg, $m_B = 90$ kg) converging on the puck at $x = 0$, $y = 0$ become entangled and fall. Before the collision, $v_A = 9\mathbf{i} + 4\mathbf{j}$ (m/s) and $v_B = -3\mathbf{i} + 6\mathbf{j}$ (m/s). If the coefficient of kinetic friction between the players and the ice is $\mu_k = 0.1$, what is their approximate position when they stop sliding?

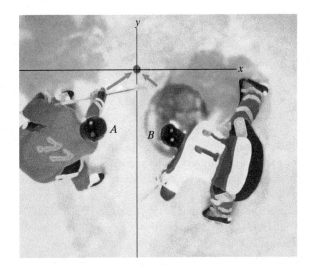

Problem 16.127

16.128 The cannon weighed 400 lb, fired a cannonball weighing 10 lb, and had a muzzle velocity of 200 ft/s. For the 10° elevation angle shown, determine (a) the velocity of the cannon after it was fired and (b) the distance the cannonball traveled. (Neglect drag.)

Problem 16.128

16.129 A 1-kg ball moving horizontally at 12 m/s strikes a 10-kg block. The coefficient of restitution of the impact is $e = 0.6$, and the coefficient of kinetic friction between the block and the inclined surface is $\mu_k = 0.4$. What distance does the block slide before stopping?

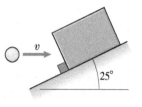

Problem 16.129

16.130 A Peace Corps volunteer designs the simple device shown for drilling water wells in remote areas. A 70-kg "hammer," such as a section of log or a steel drum partially filled with concrete, is hoisted to $h = 1$ m and allowed to drop onto a protective cap on the section of pipe being pushed into the ground. The combined mass of the cap and section of pipe is 20 kg. Assume that the coefficient of restitution is nearly zero.

(a) What is the velocity of the cap and pipe immediately after the impact?

(b) If the pipe moves 30 mm downward when the hammer is dropped, what resistive force was exerted on the pipe by the ground? (Assume that the resistive force is constant during the motion of the pipe.)

Problem 16.130

16.131 A tugboat (mass = 40,000 kg) and a barge (mass = 160,000 kg) are stationary with a slack hawser connecting them. The tugboat accelerates to 2 knots (1 knot = 1852 m/h) before the hawser becomes taut. Determine the velocities of the tugboat and the barge just after the hawser becomes taut (a) if the "impact" is perfectly plastic ($e = 0$) and (b) if the "impact" is perfectly elastic ($e = 1$). Neglect the forces exerted by the water and the tugboat's engines.

16.132 In Problem 16.131, determine the magnitude of the impulsive force exerted on the tugboat in the two cases if the duration of the "impact" is 4 s. Neglect the forces exerted by the water and the tugboat's engines during this period.

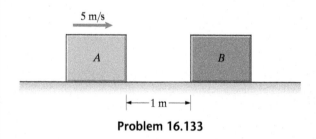

Problems 16.131/16.132

16.133 The 10-kg mass A is moving at 5 m/s when it is 1 m from the stationary 10-kg mass B. The coefficient of kinetic friction between the floor and the two masses is $\mu_k = 0.6$, and the coefficient of restitution of the impact is $e = 0.5$. Determine how far B moves from its initial position as a result of the impact.

Problem 16.133

16.134 The kinetic coefficients of friction between the 5-kg crates A and B and the inclined surface are 0.1 and 0.4, respectively. The coefficient of restitution between the crates is $e = 0.8$. If the crates are released from rest in the positions shown, what are the magnitudes of their velocities immediately after they collide?

16.135 Solve Problem 16.134 if crate A has a velocity of 0.2 m/s down the inclined surface and crate B is at rest when the crates are in the positions shown.

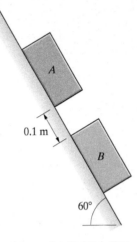

Problems 16.134/16.135

16.136 A small object starts from rest at A and slides down the smooth ramp. The coefficient of restitution of the impact of the object with the floor is $e = 0.8$. At what height above the floor does the object hit the wall?

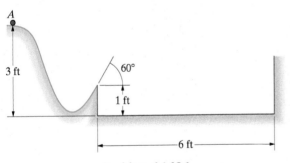

Problem 16.136

16.137 The cue gives the cue ball A a velocity of magnitude 3 m/s. The angle $\beta = 0$ and the coefficient of restitution of the impact of the cue ball and the eight ball B is $e = 1$. If the magnitude of the eight ball's velocity after the impact is 0.9 m/s, what was the coefficient of restitution of the cue ball's impact with the cushion? (The balls are of equal mass.)

16.138 What is the solution to Problem 16.137 if the angle $\beta = 10°$?

16.139 What is the solution to Problem 16.137 if the angle $\beta = 15°$ and the coefficient of restitution of the impact between the two balls is $e = 0.9$?

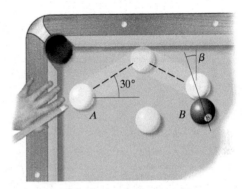

Problems 16.137–16.139

16.140 A ball is given a horizontal velocity of 3 m/s at 2 m above the smooth floor. Determine the distance D between the ball's first and second bounces if the coefficient of restitution is $e = 0.6$.

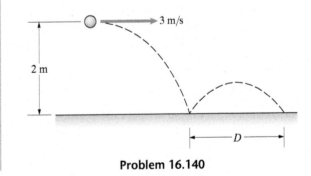

Problem 16.140

16.141* A basketball dropped on the floor from a height of 4 ft rebounds to a height of 3 ft. In the layup shot shown, the magnitude of the ball's velocity is 5 ft/s, and the angles between its velocity vector and the positive coordinate axes are $\theta_x = 42°$, $\theta_y = 68°$, and $\theta_z = 124°$ just before it hits the backboard. What are the magnitude of its velocity and the angles between its velocity vector and the positive coordinate axes just after the ball hits the backboard?

16.142* In Problem 16.141, the basketball's diameter is 9.5 in, the coordinates of the center of the basket rim are $x = 0$, $y = 0$, $z = 12$ in, and the backboard lies in the x–y plane. Determine the x and y coordinates of the point where the ball must hit the backboard so that the center of the ball passes through the center of the basket rim.

Problems 16.141/16.142

16.143 A satellite at $r_0 = 10{,}000$ mi from the center of the earth is given an initial velocity $v_0 = 20{,}000$ ft/s in the direction shown. Determine the magnitude of the transverse component of the satellite's velocity when $r = 20{,}000$ mi. (The radius of the earth is 3960 mi.)

16.144 In Problem 16.143, determine the magnitudes of the radial and transverse components of the satellite's velocity when $r = 15{,}000$ mi.

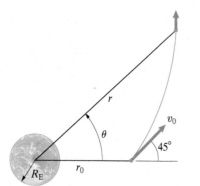

Problems 16.143/16.144

16.145 The snow is 2 ft deep and weighs 20 lb/ft³, the snowplow is 8 ft wide, and the truck travels at 5 mi/h. What force does the snow exert on the truck?

Problem 16.145

16.146 An empty 55-lb drum, 3 ft in diameter, stands on a set of scales. Water begins pouring into the drum at 1200 lb/min from 8 ft above the bottom of the drum. The weight density of water is approximately 62.4 lb/ft³. What do the scales read 40 s after the water starts pouring?

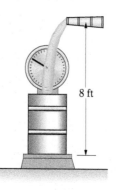

Problem 16.146

16.147 The ski boat's jet propulsive system draws water in at A and expels it at B at 80 ft/s relative to the boat. Assume that the water drawn in enters with no horizontal velocity relative to the surrounding water. The maximum mass flow rate of water through the engine is 2.5 slugs/s. Hydrodynamic drag exerts a force on the boat of magnitude $1.5v$ lb, where v is the boat's velocity in feet per second. Neglecting aerodynamic drag, what is the ski boat's maximum velocity?

16.148 The ski boat in Problem 16.147 weighs 2800 lb. The mass flow rate of water through the engine is 2.5 slugs/s, and the craft starts from rest at $t = 0$. Determine the boat's velocity at (a) $t = 20$ s and (b) $t = 60$ s.

Problems 16.147/16.148

16.149* A crate of mass m slides across a smooth floor pulling a chain from a stationary pile. The mass per unit length of the chain is ρ_L. If the velocity of the crate is v_0 when $s = 0$, what is its velocity as a function of s?

Problem 16.149

Design Project

Design and carry out experiments to measure the coefficients of restitution when (a) a table tennis ball; (b) a tennis ball; and (c) a soccer ball or basketball strike a rigid surface. Investigate the repeatability of your results. Determine how sensitive your results are to the velocity of the ball. Write a brief report describing your experiments, discussing possible sources of error, and presenting your results. Also comment on possible reasons for the differences in your results for the three kinds of balls.

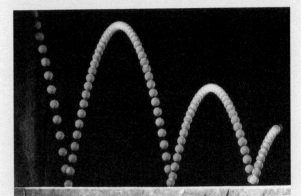

© Harold & Esther Edgerton Foundation, 2007, courtesy of Palm Press, Inc.

CHAPTER 17

Planar Kinematics of Rigid Bodies

If the external forces acting on an object are known, Newton's second law can be used to determine the motion of the object's center of mass without considering any angular motion of the object about its center of mass. In many situations, however, angular motion must also be considered. The rotational motions of some objects are central to their functions, as in the cases of gears, wheels, generators, and turbines. In this chapter we analyze motions of objects, including their rotational motions.

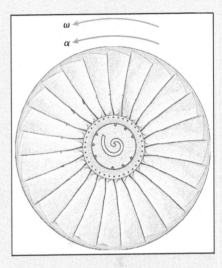

◀ In designing a jet engine, the engineers must analyze the rotational motion of its fan, compressor, and turbine.

17.1 Rigid Bodies and Types of Motion

If a brick is thrown (Fig. 17.1a), we can determine the motion of its center of mass without having to be concerned about its rotational motion. The only significant force is the weight of the brick, and Newton's second law determines the acceleration of its center of mass. But suppose that the brick is standing on the floor, it is tipped over (Fig. 17.1b), and we want to determine the motion of its center of mass as it falls. In this case, the brick is subjected to its weight and also to a force exerted by the floor. We cannot determine either the force exerted by the floor or the motion of the brick's center of mass without also considering its rotational motion.

Before we can analyze such motions, we must consider how to describe them. A brick is an example of an object whose motion can be described by treating the object as a rigid body. A *rigid body* is an idealized model of an object that does not deform, or change shape. The precise definition is that the distance between every pair of points of a rigid body remains constant. Although any object does deform as it moves, if its deformation is sufficiently small, we can approximate its motion by modeling it as a rigid body. For example, in normal use, a twirler's baton (Fig. 17.2a) can be modeled as a rigid body, but a fishing rod (Fig. 17.2b) cannot.

(a)   (b)

Figure 17.1
(a) A thrown brick—its rotation doesn't affect the motion of its center of mass.
(b) A tipped brick—its rotation and the motion of its center of mass are interrelated.

(a)   (b)

Figure 17.2
(a) A baton can be modeled as a rigid body.
(b) Under normal use, a fishing rod is too flexible to be modeled as a rigid body.

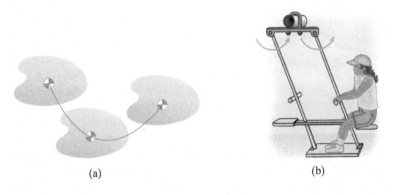

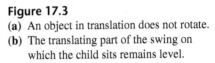

(a)   (b)

Figure 17.3
(a) An object in translation does not rotate.
(b) The translating part of the swing on which the child sits remains level.

Describing the motion of a rigid body requires a reference frame (coordinate system) relative to which the motions of the points of the rigid body and its angular motion are measured. In many situations, it is convenient to use a reference frame that is fixed with respect to the earth. For example, we would use such an *earth-fixed* reference frame to describe the motion of the center of mass and the angular motion of the brick in Fig. 17.1. In the paragraphs that follow, we discuss some types of rigid-body motions relative to a given reference frame that occur frequently in applications.

Translation

If a rigid body in motion relative to a given reference frame does not rotate, it is said to be in *translation* (Fig. 17.3a). For example, the child's swing in Fig. 17.3b is designed so that the horizontal bar to which the seats are attached is in translation. Although each point of the horizontal bar moves in a circular path, the bar does not rotate. It remains horizontal, making it easier for the child to ride safely. Every point of a rigid body in translation has the same velocity and acceleration, so we describe the motion of the rigid body completely if we describe the motion of a single point.

Rotation about a Fixed Axis

After translation, the simplest type of rigid-body motion is rotation about an axis that is fixed relative to a given reference frame (Fig. 17.4a). Each point of the rigid body on the axis is stationary, and each point not on the axis moves in a circular path about the axis as the rigid body rotates. The rotor of an electric motor (Fig. 17.4b) is an example of an object rotating about a fixed axis. The motion of a ship's propeller relative to the ship is rotation about a fixed axis. We discuss this type of motion in more detail in the next section.

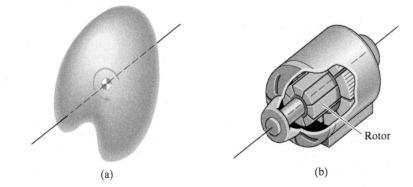

(a)   (b)

Rotor

Figure 17.4
(a) A rigid body rotating about a fixed axis.
(b) Relative to the frame of an electric motor, the rotor rotates about a fixed axis.

Planar Motion

Consider a plane that is fixed relative to a given reference frame and a rigid body intersected by the plane (Fig. 17.5a). If the rigid body undergoes a motion in which the points intersected by the plane remain in the plane, the body is said to be in *two-dimensional*, or *planar*, motion. We refer to the fixed plane as the plane of the motion. Rotation of a rigid body about a fixed axis is a special case of planar motion. As another example, when a car moves in a straight path, its wheels are in planar motion (Fig. 17.5b).

The components of an internal combustion engine illustrate these types of motion (Fig. 17.6). Relative to a reference frame that is fixed with respect to the engine, the pistons translate within the cylinders. The connecting rods are in general planar motion, and the crankshaft rotates about a fixed axis.

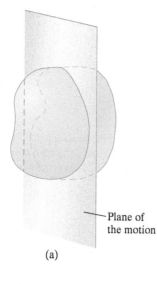

Plane of the motion

(a)

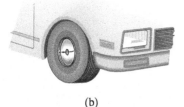

(b)

Figure 17.5
(a) A rigid body intersected by a fixed plane.
(b) A wheel undergoing planar motion.

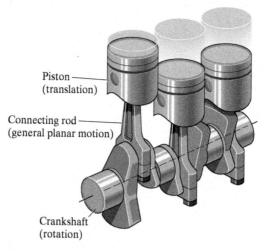

Piston (translation)

Connecting rod (general planar motion)

Crankshaft (rotation)

Figure 17.6
Translation, rotation about a fixed axis, and planar motion in an automobile engine.

RESULTS

Rigid Body
An idealized model of an object that does not deform, or change shape. The distance between every pair of points of a rigid body remains constant.

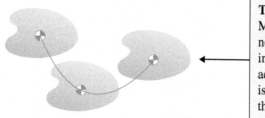

Translation
Motion of a rigid body in which it does not rotate. Every point of a rigid body in translation has the same velocity and acceleration, so the motion of the object is completely described by describing the motion of a single point.

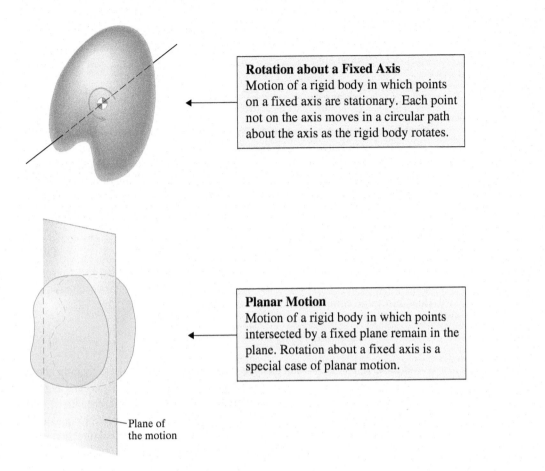

Rotation about a Fixed Axis
Motion of a rigid body in which points on a fixed axis are stationary. Each point not on the axis moves in a circular path about the axis as the rigid body rotates.

Planar Motion
Motion of a rigid body in which points intersected by a fixed plane remain in the plane. Rotation about a fixed axis is a special case of planar motion.

Plane of the motion

17.2 Rotation about a Fixed Axis

BACKGROUND

By considering the rotation of an object about an axis that is fixed relative to a given reference frame, we can introduce some of the concepts of rigid-body motion in a familiar context. In this type of motion, each point of the rigid body moves in a circular path around the fixed axis, so we can analyze the motions of points using results developed in Chapter 13.

In Fig. 17.7, we show a rigid body rotating about a fixed axis and introduce two lines perpendicular to the axis. The reference line is fixed, and the body-fixed line rotates with the rigid body. The angle θ between the reference line and

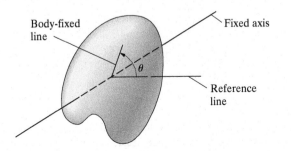

Body-fixed line

Fixed axis

θ

Reference line

Figure 17.7
Specifying the orientation of an object rotating about a fixed axis.

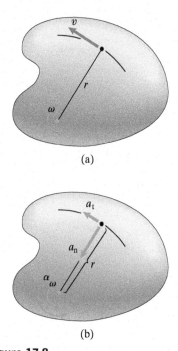

(a)

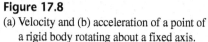

(b)

Figure 17.8
(a) Velocity and (b) acceleration of a point of a rigid body rotating about a fixed axis.

the body-fixed line describes the position, or orientation, of the rigid body about the fixed axis. The rigid body's *angular velocity*, or rate of rotation, and its *angular acceleration* are

$$\omega = \frac{d\theta}{dt}, \qquad \alpha = \frac{d\omega}{dt} = \frac{d^2\theta}{dt^2}. \tag{17.1}$$

Each point of the object not on the fixed axis moves in a circular path about the axis. Using our knowledge of the motion of a point in a circular path, we can relate the velocity and acceleration of a point to the object's angular velocity and angular acceleration. In Fig. 17.8, we view the object in the direction parallel to the fixed axis. The velocity of a point at a distance r from the fixed axis is tangent to the point's circular path (Fig. 17.8a) and is given in terms of the angular velocity of the object by

$$v = r\omega. \tag{17.2}$$

A point has components of acceleration tangential and normal to its circular path (Fig. 17.8b). In terms of the angular velocity and angular acceleration of the object, the components of acceleration are

$$a_t = r\alpha, \qquad a_n = \frac{v^2}{r} = r\omega^2. \tag{17.3}$$

With these relations, we can analyze problems involving objects rotating about fixed axes. For example, suppose that we know the angular velocity ω_A and angular acceleration α_A of the gear in Fig. 17.9 relative to a particular reference frame, and we want to determine ω_B and α_B. The velocities of the gears must be equal at P, because there is no relative motion between them in the tangential direction at P. Therefore, $r_A\omega_A = r_B\omega_B$, and we find that the angular velocity of gear B is

$$\omega_B = \left(\frac{r_A}{r_B}\right)\omega_A.$$

By taking the derivative of this equation with respect to time, we determine the angular acceleration of gear B:

$$\alpha_B = \left(\frac{r_A}{r_B}\right)\alpha_A.$$

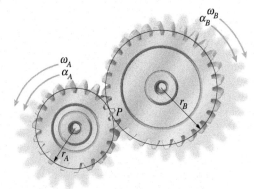

Figure 17.9
Relating the angular velocities and angular accelerations of meshing gears.

From this result, we see that the tangential components of the accelerations of the gears at P are equal: $r_A\alpha_A = r_B\alpha_B$. However, the normal components of the accelerations of the gears at P are different in direction and, if the gears have different radii, are different in magnitude as well. The normal component of the acceleration of gear A at P points toward the center of gear A, and its magnitude is $r_A\omega_A^2$. The normal component of the acceleration of gear B at P points toward the center of gear B, and its magnitude is $r_B\omega_B^2 = (r_A/r_B)(r_A\omega_A^2)$.

RESULTS

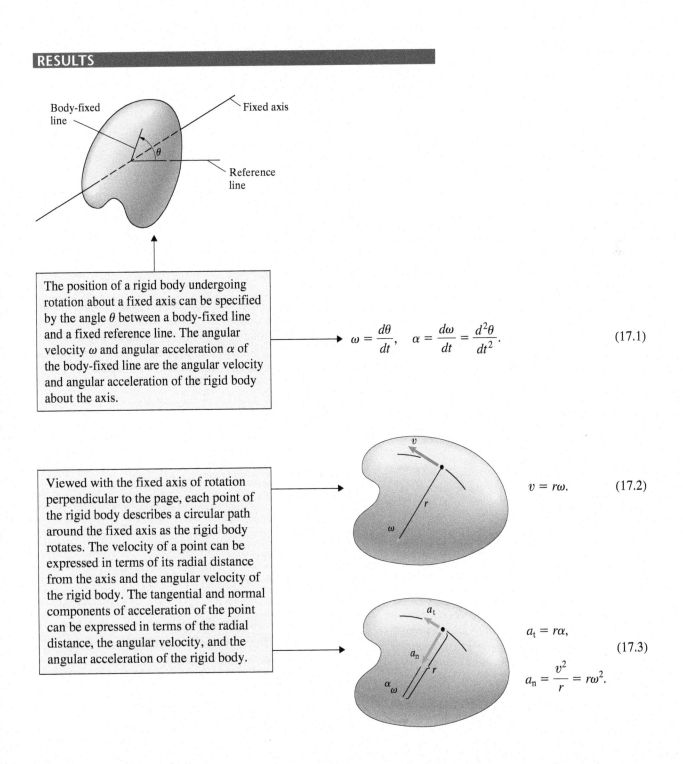

The position of a rigid body undergoing rotation about a fixed axis can be specified by the angle θ between a body-fixed line and a fixed reference line. The angular velocity ω and angular acceleration α of the body-fixed line are the angular velocity and angular acceleration of the rigid body about the axis.

$$\omega = \frac{d\theta}{dt}, \quad \alpha = \frac{d\omega}{dt} = \frac{d^2\theta}{dt^2}. \tag{17.1}$$

Viewed with the fixed axis of rotation perpendicular to the page, each point of the rigid body describes a circular path around the fixed axis as the rigid body rotates. The velocity of a point can be expressed in terms of its radial distance from the axis and the angular velocity of the rigid body. The tangential and normal components of acceleration of the point can be expressed in terms of the radial distance, the angular velocity, and the angular acceleration of the rigid body.

$$v = r\omega. \tag{17.2}$$

$$a_{\text{t}} = r\alpha,$$

$$a_{\text{n}} = \frac{v^2}{r} = r\omega^2. \tag{17.3}$$

Active Example 17.1   Objects Rotating about Fixed Axes (▶ *Related Problem 17.1*)

Gear A of the winch turns gear B, raising the hook H. Gear A starts from rest at time $t = 0$ and its clockwise angular acceleration (in rad/s^2) is given as a function of time by $\alpha_A = 0.2t$. What is the upward velocity of the hook and what vertical distance has it risen at $t = 10$ s?

Strategy

By equating the tangential components of acceleration of gears A and B at their point of contact, we can determine the angular acceleration of gear B. Then we can integrate to determine the angular velocity of gear B and the angle through which it has turned at $t = 10$ s. With that information, we can determine the velocity of the hook and the distance through which it has traveled.

Solution

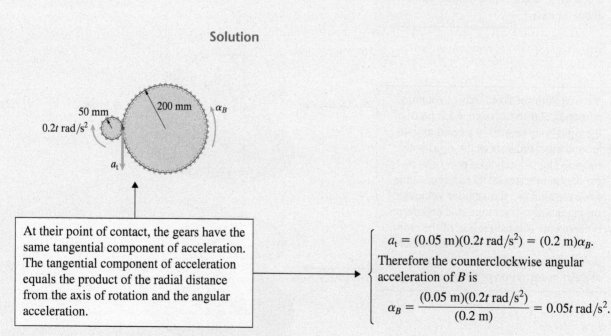

At their point of contact, the gears have the same tangential component of acceleration. The tangential component of acceleration equals the product of the radial distance from the axis of rotation and the angular acceleration.

$a_t = (0.05 \text{ m})(0.2t \text{ rad/s}^2) = (0.2 \text{ m})\alpha_B.$

Therefore the counterclockwise angular acceleration of B is

$$\alpha_B = \frac{(0.05 \text{ m})(0.2t \text{ rad/s}^2)}{(0.2 \text{ m})} = 0.05t \text{ rad/s}^2.$$

Integrate $\alpha_B = d\omega_B/dt$ to determine the angular velocity of gear B as a function of time.

$$\begin{cases} \displaystyle\int_0^{\omega_B} d\omega_B = \int_0^t 0.05t\, dt: \\[2mm] \omega_B = 0.025t^2\, \text{rad/s}. \end{cases}$$

Integrate $\omega_B = d\theta_B/dt$ to determine the angle through which gear B has turned as a function of time.

$$\begin{cases} \displaystyle\int_0^{\theta_B} d\theta_B = \int_0^t 0.025t^2\, dt: \\[2mm] \theta_B = 0.00833t^3\, \text{rad}. \end{cases}$$

Determine ω_B and θ_B at $t = 10$ s.

$$\begin{cases} \omega_B = 0.025(10)^2 \\ \quad = 2.5\ \text{rad/s}, \\ \theta_B = 0.00833(10)^3 \\ \quad = 8.33\ \text{rad}. \end{cases}$$

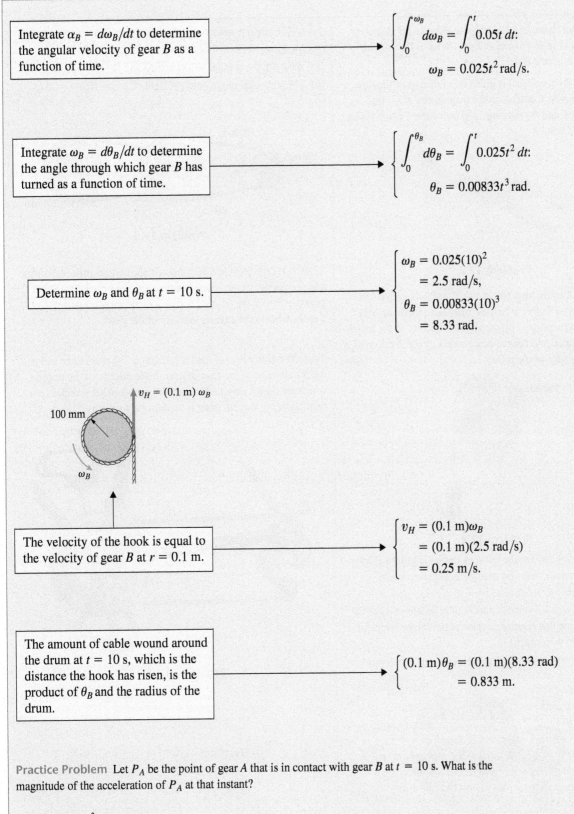

$v_H = (0.1\ \text{m})\,\omega_B$

100 mm

ω_B

The velocity of the hook is equal to the velocity of gear B at $r = 0.1$ m.

$$\begin{cases} v_H = (0.1\ \text{m})\omega_B \\ \quad = (0.1\ \text{m})(2.5\ \text{rad/s}) \\ \quad = 0.25\ \text{m/s}. \end{cases}$$

The amount of cable wound around the drum at $t = 10$ s, which is the distance the hook has risen, is the product of θ_B and the radius of the drum.

$$\begin{cases} (0.1\ \text{m})\theta_B = (0.1\ \text{m})(8.33\ \text{rad}) \\ \quad = 0.833\ \text{m}. \end{cases}$$

Practice Problem Let P_A be the point of gear A that is in contact with gear B at $t = 10$ s. What is the magnitude of the acceleration of P_A at that instant?

Answer: 5.00 m/s^2.

Problems

▶ **17.1** In Active Example 17.1, suppose that at a given instant the hook *H* is moving downward at 2 m/s. What is the angular velocity of gear *A* at that instant?

17.2 The angle θ (in radians) is given as a function of time by $\theta = 0.2\pi t^2$. At $t = 4$ s, determine the magnitudes of (a) the velocity of point *A* and (b) the tangential and normal components of acceleration of point *A*.

Problem 17.2

17.3 The mass *A* starts from rest at $t = 0$ and falls with a constant acceleration of 8 m/s². When the mass has fallen one meter, determine the magnitudes of (a) the angular velocity of the pulley and (b) the tangential and normal components of acceleration of a point at the outer edge of the pulley.

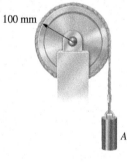

Problem 17.3

17.4 At the instant shown, the left disk has an angular velocity of 3 rad/s counterclockwise and an angular acceleration of 1 rad/s² clockwise.

(a) What are the angular velocity and angular acceleration of the right disk? (Assume that there is no relative motion between the disks at their point of contact.)

(b) What are the magnitudes of the velocity and acceleration of point *A*?

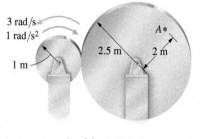

Problem 17.4

17.5 The angular velocity of the left disk is given as a function of time by $\omega_A = 4 + 0.2t$ rad/s.

(a) What are the angular velocities ω_B and ω_C at $t = 5$ s?

(b) Through what angle does the right disk turn from $t = 0$ to $t = 5$ s?

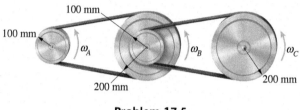

Problem 17.5

17.6 (a) If the bicycle's 120-mm sprocket wheel rotates through one revolution, through how many revolutions does the 45-mm gear turn? (b) If the angular velocity of the sprocket wheel is 1 rad/s, what is the angular velocity of the gear?

17.7 The rear wheel of the bicycle has a 330-mm radius and is rigidly attached to the 45-mm gear. If the rider turns the pedals, which are rigidly attached to the 120-mm sprocket wheel, at one revolution per second, what is the bicycle's velocity?

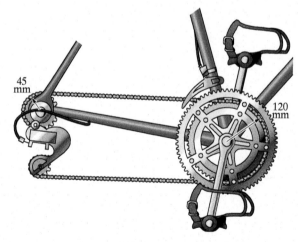

Problems 17.6/17.7

17.8 The disk is rotating about the origin with a constant clockwise angular velocity of 100 rpm. Determine the x and y components of velocity of points A and B (in in/s).

17.9 The disk is rotating about the origin with a constant clockwise angular velocity of 100 rpm. Determine the x and y components of acceleration of points A and B (in in/s^2).

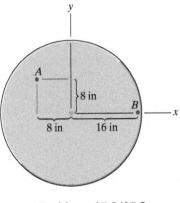

Problems 17.8/17.9

17.10 The radius of the Corvette's tires is 14 in. It is traveling at 80 mi/h when the driver applies the brakes, subjecting the car to a deceleration of 25 ft/s^2. Assume that the tires continue to roll, not skid, on the road surface. At that instant, what are the magnitudes of the tangential and normal components of acceleration (in ft/s^2) of a point at the outer edge of a tire relative to a nonrotating coordinate system with its origin at the center of the tire?

Problem 17.10

17.11 If the bar has a counterclockwise angular velocity of 8 rad/s and a clockwise angular acceleration of 40 rad/s^2, what are the magnitudes of the accelerations of points A and B?

17.12 If the magnitudes of the velocity and acceleration of point A of the rotating bar are $|\mathbf{v}_A| = 3$ m/s and $|\mathbf{a}_A| = 28$ m/s^2, what are $|\mathbf{v}_B|$ and $|\mathbf{a}_B|$?

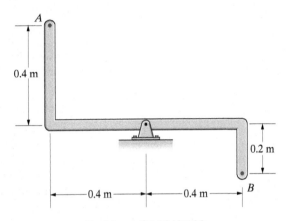

Problems 17.11/17.12

17.13 A disk of radius $R = 0.5$ m rolls on a horizontal surface. The relationship between the horizontal distance x the center of the disk moves and the angle β through which the disk rotates is $x = R\beta$. Suppose that the center of the disk is moving to the right with a constant velocity of 2 m/s.

(a) What is the disk's angular velocity?

(b) Relative to a nonrotating reference frame with its origin at the center of the disk, what are the magnitudes of the velocity and acceleration of a point on the edge of the disk?

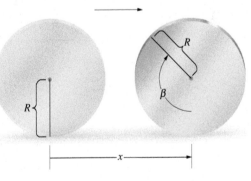

Problem 17.13

17.3 General Motions: Velocities

Each point of a rigid body in translation undergoes the same motion. Each point of a rigid body rotating about a fixed axis undergoes circular motion about the axis. To analyze more complicated motions that combine translation and rotation, we must develop equations that relate the relative motions of points of a rigid body to its angular motion.

Relative Velocities

In Fig. 17.10a, we view a rigid body from a perspective perpendicular to the plane of its motion. Points A and B are points of the rigid body contained in the latter plane, and O is the origin of a given reference frame. The position of A relative to B, $\mathbf{r}_{A/B}$, is related to the positions of A and B relative to O by

$$\mathbf{r}_A = \mathbf{r}_B + \mathbf{r}_{A/B}.$$

Taking the derivative of this equation with respect to time, we obtain

$$\mathbf{v}_A = \mathbf{v}_B + \mathbf{v}_{A/B}, \tag{17.4}$$

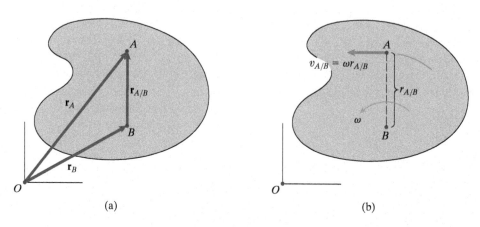

(a)

(b)

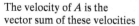

The velocity of A is the vector sum of these velocities

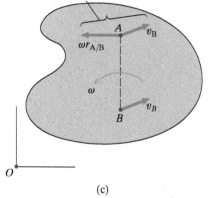

(c)

Figure 17.10
(a) A rigid body in planar motion.
(b) The velocity of A relative to B.
(c) The velocity of A is the sum of its velocity relative to B and the velocity of B.

where $\mathbf{v}_A$ and $\mathbf{v}_B$ are the velocities of A and B relative to the given reference frame and $\mathbf{v}_{A/B} = d\mathbf{r}_{A/B}/dt$ is the velocity of A relative to B. (*When we simply speak of the velocity of a point, we will mean its velocity relative to the given reference frame.*)

We can show that $\mathbf{v}_{A/B}$ is related in a simple way to the rigid body's angular velocity. Since A and B are points of the rigid body, the distance between them, $r_{A/B} = |\mathbf{r}_{A/B}|$, is constant. That means that, relative to B, A moves in a circular path as the rigid body rotates. That velocity of A relative to B is therefore tangent to the circular path and equal to the product of $r_{A/B}$ and the angular velocity ω of the rigid body (Fig. 17.10b). From Eq. (17.4), the velocity of A is the sum of the velocity of B and the velocity of A relative to B (Fig. 17.10c). This result can be used to relate velocities of points of a rigid body in planar motion when the angular velocity of the body is known.

For example, in Fig. 17.11a, we show a circular disk of radius R rolling with counterclockwise angular velocity ω on a stationary plane surface. Saying that the surface is stationary means that we are describing the motion of the disk in terms of a reference frame that is fixed with respect to the surface. By

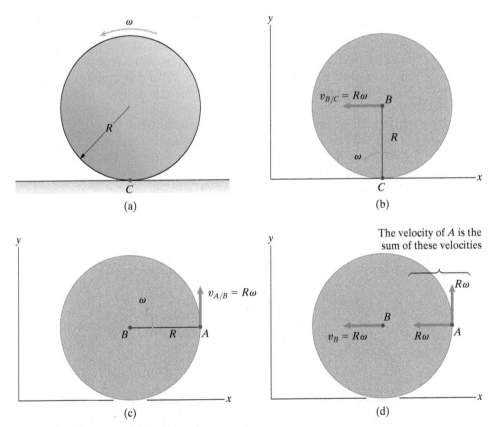

(a)

(b)

(c)

(d)

Figure 17.11
(a) A disk rolling with angular velocity ω.
(b) The velocity of the center B relative to C.
(c) The velocity of A relative to B.
(d) The velocity of A equals the sum of the velocity of B
    and the velocity of A relative to B.

rolling, we mean that the velocity of the disk relative to the surface is zero at the point of contact C. The velocity of the center B of the disk relative to C is illustrated in Fig. 17.11b. Since $\mathbf{v}_C = \mathbf{0}$, the velocity of B in terms of the fixed coordinate system shown is

$$\mathbf{v}_B = \mathbf{v}_C + \mathbf{v}_{B/C} = -R\omega\mathbf{i}.$$

This result is very useful: *The magnitude of the velocity of the center of a round object rolling on a stationary plane surface equals the product of the radius and the magnitude of the angular velocity.*

We can determine the velocity of any other point of the disk in the same way. Figure 17.11c shows the velocity of a point A relative to point B. The velocity of A is the sum of the velocity of B and the velocity of A relative to B (Fig. 17.11d):

$$\mathbf{v}_A = \mathbf{v}_B + \mathbf{v}_{A/B} = -R\omega\mathbf{i} + R\omega\mathbf{j}.$$

The Angular Velocity Vector

We can express the rate of rotation of a rigid body as a vector. *Euler's theorem* states that a rigid body constrained to rotate about a fixed point B can move between any two positions by a single rotation about some axis through B. Suppose that we choose an arbitrary point B of a rigid body that is undergoing an arbitrary motion at a time t. Euler's theorem allows us to express the rigid body's change in position relative to B during an interval of time from t to $t + dt$ as a single rotation through an angle $d\theta$ about some axis. At time t, the rigid body's rate of rotation about the axis is its angular velocity $\omega = d\theta/dt$, and the axis about which the body rotates is called the *instantaneous axis of rotation*.

The *angular velocity vector*, denoted by $\boldsymbol{\omega}$, specifies both the direction of the instantaneous axis of rotation and the angular velocity. The angular velocity vector is defined to be parallel to the instantaneous axis of rotation (Fig. 17.12a), and its magnitude is the rate of rotation, the absolute value of ω. Its direction is

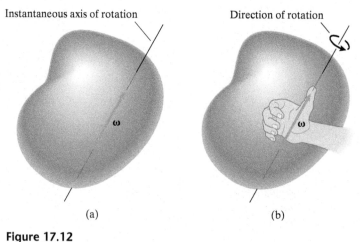

Instantaneous axis of rotation

Direction of rotation

(a)   (b)

Figure 17.12
(a) An angular velocity vector.
(b) Right-hand rule for the direction of the vector.

related to the direction of the rigid body's rotation through a right-hand rule: Pointing the thumb of the right hand in the direction of $\boldsymbol{\omega}$, the fingers curl around $\boldsymbol{\omega}$ in the direction of the rotation (Fig. 17.12b).

For example, the axis of rotation of the rolling disk in Fig. 17.11 is parallel to the z axis, so the angular velocity vector of the disk is parallel to the z axis and its magnitude is ω. Curling the fingers of the right hand around the z axis in the direction of the rotation, the thumb points in the positive z direction (Fig. 17.13). The angular velocity vector of the disk is $\boldsymbol{\omega} = \omega\mathbf{k}$.

The angular velocity vector allows us to express the results of the previous section in a convenient form. Let A and B be points of a rigid body with angular velocity $\boldsymbol{\omega}$ (Fig. 17.14a). We can show that the velocity of A relative to B is

$$\mathbf{v}_{A/B} = \frac{d\mathbf{r}_{A/B}}{dt} = \boldsymbol{\omega} \times \mathbf{r}_{A/B}. \tag{17.5}$$

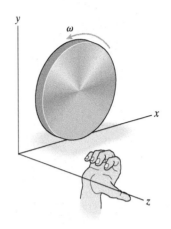

Figure 17.13
Determining the direction of the angular velocity vector of a rolling disk.

At the present instant, relative to B, point A is moving in a circular path of radius $|\mathbf{r}_{A/B}| \sin \beta$, where β is the angle between the vectors $\mathbf{r}_{A/B}$ and $\boldsymbol{\omega}$ (Fig. 17.14b). The magnitude of the velocity of A relative to B is equal to the product of the radius of the circular path and the angular velocity of the rigid body; that is, $|\mathbf{v}_{A/B}| = (|\mathbf{r}_{A/B}| \sin \beta)|\boldsymbol{\omega}|$. The right-hand side of this equation is the magnitude of the cross product of $\mathbf{r}_{A/B}$ and $\boldsymbol{\omega}$. In addition, $\mathbf{v}_{A/B}$ is perpendicular both to $\boldsymbol{\omega}$ and $\mathbf{r}_{A/B}$. But is $\mathbf{v}_{A/B}$ equal to $\boldsymbol{\omega} \times \mathbf{r}_{A/B}$ or $\mathbf{r}_{A/B} \times \boldsymbol{\omega}$? Notice in Fig. 17.14b that, pointing the fingers of the right hand in the direction of $\boldsymbol{\omega}$ and closing them toward $\mathbf{r}_{A/B}$, the thumb points in the direction of the velocity of A relative to B, so $\mathbf{v}_{A/B} = \boldsymbol{\omega} \times \mathbf{r}_{A/B}$. Substituting Eq. (17.5) into Eq. (17.4), we obtain an equation for the relation between the velocities of two points of a rigid body in terms of its angular velocity:

$$\mathbf{v}_A = \mathbf{v}_B + \underbrace{\boldsymbol{\omega} \times \mathbf{r}_{A/B}}_{\mathbf{v}_{A/B}}. \tag{17.6}$$

If the angular velocity vector and the velocity of one point of a rigid body are known, Eq. (17.6) can be used to determine the velocity of any other point of the rigid body. Returning to the example of a disk of radius R rolling with

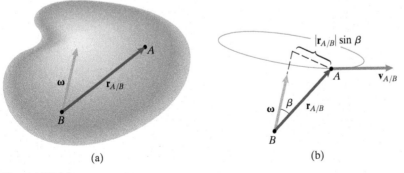

(a)   (b)

Figure 17.14
(a) Points A and B of a rotating rigid body.
(b) A is moving in a circular path relative to B.

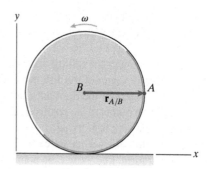

Figure 17.15
A rolling disk and the position vector of A
relative to B.

angular velocity ω (Fig. 17.15), let us use Eq. (17.6) to determine the velocity of point A. The velocity of the center of the disk is given in terms of the angular velocity by $\mathbf{v}_B = -R\omega\mathbf{i}$, the disk's angular velocity vector is $\boldsymbol{\omega} = \omega\mathbf{k}$, and the position vector of A relative to the center is $\mathbf{r}_{A/B} = R\mathbf{i}$. The velocity of point A is

$$\mathbf{v}_A = \mathbf{v}_B + \boldsymbol{\omega} \times \mathbf{r}_{A/B} = -R\omega\mathbf{i} + (\omega\mathbf{k}) \times (R\mathbf{i})$$
$$= -R\omega\mathbf{i} + R\omega\mathbf{j}.$$

Compare this result with the velocity of point A shown in Fig. 17.11d.

RESULTS

Relative Velocities

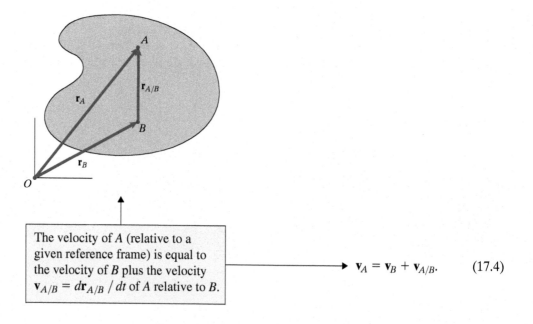

The velocity of A (relative to a given reference frame) is equal to the velocity of B plus the velocity $\mathbf{v}_{A/B} = d\mathbf{r}_{A/B}/dt$ of A relative to B.

$$\mathbf{v}_A = \mathbf{v}_B + \mathbf{v}_{A/B}. \qquad (17.4)$$

Rolling Motion

The round object of radius R is rolling on the stationary plane surface with counterclockwise angular velocity ω. The velocity of the point C in contact with the surface is zero. The velocity of the center B is $R\omega$ in the direction shown.

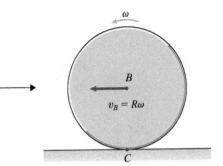

Angular Velocity Vector

The angular velocity vector $\boldsymbol{\omega}$ describes an object's rotation relative to a given reference frame. The vector is defined to be parallel to the object's axis of rotation. Its direction is specified by a right-hand rule: When the fingers of the right hand curl around the axis in the direction of the object's rotation, the thumb points in the direction of $\boldsymbol{\omega}$. The magnitude of $\boldsymbol{\omega}$ is the object's angular velocity about its axis of rotation.

Direction of rotation

Suppose that an object's axis of rotation is parallel to the z axis and the object is rotating in the counterclockwise direction with angular velocity ω. Then its angular velocity vector is $\boldsymbol{\omega} = \omega\mathbf{k}$.

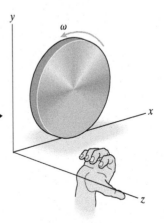

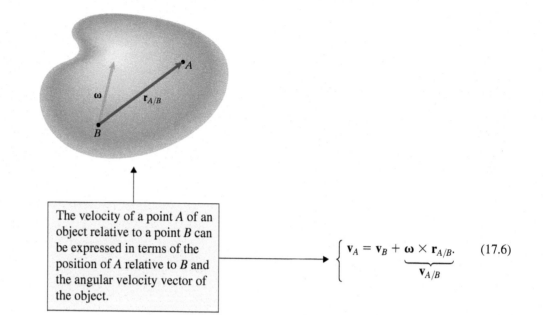

The velocity of a point A of an object relative to a point B can be expressed in terms of the position of A relative to B and the angular velocity vector of the object.

$$\left\{ \mathbf{v}_A = \mathbf{v}_B + \underbrace{\boldsymbol{\omega} \times \mathbf{r}_{A/B}}_{\mathbf{v}_{A/B}}. \right. \qquad (17.6)$$

Active Example 17.2   **Determining Velocities and Angular Velocities** (▶ *Related Problem 17.33*)

Bar AB is rotating with a clockwise angular velocity of 10 rad/s. Point C slides on the horizontal surface. At the instant shown, determine the angular velocity of bar BC and the velocity of point C.

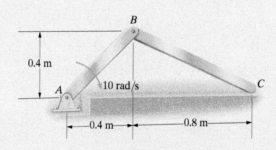

Strategy

The angular velocity of bar AB is given, and point A is stationary, so we can use Eq. (17.6) to determine the velocity of point B. Then by applying Eq. (17.6) to points B and C of bar BC, we can determine both the angular velocity of bar BC and the velocity of point C.

Solution

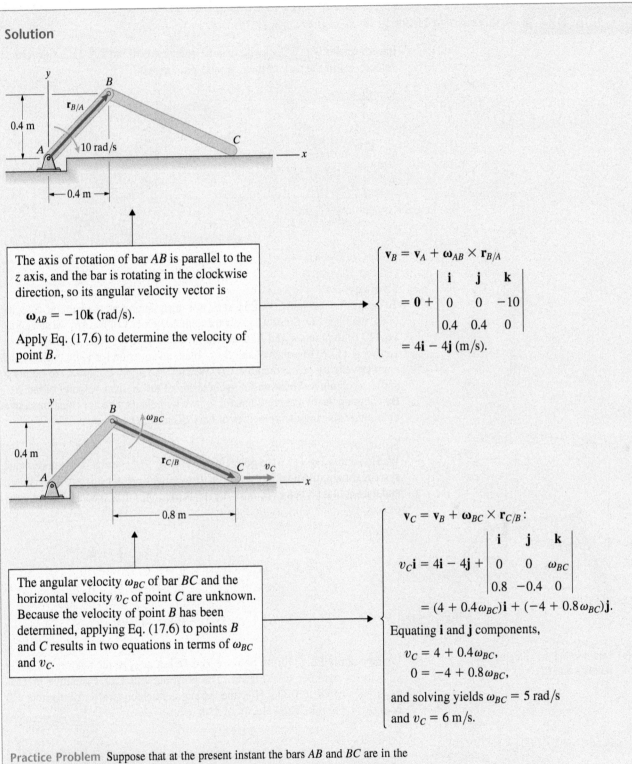

The axis of rotation of bar AB is parallel to the z axis, and the bar is rotating in the clockwise direction, so its angular velocity vector is

$\omega_{AB} = -10\mathbf{k}$ (rad/s).

Apply Eq. (17.6) to determine the velocity of point B.

$$\mathbf{v}_B = \mathbf{v}_A + \omega_{AB} \times \mathbf{r}_{B/A}$$

$$= \mathbf{0} + \begin{vmatrix} \mathbf{i} & \mathbf{j} & \mathbf{k} \\ 0 & 0 & -10 \\ 0.4 & 0.4 & 0 \end{vmatrix}$$

$$= 4\mathbf{i} - 4\mathbf{j} \text{ (m/s)}.$$

The angular velocity ω_{BC} of bar BC and the horizontal velocity v_C of point C are unknown. Because the velocity of point B has been determined, applying Eq. (17.6) to points B and C results in two equations in terms of ω_{BC} and v_C.

$$\mathbf{v}_C = \mathbf{v}_B + \omega_{BC} \times \mathbf{r}_{C/B}:$$

$$v_C\mathbf{i} = 4\mathbf{i} - 4\mathbf{j} + \begin{vmatrix} \mathbf{i} & \mathbf{j} & \mathbf{k} \\ 0 & 0 & \omega_{BC} \\ 0.8 & -0.4 & 0 \end{vmatrix}$$

$$= (4 + 0.4\omega_{BC})\mathbf{i} + (-4 + 0.8\omega_{BC})\mathbf{j}.$$

Equating $\mathbf{i}$ and $\mathbf{j}$ components,

$$v_C = 4 + 0.4\omega_{BC},$$
$$0 = -4 + 0.8\omega_{BC},$$

and solving yields $\omega_{BC} = 5$ rad/s and $v_C = 6$ m/s.

Practice Problem Suppose that at the present instant the bars AB and BC are in the positions shown. What would the angular velocity of bar AB need to be for point C to be moving toward the left at 3 m/s? What would the angular velocity of bar BC be?

Answer: Bar AB, 5 rad/s counterclockwise. Bar BC, 2.5 rad/s clockwise.

Example 17.3 | Analysis of a Linkage (▶ *Related Problem 17.36*)

Bar *AB* rotates with a clockwise angular velocity of 10 rad/s. What is the vertical velocity v_R of the rack of the rack-and-pinion gear?

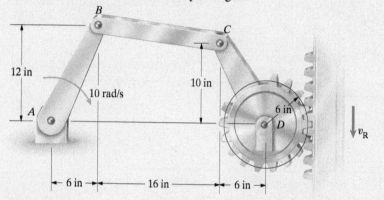

Strategy

To determine the velocity of the rack, we must determine the angular velocity of the member *CD*. Since we know the angular velocity of bar *AB*, we can apply Eq. (17.6) to points *A* and *B* to determine the velocity of point *B*. Then we can apply Eq. (17.6) to points *C* and *D* to obtain an equation for $\mathbf{v}_C$ in terms of the angular velocity of the member *CD*. We can also apply Eq. (17.6) to points *B* and *C* to obtain an equation for $\mathbf{v}_C$ in terms of the angular velocity of bar *BC*. By equating the two expressions for $\mathbf{v}_C$, we will obtain a vector equation in two unknowns: the angular velocities of bars *BC* and *CD*.

Solution

We first apply Eq. (17.6) to points *A* and *B* (Fig. a). In terms of the coordinate system shown, the position vector of *B* relative to *A* is $\mathbf{r}_{B/A} = 0.5\mathbf{i} + \mathbf{j}$ (ft), and the angular velocity vector of bar *AB* is $\boldsymbol{\omega}_{AB} = -10\mathbf{k}$ (rad/s). The velocity of *B* is

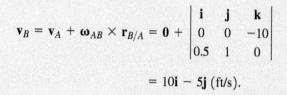

(a) Determining the velocities of points *B* and *C*.

$$\mathbf{v}_B = \mathbf{v}_A + \boldsymbol{\omega}_{AB} \times \mathbf{r}_{B/A} = \mathbf{0} + \begin{vmatrix} \mathbf{i} & \mathbf{j} & \mathbf{k} \\ 0 & 0 & -10 \\ 0.5 & 1 & 0 \end{vmatrix}$$

$$= 10\mathbf{i} - 5\mathbf{j} \ (\text{ft/s}).$$

We now apply Eq. (17.6) to points *C* and *D*. Let ω_{CD} be the unknown angular velocity of member *CD* (Fig. a). The position vector of *C* relative to *D* is $\mathbf{r}_{C/D} = -0.500\mathbf{i} + 0.833\mathbf{j}$ (ft), and the angular velocity vector of member *CD* is $\boldsymbol{\omega}_{CD} = -\omega_{CD}\mathbf{k}$. The velocity of *C* is

$$\mathbf{v}_C = \mathbf{v}_D + \boldsymbol{\omega}_{CD} \times \mathbf{r}_{C/D} = \mathbf{0} + \begin{vmatrix} \mathbf{i} & \mathbf{j} & \mathbf{k} \\ 0 & 0 & -\omega_{CD} \\ -0.500 & 0.833 & 0 \end{vmatrix}$$

$$= 0.833\omega_{CD}\mathbf{i} + 0.500\omega_{CD}\mathbf{j}.$$

Now we apply Eq. (17.6) to points B and C (Fig. b). We denote the unknown angular velocity of bar BC by ω_{BC}. The position vector of C relative to B is $\mathbf{r}_{C/B} = 1.333\mathbf{i} - 0.167\mathbf{j}$ (ft), and the angular velocity vector of bar BC is $\boldsymbol{\omega}_{BC} = \omega_{BC}\mathbf{k}$. Expressing the velocity of C in terms of the velocity of B, we obtain

$$\mathbf{v}_C = \mathbf{v}_B + \boldsymbol{\omega}_{BC} \times \mathbf{r}_{C/B} = \mathbf{v}_B + \begin{vmatrix} \mathbf{i} & \mathbf{j} & \mathbf{k} \\ 0 & 0 & \omega_{BC} \\ 1.333 & -0.167 & 0 \end{vmatrix}$$

$$= \mathbf{v}_B + 0.167\omega_{BC}\mathbf{i} + 1.333\omega_{BC}\mathbf{j}.$$

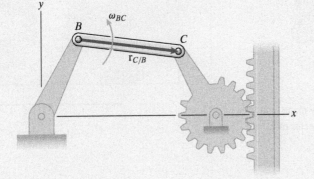

(b) Expressing the velocity of point C in terms of the velocity of point B.

Substituting our expressions for $\mathbf{v}_B$ and $\mathbf{v}_C$ into this equation, we get

$$0.833\omega_{CD}\mathbf{i} + 0.500\omega_{CD}\mathbf{j} = 10\mathbf{i} - 5\mathbf{j} + 0.167\omega_{BC}\mathbf{i} + 1.333\omega_{BC}\mathbf{j}.$$

Equating the $\mathbf{i}$ and $\mathbf{j}$ components yields two equations in terms of ω_{BC} and ω_{CD}:

$$0.833\omega_{CD} = 10 + 0.167\omega_{BC},$$
$$0.500\omega_{CD} = -5 + 1.333\omega_{BC}.$$

Solving these equations, we obtain $\omega_{BC} = 8.92$ rad/s and $\omega_{CD} = 13.78$ rad/s.

The vertical velocity of the rack is equal to the velocity of the gear where it contacts the rack:

$$v_R = (0.5 \text{ ft})\omega_{CD} = (0.5)(13.78) = 6.89 \text{ ft/s}.$$

Critical Thinking

How did we know the sequence of steps that would determine the velocity of the rack and pinion gear? The Strategy section may give you the impression that there is only one method of solution and that it should be obvious. Neither of these things is true. But most problems of this kind can be solved by repeatedly applying Eq. (17.6) until enough equations have been obtained to determine what you need to know. Just remember that Eq. (17.6) *applies only to two points of the same rigid body*. For example, in this case we could apply Eq. (17.6) to points B and C, but we could not have applied it to points A and C.

Problems

17.14 The turbine rotates relative to the coordinate system at 30 rad/s about a fixed axis coincident with the x axis. What is its angular velocity vector?

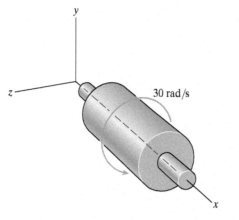

Problem 17.14

17.15 The rectangular plate swings in the x–y plane from arms of equal length. What is the angular velocity vector of (a) the rectangular plate and (b) the bar AB?

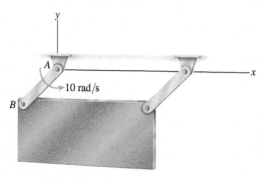

Problem 17.15

17.16 Bar OQ is rotating in the clockwise direction at 4 rad/s. What are the angular velocity vectors of the bars OQ and PQ?

Strategy: Notice that if you know the angular velocity of bar OQ, you also know the angular velocity of bar PQ.

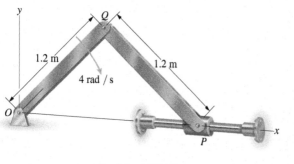

Problem 17.16

17.17 A disk of radius $R = 0.5$ m rolls on a horizontal surface. The relationship between the horizontal distance x the center of the disk moves and the angle β through which the disk rotates is $x = R\beta$. Suppose that the center of the disk is moving to the right with a constant velocity of 2 m/s.

(a) What is the disk's angular velocity?

(b) What is the disk's angular velocity vector?

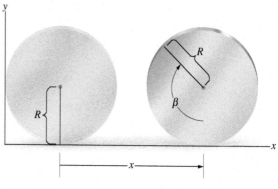

Problem 17.17

17.18 The rigid body rotates with angular velocity $\omega = 12$ rad/s. The distance $r_{A/B} = 0.4$ m.

(a) Determine the x and y components of the velocity of A relative to B by representing the velocity as shown in Fig. 17.10b.

(b) What is the angular velocity vector of the rigid body?

(c) Use Eq. (17.5) to determine the velocity of A relative to B.

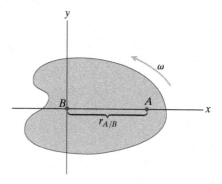

Problem 17.18

17.19 The bar is rotating in the counterclockwise direction with angular velocity ω. The magnitude of the velocity of point A is 6 m/s. Determine the velocity of point B.

17.20 The bar is rotating in the counterclockwise direction with angular velocity ω. The magnitude of the velocity of point A relative to point B is 6 m/s. Determine the velocity of point B.

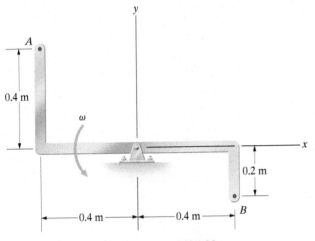

Problems 17.19/17.20

17.21 The bracket is rotating about point O with counterclockwise angular velocity ω. The magnitude of the velocity of point A relative to point B is 4 m/s. Determine ω.

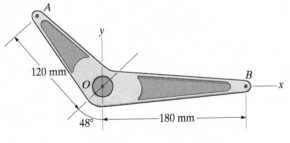

Problem 17.21

17.22 Determine the x and y components of the velocity of point A.

17.23 If the angular velocity of the bar is constant, what are the x and y components of the velocity of point A 0.1 s after the instant shown?

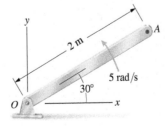

Problems 17.22/17.23

17.24 The disk is rotating about the z axis at 50 rad/s in the clockwise direction. Determine the x and y components of the velocities of points A, B, and C.

17.25 If the magnitude of the velocity of point A relative to point B is 4 m/s, what is the magnitude of the disk's angular velocity?

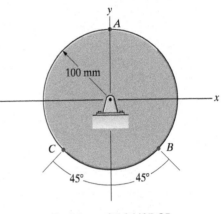

Problems 17.24/17.25

17.26 The radius of the Corvette's tires is 14 in. It is traveling at 80 mi/h. Assume that the tires roll, not skid, on the road surface.

(a) What is the angular velocity of its wheels?

(b) In terms of the earth-fixed coordinate system shown, determine the velocity (in ft/s) of the point of the tire with coordinates (−14 in, 0, 0).

Problem 17.26

17.27 Point A of the rolling disk is moving toward the right. The magnitude of the velocity of point C is 5 m/s. Determine the velocities of points B and D.

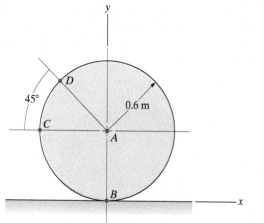

Problem 17.27

17.28 The helicopter is in planar motion in the x–y plane. At the instant shown, the position of the craft's center of mass, G, is $x = 2$ m, $y = 2.5$ m; and its velocity is $\mathbf{v}_G = 12\mathbf{i} + 4\mathbf{j}$ (m/s). The position of point T where the tail rotor is mounted is $x = -3.5$ m, $y = 4.5$ m. The helicopter's angular velocity is 0.2 rad/s clockwise. What is the velocity of point T?

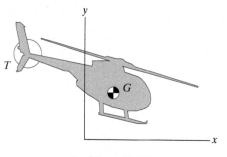

Problem 17.28

17.29 The bar is moving in the x–y plane and is rotating in the counterclockwise direction. The velocity of point A relative to the reference frame is $\mathbf{v}_A = 12\mathbf{i} - 2\mathbf{j}$ (m/s). The magnitude of the velocity of point A relative to point B is 8 m/s. What is the velocity of point B relative to the reference frame?

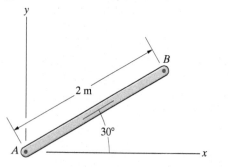

Problem 17.29

17.30 Points A and B of the 2-m bar slide on the plane surfaces. Point B is moving to the right at 3 m/s. What is the velocity of the midpoint G of the bar?

Strategy: First apply Eq. (17.6) to points A and B to determine the bar's angular velocity. Then apply Eq. (17.6) to points B and G.

Problem 17.30

17.31 Bar AB rotates at 6 rad/s in the clockwise direction. Determine the velocity (in in/s) of the slider C.

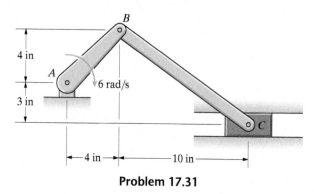

Problem 17.31

17.32 If $\theta = 45°$ and the sleeve P is moving to the right at 2 m/s, what are the angular velocities of bars OQ and PQ?

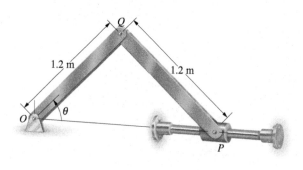

Problem 17.32

▶ 17.33 In Active Example 17.2, consider the instant when bar *AB* is vertical and rotating in the clockwise direction at 10 rad/s. Draw a sketch showing the positions of the two bars at that instant. Determine the angular velocity of bar *BC* and the velocity of point *C*.

17.34 Bar *AB* rotates in the counterclockwise direction at 6 rad/s. Determine the angular velocity of bar *BD* and the velocity of point *D*.

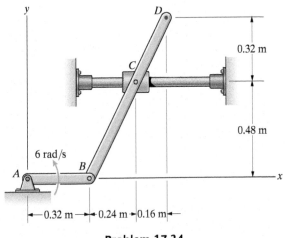

Problem 17.34

17.35 At the instant shown, the piston's velocity is $v_C = -14\mathbf{i}$ (m/s). What is the angular velocity of the crank *AB*?

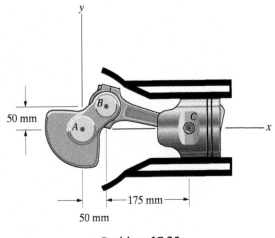

Problem 17.35

▶ 17.36 In Example 17.3, determine the angular velocity of the bar *AB* that would be necessary so that the downward velocity of the rack $v_R = 10$ ft/s at the instant shown.

17.37 Bar *AB* rotates at 12 rad/s in the clockwise direction. Determine the angular velocities of bars *BC* and *CD*.

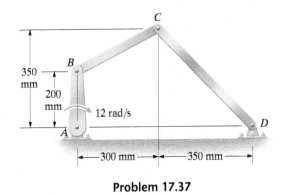

Problem 17.37

17.38 Bar *AB* is rotating at 10 rad/s in the counterclockwise direction. The disk rolls on the circular surface. Determine the angular velocities of bar *BC* and the disk at the instant shown.

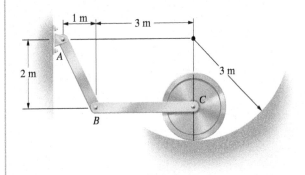

Problem 17.38

17.39 Bar *AB* rotates at 2 rad/s in the counterclockwise direction. Determine the velocity of the midpoint *G* of bar *BC*.

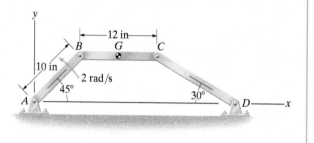

Problem 17.39

17.40 Bar *AB* rotates at 10 rad/s in the counterclockwise direction. Determine the velocity of point *E*.

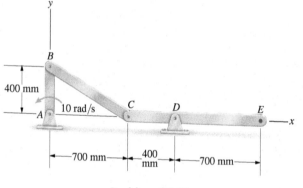

Problem 17.40

17.41 Bar *AB* rotates at 4 rad/s in the counterclockwise direction. Determine the velocity of point *C*.

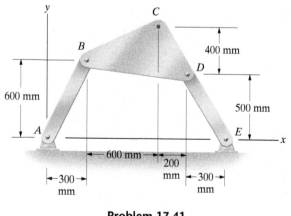

Problem 17.41

17.42 The upper grip and jaw of the pliers *ABC* is stationary. The lower grip *DEF* is rotating a 0.2 rad/s in the clockwise direction. At the instant shown, what is the angular velocity of the lower jaw *CFG*?

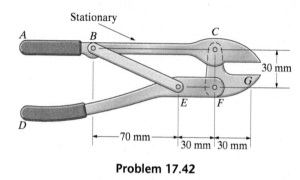

Problem 17.42

17.43 The horizontal member *ADE* supporting the scoop is stationary. If the link *BD* is rotating in the clockwise direction at 1 rad/s, what is the angular velocity of the scoop?

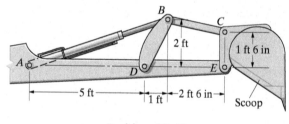

Problem 17.43

17.44 The diameter of the disk is 1 m, and the length of bar *AB* is 1 m. The disk is rolling, and point *B* slides on the plane surface. Determine the angular velocity of bar *AB* and the velocity of point *B*.

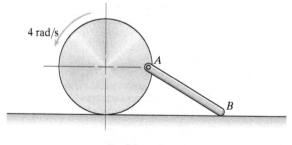

Problem 17.44

17.45 A motor rotates the circular disk mounted at A, moving the saw back and forth. (The saw is supported by a horizontal slot so that point C moves horizontally.) The radius AB is 4 in, and the link BC is 14 in long. In the position shown, $\theta = 45°$ and the link BC is horizontal. If the angular velocity of the disk is one revolution per second counterclockwise, what is the velocity of the saw?

17.46 In Problem 17.45, if the angular velocity of the disk is one revolution per second counterclockwise and $\theta = 270°$, what is the velocity of the saw?

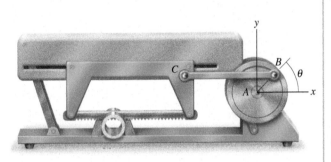

Problems 17.45/17.46

17.47 The disks roll on the plane surface. The angular velocity of the left disk is 2 rad/s in the clockwise direction. What is the angular velocity of the right disk?

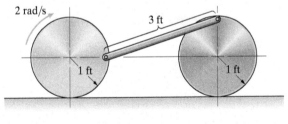

Problem 17.47

17.48 The disk rolls on the curved surface. The bar rotates at 10 rad/s in the counterclockwise direction. Determine the velocity of point A.

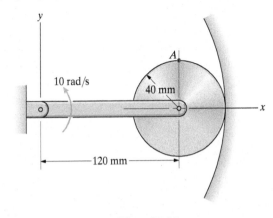

Problem 17.48

17.49 If $\omega_{AB} = 2$ rad/s and $\omega_{BC} = 4$ rad/s, what is the velocity of point C, where the excavator's bucket is attached?

17.50 If $\omega_{AB} = 2$ rad/s, what clockwise angular velocity ω_{BC} will cause the vertical component of the velocity of point C to be zero? What is the resulting velocity of point C?

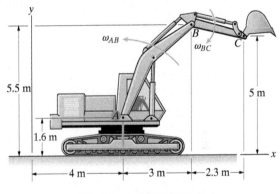

Problems 17.49/17.50

17.51 The steering linkage of a car is shown. Member DE rotates about the fixed pin E. The right brake disk is rigidly attached to member DE. The tie rod CD is pinned at C and D. At the instant shown, the Pitman arm AB has a counterclockwise angular velocity of 1 rad/s. What is the angular velocity of the right brake disk?

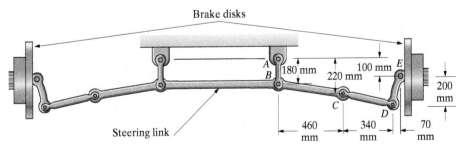

Problem 17.51

17.52 An athlete exercises his arm by raising the mass m. The shoulder joint A is stationary. The distance AB is 300 mm, and the distance BC is 400 mm. At the instant shown, $\omega_{AB} = 1$ rad/s and $\omega_{BC} = 2$ rad/s. How fast is the mass m rising?

17.53 The distance AB is 12 in, the distance BC is 16 in, $\omega_{AB} = 0.6$ rad/s, and the mass m is rising at 24 in/s. What is the angular velocity ω_{BC}?

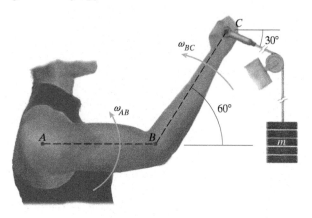

Problems 17.52/17.53

17.54 Points B and C are in the x–y plane. The angular velocity vectors of the arms AB and BC are $\mathbf{\omega}_{AB} = -0.2\mathbf{k}$ (rad/s) and $\mathbf{\omega}_{BC} = 0.4\mathbf{k}$ (rad/s). What is the velocity of point C?

17.55 If the velocity of point C of the robotic arm is $\mathbf{v}_C = -0.15\mathbf{i} + 0.42\mathbf{j}$ (m/s), what are the angular velocities of arms AB and BC?

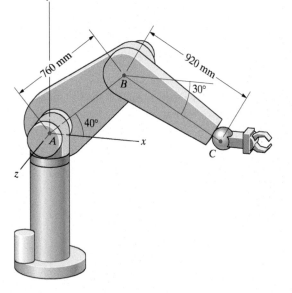

Problems 17.54/17.55

17.56 The link AB of the robot's arm is rotating at 2 rad/s in the counterclockwise direction, the link BC is rotating at 3 rad/s in the clockwise direction, and the link CD is rotating at 4 rad/s in the counterclockwise direction. What is the velocity of point D?

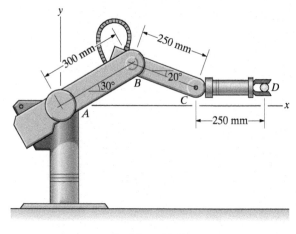

Problem 17.56

17.57 The person squeezes the grips of the shears, causing the angular velocities shown. What is the resulting angular velocity of the jaw BD?

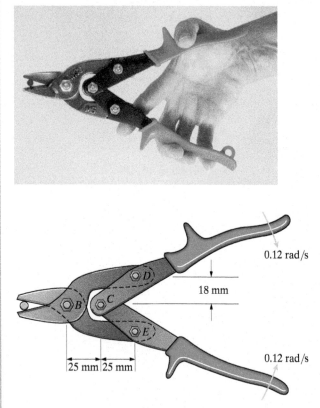

Problem 17.57

17.58 Determine the velocity v_W and the angular velocity of the small pulley.

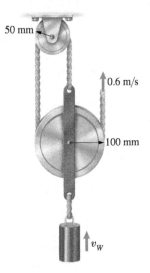

50 mm

0.6 m/s

100 mm

v_W

Problem 17.58

17.59 Determine the velocity of the block and the angular velocity of the small pulley.

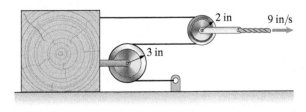

2 in

9 in/s

3 in

Problem 17.59

17.60 The device shown in the next column is used in the semiconductor industry to polish silicon wafers. The wafers are placed on the faces of the carriers. The outer and inner rings are then rotated, causing the wafers to move and rotate against an abrasive surface. If the outer ring rotates in the clockwise direction at 7 rpm and the inner ring rotates in the counterclockwise direction at 12 rpm, what is the angular velocity of the carriers?

17.61 Suppose that the outer ring rotates in the clockwise direction at 5 rpm and you want the centerpoints of the carriers to remain stationary during the polishing process. What is the necessary angular velocity of the inner ring?

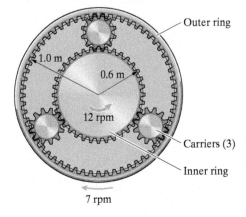

Outer ring

1.0 m

0.6 m

12 rpm

Carriers (3)

Inner ring

7 rpm

Problems 17.60/17.61

17.62 The ring gear is fixed and the hub and planet gears are bonded together. The connecting rod rotates in the counterclockwise direction at 60 rpm. Determine the angular velocity of the sun gear and the magnitude of the velocity of point A.

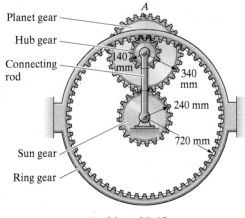

A

Planet gear

Hub gear

Connecting rod

140 mm

340 mm

240 mm

720 mm

Sun gear

Ring gear

Problem 17.62

17.63 The large gear is fixed. Bar AB has a counterclockwise angular velocity of 2 rad/s. What are the angular velocities of bars CD and DE?

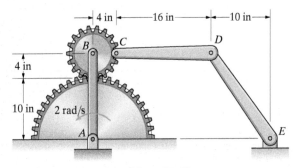

4 in | 16 in | 10 in

B C D

4 in

10 in   2 rad/s A E

Problem 17.63

17.4 Instantaneous Centers

BACKGROUND

A point of a rigid body whose velocity is zero at a given instant is called an *instantaneous center*. "Instantaneous" means that the point may have zero velocity *only* at the instant under consideration, although we also refer to a fixed point, such as a point of a fixed axis about which a rigid body rotates, as an instantaneous center.

When we know the location of an instantaneous center of a rigid body in two-dimensional motion and we know its angular velocity, the velocities of other points are easy to determine. For example, suppose that point C in Fig. 17.16a is the instantaneous center of a rigid body in plane motion with angular velocity ω. Relative to C, a point A moves in a circular path. The velocity of A relative to C is tangent to the circular path and equal to the product of the distance from C to A and the angular velocity. But since C is stationary at the present instant, the velocity of A relative to C is the velocity of A. At this instant, every point of the rigid body rotates about C (Fig. 17.16b).

The instantaneous center of a rigid body in planar motion can often be located by a simple procedure. Suppose that the directions of the motions of two points A and B are known and are not parallel (Fig. 17.17a). If we draw lines through A and B perpendicular to their directions of motion, then the point C where the lines intersect is the instantaneous center. To show that this is true, let us express the velocity of C in terms of the velocity of A (Fig. 17.17b):

$$\mathbf{v}_C = \mathbf{v}_A + \boldsymbol{\omega} \times \mathbf{r}_{C/A}.$$

The vector $\boldsymbol{\omega} \times \mathbf{r}_{C/A}$ is perpendicular to $\mathbf{r}_{C/A}$, so this equation indicates that $\mathbf{v}_C$ is parallel to the direction of motion of A. We also express the velocity of C in terms of the velocity of B:

$$\mathbf{v}_C = \mathbf{v}_B + \boldsymbol{\omega} \times \mathbf{r}_{C/B}.$$

The vector $\boldsymbol{\omega} \times \mathbf{r}_{C/B}$ is perpendicular to $\mathbf{r}_{C/B}$, so this equation indicates that $\mathbf{v}_C$ is parallel to the direction of motion of B. We have shown that the component of $\mathbf{v}_C$ perpendicular to the direction of motion of A is zero and that the component of $\mathbf{v}_C$ perpendicular to the direction of motion of B is zero, so $\mathbf{v}_C = \mathbf{0}$.

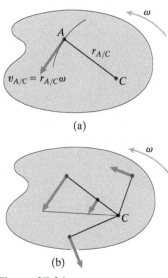

Figure 17.16
(a) An instantaneous center C and a different point A.
(b) Every point is rotating about the instantaneous center.

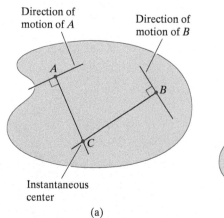

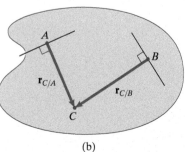

Figure 17.17
(a) Locating the instantaneous center in planar motion.
(b) Proving that $\mathbf{v}_C = \mathbf{0}$.

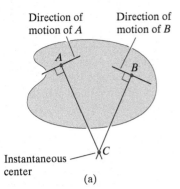

Direction of motion of A

Direction of motion of B

Instantaneous center

(a)

(b)

Figure 17.18
(a) An instantaneous center external to the rigid body.
(b) A hypothetical extended body. Point C would be stationary.

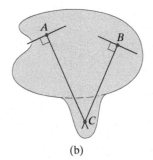

(a)

An instantaneous center may not be a point of the rigid body (Fig. 17.18a). This simply means that at the instant in question, the rigid body is rotating about an external point. It is helpful to *imagine* extending the rigid body so that it includes the instantaneous center (Fig. 17.18b). The velocity of point C of the extended body would be zero at the instant under consideration.

Notice in Fig. 17.18a that if the directions of motion of A and B are changed so that the lines perpendicular to their directions of motion become parallel, C moves to infinity. In that case, the rigid body is in pure translation, with an angular velocity of zero.

Returning once again to our example of a disk of radius R rolling with angular velocity ω (Fig. 17.19a), the point C in contact with the floor is stationary at the instant shown—it is the instantaneous center of the disk. Therefore, the velocity of any other point is perpendicular to the line from C to the point, and its magnitude equals the product of ω and the distance from C to the point. In terms of the coordinate system given in Fig. 17.19b, the velocity of point A is

$$\mathbf{v}_A = -\sqrt{2}R\omega \cos 45°\mathbf{i} + \sqrt{2}R\omega \sin 45°\mathbf{j}$$

$$= -R\omega\mathbf{i} + R\omega\mathbf{j}.$$

(b)

Figure 17.19
(a) Point C is the instantaneous center of the rolling disk.
(b) Determining the velocity of point A.

RESULTS

A point of a rigid body whose velocity is zero at a given instant is called an *instantaneous center*. "Instantaneous" means that the point may have zero velocity only at the instant under consideration, although a fixed point, such as a point of a fixed axis around which a rigid body rotates, is also referred to as an instantaneous center.

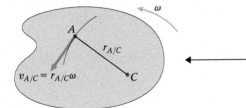

If a point C is an instantaneous center of a rigid body undergoing two-dimensional motion with angular velocity ω, the magnitude and direction of the velocity of any other point A can be determined.

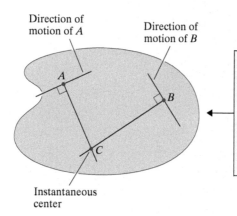

If the directions of motion of two points A and B of a rigid body in two-dimensional motion are known, the location of the instantaneous center C can be determined. It is the intersection of lines drawn through A and B perpendicular to their directions of motion. Notice that if the directions of motion of A and B are parallel, the instantaneous center is at infinity, *i.e.*, the rigid body is in translation.

Active Example 17.4 Linkage Analysis by Instantaneous Centers (▶ *Related Problem 17.70*)

Bar AB rotates with a counterclockwise angular velocity of 10 rad/s. At the instant shown, what are the angular velocities of bars BC and CD?

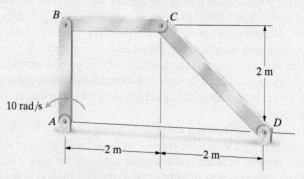

Strategy

Bars AB and CD rotate about fixed points, so we know their instantaneous centers. We also know the directions of motion of points B and C, so we can locate the instantaneous center of bar BC. Using these instantaneous centers, and the relations between the velocities of points of the bars and the angular velocities of the bars, we can determine the angular velocities.

Solution

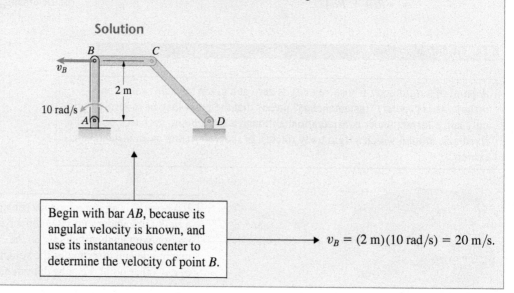

Begin with bar AB, because its angular velocity is known, and use its instantaneous center to determine the velocity of point B.

$v_B = (2\ \text{m})(10\ \text{rad/s}) = 20\ \text{m/s}.$

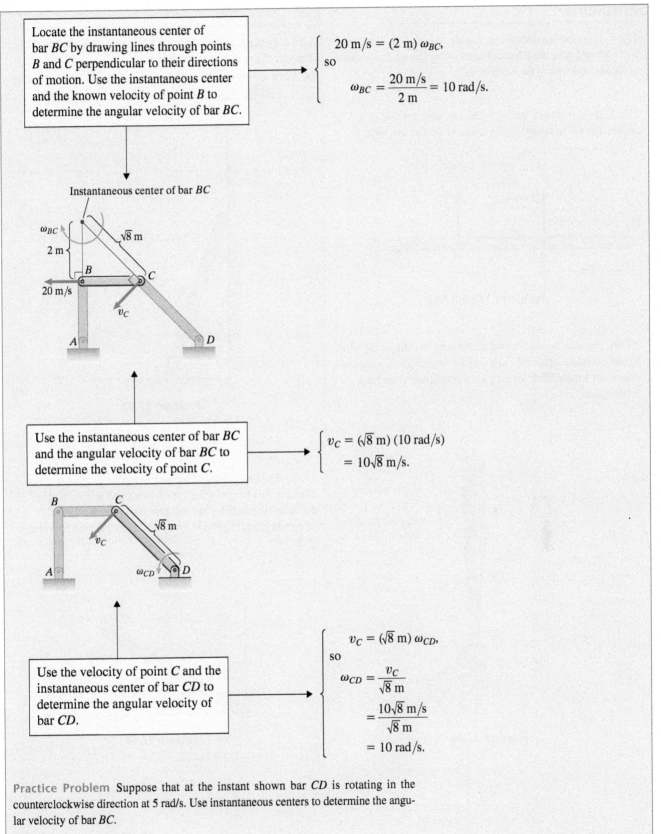

Locate the instantaneous center of bar BC by drawing lines through points B and C perpendicular to their directions of motion. Use the instantaneous center and the known velocity of point B to determine the angular velocity of bar BC.

$$20 \text{ m/s} = (2 \text{ m}) \, \omega_{BC},$$
so
$$\omega_{BC} = \frac{20 \text{ m/s}}{2 \text{ m}} = 10 \text{ rad/s}.$$

Instantaneous center of bar BC

Use the instantaneous center of bar BC and the angular velocity of bar BC to determine the velocity of point C.

$$v_C = (\sqrt{8} \text{ m})\,(10 \text{ rad/s})$$
$$= 10\sqrt{8} \text{ m/s}.$$

Use the velocity of point C and the instantaneous center of bar CD to determine the angular velocity of bar CD.

$$v_C = (\sqrt{8} \text{ m}) \, \omega_{CD},$$
so
$$\omega_{CD} = \frac{v_C}{\sqrt{8} \text{ m}}$$
$$= \frac{10\sqrt{8} \text{ m/s}}{\sqrt{8} \text{ m}}$$
$$= 10 \text{ rad/s}.$$

Practice Problem Suppose that at the instant shown bar CD is rotating in the counterclockwise direction at 5 rad/s. Use instantaneous centers to determine the angular velocity of bar BC.

Answer: 5 rad/s clockwise.

Problems

17.64 If the bar has a clockwise angular velocity of 10 rad/s and $v_A = 20$ m/s, what are the coordinates of the instantaneous center of the bar, and what is the value of v_B?

17.65 If $v_A = 24$ m/s and $v_B = 36$ m/s, what are the coordinates of the instantaneous center of the bar, and what is its angular velocity?

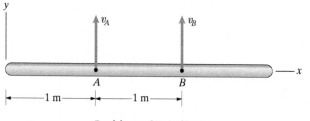

Problems 17.64/17.65

17.66 The velocity of point O of the bat is $\mathbf{v}_O = -6\mathbf{i} - 14\mathbf{j}$ (ft/s), and the bat rotates about the z axis with a counterclockwise angular velocity of 4 rad/s. What are the x and y coordinates of the bat's instantaneous center?

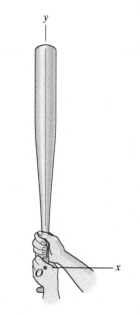

Problem 17.66

17.67 Points A and B of the 1-m bar slide on the plane surfaces. The velocity of B is $\mathbf{v}_B = 2\mathbf{i}$ (m/s).

(a) What are the coordinates of the instantaneous center of the bar?

(b) Use the instantaneous center to determine the velocity of A.

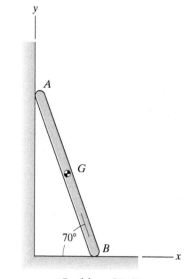

Problem 17.67

17.68 The bar is in two-dimensional motion in the x–y plane. The velocity of point A is $\mathbf{v}_A = 8\mathbf{i}$ (ft/s), and B is moving in the direction parallel to the bar. Determine the velocity of B (a) by using Eq. (17.6) and (b) by using the instantaneous center of the bar.

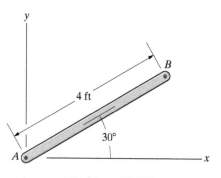

Problem 17.68

17.69 Point *A* of the bar is moving at 8 m/s in the direction of the unit vector $0.966\mathbf{i} - 0.259\mathbf{j}$, and point *B* is moving in the direction of the unit vector $0.766\mathbf{i} + 0.643\mathbf{j}$.

(a) What are the coordinates of the bar's instantaneous center?

(b) What is the bar's angular velocity?

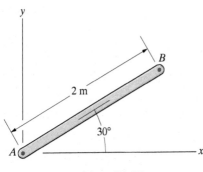

Problem 17.69

▶ **17.70** Bar *AB* rotates with a counterclockwise angular velocity of 10 rad/s. At the instant shown, what are the angular velocities of bars *BC* and *CD*? (See Active Example 17.4.)

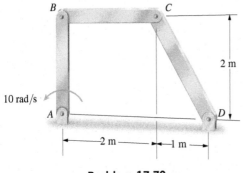

Problem 17.70

17.71 Use instantaneous centers to determine the horizontal velocity of *B*.

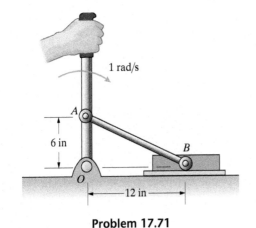

Problem 17.71

17.72 When the mechanism in Problem 17.71 is in the position shown here, use instantaneous centers to determine the horizontal velocity of *B*.

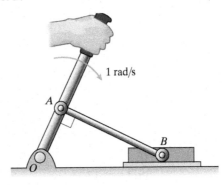

Problem 17.72

17.73 The angle $\theta = 45°$, and bar *OQ* is rotating in the counterclockwise direction at 0.2 rad/s. Use instantaneous centers to determine the velocity of the sleeve *P*.

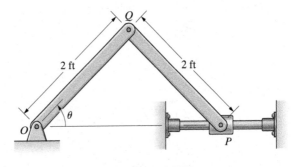

Problem 17.73

17.74 Bar *AB* is rotating in the counterclockwise direction at 5 rad/s. The disk rolls on the horizontal surface. Determine the angular velocity of bar *BC*.

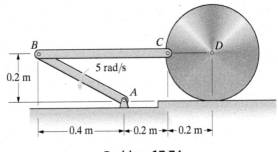

Problem 17.74

17.75 Bar AB rotates at 6 rad/s in the clockwise direction. Use instantaneous centers to determine the angular velocity of bar BC.

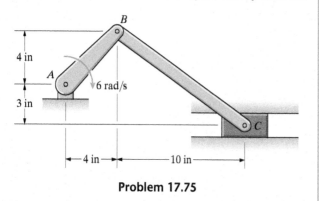

Problem 17.75

17.76 The crank AB is rotating in the clockwise direction at 2000 rpm (revolutions per minute).

(a) At the instant shown, what are the coordinates of the instantaneous center of the connecting rod BC?

(b) Use instantaneous centers to determine the angular velocity of the connecting rod BC at the instant shown.

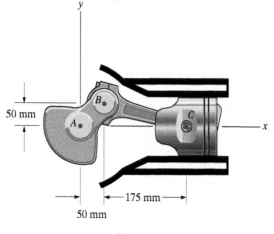

Problem 17.76

17.77 The disks roll on the plane surface. The left disk rotates at 2 rad/s in the clockwise direction. Use instantaneous centers to determine the angular velocities of the bar and the right disk.

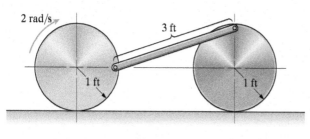

Problem 17.77

17.78 Bar AB rotates at 12 rad/s in the clockwise direction. Use instantaneous centers to determine the angular velocities of bars BC and CD.

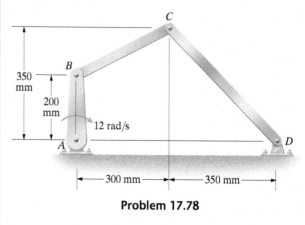

Problem 17.78

17.79 The horizontal member ADE supporting the scoop is stationary. The link BD is rotating in the clockwise direction at 1 rad/s. Use instantaneous centers to determine the angular velocity of the scoop.

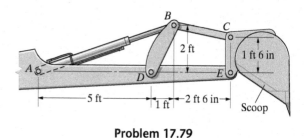

Problem 17.79

17.80 The disk is in planar motion. The directions of the velocities of points A and B are shown. The velocity of point A is $v_A = 2$ m/s.

(a) What are the coordinates of the disk's instantaneous center?

(b) Determine the velocity v_B and the disk's angular velocity.

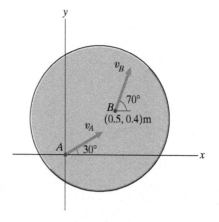

Problem 17.80

17.5 General Motions: Accelerations

BACKGROUND

In Chapter 18, we will be concerned with determining the motion of a rigid body when we know the external forces and couples acting on it. The governing equations are expressed in terms of the acceleration of the center of mass of the rigid body and its angular acceleration. To solve such problems, we need the relationship between the accelerations of points of a rigid body and its angular acceleration. In this section, we extend the methods we have used to analyze velocities of points of rigid bodies to accelerations.

Consider points A and B of a rigid body in planar motion relative to a given reference frame (Fig. 17.20a). Their velocities are related by

$$\mathbf{v}_A = \mathbf{v}_B + \mathbf{v}_{A/B}.$$

Taking the derivative of this equation with respect to time, we obtain

$$\mathbf{a}_A = \mathbf{a}_B + \mathbf{a}_{A/B},$$

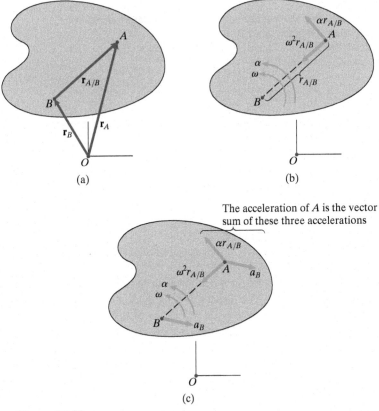

Figure 17.20
(a) Points A and B of a rigid body in planar motion and the position vector of A relative to B.
(b) Components of the acceleration of A relative to B.
(c) The acceleration of A.

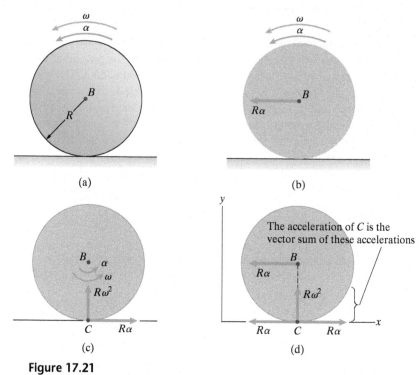

Figure 17.21
(a) A disk rolling with angular velocity ω and angular acceleration α.
(b) Acceleration of the center B.
(c) Components of the acceleration of C relative to B.
(d) The acceleration of C.

where $\mathbf{a}_A$ and $\mathbf{a}_B$ are the accelerations of A and B relative to the reference frame and $\mathbf{a}_{A/B}$ is the acceleration of A relative to B. (*When we simply speak of the acceleration of a point, we will mean its acceleration relative to the given reference frame.*) Because A moves in a circular path relative to B as the rigid body rotates, $\mathbf{a}_{A/B}$ has normal and tangential components (Fig. 17.20b). The tangential component equals the product of the distance $r_{A/B} = |\mathbf{r}_{A/B}|$ and the angular acceleration α of the rigid body. The normal component points toward the center of the circular path, and its magnitude is $|\mathbf{v}_{A/B}|^2/r_{A/B} = \omega^2 r_{A/B}$. The acceleration of A equals the sum of the acceleration of B and the acceleration of A relative to B (Fig. 17.20c).

For example, let us consider a circular disk of radius R rolling on a stationary plane surface. The disk has counterclockwise angular velocity ω and counterclockwise angular acceleration α (Fig. 17.21a). The disk's center B is moving in a straight line with velocity $R\omega$, toward the left if ω is positive. Therefore, the acceleration of B is $d/dt(R\omega) = R\alpha$ and is toward the left if α is positive (Fig. 17.21b). In other words, *the magnitude of the acceleration of the center of a round object rolling on a stationary plane surface is the product of the radius and the angular acceleration.*

Now that we know the acceleration of the disk's center, let us determine the acceleration of the point C that is in contact with the surface. Relative to B, C moves in a circular path of radius R with angular velocity ω and angular acceleration α. The tangential and normal components of the acceleration of C

relative to B are shown in Fig. 17.21c. The acceleration of C is the sum of the acceleration of B and the acceleration of C relative to B (Fig. 17.21d). In terms of the coordinate system shown,

$$\mathbf{a}_C = \mathbf{a}_B + \mathbf{a}_{C/B} = -R\alpha\mathbf{i} + R\alpha\mathbf{i} + R\omega^2\mathbf{j}$$
$$= R\omega^2\mathbf{j}.$$

The acceleration of point C parallel to the surface is zero, but C does have an acceleration normal to the surface.

Expressing the acceleration of a point A relative to a point B in terms of A's circular path about B as we have done is useful for visualizing and understanding the relative acceleration. However, just as we did in the case of the relative velocity, we can obtain $\mathbf{a}_{A/B}$ in a form more convenient for applications by using the angular velocity vector $\boldsymbol{\omega}$. The velocity of A relative to B is given in terms of $\boldsymbol{\omega}$ by Eq. (17.5):

$$\mathbf{v}_{A/B} = \boldsymbol{\omega} \times \mathbf{r}_{A/B}.$$

Taking the derivative of this equation with respect to time, we obtain

$$\mathbf{a}_{A/B} = \frac{d\boldsymbol{\omega}}{dt} \times \mathbf{r}_{A/B} + \boldsymbol{\omega} \times \mathbf{v}_{A/B}$$

$$= \frac{d\boldsymbol{\omega}}{dt} \times \mathbf{r}_{A/B} + \boldsymbol{\omega} \times (\boldsymbol{\omega} \times \mathbf{r}_{A/B}).$$

We next define the *angular acceleration vector* $\boldsymbol{\alpha}$ to be the rate of change of the angular velocity vector:

$$\boldsymbol{\alpha} = \frac{d\boldsymbol{\omega}}{dt}. \tag{17.7}$$

Then the acceleration of A relative to B is

$$\mathbf{a}_{A/B} = \boldsymbol{\alpha} \times \mathbf{r}_{A/B} + \boldsymbol{\omega} \times (\boldsymbol{\omega} \times \mathbf{r}_{A/B}).$$

Using this expression, we can write equations relating the velocities and accelerations of two points of a rigid body in terms of its angular velocity and angular acceleration:

$$\mathbf{v}_A = \mathbf{v}_B + \boldsymbol{\omega} \times \mathbf{r}_{A/B}, \tag{17.8}$$

$$\mathbf{a}_A = \mathbf{a}_B + \boldsymbol{\alpha} \times \mathbf{r}_{A/B} + \boldsymbol{\omega} \times (\boldsymbol{\omega} \times \mathbf{r}_{A/B}). \tag{17.9}$$

In the case of planar motion, the term $\boldsymbol{\alpha} \times \mathbf{r}_{A/B}$ in Eq. (17.9) is the tangential component of the acceleration of A relative to B, and $\boldsymbol{\omega} \times (\boldsymbol{\omega} \times \mathbf{r}_{A/B})$ is the normal component (Fig. 17.22). Therefore, for planar motion, we can write Eq. (17.9) in the simpler form

$$\mathbf{a}_A = \mathbf{a}_B + \boldsymbol{\alpha} \times \mathbf{r}_{A/B} - \omega^2\mathbf{r}_{A/B}. \qquad \text{planar motion} \tag{17.10}$$

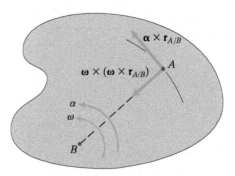

Figure 17.22
Vector components of the acceleration of A relative to B in planar motion.

Relative Velocities and Accelerations

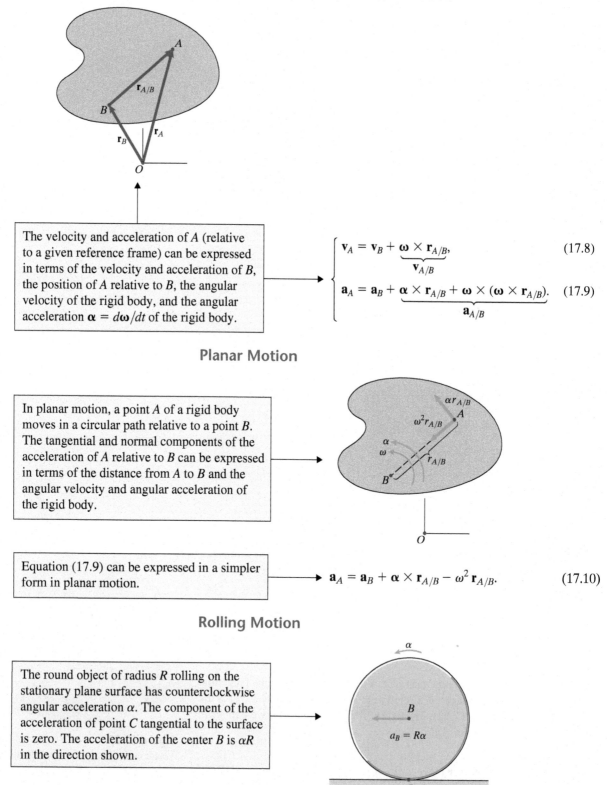

The velocity and acceleration of A (relative to a given reference frame) can be expressed in terms of the velocity and acceleration of B, the position of A relative to B, the angular velocity of the rigid body, and the angular acceleration $\boldsymbol{\alpha} = d\boldsymbol{\omega}/dt$ of the rigid body.

$$\mathbf{v}_A = \mathbf{v}_B + \underbrace{\boldsymbol{\omega} \times \mathbf{r}_{A/B}}_{\mathbf{v}_{A/B}}, \qquad (17.8)$$

$$\mathbf{a}_A = \mathbf{a}_B + \underbrace{\boldsymbol{\alpha} \times \mathbf{r}_{A/B} + \boldsymbol{\omega} \times (\boldsymbol{\omega} \times \mathbf{r}_{A/B})}_{\mathbf{a}_{A/B}}. \qquad (17.9)$$

Planar Motion

In planar motion, a point A of a rigid body moves in a circular path relative to a point B. The tangential and normal components of the acceleration of A relative to B can be expressed in terms of the distance from A to B and the angular velocity and angular acceleration of the rigid body.

Equation (17.9) can be expressed in a simpler form in planar motion.

$$\mathbf{a}_A = \mathbf{a}_B + \boldsymbol{\alpha} \times \mathbf{r}_{A/B} - \omega^2 \mathbf{r}_{A/B}. \qquad (17.10)$$

Rolling Motion

The round object of radius R rolling on the stationary plane surface has counterclockwise angular acceleration α. The component of the acceleration of point C tangential to the surface is zero. The acceleration of the center B is αR in the direction shown.

Active Example 17.5 Acceleration of a Point of a Rolling Disk (▶ *Related Problem 17.85*)

The rolling disk has counterclockwise angular velocity ω and counterclockwise angular acceleration α. What is the acceleration of point A?

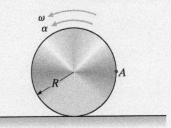

Strategy
We know the acceleration of the center of a rolling disk in terms of the radius of the disk and its angular acceleration. We can obtain the acceleration of point A by adding the acceleration of the center to the tangential and normal components of the acceleration of A relative to the center.

Solution

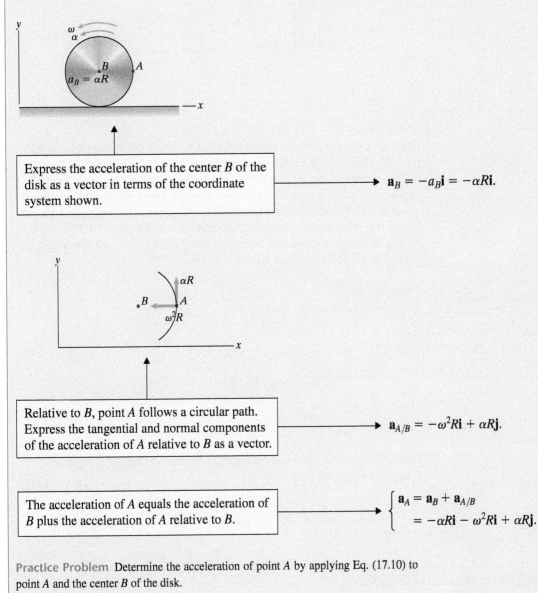

Express the acceleration of the center B of the disk as a vector in terms of the coordinate system shown.

$$\mathbf{a}_B = -a_B\mathbf{i} = -\alpha R\mathbf{i}.$$

Relative to B, point A follows a circular path. Express the tangential and normal components of the acceleration of A relative to B as a vector.

$$\mathbf{a}_{A/B} = -\omega^2 R\mathbf{i} + \alpha R\mathbf{j}.$$

The acceleration of A equals the acceleration of B plus the acceleration of A relative to B.

$$\begin{cases} \mathbf{a}_A = \mathbf{a}_B + \mathbf{a}_{A/B} \\ \quad = -\alpha R\mathbf{i} - \omega^2 R\mathbf{i} + \alpha R\mathbf{j}. \end{cases}$$

Practice Problem Determine the acceleration of point A by applying Eq. (17.10) to point A and the center B of the disk.

Answer: $\mathbf{a}_A = -\alpha R\mathbf{i} - \omega^2 R\mathbf{i} + \alpha R\mathbf{j}$.

Example 17.6 Angular Accelerations of Members of a Linkage (▶ *Related Problem 17.90*)

Bar AB has a counterclockwise angular velocity of 10 rad/s and a clockwise angular acceleration of 300 rad/s². What are the angular accelerations of bars BC and CD?

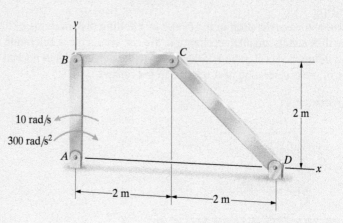

Strategy

Since we know the angular velocity of bar AB, we can determine the velocity of point B. Then we can apply Eq. (17.8) to points C and D to obtain an equation for $\mathbf{v}_C$ in terms of the angular velocity of bar CD. We can also apply Eq. (17.8) to points B and C to obtain an equation for $\mathbf{v}_C$ in terms of the angular velocity of bar BC. By equating the two expressions for $\mathbf{v}_C$, we will obtain a vector equation in two unknowns: the angular velocities of bars BC and CD. Then, by following the same sequence of steps, but using Eq. (17.10), we can obtain the angular accelerations of bars BC and CD.

Solution

The velocity of B is (Fig. a)

$$\mathbf{v}_B = \mathbf{v}_A + \boldsymbol{\omega}_{AB} \times \mathbf{r}_{B/A}$$

$$= \mathbf{0} + (10\mathbf{k}) \times (2\mathbf{j})$$

$$= -20\mathbf{i} \text{ (m/s)}.$$

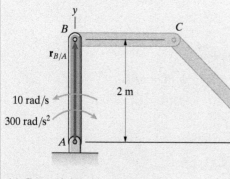

(a) Determining the motion of B.

Let ω_{CD} be the unknown angular velocity of bar CD (Fig. b). The velocity of C in terms of the velocity of D is

$$\mathbf{v}_C = \mathbf{v}_D + \boldsymbol{\omega}_{CD} \times \mathbf{r}_{C/D}$$

$$= \mathbf{0} + \begin{vmatrix} \mathbf{i} & \mathbf{j} & \mathbf{k} \\ 0 & 0 & \omega_{CD} \\ -2 & 2 & 0 \end{vmatrix}$$

$$= -2\omega_{CD}\mathbf{i} - 2\omega_{CD}\mathbf{j}.$$

Denoting the angular velocity of bar BC by ω_{BC} (Fig. c), we obtain the velocity of C in terms of the velocity of B:

$$\mathbf{v}_C = \mathbf{v}_B + \boldsymbol{\omega}_{BC} \times \mathbf{r}_{C/B}$$
$$= -20\mathbf{i} + (\omega_{BC}\mathbf{k}) \times (2\mathbf{i})$$
$$= -20\mathbf{i} + 2\omega_{BC}\mathbf{j}.$$

Equating our two expressions for $\mathbf{v}_C$ yields

$$-2\omega_{CD}\mathbf{i} - 2\omega_{CD}\mathbf{j} = -20\mathbf{i} + 2\omega_{BC}\mathbf{j}.$$

Equating the $\mathbf{i}$ and $\mathbf{j}$ components, we obtain $\omega_{CD} = 10$ rad/s and $\omega_{BC} = -10$ rad/s.

We can use the same sequence of steps to determine the angular accelerations. The acceleration of B is (Fig. a)

$$\mathbf{a}_B = \mathbf{a}_A + \boldsymbol{\alpha}_{AB} \times \mathbf{r}_{B/A} - \omega_{AB}^2 \mathbf{r}_{B/A}$$
$$= 0 + (-300\mathbf{k}) \times (2\mathbf{j}) - (10)^2(2\mathbf{j})$$
$$= 600\mathbf{i} - 200\mathbf{j} \ (\text{m/s}^2).$$

The acceleration of C in terms of the acceleration of D is (Fig. b)

$$\mathbf{a}_C = \mathbf{a}_D + \boldsymbol{\alpha}_{CD} \times \mathbf{r}_{C/D} - \omega_{CD}^2 \mathbf{r}_{C/D}$$
$$= 0 + \begin{vmatrix} \mathbf{i} & \mathbf{j} & \mathbf{k} \\ 0 & 0 & \alpha_{CD} \\ -2 & 2 & 0 \end{vmatrix} - (10)^2(-2\mathbf{i} + 2\mathbf{j})$$
$$= (200 - 2\alpha_{CD})\mathbf{i} - (200 + 2\alpha_{CD})\mathbf{j}.$$

The acceleration of C in terms of the acceleration of B is (Fig. c)

$$\mathbf{a}_C = \mathbf{a}_B + \boldsymbol{\alpha}_{BC} \times \mathbf{r}_{C/B} - \omega_{BC}^2 \mathbf{r}_{C/B}$$
$$= 600\mathbf{i} - 200\mathbf{j} + (\alpha_{BC}\mathbf{k}) \times (2\mathbf{i}) - (-10)^2(2\mathbf{i})$$
$$= 400\mathbf{i} - (200 - 2\alpha_{BC})\mathbf{j}.$$

Equating the expressions for $\mathbf{a}_C$, we obtain

$$(200 - 2\alpha_{CD})\mathbf{i} - (200 + 2\alpha_{CD})\mathbf{j} = 400\mathbf{i} - (200 - 2\alpha_{BC})\mathbf{j}.$$

Equating $\mathbf{i}$ and $\mathbf{j}$ components, we obtain the angular accelerations $\alpha_{BC} = 100$ rad/s^2 and $\alpha_{CD} = -100$ rad/s^2.

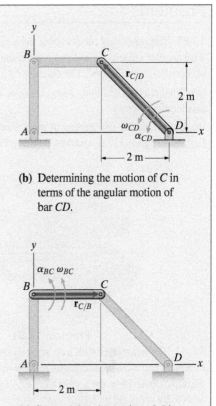

(b) Determining the motion of C in terms of the angular motion of bar CD.

(c) Determining the motion of C in terms of the angular motion of bar BC.

Critical Thinking

To determine the angular accelerations of a set of pinned rigid bodies, it is usually necessary to determine their angular velocities first, because the angular velocity appears in Eqs. (17.9) and (17.10). But as this example demonstrates, this initial step provides you with a guide for completing the solution. Once you have found a sequence of steps using Eq. (17.8) for determining the angular velocities, the same sequence of steps using Eq. (17.9) or (17.10) will determine the angular accelerations.

Problems

17.81 The rigid body rotates about the z axis with counterclockwise angular velocity $\omega = 4$ rad/s and counterclockwise angular acceleration $\alpha = 2$ rad/s^2. The distance $r_{A/B} = 0.6$ m.

(a) What are the rigid body's angular velocity and angular acceleration vectors?

(b) Determine the acceleration of point A relative to point B, first by using Eq. (17.9) and then by using Eq. (17.10).

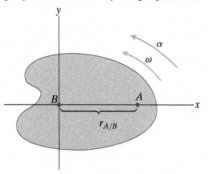

Problem 17.81

17.82 The bar rotates with a counterclockwise angular velocity of 5 rad/s and a counterclockwise angular acceleration of 30 rad/s^2. Determine the acceleration of A (a) by using Eq. (17.9) and (b) by using Eq. (17.10).

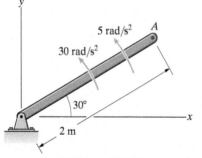

Problem 17.82

17.83 The bar rotates with a counterclockwise angular velocity of 20 rad/s and a counterclockwise angular acceleration of 6 rad/s^2.

(a) By applying Eq. (17.10) to point A and the fixed point O, determine the acceleration of A.

(b) By using the result of part (a) and applying Eq. (17.10) to points A and B, determine the acceleration of B.

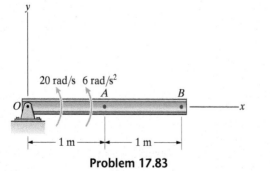

Problem 17.83

17.84 The helicopter is in planar motion in the x–y plane. At the instant shown, the position of its center of mass G is $x = 2$ m, $y = 2.5$ m, its velocity is $\mathbf{v}_G = 12\mathbf{i} + 4\mathbf{j}$ (m/s), and its acceleration is $\mathbf{a}_G = 2\mathbf{i} + 3\mathbf{j}$ (m/s^2). The position of point T where the tail rotor is mounted is $x = -3.5$ m, $y = 4.5$ m. The helicopter's angular velocity is 0.2 rad/s clockwise, and its angular acceleration is 0.1 rad/s^2 counterclockwise. What is the acceleration of point T?

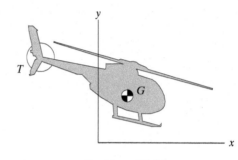

Problem 17.84

▶ **17.85** Point A of the rolling disk is moving toward the right and accelerating toward the right. The magnitude of the velocity of point C is 2 m/s, and the magnitude of the acceleration of point C is 14 m/s^2. Determine the accelerations of points B and D. (See Active Example 17.5.)

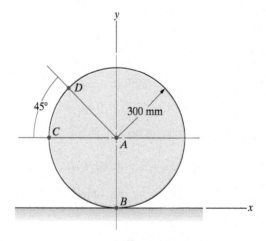

Problem 17.85

17.86 The disk rolls on the circular surface with a constant clockwise angular velocity of 1 rad/s. What are the accelerations of points *A* and *B*?

Strategy: Begin by determining the acceleration of the center of the disk. Notice that the center moves in a circular path and the magnitude of its velocity is constant.

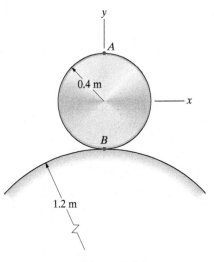

Problem 17.86

17.87 The length of the bar is *L* = 4 ft and the angle $\theta = 30°$. The bar's angular velocity is $\omega = 1.8$ rad/s and its angular acceleration is $\alpha = 6$ rad/s². The endpoints of the bar slide on the plane surfaces. Determine the acceleration of the midpoint *G*.

Strategy: Begin by applying Eq. (17.10) to the endpoints of the bar to determine their accelerations.

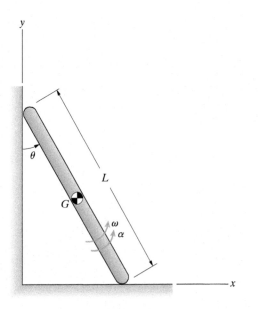

Problem 17.87

17.88 The angular velocity and angular acceleration of bar *AB* are $\omega_{AB} = 2$ rad/s and $\alpha_{AB} = 10$ rad/s². The dimensions of the rectangular plate are 12 in × 24 in. What are the angular velocity and angular acceleration of the rectangular plate?

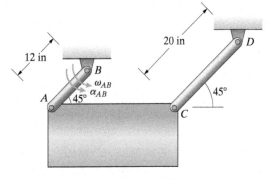

Problem 17.88

17.89 The ring gear is stationary, and the sun gear has an angular acceleration of 10 rad/s² in the counterclockwise direction. Determine the angular acceleration of the planet gears.

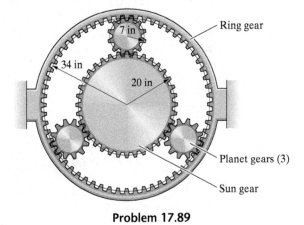

Problem 17.89

▶ **17.90** In Example 17.6, what is the acceleration of the midpoint of bar *BC*?

17.91 The 1-m-diameter disk rolls, and point *B* of the 1-m-long bar slides, on the plane surface. Determine the angular acceleration of the bar and the acceleration of point *B*.

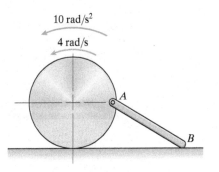

Problem 17.91

17.92 If $\theta = 45°$ and sleeve P is moving to the right with a constant velocity of 2 m/s, what are the angular accelerations of bars OQ and PQ?

17.93 If $\theta = 50°$ and bar OQ has a constant clockwise angular velocity of 1 rad/s, what is the acceleration of sleeve P?

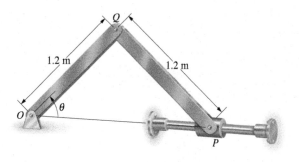

Problems 17.92/17.93

17.94 The angle $\theta = 60°$, and bar OQ has a constant counterclockwise angular velocity of 2 rad/s. What is the angular acceleration of bar PQ?

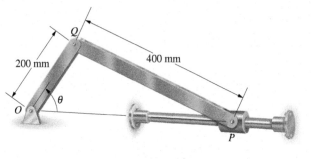

Problem 17.94

17.95 At the instant shown, the piston's velocity and acceleration are $\mathbf{v}_C = -14\mathbf{i}$ (m/s) and $\mathbf{a}_C = -2200\mathbf{i}$ (m/s^2). What is the angular acceleration of the crank AB?

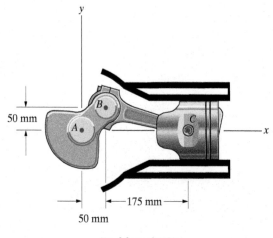

Problem 17.95

17.96 The angular velocity and angular acceleration of bar AB are $\omega_{AB} = 4$ rad/s and $\alpha_{AB} = -6$ rad/s^2. Determine the angular accelerations of bars BC and CD.

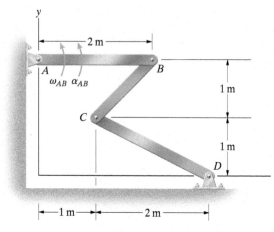

Problem 17.96

17.97 The angular velocity and angular acceleration of bar AB are $\omega_{AB} = 2$ rad/s and $\alpha_{AB} = 8$ rad/s^2. What is the acceleration of point D?

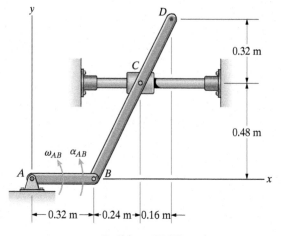

Problem 17.97

17.98 The angular velocity $\omega_{AB} = 6$ rad/s. If the acceleration of the slider C is zero at the instant shown, what is the angular acceleration α_{AB}?

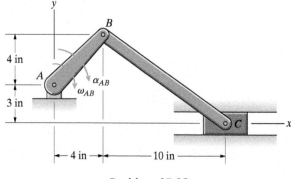

Problem 17.98

17.99 The angular velocity and angular acceleration of bar AB are $\omega_{AB} = 5$ rad/s and $\alpha_{AB} = 10$ rad/s². Determine the angular acceleration of bar BC.

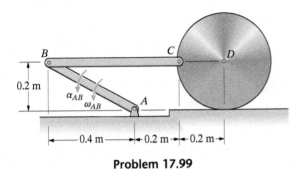

Problem 17.99

17.100 At the instant shown, bar AB is rotating at 10 rad/s in the counterclockwise direction and has a counterclockwise angular acceleration of 20 rad/s². The disk rolls on the circular surface. Determine the angular accelerations of bar BC and the disk.

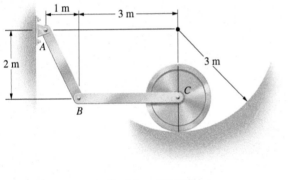

Problem 17.100

17.101 If $\omega_{AB} = 2$ rad/s, $\alpha_{AB} = 2$ rad/s², $\omega_{BC} = 1$ rad/s, and $\alpha_{BC} = 4$ rad/s², what is the acceleration of point C where the scoop of the excavator is attached?

17.102 If the velocity of point C of the excavator is $\mathbf{v}_C = 4\mathbf{i}$ (m/s) and is constant, what are ω_{AB}, α_{AB}, ω_{BC}, and α_{BC}?

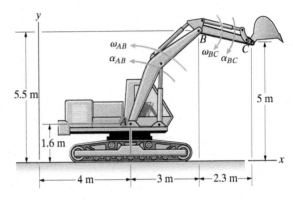

Problems 17.101/17.102

17.103 The steering linkage of a car is shown. Member DE rotates about the fixed pin E. The right brake disk is rigidly attached to member DE. The tie rod CD is pinned at C and D. At the instant shown, the Pitman arm AB has a counterclockwise angular velocity of 1 rad/s and a clockwise angular acceleration of 2 rad/s². What is the angular acceleration of the right brake disk?

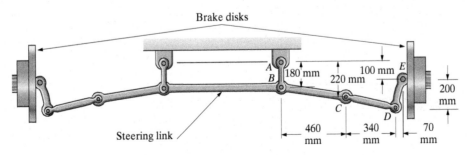

Problem 17.103

17.104 At the instant shown, bar AB has no angular velocity, but has a counterclockwise angular acceleration of 10 rad/s². Determine the acceleration of point E.

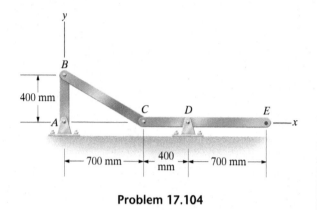

Problem 17.104

17.105 If $\omega_{AB} = 12$ rad/s and $\alpha_{AB} = 100$ rad/s², what are the angular accelerations of bars BC and CD?

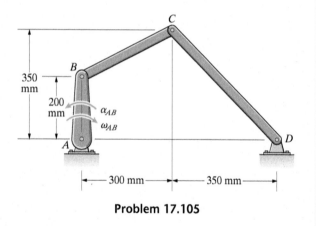

Problem 17.105

17.106 If $\omega_{AB} = 4$ rad/s counterclockwise and $\alpha_{AB} = 12$ rad/s² counterclockwise, what is the acceleration of point C?

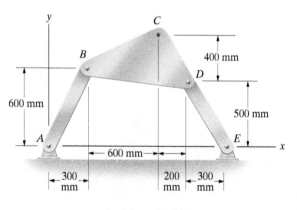

Problem 17.106

17.107 The angular velocities and angular accelerations of the grips of the shears are shown. What is the resulting angular acceleration of the jaw BD?

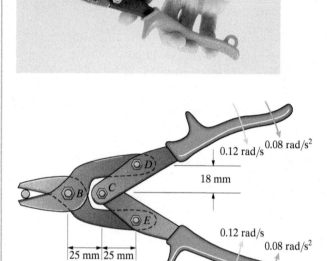

Problem 17.107

17.108 If arm *AB* has a constant clockwise angular velocity of 0.8 rad/s, arm *BC* has a constant clockwise angular velocity of 0.2 rad/s, and arm *CD* remains vertical, what is the acceleration of part *D*?

17.109 If arm *AB* has a constant clockwise angular velocity of 0.8 rad/s and you want *D* to have zero velocity and acceleration, what are the necessary angular velocities and angular accelerations of arms *BC* and *CD*?

17.110 If you want arm *CD* to remain vertical and you want part *D* to have velocity $\mathbf{v}_D = 1.0\mathbf{i}$ (m/s) and zero acceleration, what are the necessary angular velocities and angular accelerations of arms *AB* and *BC*?

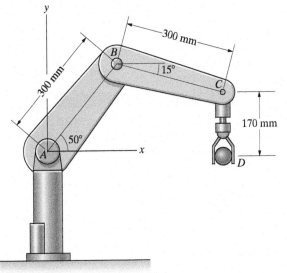

Problems 17.108–17.110

17.111 Link *AB* of the robot's arm is rotating with a constant counterclockwise angular velocity of 2 rad/s, and link *BC* is rotating with a constant clockwise angular velocity of 3 rad/s. Link *CD* is rotating at 4 rad/s in the counterclockwise direction and has a counterclockwise angular acceleration of 6 rad/s². What is the acceleration of point *D*?

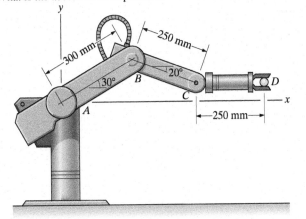

Problem 17.111

17.112 The upper grip and jaw of the pliers *ABC* is stationary. The lower grip *DEF* is rotating in the clockwise direction with a constant angular velocity of 0.2 rad/s. At the instant shown, what is the angular acceleration of the lower jaw *CFG*?

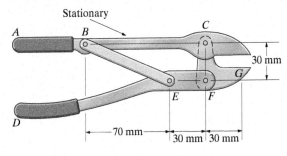

Problem 17.112

17.113 The horizontal member *ADE* supporting the scoop is stationary. If link *BD* has a clockwise angular velocity of 1 rad/s and a counterclockwise angular acceleration of 2 rad/s², what is the angular acceleration of the scoop?

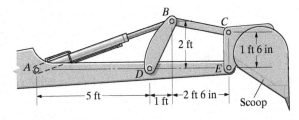

Problem 17.113

17.114 The ring gear is fixed, and the hub and planet gears are bonded together. The connecting rod has a counterclockwise angular acceleration of 10 rad/s². Determine the angular accelerations of the planet and sun gears.

17.115 The connecting rod has a counterclockwise angular velocity of 4 rad/s and a clockwise angular acceleration of 12 rad/s². Determine the magnitude of the acceleration of point *A*.

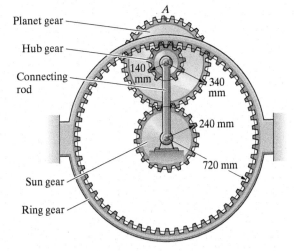

Problems 17.114/17.115

17.116 The large gear is fixed. The angular velocity and angular acceleration of bar AB are $\omega_{AB} = 2$ rad/s and $\alpha_{AB} = 4$ rad/s^2. Determine the angular accelerations of bars CD and DE.

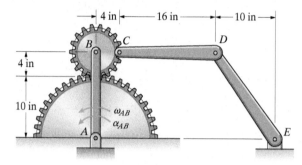

Problem 17.116

17.6 Sliding Contacts

BACKGROUND

In this section we consider a type of problem that is superficially similar to those we have discussed previously in this chapter, but which requires a different method of solution. For example, in Fig. 17.23, pin A of the bar connected at C slides in a slot in the bar connected at B. Suppose that we know the angular velocity of the bar connected at B, and we want to determine the angular velocity of bar AC. We cannot use the equation $\mathbf{v}_A = \mathbf{v}_B + \boldsymbol{\omega} \times \mathbf{r}_{A/B}$ to express the velocity of pin A in terms of the angular velocity of the bar fixed at B, because we derived that equation under the assumption that A and B are points of the same rigid body. Here pin A moves relative to the bar connected at B as it slides along the slot. This is an example of a *sliding contact* between rigid bodies. To solve this type of problem, we must rederive Eqs. (17.8), (17.9), and (17.10) without making the assumption that A is a point of the rigid body.

To describe the motion of a point that moves relative to a given rigid body, it is convenient to use a reference frame that moves with the rigid body. We say that such a reference frame is *body-fixed*. In Fig. 17.24, we introduce a body-fixed reference frame xyz with its origin at a point B of the rigid body, in addition to

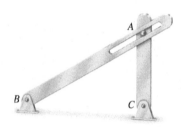

Figure 17.23
Linkage with a sliding contact.

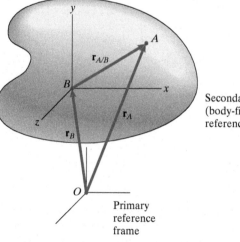

Figure 17.24
A point B of a rigid body, a body-fixed secondary reference frame, and an arbitrary point A.

the *primary reference frame* with origin O. (The primary reference frame is the reference frame relative to which we are describing the motion of the rigid body.) We do not assume A to be a point of the rigid body. The position of A relative to O is

$$\mathbf{r}_A = \mathbf{r}_B + \underbrace{x\mathbf{i} + y\mathbf{j} + z\mathbf{k},}_{\mathbf{r}_{A/B}}$$

where x, y, and z are the coordinates of A in terms of the body-fixed reference frame. Our next step is to take the derivative of this expression with respect to time in order to obtain an equation for the velocity of A. In doing so, we recognize that the unit vectors $\mathbf{i}$, $\mathbf{j}$, and $\mathbf{k}$ are not constant, because they rotate with the body-fixed reference frame:

$$\mathbf{v}_A = \mathbf{v}_B + \frac{dx}{dt}\mathbf{i} + x\frac{d\mathbf{i}}{dt} + \frac{dy}{dt}\mathbf{j} + y\frac{d\mathbf{j}}{dt} + \frac{dz}{dt}\mathbf{k} + z\frac{d\mathbf{k}}{dt}.$$

Now, what are the derivatives of the unit vectors? In Section 17.3, we showed that if $\mathbf{r}_{P/B}$ is the position of a point P of a rigid body relative to another point B of the same rigid body, $d\mathbf{r}_{P/B}/dt = \mathbf{v}_{P/B} = \boldsymbol{\omega} \times \mathbf{r}_{P/B}$. Since we can regard the unit vector $\mathbf{i}$ as the position vector of a point P of the rigid body (Fig. 17.25), its derivative is $d\mathbf{i}/dt = \boldsymbol{\omega} \times \mathbf{i}$. Applying the same argument to the unit vectors $\mathbf{j}$ and $\mathbf{k}$, we obtain

$$\frac{d\mathbf{i}}{dt} = \boldsymbol{\omega} \times \mathbf{i}, \qquad \frac{d\mathbf{j}}{dt} = \boldsymbol{\omega} \times \mathbf{j}, \qquad \frac{d\mathbf{k}}{dt} = \boldsymbol{\omega} \times \mathbf{k}.$$

Using these expressions, we can write the velocity of point A as

$$\mathbf{v}_A = \mathbf{v}_B + \underbrace{\mathbf{v}_{A\,\text{rel}} + \boldsymbol{\omega} \times \mathbf{r}_{A/B},}_{\mathbf{v}_{A/B}} \qquad (17.11)$$

where

$$\mathbf{v}_{A\,\text{rel}} = \frac{dx}{dt}\mathbf{i} + \frac{dy}{dt}\mathbf{j} + \frac{dz}{dt}\mathbf{k} \qquad (17.12)$$

is the velocity of A relative to the body-fixed reference frame. That is, $\mathbf{v}_A$ is the velocity of A relative to the primary reference frame, and $\mathbf{v}_{A\,\text{rel}}$ is the velocity of A relative to the rigid body.

Equation (17.11) expresses the velocity of a point A as the sum of three terms (Fig. 17.26): the velocity of a point B of the rigid body, the velocity $\boldsymbol{\omega} \times \mathbf{r}_{A/B}$ of A relative to B due to the rotation of the rigid body, and the velocity $\mathbf{v}_{A\,\text{rel}}$ of A relative to the rigid body.

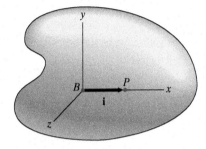

Figure 17.25
Interpreting $\mathbf{i}$ as the position vector of a point P relative to B.

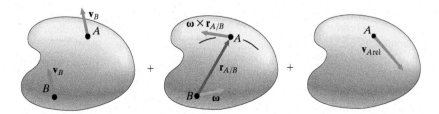

Figure 17.26
Expressing the velocity of A in terms of the velocity of a point B of the rigid body.

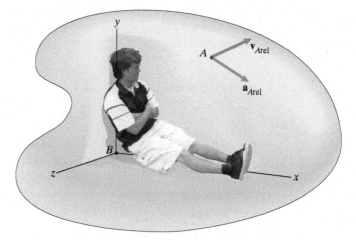

Figure 17.27
Imagine yourself to be stationary relative to
the rigid body.

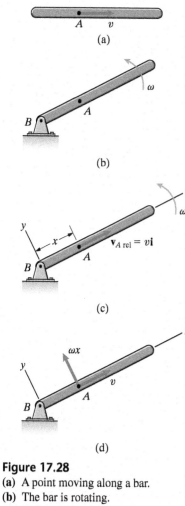

Figure 17.28
(a) A point moving along a bar.
(b) The bar is rotating.
(c) A body-fixed reference frame.
(d) Components of $\mathbf{v}_A$.

To obtain an equation for the acceleration of point A, we take the derivative of Eq. (17.11) with respect to time and use Eq. (17.12). The result is (see
Problem 17.142)

$$\mathbf{a}_A = \mathbf{a}_B + \underbrace{\mathbf{a}_{A\,\text{rel}} + 2\boldsymbol{\omega} \times \mathbf{v}_{A\,\text{rel}} + \boldsymbol{\alpha} \times \mathbf{r}_{A/B} + \boldsymbol{\omega} \times (\boldsymbol{\omega} \times \mathbf{r}_{A/B})}_{\mathbf{a}_{A/B}}, \quad (17.13)$$

where

$$\mathbf{a}_{A\,\text{rel}} = \frac{d^2x}{dt^2}\mathbf{i} + \frac{d^2y}{dt^2}\mathbf{j} + \frac{d^2z}{dt^2}\mathbf{k} \quad (17.14)$$

is the acceleration of A relative to the body-fixed reference frame. That is, $\mathbf{a}_A$
is the acceleration of A relative to the primary reference frame, and $\mathbf{a}_{A\,\text{rel}}$ is the
acceleration of A relative to the rigid body.

In the case of planar motion, we can express Eq. (17.13) in the simpler form

$$\mathbf{a}_A = \mathbf{a}_B + \underbrace{\mathbf{a}_{A\,\text{rel}} + 2\boldsymbol{\omega} \times \mathbf{v}_{A\,\text{rel}} + \boldsymbol{\alpha} \times \mathbf{r}_{A/B} - \omega^2 \mathbf{r}_{A/B}}_{\mathbf{a}_{A/B}}. \quad (17.15)$$

In summary, $\mathbf{v}_A$ and $\mathbf{a}_A$ are the velocity and acceleration of point A relative to the primary reference frame—the reference frame relative to which the
rigid body's motion is being described. The terms $\mathbf{v}_{A\,\text{rel}}$ and $\mathbf{a}_{A\,\text{rel}}$ are the velocity and acceleration of point A relative to the body-fixed reference frame.
That is, they are the velocity and acceleration measured by an observer mov-
ing with the rigid body (Fig. 17.27). If A is a point of the rigid body, then $\mathbf{v}_{A\,\text{rel}}$
and $\mathbf{a}_{A\,\text{rel}}$ are zero, and Eqs. (17.11) and (17.13) are identical to Eqs. (17.8)
and (17.9).

We can illustrate these concepts with a simple example. Figure 17.28a
shows a point A moving with velocity v parallel to the axis of a bar. (Imagine
that A is a bug walking along the bar.) Suppose that at the same time, the bar is
rotating about a fixed point B with a constant angular velocity ω relative to an
earth-fixed reference frame (Fig. 17.28b). We will use Eq. (17.11) to determine
the velocity of A relative to the earth-fixed reference frame.

Let the coordinate system in Fig. 17.28c be fixed with respect to the bar,
and let x be the present position of A. In terms of this body-fixed reference
frame, the angular velocity vector of the bar (and the reference frame) rela-
tive to the primary earth-fixed reference frame is $\boldsymbol{\omega} = \omega\mathbf{k}$. Relative to the

body-fixed reference frame, point A moves along the x axis with velocity v, so $\mathbf{v}_{A\,\text{rel}} = v\mathbf{i}$. From Eq. (17.11), the velocity of A relative to the earth-fixed reference frame is

$$
\begin{aligned}
\mathbf{v}_A &= \mathbf{v}_B + \mathbf{v}_{A\,\text{rel}} + \boldsymbol{\omega} \times \mathbf{r}_{A/B} \\
&= \mathbf{0} + v\mathbf{i} + (\omega\mathbf{k}) \times (x\mathbf{i}) \\
&= v\mathbf{i} + \omega x\mathbf{j}.
\end{aligned}
$$

Relative to the earth-fixed reference frame, A has a component of velocity parallel to the bar and also a perpendicular component due to the bar's rotation (Fig. 17.28d). Although $\mathbf{v}_A$ is the velocity of A relative to the earth-fixed reference frame, notice that it is expressed in components that are in terms of the body-fixed reference frame.

RESULTS

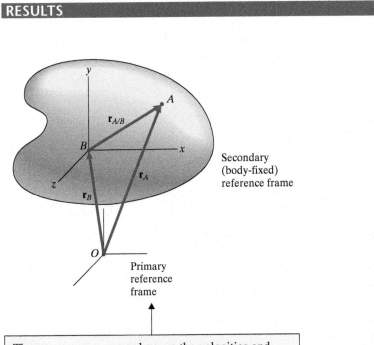

Secondary (body-fixed) reference frame

Primary reference frame

The terms $\mathbf{v}_A$, $\mathbf{v}_B$, $\mathbf{a}_A$, and $\mathbf{a}_B$ are the velocities and accelerations of points A and B relative to the primary reference frame. $\boldsymbol{\omega}$ and $\boldsymbol{\alpha}$ are the angular velocity and angular acceleration of the rigid body relative to the primary reference frame. The term

$$
\mathbf{v}_{A\,\text{rel}} = \frac{dx}{dt}\mathbf{i} + \frac{dy}{dt}\mathbf{j} + \frac{dz}{dt}\mathbf{k} \tag{17.12}
$$

is the velocity of A relative to the secondary reference frame, that is, relative to the rigid body. The term

$$
\mathbf{a}_{A\,\text{rel}} = \frac{d^2x}{dt^2}\mathbf{i} + \frac{d^2y}{dt^2}\mathbf{j} + \frac{d^2z}{dt^2}\mathbf{k} \tag{17.14}
$$

is the acceleration of A relative to the secondary reference frame.

$$
\mathbf{v}_A = \mathbf{v}_B + \mathbf{v}_{A\,\text{rel}} + \boldsymbol{\omega} \times \mathbf{r}_{A/B}, \tag{17.11}
$$

$$
\begin{aligned}
\mathbf{a}_A = \mathbf{a}_B &+ \mathbf{a}_{A\,\text{rel}} + 2\boldsymbol{\omega} \times \mathbf{v}_{A\,\text{rel}} \\
&+ \boldsymbol{\alpha} \times \mathbf{r}_{A/B} + \boldsymbol{\omega} \times (\boldsymbol{\omega} \times \mathbf{r}_{A/B}). \tag{17.13}
\end{aligned}
$$

In the case of planar motion, Eq. (17.13) can be written

$$
\begin{aligned}
\mathbf{a}_A = \mathbf{a}_B &+ \mathbf{a}_{A\,\text{rel}} + 2\boldsymbol{\omega} \times \mathbf{v}_{A\,\text{rel}} \\
&+ \boldsymbol{\alpha} \times \mathbf{r}_{A/B} - \omega^2 \mathbf{r}_{A/B}. \tag{17.15}
\end{aligned}
$$

Active Example 17.7 **Linkage with a Sliding Contact** (▶ *Related Problem 17.117*)

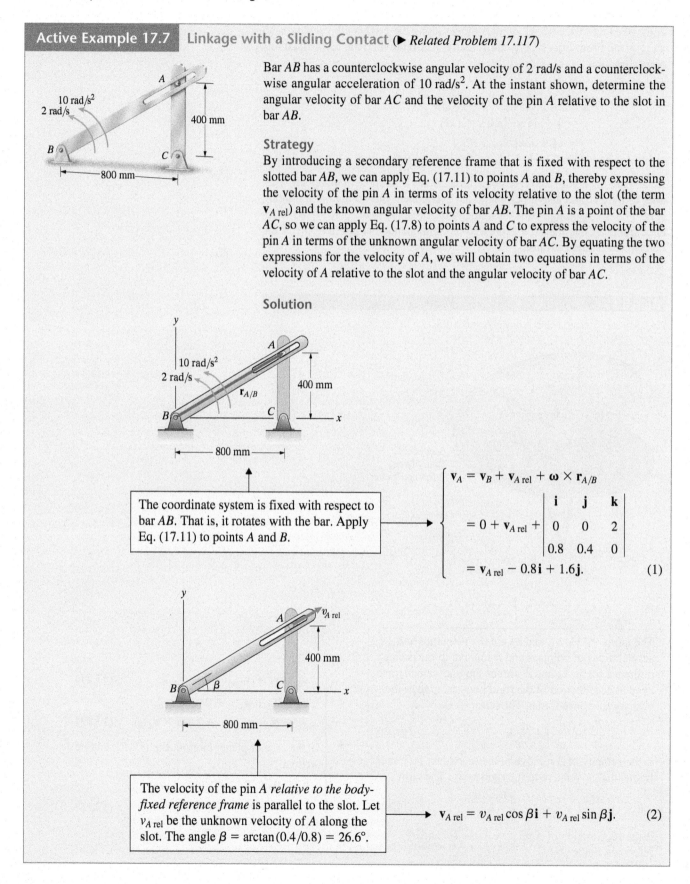

Bar AB has a counterclockwise angular velocity of 2 rad/s and a counterclockwise angular acceleration of 10 rad/s². At the instant shown, determine the angular velocity of bar AC and the velocity of the pin A relative to the slot in bar AB.

Strategy

By introducing a secondary reference frame that is fixed with respect to the slotted bar AB, we can apply Eq. (17.11) to points A and B, thereby expressing the velocity of the pin A in terms of its velocity relative to the slot (the term $\mathbf{v}_{A\,rel}$) and the known angular velocity of bar AB. The pin A is a point of the bar AC, so we can apply Eq. (17.8) to points A and C to express the velocity of the pin A in terms of the unknown angular velocity of bar AC. By equating the two expressions for the velocity of A, we will obtain two equations in terms of the velocity of A relative to the slot and the angular velocity of bar AC.

Solution

The coordinate system is fixed with respect to bar AB. That is, it rotates with the bar. Apply Eq. (17.11) to points A and B.

$$\left\{ \begin{aligned} \mathbf{v}_A &= \mathbf{v}_B + \mathbf{v}_{A\,rel} + \boldsymbol{\omega} \times \mathbf{r}_{A/B} \\ &= 0 + \mathbf{v}_{A\,rel} + \begin{vmatrix} \mathbf{i} & \mathbf{j} & \mathbf{k} \\ 0 & 0 & 2 \\ 0.8 & 0.4 & 0 \end{vmatrix} \\ &= \mathbf{v}_{A\,rel} - 0.8\mathbf{i} + 1.6\mathbf{j}. \end{aligned} \right. \quad (1)$$

The velocity of the pin A *relative to the body-fixed reference frame* is parallel to the slot. Let $v_{A\,rel}$ be the unknown velocity of A along the slot. The angle $\beta = \arctan(0.4/0.8) = 26.6°$.

$$\mathbf{v}_{A\,rel} = v_{A\,rel}\cos\beta\,\mathbf{i} + v_{A\,rel}\sin\beta\,\mathbf{j}. \quad (2)$$

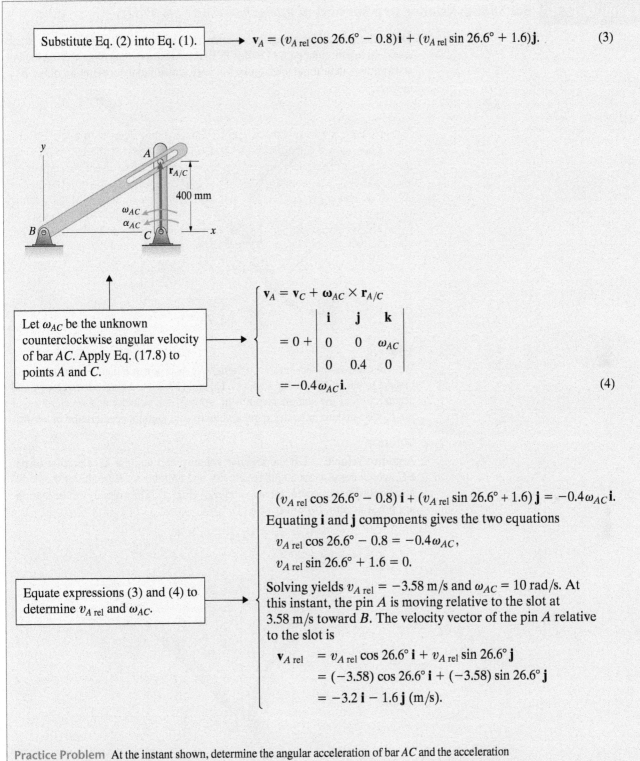

Substitute Eq. (2) into Eq. (1).

$$\mathbf{v}_A = (v_{A\,\text{rel}} \cos 26.6° - 0.8)\mathbf{i} + (v_{A\,\text{rel}} \sin 26.6° + 1.6)\mathbf{j}. \qquad (3)$$

Let ω_{AC} be the unknown counterclockwise angular velocity of bar AC. Apply Eq. (17.8) to points A and C.

$$\begin{cases} \mathbf{v}_A = \mathbf{v}_C + \boldsymbol{\omega}_{AC} \times \mathbf{r}_{A/C} \\[4pt] = 0 + \begin{vmatrix} \mathbf{i} & \mathbf{j} & \mathbf{k} \\ 0 & 0 & \omega_{AC} \\ 0 & 0.4 & 0 \end{vmatrix} \\[4pt] = -0.4\,\omega_{AC}\mathbf{i}. \end{cases} \qquad (4)$$

Equate expressions (3) and (4) to determine $v_{A\,\text{rel}}$ and ω_{AC}.

$$\begin{cases} (v_{A\,\text{rel}} \cos 26.6° - 0.8)\,\mathbf{i} + (v_{A\,\text{rel}} \sin 26.6° + 1.6)\,\mathbf{j} = -0.4\,\omega_{AC}\mathbf{i}. \end{cases}$$

Equating $\mathbf{i}$ and $\mathbf{j}$ components gives the two equations

$$v_{A\,\text{rel}} \cos 26.6° - 0.8 = -0.4\,\omega_{AC},$$

$$v_{A\,\text{rel}} \sin 26.6° + 1.6 = 0.$$

Solving yields $v_{A\,\text{rel}} = -3.58$ m/s and $\omega_{AC} = 10$ rad/s. At this instant, the pin A is moving relative to the slot at 3.58 m/s toward B. The velocity vector of the pin A relative to the slot is

$$\begin{aligned} \mathbf{v}_{A\,\text{rel}} &= v_{A\,\text{rel}} \cos 26.6°\,\mathbf{i} + v_{A\,\text{rel}} \sin 26.6°\,\mathbf{j} \\ &= (-3.58) \cos 26.6°\,\mathbf{i} + (-3.58) \sin 26.6°\,\mathbf{j} \\ &= -3.2\,\mathbf{i} - 1.6\,\mathbf{j}\ (\text{m/s}). \end{aligned}$$

Practice Problem At the instant shown, determine the angular acceleration of bar AC and the acceleration of the pin A relative to the slot in bar AB.

Answer: $\alpha_{AC} = 170$ rad/s² counterclockwise, A is accelerating at 75.1 m/s² toward B.

| Example 17.8 | Bar Sliding Relative to a Support (▶ *Related Problems 17.130, 17.131*) |

The collar at B slides along the circular bar, causing pin B to move at constant speed v_0 in a circular path of radius R. Bar BC slides in the collar at A. At the instant shown, determine the angular velocity and angular acceleration of bar BC.

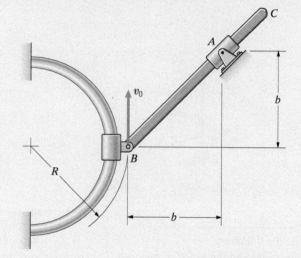

Strategy

We will use a secondary reference frame with its origin at B that is body fixed with respect to bar BC. By using Eqs. (17.11) and (17.15) to express the velocity and acceleration of the stationary pin A in terms of the velocity and acceleration of pin B, we can determine the angular velocity and angular acceleration of bar BC.

Solution

Angular Velocity   Let the angular velocity and angular acceleration of bar BC, which are also the angular velocity and angular acceleration of the body-fixed coordinate system, be ω_{BC} and α_{BC} (Fig. a). The velocity of the stationary pin A is zero. From Eq. (17.11),

$$\mathbf{v}_A = \mathbf{0} = \mathbf{v}_B + \mathbf{v}_{A\,\text{rel}} + \boldsymbol{\omega} \times \mathbf{r}_{A/B}, \tag{1}$$

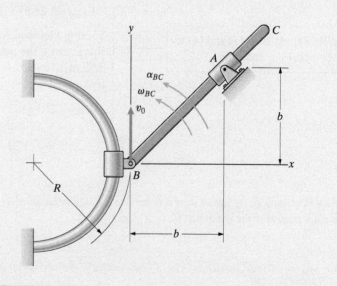

(a) A reference frame fixed with respect to bar BC.

where $\mathbf{v}_{A\,\text{rel}}$ is the velocity of A relative to the body-fixed coordinate system and $\boldsymbol{\omega} = \omega_{BC}\mathbf{k}$ is the angular velocity vector of the coordinate system. The velocity of pin B is $\mathbf{v}_B = v_0\mathbf{j}$. The velocity of the stationary pin A relative to the body-fixed coordinate system is parallel to the bar (Fig. b), so we can express it in the form

$$\mathbf{v}_{A\,\text{rel}} = v_{A\,\text{rel}}\cos 45°\mathbf{i} + v_{A\,\text{rel}}\sin 45°\mathbf{j} \qquad (2)$$

and write Eq. (1) as

$$\mathbf{0} = v_0\mathbf{j} + v_{A\,\text{rel}}\cos 45°\mathbf{i} + v_{A\,\text{rel}}\sin 45°\mathbf{j}$$

$$+ \begin{vmatrix} \mathbf{i} & \mathbf{j} & \mathbf{k} \\ 0 & 0 & \omega_{BC} \\ b & b & 0 \end{vmatrix}.$$

From the $\mathbf{i}$ and $\mathbf{j}$ components of this equation, we obtain

$$v_{A\,\text{rel}}\cos 45° - b\omega_{BC} = 0,$$
$$v_0 + v_{A\,\text{rel}}\sin 45° + b\omega_{BC} = 0.$$

Solving these equations, we determine that the velocity of pin A relative to the body-fixed coordinate system is

$$\mathbf{v}_{A\,\text{rel}} = v_{A\,\text{rel}}\cos 45°\mathbf{i} + v_{A\,\text{rel}}\sin 45°\mathbf{j}$$

$$= -\frac{v_0}{2}\mathbf{i} - \frac{v_0}{2}\mathbf{j}$$

and the angular velocity of bar BC is

$$\omega_{BC} = -\frac{v_0}{2b}.$$

Angular Acceleration The acceleration of pin A is zero. From Eq. (17.15),

$$\mathbf{a}_A = \mathbf{0} = \mathbf{a}_B + \mathbf{a}_{A\,\text{rel}} + 2\boldsymbol{\omega} \times \mathbf{v}_{A\,\text{rel}} + \boldsymbol{\alpha} \times \mathbf{r}_{A/B} - \omega^2\mathbf{r}_{A/B}. \qquad (3)$$

The acceleration of pin B is $\mathbf{a}_B = -(v_0^2/R)\mathbf{i}$. The acceleration of pin A relative to the body-fixed coordinate system is parallel to the bar (Fig. c). We can therefore express it as

$$\mathbf{a}_{A\,\text{rel}} = a_{A\,\text{rel}}\cos 45°\mathbf{i} + a_{A\,\text{rel}}\sin 45°\mathbf{j} \qquad (4)$$

and write Eq. (3) as

$$\mathbf{0} = -\frac{v_0^2}{R}\mathbf{i} + a_{A\,\text{rel}}\cos 45°\mathbf{i} + a_{A\,\text{rel}}\sin 45°\mathbf{j}$$

$$+ 2\begin{vmatrix} \mathbf{i} & \mathbf{j} & \mathbf{k} \\ 0 & 0 & \omega_{BC} \\ -v_0/2 & -v_0/2 & 0 \end{vmatrix} + \begin{vmatrix} \mathbf{i} & \mathbf{j} & \mathbf{k} \\ 0 & 0 & \alpha_{BC} \\ b & b & 0 \end{vmatrix}$$

$$- \omega_{BC}^2(b\mathbf{i} + b\mathbf{j}).$$

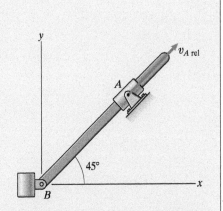

(b) Direction of the velocity of the fixed pin A relative to the body-fixed coordinate system.

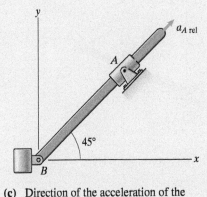

(c) Direction of the acceleration of the fixed pin A relative to the body-fixed coordinate system.

From the $\mathbf{i}$ and $\mathbf{j}$ components of this equation, we obtain

$$-\frac{v_0^2}{R} + a_{A\,\text{rel}} \cos 45° + v_0 \omega_{BC} - b\alpha_{BC} - b\omega_{BC}^2 = 0,$$

$$a_{A\,\text{rel}} \sin 45° - v_0 \omega_{BC} + b\alpha_{BC} - b\omega_{BC}^2 = 0.$$

Solving these equations, we determine that the angular acceleration of bar BC is

$$\alpha_{BC} = -\frac{v_0^2}{2b}\left(\frac{1}{R} + \frac{1}{b}\right).$$

Critical Thinking

In this example, the bar BC slides relative to its support at A. Notice that the fixed support A moves relative to the coordinate system that is body-fixed with respect to the bar BC. Because the bar BC is stationary with respect to the body-fixed coordinate system, we knew that the velocity and acceleration of the support A relative to the body-fixed coordinate system is parallel to bar BC. That was why we could express them in the forms given by Eqs. (2) and (4).

Example 17.9	Analysis of a Sliding Contact (▶ *Related Problems 17.136, 17.137*)

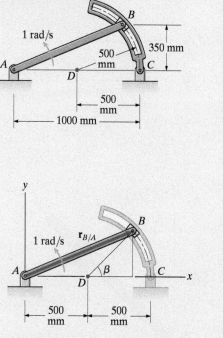

(a) Determining the velocity of point B.

Bar AB rotates with a constant counterclockwise angular velocity of 1 rad/s. Block B slides in a circular slot in the curved bar BC. At the instant shown, the center of the circular slot is at D. Determine the angular velocity and angular acceleration of bar BC.

Strategy

Since we know the angular velocity of bar AB, we can determine the velocity of point B. Because B is not a point of bar BC, we must apply Eq. (17.11) to points B and C. By equating our expressions for $\mathbf{v}_B$, we can solve for the angular velocity of bar BC. Then, by following the same sequence of steps, but this time using Eq. (17.15), we can determine the angular acceleration of bar BC.

Solution

To determine the velocity of B, we express it in terms of the velocity of A and the angular velocity of bar AB: $\mathbf{v}_B = \mathbf{v}_A + \boldsymbol{\omega}_{AB} \times \mathbf{r}_{B/A}$. In terms of the coordinate system shown in Fig. a, the position vector of B relative to A is

$$\mathbf{r}_{B/A} = (0.500 + 0.500\cos\beta)\mathbf{i} + 0.350\mathbf{j} = 0.857\mathbf{i} + 0.350\mathbf{j} \text{ (m)},$$

where $\beta = \arcsin(350/500) = 44.4°$. Therefore, the velocity of B is

$$\mathbf{v}_B = \mathbf{v}_A + \boldsymbol{\omega}_{AB} \times \mathbf{r}_{B/A} = 0 + \begin{vmatrix} \mathbf{i} & \mathbf{j} & \mathbf{k} \\ 0 & 0 & 1 \\ 0.857 & 0.350 & 0 \end{vmatrix}$$

$$= -0.350\mathbf{i} + 0.857\mathbf{j} \text{ (m/s)}. \qquad (1)$$

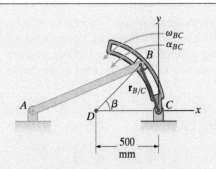

(b) A coordinate system fixed with respect to the curved bar.

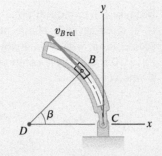

(c) The velocity of B relative to the body-fixed coordinate system.

To apply Eq. (17.11) to points B and C, we introduce a parallel secondary coordinate system that rotates with the curved bar (Fig. b). The velocity of B is

$$\mathbf{v}_B = \mathbf{v}_C + \mathbf{v}_{B\,\text{rel}} + \boldsymbol{\omega}_{BC} \times \mathbf{r}_{B/C}. \qquad (2)$$

The position vector of B relative to C is

$$\mathbf{r}_{B/C} = -(0.500 - 0.500 \cos \beta)\mathbf{i} + 0.350\mathbf{j} = -0.143\mathbf{i} + 0.350\mathbf{j}\ (\text{m}).$$

Relative to the body-fixed coordinate system, point B moves in a circular path about point D (Fig. c). In terms of the angle β, the vector

$$\mathbf{v}_{B\,\text{rel}} = -v_{B\,\text{rel}} \sin \beta\mathbf{i} + v_{B\,\text{rel}} \cos \beta\mathbf{j}.$$

We substitute the preceding expressions for $\mathbf{r}_{B/C}$ and $\mathbf{v}_{B\,\text{rel}}$ into Eq. (2), obtaining

$$\mathbf{v}_B = -v_{B\,\text{rel}} \sin \beta\mathbf{i} + v_{B\,\text{rel}} \cos \beta\mathbf{j} + \begin{vmatrix} \mathbf{i} & \mathbf{j} & \mathbf{k} \\ 0 & 0 & \omega_{BC} \\ -0.143 & 0.350 & 0 \end{vmatrix}.$$

Equating this expression for $\mathbf{v}_B$ to its value given in Eq. (1) yields the two equations

$$-v_{B\,\text{rel}} \sin \beta - 0.350\omega_{BC} = -0.350,$$

$$v_{B\,\text{rel}} \cos \beta - 0.143\omega_{BC} = 0.857.$$

Solving these equations, we obtain $v_{B\,\text{rel}} = 1.0$ m/s and $\omega_{BC} = -1.0$ rad/s.

We follow the same sequence of steps to determine the angular acceleration of bar BC. The acceleration of point B is

$$\begin{aligned}
\mathbf{a}_B &= \mathbf{a}_A + \boldsymbol{\alpha}_{AB} \times \mathbf{r}_{B/A} - \omega_{AB}^2 \mathbf{r}_{B/A} \\
&= \mathbf{0} + \mathbf{0} - (1)^2(0.857\mathbf{i} + 0.350\mathbf{j}) \\
&= -0.857\mathbf{i} - 0.350\mathbf{j}\ (\text{m/s}^2). \qquad (3)
\end{aligned}$$

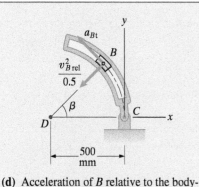

(d) Acceleration of B relative to the body-fixed coordinate system.

Because the motion of point B relative to the body-fixed coordinate system is a circular path about point D, there is a tangential component of acceleration, which we denote a_{Bt}, and a normal component of acceleration $v_{B\,rel}^2/(0.5\text{ m})$. These components are shown in Fig. d. In terms of the angle β, the vector

$$\mathbf{a}_{B\,rel} = -a_{Bt}\sin\beta\mathbf{i} + a_{Bt}\cos\beta\mathbf{j}$$

$$- (v_{B\,rel}^2/0.5)\cos\beta\mathbf{i} - (v_{B\,rel}^2/0.5)\sin\beta\mathbf{j}.$$

Applying Eq. (17.15) to points B and C, we obtain the acceleration of B:

$$\mathbf{a}_B = \mathbf{a}_C + \mathbf{a}_{B\,rel} + 2\boldsymbol{\omega}_{BC} \times \mathbf{v}_{B\,rel}$$

$$+ \boldsymbol{\alpha}_{BC} \times \mathbf{r}_{B/C} - \omega_{BC}^2\mathbf{r}_{B/C}$$

$$= \mathbf{0} - a_{Bt}\sin\beta\mathbf{i} + a_{Bt}\cos\beta\mathbf{j}$$

$$- [(1)^2/0.5]\cos\beta\mathbf{i} - [(1)^2/0.5]\sin\beta\mathbf{j}$$

$$+ 2\begin{vmatrix} \mathbf{i} & \mathbf{j} & \mathbf{k} \\ 0 & 0 & -1 \\ -(1)\sin\beta & (1)\cos\beta & 0 \end{vmatrix}$$

$$+ \begin{vmatrix} \mathbf{i} & \mathbf{j} & \mathbf{k} \\ 0 & 0 & \alpha_{BC} \\ -0.143 & 0.350 & 0 \end{vmatrix} - (-1)^2(-0.143\mathbf{i} + 0.350\mathbf{j}).$$

Equating this expression for $\mathbf{a}_B$ to its value given in Eq. (3) yields the two equations

$$-a_{Bt}\sin\beta - 0.350\alpha_{BC} + 0.143 = -0.857,$$

$$a_{Bt}\cos\beta - 0.143\alpha_{BC} - 0.350 = -0.350.$$

Solving we obtain $a_{Bt} = 0.408\text{ m/s}^2$ and $\alpha_{BC} = 2.040\text{ rad/s}^2$.

Critical Thinking

This example was distinguished by the fact that the slotted bar has the shape of a circular arc. As a result, the block B moved in a circular path relative to the coordinate system that is fixed with respect to the circular bar. The direction of the velocity $\mathbf{v}_{B\,rel}$ was tangential to the circular path. However, as a result of the curved path, we knew that $\mathbf{a}_{B\,rel}$ would have components tangential and normal to the path. Furthermore, the magnitude of the normal component could be written in terms of the magnitude of the velocity and the radius of the circular path. Therefore, once $\mathbf{v}_{B\,rel}$ was known, only the tangential component of $\mathbf{a}_{B\,rel}$ needed to be determined.

Problems

▶ **17.117** In Active Example 17.7, suppose that the distance from point C to the pin A on the vertical bar AC is 300 mm instead of 400 mm. Draw a sketch of the linkage with its new geometry. Determine the angular velocity of bar AC and the velocity of the pin A relative to the slot in bar AB.

17.118 The bar rotates with a constant counterclockwise angular velocity of 10 rad/s and sleeve A slides at a constant velocity of 4 ft/s relative to the bar. Use Eq. (17.15) to determine the acceleration of A.

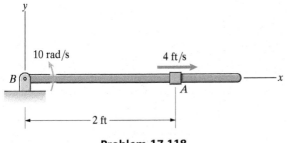

Problem 17.118

17.119 Sleeve C slides at 1 m/s relative to bar BD. Use the body-fixed coordinate system shown to determine the velocity of C.

17.120 The angular accelerations of the two bars are zero and sleeve C slides at a constant velocity of 1 m/s relative to bar BD. What is the acceleration of sleeve C?

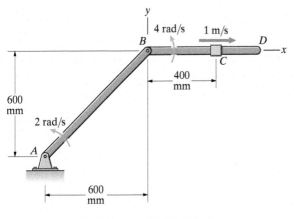

Problems 17.119/17.120

17.121 Bar AC has an angular velocity of 2 rad/s in the counterclockwise direction that is decreasing at 4 rad/s². The pin at C slides in the slot in bar BD.

(a) Determine the angular velocity of bar BD and the velocity of the pin relative to the slot.

(b) Determine the angular acceleration of bar BD and the acceleration of the pin relative to the slot.

17.122 The velocity of pin C relative to the slot is 21 in/s upward and is decreasing at 42 in/s². What are the angular velocity and angular acceleration of bar AC?

17.123 What should the angular velocity and acceleration of bar AC be if you want the angular velocity and acceleration of bar BD to be 4 rad/s counterclockwise and 24 rad/s² counterclockwise, respectively?

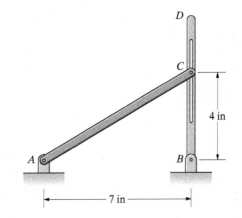

Problems 17.121–17.123

17.124 Bar *AB* has an angular velocity of 4 rad/s in the clockwise direction. What is the velocity of pin *B* relative to the slot?

17.125 Bar *AB* has an angular velocity of 4 rad/s in the clockwise direction and an angular acceleration of 10 rad/s² in the counterclockwise direction. What is the acceleration of pin *B* relative to the slot?

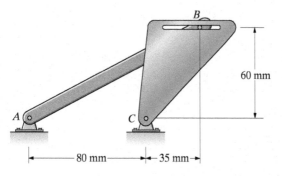

Problems 17.124/17.125

17.126 The hydraulic actuator *BC* of the crane is extending (increasing in length) at a rate of 0.2 m/s. At the instant shown, what is the angular velocity of the crane's boom *AD*?

Strategy: Use Eq. (17.8) to write the velocity of point *C* in terms of the velocity of point *A*, and use Eq. (17.11) to write the velocity of point *C* in terms of the velocity of point *B*. Then equate your two expressions for the velocity of point *C*.

17.127 The hydraulic actuator *BC* of the crane is extending (increasing in length) at a constant rate of 0.2 m/s. At the instant shown, what is the angular acceleration of the crane's boom *AD*?

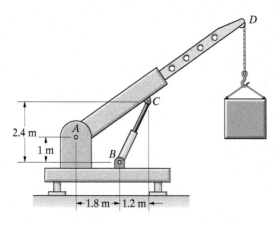

Problems 17.126/17.127

17.128 The angular velocity $\omega_{AC} = 5°$ per second. Determine the angular velocity of the hydraulic actuator *BC* and the rate at which the actuator is extending.

17.129 The angular velocity $\omega_{AC} = 5°$ per second and the angular acceleration $\alpha_{AC} = -2°$ per second squared. Determine the angular acceleration of the hydraulic actuator *BC* and the rate of change of the actuator's rate of extension.

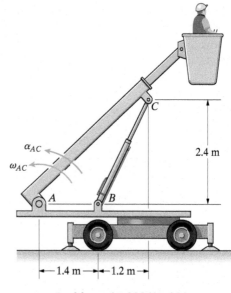

Problems 17.128/17.129

▶ **17.130** The sleeve at *A* slides upward at a constant velocity of 10 m/s. Bar *AC* slides through the sleeve at *B*. Determine the angular velocity of bar *AC* and the velocity at which the bar slides relative to the sleeve at *B*. (See Example 17.8.)

▶ **17.131** The sleeve at *A* slides upward at a constant velocity of 10 m/s. Determine the angular acceleration of bar *AC* and the rate of change of the velocity at which the bar slides relative to the sleeve at *B*. (See Example 17.8.)

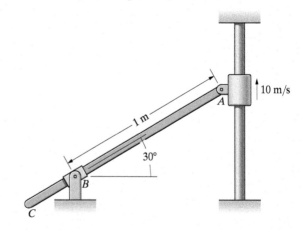

Problems 17.130/17.131

17.132 Block A slides up the inclined surface at 2 ft/s. Determine the angular velocity of bar AC and the velocity of point C.

17.133 Block A slides up the inclined surface at a constant velocity of 2 ft/s. Determine the angular acceleration of bar AC and the acceleration of point C.

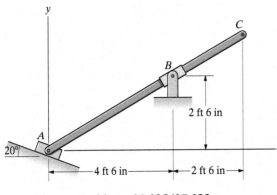

Problems 17.132/17.133

17.134 The angular velocity of the scoop is 1 rad/s clockwise. Determine the rate at which the hydraulic actuator AB is extending.

17.135 The angular velocity of the scoop is 1 rad/s clockwise and its angular acceleration is zero. Determine the rate of change of the rate at which the hydraulic actuator AB is extending.

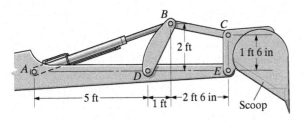

Problems 17.134/17.135

▶ **17.136** Suppose that the curved bar in Example 17.9 rotates with a counterclockwise angular velocity of 2 rad/s.
(a) What is the angular velocity of bar AB?
(b) What is the velocity of block B relative to the slot?

▶ **17.137** Suppose that the curved bar in Example 17.9 has a clockwise angular velocity of 4 rad/s and a counterclockwise angular acceleration of 10 rad/s². What is the angular acceleration of bar AB?

17.138* The disk rolls on the plane surface with a counterclockwise angular velocity of 10 rad/s. Bar AB slides on the surface of the disk at A. Determine the angular velocity of bar AB.

17.139* The disk rolls on the plane surface with a constant counterclockwise angular velocity of 10 rad/s. Determine the angular acceleration of bar AB.

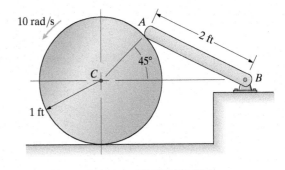

Problems 17.138/17.139

17.140* Bar BC rotates with a counterclockwise angular velocity of 2 rad/s. A pin at B slides in a circular slot in the rectangular plate. Determine the angular velocity of the plate and the velocity at which the pin slides relative to the circular slot.

17.141* Bar BC rotates with a constant counterclockwise angular velocity of 2 rad/s. Determine the angular acceleration of the plate.

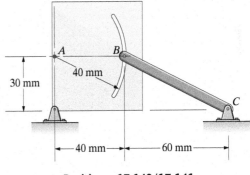

Problems 17.140/17.141

17.142* By taking the derivative of Eq. (17.11) with respect to time and using Eq. (17.12), derive Eq. (17.13).

17.7 Moving Reference Frames

BACKGROUND

In this section we revisit the subjects of Chapters 13 and 14—the motion of a point and Newton's second law. In many situations, it is convenient to describe the motion of a point by using a secondary reference frame that moves relative to some primary reference frame. For example, to measure the motion of a point relative to a moving vehicle, we would choose a secondary reference frame that is fixed with respect to the vehicle. Here we show how the velocity and acceleration of a point relative to a primary reference frame are related to their values relative to a moving secondary reference frame. We also discuss how to apply Newton's second law using moving secondary reference frames when the primary reference frame is inertial. In Chapter 14 we mentioned the example of playing tennis on the deck of a cruise ship. If the ship translates with constant velocity, we can use the equation $\Sigma\mathbf{F} = m\mathbf{a}$ expressed in terms of a reference frame fixed with respect to the ship to analyze the ball's motion. We cannot do so if the ship is turning or changing its speed. However, we can apply the second law using reference frames that accelerate and rotate relative to an inertial reference frame by properly accounting for the acceleration and rotation. We explain how this is done in this section.

Motion of a Point Relative to a Moving Reference Frame

Equations (17.11) and (17.13) give the velocity and acceleration of an arbitrary point A relative to a point B of a rigid body in terms of a body-fixed secondary reference frame:

$$\mathbf{v}_A = \mathbf{v}_B + \mathbf{v}_{A\,\text{rel}} + \boldsymbol{\omega} \times \mathbf{r}_{A/B}, \tag{17.16}$$

$$\mathbf{a}_A = \mathbf{a}_B + \mathbf{a}_{A\,\text{rel}} + 2\boldsymbol{\omega} \times \mathbf{v}_{A\,\text{rel}} + \boldsymbol{\alpha} \times \mathbf{r}_{A/B} + \boldsymbol{\omega} \times (\boldsymbol{\omega} \times \mathbf{r}_{A/B}). \tag{17.17}$$

But these equations don't require us to assume that the secondary reference frame is connected to some rigid body. They apply to any reference frame having a moving origin B and rotating with angular velocity $\boldsymbol{\omega}$ and angular acceleration $\boldsymbol{\alpha}$ relative to a primary reference frame (Fig. 17.29). The terms $\mathbf{v}_A$ and

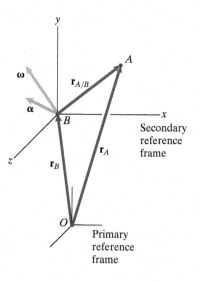

Figure 17.29
A secondary reference frame with origin B and an arbitrary point A.

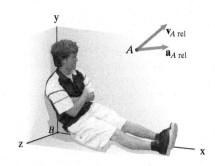

Figure 17.30
Imagine yourself to be stationary relative to the secondary reference frame.

$\mathbf{a}_A$ are the velocity and acceleration of A relative to the primary reference frame. The terms $\mathbf{v}_{A\,rel}$ and $\mathbf{a}_{A\,rel}$ are the velocity and acceleration of A relative to the secondary reference frame. That is, they are the velocity and acceleration measured by an observer moving with the secondary reference frame (Fig. 17.30).

When the velocity and acceleration of a point A relative to a moving secondary reference frame are known, we can use Eqs. (17.16) and (17.17) to determine the velocity and acceleration of A relative to the primary reference frame. There will also be situations in which the velocity and acceleration of A relative to the primary reference frame will be known and we will want to use Eqs. (17.16) and (17.17) to determine the velocity and acceleration of A relative to a moving secondary reference frame.

Inertial Reference Frames

We say that a reference frame is *inertial* if it can be used to apply Newton's second law in the form $\Sigma\mathbf{F} = m\mathbf{a}$. Why can an earth-fixed reference frame be treated as inertial in many situations, even though it both accelerates and rotates? How can Newton's second law be applied using a reference frame that is fixed with respect to a ship or airplane? We are now in a position to answer these questions.

Earth-Centered, Nonrotating Reference Frame We begin by showing why a nonrotating reference frame with its origin at the center of the earth can be assumed to be inertial for the purpose of describing motions of objects near the earth. Figure 17.31a shows a hypothetical nonaccelerating, nonrotating reference frame with origin O and a secondary nonrotating, *earth-centered reference frame*. The earth (and therefore the earth-centered reference frame)

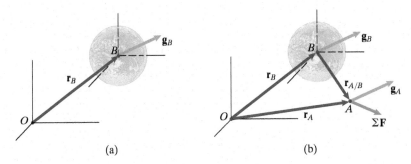

(a)   (b)

Figure 17.31
(a) An inertial reference frame and a nonrotating reference frame with its origin at the center of the earth.
(b) Determining the motion of an object A.

accelerates due to the gravitational attractions of the sun, moon, etc. We denote the earth's acceleration by the vector $\mathbf{g}_B$.

Suppose that we want to determine the motion of an object A of mass m (Fig. 17.31b). A is also subject to the gravitational attractions of the sun, moon, etc., and we denote the resulting gravitational acceleration by the vector $\mathbf{g}_A$. The vector $\Sigma\mathbf{F}$ is the sum of all other external forces acting on A, including the gravitational force exerted on it by the earth. The total external force acting on A is $\Sigma\mathbf{F} + m\mathbf{g}_A$. We can apply Newton's second law to A, using our hypothetical inertial reference frame:

$$\Sigma\mathbf{F} + m\mathbf{g}_A = m\mathbf{a}_A. \tag{17.18}$$

Here, $\mathbf{a}_A$ is the acceleration of A relative to O. Since the earth-centered reference frame does not rotate, we can use Eq. (17.17) to write $\mathbf{a}_A$ as

$$\mathbf{a}_A = \mathbf{a}_B + \mathbf{a}_{A\,\mathrm{rel}},$$

where $\mathbf{a}_{A\,\mathrm{rel}}$ is the acceleration of A relative to the earth-centered reference frame. Using this relation and our definition of the earth's acceleration $\mathbf{a}_B = \mathbf{g}_B$ in Eq. (17.18), we obtain

$$\Sigma\mathbf{F} = m\mathbf{a}_{A\,\mathrm{rel}} + m(\mathbf{g}_B - \mathbf{g}_A). \tag{17.19}$$

If the object A is on or near the earth, its gravitational acceleration $\mathbf{g}_A$ due to the attraction of the sun, etc., is very nearly equal to the earth's gravitational acceleration $\mathbf{g}_B$. If we neglect the difference, Eq. (17.19) becomes

$$\Sigma\mathbf{F} = m\mathbf{a}_{A\,\mathrm{rel}}. \tag{17.20}$$

Thus, we can apply Newton's second law using a nonrotating, earth-centered reference frame. Even though this reference frame accelerates, virtually the same gravitational acceleration acts on the object. Notice that this argument does not hold if the object is not near the earth.

Earth-Fixed Reference Frame   For many applications, the most convenient reference frame is a local, *earth-fixed reference frame.* Why can we usually assume that an earth-fixed reference frame is inertial? Figure 17.32 shows a nonrotating reference frame with its origin O at the center of the earth, and a secondary earth-fixed reference frame with its origin at a point B. Since we can assume that the earth-centered, nonrotating reference frame is inertial, we can write Newton's second law for an object A of mass m as

$$\Sigma\mathbf{F} = m\mathbf{a}_A, \tag{17.21}$$

where $\mathbf{a}_A$ is A's acceleration relative to O. The earth-fixed reference frame rotates with the angular velocity of the earth, which we denote by $\boldsymbol{\omega}_E$. We can use Eq. (17.17) to write Eq. (17.21) in the form

$$\Sigma\mathbf{F} = m\mathbf{a}_{A\,\mathrm{rel}} + m[\mathbf{a}_B + 2\boldsymbol{\omega}_E \times \mathbf{v}_{A\,\mathrm{rel}}$$
$$+ \boldsymbol{\omega}_E \times (\boldsymbol{\omega}_E \times \mathbf{r}_{A/B})], \tag{17.22}$$

where $\mathbf{a}_{A\,\mathrm{rel}}$ is A's acceleration relative to the earth-fixed reference frame. If we can neglect the terms in brackets on the right side of Eq. (17.22), we can take the earth-fixed reference frame to be inertial. Let us consider each term. (Recall from the definition of the cross product that $|\mathbf{U} \times \mathbf{V}| = |\mathbf{U}||\mathbf{V}| \sin\theta,$

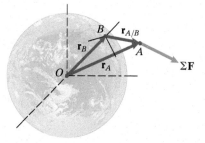

Figure 17.32
An earth-centered, nonrotating reference frame (origin O), an earth-fixed reference frame (origin B), and an object A.

where θ is the angle between the two vectors. Therefore, the magnitude of the cross product is bounded by the product of the magnitudes of the vectors.)

- The term $\boldsymbol{\omega}_E \times (\boldsymbol{\omega}_E \times \mathbf{r}_{A/B})$: The earth's angular velocity $\boldsymbol{\omega}_E$ is approximately one revolution per day $= 7.27 \times 10^{-5}$ rad/s. Therefore, the magnitude of this term is bounded by $\omega_E^2|\mathbf{r}_{A/B}| = (5.29 \times 10^{-9})|\mathbf{r}_{A/B}|$. For example, if the distance $|\mathbf{r}_{A/B}|$ from the origin of the earth-fixed reference frame to the object A is 10,000 m, this term is no larger than 5.3×10^{-5} m/s^2.

- The term $\mathbf{a}_B$: This term is the acceleration of the origin B of the earth-fixed reference frame relative to the center of the earth. B moves in a circular path due to the earth's rotation. If B lies on the earth's surface, $\mathbf{a}_B$ is bounded by $\omega_E^2 R_E$, where R_E is the radius of the earth. Using the value $R_E = 6370$ km, we find that $\omega_E^2 R_E = 0.0337$ m/s^2. This value is too large to neglect for many purposes. However, under normal circumstances, $\mathbf{a}_B$ is accounted for as a part of the local value of the acceleration due to gravity.

- The term $2\boldsymbol{\omega}_E \times \mathbf{v}_{A\,rel}$: This term is called the *Coriolis acceleration*. Its magnitude is bounded by $2\omega_E|\mathbf{v}_{A\,rel}| = (1.45 \times 10^{-4})|\mathbf{v}_{A\,rel}|$. For example, if the magnitude of the velocity of A relative to the earth-fixed reference frame is 10 m/s, this term is no larger than 1.45×10^{-3} m/s^2.

We see that in most applications, the terms in brackets in Eq. (17.22) can be neglected. However, in some cases this is not possible. The Coriolis acceleration becomes significant if an object's velocity relative to the earth is large, and even very small accelerations become significant if an object's motion must be predicted over a large period of time. In such cases, we can still use Eq. (17.22) to determine the motion, but must retain the significant terms. When this is done, the terms in brackets are usually moved to the left side:

$$\Sigma \mathbf{F} - m\mathbf{a}_B - 2m\boldsymbol{\omega}_E \times \mathbf{v}_{A\,rel} - m\boldsymbol{\omega}_E \times (\boldsymbol{\omega}_E \times \mathbf{r}_{A/B}) = m\mathbf{a}_{A\,rel}. \quad (17.23)$$

Written in this way, the equation has the usual form of Newton's second law, except that the left side contains additional "forces." (We use quotation marks because the quantities these terms represent are not forces, but arise from the motion of the earth-fixed reference frame.)

Coriolis Effects The term $-2m\boldsymbol{\omega}_E \times \mathbf{v}_{A\,rel}$ in Eq. (17.23) is called the Coriolis force. It explains a number of physical phenomena that exhibit different behaviors in the northern and southern hemispheres. The earth's angular velocity vector $\boldsymbol{\omega}_E$ points north. When an object in the northern hemisphere that is moving tangent to the earth's surface travels north (Fig. 17.33a), the cross product $\boldsymbol{\omega}_E \times \mathbf{v}_{A\,rel}$ points west (Fig. 17.33b). Therefore, the Coriolis force points east; it causes an object moving north to turn to the right (Fig. 17.33c). If the object is moving south, the direction of $\mathbf{v}_{A\,rel}$ is reversed and the Coriolis force points west; its effect is to cause the object moving south to turn to the right (Fig. 17.33c). For example, in the northern hemisphere, winds converging on a center of low pressure tend to rotate about that center in the counterclockwise direction (Fig. 17.34a).

When an object in the southern hemisphere travels north (Fig. 17.33d), the cross product $\boldsymbol{\omega}_E \times \mathbf{v}_{A\,rel}$ points east (Fig. 17.33e). The Coriolis force points west and tends to cause the object to turn to the left (Fig. 17.33f). If the object is moving south, the Coriolis force points east and tends to cause the object to turn to the left (Fig. 17.33f). In the southern hemisphere, winds converging on

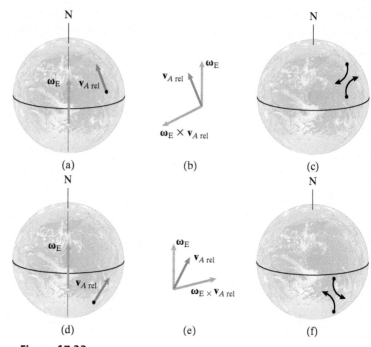

Figure 17.33
(a) An object in the northern hemisphere moving north.
(b) Cross product of the earth's angular velocity with the object's velocity.
(c) Effects of the Coriolis force in the northern hemisphere.
(d) An object in the southern hemisphere moving north.
(e) Cross product of the earth's angular velocity with the object's velocity.
(f) Effects of the Coriolis force in the southern hemisphere.

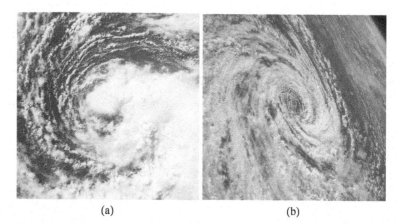

Figure 17.34
Storms in the (a) northern and (b) southern hemispheres.

a center of low pressure tend to rotate about that center in the clockwise direction (Fig. 17.34b).

Arbitrary Reference Frame   How can we analyze an object's motion relative to a reference frame that undergoes an arbitrary motion, such as a reference frame fixed with respect to a moving vehicle? Suppose that the primary reference frame with its origin at O in Fig. 17.35 is inertial and the secondary reference frame with its origin at B undergoes an arbitrary motion with angular velocity $\boldsymbol{\omega}$ and angular acceleration $\boldsymbol{\alpha}$. We can write Newton's second law for an object A of mass m as

$$\Sigma\mathbf{F} = m\mathbf{a}_A, \tag{17.24}$$

where $\mathbf{a}_A$ is A's acceleration relative to O. We use Eq. (17.17) to write Eq. (17.24) in the form

$$\Sigma\mathbf{F} - m[\mathbf{a}_B + 2\boldsymbol{\omega}\times\mathbf{v}_{A\,\text{rel}} + \boldsymbol{\alpha}\times\mathbf{r}_{A/B} + \boldsymbol{\omega}\times(\boldsymbol{\omega}\times\mathbf{r}_{A/B})] = m\mathbf{a}_{A\,\text{rel}}, \tag{17.25}$$

where $\mathbf{a}_{A\,\text{rel}}$ is A's acceleration relative to the secondary reference frame. This is Newton's second law expressed in terms of a secondary reference frame undergoing an arbitrary motion relative to an inertial primary reference frame. If the forces acting on A and the secondary reference frame's motion are known, Eq. (17.25) can be used to determine $\mathbf{a}_{A\,\text{rel}}$.

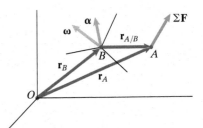

Figure 17.35
An inertial reference frame (origin O) and a reference frame undergoing an arbitrary motion (origin B).

RESULTS

Motion of a Point Relative to a Moving Reference Frame

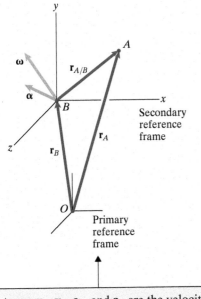

The terms $\mathbf{v}_A$, $\mathbf{v}_B$, $\mathbf{a}_A$, and $\mathbf{a}_B$ are the velocities and accelerations of points A and B relative to the primary reference frame. $\boldsymbol{\omega}$ and $\boldsymbol{\alpha}$ are the angular velocity and angular acceleration of the secondary reference frame relative to the primary reference frame. The terms $\mathbf{v}_{A\,\text{rel}}$ and $\mathbf{a}_{A\,\text{rel}}$ are the velocity and acceleration of A relative to the secondary reference frame.

$$\mathbf{v}_A = \mathbf{v}_B + \mathbf{v}_{A\,\text{rel}} + \boldsymbol{\omega}\times\mathbf{r}_{A/B}, \tag{17.16}$$

$$\mathbf{a}_A = \mathbf{a}_B + \mathbf{a}_{A\,\text{rel}} + 2\boldsymbol{\omega}\times\mathbf{v}_{A\,\text{rel}}$$
$$+ \boldsymbol{\alpha}\times\mathbf{r}_{A/B} + \boldsymbol{\omega}\times(\boldsymbol{\omega}\times\mathbf{r}_{A/B}). \tag{17.17}$$

In planar problems, Eq. (17.17) can be written

$$\mathbf{a}_A = \mathbf{a}_B + \mathbf{a}_{A\,\text{rel}} + 2\boldsymbol{\omega}\times\mathbf{v}_{A\,\text{rel}}$$
$$+ \boldsymbol{\alpha}\times\mathbf{r}_{A/B} - \omega^2\mathbf{r}_{A/B}.$$

Inertial Reference Frames

A reference frame is said to be *inertial* if it can be used to apply Newton's second law in the form $\Sigma\mathbf{F} = m\mathbf{a}$.

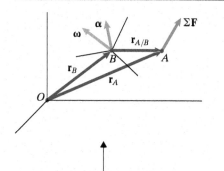

The reference frame with origin O is assumed to be inertial. The secondary reference frame with origin B is undergoing an arbitrary motion relative to the inertial reference frame. $\boldsymbol{\omega}$ and $\boldsymbol{\alpha}$ are the angular velocity and angular acceleration of the secondary reference frame relative to the inertial reference frame. Point A is the center of mass of an object of mass m.

Newton's second law for the object A can be expressed in terms of the acceleration of A relative to the inertial reference frame:

$$\Sigma\mathbf{F} = m\mathbf{a}_A. \tag{17.24}$$

By using Eq. (17.17), Newton's second law also can be expressed in terms of the acceleration of A relative to the secondary reference frame:

$$\Sigma\mathbf{F} - m[\mathbf{a}_B + 2\boldsymbol{\omega} \times \mathbf{v}_{A\text{ rel}} + \boldsymbol{\alpha} \times \mathbf{r}_{A/B} + \boldsymbol{\omega} \times (\boldsymbol{\omega} \times \mathbf{r}_{A/B})] = m\mathbf{a}_{A\text{ rel}}. \tag{17.25}$$

When Newton's second law is expressed in this way, additional "forces" appear on the left side of the equation which are artifacts arising from the motion of the secondary reference frame.

Active Example 17.10   A Rotating Secondary Reference Frame (▶ *Related Problem 17.143*)

Relative to a primary reference frame that is fixed with respect to the earth, the merry-go-round rotates with constant counterclockwise angular velocity ω. Suppose that you are in the center at B and observe a second person A using a reference frame that is fixed with respect to the merry-go-round. The person A is standing still on the ground just next to the merry-go-round. What are her velocity and acceleration relative to your reference frame at the instant shown?

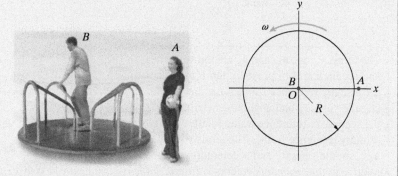

Strategy

The person A is stationary relative to the earth, so her velocity and acceleration relative to the primary reference frame are known. We can use Eqs. (17.16) and (17.17) to determine her velocity $\mathbf{v}_{A\ rel}$ and acceleration $\mathbf{a}_{A\ rel}$ relative to the secondary reference frame.

Solution

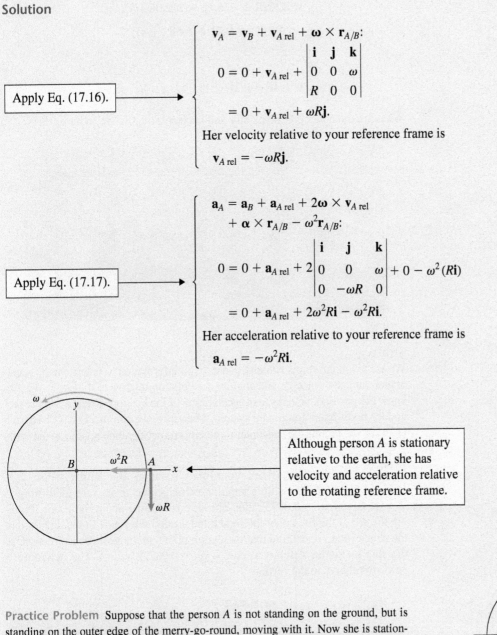

Apply Eq. (17.16).

$$\mathbf{v}_A = \mathbf{v}_B + \mathbf{v}_{A\ rel} + \boldsymbol{\omega} \times \mathbf{r}_{A/B}:$$

$$0 = 0 + \mathbf{v}_{A\ rel} + \begin{vmatrix} \mathbf{i} & \mathbf{j} & \mathbf{k} \\ 0 & 0 & \omega \\ R & 0 & 0 \end{vmatrix}$$

$$= 0 + \mathbf{v}_{A\ rel} + \omega R\mathbf{j}.$$

Her velocity relative to your reference frame is

$$\mathbf{v}_{A\ rel} = -\omega R\mathbf{j}.$$

Apply Eq. (17.17).

$$\mathbf{a}_A = \mathbf{a}_B + \mathbf{a}_{A\ rel} + 2\boldsymbol{\omega} \times \mathbf{v}_{A\ rel}$$
$$+ \boldsymbol{\alpha} \times \mathbf{r}_{A/B} - \omega^2 \mathbf{r}_{A/B}:$$

$$0 = 0 + \mathbf{a}_{A\ rel} + 2\begin{vmatrix} \mathbf{i} & \mathbf{j} & \mathbf{k} \\ 0 & 0 & \omega \\ 0 & -\omega R & 0 \end{vmatrix} + 0 - \omega^2 (R\mathbf{i})$$

$$= 0 + \mathbf{a}_{A\ rel} + 2\omega^2 R\mathbf{i} - \omega^2 R\mathbf{i}.$$

Her acceleration relative to your reference frame is

$$\mathbf{a}_{A\ rel} = -\omega^2 R\mathbf{i}.$$

Although person A is stationary relative to the earth, she has velocity and acceleration relative to the rotating reference frame.

Practice Problem Suppose that the person A is not standing on the ground, but is standing on the outer edge of the merry-go-round, moving with it. Now she is stationary relative to the secondary reference frame. Use Eqs. (17.16) and (17.17) to determine her velocity and acceleration relative to the earth.

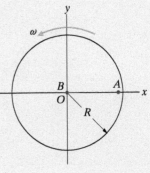

Answer: $\mathbf{v}_A = \omega R\mathbf{j}$, $\mathbf{a}_A = -\omega^2 R\mathbf{i}$.

Example 17.11 A Reference Frame Fixed with Respect to a Ship (▶ *Related Problem 17.147*)

At the instant shown, the ship B is moving north at a constant speed of 15.0 m/s relative to the earth and is turning toward the west at a constant rate of 5.0° per second. Relative to the ship's body-fixed coordinate system, its radar indicates that the position, velocity, and acceleration of the helicopter A are

$$\mathbf{r}_{A/B} = 420.0\mathbf{i} + 236.2\mathbf{j} + 212.0\mathbf{k} \ (\text{m}),$$
$$\mathbf{v}_{A \text{ rel}} = -53.5\mathbf{i} + 2.0\mathbf{j} + 6.6\mathbf{k} \ (\text{m/s}),$$

and

$$\mathbf{a}_{A \text{ rel}} = 0.4\mathbf{i} - 0.2\mathbf{j} - 13.0\mathbf{k} \ (\text{m/s}^2).$$

What are the helicopter's velocity and acceleration relative to the earth?

Strategy
We are given the ship's velocity and enough information to determine its acceleration, angular velocity, and angular acceleration relative to the earth. We also know the position, velocity, and acceleration of the helicopter relative to the secondary body-fixed coordinate system. Therefore, we can use Eqs. (17.16) and (17.17) to determine the helicopter's velocity and acceleration relative to the earth.

Solution
In terms of the body-fixed coordinate system, the ship's velocity is $\mathbf{v}_B = 15.0\mathbf{i}$ (m/s). The ship's angular velocity due to its rate of turning is $\omega = (5.0/180)\pi = 0.0873$ rad/s. The ship is rotating about the y axis. Pointing the arc of the fingers of the right hand around the y axis in the direction of the ship's rotation, we find that the thumb points in the positive y direction, so the ship's angular velocity vector is $\boldsymbol{\omega} = 0.0873\mathbf{j}$ (rad/s). The helicopter's velocity relative to the earth is

$$\mathbf{v}_A = \mathbf{v}_B + \mathbf{v}_{A \text{ rel}} + \boldsymbol{\omega} \times \mathbf{r}_{A/B}$$

$$= 15.0\mathbf{i} + (-53.5\mathbf{i} + 2.0\mathbf{j} + 6.6\mathbf{k}) + \begin{vmatrix} \mathbf{i} & \mathbf{j} & \mathbf{k} \\ 0 & 0.0873 & 0 \\ 420.0 & 236.2 & 212.0 \end{vmatrix}$$

$$= -20.0\mathbf{i} + 2.0\mathbf{j} - 30.1\mathbf{k} \ (\text{m/s}).$$

We can determine the ship's acceleration by expressing it in terms of normal and tangential components in the form given by Eq. (13.37) (Fig. a):

$$\mathbf{a}_B = \frac{dv}{dt}\mathbf{e}_t + v\frac{d\theta}{dt}\mathbf{e}_n = \mathbf{0} + (15)(0.0873)\mathbf{e}_n$$

$$= 1.31\mathbf{e}_n \ (\text{m/s}^2).$$

The z axis is perpendicular to the ship's path and points toward the convex side of the path (Fig. b). Therefore, in terms of the body-fixed coordinate system, the ship's acceleration is $\mathbf{a}_B = -1.31\mathbf{k}$ (m/s^2). The ship's angular velocity vector is constant, so $\boldsymbol{\alpha} = \mathbf{0}$. The helicopter's acceleration relative to the earth is

$$\mathbf{a}_A = \mathbf{a}_B + \mathbf{a}_{A\,\text{rel}} + 2\boldsymbol{\omega} \times \mathbf{v}_{A\,\text{rel}} + \boldsymbol{\alpha} \times \mathbf{r}_{A/B}$$

$$+ \boldsymbol{\omega} \times (\boldsymbol{\omega} \times \mathbf{r}_{A/B})$$

$$= -1.31\mathbf{k} + (0.4\mathbf{i} - 0.2\mathbf{j} - 13.0\mathbf{k}) + 2\begin{vmatrix} \mathbf{i} & \mathbf{j} & \mathbf{k} \\ 0 & 0.0873 & 0 \\ -53.5 & 2.0 & 6.6 \end{vmatrix}$$

$$+ \mathbf{0} + (0.0873\mathbf{j}) \times \begin{vmatrix} \mathbf{i} & \mathbf{j} & \mathbf{k} \\ 0 & 0.0873 & 0 \\ 420.0 & 236.2 & 212.0 \end{vmatrix}$$

$$= -1.65\mathbf{i} - 0.20\mathbf{j} - 6.59\mathbf{k} \ (\text{m/s}^2).$$

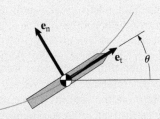

(a) Determining the ship's acceleration.

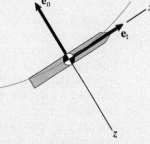

(b) Correspondence between the normal and tangential components and the body-fixed coordinate system.

Critical Thinking

Notice the substantial differences between the helicopter's velocity and acceleration relative to the earth and the values the ship measures using its body-fixed coordinate system. The ship's instruments, like those used on any moving vehicle, intrinsically make measurements relative to a body-fixed reference frame. Equations (17.16) and (17.17) must be used to transform measurements of velocity and acceleration into values relative to other reference frames.

Example 17.12 | An Earth-Fixed Reference Frame (▶ *Related Problem 17.151*)

The satellite A is in a circular polar orbit (an orbit that intersects the earth's axis of rotation). Relative to a nonrotating primary reference frame with its origin at the center of the earth, the satellite moves in a circular path of radius R with a velocity of constant magnitude v_A. At the present instant, the satellite is above the equator. The secondary earth-fixed reference frame shown is oriented with the y axis in the direction of the north pole and the x axis in the direction of the satellite. What are the satellite's velocity and acceleration relative to the earth-fixed reference frame? Let ω_E be the angular velocity of the earth.

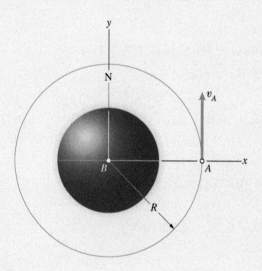

Strategy

We are given enough information to determine the satellite's velocity and acceleration $\mathbf{v}_A$ and $\mathbf{a}_A$ relative to the nonrotating primary reference frame and the angular velocity vector $\boldsymbol{\omega}$ of the secondary reference frame. We can therefore use Eqs. (17.16) and (17.17) to determine the satellite's velocity and acceleration $\mathbf{v}_{A\,\text{rel}}$ and $\mathbf{a}_{A\,\text{rel}}$ relative to the earth-fixed reference frame.

Solution

At the present instant, the satellite's velocity and acceleration relative to a nonrotating primary reference frame with its origin at the center of the earth are $\mathbf{v}_A = v_A\mathbf{j}$ and $\mathbf{a}_A = -(v_A^2/R)\mathbf{i}$. The angular velocity vector of the earth points north (confirm this by using the right-hand rule), so the angular velocity of the earth-fixed reference frame is $\boldsymbol{\omega} = \omega_E \mathbf{j}$. From Eq. (17.16),

$$\mathbf{v}_A = \mathbf{v}_B + \mathbf{v}_{A\,rel} + \boldsymbol{\omega} \times \mathbf{r}_{A/B}:$$

$$v_A \mathbf{j} = 0 + \mathbf{v}_{A\,rel} + \begin{vmatrix} \mathbf{i} & \mathbf{j} & \mathbf{k} \\ 0 & \omega_E & 0 \\ R & 0 & 0 \end{vmatrix}.$$

Solving for $\mathbf{v}_{A\,rel}$, we find that the satellite's velocity relative to the earth-fixed reference frame is

$$\mathbf{v}_{A\,rel} = v_A \mathbf{j} + R\omega_E \mathbf{k}.$$

The second term on the right side of this equation is the satellite's velocity toward the west relative to the rotating earth-fixed reference frame.
From Eq. (17.17),

$$\mathbf{a}_A = \mathbf{a}_B + \mathbf{a}_{A\,rel} + 2\boldsymbol{\omega} \times \mathbf{v}_{A\,rel} + \boldsymbol{\alpha} \times \mathbf{r}_{A/B} + \boldsymbol{\omega} \times (\boldsymbol{\omega} \times \mathbf{r}_{A/B}):$$

$$-\frac{v_A^2}{R}\mathbf{i} = 0 + \mathbf{a}_{A\,rel} + 2\begin{vmatrix} \mathbf{i} & \mathbf{j} & \mathbf{k} \\ 0 & \omega_E & 0 \\ 0 & v_A & R\omega_E \end{vmatrix} + 0 + \begin{vmatrix} \mathbf{i} & \mathbf{j} & \mathbf{k} \\ 0 & \omega_E & 0 \\ 0 & 0 & -R\omega_E \end{vmatrix}.$$

Solving for $\mathbf{a}_{A\,rel}$, we find the satellite's acceleration relative to the earth-fixed reference frame:

$$\mathbf{a}_{A\,rel} = -\left(\frac{v_A^2}{R} + \omega_E^2 R\right)\mathbf{i}.$$

Critical Thinking

In this example, we assumed that the motion of the satellite was known relative to a nonrotating primary reference frame with its origin at the center of the earth, and used Eqs. (17.16) and (17.17) to determine its velocity and acceleration relative to a secondary earth-fixed reference frame. The reverse of this procedure is more common in practice. Ground-based measuring instruments measure velocity and acceleration relative to an earth-fixed reference frame, and Eqs. (17.16) and (17.17) are used to determine their values relative to other reference frames.

Example 17.13 | Inertial Reference Frames (▶ *Related Problem 17.154*)

Suppose that you and a friend play tennis on the deck of a cruise ship, and you use the ship-fixed coordinate system with origin B to analyze the motion of the ball A. At the instant shown, the ball's position and velocity relative to the ship-fixed coordinate system are $\mathbf{r}_{A/B} = 5\mathbf{i} + 2\mathbf{j} + 12\mathbf{k}$ (m) and $\mathbf{v}_{A\,\text{rel}} = \mathbf{i} - 2\mathbf{j} + 7\mathbf{k}$ (m/s). The mass of the ball is 0.056 kg. Ignore the aerodynamic force on the ball for this example. As a result of the ship's motion, the acceleration of the origin B relative to the earth is $\mathbf{a}_B = 1.10\mathbf{i} + 0.07\mathbf{k}$ (m/s^2), and the angular velocity of the ship-fixed coordinate system is constant and equal to $\boldsymbol{\omega} = 0.1\mathbf{j}$ (rad/s). Use Newton's second law to determine the ball's acceleration relative to the ship-fixed coordinate system, (a) assuming that the ship-fixed coordinate system is inertial and (b) not assuming that the ship-fixed coordinate system is inertial, but assuming that a local earth-fixed coordinate system is inertial.

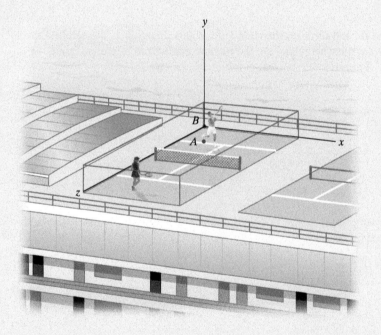

Strategy

In part (a), we know the ball's mass and the external forces acting on it, so we can simply apply Newton's second law to determine the acceleration. In part (b), we can express Newton's second law in the form given by Eq. (17.25), which applies to a coordinate system undergoing an arbitrary motion relative to an inertial reference frame.

Solution

(a) If the ship-fixed coordinate system is assumed to be inertial, we can apply Newton's second law to the ball in the form $\Sigma \mathbf{F} = m\mathbf{a}_{A\,\text{rel}}$. The only external force on the ball is its weight,

$$-(0.056)(9.81)\mathbf{j} = 0.056\mathbf{a}_{A\,\text{rel}},$$

so we obtain $\mathbf{a}_{A\,\text{rel}} = -9.81\mathbf{j}$ (m/s^2).

(b) If we treat the earth-fixed reference frame as inertial, we can solve Eq. (17.25) for the ball's acceleration relative to the ship-fixed reference frame:

$$\mathbf{a}_{A\,\text{rel}} = \frac{1}{m}\Sigma\mathbf{F} - \mathbf{a}_B - 2\boldsymbol{\omega} \times \mathbf{v}_{A\,\text{rel}} - \boldsymbol{\alpha} \times \mathbf{r}_{A/B} - \boldsymbol{\omega} \times (\boldsymbol{\omega} \times \mathbf{r}_{A/B})$$

$$= \frac{1}{0.056}[-(0.056)(9.81)\mathbf{j}] - (1.10\mathbf{i} + 0.07\mathbf{k}) - 2\begin{vmatrix} \mathbf{i} & \mathbf{j} & \mathbf{k} \\ 0 & 0.1 & 0 \\ 1 & -2 & 7 \end{vmatrix}$$

$$- \mathbf{0} - (0.1\mathbf{j}) \times \begin{vmatrix} \mathbf{i} & \mathbf{j} & \mathbf{k} \\ 0 & 0.1 & 0 \\ 5 & 2 & 12 \end{vmatrix}$$

$$= -2.45\mathbf{i} - 9.81\mathbf{j} + 0.25\mathbf{k} \ (\text{m/s}^2).$$

Critical Thinking

This example demonstrates the care you must exercise in applying Newton's second law. When we treated the ship-fixed coordinate system as inertial, Newton's second law did not correctly predict the ball's acceleration relative to the coordinate system, because the effects of the motion of the coordinate system on the ball's relative motion were not accounted for.

Problems

▶ **17.143** In Active Example 17.10, suppose that the merry-go-round has counterclockwise angular velocity ω and counterclockwise angular acceleration α. The person A is standing still on the ground. Determine her acceleration relative to your reference frame at the instant shown.

17.144 The x–y coordinate system is body fixed with respect to the bar. The angle θ (in radians) is given as a function of time by $\theta = 0.2 + 0.04t^2$. The x coordinate of the sleeve A (in feet) is given as a function of time by $x = 1 + 0.03t^3$. Use Eq. (17.16) to determine the velocity of the sleeve at $t = 4$ s relative to a non-rotating reference frame with its origin at B. (Although you are determining the velocity of A relative to a nonrotating reference frame, your answer will be expressed in components in terms of the body-fixed reference frame.)

17.145 The metal plate is attached to a fixed ball-and-socket support at O. The pin A slides in a slot in the plate. At the instant shown, $x_A = 1$ m, $dx_A/dt = 2$ m/s, and $d^2x_A/dt^2 = 0$, and the plate's angular velocity and angular acceleration are $\boldsymbol{\omega} = 2\mathbf{k}$ (rad/s) and $\boldsymbol{\alpha} = \mathbf{0}$. What are the x, y, and z components of the velocity and acceleration of A relative to a nonrotating reference frame with its origin at O?

17.146 The pin A slides in a slot in the plate. Suppose that at the instant shown, $x_A = 1$ m, $dx_A/dt = -3$ m/s, $d^2x_A/dt^2 = 4$ m/s², and the plate's angular velocity and angular acceleration are $\boldsymbol{\omega} = -4\mathbf{j} + 2\mathbf{k}$ (rad/s) and $\boldsymbol{\alpha} = 3\mathbf{i} - 6\mathbf{j}$ (rad/s²). What are the x, y, and z components of the velocity and acceleration of A relative to a nonrotating reference frame that is stationary with respect to O?

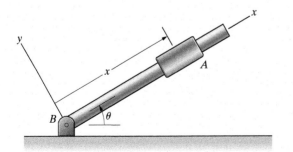

Problem 17.144

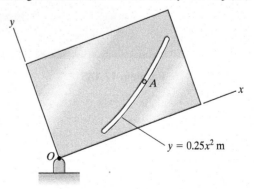

$y = 0.25x^2$ m

Problems 17.145/17.146

▶ 17.147 The coordinate system is fixed relative to the ship B. At the instant shown, the ship is sailing north at 5 m/s relative to the earth, and its angular velocity is 0.26 rad/s counterclockwise. Using radar, it is determined that the position of the airplane is $1080\mathbf{i} + 1220\mathbf{j} + 6300\mathbf{k}$ (m) and its velocity relative to the ship's coordinate system is $870\mathbf{i} - 45\mathbf{j} - 21\mathbf{k}$ (m/s). What is the airplane's velocity relative to the earth? (See Example 17.11.)

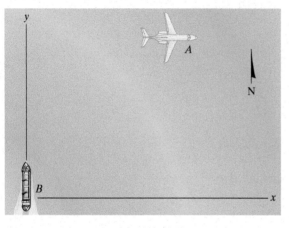

Problem 17.147

17.148 The space shuttle is attempting to recover a satellite for repair. At the current time, the satellite's position relative to a coordinate system fixed to the shuttle is $50\mathbf{i}$ (m). The rate gyros on the shuttle indicate that its current angular velocity is $0.05\mathbf{j} + 0.03\mathbf{k}$ (rad/s). The shuttle pilot measures the velocity of the satellite relative to the body-fixed coordinate system and determines it to be $-2\mathbf{i} - 1.5\mathbf{j} + 2.5\mathbf{k}$ (rad/s). What are the x, y, and z components of the satellite's velocity relative to a nonrotating coordinate system with its origin fixed to the shuttle's center of mass?

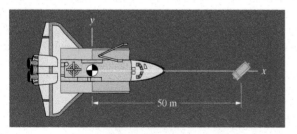

Problem 17.148

17.149 The train on the circular track is traveling at a constant speed of 50 ft/s in the direction shown. The train on the straight track is traveling at 20 ft/s in the direction shown and is increasing its speed at 2 ft/s². Determine the velocity of passenger A that passenger B observes relative to the given coordinate system, which is fixed to the car in which B is riding.

17.150 In Problem 17.149, determine the acceleration of passenger A that passenger B observes relative to the coordinate system fixed to the car in which B is riding.

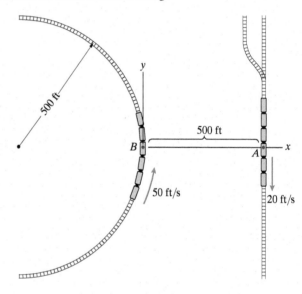

Problems 17.149/17.150

▶ 17.151 The satellite A is in a circular polar orbit (a circular orbit that intersects the earth's axis of rotation). The radius of the orbit is R, and the magnitude of the satellite's velocity relative to a nonrotating reference frame with its origin at the center of the earth is v_A. At the instant shown, the satellite is above the equator. An observer B on the earth directly below the satellite measures its motion using the earth-fixed coordinate system shown. What are the velocity and acceleration of the satellite relative to B's earth-fixed coordinate system? The radius of the earth is R_E and the angular velocity of the earth is ω_E. (See Example 17.12.)

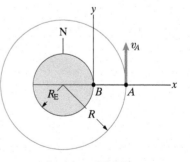

Problem 17.151

17.152 A car A at north latitude L drives north on a north–south highway with constant speed v. The earth's radius is R_E, and the earth's angular velocity is ω_E. (The earth's angular velocity vector points north.) The coordinate system is earth fixed, and the x axis passes through the car's position at the instant shown. Determine the car's velocity and acceleration (a) relative to the earth-fixed coordinate system and (b) relative to a nonrotating reference frame with its origin at the center of the earth.

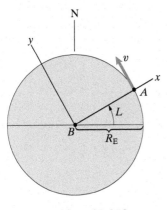

Problem 17.152

17.153 The airplane B conducts flight tests of a missile. At the instant shown, the airplane is traveling at 200 m/s relative to the earth in a circular path of 2000-m radius *in the horizontal plane*. The coordinate system is fixed relative to the airplane. The x axis is tangent to the plane's path and points forward. The y axis points out the plane's right side, and the z axis points out the bottom of the plane. The plane's bank angle (the inclination of the z axis from the vertical) is constant and equal to 20°. *Relative to the airplane's coordinate system,* the pilot measures the missile's position and velocity and determines them to be $r_{A/B} = 1000\mathbf{i}$ (m) and $\mathbf{v}_{A/B} = 100.0\mathbf{i} + 94.0\mathbf{j} + 34.2\mathbf{k}$ (m/s).

(a) What are the x, y, and z components of the airplane's angular velocity vector?

(b) What are the x, y, and z components of the missile's velocity relative to the earth?

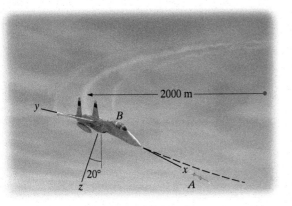

Problem 17.153

▶ 17.154 To conduct experiments related to long-term space-flight, engineers construct a laboratory on earth that rotates about the vertical axis at B with a constant angular velocity ω of one revolution every 6 s. They establish a laboratory-fixed coordinate system with its origin at B and the z axis pointing upward. An engineer holds an object stationary relative to the laboratory at point A, 3 m from the axis of rotation, and releases it. At the instant he drops the object, determine its acceleration relative to the laboratory-fixed coordinate system, (a) assuming that the laboratory-fixed coordinate system is inertial and (b) not assuming that the laboratory-fixed coordinate system is inertial, but assuming that an earth-fixed coordinate system with its origin at B is inertial. (See Example 17.13.)

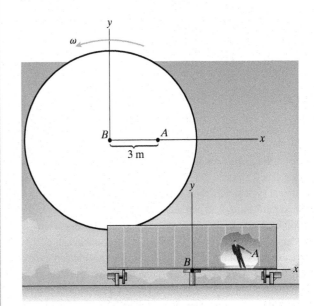

Problem 17.154

17.155 The disk rotates *in the horizontal plane* about a fixed shaft at the origin with constant angular velocity $\omega = 10$ rad/s. The 2-kg slider A moves in a smooth slot in the disk. The spring is unstretched when $x = 0$ and its constant is $k = 400$ N/m. Determine the acceleration of A relative to the body-fixed coordinate system when $x = 0.4$ m.

Strategy: Use Eq. (17.25) to express Newton's second law for the slider in terms of the body-fixed coordinate system.

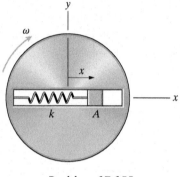

Problem 17.155

17.156* Engineers conduct flight tests of a rocket at 30° north latitude. They measure the rocket's motion using an earth-fixed coordinate system with the x axis pointing upward and the y axis directed northward. At a particular instant, the mass of the rocket is 4000 kg, the velocity of the rocket relative to the engineers' coordinate system is $2000\mathbf{i} + 2000\mathbf{j}$ (m/s), and the sum of the forces exerted on the rocket by its thrust, weight, and aerodynamic forces is $400\mathbf{i} + 400\mathbf{j}$ (N). Determine the rocket's acceleration relative to the engineers' coordinate system (a) assuming that their earth-fixed coordinate system is inertial and (b) not assuming that their earth-fixed coordinate system is inertial.

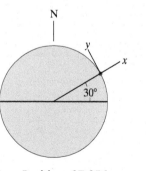

Problem 17.156

17.157* Consider a point A on the surface of the earth at north latitude L. The radius of the earth is R_E, and the earth's angular velocity is ω_E. A plumb bob suspended just above the ground at point A will hang at a small angle β relative to the vertical because of the earth's rotation. Show that β is related to the latitude by

$$\tan \beta = \frac{\omega_E^2 R_E \sin L \cos L}{g - \omega_E^2 R_E \cos^2 L}.$$

Strategy: Using the earth-fixed coordinate system shown, express Newton's second law in the form given by Eq. (17.22).

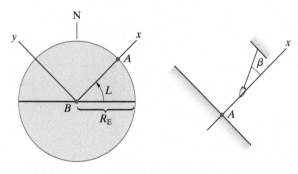

Problem 17.157

17.158* Suppose that a space station is in orbit around the earth and two astronauts on the station toss a ball back and forth. They observe that the ball appears to travel between them in a straight line at constant velocity.

(a) Write Newton's second law for the ball as it travels between the astronauts in terms of a nonrotating coordinate system with its origin fixed to the station. What is the term $\Sigma\mathbf{F}$? Use the equation you wrote to explain the behavior of the ball observed by the astronauts.

(b) Write Newton's second law for the ball as it travels between the astronauts in terms of a nonrotating coordinate system with its origin fixed to the center of the earth. What is the term $\Sigma\mathbf{F}$? Explain the difference between this equation and the one you obtained in part (a).

Review Problems

17.159 If $\theta = 60°$ and bar OQ rotates in the counterclockwise direction at 5 rad/s, what is the angular velocity of bar PQ?

17.160 If $\theta = 55°$ and the sleeve P is moving to the left at 2 m/s, what are the angular velocities of bars OQ and PQ?

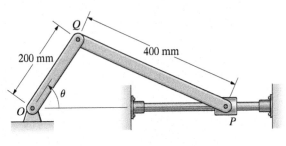

Problems 17.159/17.160

17.161 Determine the vertical velocity v_H of the hook and the angular velocity of the small pulley.

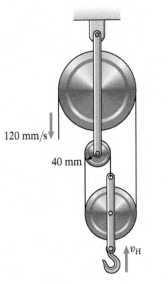

Problem 17.161

17.162 If the crankshaft AB is turning in the counterclockwise direction at 2000 rpm, what is the velocity of the piston?

17.163 If the piston is moving with velocity $\mathbf{v}_C = 20\mathbf{j}$ (ft/s), what are the angular velocities of the crankshaft AB and the connecting rod BC?

17.164 If the piston is moving with velocity $\mathbf{v}_C = 20\mathbf{j}$ (ft/s) and its acceleration is zero, what are the angular accelerations of the crankshaft AB and the connecting rod BC?

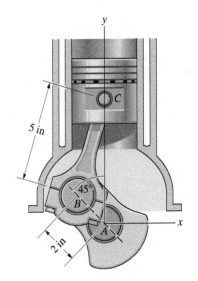

Problems 17.162–17.164

17.165 Bar *AB* rotates at 6 rad/s in the counterclockwise direction. Use instantaneous centers to determine the angular velocity of bar *BCD* and the velocity of point *D*.

17.166 Bar *AB* rotates with a constant angular velocity of 6 rad/s in the counterclockwise direction. Determine the acceleration of point *D*.

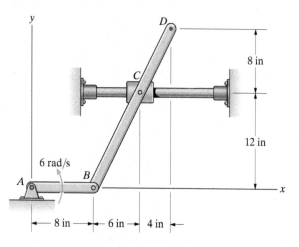

Problems 17.165/17.166

17.167 Point *C* is moving to the right at 20 in/s. What is the velocity of the midpoint *G* of bar *BC*?

17.168 Point *C* is moving to the right with a constant velocity of 20 in/s. What is the acceleration of the midpoint *G* of bar *BC*?

17.169 If the velocity of point *C* is $\mathbf{v}_C = 1.0\mathbf{i}$ (in/s), what are the angular velocity vectors of arms *AB* and *BC*?

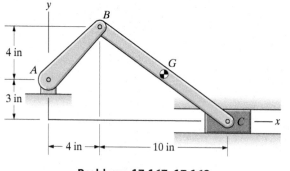

Problems 17.167–17.169

17.170 Points *B* and *C* are in the *x–y* plane. The angular velocity vectors of arms *AB* and *BC* are $\boldsymbol{\omega}_{AB} = -0.5\mathbf{k}$ (rad/s) and $\boldsymbol{\omega}_{BC} = -2.0\mathbf{k}$ (rad/s). Determine the velocity of point *C*.

17.171 If the velocity of point *C* is $\mathbf{v}_C = 1.0\mathbf{i}$ (m/s), what are the angular velocity vectors of arms *AB* and *BC*?

17.172 The angular velocity vectors of arms *AB* and *BC* are $\boldsymbol{\omega}_{AB} = -0.5\mathbf{k}$ (rad/s) and $\boldsymbol{\omega}_{BC} = 2.0\mathbf{k}$ (rad/s), and their angular acceleration vectors are $\boldsymbol{\alpha}_{AB} = 1.0\mathbf{k}$ (rad/s^2) and $\boldsymbol{\alpha}_{BC} = 1.0\mathbf{k}$ (rad/s^2). What is the acceleration of point *C*?

17.173 The velocity of point *C* is $\mathbf{v}_C = 1.0\mathbf{i}$ (m/s) and $\mathbf{a}_C = \mathbf{0}$. What are the angular velocity and angular acceleration vectors of arm *BC*?

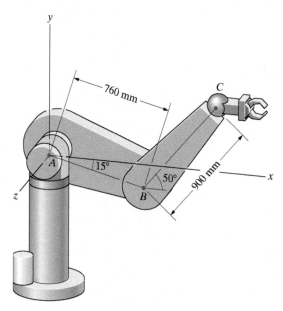

Problems 17.170–17.173

17.174 The crank *AB* has a constant clockwise angular velocity of 200 rpm. What are the velocity and acceleration of the piston *P*?

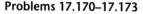
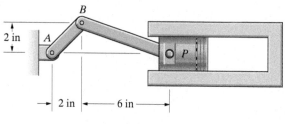

Problem 17.174

17.175 Bar AB has a counterclockwise angular velocity of 10 rad/s and a clockwise angular acceleration of 20 rad/s². Determine the angular acceleration of bar BC and the acceleration of point C.

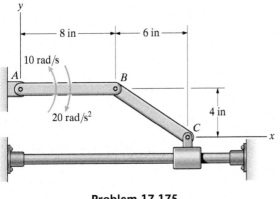

Problem 17.175

17.176 The angular velocity of arm AC is 1 rad/s counterclockwise. What is the angular velocity of the scoop?

17.177 The angular velocity of arm AC is 2 rad/s counterclockwise and its angular acceleration is 4 rad/s² clockwise. What is the angular acceleration of the scoop?

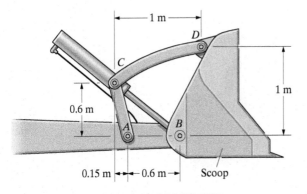

Problems 17.176/17.177

17.178 If you want to program the robot so that, at the instant shown, the velocity of point D is $\mathbf{v}_D = 0.2\mathbf{i} + 0.8\mathbf{j}$ (m/s) and the angular velocity of arm CD is 0.3 rad/s counterclockwise, what are the necessary angular velocities of arms AB and BC?

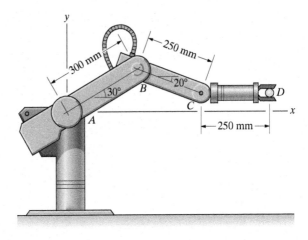

Problem 17.178

17.179 The ring gear is stationary, and the sun gear rotates at 120 rpm in the counterclockwise direction. Determine the angular velocity of the planet gears and the magnitude of the velocity of their centerpoints.

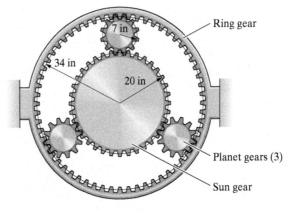

Problem 17.179

17.180 Arm *AB* is rotating at 10 rad/s in the clockwise direction. Determine the angular velocity of arm *BC* and the velocity at which the arm slides relative to the sleeve at *C*.

17.181 Arm *AB* is rotating with an angular velocity of 10 rad/s and an angular acceleration of 20 rad/s^2, both in the clockwise direction. Determine the angular acceleration of arm *BC*.

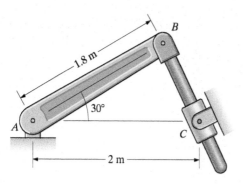

Problems 17.180/17.181

17.182 Arm *AB* is rotating with a constant counterclockwise angular velocity of 10 rad/s. Determine the vertical velocity and acceleration of the rack *R* of the rack-and-pinion gear.

17.183 The rack *R* of the rack-and-pinion gear is moving upward with a constant velocity of 10 ft/s. What are the angular velocity and angular acceleration of bar *BC*?

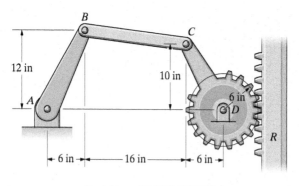

Problems 17.182/17.183

17.184 Bar *AB* has a constant counterclockwise angular velocity of 2 rad/s. The 1-kg collar *C* slides on the smooth horizontal bar. At the instant shown, what is the tension in the cable *BC*?

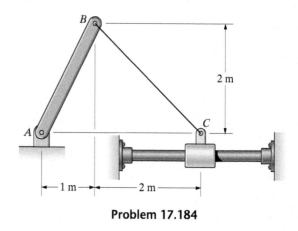

Problem 17.184

17.185 An athlete exercises his arm by raising the 8-kg mass *m*. The shoulder joint *A* is stationary. The distance *AB* is 300 mm, the distance *BC* is 400 mm, and the distance from *C* to the pulley is 340 mm. The angular velocities $\omega_{AB} = 1.5$ rad/s and $\omega_{BC} = 2$ rad/s are constant. What is the tension in the cable?

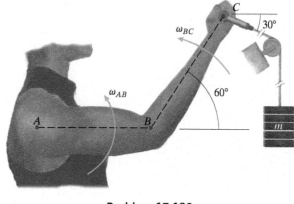

Problem 17.185

17.186 The hydraulic actuator *BC* of the crane is extending (increasing in length) at a constant rate of 0.2 m/s. When the angle $\beta = 35°$, what is the angular velocity of the crane's boom *AD*?

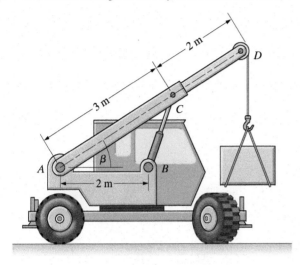

Problem 17.186

17.187 The coordinate system shown is fixed relative to the ship *B*. The ship uses its radar to measure the position of a stationary buoy *A* and determines it to be $400\mathbf{i} + 200\mathbf{j}$ (m). The ship also measures the velocity of the buoy relative to its body-fixed coordinate system and determines it to be $2\mathbf{i} - 8\mathbf{j}$ (m/s). What are the ship's velocity and angular velocity relative to the earth? (Assume that the ship's velocity is in the direction of the *y* axis.)

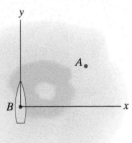

Problem 17.187

Design Project

Bar *AB* rotates about the fixed point *A* with constant angular velocity ω_0.

(a) Determine the lengths of the bars *AB* and *BC* so that as bar *AB* rotates, the collar *C* moves back and forth between the positions *D* and *E*.

(b) Draw graphs of the velocity and acceleration of the collar *C* as functions of the angular position of bar *AB*.

(c) Suppose that a design constraint is that the magnitude of the acceleration of collar *C* must not exceed 200 m/s². What is the maximum allowable value of ω_0?

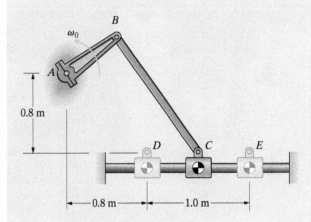

Planar Dynamics of Rigid Bodies

In Chapter 17 we analyzed planar motions of rigid bodies without considering the forces and couples causing them. In this chapter we derive planar equations of angular motion for a rigid body. By drawing the free-body diagram of an object and applying the equations of motion, we can determine both the acceleration of its center of mass and its angular acceleration in terms of the forces and couples to which it is subjected.

◄ The riders exert forces on the pedals of the exercise bicycles, resulting in planar motions of the cranks, chain, and wheels.

18.1 Momentum Principles for a System of Particles

BACKGROUND

Our derivations of the equations of motion for rigid bodies are based on principles governing the motion of a system of particles. We summarize these general and important principles in this section.

Force–Linear Momentum Principle

We begin by showing that the sum of the external forces on a system of particles equals the rate of change of the total linear momentum of the system. Let us consider a system of N particles. We denote the mass of the ith particle by m_i and denote its position vector relative to the origin O of an inertial reference frame by $\mathbf{r}_i$ (Fig. 18.1). Let $\mathbf{f}_{ij}$ be the force exerted on the ith particle by the jth particle, and let the external force on the ith particle (i.e., the total force exerted by objects other than the system of particles we are considering) be $\mathbf{f}_i^E$. Newton's second law states that the total force acting on the ith particle equals the product of its mass and the rate of change of its linear momentum; that is,

$$\sum_j \mathbf{f}_{ij} + \mathbf{f}_i^E = \frac{d}{dt}(m_i \mathbf{v}_i), \tag{18.1}$$

where $\mathbf{v}_i = d\mathbf{r}_i/dt$ is the velocity of the ith particle. Writing this equation for each particle of the system and summing from $i = 1$ to N, we obtain

$$\sum_i \sum_j \mathbf{f}_{ij} + \sum_i \mathbf{f}_i^E = \frac{d}{dt} \sum_i m_i \mathbf{v}_i. \tag{18.2}$$

The first term on the left side of this equation is the sum of the internal forces on the system of particles. As a consequence of Newton's third law $(\mathbf{f}_{ij} + \mathbf{f}_{ji} = \mathbf{0})$, that term equals zero:

$$\sum_i \sum_j \mathbf{f}_{ij} = \mathbf{f}_{12} + \mathbf{f}_{21} + \mathbf{f}_{13} + \mathbf{f}_{31} + \cdots = \mathbf{0}.$$

The second term on the left side of Eq. (18.2) is the sum of the external forces on the system. Denoting it by $\Sigma\mathbf{F}$, we conclude that the sum of the external forces on the system equals the rate of change of its total linear momentum:

$$\Sigma\mathbf{F} = \frac{d}{dt} \sum_i m_i \mathbf{v}_i. \tag{18.3}$$

Let m be the sum of the masses of the particles:

$$m = \sum_i m_i.$$

The position of the center of mass of the system is

$$\mathbf{r} = \frac{\sum_i m_i \mathbf{r}_i}{m}, \tag{18.4}$$

so the velocity of the center of mass is

$$\mathbf{v} = \frac{d\mathbf{r}}{dt} = \frac{\sum_i m_i \mathbf{v}_i}{m}.$$

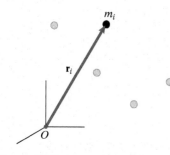

Figure 18.1
A system of particles. The vector $\mathbf{r}_i$ is the position vector of the ith particle.

By using this expression, we can write Eq. (18.3) as

$$\Sigma \mathbf{F} = \frac{d}{dt}(m\mathbf{v}).$$

The total external force on a system of particles thus equals the rate of change of the product of the total mass of the system and the velocity of its center of mass. Since any object or collection of objects, including a rigid body, can be modeled as a system of particles, this result is one of the most general and elegant in mechanics. Furthermore, if the total mass m is constant, we obtain

$$\Sigma \mathbf{F} = m\mathbf{a},$$

where $\mathbf{a} = d\mathbf{v}/dt$ is the acceleration of the center of mass. We see that the total external force equals the product of the total mass and the acceleration of the center of mass.

Moment–Angular Momentum Principles

We now obtain relations between the sum of the moments due to the forces acting on a system of particles and the rate of change of the total angular momentum of the system. In Fig. 18.2, $\mathbf{r}_i$ is the position vector of the ith particle of a system of particles, $\mathbf{r}$ is the position vector of the center of mass of the system, and $\mathbf{R}_i$ is the position vector of the ith particle relative to the center of mass. These vectors are related by

$$\mathbf{r}_i = \mathbf{r} + \mathbf{R}_i. \tag{18.5}$$

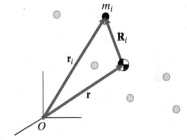

Figure 18.2
The vector $\mathbf{R}_i$ is the position vector of the ith particle relative to the center of mass.

We take the cross product of Newton's second law for the ith particle, Eq. (18.1), with the position vector $\mathbf{r}_i$ and sum from $i = 1$ to N, writing the resulting equation in the form

$$\sum_i \sum_j \mathbf{r}_i \times \mathbf{f}_{ij} + \sum_i \mathbf{r}_i \times \mathbf{f}_i^E = \frac{d}{dt} \sum_i \mathbf{r}_i \times m_i \mathbf{v}_i. \tag{18.6}$$

The first term on the left side of this equation is the sum of the moments about O due to the forces exerted on the particles by the other particles of the system. This term vanishes if we assume that the mutual forces exerted by each pair of particles are not only equal and opposite, but directed along the straight line between the particles. For example, consider particles 1 and 2 in Fig. 18.3. If the forces the particles exert on each other are directed along the line between the particles, we can write the moment about O as

$$\mathbf{r}_1 \times \mathbf{f}_{12} + \mathbf{r}_1 \times \mathbf{f}_{21} = \mathbf{r}_1 \times (\mathbf{f}_{12} + \mathbf{f}_{21}) = \mathbf{0}. \tag{18.7}$$

The second term on the left side of Eq. (18.6) is the sum of the moments about O due to external forces, which we denote by $\Sigma \mathbf{M}_O$. We write Eq. (18.6) as

$$\Sigma \mathbf{M}_O = \frac{d\mathbf{H}_O}{dt}, \tag{18.8}$$

where

$$\mathbf{H}_O = \sum_i \mathbf{r}_i \times m_i \mathbf{v}_i \tag{18.9}$$

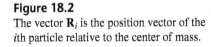
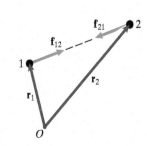

Figure 18.3
Particles 1 and 2 and the forces they exert on each other. If the forces act along the line between the particles, their total moment about O is zero.

is the total angular momentum about O. The sum of the moments about O is equal to the rate of change of the total angular momentum about O. By using Eqs. (18.4) and (18.5), we can write Eq. (18.8) as

$$\Sigma \mathbf{M}_O = \frac{d}{dt}(\mathbf{r} \times m\mathbf{v} + \mathbf{H}), \tag{18.10}$$

where

$$\mathbf{H} = \sum_i \mathbf{R}_i \times m_i \frac{d\mathbf{R}_i}{dt} \tag{18.11}$$

is the total angular momentum of the system about the center of mass.

We also need to determine the relation between the sum of the moments about the center of mass, which we denote by $\Sigma \mathbf{M}$, and $\mathbf{H}$. We can obtain this relation by letting the fixed point O be coincident with the center of mass at the present instant. In that case, $\Sigma \mathbf{M}_O = \Sigma \mathbf{M}$ and $\mathbf{r} = \mathbf{0}$, and we see from Eq. (18.10) that

$$\Sigma \mathbf{M} = \frac{d\mathbf{H}}{dt}. \tag{18.12}$$

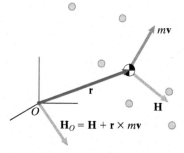

Figure 18.4
The angular momentum about O equals the sum of the angular momentum about the center of mass and the angular momentum about O due to the velocity of the center of mass.

The sum of the moments about the center of mass is equal to the rate of change of the total angular momentum about the center of mass. The angular momenta about point O and about the center of mass are related by (Fig. 18.4)

$$\mathbf{H}_O = \mathbf{H} + \mathbf{r} \times m\mathbf{v}. \tag{18.13}$$

RESULTS

The equations of motion for a rigid body can be derived from principles governing the motion of a system of particles. This section summarizes these principles.

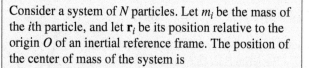

Consider a system of N particles. Let m_i be the mass of the ith particle, and let $\mathbf{r}_i$ be its position relative to the origin O of an inertial reference frame. The position of the center of mass of the system is

$$\mathbf{r} = \frac{\sum_i m_i \mathbf{r}_i}{m}, \tag{18.4}$$

where m is the total mass of the system and $\sum_i$ denotes the sum from $i = 1$ to N. The vector $\mathbf{R}_i$ is the position of the ith particle relative to the center of mass.

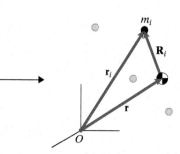

Force-Linear Momentum Principle

> The sum of the external forces on a system of particles (which can represent virtually any object, including a rigid body) is equal to the rate of change of its linear momentum. The linear momentum is the product of the total mass m and the velocity $\mathbf{v} = d\mathbf{r}/dt$ of the center of mass. The term $\mathbf{a} = d\mathbf{v}/dt$ is the acceleration of the center of mass.

$$\Sigma\mathbf{F} = \frac{d}{dt}(m\mathbf{v}).$$

If the total mass is constant,

$$\Sigma\mathbf{F} = m\frac{d\mathbf{a}}{dt}.$$

Moment-Angular Momentum Principles

> The sum of the moments about the origin O due to the forces acting on a system of particles is equal to the rate of change of the total angular momentum about O.

$$\Sigma\mathbf{M}_O = \frac{d\mathbf{H}_O}{dt}, \tag{18.8}$$

where the total angular momentum about O is

$$\mathbf{H}_O = \sum_i \mathbf{r}_i \times m_i\mathbf{v}_i. \tag{18.9}$$

> The sum of the moments about the center of mass due to the forces acting on a system of particles is equal to the rate of change of the total angular momentum about the center of mass.

$$\Sigma\mathbf{M} = \frac{d\mathbf{H}}{dt}, \tag{18.12}$$

where the total angular momentum about the center of mass is

$$\mathbf{H} = \sum_i \mathbf{R}_i \times m_i\frac{d\mathbf{R}_i}{dt} \tag{18.11}$$

> Relationship between the total angular momentum $\mathbf{H}_O$ about the origin and the total angular momentum $\mathbf{H}$ about the center of mass.

$$\mathbf{H}_O = \mathbf{H} + \mathbf{r} \times m\mathbf{v}. \tag{18.13}$$

18.2 The Planar Equations of Motion

BACKGROUND

We now derive the equations of motion for a rigid body in planar motion. We have shown that the total external force on any object equals the product of the mass of the object and the acceleration of its center of mass:

$$\Sigma\mathbf{F} = m\mathbf{a}. \tag{18.14}$$

This equation, which we refer to as Newton's second law, describes the motion of the center of mass of a rigid body. To derive the equations of angular motion, we consider first rotation about a fixed axis and then general planar motion.

Rotation about a Fixed Axis

Let O be a point that is stationary relative to an inertial reference frame, and let L_O be a nonrotating line through O. Suppose that a rigid body rotates about L_O. In terms of a coordinate system with its z axis aligned with L_O (Fig. 18.5a), we

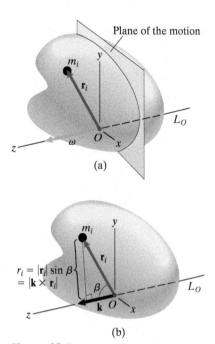

Figure 18.5
(a) A coordinate system with the z axis aligned with the axis of rotation, L_O.
(b) The magnitude of $\mathbf{k} \times \mathbf{r}_i$ is the perpendicular distance from the axis of rotation to m_i.

can express the rigid body's angular velocity vector as $\boldsymbol{\omega} = \omega\mathbf{k}$, and the velocity of the ith particle of the rigid body is

$$\frac{d\mathbf{r}_i}{dt} = \boldsymbol{\omega} \times \mathbf{r}_i = \omega\mathbf{k} \times \mathbf{r}_i.$$

Let $\Sigma M_O = \Sigma\mathbf{M}_O \cdot \mathbf{k}$ denote the sum of the moments about L_O. By taking the dot product of Eq. (18.8) with $\mathbf{k}$, we obtain

$$\Sigma M_O = \frac{dH_O}{dt}, \qquad (18.15)$$

where

$$H_O = \mathbf{H}_O \cdot \mathbf{k} = \sum_i [\mathbf{r}_i \times m_i(\omega\mathbf{k} \times \mathbf{r}_i)] \cdot \mathbf{k}$$

is the angular momentum about L_O. Using the identity $\mathbf{U} \cdot (\mathbf{V} \times \mathbf{W}) = (\mathbf{U} \times \mathbf{V}) \cdot \mathbf{W}$, we can write this expression as

$$H_O = \sum_i m_i(\mathbf{k} \times \mathbf{r}_i) \cdot (\mathbf{k} \times \mathbf{r}_i)\omega = \sum_i m_i|\mathbf{k} \times \mathbf{r}_i|^2\omega. \quad (18.16)$$

In Fig. 18.5b, we show that $|\mathbf{k} \times \mathbf{r}_i|$ is the perpendicular distance from L_O to the ith particle, which we denote by r_i. Using the definition of the moment of inertia of the rigid body about L_O,

$$I_O = \sum_i m_i r_i^2,$$

we can write Eq. (18.16) as

$$H_O = I_O\omega.$$

Substituting this expression into Eq. (18.15), we obtain the equation of angular motion for a rigid body rotating about a fixed axis. The sum of the moments about the fixed axis equals the product of the moment of inertia about the fixed axis and the angular acceleration:

$$\Sigma M_O = I_O\alpha. \qquad (18.17)$$

General Planar Motion

Figure 18.6a shows the plane of the motion of a rigid body in general planar motion. Point O is a fixed point contained in the plane, L_O is the line through O that is perpendicular to the plane, and L is the line parallel to L_O that passes through the center of mass of the rigid body. In terms of the coordinate system shown, we can express the rigid body's angular velocity vector as $\boldsymbol{\omega} = \omega\mathbf{k}$, and the velocity of the ith particle of the rigid body relative to the center of mass is

$$\frac{d\mathbf{R}_i}{dt} = \boldsymbol{\omega} \times \mathbf{R}_i = \omega\mathbf{k} \times \mathbf{R}_i.$$

By taking the dot product of Eq. (18.10) with $\mathbf{k}$, we obtain

$$\Sigma M_O = \frac{d}{dt}[(\mathbf{r} \times m\mathbf{v}) \cdot \mathbf{k} + H], \qquad (18.18)$$

where

$$H = \mathbf{H} \cdot \mathbf{k} = \sum_i [\mathbf{R}_i \times m_i(\omega\mathbf{k} \times \mathbf{R}_i)] \cdot \mathbf{k}$$

is the angular momentum about L. Using the identity $\mathbf{U} \cdot (\mathbf{V} \times \mathbf{W}) = (\mathbf{U} \times \mathbf{V}) \cdot \mathbf{W}$, we can write this equation for H as

$$H = \sum_i m_i(\mathbf{k} \times \mathbf{R}_i) \cdot (\mathbf{k} \times \mathbf{R}_i)\omega = \sum_i m_i|\mathbf{k} \times \mathbf{R}_i|^2\omega.$$

The term $|\mathbf{k} \times \mathbf{R}_i| = r_i$ is the perpendicular distance from L to the ith particle (Fig. 18.6b). In terms of the moment of inertia of the rigid body about L,

$$I = \sum_i m_i r_i^2,$$

the rigid body's angular momentum about L is

$$H = I\omega.$$

Substituting this expression into Eq. (18.18), we obtain

$$\Sigma M_O = \frac{d}{dt}[(\mathbf{r} \times m\mathbf{v}) \cdot \mathbf{k} + I\omega] = (\mathbf{r} \times m\mathbf{a}) \cdot \mathbf{k} + I\alpha. \quad (18.19)$$

With this equation we can obtain the relation between the sum of the moments about L, which we denote by ΣM, and the angular acceleration. If we let the fixed axis L_O be coincident with L at the present instant, then $\Sigma M_O = \Sigma M$ and $\mathbf{r} = \mathbf{0}$, and from Eq. (18.19) we obtain the equation of angular motion for a rigid body in general planar motion. The sum of the moments about the center of mass equals the product of the moment of inertia about the center of mass and the angular acceleration:

$$\Sigma M = I\alpha. \quad (18.20)$$

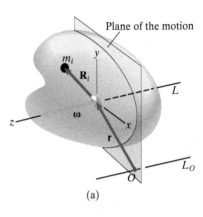

(a)

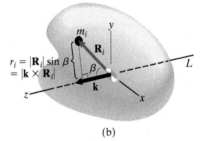

(b)

Figure 18.6
(a) A coordinate system with the z axis aligned with L.
(b) The magnitude of $\mathbf{k} \times \mathbf{R}_i$ is the perpendicular distance from L to m_i.

RESULTS

The sum of the external forces on any object equals the product of its mass and the acceleration of its center of mass relative to an inertial reference frame.

$$\Sigma\mathbf{F} = m\mathbf{a}. \quad (18.14)$$

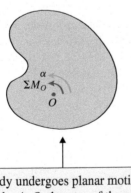

If a rigid body undergoes planar motion about a *fixed axis* O, the sum of the moments about O due to external forces and couples equals the product of the moment of inertia of the rigid body about O and the angular acceleration.

$$\Sigma M_O = I_O\alpha. \quad (18.17)$$

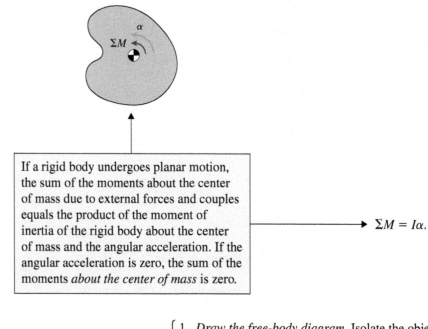

If a rigid body undergoes planar motion, the sum of the moments about the center of mass due to external forces and couples equals the product of the moment of inertia of the rigid body about the center of mass and the angular acceleration. If the angular acceleration is zero, the sum of the moments *about the center of mass* is zero.

$$\Sigma M = I\alpha. \quad (18.20)$$

The equations of planar motion can be used to obtain information about an object's motion, or determine unknown forces or couples acting on the object, or both. Doing so typically requires three steps.

1. *Draw the free-body diagram.* Isolate the object and identify the external forces and couples acting on it.
2. *Apply the equations of motion.* Write the equations of motion, choosing a suitable coordinate system for applying Newton's second law. For example, if the center of mass moves in a circular path, it may be advantageous to use tangential and normal components.
3. *Determine kinematic relationships.* If necessary, supplement the equations of motion with relationships between the acceleration of the center of mass and the angular acceleration of the object.

Active Example 18.1   **Translating Object** (▶ *Related Problem 18.4*)

The airplane weighs 830,000 lb, and the total thrust of its engines during its takeoff roll is $T = 208,000$ lb. Determine the airplane's acceleration and the normal forces exerted on its wheels by the runway at A and B during takeoff. Neglect the horizontal forces exerted on its wheels.

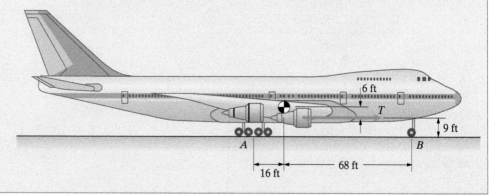

Strategy

We must draw the free-body diagram of the airplane to identify the external forces acting on it. We can apply Newton's second law to determine the horizontal acceleration. Then we will use two conditions to determine the normal forces at A and B. (1) The airplane's acceleration in the vertical direction is zero, so the sum of the forces in the vertical direction must equal zero. (2) The airplane's angular acceleration is zero during its takeoff roll, so the equation of angular motion states that the sum of the moments *about the center of mass* due to the forces on the airplane is zero.

Solution

Draw the free-body diagram of the airplane.

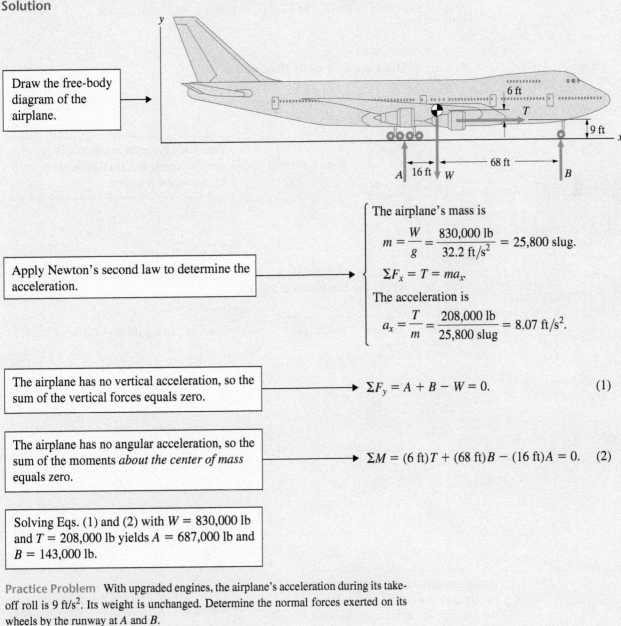

The airplane's mass is

$$m = \frac{W}{g} = \frac{830{,}000 \text{ lb}}{32.2 \text{ ft/s}^2} = 25{,}800 \text{ slug}.$$

Apply Newton's second law to determine the acceleration.

$$\Sigma F_x = T = ma_x.$$

The acceleration is

$$a_x = \frac{T}{m} = \frac{208{,}000 \text{ lb}}{25{,}800 \text{ slug}} = 8.07 \text{ ft/s}^2.$$

The airplane has no vertical acceleration, so the sum of the vertical forces equals zero.

$$\Sigma F_y = A + B - W = 0. \qquad (1)$$

The airplane has no angular acceleration, so the sum of the moments *about the center of mass* equals zero.

$$\Sigma M = (6 \text{ ft})T + (68 \text{ ft})B - (16 \text{ ft})A = 0. \quad (2)$$

Solving Eqs. (1) and (2) with $W = 830{,}000$ lb and $T = 208{,}000$ lb yields $A = 687{,}000$ lb and $B = 143{,}000$ lb.

Practice Problem With upgraded engines, the airplane's acceleration during its take-off roll is 9 ft/s². Its weight is unchanged. Determine the normal forces exerted on its wheels by the runway at A and B.

Answer: $A = 688{,}000$ lb, $B = 142{,}000$ lb.

Active Example 18.2 Rolling Motion (▶ *Related Problems 18.32, 18.33*)

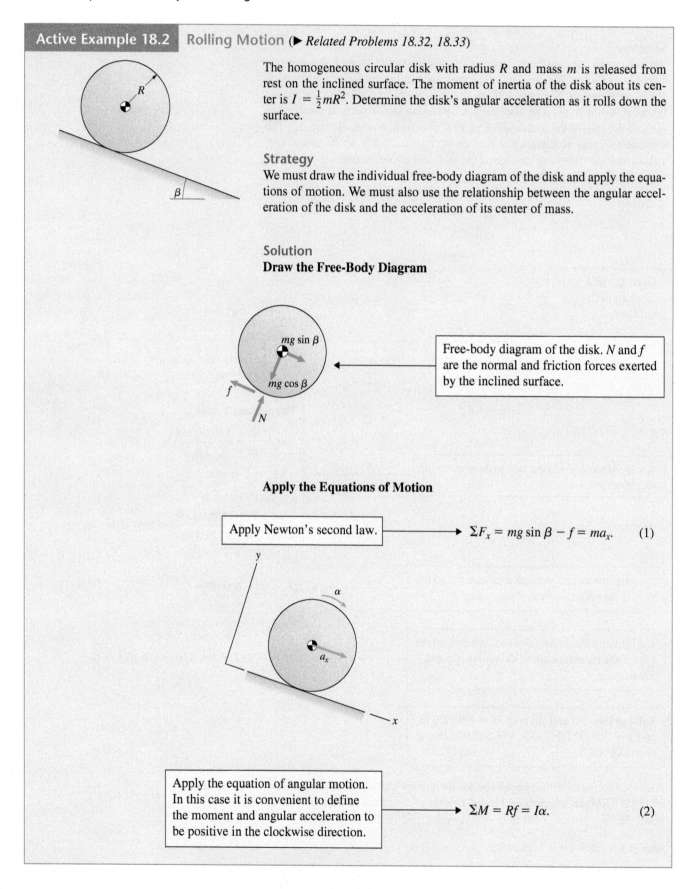

The homogeneous circular disk with radius R and mass m is released from rest on the inclined surface. The moment of inertia of the disk about its center is $I = \frac{1}{2}mR^2$. Determine the disk's angular acceleration as it rolls down the surface.

Strategy

We must draw the individual free-body diagram of the disk and apply the equations of motion. We must also use the relationship between the angular acceleration of the disk and the acceleration of its center of mass.

Solution
Draw the Free-Body Diagram

Free-body diagram of the disk. N and f are the normal and friction forces exerted by the inclined surface.

Apply the Equations of Motion

Apply Newton's second law.

$$\Sigma F_x = mg \sin \beta - f = ma_x. \quad (1)$$

Apply the equation of angular motion. In this case it is convenient to define the moment and angular acceleration to be positive in the clockwise direction.

$$\Sigma M = Rf = I\alpha. \quad (2)$$

Determine Kinematic Relationships

| Relationship between the acceleration of the center of the disk and the angular acceleration in rolling motion. | $a_x = R\alpha.$ (3) |

| Solving Eqs. (1)–(3) for the three variables a_x, α, and f and substituting $I = \frac{1}{2}mR^2$ yields the angular acceleration. | $\alpha = \dfrac{2g}{3R}\sin\beta.$ |

Practice Problem It was assumed that the disk would roll, and not slip, when it was released from rest on the inclined surface. Let μ_s be the coefficient of static friction between the disk and the surface. What is the largest value of the angle β for which the disk will roll instead of slipping?

Answer: $\beta = \arctan(3\mu_s)$.

Active Example 18.3   **Bar in General Planar Motion** (▶ *Related Problem 18.45*)

The slender bar of mass m slides on the smooth floor and wall. It has counterclockwise angular velocity ω at the instant shown. What is the bar's angular acceleration?

Strategy
After drawing the free-body diagram of the bar, we will apply Newton's second law and the equation of angular motion. This will result in three equations in terms of the two components of acceleration of the bar's center of mass, the angular acceleration, and the unknown forces exerted on the bar by the floor and wall. We must obtain two kinematic relations between the acceleration of the center of mass and the angular acceleration to complete the solution.

Solution
Draw the Free-Body Diagram

Free-body diagram of the bar. N and P are the normal forces exerted by the floor and wall.

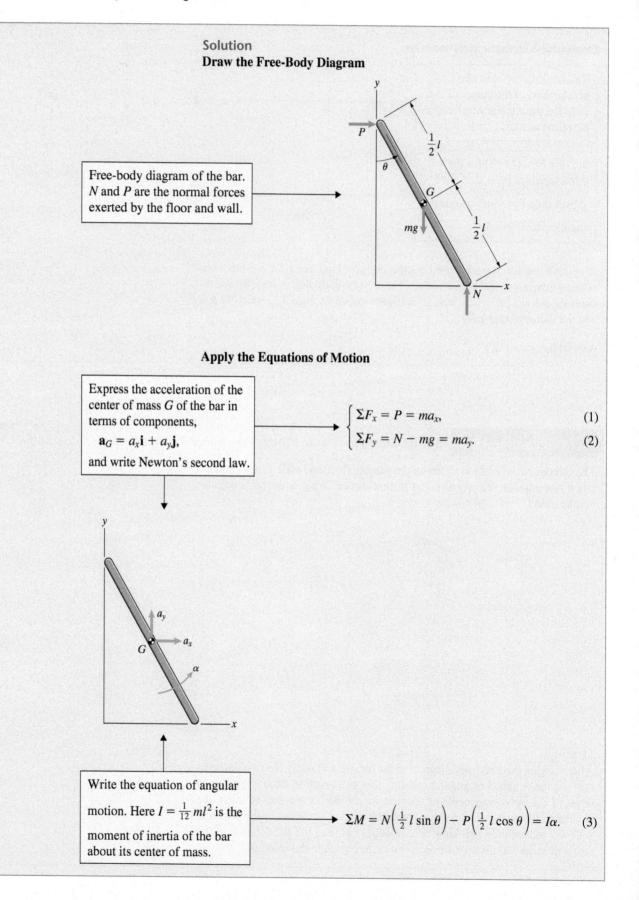

Apply the Equations of Motion

Express the acceleration of the center of mass G of the bar in terms of components,

$$\mathbf{a}_G = a_x \mathbf{i} + a_y \mathbf{j},$$

and write Newton's second law.

$$\begin{cases} \Sigma F_x = P = ma_x, & (1) \\ \Sigma F_y = N - mg = ma_y. & (2) \end{cases}$$

Write the equation of angular motion. Here $I = \frac{1}{12} ml^2$ is the moment of inertia of the bar about its center of mass.

$$\Sigma M = N\left(\frac{1}{2}l\sin\theta\right) - P\left(\frac{1}{2}l\cos\theta\right) = I\alpha. \qquad (3)$$

Determine Kinematic Relationships

Express the acceleration of the center of mass in terms of the acceleration of the bottom end A of the bar. Equating $\mathbf{j}$ components results in a relation between a_y and the angular acceleration α.

$$\mathbf{a}_G = \mathbf{a}_A + \boldsymbol{\alpha} \times \mathbf{r}_{G/A} - \omega^2 \mathbf{r}_{G/A}:$$

$$a_x\mathbf{i} + a_y\mathbf{j} = a_A\mathbf{i} + \begin{vmatrix} \mathbf{i} & \mathbf{j} & \mathbf{k} \\ 0 & 0 & \alpha \\ -\frac{1}{2}l\sin\theta & \frac{1}{2}l\cos\theta & 0 \end{vmatrix}$$

$$-\omega^2\left(-\tfrac{1}{2}l\sin\theta\,\mathbf{i} + \tfrac{1}{2}l\cos\theta\,\mathbf{j}\right).$$

Equating the $\mathbf{j}$ components yields

$$a_y = -\tfrac{1}{2}l(\alpha\sin\theta + \omega^2\cos\theta). \tag{4}$$

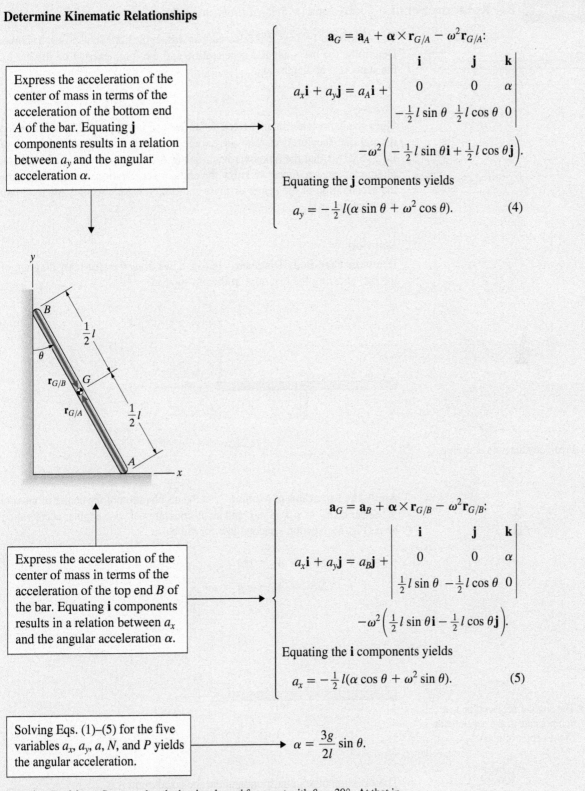

Express the acceleration of the center of mass in terms of the acceleration of the top end B of the bar. Equating $\mathbf{i}$ components results in a relation between a_x and the angular acceleration α.

$$\mathbf{a}_G = \mathbf{a}_B + \boldsymbol{\alpha} \times \mathbf{r}_{G/B} - \omega^2 \mathbf{r}_{G/B}:$$

$$a_x\mathbf{i} + a_y\mathbf{j} = a_B\mathbf{j} + \begin{vmatrix} \mathbf{i} & \mathbf{j} & \mathbf{k} \\ 0 & 0 & \alpha \\ \frac{1}{2}l\sin\theta & -\frac{1}{2}l\cos\theta & 0 \end{vmatrix}$$

$$-\omega^2\left(\tfrac{1}{2}l\sin\theta\,\mathbf{i} - \tfrac{1}{2}l\cos\theta\,\mathbf{j}\right).$$

Equating the $\mathbf{i}$ components yields

$$a_x = -\tfrac{1}{2}l(\alpha\cos\theta + \omega^2\sin\theta). \tag{5}$$

Solving Eqs. (1)–(5) for the five variables a_x, a_y, α, N, and P yields the angular acceleration.

$$\alpha = \frac{3g}{2l}\sin\theta.$$

Practice Problem Suppose that the bar is released from rest with $\theta = 30°$. At that instant, what normal forces are exerted on the bar by the floor and wall?

Answer: $N = 0.813mg$, $P = 0.325mg$.

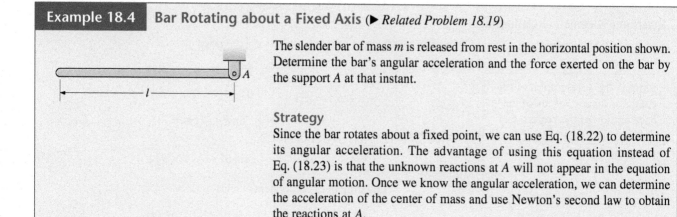

| Example 18.4 | Bar Rotating about a Fixed Axis (▶ *Related Problem 18.19*) |

The slender bar of mass m is released from rest in the horizontal position shown. Determine the bar's angular acceleration and the force exerted on the bar by the support A at that instant.

Strategy

Since the bar rotates about a fixed point, we can use Eq. (18.22) to determine its angular acceleration. The advantage of using this equation instead of Eq. (18.23) is that the unknown reactions at A will not appear in the equation of angular motion. Once we know the angular acceleration, we can determine the acceleration of the center of mass and use Newton's second law to obtain the reactions at A.

Solution

Draw the Free-Body Diagram In Fig. a, we draw the free-body diagram of the bar, showing the reactions at the pin support.

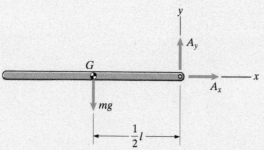

(a) Free-body diagram of the bar.

Apply the Equations of Motion Let the acceleration of the center of mass G of the bar be $\mathbf{a}_G = a_x\mathbf{i} + a_y\mathbf{j}$, and let its counterclockwise angular acceleration be α (Fig. b). Newton's second law for the bar is

$$\Sigma F_x = A_x = ma_x,$$
$$\Sigma F_y = A_y - mg = ma_y.$$

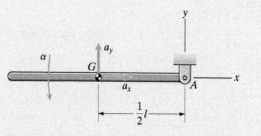

(b) The angular acceleration and components of the acceleration of the center of mass.

The equation of angular motion about the fixed point A is

$$\Sigma M_A = \left(\tfrac{1}{2}l\right)mg = I_A\alpha. \tag{1}$$

The moment of inertia of a slender bar about its center of mass is $I = \frac{1}{12}ml^2$. (See Appendix C.) From the parallel-axis theorem, the moment of inertia of the bar about A is

$$I_A = I + d^2m = \tfrac{1}{12}ml^2 + \left(\tfrac{1}{2}l\right)^2 m = \tfrac{1}{3}ml^2.$$

Substituting this expression into Eq. (1), we obtain the angular acceleration:

$$\alpha = \frac{\frac{1}{2}mgl}{\frac{1}{3}ml^2} = \frac{3}{2}\frac{g}{l}.$$

Determine Kinematic Relationships   To determine the reactions A_x and A_y, we need to determine the acceleration components a_x and a_y. We can do so by expressing the acceleration of G in terms of the acceleration of A:

$$\mathbf{a}_G = \mathbf{a}_A + \boldsymbol{\alpha} \times \mathbf{r}_{G/A} - \omega^2 \mathbf{r}_{G/A}.$$

At the instant the bar is released, its angular velocity $\omega = 0$. Also, $\mathbf{a}_A = \mathbf{0}$, so we obtain

$$\mathbf{a}_G = a_x \mathbf{i} + a_y \mathbf{j} = (\alpha \mathbf{k}) \times \left(-\tfrac{1}{2}l\mathbf{i}\right) = -\tfrac{1}{2}l\alpha\mathbf{j}.$$

Equating $\mathbf{i}$ and $\mathbf{j}$ components, we get

$$a_x = 0,$$
$$a_y = -\tfrac{1}{2}l\alpha = -\tfrac{3}{4}g.$$

Substituting these acceleration components into Newton's second law, we find that the reactions at A at the instant the bar is released are

$$A_x = 0,$$
$$A_y = mg + m\left(-\tfrac{3}{4}g\right) = \tfrac{1}{4}mg.$$

Critical Thinking
We could have determined the kinematic relationship between the bar's angular acceleration and the acceleration of the center of mass G in a less formal way. Because G describes a circular path about A, we know that the tangential component of its acceleration equals the product of the radial distance from A to G and the angular acceleration of the bar. Due to the directions in which we defined a_y and α to be positive, $a_y = -(l/2)\alpha$. The normal component of the acceleration of G is equal to the square of its velocity divided by the radius of its circular path. The velocity equals zero at the instant the bar is released, so $a_x = 0$.

Example 18.5	**Connected Rigid Bodies** (▶ *Related Problems 18.53, 18.54*)

The slender bar has mass m and is pinned at A to a metal block of mass m_B that rests on a smooth, level surface. The system is released from rest in the position shown. What is the bar's angular acceleration at the instant of release?

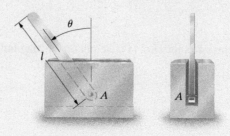

Strategy

We must draw free-body diagrams of the bar and the block and apply the equations of motion to them individually. To complete the solution, we must also relate the acceleration of the bar's center of mass and its angular acceleration to the acceleration of the block.

Solution

Draw the Free-Body Diagrams   We draw the free-body diagrams of the bar and block in Fig. a. Notice the opposite forces they exert on each other where they are pinned together.

Apply the Equations of Motion   Writing the acceleration of the center of mass of the bar as $\mathbf{a}_G = a_x\mathbf{i} + a_y\mathbf{j}$ (Fig. b), we have, from Newton's second law,

$$\Sigma F_x = A_x = ma_x$$

and

$$\Sigma F_y = A_y - mg = ma_y.$$

Letting α be the counterclockwise angular acceleration of the bar (Fig. b), its equation of angular motion is

$$\Sigma M = A_x\left(\tfrac{1}{2}l\cos\theta\right) + A_y\left(\tfrac{1}{2}l\sin\theta\right) = I\alpha.$$

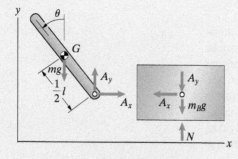

(a) Free-body diagrams of the bar and block.

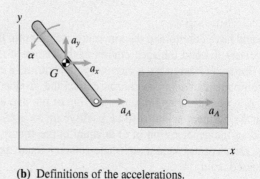

(b) Definitions of the accelerations.

We express the block's acceleration as $a_A\mathbf{i}$ (Fig. b) and write Newton's second law for the block:

$$\Sigma F_x = -A_x = m_B a_A,$$

$$\Sigma F_y = N - A_y - m_B g = 0.$$

Determine Kinematic Relationships To relate the bar's motion to that of the block, we express the acceleration of the bar's center of mass in terms of the acceleration of point A (Figs. b and c):

$$\mathbf{a}_G = \mathbf{a}_A + \boldsymbol{\alpha} \times \mathbf{r}_{G/A} - \omega^2 \mathbf{r}_{G/A},$$

$$a_x\mathbf{i} + a_y\mathbf{j} = a_A\mathbf{i} + \begin{vmatrix} \mathbf{i} & \mathbf{j} & \mathbf{k} \\ 0 & 0 & \alpha \\ -\frac{1}{2}l\sin\theta & \frac{1}{2}l\cos\theta & 0 \end{vmatrix} - \mathbf{0}.$$

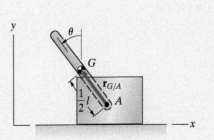

(c) Position vector of G relative to A.

Equating $\mathbf{i}$ and $\mathbf{j}$ components, we obtain

$$a_x = a_A - \tfrac{1}{2}l\alpha\cos\theta$$

and

$$a_y = -\tfrac{1}{2}l\alpha\sin\theta.$$

We have five equations of motion and two kinematic relations in terms of seven unknowns: A_x, A_y, N, a_x, a_y, α, and a_A. Solving them for the angular acceleration and using the relation $I = \frac{1}{12}ml^2$ for the bar's moment of inertia, we obtain

$$\alpha = \frac{\frac{3}{2}(g/l)\sin\theta}{1 - \frac{3}{4}\left[m/(m+m_B)\right]\cos^2\theta}.$$

Critical Thinking

This example is a simple case of a very important type of problem. The approach we used is applicable to analyzing planar motions of a broad class of machines consisting of interconnected moving parts. The free-body diagrams of the individual parts are first drawn, including the forces and couples the parts exert on each other. The equations of motion are written for each part. The solution is completed by determining kinematic relationships. In this example, the block is constrained to move horizontally. The block and bar have the same acceleration at the pin A, and the acceleration of the center of mass G of the bar is related to the bar's angular acceleration and the acceleration of point A because the bar is pinned at A. As this example illustrates, determining kinematic relationships is usually the most challenging part of the solution.

Problems

18.1 A horizontal force $F = 30$ lb is applied to the 230-lb refrigerator as shown. Friction is negligible.

(a) What is the magnitude of the refrigerator's acceleration?

(b) What normal forces are exerted on the refrigerator by the floor at A and B?

18.2 Solve Problem 18.1 if the coefficient of kinetic friction at A and B is $\mu_k = 0.1$.

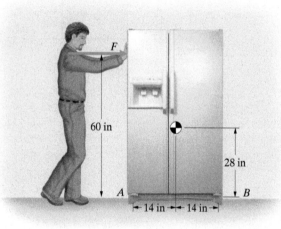

Problems 18.1/18.2

18.3 As the 2800-lb airplane begins its takeoff run at $t = 0$, its propeller exerts a horizontal force $T = 1000$ lb. Neglect horizontal forces exerted on the wheels by the runway.

(a) What distance has the airplane moved at $t = 2$ s?

(b) What normal forces are exerted on the tires at A and B?

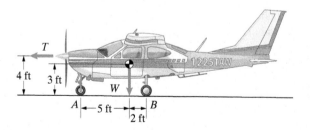

Problem 18.3

▶ **18.4** The Boeing 747 begins its takeoff run at time $t = 0$. The normal forces exerted on its tires at A and B are $N_A = 175$ kN and $N_B = 2800$ kN. If you assume that these forces are constant and neglect horizontal forces other than the thrust T, how fast is the airplane moving at $t = 4$ s? (See Active Example 18.1.)

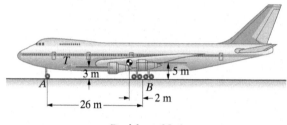

Problem 18.4

18.5 The crane moves to the right with constant acceleration, and the 800-kg load moves without swinging.

(a) What is the acceleration of the crane and load?

(b) What are the tensions in the cables attached at A and B?

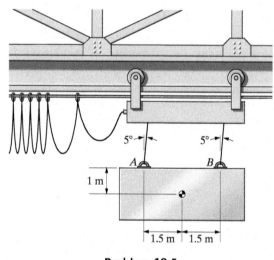

Problem 18.5

18.6 The total weight of the go-cart and driver is 240 lb. The location of their combined center of mass is shown. The rear drive wheels together exert a 24-lb horizontal force on the track. Neglect the horizontal forces exerted on the front wheels.

(a) What is the magnitude of the go-cart's acceleration?

(b) What normal forces are exerted on the tires at A and B?

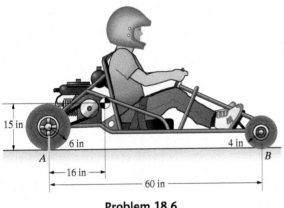

Problem 18.6

18.7 The total weight of the bicycle and rider is 160 lb. The location of their combined center of mass is shown. The dimensions shown are $b = 21$ in, $c = 16$ in, and $h = 38$ in. What is the largest acceleration the bicycle can have without the front wheel leaving the ground? Neglect the horizontal force exerted on the front wheel by the road.

Strategy: You want to determine the value of the acceleration that causes the normal force exerted on the front wheel by the road to equal zero.

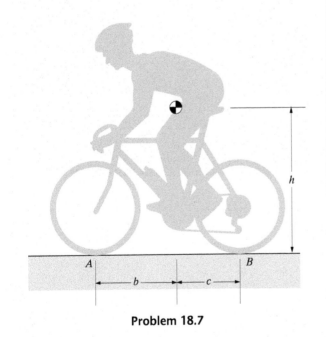

Problem 18.7

18.8 The moment of inertia of the disk about O is $I = 20$ kg-m^2. At $t = 0$, the stationary disk is subjected to a constant 50 N-m torque.

(a) What is the magnitude of the resulting angular acceleration of the disk?

(b) How fast is the disk rotating (in rpm) at $t = 4$ s?

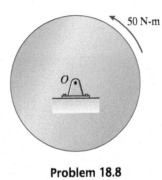

Problem 18.8

18.9 The 10-lb bar is on a smooth horizontal table. The figure shows the bar viewed from above. Its moment of inertia about the center of mass is $I = 0.8$ slug-ft^2. The bar is stationary when the force $F = 5$ lb is applied in the direction parallel to the y axis. At that instant, determine (a) the acceleration of the center of mass, and (b) the acceleration of point A.

18.10 The 10-lb bar is on a smooth horizontal table. The figure shows the bar viewed from above. Its moment of inertia about the center of mass is $I = 0.8$ slug-ft^2. The bar is stationary when the force $F = 5$ lb is applied in the direction parallel to the y axis. At that instant, determine the acceleration of point B.

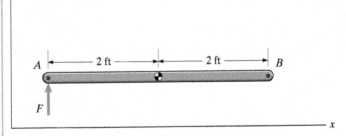

Problems 18.9/18.10

18.11 The moment of inertia of the astronaut and maneuvering unit about the axis through their center of mass perpendicular to the page is $I = 40$ kg-m^2. A thruster can exert a force $T = 10$ N. For safety, the control system of his maneuvering unit will not allow his angular velocity to exceed 15° per second. If he is initially not rotating, and at $t = 0$ he activates the thruster until he is rotating at 15° per second, through how many degrees has he rotated at $t = 10$ s?

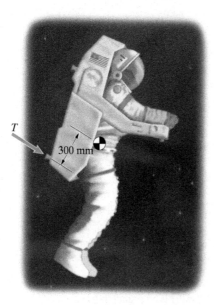

300 mm

Problem 18.11

18.12 The moment of inertia of the helicopter's rotor is 420 slug-ft^2. The rotor starts from rest. At $t = 0$, the pilot begins advancing the throttle so that the torque exerted on the rotor by the engine (in ft-lb) is given as a function of time in seconds by $T = 200t$.

(a) How long does it take the rotor to turn ten revolutions?

(b) What is the rotor's angular velocity (in rpm) when it has turned ten revolutions?

Problem 18.12

18.13 The moments of inertia of the pulleys are $I_A = 0.0025$ kg-m^2, $I_B = 0.045$ kg-m^2, and $I_C = 0.036$ kg-m^2. A 5 N-m counterclockwise couple is applied to pulley A. Determine the resulting counterclockwise angular accelerations of the three pulleys.

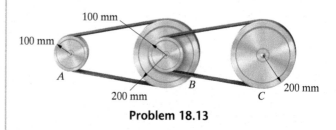

100 mm

100 mm

A

200 mm

B

C

200 mm

Problem 18.13

18.14 The moment of inertia of the wind-tunnel fan is 225 kg-m^2. The fan starts from rest. The torque exerted on it by the engine is given as a function of the angular velocity of the fan by $T = 140 - 0.02\omega^2$ N-m.

(a) When the fan has turned 620 revolutions, what is its angular velocity in rpm (revolutions per minute)?

(b) What maximum angular velocity in rpm does the fan attain?

   Strategy: By writing the equation of angular motion, determine the angular acceleration of the fan in terms of its angular velocity. Then use the chain rule:

$$\alpha = \frac{d\omega}{dt} = \frac{d\omega}{d\theta}\frac{d\theta}{dt} = \frac{d\omega}{d\theta}\omega.$$

Problem 18.14

18.15 The moment of inertia of the pulley about its axis is $I = 0.005$ kg-m². If the 1-kg mass A is released from rest, how far does it fall in 0.5 s?

Strategy: Draw individual free-body diagrams of the pulley and the mass.

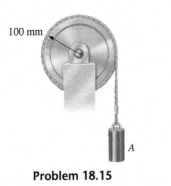

100 mm

A

Problem 18.15

18.16 The radius of the pulley is 125 mm and the moment of inertia about its axis is $I = 0.05$ kg-m². If the system is released from rest, how far does the 20-kg mass fall in 0.5 s? What is the tension in the rope between the 20-kg mass and the pulley?

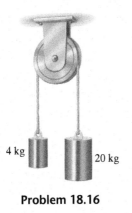

4 kg

20 kg

Problem 18.16

18.17 The moment of inertia of the pulley is 0.4 slug-ft². The coefficient of kinetic friction between the 5-lb weight and the horizontal surface is $\mu_k = 0.2$. Determine the magnitude of the acceleration of the 5-lb weight in each case.

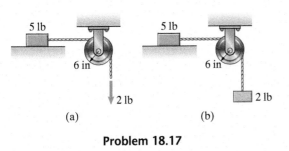

5 lb

6 in

2 lb

(a)

5 lb

6 in

2 lb

(b)

Problem 18.17

18.18 The 5-kg slender bar is released from rest in the horizontal position shown. Determine the bar's counterclockwise angular acceleration (a) at the instant it is released, and (b) at the instant when it has rotated 45°.

▶ 18.19 The 5-kg slender bar is released from rest in the horizontal position shown. At the instant when it has rotated 45°, its angular velocity is 4.16 rad/s. At that instant, determine the magnitude of the force exerted on the bar by the pin support. (See Example 18.4.)

18.20 The 5-kg slender bar is released from rest in the horizontal position shown. Determine the magnitude of its angular velocity when it has fallen to the vertical position.

Strategy: Draw the free-body diagram of the bar when it has fallen through an arbitrary angle θ and apply the equation of angular motion to determine the bar's angular acceleration as a function of θ. Then use the chain rule to write the angular acceleration as

$$\alpha = \frac{d\omega}{dt} = \frac{d\omega}{d\theta}\frac{d\theta}{dt} = \frac{d\omega}{d\theta}\omega.$$

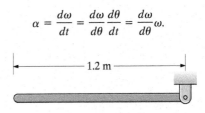

1.2 m

Problems 18.18–18.20

18.21 The object consists of the 2-kg slender bar ABC welded to the 3-kg slender bar BDE. The y axis is vertical.

(a) What is the object's moment of inertia about point D?

(b) Determine the object's counterclockwise angular acceleration at the instant shown.

18.22 The object consists of the 2-kg slender bar ABC welded to the 3-kg slender bar BDE. The y axis is vertical. At the instant shown, the object has a counterclockwise angular velocity of 5 rad/s. Determine the components of the force exerted on it by the pin support.

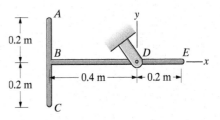

A

y

0.2 m

B

D

E

x

0.2 m

0.4 m

0.2 m

C

Problems 18.21/18.22

18.23 The length of the slender bar is $l = 4$ m and its mass is $m = 30$ kg. It is released from rest in the position shown.

(a) If $x = 1$ m, what is the bar's angular acceleration at the instant it is released?

(b) What value of x results in the largest angular acceleration when the bar is released? What is the angular acceleration?

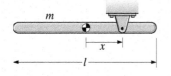

Problem 18.23

18.24 Model the arm ABC as a single rigid body. Its mass is 320 kg, and the moment of inertia about its center of mass is $I = 360$ kg-m^2. Point A is stationary. If the hydraulic piston exerts a 14-kN force on the arm at B, what is the arm's angular acceleration?

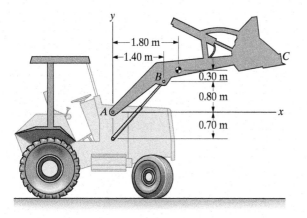

Problem 18.24

18.25 The truck's bed weighs 8000 lb and its moment of inertia about O is 33,000 slug-ft^2. At the instant shown, the coordinates of the center of mass of the bed are (10, 12) ft and the coordinates of point B are (15, 11) ft. If the bed has a counterclockwise angular acceleration of 0.2 rad/s^2, what is the magnitude of the force exerted on the bed at B by the hydraulic cylinder AB?

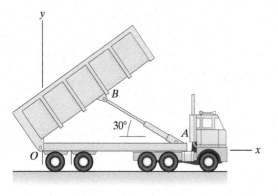

Problem 18.25

18.26 Arm BC has a mass of 12 kg and the moment of inertia about its center of mass is 3 kg-m^2. Point B is stationary and arm BC has a constant counterclockwise angular velocity of 2 rad/s. At the instant shown, what are the couple and the components of force exerted on arm BC at B?

18.27 Arm BC has a mass of 12 kg and the moment of inertia about its center of mass is 3 kg-m^2. At the instant shown, arm AB has a constant clockwise angular velocity of 2 rad/s and arm BC has a counterclockwise angular velocity of 2 rad/s and a clockwise angular acceleration of 4 rad/s^2. What are the couple and the components of force exerted on arm BC at B?

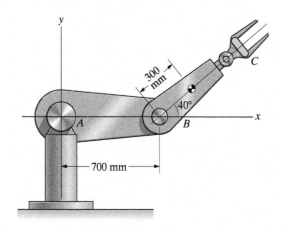

Problems 18.26/18.27

18.28 The space shuttle's attitude control engines exert two forces $F_f = 8$ kN and $F_r = 2$ kN. The force vectors and the center of mass G lie in the x–y plane of the inertial reference frame. The mass of the shuttle is 54,000 kg, and its moment of inertia about the axis through the center of mass that is parallel to the z axis is 4.5×10^6 kg-m^2. Determine the acceleration of the center of mass and the angular acceleration. (You can ignore the force exerted on the shuttle by its weight.)

18.29 In Problem 18.28, suppose that $F_f = 4$ kN and you want the shuttle's angular acceleration to be zero. Determine the necessary force F_r and the resulting acceleration of the center of mass.

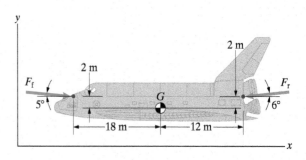

Problems 18.28/18.29

18.30 Points B and C lie in the x–y plane. The y axis is vertical. The center of mass of the 18-kg arm BC is at the midpoint of the line from B to C, and the moment of inertia of the arm about the axis through the center of mass that is parallel to the z axis is 1.5 kg-m^2. At the instant shown, the angular velocity and angular acceleration vectors of arm AB are $\omega_{AB} = 0.6\mathbf{k}$ (rad/s) and $\alpha_{AB} = -0.3\mathbf{k}$ (rad/s^2). The angular velocity and angular acceleration vectors of arm BC are $\omega_{BC} = 0.4\mathbf{k}$ (rad/s) and $\alpha_{BC} = 2\mathbf{k}$ (rad/s^2). Determine the force and couple exerted on arm BC at B.

18.31 Points B and C lie in the x–y plane. The y axis is vertical. The center of mass of the 18-kg arm BC is at the midpoint of the line from B to C, and the moment of inertia of the arm about the axis through the center of mass that is parallel to the z axis is 1.5 kg-m^2. At the instant shown, the angular velocity and angular acceleration vectors of arm AB are $\omega_{AB} = 0.6\mathbf{k}$ (rad/s) and $\alpha_{AB} = -0.3\mathbf{k}$ (rad/s^2). The angular velocity vector of arm BC is $\omega_{BC} = 0.4\mathbf{k}$ (rad/s). If you want to program the robot so that the angular acceleration of arm BC is zero at this instant, what couple must be exerted on arm BC at B?

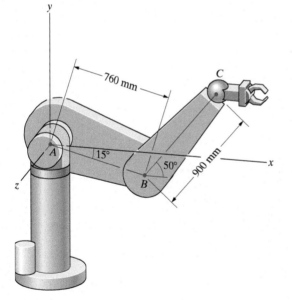

Problems 18.30/18.31

▶ **18.32** The radius of the 2-kg disk is $R = 80$ mm. Its moment of inertia is $I = 0.0064$ kg-m^2. It rolls on the inclined surface. If the disk is released from rest, what is the magnitude of the velocity of its center two seconds later? (See Active Example 18.2.)

▶ **18.33** The radius of the 2-kg disk is $R = 80$ mm. Its moment of inertia is $I = 0.0064$ kg-m^2. What minimum coefficient of static friction is necessary for the disk to roll, instead of slip, on the inclined surface? (See Active Example 18.2.)

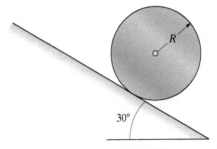

Problems 18.32/18.33

18.34 A thin ring and a homogeneous circular disk, each of mass m and radius R, are released from rest on an inclined surface. Determine the ratio $v_{\text{ring}}/v_{\text{disk}}$ of the velocities of their centers when they have rolled a distance D.

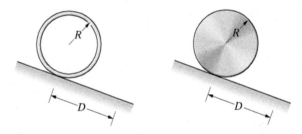

Problem 18.34

18.35 The stepped disk weighs 40 lb and its moment of inertia is $I = 0.2$ slug-ft^2. If the disk is released from rest, how long does it take its center to fall 3 ft? (Assume that the string remains vertical.)

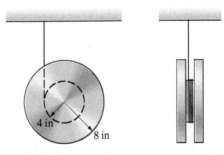

Problem 18.35

18.36 The radius of the pulley is $R = 100$ mm and its moment of inertia is $I = 0.1$ kg-m^2. The mass $m = 5$ kg. The spring constant is $k = 135$ N/m. The system is released from rest with the spring unstretched. At the instant when the mass has fallen 0.2 m, determine (a) the angular acceleration of the pulley, and (b) the tension in the rope between the mass and the pulley.

18.37 The radius of the pulley is $R = 100$ mm and its moment of inertia is $I = 0.1$ kg-m^2. The mass $m = 5$ kg. The spring constant is $k = 135$ N/m. The system is released from rest with the spring unstretched. What maximum distance does the mass fall before rebounding?

Strategy: Assume that the mass has fallen an arbitrary distance x. Write the equations of motion for the mass and the pulley and use them to determine the acceleration a of the mass as a function of x. Then apply the chain rule:

$$\frac{dv}{dt} = \frac{dv}{dx}\frac{dx}{dt} = \frac{dv}{dx}v.$$

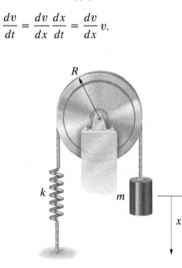

Problems 18.36/18.37

18.38 The mass of the disk is 45 kg, and its radius is $R = 0.3$ m. The spring constant is $k = 600$ N/m. The disk is rolled to the left until the spring is compressed 0.5 m and released from rest.

(a) If you assume that the disk rolls, what is its angular acceleration at the instant it is released?

(b) What is the minimum coefficient of static friction for which the disk will not slip when it is released?

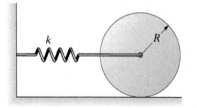

Problem 18.38

18.39 The disk weighs 12 lb and its radius is 6 in. It is stationary on the surface when the force $F = 10$ lb is applied.

(a) If the disk rolls on the surface, what is the acceleration of its center?

(b) What minimum coefficient of static friction is necessary for the disk to roll instead of slipping when the force is applied?

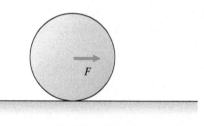

Problem 18.39

18.40 A 42-lb sphere with radius $R = 4$ in is placed on a horizontal surface with initial angular velocity $\omega_0 = 40$ rad/s. The coefficient of kinetic friction between the sphere and the surface is $\mu_k = 0.06$. What maximum velocity will the center of the sphere attain, and how long does it take to reach that velocity?

Strategy: The friction force exerted on the spinning sphere by the surface will cause the sphere to accelerate to the right. The friction force will also cause the sphere's angular velocity to decrease. The center of the sphere will accelerate until the sphere is rolling on the surface instead of slipping relative to it. Use the relation between the velocity of the center and the angular velocity of the sphere when it is rolling to determine when the sphere begins rolling.

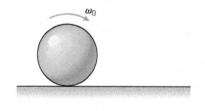

Problem 18.40

18.41 A soccer player kicks the ball to a teammate 8 m away. The ball leaves the player's foot moving parallel to the ground at 6 m/s with no angular velocity. The coefficient of kinetic friction between the ball and the grass is $\mu_k = 0.32$. How long does it take the ball to reach his teammate? The radius of the ball is 112 mm and its mass is 0.4 kg. Estimate the ball's moment of inertia by using the equation for a thin spherical shell: $I = \frac{2}{3}mR^2$.

Problem 18.41

18.42 The 100-kg cylindrical disk is at rest when the force F is applied to a cord wrapped around it. The static and kinetic coefficients of friction between the disk and the surface are 0.2. Determine the angular acceleration of the disk if (a) $F = 500$ N and (b) $F = 1000$ N.

Strategy: First solve the problem by assuming that the disk does not slip, but rolls on the surface. Determine the frictional force, and find out whether it exceeds the product of the coefficient of friction and the normal force. If it does, you must rework the problem assuming that the disk slips.

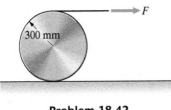

Problem 18.42

18.43 The ring gear is fixed. The mass and moment of inertia of the sun gear are $m_S = 320$ kg and $I_S = 40$ kg-m^2. The mass and moment of inertia of each planet gear are $m_P = 38$ kg and $I_P = 0.60$ kg-m^2. If a couple $M = 200$ N-m is applied to the sun gear, what is the latter's angular acceleration?

18.44 In Problem 18.43, what is the magnitude of the tangential force exerted on the sun gear by each planet gear at their points of contact when the 200 N-m couple is applied to the sun gear?

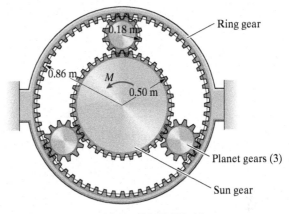

Problems 18.43/18.44

▶ **18.45** The 18-kg ladder is released from rest in the position shown. Model it as a slender bar and neglect friction. At the instant of release, determine (a) the angular acceleration of the ladder and (b) the normal force exerted on the ladder by the floor. (See Active Example 18.3.)

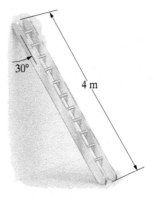

Problem 18.45

18.46 The 18-kg ladder is released from rest in the position shown. Model it as a slender bar and neglect friction. Determine its angular acceleration at the instant of release.

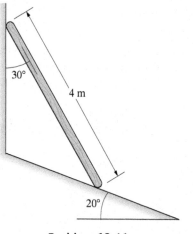

Problem 18.46

18.47 The 4-kg slender bar is released from rest in the position shown. Determine its angular acceleration at that instant if (a) the surface is rough and the bar does not slip, and (b) the surface is smooth.

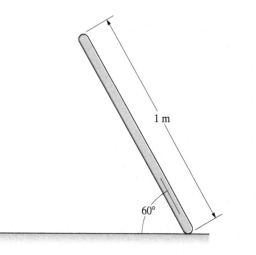

Problem 18.47

18.48 The masses of the bar and disk are 14 kg and 9 kg, respectively. The system is released from rest with the bar horizontal. Determine the bar's angular acceleration at that instant if (a) the bar and disk are welded together at A and (b) the bar and disk are connected by a smooth pin at A.

Strategy: In part (b), draw individual free-body diagrams of the bar and disk.

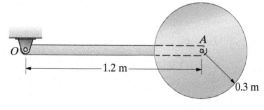

Problem 18.48

18.49 The 5-lb horizontal bar is connected to the 10-lb disk by a smooth pin at A. The system is released from rest in the position shown. What are the angular accelerations of the bar and disk at that instant?

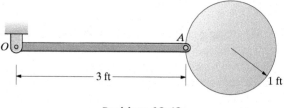

Problem 18.49

18.50 The 0.1-kg slender bar and 0.2-kg cylindrical disk are released from rest with the bar horizontal. The disk rolls on the curved surface. What is the bar's angular acceleration at the instant it is released?

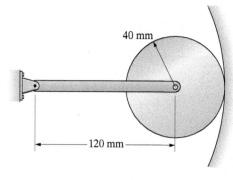

Problem 18.50

18.51 The mass of the suspended object A is 8 kg. The mass of the pulley is 5 kg, and its moment of inertia is 0.036 kg-m². If the force T = 70 N, what is the magnitude of the acceleration of A?

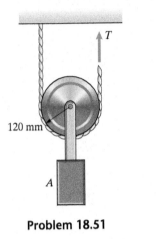

Problem 18.51

18.52 The suspended object A weighs 20 lb. The pulleys are identical, each weighing 10 lb and having moment of inertia 0.022 slug-ft². If the force T = 15 lb, what is the magnitude of the acceleration of A?

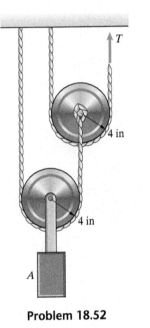

Problem 18.52

▶ **18.53** The 2-kg slender bar and 5-kg block are released from rest in the position shown. If friction is negligible, what is the block's acceleration at that instant? (See Example 18.5.)

▶ **18.54** The 2-kg slender bar and 5-kg block are released from rest in the position shown. What minimum coefficient of static friction between the block and the horizontal surface would be necessary for the block not to move when the system is released? (See Example 18.5.)

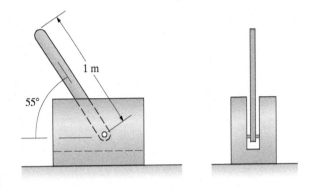

Problems 18.53/18.54

18.55 As a result of the constant couple M applied to the 1-kg disk, the angular acceleration of the 0.4-kg slender bar is zero. Determine M and the counterclockwise angular acceleration of the rolling disk.

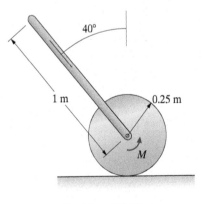

Problem 18.55

18.56 The slender bar weighs 40 lb, and the crate weighs 80 lb. At the instant shown, the velocity of the crate is zero and it has an acceleration of 14 ft/s² toward the left. The horizontal surface is smooth. Determine the couple M and the tension in the rope.

18.57 The slender bar weighs 40 lb, and the crate weighs 80 lb. At the instant shown, the velocity of the crate is zero and it has an acceleration of 14 ft/s² toward the left. The coefficient of kinetic friction between the horizontal surface and the crate is $\mu_k = 0.2$. Determine the couple M and the tension in the rope.

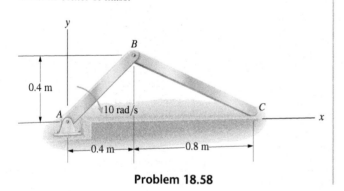

Problems 18.56/18.57

18.58 Bar AB is rotating with a constant clockwise angular velocity of 10 rad/s. The 8-kg slender bar BC slides on the horizontal surface. At the instant shown, determine the total force (including its weight) acting on bar BC and the total moment about its center of mass.

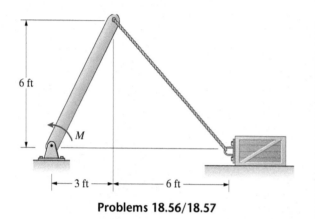

Problem 18.58

18.59 The masses of the slender bars AB and BC are 10 kg and 12 kg, respectively. The angular velocities of the bars are zero at the instant shown and the horizontal force $F = 150$ N. The horizontal surface is smooth. Determine the angular accelerations of the bars.

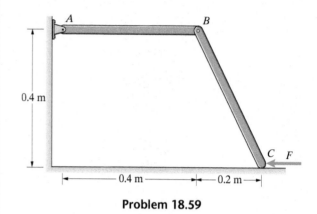

Problem 18.59

18.60 Let the total moment of inertia of the car's two rear wheels and axle be I_R, and let the total moment of inertia of the two front wheels be I_F. The radius of the tires is R, and the total mass of the car, including the wheels, is m. If the car's engine exerts a torque (couple) T on the rear wheels and the wheels do not slip, show that the car's acceleration is

$$a = \frac{RT}{R^2 m + I_R + I_F}.$$

Strategy: Isolate the wheels and draw three free-body diagrams.

Problem 18.60

18.61 The combined mass of the motorcycle and rider is 160 kg. Each 9-kg wheel has a 330-mm radius and a moment of inertia $I = 0.8$ kg-m^2. The engine drives the rear wheel by exerting a couple on it. If the rear wheel exerts a 400-N horizontal force on the road and you do *not* neglect the horizontal force exerted on the road by the front wheel, determine (a) the motorcycle's acceleration and (b) the normal forces exerted on the road by the rear and front wheels. (The location of the center of mass of the motorcycle, *not including* its wheels, is shown.)

18.62 In Problem 18.61, if the front wheel lifts slightly off the road when the rider accelerates, determine (a) the motorcycle's acceleration and (b) the torque exerted by the engine on the rear wheel.

723 mm

A ← 649 mm → B

← 1500 mm →

Problems 18.61/18.62

18.63 The moment of inertia of the vertical handle about O is 0.12 slug-ft^2. The object B weighs 15 lb and rests on a smooth surface. The weight of the bar AB is negligible (which means that you can treat the bar as a two-force member). If the person exerts a 0.2-lb horizontal force on the handle 15 in above O, what is the resulting angular acceleration of the handle?

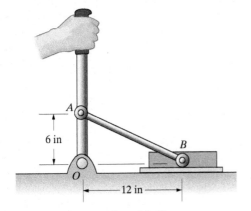

A

6 in

B

O

← 12 in →

Problem 18.63

18.64 The bars are each 1 m in length and have a mass of 2 kg. They rotate *in the horizontal plane*. Bar AB rotates with a constant angular velocity of 4 rad/s in the counterclockwise direction. At the instant shown, bar BC is rotating in the counterclockwise direction at 6 rad/s. What is the angular acceleration of bar BC?

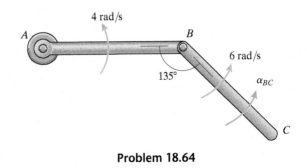

4 rad/s

A

B

135°

6 rad/s

α_{BC}

C

Problem 18.64

18.65 Bars OQ and PQ each weigh 6 lb. The weight of the collar P and the friction between the collar and the horizontal bar are negligible. If the system is released from rest with $\theta = 45°$, what are the angular accelerations of the two bars?

18.66 In Problem 18.65, what are the angular accelerations of the two bars if the collar P weighs 2 lb?

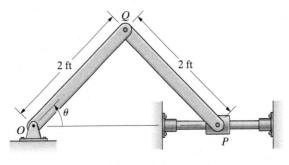

Q

2 ft   2 ft

θ

O

P

Problems 18.65/18.66

18.67 The 4-kg slender bar is pinned to 2-kg sliders at A and B. If friction is negligible and the system is released from rest in the position shown, what is the angular acceleration of the bar at that instant?

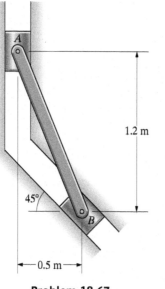

1.2 m

45°

B

0.5 m

Problem 18.67

18.68 The mass of the slender bar is m and the mass of the homogeneous disk is $4m$. The system is released from rest in the position shown. If the disk rolls and the friction between the bar and the horizontal surface is negligible, show that the disk's angular acceleration is $\alpha = 6g/95R$ counterclockwise.

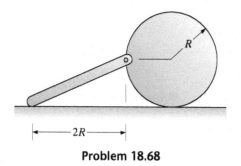

R

2R

Problem 18.68

18.69 Bar AB rotates *in the horizontal plane* with a constant angular velocity of 10 rad/s in the counterclockwise direction. The masses of the slender bars BC and CD are 3 kg and 4.5 kg, respectively. Determine the x and y components of the forces exerted on bar BC by the pins at B and C at the instant shown.

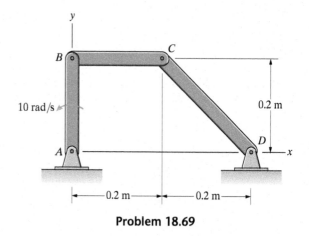

y

B   C

10 rad/s   0.2 m

                                    D

A   x

0.2 m   0.2 m

Problem 18.69

18.70 The 2-kg bar rotates *in the horizontal plane* about the smooth pin. The 6-kg collar A slides on the smooth bar. At the instant shown, $r = 1.2$ m, $\omega = 0.4$ rad/s, and the collar is sliding outward at 0.5 m/s relative to the bar. If you neglect the moment of inertia of the collar (that is, treat the collar as a particle), what is the bar's angular acceleration?

Strategy: Draw individual free-body diagrams of the bar and the collar, and write Newton's second law for the collar in polar coordinates.

18.71 In Problem 18.70, suppose that the moment of inertia of the collar about its center of mass is 0.2 kg-m². Determine the angular acceleration of the bar and compare your answer with the answer to Problem 18.70.

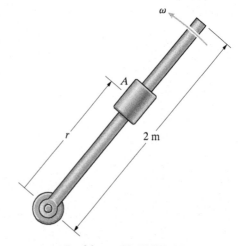

ω

A

r   2 m

Problems 18.70/18.71

Appendix: Moments of Inertia

When a rigid body is subjected to forces and couples, the rotational motion that results depends not only on the mass of the body, but also on how its mass is *distributed*. Although the two objects in Fig. 18.7 have the same mass, the angular accelerations caused by the couple M are different. This difference is reflected in the equation of angular motion $M = I\alpha$ through the moment of inertia, I. The object in Fig. 18.7a has a smaller moment of inertia about the axis L, so its angular acceleration is greater.

In deriving the equations of motion of a rigid body in Sections 18.1 and 18.2, we modeled the body as a finite number of particles and expressed its moment of inertia about an axis L_O as

$$I_O = \sum_i m_i r_i^2,$$

where m_i is the mass of the ith particle and r_i is the perpendicular distance from L_O to the ith particle (Fig. 18.8a). To calculate moments of inertia of objects, it is often more convenient to model them as continuous distributions of mass and express the moment of inertia about L_O as

$$I_O = \int_m r^2 \, dm, \tag{18.21}$$

where r is the perpendicular distance from L_O to the differential element of mass dm (Fig. 18.8b). When the axis passes through the center of mass of the object, we denote the axis by L and the moment of inertia about L by I.

The dimensions of the moment of inertia of an object are $(\text{mass}) \times (\text{length})^2$. Notice that the definition implies that its value must be positive.

Simple Objects

We begin by determining moments of inertia of some simple objects. In the next subsection we describe the parallel-axis theorem, which simplifies the task of determining moments of inertia of objects composed of combinations of simple parts.

Slender Bars   We will determine the moment of inertia of a straight slender bar about a perpendicular axis L through the center of mass of the bar (Fig. 18.9a). "Slender" means we assume that the bar's length is much greater than its width. Let the bar have length l, cross-sectional area A, and mass m. We assume that A is uniform along the length of the bar and that the material is homogeneous. Consider a differential element of the bar of length dr at a distance r from the center of mass (Fig. 18.9b). The element's mass is equal to the product of its volume and the mass density: $dm = \rho A \, dr$. Substituting this expression into Eq. (18.21), we obtain the moment of inertia of the bar about a perpendicular axis through its center of mass:

$$I = \int_m r^2 \, dm = \int_{-l/2}^{l/2} \rho A r^2 \, dr = \tfrac{1}{12} \rho A l^3.$$

The mass of the bar equals the product of the mass density and the volume of the bar ($m = \rho A l$), so we can express the moment of inertia as

$$I = \tfrac{1}{12} m l^2. \tag{18.22}$$

We have neglected the lateral dimensions of the bar in obtaining this result. That is, we treated the differential element of mass dm as if it were concentrated on

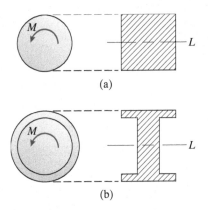

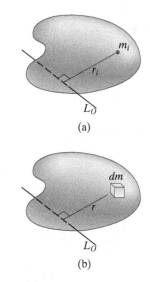

Figure 18.7
Objects of equal mass that have different moments of inertia about L.

Figure 18.8
Determining the moment of inertia by modeling an object as (a) a finite number of particles and (b) a continuous distribution of mass.

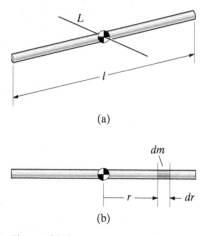

(a)

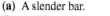

(b)

Figure 18.9
(a) A slender bar.
(b) A differential element of length dr.

the axis of the bar. As a consequence, Eq. (18.22) is an approximation for the moment of inertia of a bar. Later in this section, we will determine the moments of inertia for a bar of finite lateral dimension and show that Eq. (18.22) is a good approximation when the width of the bar is small in comparison to its length.

Thin Plates Consider a homogeneous flat plate that has mass m and uniform thickness T. We will leave the shape of the cross-sectional area of the plate unspecified. Let a cartesian coordinate system be oriented so that the plate lies in the x–y plane (Fig. 18.10a). Our objective is to determine the moments of inertia of the plate about the x, y, and z axes.

We can obtain a differential element of volume of the plate by projecting an element of area dA through the thickness T of the plate (Fig. 18.10b). The resulting volume is $T\, dA$. The mass of this element of volume is equal to the product of the mass density and the volume: $dm = \rho T\, dA$. Substituting this expression into Eq. (18.9), we obtain the moment of inertia of the plate about the z axis in the form

$$I_{z\text{ axis}} = \int_m r^2\, dm = \rho T \int_A r^2\, dA,$$

where r is the distance from the z axis to dA. Because the mass of the plate is $m = \rho T A$, where A is the cross-sectional area of the plate, the product $\rho T = m/A$. The integral on the right is the polar moment of inertia J_O of the cross-sectional area of the plate. Therefore, we can write the moment of inertia of the plate about the z axis as

$$I_{z\text{ axis}} = \frac{m}{A} J_O. \tag{18.23}$$

From Fig. 18.10b, we see that the perpendicular distance from the x axis to the element of area dA is the y coordinate of dA. Consequently, the moment of inertia of the plate about the x axis is

$$I_{x\text{ axis}} = \int_m y^2\, dm = \rho T \int_A y^2\, dA = \frac{m}{A} I_x, \tag{18.24}$$

where I_x is the moment of inertia of the cross-sectional area of the plate about the x axis. The moment of inertia of the plate about the y axis is

$$I_{y\text{ axis}} = \int_m x^2\, dm = \rho T \int_A x^2\, dA = \frac{m}{A} I_y, \tag{18.25}$$

where I_y is the moment of inertia of the cross-sectional area of the plate about the y axis.

Because the sum of the area moments of inertia I_x and I_y is equal to the polar moment of inertia J_O, the moment of inertia of the thin plate about the z axis is equal to the sum of its moments of inertia about the x and y axes:

$$I_{z\text{ axis}} = I_{x\text{ axis}} + I_{y\text{ axis}}. \tag{18.26}$$

Thus, we have expressed the moments of inertia of a thin homogeneous plate of uniform thickness in terms of the moments of inertia of the cross-sectional area of the plate. In fact, these results explain why the area integrals I_x, I_y, and J_O are called moments of inertia.

The use of the same terminology and similar symbols for moments of inertia of areas and moments of inertia of objects can be confusing, but is entrenched in engineering practice. The type of moment of inertia being referred to can be determined either from the context or from the units, $(\text{length})^4$ for moments of inertia of areas and $(\text{mass}) \times (\text{length})^2$ for moments of inertia of objects.

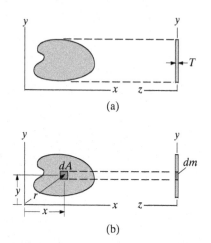

(a)

(b)

Figure 18.10
(a) A plate of arbitrary shape and uniform thickness T.
(b) An element of volume obtained by projecting an element of area dA through the plate.

Example 18.6   Moment of Inertia of an L-Shaped Bar (▶ *Related Problem 18.72*)

Two homogeneous slender bars, each of length l, mass m, and cross-sectional area A, are welded together to form an L-shaped object. Use integration to determine the moment of inertia of the object about the axis L_O through point O. The axis L_O is perpendicular to the two bars.

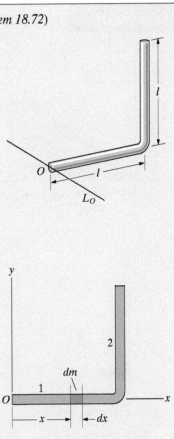

Strategy
Using the same integration procedure we used for a single bar, we can determine the moment of inertia of each bar about L_O and sum the results.

Solution
We orient a coordinate system with the z axis along L_O and the x axis collinear with bar 1 (Fig. a). The mass of the differential element of length dx of bar 1 is $dm = \rho A\, dx$. The moment of inertia of bar 1 about L_O is

$$(I_O)_1 = \int_m r^2\, dm = \int_0^l \rho A x^2\, dx = \tfrac{1}{3}\rho A l^3.$$

In terms of the mass of the bar, $m = \rho A l$, we can write this result as

$$(I_O)_1 = \tfrac{1}{3}ml^2.$$

The mass of the element of length dy of bar 2 shown in Fig. b is $dm = \rho A\, dy$. From the figure, we see that the perpendicular distance from L_O to the element is $r = \sqrt{l^2 + y^2}$. Therefore, the moment of inertia of bar 2 about L_O is

$$(I_O)_2 = \int_m r^2\, dm = \int_0^l \rho A(l^2 + y^2)\, dy = \tfrac{4}{3}\rho A l^3.$$

(a) Differential element of bar 1.

In terms of the mass of the bar, we obtain

$$(I_O)_2 = \tfrac{4}{3}ml^2.$$

The moment of inertia of the L-shaped object about L_O is

$$I_O = (I_O)_1 + (I_O)_2 = \tfrac{1}{3}ml^2 + \tfrac{4}{3}ml^2 = \tfrac{5}{3}ml^2.$$

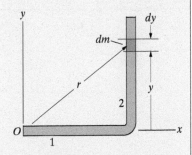

(b) Differential element of bar 2.

Critical Thinking
In this example we used integration to determine a moment of inertia of an object consisting of two straight bars. The same procedure could be applied to more complicated objects made of such bars, but it would obviously be cumbersome. Once we have used integration to determine a moment of inertia of a single bar, such as Eq. (18.22), it would be very convenient to be able to use that result to determine moments of inertia of composite objects made of bars without having to resort to integration. We show how this can be done in the next section.

Example 18.7	Moments of Inertia of a Triangular Plate (▶ *Related Problem 18.76*)

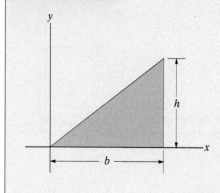

The thin, homogeneous plate is of uniform thickness and mass m. Determine its moments of inertia about the x, y, and z axes.

Strategy

The moments of inertia about the x and y axes are given by Eqs. (18.24) and (18.25) in terms of the moments of inertia of the cross-sectional area of the plate. We can determine the moment of inertia of the plate about the z axis from Eq. (18.26).

Solution

From Appendix B, the moments of inertia of the triangular area about the x and y axes are $I_x = \frac{1}{12}bh^3$ and $I_y = \frac{1}{4}hb^3$. Therefore, the moments of inertia of the plate about the x and y axes are

$$I_{x\,\text{axis}} = \frac{m}{A}I_x = \frac{m}{\frac{1}{2}bh}\left(\tfrac{1}{12}bh^3\right) = \tfrac{1}{6}mh^2$$

and

$$I_{y\,\text{axis}} = \frac{m}{A}I_y = \frac{m}{\frac{1}{2}bh}\left(\tfrac{1}{4}hb^3\right) = \tfrac{1}{2}mb^2.$$

The moment of inertia about the z axis is

$$I_{z\,\text{axis}} = I_{x\,\text{axis}} + I_{y\,\text{axis}} = m\left(\tfrac{1}{6}h^2 + \tfrac{1}{2}b^2\right).$$

Critical Thinking

As this example demonstrates, you can use the moments of inertia of areas tabulated in Appendix B to determine moments of inertia of thin homogeneous plates. For plates with more complicated shapes, you can use the methods for determining moments of inertia of composite areas.

Parallel-Axis Theorem

The parallel-axis theorem allows us to determine the moment of inertia of a composite object when we know the moments of inertia of its parts. Suppose that we know the moment of inertia I about an axis L through the center of mass of an object and we wish to determine its moment of inertia I_O about a parallel axis L_O (Fig. 18.11a). To determine I_O, we introduce parallel coordinate systems xyz and $x'y'z'$, with the z axis along L_O and the z' axis along L, as shown in Fig. 18.11b. (In this figure, the axes L_O and L are perpendicular to the page.) The origin O of the xyz coordinate system is contained in the $x'-y'$ plane. The terms d_x and d_y are the coordinates of the center of mass relative to the xyz coordinate system.

The moment of inertia of the object about L_O is

$$I_O = \int_m r^2\,dm = \int_m (x^2 + y^2)\,dm, \qquad (18.27)$$

where r is the perpendicular distance from L_O to the differential element of mass dm and x, y are the coordinates of dm in the x–y plane. The x–y coordinates of dm are related to its $x'-y'$ coordinates by

$$x = x' + d_x$$

and

$$y = y' + d_y.$$

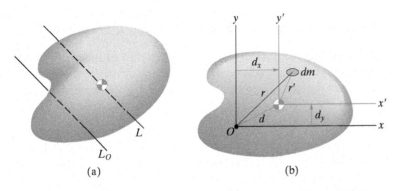

(a)   **(b)**

Figure 18.11
(a) An axis L through the center of mass of an object and a parallel axis L_O.
(b) The xyz and $x'y'z'$ coordinate systems.

Substituting these expressions into Eq. (18.27), we can write that equation as

$$I_O = \int_m [(x')^2 + (y')^2]\, dm + 2d_x \int_m x'\, dm + 2d_y \int_m y'\, dm$$

$$+ \int_m (d_x^2 + d_y^2)\, dm. \qquad (18.28)$$

Since $(x')^2 + (y')^2 = (r')^2$, where r' is the perpendicular distance from L to dm, the first integral on the right side of this equation is the moment of inertia I of the object about L. Recall that the x' and y' coordinates of the center of mass of the object relative to the $x'y'z'$ coordinate system are defined by

$$\bar{x}' = \frac{\int_m x'\, dm}{\int_m dm}, \quad \bar{y}' = \frac{\int_m y'\, dm}{\int_m dm}.$$

Because the center of mass of the object is at the origin of the $x'y'z'$ system, $\bar{x}' = 0$ and $\bar{y}' = 0$. Therefore, the integrals in the second and third terms on the right side of Eq. (18.28) are equal to zero. From Fig. 18.11b, we see that $d_x^2 + d_y^2 = d^2$, where d is the perpendicular distance between the axes L and L_O. Therefore, we obtain

$$I_O = I + d^2 m. \qquad (18.29)$$

This is the *parallel-axis theorem* for moments of inertia of objects. Equation (18.29) relates the moment of inertia I of an object about an axis *through the center of mass* to its moment of inertia I_O about any parallel axis, where d is the perpendicular distance between the two axes and m is the mass of the object.

The parallel-axis theorem makes it possible to determine moments of inertia of composite objects. Determine the moment of inertia about a given axis L_O typically requires three steps:

1. *Choose the parts.* Try to divide the object into parts whose moments of inertia can easily be determined.

2. *Determine the moments of inertia of the parts.* Determine the moment of inertia of each part about the axis through its center of mass parallel to L_O. Then use the parallel-axis theorem to determine its moment of inertia about L_O.

3. *Sum the results.* Sum the moments of inertia of the parts (or subtract in the case of a hole or cutout) to obtain the moment of inertia of the composite object.

Example 18.8   **Application of the Parallel-Axis Theorem** (▶ *Related Problems 18.82, 18.83*)

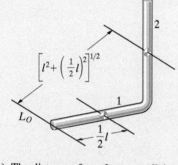

Two homogeneous, slender bars, each of length l and mass m, are welded together to form an L-shaped object. Determine the moment of inertia of the object about the axis L_O through point O. (The axis L_O is perpendicular to the two bars.)

Strategy

The moment of inertia of a straight slender bar about a perpendicular axis through its center of mass is given by Eq. (18.22). We can use the parallel-axis theorem to determine the moments of inertia of the bars about the axis L_O and sum them to obtain the moment of inertia of the composite bar.

Solution

Choose the Parts   The parts are the two bars, which we call bar 1 and bar 2 (Fig. a).

Determine the Moments of Inertia of the Parts   From Appendix C, the moment of inertia of each bar about a perpendicular axis through its center of mass is $I = \frac{1}{12}ml^2$. The distance from L_O to the parallel axis through the center of mass of bar 1 is $\frac{1}{2}l$ (Fig. a). Therefore, the moment of inertia of bar 1 about L_O is

$$(I_O)_1 = I + d^2m = \tfrac{1}{12}ml^2 + \left(\tfrac{1}{2}l\right)^2 m = \tfrac{1}{3}ml^2.$$

(a) The distances from L_O to parallel axes through the centers of mass of bars 1 and 2.

The distance from L_O to the parallel axis through the center of mass of bar 2 is $\left[l^2 + \left(\tfrac{1}{2}l\right)^2\right]^{1/2}$. The moment of inertia of bar 2 about L_O is

$$(I_O)_2 = I + d^2m = \tfrac{1}{12}ml^2 + \left[l^2 + \left(\tfrac{1}{2}l\right)^2\right] m = \tfrac{4}{3}ml^2.$$

Sum the Results   The moment of inertia of the L-shaped object about L_O is

$$I_O = (I_O)_1 + (I_O)_2 = \tfrac{1}{3}ml^2 + \tfrac{4}{3}ml^2 = \tfrac{5}{3}ml^2.$$

Critical Thinking

Compare this solution with Example 18.6, in which we used integration to determine the moment of inertia of the same object about L_O. We obtained the result much more easily with the parallel-axis theorem, but of course we needed to know the moments of inertia of the bars about the axes through their centers of mass.

Example 18.9 | Moments of Inertia of a Composite Object (▶ *Related Problems 18.102, 18.103*)

The object consists of a slender 3-kg bar welded to a thin, circular 2-kg disk. Determine the moment of inertia of the object about the axis L through its center of mass. (The axis L is perpendicular to the bar and disk.)

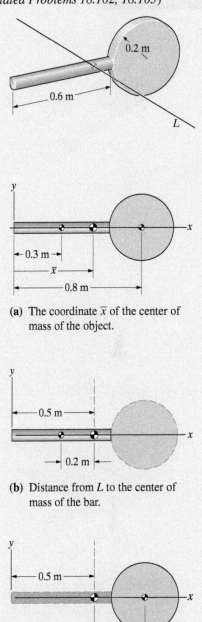

Strategy

We must locate the center of mass of the composite object and then apply the parallel-axis theorem. We can obtain the moments of inertia of the bar and disk from Appendix C.

Solution

Choose the Parts The parts are the bar and the disk. Introducing the coordinate system in Fig. a, we have, for the x coordinate of the center of mass of the composite object,

$$\bar{x} = \frac{\bar{x}_{bar}m_{bar} + \bar{x}_{disk}m_{disk}}{m_{bar} + m_{disk}}$$

$$= \frac{(0.3 \text{ m})(3 \text{ kg}) + (0.6 \text{ m} + 0.2 \text{ m})(2 \text{ kg})}{(3 \text{ kg}) + (2 \text{ kg})} = 0.5 \text{ m}.$$

(a) The coordinate $\bar{x}$ of the center of mass of the object.

Determine the Moments of Inertia of the Parts The distance from the center of mass of the bar to the center of mass of the composite object is 0.2 m (Fig. b). Therefore, the moment of inertia of the bar about L is

$$I_{bar} = \tfrac{1}{12}(3 \text{ kg})(0.6 \text{ m})^2 + (3 \text{ kg})(0.2 \text{ m})^2 = 0.210 \text{ kg-m}^2.$$

The distance from the center of mass of the disk to the center of mass of the composite object is 0.3 m (Fig. c). The moment of inertia of the disk about L is

$$I_{disk} = \tfrac{1}{2}(2 \text{ kg})(0.2 \text{ m})^2 + (2 \text{ kg})(0.3 \text{ m})^2 = 0.220 \text{ kg-m}^2.$$

(b) Distance from L to the center of mass of the bar.

Sum the Results The moment of inertia of the composite object about L is

$$I = I_{bar} + I_{disk} = 0.430 \text{ kg-m}^2.$$

Critical Thinking

This example demonstrates the most common procedure for determining moments of inertia of objects in engineering applications. Objects usually consist of assemblies of parts. The center of mass of each part and its moment of inertia about the axis through its center of mass must be determined. (It may be necessary to determine this information experimentally, or it is sometimes supplied by manufacturers of subassemblies.) Then the center of mass of the composite object is determined, and the parallel axis theorem is used to determine the moment of inertia of each part about the axis through the center of mass of the composite object. Finally, the individual moments of inertia are summed to obtain the moment of inertia of the composite object.

(c) Distance from L to the center of mass of the disk.

| **Example 18.10** | Moments of Inertia of a Homogeneous Cylinder (▶ *Related Problems 18.95, 18.96*) |

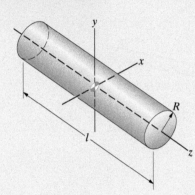

The homogeneous cylinder has mass m, length l, and radius R. Determine the moments of inertia of the cylinder about the x, y, and z axes.

Strategy
We can determine the moments of inertia of the cylinder by an interesting application of the parallel-axis theorem. We use it to determine the moments of inertia about the x, y, and z axes of an infinitesimal element of the cylinder consisting of a disk of thickness dz. Then we integrate the results with respect to z to obtain the moments of inertia of the cylinder.

Solution
Consider an element of the cylinder of thickness dz at a distance z from the center of the cylinder (Fig. a). (You can imagine obtaining this element by "slicing" the cylinder perpendicular to its axis.) The mass of the element is equal to the product of the mass density and the volume of the element: $dm = \rho(\pi R^2 \, dz)$. We obtain the moments of inertia of the element by using the values for a thin circular plate given in Appendix C. The moment of inertia about the z axis is

$$dI_{z \text{ axis}} = \tfrac{1}{2} dm \, R^2 = \tfrac{1}{2}(\rho \pi R^2 \, dz) R^2.$$

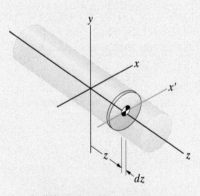

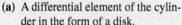

(a) A differential element of the cylinder in the form of a disk.

We integrate this result with respect to z from $-l/2$ to $l/2$, thereby summing the moments of inertia of the infinitesimal disk elements that make up the cylinder. The result is the moment of inertia of the cylinder about the z axis:

$$I_{z \text{ axis}} = \int_{-l/2}^{l/2} \tfrac{1}{2} \rho \pi R^4 \, dz = \tfrac{1}{2} \rho \pi R^4 l.$$

We can write this result in terms of the mass of the cylinder, $m = \rho(\pi R^2 l)$, as

$$I_{z \text{ axis}} = \tfrac{1}{2} m R^2.$$

The moment of inertia of the disk element about the x' axis is

$$dI_{x' \text{ axis}} = \tfrac{1}{4} dm \, R^2 = \tfrac{1}{4}(\rho \pi R^2 \, dz) R^2.$$

We use this result and the parallel-axis theorem to determine the moment of inertia of the element about the x axis:

$$dI_{x \text{ axis}} = dI_{x' \text{ axis}} + z^2 \, dm = \tfrac{1}{4}(\rho \pi R^2 \, dz) R^2 + z^2(\rho \pi R^2 \, dz).$$

Integrating this expression with respect to z from $-l/2$ to $l/2$, we obtain the moment of inertia of the cylinder about the x axis:

$$I_{x \text{ axis}} = \int_{-l/2}^{l/2} \left(\tfrac{1}{4} \rho \pi R^4 + \rho \pi R^2 z^2 \right) dz = \tfrac{1}{4} \rho \pi R^4 l + \tfrac{1}{12} \rho \pi R^2 l^3.$$

In terms of the mass of the cylinder,

$$I_{x \text{ axis}} = \tfrac{1}{4}mR^2 + \tfrac{1}{12}ml^2.$$

Due to the symmetry of the cylinder,

$$I_{y \text{ axis}} = I_{x \text{ axis}}.$$

Critical Thinking

When the cylinder is very long in comparison to its width $(l \gg R)$, the first term in the equation for $I_{x \text{ axis}}$ can be neglected, and we obtain the moment of inertia of a slender bar about a perpendicular axis, Eq. (18.22). On the other hand, when the radius of the cylinder is much greater than its length $(R \gg l)$, the second term in the equation for $I_{x \text{ axis}}$ can be neglected, and we obtain the moment of inertia for a thin circular disk about an axis parallel to the disk. This indicates the sizes of the terms you neglect when you use the approximate expressions for the moments of inertia of a "slender" bar and a "thin" disk.

Problems

▶ **18.72** The axis L_O is perpendicular to both segments of the L-shaped slender bar. The mass of the bar is 6 kg and the material is homogeneous. Use integration to determine the moment of inertia of the bar about L_O. (See Example 18.6.)

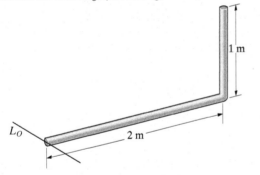

Problem 18.72

18.73 Two homogeneous slender bars, each of mass m and length l, are welded together to form the T-shaped object. Use integration to determine the moment of inertia of the object about the axis through point O that is perpendicular to the bars.

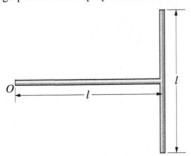

Problem 18.73

18.74 The slender bar lies in the x–y plane. Its mass is 6 kg and the material is homogeneous. Use integration to determine its moment of inertia about the z axis.

18.75 The slender bar lies in the x–y plane. Its mass is 6 kg and the material is homogeneous. Use integration to determine its moment of inertia about the y axis.

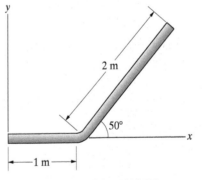

Problems 18.74/18.75

▶ 18.76 The homogeneous thin plate has mass $m = 12$ kg and dimensions $b = 1$ m and $h = 2$ m. Determine the moments of inertia of the plate about the x, y, and z axes. (See Example 18.7.)

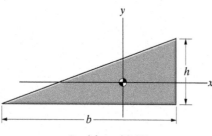

Problem 18.76

18.77 The brass washer is of uniform thickness and mass m.

(a) Determine its moments of inertia about the x and z axes.

(b) Let $R_i = 0$, and compare your results with the values given in Appendix C for a thin circular plate.

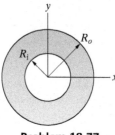

Problem 18.77

18.78 The homogeneous thin plate is of uniform thickness and weighs 20 lb. Determine its moment of inertia about the y axis.

18.79 Determine the moment of inertia of the 20-lb plate about the x axis.

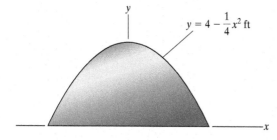

$y = 4 - \dfrac{1}{4}x^2$ ft

Problems 18.78/18.79

18.80 The mass of the object is 10 kg. Its moment of inertia about L_1 is 10 kg-m². What is its moment of inertia about L_2? (The three axes lie in the same plane.)

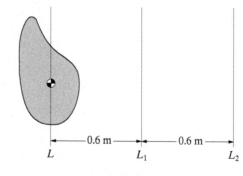

—0.6 m—   —0.6 m—

L L_1 L_2

Problem 18.80

▶ 18.81 An engineer gathering data for the design of a maneuvering unit determines that the astronaut's center of mass is at $x = 1.01$ m, $y = 0.16$ m and that her moment of inertia about the z axis is 105.6 kg-m². The astronaut's mass is 81.6 kg. What is her moment of inertia about the z' axis through her center of mass?

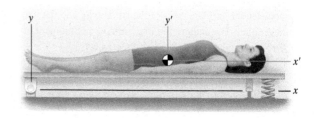

Problem 18.81

▶ 18.82 Two homogeneous slender bars, each of mass m and length l, are welded together to form the T-shaped object. Use the parallel-axis theorem to determine the moment of inertia of the object about the axis through point O that is perpendicular to the bars. (See Example 18.8.)

▶ 18.83 Use the parallel-axis theorem to determine the moment of inertia of the T-shaped object about the axis through the center of mass of the object that is perpendicular to the two bars. (See Example 18.8.)

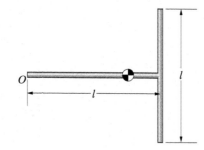

Problems 18.82/18.83

18.84 The mass of the homogeneous slender bar is 30 kg. Determine its moment of inertia about the z axis.

18.85 The mass of the homogeneous slender bar is 30 kg. Determine the moment of inertia of the bar about the z' axis through its center of mass.

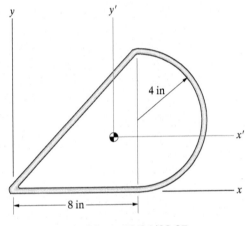

Problems 18.84/18.85

18.86 The homogeneous slender bar weighs 5 lb. Determine its moment of inertia about the z axis.

18.87 Determine the moment of inertia of the 5-lb bar about the z' axis through its center of mass.

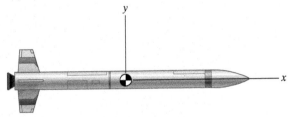

Problems 18.86/18.87

18.88 The rocket is used for atmospheric research. Its weight and its moment of inertia about the z axis through its center of mass (including its fuel) are 10,000 lb and 10,200 slug-ft^2, respectively. The rocket's fuel weighs 6000 lb, its center of mass is located at $x = -3$ ft, $y = 0$, $z = 0$, and the moment of inertia of the fuel about the axis through the fuel's center of mass parallel to the z axis is 2200 slug-ft^2. When the fuel is exhausted, what is the rocket's moment of inertia about the axis through its new center of mass parallel to the z axis?

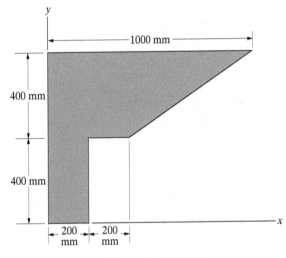

Problem 18.88

18.89 The mass of the homogeneous thin plate is 36 kg. Determine the moment of inertia of the plate about the x axis.

18.90 Determine the moment of inertia of the 36-kg plate about the z axis.

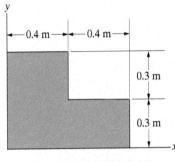

Problems 18.89/18.90

18.91 The mass of the homogeneous thin plate is 20 kg. Determine its moment of inertia about the x axis.

18.92 The mass of the homogeneous thin plate is 20 kg. Determine its moment of inertia about the y axis.

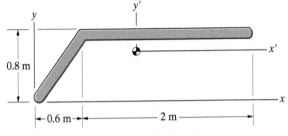

Problems 18.91/18.92

18.93 The thermal radiator (used to eliminate excess heat from a satellite) can be modeled as a homogeneous thin rectangular plate. The mass of the radiator is 5 slugs. Determine its moments of inertia about the x, y, and z axes.

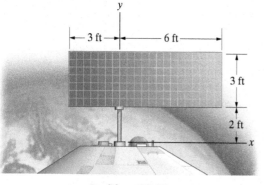

Problem 18.93

18.94 The mass of the homogeneous thin plate is 2 kg. Determine the moment of inertia of the plate about the axis through point O that is perpendicular to the plate.

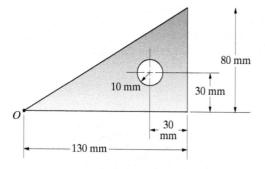

Problem 18.94

▶ 18.95 The homogeneous cone is of mass m. Determine its moment of inertia about the z axis, and compare your result with the value given in Appendix C. (See Example 18.10.)

▶ 18.96 Determine the moments of inertia of the homogeneous cone of mass m about the x and y axes, and compare your results with the values given in Appendix C. (See Example 18.10.)

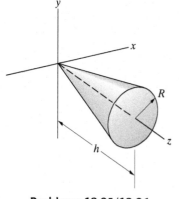

Problems 18.95/18.96

18.97 The homogeneous object has the shape of a truncated cone and consists of bronze with mass density $\rho = 8200 \text{ kg/m}^3$. Determine the moment of inertia of the object about the z axis.

18.98 Determine the moment of inertia of the object described in Problem 18.97 about the x axis.

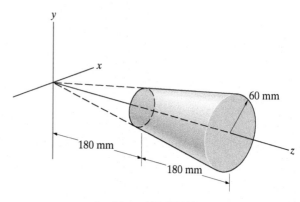

Problems 18.97/18.98

18.99 The homogeneous rectangular parallelepiped is of mass m. Determine its moments of inertia about the x, y, and z axes and compare your results with the values given in Appendix C.

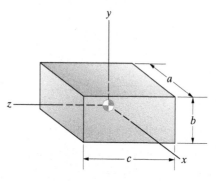

Problem 18.99

18.100 The sphere-capped cone consists of material with density 7800 kg/m³. The radius $R = 80$ mm. Determine its moment of inertia about the x axis.

18.101 Determine the moment of inertia of the sphere-capped cone described in Problem 18.100 about the y axis.

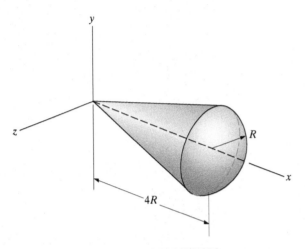

Problems 18.100/18.101

▶ **18.102** The circular cylinder is made of aluminum (Al) with density 2700 kg/m³ and iron (Fe) with density 7860 kg/m³. Determine its moment of inertia about the x' axis. (See Example 18.9.)

▶ **18.103** Determine the moment of inertia of the composite cylinder described in Problem 18.102 about the y' axis. (See Example 18.9.)

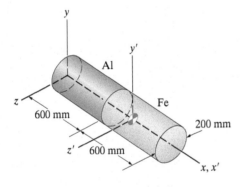

Problems 18.102/18.103

18.104 The homogeneous machine part is made of aluminum alloy with mass density $\rho = 2800$ kg/m³. Determine the moment of inertia of the part about the z axis.

18.105 Determine the moment of inertia of the machine part described in Problem 18.104 about the x axis.

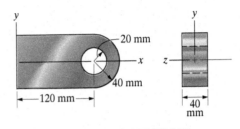

Problems 18.104/18.105

18.106 The object shown consists of steel of density $\rho = 7800$ kg/m³. Determine its moment of inertia about the axis L_O through point O.

18.107 Determine the moment of inertia of the object described in Problem 18.106 about the axis through the center of mass of the object parallel to L_O.

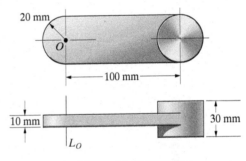

Problems 18.106/18.107

18.108 The thick plate consists of steel of density $\rho = 15$ slug/ft³. Determine the moment of inertia of the plate about the z axis.

18.109 Determine the moment of inertia of the object described in Problem 18.108 about the x axis.

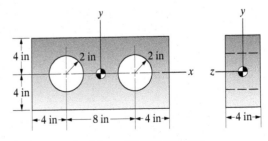

Problems 18.108/18.109

Review Problems

18.110 The airplane is at the beginning of its takeoff run. Its weight is 1000 lb, and the initial thrust T exerted by its engine is 300 lb. Assume that the thrust is horizontal, and neglect the tangential forces exerted on the wheels.

(a) If the acceleration of the airplane remains constant, how long will it take to reach its takeoff speed of 80 mi/h?

(b) Determine the normal force exerted on the forward landing gear at the beginning of the takeoff run.

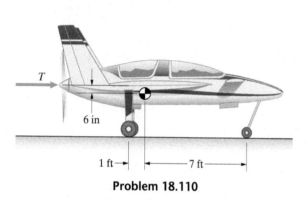

6 in

1 ft → ← 7 ft →

Problem 18.110

18.111 The pulleys can turn freely on their pin supports. Their moments of inertia are $I_A = 0.002$ kg-m², $I_B = 0.036$ kg-m², and $I_C = 0.032$ kg-m². They are initially stationary, and at $t = 0$, a constant couple $M = 2$ N-m is applied to pulley A. What is the angular velocity of pulley C and how many revolutions has it turned at $t = 2$ s?

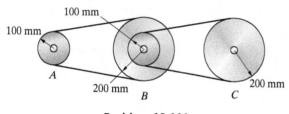

100 mm

100 mm

A

200 mm

B

200 mm

C

Problem 18.111

18.112 A 2-kg box is subjected to a 40-N horizontal force. Neglect friction.

(a) If the box remains on the floor, what is its acceleration?

(b) Determine the range of values of c for which the box will remain on the floor when the force is applied.

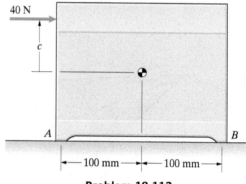

40 N

c

A

B

← 100 mm → ← 100 mm →

Problem 18.112

18.113 The slender 2-slug bar AB is 3 ft long. It is pinned to the cart at A and leans against it at B.

(a) If the acceleration of the cart is $a = 20$ ft/s², what normal force is exerted on the bar by the cart at B?

(b) What is the largest acceleration a for which the bar will remain in contact with the surface at B?

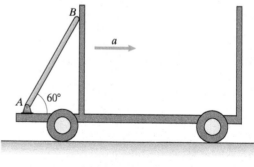

B

a

A 60°

Problem 18.113

18.114 To determine a 4.5-kg tire's moment of inertia, an engineer lets the tire roll down an inclined surface. If it takes the tire 3.5 s to start from rest and roll 3 m down the surface, what is the tire's moment of inertia about its center of mass?

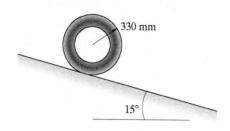

330 mm

15°

Problem 18.114

18.115 Pulley A weighs 4 lb, $I_A = 0.060$ slug-ft^2, and $I_B = 0.014$ slug-ft^2. If the system is released from rest, what distance does the 16-lb weight fall in 0.5 s?

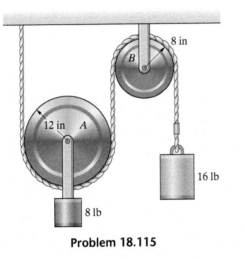

8 in

B

12 in A

16 lb

8 lb

Problem 18.115

18.116 Model the excavator's arm ABC as a single rigid body. Its mass is 1200 kg, and the moment of inertia *about its center of mass* is $I = 3600$ kg-m^2. The angular velocity of the arm is zero and its angular acceleration is 1 rad/s^2 counterclockwise. What force does the vertical hydraulic cylinder exert on the arm at B?

18.117 Model the excavator's arm ABC as a single rigid body. Its mass is 1200 kg, and the moment of inertia *about its center of mass* is $I = 3600$ kg-m^2. The angular velocity of the arm is 2 rad/s counterclockwise and its angular acceleration is 1 rad/s^2 counterclockwise. What are the components of the force exerted on the arm at A?

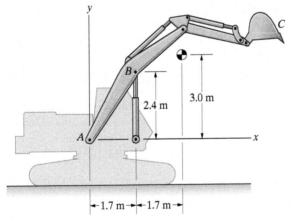

y

C

B

2.4 m

3.0 m

A

x

←1.7 m→←1.7 m→

Problems 18.116/18.117

18.118 To decrease the angle of elevation of the stationary 200-kg ladder, the gears that raised it are disengaged, and a fraction of a second later a second set of gears that lower it are engaged. At the instant the gears that raised the ladder are disengaged, what is the ladder's angular acceleration and what are the components of force exerted on the ladder by its support at O? The moment of inertia of the ladder about O is $I_O = 14,000$ kg-m^2, and the coordinates of its center of mass at the instant the gears are disengaged are $\bar{x} = 3$ m, $\bar{y} = 4$ m.

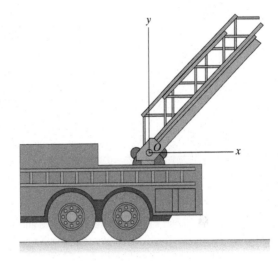

y

O x

Problem 18.118

18.119 The slender bars each weigh 4 lb and are 10 in long. The homogeneous plate weighs 10 lb. If the system is released from rest in the position shown, what is the angular acceleration of the bars at that instant?

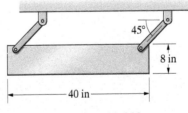

45°

8 in

←————40 in————→

Problem 18.119

18.120 A slender bar of mass m is released from rest in the position shown. The static and kinetic coefficients of friction at the floor and wall have the same value μ. If the bar slips, what is its angular acceleration at the instant of release?

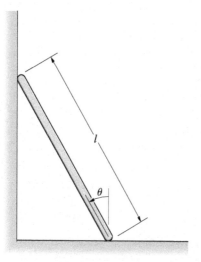

Problem 18.120

18.121 Each of the go-cart's front wheels weighs 5 lb and has a moment of inertia of 0.01 slug-ft². The two rear wheels and rear axle form a single rigid body weighing 40 lb and having a moment of inertia of 0.1 slug-ft². The total weight of the go-cart and driver is 240 lb. (The location of the center of mass of the go-cart and driver, *not including* the front wheels or the rear wheels and rear axle, is shown.) If the engine exerts a torque of 12 ft-lb on the rear axle, what is the go-cart's acceleration?

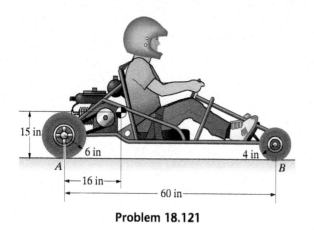

Problem 18.121

18.122 Bar AB rotates with a constant angular velocity of 10 rad/s in the counterclockwise direction. The masses of the slender bars BC and CDE are 2 kg and 3.6 kg, respectively. The y axis points upward. Determine the components of the forces exerted on bar BC by the pins at B and C at the instant shown.

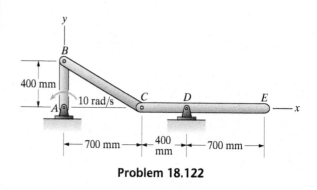

Problem 18.122

18.123 At the instant shown, the arms of the robotic manipulator have constant counterclockwise angular velocities $\omega_{AB} = -0.5$ rad/s, $\omega_{BC} = 2$ rad/s, and $\omega_{CD} = 4$ rad/s. The mass of arm CD is 10 kg, and its center of mass is at its midpoint. At this instant, what force and couple are exerted on arm CD at C?

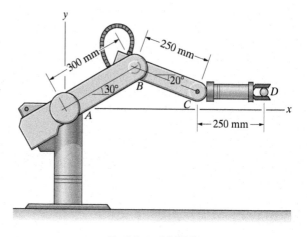

Problem 18.123

18.124 Each bar is 1 m in length and has a mass of 4 kg. The inclined surface is smooth. If the system is released from rest in the position shown, what are the angular accelerations of the bars at that instant?

18.125 Each bar is 1 m in length and has a mass of 4 kg. The inclined surface is smooth. If the system is released from rest in the position shown, what is the magnitude of the force exerted on bar OA by the support at O at that instant?

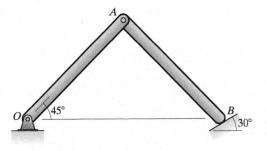

Problems 18.124/18.125

18.126* The fixed ring gear lies *in the horizontal plane*. The hub and planet gears are bonded together. The mass and moment of inertia of the combined hub and planet gears are $m_{HP} = 130$ kg and $I_{HP} = 130$ kg-m^2. The moment of inertia of the sun gear is $I_S = 60$ kg-m^2. The mass of the connecting rod is 5 kg, and it can be modeled as a slender bar. If a 1 kN-m counterclockwise couple is applied to the sun gear, what is the resulting angular acceleration of the bonded hub and planet gears?

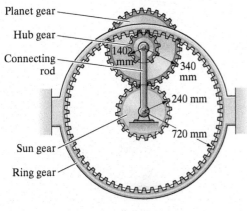

Problem 18.126

18.127* The system is stationary at the instant shown. The net force exerted on the piston by the exploding fuel–air mixture and friction is 5 kN to the left. A clockwise couple $M = 200$ N-m acts on the crank AB. The moment of inertia of the crank about A is 0.0003 kg-m^2. The mass of the connecting rod BC is 0.36 kg, and its center of mass is 40 mm from B on the line from B to C. The connecting rod's moment of inertia about its center of mass is 0.0004 kg-m^2. The mass of the piston is 4.6 kg. What is the piston's acceleration? (Neglect the gravitational forces on the crank and connecting rod.)

18.128* If the crank AB described in Problem 18.127 has a counterclockwise angular velocity of 2000 rpm at the instant shown, what is the piston's acceleration due to the couple M and the force exerted by the fuel–air mixture?

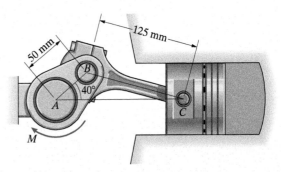

Problems 18.127/18.128

Design Project

Investigate the effects of the coefficients of friction and the dimensions b, c, and h on the bicycle rider's ability to accelerate and decelerate. Consider a range of coefficients of friction that you think might encompass dry, wet, and icy road surfaces. In studying deceleration, consider the cases in which the brakes act on the front wheel only, the rear wheel only, and both wheels. Pay particular attention to the constraint that the bicycle's front and rear wheels should not leave the ground. Write a brief report presenting your analyses and making observations about their implications for bicycle design. Notice that the rider can alter the dimensions h, b, and c to some extent by changing the position of his upper body. Comment on how the rider can thereby affect the bicycle's performance with respect to acceleration and deceleration.

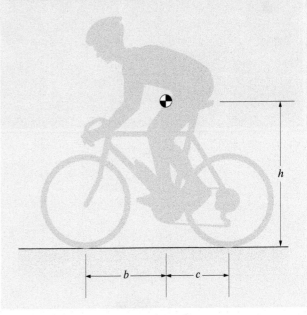

Energy and Momentum in Rigid-Body Dynamics

In Chapters 15 and 16 we demonstrated that energy and momentum methods are useful for solving particular types of problems in dynamics. If the forces on an object are known as functions of position, the principle of work and energy can be used to determine the change in the magnitude of the velocity of the object as it moves between two positions. If the forces are known as functions of time, the principle of impulse and momentum can be used to determine the change in the object's velocity during an interval of time. We now extend these methods to situations in which both the translational and rotational motions of objects must be considered.

$\int \Sigma M \, d\theta$

◀ The wind does work on the rotor of the wind turbine, causing it to rotate and power an electric generator.

19.1 Work and Energy

BACKGROUND

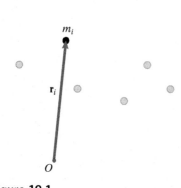

Figure 19.1
A system of particles. The vector $\mathbf{r}_i$ is the position vector of the ith particle.

We will show that the work done on a rigid body by external forces and couples as it moves between two positions is equal to the change in its kinetic energy. To obtain this result, we adopt the same approach used in Chapter 18 to obtain the equations of motion for a rigid body. We derive the principle of work and energy for a system of particles and use it to deduce the principle for a rigid body.

Let m_i be the mass of the ith particle of a system of N particles. Let $\mathbf{r}_i$ be the position of the ith particle relative to a point O that is fixed with respect to an inertial reference frame (Fig. 19.1). We denote the sum of the kinetic energies of the particles by

$$T = \sum_i \tfrac{1}{2} m_i \mathbf{v}_i \cdot \mathbf{v}_i, \tag{19.1}$$

where $\mathbf{v}_i = d\mathbf{r}_i/dt$ is the velocity of the ith particle. Our objective is to relate the work done on the system of particles to the change in T. We begin with Newton's second law for the ith particle,

$$\sum_j \mathbf{f}_{ij} + \mathbf{f}_i^{\mathrm{E}} = \frac{d}{dt}(m_i \mathbf{v}_i),$$

where $\mathbf{f}_{ij}$ is the force exerted on the ith particle by the jth particle and $\mathbf{f}_i^{E}$ is the external force on the ith particle. We take the dot product of this equation with $\mathbf{v}_i$ and sum from $i = 1$ to N:

$$\sum_i \sum_j \mathbf{f}_{ij} \cdot \mathbf{v}_i + \sum_i \mathbf{f}_i^{\mathrm{E}} \cdot \mathbf{v}_i = \sum_i \mathbf{v}_i \cdot \frac{d}{dt}(m_i \mathbf{v}_i). \tag{19.2}$$

We can express the term on the right side of this equation as the rate of change of the total kinetic energy:

$$\sum_i \mathbf{v}_i \cdot \frac{d}{dt}(m_i \mathbf{v}_i) = \frac{d}{dt} \sum_i \tfrac{1}{2} m_i \mathbf{v}_i \cdot \mathbf{v}_i = \frac{dT}{dt}.$$

Multiplying Eq. (19.2) by dt yields

$$\sum_i \sum_j \mathbf{f}_{ij} \cdot d\mathbf{r}_i + \sum_i \mathbf{f}_i^{\mathrm{E}} \cdot d\mathbf{r}_i = dT.$$

We integrate this equation, obtaining

$$\sum_i \sum_j \int_{(\mathbf{r}_i)_1}^{(\mathbf{r}_i)_2} \mathbf{f}_{ij} \cdot d\mathbf{r}_i + \sum_i \int_{(\mathbf{r}_i)_1}^{(\mathbf{r}_i)_2} \mathbf{f}_i^{\mathrm{E}} \cdot d\mathbf{r}_i = T_2 - T_1. \tag{19.3}$$

The terms on the left side are the work done on the system by internal and external forces as the particles move from positions $(\mathbf{r}_i)_1$ to positions $(\mathbf{r}_i)_2$. We see that the work done by internal and external forces as a system of particles moves between two positions equals the change in the total kinetic energy of the system.

If the particles represent a rigid body, and we assume that the internal forces between each pair of particles are directed along the straight line between them, the work done by internal forces is zero. To show that this is true, we consider two particles of a rigid body designated 1 and 2 (Fig. 19.2). The sum of the

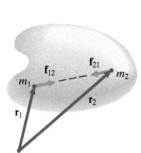

Figure 19.2
Particles 1 and 2 and the forces they exert on each other.

forces the two particles exert on each other is zero ($\mathbf{f}_{12} + \mathbf{f}_{21} = \mathbf{0}$), so the rate at which the forces do work (the power) is

$$\mathbf{f}_{12} \cdot \mathbf{v}_1 + \mathbf{f}_{21} \cdot \mathbf{v}_2 = \mathbf{f}_{21} \cdot (\mathbf{v}_2 - \mathbf{v}_1).$$

We can show that $\mathbf{f}_{21}$ is perpendicular to $\mathbf{v}_2 - \mathbf{v}_1$, and therefore the rate at which work is done by the internal forces between these two particles is zero. Because the particles are points of a rigid body, we can express their relative velocity in terms of the rigid body's angular velocity $\boldsymbol{\omega}$ as

$$\mathbf{v}_2 - \mathbf{v}_1 = \boldsymbol{\omega} \times (\mathbf{r}_2 - \mathbf{r}_1). \tag{19.4}$$

This equation shows that the relative velocity $\mathbf{v}_2 - \mathbf{v}_1$ is perpendicular to $\mathbf{r}_2 - \mathbf{r}_1$, which is the position vector from particle 1 to particle 2. Since the force $\mathbf{f}_{21}$ is parallel to $\mathbf{r}_2 - \mathbf{r}_1$, it is perpendicular to $\mathbf{v}_2 - \mathbf{v}_1$. We can repeat this argument for each pair of particles of the rigid body, so the total rate at which work is done by internal forces is zero. This implies that the work done by internal forces as a rigid body moves between two positions is zero. Notice that *if an object is not rigid, work can be done by internal forces.*

Therefore, in the case of a rigid body, the work done by internal forces in Eq. (19.3) vanishes. Denoting the work done by external forces by U_{12}, we obtain the principle of work and energy for a rigid body: *The work done by external forces and couples as a rigid body moves between two positions equals the change in the total kinetic energy of the body:*

$$U_{12} = T_2 - T_1. \tag{19.5}$$

We can also state this principle for a *system* of rigid bodies: *The work done by external and internal forces and couples as a system of rigid bodies moves between two positions equals the change in the total kinetic energy of the system.*

Kinetic Energy

The kinetic energy of a rigid body can be expressed in terms of the velocity of the center of mass of the body and its angular velocity. We consider first general planar motion and then rotation about a fixed axis.

General Planar Motion   Let us model a rigid body as a system of particles, and let $\mathbf{R}_i$ be the position vector of the ith particle relative to the body's center of mass (Fig. 19.3). The position of the center of mass is

$$\mathbf{r} = \frac{\sum\limits_i m_i \mathbf{r}_i}{m},$$

where m is the mass of the rigid body. The position of the ith particle relative to O is related to its position relative to the center of mass by

$$\mathbf{r}_i = \mathbf{r} + \mathbf{R}_i, \tag{19.6}$$

and the vectors $\mathbf{R}_i$ satisfy the relation

$$\sum\limits_i m_i \mathbf{R}_i = \mathbf{0}. \tag{19.7}$$

The kinetic energy of the rigid body is the sum of the kinetic energies of its particles, given by Eq. (19.1):

$$T = \sum\limits_i \tfrac{1}{2} m_i \mathbf{v}_i \cdot \mathbf{v}_i. \tag{19.8}$$

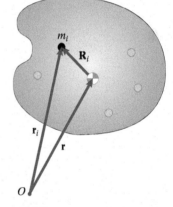

Figure 19.3
Representing a rigid body as a system of particles.

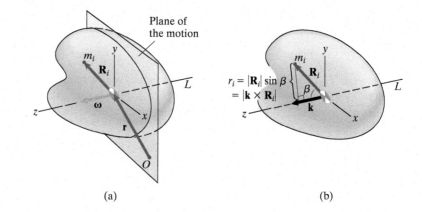

Figure 19.4
(a) A coordinate system with the z axis aligned with L.
(b) The magnitude of $\mathbf{k} \times \mathbf{R}_i$ is the perpendicular distance from L to m_i.

(a)   (b)

By taking the derivative of Eq. (19.6) with respect to time, we obtain

$$\mathbf{v}_i = \mathbf{v} + \frac{d\mathbf{R}_i}{dt},$$

where $\mathbf{v}$ is the velocity of the center of mass. Substituting this expression into Eq. (19.8) and using Eq. (19.7), we obtain the kinetic energy of the rigid body in the form

$$T = \tfrac{1}{2}mv^2 + \sum_i \tfrac{1}{2}m_i \frac{d\mathbf{R}_i}{dt} \cdot \frac{d\mathbf{R}_i}{dt}, \tag{19.9}$$

where v is the magnitude of the velocity of the center of mass.

Let L be the axis through the center of mass that is perpendicular to the plane of the motion (Fig. 19.4a). In terms of the coordinate system shown, we can express the angular velocity vector as $\boldsymbol{\omega} = \omega\mathbf{k}$. The velocity of the ith particle relative to the center of mass is $d\mathbf{R}_i/dt = \omega\mathbf{k} \times \mathbf{R}_i$, so we can write Eq. (19.9) as

$$T = \tfrac{1}{2}mv^2 + \tfrac{1}{2}\left[\sum_i m_i(\mathbf{k} \times \mathbf{R}_i) \cdot (\mathbf{k} \times \mathbf{R}_i) \right]\omega^2. \tag{19.10}$$

The magnitude of the vector $\mathbf{k} \times \mathbf{R}_i$ is the perpendicular distance r_i from L to the ith particle (Fig. 19.4b), so the term in brackets in Eq. (19.10) is the moment of inertia of the body about L:

$$\sum_i m_i(\mathbf{k} \times \mathbf{R}_i) \cdot (\mathbf{k} \times \mathbf{R}_i) = \sum_i m_i|\mathbf{k} \times \mathbf{R}_i|^2 = \sum_i m_i r_i^2 = I.$$

Thus, we obtain the kinetic energy of a rigid body in general planar motion in the form

$$T = \tfrac{1}{2}mv^2 + \tfrac{1}{2}I\omega^2, \tag{19.11}$$

where m is the mass of the rigid body, v is the magnitude of the velocity of the center of mass, I is the moment of inertia about the axis L through the center of mass, and ω is the angular velocity. We see that the kinetic energy consists of two terms, the *translational kinetic energy* due to the velocity of the center of mass and the *rotational kinetic energy* due to the angular velocity (Fig. 19.5).

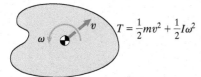

Figure 19.5
Kinetic energy in general planar motion.

Fixed-Axis Rotation   An object rotating about a fixed axis is in general planar motion, and its kinetic energy is given by Eq. (19.11). But in this case there is another expression for the kinetic energy that is often convenient. Suppose that a rigid body rotates with angular velocity ω about a fixed axis O. In terms of the distance d from O to the center of mass of the body, the velocity of the center of mass is $v = \omega d$ (Fig. 19.6a). From Eq. (19.11), the kinetic energy is

$$T = \tfrac{1}{2}m(\omega d)^2 + \tfrac{1}{2}I\omega^2 = \tfrac{1}{2}(I + d^2 m)\omega^2.$$

According to the parallel-axis theorem, the moment of inertia about O is $I_O = I + d^2 m$, so we obtain the kinetic energy of a rigid body rotating about a fixed axis O in the form (Fig. 19.6b)

$$T = \tfrac{1}{2}I_O\omega^2. \tag{19.12}$$

Work and Potential Energy

The procedures for determining the work done by different types of forces and the expressions for the potential energies of forces discussed in Chapter 15 provide the essential tools for applying the principle of work and energy to a rigid body. The work done on a rigid body by a force $\mathbf{F}$ is given by

$$U_{12} = \int_{(\mathbf{r}_p)_1}^{(\mathbf{r}_p)_2} \mathbf{F} \cdot d\mathbf{r}_p, \tag{19.13}$$

where $\mathbf{r}_p$ is the position of the point of application of $\mathbf{F}$ (Fig. 19.7). If the point of application is stationary, or if its direction of motion is perpendicular to $\mathbf{F}$, no work is done.

A force $\mathbf{F}$ is conservative if a potential energy V exists such that

$$\mathbf{F} \cdot d\mathbf{r}_p = -dV. \tag{19.14}$$

In terms of its potential energy, the work done by a conservative force $\mathbf{F}$ is

$$U_{12} = \int_{(\mathbf{r}_p)_1}^{(\mathbf{r}_p)_2} \mathbf{F} \cdot d\mathbf{r}_p = \int_{V_1}^{V_2} -dV = -(V_2 - V_1),$$

where V_1 and V_2 are the values of V at $(\mathbf{r}_p)_1$ and $(\mathbf{r}_p)_2$.

If a rigid body is subjected to a couple M (Fig. 19.8a), what work is done as the body moves between two positions? We can evaluate the work by representing the couple by forces (Fig. 19.8b) and determining the work done by the

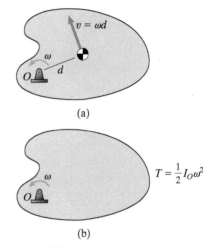

(a)

(b)

Figure 19.6
(a) Velocity of the center of mass.
(b) Kinetic energy of a rigid body rotating about a fixed axis.

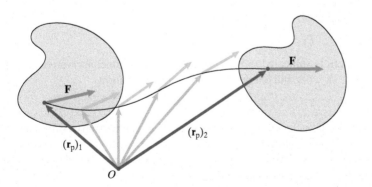

Figure 19.7
The work done by a force on a rigid body is determined by the path of the point of application of the force.

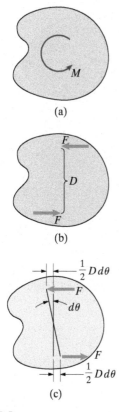

Figure 19.8
(a) A rigid body subjected to a couple.
(b) An equivalent couple consisting of two forces: $DF = M$.
(c) Determining the work done by the forces.

forces. If the rigid body rotates through an angle $d\theta$ in the direction of the couple (Fig. 19.8c), the work done by each force is $\left(\frac{1}{2}D \, d\theta\right)F$, so the total work is $DF \, d\theta = M \, d\theta$. Integrating this expression, we obtain the work done by a couple M as the rigid body rotates from θ_1 to θ_2 in the direction of M:

$$U_{12} = \int_{\theta_1}^{\theta_2} M \, d\theta. \tag{19.15}$$

If M is constant, the work is simply the product of the couple and the angular displacement:

$$U_{12} = M(\theta_2 - \theta_1) \quad \text{(constant couple)}$$

A couple M is conservative if a potential energy V exists such that

$$M \, d\theta = -dV. \tag{19.16}$$

We can express the work done by a conservative couple in terms of its potential energy:

$$U_{12} = \int_{\theta_1}^{\theta_2} M \, d\theta = \int_{V_1}^{V_2} -dV = -(V_2 - V_1).$$

For example, in Fig. 19.9, a torsional spring exerts a couple on a bar that is proportional to the bar's angle of rotation: $M = -k\theta$. From the relation

$$M \, d\theta = -k\theta \, d\theta = -dV,$$

we see that the potential energy must satisfy the equation

$$\frac{dV}{d\theta} = k\theta.$$

Integrating this equation, we find that the potential energy of the torsional spring is

$$V = \tfrac{1}{2}k\theta^2. \tag{19.17}$$

If *all* the forces and couples that do work on a system of rigid bodies are conservative, we can express the total work done as the body moves between two positions 1 and 2 in terms of the total potential energy of the forces and couples:

$$U_{12} = V_1 - V_2.$$

Combining this relation with the principle of work and energy, Eq. (19.5), we conclude that the sum of the kinetic energy and the total potential energy is constant—energy is conserved:

$$T_1 + V_1 = T_2 + V_2 \tag{19.18}$$

If a system is subjected to both conservative and nonconservative forces, the principle of work and energy can be written in the form

$$T_1 + V_1 + U_{12} = T_2 + V_2 \tag{19.19}$$

The term U_{12} includes the work done by all nonconservative forces acting on the system as it moves from position 1 to position 2. If a force is conservative, there is a choice. The work it does can be calculated and included in U_{12}, *or* the force's potential energy can be included in V.

Power

The work done on a rigid body by a force $\mathbf{F}$ during an infinitesimal displacement $d\mathbf{r}_p$ of its point of application is

$$\mathbf{F} \cdot d\mathbf{r}_p.$$

We obtain the power P transmitted to the rigid body—the rate at which work is done on it—by dividing this expression by the interval of time dt during which the displacement takes place. We obtain

$$P = \mathbf{F} \cdot \mathbf{v}_p, \qquad (19.20)$$

where $\mathbf{v}_p$ is the velocity of the point of application of $\mathbf{F}$.

Similarly, the work done on a rigid body in planar motion by a couple M during an infinitesimal rotation $d\theta$ in the direction of M is

$$M \, d\theta.$$

Dividing this expression by dt, we find that the power transmitted to the rigid body is the product of the couple and the angular velocity:

$$P = M\omega. \qquad (19.21)$$

The total work done on a rigid body during an interval of time equals the change in kinetic energy of the body, so the total power transmitted equals the rate of change of the body's kinetic energy:

$$P = \frac{dT}{dt}.$$

The average with respect to time of the power during an interval of time from t_1 to t_2 is

$$P_{av} = \frac{1}{t_2 - t_1} \int_{t_1}^{t_2} P \, dt = \frac{1}{t_2 - t_1} \int_{T_1}^{T_2} dT = \frac{T_2 - T_1}{t_2 - t_1}.$$

This expression shows that we can determine the average power transferred to or from a rigid body during an interval of time by dividing the change in kinetic energy of the body, or the total work done, by the interval of time:

$$P_{av} = \frac{T_2 - T_1}{t_2 - t_1} = \frac{U_{12}}{t_2 - t_1}. \qquad (19.22)$$

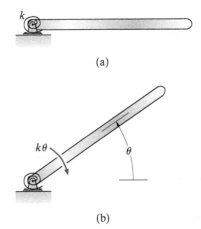

Figure 19.9
(a) A linear torsional spring connected to a bar.
(b) The spring exerts a couple of magnitude $k\theta$ in the direction opposite that of the bar's rotation.

RESULTS

Principle of Work and Energy

Let T be the total kinetic energy of a system of rigid bodies. The *principle of work and energy* states that the work done by external and internal forces and couples as the system moves between two positions equals the change in the total kinetic energy of the system. $\longrightarrow$ $U_{12} = T_2 - T_1.$ (19.5)

Kinetic Energy

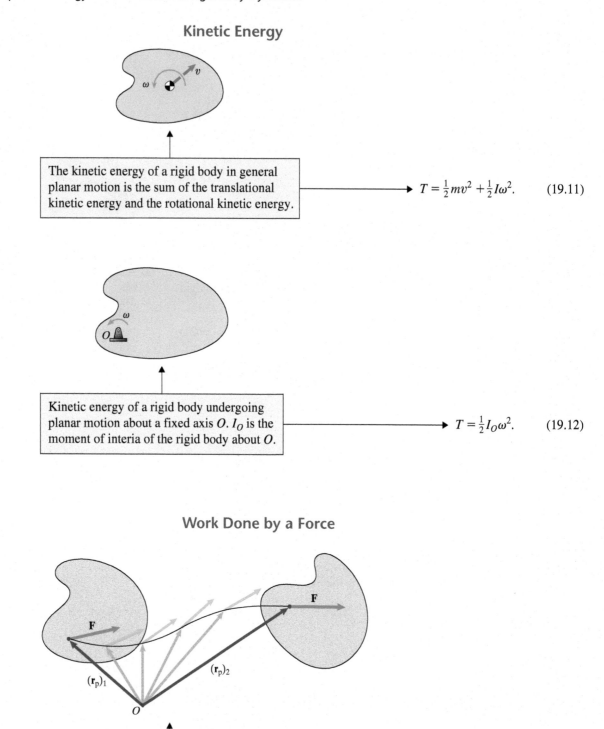

The kinetic energy of a rigid body in general planar motion is the sum of the translational kinetic energy and the rotational kinetic energy.

$$T = \tfrac{1}{2}mv^2 + \tfrac{1}{2}I\omega^2. \qquad (19.11)$$

Kinetic energy of a rigid body undergoing planar motion about a fixed axis O. I_O is the moment of interia of the rigid body about O.

$$T = \tfrac{1}{2}I_O\omega^2. \qquad (19.12)$$

Work Done by a Force

Work done on a rigid body by a force $\mathbf{F}$ as the point of application of the force moves from position $(\mathbf{r}_p)_1$ to position $(\mathbf{r}_p)_2$. A force $\mathbf{F}$ is conservative if a potential energy V exists such that

$$\mathbf{F} \cdot d\mathbf{r}_p = -dV. \qquad (19.14)$$

$$U_{12} = \int_{(\mathbf{r}_p)_1}^{(\mathbf{r}_p)_2} \mathbf{F} \cdot d\mathbf{r}_p. \qquad (19.13)$$

Work Done by a Couple

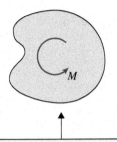

Work done on a rigid body by a couple M as the rigid body rotates from angular position (in radians) θ_1 to angular position θ_2. A couple M is conservative if a potential energy V exists such that

$$M\, d\theta = -dV. \qquad (19.16)$$

$$U_{12} = \int_{\theta_1}^{\theta_2} M\, d\theta. \qquad (19.15)$$

If M is constant,

$$U_{12} = M\,(\theta_2 - \theta_1).$$

Conservation of Energy

If all of the forces that do work on a system of rigid bodies are conservative, the sum of the total kinetic energy and the total potential energy is the same at any two positions.

$$T_1 + V_1 = T_2 + V_2. \qquad (19.18)$$

When both conservative and nonconservative forces do work on a system of rigid bodies, the principle of work and energy can be expressed in terms of the potential energy V of the conservative forces and the work U_{12} done by nonconservative forces.

$$T_1 + V_1 + U_{12} = T_2 + V_2. \qquad (19.19)$$

Applying energy methods to a rigid body or system of rigid bodies typically involves three steps.

1. *Identify the forces and couples that do work.* Use free-body diagrams to determine which external forces and couples do work on the system.

2. *Apply the principle of work and energy or conservation of energy.* Either equate the total work done during a change in position to the change in the kinetic energy, or equate the sum of the kinetic and potential energies at two positions.

3. *Determine kinematic relationships.* To complete the solution, it will often be necessary to obtain relations between velocities of points of rigid bodies and their angular velocities.

Power

| The power transmitted to a rigid body by a force. The term $\mathbf{v}_p$ is the velocity of the point of application of the force. | $\longrightarrow$ | $P = \mathbf{F} \cdot \mathbf{v}_p.$ | (19.20) |

| The power transmitted to a rigid body by a couple is the product of the couple and the angular velocity of the rigid body. | $\longrightarrow$ | $P = M\omega.$ | (19.21) |

| The average power transferred to a rigid body during an interval of time is equal to the change in kinetic energy of the body, or the total work done during that time, divided by the interval of time. | $\longrightarrow$ | $P_{av} = \dfrac{T_2 - T_1}{t_2 - t_1} = \dfrac{U_{12}}{t_2 - t_1}.$ | (19.22) |

Active Example 19.1 **Applying Work and Energy to a Rolling Disk** (▶ *Related Problems 19.19, 19.20*)

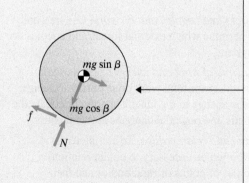

A disk of mass m and moment of inertia $I = \frac{1}{2}mR^2$ is released from rest on an inclined surface. Assuming that the disk rolls without slipping, what is the magnitude of the velocity of its center when it has moved a distance b?

Strategy
We will equate the work done as the disk rolls a distance b to the change in its kinetic energy. To determine the velocity of the center of the disk, we will need the relationship between the velocity of the center and the angular velocity of the disk.

Solution

Identify the Forces and Couples that Do Work

Free-body diagram of the disk. N and f are the normal and friction forces exerted by the inclined surface. The disk's weight does work as the disk rolls, but the normal and friction forces do not. To explain why, we can write the work done by a force $\mathbf{F}$ as

$$\int_{(\mathbf{r}_p)_1}^{(\mathbf{r}_p)_2} \mathbf{F} \cdot d\mathbf{r}_p = \int_{t_1}^{t_2} \mathbf{F} \cdot \frac{d\mathbf{r}_p}{dt}\, dt = \int_{t_1}^{t_2} \mathbf{F} \cdot \mathbf{v}_p\, dt,$$

where $\mathbf{v}_p$ is the velocity of the point of application of $\mathbf{F}$. Because the velocity of the point where the normal and friction forces act is zero, they do no work.

Apply Work and Energy

The work done by the disk's weight is the product of the component of the weight in the direction of motion and the distance b. Notice that this result is also obtained by multiplying the weight of the disk mg by the decrease in height of the center of mass.

$U_{12} = (mg \sin \beta)b.$

Equate the work to the change in the kinetic energy of the disk. The initial kinetic energy is zero. Let v be the velocity of the center of the disk and ω its angular velocity when the center has moved a distance b.

$mgb \sin \beta = \frac{1}{2}mv^2 + \frac{1}{2}I\omega^2 - 0.$ (1)

Determine Kinematic Relationships

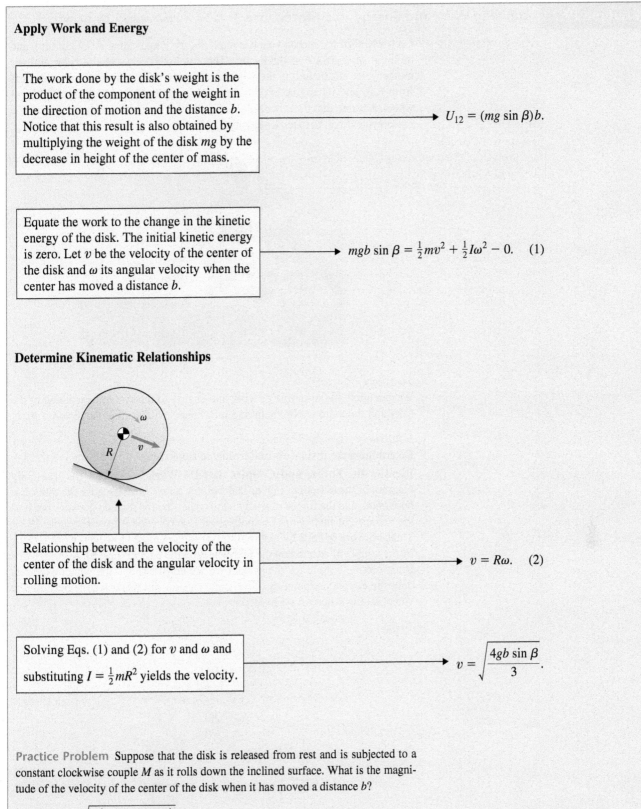

Relationship between the velocity of the center of the disk and the angular velocity in rolling motion.

$v = R\omega.$ (2)

Solving Eqs. (1) and (2) for v and ω and substituting $I = \frac{1}{2}mR^2$ yields the velocity.

$v = \sqrt{\dfrac{4gb \sin \beta}{3}}.$

Practice Problem Suppose that the disk is released from rest and is subjected to a constant clockwise couple M as it rolls down the inclined surface. What is the magnitude of the velocity of the center of the disk when it has moved a distance b?

Answer: $v = \sqrt{\dfrac{4}{3}\left(gb \sin \beta + \dfrac{Mb}{mR}\right)}.$

Example 19.2 | Applying Work and Energy to a Motorcycle (▶ *Related Problems 19.27, 19.28*)

Each wheel of the motorcycle has mass $m_W = 9$ kg, radius $R = 330$ mm, and moment of inertia $I = 0.8$ kg-m^2. The combined mass of the rider and the motorcycle, not including the wheels, is $m_C = 142$ kg. The motorcycle starts from rest, and its engine exerts a constant couple $M = 140$ N-m on the rear wheel. Assume that the wheels do not slip. What horizontal distance b must the motorcycle travel to reach a velocity of 25 m/s?

Strategy

We can apply the principle of work and energy to the system consisting of the rider and the motorcycle, including its wheels, to determine the distance b.

Solution

Determining the distance b requires three steps.

Identify the Forces and Couples that Do Work We draw the free-body diagram of the system in Fig. a. The weights do no work because the motion is horizontal, and the forces exerted on the wheels by the road do no work because the velocity of their point of application is zero. (See Active Example 19.1.) Thus, no work is done by external forces and couples! However, work is done by the couple M exerted on the rear wheel by the engine (Fig. b). Although this is an internal couple for the system we are considering—the wheel exerts an opposite couple on the body of the motorcycle—net work is done because the wheel rotates whereas the body does not.

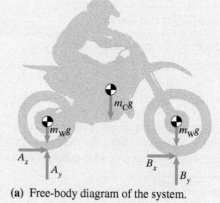

(a) Free-body diagram of the system.

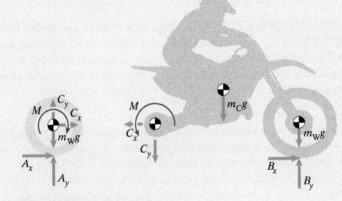

(b) Isolating the rear wheel.

Apply Work and Energy   If the motorcycle moves a horizontal distance b, the wheels turn through an angle b/R rad, and the work done by the constant couple M is

$$U_{12} = M(\theta_2 - \theta_1) = M\left(\frac{b}{R}\right).$$

Let v be the motorcycle's velocity and ω the angular velocity of the wheels when the motorcycle has moved a distance b. The work equals the change in the total kinetic energy:

$$M\left(\frac{b}{R}\right) = \frac{1}{2}m_C v^2 + 2\left(\frac{1}{2}m_W v^2 + \frac{1}{2}I\omega^2\right) - 0. \qquad (1)$$

Determine Kinematic Relationship   The angular velocity of the rolling wheels is related to the velocity v by $\omega = v/R$. Substituting this relation into Eq. (1) and solving for b, we obtain

$$b = \left(\frac{1}{2}m_C + m_W + \frac{I}{R^2}\right)\frac{Rv^2}{M}$$

$$= \left[\frac{1}{2}(142 \text{ kg}) + (9 \text{ kg}) + \frac{0.8 \text{ kg-m}^2}{(0.33 \text{ m})^2}\right]\frac{(0.33 \text{ m})(25 \text{ m/s})^2}{140 \text{ N-m}}$$

$$= 129 \text{ m}.$$

Critical Thinking

Although we drew separate free-body diagrams of the motorcycle and its rear wheel to clarify the work done by the couple exerted by the engine, notice that we treated the motorcycle, including its wheels, as a single system in applying the principle of work and energy. By doing so, we did not need to consider the work done by the internal forces between the motorcycle's body and its wheels. When applying the principle of work and energy to a system of rigid bodies, you will usually find it simplest to express the principle for the system as a whole. This is in contrast to determining the motion of a system of rigid bodies by using the equations of motion, which usually requires that you draw free-body diagrams of each rigid body and apply the equations to them individually.

Example 19.3   Applying Conservation of Energy to a Linkage (▶ *Related Problem 19.40*)

The slender bars AB and BC of the linkage have mass m and length l, and the collar C has mass m_C. A torsional spring at A exerts a clockwise couple $k\theta$ on bar AB. The system is released from rest in the position $\theta = 0$ and allowed to fall. Neglecting friction, determine the angular velocity $\omega = d\theta/dt$ of bar AB as a function of θ.

Strategy

The objective in this example—determining an angular velocity as a function of the position of the system—encourages an energy approach. We must first identify forces and couples that do work on the system. If they are conservative, we can apply conservation of energy to determine ω as a function of θ. If nonconservative forces do work on the system, we can apply the principle of work and energy.

Solution

Identify the Forces and Couples that Do Work   We draw the free-body diagram of the system in Fig. a. The forces and couples that do work—the weights of the bars and collar and the couple exerted by the torsional spring—are conservative. We can use conservation of energy and the kinematic relationships between the angular velocities of the bars and the velocity of the collar to determine ω as a function of θ.

Apply Conservation of Energy   We denote the center of mass of bar BC by G and the angular velocity of bar BC by ω_{BC} (Fig. b). The moment of inertia of each bar about its center of mass is $I = \frac{1}{12}ml^2$. Since bar AB rotates about the fixed point A, we can write its kinetic energy as

$$T_{\text{bar } AB} = \tfrac{1}{2}I_A\omega^2 = \tfrac{1}{2}\left[I + \left(\tfrac{1}{2}l\right)^2 m\right]\omega^2 = \tfrac{1}{6}ml^2\omega^2.$$

The kinetic energy of bar BC is

$$T_{\text{bar } BC} = \tfrac{1}{2}mv_G^2 + \tfrac{1}{2}I\omega_{BC}^2 = \tfrac{1}{2}mv_G^2 + \tfrac{1}{24}ml^2\omega_{BC}^2.$$

The kinetic energy of the collar C is

$$T_{\text{collar}} = \tfrac{1}{2}m_C v_C^2.$$

Using the datum in Fig. (a), we obtain the potential energies of the weights:

$$V_{\text{bar } AB} + V_{\text{bar } BC} + V_{\text{collar}} = mg(\tfrac{1}{2}l\cos\theta) + mg(\tfrac{3}{2}l\cos\theta) + m_C g(2l\cos\theta).$$

The potential energy of the torsional spring is given by Eq. (19.17):

$$V_{\text{spring}} = \tfrac{1}{2}k\theta^2.$$

We now have all the ingredients to apply conservation of energy. We equate the sum of the kinetic and potential energies at the position $\theta = 0$ to the sum of the kinetic and potential energies at an arbitrary value of θ:

$$T_1 + V_1 = T_2 + V_2:$$

$$0 + 2mgl + 2m_C gl = \tfrac{1}{6}ml^2\omega^2 + \tfrac{1}{2}mv_G^2 + \tfrac{1}{24}ml^2\omega_{BC}^2 + \tfrac{1}{2}m_C v_C^2$$
$$+ 2mgl\cos\theta + 2m_C gl\cos\theta + \tfrac{1}{2}k\theta^2.$$

To determine ω from this equation, we must express the velocities v_G, v_C, and ω_{BC} in terms of ω.

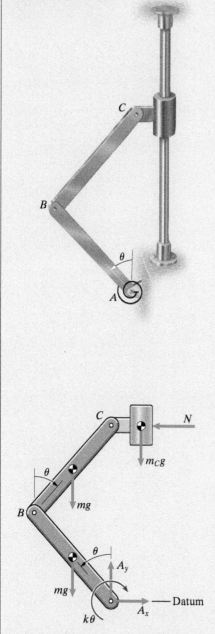

(a) Free-body diagram of the system.

Determine Kinematic Relationships We can determine the velocity of point B in terms of ω and then express the velocity of point C in terms of the velocity of point B and the angular velocity ω_{BC}.

The velocity of B is

$$\mathbf{v}_B = \mathbf{v}_A + \boldsymbol{\omega}_{AB} \times \mathbf{r}_{B/A}$$

$$= \mathbf{0} + \begin{vmatrix} \mathbf{i} & \mathbf{j} & \mathbf{k} \\ 0 & 0 & \omega \\ -l \sin \theta & l \cos \theta & 0 \end{vmatrix}$$

$$= -l\omega \cos \theta \, \mathbf{i} - l\omega \sin \theta \, \mathbf{j}.$$

The velocity of C, expressed in terms of the velocity of B, is

$$v_C \mathbf{j} = \mathbf{v}_B + \boldsymbol{\omega}_{BC} \times \mathbf{r}_{C/B}$$

$$= -l\omega \cos \theta \, \mathbf{i} - l\omega \sin \theta \, \mathbf{j} + \begin{vmatrix} \mathbf{i} & \mathbf{j} & \mathbf{k} \\ 0 & 0 & \omega_{BC} \\ l \sin \theta & l \cos \theta & 0 \end{vmatrix}.$$

Equating $\mathbf{i}$ and $\mathbf{j}$ components, we obtain

$$\omega_{BC} = -\omega, \qquad v_C = -2l\omega \sin \theta.$$

(The minus signs indicate that the directions of the velocities are opposite to the directions we assumed in Fig. b.) Now that we know the angular velocity of bar BC in terms of ω, we can determine the velocity of its center of mass in terms of ω by expressing it in terms of $\mathbf{v}_B$:

$$\mathbf{v}_G = \mathbf{v}_B + \boldsymbol{\omega}_{BC} \times \mathbf{r}_{G/B}$$

$$= -l\omega \cos \theta \, \mathbf{i} - l\omega \sin \theta \, \mathbf{j} + \begin{vmatrix} \mathbf{i} & \mathbf{j} & \mathbf{k} \\ 0 & 0 & -\omega \\ \frac{1}{2}l \sin \theta & \frac{1}{2}l \cos \theta & 0 \end{vmatrix}$$

$$= -\tfrac{1}{2} l\omega \cos \theta \, \mathbf{i} - \tfrac{3}{2} l\omega \sin \theta \, \mathbf{j}.$$

Substituting these expressions for ω_{BC}, v_C, and $\mathbf{v}_G$ into our equation of conservation of energy and solving for ω, we obtain

$$\omega = \left[\frac{2gl(m + m_C)(1 - \cos \theta) - \frac{1}{2}k\theta^2}{\frac{1}{3}ml^2 + (m + 2m_C)l^2 \sin^2 \theta} \right]^{1/2}.$$

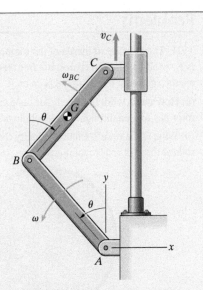

(b) Angular velocities of the bars and the velocity of the collar.

Critical Thinking

Newton's second law and the equation of angular motion for a rigid body can be applied to this example instead of conservation of energy. How can you decide what approach to use? The energy methods we have described are generally useful only when you can easily determine the work done by forces and couples acting on a system or their associated potential energies. When that is the case, an energy approach is often preferable. To apply Newton's second law and the equation of angular motion to this example, it would be necessary to draw individual free-body diagrams of the two bars and the collar C, thereby introducing into the formulation the forces exerted on the bars and collar at the pins connecting them. In contrast, we were able to apply conservation of energy to the system as a whole, greatly simplifying the solution.

Problems

19.1 The moment of inertia of the rotor of the medical centrifuge is $I = 0.2$ kg-m^2. The rotor starts from rest and the motor exerts a constant torque of 0.8 N-m on it.

(a) How much work has the motor done on the rotor when the rotor has rotated through four revolutions?

(b) What is the rotor's angular velocity (in rpm) when it has rotated through four revolutions?

Problem 19.1

19.2 The 4-lb slender bar is 2 ft in length. It started from rest in an initial position relative to the inertial reference frame. When it is in the position shown, the velocity of the end A is $2\mathbf{i} + 6\mathbf{j}$ (ft/s) and the bar has a counterclockwise angular velocity of 12 rad/s. How much work was done on the bar as it moved from its initial position to its present position?

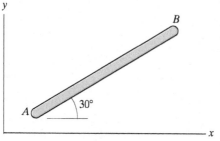

Problem 19.2

19.3 The 20-kg disk is at rest when the constant 10 N-m counterclockwise couple is applied. Determine the disk's angular velocity (in rpm) when it has rotated through four revolutions (a) by applying the equation of angular motion $\Sigma M = I\alpha$, and (b) by applying the principle of work and energy.

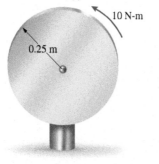

Problem 19.3

19.4 The space station is initially not rotating. Its reaction control system exerts a constant couple on it until it has rotated 90°, then exerts a constant couple of the same magnitude in the opposite direction so that its angular velocity has decreased to zero when it has undergone a total rotation of 180°. The maneuver takes 6 hours. The station's moment of inertia about the axis of rotation is $I = 1.5 \times 10^{10}$ kg-m^2. How much work is done in performing this maneuver? In other words, how much energy had to be expended in the form of reaction control fuel?

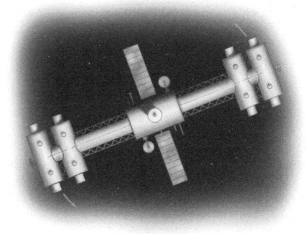

Problem 19.4

19.5 The helicopter's rotor starts from rest. Suppose that its engine exerts a constant 1200 ft-lb couple on the rotor and aerodynamic drag is negligible. The rotor's moment of inertia is $I = 400$ slug-ft^2.

(a) Use work and energy to determine the magnitude of the rotor's angular velocity when it has rotated through 5 revolutions.

(b) What average power is transferred to the rotor while it rotates through 5 revolutions?

19.6 The helicopter's rotor starts from rest. The moment exerted on it (in N-m) is given as a function of the angle through which it has turned in radians by $M = 6500 - 20\theta$. The rotor's moment of inertia is $I = 540$ kg-m^2. Determine the rotor's angular velocity (in rpm) when it has turned through 10 revolutions.

Problems 19.5/19.6

19.7 During extravehicular activity, an astronaut's angular velocity is initially zero. She activates two thrusters of her maneuvering unit, exerting equal and opposite forces $T = 2$ N. The moment of inertia of the astronaut and her equipment about the axis of rotation is 45 kg-m². Use the principle of work and energy to determine the angle through which she has rotated when her angular velocity reaches 15° per second.

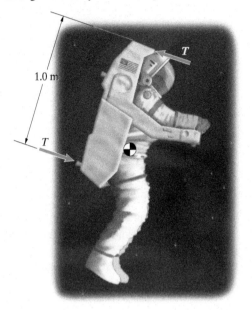

Problem 19.7

19.8 The 8-kg slender bar is released from rest in the horizontal position 1 and falls to position 2.

(a) How much work is done by the bar's weight as it falls from position 1 to position 2?

(b) How much work is done by the force exerted on the bar by the pin support as the bar falls from position 1 to position 2?

(c) Use conservation of energy to determine the bar's angular velocity when it is in position 2.

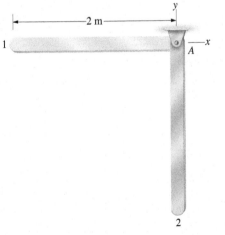

Problem 19.8

19.9 The 20-lb bar is released from rest in the horizontal position 1 and falls to position 2. In addition to the force exerted on it by its weight, it is subjected to a constant counterclockwise couple $M = 30$ ft-lb. Determine the bar's counterclockwise angular velocity in position 2.

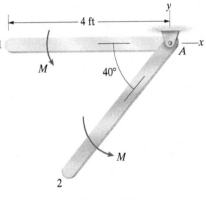

Problem 19.9

19.10 The object consists of an 8-lb slender bar welded to a circular disk. When the object is released from rest in position 1, its angular velocity in position 2 is 4.6 rad/s. What is the weight of the disk?

19.11* The object consists of an 8-lb slender bar welded to a 12-lb circular disk. The object is released from rest in position 1. Determine the x and y components of force exerted on the object by the pin support when it is in position 2.

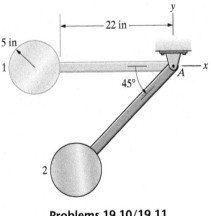

Problems 19.10/19.11

19.12 The mass of each box is 4 kg. The radius of the pulley is 120 mm and its moment of inertia is 0.032 kg-m². The surfaces are smooth. If the system is released from rest, how fast are the boxes moving when the left box has moved 0.5 m to the right?

19.13 The mass of each box is 4 kg. The radius of the pulley is 120 mm and its moment of inertia is 0.032 kg-m². The coefficient of kinetic friction between the boxes and the surfaces is $\mu_k = 0.12$. If the system is released from rest, how fast are the boxes moving when the left box has moved 0.5 m to the right?

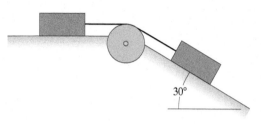

Problems 19.12/19.13

19.14 The 4-kg bar is released from rest in the horizontal position 1 and falls to position 2. The unstretched length of the spring is 0.4 m and the spring constant is $k = 20$ N/m. What is the magnitude of the bar's angular velocity when it is in position 2?

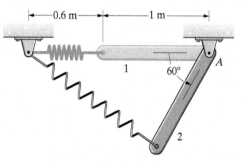

Problem 19.14

19.15 The moments of inertia of two gears that can turn freely on their pin supports are $I_A = 0.002$ kg-m² and $I_B = 0.006$ kg-m². The gears are at rest when a constant couple $M = 2$ N-m is applied to gear B. Neglecting friction, use the principle of work and energy to determine the angular velocities of the gears when gear A has turned 100 revolutions.

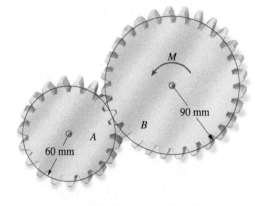

Problem 19.15

19.16 The moments of inertia of gears A and B are $I_A = 0.02$ kg-m² and $I_B = 0.09$ kg-m². Gear A is connected to a torsional spring with constant $k = 12$ N-m/rad. If gear B is given an initial counterclockwise angular velocity of 10 rad/s with the torsional spring unstretched, through what maximum counterclockwise angle does gear B rotate?

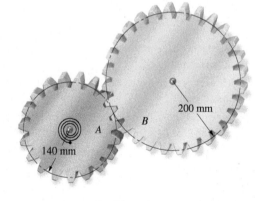

Problem 19.16

19.17 The moments of inertia of three pulleys that can turn freely on their pin supports are $I_A = 0.002$ kg-m², $I_B = 0.036$ kg-m², and $I_C = 0.032$ kg-m². The pulleys are stationary when a constant couple $M = 2$ N-m is applied to pulley A. What is the angular velocity of pulley A when it has turned 10 revolutions?

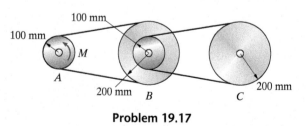

Problem 19.17

19.18 Model the arm ABC as a single rigid body. Its mass is 300 kg, and the moment of inertia about its center of mass is $I = 360$ kg-m^2. Starting from rest with its center of mass 2 m above the ground (position 1), arm ABC is pushed upward by the hydraulic cylinders. When it is in the position shown (position 2), the arm has a counterclockwise angular velocity of 1.4 rad/s. How much work do the hydraulic cylinders do on the arm in moving it from position 1 to position 2?

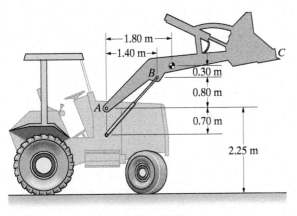

Problem 19.18

▶ **19.19** The mass of the circular disk is 5 kg and its radius is $R = 0.2$ m. The disk is stationary when a constant clockwise couple $M = 10$ N-m is applied to it, causing the disk to roll toward the right. Consider the disk when its center has moved a distance $b = 0.4$ m.

(a) How much work has the couple M done on the disk?

(b) How much work has been done by the friction force exerted on the disk by the surface?

(c) What is the magnitude of the velocity of the center of the disk?

(See Active Example 19.1.)

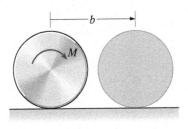

Problem 19.19

▶ **19.20** The mass of the homogeneous cylindrical disk is $m = 5$ kg and its radius is $R = 0.2$ m. The angle $\beta = 15°$. The disk is stationary when a constant clockwise couple $M = 10$ N-m is applied to it. If the disk rolls without slipping, what is the velocity of the center of the disk when it has moved a distance $b = 0.4$ m? (See Active Example 19.1.)

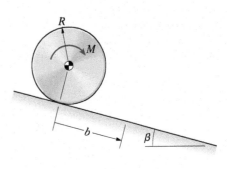

Problem 19.20

19.21 The mass of the stepped disk is 18 kg and its moment of inertia is 0.28 kg-m^2. If the disk is released from rest, what is its angular velocity when the center of the disk has fallen 1 m?

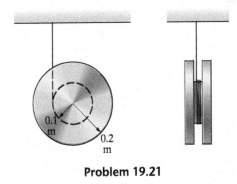

Problem 19.21

19.22 The 100-kg homogeneous cylindrical disk is at rest when the force $F = 500$ N is applied to a cord wrapped around it, causing the disk to roll. Use the principle of work and energy to determine the angular velocity of the disk when it has turned one revolution.

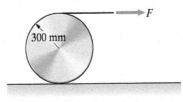

Problem 19.22

19.23 The 1-slug homogeneous cylindrical disk is given a clockwise angular velocity of 2 rad/s with the spring unstretched. The spring constant is $k = 3$ lb/ft. If the disk rolls, how far will its center move to the right?

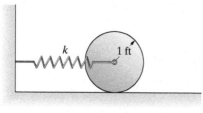

Problem 19.23

19.24 The system is released from rest. The moment of inertia of the pulley is 0.03 slug-ft^2. The slanted surface is smooth. Determine the magnitude of the velocity of the 10-lb weight when it has fallen 2 ft.

19.25 The system is released from rest. The moment of inertia of the pulley is 0.03 slug-ft^2. The coefficient of kinetic friction between the 5-lb weight and the slanted surface is $\mu_k = 0.3$. Determine the magnitude of the velocity of the 10-lb weight when it has fallen 2 ft.

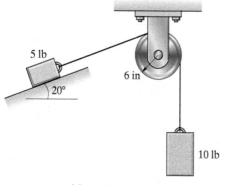

Problems 19.24/19.25

19.26 Each of the cart's four wheels weighs 3 lb, has a radius of 5 in, and has moment of inertia $I = 0.01$ slug-ft^2. The cart (not including its wheels) weighs 20 lb. The cart is stationary when the constant horizontal force $F = 10$ lb is applied. How fast is the cart going when it has moved 2 ft to the right?

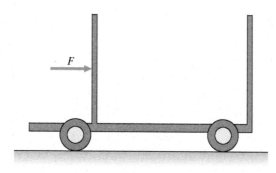

Problem 19.26

▶ **19.27** The total moment of inertia of the car's two rear wheels and axle is I_R, and the total moment of inertia of the two front wheels is I_F. The radius of the tires is R, and the total mass of the car, including the wheels, is m. The car is moving at velocity v_0 when the driver applies the brakes. If the car's brakes exert a constant retarding couple M on each wheel and the tires do not slip, determine the car's velocity as a function of the distance s from the point where the brakes are applied. (See Example 19.2.)

▶ **19.28** The total moment of inertia of the car's two rear wheels and axle is 0.24 kg-m^2. The total moment of inertia of the two front wheels is 0.2 kg-m^2. The radius of the tires is 0.3 m. The mass of the car, including the wheels, is 1480 kg. The car is moving at 100 km/h. If the car's brakes exert a constant retarding couple of 650 N-m on each wheel and the tires do not slip, what distance is required for the car to come to a stop? (See Example 19.2.)

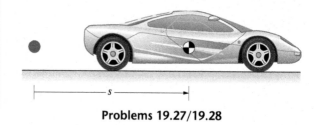

Problems 19.27/19.28

19.29 The radius of the pulley is $R = 100$ mm and its moment of inertia is $I = 0.1$ kg-m^2. The mass $m = 5$ kg. The spring constant is $k = 135$ N/m. The system is released from rest with the spring unstretched. Determine how fast the mass is moving when it has fallen 0.5 m.

Problem 19.29

19.30 The masses of the bar and disk are 14 kg and 9 kg, respectively. The system is released from rest with the bar horizontal. Determine the angular velocity of the bar when it is vertical if the bar and disk are welded together at A.

19.31 The masses of the bar and disk are 14 kg and 9 kg, respectively. The system is released from rest with the bar horizontal. Determine the angular velocity of the bar when it is vertical if the bar and disk are connected by a smooth pin at A.

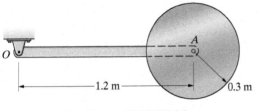

Problems 19.30/19.31

19.32 The 45-kg crate is pulled up the inclined surface by the winch. The coefficient of kinetic friction between the crate and the surface is $\mu_k = 0.4$. The moment of inertia of the drum on which the cable is being wound is $I_A = 4$ kg-m^2. The crate starts from rest, and the motor exerts a constant couple $M = 50$ N-m on the drum. Use the principle of work and energy to determine the magnitude of the velocity of the crate when it has moved 1 m.

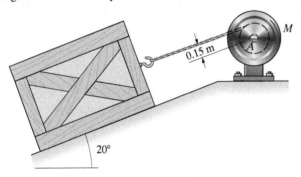

Problem 19.32

19.33 The 2-ft slender bars each weigh 4 lb, and the rectangular plate weighs 20 lb. If the system is released from rest in the position shown, what is the velocity of the plate when the bars are vertical?

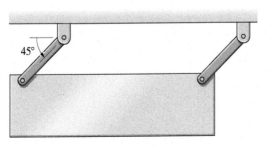

Problem 19.33

19.34 The mass of the 2-m slender bar is 8 kg. A torsional spring exerts a counterclockwise couple $k\theta$ on the bar, where $k = 40$ N-m/rad and θ is in radians. The bar is released from rest with $\theta = 5°$. What is the magnitude of the bar's angular velocity when $\theta = 60°$?

Problem 19.34

19.35 The mass of the suspended object A is 8 kg. The mass of the pulley is 5 kg, and its moment of inertia is 0.036 kg-m^2. If the force $T = 70$ N is applied to the stationary system, what is the magnitude of the velocity of A when it has risen 0.2 m?

Problem 19.35

19.36 The mass of the left pulley is 7 kg, and its moment of inertia is 0.330 kg-m^2. The mass of the right pulley is 3 kg, and its moment of inertia is 0.054 kg-m^2. If the system is released from rest, how fast is the 18-kg mass moving when it has fallen 0.1 m?

Problem 19.36

19.37 The 18-kg ladder is released from rest with $\theta = 10°$. The wall and floor are smooth. Modeling the ladder as a slender bar, use conservation of energy to determine the angular velocity of the bar when $\theta = 40°$.

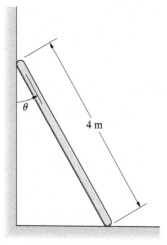

Problem 19.37

19.38 The 8-kg slender bar is released from rest with $\theta = 60°$. The horizontal surface is smooth. What is the bar's angular velocity when $\theta = 30°$.

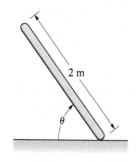

Problem 19.38

19.39 The mass and length of the bar are $m = 4$ kg and $l = 1.2$ m. The spring constant is $k = 180$ N/m. If the bar is released from rest in the position $\theta = 10°$, what is its angular velocity when it has fallen to $\theta = 20°$?

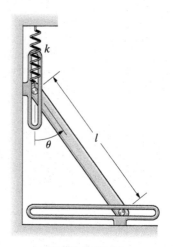

Problem 19.39

▶ **19.40** The 4-kg slender bar is pinned to a 2-kg slider at A and to a 4-kg homogeneous cylindrical disk at B. Neglect the friction force on the slider and assume that the disk rolls. If the system is released from rest with $\theta = 60°$, what is the bar's angular velocity when $\theta = 0$? (See Example 19.3.)

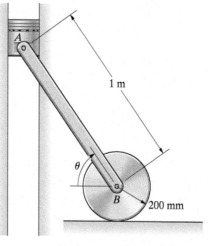

Problem 19.40

19.41* The sleeve P slides on the smooth horizontal bar. The mass of each bar is 4 kg and the mass of the sleeve P is 2 kg. If the system is released from rest with $\theta = 60°$, what is the magnitude of the velocity of the sleeve P when $\theta = 40°$?

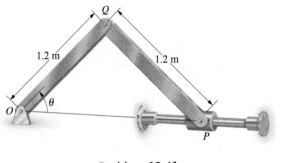

Problem 19.41

19.42* The system is in equilibrium in the position shown. The mass of the slender bar ABC is 6 kg, the mass of the slender bar BD is 3 kg, and the mass of the slider at C is 1 kg. The spring constant is $k = 200$ N/m. If a constant 100-N downward force is applied at A, what is the angular velocity of bar ABC when it has rotated 20° from its initial position?

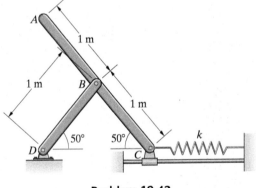

Problem 19.42

19.43* The masses of bars AB and BC are 5 kg and 3 kg, respectively. If the system is released from rest in the position shown, what are the angular velocities of the bars at the instant before the joint B hits the smooth floor?

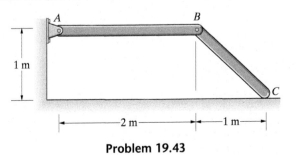

Problem 19.43

19.44* Bar AB weighs 5 lb. Each of the sleeves A and B weighs 2 lb. The system is released from rest in the position shown. What is the magnitude of the angular velocity of the bar when the sleeve B has moved 3 in to the right?

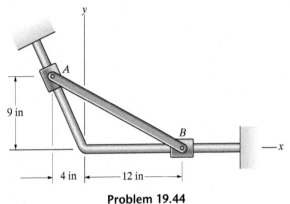

Problem 19.44

19.45* Each bar has a mass of 8 kg and a length of 1 m. The spring constant is $k = 100$ N/m, and the spring is unstretched when $\theta = 0$. If the system is released from rest with the bars vertical, what is the magnitude of the angular velocity of the bars when $\theta = 30°$?

Problem 19.45

19.46* The system starts from rest with the crank AB vertical. A constant couple M exerted on the crank causes it to rotate in the clockwise direction, compressing the gas in the cylinder. Let s be the displacement (in meters) of the piston to the right relative to its initial position. The net force toward the left exerted on the piston by atmospheric pressure and the gas in the cylinder is $350/(1 - 10s)$ N. The moment of inertia of the crank about A is 0.0003 kg-m^2. The mass of the connecting rod BC is 0.36 kg, and the center of mass of the rod is at its midpoint. The connecting rod's moment of inertia about its center of mass is 0.0004 kg-m^2. The mass of the piston is 4.6 kg. If the clockwise angular velocity of the crank AB is 200 rad/s when it has rotated 90° from its initial position, what is M? (Neglect the work done by the weights of the crank and connecting rod.)

19.47* In Problem 19.46, if the system starts from rest with the crank AB vertical and the couple $M = 40$ N-m, what is the clockwise angular velocity of AB when it has rotated 45° from its initial position?

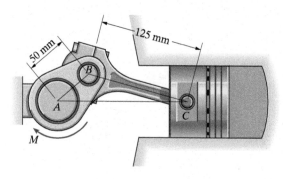

Problems 19.46/19.47

19.2 Impulse and Momentum

BACKGROUND

In this section, we review our discussion of the principle of linear impulse and momentum from Chapter 16 and then derive the principle of angular impulse and momentum for a rigid body. These principles relate time integrals of the forces and couples acting on a rigid body to changes in the velocity of its center of mass and its angular velocity.

Linear Momentum

Integrating Newton's second law with respect to time yields the principle of linear impulse and momentum for a rigid body:

$$\int_{t_1}^{t_2} \Sigma \mathbf{F} \, dt = m\mathbf{v}_2 - m\mathbf{v}_1. \tag{19.23}$$

Here, $\mathbf{v}_1$ and $\mathbf{v}_2$ are the velocities of the center of mass at the times t_1 and t_2 (Fig. 19.10). If the external forces acting on a rigid body are known as functions of time, this principle yields the change in the velocity of the center of mass of the body during an interval of time. In terms of the average of the total force from t_1 to t_2,

$$\Sigma \mathbf{F}_{\text{av}} = \frac{1}{t_2 - t_1} \int_{t_1}^{t_2} \Sigma \mathbf{F} \, dt,$$

we can write Eq. (19.23) as

$$(t_2 - t_1)\Sigma \mathbf{F}_{\text{av}} = m\mathbf{v}_2 - m\mathbf{v}_1. \tag{19.24}$$

This form of the principle of linear impulse and momentum is often useful when an object is subjected to impulsive forces. (See Section 16.1.)

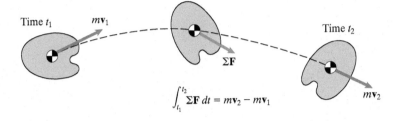

Figure 19.10
Principle of linear impulse and momentum.

If the only forces acting on two rigid bodies A and B are the forces they exert on each other, or if other forces are negligible, the total linear momentum of A and B is conserved:

$$m_A \mathbf{v}_A + m_B \mathbf{v}_B = \text{constant.} \tag{19.25}$$

Angular Momentum

When momentum principles are applied to rigid bodies, it is often necessary to determine both the velocities of their centers of mass and their angular velocities. For this task, linear momentum principles alone are not sufficient. In this section, we derive angular momentum principles for a rigid body in planar motion.

Principles of Angular Impulse and Momentum The total moment about the center of mass of a rigid body in planar motion equals the product of the moment of inertia of the body about its center of mass and the angular acceleration:

$$\Sigma M = I\alpha.$$

We can write this equation in the form

$$\Sigma M = \frac{dH}{dt}, \tag{19.26}$$

where

$$H = I\omega \tag{19.27}$$

is the rigid body's angular momentum about its center of mass. Integrating Eq. (19.26) with respect to time, we obtain one form of the principle of angular impulse and momentum:

$$\int_{t_1}^{t_2} \Sigma M \, dt = H_2 - H_1. \tag{19.28}$$

Here, H_1 and H_2 are the values of the angular momentum at the times t_1 and t_2. This equation says that angular impulse about the center of mass of the rigid body during the interval of time from t_1 to t_2 is equal to the change in the rigid body's angular momentum about its center of mass. If the total moment about the center of mass is known as a function of time, Eq. (19.28) can be used to determine the change in the angular velocity from t_1 to t_2.

We can derive another useful form of this principle: Let $\mathbf{r}$ be the position vector of the center of mass of the rigid body relative to a fixed point

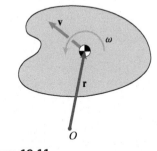

Figure 19.11
A rigid body in planar motion with velocity $\mathbf{v}$ and angular velocity ω.

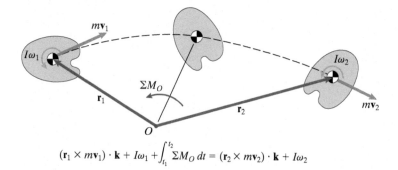

Figure 19.12
The impulse about O equals the change in the angular momentum about O.

$$(\mathbf{r}_1 \times m\mathbf{v}_1) \cdot \mathbf{k} + I\omega_1 + \int_{t_1}^{t_2} \Sigma M_O \, dt = (\mathbf{r}_2 \times m\mathbf{v}_2) \cdot \mathbf{k} + I\omega_2$$

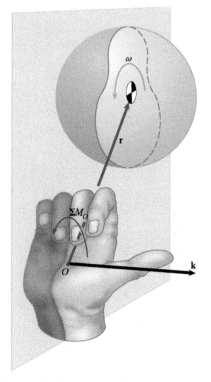

Figure 19.13
The direction of $\mathbf{k}$.

O (Fig. 19.11). In Chapter 18, we derived a relationship between the total moment about O due to external forces and couples and the rate of change of the rigid body's angular momentum about O:

$$\Sigma M_O = \frac{dH_O}{dt}, \tag{19.29}$$

where

$$H_O = (\mathbf{r} \times m\mathbf{v}) \cdot \mathbf{k} + I\omega. \tag{19.30}$$

Integrating Eq. (19.29) with respect to time, we obtain a second form of the principle of angular impulse and momentum:

$$\int_{t_1}^{t_2} \Sigma M_O \, dt = H_{O2} - H_{O1}. \tag{19.31}$$

The angular impulse about a fixed point O during the interval of time from t_1 to t_2 is equal to the change in the rigid body's angular momentum about O (Fig. 19.12).

The term $(\mathbf{r} \times m\mathbf{v}) \cdot \mathbf{k}$ in Eq. (19.30) is the rigid body's angular momentum about O due to the velocity of its center of mass. This term has the same form as the moment of a force, but with the linear momentum $m\mathbf{v}$ in place of the force. If we define ΣM_O and ω to be positive in the counterclockwise direction, the unit vector $\mathbf{k}$ points out of the page (Fig. 19.13) and $(\mathbf{r} \times m\mathbf{v}) \cdot \mathbf{k}$ is the counterclockwise "moment" of the linear momentum. The vector expression can be used to calculate this quantity, but it is often easier to use the fact that its magnitude is the product of the magnitude of the linear momentum and the perpendicular distance from point O to the line of action of the velocity. The "moment" is positive if it is counterclockwise (Fig. 19.14a) and negative if it is clockwise (Fig. 19.14b).

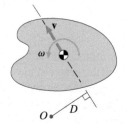
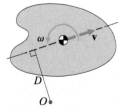

Figure 19.14
Determining the angular momentum about O by calculating the "moment" of the linear momentum.

Angular momentum
$H_O = D(m|\mathbf{v}|) + I\omega$

(a)

Angular momentum
$H_O = -D(m|\mathbf{v}|) + I\omega$

(b)

Impulsive Forces and Couples   The average of the moment about the center of mass from t_1 to t_2 is

$$\Sigma M_{av} = \frac{1}{t_2 - t_1} \int_{t_1}^{t_2} \Sigma M \, dt.$$

Using this equation, we can write Eq. (19.28) as

$$(t_2 - t_1) \Sigma M_{av} = H_2 - H_1. \tag{19.32}$$

In the same way, we can express Eq. (19.31) in terms of the average moment about point O:

$$(t_2 - t_1)(\Sigma M_O)_{av} = H_{O2} - H_{O1}. \tag{19.33}$$

When the average value of the moment and its duration are known, we can use Eq. (19.32) or Eq. (19.33) to determine the change in the angular momentum. These equations are often useful when a rigid body is subjected to impulsive forces and couples.

Conservation of Angular Momentum   We can use Eq. (19.31) to obtain an equation of conservation of total angular momentum for two rigid bodies. Let A and B be rigid bodies in two-dimensional motion in the same plane, and suppose that they are subjected only to the forces and couples they exert on each other or that other forces and couples are negligible. Let M_{OA} be the moment about a fixed point O due to the forces and couples acting on A, and let M_{OB} be the moment about O due to the forces and couples acting on B. Under the same assumption we made in deriving the equations of motion—the forces between each pair of particles are directed along the line between the particles—the moment $M_{OB} = -M_{OA}$. For example, in Fig. 19.15, A and B exert forces on each other by contact. The resulting moments about O are $M_{OA} = (\mathbf{r_p} \times \mathbf{R}) \cdot \mathbf{k}$ and $M_{OB} = [\mathbf{r_p} \times (-\mathbf{R})] \cdot \mathbf{k} = -M_{OA}$.

We apply Eq. (19.31) to A and B for arbitrary times t_1 and t_2, obtaining

$$\int_{t_1}^{t_2} M_{OA} \, dt = H_{OA2} - H_{OA1}$$

and

$$\int_{t_1}^{t_2} M_{OB} \, dt = H_{OB2} - H_{OB1}.$$

Summing these equations, the terms on the left cancel, and we obtain

$$H_{OA1} + H_{OB1} = H_{OA2} + H_{OB2}.$$

We see that the total angular momentum of A and B about O is conserved:

$$H_{OA} + H_{OB} = \text{constant}. \tag{19.34}$$

Notice that this result holds even when A and B are subjected to significant external forces and couples if the total moment about O due to the external forces and couples is zero. The point O can sometimes be chosen so that this condition is satisfied. The result also applies to an arbitrary number of rigid bodies: Their total angular momentum about O is conserved if the total moment about O due to external forces and couples is zero.

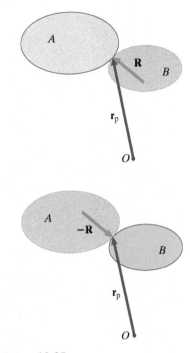

Figure 19.15
Rigid bodies A and B exerting forces on each other by contact.

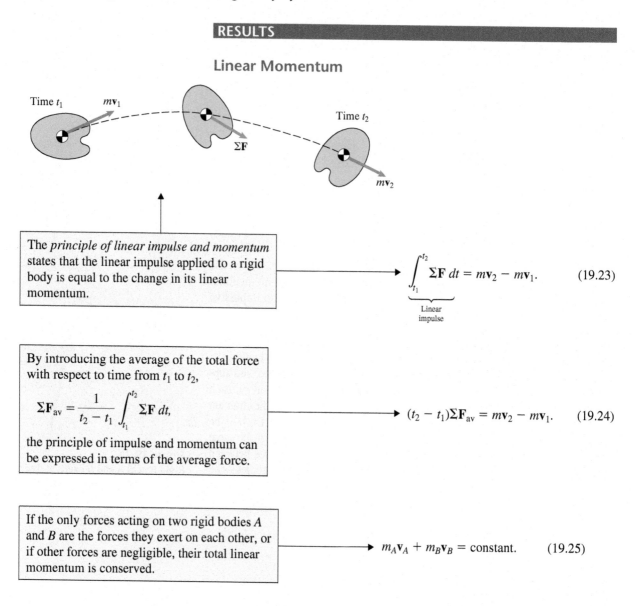

RESULTS

Linear Momentum

The *principle of linear impulse and momentum* states that the linear impulse applied to a rigid body is equal to the change in its linear momentum.

$$\int_{t_1}^{t_2} \Sigma \mathbf{F} \, dt = m\mathbf{v}_2 - m\mathbf{v}_1.$$　　(19.23)

$\underbrace{}_{\text{Linear impulse}}$

By introducing the average of the total force with respect to time from t_1 to t_2,

$$\Sigma \mathbf{F}_{av} = \frac{1}{t_2 - t_1} \int_{t_1}^{t_2} \Sigma \mathbf{F} \, dt,$$

the principle of impulse and momentum can be expressed in terms of the average force.

$$(t_2 - t_1)\Sigma \mathbf{F}_{av} = m\mathbf{v}_2 - m\mathbf{v}_1.$$　　(19.24)

If the only forces acting on two rigid bodies A and B are the forces they exert on each other, or if other forces are negligible, their total linear momentum is conserved.

$$m_A \mathbf{v}_A + m_B \mathbf{v}_B = \text{constant}.$$　　(19.25)

Angular Momentum of a Rigid Body in Planar Motion

Angular momentum about the center of mass.

$$H = I\omega.$$　　(19.27)

One form of the *principle of angular impulse and momentum* states that the angular impulse about the center of mass during an interval of time from t_1 to t_2 is equal to the change in the angular momentum about the center of mass.

$$\int_{t_1}^{t_2} \Sigma M \, dt = H_2 - H_1.$$　　(19.28)

$\underbrace{}_{\text{Angular impulse}}$

Equation (19.28) can be expressed in terms of the average moment about the center of mass.

$$(t_2 - t_1)\Sigma M_{av} = H_2 - H_1.$$　　(19.32)

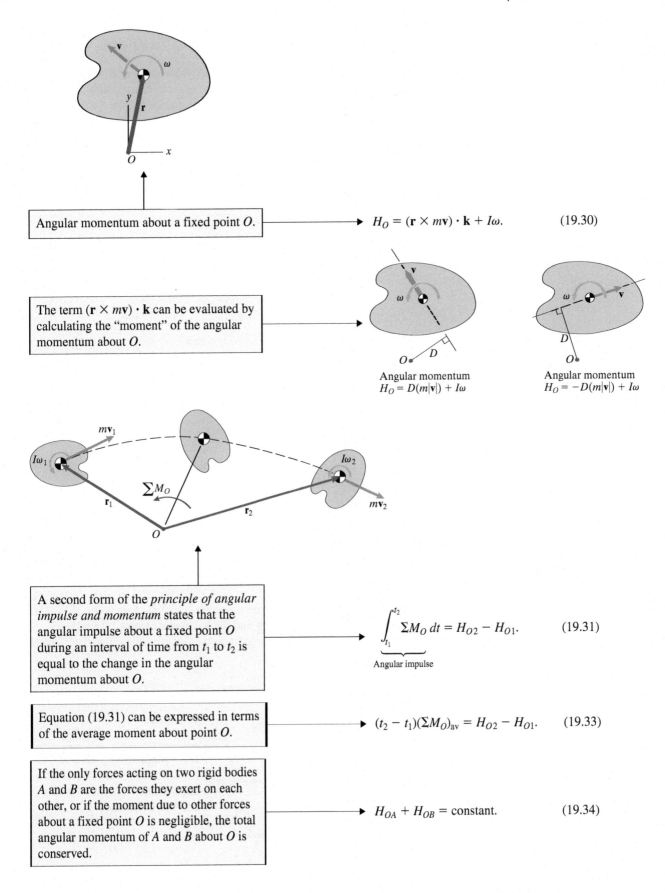

Angular momentum about a fixed point O.

$$H_O = (\mathbf{r} \times m\mathbf{v}) \cdot \mathbf{k} + I\omega. \qquad (19.30)$$

The term $(\mathbf{r} \times m\mathbf{v}) \cdot \mathbf{k}$ can be evaluated by calculating the "moment" of the angular momentum about O.

Angular momentum
$H_O = D(m|\mathbf{v}|) + I\omega$

Angular momentum
$H_O = -D(m|\mathbf{v}|) + I\omega$

A second form of the *principle of angular impulse and momentum* states that the angular impulse about a fixed point O during an interval of time from t_1 to t_2 is equal to the change in the angular momentum about O.

$$\underbrace{\int_{t_1}^{t_2} \Sigma M_O \, dt}_{\text{Angular impulse}} = H_{O2} - H_{O1}. \qquad (19.31)$$

Equation (19.31) can be expressed in terms of the average moment about point O.

$$(t_2 - t_1)(\Sigma M_O)_{\text{av}} = H_{O2} - H_{O1}. \qquad (19.33)$$

If the only forces acting on two rigid bodies A and B are the forces they exert on each other, or if the moment due to other forces about a fixed point O is negligible, the total angular momentum of A and B about O is conserved.

$$H_{OA} + H_{OB} = \text{constant}. \qquad (19.34)$$

Active Example 19.4 Principle of Angular Impulse and Momentum (▶ *Related Problem 19.55*)

Prior to their contact, disk A has a counterclockwise angular velocity ω_0 and disk B is stationary. At $t = 0$, disk A is moved into contact with disk B. As a result of friction, the angular velocity of A decreases and the clockwise angular velocity of B increases until there is no slip between the disks. What are their final angular velocities ω_A and ω_B? The moments of inertia of the disks are I_A and I_B.

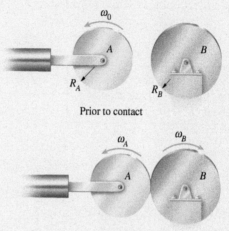

Prior to contact

Final angular velocities

Strategy

The disks rotate about fixed axes through their centers of mass while they are in contact, so we can apply the principle of angular impulse and momentum in the form given by Eq. (19.28) to each disk. When there is no longer any slip between the disks, their velocities are equal at their point of contact. With this kinematic relationship and the equations obtained from the principle of angular impulse and momentum, we can determine the final angular velocities.

Solution

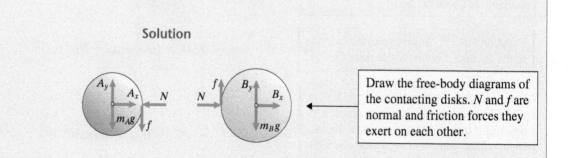

Draw the free-body diagrams of the contacting disks. N and f are normal and friction forces they exert on each other.

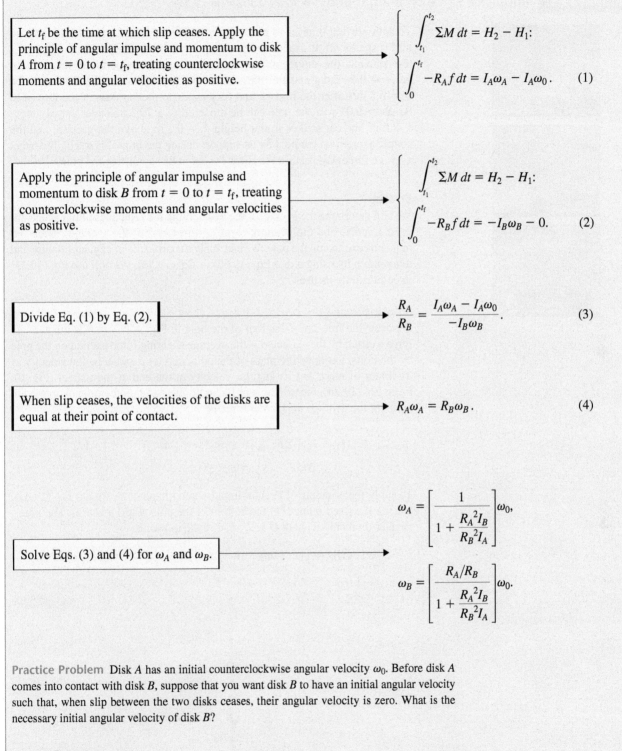

Let t_f be the time at which slip ceases. Apply the principle of angular impulse and momentum to disk A from $t = 0$ to $t = t_f$, treating counterclockwise moments and angular velocities as positive.

$$\int_{t_1}^{t_2} \Sigma M \, dt = H_2 - H_1:$$

$$\int_0^{t_f} -R_A f \, dt = I_A \omega_A - I_A \omega_0. \qquad (1)$$

Apply the principle of angular impulse and momentum to disk B from $t = 0$ to $t = t_f$, treating counterclockwise moments and angular velocities as positive.

$$\int_{t_1}^{t_2} \Sigma M \, dt = H_2 - H_1:$$

$$\int_0^{t_f} -R_B f \, dt = -I_B \omega_B - 0. \qquad (2)$$

Divide Eq. (1) by Eq. (2).

$$\frac{R_A}{R_B} = \frac{I_A \omega_A - I_A \omega_0}{-I_B \omega_B}. \qquad (3)$$

When slip ceases, the velocities of the disks are equal at their point of contact.

$$R_A \omega_A = R_B \omega_B. \qquad (4)$$

Solve Eqs. (3) and (4) for ω_A and ω_B.

$$\omega_A = \left[\frac{1}{1 + \dfrac{R_A{}^2 I_B}{R_B{}^2 I_A}} \right] \omega_0,$$

$$\omega_B = \left[\frac{R_A / R_B}{1 + \dfrac{R_A{}^2 I_B}{R_B{}^2 I_A}} \right] \omega_0.$$

Practice Problem Disk A has an initial counterclockwise angular velocity ω_0. Before disk A comes into contact with disk B, suppose that you want disk B to have an initial angular velocity such that, when slip between the two disks ceases, their angular velocity is zero. What is the necessary initial angular velocity of disk B?

Answer: $\dfrac{R_B I_A}{R_A I_B} \omega_0$ counterclockwise.

| Example 19.5 | Impulsive Force on a Rigid Body (▶ *Related Problem 19.56*) |

To help prevent injuries to passengers, engineers design a street light pole so that it shears off at ground level when struck by a vehicle. From video of a test impact, the engineers estimate the angular velocity of the pole to be $\omega = 0.74$ rad/s and the horizontal velocity of its center of mass to be $v = 6.8$ m/s after the impact, and they estimate the duration of the impact to be $\Delta t = 0.01$ s. If the pole can be modeled as a 70-kg slender bar of length $l = 6$ m, the car strikes it at a height $h = 0.5$ m above the ground, and the couple exerted on the pole by its support during the impact is negligible, what average force was required to shear off the bolts supporting the pole?

Strategy

We will determine the average force by applying the principles of linear and angular impulse and momentum, expressed in terms of the average forces and moments exerted on the pole. We can apply the principle of angular impulse and momentum by using either Eq. (19.32) or Eq. (19.33). We will use Eq. (19.33) to demonstrate its use.

Solution

We draw the free-body diagram of the pole in Fig. a, where F is the average force exerted by the car and S is the average shearing force exerted on the pole by the bolts. Let m be the mass of the pole, and let v and ω be the velocity of its center of mass and its angular velocity at the end of the impact (Fig. b). From Eq. (19.24), the principle of linear impulse and momentum expressed in terms of the average horizontal force is

$$(t_2 - t_1)(\Sigma F_x)_{av} = (mv_x)_2 - (mv_x)_1:$$
$$\Delta t(F - S) = mv - 0. \qquad (1)$$

To apply the principle of angular impulse and momentum, we use Eq. (19.33), placing the fixed point O at the bottom of the pole (Figs. a and b). The pole's angular momentum about O at the end of the impact is

$$H_{O2} = [(\mathbf{r} \times m\mathbf{v}) \cdot \mathbf{k} + I\omega]_2 = \left[\left(\tfrac{1}{2}l\mathbf{j}\right) \times m(v\mathbf{i})\right] \cdot \mathbf{k} + I\omega$$
$$= -\tfrac{1}{2}lmv + I\omega.$$

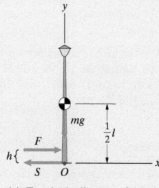

(a) Free-body diagram of the pole.

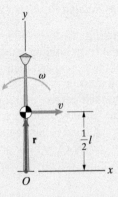

(b) Velocity and angular velocity at the end of the impact.

We can also obtain this result by calculating the "moment" of the linear momentum about O and adding the term $I\omega$. The magnitude of the "moment" is the product of the magnitude of the linear momentum (mv) and the perpendicular distance from O to the line of action of the linear momentum $\left(\frac{1}{2}l\right)$, and it is negative because the "moment" is clockwise. (See Fig. 19.14.) From Eq. (19.33), we obtain

$$(t_2 - t_1)(\Sigma M_O)_{av} = H_{O2} - H_{O1}:$$
$$\Delta t(-hF) = -\tfrac{1}{2}lmv + I\omega - 0.$$

Solving this equation together with Eq. (1) for the average shear force S, we obtain

$$S = \frac{\left(\frac{1}{2}l - h\right)mv - I\omega}{h\,\Delta t}$$

$$= \frac{\left[\frac{1}{2}(6\text{ m}) - 0.5\text{ m}\right](70\text{ kg})(6.8\text{ m/s}) - \left[\frac{1}{12}(70\text{ kg})(6\text{ m})^2\right](0.74\text{ rad/s})}{(0.5\text{ m})(0.01\text{ s})}$$

$$= 207{,}000\text{ N}.$$

Critical Thinking

This example demonstrates both the power and the limitations of momentum methods. As the car collides with the light pole, the car's structure deforms, the light pole deforms, and the bolts supporting the pole fail. The time history of the force exerted on the car and pole by the impact hinges upon the details of these complicated phenomena. Using momentum methods and information about the motion of the pole after the impact, we were able to estimate the *average* value of the force, but we cannot determine its time history. To do so would require either more elaborate experiments or an analysis of the collision in which the deformations of the car and pole are modeled. This kind of tradeoff occurs frequently in engineering. Often limited information about a phenomenon can be obtained quickly, as we do in this example, but more accurate and complete information could be obtained by investing the necessary time and resources. The question that must be answered is whether the additional investment is essential to achieve the required engineering objectives.

Example 19.6   Conservation of Angular Momentum (▶ *Related Problem 19.58*)

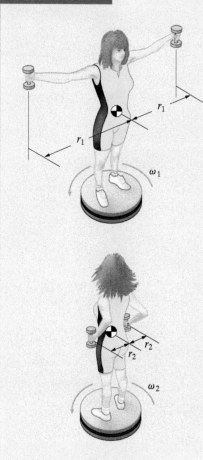

In a well-known demonstration of conservation of angular momentum, a person stands on a rotating platform holding a mass m in each hand. The combined moment of inertia of the person and platform about their axis of rotation is $I_P = 0.4$ kg-m² and each mass $m = 4$ kg. Neglect the moments of inertia of the masses she holds about their centers of mass—that is, treat them as particles. If the person's angular velocity with her arms extended to $r_1 = 0.6$ m is $\omega_1 = 1$ revolution per second, what is her angular velocity ω_2 when she pulls the masses inward to $r_2 = 0.2$ m? (You have observed skaters using this phenomenon to control their angular velocity in a spin by altering the positions of their arms.)

Strategy

If we neglect friction in the rotating platform, the total angular momentum of the person, platform, and masses about the axis of rotation is conserved. We can use this condition to determine ω_2.

Solution

The total angular momentum is the sum of the angular momentum of the person and platform and the angular momentum of the two masses. When her arms are extended, the velocity of each mass m is $r_1\omega_1$, so the "moment" of the angular momentum of each mass about the axis of rotation is $r_1(mr_1\omega_1)$. The total angular momentum is

$$H_{O1} = I_P\omega_1 + 2r_1(mr_1\omega_1).$$

When she pulls the masses inward, the total angular momentum is

$$H_{O2} = I_P\omega_2 + 2r_2(mr_2\omega_2).$$

The total angular momentum is conserved.

$$H_{O1} = H_{O2}:$$
$$(I_P + 2mr_1^2)\omega_1 = (I_P + 2mr_2^2)\omega_2,$$
$$\left[0.4 \text{ kg-m}^2 + 2(4 \text{ kg})(0.6 \text{ m})^2\right]\omega_1 = \left[0.4 \text{ kg-m}^2 + 2(4 \text{ kg})(0.2 \text{ m})^2\right]\omega_2.$$

This yields $\omega_2 = 4.56\omega_1 = 4.56$ revolutions per second.

Critical Thinking

Calculate the total kinetic energy of the person, platform, and masses when her arms are extended and when she has pulled the masses inward. You will find that the total kinetic energy is greater in the second case. It appears that conservation of energy is violated. However, she does work on the weights in pulling them inward. Her physiological energy supplies the additional kinetic energy.

Problems

19.48 The moment of inertia of the disk about O is 22 kg-m². At $t = 0$, the stationary disk is subjected to a constant 50 N-m torque.

(a) Determine the angular impulse exerted on the disk from $t = 0$ to $t = 5$ s.

(b) What is the disk's angular velocity at $t = 5$ s?

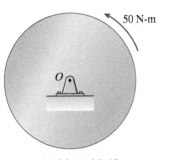

50 N-m

O

Problem 19.48

19.49 The moment of inertia of the jet engine's rotating assembly is 400 kg-m². The assembly starts from rest. At $t = 0$, the engine's turbine exerts a couple on it that is given as a function of time by $M = 6500 - 125t$ N-m.

(a) What is the magnitude of the angular impulse exerted on the assembly from $t = 0$ to $t = 20$ s?

(b) What is the magnitude of the angular velocity of the assembly (in rpm) at $t = 20$ s?

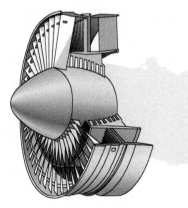

Problem 19.49

19.50 An astronaut fires a thruster of his maneuvering unit, exerting a force $T = 2(1 + t)$ N, where t is in seconds. The combined mass of the astronaut and his equipment is 122 kg, and the moment of inertia about their center of mass is 45 kg-m². Modeling the astronaut and his equipment as a rigid body, use the principle of angular impulse and momentum to determine how long it takes for his angular velocity to reach 0.1 rad/s.

19.51 The combined mass of the astronaut and his equipment is 122 kg, and the moment of inertia about their center of mass is 45 kg-m². The maneuvering unit exerts an impulsive force T of 0.2-s duration, giving him a counterclockwise angular velocity of 1 rpm.

(a) What is the average magnitude of the impulsive force?

(b) What is the magnitude of the resulting change in the velocity of the astronaut's center of mass?

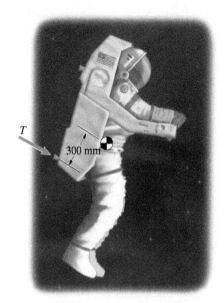

T

300 mm

Problems 19.50/19.51

19.52 A flywheel attached to an electric motor is initially at rest. At $t = 0$, the motor exerts a couple $M = 200e^{-0.1t}$ N-m on the flywheel. The moment of inertia of the flywheel is 10 kg-m².

(a) What is the flywheel's angular velocity at $t = 10$ s?

(b) What maximum angular velocity will the flywheel attain?

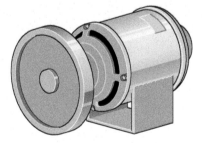

Problem 19.52

19.53 A main landing gear wheel of a Boeing 777 has a radius of 0.62 m and its moment of inertia is 24 kg-m². After the plane lands at 75 m/s, the skid marks of the wheel's tire is measured and determined to be 18 m in length. Determine the average friction force exerted on the wheel by the runway. Assume that the airplane's velocity is constant during the time the tire skids (slips) on the runway.

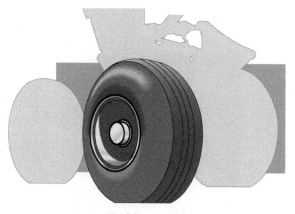

Problem 19.53

19.54 The force a club exerts on a 0.045-kg golf ball is shown. The ball is 42 mm in diameter and can be modeled as a homogeneous sphere. The club is in contact with the ball for 0.0006 s, and the magnitude of the velocity of the ball's center of mass after the ball is hit is 36 m/s. What is the magnitude of the ball's angular velocity after it is hit?

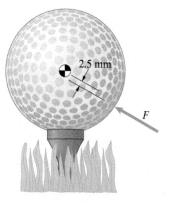

Problem 19.54

▶ **19.55** Disk A initially has a counterclockwise angular velocity $\omega_0 = 50$ rad/s. Disks B and C are initially stationary. At $t = 0$, disk A is moved into contact with disk B. Determine the angular velocities of the three disks when they have stopped slipping relative to each other. The masses of the disks are $m_A = 4$ kg, $m_B = 16$ kg, and $m_C = 9$ kg. (See Active Example 19.4.)

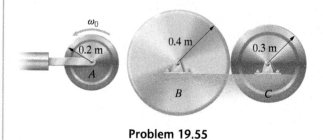

Problem 19.55

▶ **19.56** In Example 19.5, suppose that in a second test at a higher velocity the angular velocity of the pole following the impact is $\omega = 0.81$ rad/s, the horizontal velocity of its center of mass is $v = 7.3$ m/s, and the duration of the impact is $\Delta t = 0.009$ s. Determine the magnitude of the average force the car exerts on the pole in shearing off the supporting bolts. Do so by applying the principle of angular impulse and momentum in the form given by Eq. (19.32).

19.57 The force exerted on the cue ball by the cue is horizontal. Determine the value of h for which the ball rolls without slipping. (Assume that the frictional force exerted on the ball by the table is negligible.)

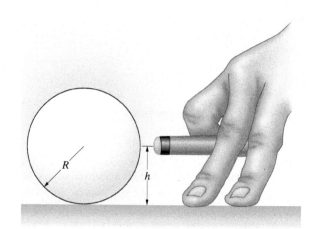

Problem 19.57

▶ 19.58 In Example 19.6, we neglected the moments of inertia of the two masses *m* about the axes through their centers of mass in calculating the total angular momentum of the person, platform, and masses. Suppose that the moment of inertia of each mass about the vertical axis through its center of mass is $I_M = 0.001$ kg-m^2. If the person's angular velocity with her arms extended to $r_1 = 0.6$ m is $\omega_1 = 1$ revolution per second, what is her angular velocity ω_2 when she pulls the masses inward to $r_2 = 0.2$ m? Compare your result to the answer obtained in Example 19.6.

19.59 Two gravity research satellites ($m_A = 250$ kg, $I_A = 350$ kg-m^2; $m_B = 50$ kg, $I_B = 16$ kg-m^2) are tethered by a cable. The satellites and cable rotate with angular velocity $\omega_0 = 0.25$ rpm. Ground controllers order satellite *A* to slowly unreel 6 m of additional cable. What is the angular velocity afterward?

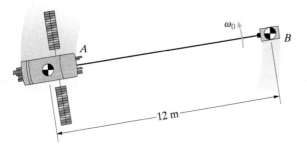

Problem 19.59

19.60 The 2-kg bar rotates *in the horizontal plane* about the smooth pin. The 6-kg collar *A* slides on the smooth bar. Assume that the moment of inertia of the collar *A* about its center of mass is negligible; that is, treat the collar as a particle. At the instant shown, the angular velocity of the bar is $\omega_0 = 60$ rpm and the distance from the pin to the collar is $r = 1.8$ m. Determine the bar's angular velocity when $r = 2.4$ m.

19.61 The 2-kg bar rotates *in the horizontal plane* about the smooth pin. The 6-kg collar *A* slides on the smooth bar. The moment of inertia of the collar *A* about its center of mass is 0.2 kg-m^2. At the instant shown, the angular velocity of the bar is $\omega_0 = 60$ rpm and the distance from the pin to the collar is $r = 1.8$ m. Determine the bar's angular velocity when $r = 2.4$ m and compare your answer to that of Problem 19.60.

19.62* The 2-kg bar rotates *in the horizontal plane* about the smooth pin. The 6-kg collar *A* slides on the smooth bar. The moment of inertia of the collar *A* about its center of mass is 0.2 kg-m^2. The spring is unstretched when $r = 0$, and the spring constant is $k = 10$ N/m. At the instant shown, the angular velocity of the bar is $\omega_0 = 2$ rad/s, the distance from the pin to the collar is $r = 1.8$ m, and the radial velocity of the collar is zero. Determine the radial velocity of the collar when $r = 2.4$ m.

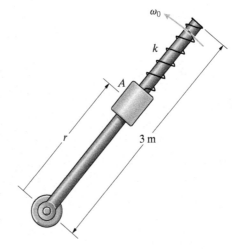

Problems 19.60–19.62

19.63 The circular bar is welded to the vertical shafts, which can rotate freely in bearings at A and B. Let I be the moment of inertia of the circular bar and shafts about the vertical axis. The circular bar has an initial angular velocity ω_0, and the mass m is released in the position shown with no velocity relative to the bar. Determine the angular velocity of the circular bar as a function of the angle β between the vertical and the position of the mass. Neglect the moment of inertia of the mass about its center of mass; that is, treat the mass as a particle.

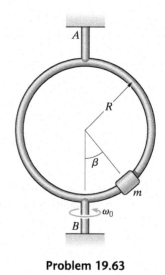

Problem 19.63

19.3 Impacts

BACKGROUND

In Chapter 16, we analyzed impacts between objects with the aim of determining the velocities of their centers of mass after the collision. We now discuss how to determine the velocities of the centers of mass *and the angular velocities* of rigid bodies after they collide.

Conservation of Momentum

Suppose that two rigid bodies A and B, in two-dimensional motion in the same plane, collide. What do the principles of linear and angular momentum tell us about their motions after the collision?

Linear Momentum If other forces are negligible in comparison to the impact forces A and B exert on each other, their total linear momentum is the same before and after the impact. But this result must be applied with care. For example, if one of the rigid bodies has a pin support (Fig. 19.16), the reactions exerted by the support cannot be neglected, and linear momentum is not conserved.

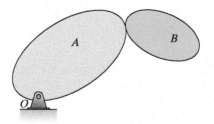

Figure 19.16
Rigid bodies A and B colliding. Because of the pin support, their total linear momentum is *not* conserved, but their total angular momentum about O is conserved.

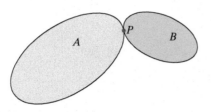

Figure 19.17
Rigid bodies A and B colliding at P. If
forces are exerted only at P, the angular
momentum of A about P and the angular
momentum of B about P are each
conserved.

Angular Momentum   If other forces and couples are negligible in comparison to the impact forces and couples that A and B exert on each other, their total angular momentum about *any* fixed point O is the same before and after the impact. [See Eq. (19.34).] If, in addition, A and B exert forces on each other only at their point of impact P and exert no couples on each other, the angular momentum about P of *each* rigid body is the same before and after the impact (Fig. 19.17). This result follows from the principle of angular impulse and momentum, Eq. (19.31), because the impact forces on A and B exert no moment about P. If one of the rigid bodies has a pin support at a point O, as in Fig. 19.16, their total angular momentum about O is the same before and after the impact.

Coefficient of Restitution

If two rigid bodies adhere and move as a single rigid body after colliding, their velocities and angular velocity can be determined by using momentum conservation and kinematic relationships alone. These relationships are not sufficient if the objects do not adhere. But some impacts of the latter type can be analyzed by using the concept of the coefficient of restitution.

Let P be the point of contact of rigid bodies A and B during an impact (Fig. 19.18). Let their velocities at P be $\mathbf{v}_{AP}$ and $\mathbf{v}_{BP}$ just before the impact and $\mathbf{v}'_{AP}$ and $\mathbf{v}'_{BP}$ just afterward. The x axis is perpendicular to the contacting surfaces at P. If the frictional forces resulting from the impact are negligible, we can show that the components of the velocities normal to the surfaces at P are related to the coefficient of restitution e by

$$e = \frac{(\mathbf{v}'_{BP})_x - (\mathbf{v}'_{AP})_x}{(\mathbf{v}_{AP})_x - (\mathbf{v}_{BP})_x}. \tag{19.35}$$

To derive this result, we must consider the effects of the impact on the individual objects. Let t_1 be the time at which they first come into contact. The objects are not actually rigid, but will deform as a result of the collision. At a time t_C, the maximum deformation will occur, and the objects will begin a "recovery" phase in which they tend to resume their original shapes. Let t_2 be the time at which they separate.

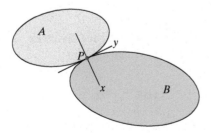

Figure 19.18
Rigid bodies A and B colliding at P. The x axis
is perpendicular to the contacting surfaces.

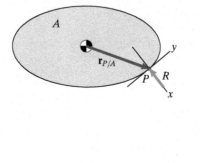

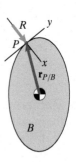

Figure 19.19
The normal force R resulting from the impact.

Our first step is to apply the principle of linear impulse and momentum to A and B for the intervals from t_1 to t_C and from t_C to t_2. Let R be the magnitude of the normal force exerted during the impact (Fig. 19.19). We denote the velocity of the center of mass of A at the times t_1, t_C, and t_2 by $\mathbf{v}_A$, $\mathbf{v}_{AC}$, and $\mathbf{v}'_A$, and denote the corresponding velocities of the center of mass of B by $\mathbf{v}_B$, $\mathbf{v}_{BC}$, and $\mathbf{v}'_B$. For A, we have

$$\int_{t_1}^{t_C} -R\,dt = m_A(\mathbf{v}_{AC})_x - m_A(\mathbf{v}_A)_x, \tag{19.36}$$

$$\int_{t_C}^{t_2} -R\,dt = m_A(\mathbf{v}'_A)_x - m_A(\mathbf{v}_{AC})_x. \tag{19.37}$$

For B, we have

$$\int_{t_1}^{t_C} R\,dt = m_B(\mathbf{v}_{BC})_x - m_B(\mathbf{v}_B)_x, \tag{19.38}$$

$$\int_{t_C}^{t_2} R\,dt = m_B(\mathbf{v}'_B)_x - m_B(\mathbf{v}_{BC})_x. \tag{19.39}$$

The coefficient of restitution is the ratio of the linear impulse during the recovery phase to the linear impulse during the deformation phase:

$$e = \frac{\displaystyle\int_{t_C}^{t_2} R\,dt}{\displaystyle\int_{t_1}^{t_C} R\,dt}. \tag{19.40}$$

If we divide Eq. (19.37) by Eq. (19.36) and divide Eq. (19.39) by Eq. (19.38), the resulting equations can be written as

$$(\mathbf{v}'_A)_x = -(\mathbf{v}_A)_x e + (\mathbf{v}_{AC})_x(1 + e),$$

$$(\mathbf{v}'_B)_x = -(\mathbf{v}_B)_x e + (\mathbf{v}_{BC})_x(1 + e). \tag{19.41}$$

We now apply the principle of angular impulse and momentum to A and B for the intervals of time from t_1 to t_C and from t_C to t_2. We denote the counterclockwise angular velocity of A at the times t_1, t_C, and t_2 by ω_A, ω_{AC}, and ω'_A and denote the corresponding angular velocities of B by ω_B, ω_{BC}, and ω'_B. We write the position vectors of P relative to the centers of mass of A and B as (Fig. 19.19)

$$\mathbf{r}_{P/A} = x_A\mathbf{i} + y_A\mathbf{j},$$
$$\mathbf{r}_{P/B} = x_B\mathbf{i} + y_B\mathbf{j}.$$
(19.42)

The moment about the center of mass of A due to the force exerted on A by the impact is $\mathbf{r}_{P/A} \times (-R\mathbf{i}) = y_A R\mathbf{k}$. From Eq. (19.28), we obtain the equations

$$\int_{t_1}^{t_C} y_A R \, dt = I_A\omega_{AC} - I_A\omega_A,$$
(19.43)

$$\int_{t_C}^{t_2} y_A R \, dt = I_A\omega'_A - I_A\omega_{AC}.$$
(19.44)

The corresponding equations for B are

$$\int_{t_1}^{t_C} -y_B R \, dt = I_B\omega_{BC} - I_B\omega_B,$$
(19.45)

$$\int_{t_C}^{t_2} -y_B R \, dt = I_B\omega'_B - I_B\omega_{BC}.$$
(19.46)

Dividing Eq. (19.44) by Eq. (19.43) and dividing Eq. (19.46) by Eq. (19.45), we can write the resulting equations as

$$\omega'_A = -\omega_A e + \omega_{AC}(1 + e),$$
$$\omega'_B = -\omega_B e + \omega_{BC}(1 + e).$$
(19.47)

By expressing the velocity of the point of A at P in terms of the velocity of the center of mass of A and the angular velocity of A, and expressing the velocity of the point of B at P in terms of the velocity of the center of mass of B and the angular velocity of B, we obtain

$$(\mathbf{v}_{AP})_x = (\mathbf{v}_A)_x - \omega_A y_A,$$
$$(\mathbf{v}'_{AP})_x = (\mathbf{v}'_A)_x - \omega'_A y_A,$$
$$(\mathbf{v}_{BP})_x = (\mathbf{v}_B)_x - \omega_B y_B,$$
$$(\mathbf{v}'_{BP})_x = (\mathbf{v}'_B)_x - \omega'_B y_B.$$
(19.48)

At time t_C, the x components of the velocities of the two objects are equal at P, which yields the relation

$$(\mathbf{v}_{AC})_x - \omega_{AC} y_A = (\mathbf{v}_{BC})_x - \omega_{BC} y_B. \tag{19.49}$$

From Eqs. (19.48),

$$\frac{(\mathbf{v}'_{BP})_x - (\mathbf{v}'_{AP})_x}{(\mathbf{v}_{AP})_x - (\mathbf{v}_{BP})_x} = \frac{(\mathbf{v}'_B)_x - \omega'_B y_B - (\mathbf{v}'_A)_x + \omega'_A y_A}{(\mathbf{v}_A)_x - \omega_A y_A - (\mathbf{v}_B)_x + \omega_B y_B}.$$

Substituting Eqs. (19.41) and (19.47) into this equation and collecting terms yields

$$\frac{(\mathbf{v}'_{BP})_x - (\mathbf{v}'_{AP})_x}{(\mathbf{v}_{AP})_x - (\mathbf{v}_{BP})_x} = e - \left[\frac{(\mathbf{v}_{AC})_x - \omega_{AC} y_A - (\mathbf{v}_{BC})_x + \omega_{BC} y_B}{(\mathbf{v}_A)_x - \omega_A y_A - (\mathbf{v}_B)_x + \omega_B y_B}\right](e + 1).$$

From Eq. (19.49), the term in brackets vanishes, and we obtain the equation relating the normal components of the velocities at the point of contact to the coefficient of restitution:

$$e = \frac{(\mathbf{v}'_{BP})_x - (\mathbf{v}'_{AP})_x}{(\mathbf{v}_{AP})_x - (\mathbf{v}_{BP})_x}. \tag{19.50}$$

In arriving at this equation, we assumed that the contacting surfaces were smooth, so that *the collision exerts no force on A or B in the direction tangential to their contacting surfaces.*

   Although we derived Eq. (19.50) under the assumption that the motions of A and B are unconstrained, the relationship also holds if they are not—for example, if one of them is connected to a pin support.

RESULTS

Suppose that two rigid bodies A and B in planar motion collide.

Linear Momentum

> If other forces are negligible in comparison to the impact forces A and B exert on each other, their total linear momentum is the same before and after the impact.

Angular Momentum

If other forces are negligible in comparison to the impact forces A and B exert on each other, their total angular momentum about *any* fixed point O is the same before and after the impact. If, in addition, A and B exert forces on each other only at their point of impact P, the angular momentum about P of *each* rigid body is the same before and after the impact.

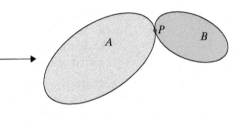

If one of the two rigid bodies has a pin support at a point O, their total angular momentum about O is the same before and after the impact.

Coefficient of Restitution

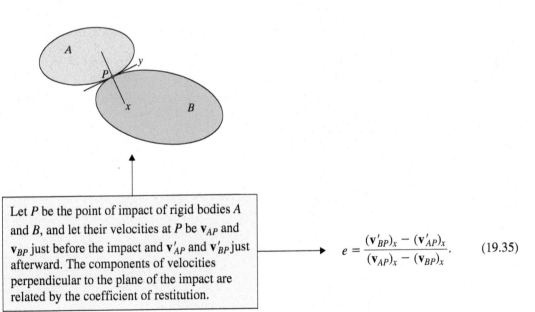

Let P be the point of impact of rigid bodies A and B, and let their velocities at P be $\mathbf{v}_{AP}$ and $\mathbf{v}_{BP}$ just before the impact and $\mathbf{v}'_{AP}$ and $\mathbf{v}'_{BP}$ just afterward. The components of velocities perpendicular to the plane of the impact are related by the coefficient of restitution.

$$e = \frac{(\mathbf{v}'_{BP})_x - (\mathbf{v}'_{AP})_x}{(\mathbf{v}_{AP})_x - (\mathbf{v}_{BP})_x}. \qquad (19.35)$$

Active Example 19.7 Impact of a Sphere and a Suspended Bar (▶ *Related Problem 19.70*)

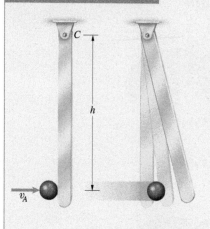

(a)

The ball of mass m_A is translating with horizontal velocity v_A when it strikes the stationary slender bar of mass m_B and length l. The coefficient of restitution of the impact is e.

(a) What is the angular velocity of the bar after the impact?

(b) If the duration of the impact is Δt, what average horizontal force is exerted on the bar by the pin support C as a result of the impact?

Strategy

(a) The total angular momentum about C of the ball and bar is the same before and after the impact. The coefficient of restitution relates the velocities of the ball and the bar *at the point of impact* before and after the impact. With these two equations and kinematic relationships, we can determine the velocity of the ball and the angular velocity of the bar after the impact.

(b) We can determine the average force exerted on the bar by the support at C by applying the principle of angular impulse and momentum to the bar.

Solution

Apply conservation of angular momentum about C. After the impact, v_A' is the velocity of the ball, v_B' is the velocity of the center of mass of the bar, and ω_B' is the angular velocity of the bar. I_B is the moment of interia of the bar about its center of mass.

$$\begin{cases} H_{CA} + H_{CB} = H'_{CA} + H'_{CB}: \\ h(m_A v_A) + 0 = h(m_A v_A') + \frac{1}{2}l(m_B v_B') + I_B \omega_B'. \quad (1) \end{cases}$$

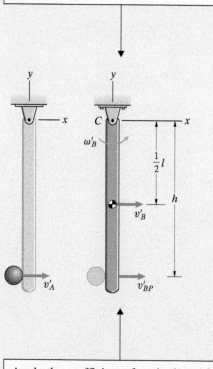

Apply the coefficient of restitution. After the impact, v'_{BP} is the velocity of the bar *at the point of impact*.

$$e = \frac{v'_{BP} - v'_A}{v_A - 0}. \quad (2)$$

Determine kinematic relationships. Because it rotates about the fixed point C, the velocities of the bar at its center of mass and at the point of impact can be expressed in terms of the bar's angular velocity.

$$\begin{cases} v'_B = \tfrac{1}{2}l\omega'_B, & (3) \\ v'_{BP} = h\omega'_B, & (4) \end{cases}$$

Solve Eqs. (1) through (4) for v'_A, v'_B, v'_{BP}, and ω'_B and use relation $I_B = \tfrac{1}{12}m_B l^2$ to obtain the bar's angular velocity.

$$\omega'_B = \frac{(1+e)hm_A v_A}{h^2 m_A + \tfrac{1}{3}m_B l^2}. \qquad (5)$$

(b)

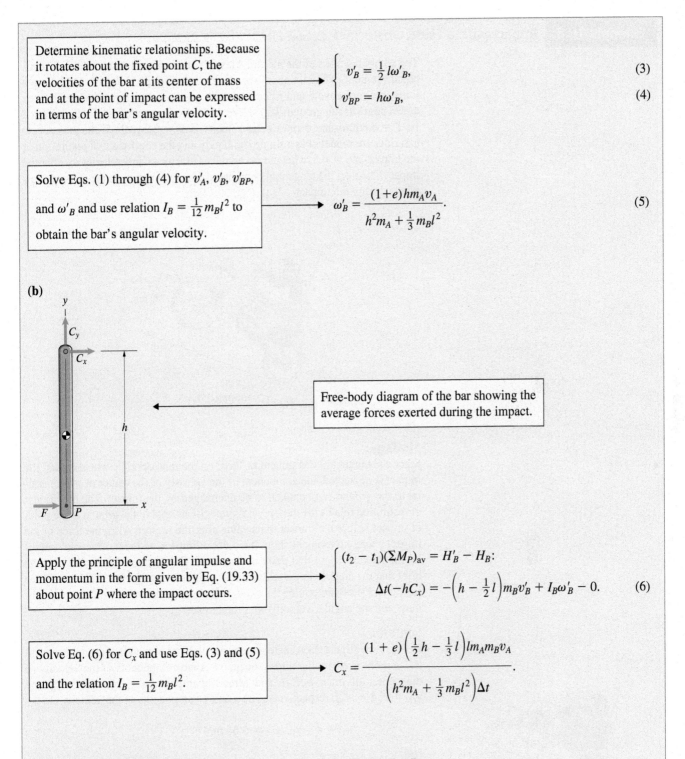

Free-body diagram of the bar showing the average forces exerted during the impact.

Apply the principle of angular impulse and momentum in the form given by Eq. (19.33) about point P where the impact occurs.

$$\begin{cases} (t_2 - t_1)(\Sigma M_P)_{\text{av}} = H'_B - H_B: \\ \Delta t(-hC_x) = -\left(h - \tfrac{1}{2}l\right)m_B v'_B + I_B \omega'_B - 0. \qquad (6) \end{cases}$$

Solve Eq. (6) for C_x and use Eqs. (3) and (5) and the relation $I_B = \tfrac{1}{12}m_B l^2$.

$$C_x = \frac{(1+e)\left(\tfrac{1}{2}h - \tfrac{1}{3}l\right)lm_A m_B v_A}{\left(h^2 m_A + \tfrac{1}{3}m_B l^2\right)\Delta t}.$$

Practice Problem Suppose that the pin support at C is removed, and the ball strikes the stationary vertical bar with horizontal velocity v_A. Assume that $m_A = m_B$ and $h = \tfrac{3}{4}l$. What is the angular velocity of the bar after the impact?

Answer: $\omega'_B = \dfrac{\tfrac{12}{11}(1+e)v_A}{l}$.

| Example 19.8 | Impact with a Fixed Obstacle (▶ *Related Problems 19.89, 19.90*) |

The combined mass of the motorcycle and rider is $m = 170$ kg, and their combined moment of inertia about their center of mass is 22 kg-m². Following a jump, the motorcycle and rider are in the position shown just before the rear wheel contacts the ground. The velocity of their center of mass is of magnitude $|\mathbf{v}_G| = 8.8$ m/s, and their angular velocity is $\omega = 0.2$ rad/s. If the motorcycle and rider are modeled as a single rigid body and the coefficient of restitution of the impact is $e = 0.8$, what are the angular velocity ω' and velocity $\mathbf{v}'_G$ after the impact? Neglect the tangential component of force exerted on the motorcycle's wheel during the impact.

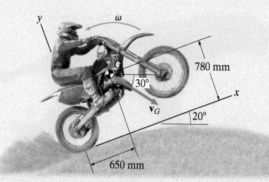

Strategy

Since the tangential component of force on the motorcycle's wheel during the impact is neglected, the component of the velocity of the center of mass parallel to the ground can be taken to be unchanged by the impact. The coefficient of restitution relates the motorcycle's velocity normal to the ground at the point of impact before the impact to its value after the impact. Also, the force of the impact exerts no moment about the point of impact, so the motorcycle's angular momentum about that point is conserved. (We assume the impact to be so brief that the angular impulse due to the weight is negligible.) With these three relations, we can determine the two components of the velocity of the center of mass and the angular velocity after the impact.

Solution

In Fig. a we align a coordinate system parallel and perpendicular to the ground at the point P where the impact occurs. Let the components of the velocity of the center of mass before and after the impact be $\mathbf{v}_G = v_x \mathbf{i} + v_y \mathbf{j}$ and $\mathbf{v}'_G = v'_x \mathbf{i} + v'_y \mathbf{j}$, respectively. The x and y components of the velocity are

$$v_x = 8.8 \cos 50° = 5.66 \text{ m/s}$$

and

$$v_y = -8.8 \sin 50° = -6.74 \text{ m/s}.$$

Because the component of the impact force tangential to the ground is neglected, the x component of the velocity of the center of mass is unchanged:

$$v'_x = v_x = 5.66 \text{ m/s}.$$

(a) Aligning the x axis of the coordinate system tangent to the ground at P.

We can express the y component of the wheel's velocity at P before the impact in terms of the velocity of the center of mass and the angular velocity (Fig. a) as

$$\mathbf{j} \cdot \mathbf{v}_P = \mathbf{j} \cdot (\mathbf{v}_G + \boldsymbol{\omega} \times \mathbf{r}_{P/G})$$

$$= \mathbf{j} \cdot \left\{ v_x\mathbf{i} + v_y\mathbf{j} + \begin{vmatrix} \mathbf{i} & \mathbf{j} & \mathbf{k} \\ 0 & 0 & \omega \\ -0.65 & -0.78 & 0 \end{vmatrix} \right\}$$

$$= v_y - 0.65\omega.$$

(Notice that this expression gives the y component of the velocity at P even though the wheel is spinning.) The y component of the wheel's velocity at P after the impact is

$$\mathbf{j} \cdot \mathbf{v}_P' = \mathbf{j} \cdot (\mathbf{v}_G' + \boldsymbol{\omega}' \times \mathbf{r}_{P/G})$$

$$= v_y' - 0.65\omega'.$$

The coefficient of restitution relates the y components of the wheel's velocity at P before and after the impact:

$$e = \frac{-(\mathbf{j} \cdot \mathbf{v}_P')}{\mathbf{j} \cdot \mathbf{v}_P} = \frac{-(v_y' - 0.65\omega')}{v_y - 0.65\omega}. \tag{1}$$

The force of the impact exerts no moment about P, so angular momentum about P is conserved:

$$H_P = H_P':$$

$$[(\mathbf{r}_{G/P} \times m\mathbf{v}_G) \cdot \mathbf{k} + I\omega] = [(\mathbf{r}_{G/P} \times m\mathbf{v}_G') \cdot \mathbf{k} + I\omega'],$$

$$\begin{vmatrix} \mathbf{i} & \mathbf{j} & \mathbf{k} \\ 0.65 & 0.78 & 0 \\ mv_x & mv_y & 0 \end{vmatrix} \cdot \mathbf{k} + I\omega = \begin{vmatrix} \mathbf{i} & \mathbf{j} & \mathbf{k} \\ 0.65 & 0.78 & 0 \\ mv_x' & mv_y' & 0 \end{vmatrix} \cdot \mathbf{k} + I\omega'.$$

Expanding the determinants and evaluating the dot products, we obtain

$$0.65mv_y - 0.78mv_x + I\omega = 0.65mv_y' - 0.78mv_x' + I\omega'. \tag{2}$$

Since we have already determined v_x', we can solve Eqs. (1) and (2) for v_y' and ω'. The results are

$$v_y' = -3.84 \text{ m/s}$$

and

$$\omega' = -14.4 \text{ rad/s}.$$

The velocity of the center of mass after the impact is $\mathbf{v}_G' = 5.66\mathbf{i} - 3.84\mathbf{j}$ m/s, and the angular velocity is 14.4 rad/s in the clockwise direction.

Critical Thinking

Although forces resulting from an impact are often so large that the effects of other forces can be neglected, that isn't always the case. In this example, we neglected the weight of the motorcycle and rider in determining their velocity and angular velocity following the impact of the rear wheel with the ground. Whenever there is doubt in engineering applications of momentum methods, such effects should be included in the analysis. In order to do so, the duration of the impact needs to be known or estimated so that the linear and angular impulses due to other forces can be evaluated.

Example 19.9	Colliding Cars (▶ *Related Problems 19.91, 19.92*)

An engineer simulates a collision between two 1600-kg cars by modeling them as rigid bodies. The moment of inertia of each car about its center of mass is 960 kg-m². The engineer assumes the contacting surfaces at P to be smooth and parallel to the x axis and assumes the coefficient of restitution to be $e = 0.2$. What are the angular velocities of the cars and the velocities of their centers of mass after the collision?

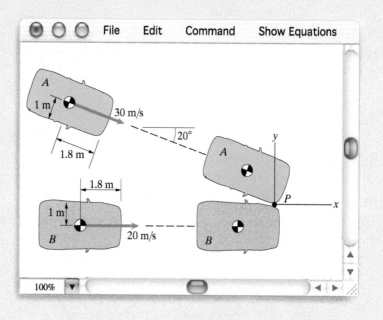

Strategy

Since the contacting surfaces are smooth, the x components of the velocities of the centers of mass are unchanged by the collision. The y components of the velocities must satisfy conservation of linear momentum, and the y components of the velocities at the point of impact before and after the impact are related by the coefficient of restitution. The force of the impact exerts no moment about P on either car, so the angular momentum of each car about P is conserved. From these conditions and kinematic relations between the velocities of the centers of mass and the velocities at P, we can determine the angular velocities and the velocities of the centers of mass after the impact.

Solution

The components of the velocities of the centers of mass before the impact are

$$\mathbf{v}_A = 30\cos 20°\mathbf{i} - 30\sin 20°\mathbf{j}$$

$$= 28.2\mathbf{i} - 10.3\mathbf{j} \ (\text{m/s})$$

and

$$\mathbf{v}_B = 20\mathbf{i} \ (\text{m/s}).$$

The x components of the velocities are unchanged by the impact:

$$v'_{Ax} = v_{Ax} = 28.2 \text{ m/s}, \qquad v'_{Bx} = v_{Bx} = 20 \text{ m/s}.$$

The y components of the velocities must satisfy conservation of linear momentum:

$$m_A v_{Ay} + m_B v_{By} = m_A v'_{Ay} + m_B v'_{By}. \tag{1}$$

Let the velocities of the two cars at P before the collision be $\mathbf{v}_{AP}$ and $\mathbf{v}_{BP}$. The coefficient of restitution $e = 0.2$ relates the y components of the velocities at P:

$$0.2 = \frac{v'_{BPy} - v'_{APy}}{v_{APy} - v_{BPy}}. \tag{2}$$

We can express the velocities at P after the impact in terms of the velocities of the centers of mass and the angular velocities after the impact (Fig. a). The position of P relative to the center of mass of car A is

$$\mathbf{r}_{P/A} = [(1.8)\cos 20° - (1)\sin 20°]\mathbf{i} - [(1.8)\sin 20° + (1)\cos 20°]\mathbf{j}$$

$$= 1.35\mathbf{i} - 1.56\mathbf{j} \text{ (m)}.$$

Therefore, we can express the velocity of point P of car A after the impact as

$$\mathbf{v}'_{AP} = \mathbf{v}'_A + \boldsymbol{\omega}'_A \times \mathbf{r}_{P/A}:$$

$$v'_{APx}\mathbf{i} + v'_{APy}\mathbf{j} = v'_{Ax}\mathbf{i} + v'_{Ay}\mathbf{j} + \begin{vmatrix} \mathbf{i} & \mathbf{j} & \mathbf{k} \\ 0 & 0 & \omega'_A \\ 1.35 & -1.56 & 0 \end{vmatrix}.$$

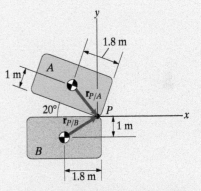

(a) Position vectors of P relative to the centers of mass.

Equating $\mathbf{i}$ and $\mathbf{j}$ components in this equation, we obtain

$$v'_{APx} = v'_{Ax} + 1.56\omega'_A,$$

$$v'_{APy} = v'_{Ay} + 1.35\omega'_A. \tag{3}$$

The position of P relative to the center of mass of car B is

$$\mathbf{r}_{P/B} = 1.8\mathbf{i} + \mathbf{j} \text{ (m)}.$$

We can express the velocity of point P of car B after the impact as

$$\mathbf{v}'_{BP} = \mathbf{v}'_B + \boldsymbol{\omega}'_B \times \mathbf{r}_{P/B}:$$

$$v'_{BPx}\mathbf{i} + v'_{BPy}\mathbf{j} = v'_{Bx}\mathbf{i} + v'_{By}\mathbf{j} + \begin{vmatrix} \mathbf{i} & \mathbf{j} & \mathbf{k} \\ 0 & 0 & \omega'_B \\ 1.8 & 1 & 0 \end{vmatrix}.$$

Equating $\mathbf{i}$ and $\mathbf{j}$ components yields

$$v'_{BPx} = v'_{Bx} - \omega'_B,$$
$$v'_{BPy} = v'_{By} + 1.8\omega'_B. \tag{4}$$

The angular momentum of car A about P is conserved:

$$H_{PA} = H'_{PA}:$$

$$[(\mathbf{r}_{A/P} \times m_A\mathbf{v}_A) \cdot \mathbf{k} + I_A\omega_A] = [(\mathbf{r}_{A/P} \times m_A\mathbf{v}'_A) \cdot \mathbf{k} + I_A\omega'_A],$$

$$\begin{vmatrix} \mathbf{i} & \mathbf{j} & \mathbf{k} \\ -1.35 & 1.56 & 0 \\ m_Av_{Ax} & m_Av_{Ay} & 0 \end{vmatrix} \cdot \mathbf{k} + 0 = \begin{vmatrix} \mathbf{i} & \mathbf{j} & \mathbf{k} \\ -1.35 & 1.56 & 0 \\ m_Av'_{Ax} & m_Av'_{Ay} & 0 \end{vmatrix} \cdot \mathbf{k} + I_A\omega'_A.$$

Expanding the determinants and evaluating the dot products, we obtain

$$-1.35m_Av_{Ay} - 1.56m_Av_{Ax}$$
$$= -1.35m_Av'_{Ay} - 1.56m_Av'_{Ax} + I_A\omega'_A. \tag{5}$$

The angular momentum of car B about P is also conserved,

$$H_{PB} = H'_{PB}:$$

$$[(\mathbf{r}_{B/P} \times m_B\mathbf{v}_B) \cdot \mathbf{k} + I_B\omega_B] = [(\mathbf{r}_{B/P} \times m_B\mathbf{v}'_B) \cdot \mathbf{k} + I_B\omega'_B],$$

$$\begin{vmatrix} \mathbf{i} & \mathbf{j} & \mathbf{k} \\ -1.8 & -1 & 0 \\ m_Bv_{Bx} & 0 & 0 \end{vmatrix} \cdot \mathbf{k} + 0 = \begin{vmatrix} \mathbf{i} & \mathbf{j} & \mathbf{k} \\ -1.8 & -1 & 0 \\ m_Bv'_{Bx} & m_Bv'_{By} & 0 \end{vmatrix} \cdot \mathbf{k} + I_B\omega'_B.$$

From this equation, it follows that

$$m_Bv_{Bx} = -1.8m_Bv'_{By} + m_Bv'_{Bx} + I_B\omega'_B. \tag{6}$$

We can solve Eqs. (1) through (6) for $\mathbf{v}'_A$, $\mathbf{v}'_{AP}$, ω'_A, $\mathbf{v}'_B$, $\mathbf{v}'_{BP}$, and ω'_B. The results for the velocities of the centers of mass of the cars and their angular velocities are as follows:

$$\mathbf{v}'_A = 28.2\mathbf{i} - 9.08\mathbf{j} \text{ (m/s)}, \qquad \omega'_A = 2.65 \text{ rad/s},$$
$$\mathbf{v}'_B = 20.0\mathbf{i} - 1.18\mathbf{j} \text{ (m/s)}, \qquad \omega'_B = -3.54 \text{ rad/s}.$$

Critical Thinking

The total angular moment of the two cars about *any point* is the same before and after their collision, because we neglected the effects of horizontal forces other than the force due to their collision. But to determine their motions after the collision, we needed to use the fact that the angular momentum of *each* car about P is the same before and after the impact. That is true because the moment about P exerted on each car by the force of the collision is zero. Notice that we could not have assumed that the angular momentum of each car about any point is the same before and after the impact.

Problems

19.64 The 10-lb bar is released from rest in the 45° position shown. It falls and the end of the bar strikes the horizontal surface at P. The coefficient of restitution of the impact is $e = 0.6$. When the bar rebounds, through what angle relative to the horizontal will it rotate?

19.65 The 10-lb bar is released from rest in the 45° position shown. It falls and the end of the bar strikes the horizontal surface at P. The bar rebounds to a position 10° relative to the horizontal. If the duration of the impact is 0.01 s, what is the magnitude of the average vertical force the horizontal surface exerted on the bar at P?

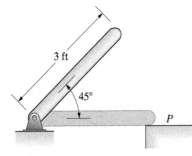

Problems 19.64/19.65

19.66 The 4-kg bar is released from rest in the horizontal position above the fixed projection at A. The distance $b = 0.35$ m. The impact of the bar with the projection is plastic; that is, the coefficient of restitution of the impact is $e = 0$. What is the bar's angular velocity immediately after the impact?

19.67 The 4-kg bar is released from rest in the horizontal position above the fixed projection at A. The coefficient of restitution of the impact is $e = 0.6$. What value of the distance b would cause the velocity of the bar's center of mass to be zero immediately after the impact? What is the bar's angular velocity immediately after the impact?

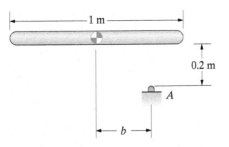

Problems 19.66/19.67

19.68 The mass of the ship is 544,000 kg, and the moment of inertia of the vessel about its center of mass is 4×10^8 kg-m². Wind causes the ship to drift sideways at 0.1 m/s and strike the stationary piling at P. The coefficient of restitution of the impact is $e = 0.2$. What is the ship's angular velocity after the impact?

19.69 In Problem 19.68, if the duration of the ship's impact with the piling is 10 s, what is the magnitude of the average force exerted on the ship by the impact?

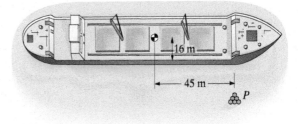

Problems 19.68/19.69

▶ **19.70** In Active Example 19.7, suppose that the ball A weighs 2 lb, the bar B weighs 6 lb, and the length of the bar is 3 ft. The ball is translating at $v_A = 10$ ft/s before the impact and strikes the bar at $h = 2$ ft. What is the angular velocity of the bar after the impact if the ball adheres to the bar?

19.71 The 2-kg sphere A is moving toward the right at 4 m/s when it strikes the end of the 5-kg slender bar B. Immediately after the impact, the sphere A is moving toward the right at 1 m/s. What is the angular velocity of the bar after the impact?

19.72 The 2-kg sphere A is moving toward the right at 4 m/s when it strikes the end of the 5-kg slender bar B. The coefficient of restitution is $e = 0.4$. The duration of the impact is 0.002 seconds. Determine the magnitude of the average horizontal force exerted on the bar by the pin support as a result of the impact.

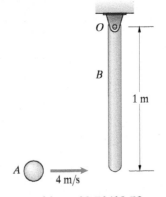

Problems 19.71/19.72

19.73 The 2-kg sphere A is moving toward the right at 10 m/s when it strikes the unconstrained 4-kg slender bar B. What is the angular velocity of the bar after the impact if the sphere adheres to the bar?

19.74 The 2-kg sphere A is moving toward the right at 10 m/s when it strikes the unconstrained 4-kg slender bar B. The coefficient of restitution of the impact is $e = 0.6$. What are the velocity of the sphere and the angular velocity of the bar after the impact?

19.75 The 5-oz ball is translating with velocity $v_A = 80$ ft/s perpendicular to the bat just before impact. The player is swinging the 31-oz bat with angular velocity $\omega = 6\pi$ rad/s before the impact. Point C is the bat's instantaneous center both before and after the impact. The distances $b = 14$ in and $\bar{y} = 26$ in. The bat's moment of inertia about its center of mass is $I_B = 0.033$ slug-ft^2. The coefficient of restitution is $e = 0.6$, and the duration of the impact is 0.008 s. Determine the magnitude of the velocity of the ball after the impact and the average force A_x exerted on the bat by the player during the impact if (a) $d = 0$, (b) $d = 3$ in, and (c) $d = 8$ in.

19.76 In Problem 19.75, show that the force A_x is zero if $d = I_B/(m_B\bar{y})$, where m_B is the mass of the bat.

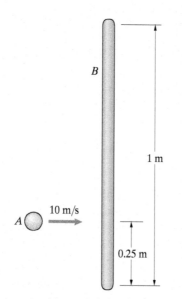

Problems 19.73/19.74

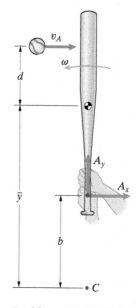

Problems 19.75/19.76

19.77 A 10-lb slender bar of length $l = 2$ ft is released from rest in the horizontal position at a height $h = 2$ ft above a peg (Fig. a). A small hook at the end of the bar engages the peg, and the bar swings from the peg (Fig. b). What is the bar's angular velocity immediately after it engages the peg?

19.78 A 10-lb slender bar of length $l = 2$ ft is released from rest in the horizontal position at a height $h = 1$ ft above a peg (Fig. a). A small hook at the end of the bar engages the peg, and the bar swings from the peg (Fig. b).

(a) Through what maximum angle does the bar rotate relative to its position when it engages the peg?

(b) At the instant when the bar has reached the angle determined in part (a), compare its gravitational potential energy to the gravitational potential energy the bar had when it was released from rest. How much energy has been lost?

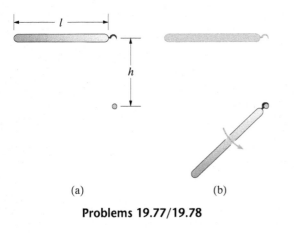

(a)   (b)

Problems 19.77/19.78

19.79 The 1-slug disk rolls at velocity $v = 10$ ft/s toward a 6-in step. The wheel remains in contact with the step and does not slip while rolling up onto it. What is the wheel's velocity once it is on the step?

19.80 The 1-slug disk rolls toward a 6-in step. The wheel remains in contact with the step and does not slip while rolling up onto it. What is the minimum velocity v the disk must have in order to climb up onto the step?

18 in

v

6 in

Problems 19.79/19.80

19.81 The length of the bar is 1 m, and its mass is 2 kg. Just before the bar hits the floor, its angular velocity is $\omega = 0$ and its center of mass is moving downward at 4 m/s. If the end of the bar adheres to the floor, what is the bar's angular velocity after the impact?

19.82 The length of the bar is 1 m, and its mass is 2 kg. Just before the bar hits the *smooth* floor, its angular velocity is $\omega = 0$ and its center of mass is moving downward at 4 m/s. If the coefficient of restitution of the impact is $e = 0.4$, what is the bar's angular velocity after the impact?

19.83 The length of the bar is 1 m, and its mass is 2 kg. Just before the bar hits the *smooth* floor, it has angular velocity ω and its center of mass is moving downward at 4 m/s. The coefficient of restitution of the impact is $e = 0.4$. What value of ω would cause the bar to have no angular velocity after the impact?

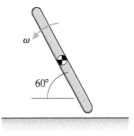

ω

60°

Problems 19.81–19.83

19.84 During her parallel-bars routine, the velocity of the 90-lb gymnast's center of mass is $4\mathbf{i} - 10\mathbf{j}$ (ft/s) and her angular velocity is zero just before she grasps the bar at A. In the position shown, her moment of inertia about her center of mass is 1.8 slug-ft^2. If she stiffens her shoulders and legs so that she can be modeled as a rigid body, what is the velocity of her center of mass and her angular velocity just after she grasps the bar?

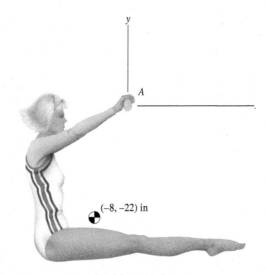

y

A

$(-8, -22)$ in

Problem 19.84

19.85 The 20-kg homogeneous rectangular plate is released from rest (Fig. a) and falls 200 mm before coming to the end of the string attached at the corner A (Fig. b). Assuming that the vertical component of the velocity of A is zero just after the plate reaches the end of the string, determine the angular velocity of the plate and the magnitude of the velocity of the corner B at that instant.

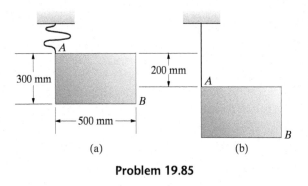

(a)   (b)

Problem 19.85

19.86* The two bars A and B are each 2 m in length, and each has a mass of 4 kg. In Fig. a, bar A has no angular velocity and is moving to the right at 1 m/s, and bar B is stationary. If the bars bond together on impact (Fig. b), what is their angular velocity ω' after the impact?

19.87* In Problem 19.86, if the bars do not bond together on impact and the coefficient of restitution is $e = 0.8$, what are the angular velocities of the bars after the impact?

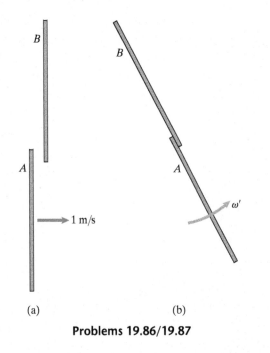

(a)   (b)

Problems 19.86/19.87

19.88* Two bars A and B are each 2 m in length, and each has a mass of 4 kg. In Fig. a, bar A has no angular velocity and is moving to the right at 1 m/s, and bar B is stationary. If the bars bond together on impact (Fig. b), what is their angular velocity ω' after the impact?

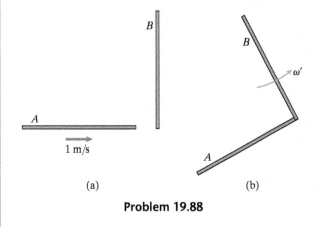

(a)   (b)

Problem 19.88

▶ **19.89*** The horizontal velocity of the landing airplane is 50 m/s, its vertical velocity (rate of descent) is 2 m/s, and its angular velocity is zero. The mass of the airplane is 12 Mg, and the moment of inertia about its center of mass is 1×10^5 kg-m^2. When the rear wheels touch the runway, they remain in contact with it. Neglecting the horizontal force exerted on the wheels by the runway, determine the airplane's angular velocity just after it touches down. (See Example 19.8.)

▶ **19.90*** Determine the angular velocity of the airplane in Problem 19.89 just after it touches down if its wheels don't stay in contact with the runway and the coefficient of restitution of the impact is $e = 0.4$. (See Example 19.8.)

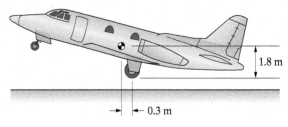

Problems 19.89/19.90

▶ **19.91*** While attempting to drive on an icy street for the first time, a student skids his 1260-kg car (A) into the university president's stationary 2700-kg Rolls-Royce Corniche (B). The point of impact is P. Assume that the impacting surfaces are smooth and parallel to the y axis and that the coefficient of restitution of the impact is e = 0.5. The moments of inertia of the cars about their centers of mass are I_A = 2400 kg-m^2 and I_B = 7600 kg-m^2. Determine the angular velocities of the cars and the velocities of their centers of mass after the collision. (See Example 19.9.)

▶ **19.92*** The student in Problem 19.91 claimed that he was moving at 5 km/h prior to the collision, but police estimate that the center of mass of the Rolls-Royce was moving at 1.7 m/s after the collision. What was the student's actual speed? (See Example 19.9.)

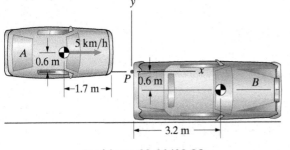

Problems 19.91/19.92

19.93 Each slender bar is 48 in long and weighs 20 lb. Bar A is released in the horizontal position shown. The bars are smooth, and the coefficient of restitution of their impact is e = 0.8. Determine the angle through which B swings afterward.

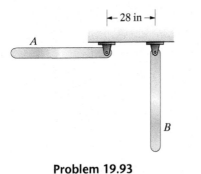

Problem 19.93

19.94* The *Apollo* CSM (A) approaches the *Soyuz* Space Station (B). The mass of the *Apollo* is m_A = 18 Mg, and the moment of inertia about the axis through its center of mass parallel to the z axis is I_A = 114 Mg-m^2. The mass of the *Soyuz* is m_B = 6.6 Mg, and the moment of inertia about the axis through its center of mass parallel to the z axis is I_B = 70 Mg-m^2. The *Soyuz* is stationary relative to the reference frame shown, and the CSM approaches with velocity $\mathbf{v}_A$ = 0.21**i** + 0.05**j** (m/s) and no angular velocity. What is the angular velocity of the attached spacecraft after docking?

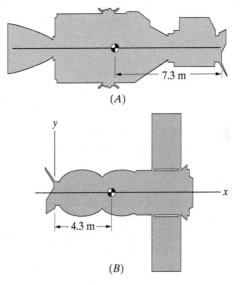

Problem 19.94

Review Problems

19.95 The moment of inertia of the pulley is 0.2 kg-m². The system is released from rest. Use the principle of work and energy to determine the velocity of the 10-kg cylinder when it has fallen 1 m.

19.96 The moment of inertia of the pulley is 0.2 kg-m². The system is released from rest. Use momentum principles to determine the velocity of the 10-kg cylinder 1 s after the system is released.

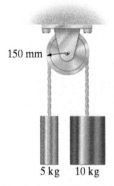

150 mm

5 kg   10 kg

Problems 19.95/19.96

19.97 Arm BC has a mass of 12 kg, and the moment of inertia about its center of mass is 3 kg-m². Point B is stationary. Arm BC is initially aligned with the (horizontal) x axis with zero angular velocity, and a constant couple M applied at B causes the arm to rotate upward. When it is in the position shown, its counterclockwise angular velocity is 2 rad/s. Determine M.

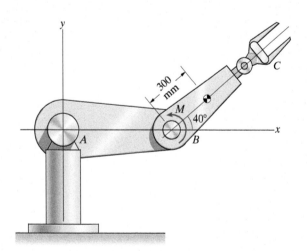

Problem 19.97

19.98 The cart is stationary when a constant force F is applied to it. What will the velocity of the cart be when it has rolled a distance b? The mass of the body of the cart is m_c, and each of the four wheels has mass m, radius R, and moment of inertia, I.

Problem 19.98

19.99 Each pulley has moment of inertia $I = 0.003$ kg-m², and the mass of the belt is 0.2 kg. If a constant couple $M = 4$ N-m is applied to the bottom pulley, what will its angular velocity be when it has turned 10 revolutions?

100 mm

M

Problem 19.99

19.100 The ring gear is fixed. The mass and moment of inertia of the sun gear are $m_S = 22$ slug and $I_S = 4400$ slug-ft^2. The mass and moment of inertia of each planet gear are $m_P = 2.7$ slug and $I_P = 65$ slug-ft^2. A couple $M = 600$ ft-lb is applied to the sun gear. Use work and energy to determine the angular velocity of the sun gear after it has turned 100 revolutions.

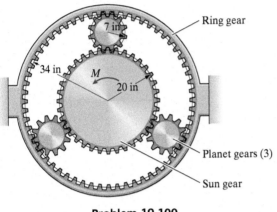

Problem 19.100

19.101 The moments of inertia of gears A and B are $I_A = 0.014$ slug-ft^2 and $I_B = 0.100$ slug-ft^2. Gear A is connected to a torsional spring with constant $k = 0.2$ ft-lb/rad. If the spring is unstretched and the surface supporting the 5-lb weight is removed, what is the velocity of the weight when it has fallen 3 in?

19.102 Consider the system in Problem 19.101.
(a) What maximum distance does the 5-lb weight fall when the supporting surface is removed?
(b) What maximum velocity does the weight achieve?

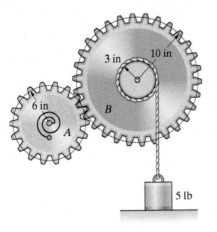

Problems 19.101/19.102

19.103 Each of the go-cart's front wheels weighs 5 lb and has a moment of inertia of 0.01 slug-ft^2. The two rear wheels and rear axle form a single rigid body weighing 40 lb and having a moment of inertia of 0.1 slug-ft^2. The total weight of the rider and go-cart, including its wheels, is 240 lb. The go-cart starts from rest, its engine exerts a constant torque of 15 ft-lb on the rear axle, and its wheels do not slip. Neglecting friction and aerodynamic drag, how fast is the go-cart moving when it has traveled 50 ft?

19.104 Determine the maximum power and the average power transmitted to the go-cart in Problem 19.103 by its engine.

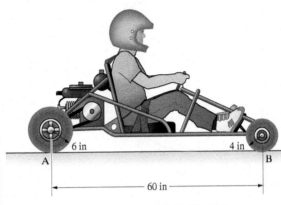

Problems 19.103/19.104

19.105 The system starts from rest with the 4-kg slender bar horizontal. The mass of the suspended cylinder is 10 kg. What is the angular velocity of the bar when it is in the position shown?

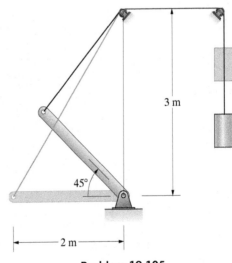

Problem 19.105

19.106 The 0.1-kg slender bar and 0.2-kg cylindrical disk are released from rest with the bar horizontal. The disk rolls on the curved surface. What is the angular velocity of the bar when it is vertical?

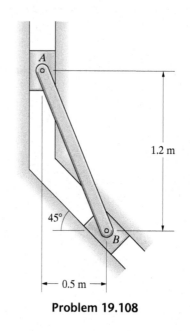

40 mm

120 mm

Problem 19.106

19.107 The slender bar of mass m is released from rest in the vertical position and allowed to fall. Neglecting friction and assuming that it remains in contact with the floor and wall, determine the bar's angular velocity as a function of θ.

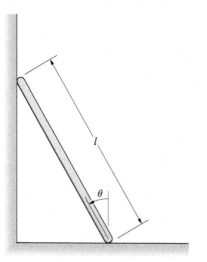

l

θ

Problem 19.107

19.108 The 4-kg slender bar is pinned to 2-kg sliders at A and B. If friction is negligible and the system starts from rest in the position shown, what is the bar's angular velocity when the slider at A has fallen 0.5 m?

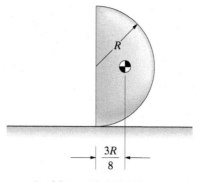

A

1.2 m

45°

B

0.5 m

Problem 19.108

19.109 The homogeneous hemisphere of mass m is released from rest in the position shown. If it rolls on the horizontal surface, what is its angular velocity when its flat surface is horizontal?

19.110 The homogeneous hemisphere of mass m is released from rest in the position shown. It rolls on the horizontal surface. What normal force is exerted on the hemisphere by the horizontal surface at the instant the flat surface of the hemisphere is horizontal?

R

$\dfrac{3R}{8}$

Problems 19.109/19.110

19.111 The slender bar rotates freely *in the horizontal plane* about a vertical shaft at O. The bar weighs 20 lb and its length is 6 ft. The slider A weighs 2 lb. If the bar's angular velocity is $\omega = 10$ rad/s and the radial component of the velocity of A is zero when $r = 1$ ft, what is the angular velocity of the bar when $r = 4$ ft? (The moment of inertia of A about its center of mass is negligible; that is, treat A as a particle.)

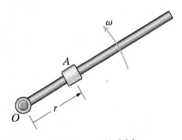

Problem 19.111

19.112 A satellite is deployed with angular velocity $\omega = 1$ rad/s (Fig. a). Two internally stored antennas that span the diameter of the satellite are then extended, and the satellite's angular velocity decreases to ω' (Fig. b). By modeling the satellite as a 500-kg sphere of 1.2-m radius and each antenna as a 10-kg slender bar, determine ω'.

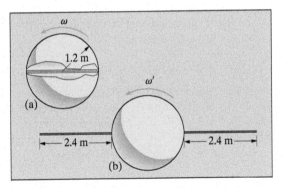

Problem 19.112

19.113 An engineer decides to control the angular velocity of a satellite by deploying small masses attached to cables. If the angular velocity of the satellite in configuration (a) is 4 rpm, determine the distance d in configuration (b) that will cause the angular velocity to be 1 rpm. The moment of inertia of the satellite is $I = 500$ kg-m^2 and each mass is 2 kg. (Assume that the cables and masses rotate with the same angular velocity as the satellite. Neglect the masses of the cables and the moments of inertia of the masses about their centers of mass.)

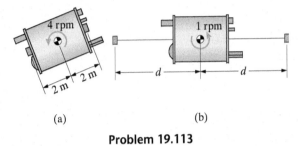

(a) (b)

Problem 19.113

19.114 The homogeneous cylindrical disk of mass m rolls on the horizontal surface with angular velocity ω. If the disk does not slip or leave the slanted surface when it comes into contact with it, what is the angular velocity ω' of the disk immediately afterward?

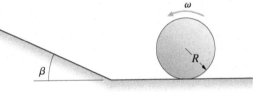

Problem 19.114

19.115 The 10-lb bar falls from rest in the vertical position and hits the smooth projection at B. The coefficient of restitution of the impact is $e = 0.6$, the duration of the impact is 0.1 s, and $b = 1$ ft. Determine the average force exerted on the bar at B as a result of the impact.

19.116 The 10-lb bar falls from rest in the vertical position and hits the smooth projection at B. The coefficient of restitution of the impact is $e = 0.6$ and the duration of the impact is 0.1 s. Determine the distance b for which the average force exerted on the bar by the support A as a result of the impact is zero.

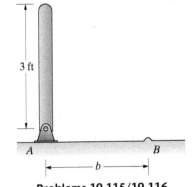

Problems 19.115/19.116

19.117 The 1-kg sphere A is moving at 2 m/s when it strikes the end of the 2-kg stationary slender bar B. If the velocity of the sphere after the impact is 0.8 m/s to the right, what is the coefficient of restitution?

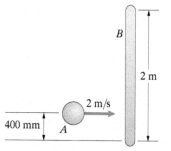

Problem 19.117

19.118 The slender bar is released from rest in the position shown in Fig. a and falls a distance $h = 1$ ft. When the bar hits the floor, its tip is supported by a depression and remains on the floor (Fig. b). The length of the bar is 1 ft and its weight is 4 oz. What is angular velocity ω of the bar just after it hits the floor?

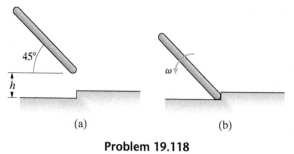

(a)   (b)

Problem 19.118

19.119 The slender bar is released from rest with $\theta = 45°$ and falls a distance $h = 1$ m onto the smooth floor. The length of the bar is 1 m and its mass is 2 kg. If the coefficient of restitution of the impact is $e = 0.4$, what is the angular velocity of the bar just after it hits the floor?

19.120 The slender bar is released from rest and falls a distance $h = 1$ m onto the smooth floor. The length of the bar is 1 m and its mass is 2 kg. The coefficient of restitution of the impact is $e = 0.4$. Determine the angle θ for which the angular velocity of the bar after it hits the floor is a maximum. What is the maximum angular velocity?

Problems 19.119/19.120

19.121 A nonrotating slender bar A moving with velocity v_0 strikes a stationary slender bar B. Each bar has mass m and length l. If the bars adhere when they collide, what is their angular velocity after the impact?

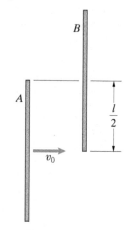

Problem 19.121

19.122 An astronaut translates toward a nonrotating satellite at $1.0\,\mathbf{i}$ (m/s) relative to the satellite. Her mass is 136 kg, and the moment of inertia about the axis through her center of mass parallel to the z axis is 45 kg-m^2. The mass of the satellite is 450 kg and its moment of inertia about the z axis is 675 kg-m^2. At the instant the astronaut attaches to the satellite and begins moving with it, the position of her center of mass is $(-1.8, -0.9, 0)$ m. The axis of rotation of the satellite after she attaches is parallel to the z axis. What is their angular velocity?

19.123 In Problem 19.122, suppose that the design parameters of the satellite's control system require that the angular velocity of the satellite not exceed 0.02 rad/s. If the astronaut is moving parallel to the x axis and the position of her center of mass when she attaches is $(-1.8, -0.9, 0)$ m, what is the maximum relative velocity at which she should approach the satellite?

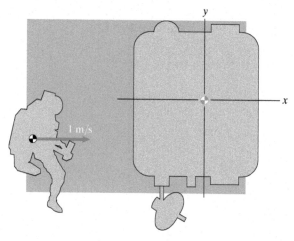

Problems 19.122/19.123

19.124 A 2800-lb car skidding on ice strikes a concrete abutment at 3 mi/h. The car's moment of inertia about its center of mass is 1800 slug-ft^2. Assume that the impacting surfaces are smooth and parallel to the y axis and that the coefficient of restitution of the impact is $e = 0.8$. What are the angular velocity of the car and the velocity of its center of mass after the impact?

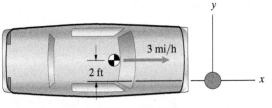

Problem 19.124

19.125 A 170-lb wide receiver jumps vertically to receive a pass and is stationary at the instant he catches the ball. At the same instant, he is hit at P by a 180-lb linebacker moving horizontally at 15 ft/s. The wide receiver's moment of inertia about his center of mass is 7 slug-ft^2. If you model the players as rigid bodies and assume that the coefficient of restitution is $e = 0$, what is the wide receiver's angular velocity immediately after the impact?

Problem 19.125

Design Project

Design and carry out experiments to determine the moments of inertia of (a) a homogeneous slender bar, such as a meter stick; and (b) a soccer ball or basketball. For the slender bar, compare your experimental values for the moments of inertia with the theoretical value $I = \frac{1}{12}ml^2$ for a slender bar of length l. For the ball, compare your experimental values with the theoretical value $I = \frac{2}{3}mR^2$ for a thin spherical shell of radius R. Investigate the repeatability of your experimental methods. Write a brief report describing your experiments, discussing possible sources of error, and presenting your results.

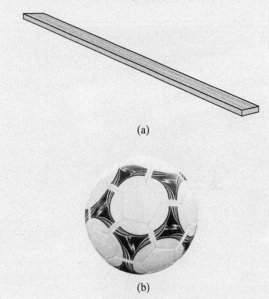

(a)

(b)

Three-Dimensional Kinematics and Dynamics of Rigid Bodies

For many engineering applications of dynamics, such as the design of airplanes and other vehicles, we must consider three-dimensional motion. After explaining how three-dimensional motion of a rigid body is described, we derive the equations of motion and use them to analyze simple motions. Finally, we introduce the Euler angles used to specify the orientation of a rigid body in three dimensions and express the equations of angular motion in terms of them.

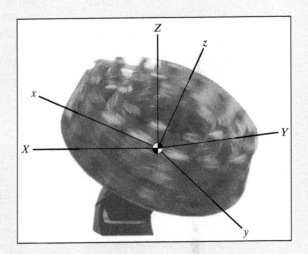

◄ The centrifuge subjects plant tissues to high accelerations for research in genetic engineering.

20.1 Kinematics

If a bicyclist rides in a straight path, the wheels undergo planar motion. But if the rider is turning, the motion of the wheels is three dimensional (Fig. 20.1a). Similarly, an airplane can remain in planar motion while it is in level flight or as it descends, climbs, or performs loops. But if it banks and turns, it is in three-dimensional motion (Fig. 20.1b). A spinning top may remain in planar motion for a brief period, rotating about a fixed vertical axis. But eventually, the top's axis begins to tilt and rotate. The top is then in three-dimensional motion and exhibits interesting, apparently gravity-defying behavior (Fig. 20.1c). In this section we begin the analysis of such motions by discussing the kinematics of rigid bodies in three-dimensional motion.

Velocities and Accelerations

We have already discussed some of the concepts needed to describe the three-dimensional motion of a rigid body relative to a given reference frame. In Chapter 17, we showed that Euler's theorem implies that a rigid body undergoing any motion other than translation has an instantaneous axis of rotation. The direction of this axis at a particular instant and the rate at which the rigid body rotates about the axis are specified by the angular velocity vector $\boldsymbol{\omega}$.

We have also shown that a rigid body's velocity is completely specified by its angular velocity vector and the velocity of a single point of the body. For the

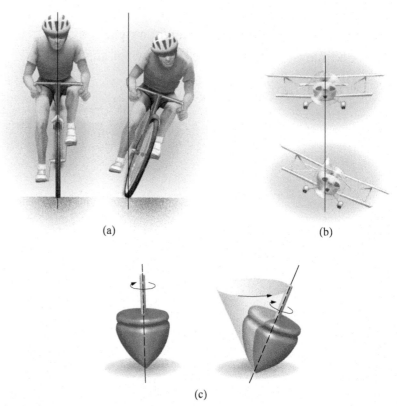

(a)

(b)

(c)

Figure 20.1
Examples of planar and three-dimensional motions.

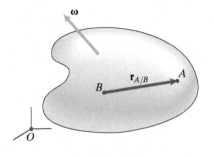

rigid body and reference frame in Fig. 20.2, suppose that we know the angular
velocity vector $\boldsymbol{\omega}$ and the velocity $\mathbf{v}_B$ of a point B. Then the velocity of *any*
other point A of the body is given by Eq. (17.8):

$$\mathbf{v}_A = \mathbf{v}_B + \boldsymbol{\omega} \times \mathbf{r}_{A/B}. \tag{20.1}$$

A rigid body's acceleration is completely specified by its angular acceleration vec-
tor $\boldsymbol{\alpha} = d\boldsymbol{\omega}/dt$, its angular velocity vector, and the acceleration of a single point
of the body. If we know $\boldsymbol{\alpha}$, $\boldsymbol{\omega}$, and the acceleration $\mathbf{a}_B$ of the point B in Fig. 20.2,
the acceleration of *any* other point A of the rigid body is given by Eq. (17.9):

$$\mathbf{a}_A = \mathbf{a}_B + \boldsymbol{\alpha} \times \mathbf{r}_{A/B} + \boldsymbol{\omega} \times (\boldsymbol{\omega} \times \mathbf{r}_{A/B}). \tag{20.2}$$

Moving Reference Frames

The velocities and accelerations in Eqs. (20.1) and (20.2) are measured relative to the reference frame indicated in Fig. 20.2, which we will refer to as the
primary reference frame. Although some situations require other choices,
the most common primary reference frame used in engineering applications is
one that is fixed relative to the earth. *When we do not state otherwise, you
should assume that the primary reference frame is earth fixed.* We also use a
secondary reference frame that moves relative to the primary reference frame.
The secondary reference frame and its motion are chosen for convenience in
describing the motion of a particular rigid body. In some situations, the secondary reference frame is defined to be fixed with respect to the rigid body.
In other cases, it is advantageous to use a secondary reference frame that moves
relative to the primary reference frame, but is not fixed with respect to the
rigid body. (See Examples 20.1–20.3.)

Figure 20.3 shows a primary reference frame, a secondary reference frame
xyz, and a rigid body. The angular velocity of the secondary reference frame

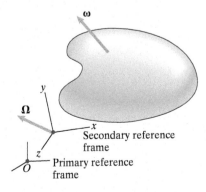

relative to the primary reference frame is specified by the vector $\boldsymbol{\Omega}$, and the angular velocity of the rigid body relative to the primary reference frame is specified by the vector $\boldsymbol{\omega}$. If the secondary reference frame is fixed with respect to the rigid body, $\boldsymbol{\Omega} = \boldsymbol{\omega}$. If we express $\boldsymbol{\omega}$ *in terms of its components in the secondary reference frame* as

$$\boldsymbol{\omega} = \omega_x \mathbf{i} + \omega_y \mathbf{j} + \omega_z \mathbf{k},$$

the rigid body's angular acceleration vector relative to the primary reference frame is

$$\boldsymbol{\alpha} = \frac{d\boldsymbol{\omega}}{dt} = \frac{d\omega_x}{dt}\mathbf{i} + \omega_x \frac{d\mathbf{i}}{dt} + \frac{d\omega_y}{dt}\mathbf{j} + \omega_y \frac{d\mathbf{j}}{dt} + \frac{d\omega_z}{dt}\mathbf{k} + \omega_z \frac{d\mathbf{k}}{dt}. \quad (20.3)$$

The derivatives of $\mathbf{i}$, $\mathbf{j}$, and $\mathbf{k}$ can be expressed in terms of the angular velocity vector $\boldsymbol{\Omega}$ as (see Section 17.5)

$$\frac{d\mathbf{i}}{dt} = \boldsymbol{\Omega} \times \mathbf{i}, \qquad \frac{d\mathbf{j}}{dt} = \boldsymbol{\Omega} \times \mathbf{j}, \qquad \frac{d\mathbf{k}}{dt} = \boldsymbol{\Omega} \times \mathbf{k}.$$

Substituting these expressions into Eq. (20.3), we obtain the angular acceleration vector of the rigid body relative to the primary reference frame in the form

$$\boldsymbol{\alpha} = \frac{d\omega_x}{dt}\mathbf{i} + \frac{d\omega_y}{dt}\mathbf{j} + \frac{d\omega_z}{dt}\mathbf{k} + \boldsymbol{\Omega} \times \boldsymbol{\omega}. \quad (20.4)$$

Notice that, in general, the derivatives $d\omega_x/dt$, $d\omega_y/dt$, and $d\omega_z/dt$ are the components of $\boldsymbol{\alpha}$ only when $\boldsymbol{\Omega} = \mathbf{0}$ or when $\boldsymbol{\Omega} = \boldsymbol{\omega}$. Otherwise, the rigid body's angular acceleration vector relative to the primary reference frame must be determined from Eq. (20.4).

When the secondary reference frame is not fixed to the rigid body, it is often convenient to express the body's angular velocity vector $\boldsymbol{\omega}$ as the sum of the angular velocity vector $\boldsymbol{\Omega}$ of the secondary reference frame and the angular velocity vector $\boldsymbol{\omega}_{\text{rel}}$ of the rigid body relative to the secondary reference frame (Fig. 20.4):

$$\boldsymbol{\omega} = \boldsymbol{\Omega} + \boldsymbol{\omega}_{\text{rel}}. \quad (20.5)$$

Figure 20.4

The vector $\boldsymbol{\omega}_{\text{rel}}$ is the rigid body's angular velocity relative to the secondary reference frame, and the vector $\boldsymbol{\Omega}$ is the angular velocity of the secondary reference frame relative to the primary reference frame. The rigid body's angular velocity vector relative to the primary reference frame is $\boldsymbol{\omega}_{\text{rel}} + \boldsymbol{\Omega}$.

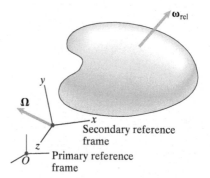

RESULTS

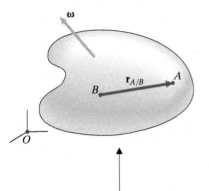

The velocity and acceleration of a point A of a rigid body (relative to a given reference frame) can be expressed in terms of the velocity and acceleration of a point B, the position of A relative to B, the angular velocity of the rigid body, and the angular acceleration $\boldsymbol{\alpha} = d\boldsymbol{\omega}/dt$ of the rigid body.

$$\begin{cases} \mathbf{v}_A = \mathbf{v}_B + \boldsymbol{\omega} \times \mathbf{r}_{A/B}, & (20.1) \\ \mathbf{a}_A = \mathbf{a}_B + \boldsymbol{\alpha} \times \mathbf{r}_{A/B} + \boldsymbol{\omega} \times (\boldsymbol{\omega} \times \mathbf{r}_{A/B}). & (20.2) \end{cases}$$

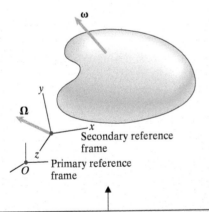

Secondary reference frame

Primary reference frame

The rate of rotation of the secondary reference frame relative to the primary reference frame is described by the angular velocity vector $\boldsymbol{\Omega}$. If the angular velocity of the rigid body relative to the primary reference frame is expressed *in terms of components in the secondary reference frame,*

$$\boldsymbol{\omega} = \omega_x \mathbf{i} + \omega_y \mathbf{j} + \omega_y \mathbf{k},$$

the angular acceleration $\boldsymbol{\alpha} = d\boldsymbol{\omega}/dt$ of the rigid body relative to the primary reference frame contains a term that arises from the rotation of the secondary reference frame.

$$\boldsymbol{\alpha} = \frac{d\omega_x}{dt}\mathbf{i} + \frac{d\omega_y}{dt}\mathbf{j} + \frac{d\omega_z}{dt}\mathbf{k} + \boldsymbol{\Omega} \times \boldsymbol{\omega}. \quad (20.4)$$

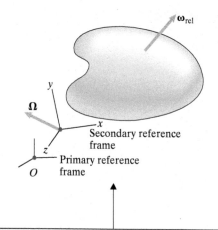

Let $\boldsymbol{\omega}_{rel}$ be the angular velocity of the rigid body *relative to the secondary reference frame.* It is sometimes convenient to express the angular velocity of the rigid body relative to the primary reference frame as the sum of the angular velocity of the secondary reference frame and the angular velocity of the rigid body relative to the secondary reference frame.

$$\boldsymbol{\omega} = \boldsymbol{\Omega} + \boldsymbol{\omega}_{rel}. \qquad (20.5)$$

Active Example 20.1 | Use of a Secondary Reference Frame (▶ *Related Problem 20.2*)

The car's tire rolls on a level surface. As the car turns, the midpoint B of the tire moves at 5 m/s in a circular path about the fixed point P (see the top view) and the tire remains perpendicular to the line from B to P. What is the tire's angular velocity vector $\boldsymbol{\omega}$ relative to an earth-fixed reference frame?

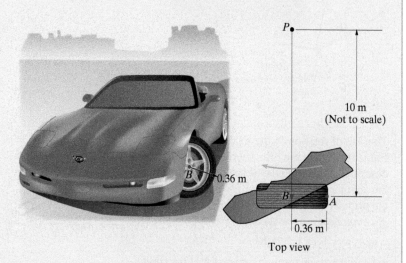

Top view

Strategy
Let us introduce a secondary reference frame with its origin at B and its y axis along the line from B to P. As the tire rolls, we assume that the y axis remains pointed toward P and the x axis *remains horizontal.* The motion of this

coordinate system is simple: It rotates about its z axis as the car turns. The motion of the tire relative to this coordinate system is also simple: It rotates about the y axis. By determining the angular velocity of the secondary coordinate system relative to an earth-fixed reference frame and the angular velocity of the tire relative to the secondary coordinate system, we can use Eq. (20.5) to determine the angular velocity of the tire relative to the earth-fixed reference frame.

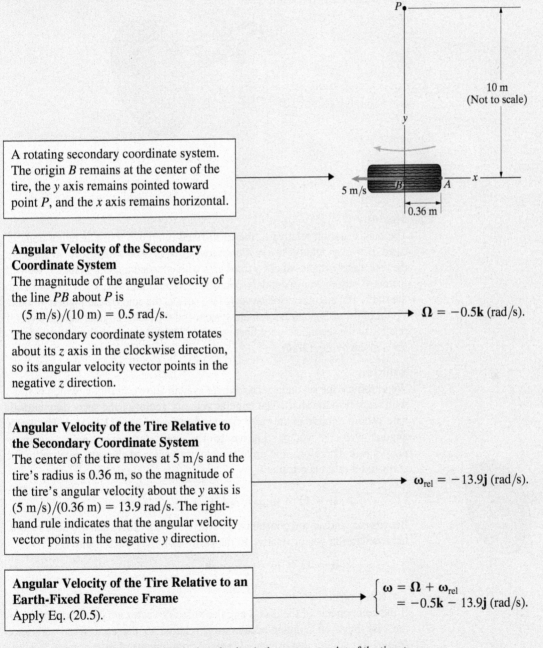

A rotating secondary coordinate system. The origin B remains at the center of the tire, the y axis remains pointed toward point P, and the x axis remains horizontal.

Angular Velocity of the Secondary Coordinate System

The magnitude of the angular velocity of the line PB about P is

$(5 \text{ m/s})/(10 \text{ m}) = 0.5 \text{ rad/s}.$

The secondary coordinate system rotates about its z axis in the clockwise direction, so its angular velocity vector points in the negative z direction.

$\boldsymbol{\Omega} = -0.5\mathbf{k} \text{ (rad/s)}.$

Angular Velocity of the Tire Relative to the Secondary Coordinate System

The center of the tire moves at 5 m/s and the tire's radius is 0.36 m, so the magnitude of the tire's angular velocity about the y axis is $(5 \text{ m/s})/(0.36 \text{ m}) = 13.9 \text{ rad/s}.$ The right-hand rule indicates that the angular velocity vector points in the negative y direction.

$\boldsymbol{\omega}_{\text{rel}} = -13.9\mathbf{j} \text{ (rad/s)}.$

Angular Velocity of the Tire Relative to an Earth-Fixed Reference Frame
Apply Eq. (20.5).

$$\begin{cases} \boldsymbol{\omega} = \boldsymbol{\Omega} + \boldsymbol{\omega}_{\text{rel}} \\ = -0.5\mathbf{k} - 13.9\mathbf{j} \text{ (rad/s)}. \end{cases}$$

Practice Problem Determine the velocity of point A, the rearmost point of the tire at the instant shown, relative to an earth-fixed reference frame.

Answer: $\mathbf{v}_A = -5\mathbf{i} - 0.18\mathbf{j} + 5\mathbf{k} \text{ (m/s)}.$

Example 20.2	Angular Velocity and Angular Acceleration of a Rotating Disk

(▶ *Related Problem 20.10*)

The disk is perpendicular to the horizontal part of the shaft and rotates relative to it with constant angular velocity ω_d. Relative to an earth-fixed reference frame, the shaft rotates about the vertical axis with constant angular velocity ω_0. Determine the angular velocity and angular acceleration vectors of the disk relative to the earth-fixed reference frame.

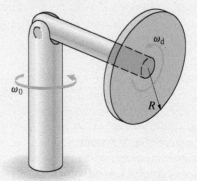

Strategy

The disk's motion relative to the earth-fixed reference frame is rather complicated. However, relative to a reference frame that is fixed with respect to the shaft, the disk simply rotates about a fixed axis with constant angular velocity. We will therefore introduce a secondary coordinate system that is fixed with respect to the shaft. The angular velocity vector we seek is the sum of the angular velocity vector of the secondary coordinate system and the disk's angular velocity vector relative to the secondary coordinate system. The disk's angular acceleration vector is given by Eq. (20.4).

Solution

We introduce the secondary coordinate system shown in Fig. a, which is fixed with respect to the shaft. The angular velocity vector of the secondary coordinate system relative to the earth-fixed reference frame is $\boldsymbol{\Omega} = \omega_0\mathbf{j}$. The disk's angular velocity vector relative to the secondary coordinate system is $\boldsymbol{\omega}_{rel} = \omega_d\mathbf{i}$. Therefore, the angular velocity vector of the disk relative to the earth-fixed reference frame is

$$\boldsymbol{\omega} = \boldsymbol{\Omega} + \boldsymbol{\omega}_{rel} = \omega_d\mathbf{i} + \omega_0\mathbf{j}.$$

Because ω_d and ω_0 are constants, we find from Eq. (20.4) that the disk's angular acceleration vector relative to the earth-fixed reference frame is

$$\boldsymbol{\alpha} = \boldsymbol{\Omega} \times \boldsymbol{\omega} = -\omega_0\omega_d\mathbf{k}.$$

Critical Thinking

If the components of the disk's angular velocity vector are constants, how can the disk have an angular acceleration relative to the earth-fixed reference frame? Remember that ω_d and ω_0 are the components of $\boldsymbol{\omega}$ *expressed in terms of the secondary coordinate system*. In this example, the magnitude and direction of the vector $\boldsymbol{\omega}$ are constant relative to the secondary coordinate system, but $\boldsymbol{\omega}$ rotates relative to the earth-fixed reference frame.

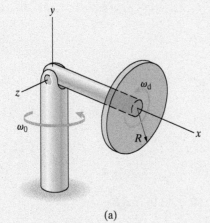

(a)

Example 20.3   Angular Velocity and Angular Acceleration of a Rolling Disk
(▶ *Related Problems 20.20–20.22*)

The bent bar is rigidly attached to the vertical shaft, which rotates with constant angular velocity ω_0. The circular disk is pinned to the bent bar and rolls on the horizontal surface.
(a) Determine the disk's angular velocity vector $\boldsymbol{\omega}_{disk}$ and angular acceleration vector $\boldsymbol{\alpha}_{disk}$.
(b) Determine the velocity of point P, which is the uppermost point of the circular disk, at the present instant.

Strategy
(a) In this example, the primary reference frame is fixed with respect to the surface on which the disk rolls. To simplify our analysis of the disk's angular motion, we will use a secondary coordinate system that is fixed with respect to the bent bar. By applying Eq. (20.1) to the bent bar, we will determine the velocity of the center of the disk. We will determine the disk's angular velocity vector by recognizing that the velocity of the point of the disk in contact with the horizontal surface is zero. We can then use Eq. (20.4) to determine the disk's angular acceleration vector.
(b) Knowing the velocity of the center of the disk and the disk's angular velocity vector, we can apply Eq. (20.1) to the disk to determine the velocity of point P.

Solution
(a) Let the coordinate system in Fig. a be fixed with respect to the bent bar. The x axis coincides with the horizontal part of the bar, and the y axis coincides with the vertical shaft. The angular velocity vector $\boldsymbol{\omega}_{bar}$ of the bar and the angular velocity vector $\boldsymbol{\Omega}$ of the coordinate system are equal:

$$\boldsymbol{\omega}_{bar} = \boldsymbol{\Omega} = \omega_0 \mathbf{j}.$$

Let point B be the stationary origin of the coordinate system, and let point A be the center of the disk (Fig. a). The position vector of A relative to B is

$$\mathbf{r}_{A/B} = (h + b \cos \beta)\mathbf{i} - b \sin \beta \mathbf{j}.$$

From Eq. (20.1), the velocity of point A is

$$\mathbf{v}_A = \mathbf{v}_B + \boldsymbol{\omega}_{bar} \times \mathbf{r}_{A/B}$$

$$= \mathbf{0} + \begin{vmatrix} \mathbf{i} & \mathbf{j} & \mathbf{k} \\ 0 & \omega_0 & 0 \\ h + b \cos \beta & -b \sin \beta & 0 \end{vmatrix}$$

$$= -\omega_0(h + b \cos \beta)\mathbf{k}.$$

Because the coordinate system is fixed with respect to the bent bar, we can write the angular velocity vector of the disk relative to the coordinate system as (Fig. b)

$$\boldsymbol{\omega}_{rel} = \omega_{rel} \cos \beta \mathbf{i} - \omega_{rel} \sin \beta \mathbf{j}.$$

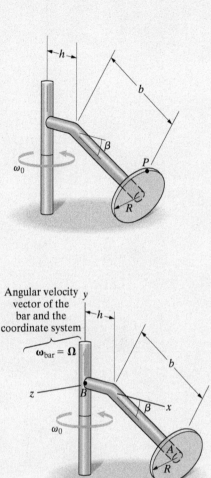

Angular velocity vector of the bar and the coordinate system

$\boldsymbol{\omega}_{bar} = \boldsymbol{\Omega}$

(a) A secondary coordinate system fixed to the bent bar.

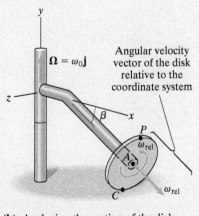

$\boldsymbol{\Omega} = \omega_0 \mathbf{j}$

Angular velocity vector of the disk relative to the coordinate system

(b) Analyzing the motion of the disk.

Let point C in Fig. b be the point of the disk that is in contact with the surface. To determine ω_{rel}, we use the condition $\mathbf{v}_C = \mathbf{0}$. The position of C relative to A is

$$\mathbf{r}_{C/A} = -R \sin \beta \mathbf{i} - R \cos \beta \mathbf{j}.$$

Therefore,

$$\mathbf{v}_C = \mathbf{v}_A + \boldsymbol{\omega}_{disk} \times \mathbf{r}_{C/A}$$

$$= -\omega_0(h + b \cos \beta)\mathbf{k} + \begin{vmatrix} \mathbf{i} & \mathbf{j} & \mathbf{k} \\ \omega_{rel} \cos \beta & \omega_0 - \omega_{rel} \sin \beta & 0 \\ -R \sin \beta & -R \cos \beta & 0 \end{vmatrix} = \mathbf{0}.$$

Solving this equation for ω_{rel}, we obtain

$$\omega_{rel} = -\omega_0\left(\frac{h}{R} + \frac{b}{R}\cos \beta - \sin \beta\right).$$

The angular velocity vector of the coordinate system is $\boldsymbol{\Omega} = \omega_0\mathbf{j}$, so the disk's angular velocity vector is

$$\boldsymbol{\omega}_{disk} = \boldsymbol{\Omega} + \boldsymbol{\omega}_{rel} = \omega_{rel} \cos \beta \mathbf{i} + (\omega_0 - \omega_{rel} \sin \beta)\mathbf{j}.$$

Even though the components of $\boldsymbol{\omega}_{disk}$ are constants, we find from Eq. (20.4) that the disk's angular acceleration is not zero:

$$\boldsymbol{\alpha}_{disk} = \boldsymbol{\Omega} \times \boldsymbol{\omega}_{disk} = \begin{vmatrix} \mathbf{i} & \mathbf{j} & \mathbf{k} \\ 0 & \omega_0 & 0 \\ \omega_{rel} \cos \beta & \omega_0 - \omega_{rel} \sin \beta & 0 \end{vmatrix}$$

$$= -\omega_0\omega_{rel} \cos \beta \mathbf{k}.$$

(b) The position vector of point P relative to the center of the disk is

$$\mathbf{r}_{P/A} = R \sin \beta \mathbf{i} + R \cos \beta \mathbf{j}.$$

Using Eq. (20.1) and our result for the velocity $\mathbf{v}_A$ of the center of the disk, we determine the velocity of point P:

$$\mathbf{v}_P = \mathbf{v}_A + \boldsymbol{\omega}_{disk} \times \mathbf{r}_{P/A}$$

$$= -\omega_0(h + b \cos \beta)\mathbf{k} + \begin{vmatrix} \mathbf{i} & \mathbf{j} & \mathbf{k} \\ \omega_{rel} \cos \beta & \omega_0 - \omega_{rel} \sin \beta & 0 \\ R \sin \beta & R \cos \beta & 0 \end{vmatrix}$$

$$= [-\omega_0(h + b \cos \beta + R \sin \beta) + \omega_{rel}R]\mathbf{k}$$

$$= -2\omega_0(h + b \cos \beta)\mathbf{k}.$$

Critical Thinking

In comparison to Example 20.2, this example was complicated by the fact that we didn't know the disk's angular velocity relative to the bar. But we knew the direction of its axis of rotation. Notice how we were able to use that information and the fact that the velocity of point C of the disk in contact with the surface is zero to determine ω_{rel}. That was the essential step in determining the disk's angular velocity vector relative to the primary reference frame.

Problems

20.1 The airplane's angular velocity relative to an earth-fixed reference frame, expressed in terms of the body-fixed coordinate system shown, is $\omega = 0.62\mathbf{i} + 0.45\mathbf{j} - 0.23\mathbf{k}$ (rad/s). The coordinates of point A of the airplane are $(3.6, 0.8, -1.2)$ m. What is the velocity of point A relative to the velocity of the airplane's center of mass?

Problem 20.1

▶ 20.2 In Active Example 20.1, suppose that the center of the tire moves at a constant speed of 5 m/s as the car turns. (As a result, when the angular velocity of the tire relative to an earth-fixed reference frame is expressed *in terms of components in the secondary reference frame,* $\omega = \omega_x\mathbf{i} + \omega_y\mathbf{j} + \omega_z\mathbf{k}$, the components ω_x, ω_y, and ω_z are constant.) What is the angular acceleration α of the tire relative to an earth-fixed reference frame?

20.3 The angular velocity of the cube relative to the primary reference frame, expressed in terms of the body-fixed coordinate system shown, is $\omega = -6.4\mathbf{i} + 8.2\mathbf{j} + 12\mathbf{k}$ (rad/s). The velocity of the center of mass G of the cube relative to the primary reference frame at the instant shown is $\mathbf{v}_G = 26\mathbf{i} + 14\mathbf{j} + 32\mathbf{k}$ (m/s). What is the velocity of point A of the cube relative to the primary reference frame at the instant shown?

20.4 The coordinate system shown is fixed with respect to the cube. The angular velocity of the cube relative to the primary reference frame, $\omega = -6.4\mathbf{i} + 8.2\mathbf{j} + 12\mathbf{k}$ (rad/s), is constant. The acceleration of the center of mass G of the cube relative to the primary reference frame at the instant shown is $\mathbf{a}_G = 136\mathbf{i} + 76\mathbf{j} - 48\mathbf{k}$ (m/s²). What is the acceleration of point A of the cube relative to the primary reference frame at the instant shown?

20.5 The origin of the secondary coordinate system shown is fixed to the center of mass G of the cube. The velocity of the center of mass G of the cube relative to the primary reference frame at the instant shown is $\mathbf{v}_G = 26\mathbf{i} + 14\mathbf{j} + 32\mathbf{k}$ (m/s). The cube is rotating relative to the secondary coordinate system with angular velocity $\omega_{rel} = 6.2\mathbf{i} - 5\mathbf{j} + 8.8\mathbf{k}$ (rad/s). The secondary coordinate system is rotating relative to the primary reference frame with angular velocity $\Omega = 2.2\mathbf{i} + 4\mathbf{j} - 3.6\mathbf{k}$ (rad/s).

(a) What is the velocity of point A of the cube relative to the primary reference frame at the instant shown?

(b) If the components of the vectors ω_{rel} and Ω are constant, what is the cube's angular acceleration relative to the primary reference frame?

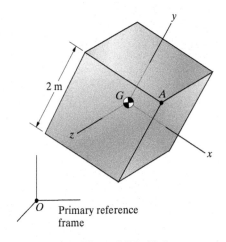

Problems 20.3–20.5

20.6 Relative to an earth-fixed reference frame, points A and B of the rigid parallelepiped are fixed and it rotates about the axis AB with an angular velocity of 30 rad/s. Determine the velocities of points C and D relative to the earth-fixed reference frame.

20.7 Relative to the xyz coordinate system shown, points A and B of the rigid parallelepiped are fixed and the parallelepiped rotates about the axis AB with an angular velocity of 30 rad/s. Relative to an earth-fixed reference frame, point A is fixed and the xyz coordinate system rotates with angular velocity $\Omega = -5\mathbf{i} + 8\mathbf{j} + 6\mathbf{k}$ (rad/s). Determine the velocities of points C and D relative to the earth-fixed reference frame.

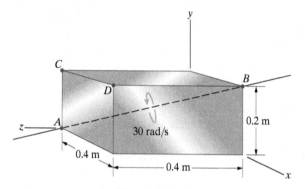

Problems 20.6/20.7

20.8 Relative to an earth-fixed reference frame, the vertical shaft rotates about its axis with angular velocity $\omega_0 = 4$ rad/s. The secondary xyz coordinate system is *fixed with respect to the shaft* and its origin is stationary. Relative to the secondary coordinate system, the disk (radius $= 8$ in) rotates with angular velocity $\omega_d = 6$ rad/s. At the instant shown, determine the velocity of point A (a) relative to the secondary reference frame, and (b) relative to the earth-fixed reference frame.

20.9 Relative to an earth-fixed reference frame, the vertical shaft rotates about its axis with constant angular velocity $\omega_0 = 4$ rad/s. The secondary xyz coordinate system is *fixed with respect to the shaft* and its origin is stationary. Relative to the secondary coordinate system, the disk (radius $= 8$ in) rotates with constant angular velocity $\omega_d = 6$ rad/s.

(a) What is the angular acceleration of the disk relative to the earth-fixed reference frame?

(b) At the instant shown, determine the acceleration of point A relative to the earth-fixed reference frame.

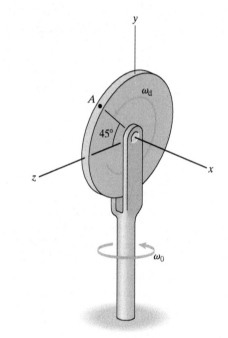

Problems 20.8/20.9

▶ 20.10 The radius of the disk is $R = 2$ ft. It is perpendicular to the horizontal part of the shaft and rotates relative to it with constant angular velocity $\omega_d = 36$ rad/s. Relative to an earth-fixed reference frame, the shaft rotates about the vertical axis with constant angular velocity $\omega_0 = 8$ rad/s.

(a) Determine the velocity relative to the earth-fixed reference frame of point P, which is the uppermost point of the disk.

(b) Determine the disk's angular acceleration vector $\boldsymbol{\alpha}$ relative to the earth-fixed reference frame.

(See Example 20.2.)

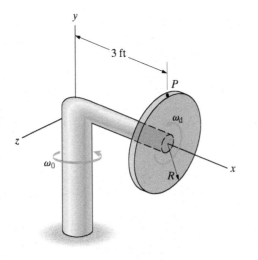

Problem 20.10

20.11 The vertical shaft supporting the disk antenna is rotating with a constant angular velocity $\omega_0 = 0.2$ rad/s. The angle θ from the horizontal to the antenna's axis is 30° at the instant shown and is increasing at a constant rate of 15° per second. The secondary xyz coordinate system shown is fixed with respect to the dish.

(a) What is the dish's angular velocity relative to an earth-fixed reference frame?

(b) Determine the velocity of the point of the antenna with coordinates $(4, 0, 0)$ m relative to an earth-fixed reference frame.

20.12 The vertical shaft supporting the disk antenna is rotating with a constant angular velocity $\omega_0 = 0.2$ rad/s. The angle θ from the horizontal to the antenna's axis is 30° at the instant shown and is increasing at a constant rate of 15° per second. The secondary xyz coordinate system shown is fixed with respect to the dish.

(a) What is the dish's angular acceleration relative to an earth-fixed reference frame?

(b) Determine the acceleration of the point of the antenna with coordinates $(4, 0, 0)$ m relative to an earth-fixed reference frame.

Problems 20.11/12.12

20.13 The radius of the circular disk is $R = 0.2$ m, and $b = 0.3$ m. The disk rotates with angular velocity $\omega_d = 6$ rad/s relative to the horizontal bar. The horizontal bar rotates with angular velocity $\omega_b = 4$ rad/s relative to the vertical shaft, and the vertical shaft rotates with angular velocity $\omega_0 = 2$ rad/s relative to an earth-fixed reference frame. Assume that the secondary reference frame shown is fixed with respect to the horizontal bar.

(a) What is the angular velocity vector ω_{rel} of the disk relative to the secondary reference frame?

(b) Determine the velocity relative to the earth-fixed reference frame of point P, which is the uppermost point of the disk.

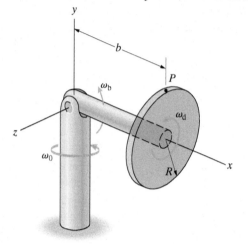

Problem 20.13

20.14 The object in Fig. a is supported by bearings at A and B in Fig. b. The horizontal circular disk is supported by a vertical shaft that rotates with angular velocity $\omega_0 = 6$ rad/s. The horizontal bar rotates with angular velocity $\omega = 10$ rad/s. At the instant shown, what is the velocity relative to an earth-fixed reference frame of the end C of the vertical bar?

20.15 The object in Fig. a is supported by bearings at A and B in Fig. b. The horizontal circular disk is supported by a vertical shaft that rotates with constant angular velocity $\omega_0 = 6$ rad/s. The horizontal bar rotates with constant angular velocity $\omega = 10$ rad/s.

(a) What is the angular acceleration of the object relative to an earth-fixed reference frame?

(b) At the instant shown, what is the acceleration relative to an earth-fixed reference frame of the end C of the vertical bar?

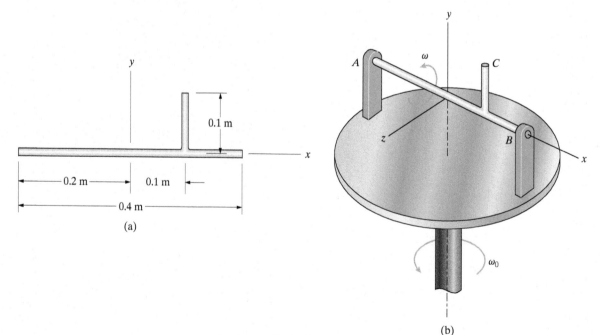

(a)

(b)

Problems 20.14/20.15

20.16 Relative to a primary reference frame, the gyroscope's circular frame rotates about the vertical axis at 2 rad/s. The 60-mm diameter wheel rotates at 10 rad/s relative to the frame. Determine the velocities of points A and B relative to the primary reference frame.

20.17 Relative to a primary reference frame, the gyroscope's circular frame rotates about the vertical axis with a constant angular velocity of 2 rad/s. The 60-mm diameter wheel rotates with a constant angular velocity of 10 rad/s relative to the frame. Determine the accelerations of points A and B relative to the primary reference frame.

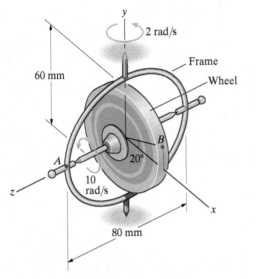

Problems 20.16/20.17

20.18 The point of the spinning top remains at a fixed point on the floor, which is the origin O of the secondary reference frame shown. The top's angular velocity relative to the secondary reference frame, $\boldsymbol{\omega}_{\text{rel}} = 50\mathbf{k}$ (rad/s), is constant. The angular velocity of the secondary reference frame relative to an earth-fixed primary reference frame is $\boldsymbol{\Omega} = 2\mathbf{j} + 5.6\mathbf{k}$ (rad/s). The components of this vector are constants. (Notice that it is expressed in terms of the secondary reference frame.) Determine the velocity relative to the earth-fixed reference frame of the point of the top with coordinates (0, 20, 30) mm.

20.19 The point of the spinning top remains at a fixed point on the floor, which is the origin O of the secondary reference frame shown. The top's angular velocity relative to the secondary reference frame, $\boldsymbol{\omega}_{\text{rel}} = 50\mathbf{k}$ (rad/s), is constant. The angular velocity of the secondary reference frame relative to an earth-fixed primary reference frame is $\boldsymbol{\Omega} = 2\mathbf{j} + 5.6\mathbf{k}$ (rad/s). The components of this vector are constants. (Notice that it is expressed in terms of the secondary reference frame.)

(a) What is the top's angular acceleration relative to the earth-fixed reference frame?

(b) Determine the acceleration relative to the earth-fixed reference frame of the point of the top with coordinates (0, 20, 30) mm.

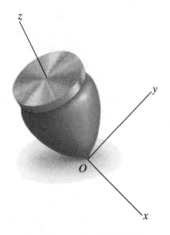

Problems 20.18/20.19

▶ **20.20*** The cone rolls on the horizontal surface, which is fixed with respect to an earth-fixed reference frame. The x axis of the secondary reference frame remains coincident with the cone's axis, and the z axis remains horizontal. As the cone rolls, the z axis rotates in the horizontal plane with an angular velocity of 2 rad/s.

(a) What is the angular velocity vector $\mathbf{\Omega}$ of the secondary reference frame?

(b) What is the angular velocity vector $\boldsymbol{\omega}_{\mathrm{rel}}$ of the cone relative to the secondary reference frame?

(See Example 20.3.)

   Strategy: To solve part (b), use the fact that the velocity relative to the earth-fixed reference frame of points of the cone in contact with the surface is zero.

▶ **20.21*** The cone rolls on the horizontal surface, which is fixed with respect to an earth-fixed reference frame. The x axis of the secondary reference frame remains coincident with the cone's axis, and the z axis remains horizontal. As the cone rolls, the z axis rotates in the horizontal plane with an angular velocity of 2 rad/s. Determine the velocity relative to the earth-fixed reference frame of the point of the base of the cone with coordinates $x = 0.4$ m, $y = 0$, and $z = 0.2$ m. (See Example 20.3.)

▶ **20.22*** The cone rolls on the horizontal surface, which is fixed with respect to an earth-fixed reference frame. The x axis of the secondary reference frame remains coincident with the cone's axis, and the z axis remains horizontal. As the cone rolls, the z axis rotates in the horizontal plane with a constant angular velocity of 2 rad/s. Determine the acceleration relative to the earth-fixed reference frame of the point of the base of the cone with coordinates $x = 0.4$ m, $y = 0$, and $z = 0.2$ m. (See Example 20.3.)

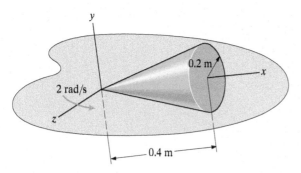

Problems 20.20–20.22

20.23* The radius and length of the cylinder are $R = 0.1$ m and $l = 0.4$ m. The horizontal surface is fixed with respect to an earth-fixed reference frame. One end of the cylinder rolls on the surface while its center, the origin of the secondary reference frame, remains stationary. The angle $\beta = 45°$. The z axis of the secondary reference frame remains coincident with the cylinder's axis, and the y axis remains horizontal. As the cylinder rolls, the y axis rotates in a horizontal plane with angular velocity $\omega_0 = 2$ rad/s.

(a) What is the angular velocity vector $\mathbf{\Omega}$ of the secondary reference frame?

(b) What is the angular velocity vector $\boldsymbol{\omega}_{\mathrm{rel}}$ of the cylinder relative to the secondary reference frame?

20.24* The radius and length of the cylinder are $R = 0.1$ m and $l = 0.4$ m. The horizontal surface is fixed with respect to an earth-fixed reference frame. One end of the cylinder rolls on the surface while its center, the origin of the secondary reference frame, remains stationary. The angle $\beta = 45°$. The z axis of the secondary reference frame remains coincident with the cylinder's axis, and the y axis remains horizontal. As the cylinder rolls, the y axis rotates in a horizontal plane with angular velocity $\omega_0 = 2$ rad/s. Determine the velocity relative to the earth-fixed reference frame of the point of the upper end of the cylinder with coordinates $x = 0.1$ m, $y = 0$, and $z = 0.2$ m.

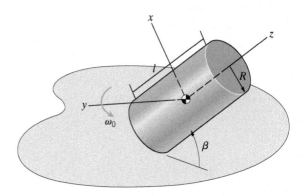

Problems 20.23/20.24

20.25* The landing gear of the P-40 airplane used in World War II retracts by rotating 90° about the horizontal axis toward the rear of the airplane. As the wheel retracts, a linkage rotates the strut supporting the wheel 90° about the strut's longitudinal axis so that the wheel is horizontal in the retracted position. (Viewed from the horizontal axis toward the wheel, the strut rotates in the clockwise direction.) The x axis of the coordinate system shown remains parallel to the horizontal axis and the y axis remains parallel to the strut as the wheel retracts. Let ω_W be the magnitude of the wheel's angular velocity when the airplane lifts off, and assume that it remains constant. Let ω_0 be the magnitude of the constant angular velocity of the strut about the horizontal axis as the landing gear is retracted. The magnitude of the angular velocity of the strut about its longitudinal axis also equals ω_0. The landing gear begins retracting at $t = 0$. Determine the wheel's angular velocity vector relative to the airplane as a function of time.

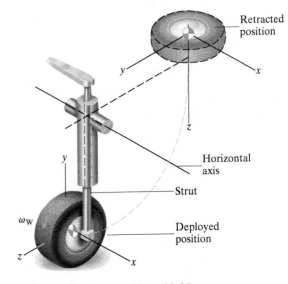

Problem 20.25

20.2 Euler's Equations

BACKGROUND

The three-dimensional equations of motion for a rigid body are called *Euler's equations*. They consist of Newton's second law,

$$\Sigma \mathbf{F} = m\mathbf{a}, \tag{20.6}$$

which states that the sum of the external forces on a rigid body equals the product of its mass and the acceleration of its center of mass, and equations of angular motion. In deriving the equations of angular motion, we consider first the special case of rotation of a rigid body about a fixed point and then general three-dimensional motion of a rigid body.

Rotation about a Fixed Point

Let m_i be the mass of the ith particle of a rigid body, and let $\mathbf{r}_i$ be its position relative to a point O that is fixed with respect to an inertial primary reference frame (Fig. 20.5). In Section 18.1, we showed that for an arbitrary system of particles, the sum of the moments about O equals the rate of change of the total angular momentum about O:

$$\Sigma \mathbf{M}_O = \frac{d\mathbf{H}_O}{dt}, \tag{20.7}$$

where the total angular momentum is

$$\mathbf{H}_O = \sum_i \mathbf{r}_i \times m_i \frac{d\mathbf{r}_i}{dt}.$$

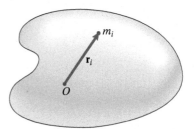

Figure 20.5
Mass and position of the ith particle of a rigid body.

If the rigid body rotates about O with angular velocity $\boldsymbol{\omega}$, the velocity of the ith particle is $d\mathbf{r}_i/dt = \boldsymbol{\omega} \times \mathbf{r}_i$, and the angular momentum is

$$\mathbf{H}_O = \sum_i \mathbf{r}_i \times m_i(\boldsymbol{\omega} \times \mathbf{r}_i). \tag{20.8}$$

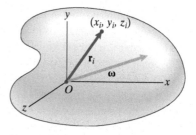

Figure 20.6
Secondary coordinate system with its
origin at O.

In Fig. 20.6, we introduce a secondary coordinate system with its origin at O. We express the vectors $\boldsymbol{\omega}$ and $\mathbf{r}_i$ in terms of their components in this coordinate system as

$$\boldsymbol{\omega} = \omega_x \mathbf{i} + \omega_y \mathbf{j} + \omega_z \mathbf{k}$$

and

$$\mathbf{r}_i = x_i \mathbf{i} + y_i \mathbf{j} + z_i \mathbf{k},$$

where (x_i, y_i, z_i) are the coordinates of the ith particle. Substituting these expressions into Eq. (20.8) and evaluating the cross products, we can write the resulting components of the angular momentum vector in the forms

$$H_{Ox} = I_{xx}\omega_x - I_{xy}\omega_y - I_{xz}\omega_z,$$

$$H_{Oy} = -I_{yx}\omega_x + I_{yy}\omega_y - I_{yz}\omega_z, \tag{20.9}$$

$$H_{Oz} = -I_{zx}\omega_x - I_{zy}\omega_y + I_{zz}\omega_z.$$

The coefficients

$$I_{xx} = \sum_i m_i(y_i^2 + z_i^2),$$

$$I_{yy} = \sum_i m_i(x_i^2 + z_i^2), \tag{20.10}$$

$$I_{zz} = \sum_i m_i(x_i^2 + y_i^2)$$

are called the *moments of inertia* about the x, y, and z axes, and the coefficients

$$I_{xy} = I_{yx} = \sum_i m_i x_i y_i,$$

$$I_{yz} = I_{zy} = \sum_i m_i y_i z_i, \tag{20.11}$$

$$I_{xz} = I_{zx} = \sum_i m_i x_i z_i$$

are called the *products of inertia*. (Evaluation of the moments and products of inertia is discussed in the appendix to this chapter.)

To obtain the equations of angular motion, we must substitute the components of the angular momentum given by Eqs. (20.9) into Eq. (20.7). The secondary coordinate system in which these components are expressed is usually chosen to be body fixed, so it rotates relative to the primary reference frame with the angular velocity $\boldsymbol{\omega}$ of the rigid body. However, we have seen in the previous section that in some situations it is convenient to use a secondary coordinate system that rotates, but is not body fixed. We denote the secondary coordinate system's angular velocity vector by $\boldsymbol{\Omega}$, where $\boldsymbol{\Omega} = \boldsymbol{\omega}$ if the coordinate system is body fixed. Expressing the angular momentum vector in terms of its components as

$$\mathbf{H}_O = H_{Ox}\mathbf{i} + H_{Oy}\mathbf{j} + H_{Oz}\mathbf{k},$$

we obtain the derivative of $\mathbf{H}_O$ with respect to time:

$$\frac{d\mathbf{H}_O}{dt} = \frac{dH_{Ox}}{dt}\mathbf{i} + H_{Ox}\frac{d\mathbf{i}}{dt} + \frac{dH_{Oy}}{dt}\mathbf{j} + H_{Oy}\frac{d\mathbf{j}}{dt} + \frac{dH_{Oz}}{dt}\mathbf{k} + H_{Oz}\frac{d\mathbf{k}}{dt}.$$

Using this expression and writing the time derivatives of the unit vectors in terms of the angular velocity $\boldsymbol{\Omega}$ of the coordinate system,

$$\frac{d\mathbf{i}}{dt} = \boldsymbol{\Omega} \times \mathbf{i}, \qquad \frac{d\mathbf{j}}{dt} = \boldsymbol{\Omega} \times \mathbf{j}, \qquad \frac{d\mathbf{k}}{dt} = \boldsymbol{\Omega} \times \mathbf{k},$$

we can write Eq. (20.7) as

$$\Sigma\mathbf{M}_O = \frac{dH_{Ox}}{dt}\mathbf{i} + \frac{dH_{Oy}}{dt}\mathbf{j} + \frac{dH_{Oz}}{dt}\mathbf{k} + \boldsymbol{\Omega} \times \mathbf{H}_O.$$

Substituting the components of $\mathbf{H}_O$ from Eq. (20.9) into this equation, we obtain the equations of angular motion (see Problem 20.60):

$$\begin{aligned}
\Sigma M_{Ox} = {} & I_{xx}\frac{d\omega_x}{dt} - I_{xy}\frac{d\omega_y}{dt} - I_{xz}\frac{d\omega_z}{dt} \\
& - \Omega_z(-I_{yx}\omega_x + I_{yy}\omega_y - I_{yz}\omega_z) \\
& + \Omega_y(-I_{zx}\omega_x - I_{zy}\omega_y + I_{zz}\omega_z), \\[6pt]
\Sigma M_{Oy} = {} & -I_{yx}\frac{d\omega_x}{dt} + I_{yy}\frac{d\omega_y}{dt} - I_{yz}\frac{d\omega_z}{dt} \\
& + \Omega_z(I_{xx}\omega_x - I_{xy}\omega_y - I_{xz}\omega_z) \\
& - \Omega_x(-I_{zx}\omega_x - I_{zy}\omega_y + I_{zz}\omega_z), \\[6pt]
\Sigma M_{Oz} = {} & -I_{zx}\frac{d\omega_x}{dt} - I_{zy}\frac{d\omega_y}{dt} + I_{zz}\frac{d\omega_z}{dt} \\
& - \Omega_y(I_{xx}\omega_x - I_{xy}\omega_y - I_{xz}\omega_z) \\
& + \Omega_x(-I_{yx}\omega_x + I_{yy}\omega_y - I_{yz}\omega_z).
\end{aligned}$$

$$(20.12)$$

In carrying out this final step, *we have assumed that the moments and products of inertia are constants.* This is so when the secondary reference frame is body fixed, but must be confirmed when it is not. Equations (20.12) can be written as the matrix equation

$$
\begin{bmatrix} \Sigma M_{Ox} \\ \Sigma M_{Oy} \\ \Sigma M_{Oz} \end{bmatrix} = \begin{bmatrix} I_{xx} & -I_{xy} & -I_{xz} \\ -I_{yx} & I_{yy} & -I_{yz} \\ -I_{zx} & -I_{zy} & I_{zz} \end{bmatrix} \begin{bmatrix} d\omega_x/dt \\ d\omega_y/dt \\ d\omega_z/dt \end{bmatrix}
$$

$$
+ \begin{bmatrix} 0 & -\Omega_z & \Omega_y \\ \Omega_z & 0 & -\Omega_x \\ -\Omega_y & \Omega_x & 0 \end{bmatrix} \begin{bmatrix} I_{xx} & -I_{xy} & -I_{xz} \\ -I_{yx} & I_{yy} & -I_{yz} \\ -I_{zx} & -I_{zy} & I_{zz} \end{bmatrix} \begin{bmatrix} \omega_x \\ \omega_y \\ \omega_z \end{bmatrix}, \quad (20.13)
$$

where

$$
\begin{bmatrix} I_{xx} & -I_{xy} & -I_{xz} \\ -I_{yx} & I_{yy} & -I_{yz} \\ -I_{zx} & -I_{zy} & I_{zz} \end{bmatrix} = [I] \qquad (20.14)
$$

is called the *inertia matrix* of the rigid body.

General Three-Dimensional Motion

Let $\mathbf{R}_i$ be the position of the *i*th particle of a rigid body relative to the center of mass of the body (Fig. 20.7). In Section 18.1, we showed that the sum of the moments about the center of mass equals the rate of change of the total angular momentum of the body relative to its center of mass; that is,

$$
\Sigma \mathbf{M} = \frac{d\mathbf{H}}{dt}, \qquad (20.15)
$$

where the total angular momentum is

$$
\mathbf{H} = \sum_i \mathbf{R}_i \times m_i \frac{d\mathbf{R}_i}{dt}.
$$

In terms of the rigid body's angular velocity $\boldsymbol{\omega}$, the velocity of the *i*th particle is $d\mathbf{R}_i/dt = \boldsymbol{\omega} \times \mathbf{R}_i$, and the angular momentum is

$$
\mathbf{H} = \sum_i \mathbf{R}_i \times m_i(\boldsymbol{\omega} \times \mathbf{R}_i). \qquad (20.16)
$$

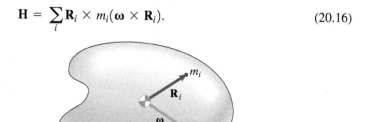

Figure 20.7
Position of the *i*th particle of a rigid body relative to the center of mass of the body.

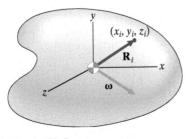

Figure 20.8
Coordinate system with its origin at
the center of mass of the body.

We introduce a secondary coordinate system with its origin at the center of mass
(Fig. 20.8) and express the vectors $\boldsymbol{\omega}$ and $\mathbf{R}_i$ in terms of their components in this
coordinate system as

$$\boldsymbol{\omega} = \omega_x \mathbf{i} + \omega_y \mathbf{j} + \omega_z \mathbf{k}$$

and

$$\mathbf{R}_i = x_i \mathbf{i} + y_i \mathbf{j} + z_i \mathbf{k}.$$

Substituting these expressions into Eq. (20.16) and evaluating the cross products, we obtain the components of the angular momentum vector in the forms

$$H_x = I_{xx}\omega_x - I_{xy}\omega_y - I_{xz}\omega_z,$$
$$H_y = -I_{yx}\omega_x + I_{yy}\omega_y - I_{yz}\omega_z, \tag{20.17}$$

and

$$H_z = -I_{zx}\omega_x - I_{zy}\omega_y + I_{zz}\omega_z,$$

where the expressions for the moments and products of inertia are again given
by Eqs. (20.10) and (20.11). Denoting the coordinate system's angular velocity
vector by $\boldsymbol{\Omega}$, and following the same steps we used to obtain Eqs. (20.12), we
obtain the equations of angular motion,

$$\Sigma M_x = I_{xx}\frac{d\omega_x}{dt} - I_{xy}\frac{d\omega_y}{dt} - I_{xz}\frac{d\omega_z}{dt}$$
$$- \Omega_z(-I_{yx}\omega_x + I_{yy}\omega_y - I_{yz}\omega_z)$$
$$+ \Omega_y(-I_{zx}\omega_x - I_{zy}\omega_y + I_{zz}\omega_z),$$

$$\Sigma M_y = -I_{yx}\frac{d\omega_x}{dt} + I_{yy}\frac{d\omega_y}{dt} - I_{yz}\frac{d\omega_z}{dt}$$
$$+ \Omega_z(I_{xx}\omega_x - I_{xy}\omega_y - I_{xz}\omega_z) \tag{20.18}$$
$$- \Omega_x(-I_{zx}\omega_x - I_{zy}\omega_y + I_{zz}\omega_z),$$

$$\Sigma M_z = -I_{zx}\frac{d\omega_x}{dt} - I_{zy}\frac{d\omega_y}{dt} + I_{zz}\frac{d\omega_z}{dt}$$
$$- \Omega_y(I_{xx}\omega_x - I_{xy}\omega_y - I_{xz}\omega_z)$$
$$+ \Omega_x(-I_{yx}\omega_x + I_{yy}\omega_y - I_{yz}\omega_z),$$

which we can write as the matrix equation

$$
\begin{bmatrix} \Sigma M_x \\ \Sigma M_y \\ \Sigma M_z \end{bmatrix} = \begin{bmatrix} I_{xx} & -I_{xy} & -I_{xz} \\ -I_{yx} & I_{yy} & -I_{yz} \\ -I_{zx} & -I_{zy} & I_{zz} \end{bmatrix} \begin{bmatrix} d\omega_x/dt \\ d\omega_y/dt \\ d\omega_z/dt \end{bmatrix}
$$
$$
+ \begin{bmatrix} 0 & -\Omega_z & \Omega_y \\ \Omega_z & 0 & -\Omega_x \\ -\Omega_y & \Omega_x & 0 \end{bmatrix} \begin{bmatrix} I_{xx} & -I_{xy} & -I_{xz} \\ -I_{yx} & I_{yy} & -I_{yz} \\ -I_{zx} & -I_{zy} & I_{zz} \end{bmatrix} \begin{bmatrix} \omega_x \\ \omega_y \\ \omega_z \end{bmatrix}.
$$

(20.19)

We have obtained equations that are identical in form to the equations of angular motion for rotation about a fixed point. Equations (20.12) and (20.13) are expressed in terms of the total moment about a fixed point O about which the rigid body rotates, and the moments and products of inertia and the components of the vectors are expressed in terms of a coordinate system with its origin at O. Equations (20.18) and (20.19) are expressed in terms of the total moment about the center of mass of the body, and the moments and products of inertia and the components of the vectors are expressed in terms of a coordinate system with its origin at the center of mass.

If the secondary coordinate system used to apply Eqs. (20.12), (20.13), (20.18), and (20.19) is body fixed, the terms $d\omega_x/dt$, $d\omega_y/dt$, and $d\omega_z/dt$ are the components of the rigid body's angular acceleration $\boldsymbol{\alpha}$. *But this is not generally the case if the secondary coordinate system rotates but is not body fixed.* [See Eq. (20.4).]

Equations of Planar Motion

Here we demonstrate how the equations of angular motion for a rigid body in planar motion can be obtained from the three-dimensional equations. Consider a rigid body that rotates about a fixed axis L_O. We introduce a body-fixed secondary coordinate system with the z axis aligned with L_O, so that the rigid body's angular velocity vector is $\boldsymbol{\omega} = \omega_z \mathbf{k}$ (Fig. 20.9). Substituting $\Omega_x = \omega_x = 0$, $\Omega_y = \omega_y = 0$, and $\Omega_z = \omega_z$ into Eqs. (20.12), we find that the third equation reduces to $\Sigma M_{Oz} = I_{zz}(d\omega_z/dt)$. Introducing the simpler notation $\Sigma M_{Oz} = \Sigma M_O$, $I_{zz} = I_O$, and $\omega_z = \omega$, we obtain

$$
\Sigma M_O = I_O \frac{d\omega}{dt}.
$$

(20.20)

This is the equation we used in Chapter 18 to analyze the rotation of a rigid body about a fixed axis. [See Eq. (18.17).] The total moment about the fixed axis equals the product of the moment of inertia about the fixed axis and the angular acceleration.

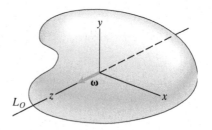

Figure 20.9

For general planar motion, we introduce a body-fixed secondary coordinate system with its origin at the center of mass of the body and the z axis perpendicular to the plane of the motion (Fig. 20.10). The rigid body's angular velocity vector is $\boldsymbol{\omega} = \omega_z \mathbf{k}$. Substituting $\Omega_x = \omega_x = 0, \Omega_y = \omega_y = 0$, and $\Omega_z = \omega_z$ into Eqs. (20.18), the third equation reduces to $\Sigma M_z = I_{zz}(d\omega_z/dt)$. With the notation $\Sigma M_z = \Sigma M, I_{zz} = I$, and $\omega_z = \omega$, we obtain

$$\Sigma M = I \frac{d\omega}{dt}. \tag{20.21}$$

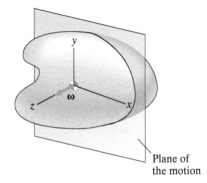

Plane of the motion

Figure 20.10

This is the equation of angular motion we used in Chapter 18 to analyze the general planar motion of a rigid body. [See Eq. (18.20).] The total moment about the center of mass of the body equals the product of the moment of inertia about the center of mass and the angular acceleration. (The term I is the moment of inertia about the axis through the center of mass that is perpendicular to the plane of the motion.)

RESULTS

The three-dimensional equations of motion for a rigid body are called *Euler's equations*. They consist of Newton's second law and equations of angular motion.

Newton's Second Law

The sum of the external forces on a rigid body is equal to the product of its mass and the acceleration of its center of mass relative to an inertial reference frame.

$\longrightarrow \quad \Sigma \mathbf{F} = m\mathbf{a}. \qquad (20.6)$

Rotation About a Fixed Point

Consider a rigid body constrained to rotate about a point O that is fixed relative to an inertial reference frame. The secondary xyz coordinate system has its origin at O. $\boldsymbol{\omega}$ is the angular velocity vector of the rigid body relative to the inertial reference frame, and $\boldsymbol{\Omega}$ is the angular velocity vector of the xyz coordinate system relative to the inertial reference frame. If the xyz coordinate system is body-fixed, $\boldsymbol{\Omega} = \boldsymbol{\omega}$.

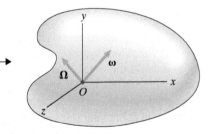

$$\Sigma M_{Ox} = I_{xx}\frac{d\omega_x}{dt} - I_{xy}\frac{d\omega_y}{dt} - I_{xz}\frac{d\omega_z}{dt}$$
$$- \Omega_z(-I_{yx}\omega_x + I_{yy}\omega_y - I_{yz}\omega_z)$$
$$+ \Omega_y(-I_{zx}\omega_x - I_{zy}\omega_y + I_{zz}\omega_z),$$

Equations of angular motion. The terms ΣM_{Ox}, ΣM_{Oy}, and ΣM_{Oz} are the components of the total external moment about point O acting on the rigid body. The terms I_{xx}, I_{yy}, and I_{zz} are the moments of inertia of the rigid body about the x, y, and z axes, and the terms $I_{xy} = I_{yx}$, $I_{yz} = I_{zy}$, and $I_{zx} = I_{xz}$ are the *products of inertia.*

$$\Sigma M_{Oy} = -I_{yx}\frac{d\omega_x}{dt} + I_{yy}\frac{d\omega_y}{dt} - I_{yz}\frac{d\omega_z}{dt}$$
$$+ \Omega_z(I_{xx}\omega_x - I_{xy}\omega_y - I_{xz}\omega_z)$$
$$- \Omega_x(-I_{zx}\omega_x - I_{zy}\omega_y + I_{zz}\omega_z),$$

(20.12)

$$\Sigma M_{Oz} = -I_{zx}\frac{d\omega_x}{dt} - I_{zy}\frac{d\omega_y}{dt} + I_{zz}\frac{d\omega_z}{dt}$$
$$- \Omega_y(I_{xx}\omega_x - I_{xy}\omega_y - I_{xz}\omega_z)$$
$$+ \Omega_x(-I_{yx}\omega_x + I_{yy}\omega_y - I_{yz}\omega_z).$$

Equations (20.12) can be written as a matrix equation.

$$\begin{bmatrix} \Sigma M_{Ox} \\ \Sigma M_{Oy} \\ \Sigma M_{Oz} \end{bmatrix} = \begin{bmatrix} I_{xx} & -I_{xy} & -I_{xz} \\ -I_{yx} & I_{yy} & -I_{yz} \\ -I_{zx} & -I_{zy} & I_{zz} \end{bmatrix} \begin{bmatrix} d\omega_x/dt \\ d\omega_y/dt \\ d\omega_z/dt \end{bmatrix}$$
$$+ \begin{bmatrix} 0 & -\Omega_z & \Omega_y \\ \Omega_z & 0 & -\Omega_x \\ -\Omega_y & \Omega_x & 0 \end{bmatrix} \begin{bmatrix} I_{xx} & -I_{xy} & -I_{xz} \\ -I_{yx} & I_{yy} & -I_{yz} \\ -I_{zx} & -I_{zy} & I_{zz} \end{bmatrix} \begin{bmatrix} \omega_x \\ \omega_y \\ \omega_z \end{bmatrix}, \quad (20.13)$$

where

$$\begin{bmatrix} I_{xx} & -I_{xy} & -I_{xz} \\ -I_{yx} & I_{yy} & -I_{yz} \\ -I_{zx} & -I_{zy} & I_{zz} \end{bmatrix} = [I] \quad (20.14)$$

is the *inertia matrix* of the rigid body.

General Three-Dimensional Motion

Consider a rigid body in general three-dimensional motion relative to an inertial reference frame. The secondary xyz coordinate system has its origin at the center of mass. ω is the angular velocity vector of the rigid body relative to the inertial reference frame, and Ω is the angular velocity vector of the xyz coordinate system relative to the inertial reference frame. If the xyz coordinate system is body-fixed, $\Omega = \omega$.

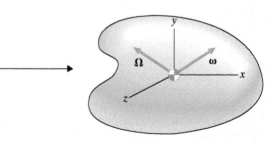

$$\Sigma M_x = I_{xx}\frac{d\omega_x}{dt} - I_{xy}\frac{d\omega_y}{dt} - I_{xz}\frac{d\omega_z}{dt}$$
$$- \Omega_z(-I_{yx}\omega_x + I_{yy}\omega_y - I_{yz}\omega_z)$$
$$+ \Omega_y(-I_{zx}\omega_x - I_{zy}\omega_y + I_{zz}\omega_z),$$

$$\Sigma M_y = -I_{yx}\frac{d\omega_x}{dt} + I_{yy}\frac{d\omega_y}{dt} - I_{yz}\frac{d\omega_z}{dt}$$
$$+ \Omega_z(I_{xx}\omega_x - I_{xy}\omega_y - I_{xz}\omega_z)$$
$$- \Omega_x(-I_{zx}\omega_x - I_{zy}\omega_y + I_{zz}\omega_z),$$

$$\Sigma M_z = -I_{zx}\frac{d\omega_x}{dt} - I_{zy}\frac{d\omega_y}{dt} + I_{zz}\frac{d\omega_z}{dt}$$
$$- \Omega_y(I_{xx}\omega_x - I_{xy}\omega_y - I_{xz}\omega_z)$$
$$+ \Omega_x(-I_{yx}\omega_x + I_{yy}\omega_y - I_{yz}\omega_z).$$

(20.18)

> Equations of angular motion. The terms ΣM_x, ΣM_y, and ΣM_z are the components of the total external moment about the center of mass of the rigid body.

> Equations (20.18) can be written as a matrix equation.

$$\begin{bmatrix} \Sigma M_x \\ \Sigma M_y \\ \Sigma M_z \end{bmatrix} = \begin{bmatrix} I_{xx} & -I_{xy} & -I_{xz} \\ -I_{yx} & I_{yy} & -I_{yz} \\ -I_{zx} & -I_{zy} & I_{zz} \end{bmatrix} \begin{bmatrix} d\omega_x/dt \\ d\omega_y/dt \\ d\omega_z/dt \end{bmatrix}$$
$$+ \begin{bmatrix} 0 & -\Omega_z & \Omega_y \\ \Omega_z & 0 & -\Omega_x \\ -\Omega_y & \Omega_x & 0 \end{bmatrix} \begin{bmatrix} I_{xx} & -I_{xy} & -I_{xz} \\ -I_{yx} & I_{yy} & -I_{yz} \\ -I_{zx} & -I_{zy} & I_{zz} \end{bmatrix} \begin{bmatrix} \omega_x \\ \omega_y \\ \omega_z \end{bmatrix}, \quad (20.19)$$

1. *Choose a coordinate system.* If an object rotates about a fixed point O, it is usually preferable to use a secondary coordinate system with its origin at O and express the equations of angular motion in the forms given by Eqs. (20.12). Otherwise, it is necessary to use a secondary coordinate system with its origin at the center of mass and express the equations of angular motion in the forms given by Eqs. (20.18). In either case, it is usually convenient to choose a coordinate system that simplifies the determination of the moments and products of inertia. Except for certain applications involving symmetric objects, a body-fixed coordinate system must be used so that the moments and products of inertia will be constant.

> Using the Euler equations to analyze three-dimensional motions of rigid bodies typically requires three steps.

2. *Draw the free-body diagram.* Isolate the object and identify the external forces and couples acting on it.

3. *Apply the equations of motion.* Use Euler's equations to relate the forces and couples acting on the object to the acceleration of the center of mass and the angular acceleration.

Active Example 20.4	Three-Dimensional Dynamics of a Bar (▶ *Related Problem 20.26*)

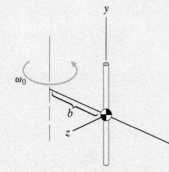

The vertical slender bar of mass m is rigidly attached to the horizontal disk. The disk is attached to a vertical shaft that is rotating with constant angular velocity ω_0. Determine the force and couple exerted on the bar by the disk.

Strategy

The external forces and couples acting on the bar are its weight and the force and couple exerted on it by the disk. We know the bar's angular velocity and angular acceleration, and we can determine the acceleration of its center of mass. So we can apply the Euler equations to determine the total force and couple exerted on the bar.

Solution
Choose a Coordinate System

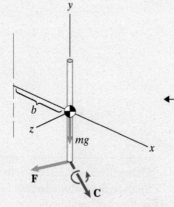

> The coordinate system is body-fixed, and its origin is at the center of mass of the bar. The y axis points upward. The x axis points radially outward from the axis of rotation of the disk.

Draw the Free-Body Diagram

> The bar is subjected to its weight and the force **F** and couple **C** exerted by the disk.

Apply Newton's Second Law

> As the disk rotates with constant angular velocity ω_0, the center of mass of the bar moves in a circular path with radius b. As a result, the center of mass has a normal component of acceleration $-b\omega_0^2\mathbf{i}$.

$$\left\{\begin{array}{c} \Sigma\mathbf{F} = m\mathbf{a}: \\[2mm] \mathbf{F} - mg\mathbf{j} = m(-b\omega_0^2\mathbf{i}). \\[2mm] \text{The force exerted on the bar by the disk is} \\[2mm] \mathbf{F} = -mb\omega_0^2\mathbf{i} + mg\mathbf{j}. \end{array}\right.$$

Apply the Equation of Angular Motion

The total moment about the center of mass of the bar is the sum of the couple $\mathbf{C}$ and the moment about the center of mass due to $\mathbf{F}$.

$$
\Sigma\mathbf{M} = \mathbf{C} +
\begin{vmatrix}
\mathbf{i} & \mathbf{j} & \mathbf{k} \\
0 & -\tfrac{1}{2}l & 0 \\
-mb\omega_0^2 & mg & 0
\end{vmatrix}
$$

$$
= \mathbf{C} - \tfrac{1}{2}mlb\omega_0^2\mathbf{k}.
$$

Moments and products of inertia for a slender bar in terms of the xyz coordinate system.

$$
\begin{bmatrix}
I_{xx} & -I_{xy} & -I_{xz} \\
-I_{yx} & I_{yy} & -I_{yz} \\
-I_{zx} & -I_{zy} & I_{zz}
\end{bmatrix}
=
\begin{bmatrix}
\tfrac{1}{12}ml^2 & 0 & 0 \\
0 & 0 & 0 \\
0 & 0 & \tfrac{1}{12}ml^2
\end{bmatrix}
$$

Apply Eqs. (20.18) with $\omega_x = \Omega_x = 0$, $\omega_y = \Omega_y = \omega_0$, and $\omega_z = \Omega_z = 0$.

$$
\Sigma M_x = I_{xx}\frac{d\omega_x}{dt} - I_{xy}\frac{d\omega_y}{dt} - I_{xz}\frac{d\omega_z}{dt}
$$
$$
- \Omega_z(-I_{yx}\omega_x + I_{yy}\omega_y - I_{yz}\omega_z)
$$
$$
+ \Omega_y(-I_{zx}\omega_x - I_{zy}\omega_y + I_{zz}\omega_z):
$$
$$
C_x = 0.
$$

$$
\Sigma M_y = -I_{yx}\frac{d\omega_x}{dt} + I_{yy}\frac{d\omega_y}{dt} - I_{yz}\frac{d\omega_z}{dt}
$$
$$
+ \Omega_z(I_{xx}\omega_x - I_{xy}\omega_y - I_{xz}\omega_z)
$$
$$
- \Omega_x(-I_{zx}\omega_x - I_{zy}\omega_y + I_{zz}\omega_z):
$$
$$
C_y = 0.
$$

$$
\Sigma M_z = -I_{zx}\frac{d\omega_x}{dt} - I_{zy}\frac{d\omega_y}{dt} + I_{zz}\frac{d\omega_z}{dt}
$$
$$
- \Omega_y(I_{xx}\omega_x - I_{xy}\omega_y - I_{xz}\omega_z)
$$
$$
+ \Omega_x(-I_{yx}\omega_x + I_{yy}\omega_y - I_{yz}\omega_z):
$$
$$
C_z - \tfrac{1}{2}mlb\omega_0^2 = 0.
$$

The couple exerted on the bar by the disk is

$$
\mathbf{C} = \tfrac{1}{2}mlb\omega_0^2\mathbf{k}.
$$

Practice Problem The bar can be regarded as rotating about any given point on the fixed axis of rotation of the disk. The origin O of the body-fixed coordinate system shown is located on the axis of rotation of the disk and the y axis points upward. The moments and products of inertia of the bar in terms of this coordinate system are

$$
\begin{bmatrix}
I_{xx} & -I_{xy} & -I_{xz} \\
-I_{yx} & I_{yy} & -I_{yz} \\
-I_{zx} & -I_{zy} & I_{zz}
\end{bmatrix}
=
\begin{bmatrix}
\tfrac{1}{3}ml^2 & -\tfrac{1}{2}mlb & 0 \\
-\tfrac{1}{2}mlb & mb^2 & 0 \\
0 & 0 & \tfrac{1}{3}ml^2 + mb^2
\end{bmatrix}.
$$

Use Newton's second law and Eqs. (20.12) to determine the force and couple exerted on the bar by the disk.

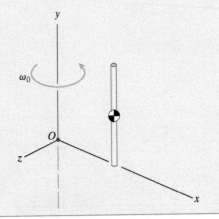

Answer: $\mathbf{F} = -mb\omega_0^2\mathbf{i} + mg\mathbf{j}$, $\mathbf{C} = \tfrac{1}{2}mlb\omega_0^2\mathbf{k}$.

Example 20.5 | **Three-Dimensional Dynamics of a Plate** (▶ *Related Problem 20.27*)

During an assembly process, the 4-kg rectangular plate is held at O by a robotic manipulator. Point O is stationary. At the instant shown, the plate is horizontal, its angular velocity vector is $\boldsymbol{\omega} = 4\mathbf{i} - 2\mathbf{j}$ (rad/s), and its angular acceleration vector is $\boldsymbol{\alpha} = -10\mathbf{i} + 6\mathbf{j}$ (rad/s²). Determine the couple exerted on the plate by the manipulator.

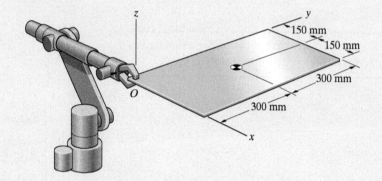

Strategy
The plate rotates about the fixed point O, so we can use Eq. (20.13) to determine the total moment exerted on the plate about O.

Solution
Draw the Free-Body Diagram We denote the force and couple exerted on the plate by the manipulator by $\mathbf{F}$ and $\mathbf{C}$ (Fig. a).

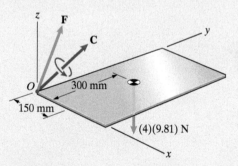

(a) Free-body diagram of the plate.

Apply the Equations of Motion The total moment about O is the sum of the couple exerted by the manipulator and the moment about O due to the plate's weight:

$$\Sigma\mathbf{M}_O = \mathbf{C} + (0.15\mathbf{i} + 0.30\mathbf{j}) \times [-(4)(9.81)\mathbf{k}]$$
$$= \mathbf{C} - 11.77\mathbf{i} + 5.89\mathbf{j} \text{ (N-m)}. \qquad (1)$$

To obtain the unknown couple $\mathbf{C}$, we can determine the total moment about O from Eq. (20.13).

We let the secondary coordinate system be body fixed, so its angular velocity $\mathbf{\Omega}$ equals the plate's angular velocity $\boldsymbol{\omega}$. We determine the plate's inertia matrix in Example 20.9, obtaining

$$[I] = \begin{bmatrix} 0.48 & -0.18 & 0 \\ -0.18 & 0.12 & 0 \\ 0 & 0 & 0.6 \end{bmatrix} \text{ kg-m}^2.$$

Therefore, the total moment about O exerted on the plate is

$$\begin{bmatrix} \Sigma M_{Ox} \\ \Sigma M_{Oy} \\ \Sigma M_{Oz} \end{bmatrix} = \begin{bmatrix} I_{xx} & -I_{xy} & -I_{xz} \\ -I_{yx} & I_{yy} & -I_{yz} \\ -I_{zx} & -I_{zy} & I_{zz} \end{bmatrix} \begin{bmatrix} d\omega_x/dt \\ d\omega_y/dt \\ d\omega_z/dt \end{bmatrix}$$

$$+ \begin{bmatrix} 0 & -\omega_z & \omega_y \\ \omega_z & 0 & -\omega_x \\ -\omega_y & \omega_x & 0 \end{bmatrix} \begin{bmatrix} I_{xx} & -I_{xy} & -I_{xz} \\ -I_{yx} & I_{yy} & -I_{yz} \\ -I_{zx} & -I_{zy} & I_{zz} \end{bmatrix} \begin{bmatrix} \omega_x \\ \omega_y \\ \omega_z \end{bmatrix}$$

$$= \begin{bmatrix} 0.48 & -0.18 & 0 \\ -0.18 & 0.12 & 0 \\ 0 & 0 & 0.6 \end{bmatrix} \begin{bmatrix} -10 \\ 6 \\ 0 \end{bmatrix}$$

$$+ \begin{bmatrix} 0 & 0 & -2 \\ 0 & 0 & -4 \\ 2 & 4 & 0 \end{bmatrix} \begin{bmatrix} 0.48 & -0.18 & 0 \\ -0.18 & 0.12 & 0 \\ 0 & 0 & 0.6 \end{bmatrix} \begin{bmatrix} 4 \\ -2 \\ 0 \end{bmatrix}$$

$$= \begin{bmatrix} -5.88 \\ 2.52 \\ 0.72 \end{bmatrix} \text{ N-m.}$$

We substitute this result into Eq. (1) to get

$$\Sigma \mathbf{M}_O = \mathbf{C} - 11.77\mathbf{i} + 5.89\mathbf{j} = -5.88\mathbf{i} + 2.52\mathbf{j} + 0.72\mathbf{k},$$

and solve for the couple $\mathbf{C}$:

$$\mathbf{C} = 5.89\mathbf{i} - 3.37\mathbf{j} + 0.72\mathbf{k} \text{ (N-m).}$$

Critical Thinking

Notice that we specified the angular velocity and angular acceleration of the plate and used the equations of angular motion to determine the couple exerted on the plate. In this chapter, we will discuss many examples in which an object's motion is specified and Euler's equations are used to determine the forces and couples acting on the object. The inverse problem, determining an object's three-dimensional motion when the forces and couples acting on it are known, is more difficult and must usually be solved by numerical methods.

Example 20.6 | **Three-Dimensional Dynamics of a Tilted Cylinder** (▶ *Related Problem 20.55, 20.56*)

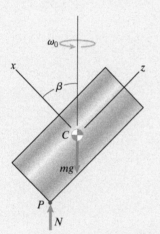

(a) Coordinate system with the z axis aligned with the cylinder axis and the y axis horizontal.

The tilted homogeneous cylinder undergoes a steady motion in which one end rolls on the floor while the center of mass of the cylinder remains stationary. The angle β between the cylinder axis and the horizontal remains constant, and the cylinder axis rotates about the vertical axis with constant angular velocity ω_0. The cylinder has mass m, radius R, and length l. What is ω_0?

Strategy

By expressing the equations of angular motion in terms of ω_0, we can determine the value of ω_0 necessary for the equations to be satisfied. Therefore, our first task is to determine the cylinder's angular velocity $\boldsymbol{\omega}$ in terms of ω_0. We can simplify this task by using a secondary coordinate system that is not body fixed.

Solution

Choose a Coordinate System We use a secondary coordinate system in which the z axis remains aligned with the cylinder axis and the y axis remains horizontal (Fig. a). The reason for this choice is that the angular velocity of the coordinate system is easy to describe—the coordinate system rotates about the vertical axis with the angular velocity ω_0—and the rotation of the cylinder relative to the coordinate system is also easy to describe. The angular velocity vector of the coordinate system is

$$\boldsymbol{\Omega} = \omega_0 \cos \beta \mathbf{i} + \omega_0 \sin \beta \mathbf{k}.$$

Relative to the coordinate system, the cylinder rotates about the z axis. Writing its angular velocity vector relative to the coordinate system as $\omega_{\text{rel}}\mathbf{k}$, we express the angular velocity vector of the cylinder as

$$\boldsymbol{\omega} = \boldsymbol{\Omega} + \omega_{\text{rel}}\mathbf{k} = \omega_0 \cos \beta \mathbf{i} + (\omega_0 \sin \beta + \omega_{\text{rel}})\mathbf{k}.$$

We can determine ω_{rel} from the condition that the velocity of the point P in contact with the floor is zero. Expressing the velocity of P in terms of the velocity of the center of mass C, we obtain

$$\mathbf{v}_P = \mathbf{v}_C + \boldsymbol{\omega} \times \mathbf{r}_{P/C}:$$

$$\mathbf{0} = \mathbf{0} + [\omega_0 \cos \beta \mathbf{i} + (\omega_0 \sin \beta + \omega_{\text{rel}})\mathbf{k}] \times [-R\mathbf{i} - \tfrac{1}{2}l\mathbf{k}]$$

$$= \left[\tfrac{1}{2}l\omega_0 \cos \beta - R(\omega_0 \sin \beta + \omega_{\text{rel}})\right]\mathbf{j}.$$

Solving for ω_{rel} yields

$$\omega_{\text{rel}} = \left[\tfrac{1}{2}\left(\frac{l}{R}\right)\cos \beta - \sin \beta\right]\omega_0.$$

Therefore, the cylinder's angular velocity vector is

$$\boldsymbol{\omega} = \omega_0 \cos \beta \mathbf{i} + \tfrac{1}{2}\left(\frac{l}{R}\right)\omega_0 \cos \beta \mathbf{k}.$$

Draw the Free-Body Diagram We draw the free-body diagram of the cylinder in Fig. a, showing the weight of the cylinder and the normal force exerted by the floor. Because the cylinder's center of mass is stationary, the floor exerts no horizontal force on the cylinder, and the normal force is $N = mg$.

Apply the Equations of Motion The moment about the center of mass due to the normal force is

$$\Sigma \mathbf{M} = \left(mgR \sin \beta - \tfrac{1}{2}mgl \cos \beta\right)\mathbf{j}.$$

From Appendix C, the moments and products of inertia are

$$\begin{bmatrix} \tfrac{1}{4}mR^2 + \tfrac{1}{12}ml^2 & 0 & 0 \\ 0 & \tfrac{1}{4}mR^2 + \tfrac{1}{12}ml^2 & 0 \\ 0 & 0 & \tfrac{1}{2}mR^2 \end{bmatrix}.$$

Although the coordinate system we are using is not body fixed, notice the moments and products of inertia are constant due to the symmetry of the cylinder. Substituting our expressions for $\mathbf{\Omega}$, $\mathbf{\omega}$, $\Sigma \mathbf{M}$, and the moments and products of inertia into the equation of angular motion, Eq. (20.19), and evaluating the matrix products, we obtain the equation

$$mg\left(R \sin \beta - \tfrac{1}{2}l \cos \beta\right) = \left(\tfrac{1}{4}mR^2 + \tfrac{1}{12}ml^2\right)\omega_0^2 \sin \beta \cos \beta$$
$$- \tfrac{1}{2}\left(\tfrac{1}{2}mR^2\right)\omega_0^2\left(\frac{l}{R}\right)\cos^2 \beta.$$

We solve this equation for ω_0^2:

$$\omega_0^2 = \frac{g\left(R \sin \beta - \tfrac{1}{2}l \cos \beta\right)}{\left(\tfrac{1}{4}R^2 + \tfrac{1}{12}l^2\right)\sin \beta \cos \beta - \tfrac{1}{4}lR \cos^2 \beta}. \tag{1}$$

Critical Thinking

If our solution yields a negative value for ω_0^2 for a given value of β, the assumed steady motion of the cylinder is not possible. For example, if the cylinder's diameter is equal to its length ($2R = l$), we can write Eq. (1) as

$$\frac{R\omega_0^2}{g} = \frac{\sin \beta - \cos \beta}{\tfrac{7}{12}\sin \beta \cos \beta - \tfrac{1}{2}\cos^2\beta}.$$

The figure shows the graph of this equation as a function of β. For values of β from approximately 40° to 45°, there is no real solution for ω_0.

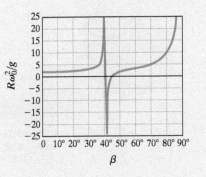

Problems

▶ **20.26** In Active Example 20.4, suppose that the shaft supporting the disk is initially stationary, and at $t = 0$ it is subjected to a constant angular acceleration α_0 in the counterclockwise direction viewed from above the disk. Determine the force and couple exerted on the bar by the disk at that instant.

▶ **20.27** In Example 20.5, suppose that the horizontal plate is initially stationary, and at $t = 0$ the robotic manipulator exerts a couple **C** on the plate at the fixed point O such that the plate's angular acceleration at that instant is $\alpha = 150\mathbf{i} + 320\mathbf{j} + 25\mathbf{k}$ (rad/s^2). Determine **C**.

20.28 A robotic manipulator moves a casting. The inertia matrix of the casting in terms of a body-fixed coordinate system with its origin at the center of mass is shown. At the present instant, the angular velocity and angular acceleration of the casting are $\omega = 1.2\mathbf{i} + 0.8\mathbf{j} - 0.4\mathbf{k}$ (rad/s) and $\alpha = 0.26\mathbf{i} - 0.07\mathbf{j} + 0.13\mathbf{k}$ (rad/s^2). What moment is exerted about the center of mass of the casting by the manipulator?

20.29 A robotic manipulator holds a casting. The inertia matrix of the casting in terms of a body-fixed coordinate system with its origin at the center of mass is shown. At the present instant, the casting is stationary. If the manipulator exerts a moment $\Sigma\mathbf{M} = 0.042\mathbf{i} + 0.036\mathbf{j} + 0.066\mathbf{k}$ (N-m) about the center of mass, what is the angular acceleration of the casting at that instant?

20.30 The rigid body rotates about the fixed point O. Its inertia matrix in terms of the body-fixed coordinate system is shown. At the present instant, the rigid body's angular velocity is $\omega = 6\mathbf{i} + 6\mathbf{j} - 4\mathbf{k}$ (rad/s) and its angular acceleration is zero. What total moment about O is being exerted on the rigid body?

20.31 The rigid body rotates about the fixed point O. Its inertia matrix in terms of the body-fixed coordinate system is shown. At the present instant, the rigid body's angular velocity is $\omega = 6\mathbf{i} + 6\mathbf{j} - 4\mathbf{k}$ (rad/s). The total moment about O due to the forces and couples acting on the rigid body is zero. What is its angular acceleration?

$$\begin{bmatrix} I_{xx} & -I_{xy} & -I_{xz} \\ -I_{yx} & I_{yy} & -I_{yz} \\ -I_{zx} & -I_{zy} & I_{zz} \end{bmatrix} = \begin{bmatrix} 4 & -2 & 0 \\ -2 & 3 & 1 \\ 0 & 1 & 5 \end{bmatrix} \text{slug-ft}^2.$$

Problems 20.30/20.31

20.32 The dimensions of the 20-kg thin plate are $h = 0.4$ m and $b = 0.6$ m. The plate is stationary relative to an inertial reference frame when the force $F = 10$ N is applied in the direction perpendicular to the plate. No other forces or couples act on the plate. At the instant F is applied, what is the magnitude of the acceleration of point A relative to the inertial reference frame?

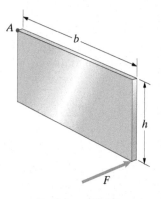

Problem 20.32

$$\begin{bmatrix} I_{xx} & -I_{xy} & -I_{xz} \\ -I_{yx} & I_{yy} & -I_{yz} \\ -I_{zx} & -I_{zy} & I_{zz} \end{bmatrix} = \begin{bmatrix} 0.05 & -0.03 & 0 \\ -0.03 & 0.08 & 0 \\ 0 & 0 & 0.04 \end{bmatrix} \text{kg-m}^2.$$

Problems 20.28/20.29

20.33 In terms of the coordinate system shown, the inertia matrix of the 6-kg slender bar is

$$
\begin{bmatrix}
I_{xx} & -I_{xy} & -I_{xz} \\
-I_{yx} & I_{yy} & -I_{yz} \\
-I_{zx} & -I_{zy} & I_{zz}
\end{bmatrix}
=
\begin{bmatrix}
0.500 & 0.667 & 0 \\
0.667 & 2.667 & 0 \\
0 & 0 & 3.167
\end{bmatrix}
\text{kg-m}^2.
$$

The bar is stationary relative to an inertial reference frame when the force $\mathbf{F} = 12\mathbf{k}$ (N) is applied at the right end of the bar. No other forces or couples act on the bar. Determine (a) the bar's angular acceleration relative to the inertial reference frame and (b) the acceleration of the right end of the bar relative to the inertial reference frame at the instant the force is applied.

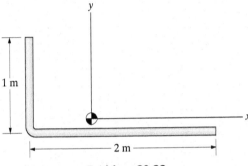

Problem 20.33

20.34 In terms of the coordinate system shown, the inertia matrix of the 12-kg slender bar is

$$
\begin{bmatrix}
I_{xx} & -I_{xy} & -I_{xz} \\
-I_{yx} & I_{yy} & -I_{yz} \\
-I_{zx} & -I_{zy} & I_{zz}
\end{bmatrix}
=
\begin{bmatrix}
2 & -3 & 0 \\
-3 & 8 & 0 \\
0 & 0 & 10
\end{bmatrix}
\text{kg-m}^2.
$$

The bar is stationary relative to an inertial reference frame when a force $\mathbf{F} = 20\mathbf{i} + 40\mathbf{k}$ (N) is applied at the point $x = 1$ m, $y = 1$ m. No other forces or couples act on the bar. Determine (a) the bar's angular acceleration and (b) the acceleration of the point $x = -1$ m, $y = -1$ m, relative to the inertial reference frame at the instant the force is applied.

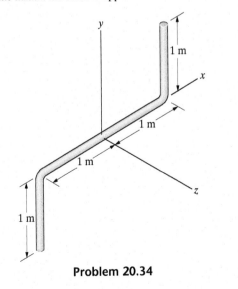

Problem 20.34

20.35 The inertia matrix of the 2.4-kg plate in terms of the given coordinate system is shown. The angular velocity vector of the plate is $\boldsymbol{\omega} = 6.4\mathbf{i} + 8.2\mathbf{j} + 14\mathbf{k}$ (rad/s), and its angular acceleration vector is $\boldsymbol{\alpha} = 60\mathbf{i} + 40\mathbf{j} - 120\mathbf{k}$ (rad/s^2). What are the components of the total moment exerted on the plate about its center of mass?

20.36 The inertia matrix of the 2.4-kg plate in terms of the given coordinate system is shown. At $t = 0$, the plate is stationary and is subjected to a force $\mathbf{F} = -10\mathbf{k}$ (N) at the point with coordinates $(220, 0, 0)$ mm. No other forces or couples act on the plate. Determine (a) the acceleration of the plate's center of mass and (b) the plate's angular acceleration at the instant the force is applied.

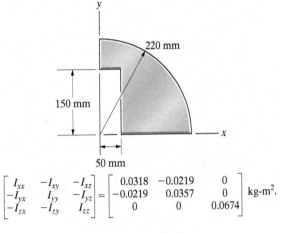

$$
\begin{bmatrix}
I_{xx} & -I_{xy} & -I_{xz} \\
-I_{yx} & I_{yy} & -I_{yz} \\
-I_{zx} & -I_{zy} & I_{zz}
\end{bmatrix}
=
\begin{bmatrix}
0.0318 & -0.0219 & 0 \\
-0.0219 & 0.0357 & 0 \\
0 & 0 & 0.0674
\end{bmatrix}
\text{kg-m}^2.
$$

Problems 20.35/20.36

20.37 A 3-kg slender bar is rigidly attached to a 2-kg thin circular disk. In terms of the body-fixed coordinate system shown, the angular velocity vector of the composite object is $\boldsymbol{\omega} = 100\mathbf{i} - 4\mathbf{j} + 6\mathbf{k}$ (rad/s) and its angular acceleration is zero. What are the components of the total moment exerted on the object about its center of mass?

20.38 A 3-kg slender bar is rigidly attached to a 2-kg thin circular disk. At $t = 0$, the composite object is stationary and is subjected to the moment $\Sigma\mathbf{M} = -10\mathbf{i} + 10\mathbf{j}$ (N-m) about its center of mass. No other forces or couples act on the object. Determine the object's angular acceleration at $t = 0$.

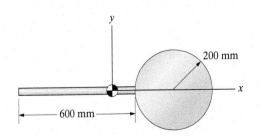

Problems 20.37/20.38

20.39 The vertical shaft supporting the dish antenna is rotating with a constant angular velocity of 1 rad/s. The angle $\theta = 30°$, $d\theta/dt = 20°/s$, and $d^2\theta/dt^2 = -40°/s^2$. The mass of the antenna is 280 kg, and its moments and products of inertia, in kg-m^2, are $I_{xx} = 140$, $I_{yy} = I_{zz} = 220$, and $I_{xy} = I_{yz} = I_{zx} = 0$. Determine the couple exerted on the antenna by its support at A at the instant shown.

Problem 20.39

20.40 The 5-kg triangular plate is connected to a ball-and-socket support at O. If the plate is released from rest in the horizontal position, what are the components of its angular acceleration at that instant?

20.41 If the 5-kg plate is released from rest in the horizontal position, what force is exerted on it by the ball-and-socket support at that instant?

20.42 The 5-kg triangular plate is connected to a ball-and-socket support at O. If the plate is released in the horizontal position with angular velocity $\boldsymbol{\omega} = 4\mathbf{i}$ (rad/s), what are the components of its angular acceleration vector at that instant?

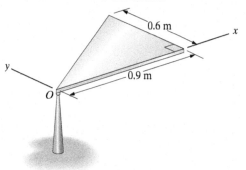

Problems 20.40–20.42

20.43 A subassembly of a space station can be modeled as two rigidly connected slender bars, each with a mass of 5000 kg. The subassembly is not rotating at $t = 0$, when a reaction control motor exerts a force $\mathbf{F} = 400\mathbf{k}$ (N) at B. What is the acceleration of point A relative to the center of mass of the subassembly at $t = 0$?

20.44 A subassembly of a space station can be modeled as two rigidly connected slender bars, each with a mass of 5000 kg. If the subassembly is rotating about the x axis at a constant rate of 1 revolution every 10 minutes, what is the magnitude of the couple its reaction control system is exerting on it?

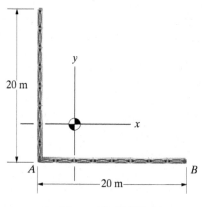

Problems 20.43/20.44

20.45 The thin circular disk of radius $R = 0.2$ m and mass $m = 4$ kg is rigidly attached to the vertical shaft. The plane of the disk is slanted at an angle $\beta = 30°$ relative to the horizontal. The shaft rotates with constant angular velocity $\omega_0 = 25$ rad/s. Determine the magnitude of the couple exerted on the disk by the shaft.

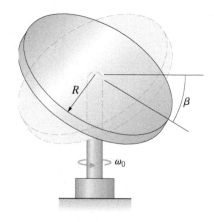

Problem 20.45

20.46 The slender bar of mass $m = 8$ kg and length $l = 1.2$ m is welded to a horizontal shaft that rotates with constant angular velocity $\omega_0 = 25$ rad/s. The angle $\beta = 30°$. Determine the magnitudes of the force **F** and couple **C** exerted on the bar by the shaft. (Write the equations of angular motion in terms of the body-fixed coordinate system shown.)

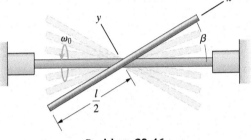

Problem 20.46

20.47 The slender bar of mass $m = 8$ kg and length $l = 1.2$ m is welded to a horizontal shaft that rotates with constant angular velocity $\omega_0 = 25$ rad/s. The angle $\beta = 30°$. Determine the magnitudes of the force **F** and couple **C** exerted on the bar by the shaft. (Write the equations of angular motion in terms of the body-fixed coordinate system shown. See Problem 20.98.)

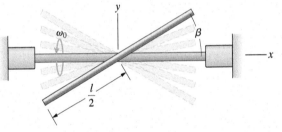

Problem 20.47

20.48 The slender bar of length l and mass m is pinned to the vertical shaft at O. The vertical shaft rotates with a constant angular velocity ω_0. Show that the value of ω_0 necessary for the bar to remain at a constant angle β relative to the vertical is

$$\omega_0 = \sqrt{3g/(2l \cos \beta)}.$$

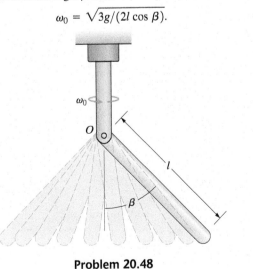

Problem 20.48

20.49 The vertical shaft rotates with constant angular velocity ω_0. The 35° angle between the edge of the 10-lb thin rectangular plate pinned to the shaft and the shaft remains constant. Determine ω_0.

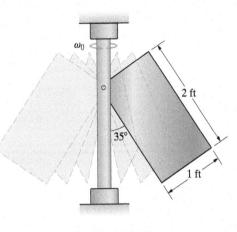

Problem 20.49

20.50 The radius of the 20-lb thin circular disk is $R = 1$ ft. The disk is mounted on the horizontal shaft and rotates with constant angular velocity $\omega_d = 10$ rad/s relative to the shaft. The horizontal shaft is 2 ft in length. The vertical shaft rotates with constant angular velocity $\omega_0 = 4$ rad/s. Determine the force and couple exerted at the center of the disk by the horizontal shaft.

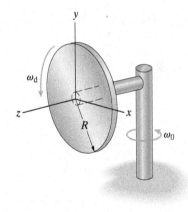

Problem 20.50

20.51 The object shown in Fig. a consists of two 1-kg vertical slender bars welded to the 4-kg horizontal slender bar. In Fig. b, the object is supported by bearings at A and B. The horizontal circular disk is supported by a vertical shaft that rotates with constant angular velocity $\omega_0 = 6$ rad/s. The horizontal bar rotates with constant angular velocity $\omega = 10$ rad/s. At the instant shown, determine the y and z components of the forces exerted on the object at A and B.

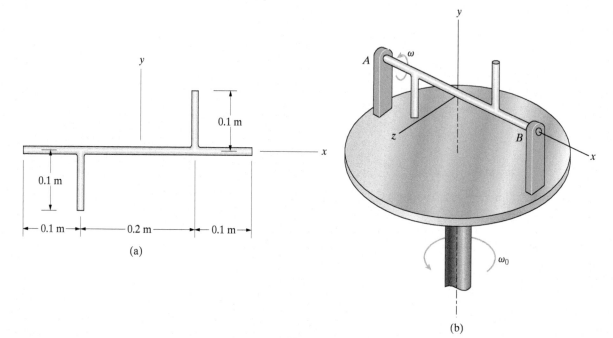

(a)

(b)

Problem 20.51

20.52 The 10-lb thin circular disk is rigidly attached to the 12-lb slender horizontal shaft. The disk and horizontal shaft rotate about the axis of the shaft with constant angular velocity $\omega_d = 20$ rad/s. The entire assembly rotates about the vertical axis with constant angular velocity $\omega_0 = 4$ rad/s. Determine the components of the force and couple exerted on the horizontal shaft by the disk.

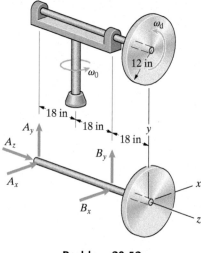

Problem 20.52

20.53 The Hubble telescope is rotating about its longitudinal axis with constant angular velocity ω_0. The coordinate system is fixed with respect to the solar panel. Relative to the telescope, the solar panel rotates about the x axis with constant angular velocity ω_x. Assume that the moments of inertia I_{xx}, I_{yy}, and I_{zz} are known, and $I_{xy} = I_{yz} = I_{zx} = 0$. Show that the moment about the x axis the servomechanisms must exert on the solar panel is

$$\Sigma M_x = (I_{zz} - I_{yy})\omega_0^2 \sin \theta \cos \theta.$$

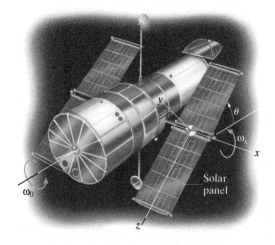

Problem 20.53

20.54 The thin rectangular plate is attached to the rectangular frame by pins. The frame rotates with constant angular velocity ω_0. Show that

$$\frac{d^2\beta}{dt^2} = -\omega_0^2 \sin \beta \cos \beta.$$

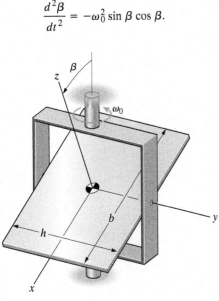

Problem 20.54

▶ **20.55*** The axis of the right circular cone of mass m, height h, and radius R spins about the vertical axis with constant angular velocity ω_0. The center of mass of the cone is stationary, and its base rolls on the floor. Show that the angular velocity necessary for this motion is $\omega_0 = \sqrt{10g/3R}$.

(See Example 20.6.)

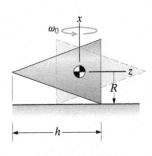

Problem 20.55

▶ **20.56** The tilted homogeneous cone undergoes a steady motion in which its flat end rolls on the floor while the center of mass remains stationary. The angle β between the axis and the horizontal remains constant, and the axis rotates about the vertical axis with constant angular velocity ω_0. The cone has mass m, radius R, and height h. Show that the angular velocity ω_0 necessary for this motion satisfies

$$\omega_0^2 = \frac{g(R \sin \beta - \frac{1}{4}h \cos \beta)}{\frac{3}{20}\left(R^2 + \frac{1}{4}h^2\right)\sin \beta \cos \beta - \frac{3}{40}hR \cos^2 \beta}.$$

(See Example 20.6.)

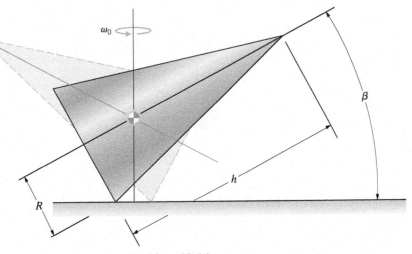

Problem 20.56

20.57* The two thin disks are rigidly connected by a slender bar. The radius of the large disk is 200 mm and its mass is 4 kg. The radius of the small disk is 100 mm and its mass is 1 kg. The bar is 400 mm in length and its mass is negligible. The composite object undergoes a steady motion in which it spins about the vertical y axis through its center of mass with angular velocity ω_0. The bar is horizontal during this motion and the large disk rolls on the floor. What is ω_0?

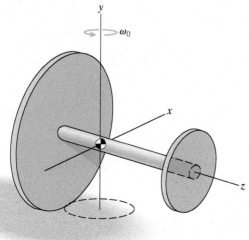

Problem 20.57

20.58 The view of an airplane's landing gear as seen looking from behind the airplane is shown in Fig. a. The radius of the wheel is 300 mm, and its moment of inertia is 2 kg-m^2. The airplane takes off at 30 m/s. After takeoff, the landing gear retracts by rotating toward the right side of the airplane, as shown in Fig. b. Determine the magnitude of the couple exerted by the wheel on its support. (Neglect the airplane's angular motion.)

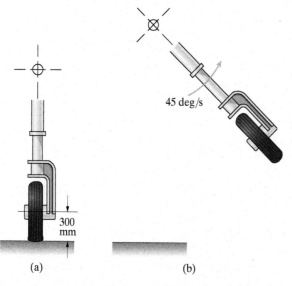

(a) (b)

Problem 20.58

20.59 If the rider turns to his left, will the couple exerted on the motorcycle by its wheels tend to cause the motorcycle to lean toward the rider's left side or his right side?

Problem 20.59

20.60* By substituting the components of $\mathbf{H}_O$ from Eqs. (20.9) into the equation

$$\Sigma \mathbf{M}_O = \frac{dH_{Ox}}{dt}\mathbf{i} + \frac{dH_{Oy}}{dt}\mathbf{j} + \frac{dH_{Oz}}{dt}\mathbf{k} + \mathbf{\Omega} \times H_O,$$

derive Eqs. (20.12).

20.3 The Euler Angles

BACKGROUND

The equations of angular motion relate the total moment acting on a rigid body to its angular velocity and acceleration. If we know the total moment and the angular velocity, we can determine the angular acceleration. But how can we use the angular acceleration to determine the rigid body's angular position, or orientation, as a function of time? To explain how this is done, we must first show how to specify the orientation of a rigid body in three dimensions.

We have seen that describing the orientation of a rigid body in planar motion requires only the angle θ that specifies the body's rotation relative to some reference orientation. In three-dimensional motion, three angles are required. To understand why, consider a particular axis that is fixed relative to a rigid body. Two angles are necessary to specify the direction of the axis, and a third angle is needed to specify the rigid body's orientation about the axis. Although several systems of angles for describing the orientation of a rigid body are commonly used, the best-known system is the one called the Euler angles. In this section we define these angles and express the equations of angular motion in terms of them.

Objects with an Axis of Symmetry

We first explain how the Euler angles are used to describe the orientation of an object with an axis of rotational symmetry, because this case results in simpler equations of angular motion.

Definitions   We assume that an object has an axis of rotational symmetry, and we introduce two reference frames: a secondary coordinate system xyz, with its z axis coincident with the object's axis of symmetry, and an inertial primary coordinate system XYZ. We begin with the object in a reference position in which xyz and XYZ are superimposed on each other (Fig. 20.11a).

Our first step is to rotate the object and the xyz system together through an angle ψ about the Z axis (Fig. 20.11b). In this intermediate orientation, we denote the secondary coordinate system by $x'y'z'$. Next, we rotate the object and the xyz system together through an angle θ about the x' axis (Fig. 20.11c). Finally, we rotate the object relative to the xyz system through an angle ϕ about the object's axis of symmetry (Fig. 20.11d). Notice that the x axis remains in the XY plane.

The angles ψ and θ specify the orientation of the secondary xyz system relative to the primary XYZ system. The angle ψ is called the *precession angle*, and θ is called the *nutation angle*. The angle ϕ specifying the rotation of the rigid body relative to the xyz system is called the *spin angle*. These three angles specify the orientation of the rigid body relative to the primary coordinate system and are called the *Euler angles*. We can obtain any orientation of the object relative to the primary coordinate system by appropriate choices of the Euler angles: We choose ψ and θ to obtain the desired direction of the axis of symmetry and then choose ϕ to obtain the desired rotational position of the object about its axis of symmetry.

Equations of Angular Motion   To analyze an object's motion in terms of the Euler angles, we must express the equations of angular motion in terms of those angles. Figure 20.12a shows the rotation ψ from the reference orientation of the xyz system to its intermediate orientation $x'y'z'$. We represent the angular velocity of the coordinate system due to the rate of change of ψ by the angular velocity vector $\dot{\psi}$ pointing in the z' direction. (We use a dot to denote the derivative with respect to time.)

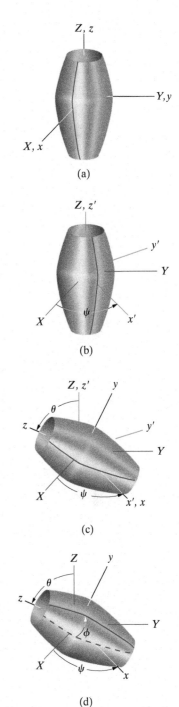

Figure 20.11
(a) The reference position.
(b) The rotation ψ about the Z axis.
(c) The rotation θ about the x' axis.
(d) The rotation ϕ of the object relative to the xyz system.

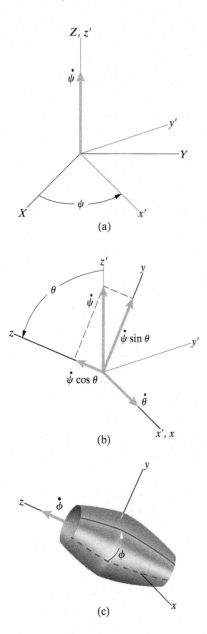

Figure 20.12

(a) The rotation ψ and the angular velocity $\dot{\psi}$.

(b) The rotation θ, the angular velocity $\dot{\theta}$, and the components of the angular velocity $\dot{\psi}$.

(c) The rotation ϕ and the angular velocity $\dot{\phi}$.

Figure 20.12b shows the second rotation θ. We represent the angular velocity due to the rate of change of θ by the vector $\dot{\theta}$ pointing in the x direction. We also resolve the angular velocity vector $\dot{\psi}$ into components in the y and z directions. The components of the angular velocity of the xyz system relative to the primary coordinate system are

$$\Omega_x = \dot{\theta},$$
$$\Omega_y = \dot{\psi}\sin\theta,$$

(20.22)

and

$$\Omega_z = \dot{\psi}\cos\theta.$$

In Fig. 20.12c, we represent the angular velocity of the rigid body relative to the xyz system by the vector $\dot{\phi}$. Adding this angular velocity to the angular velocity of the xyz system, we obtain the components of the angular velocity of the rigid body relative to the XYZ system:

$$\omega_x = \dot{\theta},$$
$$\omega_y = \dot{\psi}\sin\theta,$$

(20.23)

and

$$\omega_z = \dot{\phi} + \dot{\psi}\cos\theta.$$

Taking the derivatives of these equations with respect to time yields

$$\frac{d\omega_x}{dt} = \ddot{\theta},$$

$$\frac{d\omega_y}{dt} = \ddot{\psi}\sin\theta + \dot{\psi}\dot{\theta}\cos\theta,$$

(20.24)

and

$$\frac{d\omega_z}{dt} = \ddot{\phi} + \ddot{\psi}\cos\theta - \dot{\psi}\dot{\theta}\sin\theta.$$

As a consequence of the object's rotational symmetry, the products of inertia I_{xy}, I_{xz}, and I_{yz} are zero and $I_{xx} = I_{yy}$. The inertia matrix is of the form

$$[I] = \begin{bmatrix} I_{xx} & 0 & 0 \\ 0 & I_{xx} & 0 \\ 0 & 0 & I_{zz} \end{bmatrix}.$$

(20.25)

Substituting Eqs. (20.22)–(20.25) into Eqs. (20.18), we obtain the equations of angular motion in terms of the Euler angles:

$$\Sigma M_x = I_{xx}\ddot{\theta} + (I_{zz} - I_{xx})\dot{\psi}^2\sin\theta\cos\theta + I_{zz}\dot{\phi}\dot{\psi}\sin\theta,$$ (20.26)

$$\Sigma M_y = I_{xx}(\ddot{\psi}\sin\theta + 2\dot{\psi}\dot{\theta}\cos\theta) - I_{zz}(\dot{\phi}\dot{\theta} + \dot{\psi}\dot{\theta}\cos\theta),$$ (20.27)

$$\Sigma M_z = I_{zz}(\ddot{\phi} + \ddot{\psi}\cos\theta - \dot{\psi}\dot{\theta}\sin\theta).$$ (20.28)

To determine the Euler angles as functions of time when the total moment is known, these equations usually must be solved by numerical integration.

However, we can obtain an important class of closed-form solutions by assuming a specific type of motion.

Steady Precession The motion called *steady precession* is commonly observed in tops and gyroscopes. The object's rate of spin $\dot{\phi}$ relative to the xyz coordinate system is assumed to be constant (Fig. 20.13). The nutation angle θ, the inclination of the *spin axis* z relative to the Z axis, is assumed to be constant, and the *precession rate* $\dot{\psi}$, the rate at which the xyz system rotates about the Z axis, is assumed to be constant. The last assumption explains the name given to this motion.

With these assumptions, Eqs. (20.26)–(20.28) reduce to

$$\Sigma M_x = (I_{zz} - I_{xx})\dot{\psi}^2 \sin\theta \cos\theta + I_{zz}\dot{\phi}\dot{\psi} \sin\theta, \qquad (20.29)$$

$$\Sigma M_y = 0, \qquad (20.30)$$

and

$$\Sigma M_z = 0. \qquad (20.31)$$

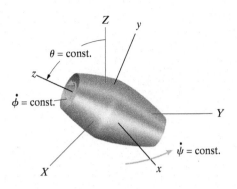

Figure 20.13
Steady procession.

We discuss two examples: the steady precession of a spinning top and the steady precession of an axially symmetric object that is free of external moments.

Precession of a Top The peculiar behavior of a top (Fig. 20.14a) inspired some of the first analytical studies of three-dimensional motions of rigid bodies. When a top is set into motion, its spin axis may initially remain vertical—a motion called *sleeping*. As friction reduces the spin rate, the spin axis begins to lean over and rotate about the vertical axis. This phase of the top's motion approximates steady precession. (The top's spin rate continuously decreases due to friction, whereas in steady precession we assume the spin rate to be constant.)

To analyze the motion, we place the primary coordinate system XYZ with its origin at the point of the top and the Z axis upward. Then we align the z axis of the xyz system with the spin axis (Fig. 20.14b). We assume that the top's point rests in a small depression so that it remains at a fixed point on the floor. The precession angle ψ and nutation angle θ specify the orientation of the spin axis, and the spin rate of the top relative to the xyz system is $\dot{\phi}$.

The top's weight exerts a moment $\Sigma M_x = mgh \sin\theta$ about the origin, and the moments $\Sigma M_y = 0$ and $\Sigma M_z = 0$. Substituting $\Sigma M_x = mgh \sin\theta$ into Eq. (20.29), we obtain

$$mgh = (I_{zz} - I_{xx})\dot{\psi}^2 \cos\theta + I_{zz}\dot{\phi}\dot{\psi}, \qquad (20.32)$$

and Eqs. (20.30) and (20.31) are identically satisfied. Equation (20.32) relates the spin rate, nutation angle, and rate of precession. For example, if we know the spin rate $\dot{\phi}$ and nutation angle θ, we can solve for the top's precession rate $\dot{\psi}$.

Moment-Free Steady Precession A spinning axisymmetric object that is free of external moments, such as an axisymmetric satellite in orbit, can exhibit a motion similar to the steady precessional motion of a top. This motion is observed when an American football is thrown in a "wobbly" spiral. To analyze it, we place the origin of the xyz system at the object's center of mass (Fig. 20.15a). Equation (20.29) then becomes

$$(I_{zz} - I_{xx})\dot{\psi} \cos\theta + I_{zz}\dot{\phi} = 0, \qquad (20.33)$$

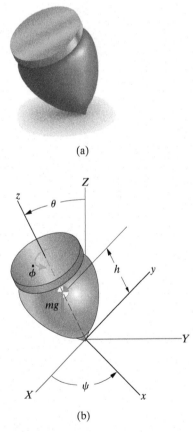

(a)

(b)

Figure 20.14
(a) A spinning top seems to defy gravity.
(b) The precession angle ψ and nutation angle θ specify the orientation of the spin axis.

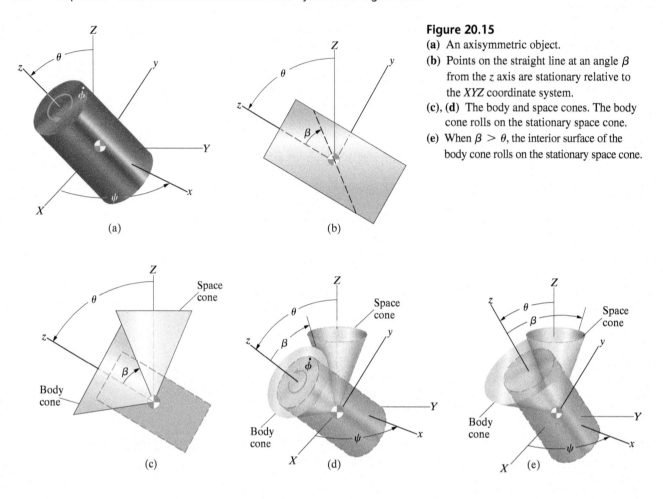

Figure 20.15

(a) An axisymmetric object.

(b) Points on the straight line at an angle β from the z axis are stationary relative to the XYZ coordinate system.

(c), (d) The body and space cones. The body cone rolls on the stationary space cone.

(e) When $\beta > \theta$, the interior surface of the body cone rolls on the stationary space cone.

(a) (b) (c) (d) (e)

and Eqs. (20.30) and (20.31) are identically satisfied. For a given value of the nutation angle, Eq. (20.33) relates the object's rates of precession and spin.

We can interpret Eq. (20.33) in a way that makes it possible to visualize the motion. We look for a point in the y–z plane at which the object's velocity relative to the center of mass is zero at the current instant. We want to find a point with coordinates $(0, y, z)$ such that

$$\boldsymbol{\omega} \times (y\mathbf{j} + z\mathbf{k}) = [(\dot{\psi} \sin \theta)\mathbf{j} + (\dot{\phi} + \dot{\psi} \cos \theta)\mathbf{k}] \times [y\mathbf{j} + z\mathbf{k}]$$
$$= [z\dot{\psi} \sin \theta - y(\dot{\phi} + \dot{\psi} \cos \theta)]\mathbf{i} = \mathbf{0}.$$

This equation is satisfied at points in the y–z plane such that

$$\frac{y}{z} = \frac{\dot{\psi} \sin \theta}{\dot{\phi} + \dot{\psi} \cos \theta}.$$

This relation is satisfied by points on the dashed black line in Fig. 20.15b, where

$$\tan \beta = \frac{y}{z} = \frac{\dot{\psi} \sin \theta}{\dot{\phi} + \dot{\psi} \cos \theta}.$$

Solving Eq. (20.33) for $\dot{\phi}$ and substituting the result into this equation, we obtain

$$\tan \beta = \left(\frac{I_{zz}}{I_{xx}}\right) \tan \theta.$$

If $I_{xx} > I_{zz}$, the angle $\beta < \theta$. In Fig. 20.15c, we show an imaginary cone of half-angle β, called the *body cone*, whose axis is coincident with the z axis. The body cone is in contact with a fixed cone, called the *space cone*, whose axis is coincident with the Z axis. If the body cone rolls on the curved surface of the space cone as the z axis precesses about the Z axis (Fig. 20.15d), the points of the body cone lying on the straight line in Fig. 20.15b have zero velocity relative to the XYZ system. That means that *the motion of the body cone is identical to the motion of the object*. The object's motion can be visualized by visualizing the motion of the body cone as it rolls around the outer surface of the space cone. This motion is called *direct precession*.

If $I_{xx} < I_{zz}$, the angle $\beta > \theta$. In this case, we must visualize the *interior* surface of the body cone rolling on the fixed space cone (Fig. 20.15e). This motion is called *retrograde precession*.

Arbitrary Objects

In our analysis of axially symmetric objects, we let the object move relative to the secondary xyz coordinate system, rotating about the z axis. As a consequence, only two angles—the precession angle ψ and nutation angle θ—are needed to specify the orientation of the xyz coordinate system, and this simplifies the equations of angular motion. The object must be axially symmetric about the z axis, so that the moments and products of inertia will not vary as it rotates. In the case of an arbitrary object, the moments and products of inertia will be constants only if the xyz coordinate system is body fixed. This means that three angles are needed to specify the orientation of the coordinate system, and the resulting equations of angular motion are more complicated.

Definitions   We begin with a reference position in which the body-fixed xyz and primary XYZ coordinate systems are superimposed on each other (Fig. 20.16a). First, we rotate the xyz system through the precession angle ψ about the Z axis (Fig. 20.16b) and denote it by $x'y'z'$ in this intermediate orientation. Then we rotate the xyz system through the nutation angle θ about the x' axis (Fig. 20.16c), denoting it now by $x''y''z''$. We obtain the final orientation of the xyz system by rotating it through the angle ϕ about the z'' axis (Fig. 20.16d). Notice that we use one more rotation of the xyz system than in the case of an axially symmetric object.

We can obtain any orientation of the body-fixed coordinate system relative to the reference coordinate system by these three rotations. We choose ψ and θ to obtain the desired direction of the z axis and then choose ϕ to obtain the desired orientation of the x and y axes.

Just as in the case of an object with rotational symmetry, we must express the components of the rigid body's angular velocity in terms of the Euler angles to obtain the equations of angular motion. Figure 20.17a shows the rotation ψ from the reference orientation of the xyz system to the intermediate orientation $x'y'z'$. We represent the angular velocity of the body-fixed coordinate system due to the rate of change of ψ by the vector $\dot\psi$ pointing in the z' direction. Figure 20.17b shows the next rotation θ that takes the body-fixed coordinate system to the intermediate orientation $x''y''z''$. We represent the angular velocity due to the rate of change of θ by the vector $\dot\theta$ pointing in the x'' direction. In this figure, we also show the components of the angular velocity vector $\dot\psi$ in the y'' and z'' directions. Figure 20.17c shows the third rotation ϕ that takes the body-fixed coordinate system to its final orientation defined by the three Euler angles. We represent the angular velocity due to the rate of change of ϕ by the vector $\dot\phi$ pointing in the z direction.

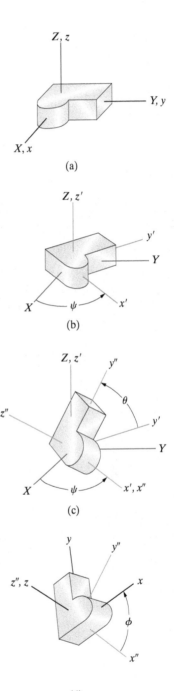

Figure 20.16
(a) The reference position.
(b) The rotation ψ about the Z axis.
(c) The rotation θ about the x' axis.
(d) The rotation ϕ about the z'' axis.

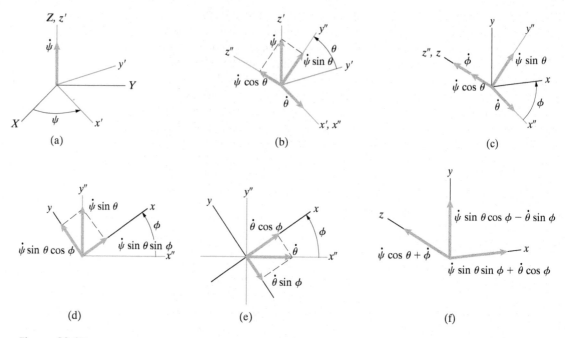

Figure 20.17

(a) The rotation ψ and the angular velocity $\dot{\psi}$.

(b) The rotation θ, the angular velocity $\dot{\theta}$, and the components of $\dot{\psi}$ in the $x''y''z''$ system.

(c) The rotation ϕ and the angular velocity $\dot{\phi}$.

(d), (e) The components of the angular velocities $\dot{\psi} \sin \theta$ and $\dot{\theta}$ in the xyz system.

(f) The angular velocities $\omega_x, \omega_y, \omega_z$.

To determine ω_x, ω_y, and ω_z in terms of the Euler angles, we need to determine the components of the angular velocities shown in Fig. 20.17c in the x-, y-, and z-axis directions. The vectors $\dot{\phi}$ and $\dot{\psi} \cos \theta$ point in the z-axis direction. In Figs. 20.17d and e, which are drawn with the z axis pointing out of the page, we determine the components of the vectors $\dot{\psi} \sin \theta$ and $\dot{\theta}$ in the x- and y-axis directions. By summing the components of the angular velocities in the three coordinate directions (Fig. 20.17f), we obtain

$$\omega_x = \dot{\psi} \sin \theta \sin \phi + \dot{\theta} \cos \phi,$$

$$\omega_y = \dot{\psi} \sin \theta \cos \phi - \dot{\theta} \sin \phi, \qquad (20.34)$$

$$\omega_z = \dot{\psi} \cos \theta + \dot{\phi}.$$

The derivatives of these equations with respect to time are

$$\frac{d\omega_x}{dt} = \ddot{\psi} \sin \theta \sin \phi + \dot{\psi}\dot{\theta} \cos \theta \sin \phi + \dot{\psi}\dot{\phi} \sin \theta \cos \phi$$

$$+ \ddot{\theta} \cos \phi - \dot{\theta}\dot{\phi} \sin \phi,$$

$$\frac{d\omega_y}{dt} = \ddot{\psi} \sin \theta \cos \phi + \dot{\psi}\dot{\theta} \cos \theta \cos \phi - \dot{\psi}\dot{\phi} \sin \theta \sin \phi$$

$$- \ddot{\theta} \sin \phi - \dot{\theta}\dot{\phi} \cos \phi, \qquad (20.35)$$

$$\frac{d\omega_z}{dt} = \ddot{\psi} \cos \theta - \dot{\psi}\dot{\theta} \sin \theta + \ddot{\phi}.$$

Equations of Angular Motion   With Eqs. (20.34) and (20.35), we can express the equations of angular motion in terms of the three Euler angles. To simplify the equations, *we assume that the body-fixed coordinate system xyz is a set of principal axes.* (See the appendix to this chapter.) Then the equations of angular motion, Eqs. (20.18), become

$$\Sigma M_x = I_{xx}\frac{d\omega_x}{dt} - (I_{yy} - I_{zz})\omega_y\omega_z,$$

$$\Sigma M_y = I_{yy}\frac{d\omega_y}{dt} - (I_{zz} - I_{xx})\omega_z\omega_x,$$

$$\Sigma M_z = I_{zz}\frac{d\omega_z}{dt} - (I_{xx} - I_{yy})\omega_x\omega_y.$$

Substituting Eqs. (20.34) and (20.35) into these relations, we obtain the equations of angular motion in terms of Euler angles:

$$\Sigma M_x = I_{xx}\ddot{\psi}\sin\theta\sin\phi + I_{xx}\ddot{\theta}\cos\phi$$
$$+ I_{xx}(\dot{\psi}\dot{\theta}\cos\theta\sin\phi + \dot{\psi}\dot{\phi}\sin\theta\cos\phi - \dot{\theta}\dot{\phi}\sin\phi)$$
$$- (I_{yy} - I_{zz})(\dot{\psi}\sin\theta\cos\phi - \dot{\theta}\sin\phi)(\dot{\psi}\cos\theta + \dot{\phi}),$$

$$\Sigma M_y = I_{yy}\ddot{\psi}\sin\theta\cos\phi - I_{yy}\ddot{\theta}\sin\phi \qquad\qquad (20.36)$$
$$+ I_{yy}(\dot{\psi}\dot{\theta}\cos\theta\cos\phi - \dot{\psi}\dot{\phi}\sin\theta\sin\phi - \dot{\theta}\dot{\phi}\cos\phi)$$
$$- (I_{zz} - I_{xx})(\dot{\psi}\cos\theta + \dot{\phi})(\dot{\psi}\sin\theta\sin\phi + \dot{\theta}\cos\phi),$$

$$\Sigma M_z = I_{zz}\ddot{\psi}\cos\theta + I_{zz}\ddot{\phi} - I_{zz}\dot{\psi}\dot{\theta}\sin\theta$$
$$- (I_{xx} - I_{yy})(\dot{\psi}\sin\theta\sin\phi + \dot{\theta}\cos\phi)(\dot{\psi}\sin\theta\cos\phi - \dot{\theta}\sin\phi).$$

If the Euler angles and their first and second derivatives with respect to time are known, Eqs. (20.36) can be solved for the components of the total moment. Or, if the total moment, the Euler angles, and the first derivatives of the Euler angles are known, Eqs. (20.36) can be solved for the second derivatives of the Euler angles. In this way, the Euler angles can be determined as functions of time when the total moment is known as a function of time, but numerical integration is usually necessary.

RESULTS

The Euler angles are a set of angles used to describe the orientation of a rigid body relative to a primary reference frame, or coordinate system.

Euler Angles for an Object with an Axis of Symmetry

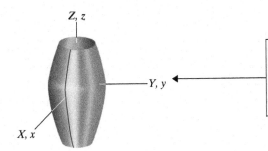

Z, z

Y, y

X, x

An object with an axis of symmetry. The primary *XYZ* coordinate system and a secondary *xyz* coordinate system are superimposed. The *Z* and *z* axes are aligned with the axis of symmetry of the object.

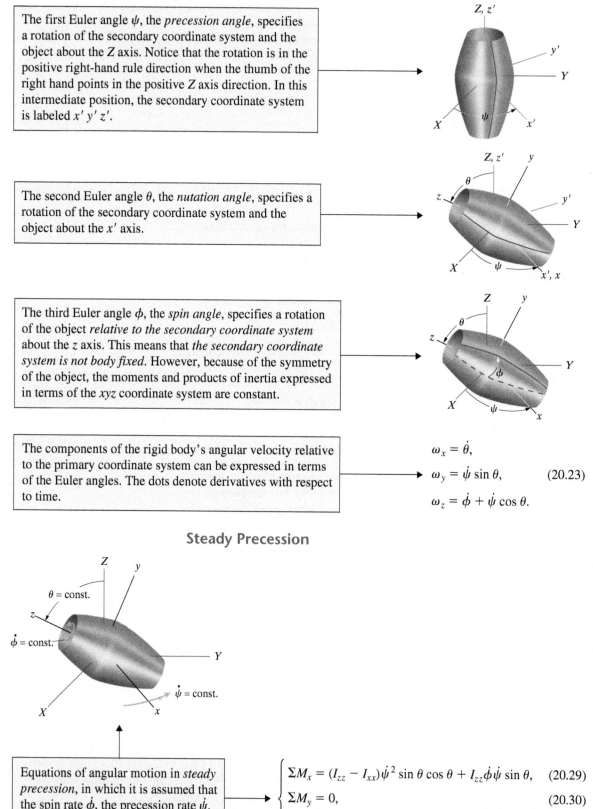

The first Euler angle ψ, the *precession angle*, specifies a rotation of the secondary coordinate system and the object about the Z axis. Notice that the rotation is in the positive right-hand rule direction when the thumb of the right hand points in the positive Z axis direction. In this intermediate position, the secondary coordinate system is labeled $x'\ y'\ z'$.

The second Euler angle θ, the *nutation angle*, specifies a rotation of the secondary coordinate system and the object about the x' axis.

The third Euler angle ϕ, the *spin angle*, specifies a rotation of the object *relative to the secondary coordinate system* about the z axis. This means that *the secondary coordinate system is not body fixed*. However, because of the symmetry of the object, the moments and products of inertia expressed in terms of the xyz coordinate system are constant.

The components of the rigid body's angular velocity relative to the primary coordinate system can be expressed in terms of the Euler angles. The dots denote derivatives with respect to time.

$$\omega_x = \dot{\theta},$$
$$\omega_y = \dot{\psi} \sin \theta, \qquad (20.23)$$
$$\omega_z = \dot{\phi} + \dot{\psi} \cos \theta.$$

Steady Precession

$\theta = $ const.

$\dot{\phi} = $ const.

$\dot{\psi} = $ const.

Equations of angular motion in *steady precession*, in which it is assumed that the spin rate $\dot{\phi}$, the precession rate $\dot{\psi}$, and the nutation angle θ are constant.

$$\begin{cases} \sum M_x = (I_{zz} - I_{xx})\dot{\psi}^2 \sin \theta \cos \theta + I_{zz}\dot{\phi}\dot{\psi} \sin \theta, & (20.29) \\ \sum M_y = 0, & (20.30) \\ \sum M_z = 0. & (20.31) \end{cases}$$

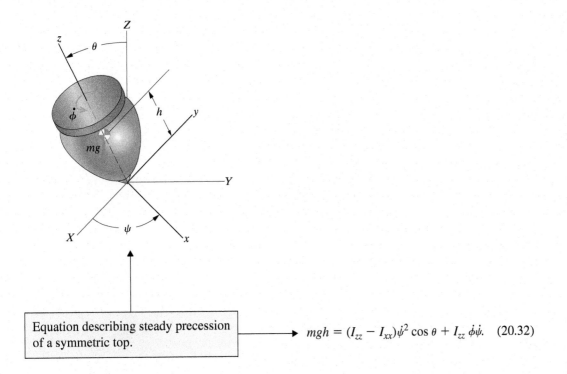

Equation describing steady precession of a symmetric top.

$$mgh = (I_{zz} - I_{xx})\dot{\psi}^2 \cos\theta + I_{zz}\,\dot{\phi}\dot{\psi}. \quad (20.32)$$

Moment-Free Steady Precession

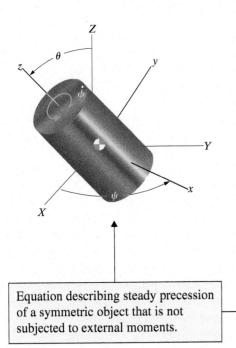

Equation describing steady precession of a symmetric object that is not subjected to external moments.

$$(I_{zz} - I_{xx})\dot{\psi}\cos\theta + I_{zz}\,\dot{\phi} = 0. \quad (20.33)$$

Space and Body Cones

A way to visualize moment-free steady precession. The space cone is fixed in space. The body cone rolls on the surface of the space cone. *The precessing object undergoes the same motion as the body cone.* The angle β is related to the nutation angle by

$$\tan \beta = \left(\frac{I_{zz}}{I_{xx}}\right) \tan \theta.$$

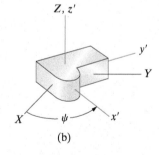

If $\beta > \theta$, the interior surface of the body cone rolls on the exterior surface of the space cone.

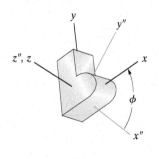

Euler Angles for an Arbitrary Object

The xyz coordinate system is body fixed and aligned with the primary XYZ coordinate system in the initial orientation (a). The precession angle ψ is a rotation of the object about the Z axis (b). The nutation angle θ is a rotation of the object about the x' axis (c). The spin angle ϕ is a rotation of the object about the z'' axis (d).

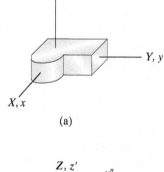

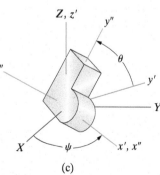

The components of the rigid body's angular velocity relative to the primary coordinate system can be expressed in terms of the Euler angles. The dots denote derivatives with respect to time.

$$\omega_x = \dot{\psi} \sin \theta \sin \phi + \dot{\theta} \cos \phi,$$
$$\omega_y = \dot{\psi} \sin \theta \cos \phi - \dot{\theta} \sin \phi, \quad (20.34)$$
$$\omega_z = \dot{\psi} \cos \theta + \dot{\phi}.$$

Equations of angular motion in terms of the Euler angles when the xyz coordinate system is a set of principal axes.

$$\Sigma M_x = I_{xx}\ddot{\psi} \sin \theta \sin \phi + I_{xx}\ddot{\theta} \cos \phi$$
$$+ I_{xx}(\dot{\psi}\dot{\theta} \cos \theta \sin \phi + \dot{\psi}\dot{\phi} \sin \theta \cos \phi - \dot{\theta}\dot{\phi} \sin \phi)$$
$$- (I_{yy} - I_{zz})(\dot{\psi} \sin \theta \cos \phi - \dot{\theta} \sin \phi)(\dot{\psi} \cos \theta + \dot{\phi}),$$

$$\Sigma M_y = I_{yy}\ddot{\psi} \sin \theta \cos \phi - I_{yy}\ddot{\theta} \sin \phi \qquad (20.36)$$
$$+ I_{yy}(\dot{\psi}\dot{\theta} \cos \theta \cos \phi - \dot{\psi}\dot{\phi} \sin \theta \sin \phi - \dot{\theta}\dot{\phi} \cos \phi)$$
$$- (I_{zz} - I_{xx})(\dot{\psi} \cos \theta + \dot{\phi})(\dot{\psi} \sin \theta \sin \phi + \dot{\theta} \cos \phi),$$

$$\Sigma M_z = I_{zz}\ddot{\psi} \cos \theta + I_{zz}\ddot{\phi} - I_{zz}\dot{\psi}\dot{\theta} \sin \theta$$
$$- (I_{xx} - I_{yy})(\dot{\psi} \sin \theta \sin \phi + \dot{\theta} \cos \phi)(\dot{\psi} \sin \theta \cos \phi - \dot{\theta} \sin \phi).$$

Active Example 20.7 **Steady Precession of a Disk** (▶ *Related Problem 20.63*)

The thin circular disk of radius R and mass m rolls along a circular path of radius r on a horizontal surface. The angle θ between the disk's axis and the vertical remains constant. (You may have seen a rolling coin exhibit this motion.) Determine the magnitude v of the velocity of the center of the disk as a function of the angle θ.

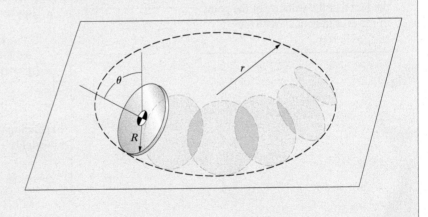

Strategy

We can determine the velocity of the center of the disk by assuming that the disk is in steady precession and determining the conditions necessary for the equations of motion to be satisfied.

Let the z axis of a secondary coordinate system be aligned with the disk's spin axis and assume that the x axis remains parallel to the horizontal surface. The angle θ is the nutation angle. The center of the disk moves in a circular path of radius $r_G = r - R \cos \theta$. Therefore, the precession rate—the rate at which the x axis rotates in the horizontal plane—is

$$\dot{\psi} = \frac{v}{r_G}. \qquad (1)$$

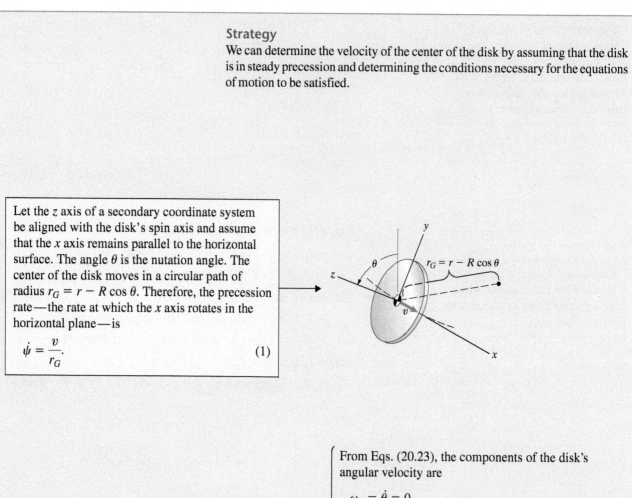

$r_G = r - R \cos \theta$

From Eqs. (20.23), the components of the disk's angular velocity are

$$\omega_x = \dot{\theta} = 0,$$

$$\omega_y = \dot{\psi} \sin \theta = \frac{v}{r_G} \sin \theta,$$

$$\omega_z = \dot{\phi} + \dot{\psi} \cos \theta = \dot{\phi} + \frac{v}{r_G} \cos \theta,$$

where $\dot{\phi}$ is the spin rate. We express the velocity of the point of the disk in contact with the surface in terms of the velocity of the center:

$$\mathbf{0} = v\mathbf{i} + \boldsymbol{\omega} \times (-R\mathbf{j})$$

$$= v\mathbf{i} + \begin{vmatrix} \mathbf{i} & \mathbf{j} & \mathbf{k} \\ 0 & \dfrac{v}{r_G} \sin \theta & \dot{\phi} + \dfrac{v}{r_G} \cos \theta \\ 0 & -R & 0 \end{vmatrix}.$$

Determine the spin rate of the disk in terms of the velocity v by using the fact that the velocity of the point of the disk in contact with the surface is zero.

Expanding the determinant and solving for the spin rate yields

$$\dot{\phi} = -v \left(\frac{1}{R} + \frac{1}{r_G} \cos \theta \right). \qquad (2)$$

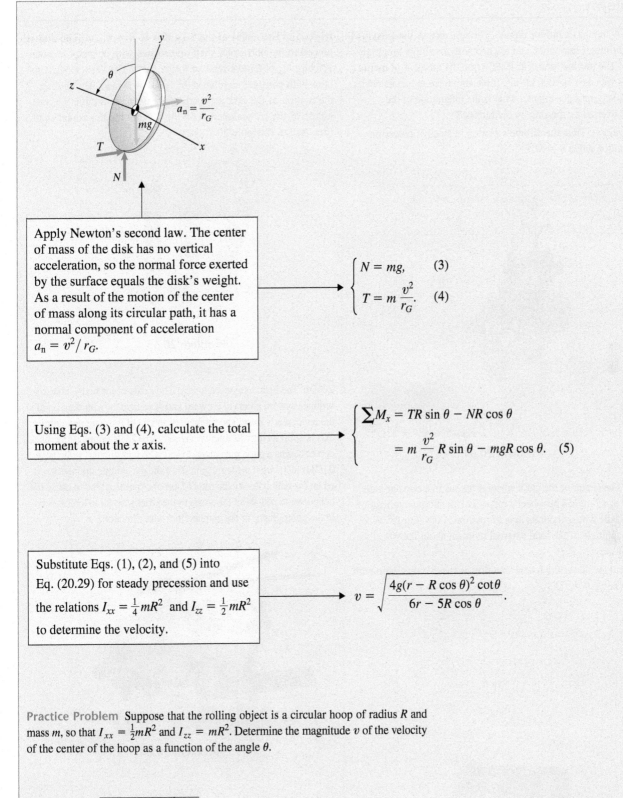

Apply Newton's second law. The center of mass of the disk has no vertical acceleration, so the normal force exerted by the surface equals the disk's weight. As a result of the motion of the center of mass along its circular path, it has a normal component of acceleration $a_n = v^2/r_G$.

$$\begin{cases} N = mg, & (3) \\ T = m\,\dfrac{v^2}{r_G}. & (4) \end{cases}$$

Using Eqs. (3) and (4), calculate the total moment about the x axis.

$$\begin{cases} \sum M_x = TR\sin\theta - NR\cos\theta \\ \qquad = m\,\dfrac{v^2}{r_G}R\sin\theta - mgR\cos\theta. \quad (5) \end{cases}$$

Substitute Eqs. (1), (2), and (5) into Eq. (20.29) for steady precession and use the relations $I_{xx} = \frac{1}{4}mR^2$ and $I_{zz} = \frac{1}{2}mR^2$ to determine the velocity.

$$v = \sqrt{\frac{4g(r - R\cos\theta)^2\cot\theta}{6r - 5R\cos\theta}}.$$

Practice Problem Suppose that the rolling object is a circular hoop of radius R and mass m, so that $I_{xx} = \frac{1}{2}mR^2$ and $I_{zz} = mR^2$. Determine the magnitude v of the velocity of the center of the hoop as a function of the angle θ.

Answer: $v = \sqrt{\dfrac{2g(r - R\cos\theta)^2\cot\theta}{4r - 3R\cos\theta}}.$

Problems

20.61 A ship has a turbine engine. The spin axis of the axisymmetric turbine is horizontal and aligned with the ship's longitudinal axis. The turbine rotates at 10,000 rpm. Its moment of inertia about its spin axis is 1000 kg-m². If the ship turns at a constant rate of 20 degrees per minute, what is the magnitude of the moment exerted on the ship by the turbine?

Strategy: Treat the turbine's motion as steady precession with nutation angle $\theta = 90°$.

Problem 20.61

20.62 The center of the car's wheel A travels in a circular path about O at 15 mi/h. The wheel's radius is 1 ft, and the moment of inertia of the wheel about its axis of rotation is 0.8 slug-ft². What is the magnitude of the total external moment about the wheel's center of mass?

Strategy: Treat the wheel's motion as steady precession with nutation angle $\theta = 90°$.

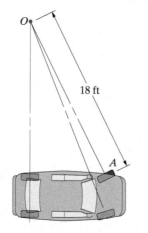

Problem 20.62

20.63 The radius of the 5-kg disk is $R = 0.2$ m. The disk is pinned to the horizontal shaft and rotates with constant angular velocity $\omega_d = 6$ rad/s relative to the shaft. The vertical shaft rotates with constant angular velocity $\omega_0 = 2$ rad/s. By treating the motion of the disk as steady precession, determine the magnitude of the couple exerted on the disk by the horizontal shaft. (See Active Example 20.7.)

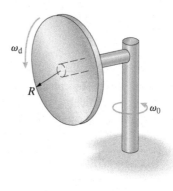

Problem 20.63

20.64 The helicopter is stationary. The z axis of the body-fixed coordinate system points downward and is coincident with the axis of the helicopter's rotor. The moment of inertia of the rotor about the z axis is 8600 kg-m². Its angular velocity is $-258\mathbf{k}$ (rpm). If the helicopter begins a pitch maneuver during which its angular velocity is $0.02\mathbf{j}$ (rad/s), what is the magnitude of the gyroscopic moment exerted on the helicopter by the rotor? Does the moment tend to cause the helicopter to roll about the x axis in the clockwise direction (as seen in the photograph) or the counterclockwise direction?

Problem 20.64

20.65 The bent bar is rigidly attached to the vertical shaft, which rotates with constant angular velocity ω_0. The disk of mass m and radius R is pinned to the bent bar and rotates with constant angular velocity ω_d relative to the bar. Determine the magnitudes of the force and couple exerted on the disk by the bar.

20.66 The bent bar is rigidly attached to the vertical shaft, which rotates with constant angular velocity ω_0. The disk of mass m and radius R is pinned to the bent bar and rotates with constant angular velocity ω_d relative to the bar. Determine the value of ω_d for which no couple is exerted on the disk by the bar.

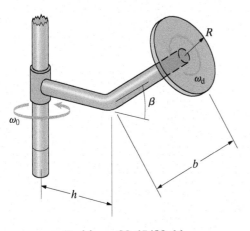

Problems 20.65/20.66

20.67 A thin circular disk undergoes moment-free steady precession. The z axis is perpendicular to the disk. Show that the disk's precession rate is $\dot{\psi} = -2\dot{\phi}/\cos\theta$. (Notice that when the nutation angle is small, the precession rate is approximately two times the spin rate.)

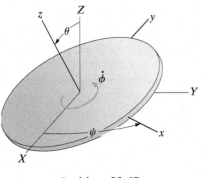

Problem 20.67

20.68 The rocket is in moment-free steady precession with nutation angle $\theta = 40°$ and spin rate $\dot{\phi} = 4$ revolutions per second. Its moments of inertia are $I_{xx} = 10,000$ kg-m^2 and $I_{zz} = 2000$ kg-m^2. What is the rocket's precession rate $\dot{\psi}$ in revolutions per second?

20.69 Sketch the body and space cones for the motion of the rocket in Problem 20.68.

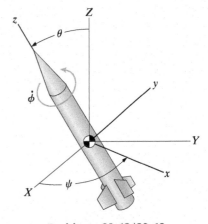

Problems 20.68/20.69

20.70 The top is in steady precession with nutation angle $\theta = 15°$ and precession rate $\dot{\psi} = 1$ revolution per second. The mass of the top is 8×10^{-4} slug, its center of mass is 1 in from the point, and its moments of inertia are $I_{xx} = 6 \times 10^{-6}$ slug-ft^2 and $I_{zz} = 2 \times 10^{-6}$ slug-ft^2. What is the spin rate $\dot{\phi}$ of the top in revolutions per second?

20.71 Suppose that top described in Problem 20.70 has a spin rate $\dot{\phi} = 15$ revolutions per second. Draw a graph of the precession rate (in revolutions per second) as a function of the nutation angle θ for values of θ from zero to 45°.

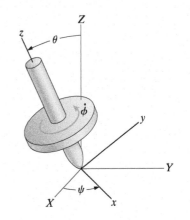

Problems 20.70/20.71

20.72 The rotor of a tumbling gyroscope can be modeled as being in moment-free steady precession. The moments of inertia of the gyroscope are $I_{xx} = I_{yy} = 0.04$ kg-m^2 and $I_{zz} = 0.18$ kg-m^2. The gyroscope's spin rate is $\dot{\phi} = 1500$ rpm and its nutation angle is $\theta = 20°$.

(a) What is the precession rate of the gyroscope in rpm?

(b) Sketch the body and space cones.

20.73 A satellite can be modeled as an 800-kg cylinder 4 m in length and 2 m in diameter. If the nutation angle is $\theta = 20°$ and the spin rate $\dot{\phi}$ is one revolution per second, what is the satellite's precession rate $\dot{\psi}$ in revolutions per second?

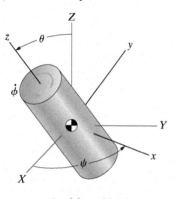

Problem 20.73

20.74* The top consists of a thin disk bonded to a slender bar. The radius of the disk is 30 mm and its mass is 0.008 kg. The length of the bar is 80 mm and its mass is negligible compared to the disk. When the top is in steady precession with a nutation angle of 10°, the precession rate is observed to be 2 revolutions per second in the same direction the top is spinning. What is the top's spin rate?

Problem 20.74

20.75 Solve Problem 20.58 by treating the motion as steady precession.

20.76* The two thin disks are rigidly connected by a slender bar. The radius of the large disk is 200 mm and its mass is 4 kg. The radius of the small disk is 100 mm and its mass is 1 kg. The bar is 400 mm in length and its mass is negligible. The composite object undergoes a steady motion in which it spins about the vertical y axis through its center of mass with angular velocity ω_0. The bar is horizontal during this motion and the large disk rolls on the floor. Determine ω_0 by treating the motion as steady precession.

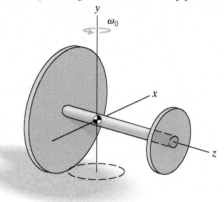

Problem 20.76

20.77* Suppose that you are testing a car and use accelerometers and gyroscopes to measure its Euler angles and their derivatives relative to a reference coordinate system. At a particular instant, $\psi = 15°$, $\theta = 4°$, $\phi = 15°$, the rates of change of the Euler angles are zero, and their second derivatives with respect to time are $\ddot{\psi} = 0$, $\ddot{\theta} = 1$ rad/s^2, and $\ddot{\phi} = -0.5$ rad/s^2. The car's principal moments of inertia, in kg-m^2, are $I_{xx} = 2200$, $I_{yy} = 480$, and $I_{zz} = 2600$. What are the components of the total moment about the car's center of mass?

20.78* If the Euler angles and their second derivatives for the car described in Problem 20.77 have the given values, but their rates of change are $\dot{\psi} = 0.2$ rad/s, $\dot{\theta} = -2$ rad/s, and $\dot{\phi} = 0$, what are the components of the total moment about the car's center of mass?

20.79* Suppose that the Euler angles of the car described in Problem 20.77 are $\psi = 40°$, $\theta = 20°$, and $\phi = 5°$, their rates of change are zero, and the components of the total moment about the car's center of mass are $\Sigma M_x = -400$ N-m, $\Sigma M_y = 200$ N-m, and $\Sigma M_z = 0$. What are the x, y, and z components of the car's angular acceleration?

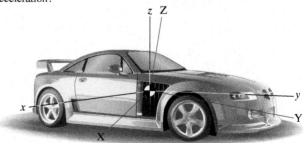

Problems 20.77–20.79

Appendix: Moments and Products of Inertia

To use the equations of motion to predict the behavior of a rigid body in three dimensions, the moments and products of inertia of the body, given by Eqs. (20.10) and (20.11), must be known. In this appendix, we demonstrate how the moments and products can be evaluated for simple objects such as slender bars and thin plates. We derive the parallel-axis theorems, which make it possible to determine the moments and products of inertia of composite objects. We also introduce the concept of principal axes, which simplifies the equations of angular motion.

Simple Objects

If we model a rigid body as a continuous distribution of mass, we can express its moments and products of inertia [Eqs. (20.10) and (20.11)] as

$$
[I] = \begin{bmatrix} I_{xx} & -I_{xy} & -I_{xz} \\ -I_{yx} & I_{yy} & -I_{yz} \\ -I_{zx} & -I_{zy} & I_{zz} \end{bmatrix}
$$

$$
= \begin{bmatrix} \displaystyle\int_m (y^2 + z^2)\, dm & -\displaystyle\int_m xy\, dm & -\displaystyle\int_m xz\, dm \\ -\displaystyle\int_m yx\, dm & \displaystyle\int_m (x^2 + z^2)\, dm & -\displaystyle\int_m yz\, dm \\ -\displaystyle\int_m zx\, dm & -\displaystyle\int_m zy\, dm & \displaystyle\int_m (x^2 + y^2)\, dm \end{bmatrix}, \quad (20.37)
$$

where x, y, and z are the coordinates of the differential element of mass dm (Fig. 20.18).

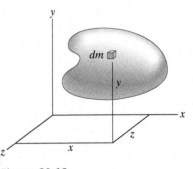

Figure 20.18
Determining the moments and products of inertia by modeling an object as a continuous distribution of mass.

Slender Bars   Let the origin of the coordinate system be at a slender bar's center of mass, with the x axis along the bar (Fig. 20.19a). The bar has length l, cross-sectional area A, and mass m. We assume that A is uniform along the length of the bar and that the material is homogeneous.

Consider a differential element of the bar of length dx at a distance x from the center of mass (Fig. 20.19b). The mass of the element is $dm = \rho A\, dx$, where ρ is the mass density. We neglect the lateral dimensions of the bar, assuming the coordinates of the differential element dm to be $(x, 0, 0)$. As a consequence of this approximation, the moment of inertia of the bar about the x axis is zero:

$$
I_{xx} = \int_m (y^2 + z^2)\, dm = 0.
$$

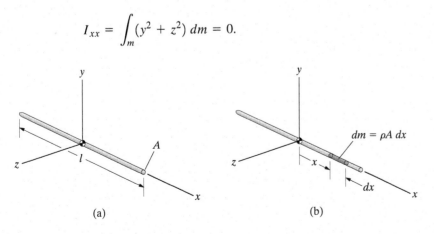

Figure 20.19
(a) A slender bar and a coordinate system with the x axis aligned with the bar.
(b) A differential element of mass of length dx.

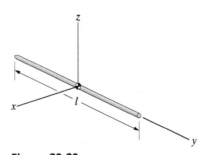

Figure 20.20
Aligning the y axis with the bar.

The moment of inertia about the y axis is

$$I_{yy} = \int_m (x^2 + z^2)\, dm = \int_{-1/2}^{1/2} \rho A x^2\, dx = \tfrac{1}{12}\rho A l^3.$$

Expressing this result in terms of the mass of the bar, $m = \rho A l$, we obtain

$$I_{yy} = \tfrac{1}{12} m l^2.$$

The moment of inertia about the z axis is equal to the moment of inertia about the y axis:

$$I_{zz} = \int_m (x^2 + y^2)\, dm = \tfrac{1}{12} m l^2.$$

Because the y and z coordinates of dm are zero, the products of inertia are zero, so the inertia matrix for the slender bar is

$$[I] = \begin{bmatrix} 0 & 0 & 0 \\ 0 & \tfrac{1}{12} m l^2 & 0 \\ 0 & 0 & \tfrac{1}{12} m l^2 \end{bmatrix}. \tag{20.38}$$

It is important to remember that the moments and products of inertia depend on the orientation of the coordinate system relative to the object. In terms of the alternative coordinate system shown in Fig. 20.20, the bar's inertia matrix is

$$[I] = \begin{bmatrix} \tfrac{1}{12} m l^2 & 0 & 0 \\ 0 & 0 & 0 \\ 0 & 0 & \tfrac{1}{12} m l^2 \end{bmatrix}.$$

Thin Plates

Suppose that a homogeneous plate of uniform thickness T, area A, and unspecified shape lies in the x–y plane (Fig. 20.21a). We can express the moments of inertia of the plate in terms of the moments of inertia of its cross-sectional area.

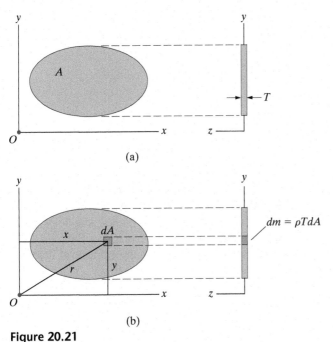

Figure 20.21
(a) A thin plate lying in the x–y plane.
(b) Obtaining a differential element of mass by projecting an
    element of area dA through the plate.

By projecting an element of area dA through the thickness T of the plate
(Fig. 20.21b), we obtain a differential element of mass $dm = \rho T\, dA$. We neglect
the thickness of the plate in calculating the moments of inertia, so the coordinates of the element dm are $(x, y, 0)$. The plate's moment of inertia about the
x axis is

$$I_{xx} = \int_m (y^2 + z^2)\, dm = \rho T \int_A y^2\, dA = \rho T I_x,$$

where I_x is the moment of inertia of the plate's cross-sectional area about the
x axis. Since the mass of the plate is $m = \rho T A$, the product $\rho T = m/A$, and
we obtain the moment of inertia in the form

$$I_{xx} = \frac{m}{A} I_x.$$

The moment of inertia about the y axis is

$$I_{yy} = \int_m (x^2 + z^2)\, dm = \rho T \int_A x^2\, dA = \frac{m}{A} I_y,$$

where I_y is the moment of inertia of the cross-sectional area about the y axis.
The moment of inertia about the z axis is

$$I_{zz} = \int_m (x^2 + y^2)\, dm = \frac{m}{A} J_O,$$

where $J_O = I_x + I_y$ is the polar moment of inertia of the cross-sectional area. The product of inertia I_{xy} is

$$I_{xy} = \int_m xy \, dm = \frac{m}{A} I_{xy}^A,$$

where

$$I_{xy}^A = \int_A xy \, dA$$

is the product of inertia of the cross-sectional area. (We use a superscript A to distinguish the product of inertia of the plate's cross-sectional area from the product of inertia of its mass.) If the cross-sectional area A is symmetric about either the x axis or the y axis, $I_{xy}^A = 0$.

Because the z coordinate of dm is zero, the products of inertia I_{xz} and I_{yz} are zero. The inertia matrix for the thin plate is thus

$$[I] = \begin{bmatrix} \dfrac{m}{A} I_x & -\dfrac{m}{A} I_{xy}^A & 0 \\[2mm] -\dfrac{m}{A} I_{xy}^A & \dfrac{m}{A} I_y & 0 \\[2mm] 0 & 0 & \dfrac{m}{A} J_O \end{bmatrix}. \tag{20.39}$$

If the moments of inertia and product of inertia of the plate's cross-sectional area are known, (Eq. 20.39) can be used to obtain the moments and products of inertia of the plate.

Parallel-Axis Theorems

Suppose that we know an object's inertia matrix $[I']$ in terms of a coordinate system $x'y'z'$ with its origin at the center of mass of the object, and we want to determine the inertia matrix $[I]$ in terms of a parallel coordinate system xyz (Fig. 20.22). Let (d_x, d_y, d_z) be the coordinates of the center of mass in the xyz coordinate system. The coordinates of a differential element of mass dm in the xyz system are given in terms of its coordinates in the $x'y'z'$ system by

$$x = x' + d_x, \qquad y = y' + d_y, \qquad z = z' + d_z. \tag{20.40}$$

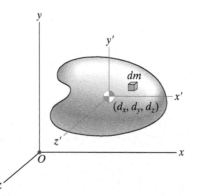

Figure 20.22
A coordinate system $x'y'z'$ with its origin at the center of mass and a parallel coordinate system xyz.

Substituting these expressions into the definition of I_{xx}, we obtain

$$I_{xx} = \int_m \left[(y')^2 + (z')^2 \right] dm + 2d_y \int_m y' \, dm$$
$$+ 2d_z \int_m z' \, dm + (d_y^2 + d_z^2) \int_m dm. \tag{20.41}$$

The first integral on the right is the object's moment of inertia about the x' axis. We can show that the second and third integrals are zero by using the definitions of the object's center of mass, expressed in terms of the $x'y'z'$ coordinate system:

$$\overline{x}' = \frac{\int_m x' \, dm}{\int_m dm}, \qquad \overline{y}' = \frac{\int_m y' \, dm}{\int_m dm}, \qquad \overline{z}' = \frac{\int_m z' \, dm}{\int_m dm}.$$

The object's center of mass is at the origin of the $x'y'z'$ system, so $\overline{x}' = \overline{y}' = \overline{z}' = 0$. Therefore, the second and third integrals on the right of Eq. (20.41) are zero, and we obtain

$$I_{xx} = I_{x'x'} + (d_y^2 + d_z^2) \, m,$$

where m is the mass of the object. Substituting Eqs. (20.40) into the definition of I_{xy}, we get

$$I_{xy} = \int_m x'y' \, dm + d_x \int_m y' \, dm + d_y \int_m x' \, dm + d_x d_y \int_m dm$$
$$= I_{x'y'} + d_x d_y m.$$

Proceeding in this way for each of the moments and products of inertia, we obtain the *parallel-axis theorems*:

$$I_{xx} = I_{x'x'} + (d_y^2 + d_z^2) \, m,$$
$$I_{yy} = I_{y'y'} + (d_x^2 + d_z^2) \, m,$$
$$I_{zz} = I_{z'z'} + (d_x^2 + d_y^2) \, m, \tag{20.42}$$
$$I_{xy} = I_{x'y'} + d_x d_y m,$$
$$I_{yz} = I_{y'z'} + d_y d_z m,$$
$$I_{zx} = I_{z'x'} + d_z d_x m.$$

If an object's inertia matrix is known in terms of a particular coordinate system, these theorems can be used to determine its inertia matrix in terms of any parallel coordinate system. They can also be used to determine the inertia matrices of composite objects.

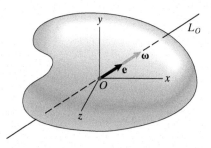

Figure 20.23
Rigid body rotating about L_O.

Moment of Inertia about an Arbitrary Axis

If we know a rigid body's inertia matrix in terms of a given coordinate system with origin O, we can determine its moment of inertia about an arbitrary axis through O. Suppose that the rigid body rotates with angular velocity $\boldsymbol{\omega}$ about an arbitrary fixed axis L_O through O, and let $\mathbf{e}$ be a unit vector with the same direction as $\boldsymbol{\omega}$ (Fig. 20.23). Then, in terms of the moment of inertia I_O about L_O, the rigid body's angular momentum about L_O is

$$H_O = I_O|\boldsymbol{\omega}|.$$

We can express the angular velocity vector as

$$\boldsymbol{\omega} = |\boldsymbol{\omega}|(e_x\mathbf{i} + e_y\mathbf{j} + e_z\mathbf{k}),$$

so that $\omega_x = |\boldsymbol{\omega}|e_x$, $\omega_y = |\boldsymbol{\omega}|e_y$, and $\omega_z = |\boldsymbol{\omega}|e_z$. Using these expressions and Eqs. (20.9), we obtain the angular momentum about L_O:

$$H_O = \mathbf{H}_O \cdot \mathbf{e} = (I_{xx}|\boldsymbol{\omega}|e_x - I_{xy}|\boldsymbol{\omega}|e_y - I_{xz}|\boldsymbol{\omega}|e_z)e_x$$

$$+ (-I_{yx}|\boldsymbol{\omega}|e_x + I_{yy}|\boldsymbol{\omega}|e_y - I_{yz}|\boldsymbol{\omega}|e_z)e_y$$

$$+ (-I_{zx}|\boldsymbol{\omega}|e_x - I_{zy}|\boldsymbol{\omega}|e_y + I_{zz}|\boldsymbol{\omega}|e_z)e_z.$$

Equating our two expressions for H_O yields

$$I_O = I_{xx}e_x^2 + I_{yy}e_y^2 + I_{zz}e_z^2 - 2I_{xy}e_xe_y - 2I_{yz}e_ye_z - 2I_{zx}e_ze_x. \quad (20.43)$$

Notice that the moment of inertia about an arbitrary axis depends on the products of inertia, in addition to the moments of inertia about the coordinate axes. If an object's inertia matrix is known, Eq. (20.43) can be used to determine the object's moment of inertia about an axis through O whose direction is specified by the unit vector $\mathbf{e}$.

Principal Axes

For *any* object and origin O, at least one coordinate system exists for which the products of inertia are zero:

$$[I] = \begin{bmatrix} I_{xx} & 0 & 0 \\ 0 & I_{yy} & 0 \\ 0 & 0 & I_{zz} \end{bmatrix}. \quad (20.44)$$

These coordinate axes are called *principal axes*, and the moments of inertia about them are called the *principal moments of inertia*.

If the inertia matrix of a rigid body is known in terms of a coordinate system $x'y'z'$ and the products of inertia are zero, then $x'y'z'$ is a set of principal axes. Suppose that the products of inertia are not zero, and we want to find a set of principal axes xyz and the corresponding principal moments of inertia (Fig. 20.24). It can be shown that the principal moments of inertia are roots of the cubic equation

$$I^3 - (I_{x'x'} + I_{y'y'} + I_{z'z'})I^2$$

$$+ (I_{x'x'}I_{y'y'} + I_{y'y'}I_{z'z'} + I_{z'z'}I_{x'x'} - I_{x'y'}^2 - I_{y'z'}^2 - I_{z'x'}^2)I \qquad (20.45)$$

$$- (I_{x'x'}I_{y'y'}I_{z'z'} - I_{x'x'}I_{y'z'}^2 - I_{y'y'}I_{z'x'}^2 - I_{z'z'}I_{x'y'}^2 - 2I_{x'y'}I_{y'z'}I_{z'x'}) = 0.$$

For each principal moment of inertia I, the vector $\mathbf{V}$ with components

$$V_{x'} = (I_{y'y'} - I)(I_{z'z'} - I) - I_{y'z'}^2,$$

$$V_{y'} = I_{x'y'}(I_{z'z'} - I) + I_{x'z'}I_{y'z'}, \qquad (20.46)$$

$$V_{z'} = I_{x'z'}(I_{y'y'} - I) + I_{x'y'}I_{y'z'}$$

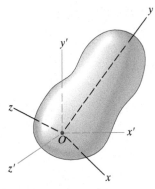

Figure 20.24
The $x'y'z'$ system with its origin at O and a set of principal axes xyz.

is parallel to the corresponding principal axis.

When the inertia matrix of an object is known in terms of a coordinate system with origin O, determining the associated principal moments of inertia and a set of principal axes involves two steps:

1. Determine the principal moments of inertia by obtaining the roots of Eq. (20.45).

2. If the three principal moments of inertia are distinct, substitute each one into Eqs. (20.46) to obtain the components of a vector parallel to the corresponding principal axis. The three principal axes can be denoted as x, y, and z arbitrarily, as long as the resulting coordinate system is right handed. If the three principal moments of inertia are equal, the moment of inertia about any axis through O has the same value, and any coordinate system with origin O is a set of principal axes. This is the case, for example, if the object is a homogeneous sphere with O at its center (Fig. 20.25a). If only two of the principal moments of inertia are equal, the third one can be substituted into Eqs. (20.46) to determine the associated principal axis. Then the moment of inertia about any axis through O that is perpendicular to the determined axis has the same value, so any coordinate system with origin O that has an axis coincident with the determined axis is a set of principal axes. This is the case when an object has an axis of rotational symmetry and O is on the axis (Fig. 20.25b).

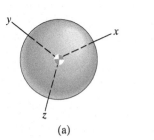

(a)

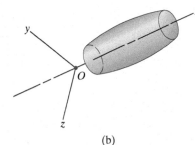

(b)

Figure 20.25
(a) A homogeneous sphere. Any coordinate system with its origin at the center is a set of principal axes.
(b) A rotationally symmetric object. The axis of symmetry is a principal axis, and any axis perpendicular to it is a principal axis.

Example 20.8 | **Parallel-Axis Theorems** (▶ *Related Problems 20.90, 20.91*)

The boom AB of the crane has a mass of 4800 kg, and the boom BC has a mass of 1600 kg and is perpendicular to AB. Modeling each boom as a slender bar and treating them as a single object, determine the moments and products of inertia of the object in terms of the coordinate system shown.

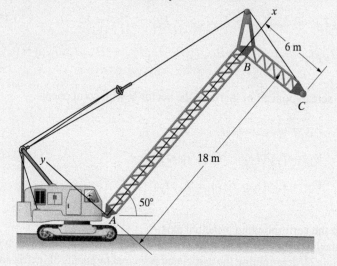

Strategy

We can apply the parallel-axis theorems to each boom to determine its moments and products of inertia in terms of the given coordinate system. The moments and products of inertia of the combined object are the sums of those for the two booms.

Solution

Boom *AB* In Fig. a, we introduce a parallel coordinate system $x'y'z'$ with its origin at the center of mass of boom AB. In terms of the $x'y'z'$ system, the inertia matrix of boom AB is

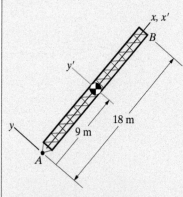

(a) Applying the parallel-axis theorems to boom AB.

$$[I'] = \begin{bmatrix} 0 & 0 & 0 \\ 0 & \frac{1}{12}ml^2 & 0 \\ 0 & 0 & \frac{1}{12}ml^2 \end{bmatrix} = \begin{bmatrix} 0 & 0 & 0 \\ 0 & \frac{1}{12}(4800)(18)^2 & 0 \\ 0 & 0 & \frac{1}{12}(4800)(18)^2 \end{bmatrix} \text{kg-m}^2.$$

The coordinates of the origin of the $x'y'z'$ system relative to the xyz system are $d_x = 9$ m, $d_y = 0$, $d_z = 0$. Applying the parallel-axis theorems, we obtain

$$I_{xx} = I_{x'x'} + (d_y^2 + d_z^2)\,m = 0,$$

$$I_{yy} = I_{y'y'} + (d_x^2 + d_z^2)\,m = \tfrac{1}{12}(4800)(18)^2 + (9)^2(4800)$$
$$= 518{,}400 \text{ kg-m}^2,$$

$$I_{zz} = I_{z'z'} + (d_x^2 + d_y^2)\,m = \tfrac{1}{12}(4800)(18)^2 + (9)^2(4800)$$
$$= 518{,}400 \text{ kg-m}^2,$$

$$I_{xy} = I_{x'y'} + d_x d_y m = 0,$$

$$I_{yz} = I_{y'z'} + d_y d_z m = 0,$$

and

$$I_{zx} = I_{z'x'} + d_z d_x m = 0.$$

Boom BC   In Fig. b we introduce a parallel coordinate system $x'y'z'$ with its origin at the center of mass of boom BC. In terms of the $x'y'z'$ system, the inertia matrix of boom BC is

$$[I'] = \begin{bmatrix} \frac{1}{12}ml^2 & 0 & 0 \\ 0 & 0 & 0 \\ 0 & 0 & \frac{1}{12}ml^2 \end{bmatrix} = \begin{bmatrix} \frac{1}{12}(1600)(6)^2 & 0 & 0 \\ 0 & 0 & 0 \\ 0 & 0 & \frac{1}{12}(1600)(6)^2 \end{bmatrix} \text{ kg-m}^2.$$

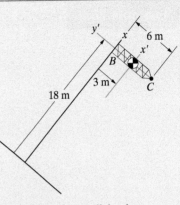

(b) Applying the parallel-axis theorems to boom BC.

The coordinates of the origin of the $x'y'z'$ system relative to the xyz system are $d_x = 18$ m, $d_y = -3$ m, $d_z = 0$. Applying the parallel-axis theorems, we obtain

$$I_{xx} = I_{x'x'} + (d_y^2 + d_z^2)\, m = \tfrac{1}{12}(1600)(6)^2 + (-3)^2(1600)$$
$$= 19{,}200 \text{ kg-m}^2,$$

$$I_{yy} = I_{y'y'} + (d_x^2 + d_z^2)\, m = 0 + (18)^2(1600) = 518{,}400 \text{ kg-m}^2,$$

$$I_{zz} = I_{z'z'} + (d_x^2 + d_y^2)\, m = \tfrac{1}{12}(1600)(6)^2 + [(18)^2 + (-3)^2](1600)$$
$$= 537{,}600 \text{ kg-m}^2,$$

$$I_{xy} = I_{x'y'} + d_x d_y m = 0 + (18)(-3)(1600) = -86{,}400 \text{ kg-m}^2,$$

$$I_{yz} = I_{y'z'} + d_y d_z m = 0,$$

and

$$I_{zx} = I_{z'x'} + d_z d_x m = 0.$$

Summing the results for the two booms, we obtain the inertia matrix for the single object:

$$[I] = \begin{bmatrix} 19{,}200 & -(-86{,}400) & 0 \\ -(-86{,}400) & 518{,}400 + 518{,}400 & 0 \\ 0 & 0 & 518{,}400 + 537{,}600 \end{bmatrix}$$

$$= \begin{bmatrix} 19{,}200 & 86{,}400 & 0 \\ 86{,}400 & 1{,}036{,}800 & 0 \\ 0 & 0 & 1{,}056{,}000 \end{bmatrix} \text{ kg-m}^2.$$

Critical Thinking

Most of the objects you will encounter in engineering will be assemblies of simpler parts, such as the crane's boom in this example. When the moments and products of inertia of the parts are known, the moments and products of inertia of the assembly can be determined using the procedure in this example. You apply the parallel-axis theorems to each part to determine its moments and products of inertia in terms of a given coordinate system, then sum the results for the parts to obtain the moments and products of inertia of the assembly in terms of that coordinate system. Notice that in order to apply the parallel-axis theorems, the moments and products of inertia of the parts must be expressed in terms of parallel coordinate systems.

| Example 20.9 | Inertia Matrix of a Plate (▶ *Related Problems 20.82, 20.83*) |

The 4-kg rectangular plate lies in the x–y plane of the body-fixed coordinate system.

(a) Determine the plate's moments and products of inertia.

(b) Determine the plate's moment of inertia about the diagonal axis L_O.

(c) If the plate is rotating about the fixed point O with angular velocity $\omega = 4\mathbf{i} - 2\mathbf{j}$ (rad/s), what is the plate's angular momentum about O?

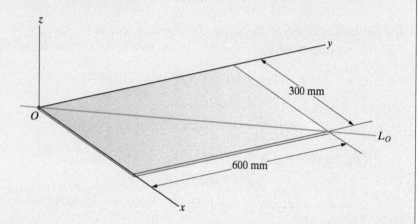

Strategy

(a) We can obtain the moments and products of inertia of the plate's rectangular area from Appendix B and use Eq. (20.39) to obtain the moments and products of inertia of the plate.

(b) Once we know the moments and products of inertia, we can use Eq. (20.43) to determine the moment of inertia about L_O.

(c) The angular momentum about O is given by Eqs. (20.9).

Solution

(a) From Appendix B, the moments of inertia of the plate's cross-sectional area are as follows (Fig. a):

$$I_x = \tfrac{1}{3}bh^3, \quad I_{xy}^A = \tfrac{1}{4}b^2h^2,$$
$$I_y = \tfrac{1}{3}hb^3, \quad J_O = \tfrac{1}{3}(bh^3 + hb^3).$$

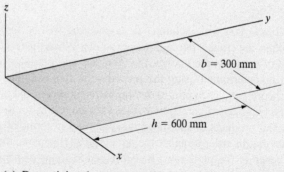

(a) Determining the moments of inertia of the plate's area.

Therefore, the moments and products of inertia of the plate are

$$I_{xx} = \frac{m}{A}I_x = \frac{(4)}{(0.3)(0.6)}\left(\frac{1}{3}\right)(0.3)(0.6)^3 = 0.48 \text{ kg-m}^2,$$

$$I_{yy} = \frac{m}{A}I_y = \frac{(4)}{(0.3)(0.6)}\left(\frac{1}{3}\right)(0.6)(0.3)^3 = 0.12 \text{ kg-m}^2,$$

$$I_{xy} = \frac{m}{A}I_{xy}^A = \frac{(4)}{(0.3)(0.6)}\left(\frac{1}{4}\right)(0.3)^2(0.6)^2 = 0.18 \text{ kg-m}^2,$$

$$I_{zz} = \frac{m}{A}J_O = \frac{(4)}{(0.3)(0.6)}\left(\frac{1}{3}\right)[(0.3)(0.6)^3 + (0.6)(0.3)^3] = 0.60 \text{ kg-m}^2,$$

and

$$I_{xz} = I_{yz} = 0.$$

(b) To apply Eq. (20.43), we must determine the components of a unit vector parallel to L_O:

$$\mathbf{e} = \frac{300\mathbf{i} + 600\mathbf{j}}{|300\mathbf{i} + 600\mathbf{j}|} = 0.447\mathbf{i} + 0.894\mathbf{j}.$$

The moment of inertia about L_O is

$$I_O = I_{xx}e_x^2 + I_{yy}e_y^2 + I_{zz}e_z^2 - 2I_{xy}e_xe_y - 2I_{yz}e_ye_z - 2I_{zx}e_ze_x$$
$$= (0.48)(0.447)^2 + (0.12)(0.894)^2 - 2(0.18)(0.447)(0.894)$$
$$= 0.048 \text{ kg-m}^2.$$

(c) The plate's angular momentum about O is

$$\begin{bmatrix} H_{Ox} \\ H_{Oy} \\ H_{Oz} \end{bmatrix} = \begin{bmatrix} I_{xx} & -I_{xy} & -I_{xz} \\ -I_{yx} & I_{yy} & -I_{yz} \\ -I_{zx} & -I_{zy} & I_{zz} \end{bmatrix}\begin{bmatrix} \omega_x \\ \omega_y \\ \omega_z \end{bmatrix}$$

$$= \begin{bmatrix} 0.48 & -0.18 & 0 \\ -0.18 & 0.12 & 0 \\ 0 & 0 & 0.6 \end{bmatrix}\begin{bmatrix} 4 \\ -2 \\ 0 \end{bmatrix}$$

$$= \begin{bmatrix} 2.28 \\ -0.96 \\ 0 \end{bmatrix} \text{ kg-m}^2/\text{s}.$$

Critical Thinking

Could you use the results of part (a) and the parallel-axis theorems to determine the moments and products of inertia of the plate in terms of any other parallel coordinate system? Remember that the parallel-axis theorems relate the moments and products of inertia of an object in terms of a coordinate system with its origin *at the center of mass* to those in terms of a parallel coordinate system. Therefore, you would first need to apply the parallel-axis theorems to the results of part (a) to determine the moments and products of inertia of the plate in terms of a parallel coordinate system with its origin at the center of mass.

| Example 20.10 | Principal Axes and Moments of Inertia (▶ *Related Problem 20.96*) |

In terms of a coordinate system $x'y'z'$ with its origin at the center of mass, the inertia matrix of a rigid body is

$$[I'] = \begin{bmatrix} 4 & -2 & 1 \\ -2 & 2 & -1 \\ 1 & -1 & 3 \end{bmatrix} \text{kg-m}^2.$$

Determine the principal moments of inertia and the directions of a set of principal axes relative to the $x'y'z'$ system.

Strategy

The principal moments of inertia are the roots of Eq. (20.45). For each principal moment of inertia, Eqs. (20.46) give the components of a vector that is parallel to the corresponding principal axis.

Solution

Substituting the moments and products of inertia into Eq. (20.45), we obtain the equation

$$I^3 - 9I^2 + 20I - 10 = 0. \tag{1}$$

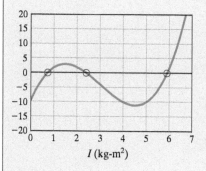

The graph shows the value of the left side of this equation as a function of I. The three roots, which are the values of the principal moments of inertia in kg-m², are $I_1 = 0.708$, $I_2 = 2.397$, and $I_3 = 5.895$.

Substituting the principal moment of inertia $I_1 = 0.708$ kg-m² into Eqs. (20.46) and dividing the resulting vector **V** by its magnitude, we obtain a unit vector parallel to the corresponding principal axis:

$$\mathbf{e}_1 = 0.473\mathbf{i} + 0.864\mathbf{j} + 0.171\mathbf{k}.$$

Substituting $I_2 = 2.397$ kg-m² into Eqs. (20.46), we obtain the unit vector

$$\mathbf{e}_2 = -0.458\mathbf{i} + 0.076\mathbf{j} + 0.886\mathbf{k},$$

and substituting $I_3 = 5.895$ kg-m² into Eqs. (20.46), we obtain the unit vector

$$\mathbf{e}_3 = 0.753\mathbf{i} - 0.497\mathbf{j} + 0.432\mathbf{k}.$$

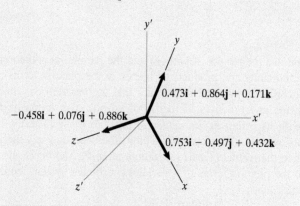

We have determined the principal moments of inertia and the components of unit vectors parallel to the corresponding principal axes. We show the principal axes, arbitrarily designating them so that $I_{xx} = 5.895$ kg-m^2, $I_{yy} = 0.708$ kg-m^2, and $I_{zz} = 2.397$ kg-m^2.

Critical Thinking

Many programmable calculators and computer programs are available that will determine roots of nonlinear algebraic equations. That is how we obtained the precise values of the roots of Eq. (1) in this example. We needed to determine all three roots of the cubic equation. Unless the program you use is designed to determine all of the roots of an Nth-order equation, you may find that the program continues to converge on a root that you have already determined instead of one you are still seeking. You can avoid this in a simple way. In this example, you would begin by asking the program to determine a root of Eq. (1). Suppose that the root obtained is 0.708. (This value is rounded off to three significant digits. In your computations, retain as much accuracy as your computer provides.) Then seek a root of the equation

$$\frac{I^3 - 9I^2 + 20I - 10}{I - 0.708} = 0.$$

In this way, you are "dividing out" the root you have determined. If the next root you obtain is 2.397, obtain the final root by seeking a root of the equation

$$\frac{I^3 - 9I^2 + 20I - 10}{(I - 0.708)(I - 2.397)} = 0.$$

Problems

20.80 The mass of the bar is 6 kg. Determine the moments and products of inertia of the bar in terms of the coordinate system shown.

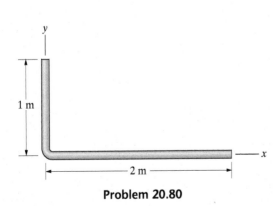

Problem 20.80

20.81 The object consists of two 1-kg vertical slender bars welded to a 4-kg horizontal slender bar. Determine its moments and products of inertia in terms of the coordinate system shown.

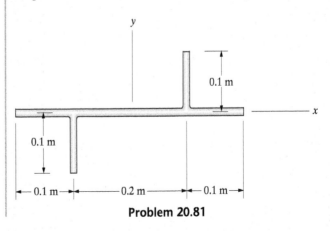

Problem 20.81

▶ **20.82** The 4-kg thin rectangular plate lies in the x–y plane. Determine the moments and products of inertia of the plate in terms of the coordinate system shown. (See Example 20.9.)

▶ **20.83** If the 4-kg plate is rotating with angular velocity $\omega = 6\mathbf{i} + 4\mathbf{j} - 2\mathbf{k}$ (rad/s), what is its angular momentum about its center of mass? (See Example 20.9.)

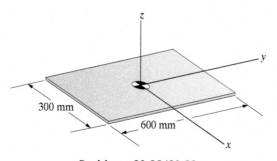

Problems 20.82/20.83

20.84 The 30-lb triangular plate lies in the x–y plane. Determine the moments and products of inertia of the plate in terms of the coordinate system shown.

20.85 The 30-lb triangular plate lies in the x–y plane.

(a) Determine its moments and products of inertia in terms of a parallel coordinate system $x'y'z'$ with its origin at the plate's center of mass.

(b) If the plate is rotating with angular velocity $\omega = 20\mathbf{i} - 12\mathbf{j} + 16\mathbf{k}$ (rad/s), what is its angular momentum about its center of mass?

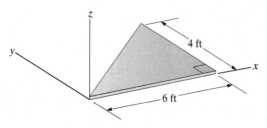

Problems 20.84/20.85

20.86 Determine the inertia matrix of the 2.4-kg steel plate in terms of the coordinate system shown.

20.87 The mass of the steel plate is 2.4 kg.

(a) Determine its moments and products of inertia in terms of a parallel coordinate system $x'y'z'$ with its origin at the plate's center of mass.

(b) If the plate is rotating with angular velocity $\omega = 20\mathbf{i} + 10\mathbf{j} - 10\mathbf{k}$ (rad/s), what is its angular momentum about its center of mass?

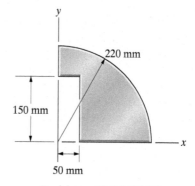

Problems 20.86/20.87

20.88 The slender bar of mass m rotates about the fixed point O with angular velocity $\omega = \omega_y\mathbf{j} + \omega_z\mathbf{k}$. Determine its angular momentum (a) about its center of mass and (b) about O.

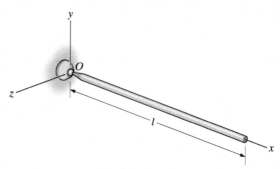

Problem 20.88

20.89 The slender bar of mass m is parallel to the x axis. If the coordinate system is body fixed and its angular velocity about the fixed point O is $\omega = \omega_y\mathbf{j}$, what is the bar's angular momentum about O?

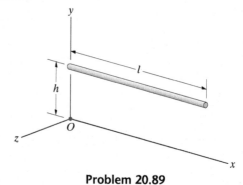

Problem 20.89

▶ 20.90 In Example 20.8, the moments and products of inertia of the object consisting of the booms *AB* and *BC* were determined in terms of the *xyz* coordinate system shown in the example. Determine the moments and products of inertia of the object in terms of a parallel coordinate system *x'y'z'* with its origin at the center of mass of the object.

▶ 20.91 Suppose that the crane described in Example 20.8 undergoes a rigid-body rotation about the vertical axis at 0.1 rad/s in the counterclockwise direction when viewed from above.

(a) What is the crane's angular velocity vector ω in terms of the body-fixed *xyz* coordinate system shown in the example?

(b) What is the angular momentum of the object consisting of the booms *AB* and *BC* about its *center of mass*?

20.92 A 3-kg slender bar is rigidly attached to a 2-kg thin circular disk. In terms of the body-fixed coordinate system shown, the angular velocity of the composite object is $\omega = 100\mathbf{i} - 4\mathbf{j} + 6\mathbf{k}$ (rad/s). What is the object's angular momentum about its center of mass?

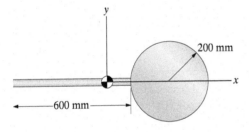

Problem 20.92

20.93* The mass of the homogeneous slender bar is *m*. If the bar rotates with angular velocity $\omega = \omega_0(24\mathbf{i} + 12\mathbf{j} - 6\mathbf{k})$, what is its angular momentum about its center of mass?

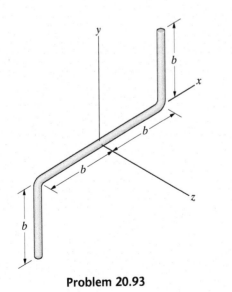

Problem 20.93

20.94* The 8-kg homogeneous slender bar has ball-and-socket supports at *A* and *B*.

(a) What is the bar's moment of inertia about the axis *AB*?

(b) If the bar rotates about the axis *AB* at 4 rad/s, what is the magnitude of its angular momentum about its axis of rotation?

20.95* The 8-kg homogeneous slender bar is released from rest in the position shown. (The *x–z* plane is horizontal.) What is the magnitude of the bar's angular acceleration about the axis *AB* at the instant of release?

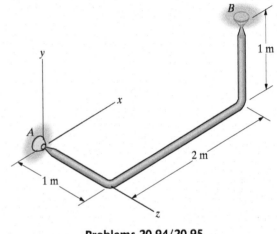

Problems 20.94/20.95

▶ 20.96 In terms of a coordinate system *x'y'z'* with its origin at the center of mass, the inertia matrix of a rigid body is

$$[I'] = \begin{bmatrix} 20 & 10 & -10 \\ 10 & 60 & 0 \\ -10 & 0 & 80 \end{bmatrix} \text{ kg-m}^2.$$

Determine the principal moments of inertia and unit vectors parallel to the corresponding principal axes. (See Example 20.10.)

20.97 For the object in Problem 20.81, determine the principal moments of inertia and unit vectors parallel to the corresponding principal axes. Draw a sketch of the object showing the principal axes.

20.98 The 1-kg, 1-m long slender bar lies in the x'–y' plane. Its inertia matrix in kg-m^2 is

$$[I'] = \begin{bmatrix} \frac{1}{12}\sin^2\beta & -\frac{1}{12}\sin\beta\cos\beta & 0 \\ -\frac{1}{12}\sin\beta\cos\beta & \frac{1}{12}\cos^2\beta & 0 \\ 0 & 0 & \frac{1}{12} \end{bmatrix}.$$

Use Eqs. (20.45) and (20.46) to determine the principal moments of inertia and unit vectors parallel to the corresponding principal axes.

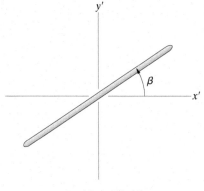

Problem 20.98

20.99* The mass of the homogeneous thin plate is 3 slugs. For the coordinate system shown, determine the plate's principal moments of inertia and unit vectors parallel to the corresponding principal axes.

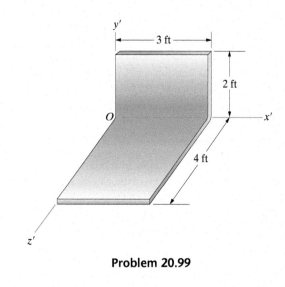

Problem 20.99

Review Problems

20.100 The disk is pinned to the horizontal shaft and rotates relative to it with angular velocity ω_d. Relative to an earth-fixed reference frame, the vertical shaft rotates with angular velocity ω_0.

(a) Determine the disk's angular velocity vector $\boldsymbol{\omega}$ relative to the earth-fixed reference frame.

(b) What is the velocity of point A of the disk relative to the earth-fixed reference frame?

20.101 The disk is pinned to the horizontal shaft and rotates relative to it with constant angular velocity ω_d. Relative to an earth-fixed reference frame, the vertical shaft rotates with constant angular velocity ω_0. What is the acceleration of point A of the disk relative to the earth-fixed reference frame?

20.102 The cone is connected by a ball-and-socket joint at its vertex to a 100-mm post. The radius of its base is 100 mm, and the base rolls on the floor. The velocity of the center of the base is $\mathbf{v}_C = 2\mathbf{k}$ (m/s).

(a) What is the cone's angular velocity vector $\boldsymbol{\omega}$?

(b) What is the velocity of point A?

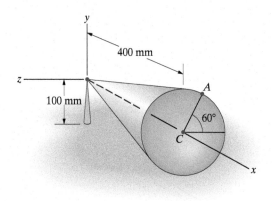

Problem 20.102

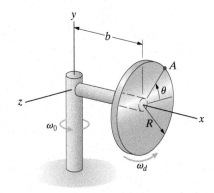

Problems 20.100/20.101

20.103 The mechanism shown is a type of universal joint called a yoke and spider. The axis L lies in the x–z plane. Determine the angular velocity ω_L and the angular velocity vector $\boldsymbol{\omega}_S$ of the cross-shaped "spider" in terms of the angular velocity ω_R at the instant shown.

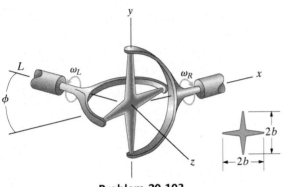

Problem 20.103

20.104 The inertia matrix of a rigid body in terms of a body-fixed coordinate system with its origin at the center of mass is

$$[I] = \begin{bmatrix} 4 & 1 & -1 \\ 1 & 2 & 0 \\ -1 & 0 & 6 \end{bmatrix} \text{kg-m}^2.$$

If the rigid body's angular velocity is $\boldsymbol{\omega} = 10\mathbf{i} - 5\mathbf{j} + 10\mathbf{k}$ (rad/s), what is its angular momentum about its center of mass?

20.105 What is the moment of inertia of the rigid body described in Problem 20.104 about the axis that passes through the origin and the point $(4, -4, 7)$ m?

Strategy: Determine the components of a unit vector parallel to the axis and use Eq. (20.43).

20.106 Determine the inertia matrix of the 0.6-slug thin plate in terms of the coordinate system shown.

20.107 At $t = 0$, the 0.6-slug thin plate has angular velocity $\boldsymbol{\omega} = 10\mathbf{i} + 10\mathbf{j}$ (rad/m) and is subjected to the force $\mathbf{F} = -10\mathbf{k}$ (lb) acting at the point $(0, 6, 0)$ in. No other forces or couples act on the plate. What are the components of its angular acceleration at that instant?

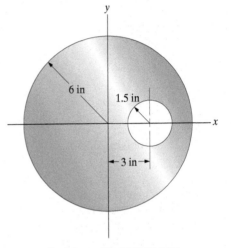

Problems 20.106/20.107

20.108 The inertia matrix of a rigid body in terms of a body-fixed coordinate system with its origin at the center of mass is

$$[I] = \begin{bmatrix} 4 & 1 & -1 \\ 1 & 2 & 0 \\ -1 & 0 & 6 \end{bmatrix} \text{kg-m}^2.$$

If the rigid body's angular velocity is $\boldsymbol{\omega} = 10\mathbf{i} - 5\mathbf{j} + 10\mathbf{k}$ (rad/s) and its angular acceleration is zero, what are the components of the total moment about its center of mass?

20.109 If the total moment about the center of mass of the rigid body described in Problem 20.108 is zero, what are the components of its angular acceleration?

20.110 The slender bar of length l and mass m is pinned to the L-shaped bar at O. The L-shaped bar rotates about the vertical axis with a constant angular velocity ω_0. Determine the value of ω_0 necessary for the bar to remain at a constant angle β relative to the vertical.

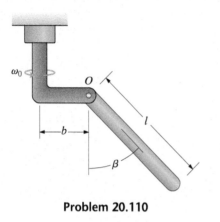

Problem 20.110

20.111 A slender bar of length l and mass m is rigidly attached to the center of a thin circular disk of radius R and mass m. The composite object undergoes a motion in which the bar rotates in the horizontal plane with constant angular velocity ω_0 about the center of mass of the composite object and the disk rolls on the floor. Show that $\omega_0 = 2\sqrt{g/R}$.

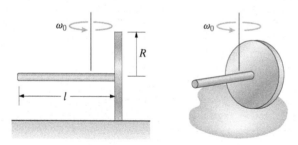

Problem 20.111

20.112* The thin plate of mass m spins about a vertical axis with the plane of the plate perpendicular to the floor. The corner of the plate at O rests in an indentation, so that it remains at the same point on the floor. The plate rotates with constant angular velocity ω_0 and the angle β is constant.

(a) Show that the angular velocity ω_0 is related to the angle β by

$$\frac{h\omega_0^2}{g} = \frac{2\cos\beta - \sin\beta}{\sin^2\beta - 2\sin\beta\cos\beta - \cos^2\beta}.$$

(b) The equation you obtained in (a) indicates that $\omega_0 = 0$ when $2\cos\beta - \sin\beta = 0$. What is the interpretation of this result?

20.113* In Problem 20.112, determine the range of values of the angle β for which the plate will remain in the steady motion described.

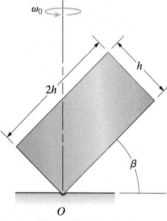

Problems 20.112/20.113

20.114 Arm BC has a mass of 12 kg, and its moments and products of inertia, in terms of the coordinate system shown, are $I_{xx} = 0.03$ kg-m^2, $I_{yy} = I_{zz} = 4$ kg-m^2, and $I_{xy} = I_{yz} = I_{xz} = 0$. At the instant shown, arm AB is rotating in the horizontal plane with a constant angular velocity of 1 rad/s in the counterclockwise direction viewed from above. Relative to arm AB, arm BC is rotating about the z axis with a constant angular velocity of 2 rad/s. Determine the force and couple exerted on arm BC at B.

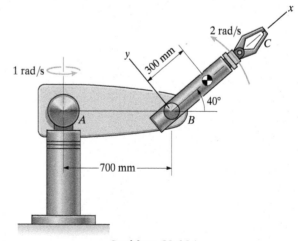

Problem 20.114

20.115 Suppose that you throw a football in a wobbly spiral with a nutation angle of 25°. The football's moments of inertia are $I_{xx} = I_{yy} = 0.003$ slug-ft^2 and $I_{zz} = 0.001$ slug-ft^2. If the spin rate is $\dot{\phi} = 4$ revolutions per second, what is the magnitude of the precession rate (the rate at which the football wobbles)?

20.116 Sketch the body and space cones for the motion of the football in Problem 20.115.

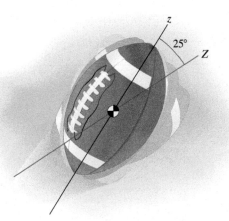

Problems 20.115/20.116

20.117 The mass of the homogeneous thin plate is 1 kg. For the coordinate system shown, determine the plate's principal moments of inertia and the directions of unit vectors parallel to the corresponding principal axes.

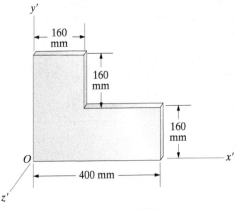

Problem 20.117

20.118 The airplane's principal moments of inertia, in slug-ft^2, are $I_{xx} = 8000$, $I_{yy} = 48,000$, and $I_{zz} = 50,000$.

(a) The airplane begins in the reference position shown and maneuvers into the orientation $\psi = \theta = \phi = 45°$. Draw a sketch showing the plane's orientation relative to the XYZ system.

(b) If the airplane is in the orientation described in (a), the rates of change of the Euler angles are $\dot{\psi} = 0$, $\dot{\theta} = 0.2$ rad/s, and $\dot{\phi} = 0.2$ rad/s, and the second derivatives of the angles with respect to time are zero, what are the components of the total moment about the airplane's center of mass?

20.119 What are the x, y, and z components of the angular acceleration of the airplane described in Problem 20.118?

20.120 If the orientation of the airplane in Problem 20.118 is $\psi = 45°$, $\theta = 60°$, and $\phi = 45°$, the rates of change of the Euler angles are $\dot{\psi} = 0$, $\dot{\theta} = 0.2$ rad/s, and $\dot{\phi} = 0.1$ rad/s, and the components of the total moment about the center of mass of the plane are $\Sigma M_x = 400$ ft-lb, $\Sigma M_y = 1200$ ft-lb, and $\Sigma M_z = 0$, what are the x, y, and z components of the airplane's angular acceleration?

Problems 20.118–20.120

CHAPTER
21

Vibrations

Vibrations have been of concern in engineering since the beginning of the industrial revolution. Beginning with the development of electromechanical devices capable of creating and measuring mechanical vibrations, engineering applications of vibrations have included the various areas of acoustics, from architectural acoustics to earthquake detection and analysis. We consider vibrating systems that have one degree of freedom—that is, the position, or configuration, of a system can be specified by a single variable. The fundamental concepts we introduce, including amplitude, frequency, period, damping, and resonance, are also used in the analysis of systems with multiple degrees of freedom.

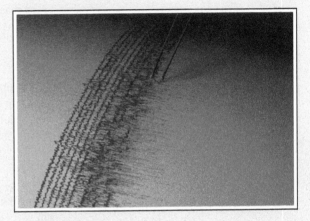

◄ Earthquakes—natural vibrations of the earth—pose a major challenge to engineering analysis and design. In this chapter we analyze the vibrations of simple mechanical systems.

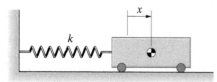

(a)

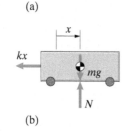

(b)

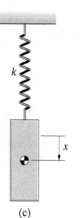

(c)

Figure 21.1
(a) The spring–mass oscillator has one degree of freedom.
(b) Free-body diagram of the mass.
(c) Suspending the mass.

21.1 Conservative Systems

BACKGROUND

We begin by presenting different examples of one-degree-of-freedom systems subjected to conservative forces, demonstrating that their motions are described by the same differential equation. We then examine solutions of this equation and use them to describe the vibrations of one-degree-of-freedom conservative systems.

Examples

The *spring–mass oscillator* (Fig. 21.1a) is the simplest example of a one-degree-of-freedom vibrating system. A single coordinate x measuring the displacement of the mass relative to a reference point is sufficient to specify the position of the system. We draw the free-body diagram of the mass in Fig. 21.1b, neglecting friction and assuming that the spring is unstretched when $x = 0$. Applying Newton's second law, we can write the equation describing the horizontal motion of the mass as

$$\frac{d^2x}{dt^2} + \frac{k}{m}x = 0. \tag{21.1}$$

We can obtain this equation by a different method that is very useful. The only force that does work on the mass, the force exerted by the spring, is conservative, which means that the sum of the kinetic and potential energies is constant:

$$\frac{1}{2}m\left(\frac{dx}{dt}\right)^2 + \frac{1}{2}kx^2 = \text{constant.}$$

Taking the derivative of this equation with respect to time, we can write the result as

$$\left(\frac{dx}{dt}\right)\left(\frac{d^2x}{dt^2} + \frac{k}{m}x\right) = 0,$$

again obtaining Eq. (21.1).

Suppose that the mass is suspended from the spring, as shown in Fig. 21.1c, and undergoes vertical motion. If the spring is unstretched when $x = 0$, it is easy to confirm that the equation of motion is

$$\frac{d^2x}{dt^2} + \frac{k}{m}x = g.$$

If the suspended mass is stationary, the magnitude of the force exerted by the spring must equal the weight $(kx = mg)$, so the equilibrium position is $x = mg/k$. (Notice that we can also determine the equilibrium position by setting the acceleration equal to zero in the equation of motion.) Let us introduce a new variable $\tilde{x}$ that measures the position of the mass relative to its equilibrium position: $\tilde{x} = x - mg/k$. Writing the equation of motion in terms of this variable, we obtain

$$\frac{d^2\tilde{x}}{dt^2} + \frac{k}{m}\tilde{x} = 0, \tag{21.2}$$

which is identical to Eq. (21.1). The vertical motion of the mass in Fig. 21.1c relative to its equilibrium position is described by the same equation that describes the horizontal motion of the mass in Fig. 21.1a relative to its equilibrium position.

Now let's consider a different one-degree-of-freedom system. If we rotate the slender bar in Fig. 21.2a through some angle and release it, it will oscillate back and forth. (An object swinging from a fixed point is called a *pendulum*.) There is only one degree of freedom, since θ specifies the bar's position. Drawing the free-body diagram of the bar (Fig. 21.2b) and writing the equation of angular motion about A yields

$$\frac{d^2\theta}{dt^2} + \frac{3g}{2l} \sin\theta = 0. \tag{21.3}$$

We can also obtain this equation by using conservation of energy. The bar's kinetic energy is $T = \frac{1}{2}I_A(d\theta/dt)^2$. If we place the datum at the level of point A (Fig. 21.2b), the potential energy associated with the bar's weight is $V = -mg(\frac{1}{2}l\cos\theta)$, so

$$T + V = \frac{1}{2}\left(\frac{1}{3}ml^2\right)\left(\frac{d\theta}{dt}\right)^2 - \frac{1}{2}mgl\cos\theta = \text{constant}.$$

Taking the derivative of this equation with respect to time and writing the result in the form

$$\left(\frac{d\theta}{dt}\right)\left(\frac{d^2\theta}{dt^2} + \frac{3g}{2l}\sin\theta\right) = 0,$$

we obtain Eq. (21.3).

Note that Eq. (21.3) does not have the same form as Eq. (21.1). However, if we express $\sin\theta$ in terms of its Taylor series,

$$\sin\theta = \theta - \frac{1}{6}\theta^3 + \frac{1}{120}\theta^5 + \cdots,$$

and assume that θ remains small enough to approximate $\sin\theta$ by θ, then Eq. (21.3) becomes identical in form to Eq. (21.1):

$$\frac{d^2\theta}{dt^2} + \frac{3g}{2l}\theta = 0. \tag{21.4}$$

Our analyses of the spring–mass oscillator and the pendulum resulted in equations of motion that are identical in form. To accomplish this in the case of the suspended spring–mass oscillator, we had to express the equation of motion in terms of displacement relative to the equilibrium position. In the case of the pendulum, we needed to assume that the motions were small. But within those restrictions, the form of equation we obtained describes the motions of many one-degree-of-freedom conservative systems.

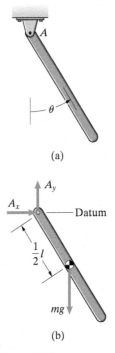

(a)

(b)

Figure 21.2
(a) A pendulum consisting of a slender bar.
(b) Free-body diagram of the bar.

Solutions

Let us consider the differential equation

$$\frac{d^2x}{dt^2} + \omega^2 x = 0, \tag{21.5}$$

where ω is a constant. We have seen that with $\omega^2 = k/m$, this equation describes the motion of a spring–mass oscillator, and with $\omega^2 = 3g/2l$, it describes small motions of a suspended slender bar. Equation (21.5) is an *ordinary differential equation*, because it is expressed in terms of ordinary (not partial) derivatives of the dependent variable x with respect to the independent variable t. Also, Eq. (21.5) is *linear*, meaning that there are no nonlinear terms in x or its derivatives, and it is *homogeneous*, meaning that each term contains x or one of its derivatives. Finally, Eq. (21.5) has *constant coefficients*, meaning that the coefficients multiplying the dependent variable x or its derivative in each term do not depend on the independent variable t. The standard approach to solving a differential equation of this kind is to assume that the solution is of the form

$$x = Ce^{\lambda t}, \tag{21.6}$$

where C and λ are constants. Substituting this expression into Eq. (21.5) yields

$$Ce^{\lambda t}(\lambda^2 + \omega^2) = 0.$$

This equation is satisfied for any value of the constant C if $\lambda = i\omega$ or $\lambda = -i\omega$, where $i = \sqrt{-1}$, so there are two nontrivial solutions of the form of Eq. (21.6), which we write as

$$x = Ce^{i\omega t} + De^{-i\omega t}. \tag{21.7}$$

By using *Euler's identity* $e^{i\theta} = \cos\theta + i\sin\theta$, we can express Eq. (21.7) in the alternative form

$$x = A\sin\omega t + B\cos\omega t, \tag{21.8}$$

where A and B are arbitrary constants.

Although in practical applications Eq. (21.8) is usually the most convenient form of the solution of Eq. (21.5), we can describe the properties of the solution more easily by expressing it in the form

$$x = E\sin(\omega t - \phi), \tag{21.9}$$

where E and ϕ are constants. To show that this solution is equivalent to Eq. (21.8), we use the identity

$$E\sin(\omega t - \phi) = E(\sin\omega t \cos\phi - \cos\omega t \sin\phi)$$
$$= (E\cos\phi)\sin\omega t + (-E\sin\phi)\cos\omega t.$$

This expression is identical to Eq. (21.8) if the constants A and B are related to E and ϕ by

$$A = E\cos\phi \quad \text{and} \quad B = -E\sin\phi. \tag{21.10}$$

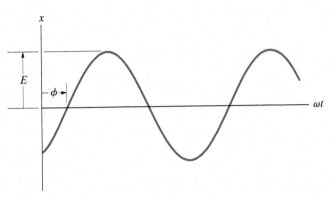

Figure 21.3
Graph of x as a function of ωt.

Equation (21.9) clearly demonstrates the oscillatory nature of the solution of Eq. (21.5). Called *simple harmonic motion*, it describes a sinusoidal function of ωt (Fig. 21.3). The positive constant E is called the *amplitude* of the vibration. By squaring Eqs. (21.10) and adding them, we obtain a relation between the amplitude and the constants A and B:

$$E = \sqrt{A^2 + B^2}. \tag{21.11}$$

Equation (21.9) can be interpreted in terms of the uniform motion of a point along a circular path. We draw a circle whose radius equals the amplitude (Fig. 21.4) and assume that the line from O to P rotates in the counterclockwise direction with constant angular velocity ω. If we choose the position of P at $t = 0$ as shown, the projection of the line OP onto the vertical axis is $E \sin(\omega t - \phi)$. Thus, there is a one-to-one correspondence between the circular motion of P and Eq. (21.9). Point P makes one complete revolution, or *cycle*, during the time required for the angle ωt to increase by 2π radians. The time $\tau = 2\pi/\omega$ required for one cycle is called the *period* of the vibration. Since τ is the time required for one cycle, its inverse $f = 1/\tau$ is the number of cycles per unit time, or *frequency* of the vibration. The frequency is usually

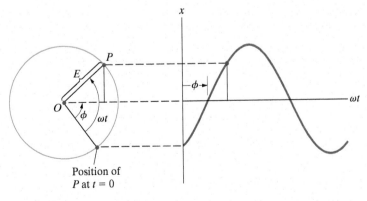

Position of
P at $t = 0$

Figure 21.4
Correspondence of simple harmonic motion with circular motion of a point.

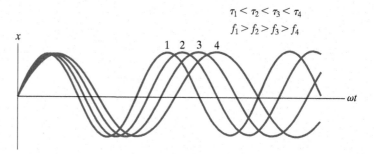

Figure 21.5
Effect of increasing the period (decreasing the frequency) of simple harmonic motion.

expressed in cycles per second, or *Hertz* (Hz). The effect of changing the period and frequency is illustrated in Fig. 21.5.

We see that the period and frequency are given by

$$\tau = \frac{2\pi}{\omega}, \tag{21.12}$$

$$f = \frac{\omega}{2\pi}. \tag{21.13}$$

A system's period and frequency are determined by its physical properties, and do not depend on the functional form in which its motion is expressed. The frequency f is the number of revolutions the point P moves around the circular path in Fig. 21.4 per unit time, so $\omega = 2\pi f$ is the number of radians per unit time. Therefore, ω *is also a measure of the frequency* and is expressed in radians per second (rad/s).

Suppose that Eq. (21.9) describes the displacement of the spring–mass oscillator in Fig. 21.1a, so that $\omega^2 = k/m$. Then the kinetic energy of the mass is

$$T = \tfrac{1}{2}m\left(\frac{dx}{dt}\right)^2 = \tfrac{1}{2}mE^2\omega^2\cos^2(\omega t - \phi), \tag{21.14}$$

and the potential energy of the spring is

$$V = \tfrac{1}{2}kx^2 = \tfrac{1}{2}mE^2\omega^2\sin^2(\omega t - \phi). \tag{21.15}$$

The sum of the kinetic and potential energies, $T + V = \tfrac{1}{2}mE^2\omega^2$, is constant (Fig. 21.6). As the system vibrates, its total energy oscillates between kinetic and potential energy. Notice that the total energy is proportional to the square of the amplitude and the square of the natural frequency.

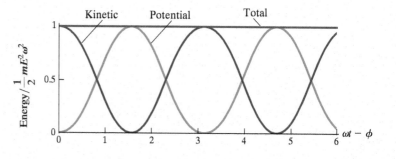

Figure 21.6
Kinetic, potential, and total energies of a spring–mass oscillator.

RESULTS

A system is said to have one *degree of freedom* if its position, or configuration, can be specified by one variable. A conservative system is one in which forces that do work are conservative.

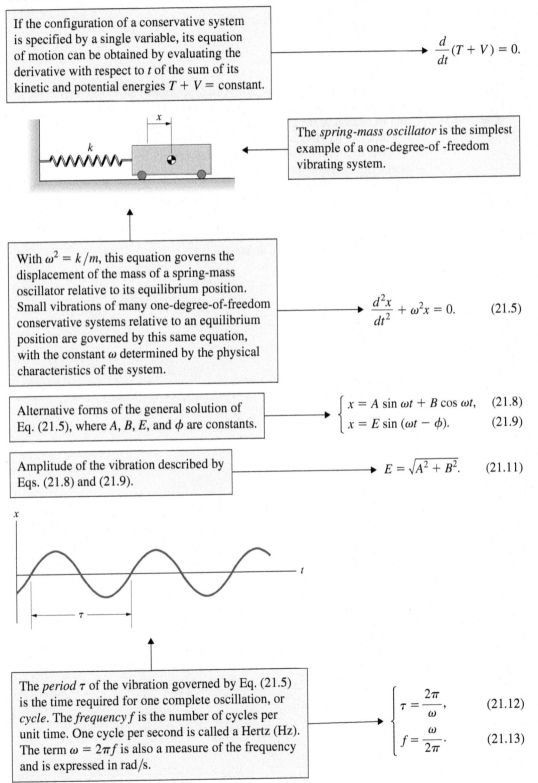

If the configuration of a conservative system is specified by a single variable, its equation of motion can be obtained by evaluating the derivative with respect to t of the sum of its kinetic and potential energies $T + V = $ constant.

$$\frac{d}{dt}(T + V) = 0.$$

The *spring-mass oscillator* is the simplest example of a one-degree-of -freedom vibrating system.

With $\omega^2 = k/m$, this equation governs the displacement of the mass of a spring-mass oscillator relative to its equilibrium position. Small vibrations of many one-degree-of-freedom conservative systems relative to an equilibrium position are governed by this same equation, with the constant ω determined by the physical characteristics of the system.

$$\frac{d^2x}{dt^2} + \omega^2x = 0. \qquad (21.5)$$

Alternative forms of the general solution of Eq. (21.5), where A, B, E, and ϕ are constants.

$$\begin{cases} x = A \sin \omega t + B \cos \omega t, & (21.8) \\ x = E \sin (\omega t - \phi). & (21.9) \end{cases}$$

Amplitude of the vibration described by Eqs. (21.8) and (21.9).

$$E = \sqrt{A^2 + B^2}. \qquad (21.11)$$

The *period* τ of the vibration governed by Eq. (21.5) is the time required for one complete oscillation, or *cycle*. The *frequency* f is the number of cycles per unit time. One cycle per second is called a Hertz (Hz). The term $\omega = 2\pi f$ is also a measure of the frequency and is expressed in rad/s.

$$\begin{cases} \tau = \dfrac{2\pi}{\omega}, & (21.12) \\[2mm] f = \dfrac{\omega}{2\pi}. & (21.13) \end{cases}$$

Vibration of a Conservative One-Degree-of-Freedom System
(▶ *Related Problems 21.1, 21.2*)

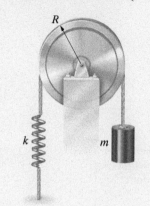

The pulley has radius R and moment of inertia I. The mass m is displaced downward from its equilibrium position and released. What is the frequency of the resulting vibrations of the system?

Strategy

A single coordinate specifying the vertical position of the mass specifies the position of the system, so it has one degree of freedom. If we can express the equation of motion of the system in the form of Eq. (21.5), we can determine the frequency of its vibrations from Eq. (21.13).

We will first obtain the equation of motion by drawing free-body diagrams and applying the rigid body equations of motion. We will then demonstrate how it can be obtained by using conservation of energy.

Determine the Equation of Motion by Applying the Rigid Body Equations of Motion

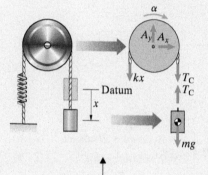

Let x be the downward displacement of the mass relative to its position when the spring is unstretched. T_C is the tension in the cable between the mass and pulley. α is the clockwise angular acceleration of the pulley.

Newton's second law for the mass is

$$mg - T_C = m\frac{d^2x}{dt^2}. \qquad (1)$$

The equation of angular motion for the pulley is

$$RT_C - R(kx) = I\alpha. \qquad (2)$$

Solving Eq. (2) for T_C, substituting the result into Eq. (1), and using the kinematic relationship

$$\alpha = \frac{1}{R}\frac{d^2x}{dt^2}$$

yields the equation of motion of the system in terms of x.

$$\left(m + \frac{I}{R^2}\right)\frac{d^2x}{dt^2} + kx = mg. \qquad (3)$$

Determine the Equation of Motion by Applying Conservation of Energy

Express the total kinetic energy in terms of x. The angular velocity ω of the pulley is related to the velocity of the mass by

$$\omega = \frac{1}{R}\frac{dx}{dt}.$$

$$\begin{cases} T = \frac{1}{2}m\left(\frac{dx}{dt}\right)^2 + \frac{1}{2}I\omega^2 \\ \qquad = \frac{1}{2}\left(m + \frac{I}{R^2}\right)\left(\frac{dx}{dt}\right)^2. \end{cases}$$

Express the total potential energy in terms of x. We place the datum for the potential energy associated with the weight of the mass at $x = 0$.

$$V = -mgx + \frac{1}{2}kx^2.$$

Obtain the equation of motion by evaluating the derivative with respect to t of the sum of the kinetic and potential energies $T + V = $ constant.

$$\begin{cases} \frac{d}{dt}(T + V) = \left(m + \frac{I}{R^2}\right)\frac{dx}{dt}\frac{d^2x}{dt^2} - mg\frac{dx}{dt} + kx\frac{dx}{dt} = 0: \\ \left(m + \frac{I}{R^2}\right)\frac{d^2x}{dt^2} + kx = mg. \qquad (3) \end{cases}$$

Express the Equation of Motion in Terms of Displacement Relative to the Equilibrium Position.

By setting $d^2x/dt^2 = 0$ in Eq. (3), we see that the equilibrium position of the system is $x = mg/k$. Let a new variable $\tilde{x} = x - mg/k$ denote the displacement of the system relative to the equilibrium position and express Eq. (3) in terms of $\tilde{x}$.

$$\begin{cases} \frac{d^2\tilde{x}}{dt^2} + \omega^2\tilde{x} = 0, \qquad (4) \\ \text{where} \\ \omega^2 = \frac{k}{m + \dfrac{I}{R^2}}. \qquad (5) \end{cases}$$

Equation (4) is of the form of Eq. (21.5). The frequency of vibration of the system is given by Eq. (21.13).

$$\begin{cases} f = \frac{\omega}{2\pi} \\ \quad = \frac{1}{2\pi}\sqrt{\frac{k}{m + \dfrac{I}{R^2}}}. \end{cases}$$

Practice Problem The mass m is displaced downward a distance h from its equilibrium position and released from rest at $t = 0$. Determine the position of the mass relative to its equilibrium position as a function of time.

Answer: $\tilde{x} = h\cos\omega t.$

Example 21.2 Frequency of a System (▶ *Related Problems 21.32, 21.33*)

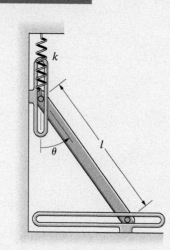

The spring attached to the slender bar of mass m is unstretched when $\theta = 0$. Neglecting friction, determine the frequency of small vibrations of the bar relative to its equilibrium position.

Strategy
The angle θ specifies the bar's position, so there is one degree of freedom. We can express the kinetic and potential energies in terms of θ and its time derivative and then take the derivative of the total energy with respect to time to obtain the equation of motion.

Solution
The kinetic energy of the bar is

$$T = \tfrac{1}{2}mv^2 + \tfrac{1}{2}I\left(\frac{d\theta}{dt}\right)^2,$$

where v is the velocity of the center of mass and $I = \tfrac{1}{12}ml^2$. The distance from the bar's instantaneous center to its center of mass is $\tfrac{1}{2}l$ (Fig. a), so $v = \left(\tfrac{1}{2}l\right)(d\theta/dt)$, and the kinetic energy is

$$T = \tfrac{1}{2}m\left[\tfrac{1}{2}l\left(\frac{d\theta}{dt}\right)\right]^2 + \tfrac{1}{2}\left(\tfrac{1}{12}ml^2\right)\left(\frac{d\theta}{dt}\right)^2 = \tfrac{1}{6}ml^2\left(\frac{d\theta}{dt}\right)^2.$$

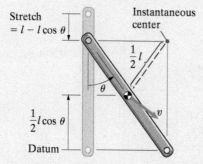

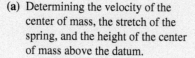

(a) Determining the velocity of the center of mass, the stretch of the spring, and the height of the center of mass above the datum.

In terms of θ, the stretch of the spring is $l - l\cos\theta$. We place the datum for the potential energy associated with the weight at the bottom of the bar (Fig. a), so the total potential energy is

$$V = mg\left(\tfrac{1}{2}l\cos\theta\right) + \tfrac{1}{2}k(l - l\cos\theta)^2.$$

The sum of the kinetic and potential energies is constant:

$$T + V = \tfrac{1}{6}ml^2\left(\frac{d\theta}{dt}\right)^2 + \tfrac{1}{2}mgl\cos\theta + \tfrac{1}{2}kl^2(1 - \cos\theta)^2 = \text{constant.}$$

Taking the derivative of this equation with respect to time, we obtain the equation of motion:

$$\tfrac{1}{3}ml^2\frac{d^2\theta}{dt^2} - \tfrac{1}{2}mgl\sin\theta + kl^2(1 - \cos\theta)\sin\theta = 0. \qquad (1)$$

To express this equation in the form of Eq. (21.5), we need to write it in terms of small vibrations relative to the equilibrium position. Let θ_e be the value of θ when the bar is in equilibrium. By setting $d^2\theta/dt^2 = 0$ in Eq. (1), we find that θ_e must satisfy the relation

$$\cos\theta_e = 1 - \frac{mg}{2kl}. \qquad (2)$$

We define $\tilde{\theta} = \tilde{\theta} - \theta_e$, and expand $\sin\theta$ and $\cos\theta$ in Taylor series in terms of $\tilde{\theta}$:

$$\sin\theta = \sin(\theta_e + \tilde{\theta}) = \sin\theta_e + \cos\theta_e\tilde{\theta} + \cdots,$$

$$\cos\theta = \cos(\theta_e + \tilde{\theta}) = \cos\theta_e - \sin\theta_e\tilde{\theta} + \cdots.$$

Substituting these expressions into Eq. (1), neglecting terms in $\tilde{\theta}$ of second and higher orders, and using Eq. (2), we obtain

$$\frac{d^2\tilde{\theta}}{dt^2} + \omega^2\tilde{\theta} = 0,$$

where

$$\omega^2 = \frac{3g}{l}\left(1 - \frac{mg}{4kl}\right).$$

From Eq. (21.13), the frequency of small vibrations of the bar is

$$f = \frac{\omega}{2\pi} = \frac{1}{2\pi}\sqrt{\frac{3g}{l}\left(1 - \frac{mg}{4kl}\right)}.$$

Critical Thinking
This example demonstrates the advantage of using conservation of energy to obtain the equation of motion of a one-degree-of-freedom conservative system. You should confirm that you can also obtain the equation of motion by drawing the free-body diagram of the bar and applying Newton's second law and the equation of angular motion. But that procedure is considerably more involved, in part because it is necessary to consider the normal forces exerted on the ends of the bar. We were able to ignore them in applying conservation of energy because they do no work on the bar.

Problems

▶ **21.1** In Active Example 21.1, suppose that the pulley has radius $R = 100$ mm and its moment of inertia is $I = 0.005$ kg-m^2. The mass $m = 2$ kg, and the spring constant is $k = 200$ N/m. If the mass is displaced downward from its equilibrium position and released, what are the period and frequency of the resulting vibration?

▶ **21.2** In Active Example 21.1, suppose that the pulley has radius $R = 4$ in and its moment of inertia is $I = 0.06$ slug-ft^2. The suspended object weighs 5 lb, and the spring constant is $k = 10$ lb/ft. The system is initially at rest in its equilibrium position. At $t = 0$, the suspended object is given a downward velocity of 1 ft/s. Determine the position of the suspended object relative to its equilibrium position as a function of time.

21.3 The mass $m = 4$ kg. The spring is unstretched when $x = 0$. The period of vibration of the mass is measured and determined to be 0.5 s. The mass is displaced to the position $x = 0.1$ m and released from rest at $t = 0$. Determine its position at $t = 0.4$ s.

21.4 The mass $m = 4$ kg. The spring is unstretched when $x = 0$. The frequency of vibration of the mass is measured and determined to be 6 Hz. The mass is displaced to the position $x = 0.1$ m and given a velocity $dx/dt = 5$ m/s at $t = 0$. Determine the amplitude of the resulting vibration.

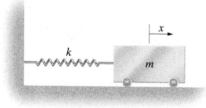

Problems 21.3/21.4

21.5 The mass $m = 4$ kg, and the spring constant is $k = 64$ N/m. For vibration of the spring–mass oscillator relative to its equilibrium position, determine (a) the frequency in Hz and (b) the period.

21.6 The mass $m = 4$ kg, and the spring constant is $k = 64$ N/m. The spring is unstretched when $x = 0$. At $t = 0$, $x = 0$, and the mass has a velocity of 2 m/s down the inclined surface. What is the value of x at $t = 0.8$ s?

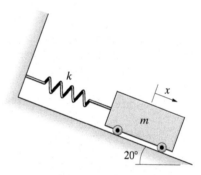

Problems 21.5/21.6

21.7 Suppose that in a mechanical design course you are asked to design a pendulum clock, and you begin with the pendulum. The mass of the disk is 2 kg. Determine the length L of the bar so that the period of small oscillations of the pendulum is 1 s. For this preliminary estimate, neglect the mass of the bar.

21.8 The mass of the disk is 2 kg and the mass of the slender bar is 0.4 kg. Determine the length L of the bar so that the period of small oscillations of the pendulum is 1 s.

    Strategy: Draw a graph of the value of the period for a range of lengths L to estimate the value of L corresponding to a period of 1 s.

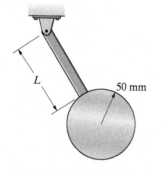

Problems 21.7/21.8

21.9 The spring constant is $k = 785$ N/m. The spring is unstretched when $x = 0$. Neglect the mass of the pulley, that is, assume that the tension in the rope is the same on both sides of the pulley. The system is released from rest with $x = 0$. Determine x as a function of time.

21.10 The spring constant is $k = 785$ N/m. The spring is unstretched when $x = 0$. The radius of the pulley is 125 mm, and moment of inertia about its axis is $I = 0.05$ kg-m^2. The system is released from rest with $x = 0$. Determine x as a function of time.

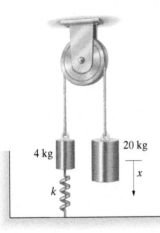

Problems 21.9/21.10

21.11 A "bungee jumper" who weighs 160 lb leaps from a bridge above a river. The bungee cord has an unstretched length of 60 ft, and it stretches an additional 40 ft before the jumper rebounds. Model the cord as a linear spring. When his motion has nearly stopped, what are the period and frequency of his vertical oscillations? (You can't model the cord as a linear spring during the early part of his motion. Why not?)

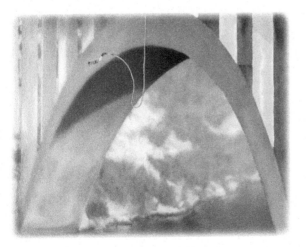

Problem 21.11

21.12 The spring constant is $k = 800$ N/m, and the spring is unstretched when $x = 0$. The mass of each object is 30 kg. The inclined surface is smooth. Neglect the mass of the pulley. The system is released from rest with $x = 0$.

(a) Determine the frequency and period of the resulting vibration.

(b) What is the value of x at $t = 4$ s?

21.13 The spring constant is $k = 800$ N/m, and the spring is unstretched when $x = 0$. The mass of each object is 30 kg. The inclined surface is smooth. The radius of the pulley is 120 mm and its moment of inertia is $I = 0.03$ kg-m^2. At $t = 0$, $x = 0$ and $dx/dt = 1$ m/s.

(a) Determine the frequency and period of the resulting vibration.

(b) What is the value of x at $t = 4$ s?

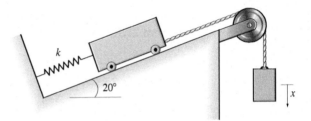

Problems 21.12/21.13

21.14 The 20-lb disk rolls on the horizontal surface. Its radius is $R = 6$ in. Determine the spring constant k so that the frequency of vibration of the system relative to its equilibrium position is $f = 1$ Hz.

21.15 The 20-lb disk rolls on the horizontal surface. Its radius is $R = 6$ in. The spring constant is $k = 15$ lb/ft. At $t = 0$, the spring is unstretched and the disk has a clockwise angular velocity of 2 rad/s. What is the amplitude of the resulting vibrations of the center of the disk?

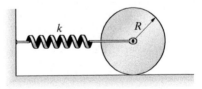

Problems 21.14/21.15

21.16 The 2-lb bar is pinned to the 5-lb disk. The disk rolls on the circular surface. What is the frequency of small vibrations of the system relative to its vertical equilibrium position?

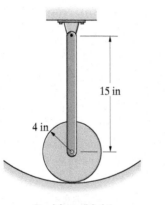

Problem 21.16

21.17 The mass of the suspended object A is 4 kg. The mass of the pulley is 2 kg and its moment of inertia is 0.018 N-m². The spring constant is $k = 150$ N/m. For vibration of the system relative to its equilibrium position, determine (a) the frequency in Hz and (b) the period.

21.18 The mass of the suspended object A is 4 kg. The mass of the pulley is 2 kg and its moment of inertia is 0.018 N-m². The spring constant is $k = 150$ N/m. The spring is unstretched when $x = 0$. At $t = 0$, the system is released from rest with $x = 0$. What is the velocity of the object A at $t = 1$ s?

Problems 21.17/21.18

21.19 The thin rectangular plate is attached to the rectangular frame by pins. The frame rotates with constant angular velocity $\omega_0 = 6$ rad/s. The angle β between the z axis of the body-fixed coordinate system and the vertical is governed by the equation

$$\frac{d^2\beta}{dt^2} = -\omega_0^2 \sin \beta \cos \beta.$$

Determine the frequency of small vibrations of the plate relative to its horizontal position.

Strategy: By writing $\sin \beta$ and $\cos \beta$ in terms of their Taylor series and assuming that β is small, show that the equation governing β can be expressed in the form of Eq. (21.5).

21.20 Consider the system described in Problem 21.19. At $t = 0$, the angle $\beta = 0.01$ rad and $d\beta/dt = 0$. Determine β as a function of time.

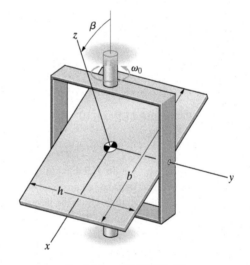

Problems 21.19/21.20

21.21 A slender bar of mass m and length l is pinned to a fixed support as shown. A torsional spring of constant k attached to the bar at the support is unstretched when the bar is vertical. Show that the equation governing small vibrations of the bar from its vertical equilibrium position is

$$\frac{d^2\theta}{dt^2} + \omega^2\theta = 0, \quad \text{where } \omega^2 = \frac{\left(k - \frac{1}{2}mgl\right)}{\frac{1}{3}ml^2}.$$

21.22* The initial conditions of the slender bar in Problem 21.21 are

$$t = 0 \begin{cases} \theta = 0 \\ \dfrac{d\theta}{dt} = \dot{\theta}_0. \end{cases}$$

(a) If $k > \frac{1}{2}mgl$, show that θ is given as a function of time by

$$\theta = \frac{\dot{\theta}_0}{\omega}\sin\omega t, \quad \text{where } \omega^2 = \frac{\left(k - \frac{1}{2}mgl\right)}{\frac{1}{3}ml^2}.$$

(b) If $k < \frac{1}{2}mgl$, show that θ is given as a function of time by

$$\theta = \frac{\dot{\theta}_0}{2h}(e^{ht} - e^{-ht}), \quad \text{where } h^2 = \frac{\left(\frac{1}{2}mgl - k\right)}{\frac{1}{3}ml^2}.$$

Strategy: To do part (b), seek a solution of the equation of motion of the form $x = Ce^{\lambda t}$, where C and λ are constants.

Problems 21.21/21.22

21.23 Engineers use the device shown to measure an astronaut's moment of inertia. The horizontal board is pinned at O and supported by the linear spring with constant $k = 12$ kN/m. When the astronaut is not present, the frequency of small vibrations of the board about O is measured and determined to be 6.0 Hz. When the astronaut is lying on the board as shown, the frequency of small vibrations of the board about O is 2.8 Hz. What is the astronaut's moment of inertia about the z axis?

21.24 In Problem 21.23, the astronaut's center of mass is at $x = 1.01$ m, $y = 0.16$ m, and his mass is 81.6 kg. What is his moment of inertia about the z' axis through his center of mass?

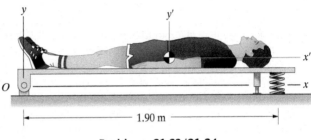

Problems 21.23/21.24

21.25* A floating sonobuoy (sound-measuring device) is in equilibrium in the vertical position shown. (Its center of mass is low enough that it is stable in this position.) The device is a 10-kg cylinder 1 m in length and 125 mm in diameter. The water density is 1025 kg/m^3, and the buoyancy force supporting the buoy equals the weight of the water that would occupy the volume of the part of the cylinder below the surface. If you push the sonobuoy slightly deeper and release it, what is the frequency of the resulting vertical vibrations?

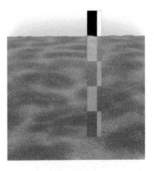

Problem 21.25

21.26 The disk rotates *in the horizontal plane* with constant angular velocity $\Omega = 12$ rad/s. The mass $m = 2$ kg slides in a smooth slot in the disk and is attached to a spring with constant $k = 860$ N/m. The radial position of the mass when the spring is unstretched is $r = 0.2$ m.

(a) Determine the "equilibrium" position of the mass, the value of r at which it will remain stationary relative to the center of the disk.

(b) What is the frequency of vibration of the mass relative to its equilibrium position?

   Strategy: Apply Newton's second law to the mass in terms of polar coordinates.

21.27 The disk rotates *in the horizontal plane* with constant angular velocity $\Omega = 12$ rad/s. The mass $m = 2$ kg slides in a smooth slot in the disk and is attached to a spring with constant $k = 860$ N/m. The radial position of the mass when the spring is unstretched is $r = 0.2$ m. At $t = 0$, the mass is in the position $r = 0.4$ m and $dr/dt = 0$. Determine the position r as a function of time.

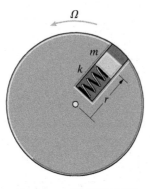

Problems 21.26/21.27

21.28 A homogeneous 100-lb disk with radius $R = 1$ ft is attached to two identical cylindrical steel bars of length $L = 1$ ft. The relation between the moment M exerted on the disk by one of the bars and the angle of rotation, θ, of the disk is

$$M = \frac{GJ}{L}\theta,$$

where J is the polar moment of inertia of the cross section of the bar and $G = 1.7 \times 10^9$ lb/ft^2 is the shear modulus of the steel. Determine the required radius of the bars if the frequency of rotational vibrations of the disk is to be 10 Hz.

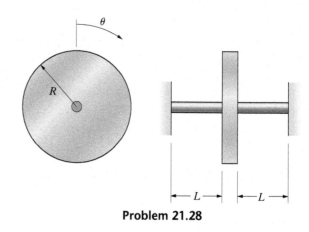

Problem 21.28

21.29 The moments of inertia of gears A and B are $I_A = 0.025$ kg-m^2 and $I_B = 0.100$ kg-m^2. Gear A is connected to a torsional spring with constant $k = 10$ N-m/rad. What is the frequency of small angular vibrations of the gears?

21.30 At $t = 0$, the torsional spring in Problem 21.29 is unstretched and gear B has a counterclockwise angular velocity of 2 rad/s. Determine the counterclockwise angular position of gear B relative to its equilibrium position as a function of time.

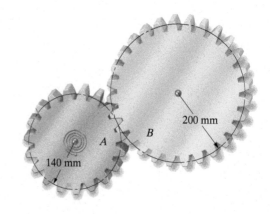

Problems 21.29/21.30

21.31 Each 2-kg slender bar is 1 m in length. What are the period and frequency of small vibrations of the system?

Problem 21.31

▶ **21.32*** The masses of the slender bar and the homogeneous disk are m and m_d, respectively. The spring is unstretched when $\theta = 0$. Assume that the disk rolls on the horizontal surface.

(a) Show that the motion of the system is governed by the equation

$$\left(\frac{1}{3} + \frac{3m_d}{2m}\cos^2\theta\right)\frac{d^2\theta}{dt^2} - \frac{3m_d}{2m}\sin\theta\cos\theta\left(\frac{d\theta}{dt}\right)^2$$

$$-\frac{g}{2l}\sin\theta + \frac{k}{m}(1 - \cos\theta)\sin\theta = 0.$$

(b) If the system is in equilibrium at the angle $\theta = \theta_e$ and $\tilde{\theta} = \theta - \theta_e$, show that the equation governing small vibrations relative to the equilibrium position is

$$\left(\frac{1}{3} + \frac{3m_d}{2m}\cos^2\theta_e\right)\frac{d^2\tilde{\theta}}{dt^2}$$

$$+ \left[\frac{k}{m}(\cos\theta_e - \cos^2\theta_e + \sin^2\theta_e) - \frac{g}{2l}\cos\theta_e\right]\tilde{\theta} = 0.$$

(See Example 21.2.)

▶ **21.33*** The masses of the bar and disk in Problem 21.32 are $m = 2$ kg and $m_d = 4$ kg, respectively. The dimensions $l = 1$ m and $R = 0.28$ m, and the spring constant is $k = 70$ N/m.

(a) Determine the angle θ_e at which the system is in equilibrium.

(b) The system is at rest in the equilibrium position, and the disk is given a clockwise angular velocity of 0.1 rad/s. Determine θ as a function of time. (See Example 21.2.)

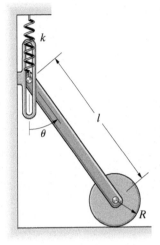

Problems 21.32/21.33

21.34 The mass of each slender bar is 1 kg. If the frequency of small vibrations of the system is 0.935 Hz, what is the mass of the object A?

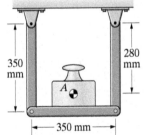

Problem 21.34

21.35* The 4-kg slender bar is 2 m in length. It is held in equilibrium in the position $\theta_0 = 35°$ by a torsional spring with constant k. The spring is unstretched when the bar is vertical. Determine the period and frequency of small vibrations of the bar relative to the equilibrium position shown.

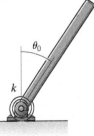

Problem 21.35

21.2 Damped Vibrations

BACKGROUND

If the mass of a spring–mass oscillator is displaced and released, it will not vibrate indefinitely. It will slow down and eventually stop as a result of frictional forces, or *damping mechanisms*, acting on the system. Damping mechanisms damp out, or *attenuate*, the vibration. In some cases, engineers intentionally include damping mechanisms in vibrating systems. For example, the shock absorbers in a car are designed to damp out vibrations of the suspension relative to the frame. In the previous section we neglected damping, so the solutions we obtained describe only motions of systems over periods of time brief enough that the effects of damping can be neglected. We now discuss a simple method for modeling damping in vibrating systems.

The spring–mass oscillator in Fig. 21.7a has a *damping element*. The schematic diagram for the damping element represents a piston moving in a cylinder of viscous fluid. The force required to lengthen or shorten a damping element is defined to be the product of a constant c, the *damping constant*, and the rate of change of the length of the element (Fig. 21.7b). Therefore, the equation of motion of the mass is

$$-c\frac{dx}{dt} - kx = m\frac{d^2x}{dt^2}.$$

By defining $\omega = \sqrt{k/m}$ and $d = c/2m$, we can write this equation in the form

$$\frac{d^2x}{dt^2} + 2d\frac{dx}{dt} + \omega^2 x = 0. \tag{21.16}$$

This equation describes the vibrations of many damped, one-degree-of-freedom systems. The form of its solution, and consequently the character of the predicted behavior of the system the equation describes, depends on whether the constant d is less than, equal to, or greater than ω. We discuss these cases in the sections that follow.

Subcritical Damping

If $d < \omega$, the system is said to be *subcritically damped*. Assuming a solution of the form

$$x = Ce^{\lambda t} \tag{21.17}$$

and substituting it into Eq. (21.16), we obtain

$$\lambda^2 + 2d\lambda + \omega^2 = 0.$$

Figure 21.7
(a) Damped spring–mass oscillator.
(b) Free-body diagram of the mass.

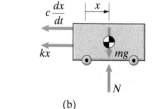

(a)

(b)

This quadratic equation yields two roots for the constant λ that we can write as

$$\lambda = -d \pm i\omega_d,$$

where

$$\omega_d = \sqrt{\omega^2 - d^2}. \tag{21.18}$$

Because we are assuming that $d < \omega$, the constant ω_d is a real number. The two roots for λ give us two solutions of the form of Eq. (21.17). The resulting general solution of Eq. (21.16) is

$$x = e^{-dt}(Ce^{i\omega_d t} + De^{-i\omega_d t}),$$

where C and D are constants. By using the Euler identity $e^{i\theta} = \cos\theta + i\sin\theta$, we can express this solution in the form

$$x = e^{-dt}(A\sin\omega_d t + B\cos\omega_d t), \tag{21.19}$$

where A and B are constants. Equation (21.19) is the product of an exponentially decaying function of time and an expression identical in form to the solution we obtained for an undamped system. The exponential function describes the expected effect of damping: The amplitude of the vibration attenuates with time. The coefficient d determines the rate at which the amplitude decreases.

Damping has an important effect in addition to causing attenuation. Because the oscillatory part of the solution is identical in form to Eq. (21.8), except that the term ω is replaced by ω_d, it follows from Eqs. (21.12) and (21.13) that the period and frequency of the damped system are

$$\tau_d = \frac{2\pi}{\omega_d}, \qquad f_d = \frac{\omega_d}{2\pi}. \tag{21.20}$$

From Eq. (21.18), we see that $\omega_d < \omega$, so *the period of the vibration is increased and its frequency is decreased as a result of subcritical damping*.

The rate of damping is often expressed in terms of the *logarithmic decrement δ*, which is the natural logarithm of the ratio of the amplitude at a time t to the amplitude at time $t + \tau_d$. Since the amplitude is proportional to e^{-dt}, we can obtain a simple relation between the logarithmic decrement, the coefficient d, and the period:

$$\delta = \ln\left[\frac{e^{-dt}}{e^{-d(t+\tau_d)}}\right] = d\tau_d.$$

Critical and Supercritical Damping

When $d \geq \omega$, the character of the solution of Eq. (21.16) is different from the case of subcritical damping. Suppose that $d > \omega$. When this is the case, the system is said to be *supercritically damped*. We again substitute a solution of the form

$$x = Ce^{\lambda t} \tag{21.21}$$

into Eq. (21.16), obtaining

$$\lambda^2 + 2d\lambda + \omega^2 = 0. \tag{21.22}$$

We can write the roots of this equation as

$$\lambda = -d \pm h,$$

where

$$h = \sqrt{d^2 - \omega^2}. \tag{21.23}$$

The resulting general solution of Eq. (21.16) is

$$x = Ce^{-(d-h)t} + De^{-(d+h)t}, \tag{21.24}$$

where C and D are constants.

When $d = \omega$, a system is said to be *critically damped*. Then the constant $h = 0$, so Eq. (21.22) has a repeated root $\lambda = -d$, and we obtain only one solution of the form (21.21). In this case, it can be shown that the general solution of Eq. (21.16) is

$$x = Ce^{-dt} + Dte^{-dt}, \tag{21.25}$$

where C and D are constants.

Equations (21.24) and (21.25) indicate that the motion of a system is not oscillatory when $d \geq \omega$. These equations are expressed in terms of exponential functions, and do not contain sines and cosines. The condition $d = \omega$ defines the minimum amount of damping necessary to avoid oscillatory behavior, which is why it is referred to as the critically damped case. Figure 21.8 shows the effect of increasing amounts of damping on the behavior of a vibrating system.

The concept of critical damping has important implications in the design of many systems. For example, it is desirable to introduce enough damping into a car's suspension so that its motion is not oscillatory, but too much damping would cause the suspension to be too "stiff."

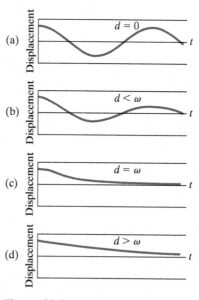

Figure 21.8
Displacement history of a vibrating system that is (a) undamped; (b) subcritically damped; (c) critically damped; (d) supercritically damped.

RESULTS

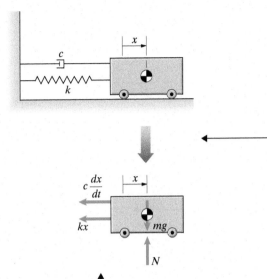

The *damped spring-mass oscillator* is the simplest example of a damped one-degree-of-freedom vibrating system. The *damping element* exerts a resisting force that is proportional to the velocity of the mass.

With $\omega^2 = k/m$ and $d = c/2m$, this equation governs the displacement of the mass of a damped spring–mass oscillator relative to its equilibrium position. Small vibrations of many damped one-degree-of-freedom systems relative to an equilibrium position are governed by this same equation, with the constants ω and d determined by the physical characteristics of the system.

$$\frac{d^2x}{dt^2} + 2d\frac{dx}{dt} + \omega^2 x = 0. \quad (21.16)$$

Subcritical Damping

When $d < \omega$, the system is said to be *subcritically damped*. The general solution of Eq. (21.16) is the product of an exponentially decaying function of time and an expression identical in form to the general solution for an undamped system.

$$\begin{cases} x = e^{-dt}(A \sin \omega_d t + B \cos \omega_d t), & (21.19) \\ \text{where} \\ \omega_d = \sqrt{\omega^2 - d^2}. & (21.18) \end{cases}$$

The period of vibration of the system is increased and the frequency is decreased as a result of subcritical damping.

$$\tau_d = \frac{2\pi}{\omega_d}, \quad f_d = \frac{\omega_d}{2\pi}. \quad (21.20)$$

The *logarithmic decrement*, a measure of attenuation, is the natural logarithm of the ratio of the amplitude at a time t to the amplitude at time $t + \tau_d$.

$$\delta = \ln\left[\frac{e^{-dt}}{e^{-d(t + \tau_d)}}\right] = d\tau_d.$$

Critical and Supercritical Damping

When $d > \omega$, the system is said to be *supercritically damped*. In this case the general solution of Eq. (21.16) does not exhibit oscillatory behavior.

$$x = Ce^{-(d-h)t} + De^{-(d+h)t}, \qquad (21.23)$$

where

$$h = \sqrt{d^2 - \omega^2}. \qquad (21.24)$$

When $d = \omega$, the system is *critically damped*. This is the minimum amount of damping for which the general solution of Eq. (21.16) does not exhibit oscillatory behavior.

$$x = Ce^{-dt} + Dte^{-dt}. \qquad (21.25)$$

Active Example 21.3   Damped Spring-Mass Oscillator (▶ *Related Problems 21.36–21.38*)

The damped spring–mass oscillator has mass $m = 2$ kg, spring constant $k = 8$ N/m, and damping constant $c = 1$ N-s/m. At time $t = 0$, the mass is released from rest in the position $x = 0.1$ m. Determine its position as a function of time.

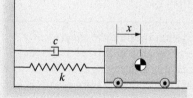

Strategy

The equation of motion for the damped spring–mass oscillator is given by Eq. (21.16). By calculating the values of ω and d, we will determine whether the damping is subcritical, critical, or supercritical and thereby choose the appropriate form of solution. Then we can use the given initial conditions to obtain the position as a function of time.

Solution

| Evaluate ω and d. The damping is subcritical. | $\begin{cases} \omega = \sqrt{\dfrac{k}{m}} = \sqrt{\dfrac{8 \text{ N/m}}{2 \text{ kg}}} = 2 \text{ rad/s}, \\[3mm] d = \dfrac{c}{2m} = \dfrac{1 \text{ N-s/m}}{2(2 \text{ kg})} = 0.25 \text{ rad/s}. \end{cases}$ |

| The general solution is given by Eq. (21.19) with $\omega_d = \sqrt{\omega^2 - d^2} = 1.98 \text{ rad/s}.$ | $\begin{cases} x = e^{-dt}(A \sin \omega_d t + B \cos \omega_d t), \\[2mm] \quad = e^{-0.25t}(A \sin 1.98t + B \cos 1.98t). \end{cases}$ (1) |

At $t = 0$, the position of the mass is $x = 0.1$ m and its velocity is $dx/dt = 0$. Use these initial conditions to solve for the constants A and B. Substituting the results into Eq. (1) determines the position of the mass as a function of time.

$\begin{cases} \dfrac{dx}{dt} = -0.25e^{-0.25t}(A \sin 1.98t + B \cos 1.98t) \\[3mm] \qquad + 1.98e^{-0.25t}(A \cos 1.98t - B \sin 1.98t). \quad (2) \end{cases}$

Substituting the initial conditions into Eqs. (1) and (2) and solving yields

$A = 0.0126$ m,
$B = 0.1$ m.

The position as a function of time is

$x = e^{-0.25t}(0.0126 \sin 1.98t + 0.1 \cos 1.98t)$ m.

The graph of the position for the first 10 s of motion clearly exhibits the attenuation of the amplitude.

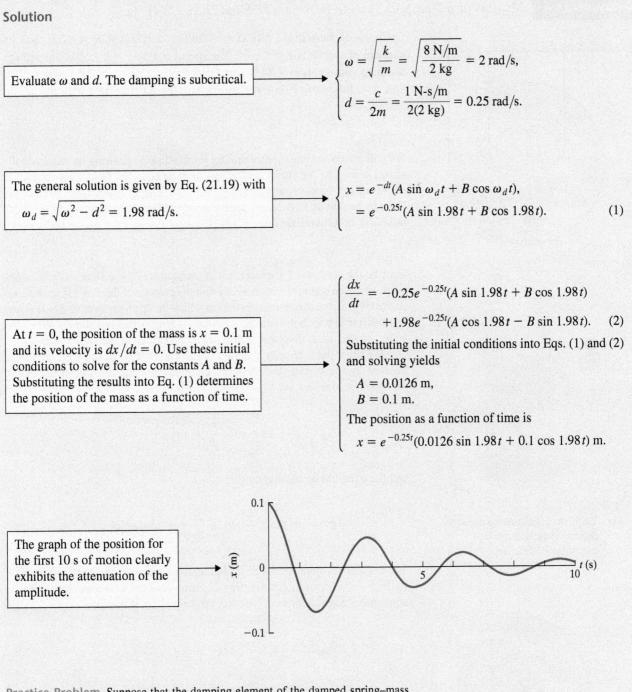

Practice Problem Suppose that the damping element of the damped spring–mass oscillator is replaced by one with damping constant $c = 12$ N-s/m. At time $t = 0$, the mass is released from rest in the position $x = 0.1$ m. Determine its position as a function of time.

Answer: $x = 0.117e^{-0.764t} - 0.0171e^{-5.24t}$ m.

Example 21.4   **Motion of a Damped System** (▶ *Related Problems 21.53, 21.54*)

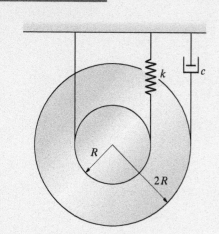

The mass of the stepped disk is $m = 20$ kg, its radius is $R = 0.3$ m, and its moment of inertia is $I = \frac{1}{2}mR^2$. The spring constant is $k = 60$ N/m and the damping constant is $c = 24$ N-s/m. Determine the position of the center of the disk as a function of time if the disk is released from rest with the spring unstretched.

Strategy

We will obtain the equation of motion for the disk by drawing its free-body diagram and using Newton's second law and the equation of angular motion. If the resulting equation can be expressed in the form of Eq. (21.16), we can analyze its motion using the same approach we applied to the damped spring–mass oscillator in Active Example 21.3.

Solution

Let x be the downward displacement of the center of the disk relative to its position when the spring is unstretched. From the position of the disk's instantaneous center (Fig. a), we can see that the rate at which the spring is stretched is $2(dx/dt)$ and the rate at which the damping element is lengthened is $3(dx/dt)$. When the center of the disk is displaced a distance x, the stretch of the spring is $2x$.

We draw the free-body diagram of the disk in Fig. b, showing the forces exerted by the spring, the damping element, and the tension in the cable. Newton's second law is

$$mg - T - 2kx - 3c\frac{dx}{dt} = m\frac{d^2x}{dt^2},$$

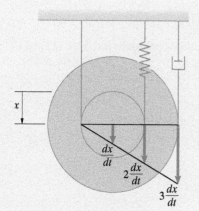

and the equation of angular motion is

$$RT - R(2kx) - 2R\left(3c\frac{dx}{dt}\right) = \left(\tfrac{1}{2}mR^2\right)\alpha.$$

(a) Using the instantaneous center to determine the relationships between the velocities.

The angular acceleration is related to the acceleration of the center of the disk by $\alpha = (d^2x/dt^2)/R$. Eliminating T from Newton's second law and the equation of angular motion, we obtain the equation of motion:

$$\frac{3}{2}m\frac{d^2x}{dt^2} + 9c\frac{dx}{dt} + 4kx = mg.$$

By setting d^2x/dt^2 and dx/dt equal to zero in this equation, we find that the equilibrium position of the disk is $x = mg/4k$. Expressing the equation of motion in terms of the position of the center of the disk relative to its equilibrium position, $\tilde{x} = x - mg/4k$, we obtain

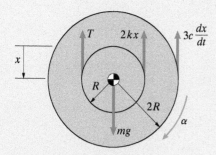

(b) Free-body diagram of the disk.

$$\frac{d^2\tilde{x}}{dt^2} + \left(\frac{6c}{m}\right)\frac{d\tilde{x}}{dt} + \left(\frac{8k}{3m}\right)\tilde{x} = 0.$$

This equation is identical in form to Eq. (21.16), where the constants are

$$d = \frac{6c}{2m} = \frac{(6)(24)}{(2)(20)} = 3.60 \text{ rad/s}$$

and

$$\omega = \sqrt{\frac{8k}{3m}} = \sqrt{\frac{(8)(60)}{(3)(20)}} = 2.83 \text{ rad/s}.$$

The damping is supercritical $(d > \omega)$ so the motion is described by Eq. (21.24) with $h = \sqrt{d^2 - \omega^2} = 2.23 \text{ rad/s}$:

$$\tilde{x} = Ce^{-(d-h)t} + De^{-(d+h)t} = Ce^{-1.37t} + De^{-5.83t}.$$

The velocity is

$$\frac{d\tilde{x}}{dt} = -1.37Ce^{-1.37t} - 5.83De^{-5.83t}.$$

At $t = 0$, $\tilde{x} = -mg/4k = -0.818$ m and $d\tilde{x}/dt = 0$. From these conditions, we obtain $C = -1.069$ m and $D = 0.252$ m, so the position of the center of the disk relative to its equilibrium position is

$$\tilde{x} = -1.069e^{-1.37t} + 0.252e^{-5.83t} \text{ m}.$$

The graph shows the position for the first 4 s of motion.

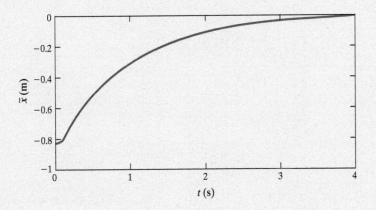

Critical Thinking

The disk in this example rotates and its center moves in the vertical direction. Why is this a one-degree-of-freedom system? As the center moves downward a distance x, the disk rotates through a clockwise angle $\theta = x/R$. Both the position of the center of the disk and the disk's angular position are specified when x is specified. The system has one degree of freedom.

Problems

▶ **21.36** The mass $m = 2$ slug, the spring constant is $k = 72$ lb/ft, and the damping constant is $c = 8$ lb-s/ft. The spring is unstretched when $x = 0$. The mass is displaced to the position $x = 1$ ft and released from rest.

(a) If the damping is subcritical, what is the frequency of the resulting damped vibrations?

(b) What is the value of x at $t = 1$ s?

(See Active Example 21.3.)

▶ **21.37** The mass $m = 2$ slug, the spring constant is $k = 72$ lb/ft, and the damping constant is $c = 32$ lb-s/ft. The spring is unstretched when $x = 0$. The mass is displaced to the position $x = 1$ ft and released from rest.

(a) If the damping is subcritical, what is the frequency of the resulting damped vibrations?

(b) What is the value of x at $t = 1$ s?

(See Active Example 21.3.)

▶ **21.38** The mass $m = 4$ slug and the spring constant is $k = 72$ lb/ft. The spring is unstretched when $x = 0$.

(a) What value of the damping constant c causes the system to be critically damped?

(b) Suppose that c has the value determined in part (a). At $t = 0$, $x = 1$ ft and $dx/dt = 4$ ft/s. What is the value of x at $t = 1$ s?

(See Active Example 21.3.)

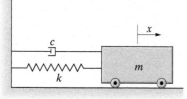

Problems 21.36–21.38

21.39 The mass $m = 2$ kg, the spring constant is $k = 8$ N/m, and the damping coefficient is $c = 12$ N-s/m. The spring is unstretched when $x = 0$. At $t = 0$, the mass is released from rest with $x = 0$. Determine the value of x at $t = 2$ s.

21.40 The mass $m = 0.15$ slugs, the spring constant is $k = 0.5$ lb/ft, and the damping coefficient is $c = 0.8$ lb-s/ft. The spring is unstretched when $x = 0$. At $t = 0$, the mass is released from rest with $x = 0$. Determine the value of x at $t = 2$ s.

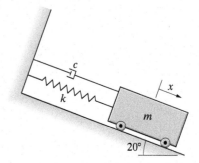

Problems 21.39/21.40

21.41 A 2570-lb test car moving with velocity $v_0 = 5$ mi/h collides with a rigid barrier at $t = 0$. As a result of the behavior of its energy-absorbing bumper, the response of the car to the collision can be simulated by the damped spring–mass oscillator shown with $k = 8000$ lb/ft and $c = 3000$ lb-s/ft. Assume that the mass is moving to the left with velocity $v_0 = 5$ mi/h and the spring is unstretched at $t = 0$. Determine the car's position (a) at $t = 0.04$ s and (b) at $t = 0.08$ s.

21.42 A 2570-lb test car moving with velocity $v_0 = 5$ mi/h collides with a rigid barrier at $t = 0$. As a result of the behavior of its energy-absorbing bumper, the response of the car to the collision can be simulated by the damped spring–mass oscillator shown with $k = 8000$ lb/ft and $c = 3000$ lb-s/ft. Assume that the mass is moving to the left with velocity $v_0 = 5$ mi/h and the spring is unstretched at $t = 0$. Determine the car's deceleration (a) immediately after it contacts the barrier; (b) at $t = 0.04$ s; and (c) at $t = 0.08$ s.

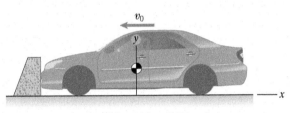

Car colliding with a rigid barrier

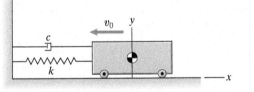

Simulation model

Problems 21.41/21.42

21.43 The motion of the car's suspension can be modeled by the damped spring–mass oscillator with $m = 36$ kg, $k = 22$ kN/m, and $c = 2.2$ kN-s/m. Assume that no external forces act on the tire and wheel. At $t = 0$, the spring is unstretched and the tire and wheel are given a velocity $dx/dt = 10$ m/s. Determine the position x as a function of time.

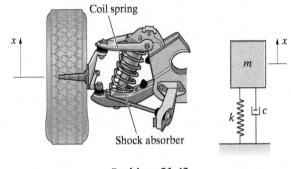

Problem 21.43

21.44 The 4-kg slender bar is 2 m in length. Aerodynamic drag on the bar and friction at the support exert a resisting moment about the pin support of magnitude $1.4(d\theta/dt)$ N-m, where $d\theta/dt$ is the angular velocity in rad/s.
(a) What are the period and frequency of small vibrations of the bar?
(b) How long does it take for the amplitude of vibration to decrease to one-half of its initial value?

21.45 The bar described in Problem 21.44 is given a displacement $\theta = 2°$ and released from rest at $t = 0$. What is the value of θ (in degrees) at $t = 2$ s?

Problems 21.44/21.45

21.46 The radius of the pulley is $R = 100$ mm and its moment of inertia is $I = 0.1$ kg-m². The mass $m = 5$ kg, and the spring constant is $k = 135$ N/m. The cable does not slip relative to the pulley. The coordinate x measures the displacement of the mass relative to the position in which the spring is unstretched. Determine x as a function of time if $c = 60$ N-s/m and the system is released from rest with $x = 0$.

21.47 For the system described in Problem 21.46, determine x as a function of time if $c = 120$ N-s/m and the system is released from rest with $x = 0$.

21.48 For the system described in Problem 21.46, choose the value of c so that the system is critically damped, and determine x as a function of time if the system is released from rest with $x = 0$.

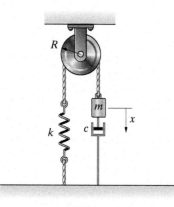

Problems 21.46–21.48

21.49 The spring constant is $k = 800$ N/m, and the spring is unstretched when $x = 0$. The mass of each object is 30 kg. The inclined surface is smooth. The radius of the pulley is 120 mm and its moment of inertia is $I = 0.03$ kg-m². Determine the frequency and period of vibration of the system relative to its equilibrium position if (a) $c = 0$ and (b) $c = 250$ N-s/m.

21.50 The spring constant is $k = 800$ N/m, and the spring is unstretched when $x = 0$. The damping constant is $c = 250$ N-s/m. The mass of each object is 30 kg. The inclined surface is smooth. The radius of the pulley is 120 mm and its moment of inertia is $I = 0.03$ kg-m². At $t = 0$, $x = 0$, and $dx/dt = 1$ m/s. What is the value of x at $t = 2$ s?

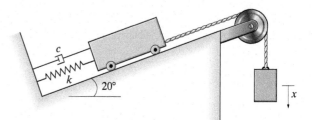

Problems 21.49/21.50

21.51 The homogeneous disk weighs 100 lb and its radius is $R = 1$ ft. It rolls on the plane surface. The spring constant is $k = 100$ lb/ft and the damping constant is $c = 3$ lb-s/ft. Determine the frequency of small vibrations of the disk relative to its equilibrium position.

21.52 In Problem 21.51, the spring is unstretched at $t = 0$ and the disk has a clockwise angular velocity of 2 rad/s. What is the angular velocity of the disk when $t = 3$ s?

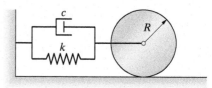

Problems 21.51/21.52

▶ 21.53 The moment of inertia of the stepped disk is I. Let θ be the angular displacement of the disk relative to its position when the spring is unstretched. Show that the equation governing θ is identical in form to Eq. (21.16), where

$$d = \frac{R^2 c}{2I} \quad \text{and} \quad \omega^2 = \frac{4R^2 k}{I}.$$

(See Example 21.4.)

▶ 21.54 In Problem 21.53, the radius $R = 250$ mm, $k = 150$ N/m, and the moment of inertia of the disk is $I = 2$ kg-m^2.

(a) At what value of c will the system be critically damped?

(b) At $t = 0$, the spring is unstretched and the clockwise angular velocity of the disk is 10 rad/s. Determine θ as a function of time if the system is critically damped.

(c) Using the result of (b), determine the maximum resulting angular displacement of the disk and the time at which it occurs.

(See Example 21.4.)

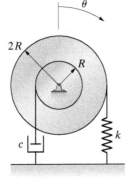

Problems 21.53/21.54

21.55 The moments of inertia of gears A and B are $I_A = 0.025$ kg-m^2 and $I_B = 0.100$ kg-m^2. Gear A is connected to a torsional spring with constant $k = 10$ N-m/rad. The bearing supporting gear B incorporates a damping element that exerts a resisting moment on gear B of magnitude $2(d\theta_B/dt)$ N-m, where $d\theta_B/dt$ is the angular velocity of gear B in rad/s. What is the frequency of small angular vibrations of the gears?

21.56 At $t = 0$, the torsional spring in Problem 21.55 is unstretched and gear B has a counterclockwise angular velocity of 2 rad/s. Determine the counterclockwise angular position of gear B relative to its equilibrium position as a function of time.

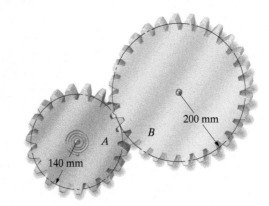

Problems 21.55/21.56

21.57 For the case of critically damped motion, confirm that the expression

$$x = Ce^{-dt} + Dte^{-dt}$$

is a solution of Eq. (21.16).

21.3 Forced Vibrations

BACKGROUND

The term *forced vibrations* means that external forces affect the vibration of a system. Until now, we have discussed *free vibration* of systems—vibration unaffected by external forces. For example, during an earthquake, a building undergoes forced vibration induced by oscillatory forces exerted on its foundations. After the earthquake subsides, the building vibrates freely until its motion damps out.

The damped spring–mass oscillator in Fig. 21.9a is subjected to a horizontal time-dependent force $F(t)$. From the free-body diagram of the mass (Fig. 21.9b), its equation of motion is

$$F(t) - kx - c\frac{dx}{dt} = m\frac{d^2x}{dt^2}.$$

Defining $d = c/2m$, $\omega^2 = k/m$, and $a(t) = F(t)/m$, we can write this equation in the form

$$\frac{d^2x}{dt^2} + 2d\frac{dx}{dt} + \omega^2x = a(t). \tag{21.26}$$

We call $a(t)$ the *forcing function*. Equation (21.26) describes the forced vibrations of many damped one-degree-of-freedom systems. It is nonhomogeneous, because the forcing function does not contain x or one of its derivatives. Its general solution consists of two parts—the homogeneous and particular solutions:

$$x = x_h + x_p.$$

The *homogeneous solution* x_h is the general solution of Eq. (21.26) with the right side set equal to zero. Therefore, the homogeneous solution is the general solution for free vibrations, which we described in Section 21.2. The *particular solution* x_p is a solution that satisfies Eq. (21.26). In the sections that follow, we discuss the particular solutions for two types of forcing functions that occur frequently in applications.

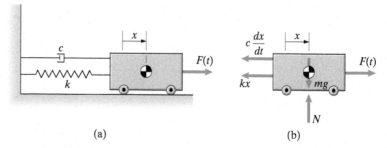

(a) (b)

Figure 21.9
(a) A damped spring–mass oscillator subjected to a time-dependent force.
(b) Free-body diagram of the mass.

Oscillatory Forcing Function

Unbalanced wheels and shafts exert forces that oscillate at their frequency of rotation. When a car's wheels are out of balance, they exert oscillatory forces that cause vibrations passengers can feel. Engineers design electromechanical devices that transform oscillating currents into oscillating forces for use in testing vibrating systems. But the principal reason we are interested in this type of forcing function is that nearly any forcing function can be represented as a sum of oscillatory forcing functions with several different frequencies or with a continuous spectrum of frequencies.

By studying the motion of a vibrating system subjected to an oscillatory forcing function, we can determine the response of the system as a function of the frequency of the force. Suppose that the forcing function is an oscillatory function of the form

$$a(t) = a_0 \sin \omega_0 t + b_0 \cos \omega_0 t, \tag{21.27}$$

where a_0, b_0, and the frequency of the forcing function ω_0 are given constants. We can obtain the particular solution to Eq. (21.26) by seeking a solution of the form

$$x_p = A_p \sin \omega_0 t + B_p \cos \omega_0 t, \tag{21.28}$$

where A_p and B_p are constants we must determine. Substituting this expression and Eq. (21.27) into Eq. (21.26), we can write the resulting equation as

$$(-\omega_0^2 A_p - 2d\omega_0 B_p + \omega^2 A_p - a_0)\sin \omega_0 t$$

$$+ (-\omega_0^2 B_p + 2d\omega_0 A_p + \omega^2 B_p - b_0)\cos \omega_0 t = 0.$$

Equating the coefficients of $\sin \omega_0 t$ and $\cos \omega_0 t$ to zero and solving for A_p and B_p, we obtain

$$A_p = \frac{(\omega^2 - \omega_0^2)a_0 + 2d\omega_0 b_0}{(\omega^2 - \omega_0^2)^2 + 4d^2\omega_0^2}$$

and

$$B_p = \frac{-2d\omega_0 a_0 + (\omega^2 - \omega_0^2)b_0}{(\omega^2 - \omega_0^2)^2 + 4d^2\omega_0^2}. \tag{21.29}$$

Substituting these results into Eq. (21.28) yields the particular solution:

$$x_p = \left[\frac{(\omega^2 - \omega_0^2)a_0 + 2d\omega_0 b_0}{(\omega^2 - \omega_0^2)^2 + 4d^2\omega_0^2} \right] \sin \omega_0 t$$

$$+ \left[\frac{-2d\omega_0 a_0 + (\omega^2 - \omega_0^2)b_0}{(\omega^2 - \omega_0^2)^2 + 4d^2\omega_0^2} \right] \cos \omega_0 t. \tag{21.30}$$

The amplitude of the particular solution is

$$E_p = \sqrt{A_p^2 + B_p^2} = \frac{\sqrt{a_0^2 + b_0^2}}{\sqrt{(\omega^2 - \omega_0^2)^2 + 4d^2\omega_0^2}}. \tag{21.31}$$

In Section 21.2, we showed that the solution of the equation describing free vibration of a damped system attenuates with time. For this reason, the particular solution for the motion of a damped vibrating system subjected to an oscillatory external force is also called the *steady–state solution*. The motion approaches the steady-state solution with increasing time. (See Active Example 21.5.)

To illustrate the effects of damping and the frequency of the forcing function on the amplitude of the particular solution, in Fig. 21.10 we plot the non-dimensional expression $\omega^2 E_p / \sqrt{a_0^2 + b_0^2}$ as a function of ω_0/ω for several values of the parameter d/ω. When there is no damping $(d = 0)$, the amplitude of the particular solution approaches infinity as the frequency ω_0 of the forcing function approaches the frequency ω. When the damping is small, the amplitude of the particular solution approaches a finite maximum value at a value of ω_0 that is smaller than ω. The frequency at which the amplitude of the particular solution is a maximum is called the *resonant frequency*. (See Problem 21.67.)

The phenomenon of resonance is a familiar one in our everyday experience. For example, when a wheel of a car is out of balance, the resulting vibrations are noticed when the car is moving at a certain speed. At that speed, the wheel rotates at the resonant frequency of the car's suspension. Resonance is of practical importance in many applications, because relatively small oscillatory forces can result in large vibrational amplitudes that may cause damage or interfere with the functioning of a system. The classic example is soldiers marching across a bridge. If their steps in unison coincide with one of the bridge's resonant frequencies, they may damage the bridge even though it can safely support their weight when they stand at rest.

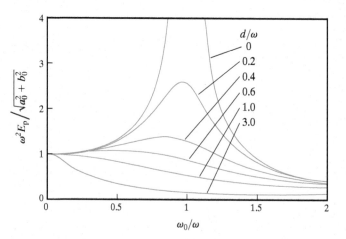

Figure 21.10
Amplitude of the particular (steady-state) solution as a function of the frequency of the forcing function.

Polynomial Forcing Function

Suppose that the forcing function $a(t)$ in Eq. (21.26) is a polynomial function of time; that is,

$$a(t) = a_0 + a_1 t + a_2 t^2 + \cdots + a_N t^N,$$

where $a_0, a_1, \ldots, a_N$ are given constants. This forcing function is important in applications because many smooth functions can be approximated by polynomials over a given interval of time. In this case we can obtain the particular solution of Eq. (21.26) by seeking a solution of the same form, namely,

$$x_p = A_0 + A_1 t + A_2 t^2 + \cdots + A_N t^N, \tag{21.32}$$

where $A_0, A_1, A_2, \ldots, A_N$ are constants to be determined.

For example, if $a(t) = a_0 + a_1 t$, Eq. (21.26) becomes

$$\frac{d^2 x}{dt^2} + 2d\frac{dx}{dt} + \omega^2 x = a_0 + a_1 t, \tag{21.33}$$

and we seek a particular solution of the form $x_p = A_0 + A_1 t$. Substituting this solution into Eq. (21.33), we can write the resulting equation as

$$(2dA_1 + \omega^2 A_0 - a_0) + (\omega^2 A_1 - a_1)t = 0.$$

This equation can be satisfied over an interval of time only if

$$2dA_1 + \omega^2 A_0 - a_0 = 0$$

and

$$\omega^2 A_1 - a_1 = 0.$$

Solving these two equations for A_0 and A_1, we obtain the particular solution:

$$x_p = \frac{a_0 - 2da_1/\omega^2 + a_1 t}{\omega^2}.$$

It can be confirmed that this is a particular solution by substituting it into Eq. (21.33).

RESULTS

The term *forced vibration* means that external forces affect the vibration of a system.

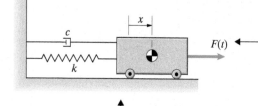

A damped spring–mass oscillator subjected to a time-dependent force $F(t)$.

With $\omega^2 = k/m$, $d = c/2m$, and $a(t) = F(t)/m$, this equation governs the displacement of the mass of the damped spring–mass oscillator relative to its equilibrium position. Small vibrations of many forced one-degree-of-freedom systems relative to an equilibrium position are governed by this same equation, with the constants ω and d and the *forcing function* $a(t)$ determined by the physical characteristics of the system and the external forces acting on it.

$$\frac{d^2x}{dt^2} + 2d\frac{dx}{dt} + \omega^2 x = a(t). \quad (21.26)$$

A homogeneous linear differential equation consists of terms that are linear in the dependent variable or its derivatives. Equation (21.26) is inhomogeneous due to the term $a(t)$. Its general solution consists of the sum of the *homogeneous solution* x_h and the *particular solution* x_p. The homogeneous solution is the general solution of Eq. (21.26) with the right side set equal to zero, which was discussed in Section 21.2. The particular solution is one that satisfies Eq. (21.26).

$$x = x_h + x_p.$$

Particular Solution for an Oscillatory Forcing Function

Particular solution of Eq. (21.26) if the forcing function is an oscillatory function of the form $a(t) = a_0 \sin \omega_0 t + b_0 \cos \omega_0 t,$ (21.27) where a_0, b_0, and ω_0 are constants.	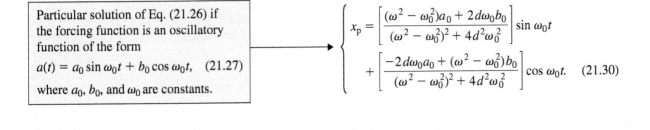

$$\begin{cases} x_p = \left[\dfrac{(\omega^2 - \omega_0^2)a_0 + 2d\omega_0 b_0}{(\omega^2 - \omega_0^2)^2 + 4d^2\omega_0^2} \right] \sin \omega_0 t \\[3mm] \quad + \left[\dfrac{-2d\omega_0 a_0 + (\omega^2 - \omega_0^2)b_0}{(\omega^2 - \omega_0^2)^2 + 4d^2\omega_0^2} \right] \cos \omega_0 t. \quad (21.30) \end{cases}$$

Amplitude of the vibration described by Eq. (21.30).	$E_p = \dfrac{\sqrt{a_0^2 + b_0^2}}{\sqrt{(\omega^2 - \omega_0^2)^2 + 4d^2\omega_0^2}}$ (21.31)

Particular Solution for a Polynomial Forcing Function

Particular solution of Eq. (21.26) if the forcing function is a polynomial function of the form $a(t) = a_0 + a_1 t + a_2 t^2 + \cdots + a_N t^N.$ where a_0, a_1, and a_N are constants. The constants $A_0, A_1, \ldots, A_N$ must be determined by substituting Eq. (21.32) into Eq. (21.26).	$x_p = A_0 + A_1 t + A_2 t^2 + \cdots + A_N t^N.$ (21.32)

Particular solution of Eq. (21.26) if the forcing function is the polynomial $a(t) = a_0 + a_1 t.$	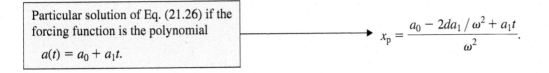 $x_p = \dfrac{a_0 - 2da_1 / \omega^2 + a_1 t}{\omega^2}.$

Active Example 21.5 | **Oscillatory Forcing Function** (▶ *Related Problem 21.61*)

An engineer designing a vibration isolation system for an instrument console models the console and isolation system as a damped spring–mass oscillator with mass $m = 2$ kg, spring constant $k = 8$ N/m, and damping constant $c = 1$ N-s/m. To determine the system's response to external vibration, she assumes that the system is initially stationary and at $t = 0$ a force $F(t) = 20 \sin 4t$ N is applied to the mass. Determine the position of the mass as a function of time.

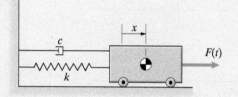

Strategy

The forcing function is $a(t) = F(t)/m = 10 \sin 4t$ m/s², which is of the form of Eq. (21.27). We can obtain the particular solution from Eq. (21.30). We must determine whether the damping is subcritical, critical, or supercritical and choose the appropriate form of the homogeneous solution. Then the initial conditions can be used to determine the unknown constants in the homogeneous solution, completing the solution.

Solution

The forcing function is of the form of Eq. (21.27) with $a_0 = 10$ m/s², $b_0 = 0$, and $\omega_0 = 4$ rad/s. The constants $\omega = \sqrt{k/m} = 2$ rad/s and $d = c/2m = 0.25$ rad/s. Substituting these values into Eq. (21.30) yields the particular solution.

$\longrightarrow$ $x_p = -0.811 \sin 4t - 0.135 \cos 4t$

Because $d < \omega$, the system is subcritically damped. The term $\omega_d = \sqrt{\omega^2 - d^2} = 1.98$ rad/s. The homogeneous solution is given by Eq. (21.19).

$\longrightarrow$ $x_h = e^{-0.25t}(A \sin 1.98t + B \cos 1.98t)$.

The solution is the sum of the homogeneous and particular solutions.

$\longrightarrow$
$$\begin{cases} x = x_h + x_p \\ \quad = e^{-0.25t}(A \sin 1.98t + B \cos 1.98t) \\ \quad - 0.811 \sin 4t - 0.135 \cos 4t. \end{cases}$$

At $t = 0$, $x = 0$ and $dx/dt = 0$. Using these conditions to determine the constants A and B determines the position of the mass as a function of time.

$$\begin{cases} x = e^{-0.25t}(1.651 \sin 1.98t + 0.135 \cos 1.98t) \\ \quad - 0.811 \sin 4t - 0.135 \cos 4t \text{ m.} \end{cases}$$

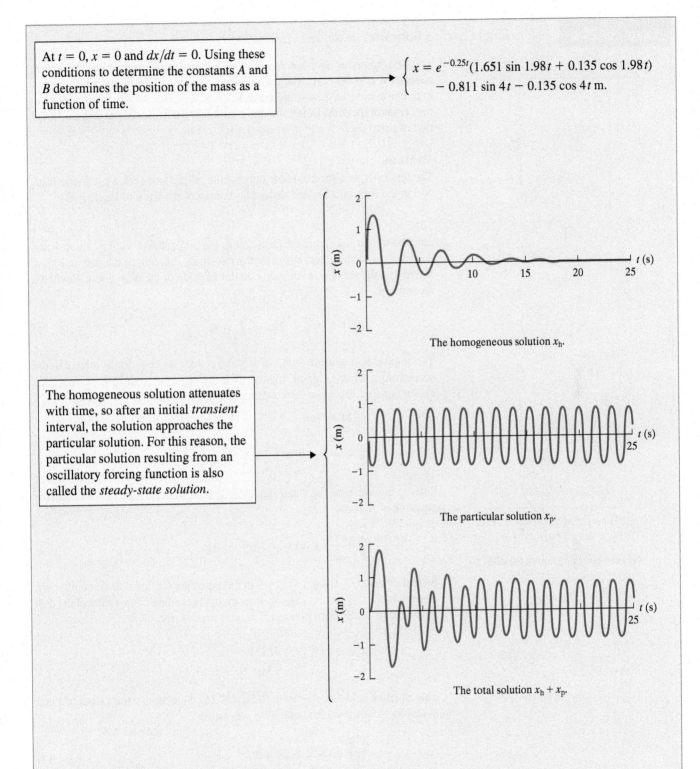

The homogeneous solution x_h.

The homogeneous solution attenuates with time, so after an initial *transient* interval, the solution approaches the particular solution. For this reason, the particular solution resulting from an oscillatory forcing function is also called the *steady-state solution*.

The particular solution x_p.

The total solution $x_h + x_p$.

Practice Problem What is the amplitude of the particular solution?

Answer: 0.822 m.

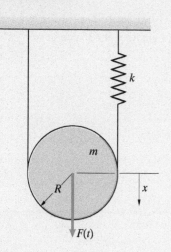

| Example 21.6 | **Polynomial Forcing Function** (▶ *Related Problem 21.63*) |

The homogeneous disk has radius $R = 2$ m and mass $m = 4$ kg. The spring constant is $k = 30$ N/m. The disk is initially stationary in its equilibrium position, and at $t = 0$ a downward force $F(t) = 12 + 12t - 0.6t^2$ N is applied to the center of the disk. Determine the position of the center of the disk as a function of time.

Strategy

The force $F(t)$ is a second-order polynomial, so we will seek a particular solution in the form of a second-order polynomial of the form of Eq. (21.32).

Solution

Let x be the displacement of the center of the disk relative to its position when the spring is unstretched. We draw the free-body diagram of the disk in Fig. a, where T is the tension in the cable on the left side of the disk. From Newton's second law,

$$F(t) + mg - 2kx - T = m\frac{d^2x}{dt^2}. \quad (1)$$

The angular acceleration of the disk in the clockwise direction is related to the acceleration of the center of the disk by $\alpha = (d^2x/dt^2)/R$. Using this expression, we can write the equation of angular motion of the disk as

$$\Sigma M = I\alpha:$$

$$TR - 2kxR = \left(\frac{1}{2}mR^2\right)\left(\frac{1}{R}\frac{d^2x}{dt^2}\right).$$

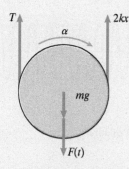

(a) Free-body diagram of the disk.

Solving this equation for T and substituting the result into Eq. (1), we obtain the equation of motion:

$$\frac{3}{2}m\frac{d^2x}{dt^2} + 4kx = F(t) + mg. \quad (2)$$

Setting $d^2x/dt^2 = 0$ and $F(t) = 0$ in this equation, we find that the equilibrium position of the disk is $x = mg/4k$. In terms of the position of the center of the disk relative to its equilibrium position $\tilde{x} = x - mg/4k$, Eq. (2) is

$$\frac{d^2\tilde{x}}{dt^2} + \frac{8k}{3m}\tilde{x} = \frac{2F(t)}{3m}.$$

This equation is identical in form to Eq. (21.26). Substituting the values of k and m and the polynomial function $F(t)$, we obtain

$$\frac{d^2\tilde{x}}{dt^2} + 20\tilde{x} = 2 + 2t - 0.1t^2. \quad (3)$$

Comparing this equation with Eq. (21.26), we see that $d = 0$ (there is no damping) and $\omega^2 = 20$ (rad/s)^2. From Eq. (21.19), the homogeneous solution is

$$\tilde{x}_h = A \sin 4.472t + B \cos 4.472t.$$

To obtain the particular solution, we seek a solution in the form of a polynomial of the same order as $F(t)$. That is,

$$\tilde{x}_p = A_0 + A_1 t + A_2 t^2,$$

where A_0, A_1, and A_2 are constants we must determine. We substitute this expression into Eq. (3) and collect terms of equal powers in t:

$$(2A_2 + 20A_0 - 2) + (20A_1 - 2)t + (20A_2 + 0.1)t^2 = 0.$$

This equation is satisfied if the coefficients multiplying each power of t equal zero, which yields

$$2A_2 + 20A_0 = 2,$$
$$20A_1 = 2,$$

and

$$20A_2 = -0.1.$$

Solving these three equations for A_0, A_1, and A_2, we obtain the particular solution:

$$\tilde{x}_p = 0.101 + 0.100t - 0.005t^2.$$

The complete solution is

$$\tilde{x} = \tilde{x}_h + \tilde{x}_p$$
$$= A \sin 4.472t + B \cos 4.472t + 0.101 + 0.100t - 0.005t^2.$$

At $t = 0$, $\tilde{x} = 0$ and $d\tilde{x}/dt = 0$. Using these conditions to determine A and B, we obtain the position of the center of the disk (in meters) as a function of time:

$$\tilde{x} = -0.022 \sin 4.472t - 0.101 \cos 4.472t + 0.101 + 0.100t - 0.005t^2.$$

The graph shows the position for the first 25 seconds of motion. The undamped, oscillatory homogeneous solution is superimposed on the slowly varying particular solution.

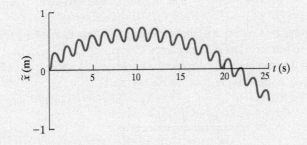

Critical Thinking

Forcing functions that arise in engineering applications can often be expressed as a Taylor series or approximated by a power series over some interval of time. When that is the case, you can analyze the response of the system during that interval of time by the approach used in this example.

Example 21.7 | **Displacement Transducers** (▶ *Related Problems 21.70–21.73*)

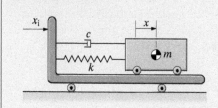

A damped spring–mass oscillator, or a device that can be modeled as a damped spring–mass oscillator, can be used to measure an object's displacement. Suppose that the base of the spring–mass oscillator shown is attached to an object and the coordinate x_i is a displacement to be measured relative to an inertial reference frame. The coordinate x measures the displacement of the mass relative to the base. When $x = 0$, the spring is unstretched. Suppose that the system is initially stationary and at $t = 0$ the base undergoes the oscillatory motion

$$x_i = a_i \sin \omega_i t + b_i \cos \omega_i t. \tag{1}$$

If $m = 2$ kg, $k = 8$ N/m, $c = 4$ N-s/m, $a_i = 0.1$ m, $b_i = 0.1$ m, and $\omega_i = 10$ rad/s, what is the resulting steady-state amplitude of the displacement of the mass relative to the base?

Strategy

We need to obtain the equation of motion for the mass, accounting for the effect of the oscillating base. To do so, we must write Newton's second law in terms of the acceleration of the mass *relative to the inertial reference frame*.

Solution

The acceleration of the mass relative to the base is d^2x/dt^2, so its acceleration relative to the inertial reference frame is $(d^2x/dt^2) + (d^2x_i/dt^2)$. Newton's second law for the mass is

$$-c\frac{dx}{dt} - kx = m\left(\frac{d^2x}{dt^2} + \frac{d^2x_i}{dt^2}\right).$$

We can write this equation as

$$\frac{d^2x}{dt^2} + 2d\frac{dx}{dt} + \omega^2 x = a(t),$$

where $d = c/2m = 1$ rad/s, $\omega = \sqrt{k/m} = 2$ rad/s, and the function

$$a(t) = -\frac{d^2x_i}{dt^2} = a_i\omega_i^2 \sin \omega_i t + b_i\omega_i^2 \cos \omega_i t. \tag{2}$$

Thus, we obtain an equation of motion identical in form to that for a spring–mass oscillator subjected to an oscillatory force. Comparing Eq. (2) with Eq. (21.27), we can obtain the amplitude of the particular (steady-state) solution from Eq. (21.31) by setting $a_0 = a_i\omega_i^2$, $b_0 = b_i\omega_i^2$, and $\omega_0 = \omega_i$:

$$E_p = \frac{\omega_i^2\sqrt{a_i^2 + b_i^2}}{\sqrt{(\omega^2 - \omega_i^2)^2 + 4d^2\omega_i^2}}. \tag{3}$$

Therefore, the steady-state amplitude of the displacement of the mass relative to its base is

$$E_p = \frac{(10)^2\sqrt{(0.1)^2 + (0.1)^2}}{\sqrt{[(2)^2 - (10)^2]^2 + 4(1)^2(10)^2}} = 0.144 \text{ m.}$$

Problems

21.58 The mass $m = 2$ slug and the spring constant is $k = 72$ lb/ft. The spring is unstretched when $x = 0$. The mass is initially stationary with the spring unstretched, and at $t = 0$ the force $F(t) = 10 \sin 4t$ lb is applied to the mass. What is the position of the mass at $t = 2$ s?

21.59 The mass $m = 2$ slug and the spring constant is $k = 72$ lb/ft. The spring is unstretched when $x = 0$. At $t = 0$, $x = 1$ ft, $dx/dt = 1$ ft/s, and the force $F(t) = 10 \sin 4t + 10 \cos 4t$ lb is applied to the mass. What is the position of the mass at $t = 2$ s?

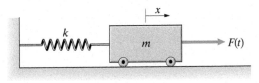

Problems 21.58/21.59

21.60 The damped spring–mass oscillator is initially stationary with the spring unstretched. At $t = 0$, a constant force $F(t) = 6$ N is applied to the mass.

(a) What is the steady-state (particular) solution?

(b) Determine the position of the mass as a function of time.

▶ **21.61** The damped spring–mass oscillator is initially stationary with the spring unstretched. At $t = 0$, a force $F(t) = 6 \cos 1.6t$ N is applied to the mass.

(a) What is the steady-state (particular) solution?

(b) Determine the position of the mass as a function of time. (See Active Example 21.5.)

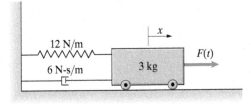

Problems 21.60/21.61

21.62 The disk with moment of inertia $I = 3$ kg-m^2 rotates about a fixed shaft and is attached to a torsional spring with constant $k = 20$ N-m/rad. At $t = 0$, the angle $\theta = 0$, the angular velocity is $d\theta/dt = 4$ rad/s, and the disk is subjected to a couple $M(t) = 10 \sin 2t$ N-m. Determine θ as a function of time.

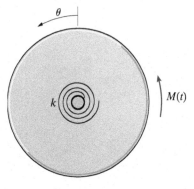

Problem 21.62

▶ **21.63** The stepped disk weighs 20 lb and its moment of inertia is $I = 0.6$ slug-ft^2. It rolls on the horizontal surface. The disk is initially stationary with the spring unstretched, and at $t = 0$ a constant force $F = 10$ lb is applied as shown. Determine the position of the center of the disk as a function of time. (See Example 21.6.)

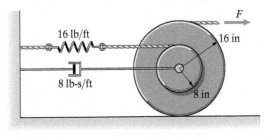

Problem 21.63

21.64* An electric motor is bolted to a metal table. When the motor is on, it causes the tabletop to vibrate horizontally. Assume that the legs of the table behave like linear springs, and neglect damping. The total weight of the motor and the tabletop is 150 lb. When the motor is not turned on, the frequency of horizontal vibration of the tabletop and motor is 5 Hz. When the motor is running at 600 rpm, the amplitude of the horizontal vibration is 0.01 in. What is the magnitude of the oscillatory force exerted on the table by the motor at its 600-rpm running speed?

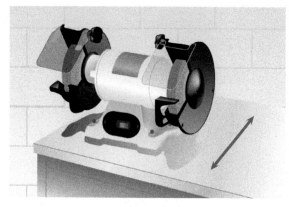

Problem 21.64

21.65 The moments of inertia of gears A and B are $I_A = 0.014$ slug-ft^2 and $I_B = 0.100$ slug-ft^2. Gear A is connected to a torsional spring with constant $k = 2$ ft-lb/rad. The system is in equilibrium at $t = 0$ when it is subjected to an oscillatory force $F(t) = 4 \sin 3t$ lb. What is the downward displacement of the 5-lb weight as a function of time?

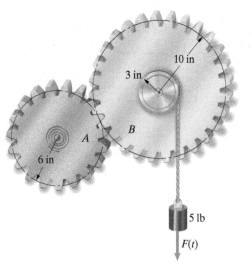

Problem 21.65

21.66* A 1.5-kg cylinder is mounted on a sting in a wind tunnel with the cylinder axis transverse to the direction of flow. When there is no flow, a 10-N vertical force applied to the cylinder causes it to deflect 0.15 mm. When air flows in the wind tunnel, vortices subject the cylinder to alternating lateral forces. The velocity of the air is 5 m/s, the distance between vortices is 80 mm, and the magnitude of the lateral forces is 1 N. If you model the lateral forces by the oscillatory function $F(t) = (1.0) \sin \omega_0 t$ N, what is the amplitude of the steady-state lateral motion of the sphere?

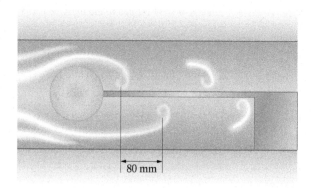

| 80 mm |

Problem 21.66

21.67 Show that the amplitude of the particular solution given by Eq. (21.31) is a maximum when the frequency of the oscillatory forcing function is $\omega_0 = \sqrt{\omega^2 - 2d^2}$.

21.68* A sonobuoy (sound-measuring device) floats in a standing-wave tank. The device is a cylinder of mass m and cross-sectional area A. The water density is ρ, and the buoyancy force supporting the buoy equals the weight of the water that would occupy the volume of the part of the cylinder below the surface. When the water in the tank is stationary, the buoy is in equilibrium in the vertical position shown at the left. Waves are then generated in the tank, causing the depth of the water at the sonobuoy's position *relative to its original depth* to be $d = d_0 \sin \omega_0 t$. Let y be the sonobuoy's vertical position relative to its original position. Show that the sonobuoy's vertical position is governed by the equation

$$\frac{d^2 y}{dt^2} + \left(\frac{A\rho g}{m}\right) y = \left(\frac{A\rho g}{m}\right) d_0 \sin \omega_0 t.$$

21.69 Suppose that the mass of the sonobuoy in Problem 21.68 is $m = 10$ kg, its diameter is 125 mm, and the water density is $\rho = 1025$ kg/m^3. If $d = 0.1 \sin 2t$ m, what is the magnitude of the steady-state vertical vibrations of the sonobuoy?

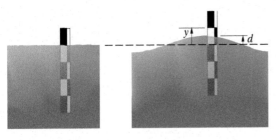

Problems 21.68/21.69

▶ **21.70** The mass weighs 50 lb. The spring constant is $k = 200$ lb/ft, and $c = 10$ lb-s/ft. If the base is subjected to an oscillatory displacement x_i of amplitude 10 in and frequency $\omega_i = 15$ rad/s, what is the resulting steady-state amplitude of the displacement of the mass relative to the base? (See Example 21.7.)

▶ **21.71** The mass is 100 kg. The spring constant is $k = 4$ N/m, and $c = 24$ N-s/m. The base is subjected to an oscillatory displacement of frequency $\omega_i = 0.2$ rad/s. The steady-state amplitude of the displacement of the mass relative to the base is measured and determined to be 200 mm. What is the amplitude of the displacement of the base? (See Example 21.7.)

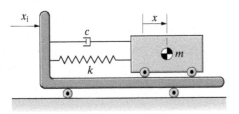

Problems 21.70/21.71

▶ **21.72** A team of engineering students builds the simple seismograph shown. The coordinate x_i measures the local horizontal ground motion. The coordinate x measures the position of the mass relative to the frame of the seismograph. The spring is unstretched when $x = 0$. The mass $m = 1$ kg, the spring constant $k = 10$ N/m, and $c = 2$ N-s/m. Suppose that the seismograph is initially stationary and that at $t = 0$ it is subjected to an oscillatory ground motion $x_i = 10 \sin 2t$ mm. What is the amplitude of the steady-state response of the mass? (See Example 21.7.)

▶ **21.73** In Problem 21.72, determine the position x of the mass relative to the base as a function of time. (See Example 21.7.)

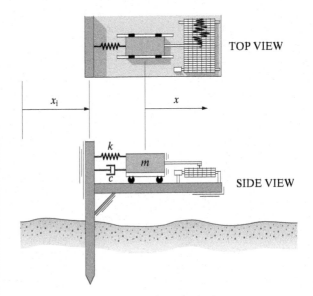

Problems 21.72/21.73

Review Problems

21.74 The coordinate x measures the displacement of the mass relative to the position in which the spring is unstretched. The mass is given the initial conditions

$$t = 0 \begin{cases} x = 0.1 \text{ m,} \\ \dfrac{dx}{dt} = 0. \end{cases}$$

(a) Determine the position of the mass as a function of time.

(b) Draw graphs of the position and velocity of the mass as functions of time for the first 5 s of motion.

21.75 When $t = 0$, the mass is in the position in which the spring is unstretched and has a velocity of 0.3 m/s to the right. Determine the position of the mass as a function of time and the amplitude of the vibration

(a) by expressing the solution in the form given by Eq. (21.8) and

(b) by expressing the solution in the form given by Eq. (21.9).

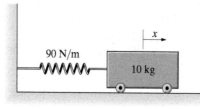

Problems 21.74/21.75

21.76 A homogeneous disk of mass m and radius R rotates about a fixed shaft and is attached to a torsional spring with constant k. (The torsional spring exerts a restoring moment of magnitude $k\theta$, where θ is the angle of rotation of the disk relative to its position in which the spring is unstretched.) Show that the period of rotational vibrations of the disk is $\tau = \pi R \sqrt{2m/k}$.

Problem 21.76

21.77 Assigned to determine the moments of inertia of astronaut candidates, an engineer attaches a horizontal platform to a vertical steel bar. The moment of inertia of the platform about L is 7.5 kg-m^2, and the frequency of torsional oscillations of the unloaded platform is 1 Hz. With an astronaut candidate in the position shown, the frequency of torsional oscillations is 0.520 Hz. What is the candidate's moment of inertia about L?

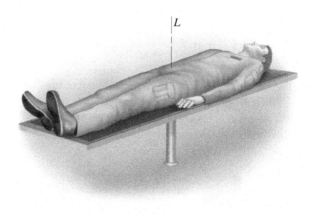

Problem 21.77

21.78 The 22-kg platen P rests on four roller bearings that can be modeled as 1-kg homogeneous cylinders with 30-mm radii. The spring constant is $k = 900$ N/m. What is the frequency of horizontal vibrations of the platen relative to its equilibrium position?

21.79 At $t = 0$, the platen described in Problem 21.78 is 0.1 m to the left of its equilibrium position and is moving to the right at 2 m/s. What are the platen's position and velocity at $t = 4$ s?

Problems 21.78/21.79

21.80 The moments of inertia of gears A and B are $I_A = 0.014$ slug-ft^2 and $I_B = 0.100$ slug-ft^2. Gear A is connected to a torsional spring with constant $k = 2$ ft-lb/rad. What is the frequency of angular vibrations of the gears relative to their equilibrium position?

21.81 The 5-lb weight in Problem 21.80 is raised 0.5 in from its equilibrium position and released from rest at $t = 0$. Determine the counterclockwise angular position of gear B relative to its equilibrium position as a function of time.

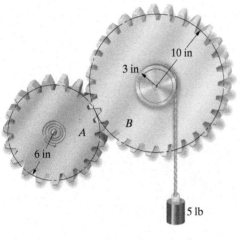

3 in

10 in

B

A

6 in

5 lb

Problems 21.80/21.81

21.82 The mass of the slender bar is m. The spring is unstretched when the bar is vertical. The light collar C slides on the smooth vertical bar so that the spring remains horizontal. Determine the frequency of small vibrations of the bar.

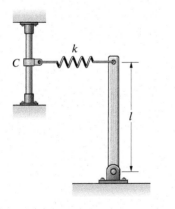

C k

l

Problem 21.82

21.83 A homogeneous hemisphere of radius R and mass m rests on a level surface. If you rotate the hemisphere slightly from its equilibrium position and release it, what is the frequency of its vibrations?

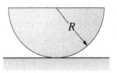

R

Problem 21.83

21.84 The frequency of the spring–mass oscillator is measured and determined to be 4.00 Hz. The oscillator is then placed in a barrel of oil, and its frequency is determined to be 3.80 Hz. What is the logarithmic decrement of vibrations of the mass when the oscillator is immersed in oil?

21.85 Consider the oscillator immersed in oil described in Problem 21.84. If the mass is displaced 0.1 m to the right of its equilibrium position and released from rest, what is its position relative to the equilibrium position as a function of time?

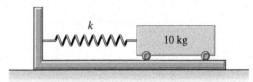

k

10 kg

Problems 21.84/21.85

21.86 The stepped disk weighs 20 lb, and its moment of inertia is $I = 0.6$ slug-ft^2. It rolls on the horizontal surface. If $c = 8$ lb-s/ft, what is the frequency of vibration of the disk?

21.87 The stepped disk described in Problem 21.86 is initially in equilibrium, and at $t = 0$ it is given a clockwise angular velocity of 1 rad/s. Determine the position of the center of the disk relative to its equilibrium position as a function of time.

21.88 The stepped disk described in Problem 21.86 is initially in equilibrium, and at $t = 0$ it is given a clockwise angular velocity of 1 rad/s. Determine the position of the center of the disk relative to its equilibrium position as a function of time if $c = 16$ lb-s/ft.

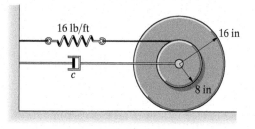

16 lb/ft   16 in

c 8 in

Problems 21.86–21.88

21.89 The 22-kg platen P rests on four roller bearings that can be modeled as 1-kg homogeneous cylinders with 30-mm radii. The spring constant is $k = 900$ N/m. The platen is subjected to a force $F(t) = 100 \sin 3t$ N. What is the magnitude of the platen's steady-state horizontal vibration?

21.90 At $t = 0$, the platen described in Problem 21.89 is 0.1 m to the right of its equilibrium position and is moving to the right at 2 m/s. Determine the platen's position relative to its equilibrium position as a function of time.

k P $F(t)$

Problems 21.89/21.90

21.91 The moments of inertia of gears A and B are $I_A = 0.014$ slug-ft^2 and $I_B = 0.100$ slug-ft^2. Gear A is connected to a torsional spring with constant $k = 2$ ft-lb/rad. The bearing supporting gear B incorporates a damping element that exerts a resisting moment on gear B of magnitude $1.5(d\theta_B/dt)$ ft-lb, where $d\theta_B/dt$ is the angular velocity of gear B in rad/s. What is the frequency of angular vibration of the gears?

21.92 The 5-lb weight in Problem 21.91 is raised 0.5 in from its equilibrium position and released from rest at $t = 0$. Determine the counterclockwise angular position of gear B relative to its equilibrium position as a function of time.

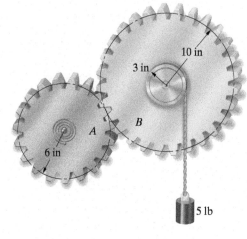

10 in

3 in

B

A

6 in

5 lb

Problems 21.91/21.92

21.93 The base and mass m are initially stationary. The base is then subjected to a vertical displacement $h \sin \omega_i t$ relative to its original position. What is the magnitude of the resulting steady-state vibration of the mass m *relative to the base*?

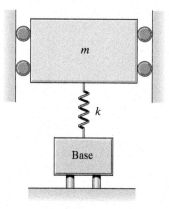

m

k

Base

Problem 21.93

21.94* The mass of the trailer, not including its wheels and axle, is m, and the spring constant of its suspension is k. To analyze the suspension's behavior, an engineer assumes that the height of the road surface relative to its mean height is $h \sin(2\pi x/\lambda)$. Assume that the trailer's wheels remain on the road and its horizontal component of velocity is v. Neglect the damping due to the suspension's shock absorbers.

(a) Determine the magnitude of the trailer's vertical steady-state vibration *relative to the road surface*.

(b) At what velocity v does resonance occur?

21.95* The trailer in Problem 21.94, not including its wheels and axle, weighs 1000 lb. The spring constant of its suspension is $k = 2400$ lb/ft, and the damping coefficient due to its shock absorbers is $c = 200$ lb-s/ft. The road surface parameters are $h = 2$ in and $\lambda = 8$ ft. The trailer's horizontal velocity is $v = 6$ mi/h. Determine the magnitude of the trailer's vertical steady-state vibration relative to the road surface, (a) neglecting the damping due to the shock absorbers and (b) not neglecting the damping.

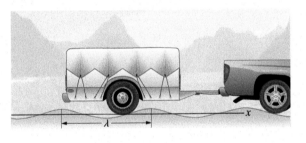

Problems 21.94/21.95

21.96* A disk with moment of inertia I rotates about a fixed shaft and is attached to a torsional spring with constant k. The angle θ measures the angular position of the disk relative to its position when the spring is unstretched. The disk is initially stationary with the spring unstretched. At $t = 0$, a time-dependent moment $M(t) = M_0(1 - e^{-t})$ is applied to the disk, where M_0 is a constant. Show that the angular position of the disk as a function of time is

$$\theta = \frac{M_0}{I} \left[-\frac{1}{\omega(1 + \omega^2)} \sin \omega t - \frac{1}{\omega^2(1 + \omega^2)} \cos \omega t \right.$$
$$\left. + \frac{1}{\omega^2} - \frac{1}{(1 + \omega^2)} e^{-t} \right].$$

Strategy: To determine the particular solution, seek a solution of the form

$$\theta_p = A_p + B_p e^{-t},$$

where A_p and B_p are constants that you must determine.

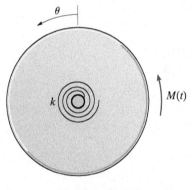

Problem 21.96

APPENDIX

A

Review of Mathematics

A.1 Algebra

Quadratic Equations

The solutions of the quadratic equation

$$ax^2 + bx + c = 0$$

are

$$x = \frac{-b \pm \sqrt{b^2 - 4ac}}{2a}.$$

Natural Logarithms

The natural logarithm of a positive real number x is denoted by $\ln x$. It is defined to be the number such that

$$e^{\ln x} = x,$$

where $e = 2.7182\ldots$ is the base of natural logarithms.

Logarithms have the following properties:

$$\ln(xy) = \ln x + \ln y,$$

$$\ln(x/y) = \ln x - \ln y,$$

$$\ln y^x = x \ln y.$$

A.2 Trigonometry

The trigonometric functions for a right triangle are

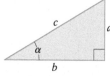

$$\sin \alpha = \frac{1}{\csc \alpha} = \frac{a}{c}, \qquad \cos \alpha = \frac{1}{\sec \alpha} = \frac{b}{c}, \qquad \tan \alpha = \frac{1}{\cot \alpha} = \frac{a}{b}.$$

The sine and cosine satisfy the relation

$$\sin^2 \alpha + \cos^2 \alpha = 1,$$

and the sine and cosine of the sum and difference of two angles satisfy

$$\sin(\alpha + \beta) = \sin \alpha \cos \beta + \cos \alpha \sin \beta,$$

$$\sin(\alpha - \beta) = \sin \alpha \cos \beta - \cos \alpha \sin \beta,$$

$$\cos(\alpha + \beta) = \cos \alpha \cos \beta - \sin \alpha \sin \beta,$$

$$\cos(\alpha - \beta) = \cos \alpha \cos \beta + \sin \alpha \sin \beta.$$

The **law of cosines** for an arbitrary triangle is

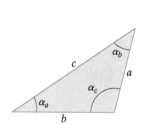

$$c^2 = a^2 + b^2 - 2ab \cos \alpha_c,$$

and the **law of sines** is

$$\frac{\sin \alpha_a}{a} = \frac{\sin \alpha_b}{b} = \frac{\sin \alpha_c}{c}.$$

A.3 Derivatives

$$\frac{d}{dx} x^n = n x^{n-1} \qquad\qquad \frac{d}{dx} \sin x = \cos x \qquad\qquad \frac{d}{dx} \sinh x = \cosh x$$

$$\frac{d}{dx} e^x = e^x \qquad\qquad \frac{d}{dx} \cos x = -\sin x \qquad\qquad \frac{d}{dx} \cosh x = \sinh x$$

$$\frac{d}{dx} \ln x = \frac{1}{x} \qquad\qquad \frac{d}{dx} \tan x = \frac{1}{\cos^2 x} \qquad\qquad \frac{d}{dx} \tanh x = \frac{1}{\cosh^2 x}$$

A.4 Integrals

$$\int x^n \, dx = \frac{x^{n+1}}{n+1} \qquad (n \neq -1)$$

$$\int x^{-1} \, dx = \ln x$$

$$\int \frac{dx}{a - bx^2} = \frac{1}{(ab)^{1/2}} \arctan \frac{(ab)^{1/2}x}{a}$$

$$\int \frac{dx}{a - bx^2} = \frac{1}{2(ab)^{1/2}} \ln \frac{a + x(ab)^{1/2}}{a - x(ab)^{1/2}}$$

$$\int \frac{x \, dx}{a - bx^2} = -\frac{1}{2b} \ln(a - bx^2)$$

$$\int (a + bx)^{1/2} \, dx = \frac{2}{3b}(a + bx)^{3/2}$$

$$\int x(a + bx)^{1/2} \, dx = -\frac{2(2a - 3bx)(a + bx)^{3/2}}{15b^2}$$

$$\int (1 + a^2x^2)^{1/2} \, dx = \frac{1}{2}\left\{ x(1 + a^2x^2)^{1/2} \right.$$

$$\left. + \frac{1}{a}\ln\left[x + \left(\frac{1}{a^2} + x^2 \right)^{1/2} \right] \right\}$$

$$\int x(1 + a^2x^2)^{1/2} \, dx = \frac{a}{3}\left(\frac{1}{a^2} + x^2 \right)^{3/2}$$

$$\int x^2(1 + a^2x^2)^{1/2} \, dx = \frac{1}{4}ax\left(\frac{1}{a^2} + x^2 \right)^{3/2}$$

$$- \frac{1}{8a^2}x(1 + a^2x^2)^{1/2} - \frac{1}{8a^3}\ln\left[x + \left(\frac{1}{a^2} + x^2 \right)^{1/2} \right]$$

$$\int (1 - a^2x^2)^{1/2} \, dx = \frac{1}{2}\left[x(1 - a^2x^2)^{1/2} + \frac{1}{a}\arcsin ax \right]$$

$$\int x(1 - a^2x^2)^{1/2} \, dx = -\frac{a}{3}\left(\frac{1}{a^2} - x^2 \right)^{3/2}$$

$$\int x^2(a^2 - x^2)^{1/2} \, dx = -\frac{1}{4}x(a^2 - x^2)^{3/2}$$

$$+ \frac{1}{8}a^2\left[x(a^2 - x^2)^{1/2} + a^2 \arcsin \frac{x}{a} \right]$$

$$\int \frac{dx}{(1 + a^2x^2)^{1/2}} = \frac{1}{a}\ln\left[x + \left(\frac{1}{a^2} + x^2 \right)^{1/2} \right]$$

$$\int \frac{dx}{(1 - a^2x^2)^{1/2}} = \frac{1}{a}\arcsin ax \quad \text{or} \quad -\frac{1}{a}\arccos ax$$

$$\int \sin x \, dx = -\cos x$$

$$\int \cos x \, dx = \sin x$$

$$\int \sin^2 x \, dx = -\frac{1}{2}\sin x \cos x + \frac{1}{2}x$$

$$\int \cos^2 x \, dx = \frac{1}{2}\sin x \cos x + \frac{1}{2}x$$

$$\int \sin^3 x \, dx = -\frac{1}{3}\cos x(\sin^2 x + 2)$$

$$\int \cos^3 x \, dx = \frac{1}{3}\sin x(\cos^2 x + 2)$$

$$\int \cos^4 x \, dx = \frac{3}{8}x + \frac{1}{4}\sin 2x + \frac{1}{32}\sin 4x$$

$$\int \sin^n x \cos x \, dx = \frac{(\sin x)^{n+1}}{n+1} \qquad (n \neq -1)$$

$$\int \sinh x \, dx = \cosh x$$

$$\int \cosh x \, dx = \sinh x$$

$$\int \tanh x \, dx = \ln \cosh x$$

$$\int e^{ax} \, dx = \frac{e^{ax}}{a}$$

$$\int xe^{ax} \, dx = \frac{e^{ax}}{a^2}(ax - 1)$$

A.5 Taylor Series

The Taylor series of a function $f(x)$ is

$$f(a + x) = f(a) + f'(a)x + \frac{1}{2!}f''(a)x^2 + \frac{1}{3!}f'''(a)x^3 + \cdots,$$

where the primes indicate derivatives.

Some useful Taylor series are

$$e^x = 1 + x + \frac{x^2}{2!} + \frac{x^3}{3!} + \cdots,$$

$$\sin(a + x) = \sin a + (\cos a)x - \frac{1}{2}(\sin a)x^2 - \frac{1}{6}(\cos a)x^3 + \cdots,$$

$$\cos(a + x) = \cos a - (\sin a)x - \frac{1}{2}(\cos a)x^2 + \frac{1}{6}(\sin a)x^3 + \cdots,$$

$$\tan(a + x) = \tan a + \left(\frac{1}{\cos^2 a}\right)x + \left(\frac{\sin a}{\cos^3 a}\right)x^2$$

$$+ \left(\frac{\sin^2 a}{\cos^4 a} + \frac{1}{3\cos^2 a}\right)x^3 + \cdots.$$

A.6 Vector Analysis

Cartesian Coordinates

The gradient of a scalar field ψ is

$$\nabla\psi = \frac{\partial\psi}{\partial x}\mathbf{i} + \frac{\partial\psi}{\partial y}\mathbf{j} + \frac{\partial\psi}{\partial z}\mathbf{k}.$$

The divergence and curl of a vector field $\mathbf{v} = v_x\mathbf{i} + v_y\mathbf{j} + v_z\mathbf{k}$ are

$$\nabla \cdot \mathbf{v} = \frac{\partial v_x}{\partial x} + \frac{\partial v_y}{\partial y} + \frac{\partial v_z}{\partial z},$$

$$\nabla \times \mathbf{v} = \begin{vmatrix} \mathbf{i} & \mathbf{j} & \mathbf{k} \\ \frac{\partial}{\partial x} & \frac{\partial}{\partial y} & \frac{\partial}{\partial z} \\ v_x & v_y & v_z \end{vmatrix}.$$

Cylindrical Coordinates

The gradient of a scalar field ψ is

$$\nabla\psi = \frac{\partial\psi}{\partial r}\mathbf{e}_r + \frac{1}{r}\frac{\partial\psi}{\partial\theta}\mathbf{e}_\theta + \frac{\partial\psi}{\partial z}\mathbf{e}_z.$$

The divergence and curl of a vector field $\mathbf{v} = v_r\mathbf{e}_r + v_\theta\mathbf{e}_\theta + v_z\mathbf{e}_z$ are

$$\nabla \cdot \mathbf{v} = \frac{\partial v_r}{\partial r} + \frac{v_r}{r} + \frac{1}{r}\frac{\partial v_\theta}{\partial\theta} + \frac{\partial v_z}{\partial z},$$

$$\nabla \times \mathbf{v} = \frac{1}{r}\begin{vmatrix} \mathbf{e}_r & r\mathbf{e}_\theta & \mathbf{e}_z \\ \frac{\partial}{\partial r} & \frac{\partial}{\partial\theta} & \frac{\partial}{\partial z} \\ v_r & rv_\theta & v_z \end{vmatrix}.$$

APPENDIX

B

Properties of Areas and Lines

B.1 Areas

The coordinates of the centroid of the area A are

$$\bar{x} = \frac{\int_A x\, dA}{\int_A dA}, \qquad \bar{y} = \frac{\int_A y\, dA}{\int_A dA}.$$

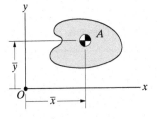

The moment of inertia about the x axis I_x, the moment of inertia about the y axis I_y, and the product of inertia I_{xy} are

$$I_x = \int_A y^2\, dA, \qquad I_y = \int_A x^2\, dA, \qquad I_{xy} = \int_A xy\, dA.$$

The polar moment of inertia about O is

$$J_O = \int_A r^2\, dA = \int_A (x^2 + y^2)\, dA = I_x + I_y.$$

Area $= bh$

$$I_x = \frac{1}{3}bh^3, \qquad I_y = \frac{1}{3}hb^3, \qquad I_{xy} = \frac{1}{4}b^2h^2$$

$$I_{x'} = \frac{1}{12}bh^3, \qquad I_{y'} = \frac{1}{12}hb^3, \qquad I_{x'y'} = 0$$

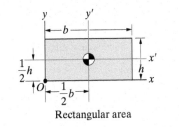

Rectangular area

$$\text{Area} = \frac{1}{2}bh$$

$$I_x = \frac{1}{12}bh^3, \qquad I_y = \frac{1}{4}hb^3, \qquad I_{xy} = \frac{1}{8}b^2h^2$$

$$I_{x'} = \frac{1}{36}bh^3, \qquad I_{y'} = \frac{1}{36}hb^3, \qquad I_{x'y'} = \frac{1}{72}b^2h^2$$

Triangular area

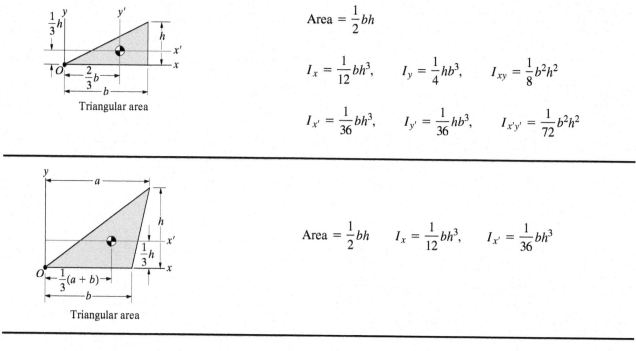

$$\text{Area} = \frac{1}{2}bh \qquad I_x = \frac{1}{12}bh^3, \qquad I_{x'} = \frac{1}{36}bh^3$$

Triangular area

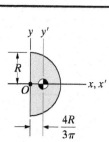

$$\text{Area} = \pi R^2 \qquad I_{x'} = I_{y'} = \frac{1}{4}\pi R^4, \qquad I_{x'y'} = 0$$

Circular area

$$\text{Area} = \frac{1}{2}\pi R^2 \qquad I_x = I_y = \frac{1}{8}\pi R^4, \qquad I_{xy} = 0$$

$$I_{x'} = \frac{1}{8}\pi R^4, \qquad I_{y'} = \left(\frac{\pi}{8} - \frac{8}{9\pi}\right)R^4, \qquad I_{x'y'} = 0$$

Semicircular area

$$\text{Area} = \frac{1}{4}\pi R^2 \qquad I_x = I_y = \frac{1}{16}\pi R^4, \qquad I_{xy} = \frac{1}{8}R^4$$

$$I_{x'} = I_{y'} = \left(\frac{\pi}{16} - \frac{4}{9\pi}\right)R^4, \qquad I_{x'y'} = \left(\frac{1}{8} - \frac{4}{9\pi}\right)R^4$$

Quarter-circular area

$$\text{Area} = \alpha R^2$$

$$I_x = \frac{1}{4}R^4\left(\alpha - \frac{1}{2}\sin 2\alpha\right), \qquad I_y = \frac{1}{4}R^4\left(\alpha + \frac{1}{2}\sin 2\alpha\right),$$

$$I_{xy} = 0$$

Circular sector

$$\text{Area} = \frac{1}{4}\pi ab$$

$$I_x = \frac{1}{16}\pi ab^3, \qquad I_y = \frac{1}{16}\pi a^3 b, \qquad I_{xy} = \frac{1}{8}a^2 b^2$$

Quarter-elliptical area

$$\text{Area} = \frac{cb^{n+1}}{n + 1}$$

$$I_x = \frac{c^3 b^{3n+1}}{9n + 3}, \qquad I_y = \frac{cb^{n+3}}{n + 3}, \qquad I_{xy} = \frac{c^2 b^{2n+2}}{4n + 4}$$

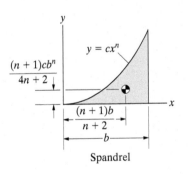

Spandrel

B.2 Lines

The coordinates of the centroid of the line L are

$$\bar{x} = \frac{\int_L x \, dL}{\int_L dL}, \qquad \bar{y} = \frac{\int_L y \, dL}{\int_L dL}, \qquad \bar{z} = \frac{\int_L z \, dL}{\int_L dL}.$$

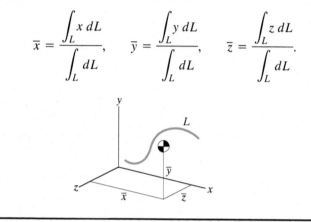

Length $= \pi R$

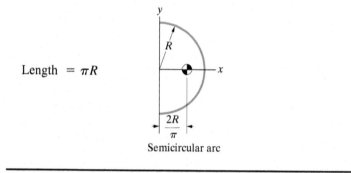

Semicircular arc

Length $= \frac{1}{2}\pi R$

Quarter-circular arc

Length $= 2\alpha R$

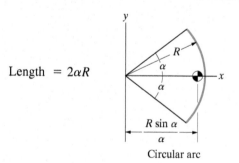

Circular arc

C

Properties of Volumes and Homogeneous Objects

The moments and products of inertia of the object in terms of the xyz coordinate system are,

$$I_{x \text{ axis}} = I_{xx} = \int_m (y^2 + z^2)\, dm,$$

$$I_{y \text{ axis}} = I_{yy} = \int_m (x^2 + z^2)\, dm,$$

$$I_{z \text{ axis}} = I_{zz} = \int_m (x^2 + y^2)\, dm,$$

$$I_{xy} = \int_m xy\, dm, \quad I_{yz} = \int_m yz\, dm,$$

$$I_{zx} = \int_m zx\, dm.$$

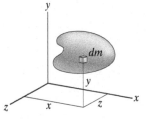

$$I_{x \text{ axis}} = 0, \qquad I_{y \text{ axis}} = I_{z \text{ axis}} = \frac{1}{3} ml^2,$$

$$I_{xy} = I_{yz} = I_{zx} = 0.$$

$$I_{x' \text{ axis}} = 0, \qquad I_{y' \text{ axis}} = I_{z' \text{ axis}} = \frac{1}{12} ml^2,$$

$$I_{x'y'} = I_{y'z'} = I_{z'x'} = 0.$$

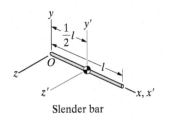

Slender bar

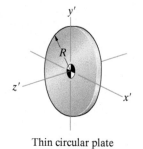

$$I_{x' \text{ axis}} = I_{y' \text{ axis}} = \frac{1}{4}mR^2, \qquad I_{z' \text{ axis}} = \frac{1}{2}mR^2,$$

$$I_{x'y'} = I_{y'z'} = I_{z'x'} = 0.$$

Thin circular plate

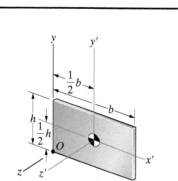

$$I_{x \text{ axis}} = \frac{1}{3}mh^2, \qquad I_{y \text{ axis}} = \frac{1}{3}mb^2, \qquad I_{z \text{ axis}} = \frac{1}{3}m(b^2 + h^2),$$

$$I_{xy} = \frac{1}{4}mbh, \qquad I_{yz} = I_{zx} = 0.$$

$$I_{x' \text{ axis}} = \frac{1}{12}mh^2, \qquad I_{y' \text{ axis}} = \frac{1}{12}mb^2, \qquad I_{z' \text{ axis}} = \frac{1}{12}m(b^2 + h^2),$$

$$I_{x'y'} = I_{y'z'} = I_{z'x'} = 0.$$

Thin rectangular plate

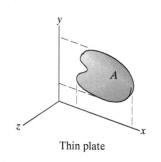

$$I_{x \text{ axis}} = \frac{m}{A}I_x, \qquad I_{y \text{ axis}} = \frac{m}{A}I_y, \qquad I_{z \text{ axis}} = I_{x \text{ axis}} + I_{y \text{ axis}},$$

$$I_{xy} = \frac{m}{A}I_{xy}^A, \qquad I_{yz} = I_{zx} = 0.$$

(The terms I_x, I_y, and I_{xy}^A are the moments and product of inertia of the plate's cross-sectional area A).

Thin plate

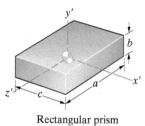

Volume $= abc$

$$I_{x' \text{ axis}} = \frac{1}{12}m(a^2 + b^2), \qquad I_{y' \text{ axis}} = \frac{1}{12}m(a^2 + c^2),$$

$$I_{z' \text{ axis}} = \frac{1}{12}m(b^2 + c^2), \qquad I_{x'y'} = I_{y'z'} = I_{z'x'} = 0.$$

Rectangular prism

Volume $= \pi R^2 l$

$$I_{x \text{ axis}} = I_{y \text{ axis}} = m\left(\frac{1}{3}l^2 + \frac{1}{4}R^2\right), \qquad I_{z \text{ axis}} = \frac{1}{2}mR^2,$$

$$I_{xy} = I_{yz} = I_{zx} = 0.$$

$$I_{x' \text{ axis}} = I_{y' \text{ axis}} = m\left(\frac{1}{12}l^2 + \frac{1}{4}R^2\right), \qquad I_{z' \text{ axis}} = \frac{1}{2}mR^2,$$

$$I_{x'y'} = I_{y'z'} = I_{z'x'} = 0.$$

Circular cylinder

Volume $= \frac{1}{3}\pi R^2 h$

$$I_{x \text{ axis}} = I_{y \text{ axis}} = m\left(\frac{3}{5}h^2 + \frac{3}{20}R^2\right), \qquad I_{z \text{ axis}} = \frac{3}{10}mR^2,$$

$$I_{xy} = I_{yz} = I_{zx} = 0.$$

$$I_{x' \text{ axis}} = I_{y' \text{ axis}} = m\left(\frac{3}{80}h^2 + \frac{3}{20}R^2\right), \qquad I_{z' \text{ axis}} = \frac{3}{10}mR^2,$$

$$I_{x'y'} = I_{y'z'} = I_{z'x'} = 0.$$

Circular cone

Volume $= \frac{4}{3}\pi R^3$

$$I_{x' \text{ axis}} = I_{y' \text{ axis}} = I_{z' \text{ axis}} = \frac{2}{5}mR^2,$$

$$I_{x'y'} = I_{y'z'} = I_{z'x'} = 0.$$

Sphere

Volume $= \frac{2}{3}\pi R^3$

$$I_{x \text{ axis}} = I_{y \text{ axis}} = I_{z \text{ axis}} = \frac{2}{5}mR^2$$

$$I_{x' \text{ axis}} = I_{y' \text{ axis}} = \frac{83}{320}mR^2, \qquad I_{z' \text{ axis}} = \frac{2}{5}mR^2$$

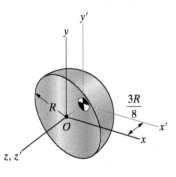

Hemisphere

APPENDIX

D

Spherical Coordinates

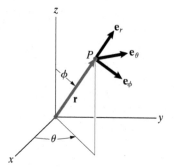

This appendix summarizes the equations of kinematics and vector calculus in spherical coordinates.

The position vector, velocity, and acceleration are

$$\mathbf{r} = r\mathbf{e}_r,$$

$$\mathbf{v} = \frac{dr}{dt}\mathbf{e}_r + r\frac{d\phi}{dt}\mathbf{e}_\phi + r\frac{d\theta}{dt}\sin\phi\,\mathbf{e}_\theta,$$

$$\mathbf{a} = \left[\frac{d^2r}{dt^2} - r\left(\frac{d\phi}{dt}\right)^2 - r\left(\frac{d\theta}{dt}\right)^2\sin^2\phi\right]\mathbf{e}_r$$

$$+ \left[r\frac{d^2\phi}{dt^2} + 2\frac{dr}{dt}\frac{d\phi}{dt} - r\left(\frac{d\theta}{dt}\right)^2\sin\phi\cos\phi\right]\mathbf{e}_\phi$$

$$+ \left[r\frac{d^2\theta}{dt^2}\sin\phi + 2\frac{dr}{dt}\frac{d\theta}{dt}\sin\phi + 2r\frac{d\phi}{dt}\frac{d\theta}{dt}\cos\phi\right]\mathbf{e}_\theta.$$

The gradient of a scalar field ψ is

$$\nabla\psi = \frac{\partial\psi}{\partial r}\mathbf{e}_r + \frac{1}{r}\frac{\partial\psi}{\partial\phi}\mathbf{e}_\phi + \frac{1}{r\sin\phi}\frac{\partial\psi}{\partial\theta}\mathbf{e}_\theta.$$

The divergence and curl of a vector field $\mathbf{v} = v_r\mathbf{e}_r + v_\theta\mathbf{e}_\theta + v_\phi\mathbf{e}_\phi$ are

$$\nabla\cdot\mathbf{v} = \frac{1}{r^2}\frac{\partial}{\partial r}(r^2v_r) + \frac{1}{r\sin\phi}\frac{\partial}{\partial\phi}(v_\phi\sin\phi) + \frac{1}{r\sin\phi}\frac{\partial v_\theta}{\partial\theta},$$

$$\nabla\times\mathbf{v} = \frac{1}{r^2\sin\phi}\begin{vmatrix} \mathbf{e}_r & r\mathbf{e}_\phi & r\sin\phi\,\mathbf{e}_\theta \\ \dfrac{\partial}{\partial r} & \dfrac{\partial}{\partial\phi} & \dfrac{\partial}{\partial\theta} \\ v_r & rv_\phi & r\sin\phi v_\theta \end{vmatrix}.$$

APPENDIX

E

D'Alembert's Principle

This appendix describes an alternative approach for obtaining the equations of planar motion for a rigid body. By writing Newton's second law as

$$\Sigma \mathbf{F} + (-m\mathbf{a}) = \mathbf{0}, \tag{E.1}$$

we can regard it as an "equilibrium" equation stating that the sum of the forces, including an *inertial force* $-m\mathbf{a}$, equals zero (Fig. E.1). To state the equation of angular motion in an equivalent way, we use Eq. (18.19), which relates the total moment about a fixed point O to the acceleration of the center of mass and the angular acceleration in general planar motion:

$$\Sigma M_O = (\mathbf{r} \times m\mathbf{a}) \cdot \mathbf{k} + I\alpha.$$

We write this equation as

$$\Sigma M_O + [\mathbf{r} \times (-m\mathbf{a})] \cdot \mathbf{k} + (-I\alpha) = 0. \tag{E.2}$$

The term $[\mathbf{r} \times (-m\mathbf{a})] \cdot \mathbf{k}$ is the moment about O due to the inertial force $-m\mathbf{a}$. We can therefore regard this equation as an "equilibrium" equation stating that the sum of the moments about any fixed point, including the moment due to the inertial force $-m\mathbf{a}$ acting at the center of mass and an *inertial couple* $-I\alpha$, equals zero.

Stated in this way, the equations of motion for a rigid body are analogous to the equations for static equilibrium: The sum of the forces equals zero and the sum of the moments about any fixed point equals zero when we properly account for inertial forces and couples. This is called *D'Alembert's principle*.

If we define ΣM_O and α to be positive in the counterclockwise direction, the unit vector $\mathbf{k}$ in Eq. (E.2) points out of the page and the term $[\mathbf{r} \times (-m\mathbf{a})] \cdot \mathbf{k}$ is the counterclockwise moment due to the inertial force. This vector operation determines the moment, or we can evaluate it by using the fact that its magnitude is the product of the magnitude of the inertial force and the perpendicular distance from point O to the line of action of the force (Fig. E.2 a). The moment is positive if it is counterclockwise, as in Fig. E.2a, and negative if it is clockwise. Notice that the sense of the inertial couple is opposite to that of the angular acceleration (Fig. E.2b).

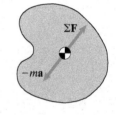

Figure E.1
The sum of the external forces and the inertial force is zero.

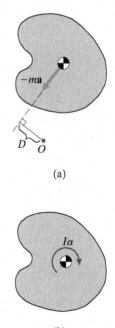

(a)

(b)

Figure E.2
(a) The magnitude of the moment due to the inertial force is $|-m\mathbf{a}|D$.
(b) A clockwise inertial couple results from a counterclockwise angular acceleration.

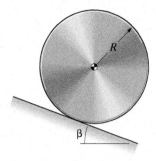

Figure E.3
Disk rolling on an inclined surface.

As an example, consider a disk of mass m and moment of inertia I that is rolling on an inclined surface (Fig. E.3). We can use D'Alembert's principle to determine the disk's angular acceleration and the forces exerted on it by the surface. The angular acceleration of the disk and the acceleration of its center are shown in Fig. E.4a. In Figure E.4b, we draw the free-body diagram of the disk showing its weight, the normal and friction forces exerted by the surface, *and the inertial force and couple.* Equation (E.1) is

$$\Sigma \mathbf{F} + (-m\mathbf{a}) = 0:$$

$$(mg \sin \beta - f)\mathbf{i} + (N - mg \cos \beta)\mathbf{j} - ma_x\mathbf{i} = 0.$$

From this vector equation, we obtain the equations

$$mg \sin \beta - f - ma_x = 0,$$

$$N - mg \cos \beta = 0. \tag{E.3}$$

We now apply Eq. (E.2). By evaluating moments about the point where the disk is in contact with the surface, we can eliminate f and N from the resulting equation:

$$\Sigma M_O + [\mathbf{r} \times (-m\mathbf{a})] \cdot \mathbf{k} + (-I\alpha) = 0:$$

$$-R(mg \sin \beta) + R(ma_x) - I\alpha = 0. \tag{E.4}$$

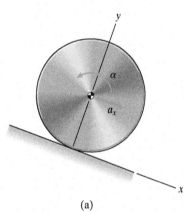

(a)

The acceleration of the center of the rolling disk is related to the counterclockwise angular acceleration by $a_x = -R\alpha$. Substituting this relation into Eq. (E.4) and solving for the angular acceleration, we obtain

$$\alpha = -\frac{mgR \sin \beta}{mR^2 + I}.$$

From this result, we also know a_x and can therefore solve Eqs. (E.3) for the normal and friction forces, obtaining

$$N = mg \cos \beta, \qquad f = \frac{mgI \sin \beta}{mR^2 + I}.$$

In Eq. (E.4) we evaluated the moment due to the inertial force by simply multiplying the magnitude of the force and the perpendicular distance from O to its line of action, but we could have evaluated it with the vector expression:

$$[\mathbf{r} \times (-m\mathbf{a})] \cdot \mathbf{k} = [(R\mathbf{j}) \times (-ma_x\mathbf{i})] \cdot \mathbf{k} = R(ma_x).$$

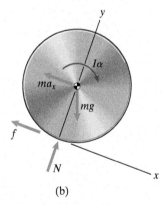

(b)

Figure E.4
(a) Acceleration of the center of the disk and its angular acceleration.
(b) Free-body diagram including the inertial force and couple.

Solutions to Practice Problems

Active Example 12.1

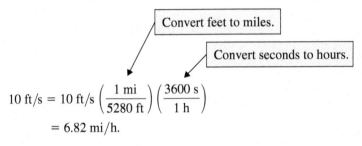

$$10\ \text{ft/s} = 10\ \text{ft/s} \left(\frac{1\ \text{mi}}{5280\ \text{ft}} \right) \left(\frac{3600\ \text{s}}{1\ \text{h}} \right)$$

$$= 6.82\ \text{mi/h}.$$

Active Example 12.4

| Use Eq. (12.6) to calculate the weight in newtons. | ⟶ | $W = mg = (0.397\ \text{kg})\,(9.81\ \text{m/s}^2) = 3.89\ \text{N}.$ |

Active Example 13.1

| Integrate the acceleration to determine the velocity as a function of time. A is an integration constant. | ⟶ | $\begin{cases} a = \dfrac{dv}{dt} = 2t, \\ v = t^2 + A. \end{cases}$ |

| Integrate the velocity to determine the position as a function of time. B is an integration constant. | ⟶ | $\begin{cases} v = \dfrac{ds}{dt} = t^2 + A, \\ s = \dfrac{1}{3}t^3 + At + B. \end{cases}$ |

| Use the known conditions at $t = 3$ s to determine A and B, obtaining $A = 5$ and $B = 6$. | ⟶ | $\begin{cases} s|_{t=3\,\text{s}} = 30 = \dfrac{1}{3}(3)^3 + A(3) + B, \\ v|_{t=3\,\text{s}} = 14 = (3)^2 + A. \end{cases}$ |

| Beginning with the x component of the position, differentiate to determine the x components of the velocity and acceleration as functions of time. | ⟶ | $\begin{cases} x = 8t^2\ \text{ft}, \\ v_x = \dfrac{dx}{dt} = 16t\ \text{ft/s}, \\ a_x = \dfrac{dv_x}{dt} = 16\ \text{ft/s}^2. \end{cases}$ |

Active Example 13.4

First apply the chain rule to express the acceleration in terms of velocity and position instead of velocity and time.

$$\frac{dv}{dt} = \frac{dv}{ds}\frac{ds}{dt} = \frac{dv}{ds}v = -0.004v^2.$$

Separate variables.

$$\frac{dv}{v} = -0.004\,ds.$$

Integrate, defining $s = 0$ to be the position at which the velocity is 80 m/s. Here v is the velocity at position s.

$$\begin{cases} \displaystyle\int_{80}^{v}\frac{dv}{v} = -0.004\int_{0}^{s}ds, \\[2ex] \left[\ln v\right]_{80}^{v} = -0.004\left[s\right]_{0}^{s}, \\[2ex] \ln v - \ln 80 = -0.004(s - 0). \end{cases}$$

Solve for s in terms of the velocity. From this equation we find that the distance required for the velocity to decrease to 10 m/s is 520 m.

$$s = 250\ln\left(\frac{80}{v}\right).$$

Active Example 13.6

Integrate the x component of the velocity to determine the x component of the position as a function of time.	$\begin{cases} v_x = \dfrac{dx}{dt} = 0.3t^2, \\[2mm] \displaystyle\int_0^x dx = 0.3\int_0^t t^2\, dt, \\[2mm] x = 0.1t^3. \end{cases}$

Evaluate the x component of the position at $t = 6$ s.	$x\big|_{t\,=\,6\,\text{s}} = 0.1(6)^3 = 21.6$ m.

Integrate the y component of the velocity to determine the y component of the position as a function of time.	$\begin{cases} v_y = \dfrac{dy}{dt} = 1.8t - 0.18t^2, \\[2mm] \displaystyle\int_0^y dy = \int_0^t (1.8t - 0.18t^2)\, dt, \\[2mm] y = 0.9t^2 - 0.06t^3. \end{cases}$

Evaluate the y component of the position at $t = 6$ s.	$y\big|_{t\,=\,6\,\text{s}} = 0.9(6)^2 - 0.06(6)^3 = 19.4$ m/s^2.

Express the position vector at $t = 6$ s in terms of its components.	$\mathbf{r}\big|_{t\,=\,6\,\text{s}} = 21.6\mathbf{i} + 19.4\mathbf{j}$ (m).

Position of the helicopter as a function of time.	

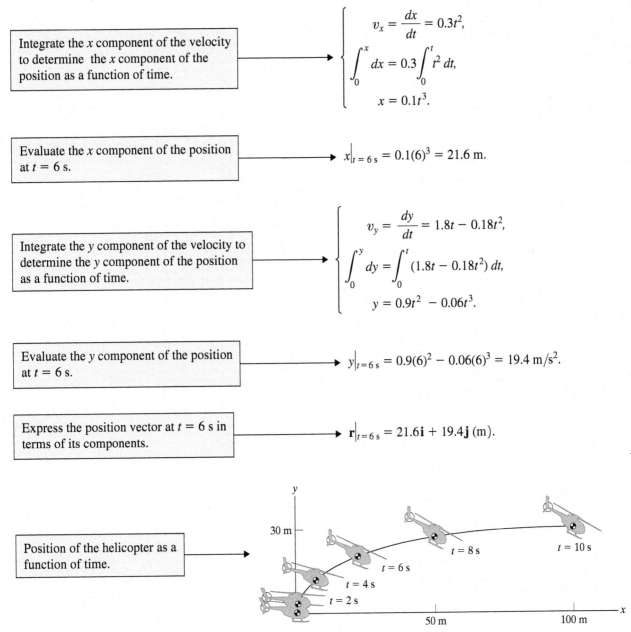

Active Example 13.8

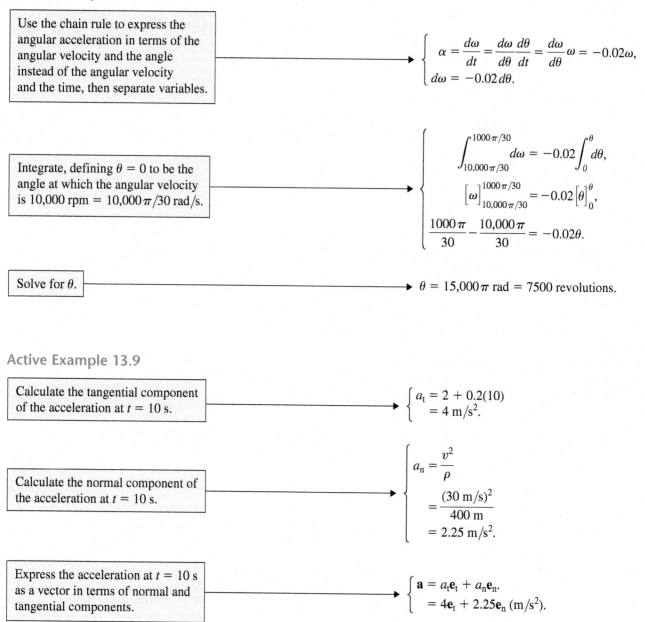

| Use the chain rule to express the angular acceleration in terms of the angular velocity and the angle instead of the angular velocity and the time, then separate variables. | $\alpha = \dfrac{d\omega}{dt} = \dfrac{d\omega}{d\theta}\dfrac{d\theta}{dt} = \dfrac{d\omega}{d\theta}\omega = -0.02\omega,$ $d\omega = -0.02\,d\theta.$ |

| Integrate, defining $\theta = 0$ to be the angle at which the angular velocity is 10,000 rpm $= 10,000\,\pi/30$ rad/s. | $\displaystyle\int_{10,000\,\pi/30}^{1000\,\pi/30} d\omega = -0.02\int_{0}^{\theta} d\theta,$ $\left[\omega\right]_{10,000\,\pi/30}^{1000\,\pi/30} = -0.02\left[\theta\right]_{0}^{\theta},$ $\dfrac{1000\,\pi}{30} - \dfrac{10,000\,\pi}{30} = -0.02\theta.$ |

| Solve for θ. | $\theta = 15,000\,\pi$ rad $= 7500$ revolutions. |

Active Example 13.9

| Calculate the tangential component of the acceleration at $t = 10$ s. | $a_t = 2 + 0.2(10)$ $= 4$ m/s^2. |

| Calculate the normal component of the acceleration at $t = 10$ s. | $a_n = \dfrac{v^2}{\rho}$ $= \dfrac{(30\ \text{m/s})^2}{400\ \text{m}}$ $= 2.25$ m/s^2. |

| Express the acceleration at $t = 10$ s as a vector in terms of normal and tangential components. | $\mathbf{a} = a_t\mathbf{e}_t + a_n\mathbf{e}_n.$ $= 4\mathbf{e}_t + 2.25\mathbf{e}_n$ (m/s^2). |

Active Example 13.13

Determine the derivatives in the expression for the acceleration.

$$\begin{cases} \dfrac{dr}{dt} = \pi \sin 2\pi t, \\[2mm] \dfrac{d^2 r}{dt^2} = 2\pi^2 \cos 2\pi t, \\[2mm] \dfrac{d\theta}{dt} = -0.4\pi \cos 2\pi t, \\[2mm] \dfrac{d^2\theta}{dt^2} = 0.8\pi^2 \sin 2\pi t. \end{cases}$$

Determine the components of the acceleration as functions of time.

$$\begin{cases} a_r = \dfrac{d^2 r}{dt^2} - r\left(\dfrac{d\theta}{dt}\right)^2 \\[2mm] \quad = 2\pi^2 \cos 2\pi t - (1 - 0.5 \cos 2\pi t)(-0.4\pi \cos 2\pi t)^2, \\[2mm] a_\theta = r\dfrac{d^2\theta}{dt^2} + 2\dfrac{dr}{dt}\dfrac{d\theta}{dt} \\[2mm] \quad = (1 - 0.5 \cos 2\pi t)(0.8\pi^2 \sin 2\pi t) \\[2mm] \quad + 2(\pi \sin 2\pi t)(-0.4\pi \cos 2\pi t). \end{cases}$$

Evaluate the acceleration at $t = 0.8$ s.

$\mathbf{a} = 5.97\mathbf{e}_r - 4.03\mathbf{e}_\theta \ (\text{m/s}^2).$

Active Example 13.16

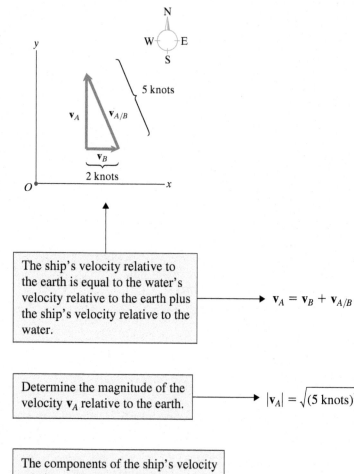

The ship's velocity relative to the earth is equal to the water's velocity relative to the earth plus the ship's velocity relative to the water.

$$\mathbf{v}_A = \mathbf{v}_B + \mathbf{v}_{A/B}$$

Determine the magnitude of the velocity $\mathbf{v}_A$ relative to the earth.

$$|\mathbf{v}_A| = \sqrt{(5 \text{ knots})^2 - (2 \text{ knots})^2} = 4.58 \text{ knots.}$$

The components of the ship's velocity relative to the water indicate that the helmsman must point the ship at arctan $(2/4.58) = 23.6°$ west of north to travel north relative to the earth.

$$\mathbf{v}_{A/B} = -2\mathbf{i} + 4.58\mathbf{j} \text{ (knots).}$$

Active Example 14.1

Apply Newton's second law to determine the crate's acceleration when there is no friction force.

$$\begin{cases} \Sigma F_x = W \sin 20° = ma_x: \\[2mm] a_x = \dfrac{mg \sin 20°}{m} = (32.2) \sin 20° = 11.0 \text{ m/s}^2. \end{cases}$$

Integrate to determine the crate's velocity as a function of time. At $t = 1$ s, the crate is moving 11.0 ft/s.

$$\begin{cases} a_x = \dfrac{dv_x}{dt} = 11.0 \text{ ft/s}^2, \\[2mm] \displaystyle\int_0^{v_x} dv_x = \int_0^t 11.0 \, dt, \\[2mm] v_x = 11.0t \text{ ft/s}. \end{cases}$$

Active Example 14.2

Beginning with the x component of the position, differentiate to determine the x components of the velocity and acceleration as functions of time.

$$\begin{cases} x = 8t^2 \text{ ft}, \\[2mm] v_x = \dfrac{dx}{dt} = 16t \text{ ft/s}, \\[2mm] a_x = \dfrac{dv_x}{dt} = 16 \text{ ft/s}^2. \end{cases}$$

Beginning with the y component of the position, differentiate to determine the y components of the velocity and acceleration as functions of time.

$$\begin{cases} y = t^3 \text{ ft}, \\[2mm] v_y = \dfrac{dy}{dt} = 3t^2 \text{ ft/s}, \\[2mm] a_y = \dfrac{dv_y}{dt} = 6t \text{ ft/s}^2. \end{cases}$$

Determine the object's mass.

$$m = \dfrac{W}{g} = \dfrac{10 \text{ lb}}{32.2 \text{ ft/s}^2} = 0.311 \text{ slug}.$$

Use Newton's second law to determine the components of the total force.

$$\begin{cases} \Sigma F_x = ma_x|_{t=4\,\text{s}} \\[1mm] \quad = (0.311 \text{ slug})(16 \text{ ft/s}^2) \\[1mm] \quad = 4.97 \text{ lb}, \\[3mm] \Sigma F_y = ma_y|_{t=4\,\text{s}} \\[1mm] \quad = (0.311 \text{ slug})[(6)(4) \text{ ft/s}^2] \\[1mm] \quad = 7.45 \text{ lb}. \end{cases}$$

Active Example 14.5

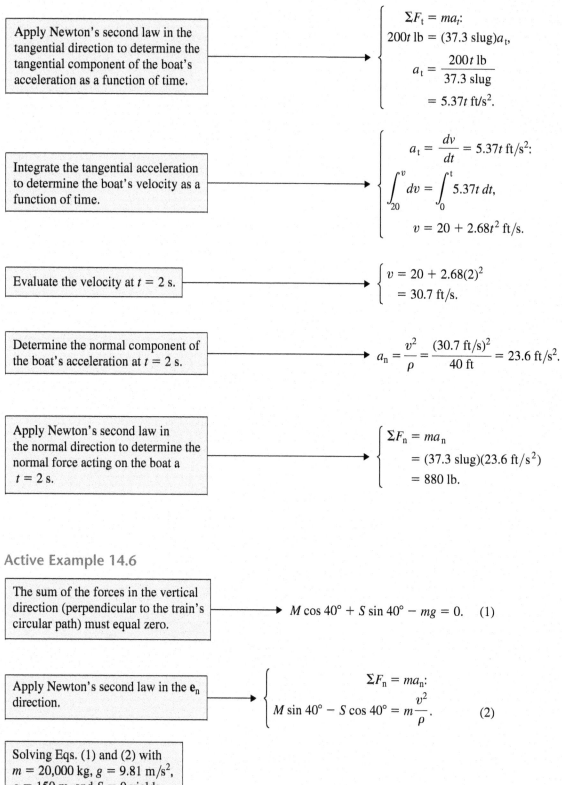

Apply Newton's second law in the tangential direction to determine the tangential component of the boat's acceleration as a function of time.

$$\Sigma F_{\mathrm{t}} = ma_t:$$
$$200t \text{ lb} = (37.3 \text{ slug})a_{\mathrm{t}},$$
$$a_{\mathrm{t}} = \frac{200t \text{ lb}}{37.3 \text{ slug}}$$
$$= 5.37t \text{ ft/s}^2.$$

Integrate the tangential acceleration to determine the boat's velocity as a function of time.

$$a_{\mathrm{t}} = \frac{dv}{dt} = 5.37t \text{ ft/s}^2:$$
$$\int_{20}^{v} dv = \int_{0}^{t} 5.37t \, dt,$$
$$v = 20 + 2.68t^2 \text{ ft/s}.$$

Evaluate the velocity at $t = 2$ s.

$$v = 20 + 2.68(2)^2$$
$$= 30.7 \text{ ft/s}.$$

Determine the normal component of the boat's acceleration at $t = 2$ s.

$$a_{\mathrm{n}} = \frac{v^2}{\rho} = \frac{(30.7 \text{ ft/s})^2}{40 \text{ ft}} = 23.6 \text{ ft/s}^2.$$

Apply Newton's second law in the normal direction to determine the normal force acting on the boat a $t = 2$ s.

$$\Sigma F_{\mathrm{n}} = ma_{\mathrm{n}}$$
$$= (37.3 \text{ slug})(23.6 \text{ ft/s}^2)$$
$$= 880 \text{ lb}.$$

Active Example 14.6

The sum of the forces in the vertical direction (perpendicular to the train's circular path) must equal zero.

$$M \cos 40° + S \sin 40° - mg = 0. \quad (1)$$

Apply Newton's second law in the $\mathbf{e}_{\mathrm{n}}$ direction.

$$\Sigma F_{\mathrm{n}} = ma_{\mathrm{n}}:$$
$$M \sin 40° - S \cos 40° = m\frac{v^2}{\rho}. \quad (2)$$

Solving Eqs. (1) and (2) with $m = 20{,}000$ kg, $g = 9.81$ m/s^2, $\rho = 150$ m, and $S = 0$ yields $M = 256$ kN and $v = 35.1$ m/s.

Active Example 14.9

Apply Newton's second law in the transverse direction. Notice that $\alpha = d\omega/dt = 0$.

$$\begin{cases} \Sigma F_\theta = ma_\theta: \\[2mm] N = m\left(r\alpha + 2\dfrac{dr}{dt}\omega\right) = 2m\omega_0 v_r. \end{cases}$$

Substitute the expression for v_r as a function of r obtained in the example to determine N as a function of r.

$$N = 2m\omega_0 \sqrt{\left(\omega_0^2 - \dfrac{k}{m}\right)(r^2 - r_0^2) + \dfrac{2k}{m}r_0(r - r_0)}.$$

Active Example 14.10

When an earth satellite in elliptic orbit is at the position of minimum radius, called its *perigee*, and when it is at the position of maximum radius, called its *apogee*, it has only a transverse component of velocity. Therefore the velocities at perigee and apogee satisfy Eq. (14.14). Let v_a be the velocity at apogee.

$$\begin{cases} r_0 v_0 = r_{max}v_a, \\[2mm] v_a = \dfrac{r_0}{r_{max}}v_0 \\[3mm] \quad = \dfrac{6600\ \text{km}}{16,000\ \text{km}}(9240\ \text{m/s}) \\[3mm] \quad = 3810\ \text{m/s}. \end{cases}$$

Active Example 15.1

In this case the normal force $N = (120\ \text{kg})(9.81\ \text{m/s}^2) = 1180\ \text{N}$.

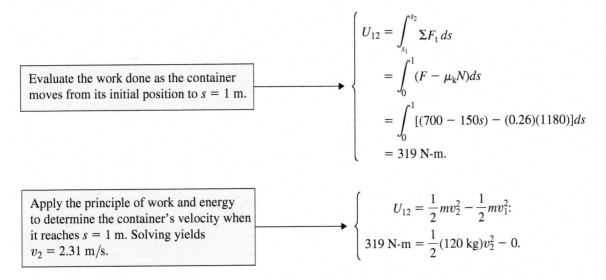

Evaluate the work done as the container moves from its initial position to $s = 1$ m.

$$\begin{cases} U_{12} = \displaystyle\int_{s_1}^{s_2} \Sigma F_t\, ds \\[3mm] \quad = \displaystyle\int_0^1 (F - \mu_k N)ds \\[3mm] \quad = \displaystyle\int_0^1 [(700 - 150s) - (0.26)(1180)]ds \\[3mm] \quad = 319\ \text{N-m}. \end{cases}$$

Apply the principle of work and energy to determine the container's velocity when it reaches $s = 1$ m. Solving yields $v_2 = 2.31$ m/s.

$$\begin{cases} U_{12} = \dfrac{1}{2}mv_2^2 - \dfrac{1}{2}mv_1^2: \\[3mm] 319\ \text{N-m} = \dfrac{1}{2}(120\ \text{kg})v_2^2 - 0. \end{cases}$$

Active Example 15.4

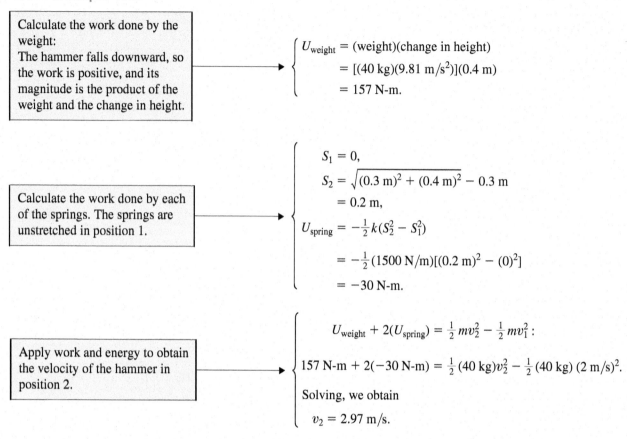

Calculate the work done by the weight:
The hammer falls downward, so the work is positive, and its magnitude is the product of the weight and the change in height.

$$U_{\text{weight}} = (\text{weight})(\text{change in height})$$
$$= [(40 \text{ kg})(9.81 \text{ m/s}^2)](0.4 \text{ m})$$
$$= 157 \text{ N-m.}$$

Calculate the work done by each of the springs. The springs are unstretched in position 1.

$$S_1 = 0,$$
$$S_2 = \sqrt{(0.3 \text{ m})^2 + (0.4 \text{ m})^2} - 0.3 \text{ m}$$
$$= 0.2 \text{ m,}$$
$$U_{\text{spring}} = -\frac{1}{2}k(S_2^2 - S_1^2)$$
$$= -\frac{1}{2}(1500 \text{ N/m})[(0.2 \text{ m})^2 - (0)^2]$$
$$= -30 \text{ N-m.}$$

Apply work and energy to obtain the velocity of the hammer in position 2.

$$U_{\text{weight}} + 2(U_{\text{spring}}) = \frac{1}{2}mv_2^2 - \frac{1}{2}mv_1^2:$$
$$157 \text{ N-m} + 2(-30 \text{ N-m}) = \frac{1}{2}(40 \text{ kg})v_2^2 - \frac{1}{2}(40 \text{ kg})(2 \text{ m/s})^2.$$

Solving, we obtain

$$v_2 = 2.97 \text{ m/s.}$$

Active Example 15.7

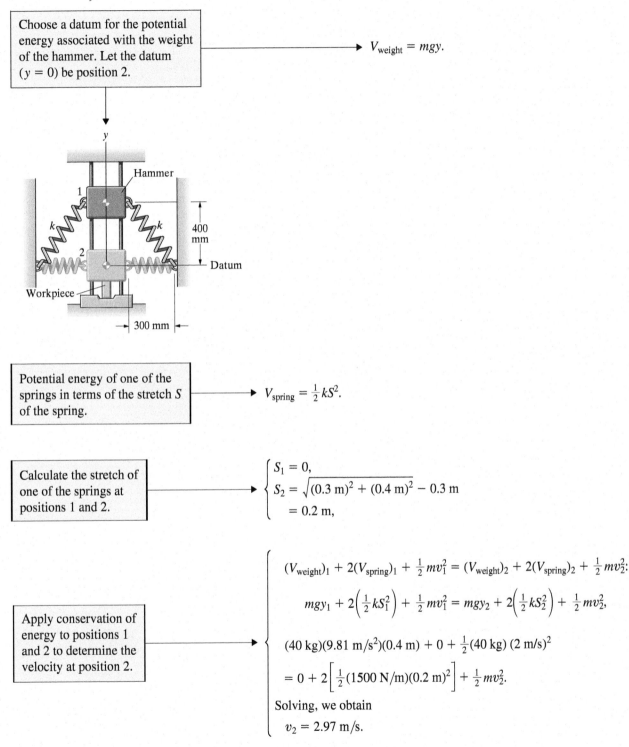

Choose a datum for the potential energy associated with the weight of the hammer. Let the datum ($y = 0$) be position 2.

$$V_{\text{weight}} = mgy.$$

Potential energy of one of the springs in terms of the stretch S of the spring.

$$V_{\text{spring}} = \tfrac{1}{2} kS^2.$$

Calculate the stretch of one of the springs at positions 1 and 2.

$$
\begin{cases}
S_1 = 0, \\
S_2 = \sqrt{(0.3 \text{ m})^2 + (0.4 \text{ m})^2} - 0.3 \text{ m} \\
\quad = 0.2 \text{ m},
\end{cases}
$$

Apply conservation of energy to positions 1 and 2 to determine the velocity at position 2.

$$
(V_{\text{weight}})_1 + 2(V_{\text{spring}})_1 + \tfrac{1}{2} mv_1^2 = (V_{\text{weight}})_2 + 2(V_{\text{spring}})_2 + \tfrac{1}{2} mv_2^2:
$$

$$
mgy_1 + 2\left(\tfrac{1}{2} kS_1^2 \right) + \tfrac{1}{2} mv_1^2 = mgy_2 + 2\left(\tfrac{1}{2} kS_2^2 \right) + \tfrac{1}{2} mv_2^2,
$$

$$
(40 \text{ kg})(9.81 \text{ m/s}^2)(0.4 \text{ m}) + 0 + \tfrac{1}{2}(40 \text{ kg})(2 \text{ m/s})^2
$$

$$
= 0 + 2\left[\tfrac{1}{2}(1500 \text{ N/m})(0.2 \text{ m})^2 \right] + \tfrac{1}{2} mv_2^2.
$$

Solving, we obtain

$$v_2 = 2.97 \text{ m/s}.$$

Active Example 15.10

The definition of a conservative force is that a potential energy exists, so the fact that the force was obtained from its potential energy guarantees that it is conservative. Alternatively, we can confirm that the force is conservative by showing that its curl is zero:

Use Eq. (15.31) to evaluate the curl of the force.

$$\left\{ \nabla \times \mathbf{F} = \frac{1}{r} \begin{vmatrix} \mathbf{e}_r & r\mathbf{e}_\theta & \mathbf{e}_z \\ \dfrac{\partial}{\partial r} & \dfrac{\partial}{\partial \theta} & \dfrac{\partial}{\partial z} \\ -\dfrac{mgR_E^2}{r^2} & 0 & 0 \end{vmatrix} \right.$$

$$= \mathbf{0}.$$

Active Example 16.1

Apply Eq. (16.2) to the interval of time from $t = 10$ s to $t = 20$ s.

$$\left\{ \begin{aligned} (t_2 - t_1)\Sigma\mathbf{F}_{av} &= m\mathbf{v}_2 - m\mathbf{v}_1: \\ (20 - 10)\Sigma\mathbf{F}_{av} &= (1200)(36\mathbf{i} + 8\mathbf{j}) - (1200)(30\mathbf{i} + 3\mathbf{j}), \\ 10\Sigma\mathbf{F}_{av} &= 7200\mathbf{i} + 6000\mathbf{j}. \end{aligned} \right.$$

Solving yields

$$\Sigma\mathbf{F}_{av} = 720\mathbf{i} + 600\mathbf{j} \text{ (N)}.$$

Active Example 16.2

The tangential component of the force as a function of time varies linearly from 300 N at $t = 0$ to $300 - 9(30) = 30$ N at $t = 30$ s, so its average value is

$$\Sigma F_{t\,av} = \frac{300 \text{ N} + 30 \text{ N}}{2} = 165 \text{ N},$$

but we can confirm this using Eq. (16.4).

Apply Eq. (16.4) to the interval of time from $t = 0$ to $t = 30$ s.

$$\left\{ \begin{aligned} (t_2 - t_1)\Sigma F_t &= mv_2 - mv_1: \\ (30 \text{ s} - 0)\Sigma F_t &= (225 \text{ kg})(22 \text{ m/s}) - (225 \text{ kg})(0). \end{aligned} \right.$$

Solving yields

$$\Sigma F_t = 165 \text{ N}.$$

Active Example 16.4

When the person is stationary in his initial position, the total linear momentum of the person and barge is zero. The only horizontal forces on the person and the barge are the forces they exert on each other, so their total linear momentum when the person is running must be zero.

> Apply conservation of linear momentum. Let v_B be the barge's velocity *toward the left.*

$\longrightarrow$

$m_P v_P + m_B(-v_B) = 0.$

The barge's velocity is

$$v_B = (m_P/m_B)v_P \text{ toward the left.}$$

Active Example 16.5

If the masses stick together after the collision, their velocity can be determined from conservation of linear momentum alone.

> Apply Eq. (16.10) (conservation of linear momentum) with $v'_A = v'_B = v'$.

$\longrightarrow$

$$m_A v_A + m_B v_B = (m_A + m_B)v':$$
$$(4 \text{ kg})(10 \text{ m/s}) + (4 \text{ kg})(-5 \text{ m/s}) = (4 \text{ kg} + 4 \text{ kg})v'.$$
Solving yields $v' = 2.5 \text{ m/s.}$

Active Example 16.7

> Express r in terms of the initial radius r_0 and the constant velocity v_0.

$\longrightarrow$

$r = r_0 - v_0 t.$

> In plane central-force motion, the product of the radial distance from the center of motion and the transverse component of the velocity is constant.

$\longrightarrow$

$$rv_\theta = r_0 v_0,$$
so
$$v_\theta = \frac{r_0 v_0}{r}$$
$$= \frac{r_0 v_0}{r_0 - v_0 t}.$$

> Express the velocity as a function of time in terms of cylindrical coordinates.

$\longrightarrow$

$$\mathbf{v} = v_r \mathbf{e}_r + v_\theta \mathbf{e}_\theta$$
$$= -v_0 \mathbf{e}_r + \frac{r_0 v_0}{r_0 - v_0 t} \mathbf{e}_\theta.$$

Active Example 16.9

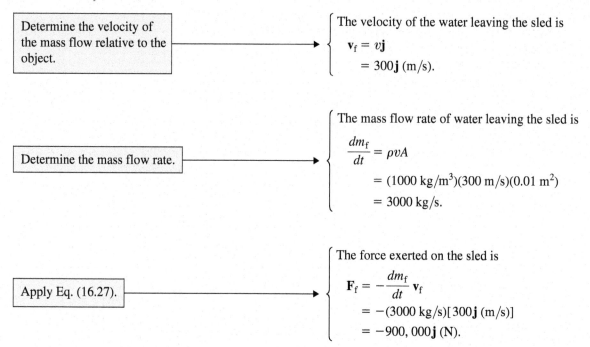

| Determine the velocity of the mass flow relative to the object. | → | The velocity of the water leaving the sled is
$$\mathbf{v}_f = v\mathbf{j}$$
$$= 300\mathbf{j} \ (\text{m/s}).$$ |

| Determine the mass flow rate. | → | The mass flow rate of water leaving the sled is
$$\frac{dm_f}{dt} = \rho v A$$
$$= (1000 \ \text{kg/m}^3)(300 \ \text{m/s})(0.01 \ \text{m}^2)$$
$$= 3000 \ \text{kg/s}.$$ |

| Apply Eq. (16.27). | → | The force exerted on the sled is
$$\mathbf{F}_f = -\frac{dm_f}{dt}\mathbf{v}_f$$
$$= -(3000 \ \text{kg/s})[\,300\mathbf{j} \ (\text{m/s})\,]$$
$$= -900,000\mathbf{j} \ (\text{N}).$$ |

Active Example 17.1

The normal and tangential components of the acceleration can be determined from Eqs. (17.3). To do so, the angular velocity and angular acceleration of gear A must be determined.

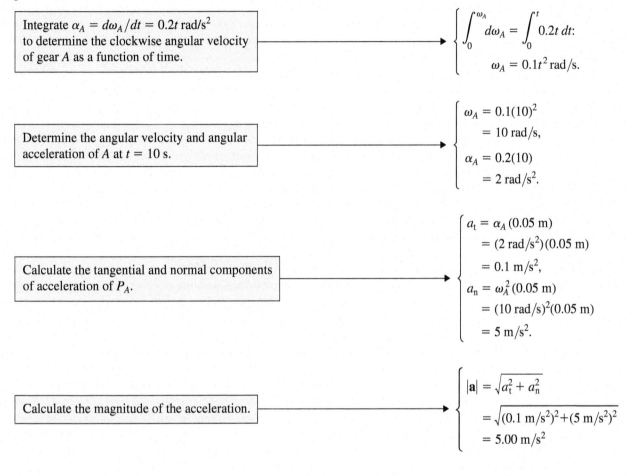

Integrate $\alpha_A = d\omega_A/dt = 0.2t$ rad/s^2 to determine the clockwise angular velocity of gear A as a function of time.

$$\int_0^{\omega_A} d\omega_A = \int_0^t 0.2t\, dt:$$

$$\omega_A = 0.1t^2 \, \text{rad/s}.$$

Determine the angular velocity and angular acceleration of A at $t = 10$ s.

$$\omega_A = 0.1(10)^2$$
$$= 10 \text{ rad/s},$$
$$\alpha_A = 0.2(10)$$
$$= 2 \text{ rad/s}^2.$$

Calculate the tangential and normal components of acceleration of P_A.

$$a_t = \alpha_A\,(0.05 \text{ m})$$
$$= (2 \text{ rad/s}^2)(0.05 \text{ m})$$
$$= 0.1 \text{ m/s}^2,$$
$$a_n = \omega_A^2\,(0.05 \text{ m})$$
$$= (10 \text{ rad/s})^2(0.05 \text{ m})$$
$$= 5 \text{ m/s}^2.$$

Calculate the magnitude of the acceleration.

$$|\mathbf{a}| = \sqrt{a_t^2 + a_n^2}$$
$$= \sqrt{(0.1 \text{ m/s}^2)^2 + (5 \text{ m/s}^2)^2}$$
$$= 5.00 \text{ m/s}^2$$

Active Example 17.2

Let ω_{AB} be the unknown angular velocity of bar AB, so that the angular velocity vector of bar AB is $\boldsymbol{\omega}_{AB} = \omega_{AB}\mathbf{k}$.

Apply Eq. (17.6) to determine the velocity of point B in terms of ω_{AB}.

$$\mathbf{v}_B = \mathbf{v}_A + \boldsymbol{\omega}_{AB} \times \mathbf{r}_{B/A}$$

$$= \mathbf{0} + \begin{vmatrix} \mathbf{i} & \mathbf{j} & \mathbf{k} \\ 0 & 0 & \omega_{AB} \\ 0.4 & 0.4 & 0 \end{vmatrix}$$

$$= -0.4\omega_{AB}\mathbf{i} + 0.4\omega_{AB}\mathbf{j}.$$

The angular velocity ω_{BC} of bar BC is also unknown, but the velocity of point C is known. Applying Eq. (17.6) to points B and C results in two equations in terms of ω_{AB} and ω_{BC}.

$$\mathbf{v}_C = \mathbf{v}_B + \boldsymbol{\omega}_{BC} \times \mathbf{r}_{C/B}:$$

$$-3\mathbf{i} = -0.4\omega_{AB}\mathbf{i} + 0.4\omega_{AB}\mathbf{j} + \begin{vmatrix} \mathbf{i} & \mathbf{j} & \mathbf{k} \\ 0 & 0 & \omega_{BC} \\ 0.8 & -0.4 & 0 \end{vmatrix}$$

$$= (-0.4\omega_{AB} + 0.4\omega_{BC})\mathbf{i} + (0.4\omega_{AB} + 0.8\omega_{BC})\mathbf{j}.$$

Equating $\mathbf{i}$ and $\mathbf{j}$ components,

$$-3 = 0.4\omega_{AB} + 0.4\omega_{BC},$$

$$0 = -0.4\omega_{AB} + 0.8\omega_{BC},$$

and solving yields $\omega_{AB} = 5$ rad/s and $\omega_{BC} = -2.5$ rad/s.

Active Example 17.4

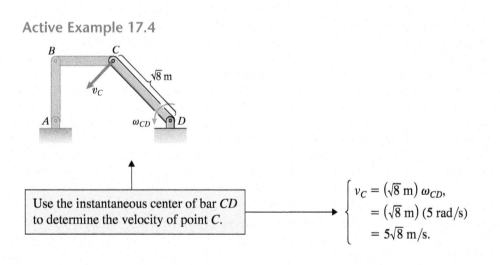

Use the instantaneous center of bar CD to determine the velocity of point C.

$$v_C = (\sqrt{8}\text{ m})\,\omega_{CD},$$

$$= (\sqrt{8}\text{ m})\,(5\text{ rad/s})$$

$$= 5\sqrt{8}\text{ m/s}.$$

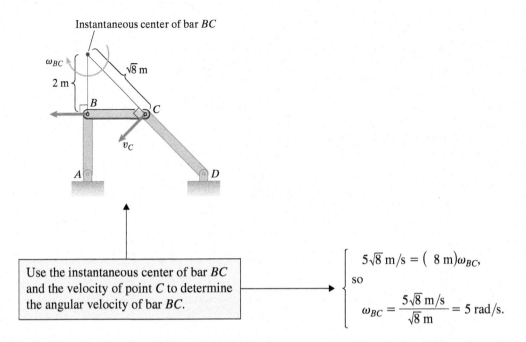

Use the instantaneous center of bar BC and the velocity of point C to determine the angular velocity of bar BC.

$$
\begin{cases}
5\sqrt{8} \text{ m/s} = (\ 8 \text{ m})\omega_{BC}, \\
\text{so} \\
\omega_{BC} = \dfrac{5\sqrt{8} \text{ m/s}}{\sqrt{8} \text{ m}} = 5 \text{ rad/s}.
\end{cases}
$$

Active Example 17.5

Apply Eq. (17.10). The angular velocity vector of the disk is $\boldsymbol{\omega} = \omega\mathbf{k}$, so its angular acceleration vector is

$$
\boldsymbol{\alpha} = \frac{d\omega}{dt} = \frac{d\omega}{dt}\mathbf{k} = \alpha\mathbf{k}.
$$

The position vector of A relative to B is $\mathbf{r}_{A/B} = R\mathbf{i}$.

$$
\begin{cases}
\mathbf{a}_A = \mathbf{a}_B + \boldsymbol{\alpha} \times \mathbf{r}_{A/B} - \omega^2 \mathbf{r}_{A/B} \\[2mm]
= -\alpha R\mathbf{i} + \begin{vmatrix} \mathbf{i} & \mathbf{j} & \mathbf{k} \\ 0 & 0 & \alpha \\ R & 0 & 0 \end{vmatrix} - \omega^2 (R\mathbf{i}) \\[4mm]
= -\alpha R\mathbf{i} + \alpha R\mathbf{j} - \omega^2 R\mathbf{i}.
\end{cases}
$$

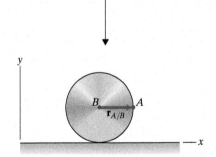

Active Example 17.7

Apply Eq. (17.15) to points A and B.

$$\begin{cases} \mathbf{a}_A = \mathbf{a}_B + \mathbf{a}_{A\,rel} + 2\boldsymbol{\omega}_{AB} \times \mathbf{v}_{A\,rel} \\ \qquad + \boldsymbol{\alpha}_{AB} \times \mathbf{r}_{A/B} - \omega_{AB}^2 \mathbf{r}_{A/B} \\[4pt] \qquad = \mathbf{0} + \mathbf{a}_{A\,rel} + 2 \begin{vmatrix} \mathbf{i} & \mathbf{j} & \mathbf{k} \\ 0 & 0 & 2 \\ -3.2 & -1.6 & 0 \end{vmatrix} \\[4pt] \qquad + \begin{vmatrix} \mathbf{i} & \mathbf{j} & \mathbf{k} \\ 0 & 0 & 10 \\ 0.8 & 0.4 & 0 \end{vmatrix} - (2)^2 (0.8\mathbf{i} + 0.4\mathbf{j}) \\[4pt] \qquad = \mathbf{a}_{A\,rel} - 0.8\mathbf{i} - 6.4\mathbf{j}. \qquad (1) \end{cases}$$

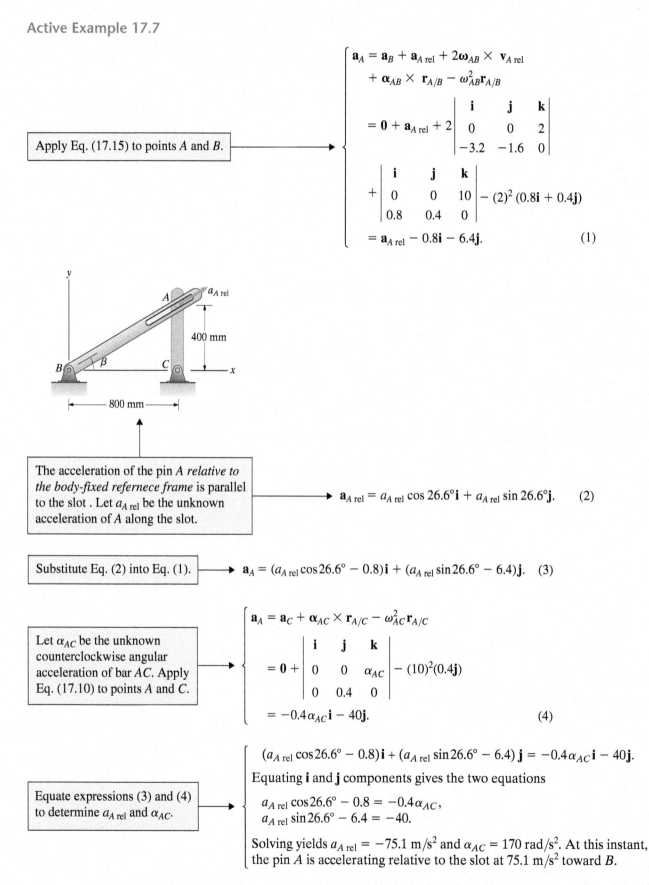

The acceleration of the pin A *relative to the body-fixed refernece frame* is parallel to the slot . Let $a_{A\,rel}$ be the unknown acceleration of A along the slot.

$$\mathbf{a}_{A\,rel} = a_{A\,rel} \cos 26.6° \mathbf{i} + a_{A\,rel} \sin 26.6° \mathbf{j}. \qquad (2)$$

Substitute Eq. (2) into Eq. (1).

$$\mathbf{a}_A = (a_{A\,rel} \cos 26.6° - 0.8)\mathbf{i} + (a_{A\,rel} \sin 26.6° - 6.4)\mathbf{j}. \qquad (3)$$

Let α_{AC} be the unknown counterclockwise angular acceleration of bar AC. Apply Eq. (17.10) to points A and C.

$$\begin{cases} \mathbf{a}_A = \mathbf{a}_C + \boldsymbol{\alpha}_{AC} \times \mathbf{r}_{A/C} - \omega_{AC}^2 \mathbf{r}_{A/C} \\[4pt] \qquad = \mathbf{0} + \begin{vmatrix} \mathbf{i} & \mathbf{j} & \mathbf{k} \\ 0 & 0 & \alpha_{AC} \\ 0 & 0.4 & 0 \end{vmatrix} - (10)^2(0.4\mathbf{j}) \\[4pt] \qquad = -0.4\alpha_{AC}\mathbf{i} - 40\mathbf{j}. \qquad (4) \end{cases}$$

Equate expressions (3) and (4) to determine $a_{A\,rel}$ and α_{AC}.

$$\begin{cases} (a_{A\,rel} \cos 26.6° - 0.8)\mathbf{i} + (a_{A\,rel} \sin 26.6° - 6.4)\,\mathbf{j} = -0.4\alpha_{AC}\mathbf{i} - 40\mathbf{j}. \\[4pt] \text{Equating } \mathbf{i} \text{ and } \mathbf{j} \text{ components gives the two equations} \\[4pt] a_{A\,rel} \cos 26.6° - 0.8 = -0.4\alpha_{AC}, \\ a_{A\,rel} \sin 26.6° - 6.4 = -40. \\[4pt] \text{Solving yields } a_{A\,rel} = -75.1 \text{ m/s}^2 \text{ and } \alpha_{AC} = 170 \text{ rad/s}^2. \text{ At this instant,} \\ \text{the pin } A \text{ is accelerating relative to the slot at 75.1 m/s}^2 \text{ toward } B. \end{cases}$$

Active Example 17.10

In this case, $\mathbf{v}_{A\,rel} = \mathbf{0}$ and $\mathbf{a}_{A\,rel} = \mathbf{0}$. Equations (17.16) and (17.17) can be used to determine her velocity $\mathbf{v}_A$ and acceleration $\mathbf{a}_A$ relative to the earth.

Apply Eq. (17.16).

$$\begin{cases} \mathbf{v}_A = \mathbf{v}_B + \mathbf{v}_{A\,rel} + \boldsymbol{\omega} \times \mathbf{r}_{A/B} \\[2mm] \quad = \mathbf{0} + \mathbf{0} + \begin{vmatrix} \mathbf{i} & \mathbf{j} & \mathbf{k} \\ 0 & 0 & \omega \\ R & 0 & 0 \end{vmatrix} \\[2mm] \quad = \omega R \mathbf{j}. \end{cases}$$

Apply Eq. (17.17).

$$\begin{cases} \mathbf{a}_A = \mathbf{a}_B + \mathbf{a}_{A\,rel} + 2\boldsymbol{\omega} \times \mathbf{v}_{A\,rel} \\[1mm] \quad\quad + \boldsymbol{\alpha} \times \mathbf{r}_{A/B} - \omega^2 \mathbf{r}_{A/B} \\[1mm] \quad = \mathbf{0} + \mathbf{0} + \mathbf{0} + \mathbf{0} - \omega^2(R\mathbf{i}) \\[1mm] \quad = -\omega^2 R \mathbf{i}. \end{cases}$$

Person A is stationary relative to the reference frame, but she has velocity and acceleration relative to the earth.

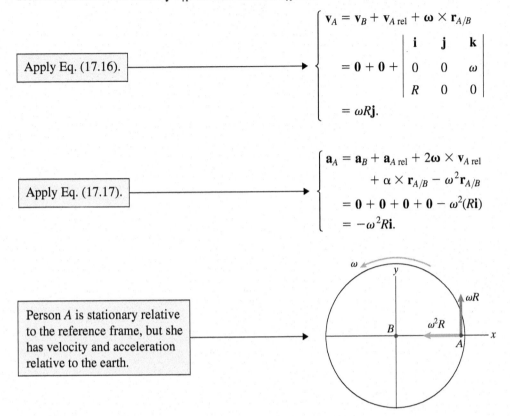

Active Example 18.1

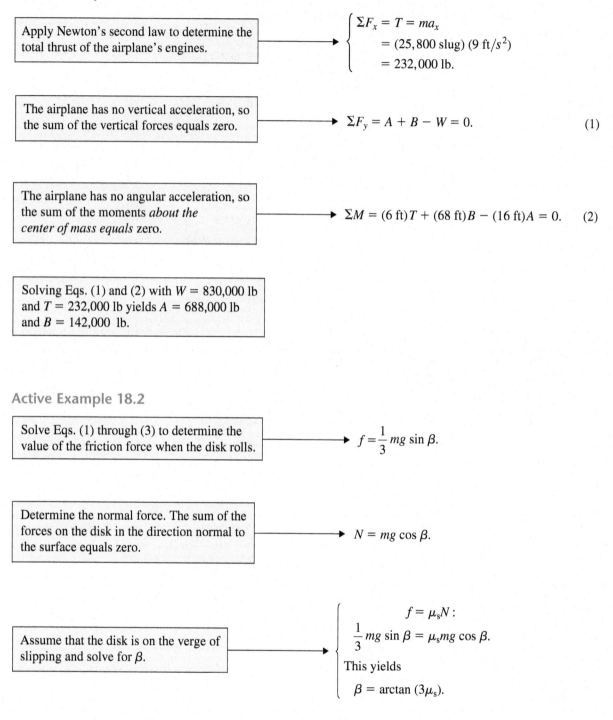

Apply Newton's second law to determine the total thrust of the airplane's engines.

$$\begin{cases} \Sigma F_x = T = ma_x \\ \quad = (25{,}800 \text{ slug}) (9 \text{ ft/s}^2) \\ \quad = 232{,}000 \text{ lb.} \end{cases}$$

The airplane has no vertical acceleration, so the sum of the vertical forces equals zero.

$$\Sigma F_y = A + B - W = 0. \qquad (1)$$

The airplane has no angular acceleration, so the sum of the moments *about the center of mass equals* zero.

$$\Sigma M = (6 \text{ ft})T + (68 \text{ ft})B - (16 \text{ ft})A = 0. \qquad (2)$$

Solving Eqs. (1) and (2) with $W = 830{,}000$ lb and $T = 232{,}000$ lb yields $A = 688{,}000$ lb and $B = 142{,}000$ lb.

Active Example 18.2

Solve Eqs. (1) through (3) to determine the value of the friction force when the disk rolls.

$$f = \frac{1}{3} mg \sin \beta.$$

Determine the normal force. The sum of the forces on the disk in the direction normal to the surface equals zero.

$$N = mg \cos \beta.$$

Assume that the disk is on the verge of slipping and solve for β.

$$\begin{cases} \quad f = \mu_s N : \\ \frac{1}{3} mg \sin \beta = \mu_s mg \cos \beta. \\ \text{This yields} \\ \quad \beta = \arctan (3\mu_s). \end{cases}$$

Active Example 18.3

At the instant the bar is released, $\omega = 0$. Equations (1) through (5) can be solved to determine P and N.

| Substitute Eq. (5) into Eq. (1) and substitute Eq. (4) into Eq. (2). | $\rightarrow$ | $\begin{cases} P = \dfrac{1}{2} ml\alpha \cos\theta, & (6) \\[2mm] N = mg - \dfrac{1}{2} ml\alpha \sin\theta. & (7) \end{cases}$ |

| Substitute Eqs (6) and (7) into Eq. (3) and solve for α, using $I = \dfrac{1}{12} ml^2$. | $\rightarrow$ | $\alpha = \dfrac{3}{2} \dfrac{g}{l} \sin\theta.$ (8) |

| Substitute Eq. (8) into Eqs (6) and (7). With $\theta = 30°$, these expressions yield $N = 0.813mg$ and $P = 0.325mg$. | $\rightarrow$ | $\begin{cases} P = \dfrac{3}{4} mg \sin\theta \cos\theta, \\[2mm] N = mg - \dfrac{3}{4} mg \sin^2\theta. \end{cases}$ |

Active Example 19.1

Apply Work and Energy

| Work is done by the disk's weight and by the couple M. As the center of the disk moves a distance b, the disk rotates through the clockwise angle (in radians) b/R, so the work done by the constant couple is Mb/R. | $\rightarrow$ | $U_{12} = (mg \sin\beta)b + M\dfrac{b}{R}.$ |

| Equate the work to the change in the kinetic energy of the disk. The initial kinetic energy is zero. Let v be the velocity of the center of the disk and ω its angular velocity when the center has moved a distance b. | $\rightarrow$ | $mgb \sin\beta + M\dfrac{b}{R} = \dfrac{1}{2}mv^2 + \dfrac{1}{2}I\omega^2 - 0.$ (1) |

Determine Kinematic Relationships

| Relationship between the velocity of the center of the disk and the angular velocity in rolling motion. | $\rightarrow$ | $v = R\omega.$ (2) |

| Solving Eqs. (1) and (2) for v and ω and substituting $I = \dfrac{1}{2}mR^2$ yields the velocity. | $\rightarrow$ | $v = \sqrt{\dfrac{4}{3}\left(gb \sin\beta + \dfrac{Mb}{mR}\right)}.$ |

Active Example 19.4

Let t_f be the time at which slip ceases. Apply the principle of angular impulse and momentum to disk A from $t = 0$ to $t = t_f$, treating counterclockwise moments and angular velocities as positive.

$$\begin{cases} \displaystyle\int_{t_1}^{t_2} \Sigma M \, dt = H_2 - H_1 : \\ \displaystyle\int_0^{t_f} -R_A f \, dt = 0 - I_A \omega_0 . \end{cases} \quad (1)$$

Let ω_{B0} be the initial counterclockwise angular velocity of disk B. Apply the principle of angular impulse and momentum to disk B from $t = 0$ to $t = t_f$, treating counterclockwise moments and angular velocities as positive.

$$\begin{cases} \displaystyle\int_{t_1}^{t_2} \Sigma M \, dt = H_2 - H_1 : \\ \displaystyle\int_0^{t_f} -R_B f \, dt = 0 - I_B \omega_{B0} - 0. \end{cases} \quad (2)$$

Divide Eq. (1) by Eq. (2), obtaining an equation for ω_{B0}.

$$\omega_{B0} = \frac{R_B I_A}{R_A I_B} \omega_0 .$$

Active Example 19.7

In this case, the total linear momentum is the same before and after the impact.

$$m_A v_A + 0 = m_A v_A' + m_B v_B'. \quad (1)$$

In this case, the total angular momentum about any fixed point is the same before and after the impact. Apply conservation of angular momentum about the center of mass of the bar.

$$\begin{cases} H_A + H_B = H_A' + H_B': \\ \left(h - \dfrac{1}{2}l\right)(m_A v_A) + 0 = \left(h - \dfrac{1}{2}l\right)(m_A v_A') + I_B \omega_B'. \end{cases} \quad (2)$$

Apply the coefficient of restitution. After the impact v_{BP}', is the velocity of the bar *at the point of impact*.

$$e = \frac{v_{BP}' - v_A'}{v_A - 0} . \quad (3)$$

Determine kinematic relationships. The velocity of the bar at the point of impact can be expressed in terms of the velocity of the center of mass.

$$
\begin{cases}
\mathbf{v}'_{BP} = \mathbf{v}'_A + \boldsymbol{\omega}'_B \times \mathbf{r}_{P/B}: \\[6pt]
v'_{BP}\mathbf{i} = v'_B\mathbf{i} + \begin{vmatrix} \mathbf{i} & \mathbf{j} & \mathbf{k} \\ 0 & 0 & \omega'_B \\ 0 & -\left(h - \dfrac{1}{2}l\right) & 0 \end{vmatrix}, \\[18pt]
v'_{BP} = v'_B + \left(h - \dfrac{1}{2}l\right)\omega'_B. \qquad (4)
\end{cases}
$$

Solving Eqs. (1)–(4) for v'_A, v'_B, and v'_{BP}, and ω'_B and use the relations $h = \dfrac{3}{4}l$, $m_A = m_B$, and $I_B = \dfrac{1}{12}m_Bl^2$ to obtain the bar's angular velocity.

$$
\omega'_B = \frac{\frac{12}{11}(1 + e)v_A}{l}.
$$

Active Example 20.1

Apply Eq. (20.1).

$$
\begin{cases}
\mathbf{v}_A = \mathbf{v}_A + \boldsymbol{\omega} \times \mathbf{r}_{A/B} \\[6pt]
= -5\mathbf{i} + \begin{vmatrix} \mathbf{i} & \mathbf{j} & \mathbf{k} \\ 0 & -13.9 & -0.5 \\ 0.36 & 0 & 0 \end{vmatrix} \\[18pt]
= -5\mathbf{i} - 0.18\mathbf{j} + 5\mathbf{k} \ (\text{m/s}).
\end{cases}
$$

Active Example 20.4

Draw the Free-Body Diagram

The bar is subjected to its weight and the force **F** and couple **C** exerted by the disk.

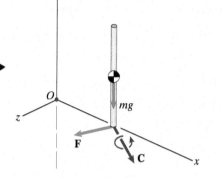

Apply Newton's Second Law

The center of mass has a normal component of acceleration $-b\omega_0^2\mathbf{i}$.

$$\begin{cases} \qquad \Sigma\mathbf{F} = m\mathbf{a}: \\[4pt] \mathbf{F} - mg\mathbf{j} = m(-b\omega_0^2\mathbf{i}). \\ \text{The force exerted on the bar} \\ \text{by the disk is} \\ \quad \mathbf{F} = -mb\omega_0^2\mathbf{i} + mg\mathbf{j}. \end{cases}$$

Apply the Equation of Angular Motion

The total moment about the point O is the sum of the couple **C** and the moment about O due to weight and the force **F**.

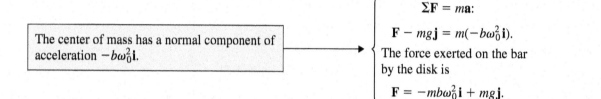

$$\begin{cases} \Sigma\mathbf{M} = \mathbf{C} + \begin{vmatrix} \mathbf{i} & \mathbf{j} & \mathbf{k} \\ b & \frac{1}{2}l & 0 \\ 0 & -mg & 0 \end{vmatrix} + \begin{vmatrix} \mathbf{i} & \mathbf{j} & \mathbf{k} \\ b & 0 & 0 \\ -mb\omega_0^2 & mg & 0 \end{vmatrix} \\ = \mathbf{C}. \end{cases}$$

$$\Sigma M_{Ox} = I_{xx}\frac{d\omega_x}{dt} - I_{xy}\frac{d\omega_y}{dt} - I_{xz}\frac{d\omega_z}{dt}$$
$$- \Omega_z(-I_{yx}\omega_x + I_{yy}\omega_y - I_{yz}\omega_z)$$
$$+ \Omega_y(-I_{zx}\omega_x - I_{zy}\omega_y + I_{zz}\omega_z):$$
$$C_x = 0.$$

$$\Sigma M_{Oy} = -I_{yx}\frac{d\omega_x}{dt} + I_{yy}\frac{d\omega_y}{dt} - I_{yz}\frac{d\omega_z}{dt}$$
$$+ \Omega_z(I_{xx}\omega_x - I_{xy}\omega_y - I_{xz}\omega_z)$$
$$- \Omega_x(-I_{zx}\omega_x - I_{zy}\omega_y + I_{zz}\omega_z),$$
$$C_y = 0.$$

$$\Sigma M_{Oz} = -I_{zx}\frac{d\omega_x}{dt} - I_{zy}\frac{d\omega_y}{dt} + I_{zz}\frac{d\omega_z}{dt}$$
$$- \Omega_y(I_{xx}\omega_x - I_{xy}\omega_y - I_{xz}\omega_z)$$
$$+ \Omega_x(-I_{yx}\omega_x + I_{yy}\omega_y - I_{yz}\omega_z):$$
$$C_z = \frac{1}{2}mlb\omega_0^2.$$

The couple exerted on the bar by the disk is

$$\mathbf{C} = \frac{1}{2}mlb\omega_0^2\mathbf{k}.$$

Apply Eqs. (20.12) with $\omega_x = \Omega_x = 0$, $\omega_y = \Omega_y = \omega_0$, and $\omega_z = \Omega_z = 0$.

Active Example 20.7

Substitute Eqs. (1), (2), and (5) into Eq. (20.29) for steady precession and use the relations $I_{xx} = \frac{1}{2}mR^2$ and $I_{zz} = \frac{1}{2}mR^2$.

$$m\frac{v^2}{r_G}R\sin\theta - mgR\cos\theta = \frac{1}{2}mR^2\frac{v^2}{r_G^2}\sin\theta\cos\theta$$
$$-mR^2\frac{v^2}{r_G}\left(\frac{1}{R} + \frac{1}{r_G}\cos\theta\right)\sin\theta.$$

Solving for v yields

$$v = \sqrt{\frac{2g(r - R\cos\theta)^2\cot\theta}{4r - 3R\cos\theta}}.$$

Active Example 21.1

Equation (4) is the equation of motion of the system in terms of the displacement of the mass relative to its equilibrium position. The general solution of this equation is given by Eq. (21.8), where A and B are constants.

$$\tilde{x} = A \sin \omega t + B \cos \omega t, \quad (5)$$

At $t = 0$, the position of the mass is $\tilde{x} = h$ and its velocity is $d\tilde{x}/dt = 0$. Use these initial conditions to solve for the constants A and B. Substituting the results into Eq. (5) determines the position of the mass as a function of time.

$$
\begin{cases}
\dfrac{d\tilde{x}}{dt} = A\omega \cos \omega t - B\omega \sin \omega t. \quad (6) \\[4pt]
\text{Substituting the initial conditions} \\
\text{into Eqs. (5) and (6) and yields} \\[2pt]
A = 0, \\
B = h. \\
\text{The position as a function of time is} \\[2pt]
\tilde{x} = h \cos \omega t.
\end{cases}
$$

Active Example 21.3

Evaluate ω and d. The damping is supercritical.

$$
\begin{cases}
\omega = \sqrt{\dfrac{k}{m}} = \sqrt{\dfrac{8 \text{ N/m}}{2 \text{ kg}}} = 2 \text{ rad/s}, \\[10pt]
d = \dfrac{c}{2m} = \dfrac{12 \text{ N-s/m}}{2(2 \text{ kg})} = 3 \text{ rad/s}.
\end{cases}
$$

The general solution is given by Eq. (21.24) with $h = \sqrt{d^2 - \omega^2} = 2.24$ rad/s.

$$
\begin{cases}
x = Ce^{-(d-h)t} + De^{-(d+h)t} \\[4pt]
\quad = Ce^{-0.764t} + De^{-5.24t}. \quad (1)
\end{cases}
$$

At $t = 0$, the position of the mass is $x = 0.1$ m and its velocity is $dx/dt = 0$. Use these initial conditions to solve for the constants C and D. Substituting the results into Eq. (1) determines the position of the mass as a function of time.

$$
\begin{cases}
\dfrac{dx}{dt} = -0.764Ce^{-0.764t} - 5.24De^{-5.24t}. \quad (2) \\[4pt]
\text{Substituting the initial conditions into Eqs. (1)} \\
\text{and (2) and solving yields} \\[2pt]
C = 0.117 \text{ m}, \\
D = -0.0171 \text{ m}. \\
\text{The position as a function of time is} \\[2pt]
x = 0.117e^{-0.764t} - 0.0171De^{-5.24t} \text{ m}.
\end{cases}
$$

Active Example 21.5

The forcing function

$$a(t) = F(t)/m = 10 \sin 4t \text{ m/s}^2$$

is of the form of Eq. (21.27) with $a_0 = 10 \text{ m/s}^2$, $b_0 = 0$, and $\omega_0 = 4$ rad/s. The constants $\omega = \sqrt{k/m} = 2$ rad/s and $d = c/2m = 0.25$ rad/s. Substituting these values into Eq. (21.31) yields the amplitude of the particular solution.

$$
\begin{cases}
E_{\mathrm{p}} = \dfrac{\sqrt{a_0^2 + b_0^2}}{\sqrt{(\omega^2 - \omega_0^2)^2 + 4d^2\omega_0^2}} \\[10pt]
\quad = 0.822 \text{ m}.
\end{cases}
$$

Answers to Even-Numbered Problems

Chapter 12

12.2 (a) $e = 2.7183$; (b) $e^2 = 7.3891$; (c) $e^2 = 7.3892$.
12.4 17.8 m^2.
12.6 The 1-in wrench fits the 25-mm nut.
12.8 (a) 267 mi/h; (b) 392 ft/s.
12.10 310 N-m.
12.12 $g = 32.2 \text{ ft/s}^2$.
12.14 (a) 0.0208 m^2; (b) 32.2 in^2.
12.16 2.07×10^6 Pa.
12.18 27.4 lb/ft.
12.20 (a) kg-m/s; (b) 2.70 slug-ft/s.
12.22 (a) 0.397 kg; (b) 0.643 N.
12.24 (a) 4.60×10^{19} slug; (b) 6.71×10^{20} kg.
12.26 163 lb.
12.28 32.1 km.
12.30 345,000 km.

Chapter 13

13.2 $v = 0$, $a = -4 \text{ in/s}^2$.
13.4 (a) $v = 13.0 \text{ m/s}$, $a = 1.28 \text{ m/s}^2$;
      (b) $v = 14.7 \text{ m/s}$ at $t = 6.67 \text{ s}$.
13.6 (a) $v = 28 \text{ m/s}$ at $t = 4 \text{ s}$; (b) $a = 0$.
13.8 (a) $v = -1.78 \text{ m/s}$, $a = -11.2 \text{ m/s}^2$;
      (b) 2.51 m/s; (c) Zero.
13.10 (a) 0.628 m/s; (b) 3.95 m/s^2.
13.12 $s = 77 \text{ m}$, $v = 66 \text{ m/s}$.
13.14 $s = 100 \text{ m}$, $v = 80 \text{ m/s}$.
13.16 4.33 m/s^2.
13.18 2.47 s, 2430 m/s.
13.20 $a = -6.66 \text{ m/s}^2$.
13.22 $s = 505 \text{ ft}$.
13.24 $s = 1070 \text{ m}$, $a = -24 \text{ m/s}^2$.
13.26 825 ft.
13.28 3630 ft.
13.30 No. 66.7 ft/s (45.5 mi/h).
13.32 51.9 solar years.
13.34 (a) 45.5 s; (b) 3390 m.
13.36 10 s.
13.38 $v = 6.73 \text{ ft/s}$, $a = -8.64 \text{ ft/s}^2$.
13.40 20 s.
13.42 $v = 3.33 \text{ m/s}$.
13.44 $v = 2.80 \text{ in/s}$.
13.46 11.2 km.
13.48 3240 ft.
13.50 2300 m.
13.52 42.5 m/s.
13.54 $v = 3.42 \text{ ft/s}$.
13.56 $c = 38,900$.

13.58 $v = -2(1 - s^2)^{1/2} \text{ m/s}$.
13.60 (a) 1.29 ft; (b) 4.55 ft/s.
13.62 $v_0 = 35,900 \text{ ft/s} = 24,500 \text{ mi/hr}$.
13.66 1266 s or 21.1 min.
13.68 $\mathbf{r} = 266.0\mathbf{i} + 75.3\mathbf{j} - 36.7\mathbf{k} \text{ (m)}$.
13.70 (a) $R = 35.3 \text{ m}$; (b) $R = 40.8 \text{ m}$; (c) $R = 35.3 \text{ m}$.
13.72 98.3 ft/s.
13.74 19.0 m.
13.76 $31.2 < v_0 < 34.2 \text{ ft/s}$.
13.78 (a) Yes; (b) No.
13.80 82.5 m.
13.82 $|\mathbf{v}| = 38.0 \text{ ft/s}$, $t = 1.67 \text{ s}$.
13.84 17.2 m.
13.86 $\mathbf{v} = 0.602\mathbf{i} - 4.66\mathbf{j} \text{ (m/s)}$.
13.90 $\mathbf{a} = -0.099\mathbf{i} + 0.414\mathbf{j} \text{ (m/s}^2)$.
13.94 (a) Positive; (b) 7.27×10^{-5} rad/s, earth rotates 15°.
13.96 430 s.
13.98 14.6 s.
13.100 (a) $\omega = 1.70 \text{ rad/s}$; (b) $\theta = 1 \text{ rad } (57.3°)$.
13.102 $d\mathbf{e}/dt = -8.81\mathbf{i} + 13.4\mathbf{j}$.
13.104 $\mathbf{v} = 12.5\mathbf{e}_t \text{ (m/s)}$, $\mathbf{a} = 3\mathbf{e}_t + 0.391\mathbf{e}_n \text{ (m/s}^2)$.
13.106 (a) 1730 rpm; (b) 3.01 rad/s^2.
13.108 1060 m/s^2 (108 g's).
13.110 (a) $\mathbf{v} = 16\mathbf{e}_t \text{ (m/s)}$, $\mathbf{a} = 16\mathbf{e}_t + 64\mathbf{e}_n \text{ (m/s}^2)$;
      (b) $s = 8 \text{ m}$.
13.112 (a) $\mathbf{v}_B = 87.3 \, \mathbf{e}_t \text{ (ft/s)}$; (b) $\mathbf{v}_B = -61.7\mathbf{i} - 61.7\mathbf{j}(\text{ft/s})$.
13.114 $v = R\sqrt{g/R_E}$.
13.116 $\mathbf{v} = 18.3\mathbf{e}_t \text{ (m/s)}$, $\mathbf{a} = 0.6\mathbf{e}_t + 6.68\mathbf{e}_n \text{ (m/s}^2)$.
13.118 $\mathbf{v} = 12.05\mathbf{e}_t \text{ (m/s)}$, $\mathbf{a} = 0.121\mathbf{e}_t + 2.905\mathbf{e}_n \text{ (m/s}^2)$.
13.120 (a) $|\mathbf{a}| = 45.0 \text{ ft/s}^2$; (b) $|\mathbf{a}| = 59.9 \text{ ft/s}^2$.
13.122 $\mathbf{v} = 6.46\mathbf{e}_t \text{ (m/s)}$.
13.124 (a) $\mathbf{v} = 15.2\mathbf{e}_t \text{ (m/s)}$, $\mathbf{a} = -1.63\mathbf{e}_t + 9.67\mathbf{e}_n \text{ (m/s}^2)$;
      (b) $\rho = 24.0 \text{ m}$.
13.126 218 m.
13.128 $\mathbf{a} = 3.59\mathbf{e}_t + 1.72\mathbf{e}_n \text{ (m/s}^2)$.
13.130 $a_n = g/\sqrt{1 + (gt/v_0)^2}$.
13.132 $dy/dt = 0.260 \text{ m/s}$, $d^2y/dt^2 = -0.150 \text{ m/s}^2$.
13.134 (a) $\rho = 16.7 \text{ ft}$; (b) $a_n = 77.3 \text{ ft/s}^2$.
13.138 $\mathbf{v} = -0.971\mathbf{e}_r - 1.76\mathbf{e}_\theta \text{ (m/s)}$.
13.140 $\mathbf{a}_A = 0.331\mathbf{e}_r + 0.480\mathbf{e}_\theta \text{ (m/s}^2)$.
13.142 (a) $\mathbf{v}_A = 3.75\mathbf{e}_r + 1.40\mathbf{e}_\theta \text{ (ft/s)}$;
      (b) $d\theta/dt = 0.468 \text{ rad/s}$.
13.144 (a) $\mathbf{v} = 0.32\mathbf{e}_r + 2.03\mathbf{e}_\theta \text{ (m/s)}$;
      (b) $\mathbf{v} = 0.32\mathbf{i} + 2.03\mathbf{j} \text{ (m/s)}$.
13.146 $\mathbf{a}_A = -58.0\mathbf{e}_r - 27.0\mathbf{e}_\theta \text{ (m/s}^2)$.
13.148 $\mathbf{a}_A = -1.09\mathbf{e}_r \text{ (m/s}^2)$.
13.152 $\mathbf{v}_C = 13.2\mathbf{e}_r + 18\mathbf{e}_\theta \text{ (m/s)}$.
13.154 $\mathbf{a} = \mathbf{0}$, $\alpha = -0.631 \text{ rad/s}^2$.

13.156 (a) $\mathbf{a} = -3.52\mathbf{e}_r + 4.06\mathbf{e}_\theta$ (m/s^2);
(b) $\mathbf{a} = -0.38\mathbf{i} - 5.36\mathbf{j}$ (m/s^2).

13.158 (a) $\mathbf{a} = -225\mathbf{e}_r - 173\mathbf{e}_\theta$ (ft/s^2);
(b) $\mathbf{a} = -108\mathbf{i} - 263\mathbf{j}$ (ft/s^2).

13.160 $\mathbf{v} = 1047\mathbf{e}_\theta + 587\mathbf{e}_z$ (ft/s),
$\mathbf{a} = -219{,}000\mathbf{e}_r$ (ft/s^2).

13.162 (a), (b) $\mathbf{v}_{A/B} = -3.66\mathbf{i} + 3.66\mathbf{j}$ (m/s).

13.164 $\mathbf{a}_{A/B} = 200\mathbf{i} + 200\mathbf{j}$ (ft/s^2).

13.166 $\mathbf{a}_{A/B} = 5\mathbf{i} - 2\mathbf{j}$ (ft/s^2).

13.168 9.93° east of north, 42.0 min.

13.170 2.34 m/s, 342 s.

13.172 (a) 7.10 m; (b) 2.22 s; (c) 11.8 m/s.

13.174 $v = 42.3$ m/s.

13.176 13.1 s.

13.178 68.6 ft/s.

13.180 $|\mathbf{v}| = 2.19$ m/s, $|\mathbf{a}| = 5.58$ m/s^2.

13.182 $\mathbf{a} = -2.75\mathbf{e}_r - 4.86\mathbf{e}_\theta$ (m/s^2).

13.184 (a) $\mathbf{v} = -2.13\mathbf{e}_r + 6.64\mathbf{e}_\theta$ (m/s);
(b) $\mathbf{v} = -5.90\mathbf{i} + 3.71\mathbf{j}$ (m/s).

Chapter 14

14.2 (a) 6.70 m/s; (b) 10.0 m.

14.4 $F = 9.91$ lb.

14.6 4.03 m/s down the inclined surface.

14.8 1.77 s.

14.10

14.12 $\Sigma\mathbf{F} = 2.40\mathbf{i} + 1.20\mathbf{j} + 2.08\mathbf{k}$ (kN).

14.14 18.8 ft/s^2.

14.16 (a) No; (b) Yes, $a_x = 0.5$ m/s^2; (c) $x = 2$ m.

14.18 $F_x = -198$ N, $F_y = 1800$ N.

14.20 $W = 88.3$ kN, $T = 61.7$ kN, $L = 66.9$ kN.

14.22 $L = 293.2$ kN, $D = 33.0$ kN.

14.24 $(66, 138, -58)$ m.

14.26 $F_x = 0.359$ N, $F_y = 19.888$ N.

14.28 2.10 m/s.

14.30 (a) 3.93 lb; (b) 19.5 ft/s.

14.32 (a) 1.01 m/s^2; (b) 6.47 m/s^2.

14.34 $F = 125$ N.

14.36 4.43 ft up the surface.

14.38 (a) 1200 N; (b) 1.84 m/s.

14.40 0.332 m/s^2 forward, 7.04 m/s^2 rearward.

14.42 $v = \pm 2\sqrt{4 - s^2}$ m/s.

14.44 (a) 773 ft/s^2 (24 g); (b) 28.0 ft/s.

14.46 (a) 11.9 ft; (b) 813 lb.

14.48 $y = -18.8$ mm.

14.50 (a) 50 s; (b) 40.8 N; (c) $4.8\mathbf{i} + \mathbf{j}$ (m/s).

14.52 2.06 m/s^2 up the bar.

14.54 $|\mathbf{a}_A| = 7.89$ ft/s^2.

14.56 $t = 0.600$ s.

14.58 $F_x = -0.544$ N.

14.60 $F_x = -73.4$ N, $F_y = 612$ N.

14.62 0.284 s.

14.64 $v_x = 0.486$ m/s, $v_y = -0.461$ m/s.

14.66 $\Sigma F_t = 0$, $\Sigma F_n = 373$ lb.

14.68 $v = 17.6$ m/s, $\Sigma F_n = 620$ N.

14.70 Tension is 4.8 N, force is 0.6 N.

14.72 (a) $\Sigma F_t = 9740$ lb, $\Sigma F_n = 28{,}800$ lb;
(b) $d\theta/dt = 3.95°$/s.

14.76 $\theta = 49.9°$, $|\mathbf{v}| = 10.8$ ft/s.

14.78 $2.62 \le v \le 3.74$ m/s.

14.80 (a) 207,000 lb; (b) 41,700 ft.

14.82 (a) $v = 140$ m/s; (b) $\rho = 815$ m.

14.86 $\theta = L_0/R - \sqrt{(L_0/R)^2 - (2v_0/R)t}$.

14.88 $\rho = 697$ m, $\mathbf{e} = 0.916\mathbf{i} - 0.308\mathbf{j} + 0.256\mathbf{k}$.

14.90 $\beta = 68.2°$

14.92 11.4 m/s^2.

14.94 $\Sigma F_r = -16.8$ N, $\Sigma F_\theta = 20.7$ N.

14.96 $9.46\mathbf{e}_r + 3.44\mathbf{e}_\theta$ (N).

14.98 $v_r = 14.8$ m/s.

14.100 $|\mathbf{v}| = 2.89$ m/s, $T = 41.6$ N.

14.102 $N = 1.02$ lb.

14.104 $\mu_s = 0.406$. The mass slips toward O.

14.106 $-1.48\mathbf{e}_r - 0.20\mathbf{e}_\theta$ (lb).

14.108 $k = 2m\omega_0^2$.

14.110 $11.5\mathbf{e}_r + 44.2\mathbf{e}_\theta + 10\mathbf{e}_z$ (N).

14.112 $|\Sigma\mathbf{F}| = 8.36$ N at $t = 4.39$ s.

14.114 (a) $|\mathbf{v}| = 1020$ m/s; (b) $t = 2.36 \times 10^6$ s (27.3 days).

14.116

14.118 $v_0 = 35{,}500$ ft/s.

14.120 (a) 34.6 ft/s^2, 3490 lb; (b) $\mu_s = 1.07$.

14.122 (a) $F_1 = 63$ kN, $F_2 = 126$ kN, $F_3 = 189$ kN.
(b) $F_1 = 75$ kN, $F_2 = 150$ kN, $F_3 = 225$ kN.

14.124 Deceleration is 1.49 m/s^2, compared with 8.83 m/s^2 on a level road.

14.126 14,300 ft (2.71 mi).

14.128 10.5 lb.

14.130 $a_A = 4.02$ ft/s^2, $T = 17.5$ lb.

14.132 29.7 ft.

14.134 9.30 N.

14.136 $\tan\alpha = v^2/\rho g$.

14.138 $\Sigma\mathbf{F} = -10.7\mathbf{e}_r + 2.55\mathbf{e}_\theta$ (N).

Chapter 15

15.2 8.06 m.

15.4 3.50 ft/s.

15.6 $P_{ave} = 1.62$ kW (kilowatts).

15.8 (a) 2.57×10^6 ft-lb/s (4670 hp).
(b) 1.57×10^6 ft-lb/s (2850 hp).

15.10 (a) 1.32×10^7 ft-lb/s (24,000 hp).
(b) 6.60×10^6 ft-lb/s (12,000 hp).

15.12 3.27 m/s.

15.14   117 ft/s (80.0 mi/h).

15.16   (a) 2160 ft-lb; (b) 7.39 lb.

15.18   (a) -12.5 ft-lb; (b) 5.67 ft/s.

15.20   Work $= 509$ N-m, $v = 2.52$ m/s.

15.22 $v = 21.8$ m/s.

15.24   4.04 m/s.

15.26   2.58 ft/s.

15.28 $v = 1.72$ m/s.

15.30

15.32   (a), (b), (c) 117 ft/s.

15.34   (a) 5.98 m/s;
        (b) 5.56 m/s.

15.36   3.55 m/s.

15.38   12.7 ft/s.

15.40 $v_0 = 15.66$ m/s, $v_{\text{top}} = 7.00$ m/s.

15.42   11.2 ft/s.

15.44   39.3 ft/s or 26.8 mi/hr.

15.46 -107 kN-m.

15.48   (a) $U_{12} = 0.210$ N-m; (b) $v_2 = 1.45$ m/s.

15.50   621 lb/ft.

15.52   (a) $k = 900$ lb/ft; (b) 274 ft/s^2.

15.54   2.18 m/s.

15.56   2.83 m/s.

15.58   (a) 5 ft; (b) 6.34 ft/s.

15.60   0.785 m.

15.62   0.385 m.

15.64 $v_2 = 7.03$ m/s.

15.66   5.77 m/s.

15.68 $k = 997$ lb/ft.

15.70   4.90 m/s.

15.72   193,000 lb.

15.74   26,600 km.

15.76 2.25×10^{10} N-m.

15.78   (a) 50 ft-lb; (b) 50 ft-lb; (c) 13.9 ft/s.

15.80   (a), (b) 6.48 m.

15.82   3.95 m/s.

15.84   (a) $\alpha = 60°$;
        (b) Before, 39.2 N; after, 58.9 N.

15.86 $\theta = 51.2°$.

15.88   (a) No; (b) 857 N.

15.90 $W[1 + \sqrt{1 + 4C/W}\,]$.

15.92   1.56 m/s.

15.94   1.99 m/s.

15.96 $V = \frac{1}{2}kS^2 + \frac{1}{4}qS^4$.

15.98 $v = 2.30$ m/s.

15.100 $v = 8.45$ ft/s.

15.102 $\mathbf{v} = 5.29\mathbf{e}_r + 4\mathbf{e}_\theta$ (m/s).

15.104 $v = 11.0$ km/s.

15.106 2880 m/s.

15.108 $v = 419$ m/s.

15.110 (a) $\mathbf{F} = -2x\mathbf{i} - 3y^2\mathbf{j}$ (N);
        (b) -2 N-m along each path.

15.112 $\mathbf{F} = -[k(r - r_0) + q(r - r_0)^3]\mathbf{e}_r$.

15.114 (a) $\mathbf{F} = (\sin\theta - 2r\cos^2\theta)\mathbf{e}_r$
              $+ (\cos\theta + 2r - \sin\theta\cos\theta)\mathbf{e}_\theta$.
        (b) The work is 2 ft-lb for any path from 1 to 2.

15.118 (a), (b) 193 ft.

15.120 He should choose (b).
        Impact velocity is 11.8 m/s, work is -251 kN-m.
        In (a), impact velocity is 13.9 m/s, work is -119 kN-m.

15.122 $v = 2.08$ m/s.

15.124 (a) 14.3×10^7 ft-lb; (b) $v = 88[1 - e^{-(F_0/88m)t}]$.

15.126 $k = 163$ lb/ft.

15.128 (a) $k = 809$ N/m; (b) $v_2 = 6.29$ m/s.

15.130 4.39 ft/s.

15.132 $h = 0.179$ m.

15.134 (a) $\theta = 27.9°$; (b) 159 lb; (c) 222 lb.

15.136 $v_1 = 4.73$ ft/s.

15.138 1.02 m.

15.140 2.00 m/s.

15.142 24.8 ft/s.

15.144 (a) $v = 0$;
        (b) $v = \sqrt{gR_E/2} = 18,300$ ft/s or 12,500 mi/hr.

15.146 (a) 11.3 MW (megawatts); (b) 9.45 MW

Chapter 16

16.2   (a) 68.8 lb-s; (b) 22.2 ft/s.

16.4   4.72 m/s (9.18 knots).

16.6 $F_0 = 0.856$ N.

16.8 $\mathbf{v} = 5.11\mathbf{i} + 5.67\mathbf{j} - 1.53\mathbf{k}$ (m/s).

16.10 $\mathbf{v} = 4.29\mathbf{i}$ (ft/s).

16.12   (a) 335 kN-s; (b) 7.84 s.

16.14   18.0 m/s.

16.16   (a) 222 N-s; (b) 1.85 m/s.

16.18   0.199 s.

16.20   17.2 ft/s.

16.22   2.01 m/s.

16.24   (a) $\mathbf{v} = 10\mathbf{i} - 12.1\mathbf{j}$ (m/s);
        (b) $-589\mathbf{j}$ (N-s);
        (c) $\mathbf{v} = 10\mathbf{i} - 12.1\mathbf{j}$ (m/s).

16.26   (a) $F_x = -32\cos 4t$ N, $F_y = -32\sin 4t$ N;
        (b) $\mathbf{v} = 3.03\mathbf{i} - 2.61\mathbf{j}$ (m/s).

16.28   5.99 s.

16.30   37.3 m/s.

16.32   105 m/s.

16.34   (a) 2070 lb; (b) 23.8 ft/s^2.

16.36   75.1 lb (approximately 600 times the watch's weight).

16.38   Horizontal force is 0.234 N, vertical force is 0.364 N.

16.40   1.54 lb.

16.42 $|\mathbf{v}_A| = 2.07$ m/s, $|\mathbf{v}_B| = 2.76$ m/s.

16.44   (a), (b) 1.63 ft/s toward the right.

16.46   (a), (b) 1.64 m/s toward the right.

16.48   (a) 6.47×10^{-4} m/s toward the left;
        (b) Zero;
        (c) 4 m to his left.

16.50   2.41 m/s.

16.52   603 m/s.

16.54   96.9 mm.

16.56   (a) $4\mathbf{i} + 30\mathbf{j}$ (mm/s); (b) $x = 6.2$ m, $y = 6$ m.

16.58 A: 0.215 ft/s toward the left. B: 1.385 ft/s toward the right.

16.60 A: 0.6 m/s toward the left. B: 0.4 m/s toward the right.

16.62 A: 228 ft/s^2 (7.09 g's). B: 155 ft/s^2 (4.81 g's).

16.64 $v_A\sqrt{mk/2}$.

16.66 $v_A = 42.2$ ft/s (28.8 mi/h).

16.68 3.00 m.

16.70 $e = 0.77$.

16.72 4.5 ft/s upward.

16.74 0.963 m/s toward the left.

16.76 33.2°.

16.78 $\mathbf{v}_A = -5.71\mathbf{i} + 4\mathbf{j} + 8\mathbf{k}$ (m/s),
$\mathbf{v}_B = 3.89\mathbf{i} - 6\mathbf{j} + 5\mathbf{k}$ (m/s).

16.80 $\mathbf{v}'_A = \mathbf{i} + \mathbf{j}$ (m/s), $\mathbf{v}'_B = -\mathbf{i} + \mathbf{j}$ (m/s).

16.84 $v_S = 35.3$ m/s.

16.86 (a) $\mathbf{v} = \frac{1}{2}t^2\mathbf{i} + 2t\mathbf{j}$ (m/s), $\mathbf{r} = \frac{1}{6}t^3\mathbf{i} + t^2\mathbf{j}$ (m);

(b), (c) $-432\mathbf{k}$ (kg-m²/s).

16.88 $|v_r| = 2490$ m/s, $|v_\theta| = 4310$ m/s.

16.90 $\omega = 3.84$ rad/s, $v_r = \pm 0.262$ m/s.

16.92 2.31 m.

16.94 6 N-m.

16.96 (a) $-1440\mathbf{k}$ (kg-m²/s); (b) 1.2 m/s.

16.100 103 lb.

16.102 750 N.

16.104 (a) $351\mathbf{i} - 849\mathbf{j}$ (N); (b) $1200\mathbf{i} - 1200\mathbf{j}$(N);
(c) $2400\mathbf{i}$ (N).

16.106 $0.406\mathbf{k}$ (N-m).

16.108 $51.9\mathbf{i} - 282.2\mathbf{j}$ (N).

16.110 1530 m/s.

16.112 $v = v_f \ln\left(\dfrac{m_0}{m_0 - \dot{m}_f t}\right) - gt$.

16.114 (a) $F = 3s + 12/32.2$ lb.
(b) 25.5 ft-lb.

16.116 $v = v_0 / [1 + (\rho_L / 2m)s]$.

16.118 6820 N.

16.120 18.5 kN.

16.122 (a) $180\mathbf{i} + 360\mathbf{j}$ (lb-s); (b) $12\mathbf{i} + 44\mathbf{j}$ (ft/s).

16.124 877 kN.

16.126 (a) 34.9 ft/s; (b) 15,700 ft-lb.

16.128 (a) 4.80 ft/s; (b) 415 ft.

16.130 (a) 3.45 m/s; (b) 18.7 kN.

16.132 (a) 8.23 kN; (b) 16.5 kN.

16.134 $|\mathbf{v}_A| = 2.46$ m/s, $|\mathbf{v}_B| = 2.90$ m/s.

16.136 1.57 ft.

16.138 $e = 0.304$.

16.140 2.30 m.

16.142 $x = -11.13$ in., $y = 6.42$ in.

16.144 $|v_r| = 11{,}550$ ft/s, $|v_\theta| = 9430$ ft/s.

16.146 867 lb (including the weight of the drum).

16.148 (a) 30.1 ft/s; (b) 46.8 ft/s.

Chapter 17

17.2 (a) 10.1 ft/s;
(b) Tangential, 2.51 ft/s²; normal, 50.5 ft/s².

17.4 (a) 1.2 rad/s clockwise, 0.4 rad/s² counterclockwise;
(b) $|\mathbf{v}_A| = 2.4$ m/s, $|\mathbf{a}_A| = 2.99$ m/s².

17.6 (a) 2.67; (b) 2.67 rad/s.

17.8 $\mathbf{v}_A = 83.8\mathbf{i} + 83.8\mathbf{j}$ (in/s), $\mathbf{v}_B = -168\mathbf{j}$ (in/s).

17.10 Tangential, 25 ft/s²; normal, 11,800 ft/s².

17.12 $|\mathbf{v}_B| = 2.37$ m/s, $|\mathbf{a}_B| = 22.1$ m/s².

17.14 $\boldsymbol{\omega} = 30\mathbf{i}$ (rad/s).

17.16 $\boldsymbol{\omega}_{OQ} = -4\mathbf{k}$ (rad/s), $\boldsymbol{\omega}_{PQ} = 4\mathbf{k}$ (rad/s).

17.18 (a) $\mathbf{v}_{A/B} = 4.8\mathbf{j}$ (m/s);
(b) $\boldsymbol{\omega} = 12\mathbf{k}$ (rad/s);
(c) $\mathbf{v}_{A/B} = 4.8\mathbf{j}$ (m/s).

17.20 $\mathbf{v}_B = 1.2\mathbf{i} + 2.4\mathbf{j}$ (m/s).

17.22 $\mathbf{v}_A = -5\mathbf{i} + 8.66\mathbf{j}$ (m/s).

17.24 $\mathbf{v}_A = 5\mathbf{i}$ (m/s), $\mathbf{v}_B = -3.54\mathbf{i} - 3.54\mathbf{j}$ (m/s),
$\mathbf{v}_C = -3.54\mathbf{i} + 3.54\mathbf{j}$ (m/s).

17.26 (a) 101 rad/s counter clockwise;
(b) $-117\mathbf{i} - 117\mathbf{j}$ (ft/s).

17.28 $\mathbf{v}_T = 12.4\mathbf{i} + 5.1\mathbf{j}$ (m/s).

17.30 $\mathbf{v}_G = 1.5\mathbf{i} - 0.546\mathbf{j}$ (m/s).

17.32 $\omega_{OQ} = 1.18$ rad/s clockwise,
$\omega_{PQ} = 1.18$ rad/s counterclockwise.

17.34 $\omega_{BD} = 8$ rad/s clockwise, $\mathbf{v}_D = 6.40\mathbf{i} - 1.28\mathbf{j}$ (m/s).

17.36 14.5 rad/s clockwise.

17.38 $\omega_{BC} = 3.33$ rad/s clockwise,
$\omega_{\text{disk}} = 20$ rad/s clockwise.

17.40 $\mathbf{v}_E = -12.3\mathbf{j}$ (m/s).

17.42 0.0857 rad/s counterclockwise.

17.44 $\omega_{AB} = 2.31$ rad/s clockwise, $v_B = 3.15$ m/s to the left.

17.46 $\mathbf{v}_C = 25.1\mathbf{i}$ in./s.

17.48 $\mathbf{v}_A = 1.2\mathbf{i} + 1.2\mathbf{j}$ (m/s).

17.50 $\omega_{BC} = 2.61$ rad/s, $\mathbf{v}_C = -9.1\mathbf{i}$ (m/s).

17.52 0.95 m/s.

17.54 $\mathbf{v}_C = 0.282\mathbf{i} + 0.202\mathbf{j}$ (m/s).

17.56 $\mathbf{v}_D = -0.557\mathbf{i} + 0.815\mathbf{j}$ (m/s).

17.58 $v_W = 0.2$ m/s, 4 rad/s counterclockwise.

17.60 35.5 rpm clockwise.

17.62 Angular velocity = 52.1 rad/s, or 497 rpm,
$|\mathbf{v}_A| = 5.21$ m/s.

17.64 $x_C = 3$ m, $y_C = 0$, $v_B = 10$ m/s.

17.66 $x = 0.35$ ft, $y = -1.5$ ft.

17.68 $\mathbf{v}_B = 6\mathbf{i} + 3.46\mathbf{j}$ (ft/s).

17.70 $\omega_{BC} = 5$ rad/s clockwise,
$\omega_{CD} = 10$ rad/s counterclockwise.

17.72 6.57 in./s.

17.74 1.67 rad/s counterclockwise.

17.76 (a) $x = 225$ mm, $y = 225$ mm;
(b) $\mathbf{v}_C = 13.5\mathbf{i}$ (m/s).

17.78 $\omega_{BC} = 5.33$ rad/s counterclockwise,
$\omega_{CD} = 4.57$ rad/s clockwise.

17.80 (a) $x = -0.425$ m, $y = 0.737$ m;
(b) $v_B = 2.31$ m/s, $\omega = 2.35$ rad/s counterclockwise.

17.82 (a), (b) $\mathbf{a}_A = -73.3\mathbf{i} + 27.0\mathbf{j}$ (m/s²).

17.84 $\mathbf{a}_T = 2.02\mathbf{i} + 2.37\mathbf{j}$ (m/s²).

17.86 $\mathbf{a}_A = -0.5\mathbf{j}$ (m/s²), $\mathbf{a}_B = 0.3\mathbf{j}$ (m/s²).

17.88 $\omega_{AC} = 0$, $\alpha_{AC} = 1.13$ rad/s² clockwise.

17.90 $500\mathbf{i} - 100\mathbf{j}$ (m/s²).

17.92 $\alpha_{OQ} = 1.39$ rad/s² clockwise,
$\alpha_{PQ} = 1.39$ rad/s² counterclockwise.

17.94 1.77 rad/s² counterclockwise.

17.96 $\alpha_{BC} = 26.8$ rad/s² counterclockwise,
$\alpha_{CD} = 12.3$ rad/s² counterclockwise.

17.98 $\alpha_{AB} = 19.0$ rad/s².

17.100 $\alpha_{BC} = 6.67$ rad/s^2 counterclockwise,
$\alpha_{\text{disk}} = 93.3$ rad/s^2 counterclockwise.

17.102 $\omega_{AB} = -0.879$ rad/s, $\alpha_{AB} = -1.06$ rad/s^2,
$\omega_{BC} = -1.15$ rad/s, $\alpha_{BC} = -2.41$ rad/s^2.

17.104 $\mathbf{a}_E = -12.3\mathbf{j}$ (m/s^2).

17.106 $\mathbf{a}_C = -7.78\mathbf{i} - 33.54\mathbf{j}$ (m/s^2).

17.108 $\mathbf{a}_D = -0.135\mathbf{i} - 0.144\mathbf{j}$ (m/s^2).

17.110 $\omega_{AB} = 3.55$ rad/s clockwise,
$\alpha_{AB} = 12.1$ rad/s^2 clockwise,
$\omega_{BC} = 2.36$ rad/s counterclockwise,
$\alpha_{BC} = 16.5$ rad/s^2 counterclockwise.

17.112 0.0571 rad/s^2 clockwise.

17.114 $\alpha_{\text{planet}} = 41.4$ rad/s^2 clockwise,
$\alpha_{\text{sun}} = 82.9$ rad/s^2 counterclockwise.

17.116 $\alpha_{CD} = 26.5$ rad/s^2 clockwise,
$\alpha_{DE} = 31.1$ rad/s^2 counterclockwise.

17.118 $\mathbf{a}_A = -200\mathbf{i} + 80\mathbf{j}$ (ft/s^2).

17.120 $\mathbf{a}_C = -8.80\mathbf{i} + 5.60\mathbf{j}$ (m/s^2).

17.122 $\omega_{AC} = 3$ rad/s counterclockwise,
$\alpha_{AC} = 6$ rad/s counterclockwise.

17.124 0.549 m/s toward the left.

17.126 0.0972 rad/s counterclockwise.

17.128 $\omega_{BC} = 6.17°$ per second counterclockwise, rate of
extension is 0.109 m/s.

17.130 $\omega_{AC} = 8.66$ rad/s counterclockwise, and bar AC slides
through the sleeve at 5 m/s toward A.

17.132 $\omega_{AC} = 0.293$ rad/s clockwise,
$\mathbf{v}_C = -0.738\mathbf{i} - 1.370\mathbf{j}$ (ft/s).

17.134 0.801 ft/s.

17.136 $\omega_{AB} = 2$ rad/s clockwise, $v_{B\,\text{rel}} = 2$ m/s toward C.

17.138 $\omega_{AB} = 5.18$ rad/s counterclockwise.

17.140 $\omega_{\text{plate}} = 2$ rad/s counterclockwise, and the velocity at
which the pin slides relative to the slot is 0.2 m/s
downward.

17.144 $\mathbf{v}_A = 1.44\mathbf{i} + 0.934\mathbf{j}$ (ft/s).

17.146 $\mathbf{v}_A = -3.5\mathbf{i} + 0.5\mathbf{j} + 4\mathbf{k}$ (m/s),
$\mathbf{a}_A = -10\mathbf{i} - 6.5\mathbf{j} - 19.25\mathbf{k}$ (m/s^2).

17.148 $-2\mathbf{i}$ (m/s).

17.150 $\mathbf{a}_{A\,\text{rel}} = -14\mathbf{i} - 2\mathbf{j}$ (ft/s^2).

17.152 (a) $\mathbf{v} = v\mathbf{j}$, $\mathbf{a} = -(v^2/R_E)\mathbf{i}$;
(b) $\mathbf{v} = v\mathbf{j} - \omega_E R_E \cos L\mathbf{k}$,
$\mathbf{a} = -(v^2/R_E + \omega_E^2 R_E \cos^2 L)\mathbf{i} +$
$\omega_E^2 R_E \sin L \cos L\mathbf{j} + 2\omega_E v \sin L\mathbf{k}$.

17.154 (a) $-9.81\mathbf{k}$ (m/s^2); (b) $3.29\mathbf{i} - 9.81\mathbf{k}$ (m/s^2).

17.156 (a) $0.1\mathbf{i} + 0.1\mathbf{j}$ (m/s^2);
(b) $0.125\mathbf{i} + 0.085\mathbf{j} + 0.106\mathbf{k}$ (m/s^2).

17.160 OQ: 9.29 rad/s counterclockwise;
PQ: 2.92 rad/s clockwise.

17.162 $\mathbf{v}_C = -32.0\mathbf{j}$ (ft/s).

17.164 $\alpha_{AB} = 13.60 \times 10^3$ rad/s^2 clockwise,
$\alpha_{BC} = 8.64 \times 10^3$ rad/s^2 counterclockwise.

17.166 $\mathbf{a}_D = -3490\mathbf{i}$ (in/s^2).

17.168 $\mathbf{a}_G = -8.18\mathbf{i} - 26.4\mathbf{j}$ (in/s^2).

17.170 $\mathbf{v}_C = -1.48\mathbf{i} + 0.79\mathbf{j}$ (m/s).

17.172 $\mathbf{a}_C = -2.99\mathbf{i} - 1.40\mathbf{j}$ (m/s^2).

17.174 Velocity is 55.9 in/s to the right;
acceleration is 390 in/s^2 to the left.

17.176 $\omega_{BD} = 0.733$ rad/s counterclockwise.

17.178 $\omega_{AB} = 0.261$ rad/s counterclockwise,
$\omega_{BC} = 2.80$ rad/s counterclockwise.

17.180 $\omega_{BC} = 1.22$ rad/s clockwise, 18.0 m/s from B toward C.

17.182 Velocity = 6.89 ft/s upward;
acceleration = 169 ft/s^2 upward.

17.184 5.66 N.

17.186 0.103 rad/s counterclockwise.

Chapter 18

18.2 (a) 0.980 ft/s^2;
(b) $N_A = 57.7$ lb, $N_B = 172$ lb.

18.4 9.23 m/s.

18.6 (a) 3.22 ft/s^2;
(b) $N_A = 182$ lb, $N_B = 58$ lb.

18.8 (a) 2.5 rad/s^2; (b) 95.5 rpm.

18.10 $-8.9\mathbf{j}$ (ft/s^2).

18.12 (a) 9.25 s; (b) 195 rpm.

18.14 (a) 565 rpm;
(b) 799 rpm.

18.16 Falls 0.721 m, tension is 80.8 N.

18.18 (a) 12.3 rad/s^2; (b) 8.67 rad/s^2.

18.20 4.95 rad/s.

18.22 $D_x = 27.5$ N, $D_y = 23.6$ N.

18.24 0.581 rad/s^2 counterclockwise.

18.26 $M_B = 27.1$ N-m counterclockwise,
$B_x = -11.0$ N, $B_y = 108.5$ N.

18.28 $\mathbf{a}_G = 0.1108\mathbf{i} - 0.0168\mathbf{j}$ (m/s^2),
$\alpha = -0.000427$ rad/s^2.

18.30 $\mathbf{F}_B = -19.1\mathbf{i} + 183.3\mathbf{j}$ (N), $\mathbf{M}_B = 62.6\mathbf{k}$ (N-m).

18.32 6.54 m/s.

18.34 $v_{\text{ring}}/v_{\text{disk}} = \sqrt{3/4}$.

18.36 (a) 14.7 rad/s^2; (b) 41.7 N.

18.38 (a) 14.8 rad/s^2 clockwise.
(b) $\mu_s = 0.227$.

18.40 Velocity = 3.81 ft/s, time = 1.97 s.

18.42 (a) It doesn't slip, $\alpha = 22.2$ rad/s^2 clockwise.
(b) It does slip, $\alpha = 53.6$ rad/s^2 clockwise.

18.44 27.9 N.

18.46 2.35 rad/s^2 counterclockwise.

18.48 (a) 9.38 rad/s^2 clockwise.
(b) 9.57 rad/s^2 clockwise.

18.50 61.3 rad/s^2 clockwise.

18.52 3.16 ft/s^2.

18.54 $\mu_s = 0.108$.

18.56 $M = 402$ ft-lb, tension is 49.2 lb.

18.58 $\Sigma\mathbf{F} = -340\mathbf{i} - 160\mathbf{j}$ (N),
$\Sigma M = 20$ N-m counterclockwise.

18.62 (a) 9.34 m/s^2; (b) 516 N-m.

18.64 $\alpha_{BC} = 17.0$ rad/s^2 counterclockwise.

18.66 $\alpha_{OQ} = 4.88$ rad/s^2 clockwise,
$\alpha_{PQ} = 4.88$ rad/s^2 counterclockwise.

18.70 $\alpha = 0.255$ rad/s^2 clockwise.

18.72 $I_0 = 14$ kg-m^2.

18.74 $I_{z\,\text{axis}} = 15.1$ kg-m^2.

18.76 $I_{x\,axis} = 2.667$ kg-m^2, $I_{y\,axis} = 0.667$ kg-m^2,
$I_{z\,axis} = 3.333$ kg-m^2.

18.78 $I_{y\,axis} = 1.99$ slug-ft^2.

18.80 20.8 kg-m^2.

18.82 $I_0 = \frac{17}{12} ml^2$.

18.84 $I_{z\,axis} = 74.0$ kg-m^2.

18.86 $I_{z\,axis} = 0.0803$ slug-ft^2.

18.88 3810 slug-ft^2.

18.90 $I_{z\,axis} = 9.00$ kg-m^2.

18.92 $I_{y\,axis} = 3.07$ kg-m^2.

18.94 $I_0 = 0.0188$ kg-m^2.

18.96 $I_{x\,axis} = I_{y\,axis} = m\left(\frac{3}{20} R^2 + \frac{3}{5} h^2\right)$.

18.98 $I_{x\,axis} = 0.844$ kg-m^2.

18.100 $I_x = 0.0535$ kg-m^2.

18.102 $I_{x'} = 0.995$ kg-m^2.

18.104 $I_{z\,axis} = 0.00911$ kg-m^2.

18.106 $I_0 = 0.00367$ kg-m^2.

18.108 $I_{z\,axis} = 0.714$ slug-ft^2.

18.110 (a) 12.1 s;
(b) 144 lb.

18.112 (a) 20 m/s^2;
(b) $c \leq 49.1$ mm.

18.114 $I = 2.05$ kg-m^2.

18.116 40.2 kN.

18.118 $\alpha = -0.420$ rad/s^2, $F_x = 336$ N, $F_y = 1710$ N.

18.120 $\alpha = (g/l)[3(1 - \mu^2)\sin\theta - 6\mu\cos\theta]/(2 - \mu^2)$
counterclockwise.

18.122 $B_x = -1959$ N, $B_y = 1238$ N, $C_x = 2081$ N,
$C_y = -922$ N.

18.124 $\alpha_{OA} = 0.425$ rad/s^2 counterclockwise,
$\alpha_{AB} = 1.586$ rad/s^2 clockwise.

18.126 $\alpha_{HP} = 5.37$ rad/s^2 clockwise.

18.128 208 m/s^2 to the left.

Chapter 19

19.2 20.7 ft-lb.

19.4 1270 N-m.

19.6 353 rpm (37.0 rad/s).

19.8 (a) 78.5 N-m; (b) Zero; (c) 3.84 rad/s.

19.10 22.2 lb.

19.12 1.39 m/s.

19.14 3.33 rad/s.

19.16 0.731 rad = 41.9°.

19.18 $U_{12} = 5630$ N-m.

19.20 2.59 m/s to the right.

19.22 16.7 rad/s clockwise.

19.24 7.52 ft/s.

19.26 5.72 ft/s.

19.28 $s = 66.1$ m.

19.30 4.33 rad/s clockwise.

19.32 0.384 m/s.

19.34 1.79 rad/s.

19.36 0.899 m/s.

19.38 2.57 rad/s counterclockwise.

19.40 4.52 rad/s counterclockwise.

19.42 2.80 rad/s counterclockwise.

19.44 2.34 rad/s.

19.46 $M = 28.2$ N-m.

19.48 (a) 250 N-m-s counterclockwise;
(b) 11.4 rad/s counterclockwise.

19.50 3 s.

19.52 (a) 126 rad/s; (b) 200 rad/s.

19.54 510 rad/s (4870 rpm).

19.56 246 kN.

19.58 $\omega_2 = 4.55$ revolutions per second.

19.60 3.94 rad/s (37.6 rpm).

19.62 1.46 m/s.

19.64 14.7°.

19.66 3.37 rad/s counterclockwise.

19.68 0.00196 rad/s counterclockwise.

19.70 1.54 rad/s.

19.72 1.27 kN.

19.74 $v_A = 1.47$ m/s, $\omega_B = 12.8$ rad/s counterclockwise.

19.78 (a) 229°; (b) 2.5 ft-lb.

19.80 5.96 ft/s.

19.82 9.6 rad/s counterclockwise

19.84 Velocity $= 5.77\mathbf{i} - 2.10\mathbf{j}$ (ft/s),
angular velocity $= 3.15$ rad/s counterclockwise.

19.86 $\omega' = 0.375$ rad/s.

19.88 $\omega' = 0.3$ rad/s.

19.90 $\omega = 0.0997$ rad/s $= 5.71$ deg/s counterclockwise.

19.92 15.0 km/hr.

19.94 0.00336 rad/s clockwise.

19.96 $v = 2.05$ m/s.

19.98 $v = \sqrt{Fb/\left[\frac{1}{2} m_c + 2(m + I/R^2)\right]}$.

19.100 $\omega_S = 12.5$ rad/s.

19.102 (a) 1.13 ft; (b) 1.09 ft/s.

19.104 $P_{max} = 580$ ft-lb/s (1.05 hp),
$P_{ave} = 290$ ft-lb/s (0.53 hp).

19.106 11.1 rad/s.

19.108 1.77 rad/s counterclockwise.

19.110 $N = (373/283)mg$.

19.112 $\omega' = 0.721$ rad/s.

19.114 $\omega' = \left(\frac{1}{3} + \frac{2}{3}\cos\beta\right)\omega$.

19.116 $b = 2$ ft.

19.118 $\omega = 8.51$ rad/s.

19.120 $\theta = 54.7°$, $\omega = 10.7$ rad/s counterclockwise.

19.122 $0.0822\mathbf{k}$ (rad/s).

19.124 0.641 rad/s clockwise, $\mathbf{v}' = -2.24\mathbf{i}$ (ft/s).

Chapter 20

20.2 $\alpha = -6.94\mathbf{i}$ (rad/s^2).

20.4 $\mathbf{a}_A = -205\mathbf{i} - 63.0\mathbf{j} - 135\mathbf{k}$ (m/s^2).

20.6 $\mathbf{v}_C = 4\mathbf{i} + 4\mathbf{k}$(m/s), $\mathbf{v}_D = 4\mathbf{i} - 8\mathbf{j}$(m/s).

20.8 (a) $-33.9\mathbf{j} + 33.9\mathbf{k}$ (in/s);
(b) $22.6\mathbf{i} - 33.9\mathbf{j} + 33.9\mathbf{k}$ (in/s).

20.10 (a) $\mathbf{v}_P = 48\mathbf{k}$ (ft/s);
(b) $\boldsymbol{\alpha} = -288\mathbf{k}$ (rad/s^2).

20.12 (a) $\boldsymbol{\alpha} = 0.0453\mathbf{i} - 0.0262\mathbf{j}$ (rad/s^2);
(b) $-0.394\mathbf{i} + 0.0693\mathbf{j} + 0.209\mathbf{k}$ (m/s^2).

20.14 $\mathbf{v}_C = 0.4\mathbf{k}$ (m/s).

20.16 $\mathbf{v}_A = 80\mathbf{i}$ (mm/s),
$\mathbf{v}_B = -102.6\mathbf{i} + 281.9\mathbf{j} - 56.4\mathbf{k}$ (mm/s).

20.18 $\mathbf{v} = -1.05\mathbf{i}$ (m/s).

20.20 (a) $\boldsymbol{\Omega} = 0.894\mathbf{i} + 1.789\mathbf{j}$ (rad/s);
(b) $\boldsymbol{\omega}_{\text{rel}} = -4.47\mathbf{i}$ (rad/s).

20.22 $\mathbf{a} = -1.28\mathbf{i} + 0.64\mathbf{j} - 3.20\mathbf{k}$ (m/s^2).

20.24 $\mathbf{v} = \mathbf{0}$.

20.26 $\mathbf{F} = mg\mathbf{j} - mb\alpha_0\mathbf{k}, \mathbf{C} = -\frac{1}{2}mlb\alpha_0\mathbf{i}$.

20.28 $\Sigma\mathbf{M} = 0.0135\mathbf{i} - 0.0086\mathbf{j} + 0.01\mathbf{k}$ (N-m).

20.30 $\Sigma\mathbf{M} = -76\mathbf{i} + 36\mathbf{j} - 60\mathbf{k}$ (ft-lb).

20.32 $|\mathbf{a}_A| = 2.5$ m/s^2.

20.34 (a) $\boldsymbol{\alpha} = 28.57\mathbf{i} + 5.71\mathbf{j} - 2.00\mathbf{k}$ (rad/s^2);
(b) $\mathbf{a} = -0.33\mathbf{i} + 2.00\mathbf{j} - 19.52\mathbf{k}$ (m/s^2).

20.36 (a) $\mathbf{a} = -4.17\mathbf{k}$ (m/s^2);
(b) $\boldsymbol{\alpha} = 49.5\mathbf{i} + 137.7\mathbf{j}$ (rad/s^2).

20.38 $\boldsymbol{\alpha} = -500.0\mathbf{i} + 24.4\mathbf{j}$ (rad/s^2).

20.40 $\boldsymbol{\alpha} = 14.5\mathbf{j}$ (rad/s^2).

20.42 $\boldsymbol{\alpha} = 14.53\mathbf{j} + 4.65\mathbf{k}$ (rad/s^2).

20.44 27.4 N-m.

20.46 $|\mathbf{F}| = 78.5$ N, $|\mathbf{M}| = 260$ N-m.

20.50 Force is $20\mathbf{j} - 19.9\mathbf{k}$ (lb), couple is $12.4\mathbf{i}$ (ft-lb).

20.52 $-10\mathbf{j} + 14.9\mathbf{k}$ (lb), $-12.4\mathbf{i}$ (ft-lb).

20.58 157 N-m.

20.62 21.5 ft-lb.

20.64 4650 N-m. Counterclockwise.

20.66 $\omega_{\text{d}} = -\frac{1}{2}\omega_0\sin\beta$.

20.68 $\dot{\psi} = 1.31$ rev/s.

20.70 29.1 rev/s.

20.72 (a) $\dot{\psi} = -2050$ rpm.
(b)

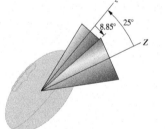

20.74 309 rad/s (49.1 revolutions per second).

20.76 $\omega_0 = 10.7$ rad/s.

20.78 $\Sigma\mathbf{M} = 2123\mathbf{i} - 155\mathbf{j} - 534\mathbf{k}$ (N-m).

20.80 $I_{xx} = 0.67$ kg-m^2, $I_{yy} = 5.33$ kg-m^2,
$I_{zz} = 6$ kg-m^2, $I_{xy} = I_{yz} = I_{zx} = 0$.

20.82 $I_{xx} = 0.12$ kg-m^2, $I_{yy} = 0.03$ kg-m^2,
$I_{zz} = 0.15$ kg-m^2, $I_{xy} = I_{yz} = I_{zx} = 0$.

20.84 $I_{xx} = 2.48$ slug-ft^2, $I_{yy} = 16.77$ slug-ft^2,
$I_{zz} = 19.25$ slug-ft^2, $I_{xy} = 5.59$ slug-ft^2, $I_{yz} = I_{zx} = 0$.

20.86 $I_{xx} = 0.0318$ kg-m^2, $I_{yy} = 0.0357$ kg-m^2,
$I_{zz} = 0.0674$ kg-m^2, $I_{xy} = 0.0219$ kg-m^2, $I_{yz} = I_{zx} = 0$.

20.88 (a) $\mathbf{H} = \frac{1}{12}ml^2(\omega_y\mathbf{j} + \omega_z\mathbf{k})$;
(b) $\mathbf{H}_O = \frac{1}{3}ml^2(\omega_y\mathbf{j} + \omega_z\mathbf{k})$.

20.90 $I_{x'x'} = 15,600$ kg-m^2, $I_{y'y'} = 226,800$ kg-m^2,
$I_{z'z'} = 242,400$ kg-m^2, $I_{x'y'} = -32,400$ kg-m^2,
$I_{y'z'} = I_{z'x'} = 0$.

20.92 $\mathbf{H} = 2.00\mathbf{i} - 1.64\mathbf{j} + 2.58\mathbf{k}$ (kg-m^2/s).

20.94 (a) $I = 3.56$ kg-m^2;
(b) 14.22 kg-m^2/s.

20.96 $I_1 = 16.15, I_2 = 62.10, I_3 = 81.75$ kg-m^2,
$\mathbf{e}_1 = 0.964\mathbf{i} - 0.220\mathbf{j} + 0.151\mathbf{k}$,
$\mathbf{e}_2 = -0.204\mathbf{i} - 0.972\mathbf{j} - 0.114\mathbf{k}$,
$\mathbf{e}_3 = 0.172\mathbf{i} + 0.079\mathbf{j} - 0.982\mathbf{k}$.

20.98 $I_1 = 0, I_2 = \frac{1}{12}, I_3 = \frac{1}{12}$ kg-m^2, $\mathbf{e}_1 = \cos\beta\mathbf{i} + \sin\beta\mathbf{j}$,
$\mathbf{e}_2 = -\sin\beta\mathbf{i} + \cos\beta\mathbf{j}, \mathbf{e}_3 = \mathbf{k}$.

20.100 (a) $\boldsymbol{\omega} = \omega_{\text{d}}\mathbf{i} + \omega_0\mathbf{j}$;
(b) $\mathbf{v}_A = -R\omega_0\cos\theta\mathbf{i} + R\omega_{\text{d}}\cos\theta\mathbf{j} +$
$(R\omega_{\text{d}}\sin\theta - b\omega_0)\mathbf{k}$.

20.102 (a) $\boldsymbol{\omega} = 20\mathbf{i} - 5\mathbf{j}$ (rad/s);
(b) $\mathbf{v}_A = 0.25\mathbf{i} + 1.00\mathbf{j} + 3.73\mathbf{k}$ (m/s).

20.104 $\mathbf{a}_B = 166\mathbf{j}$ (ft/s^2),
$\boldsymbol{\alpha} = 7.31\mathbf{i} - 5.52\mathbf{j} + 22.89\mathbf{k}$ (rad/s^2).

20.106 $\mathbf{H} = 25\mathbf{i} + 50\mathbf{k}$ (kg-m^2/s).

20.108 $I_{xx} = 0.0398$ slug-ft^2, $I_{yy} = 0.0373$ slug-ft^2,
$I_{zz} = 0.0772$ slug-ft^2, $I_{xy} = I_{yz} = I_{zx} = 0$.

20.110 $\Sigma\mathbf{M} = -250\mathbf{i} - 250\mathbf{j} + 125\mathbf{k}$ (N-m).

20.112 $\omega_0 = \left[g\sin\beta/\left(\frac{2}{3}l\sin\beta\cos\beta + b\cos\beta\right)\right]^{1/2}$.

20.114 (b) If $\omega_0 = 0$, the plate is stationary. The solution of the equation $2\cos\beta - \sin\beta = 0$ is the value of β for which the center of mass of the plate is directly above point O; the plate is balanced on one corner.

20.116 $\mathbf{F}_B = 52.72\mathbf{i} + 97.35\mathbf{j} + 9.26\mathbf{k}$ (N),
$\mathbf{M}_B = 0.05\mathbf{i} - 10.25\mathbf{j} + 30.63\mathbf{k}$ (N-m).

20.118

20.120 $\Sigma\mathbf{M} = -283\mathbf{i} - 2546\mathbf{j} - 800\mathbf{k}$ (ft-lb).

20.122 $\boldsymbol{\alpha} = 0.0535\mathbf{i} + 0.0374\mathbf{j} + 0.0160\mathbf{k}$ (rad/s^2).

Chapter 21

21.2 $\tilde{x} = 0.264\sin 3.79t$ ft.

21.4 Amplitude = 0.166 m.

21.6 $x = 0.390$ m.

21.8 $L = 0.203$ m.

21.10 $x = 0.200(1 - \cos 5.37t)$.

21.12 (a) $f = 0.581$ Hz, $\tau = 1.72$ s;
(b) $x = 0.351$ m.

21.14 $k = 36.8$ lb/ft.

21.16 0.692 Hz.

21.18 0.287 m/s downward.

21.20 $\beta = 0.01 \cos 6t$ rad.

21.24 $I_{(z' \text{ axis})} = 24.2$ kg-m^2.

21.26 (a) $r = 0.301$ m;

(b) $f = 2.69$ Hz.

21.28 0.39 in.

21.30 $\theta_B = 0.172 \sin 11.6t$ rad.

21.32 $m_A = 4.38$ kg.

21.36 (a) $f_d = 0.900$ Hz;

(b) $x = 0.0816$ ft.

21.38 (a) $c = 33.9$ lb-s/ft;

(b) $x = 0.133$ ft.

21.40 $x = 2.38$ ft.

21.42 (a) 276 ft/s^2 (8.56 g's);

(b) 67.6 ft/s^2 (2.10 g's);

(c) 15.8 ft/s^2 (0.490 g's).

21.44 (a) $\tau_d = 2.32$ s, $f_d = 0.431$ Hz;

(b) 5.28 s.

21.46 $x = e^{-2t}(-0.325 \sin 2.24t - 0.363 \cos 2.24t)$

$+ 0.363$ m.

21.48 $x = -(0.363 + 1.090t)e^{-3t} + 0.363$ m.

21.50 $x = 0.237$ m.

21.52 0.153 rad/s clockwise.

21.54 (a) 277 N-s/m;

(b) $\theta = 10te^{-4.33t}$ rad;

(c) $\theta_{\max} = 0.850$ rad at $t = 0.231$ s.

21.56 $\theta_B = 0.209e^{-6.62t} \sin 9.55t$ rad.

21.58 $x = 0.337$ ft.

21.60 (a) $x_p = 0.5$ m;

(b) $x = e^{-t}(-0.289 \sin 1.73t - 0.5 \cos 1.73t) + 0.5$ m.

21.62 $\theta = 0.581 \sin 2.58t + 1.250 \sin 2t$ rad.

21.64 11.5 lb.

21.66 0.113 mm.

21.70 16.5 in.

21.72 5.5 mm.

21.74 (a) $x = (0.1) \cos 3t$ m.

(b)

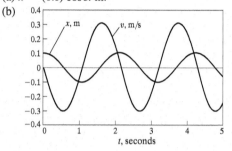

21.78 $f = 0.985$ Hz.

21.80 1.03 Hz.

21.82 $f = (1/2\pi)\sqrt{3[k/m] - (g/2l)]}$.

21.84 $\delta = 2.07$.

21.86 $f_d = 0.714$ Hz.

21.88 $x = 0.118e^{-2.68t} - 0.118e^{-14.01t}$ ft.

21.90 $\theta_B = e^{-5.05t}(0.244 \sin 3.45t + 0.167 \cos 3.45t)$ rad.

21.92 $x = 0.253 \sin 6.19t + 0.100 \cos 6.19t$

$+ 0.145 \sin 3.00t$ m.

21.94 (a) $E_p = (2\pi v/\lambda)^2 h/[(k/m) - (2\pi v/\lambda)^2]$;

(b) $v = \lambda\sqrt{k/m}/2\pi$.

Index

Properties of Areas and Lines

Areas

The coordinates of the centroid of the area A are

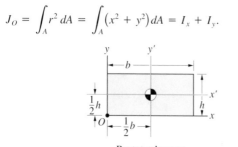

$$\bar{x} = \frac{\int_A x\, dA}{\int_A dA}, \qquad \bar{y} = \frac{\int_A y\, dA}{\int_A dA}.$$

The moment of inertia about the x axis I_x, the moment of inertia about the y axis I_y, and the product of inertia I_{xy} are

$$I_x = \int_A y^2\, dA, \qquad I_y = \int_A x^2\, dA, \qquad I_{xy} = \int_A xy\, dA.$$

The polar moment of inertia about O is

$$J_O = \int_A r^2\, dA = \int_A (x^2 + y^2)\, dA = I_x + I_y.$$

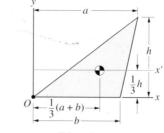

Triangular area

$$\text{Area} = \frac{1}{2}bh \qquad I_x = \frac{1}{12}bh^3, \qquad I_{x'} = \frac{1}{36}bh^3$$

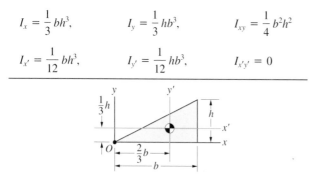

Rectangular area

$$\text{Area} = bh$$

$$I_x = \frac{1}{3}bh^3, \qquad I_y = \frac{1}{3}hb^3, \qquad I_{xy} = \frac{1}{4}b^2h^2$$

$$I_{x'} = \frac{1}{12}bh^3, \qquad I_{y'} = \frac{1}{12}hb^3, \qquad I_{x'y'} = 0$$

Circular area

$$\text{Area} = \pi R^2 \qquad I_{x'} = I_{y'} = \frac{1}{4}\pi R^4, \qquad I_{x'y'} = 0$$

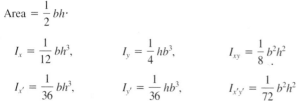

Triangular area

$$\text{Area} = \frac{1}{2}bh \cdot$$

$$I_x = \frac{1}{12}bh^3, \qquad I_y = \frac{1}{4}hb^3, \qquad I_{xy} = \frac{1}{8}b^2h^2$$

$$I_{x'} = \frac{1}{36}bh^3, \qquad I_{y'} = \frac{1}{36}hb^3, \qquad I_{x'y'} = \frac{1}{72}b^2h^2$$

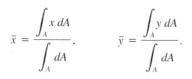

Semicircular area

$$\text{Area} = \frac{1}{2}\pi R^2 \qquad I_x = I_y = \frac{1}{8}\pi R^4, \qquad I_{xy} = 0$$

$$I_{x'} = \frac{1}{8}\pi R^4, \qquad I_{y'} = \left(\frac{\pi}{8} - \frac{8}{9\pi}\right)R^4, \qquad I_{x'y'} = 0$$